Road Vehicle

Accident Reconstruction

Road Vehicle
Accident Reconstruction

Rao V Dukkipati
PhD, PE
Fellow of ASME, CSME and SAE
Former Professor and Chairman of Mechanical Engineering
Graduate Program Director
Fairfield University
Fairfield, Connecticut
USA

Anshan
6 Newlands Road
Tunbridge Wells
Kent, UK

Co-published in the UK in 2014

ANSHAN LTD
6 Newlands Road
Tunbridge Wells
Kent, UK

Tel:+44(0) 1892 557767
e-mail: info@anshan.co.uk
Web Site: www.anshan.co.uk

ISBN: 978-184-8290-792

British Library Cataloguing in Publication Data
A Catalogue record for this book is available from the British Library

Preface

This book is a comprehensive text on road vehicle accident reconstruction. It is self-contained and the subject matter is presented in an organized and systematic manner. No previous knowledge of road vehicle accident reconstruction is assumed. This book is quite appropriate for several groups of audience including:

- senior undergraduate and graduate students in mathematics, science, engineering, lawyers, students, vehicle manufacturers and suppliers, police department and academies, law enforcement officers, insurance companies, those involved in accident reconstruction and others taking the introductory course on road vehicle accident reconstruction.
- the book can be adapted for a short professional course on road vehicle accident reconstruction.
- design and research engineers will be able to draw upon the book in selecting and developing road vehicle accident reconstruction aspects for analytical and design purposes.
- practicing engineers and managers who want to learn about the basic principles and concepts involved in accident reconstruction and how they can be applied at their own work place concerns.

Because the book is aimed at a wider audience, the level of mathematics is kept relatively low. All the principles presented in the book are illustrated by numerous practical examples. The book draws a balance between theory and practice.

A basic review of vector algebra, basic engineering mathematics, glossary of terms, units and conversion, ACTAR, miscellaneous accident reconstruction formulae, sensitivity analysis of accident reconstruction computations, graph of minimum stopping distance versus speed based on a friction coefficient of 0.7, table of minimum stopping distance versus speed, maximum safe driving speeds for various visibility conditions, speed in mph required to slow to a stop for various distances, surfaces and braking efficiency for cars are outlined in Appendices A, B, C, D, E, F, G, H, I, J, K, and L respectively. An extensive bibliography provided at the end of the book serves as helpful source for further study and research by interested readers. Both the SI and the US/English system of units have been used throughout the book.

I sincerely hope that the final outcome of this book will help the readers in developing an appreciation for the topic of road vehicle accident reconstruction.

Rao V. Dukkipati

Acknowledgements

I am grateful to all those who have had a direct impact on this work. I am greatly indebted to my coauthors of *Road Vehicle Dynamics* book. Jian Pang, Mohamad Qatu, Gang Sheng, and Shuguang Zuo and to numerous authors who have made valuable contributions to the literature on the subject of road vehicle accident reconstruction and more so to the authors of the articles listed in the bibliography of this book. The author acknowledges with thanks the editor, Victor Craig of the Accident Reconstruction Journal for permission to include a few of the example problems and solutions in Chapter 4. My sincere thanks to Mr. P.R. Naidu of Andhra University, Andhra Pradesh, India, for the excellent preparation of the complete manuscript. I sincerely thank to Martha B. Swiss, Intellectual Property Manager, and production staff for the careful processing of the manuscript. Finally, I express my heartfull thanks to my family members: Sudha, Ravi, Madhavi, Anand, Ashwin, Raghav, and Vishwa, not only for their encouragement and support but also for sharing all the pain, frustration, and fun of producing a textbook manuscript.

I would appreciate being informed of errors, or receiving other comments and helpful suggestions about the book. Please send an e-mail to *info@anshan.co.uk*

Contents

CHAPTER 1

Uncertainty in Numerical Computations and Measurements

1.1 INTRODUCTION

Errors are key and unavoidable ingredients of measurement process. Their net effect is to create an uncertainty in the value measured. When used in the context of measurements, uncertainty has a number, generally, a unit associated with it. More specifically, measurement uncertainty has the same unit as the measurand.

Uncertainty arises in two distinct situations: the process of making numerical or analytical computations and during the process of making measurement and observations. Generally speaking, measurements and observations are greatly influenced by the instruments used in the process and the inevitable presence of some uncontrolled variables. Hence, uncertainty analysis and estimation in such scenarios can provide very significant information and knowledge about the validity and acceptability of what is established or determined.

This chapter introduces the basics of uncertainty analysis and its approximation in road vehicle accident reconstruction.

1.2 ERRORS IN NUMERICAL COMPUTATIONS

The solutions of accident reconstruction problems are of two types: analytical and numerical. The analytical solutions are generally expressed in closed-form and hence these solutions are nearly error free. On the other hand, numerical methods solve problems using a computational machine (computer, calculator, etc.). Hence, for certain classes of problems, it is very difficult or even impossible to obtain an analytical solution in closed-form. For those problems, one can use numerical methods to determine the solution. Also, the solutions of non-linear ordinary differential equations cannot easily be determined by analytical methods. Such problems can easily be solved by numerical methods using modern high speed digital computers. Numerical computations are almost invariably contaminated by errors, and it is important to understand the source, propagation, magnitude, and rate of growth of these errors.

In this modern age of computers, many complex and large problems can be solved very fast and efficiently. Numerical methods seem to be more suitable for use in modern computations. Numerical methods have been highly appreciated and extensively used by engineers, scientists, accident reconstructionists, and many others.

1.3 NUMERICAL COMPUTATIONS

Numerical methods are methods for solving problems on computers by numerical calculations, often giving a table of numbers and/or graphical representations or figures. Numerical methods tend to emphasize the implementation of algorithms. The aim of numerical methods is, therefore, to provide systematic methods for solving problems in a numerical form. The process of solving problems generally involves starting from an initial data, using high precision digital computers, following the steps in the algorithms, and finally obtaining the results. Often, the numerical data and the methods used are approximate ones. Hence, the error in a computed result may be caused by the errors in the data, or the errors in the method or both.

In this section, we will describe Taylor's theorem, a few basic ideas and concepts regarding numerical computations and number representation, including binary, decimal, and hexadecimal numbers, error considerations, sources of errors, accuracy and precession, absolute and relative errors, inherent errors, round-off errors and truncation errors, error estimation, general error formulae including approximation of a function stability and condition, and uncertainty in data or noise.

1.3.1 Taylor's Theorem

Taylor's theorem allows us to represent, exactly, and fairly general functions in terms of polynomials with a known, specified, and boundable error. Taylor's theorem is stated as follows:

Let $f(x)$ has $n + 1$ continuous derivatives on $[a, b]$ for some $n \geq 0$, and let $x, x_0 \in [a, b]$. Then

$$f(x) = p_n(x) + R_n(x) \tag{1.1}$$

for

$$p_n(x) = \sum_{k=0}^{n} \frac{(x-x_0)^k}{k!} f^{(k)}(x_0) \tag{1.2}$$

and

$$R_n(x) = \frac{1}{n!} \int_{x_0}^{x} (x-t)^n f^{n+1}(t)\, dt \tag{1.3}$$

Also, there exists a point ξ_x, between x and x_0, such that

$$R_n(x) = \frac{(x-x_0)^{n+1}}{(n+1)!} f^{(n+1)}(\xi_x) \tag{1.4}$$

where $R_n(x)$ is the *remainder*. Taylor's series is an associated formula of Taylor's theorem.

Taylor's series gives us a means to predict a function value at one point in terms of the function value and its derivatives at another point.

Taylor's series expansion is defined by

$$f(x_{i+1}) = f(x_i) + f'(x_i)(x_{i+1} - x_i) + \frac{f''(x_i)}{2!}(x_{i+1} - x_i)^2$$

$$+ \frac{f^{(3)}(x_i)}{3!}(x_{i+1} - x_i)^3 + \cdots + \frac{f^n(x_i)}{n!}(x_{i+1} - x_i)^n + R_n \tag{1.5}$$

We note that Eq. (1.5) represents an infinite series. The remainder term R_n is included to account for all terms from $(n + 1)$ to infinity:

$$R_n = \frac{f^{(n+1)}(\xi)}{(n+1)!}(x_{i+1} - x_i)^{n+1} \tag{1.6}$$

where the subscript n connotes that this is the remainder for the nth order approximation and ξ is a value of x that lies somewhere between x_i and x_{i+1}.

We can rewrite the Taylor's series in Eq. (1.6) by defining a step size $h = x_{i+1} - x_i$ as

$$f(x_{i+1}) = f(x_i) + f'(x_i)h + \frac{f''(x_i)}{2!}h^2 + \frac{f^{(3)}(x_i)}{3!}h^3 + \cdots + \frac{f^n(x_i)}{n!}h^n + R_n \tag{1.7}$$

where the remainder term R_n is given by

$$R_n = \frac{f^{(n+1)}(\xi)}{(n+1)!}h^{n+1} \tag{1.8}$$

The estimation of function at a point b which is fairly close to a is desired, then the Taylor's series is written as an infinite series:

$$f(b) = f(a) + (b - a)f'(a) + \frac{(b-a)^2}{2!}f''(a) + \ldots + \frac{(b-a)^n}{n!}f^{(n)}(a) + \ldots \tag{1.9}$$

If b is very close to a, then only a few terms can give good estimation. The Taylor's series expansion for e^x, $\sin x$, and $\cos x$ are given below:

$$e^x = 1 + x + \frac{1}{2!}x^2 + \frac{1}{3!}x^3 + \ldots + \frac{1}{n!}x^n + \frac{1}{(n+1)!}x^{n+1}e^{\xi x}$$

$$= \sum_{k=0}^{n} \frac{1}{k!}x^k + R_n(x) \tag{1.10}$$

$$\sin x = x - \frac{1}{3!}x^3 + \frac{1}{5!}x^5 - \ldots + \frac{(-1)^n}{(2n+1)!}x^{2n+1} + \frac{(-1)^{n+1}}{(2n+3)!}x^{2n+3}\cos\xi x$$

$$= \sum_{k=0}^{n} \frac{(-1)^k}{(2k+1)!}x^{2k+1} + R_n(x) \tag{1.11}$$

$$\cos x = 1 - \frac{1}{2!}x^2 + \frac{1}{4!}x^4 - \ldots + \frac{(-1)^n}{(2n)!}x^{2n} + \frac{(-1)^{n+1}}{(2n+2)!}x^{2n+2}\cos\xi x$$

$$= \sum_{k=0}^{n} \frac{(-1)^k}{(2k)!}x^{2k} + R_n(x) \quad (1.12)$$

The error in Taylor's series when the series is terminated after the term containing $(x - a)^n$ will not exceed

$$\left|f^{(n-1)}\right|_{\max} \frac{[|x-a|]^{n-1}}{(n+1)!} \quad (1.13)$$

where max corresponds to the maximum magnitude of the derivative in the interval a to x. When the Taylor's series is truncated after n terms, then $f(x)$ will be accurate to $O(x - a)^n$.

Example 1.1: *Use Taylor's series expansion to approximate $f(x) = \cos x$ at $x_{i+1} = \pi/3$ with $n = 0$ to 6 on the basis of the value of $f(x)$ and its derivatives at $x_i = \pi/4$ which implies that $h = \frac{\pi}{3} - \frac{\pi}{4} = \pi/12$.*

Solution: The zero approximation is given by

$$f(x_{i+1}) \cong f(x_i) + f'(x_i)(x_{i+1} - x_i)$$

$$f(\pi/3) \cong \cos\frac{\pi}{4} = 0.70710678\,;\; f\left(\frac{\pi}{3}\right) = 0.5$$

The % relative error is

$$\in_t = \frac{0.5 - 0.70710678}{0.5}(100) = -41.4\%$$

$$f'(x) = -\sin x$$

$$f\left(\frac{\pi}{3}\right) \cong \cos\left(\frac{\pi}{4}\right) - \sin\left(\frac{\pi}{4}\right)\left(\frac{\pi}{12}\right) = 0.52198666$$

$$\in_t = -4.4\%$$

$$f''(x) = -\cos x$$

$$f\left(\frac{\pi}{3}\right) \cong \cos\left(\frac{\pi}{4}\right) - \sin\left(\frac{\pi}{4}\right)\left(\frac{\pi}{12}\right) - \frac{\cos(\pi/4)}{2}\left(\frac{\pi}{12}\right)^2 = 0.49775449$$

with $\in_t = 0.449\%$.

Table 1.1 shows the Taylor's series approximation for $n = 0$ to 6.

Table 1.1

Order n	$f^{(n)}(x)$	$F(\pi/3)$	ε_t
0	cos *x*	0.70710678	– 41.4
1	– sin *x*	0.52198666	– 4.4
2	– cos *x*	0.49775449	0.449
3	sin *x*	0.49986915	2.62×10^{-2}
4	cos *x*	0.50000755	-1.51×10^{-3}
5	– sin *x*	0.50000030	-6.08×10^{-5}
6	– cos *x*	0.49999999	2.40×10^{-6}

1.4 NUMBER REPRESENTATION

A *base-b number* is made up of individual *digits*. In *positional numbering system*, the position of a digit in the number determines that digit's contribution to the total value of the number.

For *decimal numbers*, the base (radix) is 10. Hence $(a_n\, a_{n-1} \ldots a_2\, a_1\, a_0)_b = a_n\, b^n + a_{n-1}\, b^{n-1} + \ldots + a_2\, b^2 + a_1 b + a_0 \cdot b^0$ contributes to the number's magnitude and is called the *most significant digit* (MSD). Similarly, the right most digit, a_0, contributes the least and is known as the *least significant digit* (LSD). Conversion of base-*b* fractions to base-10 is done by $(0 \cdot a_1\, a_2 \ldots a_m)_b = a_1 b^{-1} + a_2 b^{-2} + \ldots + a_m b^{-m}$. This is known as the *expansion method.*

There are two binary digits (bits) in the *binary number system*: *zero* and *one*. The left most bit is called the *most significant bit* (MSB) and the right most bit is the *least significant bit* (LSB). The rules of bit additions are: 0 + 0 = 0; 0 + 1 = 1; 1 + 0 = 1; 1 + 1 = 0 carry 1. The first ten digits 1, 2, 3, …, 10 in base 10 and their representation in base-2 are shown in Fig. 1.1.

Base 10	*Base 2*			
	2^3	2^2	2^1	2^0
1	0	0	0	1
2	0	0	1	0
3	0	0	1	1
4	0	1	0	0
5	0	1	0	1
6	0	1	1	0
7	0	1	1	1
8	1	0	0	0
9	1	0	0	1
10	1	0	1	0

Fig. 1.1 Representation of numbers in decimal and binary forms.

Most computer languages use *floating-point arithmetic*. Every number is represented using a (fixed, finite) number of binary digits, called *bits*. Each binary digit is referred to as a bit. In this method, the computer represents a number in the following form:

$$\text{Number} = \sigma\, m b^{t-p} \tag{1.14}$$

where σ = sign of the number (±), denoted by a single bit.

m = mantissa or a fraction (a value which lies between 0.1 and 1).

b = the base of the internal number system ($b = 2$ for binary, $b = 10$ for decimal or $b = 16$ for hexadecimal computers).

t = shifted exponent (the value that is actually stored).

p = shift required to recover the actual exponent. Shifting in the exponent is normally done to avoid the need for a sign bit in the exponent itself.

The number is, then, stored by storing only the values of σ, m and t. The normal way to represent and store numbers is to use a binary or base 2 number system which contains the following two digits.

$$\text{binary digits} = \{0\ 1\} \tag{1.15}$$

For positive integers, the binary form is

$$d_n 2^n + d_{n-1} 2^{n-1} + \ldots + d_1 2^1 + d_0 2^0 \tag{1.16}$$

While for positive numbers, less than one, it is

$$d_{-1} 2^{-1} + d_{-2} 2^{-2} + d_{-3} 2^{-3} + \ldots \tag{1.17}$$

with all binary digits d_i either 0 or 1. Such representations are unique.

Conversion between base 10 and base 2 is performed automatically by programming languages. Thus, conversion of an n-bit binary integer $b = b_{n-1} \ldots b_0$ to its decimal equivalent x is done as a sum of n powers of 2:

$$x = \sum_{k=0}^{n-1} b_k 2^k \tag{1.18}$$

A positive decimal integer x, in the range 0 to $2^n - 1$ is converted to its n-bit binary equivalent $b = b_{n-1} \ldots b_0$ by conducting a sequence of n divisions by decreasing powers of 2. In other words, the digits of the binary numbers are computed starting with the most significant bit, b_{n-1}, and ending with the least significant, b_0.

Noting that the hexadecimal numbers have a larger base or radix than decimal numbers, the first six letters of the alphabet are used to augment the decimal digits as follows:

$$\text{Hexadecimal digits} = \{0, 1, 2, 3, 4, 5, 6, 7, 8, 9, A, B, C, D, E, F\} \tag{1.19}$$

The conversion between binary, decimal and hexadecimal numbers can be accomplished using Table 1.2.

Table 1.2 Binary, decimal and hexadecimal numbers

Binary	*Decimal*	*Hexadecimal*	*Binary*	*Decimal*	*Hexadecimal*
0000	00	0	1000	08	8
0001	01	1	1001	09	9
0010	02	2	1010	10	*A*
0011	03	3	1011	11	*B*
0100	04	4	1100	12	*C*
0101	05	5	1101	13	*D*
0110	06	6	1110	14	*E*
0111	07	7	1111	15	*F*

Example 1.2: *Determine the decimal values of the following numbers:*

(*a*) $x = (10010110)_2$

(*b*) $x = (777)_8$

Solution:

(*a*)
$$x = \sum_{k=0}^{7} b_k 2^k \text{ using equation (1.18)}$$
$$= 2^1 + 2^2 + 2^4 + 2^7$$
$$= 2 + 4 + 16 + 128 = 150$$

(*b*)
$$x = (777)_8$$
$$x = \sum_{k=0}^{2} b_k 8^k = 7(8)^0 + 7(8)^1 + 7(8)^2 = 7 + 56 + 448 = 511$$

Example 1.3: *Convert* $(1011)_2$ *to base-10.*

Solution: $(1)\,(2)^3 + (0)\,(2)^2 + (1)\,(2)^1 + 1 = 11$

The remainder method is used to convert base-10 numbers to base-*b* numbers. Converting a base-10 fraction to base-*b* requires multiplication of the base-10 fraction and subsequent fractional parts by the base. The base-*b* fraction is formed from the integer parts of the products taken into same order in which they were determined.

The *octal (base-8) numbering system* is one of the alternatives to working with long binary numbers. Only the digits 0 to 7 are employed. For instance,

$$7 + 1 = 6 + 2 = 5 + 3 = (10)_8$$
$$7 + 2 = 6 + 3 = 5 + 4 = (11)_8$$
$$7 + 3 = 6 + 4 = 5 + 5 = (12)_8$$

Example 1.4: *Perform the following operations:*

(*a*) $(7)_8 + (6)_8$

(*b*) *Convert* $(0.14)_{10}$ *to base-8*

(*c*) *Convert* $(27.52)_8$ *to base-10.*

Solution:

(*a*) The sum of 7 and 6 in base-10 is 13. This is greater than 8. Using the remainder method, we have

$$13/8 = 1 \text{ remainder } 5$$
$$1/8 = 0 \text{ remainder } 1$$

The answer is $(15)_8$.

(*b*)
$$0.14 \times 8 = 1.12$$
$$0.12 \times 8 = 0.96$$
$$0.96 \times 8 = 7.68$$
$$0.68 \times 8 = 5.44$$
$$0.44 \times 8 = \text{etc.}$$

The answer is $(0.1075....)_8$ which is obtained from the integer parts of the products above.

(*c*) $(2)\,(8)^1 + (7)\,(8)^0 + (5)\,(8)^{-1} + (2)\,(8)^{-2} = 16 + 7 + \frac{5}{8} + \frac{2}{64} = (23.656)_{10}$

The *hexadecimal (base-16) system* is a shorthand way of representing the value of four binary digits at a time.

Example 1.5:

(*a*) *Convert $(1475)_{10}$ to base-16.*

(*b*) *Convert $(0.8)_{10}$ to base-16.*

Solution:

(*a*) Using the remainder method

$$\frac{1475}{16} = 92 \text{ remainder } 3$$

$$\frac{92}{16} = 5 \text{ remainder } 12$$

$$\frac{5}{16} = 0 \text{ remainder } 5$$

Now, $(12)_{10}$ is $(C)_{16}$ or (hex C).

Hence, the answer is $(5C_3)_{16}$.

(*b*)
$$0.8 \times 16 = 12.8$$
$$0.8 \times 16 = 12.8$$
$$0.8 \times 16 = \text{etc.}$$

Since $(12)_{10} = (C)_{16}$, we have the answer as $(0.CCCCC...)_{16}$.

Example 1.6:

(*a*) *Convert $(5431)_8$ to base-2.*

(*b*) *Convert $(1011111101111001)_2$ to base-16.*

Solution:

(*a*) First convert each octal digit to binary digits.

$$(5)_8 = (101)_2$$
$$(4)_8 = (100)_2$$
$$(3)_8 = (011)_2$$
$$(1)_8 = (001)_2$$

Hence, the answer is $(101100011001)_2$.

(*b*) Grouping the bits into fours starting at right-hand-bit, we have 1011, 1111, 0111 and 1001. Converting these groups into their hexadecimal equivalents, we get

$$(1011)_2 = (B)_{16}$$
$$(1111)_2 = (F)_{16}$$
$$(0111)_2 = (7)_{16}$$
$$(1001)_2 = (9)_{16}$$

Therefore, the answer is $(BF\ 79)_{16}$.

Example 1.7:

(*a*) *Convert the following base-2 numbers to base-10: 1011001 and 110.00101*

(*b*) *Convert the following base-8 numbers to base 10: 71563 and 3.14.*

Solution:

(*a*)
$$\begin{aligned}(1011001)_2 &= (1\times 2^6) + (0\times 2^5) + (1\times 2^4) + (1\times 2^3) + (0\times 2^2) + (0\times 2^1) + (1\times 2^0)\\ &= 1(64) + 0(32) + 1(16) + 1(8) + 0(4) + 0(2) + 1(1) = 89\\ (110.00101)_2 &= (1\times 2^2) + (1\times 2^1) + (0\times 2^0) + (0\times 2^{-1}) + (0\times 2^{-2})\\ &\quad + (1\times 2^{-3}) + (0\times 2^{-4}) + (1\times 2^{-5})\\ &= 1(4) + 1(2) + 0(1) + 0(0.5) + 0(0.25) + 1(0.125) + 0(0.0625) + (.03125)\\ &= 6.15625\end{aligned}$$

(*b*)
$$\begin{aligned}(71563)_8 &= (7\times 8^4) + (1\times 8^3) + (5\times 8^2) + (6\times 8^1) + (3\times 2^0)\\ &= 7(4096) + 1(512) + 5(64) + 6(8) + 3(1) = 29{,}555\\ (3.14)_8 &= (3\times 8^0) + (3\times 8^{-1}) + (4\times 8^{-2})\\ &= 3(1) + 1(0.125) + 4(0.015625) = 3.1875\end{aligned}$$

1.5 ERROR CONSIDERATIONS

Sources of Errors: When a computational procedure is involved in solving a scientific-mathematical problem, errors often will be involved in the process. A rough classification of the kinds of original errors that might occur is as follows:

Modeling Errors: *Mathematical modeling* is a process when mathematical equations are used to represent a physical system. This modeling introduces errors and is called *modeling errors*.

Blunders and Mistakes: Blunders occur at any stage of the mathematical modeling process and consist of all other components of error. Blunders can be avoided by sound knowledge of fundamental principles, with taking proper care in approach and design to a solution. Mistakes are due to the programming errors.

Machine Representation and Arithmetic Errors: These errors are inevitable when using floating-point arithmetic, when using computers or calculators. Examples are rounding and chopping errors.

Mathematical Approximation Errors: This error is also known as a *truncation error* or *discretization error*. These errors arise when an approximate formulation is made to a problem that otherwise cannot be solved exactly.

Accuracy and Precision: *Accuracy* refers to how closely a computed or measured value agrees with the true value. *Precision* refers to how closely individual computed or measured values agree with each other. *Inaccuracy* (also known as *bias*) is the systematic deviation from the truth. *Imprecision* (uncertainty) refers to the magnitude of the scatter. These concepts are illustrated graphically using an analogy from target practice as shown in Fig. 1.2.

Fig. 1.2 illustrates the concepts of accuracy and precision from marksmanship example (*a*) inaccurate and imprecise, (*b*) accurate and imprecise, (*c*) inaccurate and precise and (*d*) accurate and precise.

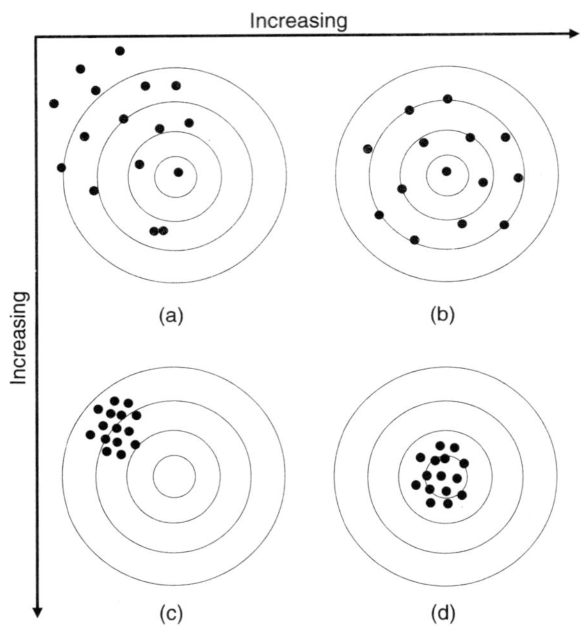

Fig. 1.2 Concepts of accuracy and precisions.

Errors are introduced by the computational process itself. Computers perform mathematical operations with only a finite number of digits. If the number x_a is an approximation to the exact result x_e, then the difference $x_e - x_a$ is called *error*. Hence

Exact value = approximate value + error

In numerical computations, we come across the following types of errors:

(*a*) Absolute and relative errors

(*b*) Inherent errors

(*c*) Round-off errors

(*d*) Truncation errors

1.5.1 Absolute and Relative Errors

If X_E is the exact or true value of a quantity and X_A is its approximate value, then $|X_E - X_A|$ is called the *absolute error* E_a. Therefore, absolute error

$$E_a = |X_E - X_A| \tag{1.20}$$

and relative error is defined by

$$E_r = \left| \frac{X_E - X_A}{X_E} \right| \tag{1.21}$$

provided $X_E \neq 0$ or X_E is not too close to zero. The percentage relative error is

$$E_p = 100E_r = 100\left| \frac{X_E - X_A}{X_E} \right| \tag{1.22}$$

Significant digits:

The concept of a significant figure, or digit, has been developed to formally define the reliability of a numerical value. The *significant digits* of a number are those that can be used with confidence.

If X_E is the exact or true value and X_A is an approximation to X_E, then X_A is said to approximate X_E to t significant digits if t is the largest non-negative integer for which

$$\frac{X_E - X_A}{|X_E|} < 5 \times 10^{-t} \tag{1.23}$$

Example 1.8: *If $X_E = e$ (base of the natural algorithm = 2.7182818) is approximated by X_A = 2.71828, what is the significant number of digits to which X_A approximates X_E?*

Solution:

$$\frac{X_E - X_A}{|X_E|} = \frac{e - 2.71828}{e} \text{ which is} < 5 \times 10^{-6}$$

Hence, X_A approximates X_E to 6 significant digits.

Example 1.9: *Let the exact or true value = 20/3 and the approximate value = 6.666.*

The absolute error is 0.000666... = 2/3000.

The relative error is (2/3000)/ (20/3) = 1/10000.

The number of significant digits is 4.

1.5.2 Inherent Errors

Inherent errors are the errors that pre-exist in the problem statement itself before its solution is obtained. Inherent errors exist because of the data being approximate or due to the limitations of the calculations using digital computers. Inherent errors cannot be completely eliminated but can be minimized if we select better data or by employing high precision computer computations.

1.5.3 Round-off Errors

Round-off error is due to the inaccuracies that arise due to a finite number of digits of precision used to represent numbers. All computers represent numbers, except for integer and some fractions, with imprecision. Digital computers use floating-point numbers of fixed word length. This type of representation will not express the exact or true values correctly. Error introduced by the omission of significant figures due to computer imperfection is called the *round-off error.*

Round-off errors are avoidable in most of the computations. When n digits are used to represent a real number, then one method is to keep the first n digits and *chop off* all remaining digits. Another method is to *round* to the nth digit by examining the values of the remaining digits. The two steps involved in rounding to n digits are as follows:

1. Add $\text{sgn}(x)\,\frac{b}{2}$ to digit $n + 1$ of x.

2. Chop x to n digits.

where x is the nonzero real number, b is the base and $\text{sgn}(x) = x/|\,x\,|$ denotes the sign of x with sgn $(0) \approx 0$. Thus, the effect of the add and chop method of rounding is to round digit n up (away from zero) if the first digit to be chopped, digit $n + 1$, is greater than or equal to $b/2$, otherwise digit n is left as it is. Errors which result from this process of chopping or rounding method are known as *round-off errors*.

Rounding to k decimal places

To round x, a positive decimal number, to k decimal places, we chop $x + 0.5 \times 10^{-5}$ after k^{th} decimal digit. Similarly, to round a negative number, we round its absolute value and then restore the sign. Table 1.3 illustrates the rounding of the numbers 234.0065792 and –234.00654983 to k decimal digits.

Table 1.3 Rounding numbers to *k* decimal digits

k	*234.0065792*	*– 234.00654983*
0	234	– 234
1	234.0	– 234.0
2	234.01	– 234.01
3	234.007	– 234.007
4	234.0065	– 234.0065
5	234.00658	– 234.00655
6	234.006579	– 234.006550
7	234.0065792	– 234.0065498
8	234.0065792	– 234.00654983

Accurate to k decimal places

When we state that Y approximates y to k decimal places provided $|\,y - Y\,| < \frac{1}{2} \times 10^{-k}$ and if both y and Y are rounded to k decimal places, then the kth decimals in the rounded versions differ by no more than one unit. Consider, for instance, the two numbers $y = 57.34$ and $Y = 57.387$ differing by $|\,y - Y\,| = 0.047 < 0.5 \times 10^{-1} = 0.05$. Hence Y approximates y to 1 decimal place. Rounding y to Y to the $k = 1$ decimal place, we find $y_r = 57.3$ and $Y_r = 57.4$, respectively. Therefore, y_r and Y_r differ in the first decimal place by no more than one unit. Also, when Y approximates y to k decimal places, then these two numbers are said to agree to k decimal places. It should be noted here that these two numbers are not necessarily the *same* when rounded to k decimal places.

Significant figures

The most significant figure in a decimal number is the leftmost nonzero digit and the least significant figure is the rightmost digit. Significant figures are all the digits that lie in between the most significant and least significant figures. However, it should be noted here that zeros on the left of the first significant figure are not treated as significant digits. For instance, in the number $Y = 0.0078560$, the first significant digit is 7 and the rightmost zero is the fifth significant digit. Table 1.4 shows the results of rounding Y to k significant figure.

Table 1.4 Significant figures

k significant digits	*Y = 0.0078560*
1	0.008
2	0.0078
3	0.00786
4	0.007856
5	0.0078560

Accurate to k significant figures

If
$$|x - X| < \frac{1}{2} \times 10^{-k} |x|$$

or
$$x - \frac{1}{2} \times 10^{-k} |x| < X < x + \frac{1}{2} \times 10^{-k} |x|$$

then we say that the floating-point number X approximates x to k significant figures. Table 1.4 shows k, the intervals $[x - d(k), x + d(k)]$, where $d(k) = \frac{1}{2} \times 10^{-k} |x|$, interval rounded, x rounded for $x = \pi = 3.141592654$. The last column in Table 1.4 shows the value of π rounded to k significant digits.

Table 1.5 Approximation of π to *K* significant figures

k	*Interval* $[x - d(k), x + d(k)]$	*Interval rounded*	*x rounded*
1	[2.984513021, 3.298672287]	[3.0, 3.3]	3.0
2	[3.125884691, 3.157300617]	[3.1, 3.2]	3.1
3	[3.140021858, 3.143163450]	[3.14, 3.14]	3.14
4	[3.141435574, 3.141749734]	[3.141, 3.142]	3.142
5	[3.141576946, 3.141608362]	[3.1416, 3.1416]	3.1416
6	[3.141591083, 3.141594225]	[3.14159, 3.14459]	3.14159
7	[3.141592497, 3.141592811]	[3.141592, 3.141593]	3.141593
8	[3.141592638, 3.141592670]	[3.1415926, 3.1415927]	3.1415927

Example 1.10: *Given the number π is approximated using n = 5 decimal digits.*

(*a*) *Determine the relative error due to chopping and express it as a percent.*

(*b*) *Determine the relative error due to rounding and express it as a percent.*

Solution:

(*a*) The relative error due to chopping is given by

$$E_r\text{ (chopping)} = \frac{3.1415 - \pi}{\pi} = 2.949 \times 10^{-5} \text{ or } 0.002949\%$$

(*b*) The relative error due to rounding is given by

$$E_r\text{ (rounding)} = \frac{3.1416 - \pi}{\pi} = 2.338 \times 10^{-6} \text{ or } 0.0002338\%.$$

Example 1.11: *If the number $\pi = 4\tan^{-1}(1)$ is approximated using 5 decimal digits, find the percentage relative error due to,*

(*a*) *chopping* (*b*) *rounding.*

Solution:

(*a*) Percentage relative error due to chopping

$$= \left(\frac{3.1415 - \pi}{\pi}\right)100 = \left(-2.949 \times 10^{-5}\right)100 \text{ or } -0.002949\%.$$

(*b*) Percentage relative error due to rounding

$$= \left(\frac{3.1416 - \pi}{\pi}\right)100 = \left(2.338 \times 10^{-6}\right)100 = 0.00023389\%.$$

Example 1.12: *Use Taylor series expansions (zero through fourth order) to predict $f(2)$ for $f(x) = \ln(x)$ with a base point at $x = 1$. Determine the true percentage relative error for each approximation.*

Solution: The true value of ln (2) = 0.693147…

Zero order:

$$f(2) \cong f(1) = 0$$

$$E_t = \left|\frac{0.693147 - 0}{0.693147}\right| 100\% = 100\%$$

First order:

$$f'(x) = \frac{1}{x} \qquad f'(1) = 1$$

$$f(2) \cong 0 + 1(1) = 1$$

$$E_t = \left|\frac{0.693147 - 1}{0.693147}\right| 100\% = 44.27\%$$

Second order:

$$f''(x) = -\frac{1}{x^2} f''(1) = -1$$

$$f(2) = 1 - 1\frac{1^2}{2} = 0.5$$

$$E_t = \left|\frac{0.693147 - 0.5}{0.693147}\right| 100\% = 27.87\%$$

Third order:

$$f^{(3)}(x) = \frac{2}{x^3} \quad f^{(3)}(1) = 2$$

$$f(2) \cong 0.5 + 2\frac{1^3}{6} = 0.833333$$

$$E_t = \left|\frac{0.693147 - 0.833333}{0.693147}\right| 100\% = 20.22\%$$

Fourth order:

$$f^{(4)}(x) = -\frac{6}{x^4} \quad f^{(4)}(1) = -6$$

$$f(2) \cong 0.833333 - 6\frac{1^4}{24} = 0.583333$$

$$E_t = \left|\frac{0.693147 - 0.583333}{0.693147}\right| 100\% = 15.84\%$$

The above results show that the series is converging at a slow rate. A smaller step size would be required to obtain more rapid convergence.

Example 1.13: *Given two numbers $a = 8.8909 \times 10^3$ and $b = 8.887 \times 10^3$. Calculate the difference between the two numbers ($a - b$) using decimal floating point approximation (scientific notation) with three significant digits in the mantissa by (a) chopping (b) rounding.*

Solution:

(*a*) In chopping, when three significant digits are in the mantissa, then

$$a = 8.890 \times 10^3 \text{ and } b = 8.887 \times 10^3$$

and $$a - b = 8.890 \times 10^3 - 8.887 \times 10^3 = 0.003 \times 10^3 = 3.$$

(*b*) In rounding, we have

$$a = 8.890 \times 10^3 \text{ and b} = 8.887 \times 10^3$$

and $$a - b = 8.890 \times 10^3 - 8.887 \times 10^3 = 0.004 \times 10^3 = 4.$$

The exact (true) difference between the numbers is 3.8 which shows that the rounding gives a value much closer to the real answer.

1.5.4 Truncation Errors

Truncation errors are defined as those errors that result from using an approximation in place of an exact mathematical procedure. Truncation error results from terminating after a finite number of terms known as *formula truncation error* or *simply truncation error*.

Let a function $f(x)$ is infinitely differentiable in an interval which includes the point $x = a$. Then the Taylor series expansion of $f(x)$ about $x = a$ is given by

$$f(x) = \sum_{k=0}^{\infty} \frac{f^{(k)}(a)\,(x-a)^k}{k!} \tag{1.23}$$

where $f^{(k)}(a)$ denotes the kth derivative of $f(x)$ evaluated at $x = a$

or
$$f^{(k)}(a) \triangleq \frac{d^k f(x)}{dx^k}\Big]_{x=0} \tag{1.24}$$

If the series is truncated after n terms, then it is equivalent to approximating $f(x)$ with a polynomial of degree $n - 1$.

$$f_n(x) \triangleq \sum_{k=0}^{n-1} \frac{f^{(k)}(a)(x-a)^k}{k!} \tag{1.25}$$

The error in approximating $E_n(x)$ is equal to the sum of the neglected higher order terms and is often called the *tail* of the series. The tail is given by

$$E_n(x) \triangleq f(x) - f_n(x) = \frac{f^{(x)}(\xi)(x-a)^n}{n!} \tag{1.26}$$

It is possible sometimes to place an upper bound on the x of $E_n(x)$ depending on the nature of function $f(x)$.

If the maximum value of $|f_n(\xi)|$ over the interval $[a, x]$ is known or can be estimated, then

$$M_n(x) \triangleq \max_{a \le \xi \le x} \left[\left|f^{(n)}(\xi)\right|\right] \tag{1.27}$$

From Eqs. (1.26) and (1.27), the worst bound on the size of the truncation error can be written as

$$|E_n(x)| \le \frac{M_n(x)\,|x-a|^n}{n!} \tag{1.28}$$

If $h = x - a$, then the truncation error $E_n(x)$ is said to be of order $O(h^n)$. In other words, as $h \to 0$, $E_n(x) \to 0$ at the same rate as h^n.

Hence
$$O(h^n) \approx ch^n \; |h| << 1 \tag{1.29}$$

where c is a nonzero constant.

The *total numerical error* is the summation of the truncation and round-off errors. The best way to minimize round-off errors is to increase the number of significant figures of the computer. It should be noted here that round-off error increases due to subtractive cancellation or due to an increase in the number of computations in an analysis. The truncation error can be reduced by decreasing the step size. In general, the truncation errors are *decreased* as the round-off errors are *increased* in numerical differentiation.

There exists no systematic and general approaches in evaluating numerical errors for all problems. In most cases, error estimates are based on experience and judgment of the engineer or scientist.

Model errors relate to bias that can be ascribed to incomplete mathematical models. Errors also enter into the analysis due to uncertainty in the physical data on which a model is based.

Example 1.14: *Given the trigonometric function $f(x) = \sin x$,*

(*a*) *expand $f(x)$ about $x = 0$ using Taylor series*

(*b*) *truncate the series to $n = 6$ terms*

(*c*) *find the relative error at $x = \pi/4$ due to truncation in (b)*

(*d*) *determine the upper bound on the magnitude of the relative error at $x = \pi/4$ and express it as a percent.*

Solution:

(*a*) Using Eq. (1.23), the Taylor series expansion is given by

$$f(x) = \sum_{k=0}^{\infty} \frac{f^{(k)}(a)(x-a)^k}{k!} = x - \frac{x^3}{3!} + \frac{x^5}{5!} - \frac{x^7}{7!} + \cdots$$

(*b*) Truncation of the Taylor series to $n = 6$ terms.

$$f_6(x) = x - \frac{x^3}{3!} + \frac{x^5}{5!}$$

(*c*) The relative error at $x = \pi/4$ due to truncation in (*b*) is given by

$$E_{r6} = \frac{f_6(\pi/4) - \sin(\pi/4)}{\sin(\pi/4)}$$

$$= \frac{\frac{\pi}{4} - \left(\frac{\pi}{4}\right)^3 \Big/ 3! + (\pi/4)^5/5! - \sin(\pi/4)}{\sin(\pi/4)} = 5.129 \times 10^{-5} \text{ or } 0.005129\%$$

(*d*) Here $f^6(x) = -\sin x$. From Eq. (1.27) using $a = 0$ and $x = \pi/4$, we obtain

$$M_6(x) \le \sin(\pi/4)$$

Now from Eq. (1.28), we have the upper bound on the truncation error given by

$$|E_6| \le \frac{\sin(\pi/4)(\pi/4)^6}{6!} = 2.305 \times 10^{-4} \text{ or } 0.02305\%$$

1.5.5 Machine Epsilon

Digital computers are *fixed-precision* devices and the number of digits the device can manipulate depends on its hardware configuration. *Machine epsilon*, $\in_M$, is the smallest positive number that the device can add to 1 while recognizing the sum as different than 1.

$\in_M$ is determined computationally by finding the smallest positive $\in$ for which $1 + \in \neq 1$. For instance, if a particular computing device computes 1.000000001 for $1 + 10^{-9}$ but 1 for $1 + 10^{-10}$, then we conclude that $10^{-10} < \in_M < 10^{-9}$ and the device in this case would be known as a *10 significant-digit device*.

1.5.6 Error Propagation

Table 1.6 summarises the errors attributed to the *round-off errors* due to the limited number of digits using fixed-precision devices. In order to illustrate these errors, we consider the following numbers: $a = 237.6581$, $b = 238.2389$, $c = 0.014789$, $d = 137469$ and $A = 238.0$, $B = 238.2$, $C = 0.01480$ and $D = 1.375 \times 10^5$.

Table 1.6 Possible types of round-off errors on a finite-precision computing device

S.No.	*Error*	*Comments*
1	Negligible addition	When two numbers of notable different magnitudes are added or subtracted, then the result rounds to the largest number.
2	Creeping round-off	Repeated rounding to k significant digits will result in accumulation of errors.
3	Error magnification	Occurs when an erroneous number is multiplied/divided by a number of large/small magnitude.
4	Subtractive cancellation	Due to the subtraction of two nearly equal numbers where the difference lies in significant digits well beyond the devices' capacity to record it.

Tables 1.7 and 1.8 show the five exact arithmetic calculations (answers rounded-off to four significant digits) and the same calculations performed on a device with four significant digits respectively.

1.6 ERROR ESTIMATION

Few computer methods are available to provide error estimates. These methods are briefly mentioned here.

1. Double precision method: In this method, the problem is solved twice, once in single precision and then in double precision. The estimate on the round-off error is then simply given by the difference between the two results obtained.
2. Interval arithmetic method: Each number in this method is represented by two machine numbers corresponding to the estimated maximum and minimum values. Two solutions are obtained at every step corresponding to the maximum and minimum values. The true solution is assumed to lie in about the center of the range. The range here is the difference between the solutions corresponding to the maximum and minimum values.
3. Significant digit arithmetic method: In this method, the digits lost due to the subtraction of two nearly equal machine numbers are tracked. Only the significant digits in a number are kept and the rest are rejected or ignored. In this way, all digits retained or kept are assumed to be significant. The results obtained with this method are considered to be very conservative.

4. Statistical approach: This method starts with the assumption that the round-off error is independent. A stochastic model for the propagation of round-off errors is then adapted in which the local errors are considered as random variables. The local round-off errors are assumed to be either uniformly or normally distributed between their extreme values. Using standard statistical analysis methods, the standard deviation, the variance, and the accumulated round-off error are estimated.
5. Backward error analysis: In this method, based on the result of a computation, the possible range of input data that could have produced it, is determined. If the result found with this approach is consistent with the input data, within the range of observational or round-off error, then some confidence is placed on the result. If this does not happen, then a major source of error is assumed to exist somewhere else, presumably within the algorithm itself.
6. Forward error analysis: The method can be illustrated by means of an example.
   Suppose the value of $A(B + C)$ is to be computed when a, b and c are the approximations to A, B, and C respectively, and the respective error amounts are e_1, e_2 and e_3.
   The true value is

$$A(B + C) = (a + e_1)(b + e_2 + c + e_3) = ab + ac + \text{error}$$

where error

$$= a(e_2 + e_3) + be_1 + ce_1 + e_1e_2 + e_1e_3$$

Now assuming the uniform bound $|e_i| \leqq e$ and that error products can be ignored, we get

$$|\,\text{error}\,| \le [2|\,a\,| + |\,b\,| + |\,c\,|]\,e$$

This procedure can be carried out for any algorithm. It is a tedious analysis. The resulting bounds are generally very conservative.

1.7 GENERAL ERROR FORMULA

1.7.1 Function Approximation

Consider the function

$$F = f(x_1, x_2, x_3, \ldots, x_n) \tag{1.30}$$

where $x_1, x_2, x_3, \ldots, x_n$ are variables.

Suppose Δx_i represents error in each x_i, so that the error in F is

$$F + \Delta F = f(x_1 + \Delta x_1, x_2 + \Delta x_2, \ldots, x_n + \Delta x_n) \tag{1.31}$$

Taylor's series expansion of the right hand side of Eq. (1.31) gives

$$F + \Delta F = f(x_1, x_2, \ldots, x_n) + \sum_{i=1}^{n} \frac{\partial f}{\partial x_i} \Delta x_i + O(\Delta x_i^2) \tag{1.32}$$

If we assume the errors in x_i as small, and $\dfrac{\Delta x_i}{x_i} \ll 1$, so that the second and higher powers of Δx_i can be ignored, Eq. (1.32) gives

$$\Delta F = \sum_{i=1}^{n} \frac{\partial f}{\partial x_i} \Delta x_i = \frac{\partial f}{\partial x_1} \Delta x_1 + \frac{\partial f}{\partial x_2} \Delta x_2 + \cdots + \frac{\partial f}{\partial x_n} \Delta x_n \tag{1.33}$$

The relative error E_r is then given by

$$E_r = \frac{\Delta f}{f} = \frac{\partial f}{\partial x_1}\frac{\partial x_1}{f} + \frac{\partial f}{\partial x_2}\frac{\partial x_2}{f} + \cdots + \frac{\partial f}{\partial x_n}\frac{\partial x_n}{f} \tag{1.34}$$

Replacing the function $f(h)$ with its approximation $\phi(h)$ and denoting the known error found as $\mu(h^n)$, where n is a positive integer, we have

$$| f(h) - \phi(h) | \leq \mu | h^n | \text{ for small } h \tag{1.35}$$

Thus, $\phi(h)$ approximates $f(h)$ with order of approximation $O(h^n)$ and we can write

$$f(h) = \phi(h) + O(h^n) \tag{1.36}$$

Example 1.15: *Determine the maximum relative error for the function*

$$F = 3x^2y^2 + 5y^2z^2 - 7x^2z^2 + 38$$

For $x = y = z = 1$ and $\Delta x = -0.05$, $\Delta y = 0.001$ and $\Delta z = 0.02$.

Solution:

$$F = 3x^2y^2 + 5y^2z^2 - 7x^2z^2 + 38$$

$$\frac{\partial F}{\partial x} = 6xy^2 - 14\,xz^2$$

$$\frac{\partial F}{\partial y} = 6x^2y + 10yz^2$$

$$\frac{\partial F}{\partial z} = 10y^2z - 14\,x^2z$$

$$(\Delta F)_{\max} = \left|\frac{\partial F}{\partial x}\Delta x\right| + \left|\frac{\partial F}{\partial y}\Delta y\right| + \left|\frac{\partial F}{\partial z}\Delta z\right|$$

$$= |(6xy^2 - 14\,xz^2)\,\Delta x| + |(6x^2y + 10yz^2)\Delta y| + |(10y^2z - 14\,x^2z)\Delta z| = 0.496$$

For $x = y = z = 1$ and $\Delta x = -0.05$, $\Delta y = 0.001$ and $\Delta z = 0.02$, we have the maximum relative error given by Eq. (1.34)

$$(E_r)_{\max} = \frac{(\Delta F)_{\max}}{F} = \frac{0.496}{39} = 0.01272$$

1.7.2 Stability and Condition

A numerical computation is said to be *numerically unstable* if the uncertainty of the input values is grossly magnified by numerical method employed.

Consider the first-order Taylor's series of a function given by

$$f(x) = f(a) + f'(a)\,(x - a) \tag{1.37}$$

The *relative error* of $f(x)$ then becomes

$$\frac{f(x)-f(a)}{f(x)} \cong \frac{f'(a)(x-a)}{f(a)} \tag{1.38}$$

The *relative error* of x becomes

$$\frac{x-a}{a} \tag{1.39}$$

A *condition number* is often defined as the ratio of the relative errors given by Eqs. (1.38) and (1.39) as

$$\text{Condition number} = \frac{a\,f'(a)}{f(a)} \tag{1.40}$$

The condition number given by Eq. (1.40) indicates the extent to which an uncertainty in x is magnified by $f(x)$.

$$\begin{aligned}&\text{Condition number} = 1 \text{ (function's relative error = relative error in } x)\\&\text{Condition number} > 1 \text{ (relative error is amplified)}\\&\text{Condition number} < 1 \text{ (relative error is attenuated)}\\&\text{Condition number} > \text{very large number (the function is ill-conditioned)}\end{aligned} \tag{1.41}$$

Example 1.16: *Compute and interpret the condition number for*

(a) $f(x) = \sin x$ *for* $a = 0.51\pi$

(b) $f(x) = \tan x$ *for* $a = 1.7$

Solution:

(*a*) The condition number is given by

$$\text{Condition number} = \frac{a\,f'(a)}{f(a)}$$

for $$a = 0.51\pi,\ f'(a) = \cos(0.51\,\pi) = -0.03141,\ f(a) = \sin(0.51\,\pi) = 0.99951$$

$$\text{Condition number} = \frac{a\,f'(a)}{f(a)} = \frac{(0.51\pi)(-0.03141)}{(0.99951)} = -0.05035$$

Since the condition number is < 1, from Eq. (1.41), we conclude that the relative error is attenuated.

(*b*) $$f(x) = \tan x,\ f'(a) = -7.6966 \text{ for } a = 1.7$$
$$f'(x) = 1/\cos^2 x,\ f'(a) = 1/\cos^2(1.7) = 60.2377$$

$$\text{Condition number} = \frac{a\,f'(a)}{f(a)} = \frac{1.7(60.2377)}{-7.6966} = -13.305$$

Thus, the function is ill-conditioned.

1.7.3 Uncertainty in Data or Noise

Uncertainty or error in the physical data on which the computation model is based can introduce errors in the analysis. This type of error is known as *noise*. The data can affect the accuracy of the numerical computations performed. The errors can exhibit both inaccuracy and imprecision. If the input data has d significant digits of accuracy, then the results obtained from the numerical computation should be reported in d significant digits of accuracy. For instance, if $a = 5.358$ and $b = 0.06437$ both have 4 significant digits of accuracy, then although $a - b = 5.29363$, we should report the correct answer as $a - b = 5.293$.

Table 1.7 Exact arithmetic rounded to four significant digits

No.	*Exact arithmetic*	*Rounded to four significant digits*
1.	$a - c = 237.6581 - 0.014897 = 237.643203$	237.6
2.	$b + d = 237.8389 + 137476 = 1377103.8389$	1.377×10^5
3.	$bd = (237.8389)(137476) = 32697140.62$	3.270×10^7
4.	$a/c = 237.6581/0.014897 = 15953.42015$	1.595×10^4
5.	$a - b = 237.6581 - 237.8389 = -0.1808$	-0.1808

Table 1.8 Calculations with a device carrying four significant digits

No.	*Calculations*
1.	$A - C = 238.0 - 0.01480 = 238.0$
2.	$B + D = 238.1 + 1.375 \times 10^5 = 1.377 \times 10^5$
3.	$BD = (238.1)(1.375 \times 10^5) = 3.274 \times 10^7$
4.	$A/C = 238.0/0.01480 = 1.608 \times 10^4$
5.	$A - B = 238.0 - 238.1 = -0.1$

The following observations can be made from the results in Tables 1.7 and 1.8.

(*a*) Negligible addition: Round-off error has crept into the fourth significant digit when we compare $a - c$ (rounded) to $A - C$.

There is a difference in the fourth significant digit when $b + d$ (rounded) is compared to $B + D$.

(*b*) Error magnification: Comparing a/c (rounded) to A/C we find a difference in the fourth significant digit when bd (rounded) and BD are compared. The two answers differ substantially by –40000.

(*c*) Creeping round-off: In the calculations of $a - c$, bd, a/c, and $b + d$ we find the result of working in four significant digits as opposed to working "exactly" and then rounding would lead to a loss of precision in the fourth significant digit. These calculations show the creeping round-off that is the gradual loss of precision as repeated rounding errors accumulate.

(*d*) Subtractive calculations: Comparing $a - b = -0.1808$ to $A - B = -0.1$, we find significant error introduced by working in fixed-precision arithmetic.

1.8 UNCERTAINTY IN MEASUREMENTS [Ref: Kline and McClintock, 1953]

Before proceeding, let us define certain terms more carefully. By *uncertainty* we mean a possible value the error might have. For a single observation, the error, which is the difference between the result and observed values, is a certain fixed number. But the uncertainty, or what one thinks the error might be, may vary considerably depending upon the particular circumstances of the observation. *Variable* will mean a basic quantity observed directly in the laboratory as opposed to the *result* which is obtained by making corrections to or calculations with the recorded values of the variables. The recorded values of the variables will be called *data*. In a few cases, of course, the results will be the same as the data. *Propagation of uncertainty* will mean the way in which uncertainties in the variables affect the uncertainty in the results.

A more precise method of estimating uncertainty in experimental results has been presented by Kline and McClintock [1953]. The method is based on a careful specification of the uncertainties in the various primary experimental measurements.

1.8.1 Uncertainty Interval in the Result

Let the result R be a function of n independent variables, $v_1, v_2, \ldots, v_n$

$$R = R\,(v_1, v_2, \ldots, v_n) \tag{1.42}$$

For small variations in the variables, this relation can be expressed in linear form as

$$\delta R = \frac{\partial R}{\partial v_1}\delta v_1 + \frac{\partial R}{\partial v_2}\delta v_2 + \ldots + \frac{\partial R}{\partial v_n}\delta v_n \tag{1.43}$$

The uncertainties in the variables v_i are represented completely by an uncertainty distribution but can be adequately described by uncertainty intervals w_i based on certain odds. Therefore, we must examine how to find the uncertainty interval for the results w_R based on essentially the same odds as the intervals for each of the variables. Certain theorems of statics concerning the way in which frequency distributions combine will be helpful in finding a reasonable value for w_R.

Theorem 1: If R is a linear function of n independent variables, and if the maximum deviation of the ith variable from its mean is $(\pm\delta v_i)_{max}$ then the maximum deviation of R from its mean value is given by

$$\delta R_{max} = \left|\frac{\partial R}{\partial v_1}\delta v_{1_{max}}\right| + \left|\frac{\partial R}{\partial v_2}\delta v_{2_{max}}\right| + \ldots + \left|\frac{\partial R}{\partial v_n}\delta v_{n_{max}}\right| \tag{1.44}$$

Equation (1.44) might be used as an approximation for calculating the uncertainty interval in the result by simply substituting w_i for v_i. This yields

$$w_R = \left|\frac{\partial R}{\partial v_1}w_1\right| + \left|\frac{\partial R}{\partial v_2}w_2\right| + \ldots + \left|\frac{\partial R}{\partial v_n}w_n\right| \tag{1.45}$$

This equation will be referred to as the linear equation. If it is employed, the odds on the uncertainty interval in the result will be much higher than the odds used in the variables. This is because of the fact that the errors in each variable can be having a range of values, and it is quite unlikely that all of them will have the most adverse values at the same time.

Theorem 2: If R is a linear function of n independent variables and each of which is distributed with a standard deviation σ_i, then the standard deviation of R is given by

$$\sigma_R = \left[\left(\frac{\partial R}{\partial v_1}\right)^2 \sigma_1^2 + \left(\frac{\partial R}{\partial v_2}\right)^2 \sigma_2^2 + ... + \left(\frac{\partial R}{\partial v_n}\right)^2 \sigma_n^2\right]^{1/2} \tag{1.46}$$

We have seen, however, that the best measure of uncertainty is neither the maximum value nor the standard deviation, but some interval based on certain odds. For the special case in which the variables are distributed normally, the distribution of the result also will be normal, and the following theorem applies:

Theorem 3: If R is a linear function of n independent variables and each of which is normally distributed, then the relation between the interval for the variables w_i, and the interval for the result w_R, which gives the same odds for each of the variables and for the result is

$$w_R = \left[\left(\frac{\partial R}{\partial v_1} w_1\right)^2 + \left(\frac{\partial R}{\partial v_2} w_2\right)^2 + ... + \left(\frac{\partial R}{\partial v_n} w_n\right)^2\right]^{1/2} \tag{1.47}$$

Equation (1.47) might be used directly as an approximation for calculating the uncertainty interval in the result. Equation (1.47) will be referred to as the second-power equation.

Suppose a set of measurements is made and the uncertainty at each measurement may be expressed with the same odds. These measurements are then used to calculate some desired result of the experiment. Suppose we like to estimate the uncertainty in the calculated result on the basis of the uncertainties in the primary measurements. The result R is a given function of the independent variables $x_1, x_2, x_3, ..., x_n$. Hence

$$R = R\,(x_1, x_2, x_3, ..., x_n)$$

Let w_R be the uncertainty in the result and $w_1, w_2, w_3, ..., w_n$ are the uncertainties in the independent variable. If the uncertainties in the independent variables are all given with the same odds, then the uncertainties in the result having these odds is given by Kline and McClintock as

$$w_R = \left[\left(\frac{\partial R}{\partial x_1} w_1\right)^2 + \left(\frac{\partial R}{\partial x_2} w_2\right)^2 + ... + \left(\frac{\partial R}{\partial x_n} w_n\right)^2\right]^{1/2} \tag{1.48}$$

1.8.2 Uncertainties for Product Functions

In majority cases, the result of function of Eq. (1.48) takes the form of a product of the respective primary variables raised to exponents and written as

$$R = x_1^{a_1} x_2^{a_2} ... x_n^{a_n} \tag{1.49}$$

Taking partial derivatives of Eq. (1.49), we obtain

$$\frac{\partial R}{\partial x_i} = x_1^{a_1} x_2^{a_2} (a_i x_1^{a_i - 1}) .... x_n^{a_n} \tag{1.50}$$

Dividing Eq. (1.50) by R from Eq. (1.49), we have

$$\frac{1}{R}\frac{\partial R}{\partial x_i} = \frac{a_i}{x_i} \tag{1.51}$$

Substituting Eq. (1.51) in Eq. (1.48), we obtain

$$\frac{w_R}{R} = \left[\sum\left(\frac{a_i w_{x_i}}{x_i}\right)^2\right]^{1/2} \tag{1.52}$$

It should be noted that Eq. (1.52) for the fractional uncertainty in the result may only be used when the result function takes the product form indicated in Eq. (1.49).

1.8.3 Uncertainties for Additive Functions

When the result function has an additive form, R can be expressed as

$$R = a_1x_1 + a_2x_2 + \ldots. + a_nx_n = \Sigma a_ix_i \tag{1.53}$$

The partial derivatives of Eq. (1.53)

$$\frac{\partial R}{\partial x_i} = a_i \tag{1.54}$$

Substituting Eq. (1.54) in Eq. (1.48), the uncertainty in the result can be expressed as

$$w_R = \left\{\sum\left[\left(\frac{\partial R}{\partial x_i}\right)^2 w_{x_i}^2\right]\right\}^{1/2} = [\Sigma(a_iw_i)^2]^{1/2} \tag{1.55}$$

Hence, Eqs. (1.52) and (1.53) can be used in combination when the result function involves both product and additive terms.

1.8.4 Propagation of Errors in Arithmetic Operations

The Errors in Sum and Difference

Consider the exact numbers $X_1, X_2, \ldots., X_n$ and their approximations respectively $x_1, x_2, \ldots., x_n$. Let $\Delta x_1, \Delta x_2, \ldots., \Delta x_n$ be the errors in $x_1, x_2, \ldots., x_n$, *i.e.*, $X_i = x_i \pm \Delta x_i$, $i = 1, 2, \ldots., n$. Also, let $X = X_1 + X_2 + \ldots. + X_n$ and $x = x_1 + x_2 + \ldots. + x_n$. Hence, the total absolute error is given by

$$|X - x| = |(X_1 - x_1) + (X_2 - x_2) + \ldots. + (X_n - x_n)|$$
$$\le |X_1 - x_1| + |X_2 - x_2| + \ldots. + |X_n - x_n|$$

Therefore, the absolute error in the sum is given by

$$|\Delta x| = |\Delta x_1| + |\Delta x_2| + \ldots. + |\Delta x_n| \tag{1.56}$$

The absolute error in sum of approximate numbers is equal to the sum of the absolute errors of the numbers.

Subtraction

Consider x_1 and x_2 are two approximate values of the corresponding exact numbers X_1 and X_2. Let $X = X_1 - X_2$ and $x = x_1 - x_2$.

Then $X_1 = x_1 \pm \Delta x_1$ and $X_2 = x_2 \pm \Delta x_2$, where Δx_1 and Δx_2 are the errors in x_1 and x_2 respectively.

or $$|X - x| = |(X_1 - x_1) - (X_2 - x_2)| \leq |X_1 - x_1| + |X_2 - x_2|$$

Hence $$|\Delta x| = |\Delta x_1| + |\Delta x_2| \tag{1.57}$$

Therefore, the absolute error in difference of two numbers is equal to the sum of individual absolute errors.

1.8.5 The Error in Product

Let us consider two exact numbers X_1 and X_2 and their approximate values x_1 and x_2. Also, let Δx_1 and Δx_2 be the errors in x_1 and x_2, *i.e.*, $X_1 = x_1 \pm \Delta x_1$ and $X_2 = x_2 \pm \Delta x_2$.

Now, $$X_1 X_2 = x_1 x_2 \pm x_1 \Delta x_2 \pm x_2 \Delta x_1 \pm \Delta x_1 \cdot \Delta x_2$$

Then $$|X_1 X_2 - x_1 x_2| \leq |x_1 \Delta x_2| + |x_2 \Delta x_1| + |\Delta x_1 \cdot \Delta x_2|$$

The last term of right hand side is small, so we discard it and dividing both sides by $|x| = |x_1 x_2|$.

Thus the relative error in the product is

$$\left|\frac{X_1 X_2 - x_1 x_2}{x_1 x_2}\right| = \left|\frac{\Delta x_2}{x_2}\right| + \left|\frac{\Delta x_1}{x_1}\right| \tag{1.58}$$

Thus, the relative errors in product of two numbers are equal to the sum of individual relative errors.

The result (1.58) can be easily extended to the product of several numbers so that, if $X = X_1 X_2 \ldots X_n$ and $x = x_1 x_2 \ldots x_n$, then

$$\left|\frac{X - x}{x}\right| = \left|\frac{\Delta x_1}{x_1}\right| + \left|\frac{\Delta x_2}{x_2}\right| + \ldots + \left|\frac{\Delta x_n}{x_n}\right| \tag{1.59}$$

Hence, the total relative error in product of n numbers is equal to the sum of individual relative errors.

A Particular Case

Let the approximate numbers $x_1, x_2, \ldots, x_n$ be all positive and $x = x_1 x_2 \ldots x_n$.

Then $$\log x = \log x_1 + \log x_2 + \ldots + \log x_n$$

or $$\frac{\Delta x}{x} = \frac{\Delta x_1}{x_1} + \frac{\Delta x_2}{x_2} + \ldots + \frac{\Delta x_n}{x_n}$$

Hence $$\left|\frac{\Delta x}{x}\right| = \left|\frac{\Delta x_1}{x_1}\right| + \left|\frac{\Delta x_2}{x_2}\right| + \ldots + \left|\frac{\Delta x_n}{x_n}\right|$$

1.8.6 The Error in Quotient

Let us consider two exact numbers X_1 and X_2 and their approximate values x_1 and x_2. Also, let $X = \frac{X_1}{X_2}$ and $x = \frac{x_1}{x_2}$.

Then $X_1 = x_1 + \Delta x_1$, $X_2 = x_2 + \Delta x_2$, where Δx_1 and Δx_2 are the errors.

Now,
$$X - x = \frac{x_1 + \Delta x_1}{x_2 + \Delta x_2} - \frac{x_1}{x_2} = \frac{x_2 \Delta x_1 - x_1 \Delta x_2}{x_2(x_2 + \Delta x_2)}$$

Dividing both sides by x and taking absolute values:

$$\left|\frac{X - x}{x}\right| = \left|\frac{x_2 \Delta x_1 - x_1 \Delta x_2}{x_1(x_2 + \Delta x_2)}\right| = \left|\frac{x_2}{x_2 + \Delta x_2}\right| \left|\frac{\Delta x_1}{x_1} - \frac{\Delta x_2}{x_2}\right|$$

The error Δx_2 is small as compared to x_2, then approximately $\frac{x_2}{x_2 + \Delta x_2} = 1$. Therefore, above relation becomes

$$\delta x = \left|\frac{\Delta x}{x}\right| = \left|\frac{X - x}{x}\right| = \left|\frac{\Delta x_1}{x_1} - \frac{\Delta x_2}{x_2}\right| \le \left|\frac{\Delta x_1}{x_1}\right| + \left|\frac{\Delta x_2}{x_2}\right| \tag{1.60}$$

i.e., $\delta x = \delta x_1 + \delta x_2$. Hence, the total relative error in quotient is equal to the sum of their individual relative errors.

The relation Eq. (1.60) can also be written as

$$\left|\frac{\Delta x}{x}\right| = \left|\frac{\Delta x_1}{x_1} - \frac{\Delta x_2}{x_2}\right| \ge \left|\frac{\Delta x_1}{x_1}\right| - \left|\frac{\Delta x_2}{x_2}\right| \tag{1.61}$$

From this relation, one can conclude that the relative error in quotient is greater than or equal to the difference of their individual relative errors.

A Particular Case

For positive approximate numbers x_1 and x_2, the Eq. (1.60) can easily be deduced. Let $x = x_1/x_2$. Then $\log x = \log x_1 - \log x_2$. Therefore,

$$\frac{\Delta x}{x} = \frac{\Delta x_1}{x_1} - \frac{\Delta x_2}{x_2} \quad i.e., \left|\frac{\Delta x}{x}\right| \le \left|\frac{\Delta x_1}{x_1}\right| + \left|\frac{\Delta x_2}{x_2}\right|$$

1.8.7 The Errors in Power and in Root

Let us consider an approximate number x_1 which has a relative error δx_1. Now, the problem is to find the relative error of $x = x_1^m$.

Then
$$x = x_1^m = \underbrace{x_1 \cdot x_1 \ldots x_1}_{m \text{ times}}$$

By (1.59), the relative error δx in the product is

$$\delta x = \underbrace{\delta x_1 + \delta x_1 + \dots + \delta x_1}_{m \text{ times}} = m\,\delta x_1 \tag{1.62}$$

Hence, when the approximate number x is raised to the power m, its relative error increases m times.

Similarly, one can calculate the relative error of the number $x = \sqrt[m]{x_1}$.

Here $$x_1 > 0$$

or $$\log x = \frac{1}{m}\log x_1$$

Thus $$\frac{\Delta x}{x} = \frac{1}{m}\frac{\Delta x_1}{x_1}$$

or $$\left|\frac{\Delta x}{x}\right| = \frac{1}{m}\left|\frac{\Delta x_1}{x_1}\right|$$

Hence, the relative error is given by

$$\delta x = \frac{1}{m}\delta x_1$$

where δx and δx_1 respectively represent the relative errors in x and x_1.

1.8.8 Error in Evaluation of a Function of Several Variables

Let $y = f(x_1, x_2, \dots, x_n)$ be a differentiable function containing n variables $x_1, x_2, \dots, x_n$ and let Δx_i be the error in x_i, for all $i = 1, 2, \dots, n$.

Then, the error Δy in y is given by

$$y + \Delta y = f(x_1 + \Delta x_1, x_2 + \Delta x_2, \dots, x_n + \Delta x_n)$$

$$= f(x_1, x_2, \dots, x_n) + \sum_{i=1}^{n}\frac{\partial f}{\partial x_i}\Delta x_i + \dots \text{ (by Taylor's series expansion)}$$

$$= y + \sum_{i=1}^{n}\frac{\partial f}{\partial x_i}\Delta x_i \text{ (neglecting second and higher powers terms of } \Delta x_i) \tag{1.63}$$

or $$\Delta y = \sum_{i=1}^{n}\frac{\partial f}{\partial x_i}\Delta x_i \tag{1.64}$$

Equation (1.64) gives the total error for computing a function containing several variables.

The relative error is given by

$$\frac{\Delta y}{y} = \sum_{i=1}^{n}\frac{\partial f}{\partial x_i}\frac{\Delta x_i}{y} \tag{1.65}$$

1.8.9 Uncertainties for Compound Data Reduction

Suppose a set of data is collected in the variables $x_1, x_2, \ldots, x_n$ and a result is calculated. At this stage, we can perturb the variables by $\Delta x_1, \Delta x_2, \ldots, \Delta x_n$ and calculate new results. Then, we will have

$$\begin{aligned} R(x_1) &= R(x_1, x_2, \ldots, x_n) \\ R(x_1 + \Delta x_1) &= R(x_1 + \Delta x_1, x_2, \ldots, x_n) \\ R(x_2) &= R(x_1, x_2, \ldots, x_n) \\ R(x_2 + \Delta x_2) &= R(x_1, x_2 + \Delta x_2, \ldots, x_n) \end{aligned} \tag{1.66}$$

For small values of Δx, the partial derivatives can be approximated by

$$\begin{aligned} \frac{\partial R}{\partial x_1} &\cong \frac{R(x_1 + \Delta x_1) - R(x_1)}{\Delta x_1} \\ \frac{\partial R}{\partial x_2} &\cong \frac{R(x_2 + \Delta x_2) - R(x_2)}{\Delta x_2} \end{aligned} \tag{1.67}$$

Now, Eqs. (1.66) and (1.67) can be substituted in Eq. (1.66) to compute the uncertainty in the result.

The procedure can be extended to evaluate the influence coefficients $\left(\frac{\partial R}{\partial x_n} w_{x_n}\right)^2$ at various points in the data reduction process. A study of the relative values of these coefficients can then indicate the data points in the experiment which contribute most to the overall uncertainty in the results. Moffat [] gave a full description of the technique.

Example 1.17: *Both tilt tale testing and the calculation of so called critical sliding velocity (CSV) have the goal of determining conditions wherein a vehicle can be tripped by sideways impact with an obstacle and roll exactly one-quarter turn. Lund and Bernard [1995] derived a formula for CSV given by*

$$CSV = \sqrt{2g(a+1)h\left(1+\left(\frac{T}{2h}\right)^2\right)\left(\sqrt{1+\left(\frac{T}{2h}\right)^2}-1\right)} \tag{E.1}$$

where a = *dimensionless constant*

g = *gravitational constant*

h = *center of gravity height*

T = *track width*

Determine the uncertainties in

(a) the track, T

(b) the center of gravity height, h

(c) the tilt table ratio TTR which is equal to $\frac{T}{2h}$.

Solution:

(*a*) and (*b*). Taking the derivative of *CSV* Eq. (E.1) with respect to track width, *T*, gives

$$\frac{dCSV}{dT}=\frac{T}{4}\sqrt{\frac{2g(a+1)}{h^3}\frac{\left(\sqrt{\left(1+\left(\frac{T}{2h}\right)^2\right)}-1\right)}{\left(\left(\frac{T}{2h}\right)^2+1\right)}}+\frac{T}{8}\sqrt{\frac{2g(a+1)}{h^3}\frac{1}{\left(\sqrt{\left(1+\left(\frac{T}{2h}\right)^2\right)}-1\right)}} \tag{E.2}$$

Using Eq. (E.1) in Eq. (E.2) and rearranging yields

$$\frac{dCSV}{CSV}=\frac{\left(\frac{T}{2h}\right)^2}{\left(1+\left(\frac{T}{2h}\right)^2\right)}\frac{dT}{T}+\frac{\frac{1}{2}\left(\frac{T}{2h}\right)^2}{\left(\sqrt{\left(1+\left(\frac{T}{2h}\right)^2\right)}-1\right)\sqrt{\left(1+\left(\frac{T}{2h}\right)^2\right)}}\frac{dT}{T} \tag{E.3}$$

Taking the derivative of *CSV* Eq. (E.1) with respect to the center of gravity height, *h*, gives

$$\frac{dCSV}{dh}=-\frac{T^2}{4}\sqrt{\frac{2g(a+1)}{h^5}\frac{\left(\sqrt{\left(1+\left(\frac{T}{2h}\right)^2\right)}-1\right)}{\left(\left(\frac{T}{2h}\right)^2+1\right)}}$$

$$+\sqrt{\frac{2g(a+1)}{4h}\left(\sqrt{\left(1+\left(\frac{T}{2h}\right)^2\right)}-1\right)\left(1+\left(\frac{T}{2h}\right)^2\right)}-\frac{T^2}{8}\sqrt{\frac{2g(a+1)}{h^5}\frac{1}{\left(\sqrt{\left(1+\left(\frac{T}{2h}\right)^2\right)}-1\right)}} \tag{E.4}$$

Using Eq. (E.1) in Eq. (E.4) yields

$$\frac{dCSV}{CSV}=\frac{1}{2}\frac{dh}{h}-\frac{\left(\frac{T}{2h}\right)^2}{\left(1+\left(\frac{T}{2h}\right)^2\right)}\frac{dh}{h}-\frac{\frac{1}{2}\left(\frac{T}{2h}\right)^2}{\left(\sqrt{\left(1+\left(\frac{T}{2h}\right)^2\right)}-1\right)\sqrt{\left(1+\left(\frac{T}{2h}\right)^2\right)}}\frac{dh}{h} \tag{E.5}$$

The sensitivities indicate that changes in *T* alone have much more effect on *CSV* than changes in *h* alone. For example, for the track width sensitivity analysis, with a *T*/2*h* of 1.10, Eq. (E.3) yields

$$\frac{dCSV}{CSV} = [1.38]\frac{dT}{T} \tag{E.6}$$

Thus, one percent change in track width T yields 1.38 percent change in CSV.

For the center of gravity height sensitivity analysis, with a $T/2h$ of 1.10, Eq. (E.5) yields

$$\frac{dCSV}{CSV} = [-0.88]\frac{dh}{h} \tag{E.7}$$

Thus, one percent change in cg height h yields negative 0.88 percent change in CSV.

(c) In this case,

$$TTR = \frac{T}{2h} \tag{E.8}$$

Differentiating with respect to track, T, yields

$$\frac{d(TTR)}{dT} = \frac{1}{2h} \tag{E.9}$$

so that

$$\frac{d(TTR)}{TTR} = \frac{dT}{T} \tag{E.10}$$

Thus a one percent change in track, T, will yield a one percent change in TTR.

Differentiating TTR of a rigid block with respect to center of gravity height, h, yields

$$\frac{d(TTR)}{dh} = \frac{T}{2h^2} \tag{E.11}$$

so that

$$\frac{d(TTR)}{TTR} = -\frac{dh}{h} \tag{E.12}$$

Thus a one percent change in cg height, h, will yield a negative one percent change in TTR.

These results indicate that, for a rigid vehicle, a small percentage change in T alone or h alone would yield the same change in tilt angle.

In practice, we expect the sensitivity of TTR to h to be dependent on the way h is changed. For example, if h is changed by changing the radius of the tires, we expect results similar to Eq. (E.12). If, on the other hand, h is changed by loading (or unloading) the sprung mass, we expect additional sensitivity of TTR due to increased (or decreased) roll gain. In any case, we do not expect CSV-like results wherein changes in track are much more important than changes in cg height.

1.8.10 Combining Standard Uncertainties

A measurand may be measured indirectly through the input quantities $x_1, x_2, \ldots, x_n$ and $y = f(x_1, x_2, \ldots, x_n)$. The standard uncertainty, $u(y)$, in y resulting from standard uncertainties $u(x_1), u(x_2), \ldots, u(x_n)$ in the input quantities is calculated using the equation

$$u^2(y) = \left(\frac{\partial y}{\partial x_1}\right)^2 u^2(x_1) + \left(\frac{\partial y}{\partial x_2}\right)^2 u^2(x_2) + \ldots + \left(\frac{\partial y}{\partial x_n}\right)^2 u^2(x_n) \quad (1.68)$$

Equation (1.68) can be written as

$$u^2(y) = \sum_{i=1}^{n} \left(\frac{\partial y}{\partial x_i}\right)^2 u^2(x_i) \quad (1.69)$$

Equation (1.69) is valid only if the x_i is mutually uncorrelated.

Example 1.18: *The velocity of a wave, v, is written in terms of the frequency, f, and the wavelength λ, as*

$$v = f\lambda$$

An ultrasonic wave has f = 41 kHz with a standard uncertainty of 0.16 kHz and λ = 0.812 cm with a standard uncertainty of 0.012 cm. Assume that there is no correlation between errors in f and λ, determine the velocity of the wave and its standard uncertainty.

Solution: The velocity $v = f\lambda$ is given by $v = (41 \times 10^3)(0.812 \times 10^{-2}) = 33.92$ m/s. Now, Eq. (1.68) can be written as

$$u^2(v) = \left(\frac{\partial v}{\partial f}\right)^2 u^2(f) + \left(\frac{\partial v}{\partial \lambda}\right)^2 u^2(\lambda) \quad (E.1)$$

and

$$v = f\lambda \quad (E.2)$$

Hence

$$\frac{\partial v}{\partial f} = \lambda, \quad \frac{\partial v}{\partial \lambda} = f$$

Substituting the values in Eq. (E.1), we get

$$u^2(v) = (0.812 \times 10^{-2})^2 (0.16 \times 10^3)^2 + (41 \times 10^3)^2 (0.012 \times 10^{-2})^2 \text{ (m/s)}^2$$

or

$$u^2(v) = 25.89 \text{ (m/s)}^2$$

Hence

$$u(v) = 5.0882 \text{ m/s}$$

Example 1.19: *The moment of inertia, I, of a solid cylinder of mass, M, rotating about its principal axis, is given by I = MR²/2, where R is the radius of the cylinder. The mean of ten values of the mass measured in an experiment is 262 g and the standard uncertainty in the mean mass is 2.6 g. The mean of the five values of the radius is 7.35 cm with a standard uncertainty in the mean radius of 0.06 cm. Determine*

(*a*) *the best estimate for the moment of inertia of the cylinder*

(*b*) *the standard uncertainty in the best estimate of the moment of inertia, assuming that the errors in the mass and radius measurements are uncorrelated.*

Solution:

(*a*) The best estimate of the moment of inertia is given by

$$I = \frac{MR^2}{2} = \frac{(262)(7.35)^2}{2} = 7076.9 \text{ g.cm}^2$$

(*b*) The variance in the best estimate is given by

$$u^2(I) = c_M^2 u^2(M) + c_R^2 u^2(R) \tag{E.1}$$

where $c_M = \dfrac{\partial I}{\partial M} = \dfrac{R^2}{2}$; $c_R = \dfrac{\partial I}{\partial R} = MR$

Substituting the values in Eq. (E.1), we have

$$u^2(I) = \left(\frac{R^2}{2}\right)^2 u^2(M) + (MR)^2 u^2(R)$$

$$= \left(\frac{7.35^2}{2}\right)^2 (2.6)^2 + (262 \times 7.35)^2 (0.06)^2$$

$$= 729.61 + 13350.1\ (\text{g.cm}^2)^2 = 14080\ (\text{g.cm}^2)^2$$

Hence $u(I) = 118.66$ g.cm^2.

1.8.11 Uncertainty and Risk Analysis

In practice, decisions are based on future estimates of costs and benefits. Estimates are likely to contain uncertainties and may not be very precise. When a single estimate of a project is desired with an analysis of how the actual amount might vary from the expected value, such an analysis is known as *uncertainty analysis* and *risk analysis*. The procedure is to combine the uncertainty of the individual components to express uncertainty in the overall project. If the uncertainty in each component is assumed as normally distributed, then the variances are additive when they are weighted for their sensitivity to the overall effect.

If the estimate for a project is Z, then the variance for Z is given by;

$$\sigma_Z^2 = \left(\frac{\partial Z}{\partial A}\right)^2 \sigma_A^2 + \left(\frac{\partial Z}{\partial B}\right)^2 \sigma_B^2 + \left(\frac{\partial Z}{\partial C}\right)^2 \sigma_C^2 \tag{1.42}$$

where $\dfrac{\partial Z}{\partial A}, \dfrac{\partial Z}{\partial B}$ and $\dfrac{\partial Z}{\partial C}$: the measures of $\dfrac{\Delta Z}{\Delta A}, \dfrac{\Delta Z}{\Delta B}$ and $\dfrac{\Delta Z}{\Delta C}$ which are the changes in Z for a unit change in A, B and C respectively. They are the individual component sensitivities.

$\sigma_A, \sigma_B, \sigma_C$: the variances of A, B, and C respectively

σ_Z : variance of Z

$\left(\dfrac{\partial Z}{\partial A}\right)^2, \left(\dfrac{\partial Z}{\partial B}\right)^2$ *and* $\left(\dfrac{\partial Z}{\partial C}\right)^2$: *weighting factors for individual variances.* If only a few estimates for a component are known, then a high value and a low value, *i.e.*, the range can often be used. The standard deviation, which is the square root of the variance, is then given by

$$\sigma = \frac{\text{High value} - \text{Low value}}{8} \tag{1.43}$$

Example 1.20: *The total cost in millions of dollars for a project is given by:*

$$C = 3X^{1.4}Y^{-0.5} + 16Y + F$$

where X is the estimated from five values: 2.5, 1.7, 1.4, 2.9 and 1.3. Y is estimated from a high value of 3 and a low value of 1. F is certain with a value of 0.4. Assuming standard normal distributions, find

(*a*) *the expected value for C*

(*b*) *sensitivity of C to X and Y*

(*c*) *variation of the actual value of C from the expected value*

(*d*) *variance of C*

Solution:

(*a*) The mean value for X is

$$\frac{2.5 + 1.7 + 1.4 + 2.9 + 1.3}{5} = \frac{9.8}{5} = 1.96$$

The variance for X is given by

$$\sigma_X^2 = \frac{(2.5 - 1.96)^2 + (1.7 - 1.96)^2 + (1.4 - 1.96)^2 + (2.9 - 1.96)^2 + (1.3 - 1.96)^2}{(5 - 1)}$$

$$= \frac{(0.2916 + 0.0676 + 0.3136 + 0.8836 + 0.4356)}{4}$$

$$= 0.498$$

The mean for Y is $\frac{(3+1)}{2} = 2$ and the variance of Y is

$$\sigma_Y^2 = \left(\frac{3-1}{8}\right)^2 = 0.0625$$

The mean value of F is 0.4 and the variance is zero.

The expected value of C is given by:

$$C = 3(1.96)^{1.4}\, 2^{-0.5} + 16(2) + 0.4 = 5.4413 + 32 + 0.4 = 37.8413$$

(*b*) The sensitivity of C to X and Y is obtained from

$$\frac{\partial C}{\partial X} = 3(1.4)X^{0.4}Y^{-0.5} = 3(1.4)(1.96)^{0.4}(2)^{-0.5} = 3.8866$$

$$\frac{\partial C}{\partial Y} = 3X^{1.4}(-0.5)Y^{-1.5} + 16 = 3(1.96)^{1.4}(-0.5)(2)^{-1.5} + 16$$

$$= 3(2.5654)(-0.5)(0.3536) + 16$$

$$= 14.6395$$

Thus, C changes 3.8866 times as fast as X for a unit change in X and 14.6395 times as fast as Y for unit change in Y. A 1% change in $X\ (\Delta x = 0.0196)$ would cause a change in C equal to

$$3.8866(0.0196) = 0.07618 \text{ or } (0.07618)(100)/37.8413 = 0.2013$$

percent. Also, a 1% change in $Y(Y = 2)$ gives 14.6395(0.02) = 0.2928 or (0.2928) × (100)/37.8413 = 0.774 percent error in C.

Hence, the total cost C is more sensitive to the estimate for Y than X.

(*c*) From equation (1.69*a*)

(*d*)

$$\sigma^2 = (3.866)^2\,(0.498) + (14.6395)^2\,(0.0625)$$
$$= 7.5226 + 13.3938$$
$$= 20.9164$$

or

$$\sigma = 4.5734$$

(*e*) Transforming the above information to a standard normal distribution, we have,

$$Z = \frac{C - 37.8413}{4.5734}$$

Assuming values for Z and from Section 1.10, (Table 1.8), the following tabulation is obtained.

Z	*Cumulative Probability for Z or less*	C
2.0	0.0228	28.6945
1.7	0.0446	30.0665
1.4	0.0808	31.4385
1.2	0.1151	32.3532
0.8	0.2119	34.1828
0.2	0.4207	36.9266
0.0	0.5000	37.8413
0.2	0.5793	38.0870
0.8	0.7881	41.5000
1.2	0.8849	43.3294
1.4	0.9192	44.2441
1.7	0.9554	45.6128
2.0	0.9773	46.9881

We find that the probability that the project will cost \$34.1828 millions or less is 0.2119 or 21.19%.

The probability of a cost of \$45.6128 million dollars or less is 0.9554 or 95.54%. The probability of a cost between \$34.1828 million dollars and \$45.6128 million dollars is

$$95.54 - 21.19 = 74.35\%$$

1.8.12 Additional Examples and Solutions

Example 1.21: *The resistance of a certain size of a metal wire is given by*

$$R = R_0[1 + \alpha(T - 18)]$$

where $R_0 = 6\ \Omega \pm 0.2$ percent is the resistance at 18°C, $\alpha = 0.004°C^{-1} \pm 1$ percent is the temperature coefficient of resistance, and the temperature of the wire is $T = 32 \pm 1°C$. Calculate the resistance of the wire and its uncertainty.

Solution:

$$R = R_0[1 + \alpha\,(T - 18)]$$

$$R_0 = 6\ \Omega \pm 0.2$$

Normal resistance:

$$R = 6[1 + 0.004(32 - 18)] = 6.335\ \Omega$$

$$\frac{\partial R}{\partial R_0} = 1 + \alpha(T - 18) = 1 + 0.004(14) = 1.056$$

$$\frac{\partial R}{\partial \alpha} = R_0(T - 18) = 84$$

$$\frac{\partial R}{\partial T} = R_0\alpha = 6(0.004) = 0.024$$

$$W_{R_0} = 6(0.002) = 0.012\ \Omega$$

$$W_\alpha = 0.004(0.01) = 4 \times 10^{-5}\ °C^{-1}$$

$$W_T = 1°C$$

Hence, the uncertainty in the resistance is

$$W_R = [(1.056)^2(0.012^2)^2 + 84^2(4 \times 10^{-5})^2 + (0.024)^2 1^2]^{1/2}$$

$$= 0.0273\ \Omega$$

Example 1.22: *A resistor has a nominal stated value of 10 Ω ±1 percent. A voltage is impressed on the resistor, and the power dissipation is to be calculated in two different ways:*

(a) *from $P = E^2/R$*

(b) *from $P = EI$*

In (a) only a voltage measurement will be made, while both current and voltage will be measured in (b).

Determine the uncertainty in the power determination in each case when the measured values of E and I are

$$E = 100\ V \pm 0.09\%\ \text{for both cases}$$

$$I = 10\ A \pm 0.09\%.$$

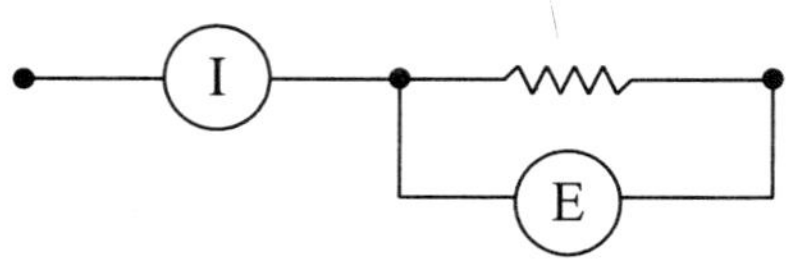

Fig. E1.22

Solution:

(*a*)

$$\frac{\partial P}{\partial E} = \frac{2E}{R}$$

$$\frac{\partial P}{\partial R} = \frac{-E^2}{R^2}$$

$$W_P = \left[\left(\frac{2E}{R}\right)^2 W_E^2 + \left(\frac{-E^2}{R^2}\right)^2 W_R^2\right]^{1/2}$$

Hence,

$$\frac{W_P}{P} = \left[4\left(\frac{W_E}{E}\right)^2 + \left(\frac{W_R}{R}\right)^2\right]^{1/2} [4(0.0009)^2 + (0.0009)^2]^{1/2}$$

$$= 2.0125 \times 10^{-3}$$

(*b*)

$$\frac{\partial P}{\partial E} = I, \quad \frac{\partial P}{\partial I} = E$$

$$\frac{W_P}{P} = \left[\left(\frac{W_E}{E}\right)^2 + \left(\frac{W_R}{R}\right)^2\right]^{1/2} [(0.0009)^2 + (0.0009)^2]^{1/2} = 1.2728 \times 10^{-3}$$

Example 1.23: *The flow rate of a certain type of flow meter is given by*

$$\dot{m} = CA\left[\frac{2g_c p_1}{RT_1}(p_1 - p_2)\right]^{1/2}$$

where A = *flow area*

C = *empirical-discharge coefficient*

p_1 *and* p_2 = *upstream and downstream pressures, respectively*

T_1 = *upstream temperature*

R = *gas constant for air*

Determine the percent uncertainty in the mass flow rate for the following conditions:

$A = 1.0\ in^2 \pm 0.0015\ in^2$

$\Delta p = p_1 - p_2 = 1.5\ psi \pm 0.005\ psi$

$T_1 = 70°F \pm 3°F;\ T_1 = 530°R$

$p_1 = 30\ psi \pm 0.5\ psi$

$C = 0.95 \pm 0.005.$

Solution:

$$\dot{m} = f(C, A, p_1, \Delta p, T_1)$$

$$\frac{\partial \dot{m}}{\partial C} = A\left(\frac{2g_c p_1}{RT_1}\Delta p\right)^{1/2}$$

$$\frac{\partial \dot{m}}{\partial A} = C\left(\frac{2g_c p_1}{RT_1}\Delta p\right)^{1/2}$$

$$\frac{\partial \dot{m}}{\partial p_1} = 0.5CA\left(\frac{2g_c}{RT_1}\Delta p\right)^{1/2} p_1^{-1/2}$$

$$\frac{\partial \dot{m}}{\partial \Delta p} = 0.5CA\left(\frac{2g_c p_1}{RT_1}\right)^{1/2} \Delta p^{-1/2}$$

$$\frac{\partial \dot{m}}{\partial T_1} = -0.5CA\left(\frac{2g_c p_1}{R}\Delta p\right)^{1/2} T_1^{-3/2}$$

The uncertainty in the mass flow rate is

$$\frac{W\dot{m}}{\dot{m}} = \left[\left(\frac{W_C}{C}\right)^2 + \left(\frac{W_A}{A}\right)^2 + \frac{1}{4}\left(\frac{Wp_1}{p_1}\right)^2 + \frac{1}{4}\left(\frac{W_{\Delta p}}{\Delta p}\right)^2 + \frac{1}{4}\left(\frac{W_{T1}}{T_1}\right)^2\right]^{1/2}$$

$$= \left[\left(\frac{0.005}{0.95}\right)^2 + \left(\frac{0.0015}{1}\right)^2 + \frac{1}{4}\left(\frac{0.5}{30}\right)^2 + \frac{1}{4}\left(\frac{0.005}{1.5}\right)^2 + \frac{1}{4}\left(\frac{3}{530}\right)^2\right]^{1/2}$$

$$= 1.05\%.$$

1.9 STATISTICAL ANALYSIS OF RANDOM UNCERTAINTIES

Statistics refers to the field or discipline of study. *Statistics* is a group of methods that are used to collect, analyze, present, and interpret data and to make decisions. *Theoretical* or *mathematical* statistics deals with the development, derivation, and proof of statistical theorems, formula, rules and laws. *Applied statistics* concerns the applications of those theorems, formulas, rules and laws to solve real-world problems. This chapter deals with *applied statistics.*

Applied statistics is divided into two branches: *descriptive statistics* and *inferential statistics. Descriptive statistics* involves methods for organizing, displaying, and describing data by using tables, graphs and summary measures. The collection of all elements of interest is called a *population* in statistics. The selection of a few elements from this population is called a *sample*. Statistics that deals with making decision, inferences, predictions, and forecasts about populations based on results obtained from samples is called descriptive statistics. Statistics that deals with decision-making procedures is called *inferential statistics*. This branch of statistics is also known as *inductive reasoning* or *inductive statistics. Inferential statistics* consists of methods that employ sample results to help make decisions or predictions about a population.

Probability, which measures the likelihood that an event will occur, is an important part of statistics. It is the basis of inferential statistics, where decisions are made under conditions of uncertainty. Probability theory is used to evaluate the uncertainty involved in those decisions. Probability statements are about occurrence or non-occurrence of a certain event under certain conditions.

1.9.1 Population and Sample

A *population* consists of all elements: individuals, items, or objects, whose characteristics are being studied. The population that is being studied is also called the *target population*. The collection of a few elements selected from a population is called a sample. Thus, a portion of the population selected for study is referred to as a *sample*. The collection of information from the elements of a population or a sample is called a survey. A *survey* that includes each and every element of the population is called a *census*. A survey conducted on a sample is called a *sample survey*. A sample that represents the characteristics of the population as closely as possible is called a *representative sample*.

A sample drawn in such a way that each element of the population has same chance of being selected is called a *random sample*. If the chance of being selected is the same for each element of the population, it is called a *simple random sample*. An *element* or *member* of a sample or population is a specific subject or object about which the information is collected. A *variable* is a characteristic under study that assumes different values for different elements. In relation to a variable, the value of a *constant* is fixed. The value of a variable for an element is called an *observation* or *measurement*. A *data set* is a collection of observations on one or more variables.

1.9.2 Types of Variables

A variable that can be measured numerically is called a *quantitative variable*. The data collected on a quantitative variable is known as *quantitative data*. Quantitative variable is classified as either *discrete variable* or *continuous variable*. A variable whose values are continuous is called a *continuous variable*. A *discrete variable* can assume only certain values with no intermediate values. On the other hand, a variable that can assume any numerical value over a certain interval or intervals is called a *continuous variable*. The time taken to complete a trip to a corner store is an example of a continuous random variable because it can assume any value, say, between one and two hours. A variable that cannot assume a numerical value but can be classified into two or more nonnumeric categories is called a *qualitative* or *categorical* variable. The data collected on such a variable are known as *qualitative data*.

1.9.3 Mean for Ungrouped Data

An *average* is a value which is typical or representative of a set of data. Since such values tend to lie centrally within a set of data arranged according to magnitude, averages are also called *measures of central tendency*.

The mean, also called the arithmetic mean or average, for ungrouped data is obtained by dividing the sum of all values by the number of values in the data set. Thus,

Mean for population data: $$\mu = \frac{\sum X}{N} \tag{1.70a}$$

Mean for sample data: $$\overline{X} = \frac{\sum X}{n} \quad (1.70b)$$

where μ is the population mean, ΣX the sum of all values, N the population size, $\overline{X}$ the sample mean, and n the sample size. Sometimes, the data may contain a few very small or a few very large values. Such values are called *outliers* or *extreme values*.

Example 1.24: *The following are the ages of 15 employees of a manufacturing company.*

65, 38, 52, 27, 24, 45, 49, 50, 26, 37, 44, 52, 41, 60, 39

Calculate the mean.

Solution:

$$\text{Mean} = \frac{\Sigma X}{n} = \frac{(65+38+52+27+24+45+49+50+26+37+44+52+41+60+39)}{15}$$

$$= \frac{649}{15} = 43.266 \text{ years.}$$

1.9.4 Measures of Dispersion for Ungrouped Data

The mean, median and mode are not sufficient measures to reveal the shape of the distribution of a data set. The measures that show the spread of a data set are called the *measures of dispersion*. There are three measures of dispersion: range, variance, and standard deviation.

Range

The range is the difference between the largest and smallest values in the data.

The range for ungrouped data = Largest value – Smallest value.

The range is not a very satisfactory measure of dispersion.

1.9.5 Variance and Standard Deviation

The standard deviation is the most used measure of dispersion. The formulas for calculating the variance and standard deviation are as follows:

Variance for ungrouped data

Population variance:

$$\sigma^2 = \frac{\sum(x-\mu)^2}{N} = \frac{\sum x^2 - \frac{\left(\sum x\right)^2}{N}}{N} \quad (1.70c)$$

Sample variance:

$$s^2 = \frac{\sum\left(x-\overline{X}\right)^2}{n-1} = \frac{\sum x^2 - \frac{\left(\sum x\right)^2}{n}}{n-1} \quad (1.71)$$

where, σ^2 is the population variance, s^2 is the sample variance, $(x - \mu)$ or $(x - \bar{x})$ is the deviation of x value from the mean, μ is the population mean, $\bar{x}$ is the sample mean, N is the population size, and n is the sample size.

$$\text{The population standard deviation } \sigma = \sqrt{\text{Population variance}} = \sqrt{\sigma^2} \tag{1.72}$$

$$\text{The sample standard deviation } s = \sqrt{\text{Sample variance}} = \sqrt{s^2} \tag{1.73}$$

A numerical value such as the mean, median, mode, range, variance, or standard deviation computed for a population data set is called a *population parameter*, or simply a *parameter*. A summary measure calculated for a sample data set is called a *sample statistic* or simply a *statistic*.

Example 1.25: *The following data refers to the pay (in thousands of dollars) of 12 employees at a manufacturing company.*

75 69 65 49 21 21 18 18 17 16 16 15

(a) calculate the mean, median and mode for this data

(b) find the range, variation and standard deviation.

Solution:

(*a*) $$\text{Mean} = \frac{\Sigma x}{n} = \frac{400}{12} = \$33{,}333$$

Median is $\left[\dfrac{n+1}{2}\right]^{th}$ term in a ranked data set. Therefore,

$$\frac{n+1}{2} = \frac{12+1}{2} = 6.5$$

$$\text{Median} = \frac{21+18}{2} = \$19{,}500$$

Mode = \$21,000, \$18,000 and \$16,000

(*b*) Range = Largest pay – Smallest pay = \$75,000 – \$15,000 = \$60,000

$$\text{Variance} = s^2 = \frac{x^2 - \dfrac{(\Sigma x)^2}{n}}{n-1} = \frac{19{,}528 - \dfrac{400^2}{12}}{12-1} = 563.1515$$

$$\text{Standard deviation} = s = \sqrt{s^2} = \sqrt{563.1515} = 23.73 = \$23{,}730.$$

1.9.6 Mean, Variance, and Standard Deviation for Grouped Data

1.9.7 Mean for Grouped Data

$$\text{Mean for population data: } \mu = \frac{\sum mf}{N} \tag{1.74}$$

$$\text{Mean for sample data: } \overline{X} = \frac{\sum mf}{n} \tag{1.75}$$

where, m is the midpoint and f is the frequency of a class.

1.9.8 Variance and Standard Deviation for Grouped Data

Population variance

$$\sigma^2 = \frac{\sum f\,(m-\mu)^2}{\mathrm{N}} = \frac{\sum m^2 f - \dfrac{(\sum mf)^2}{N}}{N} \tag{1.76}$$

Sample variance

$$s^2 = \frac{\sum f\,(m-\overline{X})^2}{n-1} = \frac{\sum m^2 f - \dfrac{(\sum mf)^2}{n}}{n-1} \tag{1.77}$$

where σ^2 is the population variance, s^2 is the sample variance, m is the midpoint of a class, N is the population size, n is the sample size, μ is the population mean, $\overline{X}$ is the sample mean, and f the class frequency.

$$\text{Population standard deviation} \quad \mathrm{s} = \sqrt{\sigma^2} \tag{1.78}$$

$$\text{Sample standard deviation} \quad s = \sqrt{s^2} \tag{1.79}$$

Example 1.26: *Calculate the mean, variance and standard deviation for the following population data.*

x	*0–3*	*4–7*	*8–11*	*12–15*	*16–19*	*20–23*
f	*7*	*4*	*18*	*12*	*7*	*6*

Solution:

Classes	*f*	*m*	*mf*	$m^2 f$
0–3	7	1.5	10.5	15.75
4–7	4	5.5	22	121
8–11	18	9.5	171	1624.5
12–15	12	13.5	162	2187
16–19	7	17.5	122.5	2143.75
20–23	6	21.5	129	2773.50
Total	54		617	8865.50

Population mean $\mu = \dfrac{\sum mf}{N} = \dfrac{617}{54} = 11.426$

Variance $= \sigma^2 = \dfrac{\sum m^2 f - \dfrac{(\sum mf)^2}{N}}{N} = \dfrac{8865.50 - \dfrac{617^2}{54}}{54} = \dfrac{8865.50 - 7049.796}{54} = 33.624$

Standard deviation $\sigma = \sqrt{\sigma^2} = \sqrt{33.624} = 5.798$

1.9.8.1 Measures of Position

A measure of position determines the position of a single value in relation to other values in a sample or a population data set. The measures of position considered here are the quartiles, percentiles, and percentile ranks.

1.9.8.2 Quartiles and Interquartile Range

The most commonly used percentiles are quartiles. Let n denote the number of observations and if we arrange the data in increasing order, then

— the first quartile is at position $\dfrac{n+1}{4}$

— the second quartile is the median, which is at position $\dfrac{n+1}{2}$

— the third quartile is at position $3\left(\dfrac{n+1}{4}\right)$.

Thus, a data set has three quartiles, and is denoted by Q_1, Q_2 and Q_3. Q_1, is the number that divides the bottom 25% of the data from the top 75%; the second quartile, Q_2, the median, is the number that divides the bottom 50% of the data from the top 50%; the third quartile, Q_3, is the number that divides the bottom 75% of the data from the top 25%. The difference between the third quartile and the first quartile of data set is called the *interquartile range* ($IQR = Q_3 - Q_1$). Interquartile range is a preferred measure of variation when the median is used as the measure of the center. *IQR* gives the range of the 50% of the observations.

1.9.8.3 Percentiles

Percentiles are the summary measures that divide ranked data set into 100 equal parts. Each ranked data set has 99 percentiles that divide it into 100 equal parts. The kth percentile is denoted by P_k, where k is an integer in the range 1 to 99.

Hence, the kth percentile, P_k, can be described as a value in a data set such that about k % of the measurements are smaller than the value of P_k and about $(100 - k)$% of the measurements are greater than the value of P_k. The value of the kth percentile is given by

$$P_k = \text{the value of the } \left(\frac{kn}{100}\right)^{\text{th}} \text{ term in a ranked data set}$$

where k denotes the number of the percentile and n represents the sample size.

The percentile rank for a particular value x_i of a data set is given by

$$\text{Percentile rank of } x_i = \frac{\text{Number of values less than } x_i}{\text{Total number of values in the data set}} \times 100$$

The percentile rank of x_i gives the percentage of values in the data set that are smaller than x_i.

Deciles: Deciles are also useful in statistics. The deciles of a data set divide it into tenths, or 10 equal parts. A typical data set has nine deciles. If we denote by $D_1, D_2 \ldots D_9$, then the first decile, D_1 is the number that divides the bottom 10% of the data from the top 90%; the second decile, D_2, is the number that divides the bottom 20% of the data from the top 80%; and so on. Hence, the first decile is the 10th percentile, the second decile is the 20th percentile, etc.

1.9.9 Random Variables

This section introduces the important concept of a probability distribution, which gives the probability for each value of a variable that is determined by chance. This section also introduces procedures for finding the mean and standard deviation for a probability distribution. A *random variable* is a variable (typically represented by x) that has a single numerical value, determined by chance, for each outcome of a procedure. In other words, a *random variable* is a quantitative variable whose value depends on chance. We defined a variable as a characteristic that varies from one member of a population to another. When one or more members are selected at random from the population, the variable, in that context, is called a *random variable*. Therefore, mathematically, a random variable is a function defined on the outcome of the sample space.

A *probability distribution* is a description that gives the probability for each value of the random variable. It is often expressed in the format of a graph, table, or formula. A random variable is classified as discrete or continuous, depending upon the range of its values. (The range of a random variable is the set of values it can assume.)

A *discrete random variable* has either a finite number of values or a countable number of values, where *countable* refers to the fact that there might be infinitely many values, but they can be associated with a counting process. Thus, *w* often says that a discrete random variable can assume at most a *countable infinite* number of values.

The number of eggs that a hen lays in a day, the salary of a new college graduate, the number of bacteria in a culture, the count of the number of students present in statistics class on a given day, the number of cavities that a child has, the price of gold on the exchange on a particular day, toss of a coin, the number of cars sold at a car dealership in a given month, color of a ball drawn from a collection of balls, etc., these are all examples of discrete random variables.

A *continuous random variable* has infinitely many values, and those values can be associated with measurements on a continuous scale without gaps or interruptions.

The maximum daily temperature, the amount of milk a cow produces in one day, the life of an electric bulb, the measure of voltage for a particular smoke detector battery, the distance by which a sharp shooter misses a target, the height of a person, the weight of a person, and the length of time a person has to wait at a bank counter, diameter of a manufactured shaft, time to failure of a machine component, repair time, the price of an automobile, the time taken to complete a medical examination, duration of a snow storm, etc. provide examples of continuous random variables.

1.9.10 Continuous Random Variables

A random variable is *continuous* if its probabilities are given by areas under a curve. The curve is called a *probability density function* for the random variable. The probability density function is sometimes called the *probability distribution*.

If the range space of the random variable X is an interval or a collection of intervals, X is called a *continuous random variable*. If a and b are real numbers and $a < b$, then the probability that X lies in the interval $[a, b]$ is defined as

$$P[a \le X \le b] = \int_a^b f(x)\,dx \tag{1.80}$$

The function $f(x)$ is known as the *probability density function* (or *pdf*) of the random variable X. The *pdf* must satisfy the following conditions:

1. $f(x) \ge 0 \qquad -\infty < x < \infty$
2. $\int f(x)\,dx = 1$ (1.81)

The probability density function (*pdf*) is shown in Fig. 1.3. The shaded area in Fig. 1.3 represents the probability that X lies in the interval $[a, b]$.

From Eq. (1.80), for any specified value of x, $p(X = x) = 0$, since

$$\int_x^x f(t)\,dt = 0 \tag{1.82}$$

Hence, we can also write

$$P(a \le X \le b) = P(a < X \le b) = P(a \le X < b) = P(a < X < b) \tag{1.83}$$

Since x is continuous

$$F(x) = \int_{-\infty}^{x} f(t)\,dt \tag{1.84}$$

The following conditions are satisfied

1. F is a non-decreasing function. If $a < b$, then $F(a) \le F(b)$.
2. $\lim_{x\to\infty} F(x) = 1$
3. $\lim_{x\to-\infty} F(x) = 0$ (1.85)

All probability questions about X can be expressed in terms of the *cdf*. For instance

$$P(a < X \leq b) = F(b) - F(a) \text{ for } a < b \tag{1.86}$$

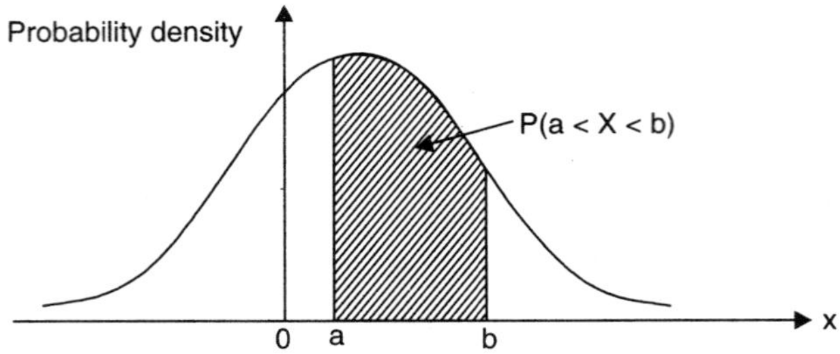

Fig. 1.3 The probability density function (*pdf*)
(The shaded area under the curve is $P(a < X < b)$.)

The cumulative distribution function of a continuous random variable X is $F(X) = P(X \leq x)$, just like for a discrete random variable. For a continuous variable, the value of $F(x)$ is obtained by integrating the probability density function.

Since $F(x) = P(X \leq x)$, we have

$$F(x) = \int_{-\infty}^{x} f(t)\, dt \tag{1.87}$$

where $f(t)$ is the probability density function.

Hence, the continuous distribution function (*cdf*) of X is the function

$$F(x) = P(X \leq x) = \int_{-\infty}^{x} f(t)\, dt \tag{1.88}$$

1.9.11 Mean and Variance for Continuous Random Variables

The population mean and variance of a continuous random variable are defined in the same way as those of a discrete random variable, except that the probability density function is used instead of the probability mass function. Specifically, if X is a continuous random variable, its population mean is defined to be the center of mass of its probability density function and its population variance is the moment of inertia around a vertical axis through the population mean.

Let X is a continuous random variable with probability density function $f(x)$. Then, the mean is given by

$$\mu_x = \int_{-\infty}^{\infty} x f(x)\, dx \tag{1.89}$$

The *mean* of X is sometimes called the *expectation* or the *expected* value of X and denoted by $E(X)$ or by μ.

The variance of X is given by

$$\sigma_X^2 = \int_{-\infty}^{\infty} (x-\mu_x)^2 f(x)\, dx \tag{1.90}$$

An alternative formula for the variance is given by

$$\sigma_X^2 = \int_{-\infty}^{\infty} x^2 f(x)\, dx - \mu_X^2 \tag{1.91}$$

The variance of X is also denoted by $V(X)$ or by σ^2.

The standard deviation is the square root of the variance

$$\sigma_x = \sqrt{\sigma_X^2} \tag{1.92}$$

Alternate Formula for the Variance

Starting with the equation

$$\sigma_X^2 = \sum_x (x-\mu_x)^2 P(X=x) \tag{1.93}$$

Expanding the above equation, we obtain

$$\sigma_X^2 = \sum_x (x^2 - 2x\mu_X + \mu_X^2) P(X=x) \tag{1.94}$$

or

$$\sigma_X^2 = \sum_x [x^2 P(X=x) - 2x\mu_X P(X=x) + \mu_X^2 P(X=x)] \tag{1.95}$$

Summing the terms separately, we have

$$\sigma_X^2 = \sum_x x^2 P(X=x) - \sum_x 2x\mu_X P(X=x) + \sum_x \mu_X^2 P(X=x) \tag{1.96}$$

Noting that

$$\sum 2x\mu_X P(X=x) = 2\mu_X \sum_x x P(X=x) = 2\mu_X \mu_X = 2\mu_X^2 \tag{1.97}$$

and

$$\sum_x \mu_X^2 P(X=x) = \mu_X^2 \sum_x P(X=x) = \mu_X^2 (1) = \mu_X^2 \tag{1.98}$$

Substituting Eqs. (1.97) and (1.98) in Eq. (1.96), we get

$$\sigma_X^2 = \sum_x x^2 P(X=x) - 2\mu_X^2 + \mu_X^2 \tag{1.99}$$

or

$$\sigma_X^2 = \sum_x x^2 P(X=x) - \mu_X^2 \tag{1.100}$$

1.9.12 Expectation

A brief summary is given here:

If X is a random variable, the expected value of X, denoted by $E(X)$, for discrete and continuous variable is defined as

$$E(X) = \sum_{all\ i} x_i p(x_i) \qquad \text{if } X \text{ is discrete} \qquad (1.101)$$

and

$$E(X) = \int_{-\infty}^{\infty} xf(x)\,dx \qquad \text{if } X \text{ is continuous} \qquad (1.102)$$

the expected value $E(X)$ of a random variable X is also referred to as the *mean* of X, μ_X (or μ), or the *first moment* of X. The quantity, $E(X^n)$, $n \geq 1$, is called the nth *moment* of X and is defined as

$$E(X^n) = \sum_{all\ i} x_i^n p(x_i) \qquad \text{if } X \text{ is discrete} \qquad (1.103)$$

and

$$E(X^n) = \int_{-\infty}^{\infty} x^n f(x)\,dx \qquad \text{if } X \text{ is continuous} \qquad (1.104)$$

The variance of a random variable, X, denoted by $V(X)$, Var(X), or σ^2, is defined by

$$V(X) = E([X - E(X)]^2)$$

A useful identity in computing $V(X)$ is given by

$$V(X) = E(X^2) - [E(X)]^2 \qquad (1.105)$$

The mean $E(X)$ is a measure of the central tendency of a random variable. The variance, $V(X)$, is a measure of the spread or dispersion of the possible values of X around the mean $E(X)$. The standard deviation, σ, is the square root of the variance.

Example 1.27: *The life of an electronic component is given by X, a continuous random variable assuming all values in the range $x \geq 0$. The pdf of the lifetime, in years, is given as*

$$f(x) = \begin{cases} \frac{1}{4} e^{-\frac{x}{4}} & x \geq 0 \\ 0 & otherwise \end{cases}$$

(*a*) *determine the probability that the life of* this component is between 2 and 4 years.

(*b*) *determine cdf of the components.*

(*c*) *determine the probability that this electronic component will last for 4 years.*

(*d*) *find the mean and variance of the life of the component.*

Solution:

(*a*)

$$P(2 \leq X \leq 4) = \frac{1}{4}\int_2^4 e^{-\frac{x}{4}}dx = -e^{-1} + e^{-\frac{1}{2}} = 0.2386$$

The probability that the laser will last between 2 and 4 years is obtained also from

$$P(2 \leq X \leq 4) = F(4) - F(2) = (1 - e^{-4/4}) - (1 - e^{-2/4})$$

$$= e^{-1/2} - e^{-1} = 0.6065 - 0.3679 = 0.2386$$

(*b*) The *cdf* of this component is given by

$$F(x) = \frac{1}{4}\int_0^x e^{-\frac{x}{4}}\,dx = 1 - e^{-\frac{x}{4}}$$

(*c*) The probability that the component will last for 4 years or less is given by

$$P(0 \le X \le 4) = F(4) - F(0) = F(4) = 1 - e^{4/4} = 0.6321$$

(*d*) The mean and variance of the life of the electronic component are determined as follows:

$$E(X) = \frac{1}{4}\int_0^\infty x e^{-\frac{x}{4}} dx = x e^{-\frac{x}{4}}\Big|_0^\infty + \int_0^\infty e^{-\frac{x}{4}} dx = 0 + \frac{1}{1/4} e^{-\frac{x}{4}}\Big|_0^\infty = 4 \text{ years}$$

To compute $V(X)$, first compute $E(X^2)$ as follows:

$$E(X^2) = \frac{1}{4}\int_0^\infty x^2 e^{-\frac{x}{4}} dx = x^2 e^{-\frac{x}{4}}\Big|_0^\infty + 8\int_0^\infty x e^{-\frac{x}{4}} dx = 32$$

Hence, variance is given by

$$V(X) = 32 - 4^2 = 16 \text{ years}^2$$

and the standard deviation is given by

$$\sigma = \sqrt{16} = 4 \text{ years.}$$

Example 1.28: *Given $f(x) = e^{-x}$ for $0 < x$. Find the following probabilities:*

(*a*) *$P(1 < X)$* (*b*) *$P(1 < X < 2.5)$*

(*c*) *$P(X = 3)$* (*d*) *$P(X < 4)$*

(*e*) *$P(3 \le X)$*

Solution:

(*a*) $P(1 < X) = \int_1^\infty e^{-x} dx = (-e^{-x})\Big|_1^\infty = e^{-1} = 0.3679$

(*b*) $P(1 < X < 2.5) = \int_1^{2.5} e^{-x} dx = (-e^{-x})\Big|_1^{2.5} = e^{-1} - e^{-2.5} = 0.2858$

(*c*) $P(X = 3) = \int_3^3 e^{-x} dx = 0$

(*d*) $P(X < 4) = \int_0^4 e^{-x} dx = (-e^{-x})\Big|_0^4 = 1 - e^{-4} = 0.9817$

$$(e)\ P(3 \le X) = \int_3^\infty e^{-x} dx = (-e^{-x})\Big|_3^\infty = e^{-3} = 0.0498$$

Example 1.29: *The probability density of the continuous random variable X is given by*

$$f(x) = \begin{cases} 1/5 & \text{for } 2 < x < 7 \\ 0 & \text{elsewhere} \end{cases}$$

Find

(a) P(3 < X < 5)

(b) area under the curve.

Solution:

$$(a)\ P(3 < X < 5) = \int_3^5 \frac{1}{5}\, dx = \frac{1}{5} x\Big|_3^5 = \frac{1}{5}(5-3) = \frac{2}{5}$$

$$(b)\ \text{Area under the curve} = \int_{-\infty}^{\infty} f(x)\, dx = \int_2^7 \frac{1}{5}\, dx = \frac{1}{5} x\Big|_2^7 = \frac{1}{5}(7-2) = 1$$

Example 1.30: *Given $f(x) = 1.5x^2$ for $-1 < x < 1$. Find the mean and variance of x.*

Solution:

$$E(X) = \int_{-1}^{1} x(1.5x^2)\, dx \int_{-1}^{1} 1.5x^3 dx = 1.5\frac{x^4}{4}\Big|_{-1}^{1} = 0$$

$$V(X) = \int_{-1}^{1} 1.5x^3(x-0)^2 dx = 1.5\int_{-1}^{1} x^5 dx = 1.5\frac{x^6}{6}\Big|_{-1}^{1} = 0$$

Example 1.31: *The pdf of the random variable X is given by*

$$f(x) = \begin{cases} \dfrac{a}{\sqrt{x}} & \text{for } 0 < x < 4 \\ 0 & \text{elsewhere} \end{cases}$$

Find (a) the value of a

(b) $P\left(X < \frac{1}{4}\right)$ and $P(X > 1)$.

Solution:

$$(a)\ \text{Area} = 1 = \int_0^4 \frac{a}{\sqrt{x}} dx = a\int_0^4 x^{-1/2} dx = a\frac{x^{1/2}}{1/2}\Big|_0^4 = 2a \cdot 2 = 4a$$

Hence, $a = 1/4$.

$$(b)\ P\left(X < \frac{1}{4}\right) = \int_0^{1/4} \frac{1}{4\sqrt{x}}\,dx = \frac{1}{4}\int_0^{1/4} x^{-1/2}\,dx = \frac{1}{4}\frac{\sqrt{x}}{1/2}\Bigg|_0^{1/4} = \frac{1}{2}\cdot\frac{1}{2} = \frac{1}{4}$$

$$P(X > 1) = 1 - \int_0^1 \frac{1}{4\sqrt{x}}\,dx = 1 - \frac{1}{2}\sqrt{x}\Bigg|_0^1 = \frac{1}{2}$$

Example 1.32: *The thickness of a conductive coating in micrometers has a density function of $f(x) = 600\,x^{-2}$ for 100 μm < x < 120 μm.*

(a) find the mean and variance of the coating thickness

(b) if the cost of coating is \$0.25 per micrometer of thickness on each component, find the average cost of the coating per component.

Solution:

(a)
$$E(X) = \int_{100}^{120} x\frac{600}{x^2}\,dx = 600\ln x\Bigg|_{100}^{120} = 109.39$$

$$V(X) = \int_{100}^{120} (x - 109.39)^2\frac{600}{x^2}\,dx = 600\int_{100}^{120} 1 - \frac{2(190.39)}{x} + \frac{(109.39)^2}{x^2}\,dx$$

$$= 600(x - 218.75\ln x - 109.39^2 x^{-1})\Big|_{100}^{120} = 33.19$$

(b) Average cost per component = \$0.25(109.39) = \$27.35.

Example 1.33: *The distribution function of the random variable X is given by*

$$F(x) = \begin{cases} 1-(1+x)e^{-x} & \text{for } x > 0 \\ 0 & \text{for } x \le 0 \end{cases}$$

Find

(a) $P(X \le 2)$ *(b) $P(1 < X < 3)$*

(c) $P(X > 4)$

Solution:

(a) $P(X \le 2) = F(2) = 1 - 3e^{-2} = 1 - 3(0.1353) = 1 - 0.4074 = 0.5926$

(b) $P(1 < X < 3) = F(3) - F(1) = 1 - 4 - e^{-3} - 1 + 2e^{-1} = 2e^{-1} - 4e^{-3}$

$= 2(0.3679) - 4(0.0498) = 0.5366$

(c) $P(X > 4) = 1 - F(4) = 5e^{-4} = 5(0.0183) = 0.0915.$

1.9.13 Permutations and Combinations

Factorial Notation: The product of the positive integers from 1 to n inclusive is denoted by $n!$ (read as "n factorial"). The value of the factorial of a number is obtained by multiplying all integers from that number to 1.

The symbol $n!$, read as "n factorial" represents the product of all integers from n to 1. In other words,

$$n! = n(n-1)\,(n-2)\,(n-3)\,\ldots\,3.2.1 \tag{1.106}$$

It is also convenient to define

$$0! = 1,\ 1! = 1 \text{ and } n! = n.((n-1)!) \tag{1.107}$$

Permutations

Any arrangement of a set of n objects in a given order is called a *permutation* of the objects (taken all at a time). Any arrangement of any $r \le n$ of those objects in a given order is called an *r-permutation* or *a permutation of the n objects taken r at a time*.

Ordered arrangement of a set of objects is called *permutations*. When the order in which they are arranged is disregarded, then the arrangements are known as *combinations*. The number of permutations of n different items is the number of different arrangements in which these items can be placed. A permutation of n different objects taken r at a time is an arrangement of r out of the n objects with attention given to the order of arrangement. The number of permutations of n objects taken r at a time is denoted by nP_r, $P(n, r)$ P_r^n, $(n)_r$ or $p_{n,r}$ and is written as

$$nP_r = n(n-1)(n-2)\ldots(n-r+1) = \frac{n!}{(n-r)!} \tag{1.108}$$

Combinations

As noted earlier, the order in which objects are arranged matters in *permutations*. However, order does not matter in *combinations*.

The number of combinations of n different objects taken r at a time is a selection of r out of the n objects with no attention given to the order of arrangement. The number of combinations of n objects taken r at a time, denoted by nC_r, $C(n, r)$, $C_{n,r}$ or $\left(\frac{n}{r}\right)$, is $r!$ times smaller than the number of permutations.

Therefore,

$$nC_r = \binom{n}{r} = \frac{nP_r}{r!} = \frac{n!}{(n-r)!\,r!} \tag{1.109}$$

It should be noted from symmetry that the identity is valid

$$nC_r = nC_{n-r} \tag{1.110}$$

Also $$\binom{n}{r} = \frac{nP_r}{r!}$$

The arrangement for dividing the number of permutations by $r!$ to get the number of combinations is that each combination gives rise to $r!$ possibilities when arranged in all possible ways, thereby giving all the permutations.

Note that in combinations, n is always greater than or equal to r. If n is smaller than r, then we cannot select r distinct elements from n.

1.10 THE NORMAL DISTRIBUTION

In everyday life, people deal with and use a wide variety of variables. Some of these variables—such as heights of people, scores in final examinations, aptitude test scores, TOEFL (Test of English as a Foreign Language), and GRE (Graduate Record Examination) share an important characteristic: their distributions have roughly the shape of a normal curve, that is, a special type of bell-shaped curve like one shown in Fig. 1.4.

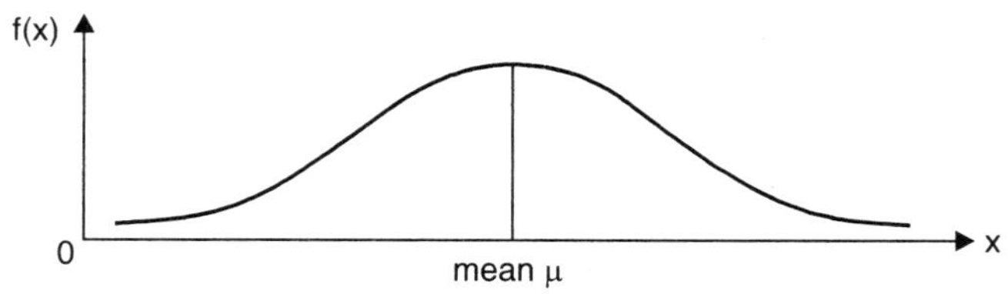

Fig. 1.4 Normal curve.

The normal distribution is in many respects the cornerstone of statistics. A random variable X is said to have a normal distribution with mean μ ($-\infty < \mu < \infty$) and variance $\sigma^2 > 0$ if it has the density function

$$f(x) = \frac{1}{\sigma\sqrt{2\pi}} e^{-(1/2)[(x-\mu)\sigma]^2} \qquad -\infty < X < \infty \tag{1.111}$$

The distribution is illustrated graphically in Fig. 1.4. The normal distribution is used so extensively that the shorthand notation $X \sim N(\mu, \sigma^2)$ is often used to indicate that the random variable X is normally distributed with mean μ and variance σ^2.

A variable is said to be a normally distributed variable or to have a *normal distribution* if its distribution has the shape of a normal curve.

If a variable of a population is normally distributed and is the only variable under consideration, the general practice is to assume that the population is *normally distributed* or that it is a *normally distributed population*. In reality, a distribution is unlikely to have exactly the shape of a normal distribution curve. If a variable's distribution is shaped roughly like a normal curve, we can consider that the variable is an *approximately normally distributed variable* or that it has *approximately a normal distribution*.

A normal distribution (and hence a normal curve) is completely determined by the mean and standard deviation. Hence, two normally distributed variables having the same mean and standard deviation must have the same distribution. We identify a normal curve by stating the corresponding mean and standard deviation and calling those as *parameters* of the normal curve.

A normal distribution is symmetric about and centered at the mean of the variable, and its spread depends on the standard deviation of the variable. The larger the standard deviation, the flatter and more spread out is the distribution.

In summary, the normal curve associated with a normal distribution is bell shaped, centered at μ, and close to the horizontal axis outside the range from $\mu - 3\sigma$ to $\mu + 3\sigma$ as shown in Figs. 1.4 and 1.4(*a*).

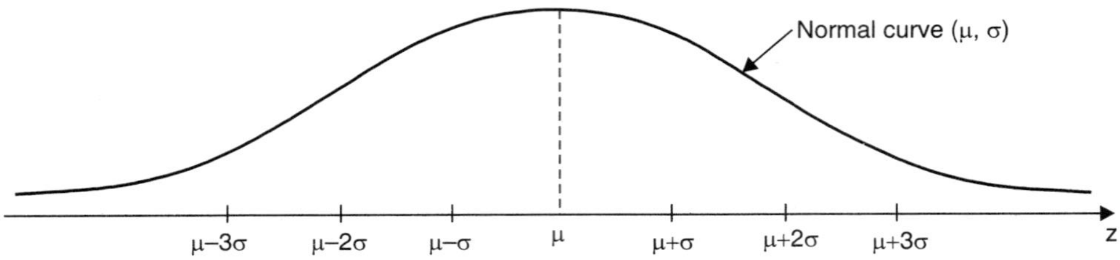

Fig. 1.4(a)

Random variables with different means and variances can be modeled by normal probability functions with appropriate choices of the center and width of the curve. The value of $E(X) = \mu$ determines the center of the probability density function and the value of $V(X) = \sigma^2$ determines the width. Figure 1.4(*b*) illustrates several normal probability density functions with selected values of μ and σ^2. Each one has the characteristic symmetric bell-shaped curve, but the centers and dispersions differ.

Definition of Normal Distribution

A random variable X with probability density function

$$f(x) = \frac{1}{\sqrt{2\pi}\sigma} e^{-\left(\frac{x-\mu}{2\sigma^2}\right)} \qquad -\infty < x < \infty$$

is a *normal random variable* with parameters μ, where $-\infty < x < \infty$, and $\sigma > 0$.

Also $\quad E(X) = \mu$ and $V(X) = \sigma^2$

and the notation $N(\mu, \sigma^2)$ is used to denote the distribution. The mean and variance of X are equal to μ and σ^2, respectively.

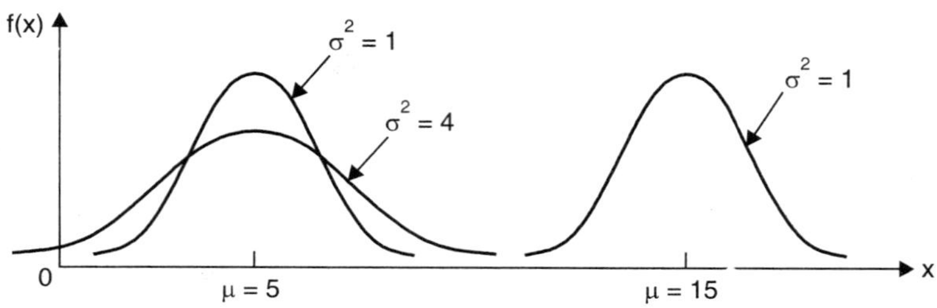

Fig. 1.4 (*b*) Normal probability density functions for selected values of the parameters μ and σ^2.

1.10.1 Properties of the Normal Distribution

The normal distribution has several important properties:

1. $\int_{-\infty}^{\infty} f(s)dx = 1$

2. $f(x) \geq 0$ for all x

3. $\lim_{x \to \infty} f(x) = 0$ and $\lim_{x \to -\infty} f(x) = 0$

4. $f[(x + \mu)] = f[-(x - \mu)]$. The density is symmetric about μ.

5. The maximum value of $f(x)$ occurs at $x = \mu$.

6. The points of inflection of $f(x)$ are at $x = \mu \pm \sigma$.

Also, $P(a \leq x \leq b) = \int_a^b f(x)\,dx$ = area under $f(x)$ from a to b for any a and b.

Property 1 may be demonstrated as follows. Let $y = (x - \mu)/\sigma$ in Eq. (3.71) and denote the integral as I. That is,

$$I = \frac{1}{\sqrt{2\pi}} \int_{-\infty}^{\infty} e^{-(1/2)y^2}\, dy \frac{1}{\sqrt{2\pi}} \int_{-\infty}^{\infty} e^{-(1/2)x^2}\, dx \tag{1.112}$$

on changing to polar coordinates with the transformation of variables $y = r \sin \theta$ and $z = r \cos \theta$, the integral becomes

$$I^2 = \frac{1}{2\pi} \int_0^{\infty} \int_0^{2\pi} r e^{-(1/2)r^2}\, d\theta\, dr = \int_0^{\infty} r e^{-(1/2)r^2}\, dr = 1 \tag{1.113}$$

1.10.2 Mean and Variance of the Normal Distribution

The mean of the normal distribution may be determined easily. Since

$$E(X) = \int_{-\infty}^{\infty} \frac{x}{\sigma\sqrt{2\pi}} e^{-(1/2)[(x-\mu)/\sigma^2]} dx \tag{1.114}$$

and if we let $z = (x - \mu)/\sigma$, we obtain

$$E(X) = \int_{-\infty}^{\infty} \frac{1}{\sqrt{2\pi}} (\mu + \sigma z) e^{-z^2/2} dz \tag{1.115}$$

$$= \mu = \int_{-\infty}^{\infty} \frac{1}{\sqrt{2\pi}} dz + \sigma \frac{1}{\sqrt{2\pi}} z e^{-z^2/2} dz \tag{1.116}$$

Since the integrand of the first integral is that of a normal density with $\mu = 0$ and $\sigma^2 = 1$, the value of the first integral is one. The second integral has value zero, that is,

$$\int_{-\infty}^{\infty} \frac{1}{\sqrt{2\pi}} z e^{-z^2/2} dz = \frac{1}{\sqrt{2\pi}} z e^{-z^2/2} \Big|_{-\infty}^{\infty} = 0$$

and thus

$$E(X) = \mu[1] = \sigma[0] = \mu \tag{1.117}$$

To find the variance, we must evaluate

$$V(X) = E[(X-\mu)^2] = \int_{-\infty}^{\infty} (x-\mu)^2 \frac{1}{\sigma\sqrt{2\pi}} e^{-(1/2)[(x-\mu)/\sigma]^2} dx$$

And letting $z = (x - \mu)/\sigma$, we obtain

$$V(X) = \int_{-\infty}^{\infty} \sigma^2 z^2 \frac{1}{\sqrt{2\pi}} e^{-z^2/2} dz = \sigma^2 \left[\int_{-\infty}^{\infty} \frac{z^2}{\sqrt{2\pi}} e^{-z^2/2} dz \right]$$

$$= \sigma^2 \left[\frac{-ze^{-z^2/2}}{\sqrt{2\pi}} \Bigg|_{-\infty}^{\infty} \right] + \int_{-\infty}^{\infty} \frac{1}{\sqrt{2\pi}} e^{-z^2/2} dz = \sigma^2 (0+1)$$

so that

$$V(X) = \sigma^2 \qquad (1.118)$$

In summary, the mean and variance of the normal density, given in Eq. (1.118), are μ and σ^2, respectively.

1.10.3 The Cumulative Normal Distribution

The distribution function F is

$$F(x) = P(X \le x) = \int_{-\infty}^{\infty} \frac{1}{\sigma\sqrt{2\pi}} e^{-(1/2)[(x-\mu)/\sigma]^2} du \qquad (1.119)$$

It is impossible to evaluate this integral without resorting to numerical methods, and even then the evaluation would have to be accomplished for each pair (μ, σ^2). However, a simple transformation of variables, $z = (X - \mu)/\sigma$, allows the evaluation to be independent of μ and σ.

That is,

$$F(X) = P(X \le x) = P\left(Z \le \frac{x-\mu}{\sigma}\right) = \int_{-\infty}^{(x-\mu)/\sigma} \frac{1}{\sqrt{2\pi}} e^{-z^2/2} dz$$

$$= \int_{-\infty}^{(x-\mu)/\sigma} \varphi(z)\, dz = \phi\left(\frac{x-\mu}{\sigma}\right) \qquad (1.120)$$

1.10.4 The Standard Normal Distribution

The probability distribution in Eq. (1.120) above,

$$\phi(z) = \frac{1}{\sqrt{2\pi}} e^{-z^2/2} \qquad -\infty < x < \infty$$

is a normal distribution with mean 0 and variance 1; that is, $Z \sim N(0, 1)$ and we say that Z has a *standard normal distribution*. A graph of the probability density function is shown in Fig. 1.5.

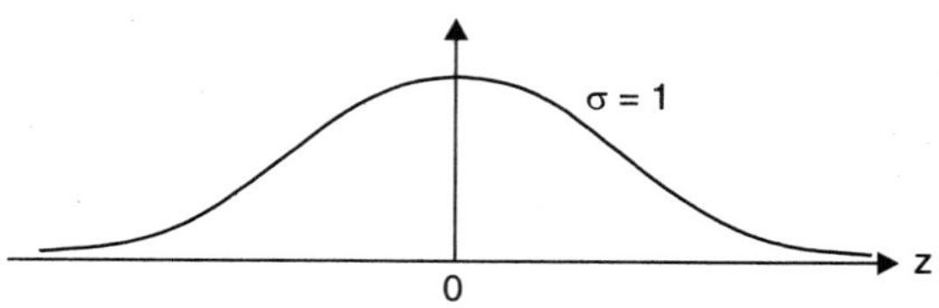

Fig. 1.5 The standard normal distribution.

The corresponding distribution function is ϕ, where

$$\phi(z) = \int_{-\infty}^{\infty} \frac{1}{\sqrt{2\pi}} e^{-z^2/2} du \quad (1.121)$$

where

$$z = \frac{x-\mu}{\sigma}$$

and this function has been well tabulated.

A few useful results for a normal distribution are summarized below and shown in Fig. 1.6(*a*). For random variable,

$$P(\mu - \sigma < X < \mu + \sigma) = 0.6827$$
$$P(\mu - 2\sigma < X < \mu + 2\sigma) = 0.9545$$
$$P(\mu - 3\sigma < X < \mu + 3\sigma) = 0.9973$$

In other words, any normally distributed variable has the following properties:

1. 68.27% of all possible observations lie within one standard deviation to either side of the mean, that is, between $\mu - \sigma$ and $\mu + \sigma$.
2. 95.45% of all possible observations lie within two standard deviation to either side of the mean, that is, between $\mu - 2\sigma$ and $\mu + 2\sigma$.
3. 99.73% of all possible observations lie within three standard deviation to either side of the mean, that is, between $\mu - 3\sigma$ and $\mu + 3\sigma$.

These properties are shown in Fig. 1.6.

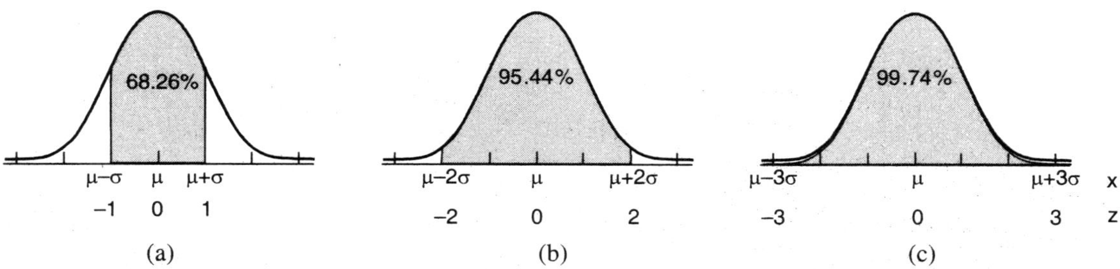

Fig. 1.6

In addition, from the symmetry of $f(x)$, $P(X > \mu) = P(X < \mu) = 0.5$. The above model assigns some probability to each interval of the real line since $f(x)$ is positive for all x. The probability density function decreases as x moves further from μ. As a result, the probability that a measurement falls far from μ is small, and at some distance from μ, the probability of an interval can be approximated as zero.

For a normally distributed variable, the percentage of all possible observations that (*i.e.*, within any specified range) equals the corresponding area under its associated normal curve, expressed as a percentage. This result holds approximately for a variable that is approximately normally distributed.

A normally distributed variable having mean 0 and standard deviation 1 is said to have the *standard normal distribution*. Its associated normal curve is called the *standard normal curve*, which is shown in Fig. 1.6(*c*).

The area under a normal probability density function beyond 3σ from the mean is quite small. Since more than 0.9973 of the probability of a normal distribution is within the interval $(\mu - 3\sigma, \mu + 3\sigma)$, 6σ is often referred to as the *width* of a normal distribution. Advanced integration methods can be used to show that the area under the normal probability density function from $-\infty < x < \infty$ is 1.

1.10.5 Problem-Solving Procedure

If X is a normal random variable with $E(X) = \mu$ and $V(X) = \sigma^2$, the random variable

$$Z = \frac{X - \mu}{\sigma}$$

is a normal random variable with $E(Z) = 0$ and $V(Z) = 1$. That is, Z is a standard normal random variable.

Suppose X is a normal random variable with mean μ and variable σ^2. Then

$$P(X \le x) = P\left(\frac{X - \mu}{\sigma} \le \frac{x - \mu}{\sigma}\right) = P(Z \le z)$$

where Z is a *standard normal variable*, and $z = \left(\frac{x - \mu}{\sigma}\right)$ is the z-value obtained by standardizing X.

The probability is obtained by using the Table 1.8 with $z = (x - \mu)/\sigma$.

Basic Properties of the Standard Normal Curve

1. The total area under the standard normal curve is 1.
2. The standard normal curve extends indefinitely in both directions, approaching, but never touching, the horizontal axis as it does so.
3. The standard normal curve is symmetric about 0. That is, the part of the curve to the left of the vertical line at the center in Fig. 1.7(*a*) is the mirror image of the curve to the right of it.
4. Almost all the area under the standard normal curve lies between –3 and 3.

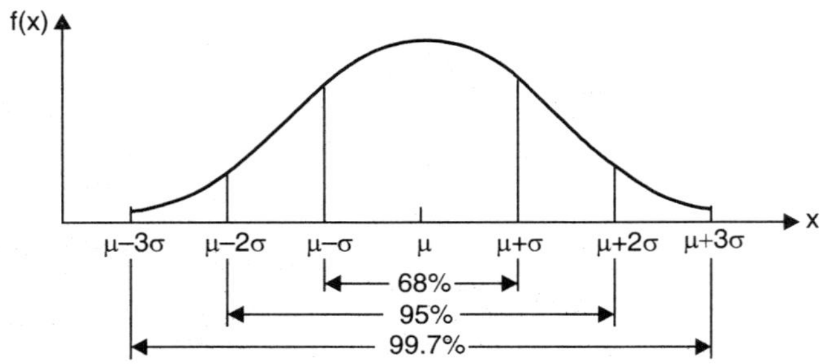

Fig. 1.7 (*a*) Probabilities associated with a normal distribution.

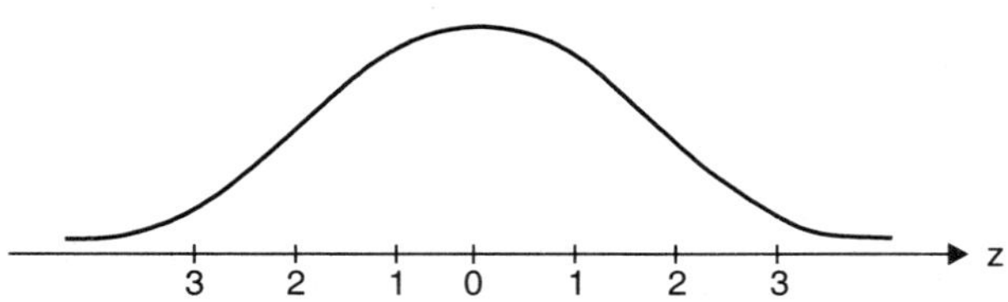

Fig. 1.7 (*b*) Standard normal distribution.

A normal random variable with $\mu = 0$ and $\sigma^2 = 1$ is called a *standard normal variable* and is denoted as Z. The cumulative distribution function of a standard normal random variable is denoted as $\phi(z) = P(Z \le z)$.

Table 1.9 provides cumulative probabilities for a standard normal random variable.

Table 1.9: Cumulative Standard Normal Distribution

$$\phi(z) = P(Z \le z) = \int_{-\infty}^{z} \frac{1}{\sqrt{2\pi}} e^{\frac{-u^2}{2}} du$$

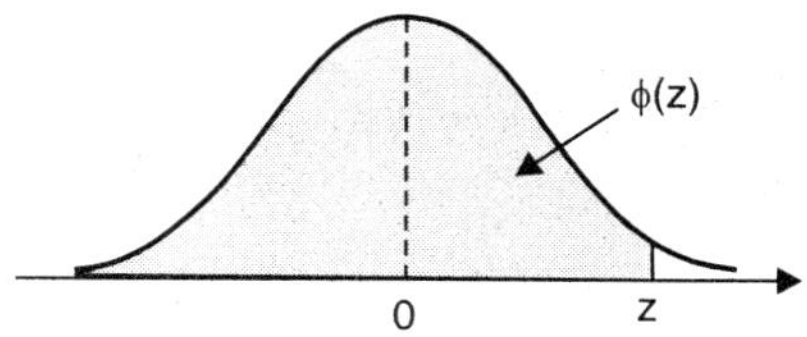

z	–0.09	–0.08	–0.07	–0.06	–0.05	–0.04	–0.03	–0.02	–0.01	–0.00	z
–3.9	0.000033	0.000034	0.000036	0.000037	0.000039	0.000041	0.000042	0.000044	0.000046	0.000048	–3.9
–3.8	0.000050	0.000052	0.000054	0.000057	0.000059	0.000062	0.000064	0.000067	0.000069	0.000072	–3.8
–3.7	0.000075	0.000078	0.000082	0.000085	0.000088	0.000092	0.000096	0.000100	0.000104	0.000108	–3.7
–3.6	0.000112	0.000117	0.000121	0.000126	0.000131	0.000136	0.000142	0.000147	0.000153	0.000159	–3.6
–3.5	0.000165	0.000172	0.000179	0.000185	0.000193	0.000200	0.000208	0.000216	0.000224	0.000233	–3.5
–3.4	0.000242	0.000251	0.000260	0.000270	0.000280	0.000291	0.000302	0.000313	0.000325	0.000337	–3.4
–3.3	0.000350	0.000362	0.000376	0.000390	0.000404	0.000419	0.000434	0.000450	0.000467	0.000483	–3.3
–3.2	0.000501	0.000519	0.000538	0.000557	0.000577	0.000598	0.000619	0.000641	0.000664	0.000687	–3.2
–3.1	0.000711	0.000736	0.000762	0.000789	0.000816	0.000845	0.000874	0.000904	0.000935	0.000968	–3.1
–3.0	0.001001	0.001035	0.001070	0.001107	0.001144	0.001183	0.001223	0.001264	0.001306	0.001350	–3.0
–2.9	0.001395	0.001441	0.001489	0.001538	0.001589	0.001641	0.001695	0.001750	0.001807	0.001866	–2.9
–2.8	0.001926	0.001988	0.002052	0.002118	0.002186	0.002256	0.002327	0.002401	0.002477	0.002555	–2.8
–2.7	0.002635	0.002718	0.002803	0.002890	0.002980	0.003072	0.003167	0.003264	0.003364	0.003467	–2.7
–2.6	0.003573	0.003681	0.003793	0.003907	0.004025	0.004145	0.004269	0.004396	0.004527	0.004661	–2.6
–2.5	0.004799	0.004940	0.005085	0.005234	0.005386	0.005543	0.005703	0.005868	0.006037	0.006210	–2.5
–2.4	0.006387	0.006569	0.006756	0.006947	0.007143	0.007344	0.007549	0.007760	0.007976	0.008198	–2.4
–2.3	0.008424	0.008656	0.008894	0.009137	0.009387	0.009642	0.009903	0.010170	0.010444	0.010724	–2.3
–2.2	0.011011	0.011304	0.011604	0.011911	0.012224	0.012545	0.012874	0.013209	0.013553	0.013903	–2.2
–2.1	0.014262	0.014629	0.015003	0.015386	0.015778	0.016177	0.016586	0.017003	0.017429	0.017864	–2.1
–2.0	0.018309	0.018763	0.019226	0.019699	0.020182	0.020675	0.021178	0.021692	0.022216	0.022750	–2.0
–1.9	0.023295	0.023852	0.024419	0.024998	0.025588	0.026190	0.026803	0.027429	0.028067	0.028717	–1.9
–1.8	0.029379	0.030054	0.030742	0.031443	0.032157	0.032884	0.033625	0.034379	0.035148	0.035930	–1.8
–1.7	0.036727	0.037538	0.038364	0.039204	0.040059	0.040929	0.041815	0.042716	0.043633	0.044565	–1.7
–1.6	0.045514	0.046479	0.047460	0.048457	0.049471	0.050503	0.050551	0.052616	0.053699	0054799	–1.6
–1.5	0.055917	0.057053	0.058208	0.059380	0.060571	0.061780	0.063008	0.064256	0.065522	0.066807	–1.5
–1.4	0.068112	0.069437	0.070781	0.072146	0.073529	0.074934	0.076359	0.077804	0.079270	0.080757	–1.4
–1.3	0.082264	0.083793	0.085343	0.086915	0.088508	0.090123	0.091759	0.093418	0.095098	0.096801	–1.3
–1.2	0.098525	0.100273	0.102042	0.103835	0.105650	0.107488	0.109349	0.111233	0.113140	0.115070	–1.2
–1.1	0.117023	0.119000	0.121001	0.123024	0.125072	0.127143	0.129238	0.131357	0.133500	0.135666	–1.1
–1.0	0.137857	0.140071	0.142310	0.144572	0.146859	0.149170	0.151505	0.153864	0.156248	0.158655	–1.0
–0.9	0.161087	0.163543	0.166023	0.168528	0.171056	0.173609	0.176185	0.178786	0.181411	0.184060	–0.9
–0.8	0.186733	0.189430	0.192150	0.194894	0.197662	0.200454	0.203269	0.206108	0.208970	0.211855	–0.8
–0.7	0.214764	0.217695	0.220650	0.223627	0.226627	0.229650	0.232695	0.235762	0.238852	0.241964	–0.7
–0.6	0.245097	0.248252	0.251429	0.254627	0.257846	0.261086	0.264347	0.267629	0.270931	0.274253	–0.6
–0.5	0.277595	0.280957	0.284339	0.287740	0.291160	0.294599	0.298056	0.301532	0.305026	0.308538	–0.5
–0.4	0.312067	0.315614	0.319178	0.322758	0.326355	0.329969	0.333598	0.337243	0.340903	0.344578	–0.4
–0.3	0.348268	0.351973	0.355691	0.359424	0.363169	0.366928	0.370700	0.374484	0.378281	0.382089	–0.3
–0.2	0.385908	0.389739	0.393580	0.397432	0.401294	0.405165	0.409046	0.412936	0.416834	0.420740	–0.2
–0.1	0.424655	0.428576	0.432505	0.436441	0.440382	0.444330	0.448283	0.452242	0.456205	0.460172	–0.1
0.0	0.464144	0.468119	0.472097	0.476078	0.480061	0.484047	0.488033	0.492022	0.496011	0.500000	0.0

Table 1.9 (Continued)

z	*–0.09*	*–0.08*	*–0.07*	*–0.06*	*–0.05*	*–0.04*	*–0.03*	*–0.02*	*–0.01*	*–0.00*	*z*
0.0	0.500000	0.503989	0.507978	0.511967	0.515953	0.519939	0.523922	0.527903	0.531881	0.535856	0.0
0.1	.0539828	0.543795	0.547758	0.551717	0.555760	0.559618	0.563559	0.567495	0.571424	0.575345	0.1
0.2	0.579260	0.583166	0.587064	0.590954	0.594835	0.598706	0.602568	0.606420	0.610261	0.614092	0.2
0.3	0.617911	0.621719	0.625516	0.629300	0.633072	0.636831	0.640576	0.644309	0.648027	0.651732	0.3
0.4	0.655422	0.659097	0.662757	0.666402	0.670031	0.673645	0.677242	0.680822	0.684386	0.687933	0.4
0.5	0.691462	0.694974	0.698468	0.701944	0.705401	0.708840	0.712260	0.715662	0.719043	0.722405	0.5
0.6	0.725747	0.729069	0.732371	0.735653	0.738914	0.742154	0.745373	0.748571	0.751748	0.754903	0.6
0.7	0.758036	0.761148	0.764238	0.767305	0.770350	0.773373	0.776373	0.779350	0.782305	0.785236	0.7
0.8	0.788145	0.791030	0.793892	0.796731	0.799546	0.802338	0.805106	0.807850	0.810570	0.813267	0.8
0.9	0.815940	0.818589	0.821214	0.823815	0.826391	0.828944	0.831472	0.833977	0.836457	0.838913	0.9
1.0	0.841345	0.843752	0.846136	0.848495	0.850830	0.853141	0.855428	0.857690	0.859929	0.862143	1.0
1.1	0.864334	0.866500	0.868643	0.870762	0.872857	0.874928	0.876976	0.878999	0.881000	0.882977	1.1
1.2	0.884930	0.886860	0.888767	0.890651	0.892512	0.894350	0.896165	0.897958	0.899727	0.901475	1.2
1.3	0.903199	0.904902	0.906582	0.908241	0.909877	0.911492	0.913085	0.914657	0.916207	0.917736	1.3
1.4	0.919243	0.920730	0.922196	0.923641	0.925066	0.926471	0.927855	0.929219	0.930563	0.931888	1.4
1.5	0.933193	0.934478	0.935744	0.936992	0.938220	0.939429	0.940620	0.941792	0.942947	0.944083	1.5
1.6	0.945201	0.946301	0.947384	0.948449	0.949497	0.950529	0.951543	0.952540	0.953521	0.954486	1.6
1.7	0.955435	0.956367	0.957284	0.958185	0.959071	0.959941	0.960796	0.961636	0.962462	0.963273	1.7
1.8	0.964070	0.964852	0.965621	0.966375	0.967116	0.967843	0.968557	0.969258	0.969946	0.970621	1.8
1.9	0.971283	0.971933	0.972571	0.973197	0.973810	0.974412	0.975002	0.975581	0.976148	0.976705	1.9
2.0	0.977250	0.977784	0.978308	0.978822	0.979325	0.979818	0.980301	0.980774	0.981237	0.981691	2.0
2.1	0.982136	0.982571	0.982997	0.983414	0.983823	0.984222	0.984614	0.984997	0.985371	0.985738	2.1
2.2	0.986097	0.986447	0.986791	0.987126	0.987455	0.987776	0.988089	0.988396	0.988696	0.988989	2.2
2.3	0.989276	0.989556	0.989830	0.990097	0.990358	0.990613	0.990863	0.991106	0.991344	0.991576	2.3
2.4	0.991802	0.992024	0.992240	0.992451	0.992656	0.992857	0.993053	0.993244	0.993431	0.993613	2.4
2.5	0.993790	0.993963	0.994132	0.994297	0.994457	0.994614	0.994766	0.994915	0.995060	0.995201	2.5
2.6	0.995339	0.995473	0.995604	0.995731	0.995855	0.995975	0.996093	0.996207	0.996319	0.996427	2.6
2.7	0.996533	0.996636	0.996736	0.996833	0.996928	0.997020	0.997110	0.997197	0.997282	0.997365	2.7
2.8	0.997445	0.997523	0.997599	0.997673	0.997744	0.997814	0.997882	0.997948	0.998012	0.998074	2.8
2.9	0.998134	0.998193	0.998250	0.998305	0.998359	0.998411	0.998462	0.998511	0.998559	0.998605	2.9
3.0	0.998650	0.998694	0.998736	0.998777	0.998817	0.998856	0.998893	0.998930	0.998965	0.998999	3.0
3.1	0.999032	0.999065	0.999096	0.999126	0.999155	0.999184	0.999211	0.999238	0.999264	0.999289	3.1
3.2	0.999313	0.999336	0.999359	0.999381	0.999402	0.999423	0.999443	0.999462	0.999481	0.999499	3.2
3.3	0.999517	0.999533	0.999550	0.999566	0.999581	0.999596	0.999610	0.999624	0.999638	0.999650	3.3
3.4	0.999663	0.999675	0.999687	0.999698	0.999709	0.999720	0.999730	0.999740	0.999749	0.999758	3.4
3.5	0.999767	0.999776	0.999784	0.999792	0.999800	0.999807	0.999815	0.999821	0.999828	0.999835	3.5
3.6	0.999841	0.999847	0.999853	0.999858	0.999864	0.999869	0.999874	0.999879	0.999883	0.999888	3.6
3.7	0.999892	0.999896	0.999900	0.999904	0.999908	0.999912	0.999915	0.999918	0.999922	0.999925	3.7
3.8	0.999928	0.999931	0.999933	0.999936	0.999938	0.999941	0.999943	0.999946	0.999948	0.999950	3.8
3.9	0.999952	0.999954	0.999956	0.999958	0.999959	0.999961	0.999963	0.999964	0.999966	0.999967	3.9

The procedure in solving problems involving calculating the cumulative normal probabilities is simple. For example, suppose that $X \sim N(50, 4)$, and we wish to find the probability that x is less than or equal to 54; that is $P(x \leq 54) = F(54)$. Since that standard normal random variable is

$$z = \frac{x - \mu}{\sigma}$$

we can *standardize* the point of interest $x = 104$ to obtain

$$z = \frac{x - \mu}{\sigma} = \frac{54 - 50}{2} = 2$$

Now the probability that the *standard* normal random variable z is less than or equal to 2 is equal to the probability that the *original* normal random variable x is less than or equal to 54. Expressed mathematically,

$$F(x) = \phi\left(\frac{x - \mu}{\sigma}\right) = \phi(z)$$

$$F(54) = \phi(2)$$

Table 1.8 contains cumulative standard normal probabilities for various values of z. From this table, we can read

$$\phi\,(2) = 0.9772$$

Note that in the relationship $z = (x - \mu)/\sigma$, the variable z measures the departure of x from the mean μ in standard deviation (σ) units. In our example, $F(54) = \phi\,(2)$ indicates that 54 is *two* standard deviations ($\sigma = 2$) above the mean. In general, $x = \mu + \sigma\, z$. In solving problems, we sometimes need to use the symmetry property of ϕ in addition to the tables.

In order to find the percentages of all possible observations of a normally distributed variable that lie within any specified range, we express the range in terms of z-scores and then determine the corresponding area under the standard normal curve. The stepwise procedure to determine a percentage or probability for a normally distributed variable is presented below:

1. Sketch the normal curve associated with the variable as shown in Fig. 1.7(*c*).
2. Shade the region of interest and mark its delimiting x-value (s).
3. Compute the z-scores for the delimiting x-value (s) found in step 2.
4. Use the Table 1.8 to find the area under the standard normal curve delimited by the z-score (s) found in step 3.

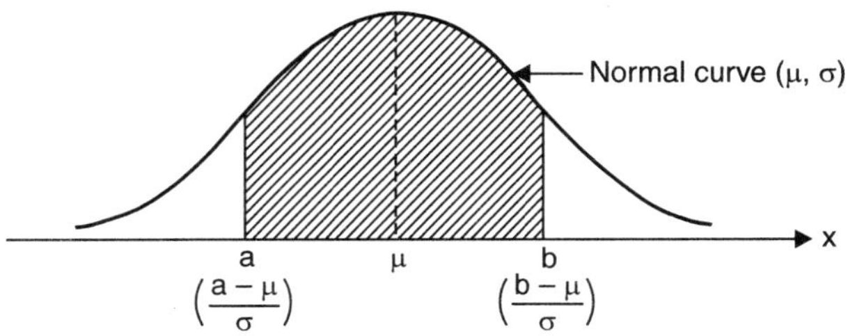

Fig. 1.7 (*c*)

Example 1.34: *The compressive strength of samples of cement can be modeled by a normal distribution with a mean of 7000 kg/cm² and a standard deviation of 100 kg/cm².*

(*a*) *find the probability of a sample's strength that is less than 7250 kg/cm².*

(*b*) *find the probability that a sample's strength is between 6800 and 6900 kg/cm².*

(*c*) *what strength is exceeded by 95% of the samples?*

Solution:

Refer to the Table 1.8.

(*a*)
$$P(X < 7250) = P\left(Z < \frac{7250-7000}{100}\right) = P(Z < 2.5) = 0.99379$$

(*b*)
$$P(6800 < X < 6900) = P\left[\frac{6800-7000}{100} < Z < \frac{6900-7000}{100}\right] = P(-2 < Z < -1)$$
$$= P(Z < -1) - P(Z < -2) = 0.13591$$

(*c*)
$$P(X > x) = P\left(Z > \frac{x-7000}{100}\right) = 0.95$$

Hence,
$$\frac{x-7000}{100} = -1.65 \text{ and } x = 6835.$$

Example 1.35:

(*a*) *A process manufactures ball bearings whose diameters are normally distributed with mean 3.505 cm and standard deviation 0.008 cm. Specifications call for the diameter to be in the interval 3.5 ± 0.01 cm. What proportion of the ball bearings will meet the specification?*

(*b*) *Suppose the process can be recalibrated so that the mean will equal to 3.5 cm, the center of the specification interval. The standard deviation of the process remains 0.008 cm. What proportion of the diameter will meet the specifications?*

Solution: Let X be the diameter of a randomly chosen ball bearing. Then, $X \sim N(3.505, 0.008^2)$. Figure E.1.35(*a*) shows the probability density function of the $N(3.505, 0.008^2)$ population. The shaded area represents $P(3.49 < X < 3.51)$, which is the proportion of ball bearings that meet the specification.

(*a*) We compute the z-scores of 3.49 and 3.51 as follows:

$$z = \frac{3.49-3.505}{0.008} = -1.88 \quad \text{and} \quad z = \frac{3.51-3.505}{0.008} = 0.63$$

The area to the left of $z = -1.88$ is 0.0301. The area to the left of $z = 0.63$ is 0.7357. The area between $z = 0.63$ and $z = -1.88$ is 0.7357 – 0.0301 = 0.7056. Hence, approximately 70.56% of the diameters of the ball bearings will meet the specifications.

(*b*) The mean is 3.5000 rather than 3.505. The calculation of area is as follows: [see Fig. 1.39(*b*)].

$$z = \frac{3.49-3.50}{0.008} = -1.25 \quad \text{and} \quad z = \frac{3.51-3.50}{0.008} = 1.25$$

The area to the left of $z = -1.25$ is 0.1056. The area to the left of $z = 1.25$ is 0.8944. The area between $z = 1.25$ and $z = -1.25$ is $0.8944 - 0.1056 = 0.7888$. Hence, recalibrating will increase the proportion of diameters that meet the specification to 78.88%.

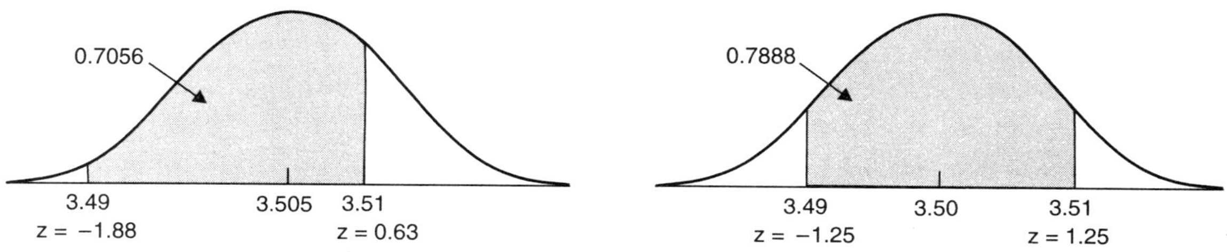

Fig. E1.35 (*a*) **Fig. E1.35** (*b*)

Example 1.36: *Refer to Example E1.35. If the process is centered so that the mean is 45 mm and the standard deviation is 0.05 mm and that 10 rods are measured and are assumed to be independent, determine*

(*a*) *the probability that all 10 rods measured are between 45.85 and 45.15 mm*

(*b*) *the expected number of the 10 rods that are between 44.85 and 45.15 mm.*

Solution:

(*a*) $P(44.85 < X < 45.15) = 0.9973$ from Example E1.35. Therefore, by independence, the probability of 10 rods that are within the given limit is $0.9973^{10} = 0.9733$.

(*b*) Let Y be the number of rods from the 10 selected that are within the given limits. Then, Y is binomial with $n = 10$ and $p = 0.9973$.

Hence, $E(Y) = 9.973$.

Example 1.37: *The weight of an electronic component is normally distributed with a mean of 6 ounces and a standard deviation of 0.25 ounce.*

(*a*) *find the probability that the electronic component weighs more than 6.5 ounces.*

(*b*) *what must the standard deviation of weight be in order for the company to state that 99% of its electronic components are less than 6.5 ounces?*

(*c*) *if the standard deviation stays at 0.25 ounces, what must the mean weight be in order for the company to state that 99.9% of its electronic components are less than 6.5 ounces?*

Solution:

(*a*) $$P(X > 6.5) = P\left(Z > \frac{6.5-6}{0.25}\right) = P(Z > 2) = 0.02275.$$

(*b*) If $P(X < 13) = 0.999$, then $P\left(Z < \frac{6.5-6}{\sigma}\right) = 0.999.$

Hence, $$\frac{0.5}{\sigma} = 3.09$$
$$\sigma = 0.1618$$

(*c*) If $P(X < 6.5) = 0.999$, then $P\left(Z < \frac{6.5-\mu}{0.25}\right) = 0.999$

Therefore, $\frac{6.5-\mu}{0.25} = 3.09$ and $\mu = 5.7275$.

Example 1.38: *The weight of a mechanical component is normally distributed with a mean of 22 oz and a standard deviation of 0.5 oz.*

(*a*) *what is the probability that a component weighs more than 23 oz?*

(*b*) *what must the standard deviation of weight be in order for the company to state that 99.9% of its components are less than 23 oz?*

(*c*) *if the standard deviation remains at 0.5 oz, what must the mean weight be in order for the company to state that 99.9% of its components are less than 23 oz?*

Solution: Refer to the Table 1.8.

(*a*)
$$P(X > 23) = P\left(Z > \frac{23-22}{0.5}\right) = P(Z > 2) = 1 - P(Z \le 2)$$
$$= 1 - 0.977250 = 0.022750.$$

(*b*)
$$\text{If } P(X < 23) = 0.999, \text{ then } P\left[Z < \frac{23-22}{\sigma}\right] = 0.999$$

Hence,
$$\frac{1}{\sigma} = 3.09 \text{ and } \sigma = \frac{1}{3.09} = 0.324$$

(*c*)
$$\text{If } P(X < 23) = 0.999, \text{ then } P\left[Z < \frac{23-\mu}{0.5}\right] = 0.999$$

Hence,
$$\frac{23-\mu}{0.5} = 3.09 \quad \text{or} \quad \mu = 21.455.$$

Example 1.39: *Refer to the Example 1.35. Assume that the process has been recalibrated so that mean diameter is now 3.5 cm. To what value must the standard deviation be lowered so that 95% of the diameters will meet the specifications?*

Solution: The specification interval is 3.49 – 3.51 cm. We must find a value for σ so that this interval spans the middle 95% of the population of ball bearing diameters, as shown in Fig. E1.39. The z-score that has 2.5% of the area to its left is $z = -1.96$. The z-score that has 2.5% of the area to its right is $z = 1.96$ (from the symmetry of the curve). It follows that the lower specification limit, 3.49, has a z-score of –1.96, while the upper limit of 3.51 has a score of 1.96. Either of these facts may be used to find σ.

Now
$$z = \frac{x-\mu}{\sigma}$$

or
$$1.96 = \frac{3.51-3.50}{\sigma}$$

or
$$\sigma = 0.005102 \text{ cm}.$$

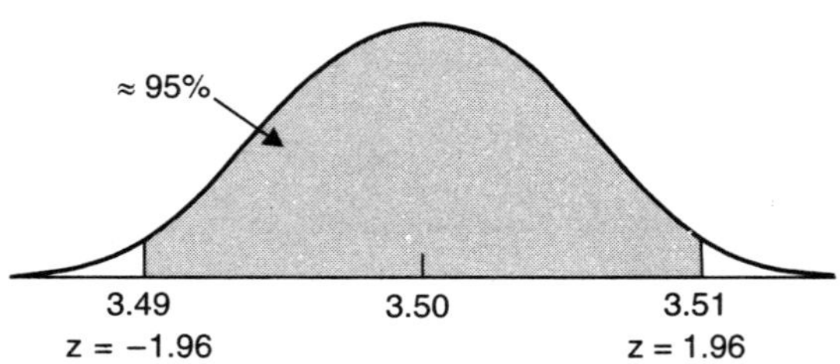

Fig. E1.39

Example 1.40: *An experiment needs a 2.41 cm diameter steel rod. Suppose that the diameter of a steel rod has a normal distribution with a mean of 2.41 cm and a standard deviation of 0.01 cm.*

(*a*) *determine the probability that a diameter is greater than 2.42 cm.*

(*b*) *what diameter is exceeded by 95% of the samples?*

(*c*) *if the specifications require that the diameter is between 2.39 cm and 2.43 cm, what proportion of the samples meets specifications?*

Solution: Refer to the Table 1.8.

(*a*) Let X denote the diameter of the steel rod.

$$X \sim N(2.41, 0.01^2)$$

$$P(X > 2.42) = 1 - P(X \leq 2.42) = 1 - \phi\left[\frac{2.42 - 2.41}{0.01}\right] = 1 - \phi(1) = 0.1587.$$

(*b*) $$P(X > x) = P\left[Z > \frac{x - 2.41}{0.01}\right] = 0.95$$

Hence, $$\frac{x - 2.41}{0.01} = 1.645 \text{ and } x = 2.3936 \text{ cm.}$$

(*c*) $$P(2.39 < X < 2.43) = \phi\left[\frac{2.43 - 2.41}{0.01}\right] - \phi\left[\frac{2.39 - 2.41}{0.01}\right] = \phi(2) - \phi(-2) = 0.9545.$$

1.10.6 Propagation of Error

Suppose $V = V(x, y)$ where V is computed from two measured quantities x and y. Then, V is in error by an amount dV as a result of the errors dx and dy in the measured quantities x and y. Generally, the error dx has no relation to the error dy, and they are said to be *independent*. When the errors dx and dy are related, then these are called *nonindependent*. Sometimes, both independent and nonindependent errors are present. In addition, there may exist a third type known as *correlated-errors*, in which the deviations of x are systematically related to those of y. Correlated errors are not an intrinsically new class of errors, but rather errors whose independent and nonindependent portions cannot be separated.

It is necessary to define the correlation coefficient ρ_{xy}, which is the average value of the sum $\Sigma\delta x_n\ \delta y_n$ for the universes of measurements of x and y, divided by $s_x s_y$. The best estimate of ρ_{xy} from a finite sample of k pairs of measurements of x and y is given by

$$\rho_{xy} = \frac{1}{(k-1)s_x s_y} \sum_{n=1}^{k} (\delta x_n \delta y_n) \tag{1.122}$$

or

$$\rho_{xy} = \frac{1}{(k-1)s_x s_y} \left[\sum_{n=1}^{k} (x_n y_n) k \bar{x}\, \bar{y}\right] \tag{1.123}$$

For completely independent errors, $\rho_{xy} = 0$ and for completely correlated errors $\rho_{xy} = \pm 1$.

1.10.7 Combining Independent and Correlated Errors

Here, we assume that we make a very large number, k, of measurements of x and y. We may, therefore, compute k values of V as follows:

$$
\begin{aligned}
V_1 &= V(x_1, y_1)\\
V_2 &= V(x_2, y_2)\\
&\vdots\\
V_n &= V(x_n, y_n)\\
&\vdots\\
V_k &= V(x_k\, y_k)
\end{aligned} \tag{1.124}
$$

Also, we assume here that the best value of V is $V(\bar{x}, \bar{y})$, where $\bar{x}$ and $\bar{y}$ are the average values of the measurements x and y respectively.

Each value V_n differs from $V(\bar{x}, \bar{y})$ by some amount δV_n, called the *deviation* of V_n or

$$\delta V_n = V_n - V(\bar{x}, \bar{y}) \tag{1.125}$$

Equation (1.125) can be written as

$$\delta V_n = \frac{\partial V}{\partial x}\delta x_n + \frac{\partial V}{\partial y}\delta y_n \tag{1.126}$$

where
$$\delta x_n = x_n - \bar{x}$$
$$\delta y_n = y_n - \bar{y}$$

The average value of $\delta V = \dfrac{\left(\sum_{n=1}^{k}\delta V_n\right)}{k} = 0$ since

$$\sum_{n=1}^{k}\delta x_n = 0 \quad \text{and} \quad \sum_{n=1}^{k}\delta y_n = 0$$

Hence
$$\bar{V} = V(\bar{x}, \bar{y}) \tag{1.127}$$

Now, the average value of the sum of the squares of the deviations from Eq. (1.126) is given by

$$\frac{\sum_{n=1}^{k}(\delta V_n)^2}{k} = \left(\frac{\partial V}{\partial x}\right)^2 \frac{\sum_{n=1}^{k}(\delta V x_n)^2}{k} + \left(\frac{\partial V}{\partial y}\right)^2 \frac{\sum_{n=1}^{k}(\delta V y_n)^2}{k} + 2\left(\frac{\partial V}{\partial x}\right)\left(\frac{\partial V}{\partial y}\right)\frac{\sum_{n=1}^{k}(\delta V x_n)(\delta V y_n)}{k} \tag{1.128}$$

Taking the square root of each side and $k \to \infty$, then Eq. (1.128) becomes

$$s_V = \sqrt{\left(\frac{\partial V}{\partial x}\right)^2 s_x^2 + \left(\frac{\partial V}{\partial y}\right)^2 s_y^2 + 2\rho_{xy}\left(\frac{\partial V}{\partial x}\right)\left(\frac{\partial V}{\partial y}\right)s_x s_y} \tag{1.129}$$

where
s_V = standard deviation of V
s_x = standard deviation of x
s_y = standard deviation of y
ρ_{xy} = correlation coefficient

Equation (1.129) gives the best estimate of s_V for a finite number of measurements of x and y. In the case of correlated errors, we have the same number of measurements of x and y. When the measurements are independent, $\rho_{xy} = 0$ and the number of measurements of x differ from the number of measurements of y.

1.10.8 Standard Deviation of an Average

By definition

$$\bar{x} = \frac{1}{k}(x_1 + x_2 + ... + x_n + ... + x_k) \tag{1.130}$$

Hence
$$\frac{\partial \bar{x}}{\partial x_1} = \frac{\partial \bar{x}}{\partial x_2} = ... = \frac{\partial \bar{x}}{\partial x_n} = ... = \frac{\partial \bar{x}}{\partial x_k} = \frac{1}{k} \tag{1.131}$$

Here, we are considering only the random errors in the measurements. Therefore, x_n's are considered independent and their mutual correlation coefficients are zero. The standard deviation of the average is given by applying Eq. (1.129).

$$s_{\bar{x}} = \sqrt{\frac{1}{k^2}\left(s_{x_1}^2 + s_{x_2}^2 + ... + s_{x_n}^2 + ... + s_{x_k}^2\right)} = \frac{s}{\sqrt{k}} \tag{1.132}$$

In Eq. (1.132), we note that $s_{x_n} = s$ and the number of them is k. Hence, the standard deviation of the average is equal to the standard deviation of an individual measurement divided by the square root of the number of independent measurements.

Let s = the rms deviation about the universe average $(\bar{x}_u) = \sum_{n=1}^{k} (\delta x_n)^2$

$\bar{x}_k$ = sample average = $\sum_{n=1}^{k} x_n$

then
$$s^2 = \frac{\sum_{n=1}^{k} x_n^2}{k} - 2\bar{x}_u \bar{x}_k + \bar{x}_u^2 \tag{1.133}$$

The square of the rms deviation about the sample average is given by

$$s'^2 = \frac{\sum_{n=1}^{k} x_n^2}{k} - \bar{x}_k^2 \tag{1.134}$$

Eq. (1.133) minus Eq. (1.134) gives

$$s^2 - s'^2 = \bar{x}_k^2 - 2\bar{x}_u \bar{x}_k + \bar{x}_u^2 = (\bar{x}_k - \bar{x}_u)^2 \tag{1.135}$$

where $(\bar{x}_k - \bar{x}_u)$ is the deviation of $\bar{x}_k$ about $\bar{x}_u$ and is equal to $s_{\bar{x}}$. Using Eq. (1.129), we can write

$$s = s^1 \sqrt{\frac{k}{k-1}} \tag{1.136}$$

Now
$$\frac{s_{\bar{x}}}{s_x} = \frac{s_{\bar{y}}}{s_y} = \frac{1}{\sqrt{k}}$$

and therefore
$$\frac{s_{\bar{v}}}{s_v} = \frac{1}{\sqrt{k}}$$

Hence
$$s_{\bar{V}} = \sqrt{\left(\frac{\partial V}{\partial x}\right)^2 s_{\bar{x}}^2 + \left(\frac{\partial V}{\partial y}\right)^2 s_{\bar{y}}^2 + 2\rho_{xy}\left(\frac{\partial V}{\partial x}\right)\left(\frac{\partial V}{\partial y}\right) s_{\bar{x}} s_{\bar{y}}} \tag{1.137}$$

1.10.9 Method of Combining Nonindependent Errors

When the errors in x and y are due to some common cause, the error of V is given by

$$dV = \frac{\partial V}{\partial x}dx + \frac{\partial V}{\partial y}dy \tag{1.138}$$

For completely correlated errors, $\rho_{xy} = 1$ and Eq. (1.129) reduces to Eq. (1.138). When both nonindependent and correlated errors are present, they can be computed from Eqs. (1.129) and (1.138). On the other hand, these groups of errors are completely independent of each other: the total error in V is obtained by computing $\sqrt{dV^2 + s_v^2}$.

1.10.10 Special Rules for Combining Errors

A few rules for commonly occurring functions as sums and products may be derived from Eqs. (1.129), (1.137) and (1.138).

(*a*) **Sum and difference**

$$V = x \pm y$$

$$\frac{\partial V}{\partial x} = 1 \quad \text{and} \quad \frac{\partial V}{\partial y} = \pm 1$$

Hence, by Eq. (1.129), we have

$$s_V = \sqrt{s_x^2 + s_y^2} \tag{1.139}$$

Hence, the absolute standard deviation of a quantity which is the sum or difference of measured quantities is equal to the square root of the sum of the squares of the absolute standard deviations of the measured quantities.

(*b*) **Product of factors raised to various powers**

Let
$$V = x^m y^q \tag{1.140}$$

where m and q are known exactly.

$$\frac{\partial V}{\partial x} = mx^{m-1}y^q \quad \text{and} \quad \frac{\partial V}{\partial y} = qx^m y^{q-1} \tag{1.141}$$

Substituting the values of Eq. (1.141) in Eq. (1.129), we obtain

$$s_V = \sqrt{m^2 x^{2(m-1)} y^{2q} s_x^2 + q^2 x^{2m} y^{2(q-1)} s_y^2} \tag{1.142}$$

Dividing Eq. (1.142) by Eq. (1.140), we get

$$\frac{s_V}{V} = \sqrt{m^2\left(\frac{s_x}{x}\right)^2 + q^2\left(\frac{s_y}{y}\right)^2} \tag{1.143}$$

Now, recalling $S_V = \frac{s_V}{V}$, $S_x = \frac{s_x}{x}$ and $S_y = \frac{s_y}{y}$, we can write

$$S_V = \sqrt{m^2 S_x^2 + q^2 S_y^2} \tag{1.44}$$

Therefore, when a calculated quantity is equal to the product of measured quantities raised to various powers, then the fractional standard deviation of the calculated quantity is equal to the square root of the sum of terms consisting of the squares of the fractional deviations of the measured quantities multiplied by the squares of their respective powers.

(*c*) **Simple product or quotient**

Here, $V = xy$ or $V = \frac{x}{y}$

Therefore, $m = 1$ and $q = \pm 1$ in (*b*) above.

Hence, Eq. (1.144) becomes

$$S_V = \sqrt{S_x^2 + S_y^2} \tag{1.145}$$

Therefore, the fractional standard deviation in a product or quotient is equal to the square root of the sum of the squares of the fractional standard deviations of the measured quantities.

(*d*) **Logarithm of quantity**

If $V = B \ln x$, where B is a constant (known), then

$$\frac{dV}{dx} = \frac{B}{x}$$

and

$$s_V = \left(\frac{B}{x}\right) s_x$$

Hence, $$s_V = BS_V \tag{1.146}$$

Therefore, the absolute standard deviation of the natural logarithm of $x (B = 1)$ is equal to the fractional standard deviation of x, while the absolute standard deviation of the logarithm to the base 10 (the case $B = 0.434$) is equal to 0.434 times the fractional standard deviation of x.

Example 1.41: *The deflection Y at the center of a horizontal beam of length L supported at its ends and loaded in the center with a weight W is given by*

$$Y = \frac{WL^3}{48\,EI}$$

where E and I are elastic modulus and the moment of inertia of the cross-section about its center of area respectively. The measurements of W and L are given by

$$W = 100 \text{ tons} \pm 1.5 \text{ ton}$$

$$L = 50 \text{ ft} \pm 0.5 \text{ ft}$$

where the ± in each case refers to the standard deviation of the mean. Determine the fractional standard deviation of Y.

Solution: We have

$$\left(\frac{\sigma_y}{Y}\right)^2 = \left(\frac{\sigma_w}{W}\right)^2 + 3^2\left(\frac{\sigma_\ell}{L}\right)^2 = (0.015)^2 + 3^2(0.01)^2$$

$$= (2.25)10^{-4} + 9(10^{-4}) = 1.125(10^{-3})$$

or

$$\left(\frac{\sigma_y}{Y}\right) = 0.03354.$$

1.11 FINITE DIFFERENCE UNCERTAINTIES

1.11.1 Finite Difference Operators

Consider a function $y = f(x)$ defined on (a, b). x and y are the independent and dependent variables respectively. If the points $x_0, x_1, \ldots, x_n$ are taken at equidistance, *i.e.*, $x_i = x_0 + ih$, $i = 0, 1, 2, \ldots, n$, then the value of y, when $x = x_i$, is denoted as y_i, where $y_i = f(x_i)$. Here, the values of x are called *arguments* and the values of y are known as *entries*. The interval h is called the *difference interval*. The differences $y_1 - y_0, y_2 - y_1, \ldots, y_n - y_{n-1}$ are called the *first differences* of the function y. They are denoted by $\Delta y_0, \Delta y_1, \ldots$, etc. That is

$$\Delta y_0 = y_1 - y_0$$
$$\Delta y_1 = y_2 - y_1$$
$$\vdots$$
$$\Delta y_n = y_n - y_{n-1} \tag{1.147}$$

The symbol Δ in Eq. 1.147 is called the *difference operator*.

1.11.2 Forward Differences

The *forward difference* or simply *difference* operator is denoted by Δ and may be defined as

$$\Delta f(x) = f(x + h) - f(x) \tag{1.148}$$

or writing in terms of y, at $x = x_i$, Eq. (1.148) becomes

$$\Delta f(x_i) = f(x_i + h) - f(x_i) \tag{1.149}$$

or

$$\Delta y_i = y_{i+1} - y_i,\ i = 0, 1, 2, \ldots, n - 1$$

The differences of the first differences are called the *second differences* and they are denoted by $\Delta^2 y_0, \Delta^2 y_1, \ldots, \Delta^2 y_n$.

Hence

$$\Delta^2 y_0 = \Delta y_1 - \Delta y_0 = (y_2 - y_1) - (y_1 - y_0) = y_2 - 2y_1 + y_0$$
$$\Delta^2 y_1 = \Delta y_2 - \Delta y_1 = (y_3 - y_2) - (y_2 - y_1) = y_3 - 2y_2 + y_1$$
$$\Delta^3 y_0 = \Delta^2 y_1 - \Delta^2 y_0 = (y_3 - 2y_2 + y_1) - (y_2 - 2y_1 + y_0)$$
$$= y_3 - 3y_2 + 3y_1 - y_0$$
$$\Delta^3 y_1 = y_4 - 3y_3 + 3y_2 - y_1, \text{ etc.}$$

Generalizing, we have

$$\Delta^{n+1} f(x) = \Delta[\Delta^n f(x)],\ i.e.,\ \Delta^{n+1} y_i = \Delta[\Delta^n y_i],\ n = 0, 1, 2, \ldots \tag{1.150}$$

Also,

$$\Delta^{n+1} f(x) = \Delta^n[f(x + h) - f(x)] = \Delta^n f(x + h) - \Delta^n f(x)$$

and

$$\Delta^{n+1} y_i = \Delta^n y_{i+1} - \Delta^n y_i,\ n = 0, 1, 2, \ldots \tag{1.151}$$

where $\Delta^0 \equiv$ identity operator, *i.e.*, $\Delta^0 f(x) = f(x)$ and $\Delta^1 = \Delta$.

Table 1.10 Forward difference table

x	y	Δy	$\Delta^2 y$	$\Delta^3 y$	$\Delta^4 y$	$\Delta^5 y$
x_0	y_0					
		Δy_0				
x_1	y_1		$\Delta^2 y_0$			
		Δy_1		$\Delta^3 y_0$		
x_2	y_2		$\Delta^2 y_1$		$\Delta^4 y_0$	
		Δy_2		$\Delta^3 y_1$		$\Delta^5 y_0$
x_3	y_3		$\Delta^2 y_2$		$\Delta^4 y_1$	
		Δy_3		$\Delta^3 y_2$		
x_4	y_4		$\Delta^2 y_3$			
		Δy_4				
x_5	y_5					

The forward differences for the arguments $x_0, x_1, \ldots, x_5$ are shown in Table 1.10. Table 1.10 is called a *diagonal difference table* or *forward difference table*. The first term in Table 1.10 is y_0 and is called the *leading term*. The differences $\Delta y_0, \Delta^2 y_0, \Delta^3 y_0, \ldots$, are called the *leading differences*. Similarly, the differences with fixed subscript are called *forward differences*.

1.11.3 Backward Differences

The *backward difference operator* is denoted by ∇ and it is defined as

$$\nabla f(x) = f(x) -- f(x - h) \tag{1.152}$$

Equation (1.152) can be written as

$$\nabla y_i = y_i - y_{i-1}, \quad i = n, n-1, \ldots, 1. \tag{1.153}$$

or

$$\nabla y_1 = y_1 - y_0, \nabla y_2 = y_2 - y_1, \ldots, \nabla y_n = y_n - y_{n-1} \tag{1.154}$$

The differences in Eq. (1.154) are called *first differences*. The *second differences* are denoted by

$$\nabla^2 y_2, \nabla^2 y_3, \ldots, \nabla^2 y_n.$$

Hence

$$\nabla^2 y_2 = \nabla(\nabla y_2) = \nabla(y_2 - y_1) = \nabla y_2 - \nabla y_1$$
$$= (y_2 - y_1) - (y_1 - y_0) = y_2 - 2y_1 + y_0.$$

Similarly, $\nabla^2 y_3 = y_3 - 2y_2 + y_1$, $\nabla^2 y_4 = y_4 - 2y_3 + y_2$, and so on.

Generalizing, we have

$$\nabla^k y_i = \nabla^{k-1} y_i - \nabla^{k-1} y_{i-1}, \qquad i = n, n-1, \ldots, k \tag{1.155}$$

where

$$\nabla^0 y_i = y_i, \nabla^1 y_i = \nabla y_i.$$

The backward differences written in a tabular form is shown in Table 1.10. In Table 1.10, the differences $\nabla^n y$ with a fixed subscript i lie along the diagonal upward sloping.

Table 1.11 Backward difference table

x	y	∇y	$\nabla^2 y$	$\nabla^3 y$	$\nabla^4 y$
x_0	y_0				
		∇y_1			
x_1	y_1		$\nabla^2 y_2$		
		∇y_2		$\nabla^3 y_3$	
x_2	y_2		$\nabla^2 y_3$		$\nabla^4 y_4$
		∇y_3		$\nabla^3 y_4$	
x_3	y_3		$\nabla^2 y_4$		
		∇y_4			
x_4	y_4				

Table 1.11 is called the *backward difference* or *horizontal table.*

1.11.4 Central Differences

The central difference operator is denoted by the symbol δ and is defined by

$$\delta f(x) = f(x + h/2) - f(x - h/2)$$

where h is the interval of differencing.

In terms of y, the first central difference is written as

$$\delta y_1 = y_{i+1/2} - y_{i-1/2} \tag{1.156}$$

where $y_{i+1/2} = f(x_i + h/2)$ and $y_{i-1/2} = f(x_i - h/2)$.

Hence $\delta y_{1/2} = y_1 - y_0,\ \delta y_{3/2} = y_2 - y_1,\ \ldots,\ \delta y_{n-1/2} = y_n - y_{n-1}.$

The second central differences are given by

$$\delta^2 y_i = \delta y_{i+1/2} - \delta y_{i-1/2} = (y_{i+1} - y_i) - (y_i - y_{i-1}) = y_{i+1} - 2y_i + y_{i-1}.$$

Generalizing

$$\delta^n y_i = \delta^{n-1} y_{i+1/2} - \delta^{n-1} y_{i-1/2}. \tag{1.157}$$

The central difference table for the seven arguments $x_0, x_1, \ldots, x_4, x_5, x_6$ and x_7 is shown in Table 1.12.

Table 1.12 Central difference table

x	y	δ	δ^2	δ^3	δ^4	δ^5	δ^6
x_0	y_0						
		$\delta y_{1/2}$					
x_1	y_1		$\delta^2 y_1$				
		$\delta y_{3/2}$		$\delta^3 y_{3/2}$			
x_2	y_2		$\delta^2 y_2$		$\delta^4 y_2$		
		$\delta y_{5/2}$		$\delta^3 y_{5/2}$		$\delta^5 y_{5/2}$	
x_3	y_3		$\delta^2 y_3$		$\delta^4 y_3$		$\delta^6 y_3$
		$\delta y_{7/2}$		$\delta^3 y_{7/2}$		$\delta^5 y_{7/2}$	
x_4	y_4		$\delta^2 y_4$		$\delta^4 y_4$		
		$\delta y_{9/2}$		$\delta^3 y_{9/2}$			
x_5	y_5		$\delta^2 y_5$				
		$\delta y_{11/2}$					
x_6	y_6						

It is noted in Table 1.11 that all odd differences have fraction suffices and all the even differences are with integral suffices.

Example 1.42:

(*a*) *Construct the forward difference table and the horizontal table for the following data:*

x	1	2	3	4	5
$y = f(x)$	4	6	9	12	17

(*b*) *Construct a forward difference table for the following data:*

x	0	10	20	30
y	0	0.174	0.347	0.518

(*c*) *Construct a difference table for* $y = f(x) = x^2 + 2x + 1$ *for* $x = 1, 2, 3, 4, 5$.

(*d*) *Obtain the backward differences for the function* $f(x) = x^3$ *from* $x = 1$ *to* 1.05 *to two decimals chopped.*

Solution:

(*a*) The forward and the horizontal or backward difference tables are shown in Tables E1.42(*a*) and E1.42 (*b*).

Table E1.42 (a) Forward difference table

x	$f(x)$	$\Delta f(x)$	$\Delta^2 f(x)$	$\Delta^3 f(x)$	$\Delta^4 f(x)$
1	4				
		2			
2	6		1		
		3		1	
3	9		0		3
		3		2	
4	12		2		
		5			
5	17				

Table E1.42 (b) Horizontal or backward difference table

x	$f(x)$	$\Delta f(x)$	$\Delta^2 f(x)$	$\Delta^3 f(x)$	$\Delta^4 f(x)$
1	4				
2	6	2			
3	9	3	1		
4	12	3	0	1	
5	17	5	2	2	3

(*b*) Table E1.46(*c*) shows the forward difference operations.

Table E1.46 (c)

x	y	Δy	$\Delta^2 y$	$\Delta^3 y$
0	0			
		0.174		
10	0.174		– 0.001	
		0.173		– 0.001
20	0.347		– 0.002	
		0.171		
30	0.518			

(*c*) Table E1.46(*d*) shows the forward difference table.

Table E1.46 (d)

x	$x = f(x)$	Δy	$\Delta^2 y$	$\Delta^3 y$
1	4			
		9		
2	13		12	
		21		6
3	34		18	
		39		6
4	73		24	
		63		
5	136			

(*d*) The following Table E1.46(*e*) shows the backward differences.

Table E1.46 (e)

x	$y = f(x)$	∇y	$\nabla^2 y$	$\nabla^3 y$	$\nabla^4 y$
1.00	1				
		0.030			
1.01	1.030		0.001		
		0.031		– 0.001	
1.02	1.061		0.000		0.002
		0.031		0.001	
1.03	1.092		0.001		– 0.001
		0.032		0.000	
1.04	1.124		0.001		
		0.033			
1.05	1.157				

Note the typical oscillations in sign and growth of the entries.

1.11.5 Error Propagation in a Difference Table

Let $y_0, y_1, y_2, \ldots, y_n$ be the true values of a function and suppose the value y_4 to be affected with an error $\in$, so that its erroneous value is $y_4 + \in$. Then the successive differences of the y are as shown in Table 1.13.

Table 1.13 Error propagation in a difference table

y	Δy	$\Delta^2 y$	$\Delta^3 y$
y_0			
	Δy_0		
y_1		$\Delta^2 y_0$	
	Δy_1		$\Delta^3 y_0$
y_2		$\Delta^2 y_0$	
	Δy_2		$\Delta^3 y_1$
y_3		$\Delta^2 y_1$	
	Δy_3		$\Delta^3 y_2 + \in$
y_4		$\Delta^2 y_3 + \in$	
	$\Delta y_4 + \in$		$\Delta^3 y_3 - 3\in$
$y_5 + \in$		$\Delta^2 y_4 - 2\in$	
	$\Delta y_5 - \in$		$\Delta^3 y_4 + 3\in$
y_6		$\Delta^2 y_5 + \in$	
	Δy_6		$\Delta^3 y_5 - \in$
y_7		$\Delta^2 y_6$	
	Δy_7		$\Delta^3 y_6$
y_8		$\Delta^2 y_7$	
	Δy_8		
y_9			

Table 1.12 shows that the effect of an error increases with the successive differences, that the coefficients of the $\in$'s are the binomial coefficients with alternating signs, and that the algebraic sum of the errors in any difference column is zero. The same effect is also true for the horizontal difference Table 1.12.

Example 1.43: *Table E1.43 gives the values of a polynomial of degree five. It is given that f (4) is in error. Correct the value of f (4).*

Table E1.43

x	1	2	3	4	5	6	7
$y = f(x)$	0.975	–0.6083	–3.5250	–5.5250	–6.3583	4.2250	36.4750

Solution:

It is given that $y = f(x)$ is a polynomial of degree five. Hence, $\Delta^5 y$ must be a constant and $f(4)$ is in error.

Let $-5.5250 + \in$ be the true or correct value. The difference table is shown in Table E1.43(*a*).

Table E1.43 (a)

x	y	Δy	$\Delta^2 y$	$\Delta^3 y$	$\Delta^4 y$	$\Delta^5 y$
1	0.975					
		–1.5833				
2	–0.6083		–1.3333			
		–2.9167		$2.25 + \in$		
3	–3.5250		$0.9167 + \in$		$-2 - 4\in$	
		$-2 + \in$		$0.25 - 3\in$		$12 + 10\in$
4	$-5.5250 + \in$		$1.1667 - 2\in$		$10 + 6\in$	
		$-0.8333 - \in$		$10.25 + 3\in$		$-10 - 10\in$
5	–6.3583		$11.4667 + \in$		$0 - 4\in$	
		10.5833		$10.25 - \in$		
6	4.2250		21.6667			
		32.2500				
7	36.4750					

Since the fifth differences of y are constant, we have

$$12 + 10\in = -10 - 10\in$$

or $$20\in = -22$$

or $$\in = -1.1$$

Hence $$f(4) = -5.5250 + \in = -5.5250 - 1.1$$

or $$f(4) = -6.6250$$

Example 1.44: *The following is a table of values of a polynomial of degrees 5. It is given that f(3) is in error. Correct the error.*

Table E1.44

2	3	4	5	6
3	2	1	3	7
3	5	0	1	7
	4	5	2	7
		4	6	7

Solution: It is given that $y = f(x)$ is a polynomial of degree 5.

Hence, $\Delta^5 y$ must be constant; $f(3)$ is in error.

Let 254 + ε be the true value, now we form the difference table as shown in Table E1.44(*a*).

Table E1.44 (a)

x	y	Δy	$\Delta^2 y$	$\Delta^3 y$	$\Delta^4 y$	$\Delta^5 y$
0	1					
		1				
1	2		30			
		31		160 + ε		
2	33		190 + ε		200 – 4ε	
		221 + ε		360 – 3ε		220 + 10ε
3	254 + ε		550 – 2ε		420 + 6ε	
		1771 – ε		1780 + 3ε		20 – 10ε
4	1054		1330 + ε		440 – 4ε	
		2101		1220 – ε		
5	3126		12550			
		4651				
6	7777					

Since the fifth differences of y are constant

$$220 + 10\varepsilon = 20 - 10\varepsilon$$

$$\Rightarrow \quad 20\varepsilon = -200$$

$$\Rightarrow \quad \varepsilon = -10$$

Hence $$f(3) = 254 + \varepsilon$$

$$\Rightarrow \quad f(3) = 244.$$

Example 1.45: Table E1.45 below shows a difference table. Find the location of error.

Table E1.45

x	$y = x^3$	Δ	Δ^2	Δ^3	Δ^3
5	125	7.651	0.306	0.006	0
5.1	132.651	7.957	0.312	0.006	0
5.2	140.608	8.269	0.318	0.006	–0.027
5.3	148.877	8.587	0.324	–0.021	0.081
5.4	157.464	8.911	***0.303***	0.060	–0.081
5.5	166.375	9.214	***0.363***	–0.021	0.027
5.6	175.616	9.577	0.342	0.006	0
5.7	185.193	9.919	0.348	0.006	
5.8	195.112	10.267	0.354		
5.9	205.379	10.621			
6	216				

Solution: The location of an error is obviously centered on 0.060 in the third difference. Instead of 9.214, one has wrongly entered it as 9.241.

1.11.6 Numerical Differentiation

Numerical differentiation deals with the following problem: *given* the function $y = f(x)$, find one of its derivatives at the point $x = x_k$. Here, the term *given* implies that we either have an algorithm for computing the function, or possess a set of discrete data points (x_i, y_i), $i = 1, 2, \ldots, n$. In other words, we have a finite number of (x, y) data points or pairs from which we can compute the derivative. Numerical differentiation is a method to compute the derivatives of a function at some values of independent variable x, when the function $f(x)$ is explicitly unknown, however, it is known only for a set of arguments.

Like the numerical interpolation discussed in Chapter 5, a number of formulae for differentiation are derived in this chapter. They are:

(*a*) Derivatives based on Newton's forward interpolation formula. This formula is used to find the derivative for some given x lying near the beginning of the data table.

(*b*) Derivatives based on Newton's backward interpolation formula. This formula is suitable to find the derivative for a point near the end of the data table.

(*c*) Derivatives based on Stirling's interpolation formula. This formula is used to find the derivative for some point lying near the middle of the tabulated value.

A method to find the maxima and minima of a given function is also discussed in this chapter.

1.11.7 Derivatives Based on Newton's Forward Integration Formula

Suppose the function $y = f(x)$ is known at $(n + 1)$ equispaced points $x_0, x_1, \ldots, x_n$ and they are $y_0, y_1, \ldots, y_n$ respectively, *i.e.*, $y_i = f(x_i)$, $i = 0, 1, \ldots, n$. Let $x_i = x_0 + ih$ and $u = \frac{x - x_i}{h}$, where h is the spacing.

Referring to Chapter 5, the Newton's forward interpolation formula is

$$y = f(x) = y_0 + u\Delta y_0 + \frac{u(u-1)}{2!}\Delta^2 y_0 + \ldots + \frac{u(u-1)\ldots(u-\overline{n-1})}{n!}\Delta^n y_0$$

$$= y_0 + u\Delta y_0 + \frac{u^2 - u}{2!}\Delta^2 y_0 + \frac{u^3 - 3u^2 + 2u}{3!}\Delta^3 y_0 + \frac{u^4 - 6u^3 + 11u^2 - 6u}{4!}\Delta^4 y_0$$

$$+ \frac{u^5 - 10u^4 + 35u^3 - 50u^2 + 24u}{5!}\Delta^5 y_0 + \ldots \qquad (1.158)$$

Differentiating Eq. (1.158) w.r.t. x, we get

$$f'(x) = \frac{1}{h}\left[\Delta y_0 + \frac{2u-1}{2!}\Delta^2 y_0 + \frac{3u^2 - 6u + 2}{3!}\Delta^3 y_0 + \frac{4u^3 - 18u^2 + 22u - 6}{4!}\Delta^4 y_0\right.$$

$$\left. + \frac{5u^4 - 40u^3 + 105u^2 - 100u + 24}{5!}\Delta^5 y_0 + \cdots\right] \qquad (1.159)$$

Note here that $\frac{du}{dx} = \frac{1}{h}$.

Differentiating Eq. (1.159) w.r.t. x, we obtain

$$f''(x) = \frac{1}{h^2}\left[\Delta^2 y_0 + \frac{6u-6}{3!}\Delta^3 y_0 + \frac{12u^2-36u+22}{4!}\Delta^4 y_0 + \frac{20u^3-120u^2+210u-100}{5!}\Delta^5 y_0 + \ldots\right] \tag{1.160}$$

and so on.

Equations (1.159) and (1.160) give the approximate derivatives of $f(x)$ at arbitrary point $x = x_0 + uh$. When $x = x_0$, $u = 0$, Eqs. (1.159) and (1.160) become

$$f'(x_0) = \frac{1}{h}\left[\Delta y_0 - \frac{1}{2}\Delta^2 y_0 + \frac{1}{3}\Delta^3 y_0 - \frac{1}{4}\Delta^4 y_0 + \frac{1}{5}\Delta^5 y_0 - \ldots\right] \tag{1.161}$$

and

$$f''(x_0) = \frac{1}{h^2}\left[\Delta^2 y_0 - \Delta^3 y_0 + \frac{11}{12}\Delta^4 y_0 - \frac{5}{6}\Delta^5 y_0 + \ldots\right] \tag{1.162}$$

and so on.

Example 1.46: *From the following table find the value of* $\frac{dy}{dx}$ *and* $\frac{d^2y}{dx^2}$ *at the point* $x = 1.0$.

x	*1*	*1.1*	*1.2*	*1.3*	*1.4*	*1.5*
y	*5.4680*	*5.6665*	*5.9264*	*6.2551*	*6.6601*	*7.1488*

Solution: The forward difference table is

x	y	Δy	$\Delta^2 y$	$\Delta^3 y$
1.0	5.4680			
		0.1985		
1.1	5.6665		0.0614	
		0.2599		0.0074
1.2	5.9264		0.0688	
		0.3287		0.0074
1.3	6.2551		0.0763	
		0.4050		0.0074
1.4	6.6601		0.0837	
		0.4887		
1.5	7.1488			

Here $x_0 = 1.0$ and $h = 0.1$. Then $u = 0$ and hence

$$\frac{dy}{dx} = y'(1.0) = \frac{1}{h}\left[\Delta y_0 - \frac{1}{2}\Delta^2 y_0 + \frac{1}{3}\Delta^3 y_0 - \ldots\right]$$

$$= \frac{1}{0.1}\left[0.1985 - \frac{1}{2}(0.0614) + \frac{1}{3}(0.0074)\right] = 1.7020$$

$$\frac{d^2y}{dx^2} = y''(1.0) = \frac{1}{h^2}\left[\Delta y_0 - \Delta^3 y_0 + ...\right]$$

$$= \frac{1}{(0.1)^2}[0.0614 - 0.0074] = 5.4040$$

Example 1.47: *Obtain the first and second derivatives of the function tabulated below at the points $x = 1.1$ and $x = 1.2$.*

x :	*1*	*1.2*	*1.4*	*1.6*	*1.8*	*2.0*
y :	*0*	*0.128*	*0.544*	*1.298*	*2.440*	*4.02*

Solution: We first construct the forward difference table as shown below.

x	y	Δy	$\Delta^2 y$	$\Delta^3 y$	$\Delta^4 y$
1.0	0				
		0.128			
1.2	0.128		0.288		
		0.416		0.05	
1.4	0.544		0.338		0
		0.754		0.05	
1.6	1.298		0.388		0
		1.142		0.05	
1.8	2.440		0.438		
		1.580			
2.0	4.02				

Since $x = 1.1$ is a non-tabulated point near the beginning of the table, we take $x_0 = 1.0$ and compute

$$p = \frac{x - x_0}{h} = \frac{1.1 - 1.0}{0.2} = 0.5$$

Hence

$$\frac{dy}{dx} = \frac{1}{h}\left[\Delta y_0 + \frac{2p-1}{2}\Delta^2 y_0 + \frac{3p^2 - 6p + 2}{6}\Delta^3 y_0\right]$$

$$= \frac{1}{0.2}\left[0.128 + 0 + \frac{3(0.5)^2 - 6(0.5) + 2}{6}(0.05)\right] = 0.62958$$

$$\frac{d^2y}{dx^2} = \frac{1}{h^2}\left[\Delta^2 y_0 + (p-1)\Delta^3 y_0\right] = \frac{1}{(0.2)^2}[0.288 + (0.5-1)0.05] = 6.575$$

Now, $x = 1.2$ is a tabulated point near the beginning of the table. For $x = x_0 = 1.2$, $p = 0$ and

$$\frac{dy}{dx} = \frac{1}{h}\left[\Delta y_0 - \frac{1}{2}\Delta^2 y_0 + \frac{1}{3}\Delta^3 y_0\right]$$

$$= \frac{1}{0.2}\left[0.416 - \frac{1}{2}(0.338) + \frac{1}{3}(0.05)\right] = 1.31833$$

$$\frac{d^2y}{dx^2} = \frac{1}{h^2}[\Delta^2 y_0 - \Delta^3 y_0] = \frac{1}{(0.2)^2}[0.338 - 0.05] = 7.2$$

Example 1.48: *Find the first and second derivatives of the functions tabulated below at the point $x = 1.1$ and $x = 1.2$.*

x	1	1.2	1.4	1.6	1.8	2.0
y	0	0.1	0.5	1.25	2.4	3.9

Solution: First, we construct the forward difference table:

x	y	Δy	$\Delta^2 y$	$\Delta^3 y$	$\Delta^4 y$
1.0	0				
		0.1			
1.2	0.1		0.3		
		0.4		0.05	
1.4	0.5		0.35		0
		0.75		0.05	
1.6	1.25		0.40		0
		1.15		0.05	
1.8	2.40		0.45		
		1.5			
2.0	3.90				

Here $x = 1.1$ is a non-tabulated point near the beginning of the table. For $x_0 = 1.0$,

$$p = \frac{x - x_0}{h} = \frac{1.1 - 1.0}{0.2} = 0.5$$

Hence

$$\frac{dy}{dx} = \frac{1}{h}\left[\Delta y_0 + \frac{2p-1}{2}\Delta^2 y_0 + \frac{3p^2 - 6p + 2}{6}\Delta^3 y_0\right]$$

$$= \frac{1}{0.2}\left[0.1 + 0 + \frac{3(0.5)^2 - 6(0.5) + 2}{6}(0.05)\right] = 0.48958$$

$$\frac{d^2y}{dx^2} = \frac{1}{h^2}\left[\Delta^2 y_0 + (p-1)\Delta^3 y_0\right] = \frac{1}{(0.2)^2}[0.3 + (0.5 - 1)0.05] = 6.875$$

For $x = 1.2$, it is a tabulated point near the beginning of the table.

Let $x = x_0 = 1.2, p = 0$

$$\frac{dy}{dx} = \frac{1}{h}\left[\Delta y_0 - \frac{1}{2}\Delta^2 y_0 + \frac{1}{3}\Delta^3 y_0\right] = \frac{1}{0.2}\left[0.4 - \frac{1}{2}(0.35) + \frac{1}{3}(0.05)\right] = 1.208$$

$$\frac{d^2y}{dx^2} = \frac{1}{h^2}[\Delta^2 y_0 - \Delta^3 y_0] = \frac{1}{(0.2)^2}[0.35 - 0.05] = 7.5$$

1.11.8 Derivatives Based on Newton's Backward Interpolation Formula

Here, we assume the function $y = f(x)$ is known at $(n + 1)$ points $x_0, x_1, \ldots, x_n$, *i.e.*, $y_i = f(x_i)$, $i = 0, 1, 2, \ldots, n$ are known. Let $x_i = x_0 + ih$, $i = 0, 1, 2, \ldots, n$ and $v = \frac{x - x_n}{h}$.

Then, the Newton's backward interpolation formula is

$$f(x) = y_n + v\nabla y_n + \frac{v(v+1)}{2!}\nabla^2 y_n + \frac{v(v+1)(v+2)}{3!}\nabla^3 y_n + \frac{v(v+1)(v+2)(v+3)}{4!}\nabla^4 y_n + \frac{v(v+1)(v+2)(v+3)(v+4)}{5!}\nabla^5 y_n + \ldots \quad (1.163)$$

When the Eq. (1.163) is differentiated w.r.t. x successively, we obtain

$$f'(x) = \frac{1}{h}\left[\nabla y_n + \frac{2v+1}{2!}\nabla^2 y_n + \frac{3v^2+6v+2}{3!}\nabla^3 y_n + \frac{4v^3+18v^2+22v+6}{4!}\nabla^4 y_n + \frac{5v^4+40v^3+105v^2+100v+24}{5!}\nabla^5 y_n + \ldots\right] \quad (1.164)$$

$$f''(x) = \frac{1}{h^2}\left[\nabla^2 y_n + \frac{6v+6}{3!}\nabla^3 y_n + \frac{12v^2+36v+22}{4!}\nabla^4 y_n + \frac{20v^3+120v^2+210v+100}{5!}\nabla^5 y_n + \ldots\right] \quad (1.165)$$

and so on.

Equations (1.164) and (1.165) can be used to determine the approximate differentiation of first, second, etc. order at any point x, where $x = x_n + vh$.

If $x = x_n$, then $v = 0$.

Equations (1.164) and (1.165) become

$$f'(x_n) = \frac{1}{h}\left[\nabla y_n + \frac{1}{2}\nabla^2 y_n + \frac{1}{3}\nabla^3 y_n + \frac{1}{4}\nabla^4 y_n + \frac{1}{5}\nabla^5 y_n + \ldots\right] \quad (1.166)$$

and

$$f''(x_n) = \frac{1}{h^2}\left[\nabla^2 y_n + \nabla^3 y_n + \frac{11}{12}\nabla^4 y_n + \frac{5}{6}\nabla^5 y_n + \ldots\right] \quad (1.167)$$

Example 1.49: *A slider in a machine moves along a fixed straight rod. Its distances x(m) along the rod are given in the following table for various values of the time t (seconds).*

t(sec.)	*1*	*2*	*3*	*4*	*5*	*6*
x(m)	*0.0201*	*0.0844*	*0.3444*	*1.0100*	*2.3660*	*4.7719*

Find the velocity and acceleration of the slider at time t = 6 sec.

Solution: The backward difference table is

t	x	∇x	$\nabla^2 x$	$\nabla^3 x$	$\nabla^4 x$	$\nabla^5 x$
1.0	0.0201					
2.0	0.0844	0.0643				
3.0	0.3444	0.2600	0.1957			
4.0	1.0100	0.6656	0.4056	0.2100		
5.0	2.3660	1.3560	0.6904	0.2847	0.0748	
6.0	4.7719	2.4059	1.0499	0.3595	0.0748	0.0000

Here $h = 1.0$

$$\frac{dx}{dt} = \frac{1}{h}\left[\nabla x + \frac{1}{2}\nabla^2 x + \frac{1}{3}\nabla^3 x + \frac{1}{4}\nabla^4 x + \frac{1}{5}\nabla^5 x + \ldots\right]$$

$$= \frac{1}{1.0}\left[2.4059 + \frac{1}{2}(1.0499) + \frac{1}{3}(0.3595) + \frac{1}{4}(0.0748) + \frac{1}{5}(0.0)\right] = 3.0694$$

$$\frac{d^2x}{dt^2} = \frac{1}{h^2}\left[\nabla^2 x + \nabla^3 x + \frac{11}{12}\nabla^5 x + \ldots\right]$$

$$= \frac{1}{(1.0)^2}\left[1.0499 + 0.3595 + \frac{11}{12}(0.0748) + \frac{5}{6}(0)\right] = 1.4780$$

1.11.9 Derivatives Based on Stirling's Interpolation Formula

Suppose $y_{\pm i} = f(x_{\pm i})$, $i = 0, 1, \ldots, n$ are given for $2n + 1$ equispaced points $x_0, x_{\pm 1}, x_{\pm 2}, \ldots, x_{\pm n}$, where $x_{\pm i} = x_0 \pm ih$, $i = 0, 1, \ldots, n$.

The Stirling's interpolation polynomial is given by

$$f(x) = y_0 + \frac{u}{1!}\left[\frac{\Delta y_{-1} + \Delta y_0}{2}\right] + \frac{u^2}{2!}\Delta^2 y_{-1} + \frac{u^3 - u}{3!}\left[\frac{\Delta^3 y_{-2} + \Delta^3 y_{-1}}{2}\right]$$
$$+ \frac{u^4 - u^2}{4!}\Delta^4 y_{-2} + \frac{u^5 - 5u^3 + 4u}{5!}\left[\frac{\Delta^5 y_{-3} + \Delta^5 y_{-2}}{2}\right] + \ldots \quad (1.168)$$

where $u = \dfrac{x - x_0}{h}$

When Eq. (1.168) is differentiated with respect to x successively, we obtain

$$f'(x) = \frac{1}{h}\left[\frac{\Delta y_{-1} + \Delta y_0}{2} + u\Delta^2 y_{-1} + \frac{3u^2 - 1}{6}\left(\frac{\Delta^3 y_{-2} + \Delta^3 y_{-1}}{2}\right)\right.$$
$$\left. + \frac{2u^3 - u}{12}\Delta^4 y_{-2} + \frac{5u^4 - 15u^2 + 4}{120}\left(\frac{\Delta^5 y_{-3} + \Delta^5 y_{-2}}{2}\right) + \ldots\right] \quad (1.169)$$

and
$$f''(x) = \frac{1}{h^2}\left[\Delta^2 y_{-1} + u\frac{\Delta^3 y_{-2} + \Delta^3 - 1}{2} + \frac{6u^2 - 1}{12}\Delta^4 y_{-2} + \frac{2u^3 - 3u}{12}\left(\frac{\Delta^5 y_{-3} + \Delta^5 y_{-2}}{2}\right) + \ldots\right] \tag{1.170}$$

At $x = x_0$, $u = 0$ and Eqs. (1.169) and (1.170) become

$$f'(x_0) = \frac{1}{h}\left[\frac{\Delta y_0 + \Delta y_{-1}}{2} - \frac{1}{6}\left(\frac{\Delta^3 y_{-1} + \Delta^3 y_{-2}}{2}\right) + \frac{1}{30}\left(\frac{\Delta^5 y_{-2} + \Delta^5 y_{-3}}{2}\right) + \ldots\right] \tag{1.171}$$

and
$$f''(x_0) = \frac{1}{h^2}\left[\Delta^2 y_{-1} - \frac{1}{12}\Delta^4 y_{-2} + \ldots\right] \tag{1.172}$$

Example 1.50: *Find* $\frac{dy}{dx}$ *and* $\frac{d^2y}{dx^2}$ *for x = 0.2 for the data given in the following table.*

x	*0*	*0.1*	*0.2*	*0.3*	*0.4*	*0.5*
y	*0*	*0.10017*	*0.20134*	*0.30452*	*0.41076*	*0.52115*

Solution: Construct the following difference table.

x	y	Δy	$\Delta^2 y$	$\Delta^3 y$	$\Delta^4 y$
0	0				
		0.10017			
0.1	0.10017		0.001		
		0.10017		0.00101	
0.2	0.20134		0.00201		0.00004
		0.10318		0.00105	
0.3	0.30452		0.00306		0.00004
		0.10624		0.00109	
0.4	0.41076		0.00415		
		0.11039			
0.5	0.52115				

Here, we use Stirling's formula. Hence, for $x = 0.2$, we have

$$\frac{dy}{dx} = \frac{1}{h}\left[\frac{\Delta y_{-1} + \Delta y_0}{2} - \frac{1}{6}\frac{\Delta^3 y_{-2} + \Delta^3 y_{-1}}{2}\right]$$

$$= \frac{1}{0.1}\left[\frac{0.10117 + 0.10318}{2} - \frac{1}{12}(0.00101 + 0.00105\right] = 1.020033$$

$$\frac{d^2y}{dx^2} = \frac{1}{h^2}\left[\Delta^2 y_{-1} - \frac{1}{12}\Delta^4 y_{-2}\right]$$

$$= \frac{1}{(0.1)^2}\left[0.00201 - \frac{1}{12}(0.00004)\right] = 0.200666$$

Example 1.51: *Compute the values of $f'(3.1)$ and $f''(3.1)$ using the following table.*

x	1	2	3	4	5
$f(x)$	0	1.4	3.3	5.6	8.1

Solution: The central difference table is

x	$y = f(x)$	Δy	$\Delta^2 y$	$\Delta^3 y$	$\Delta^4 y$
$x_{-2} = 1$	0				
		1.4			
$x_{-1} = 2$	1.4		0.5		
		1.9		–0.1	
$x_0 = 3$	3.3		0.4		–0.1
		2.3		–0.2	
$x_1 = 4$	5.6		0.2		
		2.5			
$x_2 = 5$	8.1				

Let $\quad x_0 = 3,\ h = 1,\ u = \dfrac{3.1-3}{1} = 0.1$

$$f'(3.1) = \frac{1}{h}\left[\frac{\Delta y_{-1} + \Delta y_0}{2} + u\Delta^2 y_{-1} + \frac{3u^2-1}{6}\left(\frac{\Delta^3 y_{-2} + \Delta^3 y_{-1}}{2}\right) + \frac{2u^3-u}{12}\Delta^4 y_{-2} + \ldots\right]$$

$$= \frac{1}{1}\left[\frac{1.9+2.3}{2} + 0.1(0.4) + \frac{3(0.1)^2-1}{6}\left(\frac{-0.1-0.2}{2}\right) + \frac{2(0.1)^3-0.1}{12}(-0.1)\right]$$

$$= [2.1 + 0.04 + 0.02425 + 0.00082] = 2.16507.$$

$$f''(3.1) = \frac{1}{h^2}\left[\Delta^2 y_{-1} + u\left(\frac{\Delta^2 y_{-2} + \Delta^3 y_{-1}}{2}\right) + \frac{6u^2-1}{12}\Delta^4 y_{-2} + \ldots\right]$$

$$= \frac{1}{1^2}\left[0.4 + 0.1\left(\frac{-0.1-0.2}{2}\right) + \frac{6(0.1)^2-1}{12}(-0.1)\right] = [0.4 - 0.015 + 0.00783] = 0.39283$$

1.11.10 Finite Difference Approximation of an Uncertainty Estimate

This is an alternative approach to approximate numerically the partial derivatives in the uncertainty expression.

The partial deviation term $\dfrac{\delta q}{\partial x_i}$ can be approximated numerically using the finite-difference approach as

$$\frac{\delta q}{\delta x_i} = \frac{\Delta q}{\Delta x_i} = \frac{q(x_i + \Delta x_i) - q(x_i)}{\Delta x_i} \tag{1.173}$$

Therefore, the forward finite-difference approximation can be written as

$$u_q = \sqrt{\left(\frac{\Delta q}{\Delta x_1} u_{x1}\right)^2 + \left(\frac{\Delta q}{\Delta x_2} u_{x2}\right)^2 + \ldots + \left(\frac{\Delta q}{\Delta x_k} u_{xk}\right)^2} \tag{1.174}$$

It should be noted here that the value of Δx_i is selected to be small enough so that the finite-difference expression closely approximates the actual derivative. A small value such as $\Delta x_i = 0.01x_i$ can be chosen as a good starting value and then the value of Δx_i can be decreased until approximate convergence in the value of u_q is obtained.

1.12 SUMMARY

In this chapter we described the Taylor's theorem and number representation including binary, decimal and hexadecimal numbers. We have defined absolute and relative errors, inherent errors, round-off errors, truncation errors, machine epsilon, and error propagation. Methods for the estimation were briefly outlined. General error formulae for approximating a function, stability, condition and uncertainty in data or noise were presented.

We have introduced the types of statistics, namely, the descriptive statistics and the inferential statistics. Descriptive statistics consists of methods for organizing and summarizing information. Inferential statistics consists of methods for drawing and measuring the reliability of conclusions about a population based on information obtained from a sample of the population.

The basic concepts and definitions, population and sample, types of variables, measures of central tendency for dispersion for ungrouped data (range, variance and standard deviation), mean, variance and standard deviation for grouped data.

Random variables (both discrete and continuous) and their mean and standard deviation are discussed. Permutations and combinations are defined briefly. Continuous probability distribution, namely, the normal distribution, its properties, mean and variance and the standard normal distribution are presented.

The study of uncertainties was presented based on the calculus of finite differences. Some important interpolation formulae by means of forward, backward and central differences of a function, which are frequently used in scientific and engineering calculations were also presented.

1.13 REFERENCES

Anon, *AR Software-AI Tools-Wincrash*, AR Software, Trantech Corporation, 4th ed., Edmonds, WA, 1998.

Anon, *Measurement Uncertainty*, ANSI/ASME Power Test Code 19.1-1985, American Society of Mechanical Engineers, New York, NY, 1986.

Beers, Y., *Introduction of the Theory of Error*, Addison-Wesley, Reading, MA, 1958.

Brach Raymond M., and **Matthew Brach, R.**, *Vehicle Accident Analysis and Accident Reconstruction Methods*, Product Code R-311, SAE International, Warrendale, PA, 2005.

Brach, R.M., and **Dunn, P.F.**, *Uncertainty Analysis for Forensic Science,* Lawyers and Judges Publishing Co., Tucson, AZ, 2004.

Dukkipati, R.V., and **Govindaraja, B.,** *Computer Programming and Numerical Methods through Solved Problems*, New Age International Publishers (P) Ltd., New Delhi, India, 2010.

Dukkipati, R.V., *Accident Reconstruction-Problems and Solutions,* SAE International, Warrandale, PA, 2011.

Dukkipati, R.V., and **Govindaraja, B.,** *Computer Programming and Numerical Methods*, New Age International Publishers (P) Ltd., New Delhi, India, 2010.

Dukkipati, R.V., *Applied Numerical Methods with MATLAB* , New Age International Publishers (P) Ltd., New Delhi, India, 2010.

Dukkipati, R.V., *Applied Numerical Methods with MATLAB through Solved Problems*, New Age International Publishers (P) Ltd., New Delhi, India, 2010.

Dukkipati, R.V., *Dictionary of Road Vehicle Dynamics and Accident Reconstruction,* SAE International, Warrandale, PA, 2010.

Dukkipati, R.V., *Mathematical Methods,* New Age International Publishers (P) Ltd., New Delhi, India, 2009.

Dukkipati, R.V., *Numerical Methods through Solved Problems,* New Age International Publishers (P) Ltd., New Delhi, India, 2009.

Dukkipati, R.V., *Numerical Methods,* New Age International Publishers (P) Ltd., New Delhi, India, 2009.

Dukkipati, R.V., Pang, J., Qatu, M., Sheng, G., and **Shuguang, Z.**, *Road Vehicle Dynamics*, Product Code R-366, SAE International, Warrendale, PA, 2008.

Dukkipati, R.V., Pang, J., Qatu, M., Sheng, G., and **Shuguang, Z.**, *Road Vehicle Dynamics: Problems and Solutions*, Product Code R-386, SAE International, Warrendale, PA, 2009.

Dukkipati, R.V., *Probability and Statistics for Scientists and Engineers*, New Age International Publishers, New Delhi, India, 2009.

Dukkipati, R.V., *Probability and Statistics through Solved Problems*, New Age International Publishers (P) Ltd., New Delhi, India, 2008.

Fricke, L.B., *"Traffic Accident Reconstruction", Traffic Accident Investigation Manual,* Vol. 2, Northwestern University Traffic Institute, Evanston, IL, 1990.

Goudie, D.W., Bowler, J.J., Brown, C.A., Heinrichs, B.E., and **Siegmund, G.P.,** "Tire Friction During Locked Wheel Braking", Paper No. 2000-01-1314, SAE International, Warrendale, PA, 2000.

Guttman, I., Wilks, S.S., and **Hunter, J.S.,** *Introductory Engineering Statistics,* John Wiley, New York, NY, 1982.

Hicks, C.R., *Fundamentals Concepts in the Design of Experiments*, Holt, Rinehart, and Wnston, New York, NY, 1964.

Holman, J.P., *Experimental Methods for Engineers*, 7th ed., McGraw Hill, New York, NY, 2001.

Kline, S.J. and **McClintock, F.A.**, "Describing Uncertainties in Single-Sample Experiments", Mechanical Engineering, American Society of Mechanical Engineers, New York, NY, p.3, January, 1953.

Lund, Y.L. and **Bernard, J.E.**, "Analysis of Simple Rollover Metrices", SAE paper No.950306, Society for Automotive Engineers, Warrendale, PA, 1995.

Mandel, J., *The Statistical Analysis of Experimental Data,* Dover Publications, New York, NY, 1964.

Mandel, J., *The Statistical Analysis of Experimental Data*, Interscience Publishers, Wiley, New York, NY, 1964.

Moffat, R.J., "Uncertainty Analysis in the Planning of an Experiment", J. Fluids Engr., Vol.107, pp.173-181, 1985.

Moffatt, R.J., "Contributions to the Theory of Single-Sample Uncertainty Analysis", J. Fluids Engr., Vol. 104, p. 250, June 1982.

Montgomery, D.C., and **Runger, G.C.,** *Applied Statistics and Probability for Engineers,* John Wiley, New York, 1999.

Rizenbergs, R.L., Burchett, J.L., and **Warren, L.A.,** "Relation of Accidents and Pavement Friction on Rural, Two-Lane Roads", Transportation Research Record 633, Pavement Surface Properties and Performance, National Academy of Science, Washington, DC, 1977.

Schenck, H., *Theories of Engineering Experimentation*, McGraw Hill, New York, NY, 1961.

Taylor, B.N., and **Kuyatt, C.E.,** "Guidelines for Evaluating and Expressing the Uncertainty of NIST Measurement Results", NIST Technical Note 1297, National Institute of Standards and Technology, U.S. Department of Commerce, Gaithersburg, MD, 1994.

Taylor, J.R., *An Introduction to Error Analysis*—The Study of Uncertainties in Physical Measurements, University Science Book, Mill Valley, CA, 1982.

Taylor, J.R., *An Introduction to Error Analysis,* University Science Books, Mill Valley, CA, 1982.

Wilson, E.B., *An Introduction to Scientific Research*, McGraw Hill, New York, NY, 1952.

Young, H.D., *Statistical Treatment of Experimental Data*, McGraw Hill, New York, NY, 1962.

PROBLEMS

1.1 Determine the following hyperbolic trigonometric functions to $O\ (0.9)^4$.

(*a*) sinh (0.9) (*b*) cosh (0.9)

1.2 Determine when $f(x) = 0$, given that $f\,(1.7) = -1.7781$ and $f\,'(1.7) = 4.3257$.

1.3 Determine $f\,(1.2)$, given the first order differential equation $\dfrac{df}{dx} = 2x$ with $f\,(1) = 1$.

1.4 (*a*) Convert $(327)_{10}$ to binary.

(*b*) Convert $(0.3125)_{10}$ to binary.

1.5 Represent the number 50824.6135 in the decimal system (base-10).

1.6 Find the binary and hexadecimal values of the following numbers,

(*a*) 329 (*b*) 203

1.7 Convert $(75)_{10}$ to base-2.

1.8 Perform the following operations:

(*a*) $(2)_8 + (5)_8$ (*b*) convert $(75)_{10}$ to base-8

(*c*) convert $(13)_8$ to base-10.

1.9 Convert $(4D3)_{16}$ to base-10.

1.10 Convert $(1001011)_2$ to base-8.

1.11 Show that the relative error E_{rxy} of the product where $x = x_e + \Delta x$ and $y = y_e + \Delta y$ is $E_{rxy} = E_{rx} + E_{ry}$. Assume $|E_{rx}| << 1$ and $|E_{ry}| << 1$.

1.12 Show that the relative error E_{rxy} of the quotient where $x = x_e + \Delta x$ and $y = y_e + \Delta y$ is $E_{rxy} = E_{rx} - E_{ry}$. Assume $|E_{rx}| << 1$ and $|E_{ry}| << 1$.

1.13 Determine the absolute and relative errors involved if $x = 2/3$ is represented in normalized decimal form with 6 digits by

(*a*) round-off (*b*) truncation.

1.14 Given that 5 digit chopping is used for arithmetic calculations involving x and y where $x = 1/3$ and $y = 5/7$. Determine the absolute and relative errors involved.

1.15 If $x = 3.536$, determine the absolute error and relative error when

(*a*) x is rounded (*b*) x is truncated to two decimal digits.

1.16 If the number $x = 57.46235$ is rounded off to four significant figures, find the absolute error, relative error and the percentage relative error.

1.17 If the approximate value of $\pi\left(=\frac{22}{7}\right)$ is 3.14, determine the absolute error, relative error and relative percentage error.

1.18 Determine the true error and true percentage relative error for each case.

(*a*) If the measured length of a track is approximated by 9999 cm and the true value is 10,000 cm

(*b*) If the measured length of a track is approximated by 9 cm and the true value is 10 cm.

1.19 The exponential function e^x can be computed using the Maclaurin series expansion as

$$e^x = 1 + x + \frac{x^2}{2!} + \frac{x^3}{3!} + \ldots + \frac{x^n}{n!}$$

Include six terms in the series and compute the percentage relative error and approximate estimate of the error for each term when estimating $e^{0.5}$.

1.20 Find the relative maximum error in F where $F = \frac{5x^2 y}{z^3}$. Given $\Delta x = \Delta y = \Delta z = 0.001$, where Δx, Δy and Δz denote the errors in x, y and z respectively such that $x = y = z = 1$.

1.21 Consider the trigonometric function $f(x) = \cos x$

(*a*) find the Taylor series expansion of $f(x)$ about 0.

(*b*) assuming the Taylor series is truncated to $n = 6$ terms. Determine the relative error at $x = \pi/4$ due to truncation. Express it as a percentage.

(*c*) determine an upper bound on the magnitude of the relative error at $x = \pi/4$ expressed as a percentage.

1.22 Suppose $f(x) = e^{-x}$ is to be expanded about the point $x = 1$ and truncated to $n = 6$ terms.

$$e^{-x} = 1 - x + \frac{x^2}{2} - \frac{x^3}{6} + \frac{x^4}{24} - \frac{x^5}{120}$$

Determine the upper bound on the magnitude of the absolute error due to truncation.

1.23 Determine the upper bound on the error for the function $f(x) = (x + 1)^{1/2}$ using a polynomial approximation with third-order Taylor series (computed about $x_0 = 0$) for all $x \in [0, 1]$.

1.24 Consider the power series expansion for e^x given by

$$e^x = 1 + x + \frac{x^2}{2!} + \frac{x^3}{3!} + \cdots + \frac{x^{n-1} - 1}{(n-1)!} + \frac{x^n}{n!}e^{\xi}, \; 0 < \xi < x.$$

Determine the number of terms, n such that their sum gives the value of e^x correct to 8 decimal places at $x = 1$.

1.25 Use Taylor's series expansion with $n = 0$ to 6 to approximate $f(x) = \cos x$ at $x_{i+1} = \pi/3$ on the premise that the value of $f(x)$ and its derivatives are at $x_i = \pi/4$. Assume $h = \pi/3 - \pi/4 = \pi/12$.

1.26 Compute and interpret the condition number for

$$f(x) = \tan x \text{ for } a = \frac{\pi}{2} + 0.1\left(\frac{\pi}{2}\right)$$

$$f(x) = \tan x \text{ for } a = \frac{\pi}{2} + 0.01\left(\frac{\pi}{2}\right)$$

1.27 Evaluate and interpret the condition numbers for

(*a*) $f(x) = (x^2 - 1)^{1/2} - x$ for $x = 200$ (*b*) $f(x) = \dfrac{e^x + 1}{x}$ for $x = 0.01$

(*c*) $f(x) = \dfrac{\cos x}{1 + \sin x}$ for $x = 0.001\pi$ (*d*) $f(x) = e^{-2x}$ for $x = 5$

1.28 The two resistors R and R_s are connected in series as shown in Fig. P1.28. The voltage drops across each resistor are measured as

$$E = 10 \text{ V} \pm 0.11 \text{ V } (1.1\%)$$
$$E_s = 1.2 \text{ V} \pm 0.005 \text{ V } (0.467\%)$$

along with a value of $R_s = 0.006\ \Omega \pm 1/4\%$

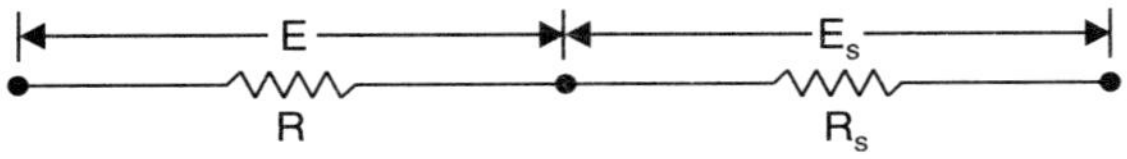

Fig. P1.28

Determine the power dissipated in resistor R and its uncertainty.

1.29 The power measurement was conducted by measuring voltage and current across the resistor with the circuit shown in Fig. P1.29. The voltmeter has an internal resistance R_m, and the value of R is known only approximately. Calculate the nominal value of the power dissipated in R and the uncertainty for the following conditions:

$$R = 100\ \Omega$$
$$R_m = 1000\ \Omega \pm 4\%$$
$$I = 5 \text{ A} \pm 1\%$$
$$E = 500 \text{ V} \pm 1.1\%$$

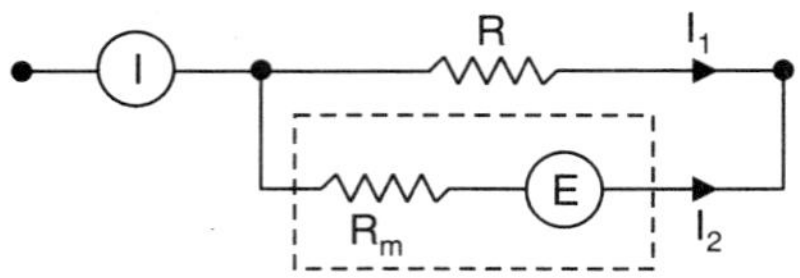

Fig. P1.29

1.30 A certain resistor draws 110.3 V and 5.4 A. The uncertainties in the measurements are ± 0.2 V and ± 0.05 A, respectively. Determine the power dissipated in the resistor and the uncertainty in the power.

1.31 A small mechanical component has dimensions 50 by 160 mm. The uncertainty in the 50 mm dimension is ± 0.015 mm. Determine the uncertainty with which the 160 mm dimension must be measured to ensure that the total resistance in the area is not greater than 150 percent of that value it would have if the 160 mm dimension were exact.

1.32 A resistance arrangement of 50 Ω is desired. Two resistances of 100 ± 0.11 Ω and two resistances of 25 ± 0.025 Ω are available. Which should be used, a series arrangement with the 25 Ω resistors or a parallel arrangement with the 100 Ω resistors? Determine the uncertainty for each arrangement.

1.33 Reynolds numbers for the pipe flow may be expressed as

$$\text{Re} = \frac{4\dot{m}/\pi d}{\mu}$$

where d = pipe diameter, m

$\dot{m}$ = mass flow, kg/s

μ = viscosity, kg/m s

In a certain system the flow rate is 10 lb m/min, ± 0.5 percent, through a 0.5 in diameter (± 0.004 in) pipe. The viscosity is 4.6×10^{-4} lb m/h ft, ± 1 percent. Determine the value of the Reynolds number and its uncertainty. Use both the analytical and numerical methods.

1.34 In a certain dynamic measurement application, the output (y) is expressed in terms of the input (x) as

$$y = 1 + (1 + x)\, e^{-x}$$

With what precision must x be determined to give a precision of ± 0.09 percent for y at a value of $x = 45$?

1.35 Determine the uncertainty of critical speed formula given by

$$V_{er} = \sqrt{gR\frac{(f+e)}{1-fe}}$$

where g = acceleration due to gravity, 32.2 ft/sec^2

f = tire-roadway frictional coefficient

R = radius of circular path, ft

$e = \tan\gamma$

γ = bank angle (super elevation)

1.36 The velocity for a vehicle that slows to rest uniformly with deceleration, fg, over a distance, d, is given by

$$v = \sqrt{2fgd}$$

where f is a constant frictional drag coefficient. Given that f is normally distributed with a mean of, $\mu_f = 0.42$ and a standard deviation, $\sigma_f = 0.75$. Determine the mean of the initial speed if the stopping distance d is 18 m.

1.37 Two resistors R_1 and R_2 are connected in series and parallel. The values of the resistances are

$$R_1 = 100 \pm 0.12\ \Omega$$
$$R_2 = 50 \pm 0.02\ \Omega$$

Calculate the uncertainty in the combined resistance of the both the series and the parallel arrangements.

1.38 Two resistors are connected in series and have the following values:

$$R_1 = 10{,}000\ \Omega \pm 6\%$$
$$R_2 = 1\text{M}\Omega \pm 9\%$$

Calculate the percent uncertainty for the series total resistance.

1.39 Four resistances having nominal values of 1, 2, 3 and 2.5 kΩ are connected in parallel. The uncertainties are ± 9 percent. A voltage of 100 V ± 1.0 V is impressed on the combination. Calculate the power drawn and its uncertainty. Use the technique of evaluation of uncertainties by result perturbation.

1.40 A model car is placed on a tethered circular track having a diameter of 11 m ± 1 cm. The speed of the car is obtained by measuring the time required for travelling each lap. A stopwatch is employed for the measurement and the estimated uncertainty in both starting and stopping the watch is ± 0.1 seconds. For a nominal speed of 95 miles per hour, determine the uncertainty in the speed measurement when made over 1, 2, 3, and 4 laps.

1.41 The following data gives the number of hours spent exercising in a gymnasium by 10 randomly selected members of a university during the past week.

8 13 6 0 1 6 11 4 0 7

Compute the mean, median and mode.

1.42 The following data gives the number of vacuum cleaners produced by a manufacturing company for a sample of 12 days.

25 33 28 24 36 34 30 22 24 29 31 23

Calculate the mean, median and mode.

1.43 The following data gives the number of heavy machinery sold by a large manufacturing company for the 12 months in the year 2008.

3 2 47 118 204 97 68 86 62 57 98 99

Find the mean, median and mode.

1.44 The following data are the test scores (in points) of 9 students in a class out 6 points:

5.8 5.7 5.9 5.7 5.5 5.7 5.7 5.7 5.6

Find the mean, median and mode for this data.

1.45 The following data gives the number of ball bearings manufactured by a company per day for a sample of 10 days.

24 32 27 23 35 33 29 21 23 28

Calculate the mean median, and mode for these data.

1.46 The following data gives the number of automobile accidents in a city during the past 12 days.

3 7 11 5 6 3 8 7 6 9 13 2

Calculate the range, variance and standard deviation.

1.47 Compute the range, variance and standard deviation for the data given in Problem 1.24.

1.48 The following data gives the number of hours spent doing the homework and attending classes by 10 randomly selected college students during the past week.

18 16 27 34 39 35 42 29 31 24

Compute the range, variance and standard deviation.

1.49 The following data gives the number of new automobiles sold at a dealership during a 15-day period.

13 5 9 7 6 10 11 14 12 8 4 8 16 10 8

Find the range, mean, variance and standard deviation.

1.50 Following are the 2008 earnings (in thousands of dollars) before taxes of all 10 employees of a small company:

30 35 22 28 19 47 52 43 39 32

Calculate the variance and standard deviation of these data.

1.51 The following data gives the number of automobiles that stopped at a service station during each of the 10 hours observed:

30 32 35 29 28 17 19 30 18 25

Find the range, variance and standard deviation.

1.52 Consider the following two data sets:

Data set 1:	8	12	25	37	41
Data set 2:	15	19	32	44	48

Calculate the standard deviation of each data set. Comment on the results obtained.

1.53 The following table gives the frequency distribution of the amounts of long distance telephone bills for the month of August 2008 for a sample of 60 families in a city.

Amount of long distance telephone bill (dollars)	*Number of families*
25 to less than 50	15
50 to less than 75	12
75 to less than 100	20
100 to less than 125	8
125 to less than 150	5

Calculate the mean, variance and standard deviation.

1.54 Using the sample formulas, find the mean, variance and standard deviation for the grouped data given in the following table:

x	f
0 to less than 4	16
4 to less than 8	23
8 to less than 12	14
12 to less than 16	10
16 to less than 20	8
20 to less than 24	7

1.55 The following table gives the frequency distribution of entertainment expenditure (in dollars) incurred by 50 families during the past week.

Entertainment expenditure (dollars)	*Number of families*
0 to less than 10	5
10 to less than 20	10
20 to less than 30	14
30 to less than 40	10
40 to less than 50	6
50 to less than 60	5

Find the mean, variance and standard deviation.

1.56 The following table gives the grouped data on the weights of all 100 babies born at a hospital in 2005.

Weight (kg)	*Number of babies*
1 to less than 2	5
2 to less than 3	30
3 to less than 4	40
4 to less than 5	20
5 to less than 6	5

Find the mean, variance and standard deviation.

1.57 The following table gives the frequency distribution of the number of personal computers sold during the past month at 40 computer stores in a big city. Calculate the mean, variance and standard deviation.

Computers sold	*Number of stores*
4 to 12	6
13 to 21	10
22 to 30	13
31 to 39	6
40 to 48	5

1.58 The following table gives the number television sets owned by 80 households.

Number of television Sets owned	*Number of households*
0	4
1	32
2	27
3	11
4	6

Find the mean, variance and standard deviation. Use the number of television sets owned as the values of m in the formulas.

1.59 The inside diameter of a finished shaft of uniform diameter is normally distributed with a mean of 4.50 cm and a standard deviation of 0.01 cm. What is the probability of obtaining a diameter exceeding 4.52 cm?

1.60 The resistance of a foil strain gauge is normally distributed with a mean of 100 ohms and a standard deviation of 0.8 ohm. The specification limits are 100 ± 1.0 ohms. What percentage of gauges will be defective?

1.61 The measurement of the diameter of a special steel pipe is normally distributed with a mean of 5.01 cm and standard deviation of 0.03 cm. The specification limits are 5.00 ± 0.05 cm. What percentage of pipes is not acceptable?

1.62 The yield strength of metal component manufactured is N (192.36). A purchaser of the metal components requires strength of at least 180 psi. The probability that a metal component will meet or exceed the specifications is given by $P(x \geq 180)$. Determine the probability that a metal component meets or exceeds the specifications.

1.63 Find the probability that the yield strength of the metal components in Problem 1.64 is between 180 and 198 psi.

1.64 The diameter of a machine shaft produced in a manufacturing company is normally distributed with a mean diameter of 0.001 inches and a standard deviation of 0.0002 inches.

(*a*) what is the probability that the diameter of the shaft exceeds 0.0013 inches?

(*b*) what is the probability that a diameter of the shaft is between 0.0007 and 0.0013 inches?

(*c*) what standard deviation of diameters is needed so that the probability in part (*b*) is 0.995?

1.65 The mass, m, of a particular machine part is normally distributed with a mean of 66 kg and a standard deviation of 5 kg.

(*a*) what percentage of the parts will have a mass less than 72 kg?

(*b*) what percentage of parts will have a mass in excess of 72 kg?

(*c*) what percentage of the parts will have a mass between 61 and 72 kg?

1.66 It was determined experimentally that the load X required to break a plate is normally distributed with mean 2.5 and standard deviation 0.24. Determine the probability that

(*a*) the plate breaks at a load of 2.61 or less (*b*) the plate breaks at a load of more than 2.39

(*c*) the breaking load is at least 2.86 (*d*) the breaking load is in the interval (2.61, 2.86).

1.67 Use the Table in Appendix-E to determine the following probabilities for the standard normal random variable z.

(*a*) $P(z < 3.0)$ (*b*) $P(z > -2.15)$

(*c*) $P(-3 < z < 3)$ (*d*) $P(0 < z < 1)$

(*e*) $P(z > 3)$

1.68 Assume z has a standard normal distribution. Use Appendix-*E* table to determine the value for z that solves each of the following:

(*a*) $P(Z < z) = 0.5$ (*b*) $P(Z > z) = 0.9$

(*c*) $P(-z < Z < z) = 0.99$ (*d*) $P(-z < Z < z) = 0.68$

(*e*) $P(-1.24 < Z < z) = 0.8$

1.69 The diameter of a component in a machine system is normally distributed with mean 0.2508 cm and standard deviation 0.0005 cm. The specifications on the component are 0.25 ± 0.0015 cm. Determine the proportion of components confirms to specifications.

1.70 The mass, m, of a particular electronic component is normally distributed with a mean of 66 g and a standard deviation of 5 g. Determine

(*a*) the percent of components that will have a mass less than 72 g

(*b*) the percent of components that will have a mass in excess of 72 kg

(*c*) the percent of components that will have a mass between 61 and 72 g.

1.71 Customers buying copper rods supplied by a certain manufacturer require that the rods be between 9.9 cm and 10.5 cm, inclusive. The manufacturing process is such that the actual rod lengths are well approximated by a normal distribution with mean 10.1 cm and standard deviation 0.20 cm. Determine the percentage of the manufacturer's production which is acceptable to the customer.

1.72 Refer to Problem 1.71 and determine what rod length is exceeded by 95% of the manufacturer's product.

1.73 Construct a forward difference table for the following data:

(*a*)

x	45	55	65	75
$y = f(x)$	20	60	120	180

(*b*)

x	40	50	60	70	80	90
$y = f(x)$	204	224	246	270	296	324

1.74 Construct a difference table for $y = x^3 + 2x + 3$ for $x = 1, 2, 3, 4, 1.7$.

1.75 Given $u_0 = 1$, $u_1 = 5$, $u_2 = 10$, $u_3 = 30$, $u_4 = 30$, find $\Delta^4 u_0$.

1.76 Given $u_0 = 5$, $u_1 = 24$, $u_2 = 81$, $u_3 = 200$, $u_4 = 100$ and $u_5 = 8$, find $\Delta^5 u_0$.

1.77 Estimate the missing term in the following tables:

(*a*)

x	1	2	3	4	5
$y = f(x)$	5	14	?	74	137

(*b*)

x	1	2	3	4	5
$y = f(x)$	8	17	38	?	140

(*c*)

x	0	1	2	3	4
$y = f(x)$	3	2	3	?	11

1.78 From the following table of values, estimate $y'(1.0)$ and $y''(1.0)$:

(*a*)

x	1	2	3	4	5	6
y	–4	+3	22	59	120	211

(*b*)

x	1	1.5	2	2.5	3	3.5
y	5	6.125	9	14.375	23	35.625

(*c*)

x	1	2	3	4	5	6
y	6	25	104	309	730	1481

(*d*)

x	1	1.5	2	2.5	3	3.5
y	2.7	–5.5188	–27.8	–75.4688	–163.3	–309.5188

(*e*)

x	1	2	3	4	5	6
y	2.9	–26.2	–157.9	–523	–1307.5	–2752.6

(*f*)

x	1	3	5	7	9	11
y	5.2	28.4	130	367.6	798.8	1481.2

(*g*)

x	1	1.5	2	2.5	3	3.5
y	–48	–41.0068	–24.362	8.1098	64.084	152.7363

1.79 Find the values of $y'(3.5)$ and $y''(3.5)$ from the following table:

(*a*)

x	1	1.5	2	2.5	3	3.5
y	–49	–46.6250	–42	–34.3750	–23	–7.1250

(*b*)

x	1	1.5	2	2.5	3	3.5
y	5	6.1250	9	14.3750	23	35.6250

(*c*)

x	1	1.5	2	2.5	3	3.5
y	0.5470	0.4536	0.2020	–0.3284	–1.2930	–2.8814

(*d*)

x	1	1.5	2	2.5	3	3.5
y	0.2377	0.3108	0.4829	0.7941	1.2849	1.9953

1.80 Find the values of $y'(6.0)$ and $y''(6.0)$ from the following table:

(*a*)

x	1	2	3	4	5	6
y	–0.5530	–0.7740	–1.2490	–2.0380	–3.1770	–4.6780

(*b*)

x	1	2	3	4	5	6
y	0.0858	–0.0099	–0.3242	–0.9827	–2.1094	–3.8270

1.81 A particle is moving along a straight line. The displacements x at some time instances t are given below. Find the velocity and acceleration of the particle at $t = 4$.

t	1	3	5	7	9	11
x	0.1405	0.7676	3.5135	9.9351	21.5892	40.0324

1.82 Find the values of $y'(4)$ and $y''(4)$ from the following table:

x	0	1	2	3	4
y	5	8	12	17	26

1.83 Find the values of $y'(2)$ and $y''(2)$ from the following table:

x	1.5	1.6	1.7	1.8	1.9	2.0
y	0.3328	0.5312	0.7651	1.0384	1.3552	1.7198

1.84 Compute the values of $y'(3)$ and $y''(3)$ from the following table:

(*a*)

x	1	2	3	4	5
y	0	1.4	1.65	5.673	8.0978

(*b*)

x	1	2	3	4	5
y	0.4	0.65	0.75	0.87	0.98

1.85 Compute the values of $y'(2)$ and $y''(2)$ from the following table:

(*a*)

x	1	1.5	2	2.5	3
y	0	0.5	1.1	3.2	5.3

(*b*)

x	1	1.5	2	2.5	3
y	–2	2	3.456	5.674	8.4592

1.86 Compute the values of $y'(1.2)$ and $y''(1.2)$ from the following table:

(*a*)

x	1	1.1	1.2	1.3	1.4
y	0.1	0.34	0.42	0.53	0.62

(*b*)

x	1	1.1	1.2	1.3	1.4
y	0.0254	0.0437	0.0587	0.0670	0.0780

(*c*)

x	1	1.1	1.2	1.3	1.4
y	0.0012	0.2342	0.5786	0.7693	0.8934

LIST OF SYMBOLS

b	base of the internal number system ($b = 2$ for binary, $b = 10$ for decimal or $b = 16$ for hexadecimal computers)
c	nonzero constant
$E(X)$	expected value of random variable X
E_a	absolute error
E_p	percentage relative error
E_r	relative error
f	frequency
$f'(x)$	derivative of function $f(x)$
$f''(x)$	second derivative of function $f(x)$
$f(x)$	function
$F(x)$	normal distribution function
I	moment of inertia of a solid cylinder
m	mantissa or a fraction (a value which lies between 0.1 and 1)
$n!$	n factorial
$N(\mu, \sigma^2)$	normal distribution with μ and σ

nC_r	number of combinations of n objects taken r at a time
nP_r	number of permutations of n objects taken r at a time
p	shift required to recover the actual exponent
R	linear function of n independent variables
$R_n(x)$	remainder
s	sample standard deviation
s	the rms deviation about the universe average $\bar{X}_u$
s^1	the rms deviation about the sample average x_k
s^2	sample variance
s_v	standard deviation of v
s_x	standard deviation of x
s_y	standard deviation of y
t	shifted exponent (the value that is actually stored)
u_q	forward finite-difference approximation
v	velocity of a wave
$V(X)$	variance of random variable X
z	standard random variable for standard normal distribution
$\frac{dy}{dx}$	derivative of y w.r.t. x
$\bar{X}$	mean for sample data
$\left\lvert \frac{\Delta x}{x} \right\rvert$	relative error in quotient
$\lvert \Delta x \rvert$	absolute error
$\frac{d^2y}{dx^2}$	second derivative of y w.r.t. x
$s_{\bar{x}}$	standard deviation of the average $\bar{x}$
$\left\lvert \frac{X - x}{x} \right\rvert$	total relative error
δ	central difference operator
δx	relative error in x
Δ	forward difference operator
Δ^o	identity operator
ε_M	machine epsilon
$\phi(x)$	standard normal distribution
λ	wavelength
μ	mean for population data
ρ_{xy}	correlation coefficient for x and y
σ	population standard deviation
σ	sign of the number (±), denoted by a single bit

σ^2	population variance
σ_R	standard deviation of R
σ_x	standard deviation of random variable x
ω_i	interval for the variable
ω_R	interval for the result
∇	backward difference operator

GLOSSARY OF TERMS

Accuracy Error: Accuracy error is always present when the numerical average of successive readings deviates from the known correct reading and continues to deviate no matter how many successive readings are made.

Accuracy of an Instrument: The accuracy of an instrument indicates the deviation of the reading from a known input. Accuracy is often expressed as a percentage of full-scale reading, so that a 100 kPa pressure gage having an accuracy of 1 percent would be accurate within ± 1 kPa over the entire range of the gage.

Accuracy: A value obtained through measurement may or may not be close to the true value. In situations, where we believe that the measured value is close to the true value, we say that the measured value is accurate. Accuracy is a qualitative term.

Actual Value: See True Value.

Array: An array is an arrangement of raw numerical data in ascending or descending order of magnitude.

ASME: The American Society of Mechanical Engineers (ASME).

Base Units: Base units are *dimensionally independent*.

Best Estimate of the True Value: When only random sources act to influence the values obtained, the best estimate of the true value is usually taken to be the mean, $\bar{x}$ of n values, where

$$\bar{x} = \frac{\sum_{i=1}^{n} x_i}{n}$$

and x_i is the ith value obtained through measurement.

Calibration: Calibration is the comparison with a true or known value. During calibration, a value is measured by an instrument or provided by an artifact. If there is a discrepancy between the value as indicated by the instrument or artifact and the corresponding standard, then the difference between the two is quoted as a correction to the instrument or artifact. This process is referred to as *calibration*.

CIPM: The International Committee on Weights and Measures (CIPM).

Coefficient Interval: All the values for a parameter that would not be rejected if they were used as the null hypothesis in a statistical test of that parameter.

Coefficient of Variation: A measure of relative variability that expresses standard deviation as a percentage of the mean.

Completely Correlated Errors: Completely correlated errors are associated with the individual measurements of x and y and could be detected by calculating the correlation coefficient from the deviations and finding that it equals unity.

Continuous Random Variable: A random variable that can assume any value in one or more intervals. A random variable whose value is countable.

Control: Control or the isolation of a test from its surroundings is one of the great, basic conceptions of western scientific man.

Controlled Experiment: A controlled experiment is one in which the effects of extraneous variables have been eliminated and in which the independent variables can be varied exactly as the investigator chooses.

Correlated Errors: Correlated errors are not an intrinsically new class of errors, but rather errors whose independent and non-independent portions cannot be separated. The method of their addition is intermediate between the methods for independent and non-dependent errors.

Data or Data Set: Collection of observations or measurements on a variable.

Data: Data applies to all the *symbolic* products of an experiment. That is, data might be photographic images, magnitude impulses on tape, figures on sheets of paper, readings on mechanical contents, a single yes-or-no answer in the mind of an observer, and so on.

Debugging: The debugging process involves modifying the rig or the data-reduction program to remove, or account for, errors introduced by the apparatus and procedure. This phase ends when the experiment has acceptably small fixed errors over its entire operating range.

Degrees of Freedom: The number of observations that can be chosen freely.

Dependent Variable: See Variable.

Derived Units: Derived units literally are derived from the base and supplementary units and, therefore are comprised of products, quotients and powers of base and supplemental units. Other units, created by combining SI base units, are referred to as derived units.

Designed Experiment: A study in which the experiment controls the assignment of elements to different treatment groups.

Detailed Uncertainty Analysis: Detailed uncertainty analysis is a more thorough approach that identifies the systematic and random errors contributing to each measurand's overall uncertainty. The propagation of systematic and random errors into the final result is computed in parallel.

Dimension: A dimension is a physical variable used to specify the behavior or nature of a particular system.

Discrepancy: Discrepancy is the difference between two measured values of the same quantity.

Discrete Random Variable: A random variable whose values are countable.

Discrete Variable: A variable that can have only discrete values.

Dispersion: The degree to which numerical data tends to spread about an average value is called the variation or dispersion of the data. The dispersion of data is characterized numerically by a standard deviation.

Element or Member: A specific subject or object included in a sample or population.

Elemental Error: Elemental error is an error that can be associated with a single uncertainty source.

Error: In the context of measurement, error is defined as the difference between the measured value and the true value:

$$\text{Error} = \text{Measured value} - \text{True value}$$

Error: The amount of variation in the response caused by factors other than controllable factors included in the experiment.

Estimation: Estimation is the process of analyzing a sample result in order to predict the corresponding value of the population parameter. The sample statistic is used to estimate a population parameter.

Experiment: A process with well-defined outcomes that, when performed, results in one and only one of the outcomes per repetition.

Experimental Plain: The *experimental plain* is a general term relating to any specific set of test operating instructions that give sequence of test work, kinds and amounts of variable changes, and any replication decisions.

Gaussian Distribution: Another name for *normal distribution.*

General Uncertainty Analysis: General uncertainty analysis is a simplified approach that considers each measurand's overall uncertainty and its propagation into the final result. It does not consider the specific systematic and random errors that contribute to the overall uncertainty.

Hysteresis Error: Hysteresis error can be ascertained by performing first an upscale, then a down scale calibration. It usually arises from having a physical change in part of the measurement system on reversing the system's input.

Independent Variable: See Variable.

Instrument Error: Instrument errors (elemental errors) are identified through calibration.

Instrument Uncertainty: The instrument uncertainty usually is stated by the manufacturer and results from a number of possible elemental instrument uncertainties. Examples are linearity, hysteresis, sensitivity, zero-shift, repeatability, stability and thermal-drift errors.

Instrument: Instruments sense, count, read, measure, observe and then record, store, correct, read out what they have been set to *see*.

ISO: International Standardization Organization.

Least Significant Digit: The right most *non-zero* digit is called the least significant digit.

Least Squares Method: The method used to fit a regression line through a scatter diagram such that the error sum of squares is a minimum.

Least Squares Model: The least squares model that forms the basis is written as follows:

Measured values = Function of one or more parameters to be estimated + Scatter.

Least Squares Regression Line: A regression line obtained by using the least squares method.

Least-Count: The *lease count* is the smallest difference between two indications that can be detected on the instrument scale.

Linear Correlation Coefficient: A measure of the strength of the linear relationship between two variables.

Linearity Error: Linearity error is a measure of how linear is the best-fit of the instrument's calibration data.

Mean: The average of the data values, that is, the sum of all of the data values divided by the number of data values.

Measurand: A particular quantity determined through measurement is called a measurand. Often, a value of a measurand can be established directly with an instrument.

Measurement Error: Measurement error is the difference between the true value and the measured value. It is a specific value.

Measurement Uncertainty: Measurement uncertainty is an estimate of the error in a measurement. It represents a range of possible values that the error might assume for a specific measurement.

Measurement: Measurement is a process by which a value of a particular quantity such as temperature of a water bath or the pH of a solution is obtained.

Measurements: Measurements are made of the independent, dependent, and often, of the extraneous variables by the instruments.

Metrology: The branch of science concerned with maintaining and increasing the accuracy of measurement is known as metrology.

Modeling Error: Modeling error is defined as a *recognizable* deficiency not due to lack of knowledge.

Modeling Uncertainty: Modeling uncertainty is defined as a *potential* deficiency due to a lack of knowledge.

Most Significant Digit: The left most *non-zero* digit is called the most significant digit.

Non-independent Errors: Non-independent errors are due to external causes and cannot be detected by a study of the deviations.

Normal Distribution: The probability distribution that describes or approximates the distribution of many variables. Most observations in a normal distribution occur near the mean.

Normal Probability Distribution: The probability distribution of a continuous random variable that, when plotted, gives a specific bell-shaped curve. The parameters of the normal distribution are the mean μ and the standard deviation, σ.

Observation or **Measurement:** The value of a variable for an element. A person, place, or thing for which data value is available or is measured.

Parameter: An unknown value in a model that is estimated from the data.

Population: A population consists of all of the individuals or objects in a well-defined group about which information is desired to answer a question. In statistics, the term population refers to the number of possible, but not necessarily actual, measured values.

Precision Error: Precision error is always present when successive measurements of an unchanged quantity yield different numerical values.

Precision Error: Precision error is not a fixed quantity like an accuracy error, but is different for each replication of a reading.

Precision of an Instrument: The precision of an instrument indicates its ability to reproduce a certain reading with a given accuracy.

Precision: When values obtained by repeated measurements of a particular quantity exhibit little variability, we say that those values are precise. Precision, like accuracy, is a qualitative term.

Primary Standard: One of the duties of National Measurement Institute of Australia in Sydney (NMI) is to house, safeguard and maintain the highest quality standards, referred to as *primary standard*.

Process Data: Process data is the symbolic materials after some additional mathematical operations have been performed on them, such as corrections by a calibration curve or plotting on a graph.

Random Errors: In some cases, the influences that affect the measurement process, or the quantity being measured, cause values to be randomly distributed above and below the true value. The errors caused by such influences are termed random errors.

Random Sample: A sample drawn in such a way that each element of the population has same chance of being included in the sample.

Random Variable: A variable whose value is determined by the outcome of a random experiment. Also called a *chance variable*.

Raw Data: Data recorded in the sequence in which they are collected and before they are processed.

Raw Data: Raw data is the symbolic materials obtained directly from the test instruments.

Readability of an Instrument: The readability of an instrument is the closeness with which the scale of the instrument is read by an experimenter. This is a subjective value. Readability does not enter into assessing the uncertainty of the instrument.

Readability of an Instrument: The term indicates the closeness with which the scale of the instrument may be read; an instrument with a 12-in scale would have a higher readability than an instrument with 6-in scale and the same range.

Repeatability Error: Repeatability error is related to the precision of the calibration. This is determined by repeating the calibration many times for the same input values.

Repeatability: In many circumstances, measurements are made under identical conditions. When this happens, it is possible for the values obtained to exhibit little variation or scatter. In such cases, we speak of measurement being repeatable.

Replication: Replication means, in general, repetition, but more specifically a return to a previous condition.

Reproducibility: If there is consistency between values obtained by different experimenters, we say that the value is reproducible.

Resolution of an Instrument: The resolution of an instrument is the smallest, physically indicated division that the instrument displays or is marked.

Rounding: When a number has too many significant figures for a particular purpose, the number of significant figures can be reduced by a simple procedure known as rounding. For example, if a length has been measured as 3.2451 m, that is to five significant figures, but three significant figures are considered sufficient, that value can be rounded to 3.25 m. The value has been rounded to three significant figures.

Rules of Decision: See Tests of Hypothesis.

Sample Point: An outcome of an experiment.

Sample Statistics: In almost every case of measurement we sample a population, and the quantities of interest obtained from the sample (some times called sample statistics) should reliably represent corresponding parameter in the population (the population parameters).

Sample: A selection of observation from a larger population of all possible observations. Ideally, the sample is selected randomly from the population.

Sampling Error: The difference between the values of a sample statistic calculated from a random sample and the value of the corresponding population parameter. This type of error occurs due to chance.

Scatter Diagram: A two-variable plot in which each point represents the levels of an observation on each variable. Often the variable on the horizontal axis is to be used to predict the variable on the vertical axis. Scatter plots are useful for viewing *correlations* and for *simple regression.*

Scatter: The term *scatter* is a quantitative descriptor of the variance in data, the extent to which the individual values differ from mean.

Selection Error: The error that occurs because the sampling frame is not representative of the population.

Sensitivity Error: Sensitivity error is characterized by the greatest change in the slope (static sensitivity) of the calibration fit.

Sensitivity of an Instrument: The sensitivity of an instrument is the ratio of the linear movement of the pointer on an analog instrument to the change in the measured variable causing this motion.

Sequence of Testing: Sequence of testing refers to the order in which various operating changes are made in the test apparatus.

SI: System International.

Stability Error: Stability error is a measure of how much the output can drift over a period of time for the same input.

Standard Deviation of a Discrete Random Variable: A measure of spread for the probability distribution of a discrete random variable.

Standard Deviation of Errors: A measure of spread for the random errors.

Standard Deviation: A measure of spread that is given by the positive square root of variance.

Standard Error of Mean: The standard deviation of the sampling distribution of the mean.

Standard Error: The *standard deviation* of the *sample distribution* for any given statistic.

Standard Normal Distribution: The normal distribution with $\mu = 0$ and $\sigma = 1$, the units of the standard normal distribution are denoted by z.

Static Measurement: A *static* measurement of a physical quantity is performed when the quantity is not changing with time.

Statistic: A summary measure calculated for sample data.

Statistics: Statistics deals with scientific methods for collecting, organizing, summarizing, presenting and analyzing data, as well as drawing valid conclusions and making reasonable decisions based on such analysis.

Survey: Collecting data on the elements of a population or sample.

Systematic Error: Systematic error causes all measured values to be consistently under or over estimated. Systematic errors sometimes can be difficult to detect and can be found and minimized through the process of calibration.

Test Apparatus: The test apparatus is usually taken to be everything required to perform a test including the instruments and the test piece.

Test Piece: The test piece is the specific hardware item that is undergoing test and may be replaced at will.

Test Run: To make a run, or test run means simply to put the test apparatus in a certain fixed condition or configuration and record all the instrument measurements.

Tests of Hypothesis: Procedures, which will enable to decide whether to accept or reject hypothesis or to determine whether observed samples differ significantly from expected results are called tests of hypothesis; tests of significance, or rules of decision.

Tests of Significance: See Tests of Hypothesis.

Thermal-Drift Error: The thermal-drift error e_T is given by

$$e_T = \left(\frac{e_{T,\max}}{FSO}\right)$$

where FSO is the full-scale-output and $e_{T,\max}$ is the greatest deviation in the output value for affixed input value that could occur during operation because of variations in the environmental temperature. The stability and thermal-drift errors are similar in behavior to the zero-shift error.

Traceability: The result of a measurement is said to be traceable if, through an unbroken chain of comparisons often involving working and secondary standards, the result can be compared with a primary standard.

True Value: In most cases, we assume that there is an *actual* or *true* value of a quantity, such as the value of the charge of an electron. It is the true value that one would like to establish through measurement.

Uncertainty Analysis: Uncertainty analysis is the prediction of the uncertainty interval which should be associated with an experimental result, based on observations of the scatter in the raw data used in calculating the result.

Uncertainty: Errors are key and unavoidable ingredients of the measurement process. Their net effect is to create an uncertainty in the value measured. When used in the context of measurement, uncertainty has a number and generally a unit associated with it. More specifically, measurement uncertainty has the same unit as the measurand. Uncertainty, like error, is a number, but an estimated one. We define uncertainty as *what we think the error would be if we could and did measure it by calibration.*

Value: The process of measurement yields a value of a particular quantity.

Variable: A characteristic, trait, attribute, or measurement that can take on different values. A variable must vary, having at least one value for some observations and another value for other observations.

Variable: Variable is used in a general sense to imply any physical quantity that undergoes change. If we control this change independent of other quantities we have an *independent variable*. If a physical quantity changes in response to the variation of one or more independent variables, we have a *dependent variable*.

Variables, Factors, or Parameters: These words are used synonymously to indicate the controllable factors in an experiment. In case of a plastic molding experiment, molding temperature, injection pressure, set time, etc., are factors.

Variance: The typical or average squared error. For a population, the sum of squared error is divided by n, the number of observations. More often, interest is in sample estimate of the variance in which case the sum of squared errors is divided by $n - 1$.

Variation: See Dispersion.

Zero-Centered Experiment: The zero-centered experiment is a goal, not often a reality experiments.

Zero-Shift Error: Zero-shift error refers to the greatest possible shift that occurs in the intercept of the calibration fit.

z-Score Formula: The difference between the observations and the mean, divided by the standard deviation.

CHAPTER 2

Basic Vehicle Motion Analysis

2.1 INTRODUCTION

In this Chapter we present an introduction to Newtonian mechanics, basic terms and definitions, Newton's laws of motion, basic equations of motion, coefficient of friction and drag factor, laws of friction, drag factor on a grade, unequal drag factors on axles, work, energy and laws of conservation of energy, potential energy, kinetic energy, law of conservation of energy, linear momentum, time distance equations, velocity equations, time equations, distance equations, acceleration factor equations, special equations, kinetic energy, combined speed equations, slide-to-stop equation, slide-to-stop equation with deceleration (drag factor), slide-to-stop equation with grade and percentage of braking, drag factor equation, braking distance equation, weight shift, horizontal center of mass (CM) equation, height of center of mass (CM) equation, weight shift equation, speed equation with weight shift, speed equation with any wheel braking with weight shift, weight shift equation with grade, critical speed yaw, critical speed yaw equation, airborne equations, fall speed equations, level takeoff, uphill takeoff, downhill takeoff, flips and vaults, vehicle lands lower than takeoff, vehicle lands higher than takeoff, maximum height, driver perception and response, estimation of vehicle critical speed from yaw marks, speed estimates from yaw marks, radius equation, lane passing with changing velocity, lane changing maneuver model, engineering models and animations, function of accident scene models, model application, reconstruction animations, and example problems and solutions.

2.2 NEWTONIAN MECHANICS

Engineering mechanics is the science that considers the motion of bodies under the action of forces and the effects of forces on that motion. Mechanics includes *statics* and *dynamics.* Statics deals with the special case of a body at rest or a body that moves with constant velocity. A body at rest or moving with constant velocity is said to be in *equilibrium.* This is sometimes also called as *static equilibrium.* When the body moves with finite velocity or acceleration, the principles of statics are no longer applicable. The mechanics of such a system is called *dynamics.* When the body has no rotational motion, it is called as *particle*. Usually, each point on a rigid body is always at a constant distance from any other point in the body. Dynamics is further divided into *kinematics* and *kinetics. Kinematics* defines the relationships among displacement, velocity, and acceleration of a moving body. *Kinetics* defines the relationship between the forces that act on a body and the motion of the body. Analysis of the bodies can either be conducted in plane or in three-dimensions. A rigid-body in space has six degree of freedom.

Collectively, the study of statics and dynamics is called *classical mechanics.* Classical mechanics treats the motion of bodies of ordinary size that move at speeds that are small compared to the speed of light. Newton (1642–1727) formulated the law of universal gravitation and the mathematics of calculus. Newton introduced the concepts of force and mass. Newton formulated the three laws of motion that are the basis of engineering applications of mechanics. The classical mechanics is often called *Newtonian mechanics.*

2.3 BASIC TERMS AND DEFINITIONS

Acceleration: The mathematical opposite of deceleration. Acceleration is the rate change of an object's speed or velocity with respect to time with speed or velocity becoming greater over time.

Acceleration Factor: A percentage or fraction of the earth's gravity constant (32.2 feet per second per second or 9.81 meters per second per second) used to describe the magnitude of an acceleration or deceleration. Acceleration factor is described in *g numbers* such that a fractional g-value is equal to the equivalent fraction of the earth's gravity constant.

Braking Distance: The distance required for a vehicle to come to a complete stop from the point at which the actual braking process begins.

Coefficient of Friction: The ratio of lateral force required to move an object across another object to the weight of the object being moved. The term coefficient of friction (CoF) will refer to the dynamic coefficient of friction.

Constant: A mathematical quantity or value that does not change.

Conversion Factor: A constant by which a value is multiplied or divided to yield an equal value with a different unit of measure.

Deceleration: The mathematical opposite of acceleration. Deceleration is the rate of change of an object's speed or velocity with respect to time with speed or velocity becoming less over time.

Density: The ratio of an object's mass to its volume.

Displacement: The degree or amount of separation between two points or objects, measured along the shortest path between them. Displacement is thus a vector quantity, having both magnitude and direction.

Distance: The degree or amount of separation between two points or objects, measured along the path between them. Distance is a scalar quantity, having magnitude only.

Force: Force is mass times acceleration. This means that weight (which is mass times acceleration of gravity) is a force.

Friction: Friction is the resistance to motion that arises when one object is moved across another. Friction is that force which acts between an object and the surface across which the object is being moved to resist the sliding motion between the two objects.

Gravitational Acceleration: The acceleration of objects resulting from the effect of earth's gravity upon the object.

Kinetic Energy: This is energy associated with movement; the energy of motion.

Mass: The mass of any object is its natural weight (W) divided by the acceleration of gravity (g).

Mechanical Work: Mechanical work is the product of the force F, which acts to produce displacement of a body, and the magnitude of the displacement, d, through which this force acts (work $= Fd$).

Newton's Second Law of Motion: The acceleration of any body (time rate of change of velocity) is directly proportional to the force acting on the body and inversely proportional to the mass ($a = F/m$).

Perception/Reaction Distance: The distance through which a vehicle travels while the driver is processing information that will give rise to a decision regarding a response to a given stimulus or set of stimuli.

Perception/Reaction Time: The time taken by a vehicle driver while the driver is processing information that will give rise to a decision regarding a response to a given stimulus or set of stimuli.

Speed: Time rate of change of position of a body *without* regard to direction.

Stopping Distance: The total distance required to bring a vehicle to a stop, including the distance traveled during the perception/reaction phase of stopping and the braking phase of stopping.

Variable: A mathematical quantity or value that changes or is permitted to be changed to yield a desired result.

Velocity: Time rate of change of position of a body *with* regard to direction.

2.4 NEWTON'S LAWS OF MOTION

Newton's laws of motion are fundamentally applicable directly to a particle that is a body which may be treated as having point mass. Newton's laws may be stated in terms of particle parameters as follows:

First Law: In the absence of applied forces, a particle originally at rest or moving with constant speed in a straight line will remain at rest or continue to move with constant speed in a straight line.

Second Law: If a particle is subjected to a force, the particle will accelerate. The acceleration of the particle will be in the direction of the force, and the magnitude of the acceleration will be proportional to the force and inversely proportional to the mass of the particle.

Newton's second law can be expressed in the form

$$a = \frac{F}{m} \tag{2.1}$$

where F = force

m = mass

a = acceleration

Newton's second law is the basis for the study of kinetics of a particle.

Third Law: For every action, there is an equal and opposite reaction, or the mutual forces exerted by two particles on each other are always equal and oppositely directed.

2.5 BASIC EQUATIONS OF MOTION

Nomenclature:

Variable	*Name*	*US units*	*SI units*
s	speed	mph	km/h
d	distance	ft	m
t	time	sec	sec
f	acceleration factor	dimensionless	dimensionless
	drag factor = a/g		
	start (initial) speed		
s_0	final (end) speed	mph	km/h
s_f	start (initial) speed	mph	km/h
v_0	final (end) speed	mph	km/h
v_f	force	mph	km/h
F	tire-road friction coefficient	lb	kg
μ	% braking	dimensionless	dimensionless
n	middle ordinate	dimensionless	dimensionless
m, M	mass	1b	kg
R	curve radius	ft	m
h	height	ft	m
KE	kinetic energy	lb ft^2/s^2	J
W	weight	lb ft/s^2	N
V	speed	ft/s	m/s
d_0	total distance	ft	m
f_s	lateral drag factor	dimensionless	dimensionless
g	acceleration due to gravity	ft/s^2	m/s^2
Q	take off angle	deg	deg
a	acceleration	ft/s^2	m/s^2

The basic vehicle motion equations are summarized in Table 2.1. As shown in Table 2.1, these equations are useful in finding the acceleration, distance, final velocity, initial velocity and time. Speed equations are summarized in Table 2.2. Time equations are given in Table 2.3. Table 2.4 gives the distance equations.

Table 2.1 Basic motion equations

No.	*To find*	*Given*	*Equation to use*
1.	Acceleration, a (ft/s^2 or m/s^2)	t, v_0, v_f	$a = \dfrac{v_f - v_0}{t}$
		t, v_0, d	$a = \dfrac{2d - 2v_0 t}{t^2}$

		v_0, v_f, d	$a = \frac{v_f^2 - v_0^2}{2d}$
2.	Distance, d (ft or m)	t, a, v_0	$d = v_0 t + at^2/2$
		a, v_0, v_f	$d = \left(v_f^2 - v_0^2\right)/2a$
		t, v_0, v_f	$d = t(v_0 + v_f)/2$
3.	Final velocity, v_f (ft/s or m/s)	t, a, v_0	$v_f = v_0 + at$
		a, v_0, d	$v_f = \sqrt{v_0^2 + 2ad}$
4.	Initial velocity, v_0 (ft/sec or m/s)	t, a, v_f	$v_0 = v_f - at$
		t, a, d	$v_0 = \frac{d}{t} - \frac{at}{2}$
		a, v_f, d	$v_0 = \sqrt{v_f^2 - 2ad}$
5.	Time, t (seconds)	a, v_0, v_f	$t = \frac{v_f - v_0}{a}$

Table 2.2 Speed equations

No.	*US units*		*SI units*	
	Speed	*Velocity*	*Speed*	*Velocity*
1.	$s = \sqrt{30df}$	$v = \sqrt{64.4\ df}$	$s = 15.9\sqrt{df}$	$v = 4.42\sqrt{df}$
2.	$s = \frac{d}{1.467t}$	$v = \frac{d}{t}$	$s = \frac{d}{0.278t}$	$v = \frac{d}{t}$
3.	$s = 21.95$ ft	$v = 32.2$ ft	$s = 35.3$ ft	$v = 9.81$ ft
4.	$s_f = s_0 + 21.95$ ft	$v_f = v_0 + 32.2$ ft	$s_f = s_0 + 35.3$ ft	$v_f = v_0 + 9.81$ ft
5.	$s_f = \sqrt{s_0^2 + 30df}$	$v_f = \sqrt{v_0^2 + 64.4\ fd}$	$s_f = \sqrt{s_0^2 + 254df}$	$v_f = \sqrt{v_0^2 + 19.62\ fd}$
6.	$s_0 = s_f - 21.95$ ft	$v_0 = v_f - 32.2$ ft	$s_0 = s_f - 35.3$ ft	$v_0 = v_f - 9.81$ ft
7.	$s_0 = \sqrt{s_f^2 - 30\ df}$	$v_0 = \sqrt{v_f^2 - 64.4\ df}$	$s_0 = \sqrt{s_f^2 - 254\ df}$	$v_0 = \sqrt{v_f^2 - 19.62\ df}$

Table 2.3 Time equations

No.	US units		SI units	
	Time	Time	Time	Time
1.	$t = \frac{d}{1.466s}$	$t = \frac{d}{v}$	$t = \frac{d}{0.278s}$	$t = \frac{d}{v}$
2.	$t = \frac{s}{21.95f}$	$t = \frac{v}{32.2f}$	$t = \frac{s}{35.3f}$	$t = \frac{v}{9.81f}$
3.	$t = \frac{s_f - s_0}{21.95f}$	$t = \frac{v_f - v_0}{32.2f}$	$t = \frac{s_f - s_0}{35.3f}$	$t = \frac{v_f - v_0}{9.81f}$
4.	$t = 0.249\sqrt{\frac{d}{f}}$	$t = 0.249\sqrt{\frac{d}{f}}$	$t = 0.45\sqrt{\frac{d}{f}}$	$t = 0.45\sqrt{\frac{d}{f}}$

Table 2.4 Distance equations

No.	US units		SI units	
	Distance	Distance	Distance	Distance
1.	$d = \frac{s^2}{30f}$	$d = \frac{v^2}{64.4f}$	$d = \frac{s^2}{254f}$	$d = \frac{v^2}{19.62f}$
2.	$d = 1.466\ st$	$d = vt$	$d = 0.278\ st$	$d = vt$
3.	$d = \frac{s_f^2 - s_0^2}{30f}$	$d = \frac{v_f^2 - v_0^2}{64.4f}$	$d = \frac{s_f^2 - s_0^2}{254f}$	$d = \frac{v_f^2 - v_0^2}{19.62f}$
4.	$d = 1.466\ s_0t + 16.1$ ft	$d = v_0t + 16.1$ ft^2	$d = 0.278\ s_0t + 4.9$ ft^2	$d = v_0t + 4.9$ ft^2

Example 2.1: *A road vehicle skids to a slop over two surfaces. The vehicle has skidded 27.43 m (90 ft) over the first surface and 15.24 m (50 ft) over the second surface. Assuming the first surface has a drag factor of 0.80 and the second surface has a drag factor of 0.70, determine*

(*a*) *the vehicle's speed when the skidding first occurred*

(*b*) *the time taken for the vehicle to skid over both surfaces*

(*c*) *the vehicle's speed after skidding 33.53 m (110 ft).*

Solution:

(*a*) From Table 2.1, the initial velocity v_0 is obtained from

$$v_0 = \sqrt{v_f^2 - 2ad} = \sqrt{0^2 - 2(-6.87)(15.24)} = 14.47 \text{ m/s } (47.48 \text{ ft/sec})$$

where $a = fg = -0.70(9.81) = -6.87$ m/s^2 (-22.54 ft/sec^2)

Here, 14.47 m/s (47.48 ft/sec) is the velocity where the vehicle starts to skid on the second surface. It is the end velocity after skidding on the first surface. Therefore, on the first surface $v_f = 14.47$ m/s (47.48 ft/sec), $d = 27.43$ m (90 ft), and $a = fg = -0.80(9.81) = -7.85$ m/s^2 (-25.76 ft/sec^2).

Now $$v_0 = \sqrt{v_f^2 - 2ad} = \sqrt{14.47^2 - 2(-7.85)(27.43)} = 25.30 \text{ m/s (83.01 ft/sec)}$$

$$= 90.86 \text{ km/h (56.47 mph)}$$

(*b*) Surface 1:

$v_0 = 25.30$ m/s (83.01 ft/sec),

$v_f = 14.47$ m/s (47.48 ft/sec)

and $a = -7.85$ m/s^2 (25.76 ft/sec^2)

From Table 2.1, we have

$$t = \frac{v_f - v_0}{a} = \frac{14.47 - 25.30}{-7.85} = 1.38 \text{ sec}$$

Surface 2: $v_0 = 14.47$ m/s (47.48 ft/sec),

$v_f = 0$

and $a = -6.87$ m/s^2 (–22.54 ft/sec^2)

$$t = \frac{v_f - v_0}{a} = \frac{0 - 14.47}{-6.87} = 2.11 \text{ sec}$$

Total skidding time = 1.38 + 2.11 = 3.49 sec

(*c*) The end velocity can be obtained by knowing the initial velocity at the beginning of the second surface. An additional distance of 6.10 m (20 ft) is required to reach a total skidding distance of 33.53 m (110 ft).

Here $v_0 = 14.47$ m/s (47.48 ft/sec),

$a = -6.87$ m/s^2 (–22.54 ft/sec^2)

and $d = 6.10$ m (20 ft).

$$v_f = \sqrt{v_0^2 - 2ad} = \sqrt{14.47^2 + 2(-6.87)(6.10)} = 17.12 \text{ m/s (56.18 ft/sec)}$$

$$= 64.50 \text{ km/h (38.22 mph)}$$

2.6 COEFFICIENT OF FRICTION AND DRAG FACTOR

A vehicle's acceleration or deceleration is related to its drag factor, f, and the coefficient of friction, μ. The drag factor is related to the acceleration by the following equation:

$$a = fg \tag{2.2}$$

where a = acceleration, m/s^2 (ft/s^2)

f = drag factor (no dimension)

g = acceleration due to gravity, 9.81 m/s^2 (32.2 ft/s^2)

Drag factors and coefficients of friction are generally similar for all types of four-tired vehicles. The drag factors for motorcycles and heavy vehicles can often differ from those of the four-tired vehicles.

Acceleration factor is the ratio of an acceleration rate to the acceleration of gravity. The deceleration rate, related to friction between a skidding or sliding tire and the roadway surface due to the force of gravity, is usually less than the acceleration rate of gravity and therefore is usually expressed as a decimal fraction of gravity. This fraction is known as the drag factor (f) or, on a level surface, the coefficient of friction value.

Friction can be thought of as the resisting force to motion between two surfaces at their interface (contact). *Static friction* between two bodies is the tangential force that opposes the sliding of one body relative to the other. *Limiting friction*, f', is the maximum value of static friction that occurs when motion is impending. *Kinetic friction* is the tangential force between two bodies after motion begins. It is less than static friction. *Angle of friction* is the angle between the action of the line of the total reaction of one body on another and the normal to the common tangent between the bodies when the motion is impending. *Coefficient of static friction* is the ratio of the limiting friction f' to the normal force N.

$$\mu = \frac{f'}{N} \tag{2.3}$$

Coefficient of kinetic friction is the ratio of the kinetic friction to the normal force. *Rolling friction (rolling resistance)* refers to the resisting forces that come into play when a vehicle is rolling with no braking. Generally, these values are very low and are assumed to be insignificant in most accident reconstruction problems. Rolling friction is, however, important in tire design and other vehicle design considerations.

2.6.1 Laws of Friction

1. The coefficient of friction is independent of the normal force; however, the limiting friction and kinetic friction are proportional to the normal force.
2. The coefficient of friction is independent of the area of contact.
3. The coefficient of kinetic friction is less than that of static friction.
4. At low speeds, friction is independent of the speed. At higher speeds, a decrease in friction has been noticed.

Drag factor, f, is defined as the force required for acceleration (or deceleration) in the direction of the acceleration (or deceleration) divided by the object's (vehicle's) weight. Thus, we have

$$f = F/W \tag{2.4}$$

where f = drag factor (no dimension)

F = force, N (lbs)

W = weight of the object (vehicle) kg (lbs)

For coefficient of friction, μ, the object must be sliding across the surface. This is not the case for drag factor, f. Drag factor and coefficient of friction will be equal only in cases where all wheels get locked and sliding on a level surface. *Drag factors* refer to conditions where a vehicle or object is slowing. The word drag implies slowing. There are several methods available for determining the drag factor/ coefficient of friction for an actual accident. They are:

1. test-skit the accident vehicle or an exemplar vehicle.
2. slide a test tire (not the whole vehicle) to obtain the friction coefficient.
3. use existing highway department skid numbers for the road in question.
4. look-up coefficients of friction in a table and apply the appropriate adjustments to them for the case at hand.

2.6.2 Drag Factor on a Grade

The drag factor on a grade is given by the following formula:

$$f_G = \frac{\mu + G}{\sqrt{1+G^2}} \tag{2.5}$$

where f_G = drag factor on a grade

μ = coefficient of friction between the tires and pavement

G = the percent grade expressed as a decimal [positive (+) for an upgrade and negative (–) for a downgrade].

Equation (2.5) applies for a case where a vehicle is skidding with all wheels locked on a non-level surface (that is, up or down grade); the drag factor is not equal to the coefficient of friction.

Table 2.5 gives a range of coefficients of friction called Northwestern drag factors for several road surface descriptions for passenger cars and pick up trucks equipped with typical tires. It can be seen from Table 2.5 that higher friction coefficients are suggested for the lower speed range. Generally, it is very rare to see a friction coefficient much over 0.90. Tables 2.6, 2.7 and 2.8 give Collin's drag factors, some unequal coefficient of friction factors, and motor cycle drag factors respectively.

Table 2.5 Northwestern drag factors

Description of road surface	*Dry under 30 mph*	*Dry over 30 mph*	*Wet under 30 mph*	*Wet over 30 mph*
PORTLAND CEMENT				
New, Sharp	0.80–1.20	0.70–1.00	0.50–0.80	0.40–0.75
Traveled	0.60–0.80	0.60–0.75	0.45–0.70	0.45–0.65
Traffic polished	0.55–0.75	0.50–0.65	0.45–0.65	0.45–0.60
ASPHALT, TAR				
New, Sharp	0.80–1.20	0.65–1.00	0.50–0.80	0.45–0.75
Traveled	0.60–0.80	0.55–0.70	0.45–0.70	0.40–0.65
Traffic polished	0.55–0.75	0.45–0.65	0.45–0.65	0.40–0.60
Excess tar	0.50–0.60	0.35–0.60	0.30–0.60	0.25–0.55
GRAVEL				
Packed, Oiled	0.55–0.85	0.50–0.80	0.40–0.80	0.40–0.60
Loose	0.40–0.70	0.40–0.70	0.45–0.75	0.45–0.75
CINDERS				
Packed	0.50–0.70	0.50–0.70	0.65–0.75	0.65–0.75
ROCK				
Crushed	0.55–0.75	0.55–0.75	0.55–0.75	0.55–0.75
ICE				
Smooth	0.10–0.25	0.07–0.20	0.05–0.10	0.05–0.10

SNOW				
Packed	0.30–0.55	0.32–0.55	0.30–0.60	0.30–0.60
Loose	0.10–0.25	0.10–0.20	0.30–0.60	0.30–0.60

Sources: 1. **Warner, C.Y., Smith, G.C., James, M.B., and Germane, G.J.,** "Friction Applications in Accident Reconstruction", SAE paper No. 830612, Society of Automotive Engineers, Warrendale, PA, 1983.

2. **Baker, J.S.,** *Traffic Accident Investigation Manual*, Northwestern University, Evanston, IL, 1975.
3. **Collins, J.C.,** *Accident Reconstruction*, C.C. Thomas Publisher, Springfield, IL, 1979.

Table 2.6 Collins drag factors

Description of road surface	*Automobile tire*	*Truck tire*
Dry concrete	0.85	0.65
Dry asphalt	0.80	0.60
Wet concrete	0.70–0.80	0.50
Wet asphalt	0.45–0.80	0.30
Packed snow	0.15	0.15
Ice	0.05	0.11 (dry)
		0.07 (wet)
Dry dirt	0.65	—
Mud	0.40–0.50	—
Grave or sand	0.55	—
Wet, oily, smooth concrete	—	0.25
Hard-packed snow w/chains	—	0.60
Dry ice w/chains	—	0.25

Source: **Collins, J.C.,** Accident Reconstruction, C.C. Thomas Publisher, Springfield, IL, 1979.

Table 2.7 Unusual coefficient of friction factors

Unusual friction coefficients		*Drag*
Passenger car skidding	on top on concrete	0.3
	on rough asphalt	0.4
	on gravel	0.5–0.7
	on dry grass	0.5
Body sheet metal on asphalt		0.4
Body sheet metal on dirt		0.2
Body sheet metal on body sheet metal (sideswipe)		0.6

Car to car		0.55
Engine braking	High gear	0.1
	Low gear	0.1–0.2
Rolling resistance	Normal inflation	0.01
	Partial inflation	0.013
	Flat tire	0.017
Skidding on packed snow		0.15
Skidding on ice or sleet		0.07
Human body skidding		1.1
Human body tumbling		0.8

Source: 1. **Collins, J.C.,** *Accident Reconstruction*, Charles C. Thomas, Springfield, IL, 1979.
2. **Collins, J.C. and Morris, J.L.,** *Highway Collision Analysis*, Charles C. Thomas, Springfield, IL, 1967.
3. **Emori, R.I.,** "Analytical Approach to Automobile Collisions", SAE Paper #680016, Warrendale, PA, 1968.
4. **McHenry, R.R., Lynch, J.P.,** "CRASH2 User's Manual", U.S. DOT Research Report, Contract No. DOT-HS-5-01124, Sept. 1976.

Table 2.8 Motorcycle drag factors

Motorcycles: Sliding on side	
Situation	*Drag factor*
Low drop test: fine textured surface	0.25–0.35
Low drop test: coarse textured surface	0.33–0.65
Towed upright and dropped: fine textured surface	0.32–0.53
Towed upright and dropped: coarse textured surface	0.39–0.79

Source: **Lambourne, R.F.,** "The Calculation of Motorcycle Speeds from Sliding Distances", Paper No. 910125, SAE International, Warrendale, PA, 1991.

2.6.3 Unequal Drag Factors on Axles

For passenger cars during hard braking, load is shifted to the front axle. Because of this load shift, no correction needs to be made when all the wheels are locked. If, however, one axle has a drag factor different from that of other axle, then the drag factor for the vehicle is given by

$$f_R = \frac{f_f - x_f(f_f - f_r)}{1 - z(f_f - f_r)} \tag{2.6}$$

where f_R = drag factor on the vehicle

f_f = drag factor on the front axle

f_r = drag factor on the rear axle

x_f = horizontal distance of the center of mass from the front axle as a decimal fraction of the wheel base

z = height of the center of mass as a decimal fraction of the wheel base

Table 2.9 Acceleration factors

Acceleration or Deceleration	*Speed range*	*Drag*	*m/s²*	*ft/sec²*
Free fall	Less than 20 mph (30 km per hr)	+1.00	+9.81	+32.2
Passenger cars,		+0.15	+1.47	+4.8
normal acceleration	20 to 40 mph (30 to 60 km per hr)	+0.10	+0.98	+3.2
		+0.05	+0.48	+1.6
Passenger cars,		+0.30	+2.94	+9.7
rapid acceleration	More than 40 mph (60 km per hr)	+0.15	+1.47	+4.8
		+0.10	+0.98	+3.2
		+0.10	+0.98	+3.2
		+0.05	+0.48	+1.6
Medium trucks,		+0.03	+0.29	+1.0
normal acceleration	Less than 20 mph (30 km per hr)	+0.05	+0.48	+1.6
	20 to 40 mph (30 to 60 km per hr)	+0.03	+0.29	+1.0
		+0.01	+0.10	+0.3
Big trucks loaded,		+0.05	+0.48	+1.6
normal acceleration	Morc than 40 mph (60 km per hr)	+0.10	+0.98	+3.2
	Less than 20 mph (30 km per hr)	–0.01	–0.10	–0.3
	20 to 40 mph (30 to 60 km per hr)	–0.02	–0.20	–0.6
	More than 40 mph (60 km per hr)	–0.04	–0.39	–1.3
Pedestrians, normal acceleration		–0.04	–0.39	–1.3
in walking		–0.05	–0.48	–1.6
Pedestrians in a hurry	Less than 20 mph (30 km per hr)	–0.08	–0.78	–2.6
Passenger cars, coasting out of	20 to 40 mph (30 to 60 km per hr)	–0.10	–0.98	–3.2
gear	More than 40 mph (60 km per hr)	–0.20	–1.96	–6.4
Passenger cars, engine braking		–0.40	–3.92	–12.9
high gear		–0.65	–6.38	–20.7
Gradual slowing, light braking	Less than 20 mph (30 km per hr)	–0.95	–9.32	–30.6
Normal braking, no skidding	20 to 40 mph (30 to 60 km per hr)	–5.00	—	—
Quick stop, skids on ice or snow	More than 40 mph (60 km per hr)	—	49.01	161.0
Hard braking, skids on most		20.00	—	—
surfaces			196.0	644.0
Maximum braking on high-friction	Less than 20 mph (30 km per hr)			
surfaces				
Car crash into standing car	20 to 40 mph (30 to 60 km per hr)			
Crash into solid fixed objects	More than 40 mph (60 km per hr)			

Source: **Baker, J.S.,** *Traffic Accident InvestigatSource:* **Baker, J.S.,** *Traffic Accident Investigation Manual*, Northwestern University, Evanston, IL, 1975.

Example 2.2: *A vehicle is sliding down as 5 percent grade and the coefficient of friction is 0.75. Determine the effective drag factor on the grade.*

Solution: The drag factor on grade is obtained from Eq. (2.5).

$$F_G = \frac{\mu + G}{\sqrt{1+G^2}} = \frac{0.75+(-0.05)}{\sqrt{1+(-0.05)^2}} = \frac{0.70}{1.00125} = 0.699 \approx 0.70$$

Table 2.10 Acceleration factor equations

No.	US units		SI units	
	Acceleration factor	*Acceleration factor*	*Acceleration factor*	*Acceleration factor*
1.	$f = \frac{s^2}{30d}$	$f = \frac{v^2}{64.4d}$	$f = \frac{s^2}{254d}$	$f = \frac{v^2}{19.62d}$
2.	$f = \frac{s}{21.95t}$	$f = \frac{v}{32.2t}$	$f = \frac{s}{35.3t}$	$f = \frac{v}{9.81t}$
3.	$f = \frac{s_f - s_0}{21.95t}$	$f = \frac{v_f - v_0}{32.2\,t}$	$f = \frac{s_f - s_0}{35.3t}$	$f = \frac{v_f - v_0}{9.81\,t}$
4.	$f = \frac{s_f^2 - s_0^2}{30d}$	$f = \frac{v_f^2 - v_0^2}{64.4\,d}$	$f = \frac{s_f^2 - s_0^2}{254d}$	$f = \frac{v_f^2 - v_0^2}{19.62\,d}$
5.	$f = \frac{d}{16.1t^2}$	$f = \frac{d}{16.1t^2}$	$f = \frac{d}{4.9t^2}$	$f = \frac{d}{4.9t^2}$
6.	$f = \frac{d - 1.466 s_0 t}{16.1t^2}$	$f = \frac{d - v_0 t}{16.1t^2}$	$f = \frac{d - 0.278 s_0 t}{4.9t^2}$	$f = \frac{d - v_0 t}{4.9t^2}$
7.	$f = \mu n + m$	$f = \mu n + m$	$f = \mu n + m$	$f = \mu n + m$
8.	$f = \frac{F}{W}$	$f = \frac{F}{W}$	$f = \frac{F}{W}$	$f = \frac{F}{W}$

Table 2.9 gives the acceleration data taken from Baker (1975). The acceleration factor equations are summarized in Table 2.10. It is common in collision reconstruction problems to need to assess the non-collision performance of vehicles, such as preimpact acceleration or *normal* braking. For analyses such as these, it is useful to know the normal capability of the vehicle in question. If circumstances warrant, independent testing can identify a vehicle's acceleration capability at various speeds.

2.6.4 Skid Marks

Skid Marks: *Marks left on the road surface when the tire stops rotating, locks up, and skids.* A friction mark left on road or any surface by a braked or locked tire. Skid marks are divided into "acceleration marks" created on acceleration, if the engine provides more power than the tire can transmit, "braking marks," if the brakes "lock up" and cause the tire to slide, or "yaw marks", if the tire slides sideways. Each has characteristic appearance, and an experienced accident reconstructor or forensic engineer can often determine what the vehicle was doing by examining the marks left by the tire. Skid marks are marks made by tires that are not rolling. The simplest form of computing vehicle speed is that of a vehicle coming to a complete stop at the end of the braking maneuver while the deceleration remains constant.

Acceleration Marks: Marks that are just the opposite of skid marks. The tires are being rotated by the axle and when done fast enough, the outside of the tire, the tread, takes time to catch up to the rest of the tire, which is being held by the coefficient of friction between the tires and the road surface. Therefore, the acceleration marks are heavy at the beginning and lighten up as the tire tread catches up.

Critical Speed Scuff: A tire mark left by a rotating wheel that is sliding or slipping sideways due to the centrifugal force exceeding peak lateral friction. The skid mark will normally have striation marks. Also referred to as yaw, centrifugal skid, sideslip or inertial mark.

Curved Skid Marks: Measure a curved skid mark by letting the measuring device follow the path of the skid mark around its curve. Do not confuse a curved skid mark with a yaw mark as shown in the Fig. 2.1(*a*).

Fig. 2.1 (*a*) Curved skid marks.

Dual-Wheel Skid Marks: Two tires on one wheel (a dual wheel) leave two side-by-side skid marks. They are to be counted as one skid mark. In the case of dual-wheel skid marks, if one of the marks is longer than the other, the longer mark should be used as the skid distance. But if the two marks do not begin and end at the same place, the length of the skid for that dual wheel will be measured from the first indication of a skid to the last indication of a skid regardless of which of the two side-by-side tires left the first and last such indications. See Fig. 2.1(*b*).

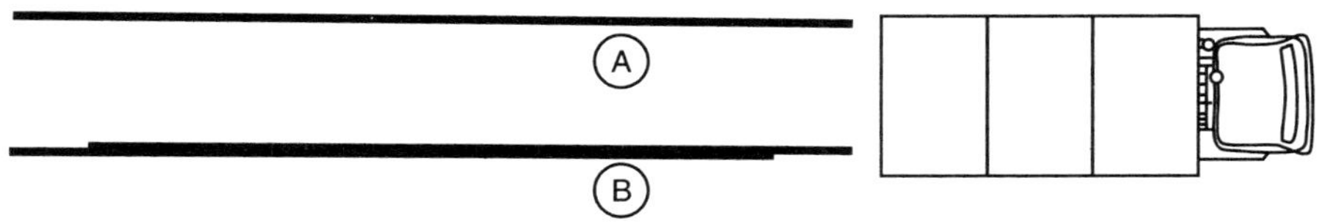

Fig. 2.1 (*b*) Dual-wheel skid marks.

Gap or Intermittent Skid Marks: Intermittent skid marks appear with gaps between them as shown in the figure below. These gaps occur when wheels are locked up, released and then relocked through braking action. Each blank space or gap between intermittent skid marks is usually a minimum of 15 ft to 20 ft (5 m to 7 m) in length, depending upon the speed of the vehicle and the driver's reaction time.

Measure intermittent skid marks separately and in the same manner as a single skid. Use heir sum (excluding the blank spaces or gaps) as the total skid distance. See Fig. 2.1(*c*).

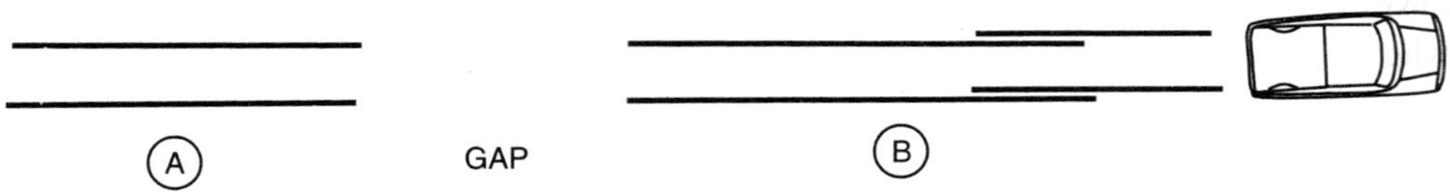

Fig. 2.1 (*c*) Intermittent of gap skid marks. Total skid distance = *A* + *B*.

Gap Skid: Skid mark in which there exists a gap of ten feet or more between segments from the release and reapplication of brakes. Each segment is measured separately .

Impending Skid mark: Left by a braked wheel rotating slower that the forward motion of the vehicle. Also referred to as spin-down.

Locked-Wheel Skid mark: A skid mark left by a braked, non rotational wheel sliding in contact with a road surface.

Offset Skid Marks: Measure the entire length of the skid mark, including the offset. When using the length of a skid mark for speed calculation purposes and when the skid mark is offset because the vehicle was struck while sliding, the vehicle's energy loss during its skid may not necessarily include that portion involving the offset mark. Under these circumferences, measure the length of the actual skid mark and the offset separately. See Fig. 2.1(*d*).

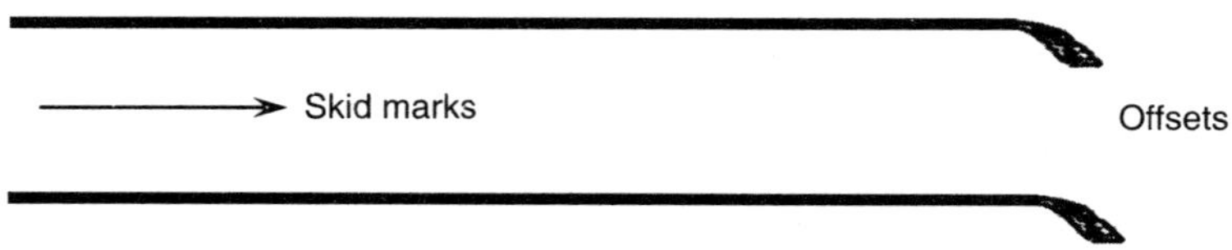

Fig. 2.1 (*d*) Offset skid marks.

Overlapping Skid Marks: Measure the total length of overlapping skid marks (including the segment at the start made by the rear tires alone and the segment at the end made by the front tires alone). The actual skid distance will be this length minus the length of the vehicle's wheelbase. See Fig. 2.1(*e*).

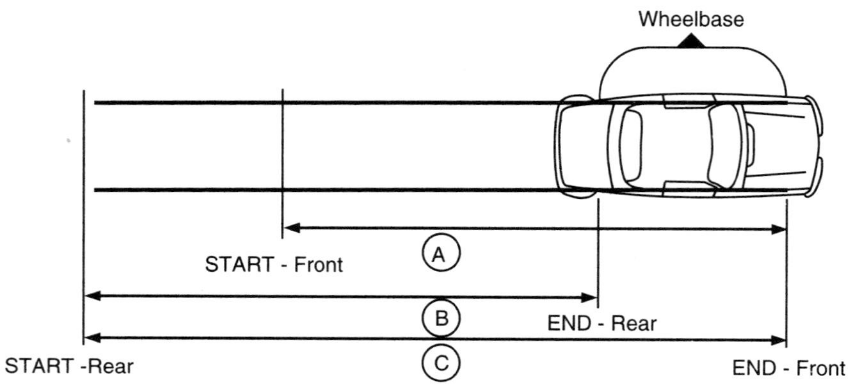

Fig. 2.1 (*e*) Overlapping skid marks.

Positive Skid Mark: A positive skid mark is the long black smear left on the pavement by a locked wheel as shown in the figure below:

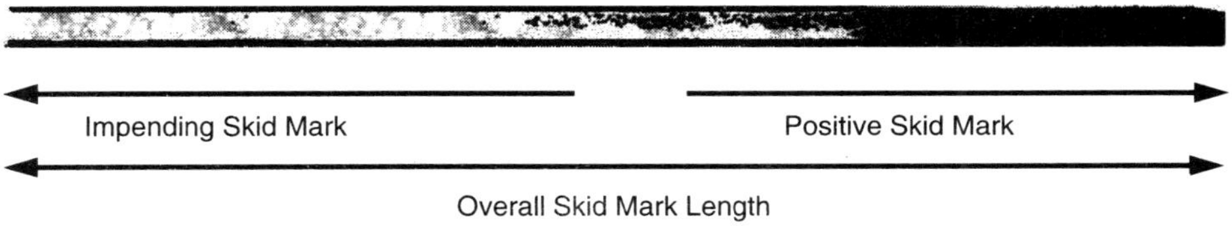

Fig. 2.1 (*f*) Impending and positive skid marks.

Puddle: A wet area on the road or roadside where dribble accumulates after a *vehicle* has come to rest. A puddle often marks the final position of a vehicle after an accident.

Scar: Any mark on a road, on the ground, or on a fixed object such that surface material is removed or displaced; durable sign on the road of the location of a traffic accident.

Scrub: Skid marks left by tires pushed sideways or kept from rotating by forces of a collision.

Skip Skid Marks: Skip skid marks occur when a locked wheel bounces on the roadway. The blank spaces between the skid marks are usually not longer than 3 ft (1 m). Bouncing is usually caused by such actions and conditions as:

(1) A locked wheel striking a hole or bump, or

(2) A locked wheel on an unloaded or lightly loaded trailer, particularly a semi-trailer.

Measure skip skid marks as one continuous skid mark from the beginning of the first skid mark to the end of the last skid mark for each wheel. See Fig. 2.1(*g*).

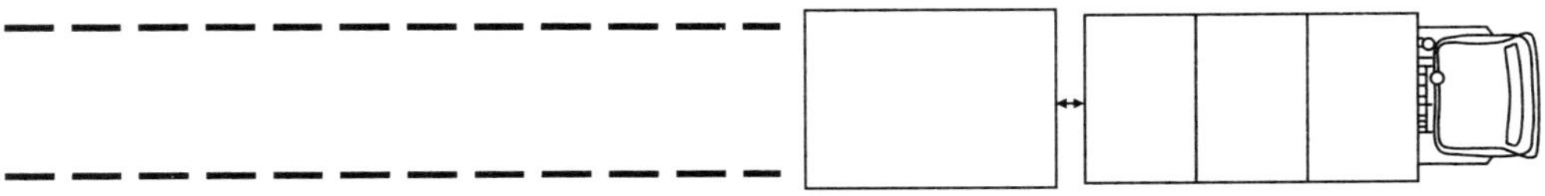

Fig. 2.1 (*g*) Skip skid marks.

Spin Skid Marks: The length and exact position of each skid mark in relation to all other skid marks on the roadway must be measured. An average of the skid-mark distances may be used as the skid distance, or the center-of-mass path may be plotted and its length used as the skid distance. See Fig. 2.1(*h*).

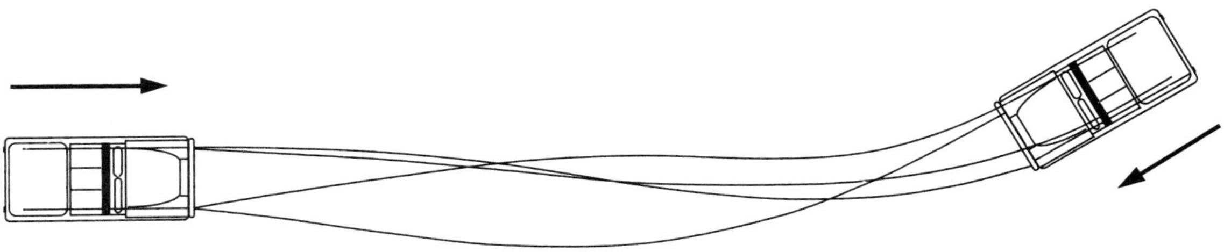

Fig. 2.1 (*h*) Spin skid marks.

Various Surface Skid Marks: When a skid mark traverses different kinds of roadway surfaces, measure the skid distance on each surface separately because of the different drag factor (*f*) values. See Fig. 2.1(*i*).

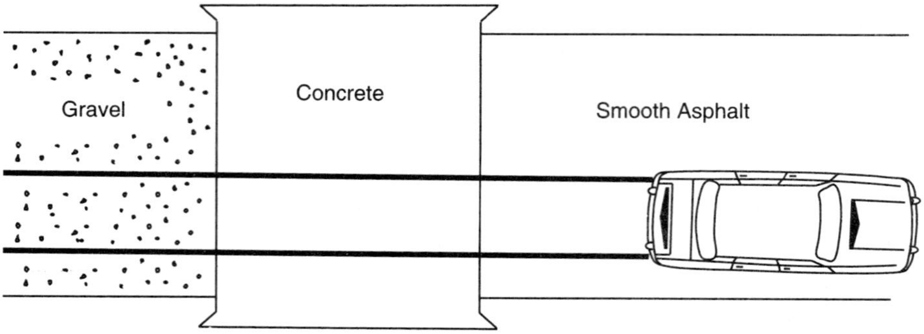

Fig. 2.1 (*i*) Skid marks on various surfaces.

Yaw Mark: Very distinctive marks left by a tire that is rotating and turning at the same time. Instead of the tire leaving dark marks parallel to the direction of vehicle movement, there are striations in the marks at various angles to the direction of travel. The direction of the marks is dependent on the amount of "rolling under" the tread does during the maneuver. In addition, the entire mark is curved. The width of the yaw mark as well as the number and angle of the striations are highly dependent on the speed of the vehicle.

2.6.5 Gouges

Gouge: A pavement scar deep enough to be easily felt with the fingers; chip, chop, or groove. See Gouge Mark.

Highway Marks: Highway marks including gouges in the pavement, oil, gasoline, water from a broken radiator and skidmarks.

Scrape Marks: Scrape marks on vehicles as marks which are wide than deep, and longer than wide. Scrape and gouge are often used interchangeably and there is usually no harm in this as long as the dimensions of the mark are given. A scrape runs parallel to the scraped surface. It does not penetrate deeply and does not come to a sudden, deep stop. The length, width and position of scrapes should be noted, and the area examined for rust or dirt to determine whether the scrape is new or old. It should also be examined for the presence of different colored paint which may be used to prove that the scrape mark was made by a particular vehicle. Note carefully the direction of the scrape marks; are they parallel to the ground (indicating sidewipe)? do they pitch downward (indicating the other vehicle was violently decelerated)? or do they pitch upward (indicating this vehicle was violently decelerated)? See Fig. 2.1(*j*).

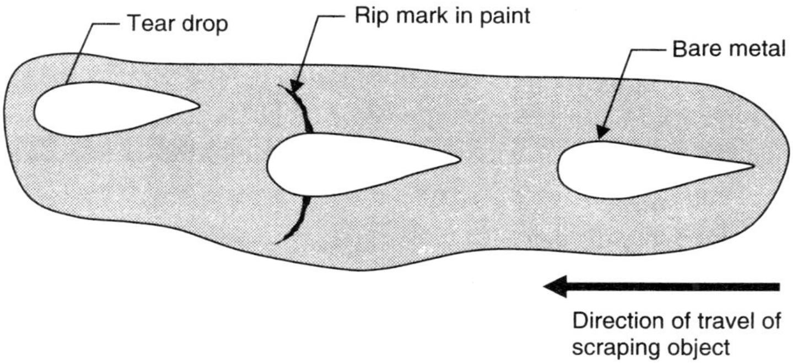

Fig. 2.1 (*j*) Scrape marks as they appear in the automobile body paint.

Scrape: A broad area of a hard surface covered with many *scratches* or *striations* made by a sliding metal part without great pressure.

Chip: A short, deep *gouge;* a hole in *pavement* made by a strong, sharp, pointed metal object under great pressure, usually without *striations.*

Chop Mark: A broad shallow gauge even and regular on the deeper side and terminating in scratches and striations on the opposite side.

Chop: A broad shallow gouge in a road surface, beginning with an even, regular, deeper side and terminating in scratches and striations on the opposite shallower side; a depression in pavement made by strong, sharp metal edge moving under heavy pressure, more commonly occurring at an impact event as opposed to post impact trajectory.

Gouges: Gouge as a mark which is deeper than it is wide, and which bends or tears the metal in the local area. A gouge is made by a sturdy piece of metal sliding into a lighter piece; for example, the mark left by a door handle on a fender during a sideswipe. The fender gouge resembles a mold of the profile of the door handle. Gouge marks should be examined for the same kind of information that a scrape mark would reveal; and, in addition, gouge marks can be used to determine other things. A gouge mark usually becomes deeper the further along it proceeds. This shows the direction of travel of the gouging object. Further, the gouging object may tear the fender or door and leave upturned metal, or chips, along its path. Such chips indicate the direction of travel. Assuming one to be looking sideways into the two gouges drawn in Fig. 2.1(*k*), the direction of travel of the gouging object is shown by the arrows. The metal chip always curls up towards the direction of travel of the gouging object.

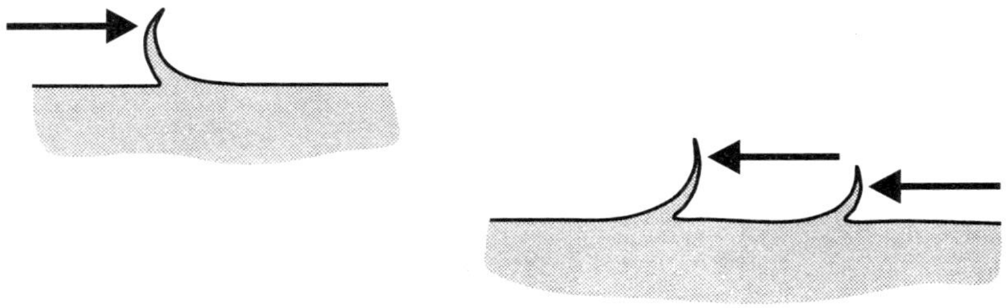

Fig. 2.1 (*k*) Cross-section of a gouge mark showing upturned metal chips.

Gouge Marks: Gouge marks are made by metal components that are forced into the pavement during impact or as the vehicle travels to its rest position. Gouge marks are extremely important since they to a large extent determine the point or area of impact between two vehicles. Gouge marks made during impact generally are deeper at the leading end of the mark than at the end of the mark. Gouge marks in the pavement must be treated with great care. A careful check must be made to determine whether or not the gouge was a result of the accident under investigation. Gouges are usually made by automobile bumpers pitching into the pavement. The gouge usually has the shape of the gouging object; hence, a particular gouge can be identified with a particular part if they fit together fairly well. A careful examination of the gouge will determine which direction the gouging object was traveling when it made the mark. Asphalt and concrete gouges usually look (in cross-section). See Fig. 2.1(*l*).

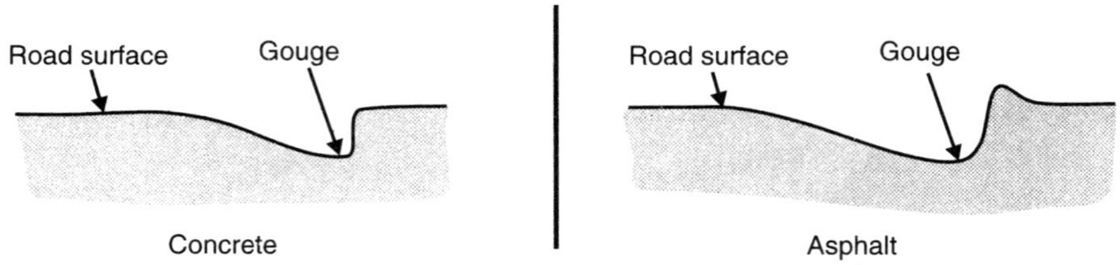

Fig. 2.1 (*l*) Cross-section of bumper gouges in concrete and asphalt pavements.

2.7 WORK, ENERGY AND LAW OF CONSERVATION OF ENERGY

2.7.1 Work

Work (W) for a force, F, constant in magnitude and with its direction along a straight line is given by

$$W = Fd \tag{2.7}$$

where W = work done, $N\,m$ (ft-lb)

F = constant force, N (lb)

d = displacement during the motion along the straight line, m (ft)

If the force and displacement (distance) are not in the same direction (collinear), then

$$W = Fd \cos \theta \tag{2.8}$$

where θ = angle between the force vector and the displacement vector of travel.

Work is a scalar quantity.

There are two forms of mechanical energy that are important in the understanding of momentum and speed analysis: potential energy and kinetic energy. Potential energy is the capacity of a mass or body to do work due to the body's position.

2.7.2 Potential Energy (V)

The potential energy of a weight w close to the surface of the earth and at a height y above a given datum is

$$V = wy \tag{2.9}$$

where w = weight, N (lbs)

y = height, m (ft)

2.7.3 Kinetic Energy (KE)

The Kinetic Energy (KE) of a body with mass m and moving with speed v is given by

$$KE = \frac{1}{2}mv^2 = \frac{1}{2}\left(\frac{w}{g}\right)v^2 \tag{2.10}$$

where KE = kinetic energy, Nm (ft-lbs)

w = weight, N (lbs)

v = velocity, m/s (ft/s)

g = acceleration due to gravity, 9.81 m/s^2 (32.2 ft/s^2)

The amount of kinetic energy that must be converted to another form in order to bring a vehicle to a stop is

$$W = \text{work, Nm (ft-lbs)} = w\,f\,d \tag{2.11}$$

where w = weight, N (lbs)

f = drag factor

d = distance, m (ft)

The work done to compress a spring is

$$W = \text{ work} = \frac{1}{2}kx^2 \tag{2.12}$$

where W = work, N m (ft-lbs)

k = spring constant, N/m (lb/ft)

x = distance compressed/stretched, m (ft)

2.7.4 Law of Conservation of Energy

When a body moves under the action of conservative forces, the sum of the kinetic energy and of the potential energy of the body remains constant. The sum $KE + V$ is called the mechanical energy of the body and is denoted by E.

$$\text{Mechanical energy } (E) = KE + V = \text{constant} \tag{2.13}$$

2.7.5 Linear Momentum

Replacing the acceleration a by the derivative $\frac{dV}{dt}$, we can write Newton's second law of motion as

$$\Sigma F = m\frac{dV}{dt} \tag{2.14}$$

Since the mass m of a body is constant, we have

$$\Sigma F = \frac{d}{dt}(mV) \tag{2.15}$$

The vector mV is called the *linear momentum*, or simply the *momentum* of the body. Eq. (2.15) expresses that the resultant of the forces acting on the body is equal to the rate of change of linear momentum of the body. Denoting by L the linear momentum of the body

$$L = mV \tag{2.16}$$

and by $\dot{L}$ the derivative of L with respect to t, we can write Eq. (2.15) as

$$\Sigma F = \dot{L} \tag{2.17}$$

It follows from Eq. (2.15) that the rate of change of the linear momentum mV is zero when $\Sigma F = 0$. Hence, if the resultant force acting on a body is zero, the linear momentum of the body remains constant, in both magnitude and direction. This is the principle of conservative of linear momentum for a body, which can be recognized as an alternative statement of Newton's first law of motion.

The energy equations are summarized in Table 2.11.

Table 2.11 Energy equations

No.	*US units*		*SI units*	
	To find	*To find*	*To find*	*To find*
1.	$s = \sqrt{\dfrac{30\ KE}{W}}$	$v = \sqrt{\dfrac{64.4\ KE}{W}}$	$s = \sqrt{\dfrac{25.92\ KE}{W}}$	$v = \sqrt{\dfrac{2\ KE}{W}}$
2.	$KE = \dfrac{Ws^2}{30}$	$KE = \dfrac{Wv^2}{64.4}$	$KE = \dfrac{Ws^2}{25.92}$	$KE = \dfrac{Wv^2}{2}$
3.	$F = \dfrac{Ws^2}{30\ d}$	$F = \dfrac{Wv^2}{64.4\ d}$	$F = \dfrac{Ws^2}{254\ d}$	$F = \dfrac{Wv^2}{19.62\ d}$
4.	$d = \dfrac{Ws^2}{30\ F}$	$d = \dfrac{Wv^2}{64.4\ F}$	$d = \dfrac{Ws^2}{254\ F}$	$d = \dfrac{Wv^2}{19.62\ F}$

2.8 SPECIAL EQUATIONS

In this section, we derive the various slide-to-stop equations. The basic slide-to-stop equation is of the form $S = \sqrt{30\ df}$, where d is the skidding distance, f is the drag factor, and 30 is a constant obtained from the derivation of the equation. The analytical basis for the above equation and for other equations such as the speed equation, and weight-shift equation is the *Work-Energy Theorem*. The work-energy theorem simply states that

$$\text{Work} = \text{Kinetic Energy} \tag{2.18}$$

or

$$Fd = \frac{1}{2}Mv^2 \tag{2.18a}$$

where F = force required to stop the vehicle.

d = the skidding distance.

M = mass of the vehicle.

v = velocity of the vehicle.

2.8.1 Kinetic Energy

Kinetic energy is the energy possessed by an object due to its motion. In inelastic conditions (such as the impact involving two lumps of clay), kinetic energy is not conserved as kinetic energy. Here we derive an equation for kinetic energy. Recalling that

$$v = \frac{d}{t},\ vt = d, \text{ and } a = \frac{v_f - v_0}{t} \tag{2.19}$$

Multiply both sides of Eq. (2.19) by t, we have

$$v_f = v_0 + at \tag{2.19a}$$

or

$$v_f = at$$

if

$$v_0 = 0$$

The average velocity, $\bar{v}$ for a given time t is given by

$$\bar{v} = \frac{v_0 + v_f}{2} = \frac{0 + v_f}{2} = \frac{v_f}{2} \tag{2.19b}$$

Since $v_f = at$, Eq. (2.19b) becomes

$$\bar{v} = \frac{at}{2}$$

also

$$\bar{v} = \frac{d}{t} \tag{2.19c}$$

Hence

$$\frac{d}{t} = \frac{at}{2}$$

or

$$d = \frac{1}{2}at^2 \tag{2.19d}$$

If the body has initial velocity, then equation (2.19d) becomes

$$d = v_0 t + \frac{1}{2}at^2$$

Squaring both sides of Eq. (2.19a), we get

$$v_f^2 = (v_0 + at)(v_0 + at)$$

or

$$v_f^2 = v_0^2 + 2v_0 at + a^2t^2$$

or

$$\frac{v_f^2}{2a} = \frac{v_0^2}{2a} + v_0 t + \frac{at^2}{2} \tag{2.20}$$

Substituting

$$d = v_0 t + \frac{1}{2}at^2 = v_0 t + \frac{at^2}{2}$$

into Eq. (2.20), we obtain

$$\frac{v_f^2}{2a} = \frac{v_0^2}{2a} + d \tag{2.20a}$$

or

$$\frac{v_f^2 - v_0^2}{2d} = a \tag{2.20b}$$

Now, work = energy = force × distance

or E = energy = $m\,a\,d$

or

$$E = m\left[\frac{v_f^2 - v_0^2}{2d}\right] d = m\left[\frac{v_f^2 - v_0^2}{2}\right] \tag{2.21}$$

or

$$KE = \frac{1}{2}m\left(v_f^2 - v_0^2\right) \tag{2.21a}$$

If $v_0 = 0$, then

$$KE = \frac{1}{2}mv^2 \tag{2.22}$$

2.8.2 Combined Speed Equations

Since we are concerned here with deceleration, the kinetic energy equation (2.21*a*) can be written as

$$KE = \frac{1}{2}m\left(v_0^2 - v_f^2\right) \tag{2.23}$$

$$= Fd = wfd \tag{2.23a}$$

where F = force, w = weight and f = the drag factor.

Hence, Eq. (2.23) may be written as

$$wfd = \frac{w}{2g}\left(v_0^2 - v_f^2\right)$$

or

$$2gfd = \left(v_0^2 - v_f^2\right) \tag{2.23b}$$

Noting that $g = 32.2$ ft/sec^2, Eq. (2.23*b*) becomes

$$64.4fd = v_0^2 - v_f^2 \tag{2.23c}$$

or $v = 1.466\ S$

and

$$v^2 = (1.466\ S)^2 = 2.14\ S^2 \tag{2.23d}$$

From Eq. (2.23*c*) and (2.23*d*), we have

$$\frac{64.4}{2.14}fd = S_0^2 - S_f^2 \tag{2.23e}$$

or

$$30fd = S_0^2 - S_f^2 \tag{2.24}$$

Eq. (2.24) can be rewritten as

$$S_0^2 = S_f^2 + 30\,fd \tag{2.24a}$$

or

$$S_0 = \sqrt{S_f^2 + 30\,fd} \tag{2.24b}$$

or

$$S_0 = \sqrt{S_1^2 + S_2^2} \tag{2.24c}$$

or

$$S_0 = \sqrt{30(f_1 d_1 + f_2 d_2 + ...)} \tag{2.25}$$

Eq. (2.25) is known as the combined speed equation.

2.8.3 Slide-to-Stop Equation

Recall that Work = Kinetic Energy (Work-Energy Theorem)

or

$$Fd = \frac{1}{2} m v^2 \tag{2.26}$$

where W = work

F = force

d = distance

m = mass

and v = velocity

Also $F = wf$

where f = deceleration factor (coefficient of friction)

and w = vehicle weight ($w = mg$, where g = gravitational acceleration = 32.2 ft/sec^2).

Eq. (2.26) becomes

$$wfd = \frac{w}{2g} v^2 \tag{2.26a}$$

or

$$2gfd = v^2 \tag{2.26b}$$

or

$$v^2 = 64.4\,fd \text{ (noting that } g = 32.2 \text{ ft/sec}^2) \tag{2.26c}$$

We also know that $v = 1.466\,S$ or $v^2 = 2.14\,S^2$.

Hence, Eq. (2.26b) becomes

$$S^2 = 30.09\,fd \approx 30\,fd \tag{2.26d}$$

or

$$S = \sqrt{30\,fd} \tag{2.26e}$$

Equation (2.26e) is known as the *slide-to-stop equation*. Equation (2.26e) may also be used to compute the speed of a vehicle that accelerates from a stop with a known acceleration factor over a known distance.

2.8.4 Slide-to-Stop Equation with Deceleration (Drag Factor)

(*a*) Uphill grades of 10% or less

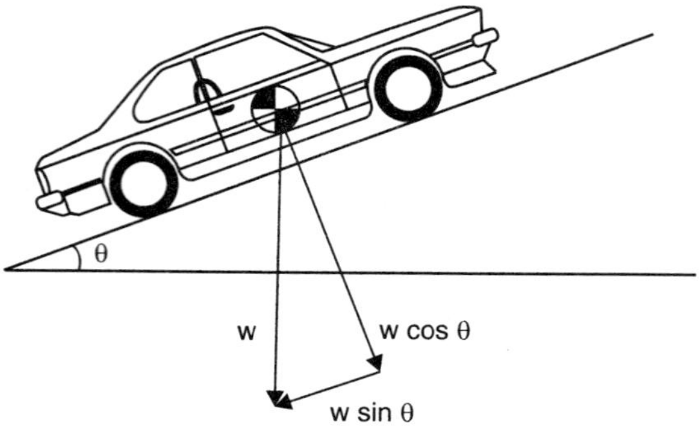

Fig. 2.2 (*a*)

Consider a vehicle that is stopping uphill as shown in Fig. 2.2(*a*). Summing up the forces parallel to the road surface, we obtain

$$(w \cos \theta) f + w \sin \theta = F \tag{2.27}$$

For θ ≈< 6 degrees, or a grade up to 10%, we have $\cos \theta \approx 1$, $\sin \theta = \tan \theta = m =$ slope.

Equation (2.27) can now be written as

$$wF + wm = F \tag{2.27a}$$

or

$$w\,(f + m) = F \tag{2.27b}$$

Substituting Eq. (2.27*b*) into the work-energy equation, we get

$$w\,(f + m)\,d = \frac{w}{2g} v^2 \tag{2.27c}$$

Noting $g = 32.2$ ft/sec^2, Eq. (2.27*c*) can be simplified as

$$v^2 = 64.4\,d\,(f + m) \tag{2.27d}$$

Now

$$v = 1.466\,S$$

or

$$v^2 = 2.14\,S^2 \tag{2.27e}$$

Equation (2.27*d*) becomes

$$2.14\,S^2 = 64.4\,d\,(f + m) \tag{2.27f}$$

or

$$S^2 = 30.09\,d\,(f + m) \approx 30\,d\,(f + m) \tag{2.27g}$$

or

$$S = \sqrt{30d\,(f + m)} \tag{2.27h}$$

Equation (2.27*h*) is the slide-to-stop equation for uphill grads of 10% or less.

(*b*) Downhill grades of 10%

The slide-to-stop equation for downhill grades of 10% or less can be derived by following a similar procedure as in (*a*) above. The slide-to-stop equation for downhill grades of 10% or less is given by

$$S = \sqrt{30d\,(f - m)} \tag{2.27i}$$

(*c*) Uphill on a grade steeper than 10%

Here Eq. (2.27) can be rewritten as

$$F = (w \cos \theta) f + w \sin \theta \tag{2.27j}$$

or

$$F = w\,(f \cos \theta + \sin \theta) \tag{2.27k}$$

Consider the work-energy theorem, work = kinetic energy.

$$wd(f \cos \theta + \sin \theta) = \frac{w}{2g} v^2$$

or

$$d\,(\cos \theta + \sin \theta) = \frac{1}{2g} v^2 \tag{2.27l}$$

Noting $g = 32.2$ ft/sec^2, and $v = 1.466$ S, Eq. (2.27*l*) becomes

$$64.4\, d\,(f \cos \theta + \sin \theta) = v^2$$

or

$$64.4\, d\,(f \cos \theta + \sin \theta) = 2.14\, S^2$$

or

$$S^2 = 30.09\, d\,(\approx 30)\,(f \cos \theta + \sin \theta) \tag{2.27m}$$

or

$$S = \sqrt{30 d (f \cos \theta + \sin \theta)} \tag{2.27n}$$

Equation (2.27*m*) is the slide-to-stop equation for uphill grades greater than 10%.

2.8.5 Slide-to-Stop Equation with Grade and Percentage of Braking

Here, we consider a vehicle that has defective brakes. When the percentage is, *n* of the total vehicle weight being carried out on the braked wheels, then we can compute the force available to stop the vehicle.

Consider the Work-Energy Theorem:

$$Fd = \frac{1}{2} mv^2 \tag{2.28}$$

Here

$$F = wf \pm wm \qquad \text{(– sign for downhill and + sign for uphill)} \tag{2.28a}$$

or

$$F = wnf \pm wm \tag{2.28b}$$

or

$$F = w\,(fn \pm m) \tag{2.28c}$$

From Eqs. (2.28) and (2.28*c*) with $m = w/g$, we obtain

$$w\,(fn \pm m)\, d = \frac{w}{2g} v^2$$

or

$$(fn \pm m) d = \frac{1}{2g} v^2 \tag{2.28d}$$

Noting that $g = 32.2$ ft/sec^2, Eq. (2.28*d*) becomes

$$64.4 (fn \pm m)\, d = v^2 \tag{2.28e}$$

Also, we have $v = 1.466\, S$ or $v^2 = 2.14\, S^2$.

Hence, Eq. (2.28*e*) becomes

$$2.14\, S^2 = 64.4\,(fn \pm m)\, d$$

or

$$S^2 = 30.09\,(\approx 30)\, d\,(fn \pm m) \tag{2.28f}$$

or
$$S = \sqrt{30d(fn \pm m)}$$

Equation (2.28*f*) represents the slide-to-stop equation accounting for both grade and percentage of braking. John Daily (1988).

Passenger Cars: $n = 60\%$ for both front wheels, 40% for both rear wheels, 70% for one front and both rear wheels, 80% for both front and one rear wheel, 30% for each front wheel, and 20% for each rear wheel.

Bobtail Tractors: $n = 30\text{–}40\%$.

Loaded Tractor Trailers: $n = 67\%$ (with all but steering axle bearing).

2.8.6 Drag Factor Equation

The slide-to-stop equation is given by Eq. (2.26*e*)

$$S = \sqrt{30\,fd} \tag{2.29}$$

or
$$S^2 = 30\,fd \tag{2.29a}$$

or
$$f = \frac{S^2}{30\,d} \tag{2.29b}$$

Equation (2.29*b*) is known as the *drag factor equation*.

2.8.7 Braking Distance Equation

We begin with the slide-to-stop equation (2.26*e*)

$$S = \sqrt{30\,fd} \tag{2.30}$$

or
$$S^2 = 30\,fd \tag{2.30a}$$

or
$$d = \frac{S^2}{30\,f} \tag{2.30b}$$

Equation (2.30*b*) is known as the *braking distance equation*.

2.9 WEIGHT SHIFT

Weight shift happens when the weight of a vehicle shifts in reaction to a change in direction speed. When a vehicle stops, weight is transferred to the front axle and there is a corresponding reduction of weight on the rear wheels.

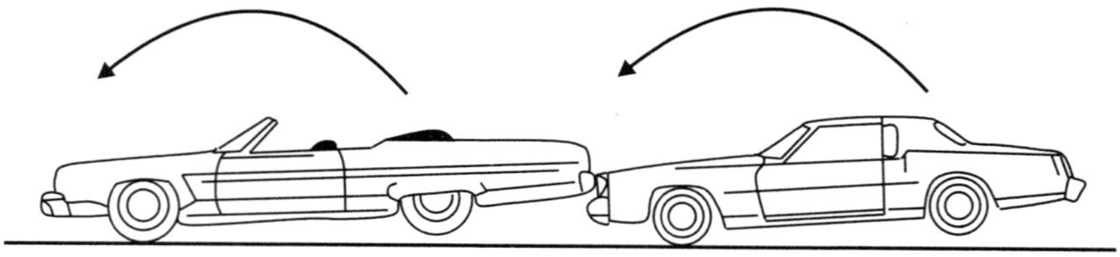

Fig. 2.2 (*b*) Weight shift causes the front end to dip because of the extra weight and the rear end to rise because of the loss in weight.

Here we present formulae for the analysis of the following:

(*a*) Horizontal Center of Mass (CM) equation.

(*b*) Height of Center of Mass (CM).

(*c*) Weight shift equation.

(*d*) Speed equation with weight shift.

(*e*) Weight shift equation with grade.

2.9.1 Horizontal Center of Mass (CM) Equation

The following notation is used:

CM = center of mass (CM) distance from rear axle, in.

W_b = wheel base, in.

W_f = weight of front axle, lbs

W_t = total weight, lbs

Figure 2.2(*c*) shows the vehicle in equilibrium.

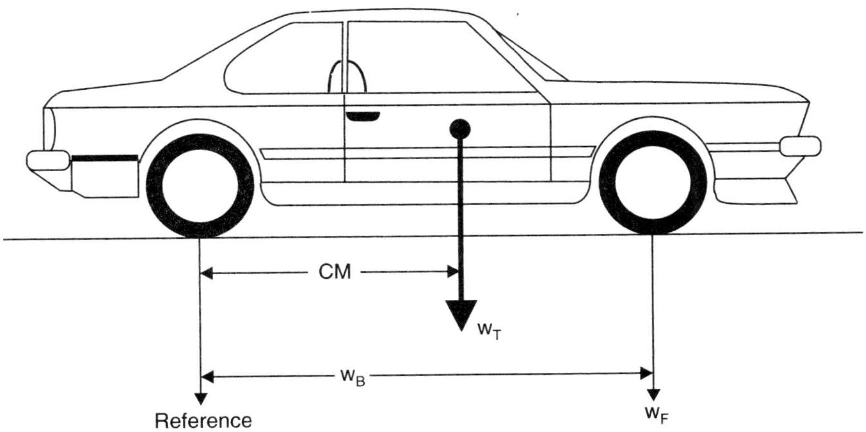

Fig. 2.2 (*c*) Vehicle in equilibrium.

Since the vehicle is in equilibrium, the sum of the torques acting on the vehicle must be equal to zero.

Taking the center of the rear axle as the pivot, we can write

$$W_f (W_b) = CM\ (W_t) \tag{2.31}$$

or

$$CM = \frac{W_f W_b}{W_t} \tag{2.32}$$

Equation (2.32) is the horizontal center of mass (CM) equation.

2.9.2 Height of Center of Mass (CM) Equation

The height of center of mass (CM) is given by

$$h = h_2 + \frac{(W_w - W_f)l\sqrt{l^2 - (h_1 - h_2)^2}}{W_t(h_1 - h_2)} \tag{2.33}$$

where h = height of the center of mass (CM) above the road surface, in.
h_1 = height of the rear wheels raised, in. (in minimum 30 in.)
h_2 = height of center of front wheels (axle center to ground), in.
l = wheelbase length, in.
W_f = weight on front wheels (on level), lbs.
W_t = total weight of vehicle, lbs.
W_w = weight on front wheels with rear axle raised, lbs.

2.9.3 Weight Shift Equation

The weight shift to front axle (lost by rear) is given by

$$\Delta W = \frac{\frac{h}{l}(W_f f_f + W_r f_r)}{\left[1-\left(\frac{h}{l} f_f\right)\right]+\left(\frac{h}{l} f_r\right)} \tag{2.34}$$

where f_f = drag factor for front wheels, dimensionless.
f_r = drag factor for rear wheels, dimensionless.
h = height of center of mass (CM), in.
l = length of wheelbase, in.
W_f = weight on front wheels (at rest), lbs.
W_r = weight on rear wheels (at rest), lbs.
ΔW = weight shift to front axle (lost by rear), lbs.

2.9.4 Speed Equation with Weight Shift

The vehicle speed accounting for weight shift is given by

$$S = \sqrt{30\left[\frac{f_f d_f (W_f + \Delta W) + f_r d_r (W_r - \Delta W)}{W_t}\right]} \tag{2.35}$$

where d_f = skid distance of front tires, in.
d_r = skid distance of rear tires, in.
f_f = drag factor for front wheels, dimensionless.
f_r = drag factor for rear wheels, dimensionless.
S = vehicle speed accounting for weight shift, mph.
W_f = weight on front wheels (at rest), lbs.
W_r = weight on rear wheels (at rest), lbs.
W_t = total weight of vehicle, lbs.
ΔW = weight shift to front axle (lost by rear), lbs.

2.9.4.1 Derivation of Speed Equation with Weight Shift

Here we present the derivation of the speed equation (2.35) accounting for weight shift.

The speed equation can be written as (work done = kinetic energy):

$$Fd = \frac{1}{2}MV^2 \tag{2.36}$$

where F = force, lbs.

d = distance, in.

M = mass of the vehicle (W/g), where W = weight of the vehicle (lbs) and g = acceleration due to gravity (32.2 ft/sec^2 or 9.81 m/sec^2).

V = vehicle velocity, ft/sec.

Now F can be written as

$$F = F_f f_f + F_r f_r \tag{2.37}$$

where F_f = downward force on the front wheels, lbs.

F_r = downward force on the rear wheels, lbs.

f_f = drag factor on front wheels, dimensionless.

f_r = drag factor on rear wheels, dimensionless.

When a vehicle stops, a certain amount of weight (ΔW) is shifted to the front wheels. This weight, ΔW, is added to the front wheels and subtracted from the rear wheels.

Hence,
$$F_f = W_f + \Delta W \tag{2.38}$$
and
$$F_r = W_r - \Delta W \tag{2.39}$$

where W_f = weight on front wheels, lbs.

W_r = weight on rear wheels, lbs.

Now, Eq. (2.37) can be written as

$$F = [(W_f + \Delta W)\, f_f + (W_r - \Delta W)\, f_r] \tag{2.40}$$

The work done by the front and rear wheels can be written respectively as

$$(W_f + \Delta W)\, f_f\, d_f \tag{2.41}$$
and
$$(W_r + \Delta W)\, f_r\, d_r \tag{2.42}$$

where d_f = skidding distance of front wheels, in.

d_r = skidding distance of rear wheels, in.

Hence, the total work done from the start of the skid to stop can be written as

$$Fd = [(W_f + \Delta W)\, f_f\, d_f + (W_r - \Delta W)\, f_r\, d_r] \tag{2.43}$$

From Eqs. (2.36) and (2.43), we can write

$$[(W_f + \Delta W) f_f\, d_f + (W_r - \Delta W) f_r\, d_r\,] = \frac{1}{2g} W_t V^2 \tag{2.44}$$

where M in Eq. (2.36) is replaced with W_t/g and W_t = total vehicle weight (lbs).

Noting that g = 32.2 ft/sec^2 and $V^2 = 2.14\ S^2$, Eq. (2.44) can be simplified as

$$S^2 = 30\left[\frac{f_f d_f (W_f - \Delta W) + f_r d_r (W_r - \Delta W)}{W_t}\right] \tag{2.45}$$

or

$$S = \sqrt{30\left[\frac{f_f d_f (W_f + \Delta W) + f_r d_r (W_r - \Delta W)}{W_t}\right]} \tag{2.46}$$

where S = speed of the vehicle accounting for weight shift, mph.

2.9.5 Speed Equation with Any Wheel Braking with Weight Shift

The vehicle speed equation with weight shift and with any wheel braking is given by

$$S = \sqrt{30\left[\frac{f_{rf} d_{rf}\left(W_{rf} + \frac{\Delta W}{2}\right) + f_{lf} d_{lf}\left(W_{lf} + \frac{\Delta W}{2}\right) + f_{rr} d_{rr}\left(W_{rr} - \frac{\Delta W}{2}\right) + f_{lr} d_{lr}\left(W_{lr} - \frac{\Delta W}{2}\right)}{W_t}\right]} \tag{2.47}$$

where d_{lf} = left front skid distance, in.
d_{lr} = left rear skid distance, in.
d_{rf} = right front skid distance, in.
d_{rr} = right rear skid distance, in.
f_{lr} = left rear drag factor, dimensionless.
f_{rr} = right rear drag factor, dimensionless.
f_{lf} = left front drag factor, dimensionless.
f_{lr} = left rear drag factor, dimensionless.
W_{lf} = left front vehicle weight, lbs.
W_{lr} = left rear vehicle weight, lbs.
W_{rf} = right front vehicle weight, lbs.
W_{rr} = right rear vehicle weight, lbs.

2.9.6 Weight Shift Equation with Grade

The vehicle will stop going uphill even without braking and the vehicle would not stop going downhill with no braking.

The vehicle speed equation with weight shift and grade is given by

$$S = \sqrt{30\left[\left(\frac{f_f d_f (W_f + \Delta W) + f_r d_r (W_r - \Delta W)}{W_t}\right) \pm dm\right]} \tag{2.48}$$

where d = total distance moved by the vehicle center of mass (CM), in.
d_f = skid distance of front tires, in.
d_r = skid distance of rear tires, in.
f_f = drag factor for front wheels, dimensionless.
f_r = drag factor for rear wheels, dimensionless.

W_f = weight on front wheels (at rest), lbs.

W_r = weight on rear wheels (at rest), lbs.

W_t = total weight of vehicle, lbs.

ΔW = weight shift to front axle (lost by rear), lbs.

m = grade, percentage.

Example 2.3: *A 4000 lb four wheeled vehicle skids to a stop on its rear wheels. The drag factor on the roadway is 0.65 and the skidding distance is 250 ft. The driver of the vehicle steers the vehicle to stop straight. The radius of the loaded front tire measured from the road surface to the center of the axle is 15 inches. The vehicle wheelbase is 102 inches. The weight on the front wheels is 2500 lbs and the weight on the rear wheels is 1500 lbs. When the rear axle is raised 35 inches, the front axle weighed 2550 lbs. Determine*

(*a*) *height of the center of mass (CM).*

(*b*) *the weight to the front axle.*

(*c*) *the vehicle speed at the beginning of the skid.*

Solution:

(*a*) The height of the center of mass is given by Eq. (2.33)

$$h = h_2 + \frac{(W_w - W_f)l\sqrt{l^2 - (h_1 - h_2)^2}}{W_t(h_1 - h_2)}$$

Here, $h_1 = 35$ in, $h_2 = 15$ in, $l = 102$ in, $W_f = 2500$ lbs, $W_w = 2550$ lbs, and $W_t = 4000$ lbs.

$$h = 15 + \frac{(2550 - 2500)(102)\sqrt{(102)^2 - (35-15)^2}}{4000(35-15)} = 21.376 \text{ inches}$$

(*b*) The weight shift to the front axle is given by Eq. (2.34)

$$\Delta W = \frac{\frac{h}{l}(W_f f_f + W_r f_r)}{\left[1 - \left(\frac{h}{l} f_f\right)\right] + \left(\frac{h}{l} f_r\right)}$$

$h = 21.376$ in. from part (*a*), $f_f = 0$, $f_r = 0.65$, and $W_r = 1500$ lbs.

$$\Delta W = \frac{21.376}{102}\left[\frac{2500(0) + (1500)(0.65)}{[1-0] + \frac{21.376}{102}(0.65)}\right] = \frac{0.2096(1500)(0.65)}{1 + 0.2096(0.65)} = 179.86 \text{ lbs.}$$

(*c*) The vehicle speed at the beginning of the skid is given by Eq. (2.35)

$$S = \sqrt{30\left[\frac{f_f d_f (W_f + \Delta W) + f_r d_r (W_r - \Delta W)}{W_t}\right]}$$

Here $d_f = 0$, $d_r = 250$ ft, $W_f = 2500$ lbs, $W_r = 1500$ lbs, $f_f = 0$, $f_r = 0.65$, $\Delta W = 179.86$ lbs (from part (*b*)), and $W_t = 4000$ lbs.

$$S = \sqrt{30\left[\frac{0 + 0.65(250)(1500 - 179.86)}{4000}\right]} = 40.11 \text{ mph.}$$

Example 2.4: *Refer to Example 2.1. Determine the horizontal center of mass (CM) of the vehicle.*

Solution: The horizontal center of mass (CM) of the vehicle is given by Eq. (2.32).

$$CM = \frac{W_f W_b}{W_t}$$

Here $W_f = 2500$ lbs, $W_b = 102$ inches, and $W_t = 4000$ lbs.

Hence, $$CM = \frac{2500(102)}{4000} = 63.75 \text{ inches.}$$

2.10 CRITICAL SPEED YAW

A critical speed yaw occurs when a vehicle is running in a radius that is as tight as possible for the given speed. In this situation the tires cannot grip the road and they slide sideways while still rolling. Under these circumstances, the rear tires track outside the corresponding front tires. Without yaw, the rear tires track inside the leading front tires.

2.10.1 Critical Speed Yaw Equation

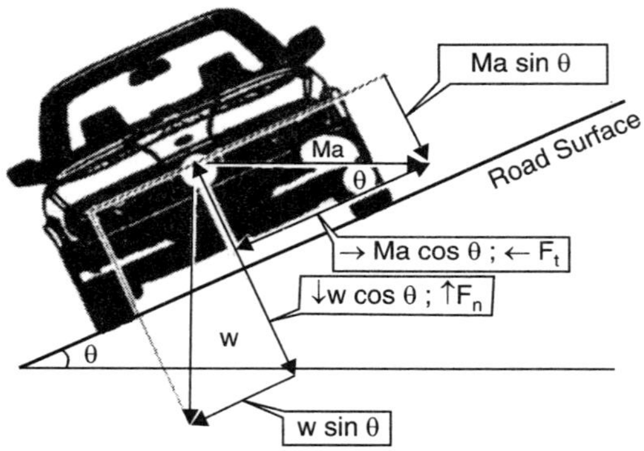

Fig. 2.3

Consider a vehicle traveling in a circular path at a constant speed as shown in Fig. 2.3. The magnitude of the vehicle remains unchanged while the direction of the velocity undergoes change which will result in lateral acceleration.

(*a*) Forces Normal to the Road Surface (F_n)

When the vehicle is in equilibrium, we have from Newton's first law of motion

$$\Sigma F_n = 0 \tag{2.49}$$

or $$F_n - w\cos\theta - Ma\sin\theta = 0$$

or $$F_n = w\cos\theta + Ma\sin\theta \tag{2.50}$$

where M = mass

a = acceleration

w = weight

F_n = normal force

(*b*) Forces Parallel to the Road Surface (F_t)

$$\Sigma F_t = 0 \tag{2.51}$$

When the vehicle is not sliding up or down the superelevation, we can write

$$F_t + w \sin \theta - Ma \cos \theta = 0$$

or

$$F_t + w \sin \theta = Ma \cos \theta \tag{2.52}$$

Since $F_t = f F_n$

From Eqs. (2.50) and (2.52), we have

$$F_t = f(w \cos \theta + Ma \sin \theta) \tag{2.53}$$

or $f(w \cos \theta + Ma \sin \theta) + w \sin \theta = Ma \cos \theta$

or

$$wf \cos \theta + Ma f \sin \theta + w \sin \theta = Ma \cos \theta \tag{2.54}$$

From Newton's second law of motion, we have

$$a = \frac{v^2}{R}$$

and

$$M = \frac{w}{g} \tag{2.55}$$

Using Eq. (2.55), Eq. (2.54) becomes

$$wf \cos\theta + \frac{w}{g}\left(\frac{v^2}{r}\right) f \sin\theta + w\sin\theta = \frac{w}{g}\left(\frac{v^2}{r}\right)\cos\theta \tag{2.56}$$

or

$$f + \left(\frac{v^2}{gr}\right) f \tan\theta + \tan\theta = \frac{v^2}{gr} \tag{2.57}$$

where r = radius of the path of the center of mass of a turning vehicle.

Now $\tan \theta = m$ = slope and Eq. (2.57) can be written as

$$f + \left(\frac{v^2}{gr}\right) f\, m + m = \frac{v^2}{gr} \tag{2.58}$$

Eq. (2.58) can be rearranged as

$$v^2 = \frac{gr(f + m)}{(1 - f\,m)}$$

or

$$v = \frac{\sqrt{g}\sqrt{r(f + m)}}{\sqrt{1 - f\,m}} \tag{2.59}$$

Noting that $g = 32.2$ ft/sec^2 and $v = 1.466\ S$, Eq. (2.59) becomes

$$S = \frac{3.86\sqrt{r(f + m)}}{\sqrt{(1 - f\,m)}} \tag{2.60}$$

Eq. (2.60) gives the critical speed on a curve with positive superelevation.

When $m = 0$, Eq. (2.60) becomes

$$S = 3.86\sqrt{r\,f} \tag{2.61}$$

Eq. (2.61) gives the critical speed on a level curve.

For a negative superelevation, the equation for critical speed on a curve is given by

$$S = \frac{3.86\sqrt{r(f-m)}}{\sqrt{1+f\,m}} \tag{2.62}$$

2.11 AIRBORNE EQUATIONS

The airborne equations, including fall and vault speeds are derived from the application of Newtonian mechanics based on a concept called uniform project motion.

According to the concept, the horizontal motion and vertical motion are independent of each other. Also, the center of mass (CM) of a body also known as center of gravity (*CG*), may be defined as the point at which the center weight of the body appears to be concentrated and could be represented as a single force, regardless of the angle at which the body might be turned. The vehicle is considered as a point mass. In other words, we consider that all the weight of the vehicle is concentrated at the center of mass (CM) of the vehicle. Hence, all measurements will be made with respect to takeoff and landing points of the center of mass. We also ignore the aerodynamic forces acting on the vehicle due to the fact that the distances involved in vehicle fall and the airborne times are short.

When vehicles travel through the air in accidents, the motion can be described as a fall, flip, or vault. A fall occurs when a vehicle is traveling forward and is no longer in contact with the surface it is moving over (see Fig. 2.4). Flip is the movement of a vehicle from a place where the forward velocity of a part of the vehicle suddenly is stopped by an object below its center of gravity such as curb, rail, or furrow, with the result that the ensuing rotational motion lifts the vehicle from the ground. Flips occur when vehicles are moving sideways and the resistance at the tires is sufficient to cause the vehicle to rise and move through the air. Vault is a roll or pitch motion of a vehicle made following loss of ground contact. Vaults are similar to flips; except they are end-for-end flips (see Fig. 2.5 to 2.7).

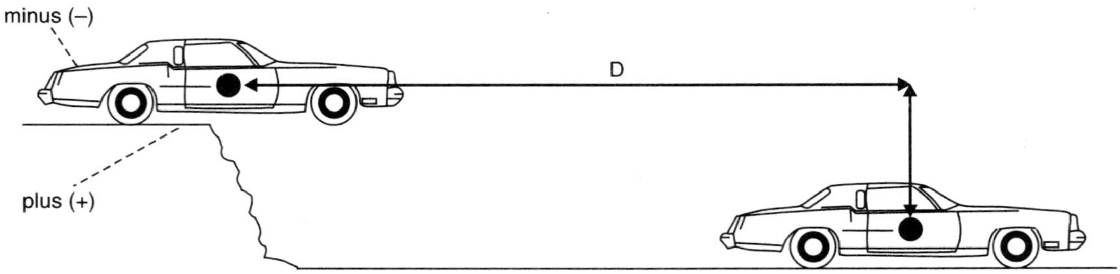

Fig. 2.4 A fall.

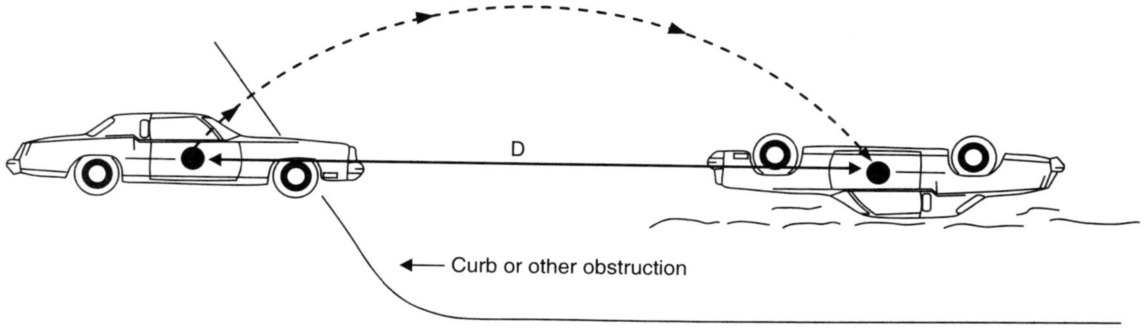

Fig. 2.5 A flip and vault (level takeoff and landing).

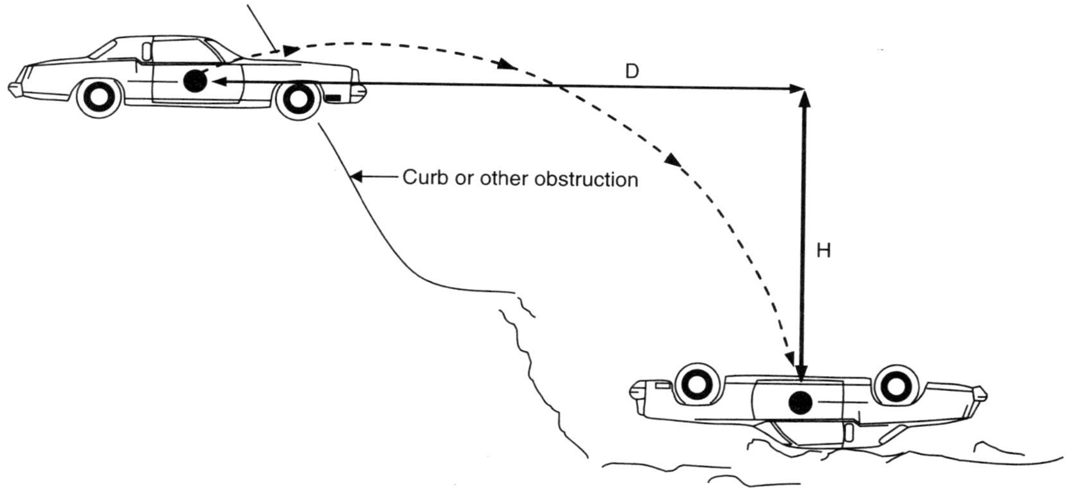

Fig. 2.6 A flip and vault (landing lower than the point of takeoff).

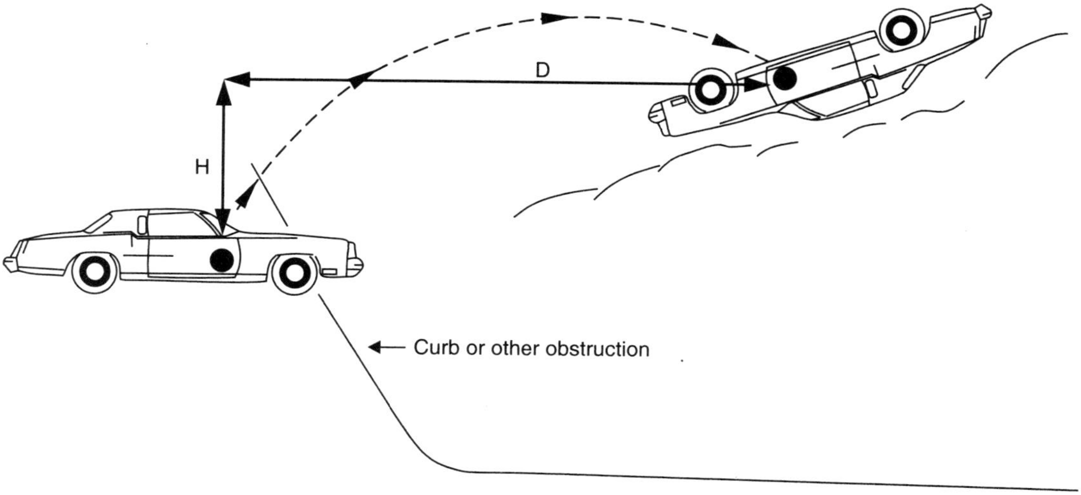

Fig. 2.7(*a*) A flip and vault (landing higher than the point of takeoff).

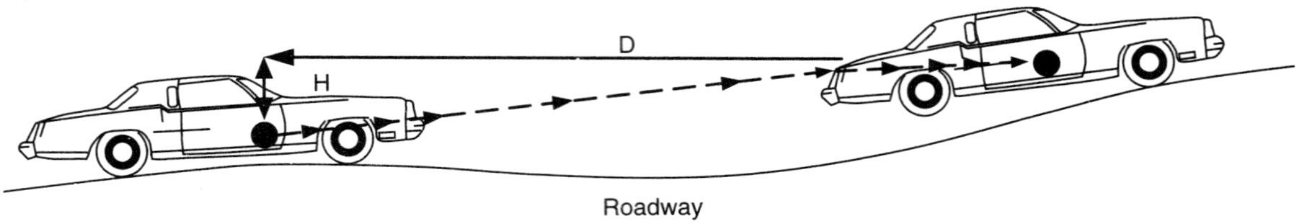

Fig. 2.7 (*b*) A vault (landing higher than the point of takeoff).

2.11.1 Fall Speed Equations

A *fall* is a situation where a vehicle leaves the roadway and falls to a landing some distance away from and always below its point of takeoff or takeoff point.

Point of takeoff or takeoff point is the location of the vehicle's center of mass when the vehicle begins its fall, vault, or flip and vault. While traveling at an excessive speed over a dip in the roadway, a vehicle may become temporarily airborne and land once again on the roadway. Also, a vehicle may for any number of reasons drive off the highway and fall over an embankment.

Fall is to drop under the force of gravity from a higher to a lower place or position. This applies to a vehicle that falls from a highway and lands some distance away from and below its point of takeoff. A fall also occurs when, for example, a vehicle fails to negotiate a curve and is projected off the highway, frequently over an embankment. In such cases, we must take into consideration any grade, plus or minus, of the roadway or area leading up to the point of takeoff.

The takeoff angle is defined as the angle between the level plane and the plane of the grade on which the vehicle is positioned at the point of takeoff. The area leading up to the point of takeoff may be level or have a positive (+) or negative (–) grade. A positive grade will give a positive or uphill or upward takeoff angle; a negative grade will give a negative or downhill or downward takeoff angle. The angle will always be positive for a vault.

In the direction of the fall speed equations, we assume that the vehicle may takeoff either level, or in an upward direction, or in a downward direction. The vehicle must land lower than its takeoff point in any of the three possible cases.

2.11.1.1 Level Takeoff

We employ here two time-dependent equations describing a falling vehicle. They are

1. The horizontal distance assuming a constant velocity throughout the fall:

$$d = vt \tag{2.63}$$

2. The vertical distances resulting from the acceleration due to gravity:

$$h = \frac{1}{2} g t^2 \tag{2.64}$$

Since t in Eqs. (2.63) and (2.64) must be the same when the vehicle strikes the ground, we have from Eq. (2.64)

$$t = \sqrt{\frac{2h}{g}} \tag{2.65}$$

Substituting t from Eq. (2.65) in Eq. (2.63), we get

$$d = v\sqrt{\frac{2h}{g}} \tag{2.66}$$

or

$$v = \frac{d}{\sqrt{\frac{2}{g}}\sqrt{h}} \tag{2.67}$$

For $g = 32.2$ ft/sec^2, we have from Eq. (2.67)

$$v = \frac{4.01d}{\sqrt{h}} \text{ ft/sec} \tag{2.68}$$

Since $v = 1.466\ S$, where S is in mph, Eq. (2.68) becomes

$$S = \frac{2.73d}{\sqrt{h}} \text{ mph} \tag{2.69}$$

In SI units

$$S = \frac{7.97d}{\sqrt{h}} \text{ km/h} \tag{2.70}$$

Example 2.5: *[Level takeoff area] A road vehicle left the highway on a curve. The area leading up to the takeoff point was level or zero grades. The vehicle first touched the ground at a horizontal distance of 50 ft (15 m) and a vertical distance of 12 ft (3.6 m) below the takeoff point. All measurements were taken to vehicle's center of mass (CM) positions. Determine the speed of the vehicle at takeoff.*

Solution: The vehicle speed is obtained from Eq. (2.69) in US units.

$$S = \frac{2.73d}{\sqrt{h}} = \frac{2.73(50)}{\sqrt{12}} = 39.40 \text{ mph}$$

In SI units: The speed of vehicle is obtained from Eq. (2.70).

$$S = \frac{7.97d}{\sqrt{h}} = \frac{7.97(15)}{\sqrt{3.6}} = 63 \text{ km/h}$$

Fall: The velocity of a vehicle at the point of takeoff is given by

$$v = d\sqrt{\frac{g}{2\cos\theta(d\sin\theta - h\cos\theta)}} \tag{2.71}$$

where v = velocity of the vehicle at takeoff (ft/s or m/s)

d = distance traveled horizontally by the vehicle's center of mass from takeoff to landing (ft or m)

g = 32.2 ft/s^2 or 9.81 m/s^2

h = distance traveled vertically by the vehicle's center of mass from takeoff to landing (ft or m). h is positive if the landing point is above the takeoff point, negative if the landing point is below the takeoff point

G = the percent grade

θ = the angle of takeoff as measured relative to a horizontal plane.

For small angles, Eq. (2.71) becomes

$$v = d\sqrt{\frac{g}{2(dG - h)}} \tag{2.72}$$

where $G = \sin\theta$ and $\cos\theta \approx 1$.

Figures. 2.8 to 2.11 show how the horizontal distance d and the vertical distance h are measured for different situations. Figure 2.8 shows a vehicle falling through the air and landing at a higher ground than takeoff. Figure 2.9 shows the measurements required for speed estimate. They are the horizontal distance that the center of mass moves from takeoff to landing, d. The vertical distance, the center of mass falls, h, and the slope, up or down of the takeoff. Figure 2.10 shows the landing position after a fall, which may not be the point where the vehicle finally comes to rest. Figure 2.11 shows the vehicle landing on a slope. A connection should be made on the basis of marks made where the vehicle first struck the ground after a fall. Figure 2.12 shows the takeoff slopes where an adjustment must be made to the vertical distance of the fall. This adjustment is equal to the slope times the horizontal distance. This adjustment is added to the vertical fall for a slope up and subtracted for a slope down.

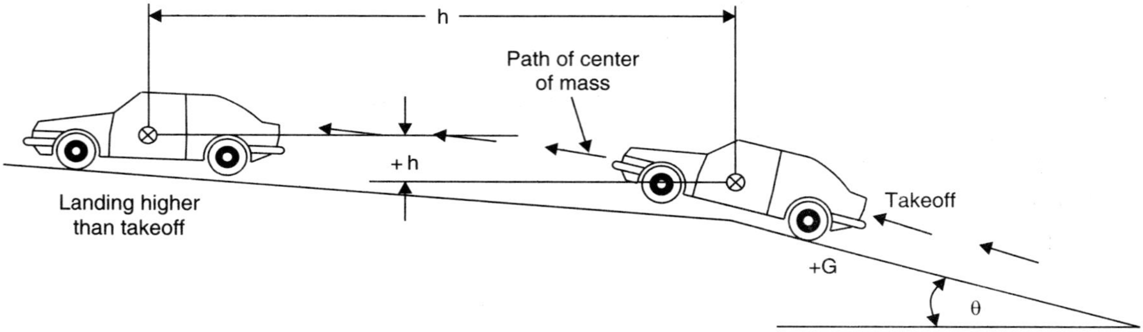

Fig. 2.8 Vehicle landing at a higher ground than takeoff.

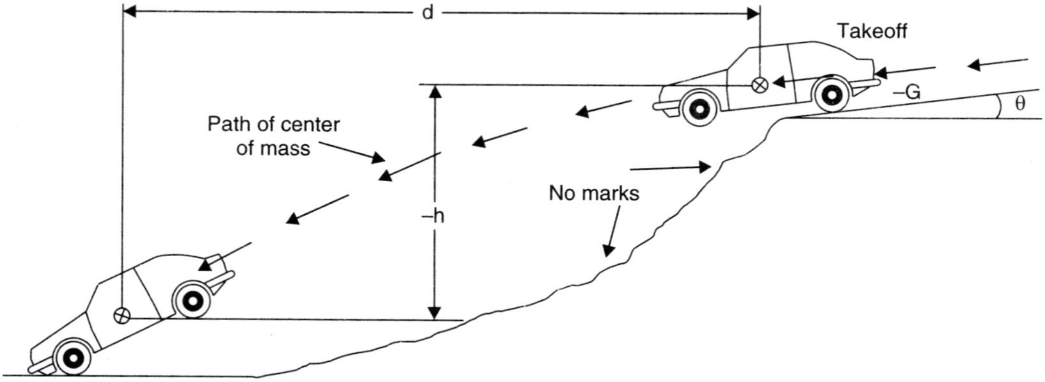

Fig. 2.9 Measurements required for speed estimate.

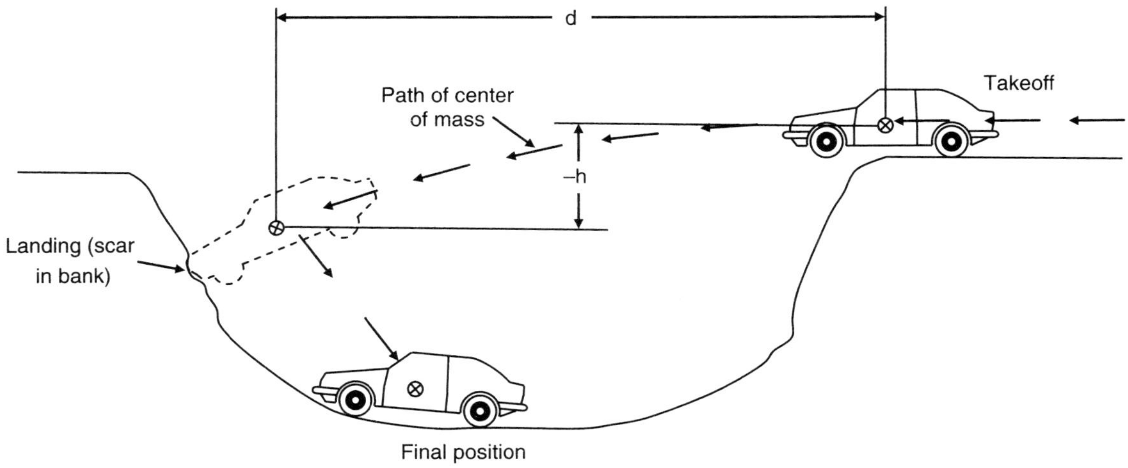

Fig. 2.10 Landing position after a fall.

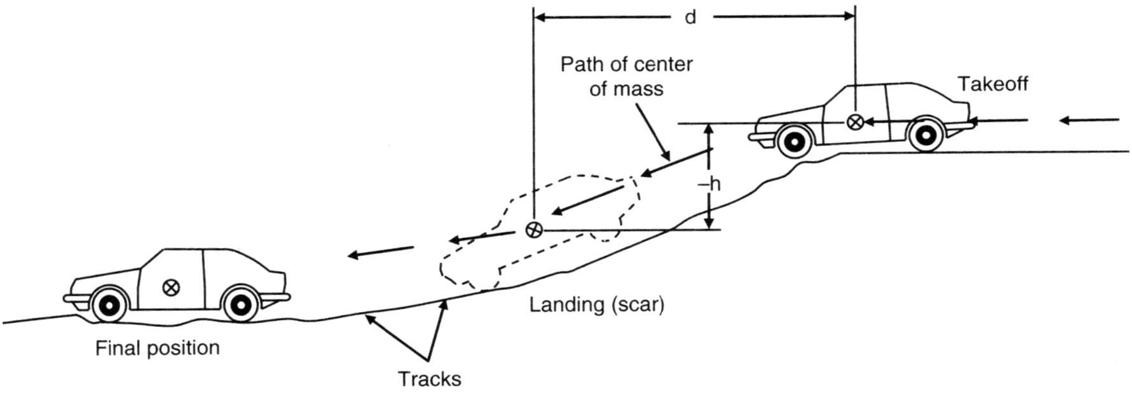

Fig. 2.11 Vehicle landing on a slope.

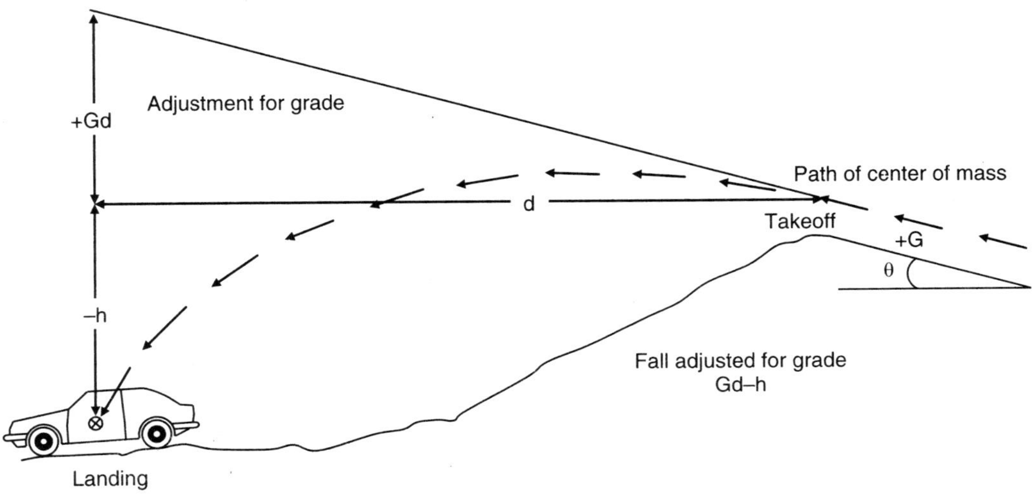

Fig. 2.12 Takeoff slopes.

2.11.1.2 Uphill takeoff

For the uphill takeoff case, the positive y direction is in the same sense of gravity, or downward.

The horizontal distance traveled in time t is

$$d = vt \cos \theta \tag{2.73}$$

The height reached in time t is

$$h = -vt \sin\theta + \frac{1}{2} gt^2 \tag{2.74}$$

In Eq. (2.74), $-v \sin \theta$ is the vertical component of the initial velocity and g is the gravitational acceleration.

From Eq. (2.73), we have for t

$$t = \frac{d}{v\cos\theta} \tag{2.75}$$

Substituting the value of t and t^2 from Eq. (2.75) in Eq. (2.74) and solving for v^2, we get

$$v^2 = \frac{g}{2}\left(\frac{1}{\cos^2\theta}\right)\frac{d^2}{(h + d\tan\theta)} \tag{2.76}$$

or

$$v = \sqrt{\frac{g}{2}}\left(\frac{1}{\cos\theta}\right)\frac{d}{\sqrt{h + d\tan\theta}} \tag{2.77}$$

Noting that for θ up to 6° (10 % slope), $\cos \theta \approx 1$. Also $\tan \theta = m$ and $g = 32.2$ ft/sec^2.

Eq. (2.77) can be written as

$$v = \sqrt{\frac{32.2}{2}\frac{d}{(h + d\tan\theta)}} \text{ ft/sec.} \tag{2.78}$$

Eq. (2.78) is good for fall velocity for uphill takeoff no greater than 10%. Also noting that $v = 1.466\, S$ where S is in mph, Eq. (2.78) becomes

$$S = \frac{2.73d}{\sqrt{h + dm}} \text{ mph} \tag{2.79}$$

In SI units

$$S = \frac{7.97\, d}{\sqrt{h + dm}} \text{ km/h} \tag{2.80}$$

Example 2.6: *[Positive grade in takeoff area] Refer to Example E2.5. The takeoff area had a grade of +7%. Determine the vehicle speed at takeoff.*

Solution: The vehicle speed is given by Eq. (2.79)

$$S = \frac{2.73d}{\sqrt{h + dm}} \text{ mph}$$

Here $d = 50$ ft, $h = 12$ ft, and $m = 0.07$.

$$S = \frac{2.73(50)}{\sqrt{12 + 50(0.07)}} = 34.67 \text{ mph}$$

In SI units, the vehicle speed is given by Eq. (2.80)

Here $d = 15$ m, $h = 3.6$ m, and $m = 0.07$.

$$S = \frac{7.97\,d}{\sqrt{h+dm}} \text{ km/h} = \frac{7.97(15)}{\sqrt{3.6+15(0.07)}} = 55.44 \text{ km/h}$$

2.11.2 Downhill Takeoff

For the downhill takeoff case also, the positive y direction is in the same sense of gravity, or downward.

The horizontal distance traveled in time t is

$$d = vt\cos\theta \tag{2.81}$$

The height reached in time t is

$$h = vt\sin\theta + \frac{1}{2}gt^2 \tag{2.82}$$

In Eq. (2.82), $v \sin\theta$ is the vertical component of the initial velocity and g is the gravitational acceleration.

Solving for t in Eq. (2.81), we have

$$t = \frac{d}{v\cos\theta} \tag{2.83}$$

Substituting the values of t and t^2 from Eq. (2.83) into Eq. (2.82), and solving for v, we get

$$v = \sqrt{\frac{g}{2}}\left(\frac{1}{\cos\theta}\right)\left(\frac{d}{\sqrt{h-d\tan\theta}}\right) \text{ ft/sec.} \tag{2.84}$$

Noting that $g = 32.2$ ft/sec^2 and $1.466\,S = v$, where S is in mph, Eq. (2.84) can be written as

$$S = \frac{2.73\,d}{\cos\theta\sqrt{h-d\tan\theta}} \tag{2.85}$$

Eq. (2.85) is the exact equation for any downhill takeoff angle and d and h are measured from the vehicle's center of mass (CM).

For $\theta \le 6°$, $\cos\theta \approx 1$. Also, that $\tan\theta = m$. Substituting these values in Eq. (2.85), we obtain

$$S = \frac{2.73d}{\sqrt{h-dm}} \text{ mph} \tag{2.86}$$

Similarly, in SI units, Eq. (2.84) becomes

$$S = \frac{2.214d}{\cos\theta\sqrt{h-d\tan\theta}} \text{ m/sec.} \tag{2.86a}$$

or

$$S = \frac{7.97\,d}{\sqrt{h-dm}} \text{ km/h} \tag{2.87}$$

$$S = \frac{7.97\,d}{\cos\theta\sqrt{h-d\tan\theta}} \text{ km/h} \tag{2.88}$$

Example 2.7: *[Minus grade in takeoff area] Refer to Example E2.5. The takeoff area had a grade of –7%. Determine the vehicle speed at takeoff.*

Solution: In US units the vehicle speed is obtained from Eq. (2.86)

$$S = \frac{2.73d}{\sqrt{h - dm}} \text{ mph}$$

Here $d = 50$ ft, $h = 12$ ft, and $m = 0.07$.

$$S = \frac{2.73(50)}{\sqrt{12 - 50(0.07)}} = 46.82 \text{ mph}$$

In SI units, the vehicle speed is given by Eq. (2.87)

$$S = \frac{7.97d}{\sqrt{h - dm}} \text{ km/h}$$

Here $d = 15$ m, $h = 3.6$ m, and $m = 0.07$.

$$S = \frac{7.97(15)}{\sqrt{3.6 - 15(0.07)}} = 74.86 \text{ km/h}$$

Example 2.8: *[Negative grade or slope in excess of 10%] A vehicle failed to make a curve in the roadway. The point where it left the roadway had a minus grade or stops of 20 degrees. After leaving the roadway, the vehicle traveled a horizontal distance of 35 ft (10.5 m) and landed 20 ft (6 m) below the takeoff point. Determine the vehicle speed at the point of takeoff.*

Solution: In US units, the vehicle speed at the point of takeoff is given by Eq. (2.85).

$$S = \frac{2.73\, d}{\cos\theta\sqrt{h - d\tan\theta}} \text{ mph}$$

Here $d = 35$ ft, $h = 20$ ft, and slope = 20 degrees.

$$S = \frac{2.73(35)}{\cos 20\sqrt{20 - 35\tan 20^\circ}} = 37.74 \text{ mph}$$

In SI units, the vehicle speed at the point of takeoff is given by Eq. (2.88)

$$S = \frac{7.97\, d}{\cos\theta\sqrt{h - d\tan\theta}} \text{ km/h}$$

Here $d = 10.5$ m, $h = 6$ m, and slope = 20 degrees.

$$S = \frac{7.97(10.5)}{\cos 20^\circ\sqrt{6 - 10.5(\tan 20^\circ)}} = 60.34 \text{ km/h}$$

2.11.3 Flips and Vaults

A flip and vault is a situation where a moving vehicle strikes a curb or some other object that has a height less than the height of the vehicle's center of mass (CM) and the vehicle stops suddenly, pivots at the point of impact, flips and then vaults through the air to land some distance away. To flip is to toss or put in motion with a sudden impulse as to cause a vehicle to turn over in the air. This applies to a vehicle in motion that strikes a substantial object at a point on the vehicle that is lower than its center of mass (CM). To vault is to leap or spring over. A vault occurs when a vehicle leaves the travel surface at a positive (+) angle and, without having flipped, lands some distance away. The landing point can be either level with or higher or lower than the takeoff point or point of takeoff.

A *vault* is an end-for-end flip. A vault occurs when the front of a vehicle strikes an object in which further movement at the ground level stops. As with falls and flips, a vehicle has vaults which will likely continue moving after it lands. It may roll or slide to its final position. Such flips and vaults can occur when a vehicle strikes an object head-on or sideways and the landing point can be either level with or higher or lower than the point of takeoff.

Level Takeoff and Landing

In US units, the speed of the vehicle on a level takeoff and landing is given by

$$S = 3.86\sqrt{d} \text{ mph} \tag{2.89}$$

where d = distance, ft

S = vehicle speed at takeoff, mph.

In SI units:

$$S = 11.27\sqrt{d} \tag{2.90}$$

where d = distance, m

S = vehicle speed at takeoff, km/h.

Flip: The general equation for a flip is given by Eq. (2.71). The distance is obtained by taking the shortest distance between any sign of where the vehicle took off (a curb or dig-in at the end of a furrow) and any sign of where it first landed, and adding the width of the vehicle (sec Fig. 2.14). Figure (2.13) shows the slope of the takeoff in a fall. It is the slope of the exact path that is followed by the vehicle. A slope up is positive and a slope down is negative.

The minimum speed is obtained when θ (the take off angle) is approximately 45 degrees. The angle for minimum speed estimate is given by

$$\theta = \frac{1}{2}\cos^{-1}\left[\frac{-h}{\sqrt{d^2 - h^2}}\right] \tag{2.91}$$

where θ = 45°, Eq. (2.88) gives the minimum speed at takeoff point.

$$v = d\sqrt{\frac{g}{d - h}} \tag{2.92}$$

Vault: A vault is an end-for-end flip. Figure 2.14 shows the vault of a vehicle. A vault occurs when the front of a vehicle strikes an object in which further movement at the ground level stops. As with falls and flips, a vehicle hat vaults will likely continue moving after it lands. It may roll or slide to its final position. The three equations that are used for a vault are given by Eqs. (2.71), (2.91) and (2.92). The variables v, d, g, θ, and h are the same variables used for a flip. Figure 2.15 shows the path of center of mass. The airborne equations are summarized in Table 2.12.

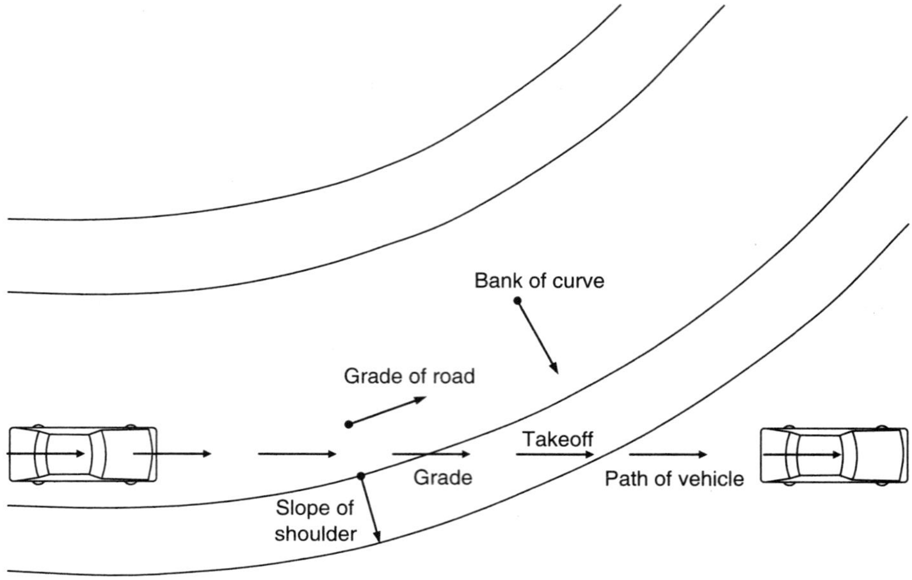

Fig. 2.13 Slope of the takeoff in a fall.

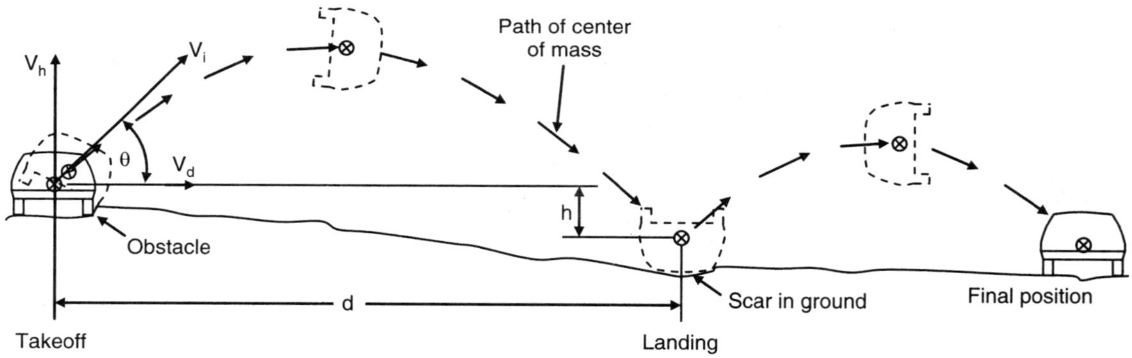

Fig. 2.14 Vault of a vehicle.

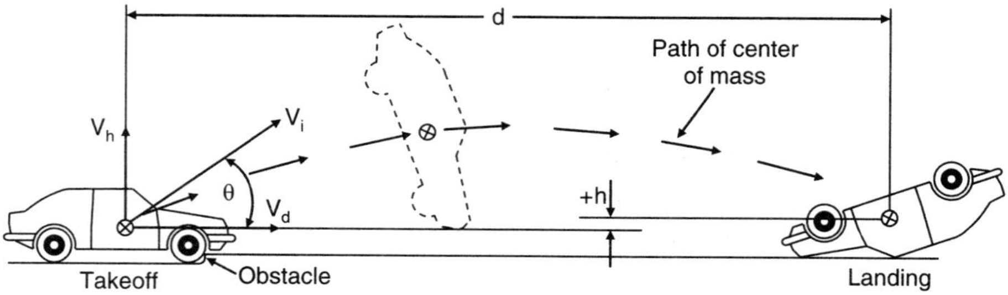

Fig. 2.15 Path of center of mass.

Example 2.9: *[Level takeoff and landing] A vehicle traveled into an intersection, struck a curb, flipped and vaulted through the air and landed on its roof on a nearby ground. The positions of the vehicle's center of mass (CM) at takeoff and landing were level and measured a horizontal distance of 50 ft (15.217 m) apart. Determine the speed of the vehicle at takeoff.*

Solution: The vehicle speed at takeoff where takeoff and landing are level is given by Eq. (2.89).

$$S = 3.86\sqrt{d} \text{ mph}$$

Here $d = 50$ ft.

Hence $S = 3.86\sqrt{50} = 27.29$ mph

In SI units, the vehicle speed at takeoff where the takeoff and landing are level is given by Eq. (2.90).

$$S = 11.27\sqrt{d} \text{ km/h}$$

Here $d = 15.217$ m.

Therefore $S = 11.27\sqrt{15.217} = 43.96$ km/h.

2.11.3.1 Vehicle lands lower than takeoff

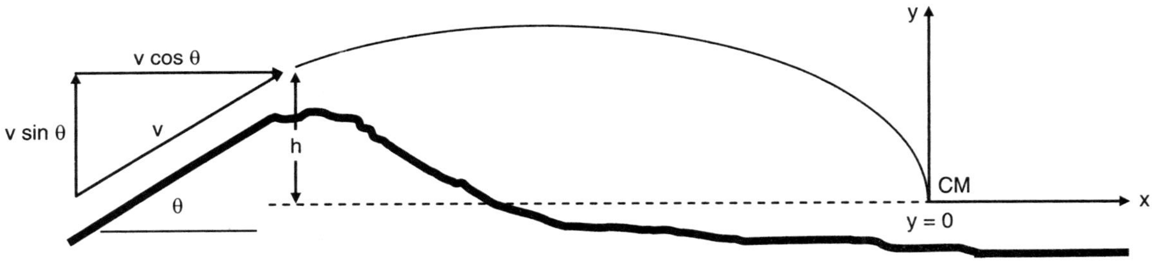

Fig. 2.16

Referring to Fig. 2.16, we can write the vehicle position equations as a function of time as follows:

In the x-direction:

$$x = d = vt\cos\theta \tag{2.93}$$

In the y-direction:

$$y = vt\sin\theta - \frac{1}{2}gt^2 + h \tag{2.94}$$

From Eq. (2.93), solving for t, we obtain

$$t = \frac{d}{v\cos\theta} \tag{2.95}$$

Substituting the value of t from Eq. (2.95) in Eq. (2.94) and noting that $y = 0$ at ground contact, and $g = 32.2$ ft/sec^2, we obtain v as

$$v = \frac{4.01\,d}{\sqrt{d\sin\theta\cos\theta + h\cos^2\theta}} \text{ vault velocity in ft/sec.} \tag{2.96}$$

Converting to speed with 1.466 $S = v$, where S is in mph, Eq. (2.96) becomes

$$S = \frac{2.73\, d}{\sqrt{d \sin\theta \cos\theta + h \cos^2\theta}} \tag{2.97}$$

Figure (2.97) gives the vault speed S in mph, where the vehicle takes off uphill and lands lower than takeoff.

In SI units

$$S = \frac{7.97\, d}{\sqrt{d \sin\theta \cos\theta + h \cos^2\theta}} \tag{2.98}$$

For level takeoff, the vehicle speed in US units (in mph) is given by

$$S = \frac{3.86\, d}{\sqrt{d + h}} \text{ mph} \tag{2.99}$$

In SI units, the vehicle speed for level takeoff is given by

$$S = \frac{11.27\, d}{\sqrt{d + h}} \text{ km/h.} \tag{2.100}$$

Example 2.10: *[Landing lower than takeoff point] Refer to Example E2.5. The vehicle landed 12 ft (3.6 m) below the level of the takeoff point measured to and from the center-of-mass (CM) position. Determine the vehicle speed at takeoff point.*

Solution: For level takeoff, the vehicle speed at takeoff is given by Eq. (2.99).

$$S = \frac{3.86\, d}{\sqrt{d + h}} \text{ mph}$$

Here $d = 50$ ft, and $h = 12$ ft.

$$S = \frac{3.86(50)}{\sqrt{50 + 12}} = 24.51 \text{ mph}$$

In SI units, the vehicle speed at takeoff is given by Eq. (2.100)

$$S = \frac{11.27\, d}{\sqrt{d + h}} \text{ km/h}$$

Here $d = 15.217$ m, and $h = 3.6$ m.

$$S = \frac{11.27(15.217)}{\sqrt{15.217 + 3.6}} = 39.53 \text{ km/h.}$$

2.11.3.2 Vehicle Lands Higher Than Takeoff

Referring to Fig. 2.17, we can write the vehicle position equations as a function of time as follows:

In the x-direction:

$$x = vt \cos\theta \tag{2.101}$$

In the y-direction:

$$y = vt \sin\theta - \frac{1}{2} g t^2 - h \tag{2.102}$$

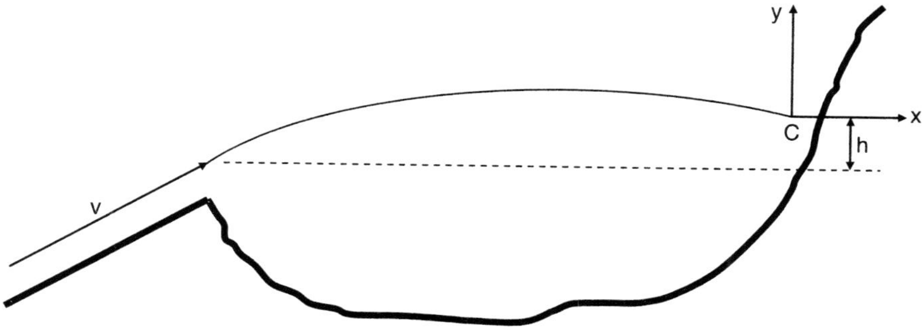

Fig. 2.17 Vehicle landing higher than takeoff.

Note here that since the landing point is higher than the takeoff point and since $y = 0$ at the landing point, h will have a negative (–ve) y value. Also, at $t = 0$, or takeoff, $y = h$. From Eq. (2.101), solving for t, we obtain

$$t = \frac{d}{v\cos\theta} \tag{2.103}$$

Substituting the value of t from Eq. (2.103) in Eq. (2.102) and noting that $y = 0$ at ground contact and $g = 32.2$ ft/sec^2, we obtain v as

$$v = \frac{4.01d}{\sqrt{d\sin\theta\cos\theta - h\cos^2\theta}} \text{ ft/sec} \tag{2.104}$$

Equation (2.104) is the vault equation in ft/sec.

Converting to speed with 1.466 S = v, where S is in mph, Eq. (2.104) becomes

$$S = \frac{2.73d}{\sqrt{d\sin\theta\cos\theta - h\cos^2\theta}} \text{ mph} \tag{2.105}$$

Equation (2.105) gives the vault speed in mph where vehicle takes off uphill and lands higher than the takeoff point.

Vault speed for $\theta = 45°$:

In substituting $\theta = 45°$ in Eq. (2.104), we obtain

$$S = \frac{3.86\,d}{\sqrt{d \pm h}} \text{ mph} \tag{2.106}$$

In SI units:

$$S = \frac{11.27\,d}{\sqrt{d \pm h}} \text{ km/h} \tag{2.107}$$

Example 2.11: *[Landing higher than the level of takeoff] Refer to Example E2.5. The vehicle landing 12 ft (3.6 m) above the level of the takeoff point measured to and from the center of mass (CM) position on the vehicle. Determine the vehicle speed at takeoff point.*

Solution: The vehicle speed at takeoff is given by Eq. (2.106)

$$S = \frac{3.86\,d}{\sqrt{d \pm h}} \text{ mph}$$

When a flip and vault landing is higher than the level of takeoff, the vertical distance h is subtracted (–) in the formula. Therefore,

$$S = \frac{3.86\,d}{\sqrt{d-h}}\ \text{mph}$$

Here $d = 50$ ft, and $h = 12$ ft.

$$S = \frac{3.86(50)}{\sqrt{50-12}} = 31.31\ \text{mph}$$

In SI units, we have from Eq. (2.107)

$$S = \frac{11.27\,d}{\sqrt{d-h}}\ \text{km/h}$$

Here we have $d = 15.217$ m and $h = 3.6$ m.

$$S = \frac{11.27(15.217)}{\sqrt{15.217-3.6}} = 50.32\ \text{km/h}.$$

Example 2.12: *[Vehicle landing lower] A vehicle was traveling at an excessive speed along a highway. Upon approaching a dip, the vehicle became momentarily airborne and landed on the highway some distance away. From the takeoff point to the landing point, the measurements (mass center to mass center) were given as follows:*

Horizontal distance = 90 ft (27 m)

Vertical distance = 4 ft (1.2 m)

The departure angle = 6 degrees

Determine the vehicle speed at takeoff.

Solution: The vehicle speed at takeoff is given by Eq. (2.105)

$$S = \frac{2.73d}{\sqrt{d\sin\theta\cos\theta - h\cos^2\theta}}\ \text{mph}$$

Here $d = 90$ ft, $h = 4$ ft and $\theta = 6$ degrees.

$$S = \frac{2.73(90)}{\sqrt{90\sin 6^\circ\cos 6^\circ + 4\cos^2 6^\circ}} = 67.49\ \text{mph}$$

In SI units, the vehicle speed at takeoff is given by Eq. (2.106)

$$S = \frac{7.97\,d}{\sqrt{d\sin\theta\cos\theta + h\cos^2\theta}}\ \text{km/h}$$

Here $d = 27$ m, $h = 1.2$ m, and $\theta = 6$ degrees.

$$S = \frac{7.97(27)}{\sqrt{27\sin 6^\circ\cos 6^\circ + 1.2\cos^2 6^\circ}} = 107.91\ \text{km/h}$$

2.11.3.3 Maximum Height (Uphill Takeoff)

We start with the equation (see derivation of Kinetic Energy (KE) equation)

$$v_f = \sqrt{v_0^2 - 2ad} \tag{2.108}$$

or $$v_f^2 = v_0^2 - 2ad \tag{2.109}$$

When the vehicle has reached its peak trajectory height, its final velocity along the y-axis will be zero ($v_f = 0$). Also, the acceleration will be the force of gravity ($a = g$). The distance the vehicle has traveled along the y-axis when $v_f = 0$ will be the peak trajectory height or $d = h$.

Substituting for v_f, a and d the values o, g, and h respectively in Eq. (2.109), we obtain

$$o = v_0^2 - 2gh \tag{2.110}$$

In the y-direction,

$$v_0 = v \sin \theta \tag{2.111}$$

where v in Eq. (2.111) is the takeoff velocity.

Substituting v_0 from Eq. (2.111) in Eq. (2.110) and solving for h, we get

$$h = \frac{v^2 \sin^2 \theta}{2g} \tag{2.112}$$

Substituting $g = 32.2$ ft/sec², and $1.466\ S = v$ in Eq. (2.112), we obtain h as

$$h = 0.033\ S^2 \sin^2 \theta \tag{2.113}$$

Equation (2.113) gives the maximum height reached by a vehicle with uphill takeoff. In Eq. (2.113), S is in mph. The airborne equations are summarized in Table 2.12.

Table 2.12 Airborne equations

No.	*US units*		*SI units*	
	Speed	*Velocity*	*Speed*	*Velocity*
1.	$S = \frac{2.73d}{\sqrt{-h}}$	$v = \frac{4.01d}{\sqrt{-h}}$	$S = \frac{7.97d}{\sqrt{-h}}$	$v = \frac{2.215d}{\sqrt{-h}}$
2.	$S = \frac{2.73d}{\sqrt{dm - h}}$	$v = \frac{4.01d}{\sqrt{dm - h}}$	$S = \frac{7.97d}{\sqrt{dm - h}}$	$v = \frac{2.215d}{\sqrt{dm - h}}$
3.	$S = \frac{2.73d}{\cos\theta\sqrt{d\tan\theta - h}}$	$v = \frac{4.01d}{\cos\theta\sqrt{d\tan\theta - h}}$	$S = \frac{7.97d}{\cos\theta\sqrt{d\tan\theta - h}}$	$v = \frac{2.215d}{\cos\theta\sqrt{d\tan\theta - h}}$

2.12 DRIVER PERCEPTION AND RESPONSE

In traffic accident reconstruction, perception and response values are primarily used to

(*a*) estimate possible stopping distance of vehicles, generally to determine whether a hazard could have been avoided under specified circumferences

(*b*) compare actual points of perception with possible points of perception to estimate perception delay as an indication of attentiveness

(*c*) determine whether a roadway users' response to a situation hazard was the best possible under existing circumferences.

Perception is a human factor, which means the general process of detecting some object or situation and comprehending its significance. Reaction or response is a person's voluntary or involuntary response to a hazard of other situation that has been perceived, the response to a sensory stimulus. Reaction time is the time required from perception to start of vehicle control for tactical or strategic operations.

In traffic accident reconstruction, perception and reaction time are mainly for two significant uses:

(*a*) to compare the least possible reaction under the circumferences with evasive tactics actually undertaken in order to evaluate the knowledge and skill of a driver or pedestrian

(*b*) to establish the possible slowing or stopping distance of a vehicle at a stipulated speed.

McGee et al. [1983] attempted to deduce perception-response time by listing its individual components, surveying to literature to find appropriate values for each, summing these values to obtain totals. Their results are shown in Table 2.13.

Table 2.13 Results of an effort to deduce driver perception-response time, based on summation of assumed components taken from the research literature

	Element	*Percentile of drivers*					
		50	*75*	*85*	*90*	*95*	*99*
1.	Perception						
	(*a*) Latency	0.24	0.27	0.31	0.33	0.35	0.45
	(*b*) Eye moment	0.09	0.09	0.09	0.09	0.09	0.09
	(*c*) Fixation	0.20	0.20	0.20	0.20	0.20	0.20
	(*d*) Recognition	0.40	0.45	0.55	0.55	0.60	0.65
2.	Decision	0.50	0.75	0.90	0.90	0.95	1.00
3.	Brake reaction	0.85	1.11	1.42	1.42	1.63	2.16
	Total *A* (1*a*–*d* + 2 + 3)	2.3	2.9	3.5	3.5	3.8	4.6
	Total *B* (1*c*–*d* + 2 + 3)	2.0	2.5	2.8	3.1	3.4	4.1
	Total *C* (1*a*–*d* + 3)	1.8	2.1	2.3	2.6	2.9	3.6

From the *proceeding* discussion, it is extremely difficult to determine the time required for an individual to actually perceive a hazard. Tests conducted to assess the combined time lapse for perception and reaction time in altered subjects indicated a range of 0.4 to 1.7 seconds. Using the result from Rumar's experiments (Olson [1989]) and adding 0.3 seconds to the test results, the perception reaction range becomes 0.7 to 2.0 seconds. Given Rumar's correction factor, the mean for perception/reaction time would be in the area of 1.5 seconds.

The safe stopping distance for vehicles knowing the actual time required for the driver to perceive and react to a hazard is given by

$$SSD = vt_p + \frac{v^2}{2a}$$

where SSD = safe stopping distance for the vehicle (ft or m)

t_p = perception reaction time (seconds)

v = velocity, m/s (ft/s)

a = acceleration factor (the effective drag factor times the gravitational constant)

Most design standards for rural and urban roadways use a 2.5 second perception/reaction time to include nearly all drivers and conditions for computing a safe stopping sight distance.

The time required by a vehicle operator to perceive and respond to a hazard is a subject. A short table of values cannot hope to convey all the background information needed for full understanding. An excellent paper on the subject is reference Paul Olson [1989]. The data included here in Table 2.14 is drawn from this reference.

Table 2.14 Perception-response times

Situation	*PR Time*
Alerted test subjects pressing the brake pedal with foot originally on the accelerator in response to a clear stimulus such as a light on the hood.	0.45 to 0.85 sec 85th %ile value = 0.7 sec
"Alerted" test subjects (with prior knowledge of hazard) encountering an obstacle in road as they crest a hill and responding by braking.	0.05 to 1.3 sec 50th %ile value = 0.7 sec
"Surprised" test subjects (with no prior knowledge of hazard) encountering an obstacle in the road as they crest a hill and responding by braking.	0.9 to 1.8 sec 50th %ile value = 1.1 sec
Australian data of drivers cresting a hill and responding to a protruding vehicle with tire change.	1.5 sec
Australian data of drivers cresting a hill and responding to the presence of an Amphometer, a speed measuring device.	3.4 to 3.6 sec
Most drivers (about 85%) given a "reasonably clear stimulus and a fairly straightforward situation".	about 1.5 sec

Source: Olson, P.L., "Driver Perception Response Times", Paper No. 890731, SAE International, Warrendale, PA, 1989.

2.13 ESTIMATION OF VEHICLE CRITICAL SPEED FROM YAW MARKS

2.13.1 Speed Estimates from Yaw Marks

Yaw is the action of a vehicle revolving around its center of mass. This action is a sideslip movement and is commonly observed in a vehicle traveling around a curve whose rear end breaks away and appears to want to move in a direction other than that in which the vehicle is headed. Yaw commences at the time the rear tires start to sideslip more than the front tires and develops as the rear tires depart from their normal tracking path and cross over the front tire paths.

Yaw mark is a tire mark caused by a slide-slipping tire often showing a striped pattern called striations. Yaw marks are the result of an avoidance maneuver, where a driver perceives an object or potential hazard in front of the vehicle and swerves suddenly to the left or right or from a sudden yaw velocity. During extreme steering maneuvers the available tire friction can be saturated by the side forces on the tire, and the vehicle will then yaw and leave a peculiar type of tire mark. The speed at the beginning of the maneuver can often be calculated from yaw marks and is often referred to as critical speed [see Figs. 2.17(*a*) and (*b*)].

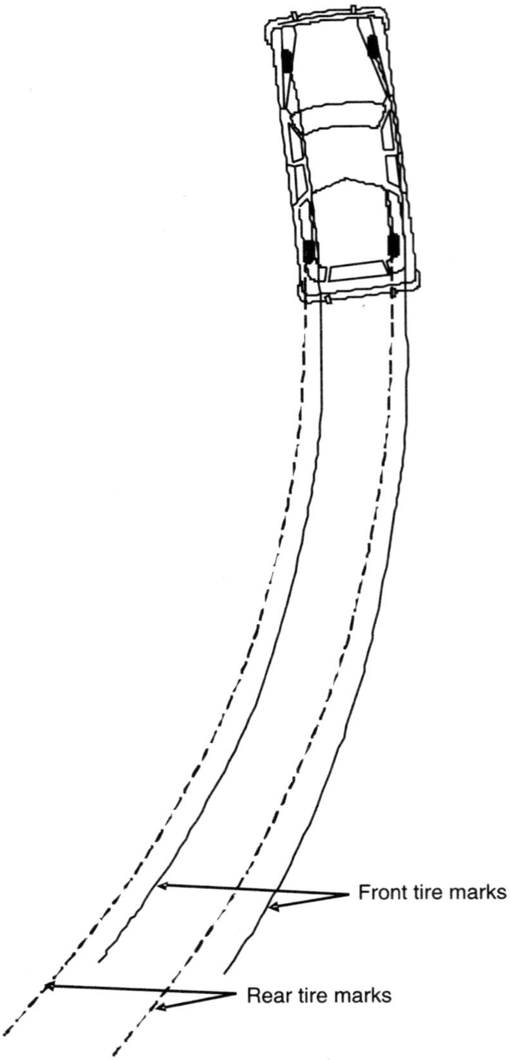

Fig. 2.17 (*a*) Under normal conditions, rear tires track inside front tires on a curve, *i.e.*, each rear tire tracks inside its respective front tire. [Rivers (1997)].

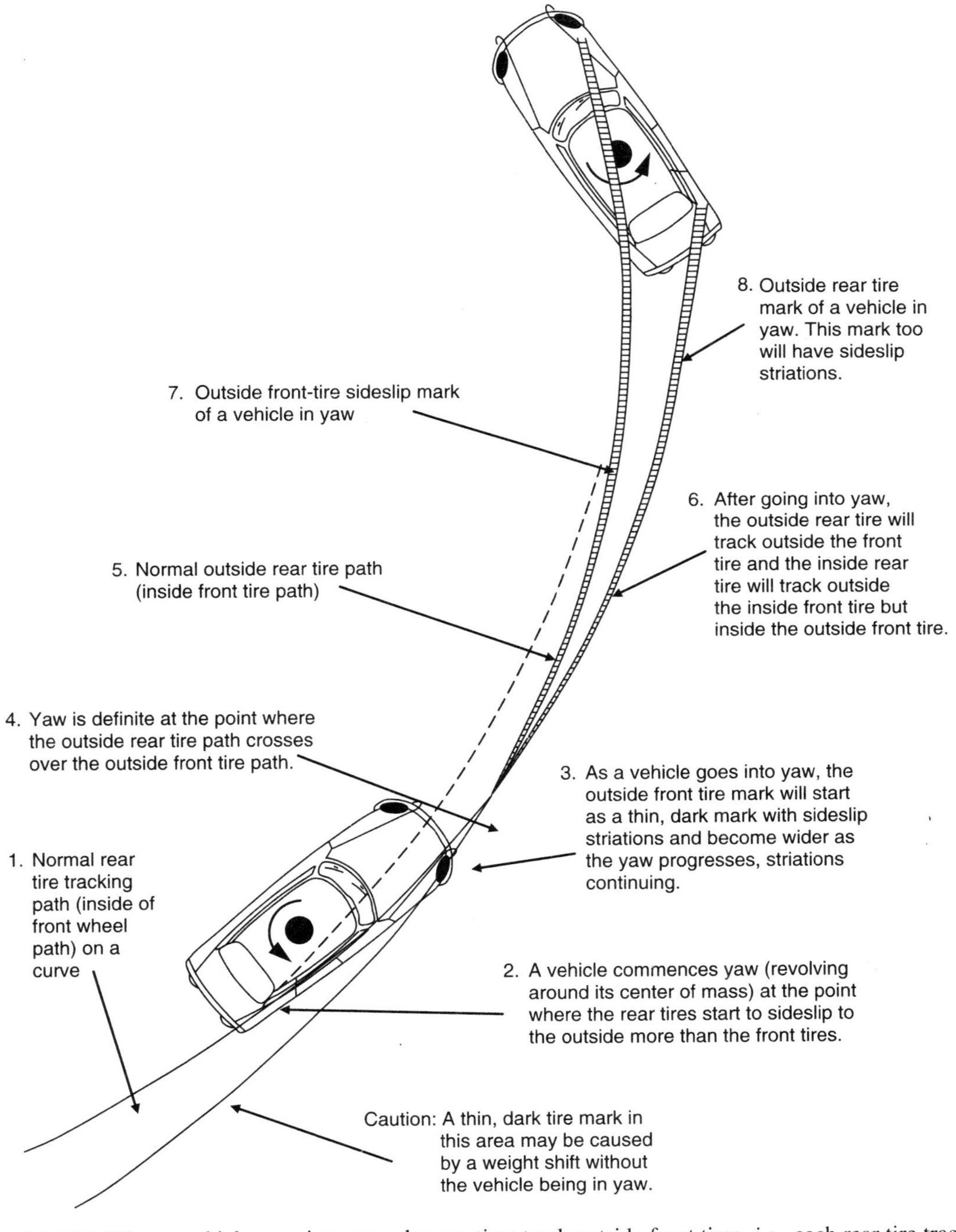

Fig. 2.17 (*b*) When a vehicle goes into yaw, the rear tires track outside front tires, *i.e.*, each rear tire tracks outside its respective front tire. (Courtesy of Rivers Traffic Consultants Ltd.) [Rivers (1997)]

In order to calculate the speed to slide slip a vehicle, one must know the radius of the curved path followed by the vehicle. Here we derive an equation that will give us the radius of a circle if we know the chord and middle ordinate.

2.13.2 Radius Equation

In Fig. 2.18, M denotes the middle ordinate (equals h), C is the chord and R is the radius.

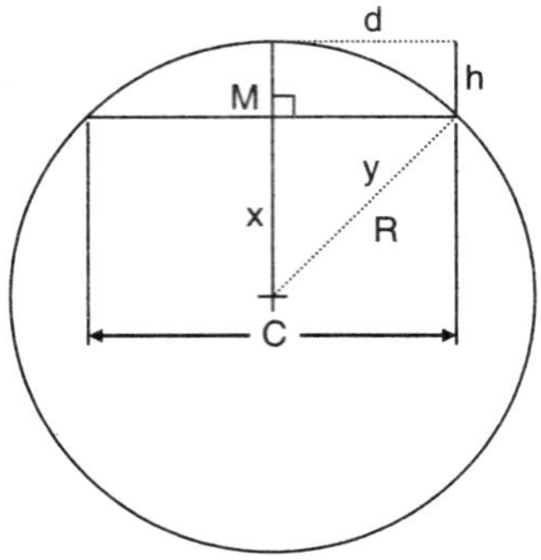

Fig. 2.18

Now $$y = \frac{1}{2}C = d$$

and $$x = R - M$$

From the basic geometry, we have

$$x^2 + y^2 = R^2 \tag{2.114}$$

or $$(R - M)^2 + \left(\frac{C}{2}\right)^2 = R^2 \tag{2.115}$$

Equation (2.115) simplifies to

$$M^2 + \frac{C^2}{4} = 2RM \tag{2.116}$$

From Eq. (2.116), we have

$$R = \frac{C^2}{8M} + \frac{M}{2} \tag{2.117}$$

From Eq. (2.114), h can also be written as

$$h = R - \sqrt{R^2 - d^2} \tag{2.118}$$

Equation (2.118) is known as the *tangent-offset equation*.

A chord measurement of 30.48 m (100 ft) is recommended. The vehicle's speed is higher at the beginning of the yaw mark since the vehicle will decelerate as it side slips. The critical speed is given by

$$V_{cr} = \sqrt{gRf} \tag{2.119}$$

where V_{cr} = critical speed, m/s (ft/sec)

f = coefficient of friction.

Equation (2.119) can also be written as

$$S = \sqrt{15Rf}$$

where S = speed (mph)

R = radius, ft

f = coefficient of friction.

Also in SI units,

$$S = \sqrt{127\,Rf}$$

where S = speed (km/h)

R = radius, m

f = coefficient of friction.

For a roadway with super elevation, the critical speed is given by

$$V_{cr} = \sqrt{\frac{gR(f + \tan\gamma)}{(1 - f\tan\gamma)}} \tag{2.120}$$

where g = acceleration due to gravity, 9.81 m/s^2 (32.2 ft/s^2)

f = coefficient of friction

R = radius, m (ft)

g = bank angle (super elevation)

Example 2.13: *The measured chord and middle ordinates from a test of a passenger road vehicle of a sudden steer maneuver are 15.24 m (50 ft) and 0.37 m (1.2 ft). The test site has a zero grade and zero super elevation. Determine*

(*a*) *the speed of the vehicle over the measured arc of yaw marks. Assume an average frictional drag coefficient of 0.8*

(*b*) *the frictional drag coefficient that gives the measured speed of 80.45 km/h (50 mph).*

Solution:

(*a*) From Eq. (2.96), the radius of the circular arc is

$$R = \frac{C^2}{8M} + \frac{M}{2} = \frac{15.24^2}{8(0.37)} + \frac{0.37}{2} = 78.65 \text{ m (258.04 ft)}$$

The speed of the vehicle is given by Eq. (2.98)

$$V_{cr} = \sqrt{127Rf} = \sqrt{127(78.65)(0.8)} = 89.39 \text{ km/h (55.56 mph)}$$

(*b*) From Eq. (2.98)

$$f = \frac{V_{cr}^2}{127R} = \frac{89.39^2}{127(78.65)} = 0.80$$

A few special equations useful in accident reconstruction are summarized in Table 2.15. In Table 2.15, e is the superelevation.

Table 2.15 Special equations

No.	*US units*		*US units*	
	To find	*To find*	*To find*	*To find*
1.	$h = R - \sqrt{R^2 - d^2}$	$h = R - \sqrt{R^2 - d^2}$	$h = R - \sqrt{R^2 - d^2}$	$h = R - \sqrt{R^2 - d^2}$
2.	$R = \frac{C^2}{8m} + \frac{M}{2}$	$R = \frac{C^2}{8m} + \frac{M}{2}$	$R = \frac{C^2}{8m} + \frac{M}{2}$	$R = \frac{C^2}{8m} + \frac{M}{2}$
3.	$s = \frac{3.86\sqrt{R(f+e)}}{\sqrt{1-fe}}$	$v = \frac{5.67\sqrt{R(f+e)}}{\sqrt{1-fe}}$	$s = \frac{11.27\sqrt{R(f+e)}}{\sqrt{1-fe}}$	$v = \frac{3.13\sqrt{R(f+e)}}{\sqrt{1-fe}}$
	e = superelevation	e = superelevation	e = superelevation	e = superelevation
4.	$s = \sqrt{s_0^2 + 30(d_1 f_1 + d_2 f_2 + d_3 f_3)}$	$v = \sqrt{v_0^2 + 64.4(d_1 f_1 + d_2 f_2 + d_3 f_3)}$	$s = \sqrt{s_0^2 + 254.3(d_1 f_1 + d_2 f_2 + d_3 f_3)}$	$v = \sqrt{v_0^2 + 19.62(d_1 f_1 + d_2 f_2 + d_3 f_3)}$
5.	$s_f = \sqrt{s_0^2 + s_1^2 + s_2^2 + s_3^2 + s_4^2}$	$v_f = \sqrt{v_0^2 + v_1^2 + v_2^2 + v_3^2 + v_4^2}$	$s_f = \sqrt{s_0^2 + s_1^2 + s_2^2 + s_3^2 + s_4^2}$	$v_f = \sqrt{v_0^2 + v_1^2 + v_2^2 + v_3^2 + v_4^2}$

Example 2.14: *[Level takeoff] In an accident a vehicle goes off a level road and travels through the air 25 ft horizontally and 12 inches vertically. How fast is the vehicle going when it becomes airborne?*

Solution:

$$S = \frac{2.73d}{\sqrt{h}}$$

Here d = 25 ft and h = 12/12 = 1 foot.

$$S = \frac{2.73(25)}{\sqrt{1}} = 68.25 \text{ mph.}$$

Example 2.15: *[Downhill takeoff with a takeoff angle greater than 10%] A vehicle drives down a 30 degree slope and then goes airborne, traveling 25 ft horizontally while falling 10 feet. Determine the vehicle speed when it becomes airborne.*

Solution: Here d = 25 ft, h = 10 ft, and θ = 30 degrees.

$$S = \frac{2.73d}{\cos\theta\sqrt{h - d\tan\theta}} = \frac{2.73(25)}{\cos 30^\circ \sqrt{10 - 25\tan 30^\circ}} = 83.05 \text{ mph.}$$

Example 2.16: *[Vehicle lands higher than takeoff] A vehicle goes airborne off a 25 degrees slope and travels 50 ft horizontally. The vehicle lands 2 ft higher than takeoff. Determine the speed at takeoff.*

Solution: Here d = 50 ft, h = 2 ft (higher), and θ = 25 degrees.

$$S = \frac{2.73\,d}{\sqrt{d\sin\theta\cos\theta - h\cos^2\theta}} = \frac{2.73(50)}{\sqrt{50\sin 25°\cos 25° - 2(\cos 25°)^2}}$$

$$= 32.62 \text{ mph.}$$

Example 2.17: *Test skids were done on a level road and gave a drag factor of 0.70 for the road surface. The drag factor for the shoulder was calculated as 0.50. The accident skid is on a 6% grade. The accident vehicle's right front wheel, which is on the shoulder, is not working. The accident skid distance is 150 ft. How fast was the vehicle traveling?*

Solution: Here $f_1 = 0.70, f_2 = 0.50, n_1 = 50\%$ (one front and one rear wheel braking), $n_2 = 20\%$ (one rear wheel braking), $m = +0.06$, and $d = 150$ ft.

$$S = \sqrt{30[(f_1 n_1 + f_2 n_2) \pm m]d}$$

$$= \sqrt{30[(0.70(0.50) + 0.50(0.20)) + 0.06]150} = 47.91 \text{ mph.}$$

Example 2.18 *[Slide-to-stop with percentage of braking and grade] A vehicle skids uphill to a stop on a +9% grade. The skidding distance is 150 ft. Only the front wheels brake. The level road drag factor is 0.70. How fast was the vehicle traveling?*

Solution: Here $d = 150$ ft, $f = 0.70$, $n = 0.60$ and $m = +0.09$.

$$S = \sqrt{30d(fn + m)} = \sqrt{30(150)[0.70(0.60) + 0.09]} = 47.91 \text{ mph}$$

2.14 LANE PASSING WITH CHANGING VELOCITY

The lane passing with changing velocity by a vehicle is shown schematically in Fig. 2.19.

Notation:

a_1 = acceleration of vehicle during first lane change, ft/sec^2

a_2 = acceleration during the second phase, the actual passing segment while in the passing lane, ft/sec^2

a_3 = vehicle acceleration during return lane change, ft/sec^2

a_y = acceleration of the vehicle before start of lane change, ft/sec^2

D = total passing distance, ft

D_1 = distance of the first lane change, ft

D_2 = distance required for the actual passing phase, ft

D_3 = distance required for the return lane change, ft

l = vehicle length, ft

l_1 = first lane length, ft

l_2 = actual passing lane length, ft

l_3 = return lane length, ft

t_2 = acceleration time while passing lane (second phase), sec

V_0 = initial speed of passing vehicle and speed of vehicle being passed, ft/sec

V_1 = speed of passing vehicle when first lane change is completed, ft/sec

V_2 = passing vehicle speed where passing phase is completed, ft/sec

V_3 = passing vehicle speed at the completion of the return lane change, ft/sec

The passing maneuver distances are obtained by calculations in three phases: (*a*) lane change phase, (*b*) passing phase and (*c*) return lane phase. Knowing the accelerations of the passing vehicle in the three phases, the velocity at the end of each passing segment and the distances for each phase can be determined. Here, it is assumed that both vehicles travel at the same speed before starting of passing.

The lane passing with changing velocity by a vehicle is shown schematically in Fig. 2.19.

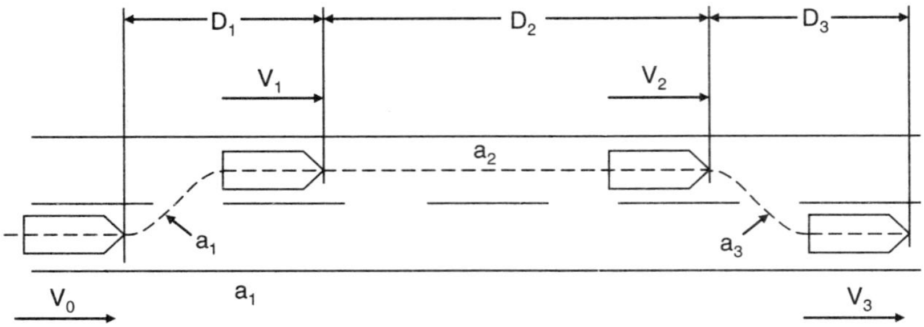

Fig. 2.19 Lane passing with acceleration.

The distance of the first lane change is given by

$$D_1 = 0.45\sqrt{\frac{l}{a_y}}\left[\frac{V_0 + V_1}{2}\right] \text{ ft} \tag{2.121}$$

The speed V_1 is given by

$$V_1 = 0.45a_1\sqrt{\frac{l}{a_y}} + V_0 \text{ ft/sec} \tag{2.122}$$

The distance D_2 required for the actual passing phase is

$$D_2 = \left[\frac{l_1 + l_2 + l_3}{\frac{V_1 + V_2}{2} - V_0}\right]\left[\frac{V_1 + V_2}{2}\right] \text{ft/sec} \tag{2.123}$$

The speed V_2 is given by

$$V_2 = \sqrt{V_1^2 + 2a_2D_2} \quad \text{ft/sec} \tag{2.124}$$

The speed V_2 can also be obtained from

$$V_2 = V_1 + a_2t_2 \text{ ft/sec} \tag{2.125}$$

Eqs. (2.123), (2.124), and (2.125) are solved simultaneously by trial and error in order to obtain the motion analysis of the passing phase.

The distance D_3 required for the return lane is given by

$$D_3 = 0.45\sqrt{\frac{l}{a_y}}\left[\frac{V_2 + V_3}{2}\right] \quad \text{ft} \tag{2.126}$$

The speed V_3 is obtained from

$$V_3 = 0.45 a_3 \sqrt{\frac{l}{a_y}} + V_2 \text{ ft/sec} \tag{2.127}$$

Example 2.19: *Compute the passing maneuver distance and time for the following given data: $V_0 = 25$ mph (36.625 ft/sec), $l_1 = 18$ ft, $l_2 = 65$ ft, $l_3 = 45$ ft, $a_1 = 6$ ft/sec², $a_2 = 4$ ft/sec², $a_3 = 0$, $l = 12$ ft and $a_y = 0.22$ g.*

Solution:

$$l_1 + l_2 + l_3 = 18 + 65 + 45 = 128 \text{ ft}$$

The speed at the end of the first lane change is given by Eq. (2.122).

$$V_1 = 0.45 a_1 \sqrt{\frac{l}{a_y}} + V_0 \text{ ft/sec}$$

$$V_1 = (0.45)(6)\sqrt{\frac{12}{0.22}} + 36.625 = 56.57 \text{ ft/sec}$$

The distance of the first lane change is obtained from Eq. (2.121).

$$D_1 = 0.45\sqrt{\frac{l}{a_y}}\left[\frac{V_0 + V_1}{2}\right] \text{ ft}$$

$$= 0.45\sqrt{\frac{12}{0.22}}\left[\frac{36.325 + 56.57}{2}\right] = 162.42 \text{ ft}$$

The distance needed for the actual passing phase is computed initially by assuming a value for V_2 in Eq. (2.123) and then computing it with Eq. (2.124).

Assume a value of $V_2 = 75$ ft/sec. Now Eq. (2.123) gives

$$D_2 = \left[\frac{l_1 + l_2 + l_3}{\frac{V_1 + V_2}{2} - V_0}\right]\left[\frac{V_1 + V_2}{2}\right] \text{ ft}$$

$$= \left[\frac{128}{\left[\frac{56.57 + 75}{2}\right] - 36.625}\right]\left[\frac{56.57 + 75}{2}\right] = 288.768 \text{ ft}$$

Now, we compute V_2 using Eq. (2.124)

$$V_2 = \sqrt{V_1^2 + 2a_2 D_2} = \sqrt{56.57^2 + 2(4)(288.768)} = 74.23 \text{ ft/sec.}$$

Therefore, the computed value of $V_2 = 74.23$ ft/sec is acceptable. The velocity V_3 at the completion of the return lane change maneuver is given by Eq. (2.127).

$$V_3 = 0.45 a_3 \sqrt{\frac{l}{a_y}} + V_2 \text{ ft/sec}$$

$$= 0.45(0)\sqrt{\frac{12}{0.22}} + 75 = 75 \text{ ft/sec.}$$

The return lane change distance is given by Eq. (2.126)

$$D_3 = 0.45\sqrt{\frac{l}{a_y}}\left[\frac{V_2+V_3}{2}\right] \text{ ft}$$

$$= 0.45\sqrt{\frac{12}{0.22}}[75] = 249.26 \text{ ft.}$$

The total passing distance is

$$D = D_1 + D_2 + D_3 = 162.42 + 288.768 + 249.26 = 700.45 \text{ ft.}$$

The time required for the complete passing maneuver is computed from $t = \dfrac{2S}{V_0 + V_t}$ for each segment. The time Δt_2 for the second phase (the actual passing segment) is

$$\Delta t_2 = \frac{2D_2}{V_1+V_2} = \frac{2(288.768)}{(56.57+75)} = 4.39 \text{ sec.}$$

Similarly,

$$\Delta t_1 = \frac{2D_1}{V_0+V_1} = \frac{2(162.42)}{(36.625+56.57)} = 3.48 \text{ sec.}$$

and

$$\Delta t_3 = \frac{2D_3}{V_2+V_3} = \frac{2(249.26)}{(74.23+75)} = 3.34 \text{ sec.}$$

The total time required for the passing maneuver is

$$\Delta t_1 + \Delta t_2 + \Delta t_3 = 4.39 + 3.48 + 3.34 = 11.21 \text{ sec.}$$

2.15 LANE CHANGE MANEUVER MODEL

Araszewski, Toor, Overgoard and Johal [2002] developed a lane change model for determining the expected vehicle dynamics including lateral, longitudinal and angular vehicle displacement during the lane change maneuver. This model is based on the vehicle speed, peak and average accelerations, and driver steering input. The model uses vehicle handling parameters during steady state handling conditions by incorporating tire, suspension, and handling characteristics.

A vehicle may perform a range of varying vehicle speeds, accelerations, and corresponding times over a range of lateral and longitudinal distances. Ultimately, the parameters of the lane change maneuver will be dependent on the driver input and specifically how the steering wheel is turned during a maneuver. Therefore, there is a significant human factor involved in a lane change analysis.

Considering the lateral lane change distance that corresponds to a constant vehicle speed and the average vehicle acceleration over the first half and latter half of the lane change maneuver derived the lane change model. This resulted in formulae for the longitudinal lane change distance and the lane change time as follows:

$$d = 2V\sqrt{\frac{S}{a_{\text{AVG}}}} \tag{2.128}$$

$$t = 2\sqrt{\frac{S}{a_{\text{AVG}}}} \tag{2.129}$$

During a real world lane change, the acceleration profile varies and is not constant. Since the largest unknown factor of the equations is the acceleration profile and thus the effective average lateral acceleration during a lane change maneuver, the model was adapted to consider different acceleration profiles. Two of the considered acceleration profiles included those obtained from a sinusoidal and triangular steer input by the vehicle operator. These adjustments allowed the peak acceleration to be an input parameter such that a theoretical upper limit could be obtained. For a sinusoidal steer input, the lane change longitudinal distance is given as follows:

$$d_{\text{Sine}} = V\sqrt{\frac{2\pi s}{a_{\text{PEAK}}}} \tag{2.130}$$

where a_{PEAK} is the peak lateral acceleration in [m/s^2].

For a triangular steer input, the lane change longitudinal distance is given by

$$d_{\text{Tri}} = 2V\sqrt{\frac{2s}{a_{\text{PEAK}}}} \tag{2.131}$$

The vehicle angle during the lane change maneuver is also derived here. The course angle may be expressed as the sum of the heading angle due to the travel path and the sideslip angle

$$\Theta = \Psi + \beta \tag{2.132}$$

where Θ = course angle
Ψ = heading angle
β = sideslip angle

The heading angle component derivation first requires a theoretical average radius during the lane change maneuver to be calculated. The theoretical average radius has not real world or practical reference, but is a useful analysis too. The theoretical average radius is given by

$$R_{\text{AVG}} = \frac{V^2}{a_{\text{AVG}}} \tag{2.133}$$

Using a geometrical relation for the ratio between the traveled distance over half a lane change and the average turn radius, the heading angle at the middle point of the lane change may be determined as follows:

$$\Psi_{\text{Middle}} = \left(\frac{180^\circ}{\pi}\right)\frac{Vt}{R_{\text{AVG}}} \tag{2.134}$$

For a symmetrical lane change maneuver, the maximum course angle occurs halfway through the lane change. At this point, the lateral acceleration is nominally zero; therefore, the sideslip effects are absent. For a sinusoidal steer input during the lane change maneuver, the maximum course angle is given as

$$\Theta_{\text{Sine(MAX)}} = \left(\frac{180^\circ}{\pi^2}\right)\frac{a_{\text{PEAK}}t}{V} \tag{2.135}$$

For a triangular steer input, the maximum course angle is given by

$$\Theta_{\text{Tri(MAX)}} = \left(\frac{45^\circ}{\pi}\right)\frac{a_{\text{PEAK}}t}{V} \tag{2.136}$$

The maximum sideslip angle will occur when the peak lateral acceleration is experienced. For a symmetrical lane change maneuver, this will occur at the quarter and three quarter points of the lane change. At these points, the heading angle will have an absolute value of half the maximum course angle and the maximum sideslip angle is defined as

$$\beta_{MAX} = a_{PEAK}\left[k_\beta \frac{b}{V^2}\right] \tag{2.137}$$

For most accident reconstruction investigations, further analyses of the input parameters may not be required. However, in some cases the steering angle for a lane change maneuver may be useful. This allows the required driver input to be calculated and assessed. The evaluation of the steering wheel input requires the knowledge of the steering wheel to tire angle ratio for the specific vehicle and the under steer gradient of the vehicle to be known. Although the steering wheel to vehicle angle ratio is often part of the vehicle specification data list, the under steer gradient is not readily available.

Additional relationships were derived for the peak steering angle required by the vehicle operator to conduct a lane change maneuver. The peak steering angle occurs when the lateral acceleration is also a peak value. Therefore, the peak steering angle is given by

$$\delta_{SW(MAX)} = G\left[k_U a_{PEAK} + \arctan\left(\frac{a_{PEAK} L}{V^2}\right)\right] \tag{2.138}$$

Other formulae that may be useful for accident reconstruction purposes are the lateral acceleration profile, the longitudinal and lateral displacement profile, and the steering angle profile. These theoretical relationships are presented as follows:

Time: t_i = instantaneous time

t_{Tot} = total lane change maneuver time.

Instantaneous lateral acceleration:

$$a_i(\text{sne}) = a_{PEAK}\left[\sin\left(\frac{2\pi t_i}{t_{TOT}}\right)\right] \quad \text{sinusoidal steer input } (a_{AVG} = 0.64 a_{PEAK}) \tag{2.139}$$

$$a_{i(\text{Tri})} = \begin{cases} a_{PEAK}\left[\dfrac{4t_i}{t_{TOT}}\right] & 0 \le t_i < \dfrac{1}{4}t_{TOT} \\ a_{PEAK}\left[2 - \dfrac{4t_i}{t_{TOT}}\right] \quad \text{for} & \dfrac{1}{4}t_{TOT} \le t_i < \dfrac{3}{4}t_{TOT} \\ a_{PEAK}\left[\dfrac{4t_i}{t_{TOT}} - 4\right] & \dfrac{3}{4}t_{TOT} \le t_i \le t_{TOT} \end{cases} \tag{2.140}$$

triangular steer input ($a_{AVG} = 0.5 a_{PEAK}$).

where a_{PEAK} = peak lateral acceleration

a_i = instantaneous lateral acceleration

Instantaneous turn radius is

$$r_i = \frac{V_2}{a_i} \tag{2.141}$$

Instantaneous steer angle:

$$\delta_{Ki} = \arctan\left(\frac{L}{r_i}\right) \quad \text{where } L \text{ is the wheelbase.}$$

$$\delta_{Ui} = K_U a_i \quad \text{where } k_U \text{ is the under-steer gradient} \tag{2.142}$$

$$\delta_i = \delta_{Ki} + \delta_{Ui}$$

Instantaneous steering wheel angle:

$$\delta_{SWi} = G\delta_i \quad \text{where } G \text{ is the steering to tire angle ratio} \tag{2.143}$$

Longitudinal displacement: Step-wise incremental calculation

$$d_{\text{long}} = d_{Vi} = d_{V(i-1)} + V[t_i - t_{(i-1)}]\cos(\psi_i) \tag{2.144}$$

Lateral displacement: Step-wise incremental calculation

$$d_{\text{lat}} = d_{Xi} = d_{X(i-1)} + V[t_i - t_{(i-1)}]\sin(\Psi_i) \tag{2.145}$$

Instantaneous heading angle: Step-wise incremental calculation

$$\Psi_i = \Psi_{(i-1)} + \left(\frac{180^\circ}{\pi}\right)\frac{V\left|t_i - t_{(i-1)}\right|}{r_i} \tag{2.146}$$

The staged lane change tests resulted in limited data over a range of vehicle speeds, lateral lane change distances, longitudinal lane change distances and lane change times. Correspondingly, the peak and average accelerations of the staged lane change maneuvers also varied. The acceleration curve during the lane change maneuver was a direct relation of the steer input angle of the vehicle operator. Thus, as expected, the resulting acceleration profile varied from run to run and significantly varied from operator to operator. This was observed to be a factor of the driving style of the operator. Assessment of the steering wheel angle during the lane change maneuver suggests that the typical lane change maneuver likely falls within the range of results defined by the sinusoidal and triangular steer input. However, this assessment was made on a limited number of staged tests with a limited number of vehicle operators.

The test results were compared to the Limpert formulae (1994) that define the lane change distance and time in terms of average lateral acceleration. The Limpert formulae are summarized here:

Limpert (Imperial):

$$d_{[ft]} = 0.458\, V_{[\text{ft/sec}]}\sqrt{\frac{s_{[ft]}}{a_{[g]}}}$$

$$d_{[ft]} = \frac{d_{[m]}}{0.3048}$$

$$s_{[ft]} = \frac{s_{[m]}}{0.3048}$$

$$V_{[\text{ft/sec}]} = \frac{V_{[\text{m/s}]}}{0.3048} \tag{2.147}$$

$$a[g] = \frac{a_{[\text{m/s}^2]}}{9.81}$$

$$\left(\frac{d_{[m]}}{0.3048}\right) = 0.458\left(\frac{V_{[m/s]}}{0.3048}\right)\sqrt{\frac{\left(\frac{s_{[m]}}{0.3048}\right)}{\left[\frac{a_{[m/s^2]}}{9.81}\right]}}$$

Limpert (Metric):

$$\Rightarrow \qquad d_{[m]} = 2.60\,V_{[m/s]}\sqrt{\frac{s_{[m]}}{a_{[m/s^2]}}} \tag{2.148}$$

where d = longitudinal lane change distance
V = speed
s = lateral lane change distance
a = average lateral acceleration

Comparison of the staged lane change maneuvers with the formulae presented in the study by Limpert [1989], suggests that the Limpert formulae consistently over-estimates the lane change time and lane change distance. This resulted in an average over-estimate of about 29% when the Limpert formula was applied using the average acceleration as recorded during the lane change maneuver.

The formulae provided by Daily [1988] were also used in a comparison with the staged lane change maneuvers. The Daily formulae assume a constant turn radius and therefore a constant lateral acceleration for each half of the lane change. The Daily formulae are given here:

Daily (Imperial):

$$d_{[ft]} = 0.732V_{[mph]}\sqrt{\frac{m_{[ft]}}{a_{[g]}}}$$

$$m_{[ft]} = \left[\frac{1}{2}\right]s_{[ft]} = \left[\frac{1}{2}\right]\frac{s_{[m]}}{0.3048}$$

$$d_{[ft]} = \frac{d[m]}{0.3048}$$

$$V_{[mph]} = \frac{V_{[km/h]}}{1.609} = \left(\frac{3.6}{1.609}\right)V_{[m/s]} \tag{2.149}$$

$$a_{[g]} = \frac{a_{[m/s^2]}}{9.81}$$

$$\left(\frac{d_{[m]}}{0.3048}\right) = 0.732\left(\frac{3.6}{1.609}\right)V_{[m/s]}\sqrt{\frac{\left[\frac{1}{2}\right]\left(\frac{s_{[m]}}{0.3048}\right)}{\left(\frac{a_{[m/s^2]}}{9.81}\right)}}$$

Limpert (Metric):

$$\Rightarrow \qquad d_{[m]} = 2.00\, V_{[\mathrm{m/s}]} \sqrt{\frac{s_{[m]}}{a_{[\mathrm{m/s^2}]}}} \qquad (2.150)$$

where d = longitudinal lane change distance.
V = speed.
m = half of the lateral lane change distance.
a = average lateral acceleration.

It is noted that although the Daily formula accurately predicts the lane change distance and time if the average acceleration is known, it does not accurately model the travel path, peak acceleration, or the acceleration profile. Due to the constant turn radius model, the vehicle path may deviate as much as about 30% at the quarter point of the full lane change maneuver. It is also noted that the Daily formula is likely not accurate if peak or threshold acceleration values are used.

As expected, the average acceleration of the tests could be used to accurately evaluate the lane change distance, time, and angle of the test vehicle; the test results were predicted with nominally zero percentage average difference, and a standard deviation of about 5%. However, for most applications, the peak lateral acceleration assessment of the collision events.

The application of the peak acceleration values resulted in calculated value that were near the actual value. The sinusoidal input condition yielded results that were on average about 5% below the actual value, with a standard deviation of about 6%. The triangular input condition yielded results that were on average about 7% above the actual value, with a standard deviation of about 7%. Therefore, the sinusoidal and triangular models appear to provide an envelope of the expected results. A significant portion of the resulting deviation in the calculated values likely originated from the operator input effect and the fact that the input was not perfectly symmetrical or smooth. However, this effect would likely also exist in real world driving conditions.

Evaluation of the test data results yielded empirical coefficients to be determined. Considering the peak acceleration values generated during the lane change maneuvers, the empirical coefficient was determined to be 2.7, with a standard deviation of 0.2. Considering the average acceleration values generated during the lane change maneuvers, the empirical coefficient was determined to be 2.0, with a standard deviation of 0.1. Therefore, the peak and average accelerations can be used in the following empirical relations for a lane change maneuver.

$$d = 2.7V\sqrt{\frac{s}{a_{\mathrm{PEAK}}}} \qquad (2.151)$$

where a_{AVG} = average acceleration in [m/s^2].
a_{PEAK} = peak acceleration in [m/s^2].

The ratio of the average acceleration to the peak acceleration during the test was about 0.58, with a standard deviation of about 0.04. Triangular and sinusoidal steering inputs result in average to peak acceleration ratios of 1.5 and about 0.64, respectively. This implies that the typical lane change maneuver was conducted with a steer input in between a sinusoidal and triangular steer input.

The formulae presented here allow manual calculations to be conducted for lane change modeling. Alternatively, similar results may be possible if a validated simulation program is available and if data is correctly applied.

The models presented here appear to model the lane change maneuver with adequate accuracy and more completely define the lane change maneuver as compared to other manual techniques.

2.16 ENGINEERING MODELS AND ANIMATIONS

Fay, Robinette and Larson (1988) have emphasized the importance of engineering models and animations in vehicular accident studies. Vehicle accident analysis relies heavily on mathematics and the principles of conservation of energy and momentum and Newton's laws of motion. In order to apply these principles, it is first necessary to know the approximate vehicle motions. The analytical procedure is interactive using a combination of model analysis to determine linear and angular velocities and accelerations. Scale accident scene models combined with aerial photography to enhance realism have been extensively utilized in evaluation, analysis and presentation of vehicular accident reconstructions to non technical audience. Slide and video accident animations have been produced directly from aerial photograph enhanced models and have been used successfully in courtroom presentations since the 1970's.

2.16.1 Function of Accident Scene Models

Motions, which occur in vehicle accidents, are sometimes extremely complex because they involve collisions between vehicles, collisions between vehicles and fixed structures, collisions between vehicles and pedestrians, and sometimes include vaulting, flipping and rolling. Finding out what happened in a vehicle accident requires an engineering analysis of vehicle dynamics and kinematics and frequently an additional analysis of the dynamics and kinematics of vehicle occupant and pedestrians. Models are useful at all phases of the engineering analysts and have an important role in presenting the engineering results in ways that are easy for people to understand. Actual vehicle impacts have been conducted with scale models to study vehicle motions and interactive forces.

A reconstructionist will usually have to expend a significant effort studying the physical evidence such as accident scene photos, the accident site, the vehicles and driver and witness statements, before the mathematical engineering analysis can be started. One of the most important steps is the proper evaluation and interpretation of the information available. Models have an important role when evaluating this information.

The base model for vehicle accident reconstruction is a model of the accident scene. The simplest possible model is a line drawing of a section of roadway or an intersection. The use of an aerial photograph is preferred over the line drawing because of the introduction of realism, detail, relative scale and credibility. The aerial photograph can be enlarged to an appropriate scale for use with commercially available scaled vehicle models. If the accident scene is flat, the aerial photograph can be mounted on a flat surface. Surroundings landmarks such as trees, bushes, buildings and traffic control devices are all directly displayed in the photo. Superimposing three-dimensional models at the appropriate locations can highlight any of these. If the accident scene is hilly or if specific elevation changes are important to the analysis or need to be emphasized, a scale three-dimensional terrain model can be constructed from an accident scene engineering survey. The aerial photograph can be superimposed over all or parts of the three-dimensional terrain base. Frequently, it is sufficient to mount only the road surface from the aerial photograph onto a plaster of paris model and construct the surrounding landscape with commercially available model supplies, or plant.

If possible, the aerial photograph is flown shortly after the accident to show vehicle skid marks, gouges, furrow marks, fluid strains, etc. Usually there is a time delay between the accident and the aerial photograph and the reconstructionist will have to superimpose this information of the accident onto the model. Most experienced reconstructionists have developed techniques and special equipment to measure or reconstruct such information, which can then be plotted onto the model. In addition, there are special tools and computer programs commercially available to assist with this process.

2.16.2 Model Application

The first step in analyzing a vehicular accident is to catalog the available information about the accident and then to superimpose or plot this information on the scene model and the vehicle models. In some cases, features of the terrain influence vehicle motion. Sometimes full-size models are used in accident studies.

Most accidents can be split into three phases: pre-impact, impact and post-impact. Each phase is analyzed separately and collectively. This activity is accomplished with the assistance of accident reconstruction software.

2.16.3 Reconstruction Animations

Accurate physical and mathematical construction of the accident is of critical importance to the reconstructionist. Of equal importance is the process of communicating the reconstruction to a non-technical client or jury in such a manner that they can comprehend it. This step is frequently a central part of the litigation team effort in which the attorney must rely upon the reconstructionist to present and defend the reconstruction. The model is an important demonstrative tool in effecting this goal. A logical extension of the model is a visual display of the vehicle motions presented over small increments of time and edited together into a smooth animation. Aerial photograph enhanced models were first used in the mid 1970's. Video animations of the reconstruction in real time have been used since the early 1980's.

To produce an animation, the motions of the vehicles and occupants are determined by computer analysis and plotted on the model. The vehicles are moved to the plotted positions and photographed with video equipment. The frames are edited together to produce a smooth running animation, which shows the accident in real time. A sequence of slides can be made with fewer positions to produce a step view' of the accident.

Introduction of terrain elevation changes and vehicle rollover motions can significantly increase the complexity of the mathematical motion analysis. Consider the example of a single vehicle which left a divided highway and rolled down a hill high embankment. The vehicle exhibits multiple types of motion; linear displacement, yawing, vehicle rollover and vaulting down the embankment. The mathematic motion analysis becomes three-dimensional with multiple degrees of freedom constrained by the physical evidence associated with the accident, *i.e.* skid marks, points of embankment, rollover path, location and position of the impacts between the vehicle and the ground and the final point and orientation at rest. Evaluation and identification of these constraints constitute a major effort in itself and employ research data.

Summarizing, engineering models are useful in understanding the motions of vehicles during accident sequences and are an important part of the interactive analysis, which involves accident data, and the principles of engineering and physics. The aerial photographs and models are useful in producing slide sequences and video animations, which portray realistic and physically accurate vehicle motions.

2.17 EXAMPLE PROBLEMS AND SOLUTIONS

Example P. 2.1: *From test skids, an investigator determines that the drag factor of a roadway surface is 0.65. Using the force required formula, he calculates the force that will be required to pull his drag sled weighing 30 lbs (13.6 kg) over the same roadway surface. The required force is 19.5 lbs (8.8 kg). He then conducts test pulls with the drag sled until the force consistently registers 19.5 lbs (8.8 kg) on the scale. Verify whether the test pulls were properly done.*

Solution:

$F = W \times f$

F = force required to move unit at a constant speed

W = weigh of unit

F = drag factor

$F = 30 \times 0.65 = 19.5$ lbs U.S.

$F = 13.6 \times 0.65 = 8.8$ kg S.I.

Example P. 2.2: *A vehicle skidded on a roadway having a minus (–) 0.06% grade. Drag factor tests conducted on a nearby level surface yielded a drag factor (f) value, of 0.62 for the level surface. Determine the actual or adjusted drag factor at the place where the accident vehicle skidded.*

Solution: $f = 0.62 - 0.06 = 0.56$

Example P. 2.3: *A vehicle skidded on a roadway having an upgrade (+) of 0.06%. Drag factor tests conducted on a nearby level surface yielded a drag factor (f) value of 0.62 for the level surface. Determine the actual or adjusted drag factor at the place where the accident vehicle skidded.*

Solution: $f = 0.62 + 0.06 = 0.68$

Example P. 2.4: *A vehicle skidded 164 ft (50 m) to a stop on an asphalt surface having a drag factor (f) value of 0.70. Determine the vehicle's minimum speed at the beginning of the skid mark.*

Solution:

U.S.	*S.I.*
$S = \sqrt{30 \times D \times f}$	$S = \sqrt{254 \times D \times f}$
$S = \sqrt{30 \times 164 \times 0.70}$	$S = \sqrt{254 \times 50 \times 0.70}$
$S = \sqrt{3444}$	$S = \sqrt{8890}$
$S = 58$(mph)	$S = 94$(km/h)

Example P. 2.5: *A vehicle skidded 95 ft (29 m) on a concrete surface having a drag factor (f) value of 0.80 and collided with a utility pole, which brought the vehicle to a stop. The impact was great enough to cause extensive damage to both the pole and the vehicle. Calculate the actual minimum speed at the beginning of the skid marks.*

Solution:

U.S.	*S.I.*
$S = \sqrt{30 \times D \times f}$	$S = \sqrt{254 \times D \times f}$
$S = \sqrt{30 \times 95 \times 0.80}$	$S = \sqrt{254 \times 29 \times 0.80}$
$S = \sqrt{2280}$	$S = \sqrt{5893}$
$S = 47$(mph)	$S = 76$(km/h)

Example P. 2.6: *A driver applied the brakes, causing his car to skid to a stop. The car had brakes on only front wheels. An investigation revealed the following information:*

Skid distance = 110 ft (33.5 m) on level surface.

Drag factor = 0.80 on level surface.

Weight carried on front wheels during braking action = 65 % of total vehicle weight.

Calculate the actual minimum speed at the beginning of the skid marks.

Solution:

U.S.	*S.I.*
$S = \sqrt{30 \times D \times f \times n}$	$S = \sqrt{254 \times D \times f \times n}$
$S = \sqrt{30 \times 110 \times 0.80 \times 0.65}$	$S = \sqrt{254 \times 33.5 \times 0.80 \times 0.65}$
$S = \sqrt{1716}$	$S = \sqrt{4425}$
$S = 41(\text{mph})$	$S = 66(\text{km/h})$

Example P. 2.7: *A car with braking on all four wheels was towing a utility trailer that was not equipped with brakes. The car weighed 3200 lbs (1450 kg). The car and trailer combined weighed 4500 lbs (2040 kg). In an attempt to avoid an accident, the driver applied the brakes of the car. An investigation revealed the following:*

Skid distance = 60 ft (18m); Drag factor = 0.65. Calculate the actual minimum speed at the beginning of the skid marks.

Solution: The n value can be calculated as:

$$n = 3200/4500 = 0.711$$

U.S.	*S.I.*
$S = \sqrt{30 \times D \times f \times n}$	$S = \sqrt{254 \times D \times f \times n}$
$S = \sqrt{30 \times 60 \times 0.65 \times 0.71}$	$S = \sqrt{254 \times 18 \times 0.65 \times 0.71}$
$S = \sqrt{821}$	$S = \sqrt{2110}$
$S = 28(\text{mph})$	$S = 46(\text{km/h})$

Example P. 2.8: *A motorcycle operator caused his motorcycle to skid to a stop by applying only the rear wheel brake. Investigation revealed the following:*

Skid distance = 65 ft (20m)

Drag factor = 0.7

Combined weight of motorcycle and rider static = 850 lbs (386 kg)

Weight carried on front wheel (static) = 350 lbs (159 kg)

Weight carried on rear wheel (static) = 500 lbs (227 kg)

20 % forward shift during braking = 100 lbs (45.4 kg)

Total weight on rear wheel during braking action = 500 – 100 = 400 lbs (182 kg)

Determine the speed.

Solution: The n value can be calculated as:

$$n = 400/850 = 0.47$$

U.S.	*S.I.*
$S = \sqrt{30 \times D \times f \times n}$	$S = \sqrt{254 \times D \times f \times n}$
$S = \sqrt{30 \times 65 \times 0.7 \times 0.47}$	$S = \sqrt{254 \times 20 \times 0.7 \times 0.47}$
$S = \sqrt{641}$	$S = \sqrt{1671}$
$S = 25(\text{mph})$	$S = 41(\text{km/h})$

Example P. 2.9: *A vehicle skidded to a stop on a roadway. The first part of the skid was on an asphalt surface. The second part was on a concrete bridge surface. An investigation provided the following information:*

Drag factor (f1) for the asphalt surface = 0.65
Drag factor (f2) for the concrete surface = 0.85
Length of skid (D1) on the asphalt surface = 70 ft (21 m)
Length of skid (D2) on the concrete surface = 78 ft (24 m)

Determine the speed.

Solution:

U.S.	*S.I.*
$S = \sqrt{30 \times (D1 \times f1 + D2 \times f2)}$	$S = \sqrt{254 \times (D1 \times f1 + D2 \times f2)}$
$S = \sqrt{30 \times (70 \times 0.65 + 78 \times 0.85)}$	$S = \sqrt{254 \times (21 \times 0.65 + 24 \times 0.85)}$
$S = \sqrt{3405}$	$S = \sqrt{8484}$
$S = 58(\text{mph})$	$S = 93(\text{km/h})$

Example P. 2.10: *In an attempt to avoid an accident, the driver applied the brakes of his vehicle and skidded to a stop. During the entire skid, the vehicle's left wheels (Side 1) were on an asphalt surface and its right wheels (Side 2) were on a gravel surface.*

The investigation revealed:

Length of skid = 100 ft (30m)

Drag factor for Side 1 (asphalt surface) = 0.70

Drag factor for Side 2 (gravel surface) = 0.4

Determine the speed.

Solution:

U.S.	*S.I.*
$S = \sqrt{15 \times (f1 + f2) \times D}$	$S = \sqrt{127 \times (f1 + f2) \times D}$
$S = \sqrt{15 \times (0.7 + 0.4) \times 100}$	$S = \sqrt{127 \times (0.7 + 0.4) \times 30}$
$S = \sqrt{1650}$	$S = 4200$
$S = 41(\text{mph})$	$S = 65(\text{km/h})$

Example P. 2.11: *A pedestrian crossing a roadway was struck by and carried on the hood of a car. The driver of the car applied the brakes of the vehicle. This sudden vehicle deceleration caused the body disengage the hood, fall onto the roadway and slide along the roadway surface. The investigator was able to locate the beginning and the end of the slide. To duplicate the slide taken by the body, the investigator used a spring scale to pull an assistant along the roadway. The assistant was clothed in material similar to that worn by the accident victim. The investigation revealed the following information:*

Distance pedestrian's body slid across the roadway = 125 ft (38 m)

Average force (pull) to slide the assistant along the roadway F = 125 lbs (57 kg)

Assistant's weight W = 175 lbs (79.4 kg)

Determine the speed.

Solution: From the $F = f \times W$ formula $f = F/W$ (drag factor)

$$f = 125/175$$
$$f = 0.71$$

U.S.	*S.I.*
$S = \sqrt{254 \times D \times f}$	$S = \sqrt{254 \times D \times f}$
$S = \sqrt{30 \times 125 \times 0.71}$	$S = \sqrt{254 \times 38 \times 0.71}$
$S = \sqrt{2662}$	$S = \sqrt{6853}$
$S = 52(\text{mph})$	$S = 83(\text{km/h})$

Example P. 2.12: *A vehicle traveled into a curve at a high rate of speed and side slipped, leaving a yaw mark which measured 320 ft (96 m) in length. The first one-third (100 ft or 30 m) of the yaw mark was used and a chord 80 ft (24 m) and a middle ordinate of 6 ft (2m) (M) were measured. A 0.75 (f) drag factor was obtained by conducting test skids on a nearby level section of the roadway. The curve, however, had a super elevation of +05% (e), so that an adjustment to the drag factor obtained by test skids was required. Determine the critical vehicle curve speed from the yaw marks.*

Solution: Radius calculation

R = radius

C = Chord

M = middle ordinate

3.86 and 11.27 constants

$$R = \frac{C^2}{8M} + \frac{M}{2}$$

$$R = \frac{80^2}{8 \times 6} + \frac{6}{2}$$

U.S.	*S.I.*
$R = 135(ft)$	$S = 41(m)$

Critical vehicle curve speed from the yaw mark:

$S = 3.86\sqrt{R \times (f \pm e)}$	$S = 11.27\sqrt{R \times (f \pm e)}$
$S = 3.86\sqrt{135 \times (0.75 + 0.05)}$	$S = 11.27\sqrt{41 \times (0.75 + 0.05)}$

$S = 3.86\sqrt{108}$ $\qquad$ $S = 11.27\sqrt{33}$

$S = 40(\text{mph})$ $\qquad$ $S = 65(\text{km/h})$

Example P. 2.13: *A vehicle went into a curve at a high rate of speed; side slipped and went off the roadway. The yaw mark was not adequate for obtaining the radius needed to make a calculation. The investigator therefore measured a chord between the points on the center line of the vehicle's proper lane of travel and calculated a radius of 220 ft (67 m). The roadway had a drag factor of 0.75 and super elevation of 5%. Calculate the critical curve speed.*

Solution:

U.S. $\qquad$ *S.I.*

$$S = 3.86\sqrt{R\times(f \pm e)} \qquad S = 11.27\sqrt{R\times(f \pm e)}$$

$$S = 3.86\sqrt{220\times(0.75+0.05)} \qquad S = 11.27\sqrt{67\times(0.75+0.05)}$$

$$S = 3.86\sqrt{176} \qquad S = 11.27\sqrt{53.6}$$

$$S = 51(\text{mph}) \qquad S = 82(\text{km/h})$$

Example P. 2.14: *In Problem 2.13 it was possible to use the center line of the proper lane of travel and determine the radius. If, however, this had not been possible, the investigator could have measured a chord between two points on the right-hand edge (A) of the roadway, resulting in a right-hand edge radius of 214 ft (65 m). The radius of the center of the lane of travel could have been calculated indirectly by adding half the 12-foot width of the lane of travel to that radius.*

Solution:

$$R = 214 + 6 = 220 \text{ (ft)}$$
$$R = 65 + 2 = 67 \text{ (m)}$$

Note that if the center line had been used to determine the radius of the center line of the lane of travel, and that radius had been 226 ft, then the radius can be calculated subtracting half width of the lane.

$$R = 226 - 6 = 220 \text{ (ft)}$$
$$R = 69 - 2 = 67 \text{ (m)}$$

Example P. 2.15: A *vehicle left the highway on a curve. The area leading up to the point of takeoff was level or zero grades. The vehicle first struck the ground at a horizontal distance (D) of 46 ft (14 m) beyond and a vertical distance (h) of 16 ft (5 m) below the takeoff point. Measurements were taken from center of mass to center of mass.*

Solution:

U.S. $\qquad$ *S.I.*

$$S = \frac{2.73\times D}{\sqrt{h}} \qquad S = \frac{7.97\times D}{\sqrt{h}}$$

$$S = \frac{2.73\times 46}{\sqrt{16}} \qquad S = \frac{7.97\times 14}{\sqrt{5}}$$

$$S = 31(\text{mph}) \qquad S = 50(\text{km/h})$$

Example P. 2.16: *A vehicle traveling on a level road struck a power pole head-on. The base of the pole was a road level. At the time of impact, a loose piece of luggage in a roof rack was projected forward and landed on the roadway in front of the vehicle. The height of the piece of luggage at the point of takeoff was 6 ft (1.8m). The horizontal distance the piece was projected was 33 ft (10m). All measurements were taken from center of mass to center of mass. Determine the speed of the vehicle.*

Solution:

U.S.	*S.I.*
$S = \dfrac{2.73 \times D}{\sqrt{h}}$	$S = \dfrac{7.97 \times D}{\sqrt{h}}$
$S = \dfrac{2.73 \times 33}{\sqrt{6}}$	$S = \dfrac{7.97 \times 10}{\sqrt{1.8}}$
$S = 37(\text{mph})$	$S = 59(\text{km/h})$

Example P. 2.17: *A vehicle left the highway on a curve. The area leading up to the point of takeoff had a grade (m) of minus 7%. The vehicle first struck the ground at a horizontal distance (D) of 46 ft (14 m) beyond and a vertical distance (h) of 10 ft (3m) below the point of takeoff. Measurements were taken from center of mass to center of mass at the points of takeoff and landing. Determine the speed of the vehicle.*

Solution:

U.S.	*S.I.*
$S = \dfrac{2.73 \times D}{\sqrt{h \pm (Dm)}}$	$S = \dfrac{7.97 \times D}{\sqrt{h \pm (Dm)}}$
$S = \dfrac{2.73 \times 46}{\sqrt{10 - 46 \times 0.07}}$	$S = \dfrac{7.97 \times 14}{\sqrt{3 - 14 \times 0.07}}$
$S = 48(\text{mph})$	$S = 78(\text{km/h})$

Example P. 2.18: *A vehicle traveled into an intersection and upon striking a curb with its front end, vaulted through the air and landed on its roof on a lawn. A 45° takeoff angle is assumed. The center of mass at landing was at the same level as at takeoff. The distance between the two center-of-mass positions was 50 ft (15 m). Calculate the speed at takeoff. Determine the speed of the vehicle.*

Solution:

U.S.	*S.I.*
$S = \sqrt{15 \times D}$	$S = \sqrt{127 \times D}$
$S = \sqrt{15 \times 50}$	$S = \sqrt{127 \times 15}$
$S = 27(\text{mph})$	$S = 44(\text{km/h})$

Example P. 2.19: *A vehicle was traveling at a high rate of speed along a roadway. Upon approaching a dip, the vehicle became momentarily airborne and landed on the roadway some distance away. From the point of takeoff to the point of landing, the measurements (center-of-mass to center-of-mass) were:*

Horizontal distance = 85 ft (26m)

Vertical distance (lower) = 4 ft (1.2m)

Determine the speed of the vehicle.

Solution:

U.S.	*S.I.*
$S = \dfrac{2.73 \times D}{\sqrt{D \times \sin\theta \times \cos\theta + [h \times (\cos\theta)^2]}}$	$S = \dfrac{7.97 \times D}{\sqrt{D \times \sin\theta \times \cos\theta + [h \times (\cos\theta)^2]}}$
$S = \dfrac{2.73 \times 85}{\sqrt{85 \times \sin 6° \times \cos 6° + [4 \times (\cos 6°)^2]}}$	$S = \dfrac{7.97 \times 26}{\sqrt{26 \times \sin 6° \times \cos 6° + [1.2 \times (\cos 6°)^2]}}$
$S = 65(\text{mph})$	$S = 105(\text{km/h})$

Example P. 2.20: *A driver locked the car's brakes and the vehicle stopped in a straight skid over a concrete surface first and then Asphalt surface. Calculate:*

(*a*) *the vehicle's slide to stop speed (S1) based on the skid marks left on the first surface (concrete).*

(*b*) *the vehicle's slide to stop speed (S2) based on the skid marks left on the second surface (Asphalt).*

(*c*) *combine the first and second surface and calculate the speed at the beginning of the skid marks on the skid marks in the first surface.*

Concrete: *Skid distance = 99 ft (30 m); Drag factor = 0.8*

Asphalt: *Skid distance = 76 ft (23 m); Drag factor = 0.7*

Solution:

	U.S.	*S.I.*
(*a*) Concrete		
	$S1 = \sqrt{30 \times D \times f}$	$S1 = \sqrt{254 \times D \times f}$
	$S1 = \sqrt{30 \times 99 \times 0.8}$	$S1 = \sqrt{254 \times 30 \times 0.8}$
	$S1 = 49(\text{mph})$	$S1 = 78(\text{km/h})$
(*b*) Asphalt		
	$S2 = \sqrt{30 \times D \times f}$	$S2 = \sqrt{254 \times D \times f}$
	$S2 = \sqrt{30 \times 75 \times 0.7}$	$S2 = \sqrt{254 \times 23 \times 0.7}$
	$S2 = 40(\text{mph})$	$S2 = 64(\text{km/h})$
(*c*) Combined		
	$Sc = \sqrt{S1^2 + S2^2}$	$Sc = \sqrt{S1^2 + S2^2}$
	$Sc = \sqrt{49^2 + 40^2}$	$Sc = \sqrt{78^2 + 64^2}$
	$Sc = 63(\text{mph})$	$Sc = 101(\text{km/h})$

Example P. 2.21: *A vehicle traveling on a roadway skidded 66 ft (20 m) before striking a solid barrier. The vehicle stopped abruptly at the time of the impact. Evidence from a witness and resultant damage indicated a collision speed of approximately 45 mph (72 km/h) (S2). The roadway drag factor was 0.75. Calculate:*

(*a*) *the vehicle's speed (S1) based on the skid marks.*

(*b*) *calculate the initial speed combining S1 and S2.*

Solution:

	U.S.	*S.I.*
(*a*) *S*1 speed		
	$S1 = \sqrt{30 \times D \times f}$	$S1 = \sqrt{254 \times D \times f}$
	$S1 = \sqrt{30 \times 66 \times 0.75}$	$S1 = \sqrt{254 \times 20 \times 0.75}$
	$S1 = 38$(mph)	$S1 = 61$(km/h)
(*b*) Combined		
	$Sc = \sqrt{S1^2 + S2^2}$	$Sc = \sqrt{S1^2 + S2^2}$
	$Sc = \sqrt{38^2 + 45^2}$	$Sc = \sqrt{61^2 + 72^2}$
	$Sc = 59$(mph)	$Sc = 94$(km/h)

Example P. 2.22: *A vehicle traveling into a T intersection, braked, skidded on an asphalt surface and then went into a fall from the highway.*

Distance of skid on asphalt = 75 ft (23 m)

Horizontal distance of fall = 85 ft (26 m)

Vertical distance of fall = 33 ft (10 m)

Roadway drag factor = 0.75

Grade at point of takeoff = 0°

Calculate:

(a) the vehicle's speed (S1) based on the skid marks.

(b) fall speed (S2).

(c) combined speed (Sc).

Solution:

	U.S.	*S.I.*
(*a*) *S*1 speed		
	$S1 = \sqrt{30 \times D \times 0.75}$	$S1 = \sqrt{254 \times D \times f}$
	$S1 = \sqrt{30 \times 75 \times 0.75}$	$S1 = \sqrt{254 \times 23 \times 0.75}$
	$S1 = 41$(mph)	$S1 = 66$(km/h)
(*b*) *S*2 Speed (fall speed)		
	$S2 = \frac{2.73 \times D}{\sqrt{h}}$	$S2 = \frac{7.97 \times D}{\sqrt{h}}$
	$S2 = \frac{2.73 \times 85}{\sqrt{33}}$	$S2 = \frac{7.97 \times 26}{\sqrt{10}}$
	$S2 = 40.5$(mph)	$S2 = 65$(km/h)

(*c*) Combined

U.S.	S.I.
$Sc = \sqrt{S1^2 + S2^2}$	$Sc = \sqrt{S1^2 + S2^2}$
$Sc = \sqrt{41^2 + 40.5^2}$	$Sc = \sqrt{66^2 + 65^2}$
$Sc = 58$(mph)	$Sc = 93$(km/h) .

Example P. 2.23: *A vehicle accelerating from a stop sign at a rate of 5 ft/s² covers an unknown distance during each of the first 3 seconds of acceleration. Find the total distance covered by the car.*

Solution:

Dt = Total distance

Vo = Initial velocity = 0

a = acceleration = 5 ft/s^2

t = time in seconds.

Total distance covered:

U.S.	*S.I.*
$D = V_0 \times t + \frac{1}{2} \times a \times t^2$	$D = V_0 \times t + \frac{1}{2} \times a \times t^2$
$Dt = \frac{1}{2} \times 5 \times 3^2$	$Dt = \frac{1}{2} \times 1.52 \times 3^2$
Dt = 22.5 ft	Dt = 6.9 (m)

Example P. 2.24: *From Problem 2.23, calculate the final speed in mph and km/h after each second.*

Solution:

	U.S.	*S.I.*
1st second		
	$S = a \times t$	$S = a \times t$
	$S = 5 \times 1 \times 0.682$	$S = 1.52 \times 1 \times 3.6$
	S = 3.42 (mph)	S = 5.5 (km/h)
2nd second		
	$S = 5 \times 2 \times 0.682$	$S = 1.52 \times 2 \times 3.6$
	S = 6.82 (mph)	S = 10.95 (km/h)
3rd second		
	$S = 5 \times 3 \times 0.682$	$S = 1.52 \times 3 \times 3.6$
	S = 10.23 (mph)	S = 16.4 (km/h)

Example P. 2.25: *A vehicle accelerated from a stop sign at a rate of 10 ft/s² (3m/s²) over a distance of 50 ft (15.2m). Calculate the speed.*

Solution:

U.S.	*S.I.*
$V = \sqrt{2 \times a \times D}$	$V = \sqrt{2 \times a \times D}$
$V = \sqrt{2 \times 10 \times 50}$	$V = \sqrt{2 \times 3 \times 15.2}$

$V = 31.6$ (ft/s) $\quad V = 9.5$(m/s)

$S = 21.6$ (mph) $\quad V = 34.2$(km/h)

Example P. 2.26: *From Problem 2.25, calculate the total time required for the car to reach 30 ft/s (9.14 m/s).*

Solution:

$$t = \frac{V}{a}$$

$$t = \frac{30}{10}$$

$$t = 3(s)$$

Example P. 2.27: *As part of an accident investigation that occurred within an intersection, the investigator measured the length of time it took each of five vehicles to travel a distance of 50 ft (15.24m) from a stop at a stop sign to the point in the intersection where a collision had taken place. The average time for the five vehicles to travel the distance was 5 s.*

Solution:

U.S.	*S.I.*
$a = \frac{2 \times D}{t^2}$	$a = \frac{2 \times D}{t^2}$
$a = \frac{2 \times 50}{5^2}$	$a = \frac{2 \times 15.24}{5^2}$
$a = 4(\text{ft/s}^2)$	$a = 1.22(\text{m/s}^2)$

Example P. 2.28: *Vehicle A stopped at a stop sign and then accelerated into an intersection, striking Vehicle B. The investigation revealed that vehicle A accelerated at the rate of 6 ft/s² (1.82 m/s²). Calculate the acceleration factor (fa).*

fa = *acceleration factor*

g = *acceleration of gravity*

a = *acceleration rate*

Solution:

U.S.	*S.I.*
$fa = \frac{a}{g}$	$fa = \frac{a}{g}$
$fa = \frac{6}{32.2}$	$fa = \frac{1.82}{9.81}$
$fa = 0.186$	$fa = 0.186$

Example P. 2.29: *Vehicle A entered an intersection where a stop sign was in place for vehicles traveling in Vehicle's 1 direction. After a point 20 ft (6m) into de intersection, Vehicle 1 struck Vehicle 2. Investigation revealed that at the time of collision, Vehicle 1 was traveling 40 mph (64 km/h). Test conducted with Vehicle 1 revealed that its maximum acceleration rate was 11.5 ft/s² (3.5 m/s²).*

(*a*) *determine if Vehicle 1 stopped at the stop sign.*

(*b*) *calculate the maximum speed attainable by vehicle A, based on its acceleration factor, over the distance it traveled from the stop sign to the point of collision.*

fa = acceleration factor

g = acceleration of gravity

a= acceleration rate

Solution:

	U.S.	S.I.
(*a*)	$fa = \frac{a}{g}$	$fa = \frac{a}{g}$
	$fa = \frac{11.50}{32.2}$	$fa = \frac{3.5}{9.81}$
	$fa = 0.36$	$fa = 0.36$
(*b*)	$S = \sqrt{30 \times D \times fa}$	$S = \sqrt{254 \times D \times fa}$
	$S = \sqrt{30 \times 20 \times 0.36}$	$S = \sqrt{254 \times 6 \times 0.36}$
	$S = 14.5$ (mph)	S = 23.4(km/h)

The investigator revealed that at the time of collision, Vehicle 1 was traveling 40 mph (64 km/h). Therefore, one must conclude that Vehicle 1 did not stop at the stop sign.

Example P. 2.30: *Vehicle 1 and 2 were traveling side-by-side at 25 mph (40 km/h) (reported by vehicle 1). After passing a traffic light, Vehicle 2 accelerated sharply. At the end of the block, a distance of 328 ft (100 m) from the traffic light, Vehicle 2 struck a pedestrian who was in a crosswalk. Calculate the possible velocity of Vehicle 2 when it struck the pedestrian. Assuming the maximum acceleration rate of 10 ft/s*^{2}*(3m/s*^{2}*) for that type of vehicle. Determine the speed of Vehicle 2 when it struck the pedestrian.*

Solution:

U.S.	S.I.
$V = S \times 1.466$	$V = S \times 0.278$
$V = 25 \times 1.466$	$V = 40 \times 0.278$
$V = 37$(ft/s)	$V = 11.1$(m/s)
$Vf = \sqrt{V_0^2 + (2aD)}$	$Vf = \sqrt{V_0^2 + (2aD)}$
$Vf = \sqrt{37^2 + (2 \times 10 \times 328)}$	$Vf = \sqrt{11.1^2 + (2 \times 3 \times 100)}$
$Vf = 89$(ft/s)	$Vf = 27$(m/s)
$S = \frac{Vf}{1.466}$	$S = \frac{Vf}{.278}$
$S = 61$(mph)	$S = 97$(km/s)

Example P. 2.31: *A vehicle left skid marks on a roadway. The drag factor was 0.70. Calculate the deceleration rate of the vehicle during the skid.*

Solution:

U.S.	S.I.
$a = f \times g$	$a = f \times g$
$a = .70 \times 32.2$	$a = 0.7 \times 9.81$
$a = 22.5(\text{ft/s^2})$	$a = 6.9(\text{m/s})$

Example P. 2.32: *A driver of a vehicle traveling 50 mph (80 km/h) applied his brakes in an unsuccessful attempt to avoid striking a pedestrian who had stepped out into the roadway. After skidding 49 ft (13 m), the vehicle struck the pedestrian, and then continued to skid to a stop for a total skid distance of 85 ft (26m). The roadway drag factor was 0.80, giving a deceleration rate of 25.76 ft/s² (7.85 m/s²). The speed of the vehicle at the time it struck the pedestrian was:*

Solution:

U.S.	S.I.
$Vf = \sqrt{V_0^2 - (2aD)}$	$Vf = \sqrt{V_0^2 - (2aD)}$
$Vf = \sqrt{(50 \times 1.466)^2 - (2 \times 25.76 \times 49)}$	$Vf = \sqrt{(80/3.6)^2 - (2 \times 7.85 \times 13)}$
$S = 53(\text{ft/s})$	$Vf = 16.5(\text{m/s})$
$S = \dfrac{Vf}{1.466}$	$S = \dfrac{Vf}{.278}$
$S = 36.4\ (\text{mph})$	$S = 59(\text{km/h})$

Example P. 2.33: *From Problem 2.32, calculate the time it took the vehicle to travel from the beginning of the skid to where it struck the pedestrian.*

Solution:

$$t = \frac{V_0 - V_f}{a}$$

$$t = \frac{(50 \times 1.466) - 53}{25.76}$$

$$t = 0.78(s)$$

Example P. 2.34: *Calculate the time it took a vehicle to travel from the beginning of the skid to a complete stop. V = 65 ft/s²; a = 27.37 ft/s².*

Solution:

$$t = \frac{V}{a}$$

$$= \frac{65}{27.37}$$

$$t = 2.4(s)$$

Example P. 2.35: *From Problem 2.34, calculate the distance traveled during each second.*

Solution:

U.S.	S.I.
$D=(V_0\times t)-(0.5\times a\times t^2)$	$D=(V_0\times t)-(0.5\times a\times t^2)$

1st second

$D=(65\times 1)-(0.5\times 27.37\times 1^2)$	$D=(19.74\times 1)-(0.5\times 8.3\times 1^2)$
$D=51.31(\text{ft})$	$D=15.59(\text{m})$

2nd seconds

$D=(65\times 2)-(0.5\times 27.37\times 2^2)$	$D=(19.74\times 2)-(0.5\times 8.3\times 2^2)$
$D=75.26(\text{ft})(\text{total})$	$D=22.88(\text{m})(\text{total})$

Distance traveled in 2nd seconds

$D=75.26-51.31$	$D=22.88-15.59$
$D=23.95(\text{ft})$	$D=7.29(\text{m})$

2.4 seconds

$D=(65\times 2.4)-(0.5\times 27.37\times 2.4^2)$	$D=(19.74\times 2.4)-(0.5\times 8.3\times 2.4^2)$
$D=80(\text{ft})(\text{total})$	$D=24(\text{m})(\text{total})$

Distance traveled in 2.4 seconds

$D=80-75.26$	$D=24-22.88$
$D=4.74(\text{ft})(\text{total})$	$D=1.12(\text{m})(\text{total})$

Example P. 2.36: *Calculate the stop distance required for a vehicle traveling at 70mph (113 km/h) on a roadway having a drag factor of 0.80.*

Solution:

U.S.	S.I.
$D=\dfrac{S^2}{30f}$	$D=\dfrac{S^2}{254f}$
$D=\dfrac{70^2}{30\times 0.8}$	$D=\dfrac{113^2}{254\times 0.8}$
$D = 204\ (\text{ft})$	$D = 62.5(\text{m})$

Example P. 2.37: *A vehicle that weighed 1500 lbs (675 kg) was traveling along a roadway at 44ft/s (13 m/s). Calculate the vehicle's momentum.*

Mom = Momentum

M = Vehicle's mass.

Solution:

U.S.	S.I.
$M_{om}=M\times V$	$M_{om}=M\times V$
$M_{om}=\dfrac{1500}{32.2}\times 44$	$M_{om}=\dfrac{675}{9.81}\times 13$
$M_{om}=2049.52(\text{lb-ft/s})$	$M_{om}=894.4(\text{kg-m/s})$

Example P. 2.38: (*a*) *A motorcycle skids 55 feet [16.8m] with only the rear wheel skidding (drag factor = 0.35). It then skids 34 feet [10.4 m] with both wheels sliding (drag factor = 0.88). It then slides 83 feet [25.3 m] on its side (drag factor = 0.52). Determine the speed of the bike at the start of the one-wheel skid.*

(*b*) *V-l is travelling 32 mph [51.5 km/h] when it hits another car. V-l Skidded 41 feet [12.5 m] at a drag factor of 0.77 just prior to impact. What was the speed of V-l at the start of skid?*

Solution: (*a*) We have a combined speed problem on our hands. The variables

S = Speed of motorcycle at start of precrash skid

d_1 = One-wheel skid distance = 55 feet

f_1 = One-wheel skid drag factor = 0.35

d_2 = Two-wheel skid distance = 34 feet

f_2 = Two-wheel skid drag factor = 0.88

d_s = Side slide distance = 83 feet

f_s = Side slide drag factor = 0.52

The speed of V-l at the start of precrash braking

$$S = \sqrt{30 \times d_1 \times f_1 + 30 \times d_2 \times f_2 + 30 \times d_s \times f_s}$$

$$S = \sqrt{30 \times 55 \times 0.35 + 30 \times 34 \times 0.88 + 30 \times 83 \times 0.52}$$

$$S = \sqrt{577.5 + 897.6 + 1294.8} = \sqrt{2870}$$

S = 53.57 or 53 mph [85 km/h]

(*b*) This problem is solved by using the combined speed formula. The variables

S_1 = Speed of V-1 at impact = 32 mph

d = Pre-impact skid distance = 41 feet

f = Drag factor of truck = 0.79

$$S = \sqrt{S_1^2 + 30 \times d \times f}$$

$$S = \sqrt{32^2 + 30 \times 41 \times 0.79}$$

$$S = \sqrt{1024 + 972} = \sqrt{1996}$$

S = 44.67 or 45 mph [72 km/h]

Example P. 2.39: *A northbound train is stopped on a siding with its last car a short distance from a highway grade crossing as shown in Fig. 2.39. A southbound train is approaching the crossing, but is screened by the other train. Meanwhile, two eastbound autos are stopped for the flashing crossing light and gate. A third auto stops behind the other two. The third car accelerates around the other two cars, attempts to maneuver around the crossing the crossing gates, and is struck by the southbound train.*

(*a*) *The car accelerated from a stop to 25 mph [40.2 km/h], and maintained constant speed until struck by the train. If its acceleration rate was 0.30g, how much time and how much distance was required for it to reach 25 mph?*

(*b*) *The car travelled 103 feet [31.4 m] from where it was stopped to where it was Struck. How much time did it require to travel this distance?*

(*c*) *The southbound train was travelling at a constant speed of 62 mph [100 km/h]. How far from the point of impact was it when the car started accelerating?*

(*d*) *The car needed to travel an additional 12 feet [3.7 m] to avoid the train. At 6.5 mph, how many more seconds would this require?*

(*e*) *After hitting the car, the train takes 5075 feet [1547 m] to stop. What was the average post impact drag factor of the train.*

(*f*) *What was the kinetic energy of the 3280-lb [1488 kg] car at 25 mph? What was the kinetic energy of the entire train at 62 mph if the train has a total weight of 5000 tons?*

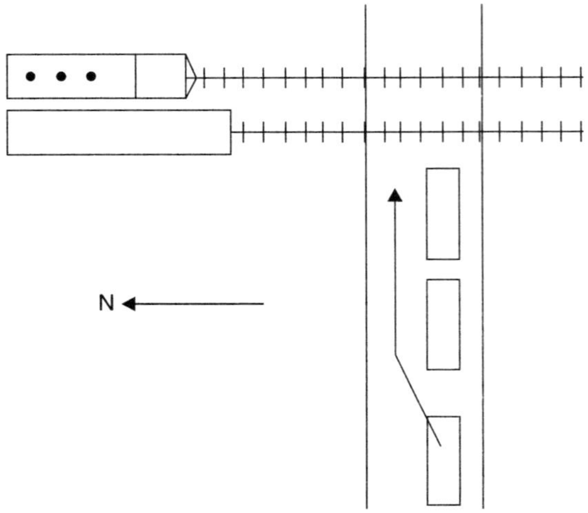

Fig. P. 2.39

Solution: (*a*) First we must find the velocity of the car (after acceleration) in feet per sec:

$$V = 25 \times 1.467 = 36.7 \text{ ft/sec}$$

We know that one '*g*' equals 32.2 ft/sec/sec, an acceleration of 0.3 g equals:

$$0.3 \times g = 0.3 \times 32.2 = 9.66 \text{ ft/sec/sec}$$

The time required to accelerate to 25 mph:

$$t = V/a = 36.7/9.66 = 3.8 \text{ seconds}$$

The distance required to accelerate to 25 mph:

$$d = V_0 \times t + 0.5 \times a \times t^2 = 0 \times 3.8 + 0.5 \times 9.66 \times 3.8^2$$

$$d = 0 + 0.5 \times 9.66 \times 14.4 = 69.7 \text{ feet [21.3 m]}$$

(*b*) We know how much time/distance it took for the car to hit 25 mph. Now we must determine how much additional time/distance it took to reach point of impact. Total distance from start to point of impact (d_T) was 103 feet.

The distance traveled at a constant 25 mph (d_C) was:

$$d_C = d_T - d = 103 - 69.7 = 33.3$$

Time required covering the final 33.3 feet:

$$t = d_C/V = 33.3/36.7 = 0.9 \text{ seconds}$$

Total elapsed Time (t_T):

$$t_T = 3.8 + 0.9 = 4.7 \text{ seconds}$$

(*c*) We have another time/distance problem on our hands. We know that the time from when the car starts accelerating until it reaches the point of impact is 4.7 seconds. We also know the train's speed (S_T) equals 62 mph. If we next determine the train's velocity (V_T) in feet per second we can cover the distance in that 4.7 sec. timeframe:

$$V_T = S_T \times 1.467 = 62 \times 1.467 = 91.0 \text{ ft/sec}$$

$$d_T = V \times t = 91.0 \times 4.7 = 428 \text{ feet}$$

(*d*) From Problem 1, we learned a constant 25 mph converts to 36.7 ft/sec. The time required to go 12 feet:

$$t = d/V = 12/\,36.7 = 0.33 \text{ seconds}$$

(*e*) We can find the train's average drag factor by simply using a variant of the basic skid formula:

$$f = S^2/\,(30 \times d)$$

We can find the train's average drag factor by simply using a variant of the basic skid formula:

$$f = 62^2/(30 \times 5075) = 3844/152250 = 0.025$$

(*f*) The same method is used for both the train and the car. To determine a moving object's kinetic energy, we use the classic physics formula:

$$KE = 0.5 \times M \times V^2 = 0.5 \times (W/g) \times V^2$$

where KE = Kinetic Energy (ft × lb)

M = Mass of the object = (lb × sec²/ft)

V = Velocity of the object (ft/sec) (36.7 ft/sec for car; 91.0 ft/sec for train)

W = Weight of the object (5000 tons × 2000 lb/ton = 10,000,000 lb for train)

g = Acceleration of gravity = 32.2 ft/sec²

Calculating for the car (weight = 3280 lb):

$$KE = 0.5 \times (W/g) \times V^2 = 0.5 \times 3280 \times 36.7^2 = 2{,}210{,}000 \text{ ft} \times \text{lb}$$

Calculating for the train:

$$KE = 0.5 \times (W/g) \times V^2 = 0.5 \times 10{,}000{,}000 \times 98.3^2$$

$$= 48{,}300{,}000{,}000 \text{ ft} \times \text{lb}$$

Example P. 2.40: *V-l is travelling 36 mph [57.9 km/h] when it hits another car. V-l skidded 52 feet [15.8 m] at a drag factor of 0.77 just prior to impact. What was the speed of V-l at the start of skid?*

Solution: This problem is solved by using the combined speed formula. The variables:

S_1 = Speed of V-l at impact = 36 mph

d = Pre-impact skid distance = 52 feet

f = Drag factor of truck = 0.77

$$S = \sqrt{S_1^2 + 30 \times d \times f}$$

$$S = \sqrt{36^2 + 30 \times 52 \times 0.72}$$

$$S = \sqrt{1296 + 1201} = \sqrt{2497}$$

$$S = 49.97 \text{ or } 50 \text{ mph } [80 \text{ km/h}]$$

Example P. 2.41: *(a) The acceleration of a car was tested. It was found that the car went 175 feet [53.3 m] in 4.2 seconds. What was the average acceleration of the car? Approximately how fast was the car going at the end of the test?*

(b) A motorcycle goes down on its side and slides 186 feet [56.7 m]. The LEVEL friction between the bike and the dry asphalt surface is 0.48. The grade over the slide was measured and found to have an average of 2% downward. Determine the speed of the bike at the start of the slide.

(c) A jeep runs off an embankment and goes airborne. The horizontal distance of the jump is 42 feet [12.8 m]. The vertical drop is 7.5 feet [2.3 m]. The takeoff grade is 5% downward. Determine the speed of the jeep at takeoff.

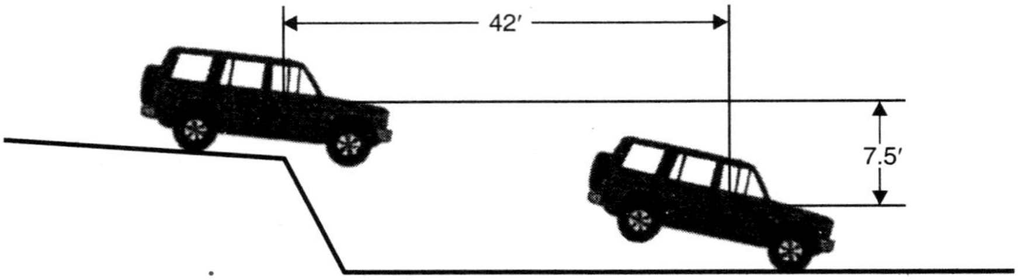

Fig. P. 2.41

Solution: (*a*) To calculate the average, we need to utilize the equation that gives the relation between acceleration, initial velocity and distance.

$$d = d_0 + V_0 \times t + 0.5 \times a \times t^2$$

where d = Total distance = 175 feet

d_0 = Starting point = 0 feet

V_0 = Initial velocity = 0 feet/sec

a = acceleration rate

t = time = 4.2 seconds

Substituting and solving:

$$175 = 0 + 0 \times 4.2 + 0.5 \times a \times 4.2^2$$

$$175 = 0.5 \times a \times 17.6 = 8.82 \times a$$

$$a = 175/8.82 = 19.8 \text{ ft/sec/sec } [6.0 \text{ m/s}^2]$$

Velocity at the end of the test can be estimated by multiplying the acceleration by the time elapsed:

$$V = a \times t = 19.8 \times 4.2 = 83.16 \text{ ft/sec}$$

Converting to miles per hour:

$$S = V/1.467 = 83.16/1.467 = 56.7 \text{ or } 57 \text{ mph } [96 \text{ km/h}]$$

(*b*) To find speed at start of the skid, skid distance, the friction and the basic skid formula. The variables:

S = Speed of the vehicle

d = Skid distance = 186 ft.

m = Slope of the surface = –2% = –2/100 = –0.02 (downward)

f_L = Level friction coefficient = 0.48

f = Drag factor = $f_L + m = 0.48 - 0.02 = 0.46$

Substituting into the formula:

$$S = \sqrt{30 \times d \times f} = \sqrt{30 \times 186 \times 0.46}$$

$$S = \sqrt{2567} = 50.67 \text{ or } 51 \text{ mph [82 km/h]}$$

(c) The variables:

S = Speed of vehicle at takeoff

d = Horizontal flight distance = 42 feet

H = Height change = –7.5 feet (downward)

m = Takeoff slope = –5% = –5/100 = –0.05 (downward)

To find speed at takeoff, we use the free fall formula:

$$S = \frac{2.74 \times d}{\sqrt{(m \times d - H)}} = \frac{2.74 \times 42}{\sqrt{(-0.05 \times 42 - -7.5)}} = \frac{2.74 \times 42}{\sqrt{(-2.1 + 7.5)}}$$

$$S = 115.1/\sqrt{(5.4)} = 115.1/2.32$$

$$S = 49.53 \text{ or } 50 \text{ mph [80 km/h]}$$

Example P. 2.42: (a) *A dump truck skids 103 feet [31.4 m] on a dry asphalt surface. The truck/ surface level coefficient of friction is 0.45, but the surface has average upslope of 2%. Calculate the speed of the truck at the start of skid.*

(b) *A truck accelerates at 0.14 g's. How much time will it take to go from a standing start to 60 mph [97 km/h]? How much distance did it require to go from 0 to 60?*

(c) *A car is travelling at 85 mph [137 km/h]. How much distance is required to brake to a stop if the drag factor is 0.74? How much time would be required?*

Solution: (a) To find speed at start of the skid, skid distance, the friction and the basic skid formula. The variables:

S = Speed of the vehicle

d = Skid distance = 103 ft.

f = Drag factor = 0.45

m = cross slope = +2% (upward) = +2/100 = +0.02

Substituting into the formula:

$$S = \sqrt{30 \times d \times (f + m)} = \sqrt{30 \times 103 \times (.045 + 0.02)} = \sqrt{30 \times 103 \times 0.47}$$

$$S = \sqrt{1452} = 38.11 \text{ or } 38 \text{ mph [61 km/h]}$$

(b) The variables:

d = Total distance

d_0 = Starting point = 0 feet

S = Final speed = 60 mph

V_0 = Initial velocity = 0 feet/sec

V = Final velocity (in feet/sec)

g = acceleration of gravity = 32.2 ft/sec/sec

a = acceleration rate = 0.14 × g = 4.5 ft/sec/sec

t = time

Converting miles per hour to feet per second:

$$V = S \times 1.467 = 60 \times 1.467 = 88.0 \text{ ft/sec}$$

Time required can be found by dividing Velocity by acceleration:

$$t = V/a = 88.0/4.5 = 19.5 \text{ sec}$$

To calculate the distance required, we need to utilize the equation that gives the relation between acceleration, initial velocity and distance.

$$d = d_0 + V_0 \times t + 0.5 \times a \times t^2$$

$$d = 0 + 0 \times 19.5 + 0.5 \times 4.5 \times 19.5^2$$

$$d = 0.5 \times 4.5 \times 381 = 856 \text{ feet [261 m]}$$

(*c*) To find the total distance needed to stop, we use a variant of basic skid formula. The variables:

S = Speed of the car = 85 mph

d = Distance required to brake to stop

f = Drag factor = 0.74

Substituting into the equation:

$$d = S^2/ (30 \times f) = 85^2/ (30 \times 0.74)$$

$$d = 7225/22.2$$

$$d = 325.4 \text{ or } 325 \text{ feet [99 m]}$$

Example P. 2.43: *A motorcycle goes down on its side and slides 41 feet [12.5 m]. The LEVEL friction between the bike and the dry asphalt surface is 0.52. The grade over the slide was measured and found to have an average of 2% downward. Determine the speed of the bike at the start of the slide.*

Solution: To find speed at start of the skid, skid distance, the friction and the basic skid formula. The variables:

S = Speed of the vehicle d = Skid distance = 41 ft.

m = Slope of the surface = –2% = –2 / 100 = –0.02 (downward)

f_L = Level friction coefficient = 0.52

f = Drag factor = $f_L + m$ = 0.52 + –0.02 = 0.50

Substituting into the formula:

$$S = \sqrt{30 \times d \times f} = \sqrt{30 \times 41 \times 0.50}$$

$$S = \sqrt{615} = 24.8 \text{ or } 25 \text{ mph [40 km/h]}$$

Example P. 2.44: *A motorcycle skids 75 feet [23.0 m] with only the rear wheel skidding (drag factor = 0.35). It then skids 28 feet [8.5 m] with both wheels sliding (drag factor = 0.88). It then slides 103 feet [31.4 m] on its side [drag factor = 0.52). Determine the speed of the bike at the start of the one-wheel skid.*

Solution: We have a combined speed problem on our hands. The variables:

S = Speed of motorcycle at start of precrash skid

d_1 = One-wheel skid distance = 75 feet

f_1 = One-wheel skid drag factor = 0.35

d_2 = Two-wheel skid distance = 28 feet

f_2 = Two-wheel skid drag factor = 0.88

d_s = Side slide distance = 103 feet

f_s = Side slide drag factor = 0.52

The speed of V-l at the start of precrash braking:

$$S = \sqrt{30 \times d_1 \times f_1 + 30 \times d_2 \times f_2 + 30 \times d_s \times f_s}$$

$$S = \sqrt{30 \times 75 \times 0.35 + 30 \times 28 \times 0.88 + 30 \times 103 \times 0.52}$$

$$S = \sqrt{787.5 + 739.2 + 1606.8} = \sqrt{3133.5}$$

$$S = 55.97 \text{ or } 56 \text{ mph [90 km/h]}$$

Example P. 2.45: *A jeep runs off an embankment and goes airborne. The horizontal distance of the jump is 31 feet [9.4 m]. The vertical drop is 5.5 feet [1.7 m]. The takeoff grade is 3% downward. Determine the speed of the jeep at takeoff.*

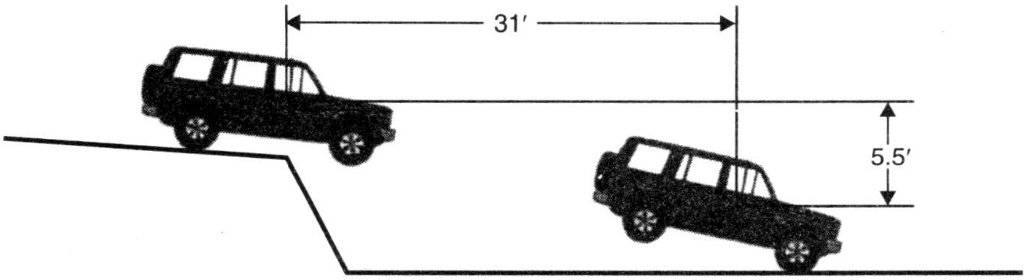

Fig. P. 2.45

Solution: The variables:

S = Speed of vehicle at takeoff

d = Horizontal flight distance = 31 feet

H = Height change = –5.5 feet (downward)

m = Takeoff slope = –3% = –3/100 = –0.03 (downward)

To find speed at takeoff, we use the free fall formula:

$$S = \frac{2.74 \times d}{\sqrt{(m \times d - H)}} = \frac{2.74 \times 31}{\sqrt{(-0.03 \times 31 - (-5.5))}} = \frac{2.74 \times 31}{\sqrt{(-0.93 + 5.5)}}$$

$$S = 84.9/\sqrt{4.57} = 84.9/2.14$$

$$S = 39.71 \text{ or } 40 \text{ mph [64 km/h]}$$

Example P. 2.46: *An automobile begins to yaw on a banked curve. The critical speed scuff mark laid down by the outside front tire was measured with a 60 foot [18.3 m] chord and found to have a middle ordinate of 20.5 inches [52 cm]. The LEVEL coefficient of friction is 0.79. Along the scuff mark the surface is banked 1%. Determine the speed of the vehicle.*

Solution: Solution of this problem has two main steps. We will ultimately utilize the critical speed to sideslip equation (banked surface). First we must determine the radius of curvature of the tire mark used to estimate the radius of curvature of the center of mass. The equation:

$$R = \frac{C^2}{8 \times m_O} + \frac{m_O}{2}$$

where R = Radius of curvature, feet

C = Chord length = 60 feet

m_O = Middle ordinate length, feet = 20.5"/12 = 1.71 feet

Substituting and solving:

$$R = \frac{60^2}{8 \times 1.71} + \frac{1.71}{2} = 264 \text{ ft.}$$

We now plug the radius of curvature, lateral friction coefficient and cross slope into the critical speed to sideslip formula. The equation:

$$S = \frac{\sqrt{15 \times R \times (f + m)}}{\sqrt{(1 - f \times m)}}$$

where S = Speed in miles per hour

f = Lateral friction coefficient = .79

R = Radius of curvature = 264 ft

m = cross slope = 1% = 1/100 = 0.01

Substituting and solving:

$$S = \frac{\sqrt{15 \times 264 \times (0.79 + 0.01)}}{\sqrt{(1 - 0.79 \times 0.01)}} = \frac{\sqrt{3168}}{\sqrt{0.992}}$$

$$S = 56.28/0.996 = 56.51 \text{ or } 57 \text{ mph } [91 \text{ km/h}]$$

Example P. 2.47: *A Camaro is drag racing another vehicle when it goes out of control, hits a curb, and goes airborne. As it leaves the ground it scrapes the top off a small dirt mound. The scrape is measured and found to have a take-off slope of 8%. The horizontal distance of the jump is 109 feet [33.2 m]. There is no change in elevation from take-off to landing. Determine the take-off speed of the Camaro.*

Solution: The variables:

S = Speed of vehicle at takeoff

d = Horizontal flight distance = 109 feet

H = Height change = 0 feet

m = Takeoff slope = +8% = +8/100 = +0.08 (upward)

To find speed at takeoff, we use the free fall formula:

$$S = \frac{2.74 \times d}{\sqrt{(m \times d - H)}} = \frac{2.74 \times 109}{\sqrt{(0.08 \times 109 - 0)}} = \frac{2.74 \times 109}{\sqrt{(8.7 - 0)}}$$

$$S = 298.7/\sqrt{(8.7)} = 298.7/2.95$$

$$S = 101.15 \text{ or } 101 \text{ mph } [163 \text{ km/h}]$$

Example P. 2.48: *A pickup truck skids 50 feet [15.2 m] on a dirt surface (drag factor of 0.63), skids another 62 feet [18.9 m] on a grass surface (drag factor of 0.42) and hits a one foot [0.3 m] diameter tree head-on. The impact leaves a 27 inch [68.5 cm] deep dent in the front of the pickup. Using one of the equations below,*

S in mph, C_{MAX} in inches: $S = 0.84 \times C_{MAX} + 4.0$

S in km/h, C_{MAX} in cm: $S = 0.53 \times C_{MAX} + 6.4$

calculate the speed of the pickup at impact. Also calculate the speed of the pickup at start of skid on the dirt.

Solution: Since there is no significant post impact movement to consider, finding impact speed (S_1) is easy. Simply substitute the 27 inches for C_{MAX} in the equation and crunch the numbers:

$$S = 0.84 \times C_{MAX} + 4.0$$

$$S = 0.84 \times 27 + 4.0 = 22.7 + 4.0$$

$$S = 26.7 \text{ or } 27 \text{ mph } [43 \text{ km/h}]$$

The speed at the start of skid can be solved by using the combined speed formula. The variables:

S = Speed of V-l at impact = 26.7 mph

d_A = Skid distance on dirt = 50 feet

f_A = Drag factor on dirt = 0.63

d_G = Skid distance on grass = 62 feet

f_G = Drag factor on grass = 0.42

$$S = \sqrt{S_1^2 + 30 \times d_D \times f_D + 30 \times d_G \times f_G}$$

$$S = \sqrt{26.7^2 + 30 \times 50 \times 0.63 + 30 \times 62 \times 0.42}$$

$$S = \sqrt{713 + 945 + 781} = \sqrt{2439}$$

$$S = 49.45 \text{ or } 49 \text{ mph } [79 \text{ km/h}]$$

Example P. 2.49: *An automobile begins to yaw on a banked curve. The critical speed scuff mark laid down by the outside front tire was measured with a 60 foot [18.3 m] chord and found to have a middle ordinate of 17.5 inches [44 cm]. The LEVEL coefficient of friction is 0.81. Along the scuff mark the surface is banked 4%. Determine the speed of the vehicle.*

Solution: Solution of this problem has two main steps. We will ultimately utilize the critical speed to sideslip equation (banked surface). First we must determine the radius of curvature of the tire mark used to estimate the radius of curvature of the center of mass. The equation:

$$R = \frac{C^2}{8 \times m_O} + \frac{m_O}{2}$$

where R = Radius of curvature, feet

C = Chord length = 60 feet

m_O = Middle ordinate length, feet = 17.5" / 12 = 1.46 feet

Substituting and solving:

$$R = \frac{60^2}{8 \times 1.46} = \frac{1.46}{2} = 310 \text{ ft.}$$

We now plug the radius of curvature, lateral friction coefficient and cross slope into the critical speed to sideslip formula. The equation:

$$S = \frac{\sqrt{15 \times R \times (f + m)}}{\sqrt{(1 - f \times m)}}$$

where S = Speed in miles per hour
f = Lateral friction coefficient = 0.81
R = Radius of curvature = 310 ft
m = Cross slope = 4% = 4/100 = 0.04

Substituting and solving:

$$S = \frac{\sqrt{15 \times 310 \times (0.81 + 0.04)}}{\sqrt{(1 - 0.81 \times 0.04)}} = \frac{\sqrt{3952}}{\sqrt{0.968}}$$

$$S = 62.87/0.984 = 63.91 \text{ or } 64 \text{ mph } [103 \text{ km/h}]$$

Example P. 2.50: *An automobile begins to yaw on a banked curve. The critical speed scuff mark laid down by the outside front tire was measured with a 60 foot [18.3 m] chord and found to have a middle ordinate of 18.5 inches [47 cm]. The LEVEL coefficient of friction is 0.76. Along the scuff mark the surface is banked 2%. Determine the speed of the vehicle.*

Solution: Solution of this problem has two main steps. We will ultimately utilize the critical speed to sideslip equation (banked surface). First we must determine the radius of curvature of the tire mark used to estimate the radius of curvature of the center of mass. The equation

$$R = \frac{C^2}{8 \times m_O} + \frac{m_O}{2}$$

where R = Radius of curvature, feet
C = Chord length = 60 feet
m_O = Middle ordinate length, feet = 18.5"/12 = 1.54 feet

Substituting and solving:

$$R = \frac{60^2}{8 \times 1.54} + \frac{1.54}{2} = 293 \text{ ft.}$$

We now plug the radius of curvature, lateral friction coefficient and cross slope into the critical speed to sideslip formula. The equation:

$$S = \frac{\sqrt{15 \times R \times (f + m)}}{\sqrt{(1 - f \times m)}}$$

where S = Speed in miles per hour
f = Lateral friction coefficient = 0.76
R = Radius of curvature = 293 ft
m = cross slope = 2% = 2/100 = 0.02

Substituting and solving:

$$S = \frac{\sqrt{15 \times 293 \times (0.76 + 0.2)}}{\sqrt{(1 - 0.76 \times 0.2)}} = \frac{\sqrt{3428}}{\sqrt{0.985}}$$

$$S = 58.55/0.985 = 58.96 \text{ or } 59 \text{ mph } [95 \text{ km/h}]$$

Example P. 2.51: *An automobile begins to yaw on a banked curve. The critical speed scuff mark laid down by the outside front tire was measured with a 60 foot [18.3 m] chord and found to have a middle ordinate of 15.5 inches [39.4 cm]. The level coefficient of friction is 0.73. Along the scuff mark the surface is banked 2%. Determine the speed of the vehicle.*

Solution: Solution of this problem has two main steps. We will ultimately utilize the critical speed to sideslip equation (banked surface). First we must determine the radius of curvature of the tire mark used to estimate the radius of curvature of the center of mass. The equation:

$$R = \frac{C^2}{8 \times m_O} + \frac{m_O}{2}$$

where R = Radius of curvature, feet

C = Chord length = 60 feet

M_O = Middle ordinate length, feet = 15.5″/12 = 1.29 feet

Substituting and solving:

$$R = \frac{60^2}{8 \times 1.29} + \frac{1.29}{2} = 348 \text{ ft}$$

We now plug the radius of curvature, lateral friction coefficient and cross slope into the critical speed to sideslip formula. The equation:

$$S = \frac{\sqrt{15 \times R \times (f + m)}}{\sqrt{(1 - f \times m)}}$$

where S = Speed in miles per hour

f = Lateral friction coefficient = 0.73

R = Radius of curvature = 348 ft

m = Cross slope = 2% = 2/100 = 0.02

Substituting and solving:

$$S = \frac{\sqrt{15 \times 348 \times (0.73 + 0.02)}}{\sqrt{(1 - 0.73 \times 0.02)}} = \frac{\sqrt{3919}}{\sqrt{0.985}}$$

$$S = 62.60/0.993 = 63.04 \text{ or } 63 \text{ mph } [101 \text{ km/h}]$$

Example P. 2.52: *An automobile begins to yaw on a banked curve. The critical speed scuff mark laid down by the outside front tire was measured with a 60 foot [18.3 m] chord and found to have a middle ordinate of 16.5 inches [42 cm]. The level coefficient of friction is 0.78. Along the scuff mark the surface is banked 2%. Determine the speed of the vehicle.*

Solution: Solution of this problem has two main steps. We will ultimately utilize the critical speed to sideslip equation (banked surface). First we must determine the radius of curvature of the tire mark used to estimate the radius of curvature of the center of mass. The equation:

$$R = \frac{C^2}{8 \times m_O} + \frac{m_O}{2}$$

where R = Radius of curvature, feet

C = Chord length = 60 feet

m_O = Middle ordinate length, feet = 16.5"/12 = 1.375 feet

Substituting and solving:

$$R = \frac{60^2}{8 \times 1.375} + \frac{1.375}{2} = 328 \text{ ft.}$$

We now plug the radius of curvature, lateral friction coefficient and cross slope into the critical speed to sideslip formula. The equation:

$$S = \frac{\sqrt{15 \times R \times (f + m)}}{\sqrt{(1 - f \times m)}}$$

where S = Speed in miles per hour

f = Lateral friction coefficient = 0.78

R = Radius of curvature = 328 ft

m = cross slope = 2% = 2/100 = 0.02

Substituting and solving:

$$S = \frac{\sqrt{15 \times 328 \times (0.78 + 0.2)}}{\sqrt{(1 - 0.78 \times 0.02)}} = \frac{\sqrt{3936}}{\sqrt{0.984}}$$

S = 62.74/0.991 = 63.24 or 63 mph [102 km/h]

Example P. 2.53: *A train consists of 3 locomotives (weight: 140 tons [1127,000 kg] each), 42 loaded hopper cars (weight: 85 tons [77,000 kg] each), and 24 loaded flatcars (weight: 73 tons [66,000 kg] each). When the brake lines are fully pressurized, each locomotive generates 10,800 pounds [48,000 kN] of stopping force, each hopper car generates 5900 pounds [26,000 kN] of stopping force, and each flatcar generates 5200 pounds [23,000 kN] of stopping force. Determine the deceleration of each locomotive and car type as if it were sliding independently. Determine the deceleration of the entire train.*

Solution: To solve this problem we use the basic physics equation drag factor (f) equals friction force (F) (sometimes called slopping force) divided by weight (W). In order for the equation to work, all units of measure must be consistent. Therefore, we will convert stopping force into tons. One ton equals 2000 pounds.

Locomotive stopping force: 10,800 lb/2000 lb per ton = 5.4 tons

Hopper car stopping force: 5900 lb/2000 lb per ton = 2.95 tons

Flat car stopping force: 5200 lb/2000 lb per ton = 2.6 tons

Once drag factor for each car is determined, it can be multiplied by the acceleration of gravity (g), which at the earth's surface is 32.2 ft/sec per sec, to get each vehicle's deceleration.

Locomotive: $f = F/W$ = 5.4/140 = 0.0386

$a = f \times g$ = 0.0386 × 32.2 = 1.24 ft/sec^2

Hopper car: $f = F/W = 2.92/85 = 0.0343$

$a = f \times g = 0.0347 \times 32.2 = 1.11$ ft/sec^2

Flat car: $f = F/W = 2.6/73 = 0.0356$

$a = f \times g = 0.0356 \times 32.2 = 1.15$ ft/sec^2

We do the same thing to get the deceleration of the entire train. But first we must add up the total weights of each locomotive and car, and the total stopping force on each locomotive and car (3 locomotives, 42 hopper cars, 24 flat cars).

Train weight: $W = 3 \times 140 + 42 \times 85 + 24 \times 33 = 5742$ tons

Train stopping force: $F = 3 \times 5.4 + 42 \times 2.95 + 24 \times 2.6 = 202.5$ tons

Train drag factor: $f = F/W = 202.5/5742 = 0.0352$

Train deceleration: $a = f \times g = 0.0352 \times 32.2 = 1.14$ ft/sec^2

Alternatively, we could have converted locomotive and car weights to pounds, divided that into stopping forces (in pounds) and gotten the same results.

2.17.1 Additional Example Problems and Solutions

Example 2.1: *A vehicle accelerates from a stop over a distance of 24.38 m (80 ft) in 7 seconds. Determine*

(*a*) *the vehicle's speed after 9 seconds of acceleration from a stop, if the acceleration is uniform over the entire 9 seconds*

(*b*) *the distance traveled in the first 9 seconds*

(*c*) *the vehicle's speed after the first 4 seconds of acceleration*

(*d*) *the time required to accelerate over the first 12.19 m (40 feet).*

Solution: Here $v_i = 0$ ft/sec, $d = 80$ ft (24.38 m) and $t = 7$ sec.

From Table 2.1,

$$a = \frac{2d - 2v_i t}{t^2} = \frac{2(80) - 2(0)(7)}{49} = \frac{160}{49} = 3.2 \text{ ft/sec}^2 \ (0.98 \text{ m/s}^2)$$

(*a*) Now $a = 3.2$ ft/sec^2 (0.98 m/s^2), $v_i = 0$ ft/sec, and $t = 9$ sec.

From Table 2.1,

$$v_e = v_i + at = 0 + (3.2)(9) = 28.8 \text{ ft/sec } (8.8 \text{ m/s})$$

(*b*) $v_i = 0$ ft/sec, $a = 3.2$ ft/sec^2, $t = 9$ sec

From Table 2.1,

$$d = v_i t + \frac{1}{2} at^2 = 0(9) + \frac{1}{2}(3.2)(9^2) = 129.6 \text{ ft } (39.04 \text{ m})$$

(*c*) $v_i = 0$ ft/sec, $a = 3.2$ ft/sec^2, $t = 4$ sec

From Table 2.1,

$$v_e = v_i + at = 0 + (3.2)(4) = 12.8 \text{ ft/sec} = 3.90 \text{ m/sec. } (14.04 \text{ km/h})$$

(*d*) $d = 40$ ft, $v_i = 0$ ft/sec, $a = 3.2$ ft/sec^2

From Table 2.1,

$$t = \frac{v_e - v_i}{a} \text{ and } v_e = \sqrt{v_i^2 + 2ad} = \sqrt{0^2 + 2(3.2)(40)}$$

$$= 16 \text{ ft/sec } (4.87 \text{ m/s})$$

Therefore, $t = \frac{v_e - v_i}{a} = \frac{16 - 0}{3.2} = 5$ sec.

Example 2.2: *A vehicle skids for 45.72 m (150 ft) and hits a pedestrian. The vehicle continues to skid another 30.48 m (100 ft) before coming to a stop. The vehicle's drag factor during skidding was 0.8. The pedestrian walked from the pavement edge northbound a distance of 4.57 m (15 ft) and was then struck by the vehicle. The pedestrian's working speed was 1.524 m/s (5 ft/sec). Determine*

(*a*) *the initial speed of the vehicle*

(*b*) *the vehicle's velocity when the pedestrian was struck*

(*c*) *the distance the vehicle is from its first contact position when the pedestrian first steps on to the pavement.*

Solution: (*a*) $v_e = 0$, $a = fg = -0.8(32.2) = -25.76$ ft/sec^2 (-7.85 m/s^2)

and $d = 150 + 100 = 250$ ft (76.2 m).

From Table 2.1,

$$v_i = \sqrt{v_e^2 + 2ad} = \sqrt{0^2 - 2(-25.76)(250)}$$

$$= 113.49 \text{ ft/sec.} = 77.20 \text{ mph } (124.21 \text{ km/h})$$

(*b*) $v_e = 0$ ft/sec, $a = -25.76$ ft/sec^2 and $d = 100$ (30.48 m).

From Table 2.1,

$$v_i = \sqrt{v_e^2 - 2ad} = \sqrt{0^2 - 2(-25.76)(100)}$$

$$= 71.78 \text{ ft/sec.} = 48.83 \text{ mph } (78.57 \text{ km/h})$$

(*c*) The time the pedestrian walks to the collision point is the same as the vehicle towards to the collision point. Hence, we first calculate the time for the pedestrian to walk to the collision point for a constant speed of 5 ft/sec and a distance of 15 ft, $d = 15$ ft, and $v = 5$ ft/sec.

From Table 2.1,

$$t = d/v = \frac{15}{5} = 3 \text{ sec.}$$

Now we determine where the vehicle was 3 sec before the collision. We will compute the time it took to skid 150 ft before the collision. Now $v_i = 113.49$ ft/sec, $v_e = 71.78$ ft/sec and $a = -25.76$ ft/sec^2 (-7.85 m/s^2). From Table 2.1,

$$t = \frac{v_e - v_i}{a} = \frac{71.78 - 113.49}{-25.76} = 1.62 \text{ sec.}$$

Hence, of the 3 sec. that the pedestrian walks towards the collision, the vehicle is skidding for 1.62 sec. Hence, for the remaining time of $3 - 1.62 = 1.38$ sec., the vehicle travels at the speed calculated at the beginning of the skid (113.49 ft/sec.). Hence, the distance traveled for 1.38 sec. is $d = vt =$ (113.49) (1.38) = 156.62 ft (47.74 m). The distance the vehicle is from its first contact position, when the pedestrian first steps on to the pavement, is given by the sum of 156.62 ft plus 150 ft. (distance skidded before impact).

Hence, total distance = 156.62 + 150 = 306.32 ft (93.46 m).

Example 2.3: *The height of a vehicle's center of mass is 0.64 m (2.1 ft) and is located 1.31 m (4.3 ft) behind the front axle. The wheelbase of the car is 2.74 m (9.5 ft). Assume there is no braking on the front axle, resulting in a drag factor due to rolling resistance to be 0.01. Also, the rear wheels are locked by brakes and slide on a pavement with a coefficient of friction of 0.70. Determine the drag factor for the whole vehicle.*

Solution: The drag for the whole vehicle can be obtained from

$$f_R = \frac{f_f - x_f(f_f - f_r)}{1 - z(f_f - f_r)} \tag{1}$$

Here,

$$z = \frac{2.1}{9.5} = 0.221$$

$$x_f = \frac{4.3}{9.5} = 0.453$$

$$f_f = 0.01 \text{ and } f_r = 0.70.$$

Substituting the above values in Eq. (1), we have

$$f_R = \frac{0.01 - 0.453(0.01 - 0.70)}{1 - 0.221(0.01 - 0.70)} = \frac{0.323}{1.152} = 0.28.$$

Example 2.4: *The measured values of chord and middle ordinate from a test of a sudden maneuver of a passenger road vehicle are 14.63 m (48 ft) and 0.46 m (1.5 ft). Assume an average frictional drag coefficient of 0.75. The test site has a zero grade and zero super elevation. Determine,*

(*a*) *the speed over the measured arc of the yaw marks*

(*b*) *the frictional drag coefficient that gives a speed of 96.54 km/h (60 mph).*

Solution: The radius of the circular arc is given by

(*a*)

$$R = \frac{c^2}{8M} + \frac{M}{2} = \frac{48^2}{8(1.5)} + \frac{1.5}{2} = 192.75 \text{ ft (58.75 m)}.$$

The speed of the vehicle is given by

$$V_{cr} = \sqrt{15Rf} = \sqrt{15(192.75)(0.75)} = 46.57 \text{ mph (74.93 km/h)}.$$

(*b*) The frictional drag coefficient is given by

$$f = \frac{V_{cr}^2}{15R} = \frac{48^2}{15(192.75)} = 0.797.$$

Example 2.5: *Determine highest speed of a road vehicle that can travel with a circular path on a level unbanked road with 335.28 m (1100 ft) radius under icy conditions, with $f = 0.08$.*

Solution: With $e = \tan \gamma = 0.$

$$V_{cr} = \sqrt{gRf} = \sqrt{32.2(1100)(0.08)} = 53.23 \text{ ft/s (85.65 km/h)}.$$

Example 2.6: *A four-wheeled vehicle had brakes on only its front wheels. When the brakes were applied it caused the vehicle to skid to a stop.*

(*a*) *Determine the vehicle's speed with the following information:*

Skid distance = 33.53 m (110 ft) on level surface.

Drag factor = 0.80 (level surface).

Weight on front wheels during braking = 65% of total vehicle weight.

(*b*) *If the vehicle is traveling on a downhill grade of 5% and the drag factor is 0.82, calculate the vehicle's speed.*

Solution:

(*a*) English S.I.

$$S = \sqrt{30\,D\,f\,n} \qquad S = \sqrt{254\,D\,f\,n} = \sqrt{30(110)(0.80)(0.65)}$$

(*a*) English S.I.

$$S = \sqrt{30\,D(\mu n \pm m)} \qquad S = \sqrt{254\,D(\mu\; n \pm m)}$$

$$= \sqrt{30(110)[(0.80)(0.65) - 0.05]} \qquad = \sqrt{254(33.53)[(0.80)(0.65) - 0.05]}$$

$$= 39.38 \text{ mph.} \qquad = 63.27 \text{ km/h.}$$

Example 2.7: *A vehicle with braking on all four wheels was towing a trailer that was not equipped with brakes. In an attempt to avoid a collision with another vehicle, the driver applied the brakes of the vehicle. Determine the speed of the vehicle and trailer at the beginning of the vehicle's skid marks with the following information:*

Vehicle weight = 1587.60 kg (3500 lbs)

Trailer weight = 680.40 kg (1500 lbs)

Skid distance = 19.81 m (65 ft)

Drag factor = 0.65

Solution: English

S.I.

$$S = \sqrt{30\,D\,f\,n}$$

$$S = \sqrt{254\,D\,f\,n} = \sqrt{30(65)(0.65)(0.7)} = \sqrt{254(19.81)(0.65)(0.7)}$$

$$= 29.79 \text{ mph } (47.85 \text{ km/h})$$

where $n = \dfrac{3500}{5000} = 0.7$

Example 2.8: *The operator of a motorcycle applied only the rear wheel brake and caused the motorcycle to skid to a stop. Determine the motorcycle's speed at the beginning of its skid mark with the following information:*

Weight of motorcycle and rider = 358.56 kg (850 lbs)

Weight on front wheel of motorcycle = 158.76 kg (350 lbs)

Weight on rear wheel of motorcycle = 226.80 kg (500 lbs)

20% forward shift during braking = 49.90 kg (110 lbs)

Total weight on rear wheel during braking action = 500 – 110 = 176.90 kg (390 lbs)

Skid distance = 19.81 m (65 ft)

Drag factor = 0.65

Solution:

$$n = \frac{390}{850} = \frac{176.90}{385.56} = 0.46$$

English

S.I.

$$S = \sqrt{30\,D\,f\,n}$$

$$S = \sqrt{254\,D\,f\,n} = \sqrt{30(65)(0.65)(0.46)} = \sqrt{254(19.81)(0.65)(0.46)}$$

$$= 24.15 \text{ mph } (38.78 \text{ km/h})$$

Example 2.9: *The drag of a vehicle applied the brakes of the vehicle in order to avoid an accident and caused the vehicle to skid to a stop. The vehicle's left wheels were on an asphalt surface and its right wheels were on a gravel surface during the entire skid. Determine the vehicle's speed at the beginning of its skid marks with the following information:*

Skid distance = 30.48 m(100 ft)

Drag factor for asphalt surface = 0.70

Drag factor for gravel surface = 0.45

Solution:

$$S = \sqrt{15(f_1 + f_2)D}$$

$$S = \sqrt{127(f_1 + f_2)D} = \sqrt{15(0.70 + 0.45)(100)} = \sqrt{127(0.7 + 0.45)(30.48)}$$

$$= 41.53 \text{ mph } (66.72 \text{ km/h})$$

Example 2.10: *A vehicle traveling on a curve side slipped leaving a yaw mark of 100.58 m (330 ft) in length. The first one-third 33.53 m (110 ft) of the yaw mark was used and a chord of 21.34 m (70 ft) and a middle ordinate of 1.07 m (3.5 ft) were measured. The drag factor is known to be 0.80. The curve had a super elevation of +06%. Determine:*

(a) the radius of the yaw mark

(b) the critical vehicle curve speed from the yaw mark.

Solution:

(*a*)

$$R = \frac{C^2}{8M} + \frac{M}{2}$$

$$R = \frac{C^2}{8M} + \frac{M}{2} = \frac{70^2}{8(3.5)} + \frac{3.5}{2} = \frac{21.34^2}{8(1.07)} + \frac{1.07}{2}$$

$$= 176.75 \text{ ft } (53.74 \text{ m}).$$

(*b*)

$$S = 3.86\sqrt{R(f \pm e)}$$

$$S = 11.27\sqrt{R(f \pm e)} = 3.86\sqrt{176.75(0.80+0.06)}$$

$$= 11.27\sqrt{53.75(0.80+0.06)}$$

$$= 47.59 \text{ mph } (76.62 \text{ km/h}).$$

Example 2.11: *A road vehicle was traveling on a level asphalt road when it struck a utility pole head-on. At the time of impact, a carton in roof rack was projected forward and landed on the road in front of the vehicle. The base of the pole was assumed to be at the road level. The height of the carton at the point of takeoff was 1.83 m (6 ft). The horizontal distance the carton was projected was 9.14 m (30 ft). All the measurements were taken from center of mass to center of mass. Determine the vehicle speed.*

Solution:

$$S = \frac{2.73D}{\sqrt{h}}$$

$$S = \frac{7.97\,D}{\sqrt{h}} = \frac{2.73(30)}{\sqrt{6}} = \frac{7.97(9.14)}{\sqrt{1.83}}$$

$$= 33.44 \text{ mph } (53.85 \text{ km/h}).$$

Example 2.12: *A vehicle upon striking an unforeseen object with its front end vaulted through the air landed on its roof on a lawn. The takeoff angle was assumed as 45 degrees and the center of mass at landing was at the same level as at takeoff. The distance between the two center-of-mass positions was 15.24 m (50 ft). Determine the speed at takeoff.*

Solution:

$$S = \sqrt{15\,D} \text{ mph}$$

$$S = \sqrt{127\,D} \text{ km/h}$$

$$= \sqrt{15(50)}$$

or

$$= \sqrt{127(15.24)}$$

$$= 27.39 \text{ mph } (43.99 \text{ km/h}).$$

Example 2.13: *A vehicle traveling on a highway and upon approaching dip, became momentarily airborne and landed on the roadway at some distance. Determine the vehicle's speed at the point of takeoff with the following information:*

Horizontal distance = 27.43 m (90 ft)

Vertical distance (lower) = 0.91 m (3 ft)

Angle of departure = 5 degrees

Solution:

$$S = \frac{2.73\,D}{\sqrt{D\sin\theta\cos\theta + h\cos^2\theta}}$$

$$S = \frac{7.97\,D}{\sqrt{D\sin\theta\cos\theta + h\cos^2\theta}} = \frac{2.73(90)}{\sqrt{90(0.087)(0.996) + 3(0.996)^2}}$$

$$= \frac{7.97(27.43)}{\sqrt{27.43(0.087)(0.996) + (0.91)(0.996)^2}}$$

$$= 74.86 \text{ mph } (120.72 \text{ km/h})$$

where $\cos\theta = \cos 5° = 0.996$, $\sin\theta = \sin 5° = 0.087$.

Example 2.14: *A four-wheeled vehicle entered into a T intersection, braked, skidded on an asphalt surface and then went into a fall from the highway. Determine,*

(*a*) *the speed from skid marks.*

(*b*) *the fall speed.*

(*c*) *the combined speed.*

Given: Skid distance on asphalt surface = 24.38 m (80 ft)

Drag factor = 0.65

Grade at takeoff point = 0 degrees

Horizontal distance of fall = 27.43 m (90 ft)

Vertical distance of fall = 10.67 m (35 ft)

Solution:

(*a*) In English units:

$$S_1 = \sqrt{30\,D\,f}$$

$$S_1 = \sqrt{254\,D\,f} = \sqrt{30(80)(0.65)} = \sqrt{254(24.38)(0.65)}$$

$$= 39.49 \text{ mph } (63.44 \text{ km/h}).$$

(*b*)

$$S_2 = \frac{2.73\,D}{\sqrt{h}}$$

$$S_2 = \frac{797\,D}{\sqrt{h}} = \frac{2.73(80)}{\sqrt{35}} = \frac{7.97(24.38)}{\sqrt{10.67}}$$

$$= 36.92 \text{ mph } (59.49 \text{ km/h}).$$

(*c*)

$$S_c = \sqrt{S_1^2 + S_2^2}$$

$$S_c = \sqrt{S_1^2 + S_2^2} = \sqrt{39.49^2 + 36.92^2} = \sqrt{63.44^2 + 59.49^2}$$

$$= 54.06 \text{ mph } (86.97 \text{ km/h}).$$

Example 2.15: *Two vehicles 1 and 2 were traveling side-by-side lines at 40.23 km/h (25 mph). After passing through an intersection traffic light, vehicle 2 accelerated sharply and at a distance of 106.68 m (350 ft) from the traffic light, vehicle 2 struck a pedestrian who was in a crosswalk. Determine:*

(*a*) *the speed of vehicle 2 when it struck the pedestrian.*

(*b*) *the time it took for vehicle 1 to accelerate from its initial and final speeds.*

Assume the maximum acceleration rate of 3.66 m/s^2 *(12 ft/s*^{2}*).*

Solution:

(*a*) In English units:

$$V = S(1.466)$$
$$V = S(0.278) = 25(1.466)$$
$$= 40.23(0.278) = 36.65 \text{ ft/s}$$
$$= 11.18 \text{ m/s}$$

Speed of vehicle 2 when it struck the pedestrian

$$V_f = \sqrt{V_0^2 + 2\,a\,D}$$
$$V_f = \sqrt{V_0^2 + 2\,a\,D} = \sqrt{36.65^2 + 2(12)(350)} = \sqrt{11.18^2 + 2(3.66)(106.68)}$$
$$= 98.71 \text{ ft/s } (30.09 \text{ m/s}).$$

or $$V_f = \frac{98.71}{1.466} = 67.33 \text{ mph } (108.27 \text{ km/h}).$$

(*b*) Time it took vehicle 1 to travel 350 ft (106.68 m) under acceleration is given by

In English units:

$$t = \frac{V_f - V_0}{a}$$
$$t = \frac{V_f - V_0}{a} = \frac{98.71 - 36.65}{12} = \frac{30.09 - 11.18}{3.66}$$
$$= 5.17 \text{ sec.}$$

Example 2.16: *A vehicle traveling at 80.45 km/h (50 mph) skidded a distance of 30.48 m (100 ft) to a stop on a roadway having a drag factor of 0.80. Determine the time it took the vehicle to skid the 100 ft distance.*

Solution: The time it took the vehicle to skid the 30.48 m (100 ft) is given by

In English units:

$$t = 0.249\sqrt{D/f}$$
$$t = 0.45\sqrt{D/f} = 0.249\sqrt{100/0.80} = 0.45\sqrt{30.48/0.80}$$
$$= 2.78 \text{ sec.}$$

2.18 SUMMARY

In this Chapter we presented an introduction to Newtonian mechanics, basic terms and definitions, Newton's laws of motion, basic equations of motion, coefficient of friction and drag factor, laws of friction, drag factor on a grade, unequal drag factors on axles, work, energy and laws of conservation of energy, potential energy, kinetic energy, linear momentum, time distance equations, velocity equations, time equations, distance equations, acceleration factor equations, kinetic energy, combined speed equations, slide-to-stop equation, slide-to-stop equation with deceleration (drag factor), slide-to-stop equation with grade and percentage of braking, drag factor equation, braking distance equation, weight shift, horizontal center of mass (CM) equation, height of center of mass (CM) equation, weight shift equation, speed equation with weight shift, speed equation with any wheel braking with weight shift, weight shift equation with grade, critical speed yaw, critical speed yaw equation, airborne equations, fall speed equations, level takeoff, uphill takeoff, downhill takeoff, flips and vaults, vehicle lands lower than takeoff, vehicle lands higher than takeoff, maximum height, driver perception and response, estimation of vehicle critical speed from yaw marks, speed estimates from yaw marks, radius equation, lane passing with changing velocity, lane changing maneuver model, engineering models and animations, function of accident scene models, model application, reconstruction animations, and example problems and solutions.

2.19 REFERENCES

Araszewski, M., Toor, A., Overgaard, R. and **Johal., R.,** "Lane Change Maneuver Modeling for Accident Reconstruction Applications", Paper No. 2002-01-0817, SAE International, Warrendale, PA, 2002.

Atkinson, D.R., Hennessy, C.J., and **Watts, A.J.,** "Low Speed Automobile Accidents and Occupant Kinematics, Dynamics, and Biomechanics", Lawyers and Judges Publishing Company, 1996.

Baker, J.S., *Traffic Accident Investigation Manual*, Northwestern University, Evanston, IL, 1975.

Baker, J.S., and **Fricke, L.B.,** "The Traffic Accident Investigation Manual", Northwestern University Traffic Institute, Evanston, IL, 1986.

Brach, R.M., "An Analytical Assessment of the Critical Speed Formula", Paper 970857, SAE International, Warrendale, PA, 1997.

Brach, Raymond M., and **Brach, R. Matthew.,** *Vehicle Accident Analysis and Reconstruction Methods*, The Society of Automotive Engineers, Warrendale, 2005.

Baxter, A.T, and **Mentzer, J.R.,** "Critical Speed Field Testing of a Passenger Vehicle", Annual Meeting of the National Association of Traffic Accident Reconstructionists and Investigators, Atlantic City, NJ, 1991.

Collins, J.C., *Accident Reconstruction,* Charles C. Thomas Publisher, Springfield, MA, November 1979.

Daily, J., *Fundamentals of Traffic Accident Reconstruction,* Institute of Police Technology and Management, University of North Florida, 1988.

Dewar, R.E., and **P.L. Olson.,** *Human Factors in Traffic Safety,* Lawyers and Judges Publishing, Tucson, AZ, 2002.

Dickerson, C.P., Arndt, M.W., Arndt, S.M., and **Mowry, G.A.,** "Evaluation of Vehicle Velocity Predictions Using the Critical Speed Formula", Paper 950137, SAE International, Warrendale, PA, 1995.

Dukkipati, R.V., Pang, J., Qatu, M., Sbeng, G., and **Shuguang, Z.,** *Road Vehicle Dynamics: Problems and Solutions Set, R-386,* SAE International, Warrandale, PA, 2009.

Dukkipati, R.V., Pang, J., Qatu, M., Sheng, G., and **Shuguang, Z.,** *Road Vehicle Dynamics*, Product Code R-366, SAE International, Warrandale, PA, 2008.

Dukkipati. R.V., *Vehicle Dynamics*, Narosa Publishing House, New Delhi, India, 2000.

Ellis, J.R., *Road Vehicle Dynamics*, John R. Ellis, Inc., Akron, OH, 1988.

Ellis, J.R., *Vehicle Dynamics*, Business Books Limited, London, 1969.

Fay, R.J., and **Scott, J.D.,** "New Dimensions in Rollover Analysis," Paper 1999-01-0448, SAE International, Warrendale, PA, 1999.

Fay, R.J., "Corrugated Metal Arch Barrier Phase I, Scale Model Study", University of Denver, Colorado, Department of Highways, December 1971.

Fay, R.J., "Models in Vehicle Accident Studies", 1987 American Engineering Model Society Conference, Denver, CO, May 11-12, 1987.

Fay, R.J., Robinette, R.D., and **Larson, V.D.,** "Engineering Models and Animations in Vehicular Accident Studies", SAE Paper: 880719, SAE International, Warrendale, PA, 1988.

Fay, R.J., "Scale Model Tests of an Energy Absorbing Highway Buffer", Highway Record No. 343, University of Denver, Colorado, CO, Department of Highways, 1971.

Fricke, L.B., *Traffic Accident Reconstruction,* Traffic Accident Investigation Manual, vol. 2, Northwestern University Traffic Institute, Evanston, IL, 1990.

Gillespie, T., *Fundamentals of Vehicle Dynamics,* SAE International, Warrendale, PA, 1992.

Lambourn, R.F., "The Calculation of Motor Car Speeds from Curved Tire Marks", Journal of the Forensic Science Society (U.K.), 29, 1989, pp. 371-386.

Limpert, R., *Motor Vehicle Accident Reconstruction and Cause Analysis*, 4th ed, The Michie Company, Charlottesville, VA, 1994.

Martinez, L., "Estimating Speed from Yawmarks - An Empirical Study", Accident Reconstruction Journal, Waldorf, MD, May/June 1993.

McGee, H.W., Hooper, K.G., Hughes, W.E. and **Benson, W.,** "Highway Design and Operations Standards Affected by Driver Characteristics", Volume II, Final Report, Bellomo-McGee, Inc, Report No. FHWA-RD-83-015, Vienna, Virginia, 1983.

Milliken, W.F., and **Milliken, D.L.,** *Race Car Vehicle Dynamics*, Society of Automotive Engineers Inc., Warrendale, PA, 1995.

Muttart, J.W, "Development and Evaluation of Driver Response Time Predictors Based Upon Meta Analysis", Paper 2003-01-0885, SAE International, Warrendale, PA, 2003.

Noon, R.K., "Engineering Analysis of Vehicle Accidents", CRC Press, FL, 1994.

Olson, PL., *Forensic Aspects of Driver Perception and Response,* Lawyers and Judges Publishing, Tucson, AZ, 1996.

Prasad, A.K., "Energy Absorbed by Vehicle Structures in Side Impacts", Paper 910599, SAE International, Warrendale, PA, 1991.

Rivers, R.W., *Speed Analysis for Traffic Accident Investigation Manual*, Institute of Police Technology and Management (IPTM), University of North Florida, Jacksonville, FL, 1997.

Semon, M., "Determination of Speed from Yaw Marks", Chapter 4 in Forensic Accident Investigation: Motor Vehicles, Bohan and Damask [ed.], Michie Butterworth, Charlottesville, VA, 1995.

Shelton, Sgt. Thomas, "Validation of the Estimation of Speed from Critical Speed Scuffmarks", Accident Reconstruction Journal, Waldorf, MD, January/February 1995.

Sledge, N.H., and **Marshek, K.M.,** "Formulas for Estimating Vehicle Critical Speed from Yaw Marks-A Review", Paper 971147, SAE International, Warrendale, PA, 1997.

Sledge, N.H., and **Marshek, K.M.,** "Vehicle Critical Speed Formula-Values for the Coefficient of Friction - A Review", Paper 971148, SAE International, Warrendale, PA, 1997.

Warner, C.Y., Smith, G.C., James, M.B., and **Germane, G.J.,** "Friction Applications in Accident Reconstruction", SAE paper No. 830612, Society of Automotive Engineers, Warrendale, PA, 1983.

Wong, J.Y., *Theory of Ground Vehicles*, 3rd ed., Wiley, New York, NY, 2001.

Tire Forces

CHAPTER 3

3.1 INTRODUCTION

Tires or *tyres* (in American and British English, respectively), are ring-shaped parts, either pneumatic or solid (including rubber, metals and plastic composites), that fit around wheels to protect them and enhance their function.

Pneumatic tires are used on many types of vehicles, such as bicycles, motorcycles, cars, trucks, earthmovers and aircraft. Tires enable better vehicle performance by providing traction, braking, steering, and load support. Tires form a flexible cushion between the vehicle and the road, which smoothes out shock and makes for a comfortable ride.

As a significant part of the vehicle, pneumatic tires perform the following functions:

- to support the weight of the vehicle.
- to cushion the vehicle over surface irregularities.
- to provide sufficient traction for driving and braking.
- to provide adequate steering control and direction stability.

The primary forces acting on the vehicle are from pneumatic tires. The critical performances of vehicle (such as driving, braking, stability, ride comfort, and traveling) are related to pneumatic tires. Therefore, analyses of tire dynamics are the precondition and basic for the studies and analyses of vehicle dynamics.

The ground forces affecting the vehicle are applied through the pneumatic tires. When the vehicle runs forward directly without side force, the forces exerted on the tire include normal force F_z, rolling resistance moment M_y, tractive force F_x and tractive moment $F_x \cdot R$. When trailing wheel puts on brake, the braking moment will increase the longitudinal resistance force. When the driving wheel moves under the driving moment, the direction of the ground longitudinal force is opposite to its direction under braking.

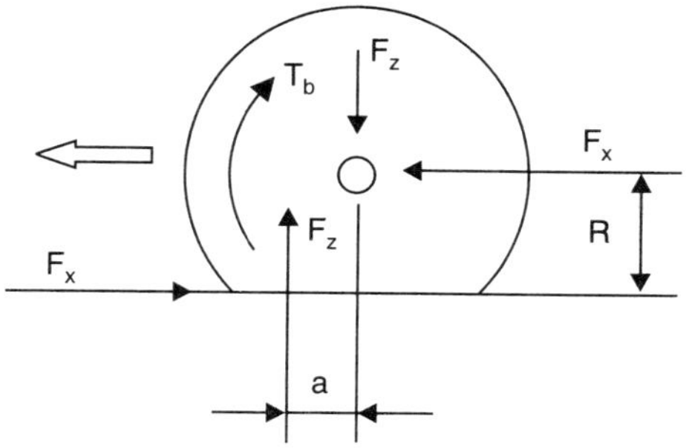

Fig. 3.1 Forces exerted on the tire without lateral under braking.

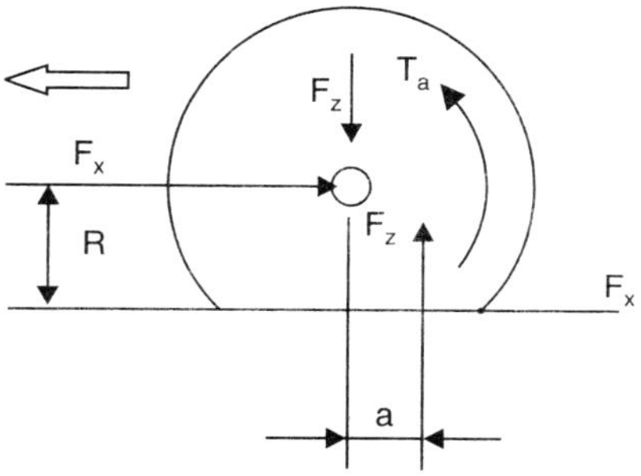

Fig. 3.2 Forces exerted on the tire without lateral force under driving.

When the vehicle changes the direction or there is a side force that is applied to the vehicle, a lateral force and moment will be developed on the tire. Because the lateral elasticity of the tire increases gradually as well as the lateral deformation at the tire-ground contact patch when tire is rolling, there is a distance *e* which is called the pneumatic trail between the resultant of force of the lateral forces and the center of contact patch. The product, $F_y e$ of the ground lateral forces applied to the tire and the pneumatic trail, determines the self-aligning moment of tire.

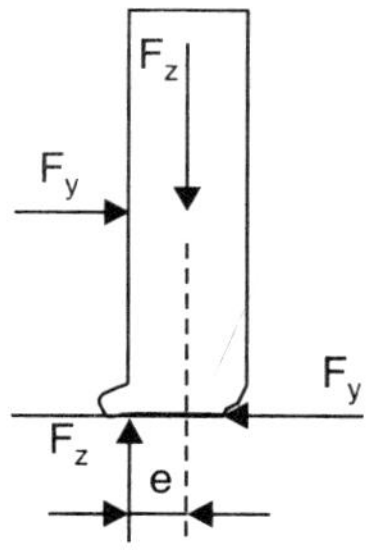

Fig. 3.3 Lateral forces exerted on the tire.

For a better analysis of the ground vehicle dynamics and its dynamics performance, it is a must to analyze the characteristics of tire dynamics firstly, and gain three tire forces and three tire moments.

To describe the characteristics of a tire and the forces and moments acting on it, it is necessary to define an axis system that serves as a reference for the definition of various parameters. One of the commonly used axis systems recommended by the Society of Automotive Engineers is shown in Fig. 3.5. The origin of the axis system is the center of tire contact. The *X* axis is the intersection of the wheel plane and the ground plane with a positive direction forward.

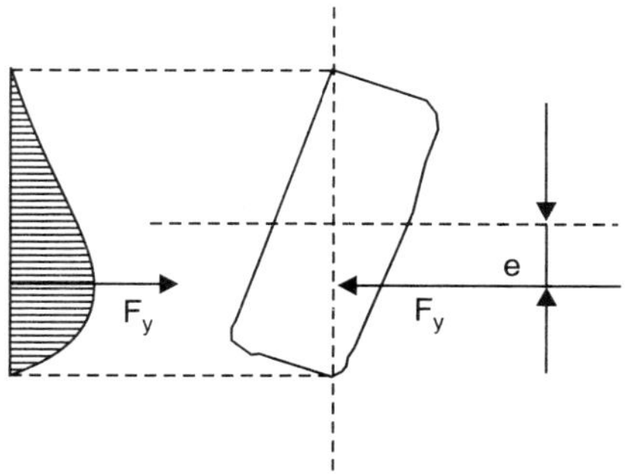

Fig. 3.4 Distribution of lateral forces at the contact patch.

The *Z* axis is perpendicular to the ground plane with a positive direction downward. The *Y* axis is in the ground plane, and its direction is chosen to make the axis system orthogonal and right hand.

There are three forces and three moments acting on the tire from the ground. Tractive force (or longitudinal force) F_x is the component in the *X* direction of the resultant force exerted on the tire by the road. Lateral force F_y is the component in the *Y* direction, and normal force F_z is the component in the *Z* direction. Overturning moment M_x is the moment about the *X* axis exerted on the tire by the road. Rolling resistance moment M_y is the moment about the *Y* axis, and aligning torque M_z is the moment about the *Z* axis.

Tire is an elastic body, which can be equivalent to a subsystem consisted of mass elements, elastic elements and damping elements in terms of mechanics. The dynamics analysis and modeling of tire are always the important part of vehicle dynamics analyses. This chapter presents mechanics characteristics of tire chiefly at the direction vertical, longitudinal, and lateral, respectively, and introduces related dynamics models and concepts of tire.

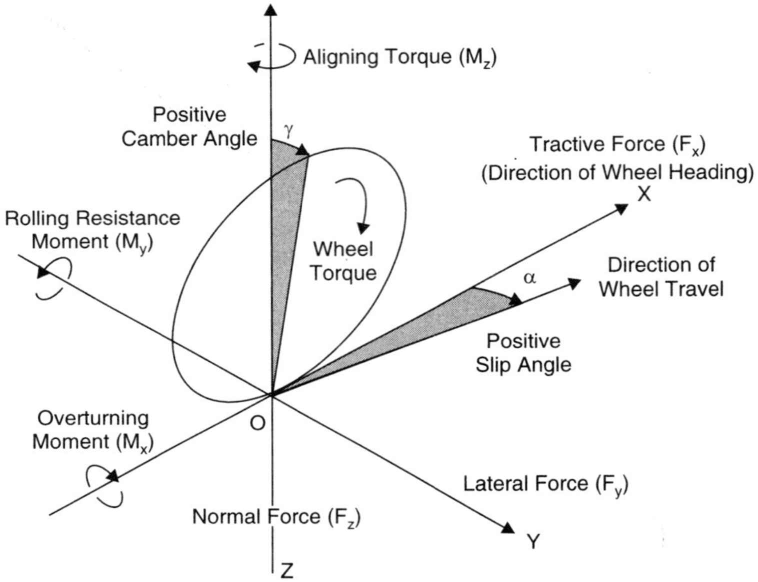

Fig. 3.5 The commonly used axis systems recommended by SAE.

3.2 VERTICAL DYNAMICS OF TIRES

Vertical dynamics of tires is the base of research on lateral and longitudinal dynamics of tires. Vertical stiffness/damping characteristics, classical vertical dynamic model of tires, and advantage and disadvantage of tires models are to follow.

3.2.1 Vertical Stiffness/Damping Characteristics of Tires

Pneumatic tire can cushion the vehicle over surface irregularities. The cushioning characteristics of tire have direct relationship with vertical stiffness and damping of tires. The static stiffness of tire K_s is determined by the slope of the static load-deflection curves. Basically, the static stiffness of the tire is a constant. The dynamic stiffness of the tire, K_d which varies with the frequency of the dynamic load, is determined by the transfer characteristics of the dynamic load to the deflection when the tire is rolling. There is a distinct difference between static stiffness and dynamic stiffness of a tire.

In the experiment of measuring the tire stiffness, it was found that the force required to make the tire produce the same deflection is different from the force that makes the tire restore from the same deflection whether under static or dynamic load. Fig. 3.6 shows the relationship between the tire load and the deflection. The area of close part in Fig. 3.6 represents the dissipative power of the rolling tire, which manifests the damping characteristics of the tire. With the increase of the frequency of excitation, the damping characteristics decrease. This is opposite to the state of the decrease in tire inflation pressure, where damping characteristics increase.

Figure 3.7 shows the variation of the dynamic stiffness of a tire with the excitation frequency: when the excitation frequency is zero, the corresponding stiffness is the static tire stiffness; the dynamic stiffness of tires decreases sharply with the increase of excitation frequency; however, beyond a frequency, the influence of excitation becomes less important. Generally, the dynamic stiffness may be 10 – 15% less than the static value. Under the same tire size, the dynamic stiffness of radial ply tires is lower than the counterpart of bias ply tires. Inflation pressure has a noticeable influence on tire stiffness. Tire stiffness will sharply decrease as soon as the pressure varies in the same direction, which is because the compressed air supports 85% of tire load; the rest is exerted on the tire body.

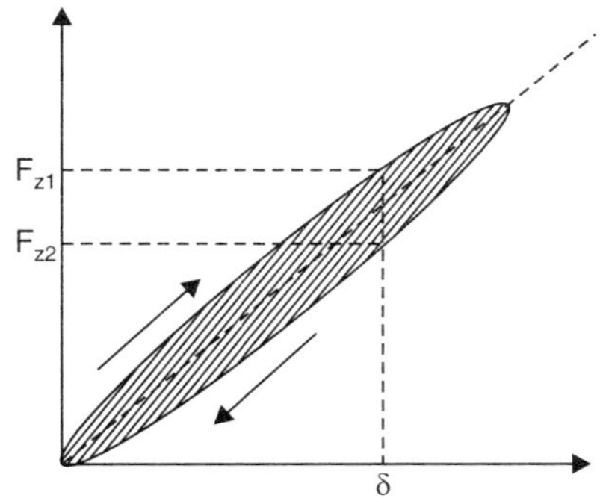

Fig. 3.6 Load-deflection relationship of a tire.

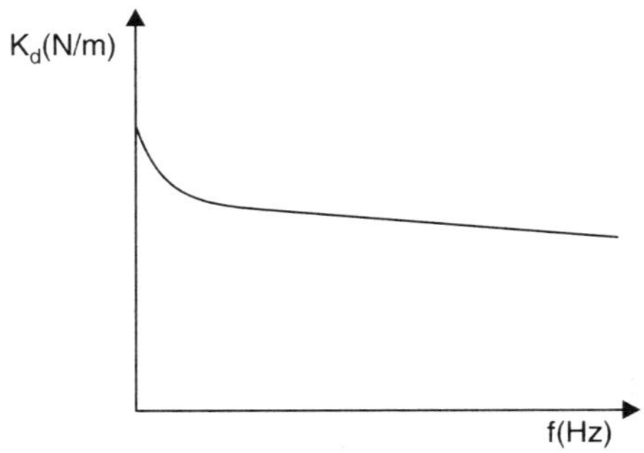

Fig. 3.7 Dynamic stiffness-excitation frequency relationship of a tire.

3.2.2 Vertical Vibration Mechanics Models of Tires

Dynamic vertical forces exerting on the tires must be obtained exactly, which is the important content of the analysis of the entire vehicle dynamics. The vertical dynamic model of tires is the basis of analyzing the vertical forces of tires. The mechanics characteristics of tire models not only embody the vertical elastic characteristic of tires, but also manifest tire-ground contact relationship as well as the influence derived from the surface irregularities.

When the influence of surface irregularities is less important with a constant velocity, point contact tire model is suitable for the analysis of vehicle vibration dynamics under the steady state condition; if the influence of surface irregularities is greater, the point contact model cannot represent the contact relationship between the tire and the road even with a constant velocity, thus, the fixed contact patch tire model is required; with the obvious variation of surface irregularities and unsteady velocity, the models introduced above are incompetent to be used in the analysis of the vertical dynamics of vehicle. Here, the time-varying contact patch tire model is needed to exactly embody the vertical dynamics of the vehicle. The models above will be introduced as follows.

3.2.2.1 Point Contact Model of Tires

An idealized tire can be equivalent to a spring-damping system, in which the elasticity characteristic of tires can be defined as the nominal stiffness and the resistance characteristic can be defined as the damping coefficient. It is assumed in this model that the contact region between the tire and ground is a point which is the projection of the center of tire onto the ground. Therefore, the vertical dynamic force exerted on the tire can be defined as:

$$F_Z = K\left(h(X) - \zeta(X)\right) + C\left(\dot{h}(X) - \dot{\zeta}(X)\right) \tag{3.1}$$

where $h(X)$ is the surface irregularity just under the center of the tire when it is rolling to the site X, $\zeta(X)$ is the displacement of the center of axis arm $\dot{h}(X)$, is the slope of the surface irregularity just under the center of the tire $\dot{\zeta}(X)$, is the velocity of the center of axis arm, is the nominal stiffness of the tire, and C is the damping coefficient of the tire.

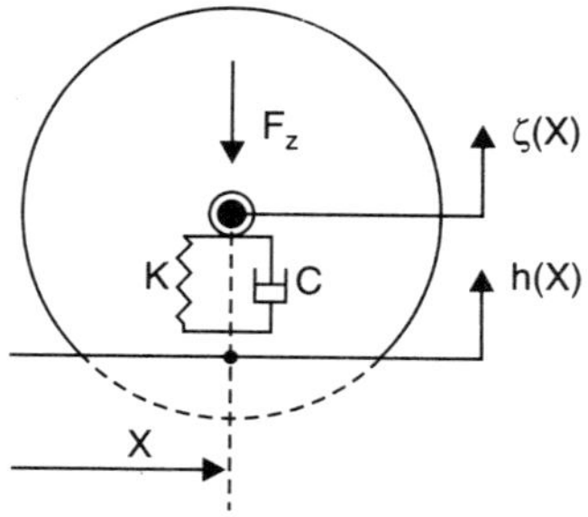

Fig. 3.8 Point contact model of tires.

This tire model can be established and applied with the conditions as follows:

1. Tire stiffness is an equivalent concentrated stiffness coefficient, which does not vary with the vertical dynamic load of tire. As a result, the model cannot manifest nonlinear characteristics of tires with massive deformation under the unsteady maneuvers such as sudden accelerating or severe braking.
2. $h(X) - \zeta(X)$ presents the equivalent deformation of the tire just under its center, which is only related to the displacement of the center of the axis arm and the surface irregularity under the contact point. When the influence of surface irregularities under the contact region is severe,

there will be a massive error between the vertical deformation calculated and the practical tire deformation, which is reasonable only over a good road surface.

As the results 1 and 2 indicate, the point contact tire model is the most cursory and simplest one and is applied primarily to the analyses of vehicle dynamics in a state with little surface irregularities, constant velocity and light variation of the tire load. The advantages of this model are simple mechanics characteristics and easy usage.

3.2.2.2 Fixed Contact Patch Model of Tires

Fixed contact patch length tire model is created on the basis of the point contact tire model by converting equivalent concentrated stiffness of tires into the equivalent distributed stiffness within the fixed contact patch length of tires.

$$F_z = \int_l k_i\left(h(X-x)-\zeta(X)\right)dx + C\left(\dot{h}(X)-\dot{\zeta}(X)\right) \tag{3.2}$$

where $h(X-x)$ is the surface irregularity of the road with a distance of x apart from the contact point when the tire is rolling to the site X, $\zeta(X)$ is the displacement of the center of the axis arm, $\dot{h}(X)$ is the slope of the surface irregularity just under the center of tire, $\dot{\zeta}(X)$ is the velocity of the center of the axis arm, k_i is the distributed stiffness coefficient of the tire corresponding to the site x, and C is the damping coefficient of the tire.

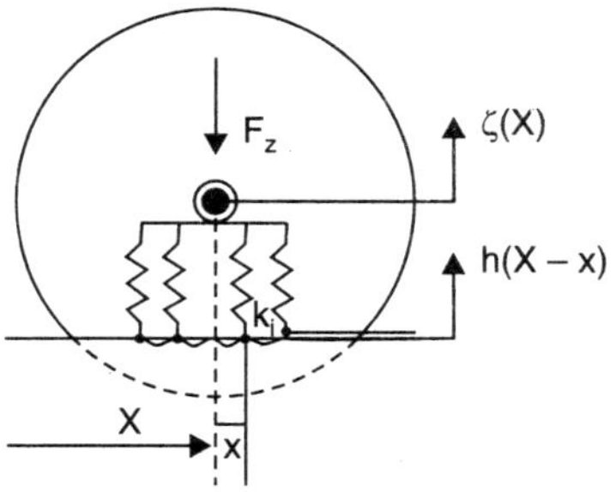

Fig. 3.9 Fixed contact patch model of tires.

Compared to the point contact tire model, the fixed contact patch length tire model embodies the comprehensive effect over the tire force from the surface irregularities within contact patch, from which the dynamic tire force derived is closer to the practical value. As it is assumed that the length of contact patch is unvaried, the equivalent stiffness of the tires is still a constant value. Therefore, this model is not available for the analyses of vehicle dynamics in the maneuvers with strong variation of contact patch, such as the sudden accelerating or severe braking.

3.2.2.3 Time-Varying Contact Patch Model of Tires

The time-varying contact patch length tire model is employed to offset the imperfections due to the models above. There are several preconditions required for the model:

1. The distributions with respect to the elasticity and damping of the tire are uniform in the radial direction.
2. The influence of surface irregularities is neglected along the width of the tire.

3. The elastic deformations turn out only within the contact region of the tire.

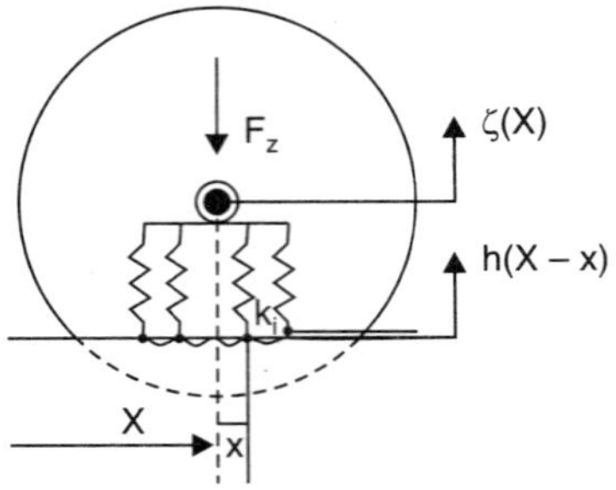

Fig. 3.10 Time-varying contact patch model of tires.

According to the above conditions, a brief tire model is illustrated as Fig. 3.10. There are many stiffness and damping elements which distribute equally in the radial direction within the contact patch. They deform in accordance with the variation of surface irregularities. Thus, the radial deformation and the corresponding velocity at the location which has an angle θ with the vertical direction can be expressed as follows:

$$\delta(X - x) = R - \frac{\left[h(X - x) - \zeta(X)\right]}{\cos\theta} \tag{3.3}$$

$$\dot{\delta}(X + x) = \frac{v\dfrac{dh(X + x)}{dx} - w}{\cos\theta} \tag{3.4}$$

where R is the free radius of the tire, v and w represent the horizontal and vertical velocities, respectively, $h(X - x)$ is the surface irregularity of the location which has a distance of x from the projection of the tire center, $\dfrac{dh(X + x)}{dx}$ is the slope of surface irregularity with a distance of x from the tire centre.

According to the geometric relationship in Fig. 3.10, the angle θ is calculated as:

$$\theta = \operatorname{arctg}\left(\frac{x}{h(X + x) - \zeta(X)}\right) \tag{3.5}$$

Then, the $F_c(X)$ radial force acting on the spring and damping elements can be written as:

$$F_c(X + x) = \begin{cases} k\delta(X + x) + c\dot{\delta}(X + x) & \delta(X + x) \geq 0 \\ 0 & \delta(X + x) < 0 \end{cases} \tag{3.6}$$

when $\delta(X + x)$ is less than zero, it means that the tire does not deform at the contact point, $X + x$, or that there turns out to be breakaway between the tire and the ground.

Thus, the vertical dynamic load acting on the tire is equal to the vector sum of the radial force derived from each element at the vertical direction, which is expressed as:

$$F_z = \int \cos\theta F_c(X + x)dx \tag{3.7}$$

Compared to the other tire models, the time-varying contact tire model obtains various characteristics as follows:

1. The dynamic vertical load acting on the tire is the comprehensive contribution of the surface irregularities on the contact patch rather than the single contact point just under the tire center, which complies with the practical filterability of road surface. The load which is derived from the model is much closer to the experimental test data.
2. The equivalent concentrated stiffness coefficient of the tire varies with contact patch. For instance, under the driving conditions with sudden accelerating or severe braking, if the tire load rises dramatically, then the contact patch will vary in the same way. Thus, more elements will contribute to the deformation of the tire, so the stiffness will increase; on the contrary, the stiffness will decrease. Therefore, this model can be also called *a nonlinear time-varying model.*

3.2.3 Enveloping Characteristics of Tires

A tire obtains enveloping characteristics which are defined as that the tire can absorb the vertical influence of the surface irregularities when it is rolling on the irregular road. The surface irregularities and vertical elasticity of the tires are the fundamental reasons for the generation of enveloping characteristics of the tires.

According to the tire models presented above, the second and the last bear the enveloping characteristics, but the first does not. By the way, when the point contact model is applied to the analysis and calculation of vehicle dynamics, the road input to compute the tire vertical dynamic load is not the practical road input, but the equivalent enveloping road input, and the formula for the force should be rewritten as:

$$F_z = K\left(h_q(X) - \zeta(X)\right) + C\left(\dot{h}_q(X) - \dot{\zeta}(X)\right) \tag{3.8}$$

where $h_q(X)$ is the

$$h_q(X) = h(X + x) - \left[R - \sqrt{R^2 - x^2}\right] \tag{3.9}$$

Adding a rigid wheel with a radius R to a point contact tire model to simulate the motion of the tire over irregular road, the practical contact point between the tire and the surface is not the projection of the tire centre on the road, but it will offset a distance of x from the projection of the tire centre on the road according to the geometric relationship related to the variation of surface irregularities and the rigid wheel. Thus, the trail of rigid wheel centre is equal to the effective surface irregularities, as shown in Fig. 3.13.

The relationship between the effective surface irregularities and the practical surface irregularities is defined as:

$$h_q(X) = h(X + x) - \left[R - \sqrt{R^2 - x^2}\right] \tag{3.9}$$

It must be noted that the point contact model with a rigid wheel can only represent the geometric enveloping characteristics of tires.

In fact, the tire also owns elastic enveloping characteristics. For instance, when a tire is slowly rolling over a real road with a semi-sine wave heave or hollow, there will be a big difference from the trail of the tire centre and the real surface irregularities due to the elasticity of the tires. Moreover, the amplitude of the former decreases with the increase of wavelength. The following figures show the relationship of the trail of the tire centre with the real surface irregularities when a time-varying contact tire is rolling over a surface with a semi-sine wave. It is noted that the shorter the wavelength is, the more dramatic the enveloping properties are. The fixed contact patch tire model has similar properties.

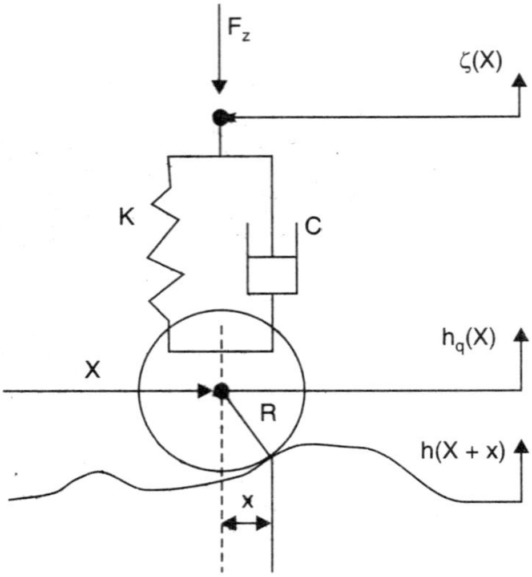

Fig. 3.11 Point contact model with a rigid wheel.

3.3 TIRE LONGITUDINAL DYNAMICS

Tire longitudinal dynamics mainly presents tire rolling resistance, the slip coefficient and the relationship between the tire traction and brake forces and the road surface condition.

3.3.1 Tire Rolling Resistance

The rolling resistance of the tire, which is applied to the wheel, is due to the deformation of the tire at the tire/road interface. As is known from the tire experiment, tire deformation needs to consume energy, and the normal pressure distribution over the tire/road contact patch is not uniform because of the unequal force needed during the compression and elastic recovery, as shown in the following figure. In the direction of travel the normal force in the leading half of the contact patch is higher than that in the trailing half. The center point of the sum of normal forces offsets a distance a to the center line in the direction of rolling. The normal force produces a moment about the axis of rotation of the tire, which is the rolling resistance moment, expressed as $M_f = F_z a$. While the driving force F_{ax} applied to the wheel produces a moment to balance the rolling resistance moment, the equation can be expressed as follows:

Fig. 3.12 Point contact model with a rigid wheel.

$$F_{ax} r = M_f \tag{3.10}$$

$$F_{ax} = \frac{M_f}{r} = F_z \frac{a}{r} \tag{3.11}$$

set
$$f = \frac{a}{r}$$

then
$$F_{ax} = F_z f \tag{3.12}$$

or:
$$f = \frac{F_{ax}}{F_z} \tag{3.13}$$

The nondimensional coefficient f is defined as the rolling resistance coefficient; it is the ratio of the driving force and the normal force on the rolling wheel in certain condition. Usually, tire rolling resistance changes linearly with the normal force of the wheel, that is

$$F_f = fF_Z \tag{3.14}$$

Experiments show the rolling resistance coefficient is proportional to the tire deformation, while inversely proportional to the radius of a loaded tire.

Table 3.1 lists the value of rolling resistance coefficient of different road surface recorded in the literature. According to USA standard, if $v < 50$ km/h, f will be 0.0165, if $v > 50$ km/h, f may be expressed by

$$f = 0.0165\left[1 + 0.01(v - 50)\right] \tag{3.15}$$

Table 3.1 Coefficient of Rolling Resistance

Road surface	*Rolling resistance coefficient*
Pneumatic tire of a car rolls	
on the coarse stone road surface	0.015
on the fine stone road surface	0.015
on the rolled asphalt mixed aggregate road surface	0.015
on the impacted coarse gravel road surface	0.012
on the mixed bituminous macadam road surface	0.025
on the soil road surface	0.05
on the field road surface	0.1~0.35
Pneumatic truck tire rolls on the asphalt mixed aggregate road surface	0.006~0.01
Steel wheel rolls on the field road surface	0.14~0.24
Steel wheel on the rail road surface	0.001~0.004

3.3.2 Rolling Resistance of the Tire with Toe-in

In actual vehicle structure there is a toe-in angle of the front wheel. This produces a toe-in resistance acting on the front wheel in the opposite direction of traveling, as shown in the Fig. 3.13.

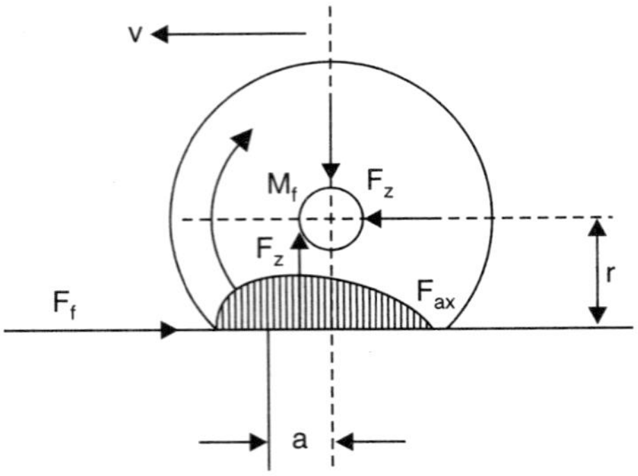

Fig. 3.13 Distribution of the forces and moments of a rolling tire.

In the Fig., δ_{v0} is the toe-in angle of the front wheel on one side; $F_{\delta v}$ is the side force due to the tire lateral deformation caused by the angle $\delta_{v0} \cdot F_{\delta v} = C_r \cdot \delta_{v0}$, where, C_r is the cornering stiffness of the tire, when the angle δ_{v0} is small enough, the value of sin δ_{v0} approximately equals δ_{v0}, that is $\sin \delta_{v0} \approx \delta_{v0}$. Hence, the component of $F_{\delta v}$ of the wheel is the toe-in resistance, as follows

$$F_v = 2F_{\delta v} \sin \delta_{v0} = 2C_r \delta_{v0}^2 \tag{3.16}$$

f_δ is defined as toe-in resistance coefficient

$$f_\delta = \frac{C_r}{F_z} \delta_{v0}^2 \tag{3.17}$$

then toe-in resistance will be expressed as follow

$$F_v = 2f_\delta F_z \tag{3.18}$$

3.3.3 Rolling Resistance of Turning Wheel

The value of additional rolling resistanc when the wheel is turning depends on the vehicle volecity, turning radius and the vehicle parameters. The rolling resistance coefficent f_R during a turn can be expressed as follow:

$$f_R = F + \Delta f \tag{3.19}$$

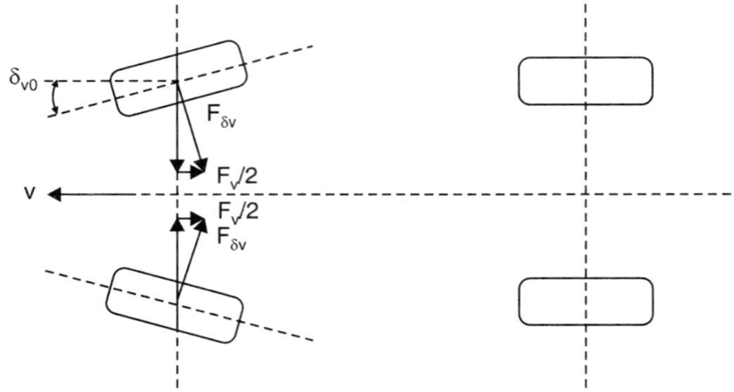

Fig. 3.14 Toe-in resistance acting on the front wheel.

α_f and α_r are the slip angles of the front and rear tire respectively in the conditon of the steering angle δ_0, correspondingly F_{yf} and F_{yr} are the cornering forces to balance the centrifugal force of the vehicle when steering, as shown in the figure, according to the equilibrium of forces, F_{yf} and F_{yr} can be expressed as:

$$F_{yf} = \frac{\cos\gamma F}{\sin\delta_0} = m\left(\frac{v^2}{R}\right)\left(\frac{\sin(\beta-\alpha_r)}{\sin\delta_0}\right) \tag{3.20}$$

$$F_{yr} = \sin\gamma F - F_{yf}\cos\delta_0 \tag{3.21}$$

$$= m\frac{v^2}{R}\left[\cos(\beta-\alpha_r) - \sin(\beta-\alpha_r)ctag\delta_0\right]$$

According to relationships of trigonometric function, then

$$\beta = \arcsin\left(\frac{l_b\cos\alpha_r}{R}\right) \tag{3.22}$$

$$\gamma = \frac{\pi}{2} - \beta + \alpha_r \tag{3.23}$$

where m is the mass of the vehicle, v is the vehicle velocity when turning, R is the turning radius.

Then the additional resistance respectively applied on the front and rear wheel can be obtained by the following equations:

$$\Delta F_f = \sin\alpha_f F_{yf} = m\left(\frac{v^2}{R}\right)\left(\frac{\sin(\beta-\alpha_r)}{\sin\delta_0}\right)\sin\alpha_f \tag{3.24}$$

$$\Delta F_r = \sin\alpha_r F_{yr} = m\frac{v^2}{R}\left[\cos(\beta-\alpha_r) - \sin(\beta-\alpha_r)\cdot ctag\delta_0\right]\cdot\sin\alpha_r \tag{3.25}$$

The additional resistance coefficient under the condition of vehicle steering can be described as follows

$$\Delta f = \frac{\Delta F_f + \Delta F_r}{mg} = \frac{v^2}{Rg}\left\{\frac{\sin(\beta-\alpha_r)}{\sin\delta_0}\sin\alpha_f + \left[\cos(\beta-\alpha_r) - \sin(\beta-\alpha_r)ctag\delta_0\right]\sin\alpha_r\right\} \tag{3.26}$$

It can be seen from the equation (3.26) that additional rolling resistance coefficient increases with both the vehicle velocity and the steering angle increase, but decreases with the turning angle increase, and also depends on the vehicle parameters.

3.3.4 Longitudinal Adhesion Coefficient

When a driving (or braking) torque is applied to a pneumatic tire, a tractive force (or braking force), which is limited to the critical coefficient of the road adhesion, is developed at the tire-ground contact patch. The maximum force F_ϕ of a tire on hard surfaces is defined as adhesion force and is proportional to the normal force applied on the wheel during driving or braking, as follows

$$F_\varphi = F_z\varphi \tag{3.27}$$

where, ϕ is the adhesion coefficient which varies with the state of the tire rolling or sliping.

When the driving torque of the wheel M_t is greater than the torque $F_\varphi r_d$ that is produced by the adhesion force around the wheel center, that is $M_t > F_\varphi r_d$,(where, r_d is the effective rolling radius of a tire, the vertical distance from the center of the wheel to tire-ground contact point) the tire will be sliping; while if the braking torque of the wheel M_b is more than the torque $F_\varphi r_d$, that is $M_b > F_\varphi r_d$, the tire will be skiding. And also the tire will be in the state of rolling-sliping with the variation of the driving torque or braking torque of the wheel. If ω is the angular speed of a rolling tire, then the longitudinal speed of the tire-ground contact point may be expressed as ωr_d, and if v_x is the linear speed of the tire relative to the ground, then the relationship of ωr_d and v_x can be used to describe the rolling or sliping state of the tire. When $\omega r_d = v_x$, there will be no relative motion at the tire-ground contact point. The tire is then at the state of pure rolling. When $\omega r_d > v_x$, there will be a negative linear velocity at the tire-ground contact point. Then the tire is rolling and slipping and will develop a longitudinal tractive force. When $\omega r_d < v_x$, there is positive linear velocity at the tire-ground contact point. The tire is rolling and sliding and will develop a longitudinal braking force.

In order to describe the tire slip accurately in braking maneuver, longitudinal skid s_b is defined as:

$$s_b = \frac{v_x - \omega r_d}{v_x} \times 100\% \tag{3.28}$$

Braking maneuver: if $s_b = 0$, the tire is purely rolling; if $s_b = 100\%$, the tire is purely skidding; if $0 < s_b < 100\%$, the tire is rolling and skidding, the slip becomes serious with the longitudinal skid s_b increase.

In order to describe the tire slip accurately in tractive maneuver, longitudinal slip s_a is defined as:

$$s_a = \frac{\omega r_d - v_x}{\omega r_d} \times 100\% \tag{3.29}$$

Driving maneuver: if $s_a = 0$, the tire is purely rolling; if $s_a = 100\%$, the tire is purely spinning; if $0 < s_a < 100\%$, the tire is rolling and slipping, the slip becomes serious as the longitudinal slip s_a increases.

Both the longitudinal slip and skip describe the relationship of the rolling and sliding, the value of them vary from 0 to 100%. For driving and braking are opposite in the longitudinal direction, one single index , slip ratio s, can be used to express both longitudinal slip and longitudinal skid. 0 is the division value, when $0 < s < 100\%$, it expresses braking maneuver, when $s = 100\%$, the wheel will lock completely; when $-100\% < s < 0$, it expresses driving maneuver, and when $s = -100\%$ the wheels are spinning at a high angular speed, but the vehicle does not move forward.

Based on available experimental data, there is closed relationship between the coefficient of road adhesion and the longitudinal slip, as shown in the Fig. as follows:

1. If $|s| = 0 \sim 15\%$, the value of ϕ increases linearly with s
2. If $|s| = 15\% \sim 30\%$, the value of ϕ reaches the maximum and is defined as peak coefficient of road adhesion.
3. If $|s| = 30\% \sim 100\%$, the value of ϕ gradually falls with s increase and is defined as sliding coefficeint of road adhesion when $|s| = 100\%$.

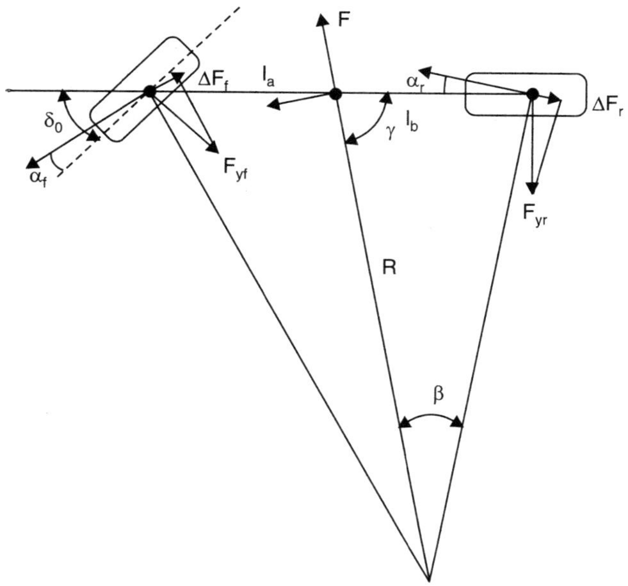

Fig. 3.15 Rolling resistance of turning wheel.

The tractive effort or braking effort will reach the maximum when the tire is rolling and slipping. Correspondingly, the value of $|s|$ is between 15% and 30%. Therefore, it is very important to avoid wheel lock-up during braking or wheel spinning during acceleration. This is one of the impetuses to the development of antilock brake systems and traction control systems for road vehicles. Further detailed discussion of braking and accelerating is provided in Chapters 6 and 7 of this book.

The peak value of the coefficient of road adhesion is approximately 1.2 times the sliding value on a dry surface; and is approximately 1.3 times on a wet surface. Average peak and sliding values of the coefficient of road adhesion ϕ_p and ϕ_s on various surfaces are given in Table 3.2.

Table 3.2 Peak and Sliding Values of the Coefficient of Road Adhesion

p	*Peak value* ϕ_p	*Sliding value* ϕ_s
Asphalt and concrete(dry)	0.8~0.9	0.75
Asphalt(wet)	0.5~0.7	0.45~0.6
Concrete(wet)	0.8	0.7
Gravel	0.6	0.55
Earth road(dry)	0.68	0.65
Earth road(wet)	0.55	0.4~0.5
Snow(hard-packed)	0.2	0.15
Ice	0.1	0.07

The coefficient of road adhesion mainly depends on road texture and suface, tire structure, tread pattern, material and inflation pressure, normal loading on the wheel and the travel speed of the vehicle, etc. Generally speaking, on a given road the larger area of the tire/road contact patch, the larger limited adhesion force will produce, then the high value of adhesion coefficient. On asphalt or concrete, the tire adhesion coefficient is considerably higher than that on field or ice. On wet surfaces, a lower tire adhesion coefficient than on dry surfaces is usually observed. The tire /road contact path of the tire with wide tread is larger than that of the tire with narrow tread, so the adhesion coefficient of wide tread tire is better than narrow tread tire. The experiment data show the adhesion coefficient of radial tire is higher than that of bias tire and the tire with high inflation pressure is lower than the tire with low inflation pressure. In addition, vehicle speed has an effect on the adhesion coefficient, a low travel speed can produce a relatively high adhesion coefficient of the tire.

3.3.5 Theoretical Model of Tire Longitudinal Force under Driving and Braking

In fact, it is difficult to obtain an accurate longitudinal friction force of the tire through the equation $F_x = F_z\varphi$. In Julien's theory, it is assumed that the tire tread can be regarded as an elastic band, and that the contact patch is rectangular and the normal pressure is uniformly distributed. It is further assumed that the contact patch can be divided into an adhesion region and a sliding region. For a pure rolling tire only adhesion region exists. The adhesion region will become smaller and smaller when the tire slip becomes more and more serious. When the tire is sliding the adhesion region will disappear, then there exists only sliding region.

It is assumed that in the adhesion region the longitudinal stresses depend on the longitudinal stiffness and the longitudinal elastic deformation of the tire, whereas in the sliding region the longitudinal stresses depend upon the adhesive properties of the tire-ground interface such as contact pressure and road/tire friction coefficient, then road/tire friction coefficient can be approximately expressed linearly as follows

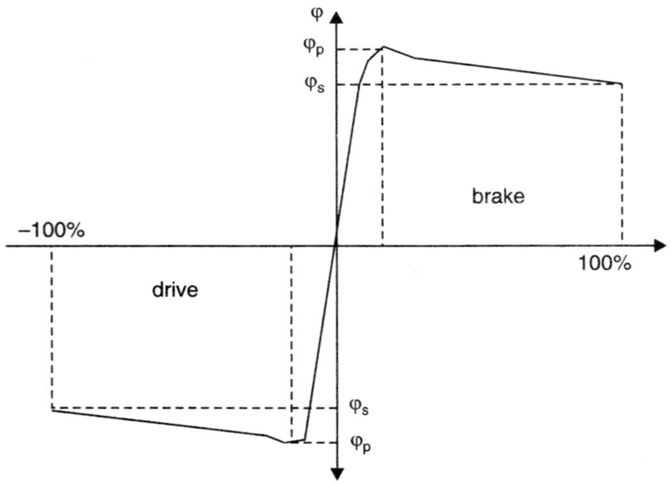

Fig. 3.16 Variation of the coefficient of road adhesion with longitudinal slip.

$$\varphi_x = \varphi_0(1 - As) \tag{3.30}$$

$$A = \frac{\left(1 - \dfrac{\phi_1}{\phi_0}\right)}{s} \tag{3.31}$$

where, ϕ_0 is the road/tire friction coefficient when no sliding takes place.

ϕ_1 is the road/tire friction coefficient when slip ratio is s_1.

A is the parameter that can be obtained through the two points $(0, \phi_0),(S_1, \phi_1)$.

The normal force of the tire on contact patch can be expressed as:

$$F_z = k_z\delta \tag{3.32}$$

where, k_z is the radial stiffness of the tire.

δ is the radial deformation of the tire.

The longitudinal stress can be expressed as

$$\sigma_\xi = \begin{cases} k_x s\xi & 0 \le \xi \le l_a \\ \varphi_x p & l_a \le \xi \le l \end{cases} \tag{3.33}$$

where, σ_ξ is the Longitudinal stress of the tire tread.

ϕ_x is the tire/road longitudinal friction coefficient.

l is the tire contact patch length.

p is the tire/road contact pressure.

k_x is the tire longitudinal stiffness per unit area in the contact patch.

s is the absolute value of longitudinal slip ratio.

ξ is the tire longitudinal deformation.

l_a is the Length of tire contact patch.

The nondimensional tire contact patch length is expressed as: $S_n = s/S_c$; $l_n = l - S_n$, then, the longitudinal stress of the tire tread is integrated in the longitudinal direction, the longitudinal friction force can be expressed as follows:

$$F_x = \begin{cases} K_x s l_n^2 + \varphi_x F_z\left(1 - 3l_n^2 + 2l_n^3\right) & s < S_c \\ \varphi_x F_z & s \ge S_c \end{cases} \tag{3.34}$$

where, S_c is the critical value to divide adhesion region and sliding region expressed as:

$$S_c = 3\phi_x F_z / K_x \tag{3.35}$$

where, K_x is the total longitudinal stiffness.

3.4 TIRE LATERAL DYNAMICS

Tire lateral dynamics mainly discuss the cornering characteristic, roll characteristic, lateral forces at the tire-ground contact patch and self-aligning torque caused by slipping. It also analyzes the relationships between slip angle, slip stiffness, cornering force and self-aligning torque to obtain the precise lateral dynamics of the tire.

3.4.1 Tire Cornering Characteristics

Due to the road bank, lateral wind or the centrifugal force when driving, a side force F_y will be applied to the center of the tire, correspondingly a lateral reaction force will also be developed at the contact patch. And it is called the cornering force F_y. The phenomenon of side slip is mainly due to the tire lateral elasticity, which makes the driving direction deviate from the direction of the tire plane. When tire begins to slip, the center line which is connected by the points A_0, A_1, A_2, A_3, A_4 on the wheel tread will be distorted. Then it can be seen at the following picture that the points A_0, A_1, A_2, A_3, A_4 will be located on the points A_0', A_1', A_2', A_3', A_4' during the wheel rolling. The angle between the path line $a-a$ connected by those points and the line $c-c$ of intersection of the wheel plane with the road surface will be formed. It is usually referenced as slip angle α and the wheel rolls along the direction of $a-a$.

The value of the slip angle is related to that of the cornering forces. Experiments indicate that, if the slip angle is not more than 4 ~ 5 degrees, the lateral force increases linearly with the slip angle. When the vehicle is traveling normally, the lateral acceleration is not more than 0.4g and the slip angle is less than 4 ~ 5 degrees. Subsequently, the relationship between cornering force and slip angle in practice can be shown as:

$$F_y = K_y \alpha \tag{3.36}$$

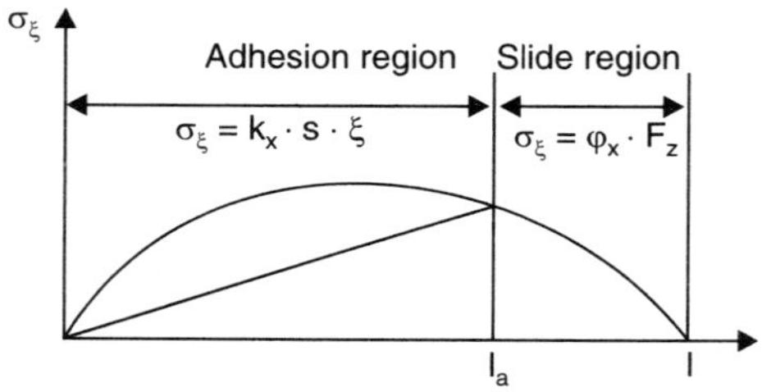

Fig. 3.17 Variation of longitudinal stiffness with tire contact patch length.

K_y which can be referenced to the slip stiffness of tire is the slope of $F_y \sim \alpha$ curve at $\alpha = 0$. According to right hand rule, the positive cornering force generates the negative slip angle. The tire slip stiffness of the passenger car usually ranges from – 28000 – 80000 N/rad and it is an important parameter of automobile stability. The absolute value of the tire slip stiffness should be large in order to keep better manipulating stability. With the large cornering force, the slip angle increases at the higher rate. Partial sideslip has occurred on the tire-ground contact patch. When the cornering force comes up to the ultimate adhesion, the whole tire will begin to sideslip. So the maximum value of the cornering force is determined by the adhesion conditions. Generally speaking, the larger the tire slip force is, the better the automobile limit performance.

Size, model and structure parameters of tire have great effect on the slip stiffness. Big size and low profile tire have high slip stiffness. Due to the wide contact patch of the radial ply tire, its slip stiffness is also very high. Moreover, the normal load on the tire strongly influences the cornering characteristics. The cornering force generally increases with an increase of the normal load. However, it will be decreased, if the perpendicular loads are too large, which makes the pressure of ground-tire contact an inhomogeneous distribution. Inflation pressure also has a moderate effect on the cornering properties of a tire. In general, the tires cornering stiffness, which increases with the increase of the inflation pressure, does not change when the tire pressure arrives to a certain high level. But the velocity of vehicle has little influence on the slip stiffness.

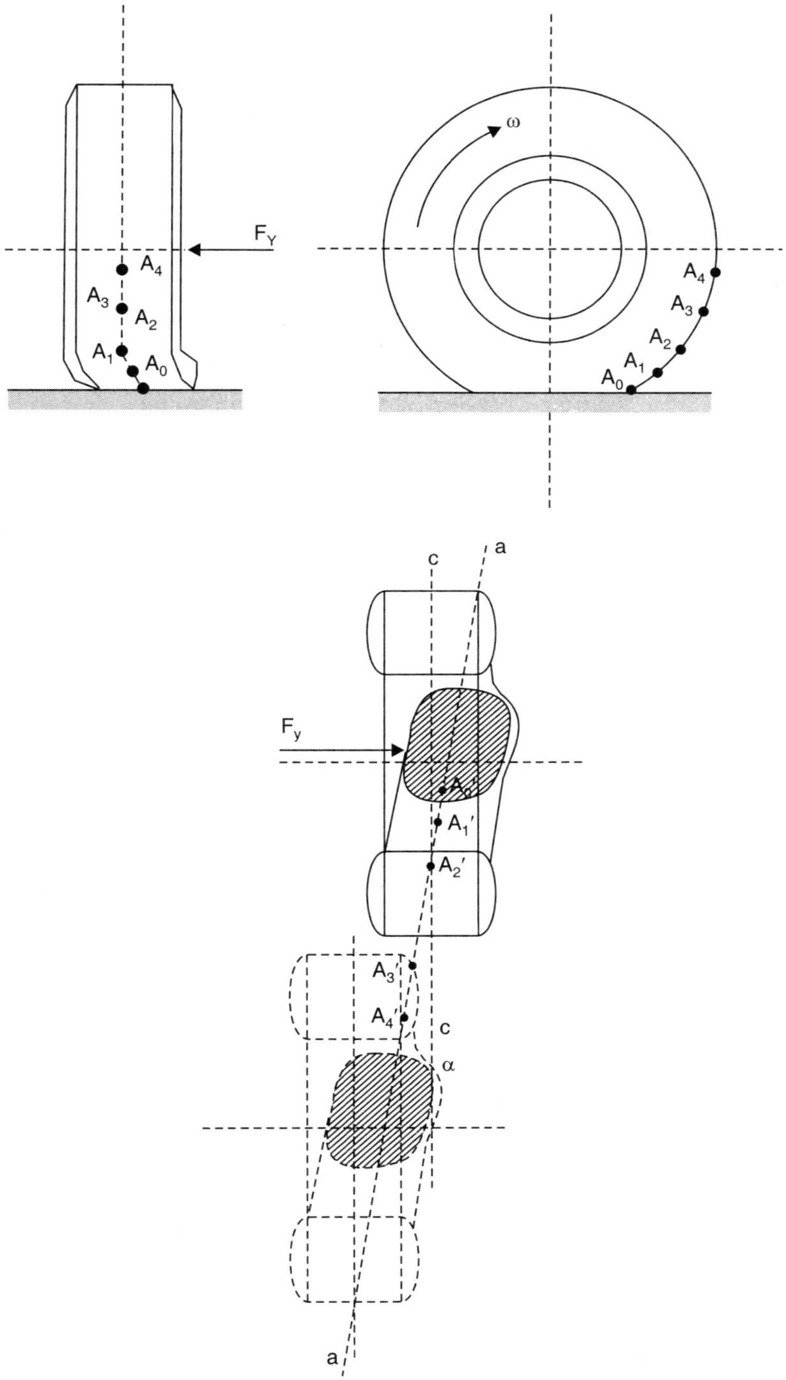

Fig. 3.18 Cornering characteristics of rolling tire.

When the tire is rolling, there is distance and slip angle between the patch line and the wheel plane. The forepart of the patch is close to the wheel plane, which results in the small lateral deformation while the rearward is far from it, which results in large lateral deformation. The patch distribution force is consistent with the deformation. When the tire is under cornering, the slip force is not added to the center of the patch and the distance which is called pneumatic trail is multiplied with the slip forces to form a torque (or couple), which tends to align the wheel plane with the direction of motion. This torque is called the aligning or self-aligning torque, and is one of the primaries restoring moments which help the steered tire return to the original position after negotiating a turn.

The rearward of the contact patch will reach the ultimate adhesion if the cornering force is increased to a certain level. The ultimate adhesion area increases with the increase of the cornering force. Tire will slide if the area comes up to the whole contact patch area. The aligning torque first increases with an increase of the slip angle. It reaches a maximum at the slip angle of 4 ~ 6 degrees, and then decreases with a further increase. It reaches zero at the slip angle of 10 ~ 16 degrees and decrease to a negative value with a further increase in the skip angle. The pneumatic trail has the same trend of the aligning torque. It first increases with an increase in the cornering forces and then decreases with a further increase.

At the same slip angle, a big tire has always high aligning torque. Also, the aligning torque of radial-ply tires is higher than that of bias-ply tires. Lower inflation pressure results in longer tire contact length, and hence pneumatic trail. This causes an increase in the aligning torque.

Longitudinal force affects the aligning torque significantly. When the tire is exposed to a longitudinal force, it will spend some adhesion, which decreases the adhesion that the slide forces can use. Generally speaking, the effect of a driving torque is to increase the aligning torque for a given slip angle. It reaches a maximum at a particular value, and then decreases with a further increase. The braking torque has the opposite effect. The variation of aligning torque is shown in the following figure.

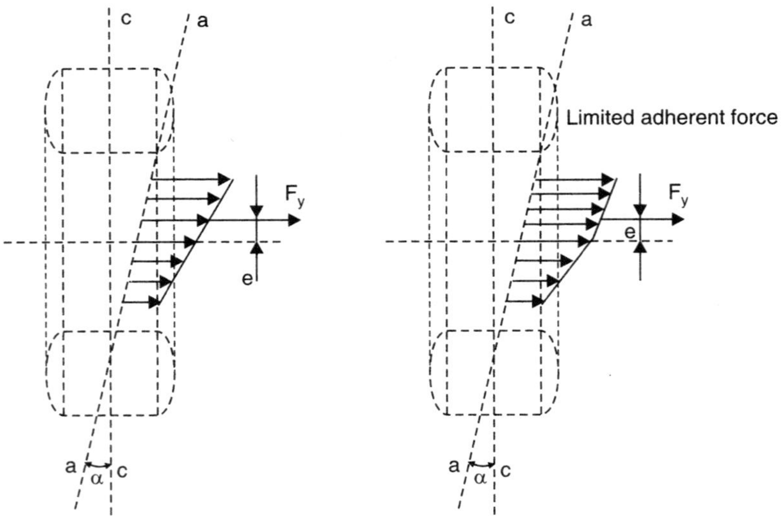

Fig. 3.19 Distribution of lateral ground force in the contact patch.

3.4.2 Mathematical Model of the Tire Cornering Characteristic

Tire cornering characteristic has great effect on the maneuver, stability and safety of the automobile and it is also affected by many factors. Because it is difficult to make research in which the only factor to depend on is tire performance, it is necessary to build up mathematic model to pursue a theoretical model. The purpose of building the model is to acquire the mathematical relationship between the tire structure factor and cornering characteristic and second to provide useful reference for the automobile design and automobile dynamics research.

3.4.2.1 Simplified Mathematical Model of the Tire Cornering Characteristic

The mathematical model is based on the following assumptions:

1. Tire case is rigid, and the whole tire elasticity is applied on the layer of tread.
2. The effect of longitudinal force and skid is ignored. The tire can roll freely.
3. Tire slip angle is zero.
4. The friction coefficient between patch and road is constant.

According to the above assumptions, we can draw the picture of tire force situation as following. When tire rolls at the given slip angle α, we build a coordinate system on the front point of the patch. So the whole patch length is $2a$. With the condition of side force, the tire lateral deformation increases with the increase of x and forms an angle α between the patch line and tire center.

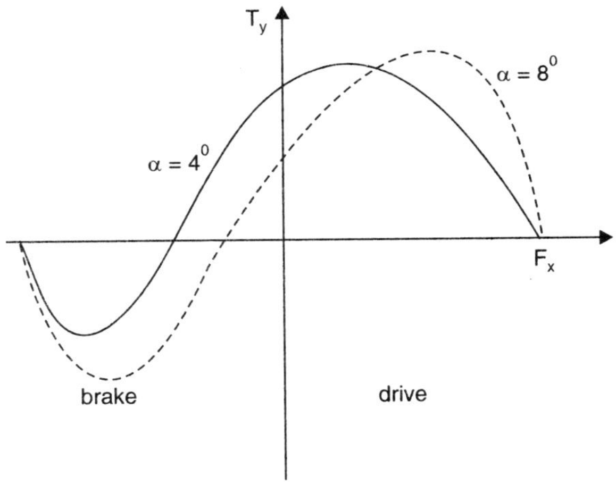

Fig. 3.20 Variation of aligning torque with longitudinal force.

Distribution of tire vertical load

Experiments indicate that the distribution of vertical load is not symmetrical at the patch length (Fig. 3.21). It can be expressed as:

$$q_z = \frac{F_z}{2a}\eta(\mu) \tag{3.37}$$

Where, $\mu = x/a$, $\eta(\mu)$ is the load distribution function. It has two constraints:

$$\eta(\mu) = 0 \quad \mu < 0 \text{ or } \mu > 2 \tag{3.38}$$

According to the balance of force and moments in the z direction, we can get the equations as :

$$\begin{cases} F_z = \int_0^{2a} q_z dx = \int_0^2 aq_z d\mu \\ F_z \cdot (a - \Delta) = \int_0^{2a} xq_z dx = \int_0^2 \mu a^2 q_z d\mu \end{cases} \tag{3.39}$$

where, Δ is defined as the distance between normal force F_z and wheel center line.

Substitute Eq. (3.36) and Eq. (3.38), then

$$\begin{cases} F_z = \int_0^2 a\left(\dfrac{F_z}{2a}\right)\eta(\mu)\,d\mu \\ F_z(a-\Delta) = \int_0^2 (a^2\mu)\left(\dfrac{F_z}{2a}\right)\eta(\mu)\,d\mu \end{cases} \tag{3.40}$$

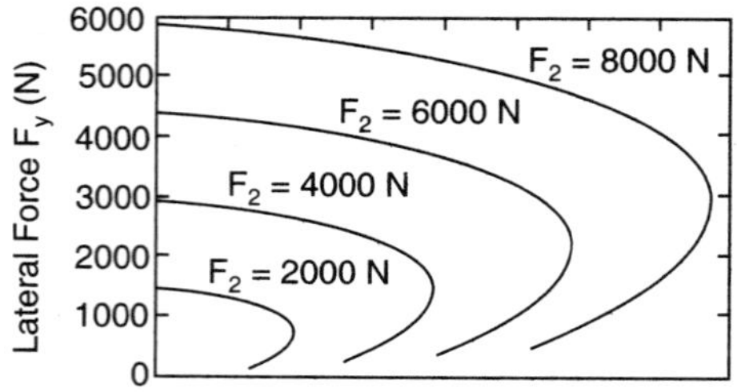

Fig. 3.21 Vertical forces acting on the rolling tire within the contact patch.

After rearranging terms, we can get :

$$\begin{cases} \int_0^2 \eta(\mu)\,d\mu = 2 \\ \int_0^2 \mu\eta(\mu)\,d\mu = 2\left(1-\dfrac{\Delta}{a}\right) = 2(1-\theta) \qquad \theta = \dfrac{\Delta}{a} \end{cases} \tag{3.41}$$

Tire cornering force and aligning torque

When the tire is rolling, the lateral deformation of patch is proportional to the value of $tg\,\alpha$. So the lateral load distribution is:

$$q_y = k_y\, x\, tg\,\alpha \tag{3.42}$$

Where, k_y is the lateral distribution stiffness of the rubber layer, which is defined as a constant. So the equation of the cornering force and aligning torque can be written as:

$$F_y = \int_0^{2a} q_y dx = \int_0^{2} a^2\mu k_y tg\alpha\, d\mu = 2a^2 k_y \cdot tg\alpha \tag{3.43}$$

$$T_y = \int_{-a}^{a} x q_y dx = \int_{-a}^{a} k_y tg\alpha x^2\, dx = \frac{2}{3}a^3 k_y \cdot tg\alpha = \frac{a}{3}F_y \tag{3.44}$$

Define $K_y = 2ak_y$ and substitute it in the Eq. 3.42, we can get the equation $F_y = K_y\, tg\alpha$ Compared with the experiment results, when the value of α is small, $tg\,\alpha \approx \alpha$, and $F_y = K_y\alpha$. Note that K_y is the slip stiffness of tire.

Define $e = \dfrac{a}{3}$, and substitute it in the Eq. (3.43); we can get the equation $T_y = F_y e$, where e is the tire trail.

Defining the tire cornering force and aligning torque when partial sideslip occurs on the tire-ground contact patch.

With the large cornering force and the slip angle, experiment shows that partial sideslip has occurred on the tire, because of $q_y \geq \varphi q_z$ at the rearward patch. ϕ is the friction coefficient, and the origin position can be defined by $q_y = \varphi q_z$.

Substitute Eq. (3.36) and (3.42), one gets

$$k_y x tg\alpha = \varphi \frac{F_z}{2a}\eta(\mu) \tag{3.45}$$

Substitute the equations $x = \mu a$ and $K_y = 2ak_y$ one gets

$$K_y \mu tg\alpha = \varphi F_z \eta(\mu) \tag{3.46}$$

Define the skid ratio as :

$$\phi_y = \frac{K_y tg\alpha}{\varphi F_z} \tag{3.47}$$

Then:

$$\varphi_y = \frac{\eta(\mu')}{\mu'} \tag{3.48}$$

Where, μ' is referenced as the original position of skidding and the value of it at different slip angle or slip ratio can be acquired. The whole contact patch can be divided into two parts: $0 \le \mu' \le \mu$ is the part of adhesion and its side force distribution is $q_y = k_y x tg\alpha$; $\mu \le \mu' \le 2$ is the part of skidding. Its side force distribution is $q_z = \frac{F_z}{2a}\eta(\mu)$. The total side force can be expressed as:

$$\begin{aligned} F_y &= \int_0^{\mu'} a^2 \cdot k_y \cdot tg\alpha \cdot \mu \cdot d\mu + \int_{\mu'}^{2} a \cdot \phi \cdot \frac{F_z}{2a} \cdot \eta(\mu) \cdot d\mu \\ &= \frac{\mu'^2}{4} \cdot K_y \cdot tg\alpha + \frac{\phi \cdot F_z}{2} \int_{\mu'}^{2} \eta(\mu)\, d\mu \\ &= \frac{\mu'^2}{4} \varphi_y \phi \cdot F_z + \frac{\phi \cdot F_z}{2} \left[\int_0^2 \eta(\mu)\, d\mu - \int_0^{\mu'} \eta(\mu)\, d\mu \right] \\ &= \left(\frac{\phi \cdot F_z}{4} \right) \cdot \eta(\mu') \cdot \mu' + \frac{\phi \cdot F_z}{2} \left[\int_0^2 \eta(\mu)\, d\mu - \int_0^{\mu'} \eta(\mu)\, d\mu \right] \end{aligned} \tag{3.49}$$

Define the zero order and first order moment of $\eta(\mu)$ at the range of $0 \sim \mu$ as:

$$m_0(\mu) = \int_0^{\mu} \eta(\mu)\, d\mu \tag{3.50}$$

$$m_1(\mu) = \int_0^{\mu} \mu\eta(\mu)\, d\mu \tag{3.51}$$

when, $\mu = 2$

$$m_0(2) = 2, \quad m_1(2) = 2(1 - \theta) \tag{3.52}$$

The total cornering force with no dimension can be defined as :

$$\frac{F_y}{\varphi F_z} = \begin{cases} \frac{\phi_y}{4} \times \mu'^2 + 1 - \frac{m_0(\mu')}{2} & 0 \le \mu' < 2 \\ \phi_y & \mu' = 2 \end{cases} \tag{3.53}$$

According to Eq. (3.52), the nondimensional cornering force $\frac{F_y}{\varphi F_z}$ is a single variable function.

The total aligning torque is :

$$T_z = \int_0^{\mu'} a^3 tg\alpha k_y \mu^2 d\mu + \int_{\mu'}^{2} a\mu\left[\varphi\frac{F_z}{2}\right]\eta(\mu)d\mu - F_y a \tag{3.54}$$

$$= \frac{a}{6}K_y tg\alpha\mu'^3 + a\varphi F_z\left(1-\theta-\frac{m_1(\mu')}{2}\right) - F_y a$$

The nondimensional total aligning torque can be defined as:

$$\frac{T_z}{\varphi F_z a} = \begin{cases} \frac{\mu'^2}{2}\left(\frac{\mu'}{3}-\frac{1}{2}\right)\times\phi_y - \theta + \frac{m_0(\mu')-m_1(\mu')}{2} & 0\le\mu'<2 \\ \frac{\phi_y}{3} & \mu'=2 \end{cases} \tag{3.55}$$

With the condition of vertical load distribution function, the procedure of acquiring nondimensional cornering characteristics is given below:

1. Given nondimensional side skidding ratio, the original position of skidding can be obtained according to Eq. (3. 47).
2. With the load distribution function, the value of $m_0(\mu)$ and $m_1(\mu)$ at $\mu = \mu'$ can be calculated by Eqs. (3.49) and (3.50). Then calculate the value of θ by Eq. (3.40).
3. Substitute the value of φ_y, μ', $m_0(\mu)$, $m_1(\mu)$, θ into Eqs. (3.52) and (3.54) to get the nondimensional cornering force $\frac{F_y}{\varphi F_z}$ and aligning torque $\frac{T_z}{\varphi F_z a}$.

This kind of tire model is a rigid case and flexible tread. Its nondimensional cornering characteristic (including $\frac{F_y}{\varphi F_z} \sim \phi_y$ and $\frac{T_z}{\varphi F_z a} \sim \phi_y$) is only related to the vertical load distribution function. If the $\eta(\mu)$ function doesn't change when the vertical load F_z changes, the result of $\frac{F_y}{\varphi F_z} \sim \phi_y$ and $\frac{T_z}{\varphi F_z a} \sim \phi_y$ at different F_z can be drawn only by one curve.

Experiments results indicate that the vertical load distribution function $\eta(\mu)$ always changes with the load F_z. When the load is low, the middle results of the $\eta(\mu)$ curve are large. And it tends to two sides of coordination with the increase in F_z (Figure 3.22).

3.4.2.2 Cornering Characteristic with Lateral Bending Deformation of Tire Case

Cornering characteristic with lateral bending deformation of the tire case is more close to realism. Consider the deformation of tire case and tread with the function of cornering forces, the relationship between the contact patch and wheel centric plane is shown as:

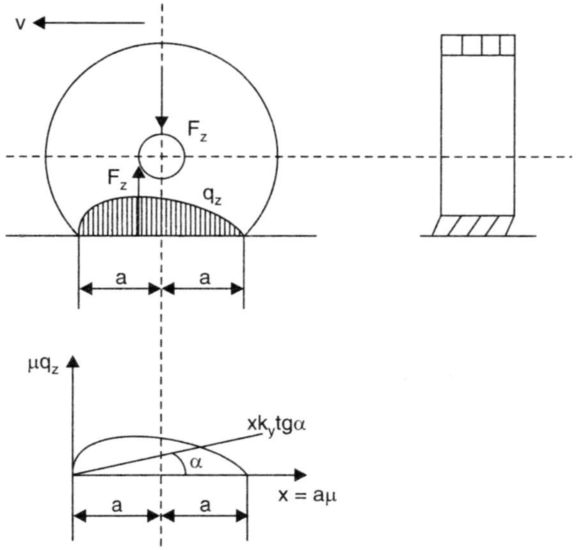

Fig. 3.22 Variation of load distribution function with load and μ.

OX is the wheel centric plane with no cornering force. With the cornering force, the centric plane will be moved from line *OX* to line *AA* and the tire case deforms to the line *AEC*. The contact patch rolls from point *A*, along the direction of line *APB*. The angle between line *APB* and line *AA* is the slip angle α. *T* is time, point *A* rolls to point *P*. The distance between point *P* and *E* is the tread deformation y_r and the distance between point *E* and point *X* is the tire case deformation. It is made of two parts: lateral deformation y_0 and lateral bending deformation y_b and they are related to the cornering forces.

$$\begin{cases} y_0 = \dfrac{F_y}{k_{y0}} \\ y_b = F_y \dfrac{\xi(\mu)}{k_{yb}} \end{cases} \tag{3.56}$$

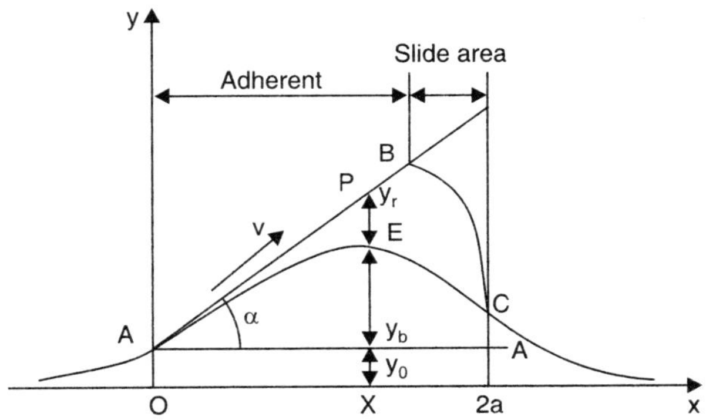

Fig. 3.23 Relationship between the contact patch and the wheel centric plane.

where, k_{y0}, and k_{yb} are defined as two constants of the tire case deformation. $\mu = x/a$ is defined as the relative coordinate to point *P*; $\xi(\mu)$ is the shape function of the tire case deformation.

We can easily get the following equation from the drawings

$$y_r + y_b = xtg\alpha \tag{3.57}$$

Respectively, define the slip ratio of the tire tread and case as $S_{yr} = y_r/x$, $S_{yb} = y_b/x$, then

$$y_r = a\mu tg\alpha - a\mu S_{yr} \tag{3.58}$$

$$S_{yb} = F_y\xi(\mu)/k_{yb}a\mu \tag{3.59}$$

Subsequently, the lateral shear stress distribution on the whole patch can be written as:

$$q_y = k_{yr}y_r = k_{yr}\left(a\mu tg\alpha - q\mu S_{yb}\right) \tag{3.60}$$

where, k_{yr} is the lateral stiffness of the tire tread.

With small slip angle α, there is not side skidding on the whole patch. So the total cornering force can be expressed as:

$$F_y = \int_0^{2a} q_r dx = \int_0^2 a^2\mu k_{yr} tg\alpha d\mu - \int_0^2 a^2\mu k_{yr}\frac{F_y}{k_{yb}a\mu}\xi(\mu)d\mu \tag{3.61}$$

After integrating:

$$F_y = 2a^2k_{yr}tg\alpha - a\frac{k_{yr}}{k_{yb}}F_y\int_0^2\xi(\mu)d\mu \tag{3.62}$$

Define the zero order moment of the relative deformation of the tire case as

$$D(\mu) = \left(ak_{yr}/k_{yb}\right)\int_0^u\xi(\mu)d\mu \tag{3.63}$$

When $\mu = 2$, $D(2) = 2ak_{yr}/k_{yb}$

Not considering the deformation of the tire case, $K_{yr} = K_y = 2a^2k_{yr}$ leads to $\xi(\mu) = 0$ and the tire cornering force can be expressed as

$$F_y = 2a^2k_{yr}tg\alpha = K_{yr}tg\alpha \tag{3.64}$$

Considering the tire deformation, then

$$F_y = K_{yr}tg\alpha - F_yD(2) = K_{yr}tg\alpha/\left[1 + D(2)\right] = K_ytg\alpha \tag{3.65}$$

where, $K_y = K_{yr}/[1 + D(2)]$, which includes the rigidity of the tire case and tread. So it is referenced as general slip stiffness.

Define $K_{yb} = ak_{yb}$, then we can get the following equation

$$K_y = \frac{K_{yr}K_{yb}}{K_{yr} + K_{yb}} \tag{3.66}$$

General slip stiffness is just like two springs in series connected by slip stiffness of tire case and tread. Due to the elasticity of tire case, the whole tire slip stiffness decreases $[1 + D(2)]$ times.

With small slip angle, the aligning torque is expressed as

$$T_z = \int_0^2 a^2q_y\mu d\mu - F_ya$$

$$= \int_0^2 a^3\mu^2 k_{yr}\left[tg\alpha - F_y/\left(ak_{yb}\right)\xi(\mu)/\mu\right]d\mu - F_y a \qquad (3.67)$$

Define the first order moment of the relative deformation function of tire case as:

$$D_1(\mu) = \left(\frac{ak_{yr}}{k_{yb}}\right)\int_0^u \xi(\mu)\mu d\mu \qquad (3.68)$$

Finally, the aligning torque is given by

$$T_z = F_y a\left[\frac{1}{3} + 4D(2)/3 - D_1(2)\right] \qquad (3.69)$$

According to the cornering properties of tire model that considers lateral elasticity of the tire case. The main factors that have great effect on the tire cornering properties are listed as follows:

1. **Tire load:** the slip stiffness of the tire tread is proportional to the load while the slip stiffness of the tire case is opposite. Subsequently with the increase of load, K_{yr} increases and K_{yb} decreases, which result that $D(2)$ increases sharply and the general slip stiffness decreases rapidly.
2. **Tire abrasion:** when the tire begins to abrase, the shear stiffness of tread increases, which increases the general slip stiffness. When the tire treads have abrased 1/2, the general slip stiffness will increase by 20%.
3. **Tire structure:** passenger car tire usually has few cord layers and its inflation pressure is low, which results in the small stiffness of the passenger car tire case. So the slip stiffness proportion $D(2) = K_{yr}/K_{yb}$ of the passenger car tire is larger than that of the truck. As a result, the tire slip stiffness of the passenger car is obviously affected by the variation of the load. When the rear axle is over-loaded, it is easy for a passenger car to oversteer with its tire trail also large. Hence, it is better for the passenger car to use small kingpin caster angle, especially the negative angle. The slip stiffness of truck tire case is high and its general slip stiffness is not sensitive to the variation of load.

3.4.3 Rolling Properties of Tires

There is an angle between the wheel plane and the line perpendicular to the ground plane in the modern vehicle, which is called the camber angle of tire. If the upper part of wheel inclines outside the car body, the angle is positive. The front wheels of the car usually contain a tiny positive camber angle $\gamma = 5' \sim 10'$ with zero load, which makes the tire perpendicular to the ground surface with a little vaulting and abrades uniformly with decrease of rolling resistance. The angle turns out to be tinily minus under full load.

A free-rolling tire with a camber angle would revolve about point O, as shown in the figure. However, the cambered tire in a vehicle is constrained to move in a straight line. There must be a side force F_{y0} acting on the tire centre. A lateral force in the direction of the camber is, therefore, developed at the contact point, which is referred to as the camber thrust F_{yr}. The relationship of the camber thrust with the camber angle is linear, and is defined as:

$$F_{yr} = K_r\gamma \qquad (3.70)$$

where K_r is defined as the camber stiffness. It is noted that minus camber angle generates positive camber thrust.

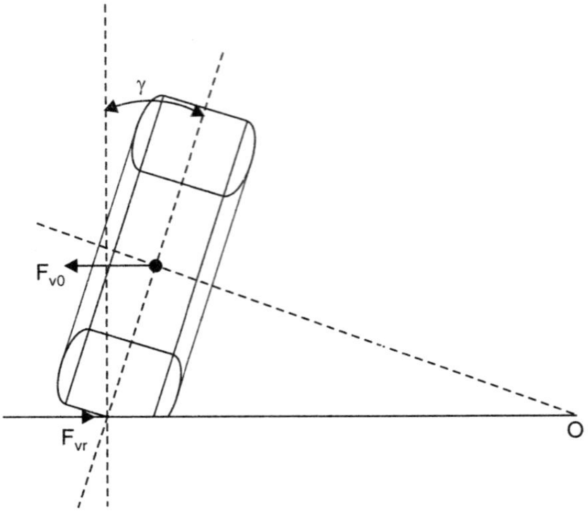

Fig. 3.24 Side forces of tires with camber angle.

The relationship between the camber thrust and the camber angle (at zero slip angles) is illustrated in Fig. 3.25. Three lines present the cornering properties under a little slip angle corresponding to a camber angle which is positive, zero and negative, respectively.

1. The ground lateral force with a zero slip angle is the camber thrust.
2. When there is a camber angle, the relationship between the cornering force and the camber angle as well as the cornering angle is given as:

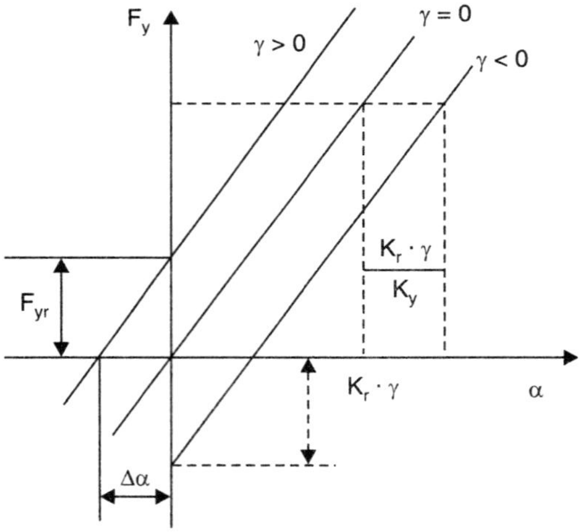

Fig. 3.25 The relationship between cambers thrust and camber angle (at zero slip angle).

$$F_y = K_y\alpha + K_r\gamma \tag{3.71}$$

where K_y is the cornering stiffness of tires without inclination, K_r is the camber stiffness of tires, α is the slip angle of tires without inclination, and γ is the camber angle of tires.

It should be noted that with the increase of the camber angle, the contact state of the tire and the road surface would be worse, which would influence the maximum ground lateral force and degrade the ultimate performance of the vehicle. Therefore, it should make sure that the front wheel is perpendicular to the ground plane as much as possible under the steering driving condition.

According to the tire axis system, the positive slip angle corresponds to the negative cornering force and positive self-aligning moment. Also, positive camber angle corresponds to negative camber thrust and camber self-aligning moment.

3.4.3.1 Cambered Tire Models

A simplified cambered tire models

There are some assumptions as follows:

1. the whole elasticity of tire is concentrated on the tire tread, without considering the elasticity of the tire carcass, and the tire is equivalent to be a brush.
2. the width of tire is neglected.
3. the form of contact patch is rectangle.

The broken line shown in the figure represents the deformed part of tire. A coordinate system Oxz is located at the left side of the contact patch with a length of $2a$. The vertical deformation of the tire is a function of x, which can be viewed as a quadratic function.

$$\Delta_z = \Delta z_m \mu(2-\mu) = \frac{a^2}{2R}\mu(2-\mu) \tag{3.72}$$

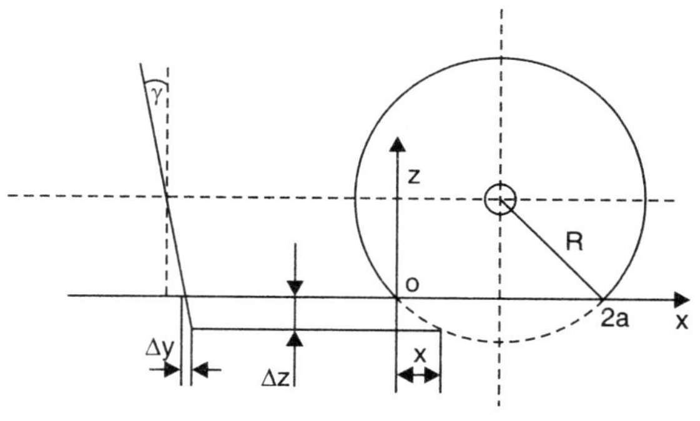

Fig. 3.26 Vertical deformation of rolling tires with camber angle in the contact patch.

where $\Delta z = \dfrac{a^2}{2R}$ presents the maximun deformation in the middle of the contact patch. Corresponding to x, the lateral deformation is given as:

$$\Delta y = \Delta z tg\gamma = \frac{a^2}{2R}\mu(2-\mu)tg\gamma \tag{3.73}$$

If it is assumed that the lateral distributed stiffness per length of tire tread is k_y, then the lateral stress at the site x can be defined by:

$$q_y = k_y \Delta y = k_y a^2 \mu(2-\mu) tg\gamma / 2R \tag{3.74}$$

Therefore, the lateral force with camber is written as:

$$F_{yr} = \left(\frac{k_y a^2}{2R}\right) tg\gamma \int_0^{2a} \mu(2-\mu)\,dx = \left(\frac{2k_y a^3}{3R}\right) tg\gamma \tag{3.75}$$

Set the cornering stiffness of tire as $K_y = 2a^2 k_y$, the lateral force with camber can be given by:

$$F_{yr} = \left(\frac{a}{3R}\right) K_y tg\gamma = K_r tg\gamma \tag{3.76}$$

Where $K_r = (a/3R)K_y$ presents camber stiffness. The camber stiffness is in proportion to the cornering stiffness and the contact patch is in inverse proportion to the radius of tire.

3.4.3.2 Cambered Tire Model with Roll Elastic Deformation of Tire Carcass

With the consideration of roll elastic deformation of tire in the modeling of cambered tire, there will be an angle, called the roll angle γ_c, formed between the tire carcass and felloe under the action of lateral force with camber. Especially, for the radial ply tire, the influence of the roll of tire cannot be neglected due to the biggish elasticity of tire carcass.

With the roll force acting on the tire, the relationship between the roll force and the roll angle of the tire carcass and felloe is expressed as:

$$tg\gamma_c = F_{ry}/K_c \tag{3.77}$$

where K_c is the camber stiffness of tire carcass.

There is a part of lateral deformation of tire tread absorbed by the roll of tire carcass, and then the new deformation is given by:

$$\Delta y' = \Delta y - \Delta z tg\gamma_c \tag{3.78}$$

According to the Eq. (3.74), the roll force can be obtained by:

$$F_{yr} = \left(\frac{a}{3R}\right) K_y \left(tg\gamma - \frac{F_{yr}}{K_c}\right) \tag{3.79}$$

Finally, the roll force of tire owing to the roll of tire can be gained as:

$$F_{yr} = K_r' tg\gamma \tag{3.80}$$

where:

$$K_r' = \frac{K_r K_c}{K_r + K_c} \tag{3.81}$$

The Eq. (3.81) indicates that the camber stiffness of tire is exactly formed in series with the camber stiffness of tire tread K_r, and the camber stiffness of tire carcass K_c. The camber stiffness of tire carcass of radial ply tire is less than the counterpart of the bias ply tire, so the former obtains a smaller camber stiffness. With the same camber angle, there should be a less toe-in angle for the former.

3.5 TIRE MECHANICS MODEL CONSIDERING LONGITUDINAL SLIP AND CORNERING CHARACTERISTICS

Longitudinal force and lateral force acting on the wheel when the vehicle is in motion, result from many factors such as tire toe-in, camber, road surface roughness, lateral wind drag and wheel's steering, thus, in most cases each tire moves under combined slip condition. To establish a tire model under the condition of combined silp is, therefore, very important to analyze the vehicle dynamics. Several main longitudinal slip and cornering characteristics tire models are introduced as follows.

3.5.1 C. G. Gim Theoretical Model

In view of the difference between the case of braking and driving, first, define the longitudinal and lateral slip ratio as follows:

$$S_s = \begin{cases} \dfrac{v_x - v_c}{v_x} > 0 & \text{brake} \\ \dfrac{v_c - v_x}{v_c} < 0 & \text{drive} \end{cases} \tag{3.82}$$

$$S_\alpha = \begin{cases} |\tan\alpha| & \text{brake} \\ (1-|S_s|)|\tan\alpha| & \text{drive} \end{cases} \tag{3.83}$$

where v_c is circumference speed of wheel.
v_x is longitudinal velocity of vehicle.
v_y is lateral velocity of vehicle.
α is slip angle expressed as $\alpha = \text{arctg}\,(v_y/v_x)$.

Due to the influence of longitudinal and lateral force on tire, the adhesion coefficient is defined as:

$$\varphi = \varphi_0\,(1 - AS_{sa}) \tag{3.84}$$

where,
$$A = \frac{\left(1 - \dfrac{\phi_1}{\phi_0}\right)}{S_1} \tag{3.85}$$

$$S_{sa} = \sqrt{S_s^2 + S_a^2} \tag{3.86}$$

ϕ_0 is the static friction coefficient between tire and road surface;
S_{sa} is the associated parameter of slip ratio.

The longitudinal and lateral and comprehensive adhesion coefficients of the tire are given by:

$$\varphi_x = \varphi\frac{S_s}{S_{sa}} = \varphi\cos\alpha \tag{3.87}$$

$$\phi_y = \varphi\frac{S_a}{S_{sa}} = \varphi\sin\alpha \tag{3.88}$$

$$\phi = \sqrt{\phi_x^2 + \phi_y^2} \tag{3.89}$$

Rolling and slip critical point in the contact patch is

$$S_n = \frac{1}{3\phi F_z}\sqrt{(K_s S_s)^2 + (K_a S_a)^2} \tag{3.90}$$

The slip critical point is:

$$S_{sc} = 3\phi F_z/K_s \tag{3.91}$$

The cornering critical point is:

$$S_{ac} = \frac{K_s}{K_a}\sqrt{S_{sc}^2 - S_s^2} \tag{3.92}$$

where K_s is the tire longitudinal stiffness.

K_a is the tire lateral stiffness.

Set $l_n = l - S_n$, thus the longitudinal force between the tire and road surface is expressed as:

$$F_x = \begin{cases} K_s S_s l_n^2 + \phi_x F_z\left(1 - 3l_n^2 + 2l_n^3\right) & S_s < S_{sc} \\ \phi_x F_z & S_s \geq S_{sc} \end{cases} \tag{3.93}$$

The lateral force between the tire and road surface is written as:

$$F_y = \begin{cases} K_a S_a l_n^2 + \phi_y F_z\left(1 - 3l_n^2 + 2l_n^3\right) & S_a < S_{ac} \\ \phi_y F_z & S_a \geq S_{ac} \end{cases} \tag{3.94}$$

The self-aligning moment can be formulated as:

$$T_z = \begin{cases} \left[K_a S_a\left(-\frac{1}{2} + \frac{2}{3}l_n\right) + \frac{3}{2}\phi_y F_z S_n^2\right] \cdot l \cdot l_n^2 & S_a < S_{ac} \\ 0 & S_a \geq S_{ac} \end{cases} \tag{3.95}$$

where l is the contact length of the tire.

The parameters required by the model mentioned above are specific, which can be obtained by experiment, without fitting a large quantity of experimental data of mechanical characteristics. The model is of high precision in the fitting of the cornering and longitudinal forces. However, due to using a simple physical mechanism in the tire model, there is a big difference between the theoretical calculation of the self-aligning moment and testing data. The self-aligning moment derived from the model will be zero with a large slip angle, which can not reflect the fact that there will be a minus self-aligning moment under a large slip angle.

3.5.2 K. H. Guo Tire Model

3.5.2.1 Steady-State Simplified Theoretical Tire Model

By assuming that elastic translation can take place only in y direction on carcass and angle about z-axis and the displacement along x-axis all can be neglected, therefore, the deformation of the tire contact patch under the condition of braking (or driving) and cornering is shown in the Fig.: α is the slip angle in the Fig., ABC is the tire centerline after cornering, the total length of the tire contact patch is $2a$. The translation deformation of the tire carcass resulting from the influence of cornering force F_y is:

$$y_b = F_y/K_{by} \tag{3.96}$$

Where K_{by} is the lateral carcass stiffness. R is tire rolling radius, ω is angular speed of tire, the initial contact point A on the tire tread is rolling to the point P, through a time t, thus, the counterpart on the tire plane turns from point O to point X.

$$X = \omega R t \tag{3.97}$$

The deformations of the tire tread in the contact patch along x and y direction are written by:

$$\begin{cases} \Delta x = vt\cos\alpha - x \\ \Delta y = vt\sin\alpha \end{cases} \tag{3.98}$$

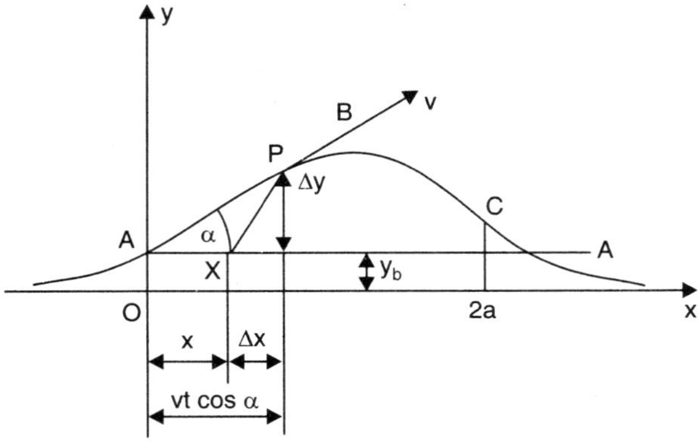

Fig. 3.27 Deformation of tire contact patch under the condition of braking (or driving) and cornering.

Define braking slip ratio S_b and driving slip ratio S_d as:

$$\begin{cases} S_b = \dfrac{\Delta x}{vt\cos\alpha} \\ S_d = -\dfrac{\Delta x}{x} \end{cases} \tag{3.99}$$

hence

$$\begin{cases} \Delta x = S_b x/(1-S_b) = S_x x \\ \Delta y = xtg\alpha/(1-S_b) = S_y x \end{cases} \tag{3.100}$$

Define the longitudinal slip ratio and cornering slip ratio as:

$$\begin{cases} S_x = S_b/(1-S_b) = -S_d \\ S_y = tg\alpha/(1-S_b) = (1+S_x)tg\alpha \end{cases} \tag{3.101}$$

Assuming that the stiffnesses of the tire tread in the x and y directions are respectively k_x and k_y, correspondingly the shear stresses in the x and y directions at point P in contact patch are respectively expressed as:

$$\begin{cases} q_x = k_x\Delta x = k_x S_x x \\ q_y = k_y\Delta y = k_y S_y x \end{cases} \tag{3.102}$$

Tire Without Slip

The resultant shear stress in the contact patch is given by:

$$q = \sqrt{q_x^2 + q_y^2} = \sqrt{(k_x S_x)^2 + (k_y S_y)^2}\, x \tag{3.103}$$

Define the direction of resultant shear stress as:

$$tg\theta = \frac{q_y}{q_x} = \frac{k_y S_y}{(k_x S_x)} \tag{3.104}$$

Integrate the shear stress over the contact patch and obtain the total longitudinal force and cornering force as follows:

$$\begin{cases} F_x = 2a^2 k_x S_x = K_x S_x \\ F_y = 2a^2 k_y S_y = K_y S_y \end{cases} \tag{3.105}$$

where the longitudinal slip stiffness is defined as $K_x = 2ak_x$, the cornering stiffness as $K_y = 2ak_y$. The resultant tangential force in contact patch can be described as:

$$F = \sqrt{F_x^2 + F_y^2} = \sqrt{(K_x S_x)^2 + (K_y S_y)^2} \tag{3.106}$$

The nondimensional longitudinal force $\overline{F}_x$, lateral force $\overline{F}_y$ and relative resultant tangential force $\overline{F}$ are defined respectively as:

$$\begin{cases} \overline{F}_x = F_x/(\varphi F_z) \\ \overline{F}_y = F_y/(\varphi F_z) \\ \overline{F} = F/(\varphi F_z) \end{cases} \tag{3.107}$$

Define the nondimensional longitudinal slip ratio ϕ_x, lateral slip ratio ϕ_y and relative resultant slip ratio ϕ as follows:

$$\begin{cases} \phi_x = K_x S_x/(\varphi F_z) \\ \phi_y = K_y S_y/(\varphi F_z) \\ \phi = \sqrt{\phi_x^2 + \phi_y^2} \end{cases} \tag{3.108}$$

where F_z is the tire vertical load, ϕ is friction coefficient between the tire and road surface.

Then the nondimensional formulas of the tire tangential forces in respective directions under condition without slip can be obtained as:

$$\begin{cases} \overline{F}_x = \phi_x = \overline{F}\phi_x/\phi \\ \overline{F}_y = \phi_y = \overline{F}\phi_y/\phi \\ \overline{F} = \sqrt{\overline{F}_x^2 + \overline{F}_y^2} = \Phi \\ tg\theta = \phi_x/\phi_y \end{cases} \tag{3.109}$$

assume that the nondimensional relative shear stress and nondimensional coordinates are as follows

$$\begin{cases} \overline{q} = 2aq/(\varphi F_z) \\ u = x/a \end{cases} \tag{3.110}$$

thus under condition without slip:

$$\begin{cases} q_x = \phi_x u\,\varphi F_z/(2a) \\ q_y = \phi_y u\,\varphi F_z/(2a) \\ q = \phi u\,\varphi F_z/(2a) \end{cases} \tag{3.111}$$

so the aligning moment can be obtained according to the shear stress in the contact patch:

$$T_z = -F_x\left(y_b + \frac{4a}{3}S_y\right) + F_y\frac{a}{3} = F_y D_x - F_x\left(D_y + y_b\right) \tag{3.112}$$

where longitudinal trail is $D_x = 1/3$, the lateral trail is $D_y = \frac{4a}{3} \cdot S_y$.

The nondimensional aligning moment finally can be described as:

$$\overline{T}_z = T_z/(a\varphi F_z) = \overline{F}_y \overline{D}_x - \overline{F}_x\left(\overline{D}_y + y_b/a\right) \tag{3.113}$$

where nondimensional trails are

$$\begin{cases} \bar{D}_x = \dfrac{1}{3} \\ \bar{D}_y = \dfrac{4}{3} S_y \end{cases} \tag{3.114}$$

Tire under condition with slip.

The resultant tangential stress in slide region is $q(x) = \varphi q_z$, on condition that the vertical pressure distribution q_z is known, the coordinate of slide start point $x = x'$ or relative coordinate $u = x/a = u' = x'/a$ can be obtained. Give the common formula of q_z as:

$$q_z = \frac{F_z}{2a}\eta(u) \tag{3.115}$$

where $\eta(u)$ is the nondimensional formula of distribution form of vertical load.
the tenable condition for slip point is

$$x\sqrt{(k_x S_x)^2 + (k_y S_y)^2} = \varphi \frac{F_z}{2a}\eta(u) \tag{3.116}$$

then the nondimensional slipping conditions are derived as:

$$\begin{cases} \varphi = \eta(u)/u \\ u \le 2 \end{cases} \tag{3.117}$$

By assuming that comprehensive relative slip ratio ϕ is known, the position of slide start point $u = u'$ can be derived from formula (3.116). When $u < u'$, in the region of adhesion,

$$\begin{cases} q_x = \phi_x u \varphi F_z/(2a) \\ q_y = \phi_y u \varphi F_z/(2a) \\ q = \phi u \varphi F_z/(2a) \end{cases} \tag{3.118}$$

else $u \ge u'$, in the slide region:

$$q = \varphi q_z \tag{3.119}$$

Therefore, the total tangential force can be formulated as:

$$F = a\int_0^{u'} \varphi F_z \phi u/(2a)\,du + a\int_{u'}^{u} \varphi F_z\, \eta(u)/(2a)\,du \tag{3.120}$$

The nondimensional formula of the total tangential force is

$$\bar{F} = \frac{1}{4}\phi u'^2 + 1 - \frac{m_0}{2} \tag{3.121}$$

where

$$m_0 = \int_0^{u'} \eta(u)\,du \tag{3.122}$$

With the assumption that the tangential stresses in the slide and in adhesive regions are in the same direction, the nondimensional formula of the longitudinal and cornering forces can be expressed as:

$$\begin{cases} \bar{F}_x = \bar{F}\dfrac{\phi_x}{\phi} = \left(\dfrac{\phi_x}{4}\right)u'^2 + \left(1 - \dfrac{m_0}{2}\right)\dfrac{\phi_x}{\phi} \\ \bar{F}_y = \bar{F}\dfrac{\phi_y}{\phi} = \left(\dfrac{\phi_y}{4}\right)u'^2 + \left(1 - \dfrac{m_0}{2}\right)\dfrac{\phi_y}{\phi} \end{cases} \tag{3.123}$$

Likewise, the aligning moment can be formulated as:

$$\bar{T}_z = \phi_y \frac{u'^3}{6} + \frac{\phi_y}{\phi}\left(1 - \frac{\Delta}{a} - \frac{m_1}{2}\right) - \bar{F}_y - \frac{S_y}{6}\phi_x u'^3 - \frac{S_y}{2} E \frac{\phi_x}{\phi^2} - \bar{F}_x \frac{y_b}{a} \tag{3.124}$$

where Δ is the eccentric moment generated by vertical load.

$$m_1 = \int_0^{u'} u\eta(u)\,du \tag{3.125}$$

$$E = \int_{u'}^{2} \eta^2(u)\,du \tag{3.126}$$

The generalized theoretical tire model gives a clear physical concepts and is primarily applied to the description of a steady-state tire dynamics when the vehicle is traveling over a good road surface under a normal driving condition. However, it is not suitable for the analysis of the non-steady state tire dynamics characteristics with the starting and emergency braking and severe steering conditions. Thus, the tire cornering and longitudinal slip characteristics on the non-steady state condition are available.

3.5.2.2 Unsteady State Semi-Empirical Tire Mechanics Model

The figure describes the state of a tire in the unsteady state of motion on road, where *XOY* is the spatial fixed coordinate system, xoy is the tire coordinate system, *o* is the center of the tire contact patch with the coordinates (*X*, *Y*) in the spatial fixed coordinate system, the angle formed between the intersection of the wheel plane and the ground plane and *OX* axis of the spatial fixed coordinate system is the angle Ψ, called the tire steering angle. The length of tire contact patch is $2a$.

$$V_{\text{syn}} = V_r\left[\psi - \left(\frac{dY}{dX} + r\frac{d\gamma}{dX}\right) - a\left(\frac{d\psi}{dX} + \eta_\gamma \frac{\sin\gamma}{r}\right)\right] \tag{3.127}$$

where $V_r = |\omega r|$

ω is the angular speed of tire.

r is effective rolling radius of tire.

γ is camber angle of tire.

and η_γ is influence coefficient of camber angle that represents the influence resulting from the deformation of carcass on the tire effective camber angle.

The nominal longitudinal slip velocity of the center of the tire contact patch is written as:

$$V_{\text{sxn}} = V_x - V_r \tag{3.128}$$

Where V_x is the velocity of the tire center in longitudinal direction, V_r the component speed of the tire center relative to ground in *X*-axis in the tire coordinate system.

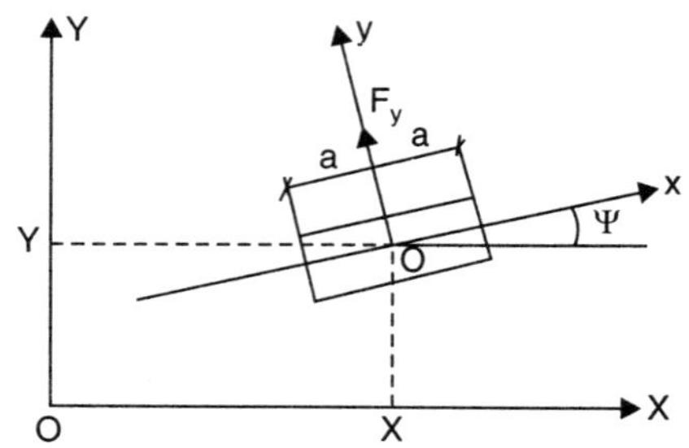

Fig. 3.28 State of a tire in unsteady state motion.

With the assumption that lateral displacement Y and the tire steering angle ψ are lesser, the longitudinal slip and cornering characteristics of the physical model with unsteady state can be described as in Fig. 3.29. Thus the effective longitudinal slip ratio S_x and effective lateral slip ratio S_y on the unsteady state derived from this model can be given by:

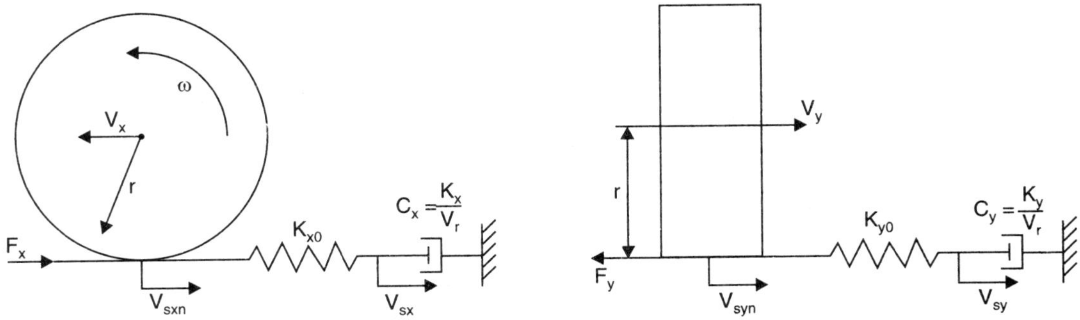

Fig. 3.29 Longitudinal slip and cornering characteristics physical model with unsteady state: (*a*) non-steady state longitudinal slip characteristic physical model, and (*b*) non-steady state cornering characteristics physical model.

$$S_x = \frac{V_{sx}}{V_r} \tag{3.129}$$

$$S_y = \frac{V_{sy}}{V_r} \tag{3.130}$$

$$V_{sx} + \frac{l_x}{V_r}\dot{V}_{sx} = V_{\text{sxn}} \tag{3.131}$$

$$V_{sy} + \frac{l_y}{V_r}\dot{V}_{sy} = V_{\text{syn}} \tag{3.132}$$

Where V_{sx} and V_{sy} are respectively effective longitudinal and lateral slip velocity.

The longitudinal relaxation length l_x and lateral relaxation length l_y are defined respectively as:

$$l_x = \frac{K_x}{K_{x0}} \tag{3.133}$$

$$l_y = \frac{K_y}{K_{y0}} \tag{3.134}$$

where K_{x0} and K_{y0} are respectively the longitudinal and lateral carcass deformation stiffness, K_x and K_y are the tire longitudinal and cornering stiffness corresponding to the effective slip ratio S_x and S_y.

$$K_x = \frac{\partial F_x}{\partial s_x} \tag{3.135}$$

$$K_y = \frac{\partial F_y}{\partial s_y} \tag{3.136}$$

The relation obtained from experiments between the longitudinal friction coefficient ϕ_x as well as lateral friction coefficient and the vertical load F_z can be formulated as:

$$\begin{cases} \phi_x = a_1 + a_2 F_z + a_3 F_z^2 \\ \phi_y = b_1 + b_2 F_z + b_3 F_z^2 \end{cases} \tag{3.137}$$

Where a_1, a_2, a_3, b_1, b_2, b_3 are experimental coefficients.

The relative longitudinal slip ratio φ_x, lateral slip ratio φ_y and total slip ratio φ are :

$$\begin{cases} \phi_x = K_x S_x / (\varphi_x F_z) \\ \phi_y = K_y S_y / (\varphi_y F_z) \\ \phi = \sqrt{\phi_x^2 + \phi_y^2} \end{cases} \tag{3.138}$$

The nondimensional total tangential force $\bar{F}$ is written as:

$$\bar{F} = 1 - \exp\left[-\varphi - E_1 \varphi^2 - \left(E_1^2 + \frac{1}{12} \right) \varphi^3 \right] \tag{3.139}$$

where coefficient E_1 is

$$E_1 = \frac{0.5}{1 + \exp\left(-\frac{F_z - b_4}{b_5} \right)} \tag{3.140}$$

Where b_4, b_5 are the experimental coefficients.

The tire longitudinal force F_x, lateral force F_y and self-aligning moment T_z are respectively expressed as:

$$\begin{cases} F_x = -\frac{\phi_x}{\phi} \varphi_x F_z \bar{F} \\ F_y = -\frac{\phi_y}{\phi} \varphi_y F_z \bar{F} \\ T_z = -F_y D_x - F_x y_0 \end{cases} \tag{3.141}$$

where D_x is pneumatic trail formulated as:

$$D_x = (D_{x0} + D_e) \exp\left(-D_1 \phi - D_2 \phi^2 \right) - D_e \tag{3.142}$$

The relation between the initial value of the pneumatic trail D_{x0} and the vertical load F_z is determined by the experimental data formulated as:

$$D_{x0} = c_1 + c_2 F_z + c_3 F_z^2 \tag{3.143}$$

The relation between the total value of the pneumatic trail D_e and the vertical load F_z is determined by the experimental data formulated as:

$$D_e = c_4 + c_5 F_z + c_6 F_z^2 \tag{3.144}$$

D_1 is a coefficient

$$D_1 = c_7 \exp\left(-\frac{F_z}{c_8} \right) \tag{3.145}$$

D_2 is a coefficient

$$D_2 = c_9 \exp\left(\frac{F_z}{c_{10}}\right) \tag{3.146}$$

y_0 is the lateral translation deformation written by:

$$y_0 = \frac{F_y}{K_{y0}} \tag{3.147}$$

C_{y0} is the lateral deformation stiffness expressed by:

$$C_{y0} = d_1 F_z + d_2 F_z^2 \tag{3.148}$$

c_1, c_{10}, d_1, d_2 in formulas mentioned above are experimental coefficients.

3.5.3 H.B.Pacejka Magic Formula Model

Magic formula is presented by professor Pacejka in Holland Delft Industrial University, which has been shown to suitably match the experimental data of the dynamic tire characteristics by combinational formula of trigonometric function. Pacejka obtained a set of formulas of tire models which can describe the longitudinal and lateral forces and self-aligning moment comprehensively with same form. The formula is called "magic formula".

In pure cornering and pure longitudinal slip conditions, the longitudinal and lateral force and self-aligning moment on tire can be written as:

lateral force:

$$F_x = D\sin\left(C\arctan\left\{B(1-E)(\sigma+S_h)+E\arctan\left[B(\sigma+S_h)\right]\right\}\right)+S_v \tag{3.149}$$

longitudinal slip ratio:

$$\sigma = \frac{\omega}{(V/R_e)} - 1 \tag{3.150}$$

where . ω is the angular speed of tire.

V is the linear speed of tire center.

R_e is the effective rolling radius of tire.

$$C = b_0 \tag{3.151}$$

$$D = \mu p F_z \tag{3.152}$$

$$\mu_p = b_1 F_z + b_2 \tag{3.153}$$

$$\text{B.C.D} = \left(b_3 F_z^2 + b_4 F_z\right) e^{-b_5 F_z} \tag{3.154}$$

$$E = b_6 F_z^2 + b_7 F_z + b_8 \tag{3.155}$$

$$S_h = b_9 F_z + b_{10} \tag{3.156}$$

$$S_v = 0 \tag{3.157}$$

F_z is vertical load on tire, unit: kN

The lateral force is given by:

$$F_y = D\sin\left(C\arctan\left\{B(1-E)(\alpha+S_h)+E\arctan\left[B(\alpha+S_h)\right]\right\}\right)+S_v \tag{3.158}$$

where α is the slip angle, degree.

γ is the camber angle, degree.

$$C = a_0 \tag{3.159}$$

$$D = \mu_{yp}F_z \tag{3.160}$$

$$\mu_{yp} = a_1F_z + a_2 \tag{3.161}$$

$$\text{B.C.D.} = a_3\sin\left[2\arctan\left(\frac{F_z}{a_4}\right)\right](1-a_5|\gamma|) \tag{3.162}$$

$$E = a_6F_z + a_7 \tag{3.163}$$

$$S_h = a_8\gamma + a_9F_z + a_{10} \tag{3.164}$$

$$S_v = a_{11}\gamma F_z + a_{12}F_z + a_{13} \tag{3.165}$$

$$a_{11} = a_{111}F_z + a_{112} \tag{3.166}$$

F_z is tire vertical load, kN.

Self-aligning moment is formulated as

$$M_z = D\sin\left(C\arctan\left\{B(1-E)(\alpha+S_h)+E\arctan\left[B(\alpha+S_h)\right]\right\}\right)+S_v \tag{3.167}$$

where α is the slip angle, degree.

γ is the camber angle, degree.

$$C = c_0 \tag{3.168}$$

$$D = c_1F_z^2 + c_2F_z \tag{3.169}$$

$$\text{B.C.D.} = \left(c_3F_z^2 + c_4F_z\right)(1-c_6|\gamma|)e^{-c_5F_z} \tag{3.170}$$

$$E = \left(c_7F_z^2 + c_8F_z + c_9\right)\left((1-c_{10}|\gamma|)\right) \tag{3.171}$$

$$S_h = c_{11}\gamma + c_{12}F_z + c_{13} \tag{3.172}$$

$$S_v = \left(c_{14}F_z^2 + c_{15}F_z\right)\gamma + c_{16}F_z + c_{17} \tag{3.173}$$

F_z is the tire vertical load, unit: kN.

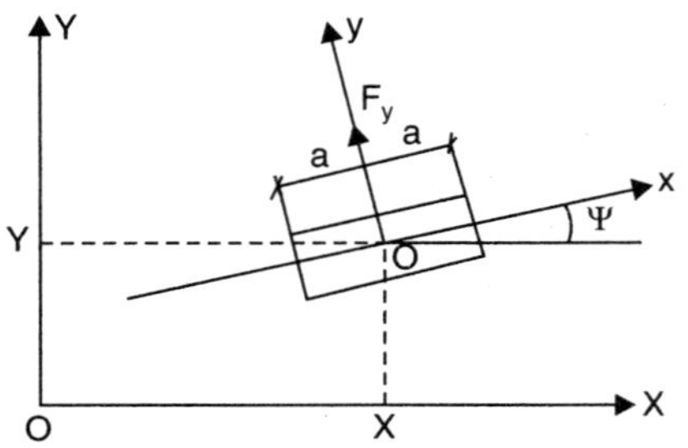

Fig. 3.30 Vector diagram of velocity in combined condition of steering and braking.

In the case of combined condition of steering and braking, the vector diagram of velocity is shown in Fig. 3.31.

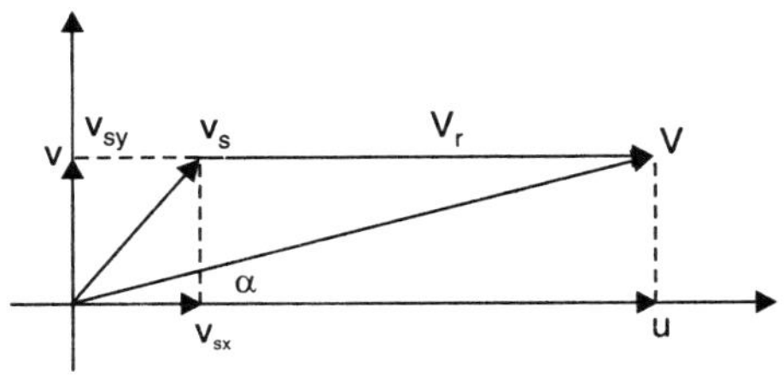

Fig. 3.31 Vector diagram of velocity in combined condition of steering and braking.

Set
$$\begin{cases} \sigma_x = \dfrac{v_{sx}}{V_r} \\ \sigma_y = \dfrac{v_{sy}}{V_r} \end{cases} \tag{3.174}$$

When $\alpha = 0$, define $\sigma = \dfrac{-v_{sx}}{u}$, thus longitudinal slip ratio and lateral slip ratio can be described as

$$\begin{cases} \sigma_x = \dfrac{v_{sx}}{u - v_{sx}} = \dfrac{-\sigma}{1+|\sigma|} \\ \sigma_y = \dfrac{v_{sy}}{u - v_{sx}} = \dfrac{\tan(\alpha)}{1+|\sigma|} \end{cases} \tag{3.175}$$

The correct values of longitudinal and lateral slip ratios are given as following equations:

$$\begin{cases} \sigma_{xtot} = \dfrac{-\sigma}{1+|\sigma|} - \delta\sigma \\ \sigma_y = \dfrac{\tan(\alpha)}{1+|\sigma|} + \delta\alpha \end{cases} \tag{3.176}$$

where:
$$\begin{cases} \delta\sigma = -S_h \\ \delta\alpha = -S_h - \dfrac{S_v}{BCD} \end{cases} \tag{3.177}$$

define
$$\sigma_{tot} = \sqrt{\sigma_{xtot}^2 + \sigma_{ytot}^2}$$

Thus the tire mechanics characteristics are as follows :

$$\begin{cases} F_x = \dfrac{\sigma_{xtot}}{\sigma_{tot}} F_x(\sigma_{tot}) \\ F_y = \dfrac{\sigma_{ytot}}{\sigma_{tot}} F_y(\sigma_{tot}) \\ M_z = \dfrac{\sigma_{ytot}}{\sigma_{tot}} M_z(\sigma_{tot}) \end{cases} \tag{3.178}$$

The magic formula presents generalized tire mechanics property and high fitting precision, but requires a massive quantity of calculation, which is suitable for the product design, vehicle dynamic simulation and test comparison in the field of high-precision tire mechanics properties analysis.

Magic model: The Magic model is empirical, requiring specification of a number of parameters determined from experimental measurements of tire forces and moments. Such measurements require sophisticated test equipment which makes the Magic model impractical for organizations with modest tire-testing capabilities. Also, the Magic model is not well suited to parametric studies of the impacts of snow and ice induced changes in the surface friction on vehicle performance. A static friction coefficient appears in the model, but it determines the peak forces and moments only. Its adjustment alone does not properly account for the variation in the surface friction commonly observed at different tire/road slip conditions. A full compliment of force and moment measurements seems to be required instead for each road surface condition examined. Hence, an alternative tire model is needed that: (1) only requires a limited set of easily measured tire input parameters, (2) is analytical in nature, with physically meaningful parameters that directly relate to tire characteristics and road surface conditions, (3) realistically accounts for changes in surface friction and (4) is numerically simple and practical for use in complex vehicle simulations. A steady-state form of solution is adequate since the problem being addressed is low frequency vehicle maneuvering on smooth non-deformable road surfaces.

3.6 EXAMPLE PROBLEMS AND SOLUTIONS

Example 3.1: *The mass of a vehicle is $m = 1820$ kg, the distance between the centre of the mass and the front axle is $l_a = 1.463$m, distance to the rear axle is $l_b = 1.585$m. Assume that the rolling resistance coefficient between the road and the tire is $f_0 = 0.0165$, when the vehicle is running at a speed of $v = 36$km/h and a steering angle of $\delta_0 = 20°$, the slip angle of the front and rear tire are both 4°, try to calculate the rolling resistance of the tire.*

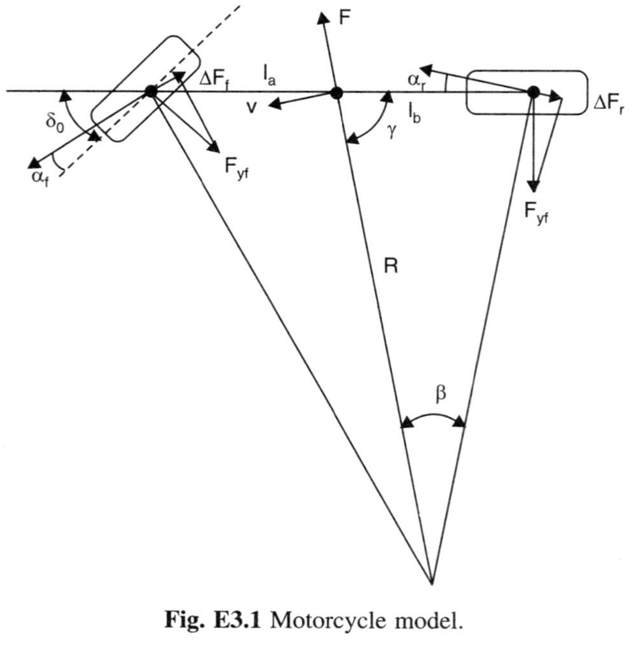

Fig. E3.1 Motorcycle model.

Solution: According to the relationships of trigonometric function.

Then $$\frac{R'}{\sin(90° - \delta_0 + \alpha_f)} = \frac{l_a + l_b}{\sin(\delta_0 - \alpha_f + \alpha_r)} \tag{1}$$

$$R' = \frac{(l_a + l_b)\sin(90° - \delta_0 + \alpha_f)}{\sin(\delta_0 - \alpha_f + \alpha_r)} = \frac{(1.463 + 1.585) \times \sin(90° - 20° + 4°)}{\sin(20° - 4° + 4°)} = 8.567\text{m} \tag{2}$$

$$R = \sqrt{l_b^2 + R'^2 - 2l_b \cdot R' \cos(90° - \alpha_r)}$$
$$= \sqrt{1.585^2 + 8.567^2 - 2 \times 1.585 \times 8.567 \times \cos(90° - 4°)} \tag{3}$$
$$= 8.603\text{m}$$

$$\beta = \arcsin\left(\frac{l_b \cos\alpha_r}{R}\right) = \arcsin\left(\frac{1.585 \times \cos 4°}{8.603}\right) = 10.6° \tag{4}$$

$$\gamma = 90° - \beta + \alpha_r = 90° - 10.6° + 4° = 83.4° \tag{5}$$

Cornering force of the front wheel

$$F_{yf} = m\frac{v^2}{R}\frac{\sin(\beta - \alpha_r)}{\sin\delta_0} = 1820 \times \frac{10^2}{8.603}\frac{\sin(10.6° - 4°)}{\sin 20°} = 7109\text{N} \tag{6}$$

Cornering force of the rear wheel

$$F_{yr} = m\frac{v^2}{R}\left[\cos(\beta - \alpha_r) - \sin(\beta - \alpha_r) c\tan\delta_0\right] \tag{7}$$

$$= 1820 \times \frac{10^2}{8.603}\left[\cos(10.6° - 4°) - \sin(10.6° - 4°)\, c\tan 20°\right] = 14335\text{N}$$

The additional resistance respectively applied on the front and rear wheel

$$\begin{cases} \Delta F_f = \sin\alpha_f F_{yf} = \sin 4° \times 7109 = 496\text{N} \\ \Delta F_r = \sin\alpha_r F_{yr} = \sin 4° \times 14335 = 1000\text{N} \end{cases} \tag{8}$$

The additional resistance coefficient under the condition of vehicle steering

$$\Delta f = \frac{\Delta F_f + \Delta F_r}{mg} = \frac{496 + 1000}{1820 \times 9.8} = 0.0839 \tag{9}$$

The rolling resistance coefficient during a turn

$$f_R = f_0 + \Delta f = 0.0165 + 0.0839 = 0.1004 \tag{10}$$

The rolling resistance of the tire

$$f = mgf_R = 1820 \times 9.8 \times 0.1004 = 1791\text{N} \tag{11}$$

Example 3.2: *Based on brush model, the load of the tire is 4500N, contact patch length of tires is 8 cm, when slip ratio is 0.15, longitudinal stiffness of the tire is* $c_{ex} = 1300\text{N/cm}^2$. *Calculate the longitudinal force under this slip ratio.*

Solution: Assume vertical force distribute as Quadratic function

So $$F_z = \int_{-4}^{4} \lambda(4^2 - x^2)dx = 4500\text{ N} \tag{1}$$

We get $\lambda = 52.7$

So $F_{ez}(x) = 52.7 \times (16 - x^2)$ (2)

The length of sliding region

$$l_a = c_{ex}\sigma/(\mu\lambda) = 1300 \times 0.15/(1.15 \times 0.8 \times 52.7) = 4 \text{ cm} \quad (3)$$

Hence $$F_x = \int_{-4}^{0} \mu\lambda(4^2 - x^2)dx + \int_{0}^{4} c_{ex}\sigma(x-4)dx = 3450.83 \text{ N} \quad (4)$$

Example 3.3: *The mass of one car is 1600kg. If it is front-engine front-wheel-drive (FF), the road of the Front shaft is 61% of the whole car. If it is front-engine rear-wheel-drive (FR), the load of the front shaft is 55% of the whole car. The coefficient of road adhesion $\varphi = 0.7$, calculate the driving force of FF and FR respectively.*

Solution: For front wheel drive

$$G = mg = 1600\text{kg} \times 9.8\frac{\text{N}}{\text{kg}} = 15680\text{N} \quad (1)$$

$$F_z = 0.61\text{G} = 0.61 \times 15680\text{N} = 9564.8\text{N} \quad (2)$$

$$F_x = F_z\varphi = 9564.8\text{N} \times 0.7 = 6695.36\text{N} \quad (3)$$

For rear wheel drive

$$G = mg = 1600\text{kg} \times 9.8\frac{\text{N}}{\text{kg}} = 15680\text{N} \quad (4)$$

$$F_z = (1-0.55)G = (1-0.55) \times 15680\text{N} = 7056\text{N} \quad (5)$$

$$F_x = F_z\varphi = 7056\text{N} \times 0.7 = 4939.2\text{N} \quad (6)$$

Example 3.4: *The load of the tire is F_z = 4500N. The coefficient of road adhesion φ_p = 1.0, φ_s = 0.7. When slip ratio s = 0.04, the longitudinal force F_x = 1000 N, and when s = 0.2, the F_x reachs the max value. Basic on H.B. Pacejka magic formula model, analyze relationships between longitudinal force and slip ratio.*

Solution: MATLAB Program:

```
f_z=4500,mp=1.0,ms =0.7,s1=0.04,f1=1000,sp = 0.2;
f_p=z*mp;
f_s=fz*ms;
D=fp;
C=1 + (1 - 2*a sin(fs/D)/pi);
B=(f1/s1)/(C*D);
E=(sp*B-tan(pi/(2*C)))/(B*sp-atan(B*sp));
x=0:0.01:1;
y=D*sin(C*a tan(B*x-E*(B*x-a tan(B*x))));
plot(x, y)
```

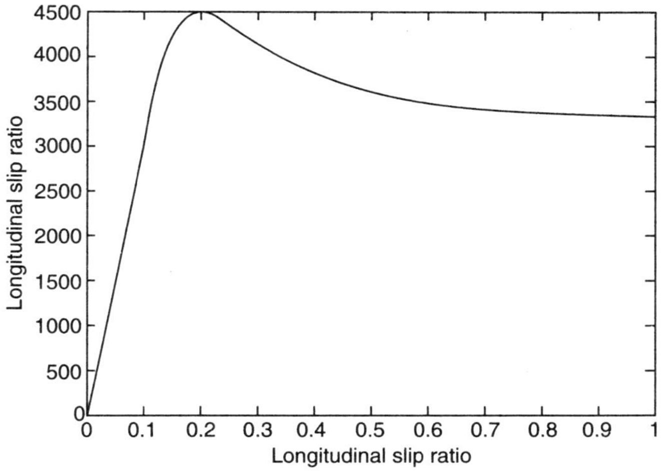

Fig. E3.4 Relationships between longitudinal force and slip ratio.

Example 3.5: *The distribution function of tire load is given as follows:*

$$\eta(u) = \begin{cases} 1 & 0 < u < 2 \\ 0 & u \le 0 \ \ or \ \ u \ge 2 \end{cases}$$

Determine the tire-cornering characteristic under simple distribution load of tire

$$\left(\frac{F_y}{\varphi \cdot F_z} \sim \phi_y \text{ and } \frac{T_z}{\varphi F_z a} \sim \phi_y \right).$$

Solution: When $\mu' < 2$, we have

$$\phi_y = \frac{\eta(\mu')}{\mu'} = \frac{1}{\mu'}, \quad \mu' = \frac{1}{\phi_y}, \quad m_0(\mu') = \int_0^{\mu'} \eta(\mu) d\mu = \int_0^{\mu'} 1 d\mu = \mu'$$

Hence $$\frac{F_y}{\varphi F_z} = \frac{\phi_y}{4} \times \mu'^2 + 1 - \frac{m_0(\mu')}{2} = \frac{\mu'}{4} + 1 - \frac{\mu'}{2} = 1 - \frac{1}{4\phi_y} \tag{1}$$

When $\mu' = 2$

$$\frac{F_y}{\varphi F_z} = \phi_y \tag{2}$$

When $\mu' > 2$

$$\frac{F_y}{\varphi F_z} = \phi_y \tag{3}$$

$\frac{F_y}{\varphi F_z} \sim \phi_y$ Fig. as follows

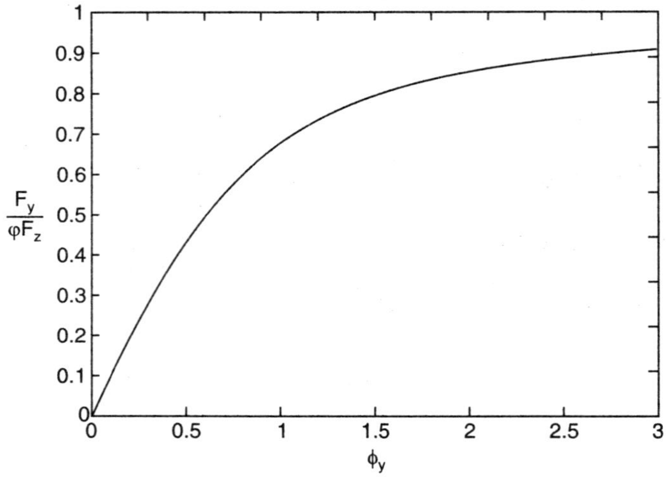

Fig. E3.5 (*a*) Relationships between lateral force and relative slip ratio.

Aligning torque:

$$\mu' = \frac{1}{\phi_y}, \quad m_0(\mu') = \int_0^{\mu'} \eta(\mu)\, d\mu = \mu' = \frac{1}{\phi_y} \tag{4}$$

$$m_1(\mu') = \int_0^{\mu'} \mu\eta(\mu)\, d\mu = \frac{1}{2}\mu'^2 = \frac{1}{2\phi_y{}^2} \tag{5}$$

Hence $\mu' < 2$,

$$\frac{T_z}{\varphi F_z a} = \frac{\mu'^2}{2}\left(\frac{\mu'}{3} - \frac{1}{2}\right) \times \phi_y - \theta + \frac{m_0(\mu') - m_1(\mu')}{2} = \frac{1}{4\phi_y}\left(1 - \frac{1}{3\phi_y}\right) \tag{6}$$

$\mu' \geq 2$

$$\frac{T_z}{\varphi F_z a} = \frac{\phi_y}{3} \tag{7}$$

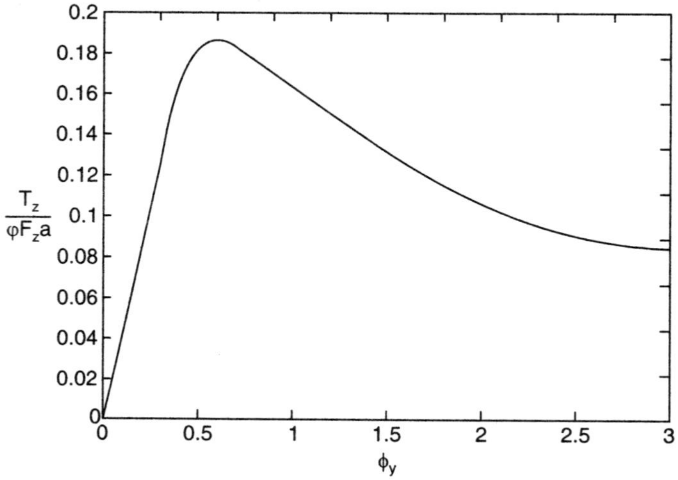

Fig. E3.5 (*b*) Relationships between aligning torque and relative slip ratio.

Example 3.6: *Define slip ratio. Draw a curve to show the relationship between the coefficient of road adhesion and the longitudinal slip ratio, then describe the characteristics of the curve, analyze the affections of different factors to the coefficient of road adhesion.*

Solution: Slip ratio

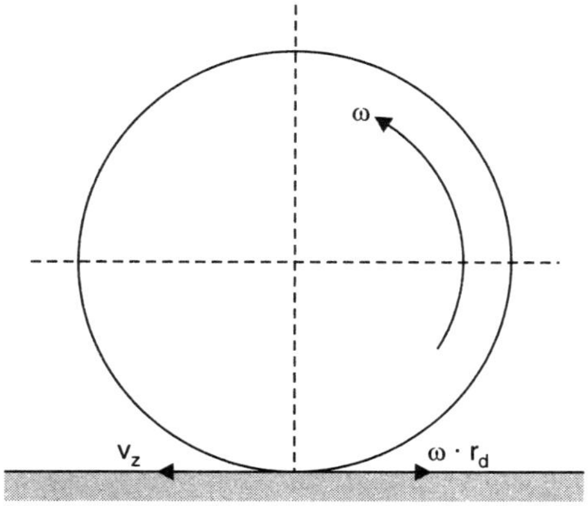

Fig. E3.6 (*a*) Tire model.

Braking maneuver longitudinal skid S_b

$$S_b = \frac{v_x - \omega r_d}{v_x} \times 100\% \tag{1}$$

Tractive maneuver longitudinal slip S_a

$$S_a = \frac{\omega r_d - v_x}{\omega r_d} \times 100\% \tag{2}$$

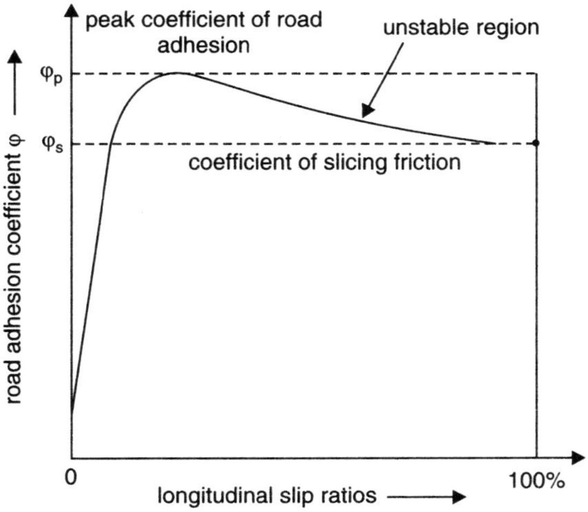

Fig. E3.6 (*b*) Free body diagram

$\phi \sim s$ curve has following characteristics:

1. If $|s| = 0 \sim 15\%$, the value of φ increases linearly with s.

2. If $| s | = 15\% \sim 30\%$, the value of φ reaches the maximum and is defined as peak coefficient of road adhesion.
3. If $| s | = 30\% \sim 100\%$, the value of φ gradually falls with s increase and is defined as sliding coefficeint of road adhesion when $| s | = 100\%$.

On a dry surface: $\varphi_{max} = 1.2\varphi_s$;

On a wet surface: $\varphi_{max} = 1.3\ \varphi_s$

The coefficient of road adhesion mainly depends on road texture and suface, tire structure, tread pattern, material and inflation pressure, normal loading on the wheel, the travel speed of the vehicle, etc. The experimental data shows the adhesion coefficient of radial tire is higher than that of bias tire and the tire with high inflation pressure is lower than the tire with low inflation pressure. In addition, a low travel speed can produce a relatively high adhesion coefficient of the tire.

Example 3.7: *Describe the Tire Cornering Characteristics. Analyze the main affecting factors. Draw a curve to show the relationship between the self-aligning torque and the slip angle.*

Solution:

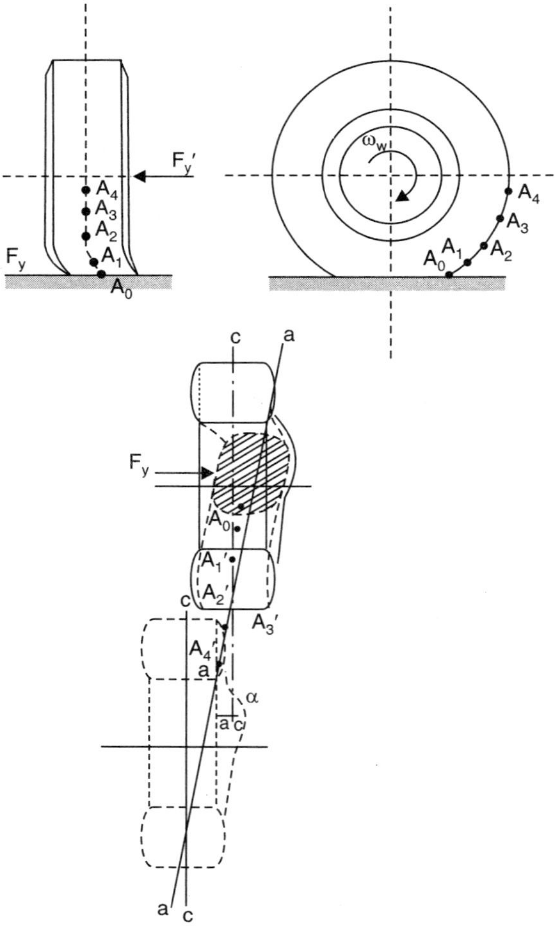

Fig. E3.7 (*a*) Tire Cornering Characteristics.

Due to the lateral wind or the centrifugal force when driving, a side force F_y will be applied to the center of the tire, correspondingly a lateral reaction force will also be developed at the contact patch. And it is called the cornering force F_y. Because of the tire lateral elasticity, tire begins to slip, the center line which is connected by the points A_0, A_1, A_2, A_3, A_4 on the wheel tread will be distorted. The slip angle between the path line will be formed,. That phenomenon is called the Tire Cornering Characteristics.

The principle factors affecting the tire corning characteristics: tire structure, working condition, and road state.

1. Tire structure: Big size, low profile and small flat rate tire, has high slip stiffness.
2. Work condition: The tire slip stiffness generally increases with the increase of the normal force. And also increases with an increase of the inflation pressure, does not change when the tire pressure arrives to a certain high level.

Example: Data from experiment

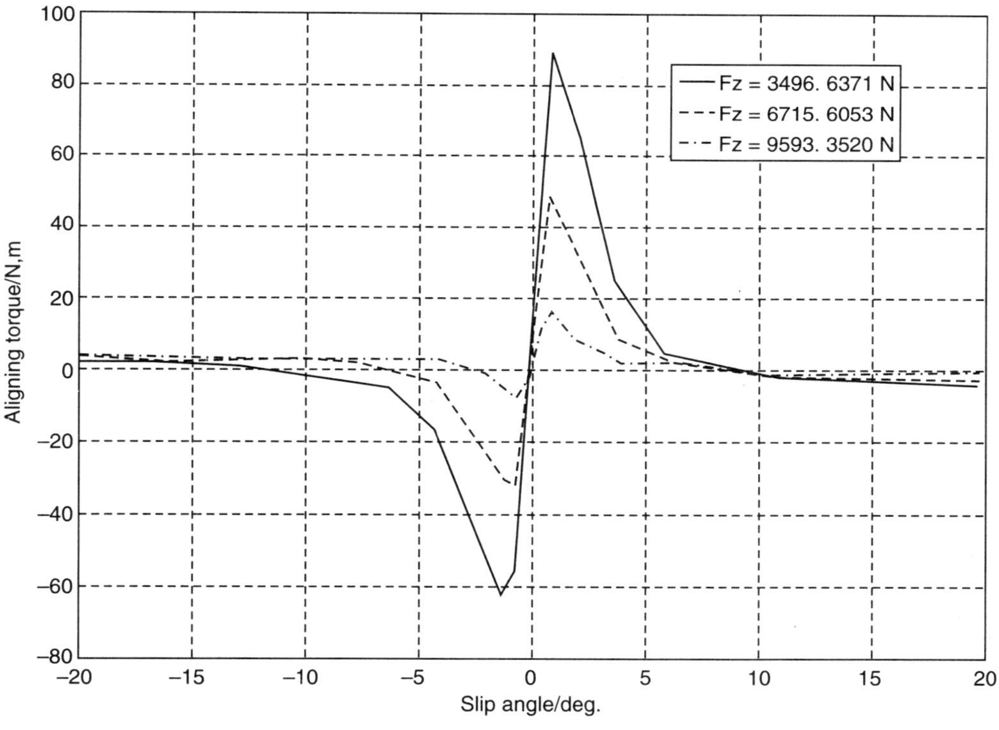

Fig. E3.7 (*b*) Relationship between aligning torque and slip angle.

The experiment data shows that:

The aligning torque first increases with an increase of the slip angle. It reaches a maximum at the slip angle of 4 ~ 6 degrees, and then decreases with a further increase. It reaches zero at the slip angle of 10 ~ 16 degrees and decreases to a negative value with a further increase in the skip angle. And also we can see that the aligning torque increases with an increase of the normal load.

Example 3.8: *Distribution function of tire load is as follows:*

$$h(u)=\begin{cases}\frac{3}{2}u(2-u), & 0<u<2\\ 0, \ u\le 0 \ or \ u\ge 2\end{cases}$$

Calculate the tire-cornering characteristic under a parabola distribution load of tire $\left(\frac{F_y}{\varphi F_z}\sim\phi_y \text{ and } \frac{T_z}{\varphi F_z a}\sim\phi_y\right)$.

Solution: The distribution of normal force is

$$\eta(u)=\begin{cases}\frac{3}{2}u(2-u), & 0<u<2\\ 0, & u\le 0 \ \text{ or } \ u\ge 2\end{cases} \tag{1}$$

And this function complies with the two constrains

$$\begin{cases}\int_0^2 u(u)du=2\\ \int_0^2 aq(u)du=Pa\end{cases} \tag{2}$$

$$\int_0^2 u\eta(u)du = 2\left(1-\frac{\Delta}{a}\right)=2(1-\theta), \qquad \theta=\frac{\Delta}{a} \tag{3}$$

Substitute Eq. (1) and Eq. (3), then can get

$$\begin{cases}\int_0^2 u\eta(u)du=2\\ \theta=0\end{cases} \tag{4}$$

With the large cornering force and the slip angle, experiment shows that partial sideslip has occurred on the tire, because of $q_y\ge\varphi\cdot q_z$ at the rearward patch. φ is the friction coefficient, and the origin position can be defined by $q_y\ge\varphi\cdot q_z$.

One gets
$$k_y x tg\alpha = \varphi\frac{F_z}{2a}\eta(\mu) \tag{5}$$

Define the skid ratio as

$$\varphi_y=\frac{K_y\cdot tg\alpha}{\phi\cdot F_z} \tag{6}$$

Then
$$\phi_y=\frac{\eta(\mu')}{\mu'} \tag{7}$$

Substitute Eq. (1) and Eq. (8), then can get

$$\phi_y = 3/2(2-u^*), (0\le u^*\le 2) \tag{8}$$

or
$$u=\begin{cases}2-\frac{2}{3}\phi_y & (\phi_y\le 3)\\ 0 & (\phi_y\ge 3)\end{cases} \tag{9}$$

In order to get non-dimensional lateral force expression.

Substituted Eq. (1), the zero order and first order moment of $\eta(\mu)$ at the range of $0 \sim \mu$ as

$$m_0(\mu) = \int_0^{\mu} \eta(\mu)\, d\mu \tag{10}$$

$$m_1(\mu) = \int_0^{\mu} \mu\eta(\mu)\, d\mu \tag{11}$$

And then we can get equation

$$m_0(u^*) = 3/2 * \int_0^{u^*} u(2-u)\, du = 3/2 * u^{*2} - 1/2 * u^{*3}$$

Substitute the result of Eq. (10) and Eq. (11) for the equation

$$\frac{F_y}{\varphi F_z} = \begin{cases} \dfrac{\phi_y}{4} \times u^{*2} + 1 - \dfrac{m_0(u^*)}{2} & 0 \le u^* < 2 \\ \phi_y & u^* = 2 \end{cases} \tag{12}$$

Non-dimensional lateral force expression is a single variable function.

We can get
$$\frac{F_y}{\varphi F_z} = \begin{cases} 1 - (1 - \phi_y/3)^3 & \phi_y \le 3 \\ 1 & \phi_y \ge 3 \end{cases} \tag{13}$$

Then use the software (MATLAB) to draw a curve to show the relationship between $\dfrac{F_y}{\varphi \cdot F_z}$ and ϕ_y.

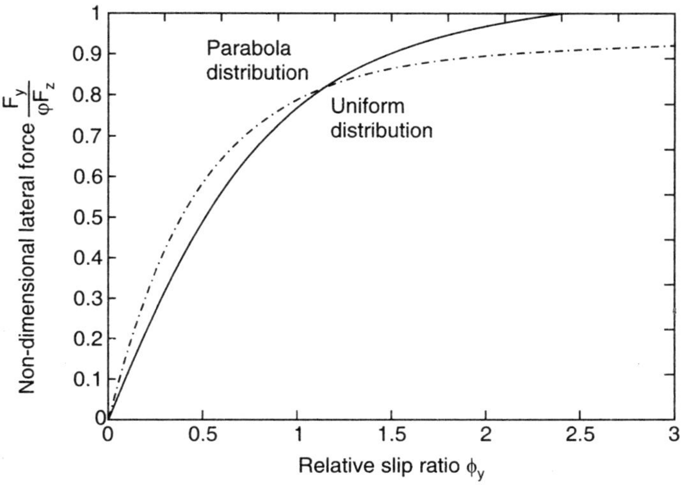

Fig. E3.8 (*a*) Relationship between non-dimensional lateral force and relative slip ratio.

From the fig. we can see that when $\phi_y \to 0$ or $\phi_y \to \infty$, the Parabola distribution and the Uniform distribution have the same result. And two curves intersect around $\phi_y = 1.24.$, which shows when

$\phi_y = 0 - 1.24$, the value of $\dfrac{F_y}{\varphi \cdot F_z}$ in Parabola distribution condition is smaller than in Uniform distribution condition. And this result totally complies with the result of Falia model.

Using the same method, we can get the expression of the non-dimensional total aligning torque.

$$\frac{T_z}{\varphi F_z a} = \begin{cases} 1/3 * \phi_y (1 - \phi_y / 3)^3 & \phi_y \leq 3 \\ 0 & \phi_y \geq 3 \end{cases} \tag{14}$$

Also we can get the figure as follows

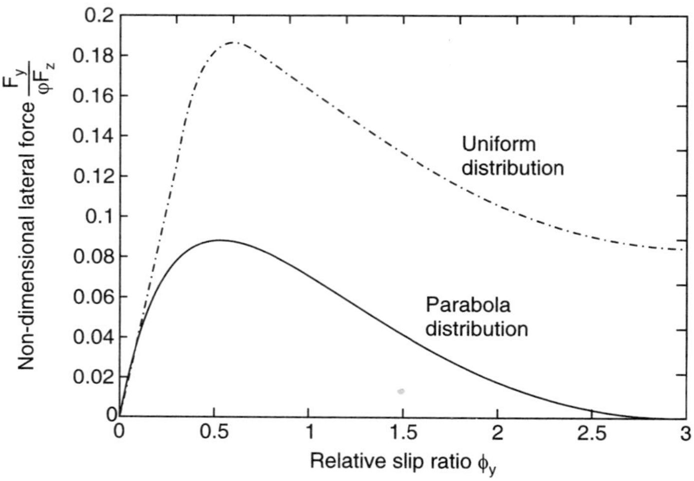

Fig. E3.8 (*b*) Relationship between non-dimensional aligning torque and relative slip ratio.

Example 3.9: *Establish a model and then calculate the longitudinal force. (Take brush model for example)*

Solution: Brush model is a simplified Theoretical Model, it takes tire as a rigid base connected with a lot of bristles, and through the deformation of these bristles to assume the longitudinal force and the lateral force. The brush model states that the contact patch between the tire and the road is divided into an adhesion region where the rubber is gripping the road and a sliding region where the rubber slides on the road surface. The total force generated by the tire is then composed of components from these two regions.

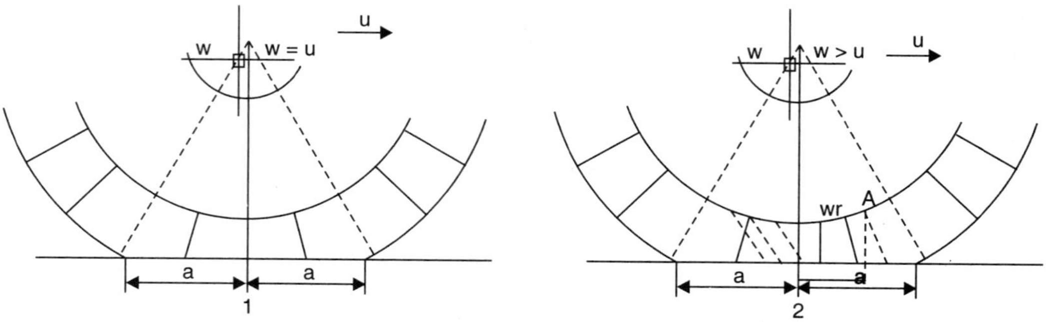

Fig. E3.9 (*a*) Brush model.

In the state of pure rolling, just as shown in the figure1, the tire contact patch length is '2a' in the adhesion region. Because of the friction function of the ground, when $wr > u$, the bristle element at the end of the tire will be deformed, which causes the speed difference between the two ends. As shown in the picture 2. And it is assumed that the radius of the wheel is much larger than the length of the contact patch ($r \gg a$). So the longitudinal deformation of the bristle element can be expressed as follows:

$$\xi = (wr - u)^{*}\ \Delta t \tag{1}$$

or

$$\xi = (wr - u)^{*}\frac{\Delta t}{\Delta x}\Delta x = \frac{(wr - u)}{wr}\Delta x \tag{2}$$

Define $\sigma_x = (wr - u)/wr$ as slip ratio, and also it is called amended slip ratio.

The deformation of the bristle element can be expressed

$$\xi = \sigma_x \Delta x.$$

Assume the longitudinal force is proportional to the deformation of the bristle element.

So the force of the flexible unit can be expressed

$$F_{ex} = cex\xi = c_{ex}\sigma_x(a - x) \tag{3}$$

c_{ex}: the stiffness of the bristle element.

In the whole adhesion region, the longitudinal force can be got through integrating

$$F_x = \int_{-a}^{a} c_{ex}\xi dx = 2c_{ex}a^2\sigma_x \tag{4}$$

When the slip ratio is very little, we can get the equation $\omega_r \approx u$, $\sigma_x \approx s$.

So the longitudinal force can be expressed as

$$d = |a| - |x_A| = c_{ex}\sigma_x/(\mu\lambda) \tag{5}$$

Eq. (3) shows that, the longitudinal force is proportional to the slip ratio of the wheel, however, this model hasn't concerned about the limit condition that for example, when the longitudinal force has reached or exceeded the limits of the ground adhesion, so the model needs to be amended for further.

The distribution of the normal force can be expressed as

$$F_{ez}(x) = \lambda(a^2 - x^2) \tag{6}$$

Through the equation $F_z = \int_{-a}^{a} \lambda(a^2 - x^2)dx$, we can get the value of λ.

When $F_{ex} \leq \mu\ F_{ez}(x)$ the critical point A has divided the adhesion region into two parts, one is adhesion region, the other is slipping region.

$$d = |a| - |x_A| = c_{ex}\sigma_x/(\mu\lambda) \tag{7}$$

So the whole longitudinal force can be expressed through following equation

$$F_x = \mu\lambda\int_{-a}^{x_A}(a^2 - x^2)dx + \int_{x_A}^{a} c_{ex}\sigma_{xc}(x - a)dx$$

$$= 1/3\mu\lambda d^2(3a - d) + 1/2\mu_{st}\lambda d(2a - d)^2 \tag{8}$$

From the Fig. we can know that the premise condition is $d \geq 2a$, if the wheel is in the pure slipping state. Then front this premise, we can get the critical slip ratio as: $\sigma_{x,c} = 2a\mu\lambda/c_{ex}$.

Generally, $\mu_{st} > \mu_{sd}$, so we can change the former equation into the following one

$$F_x = 1/3\mu_{sd}\lambda d^2(3a-d) + 1/2\mu_{st}\lambda d(2a-d)^2 \tag{9}$$

And now, the critical slip ratio is

$$\sigma_{x,c} = 2a\mu\lambda / c_{ex} \tag{10}$$

At last, use the software (MATLAB) to calculate the equation to draw a curve as follows which has shown the relationship between the longitudinal force and the slip ratio.

Use the following data: $F_z = 4000\text{N}$, $\mu_{st} = 1.0$; $\mu_{sd} = 0.6$; $a = 4\text{cm}$; $\sigma_{xc} = 0.259$;

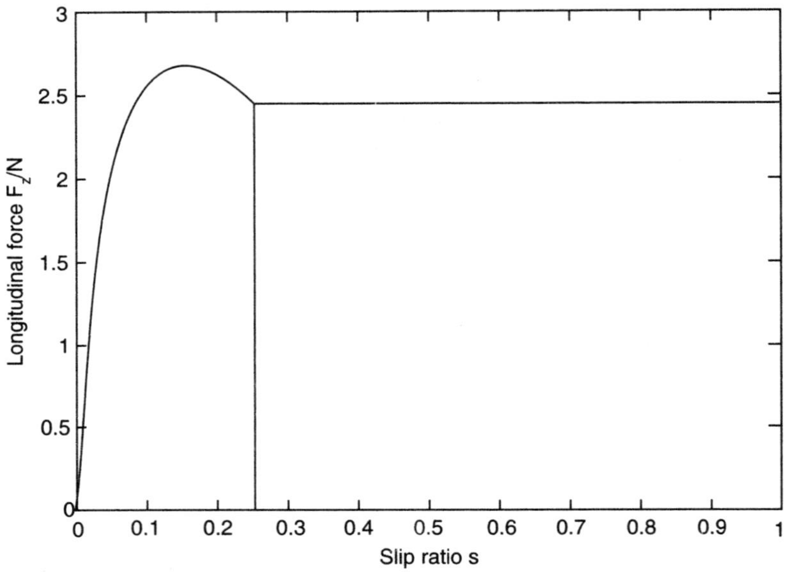

Fig. E3.9 (*b*) Relationship between lateral force and slip ratio.

MATLAB Program:

```
 Fz=4000,Us=1.0,Ud=0.6,a=0.004;
 k=ones(1,1/0.01+1);
 e=4.6875e+007;
 s=0:0.01:1;
 Ce=1101100;
 C=8.6873e+005;
 d=C.*s/(Ud*4.6875e+007);
 m=length(d);
for i=1:1:m;
   if d(i)>2*a
 d(i)=2*a;
end;
end;
```

```
Fx=1/3*Ud*e.*d.^2.*(3*a.*k-d)+1/2*Us*e.*d.*(2*a.*k-d).^2;
plot(s, Fx)
```

Example 3.10: *Please simply analyze the influence of the rolling resistance on the axle-load.*

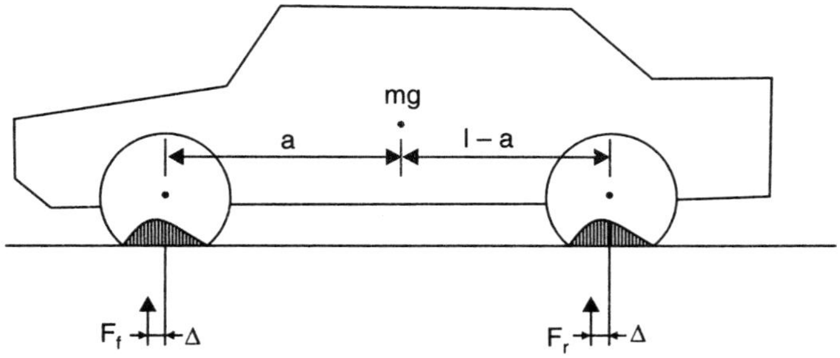

Fig. E3.10 The influence of the rolling resistance to the axle-load.

Solution: If the wheelbase is l, and the distance from the front axle to the centroid is a, then the distance from the front axle to the centroid is $l - a$.

Thus, when the vehicle is static, we can calculate the load on each axle as follows:

The front axle

$$F_{rs} = \frac{l-a}{l}mg \tag{1}$$

And the rear axle

$$F_{rs} = \frac{a}{l}mg \tag{2}$$

When the vehicle is running and the wheels are rolling, because of the deformation of the tires, these axle-loads will be changed. As shown in Fig. 1, we can get the new result

The front axle

$$F_{fd} = \frac{l-a-\Delta}{l}mg = F_{fs} - \frac{\Delta}{l}mg \tag{3}$$

The rear axle

$$F_{fd} = \frac{a+\Delta}{l}mg = F_{rs} + \frac{\Delta}{l}mg\ . \tag{4x}$$

Where $\Delta = fR$ the distance between normal force F and wheel center line, f is the rolling resistance coefficient and R is the rolling radius of the tires.

Substitute $\Delta = fR$ into the equations above, then

$$F_{fd} = F_{fs} - \frac{fR}{l}mg \tag{5}$$

$$F_{rd} = F_{rs} + \frac{fR}{l}mg \tag{6}$$

So, from the result above, we know that, because of the rolling resistance, the axle-load on both front axle and rear axle are changed. The loads on the front axle reduce, and the loads on the rear axle increase, that means some of the loads are transferred from front axle to the rear axle. The magnitude of the transferred loads is approximately $\frac{fR}{l}mg$, the bigger the rolling resistance coefficient is, the smaller the front axle-load will become, and the rear axle-load will become bigger at the same magnitude.

Example 3.11: *The load distribution function can be represented in this form:* $\eta(u) = \frac{n+1}{n}\left[1-(u-1)^n\right]$, *here, different n represents different types of the tires. But when the load on the tire is not heavy, we can use another equation :* $\eta(u) = c_0 + c_1 u + c_2 u^2 + c_3 u^3$, *it's much easy to calculate, as long as the coefficients* c_i *are known, the load distribution function* $\eta(u)$ *is also known. So, please determine the coefficients* c_i, *if the tires are rolling on good condition road and the following parameters are given: the rolling resistance coefficient f, the whole patch length 2a and the rolling radius R.*

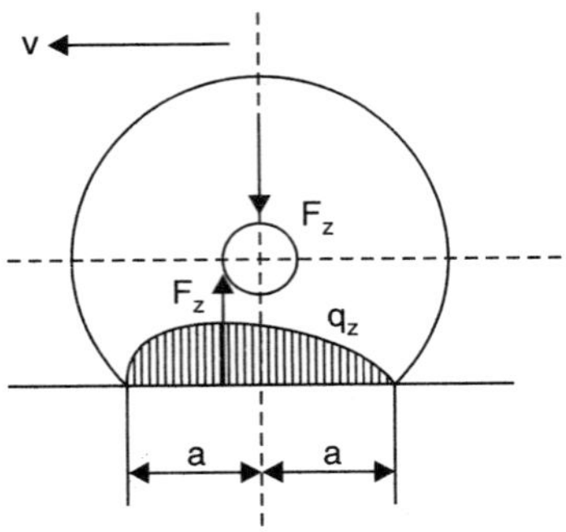

Fig. E3.11 The load distribution of the rolling tire.

Solution: Because the tires are rolling on good road, so we can think that, the rolling resistance coefficient is caused only by the deformation of the tires. Then

$$f = \frac{\Delta}{R}, \text{ that is } \Delta = fR.$$

From the Eq. (3-40) we get

$$\begin{cases} \int_0^2 (c_0 + c_1 u + c_2 u^2 + c_3 u^3)du = 2 \\ \int_0^2 u\left(c_0 + c_1 u + c_2 u^2 + c_3 u^3\right)d\mu = 2\left(1-\frac{\Delta}{a}\right) = 2\left(1-\frac{fR}{a}\right) \end{cases} \quad (1)$$

Additional, when $u = 0$, $\eta(u) = 0$ and when $u = 2$, $\eta(u) = 0$, so

$$\begin{cases} c_0 = 0 \\ c_0 + c_1 2 + c_2 2^2 + c_3 2^3 = 0 \end{cases} \quad (2)$$

Solve these four equations, we can get the coefficients c_i

$$\begin{cases} c_0 = 0 \\ c_1 = 3 + 15\dfrac{fR}{a} \\ c_2 = -\dfrac{3}{2} - \dfrac{45}{2}\dfrac{fR}{a} \\ c_3 = \dfrac{15}{2}\dfrac{fR}{a} \end{cases} \tag{3}$$

So the load distribution function is

$$\eta(u) = (3 + 15\frac{fR}{a})u + (-\frac{3}{2} - \frac{45}{2}\frac{fR}{a})u^2 + (\frac{15}{2}\frac{fR}{a})u^3 \tag{4}$$

Example 3.12: *When the vehicle is running on wet surface, it's easy for waterskiing phenomenon to occur when the velocity of the vehicle reaches a certain values, and it's always very dangerous. So please simply analyze the variation of the tire side force and aligning torque when the waterskiing phenomenon occurs (suppose the slip angle α is constant).*

Solution: The distribution of vertical load can be expressed as

$$q_z = \frac{F_z}{2a}\eta(u) \tag{1}$$

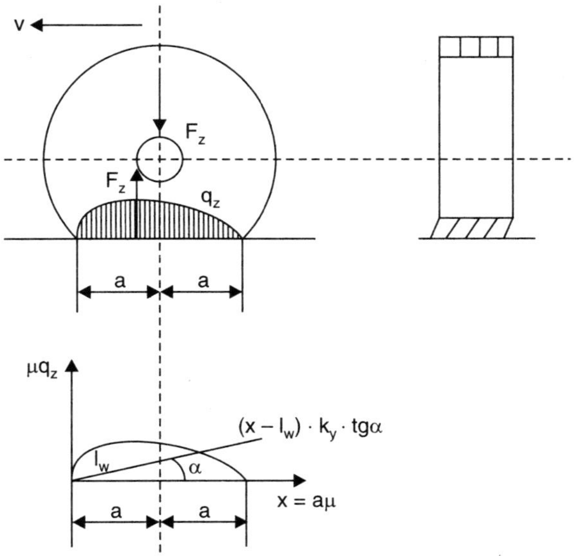

Fig. E3.12 (*a*) The Tire Cornering Characteristic under waterskiing.

If the length of the waterskiing area is l_w, and the adhesion coefficient in the waterskiing area is supposed to be zero, then the lateral load distribution at the frontward patch is

$$\begin{cases} q_y = k_y(x - l_w)tg\alpha, & l_w \le x \le u'a \\ q_y = 0, & x \le l_w \end{cases} \tag{2}$$

Partial sideslip has occurred on the tire, because of $q_y \geq \varphi \cdot q_z$ at the rearward patch. The origin position can be defined by $q_y = \varphi q_z$. That is

$$k_y(x-l_w)tg\alpha = \phi \frac{F_z}{2a}\eta(u) \tag{3}$$

If u' is the answer of the above function, then the total side force can be expressed as

$$F_y = \int_{\frac{l_w}{a}}^{\mu'} a^2 k_y (u - \frac{l_w}{a}) tg\alpha \cdot u \cdot du + \int_{\mu'}^{2} a \cdot \varphi \cdot \frac{F_z}{2a} \cdot \eta(u) \cdot du \tag{4}$$

And the total aligning torque is

$$M_z = \int_{\frac{l_w}{a}}^{\mu'} a^3 k_y (u - \frac{l_w}{a}) tg\alpha u^2 \cdot du + \int_{\mu'}^{2} a^2 u \varphi \frac{F_z}{2a} \eta(\mu) d\mu - F_y a \tag{5}$$

If all the parameters are given, it's easy to calculate the side force and the total aligning torque. Following figures are the result of the calculation by computer.

The parameter used in the calculation is the following:

The vertical load F_z = 4000N; the rolling radius of the tire R = 0.03m;

The whole patch length $2a$ = 0.16 m;

The tire deformation rolling resistance coefficient f = 0.015;

The lateral distribution stiffness of the rubber layer $k_y = 3.5 \times 10^6$ N/m^2;

The slip angle α = 4; the friction coefficient φ = 0.6;

Distribution function $\eta(u) = c_0 + c_1 u + c_2 u^2 + c_3 u^3$.

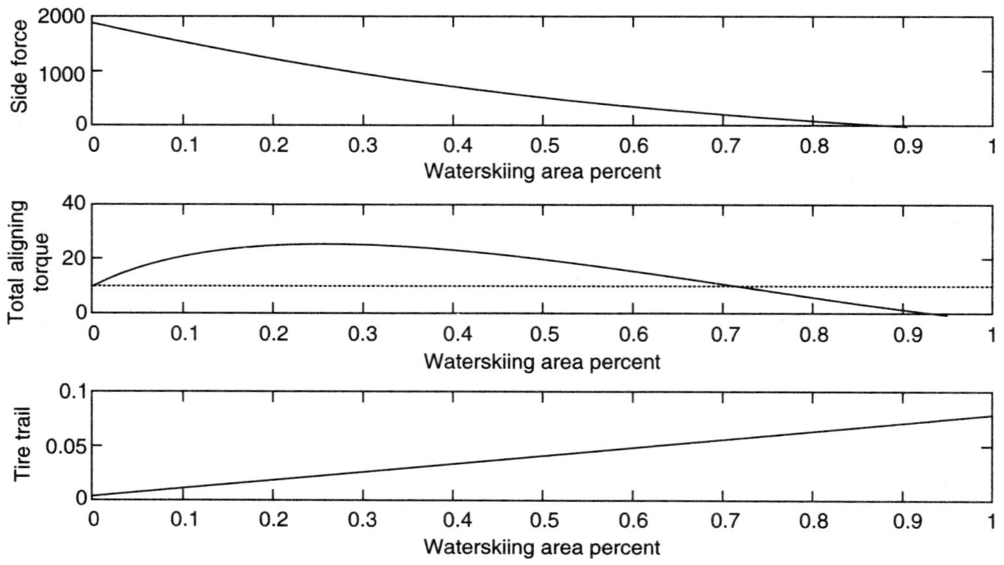

Fig. E3.12 (*b*) The result of the calculation.

From the Fig. we know that, the side force reduces when the waterskiing area becomes longer, and when the waterskiing area covers the whole patch length, the tire will lose side force. But the variation of the total aligning torque is not so simply. From the Fig., we get the following information: the aligning torque does not reduce but rise when the waterskiing area becomes longer under a certain value, and until the percent of the waterskiing reaches 30%, the aligning torque stops to rise and begins to fall down; but the aligning torque is still larger than non waterskiing happens till the percent of the waterskiing area goes up to 70%.

It's all because the tire trail increases rapidly when the waterskiing area expands.

It's very dangerous when waterskiing phenomenon occurs, not only because of the decrease of the side force, but also because of the increase of the aligning torque. The drivers always judge the situation of the tires through the aligning torque, so when the aligning torque rises, they often think the road condition is very good, but in fact, they are in danger.

Example 3.13: *When considering the cornering characteristic with Lateral Bending Deformation Of Tire Case, the shape function of the tire case deformation is needed. The shape function of the tire case deformation can be represented as:* $\xi(u) = 2u(2 - X_e u)/(4 - 8/3X_e)$, *and* X_e *is a constant determined by tire types. Please simply analyze the influence of the tire case bending stiffness to the tire side force and aligning torque (suppose the slip angle* α *is little).*

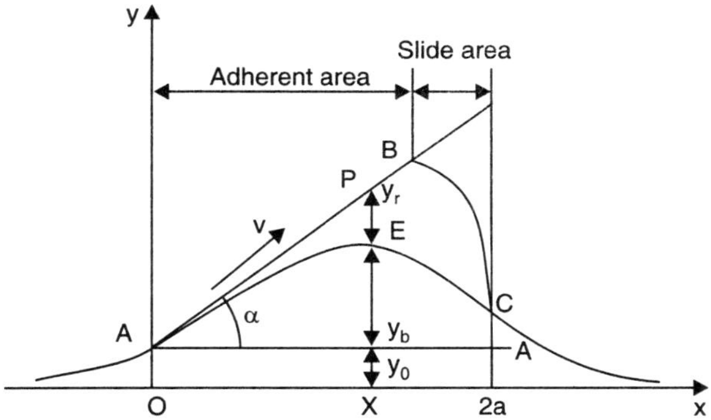

Fig. E3.13 (*a*) Cornering characteristic with lateral bending deformation of tire case.

Solution: With small slip angle α, there is no side skidding on the whole patch. So the total cornering force can be expressed as

$$F_y = \int_0^{2a} q_r dx = \int_0^2 a^2 u k_{yr} tg\alpha du - \int_0^2 a^2 u k_{yr} \frac{F_y}{k_{yb} a u} \xi(u) du \tag{1}$$

After integrating

$$F_y = 2a^2 k_{yr} tg\alpha - a \frac{k_{yr}}{k_{yb}} F_y \int_0^2 \xi(u) du \tag{2}$$

Solve this function, one gets

$$F_y = \frac{2a^2 k_{yr} tg\alpha}{1 + a \dfrac{k_{yr}}{k_{yb}} \displaystyle\int_0^2 \xi(u) du} \tag{3}$$

$\xi(u)$ can be represented as

$$\xi(u) = 2u(2 - X_e u)/(4 - 8/3X_e) \tag{4}$$

Then $$\int_0^2 \xi(u)\,du = 2 \text{ and } \int_0^2 u\xi(u)\,du = \frac{6X_e - 8}{2X_e - 3}$$

Substitute $\int_0^2 \xi(u)\,du = 2$, $K_{yr} = 2a^2 k_{yr}$ and $K_{yb} = ak_{yr}$ into the above equation, then

$$F_y = \frac{K_{yr}}{1 + \dfrac{K_{yr}}{K_{yb}}} tg\alpha = \frac{K_{yb}K_{yr}}{K_{yb} + K_{yr}} tg\alpha \tag{5}$$

And the aligning torque can be expressed as:

$$\begin{aligned} T_z &= \int_0^2 a^2 q_y u\,du - F_y a \\ &= \int_0^2 a^3 u^2 k_{yr}\left[tg\alpha - F_y/(ak_{yb})\xi(u)/u\right]du - F_y a \\ &= \frac{4}{3}aK_{yr}tg\alpha - \frac{a}{2}\frac{K_{yr}}{K_{yb}}F_y\int_0^2 u\xi(u)du - a\frac{K_{yr}K_{yb}}{K_{yr} + K_{yb}}tg\alpha \qquad (6) \\ &= \frac{4}{3}aK_{yr}tg\alpha - \frac{a}{2}\frac{K_{yr}^2}{K_{yb} + K_{yr}}\frac{6X_e - 8}{2X_e - 3}tg\alpha - a\frac{K_{yr}K_{yb}}{K_{yr} + K_{yb}}tg\alpha \\ &= \left(\frac{4}{3}K_{yr} - \frac{1}{2}\frac{K_{yr}^2}{K_{yb} + K_{yr}}\frac{6X_e - 8}{2X_e - 3} - \frac{K_{yr}K_{yb}}{K_{yr} + K_{yb}}\right)atg\alpha \end{aligned}$$

If all the parameters needed are given, it's easy to calculate the side force and the total aligning torque. Following Figures are the result of the calculation by computer.

The parameter used in the calculation is the following

The whole patch length $2a = 0.16$m;

The slip stiffness of tire tread $K_{yr} = 45000$;

The bend stiffness of tire case K_{yb} changes from 20000 ~ 40000;

The slip angle $\alpha = 4°$; shape function coefficient $X_e = 0.8$;

From the Figures we know that, the side force increases when the bending stiffness of the tire case K_{yb} increases, and the aligning torque has the same trend. However, the tire trail has the opposite trend. When the bending stiffness of the tire case K_{yb} increases, the tire trail decreases. It means that, when the side force is at the same magnitude, the tire with softer tire case will have larger aligning torque, that's also the reason why the radial ply tire always gets more aligning torque than bias ply tire.

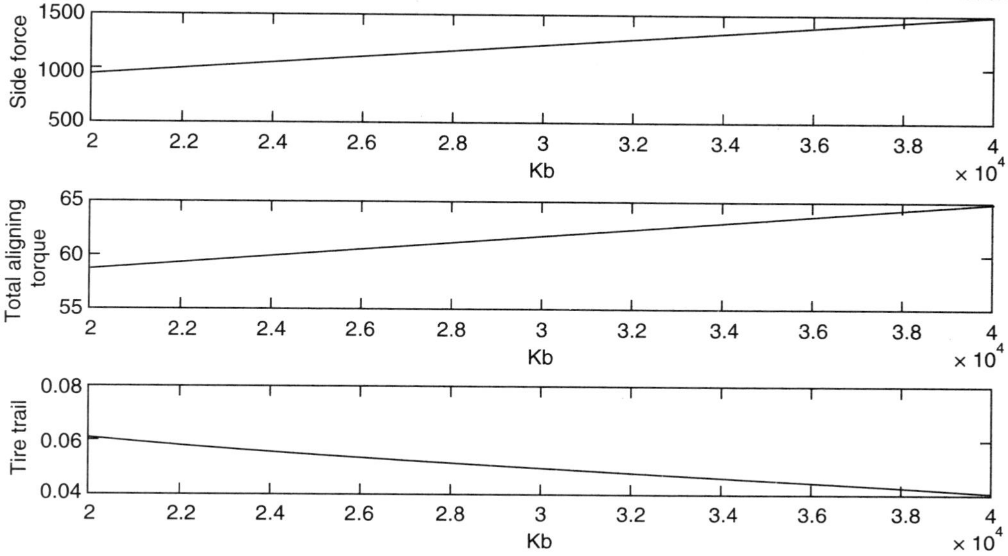

Fig. E3.13 (*b*) The result of the calculation.

Example 3.14: *Based on the parameters given in the last exercise, try to write a piece of MATLAB program which can generate a two-dimensional curve representing the relationship between the steering angle and the rolling resistance of the tire. (The steering angle ranges between 0°and 30°.)*

Solution: MATLAB Program:

```
m=1820,la = 1.463,lb = 1.585,f0 = 0.0165,g = 9.8,v0 = 36;
afdeg=4;
ardor=4;
xdeg=0:0.001:30;
af=afdeg*pi/180;
ar=ardeg*pi/180;
x=xdeg*pi/180;
v=v0/E3.6;
r0=(la+lb).*sin(pi/2-x+af)./sin(x-af+ar);
r=sqrt(lb.^2+r0.^2-2*lb.*r0.*cos(pi/2-ar));
b=asin(lb.*cos(ar)./r);
y=pi/2-b+ar;
fyf=m.*v.^2.*sin(b-ar)./(r.*sin(x));
fyr=m.*v.^2.*(cos(b-ar)-sin(b-ar)./tan(x))./r;
dff=sin(af).*fyf;
dfr=sin(ar).*fyr;
df=(dff+dfr)./(m.*g);
fr=f0+df;
```

```
f=m.*g.*fr;
plot(xdeg, f);
```

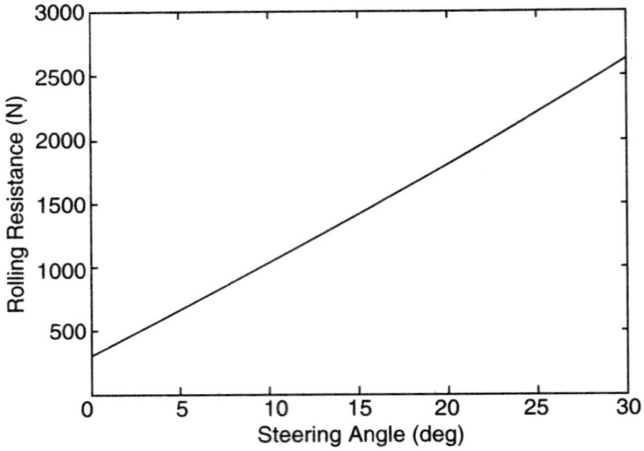

Fig. E3.14 Rolling resistance changes with the steering angle.

Example 3.15: *The radius of a vehicle's tire is R = 0.03m, the camber-angle is* $\gamma = 1°$ *camber stiffness of tire carcass is* $K_c = 730800$ *N/m, the length of the contact patch is 2a = 0.06 m, the lateral distributed stiffness per length of tire tread is* $k_y = 4\times10^8\,\text{N/m}$. *Try to calculate the lateral force with camber.*

Solution: The broken line shown in the Fig. represents the deformed part of tire. A coordinate system *Oxz* is located at the left side of the contact patch with a length of 2*a*.

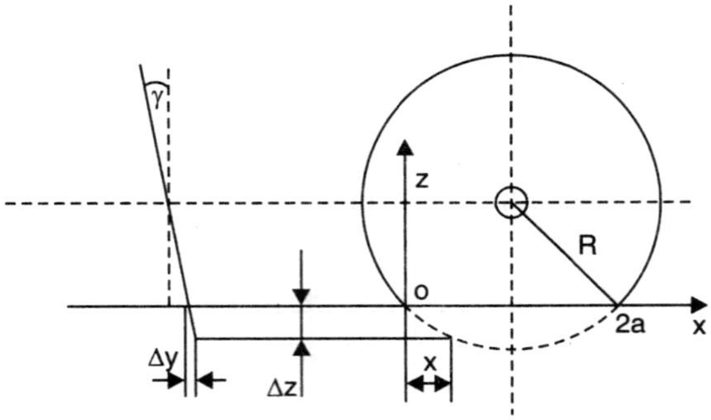

Fig. E3.15 Lateral force model.

So the vertical deformation of the tire is

$$\Delta z = \Delta z_m \mu(2-\mu) = \frac{a^2}{2R}\mu(2-\mu) = 0.0015\mu(2-\mu) \tag{1}$$

Corresponding to *x*, the lateral deformation is

$$\Delta y = \Delta z tg\gamma = 2.618\times10^{-5}\mu(2-\mu) \tag{2}$$

The lateral stress at the site x is

$$q_y = k_y \Delta y = 10472\mu(2-\mu) \tag{3}$$

The lateral force with camber is

$$F_{yr0} = \int_0^{2a} q_y dx = 10472 \int_0^{2a} \mu(2-\mu)dx = 10472 \times \frac{4}{3} \times 0.03 = 419\text{N} \tag{4}$$

The cornering stiffness of the tire is

$$K_y = 2a^2 k_y = 2 \times 0.03^2 \times 4 \times 10^8 = 720000\text{N/m} \tag{5}$$

The camber stiffness is

$$K_r = \frac{a}{3R} K_y = \frac{0.03 \times 720000}{3 \times 0.3} = 24000\text{N/m} \tag{6}$$

With the consideration of roll elastic deformation of tire in the modeling of cambered tire, finally, the roll force of tire owing to the roll of tire should be

$$F_{yr} = \frac{K_r K_c}{K_r + K_c} tg\gamma = \frac{24000 \times 730800}{24000 + 730800} \times tg(1^0) = 406N \tag{7}$$

Example 3.16: *A vehicle is in drive state, the circumference speed of wheel is v_c = 70 km/h, the longitudinal stiffness of the tire is K_s = 1001100N/m, the lateral stiffness is K_a = 975000N/m, the longitudinal velocity of the vehicle is v_x = 65km/h, the lateral velocity is v_y = 4km/h, the static friction coefficient between the tire and road surface is φ_0 = 0.9, the friction coefficient is φ_1 = 0.6 when the slip ratio is S_1 = 8%, the vertical load is F_z = 4000N, the length of the tire's contact patch is l = 0.06m. Try to calculate the longitudinal force, the lateral force and the self-aligning moment of the driving tire by taking advantage of the C.G. Gim theoretical model.*

Solution: The slip angle of the tire

$$\alpha = \arctan \frac{v_y}{v_x} = \arctan \frac{4}{65} = 3.52° \tag{1}$$

The longitudinal slip ratio

$$S_s = \frac{v_c - v_x}{v_c} = \frac{70-65}{70} \times 100\% = 7.14\% \tag{2}$$

The lateral slip ratio

$$S_\alpha = (1 - S_s)|\tan\alpha| = (1 - 7.14\%) \times |\tan 3.52°| = 5.71\% \tag{3}$$

Associated parameter of slip ratio

$$S_{s\alpha} = \sqrt{S_s^2 + S_\alpha^2} = \sqrt{(7.14\%)^2 + (5.71\%)^2} = 9.15\% \tag{4}$$

Then get the comprehensive adhesion coefficient of the tire

$$\varphi = \varphi_0 + (\varphi_1 - \varphi_0) \times S_{sa} = 0.95 + (0.75 - 0.95) \times 9.15\% = 0.9317 \tag{5}$$

So, the longitudinal adhesion coefficient of the tire

$$\varphi_x = \varphi \frac{S_s}{S_{s\alpha}} = \varphi\cos\alpha = 0.9317 \times \cos 3.52° = 0.9299 \tag{6}$$

The lateral adhesion coefficient of the tire

$$\varphi_y = \varphi \frac{S_\alpha}{S_{s\alpha}} = \varphi \sin\alpha = 0.9317 \times \sin 3.52° = 0.0572 \tag{7}$$

The critical point of rolling and slip in the contact patch is

$$S_n = \frac{1}{3\varphi F_z}\sqrt{(K_s S_s)^2 + (K_\alpha S_\alpha)^2} = \frac{\sqrt{(79981 \times 7.14\%)^2 + (65836 \times 5.71\%)^2}}{3 \times 0.9317 \times 4000} = 0.6118 \tag{8}$$

The slip critical point is

$$S_{sc} = \frac{3\varphi F_z}{K_s} = \frac{3 \times 0.9317 \times 4000}{79981} = 0.1398 \tag{9}$$

The cornering critical point is

$$S_{ac} = \frac{K_s}{K_\alpha}\sqrt{S_{sc}^2 - S_s^2} = \frac{79981}{65836}\sqrt{0.1398^2 - 0.0714^2} = 0.1460 \tag{10}$$

For $\quad S_s < S_{sc}, \quad S_a < S_{ac}$

Set $\quad l_n = 1 - S_n = 1 - 0.6118 = 0.3882$

Thus the longitudinal force between the tire and road surface is

$$\begin{aligned} F_x &= K_s S_s l_n^2 + \varphi_x F_z \left(1 - 3l_n^2 + 2l_n^3\right) \\ &= 79981 \times 7.14\% \times 0.3882^2 + 0.9299 \times 4000 \times \left(1 - 3 \times 0.3882^2 + 2 \times 0.3882^3\right) \\ &= 3333.8\text{N} \end{aligned} \tag{11}$$

The lateral force between the tire and road surface is

$$\begin{aligned} F_y &= K_a S_a l_n^2 + \varphi_y F_z \left(1 - 3l_n^2 + 2l_n^3\right) \\ &= 65836 \times 5.71\% \times 0.3882^2 + 0.0572 \times 4000 \times \left(1 - 3 \times 0.3882^2 + 2 \times 0.3882^3\right) \\ &= 719.1\ \text{N} \end{aligned} \tag{12}$$

The self-aligning moment is

$$\begin{aligned} T_z &= \left[K_a S_a \left(-\frac{1}{2} + \frac{2}{3} l_n \right) + \frac{3}{2} \varphi_y F_z S_n^2 \right] \cdot l \cdot l_n^2 \\ &= \left[65836 \times 5.71\% \times \left(-\frac{1}{2} + \frac{2}{3} \times 0.3882 \right) + \frac{3}{2} \times 0.0572 \times 4000 \times 0.6118^2 \right] \times 0.06 \times 0.3882^2 \\ &= -704.\text{N.m} \end{aligned} \tag{13}$$

3.7 REFERENCES

"Statement of Work, Truck Tire Characterization Phase 1 Part 2", Cooperative Research, Funding by NHTSA (National Highway Traffic Safety Administration) Contract No. DTNH22-92-C-17189, SAE International, Warrendale, PA, 1995.

Adler, Ulrich [ed.], *Automotive Handbook,* 3rd edition, Robert Bosch, Stuttgart, Germany, 1993.

Adler, Ulrich [ed.], *Automotive Handbook,* 3rd edition, Robert Bosch, Stuttgart, Germany, 1993.

Allen, W.R., Rosenthal, T.J., and **Chrstos, J.P.,** "Vehicle Dynamics Tire Model for Both Pavement and Off-Road Conditions", *SAE Special Publications*, Vol. 1228, *Research into Vehicle Dynamics and Simulation*, 970559, p 27-38, 1997.

Ammon, D., "Vehicle Dynamics Analysis Tasks and Related Tire Simulation Challenges", *Vehicle System Dynamics*, Vol. 43, p 30-47, 2005.

Bakker, E., Nyborg, L., and **Pacejka, H.B.,** "Tyre Modeling for Use in Vehicle Dynamic Studies", Paper 870421, SAE International, Warrendale, PA, 1987.

Blythe, W., and **Day, T.D.,** "Single Vehicle Wet Road Loss of Control; Effects of Tire Tread Depth and Placement", Paper 2002-01-0553, SAE International, Warrendale, PA, 2002.

Bohm, F., "Tire Models for Computational Car Dynamics in the Frequency Range up to 1000 Hz", *Vehicle System Dynamics*, Vol. 21, pp. 82-91, 1993.

Brach, Matthew., R., and **Brach, R.M.,** "Modeling Combined Braking and Steering Forces", Paper 2000-01-0357, SAE International, Warrendale, PA, 2000.

Browne, A.L., "Mathematical Analysis for Pneumatic Tire Hydroplaning", ASTM STP 583, American Society for Testing Materials, West Conshohocken, PA, 1975, pp. 75-94.

Cabrera, J.A., Ortiz, A., Carabias, E., and **Simon, A.,** "An Alternative Method to Determine the Magic Tire Model Parameters Using Genetic Algorithms", *Vehicle System Dynamics*, Vol. 41, No. 2, p 109-127, 2004.

Clark, S.K. [ed.], "Mechanics of Pneumatic Tires", DOT HS 805 952, U.S. Department of Transportation, National Highway Traffic Safety Administration, Washington, DC, 1981.

Clover, C.L., and **Bernard, J.E.,** "Longitudinal Tire Dynamics", *Vehicle System Dynamics*, Vol. 29. No. 4, p 231-259, 1998.

Cui, S., and **Yu, Q.,** *Traveling Performance and Test of Vehicle Tire*, Mechanical Industry Press, in China, 1995.

d'Entremont, K.L., "The Behavior of Tire-Force Model Parameters Under Extreme Operating Conditions", Paper 970558, SAE International, Warrendale, PA, 1997.

Dear, J., Asgari, J., and **Hrovat, D.A.,** "3D Brush-Type Dynamic Tire Friction Model", *Vehicle System Dynamics*, Vol. 42, No. 3, p o133-173, 2004.

Deur, J., "Modeling and Analysis of Longitudinal Tire Dynamics Based on the LuGre Friction Model", *Advances in Automotive Control 2001, Proceedings volume from the 3rd IFAC Workshop*, p 91-96, 2001.

Dong, P., Liu, Q., and **Guo, K.,(In Chinese).,** "Analysis of the E-Function for Non-Steady State Tire Cornering Mechanics", *Qinghua Daxue Xuebao/Journal of Tsinghua University,* Vol. 42, No. 2, p 251-253, 2003.

Dukkipati, R.V., Pang, J., Qatu, M., Sheng, G., and **Shuguang, Z.,** *Road Vehicle Dynamics: Problems and Solutions Workbook,* Product Code R-386, SAE International, Warrandale, PA, 2009.

Dukkipati, R.V., Pang, J., Qatu, M., Sheng, G., and **Shuguang, Z.,** *Road Vehicle Dynamics,* Product Code R-366, SAE International, Warrandale, PA, 2008.

Dukkipati, R.V., *Vehicle Dynamics*, Narosa Publishing House, New Delhi, India, 2000.

Fiala, E., "Seitenkrafte am rollenden Luftreifen" (Lateral Forces at the Rolling Pneumatic Tire), no. 96, VDI-Zeitschrift, 1954, p. 973.

Fiala, E., "Seitenkrafte am rollenden Luftreifen" (Lateral Forces at the Rolling Pneumatic Tire), no. 96, VDI-Zeitschrift, 1954, p. 973.

Gim, G., and **Nikravesh, P.E.,** "An Analytical Model of Pneumatic Tyres for Vehicle Dynamic Simulations. Part 2: Comprehensive Slips", *International Journal of Vehicle Design,* vol. 12, no. 1, Inderscience, Geneva, Switzerland, 1991.

Gim, G., and **R.E. Nikravesh.,** "An Analytical Model of Pneumatic Tyres for Vehicle Dynamic Simulations. Part 1: Pure Slips", *International Journal of Vehicle Design,* vol. 11, no. 6, Inderscience, Geneva, Switzerland, 1990.

Gim, G., and **R.E. Nikravesh.,** "An Analytical Model of Pneumatic Tyres for Vehicle Dynamic Simulations. Part 3: Validation Against Experimental Data", *International Journal of Vehicle Design,* vol. 12, no. 2, Inderscience, Geneva, Switzerland. 1991.

Gim, G., Choi, Y., and **Kim, S.,** "A Semi Physical Tire Model for Vehicle Dynamics Analysis of Handling and Braking", *Vehicle System Dynamics*, Vol. 43, p 267-280, 2005.

Guan, D., Fan, C., and **Xie, X.,** "A Dynamic Tyre Model of Vertical Performance Rolling Over Cleats", *Vehicle System Dynamics*, Vol. 43, p 209-222, 2005.

Guo, K., and **Liu, Q.,** "Generalized Theoretical Model of Tire Cornering Properties in Steady State Condition", *SAE Special Publications,* Vol. 1308, *Heavy Vehicle and Highway Dynamics,* 973191, p 31-38, 1997.

Guo, K., and **Liu, Q.,** "Model of Tire Enveloping Properties and its Application on Modeling of Automobile Vibration Systems", *SAE Special Publications, Vol. 1338, 1998, Developments in Tire, Wheel, Steering, and Suspension Technology*, 980253, p 21-27.

Guo, K., and **Liu, Q.,** "Tire Models for Vehicle Dynamics Analysis in Steady State Condition", *Qiche Gongcheng/Automotive Engineering*, Vol. 20, No. 3, p 129-134, 1998.

Guo, K., Lu, D., Chen, S.K., Lin, W.C., and **Lu, X.P.,** "The UniTire model: A Nonlinear and Non-Steady-State Tire Model For Vehicle Dynamics Simulation", *Vehicle System Dynamics*, Vol. 43, p 341-358, 2005.

Guo, K.H., Yuan, Z.C., Lu, D., and **Lin, B.Z.,** "Speed Prediction Ability of UniTire Steady State Model", *Jilin Daxue Xuebao (Gongxueban)/Journal of Jilin University (Engineering and Technology Edition),* Vol. 35, No. 5, p 457-461, 2005.

Holtschulze, J., Goertz, H., and **Husemann, T.,** "A Simplified Tire Model For Intelligent Tires," *Vehicle System Dynamics*, Vol. 43., p 305-316, 2005.

Kao, B.G., and **Warholic, T.,** "Tire Lateral Force Modeling and the Bushing Analogy Tire Model", *Tire Science and Technology,* Vol. 33, No. 1, p 18-37, 2005.

Konghui, Guo., *Automobile Handling Dynamics,* Jilin Press of Science and Technology, China, December 1991.

Liu, Q., and **Guo, K., (In Chinese),** "Numerical Simulation Algorithms of Tire Cornering Properties in Non Steady State Conditions", *Qinghua Daxue Xuebao/Journal of Tsinghua University,* Vol. 39, No. 11, p 69-72 , 1999.

Lugner, P., Pacejka, H., and **Plochl, M.,** "Recent Advances in Tire Models and Testing Procedures", *Vehicle System Dynamics*, Vol. 43, No. 6-7, p 413-436, 2005.

Miege, A.J.P., and **Popov, A.A.,** "Truck Tire Modeling For Rolling Resistance Calculations Under a Dynamic Vertical Load", *Proceedings of the Institution of Mechanical Engineers, Part D: Journal of Automobile Engineering,* Vol. 219, No. 4, p 441-454, 2005.

Miege, Arnaud J.P., and **Popov, Atanas A.,** "The Rolling Resistance of Truck Tires Under a Dynamic Vertical Load", *Vehicle System Dynamics*, Vol. 43, p 135-144, 2005.

Milliken, W.F., and **D.L. Milliken.,** *Race Car Vehicle Dynamics,* SAE International, Warrendale, PA, 1995.

Miyashita, N., and **Kabe, K.,** "A Study of the Cornering Force by use of the Analytical Tire Model," *Vehicle System Dynamics,* Vol. 43, p 123-134, 2005.

Nguyen, P.K., and **E.R. Case.,** "Tire Friction Models and Their Effect on Simulated Vehicle Dynamics", Proceedings of a Symposium on Commercial Vehicle Braking and Handling, UM-HSRI-PF-75-6, Ann Arbor, MI, 1975.

Nicolas, V.T., and **Comstock, T.R.,** "Predicting Directional Behavior of Tractor Semitrailers When Wheel Anti-Skid Brake Systems are Used", Paper No. 72—WA/Aut-16, Winter Annual Meeting, November 26-30, ASME, New York, NY, 1972.

Nielsen, L., and **Sandberg, T.,** "A New Model for Rolling Resistance of Pneumatic Tires", Paper 2002-01-1200, SAE International, Warrendale, PA, 2002.

Pacejka, H.B., "Spin: Camber and Turning", *Vehicle System Dynamics,* Vol. 43, p 3-17, 2005.

Pacejka, H.B., *Tire and Vehicle Dynamics*, 2nd ed, SAE International, Warrandale, PA, 2006.

Palkovics, L., El-Gindy, M., and **Pacejka, H.B.,** "Modeling of the Cornering Characteristics of Tires on an Uneven Road Surface: a Dynamic Version of the 'Neuro-Tyre'", *International Journal of Vehicle Design*, Vol. 15, No. 1-2, p 189-215, 1994.

Poliey, M. Alleyne., "*A.G. Dimensionless* Analysis of Tire Characteristics for Vehicle Dynamics Studies", *Proceedings of the American Control Conference, Vol. 4, Proceedings of the 2004 American Control Conference (AAC)*, p 3411-3416, 2004.

SAE., "Vehicle Dynamics Terminology, SAE Recommended Practice", SAE Standard J670e, SAE International, Warrendale, PA, 1976.

Schmeitz, A.J.C., Jansen, S.T.H., Pacejka, H.B., Davis, J.C., Kota, N.M., Liang, C.G., and **Lodewijks, G..,** "Application of a Semi-Empirical Dynamic Tire Model for Rolling over Arbitrary Road Profiles", *International Journal of Vehicle Design*, Vol. 36, No. 2-3, p 194-215, 2004.

Schuring, D.J., Pelz,W., and **Pottinger, M.G.,** "The BNPS Model–An Automated Implementation of the 'Magic Formula' Concept", Paper 931909, SAE International. Warrendale, PA 1993.

Schuring, D.J., Pelz,W., and **Pottinger, M.G.,** "A Model for Combined Tire Cornering and Braking Forces", Paper 960180, SAE International, Warrendale, PA, 1996.

Shang, J., Guan, D., and **Yam, L.H.,** "Study on Tire Dynamic Cornering Properties Using Experimental Model Parameters", *Vehicle System Dynamics*, Vol. 37, No. 2, p 129-144, 2003.

Velenis, E., Tsiotras, P., Canudas, D.W., and **Sorine, M.,** "Dynamic Tire Friction Models for Combined Longitudinal and Lateral Vehicle Motion", *Vehicle System Dynamics*, Vol. 41, No. 1, p 3-29, 2005.

Velumot, H.P., *Method and Simulation of Vehicle Dynamics*, Press of Peking University of Science and Technology, 1998.

Warner, C. Y., Smith, G.C., James, M.B., and **Germane, G.J.,** "Friction Applications in Accident Reconstruction", Paper 830612, SAE International, Warrendale, PA, 1983.

Wille, R., Bohm, F., and **Duda, A.,** "Calculation of the Rolling Contact Between a Tire and Deformable Ground", *Vehicle System Dynamics*, Vol. 43, p 483-492, 2005.

Wong, J.Y., *Theory of Ground Vehicles,* John Wiley and Sons, Inc., New York, NY, 1993.

Yi, J., Alvarez, L., Claeys, X., Horowitz, R., and **Canudas De Wit, C.,** "Emergency Braking Control with an Observer-Based Dynamic Tire/Road Friction Model and Wheel Angular Velocity Information", *Proceedings of the American Control Conference*, Vol. 1, p 19-24, 2001.

Yu, Z., *Automobile Theory*, Mechanical Industry Press, 3rd ed., Beijing, China, October 2000.

Zhang, H., *Automobile System Dynamics*, Tongji University Press, Tongji, China, 1996.

Zuo, S., *Influence of Road Surface Roughness on Braking Performance of Tractor-trailing*, Dissertation Submitted to Jilin University of Technology for Ph.D., June 1996.

Zuo, S., Liu, D., and **Zhang J.,** "Tire Model Due To Braking Process For Vehicles On Uneven Road Surface", *Transactions of Nanjing University of Aeronautics & Astronautics*, 1997.

3.8 PROBLEMS

Problem 3.1: *The whole mass of a four-wheel-drive car is 1500kg, the drive ratio is 1:1, the wheelbase is 2.6m, the height of the centroid is 0.5m, the load of front shaft is 55%, the coefficient of load adhesion $\varphi = 0.8$, rolling resistance coefficient $f = 0.02$, calculate the max acceleration.*

Solution: Total tire rolling resistance:

$$E_f = mgf = 1500 \times 9.81 \times 0.02 = 294.3\text{N} \tag{1}$$

The load transfer

$$\Delta F = \frac{mgha}{L} = \frac{1500 \times 9.81 \times 0.5}{2.6} a = 2829.8a \tag{2}$$

Analyze which wheel will first achieve the utmost, assume the load transfer $\Delta F = 0.05mg$, then:

$$a = \frac{0.05\text{mg}}{2829.8} = \frac{0.05 \times 1500 \times 9.81}{2829.8} = 0.26\text{m/s}^2 \tag{3}$$

The drive force:

$$F = ma = 1500 \times 0.26 = 390\text{N} < F_{\max} = m \cdot g \cdot \varphi = 11772\text{N} \tag{4}$$

So the load transfer is larger than $0.05mg$, so we can make sure that the front wheel first achieves the utmost, then:

$$2 \times (0.55mg - \Delta F) \times 0.8 - F_f = ma \tag{5}$$

Solve the formula, get the max acceleration:

$$a_{\max} = 2.0995\text{m/s}^2 \tag{6}$$

Problem 3.2: *A car is driving with a constant velocity around a circle($r = 30m$), the mass of the car is 1600kg; the wheelbase is 2.6m, the load of front shaft is 60%, assume the coefficient of load adhesion $\varphi = 0.9$, rolling resistance coefficient $f = 0.03$, front wheel turn angle is 15°, the relationship between longitudinal force F_x and lateral force F_y is $\sqrt{F_x^2 + F_y^2} = F_z \cdot \varphi$, ignore the lateral stiffness, calculate the max velocity without sideslip.*

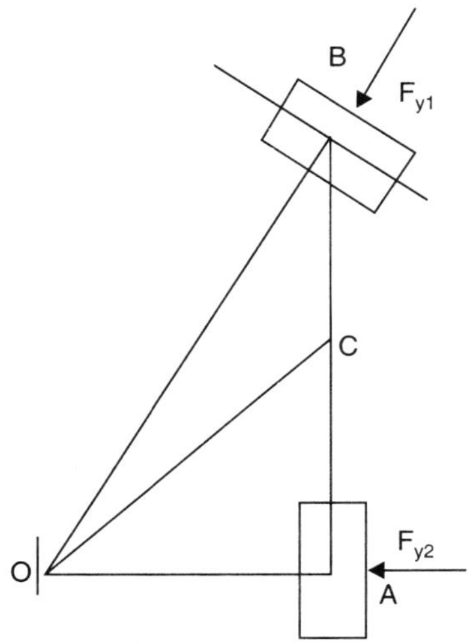

Fig. P. 3.2 Two-DOF car model.

Solution: First, we analyze the relationship between lateral force F_y and F_{y1}, F_{y2}.

$$|OA| = \sqrt{|OC|^2 - |AC|^2} \tag{1}$$

$$|OB| = \sqrt{|OA|^2 + |AB|^2} \tag{2}$$

$$F_y \cdot \frac{|OA|}{|OC|} = F_{y2} + F_{y1} \cdot \frac{|OA|}{|OB|} \tag{4}$$

$$F_y \cdot \frac{|AC|}{|OC|} = F_{y1} \cdot \frac{|AB|}{|OB|} \tag{4}$$

From the formula above, we get:

$$\frac{F_{y2}}{F_{y1}} = 0.6 < \frac{F_{z2}}{F_{z1}} = 0.67 \tag{5}$$

So the front wheel first achieves the utmost. Then the max lateral force of front wheel is as follows:

$$F_{y1} = \sqrt{(F_{z1} \cdot \varphi)^2 - (F_{z1} \cdot f)^2} = 1600 \times 9.81 \times 0.6 \times \sqrt{0.8^2 - 0.03^2} = 7528.8\text{N} \tag{6}$$

The max lateral force is:

$$F_y = F_{y1} \cdot \frac{|AB|}{|OB|} \cdot \frac{|OC|}{|AC|} = 7528.8 \times 1.6627 = 12518\text{N} \tag{7}$$

The max velocity without sideslip is:

$$V_{\max} = \sqrt{\frac{F_y \cdot r}{m}} = \sqrt{\frac{12518 \times 30}{1600}} = 15.32\text{m/s} \tag{8}$$

Problem 3.3: *When adding a rigid wheel with a radius R to a point contact tire model, it obtains enveloping characteristics. So, please calculate the effective surface irregularities $h_q(X)$ when the practice surface irregularities are h(X) use point contact tire model with a rigid wheel.*

Here, h(X) = 0.05 sin (15X)m, R = 0.3m, assume that the wheel rotates slowly and maintain contact with the road surface.

Solution: From the equation (3.9) we know the effective surface irregularities:

$$h_q(X) = h(X + x) - \left[R - \sqrt{R^2 - x^2}\right] \tag{1}$$

Here, $h(X)$ and R is given, so if x is known, the answer will be clear. Maybe it's hard to calculate the $h_q(X)$ by hand, but it's easy for computer to obtain the result. So with the help of computer, we can get the effective surface irregularities $h_q(X)$ as follow Fig.:

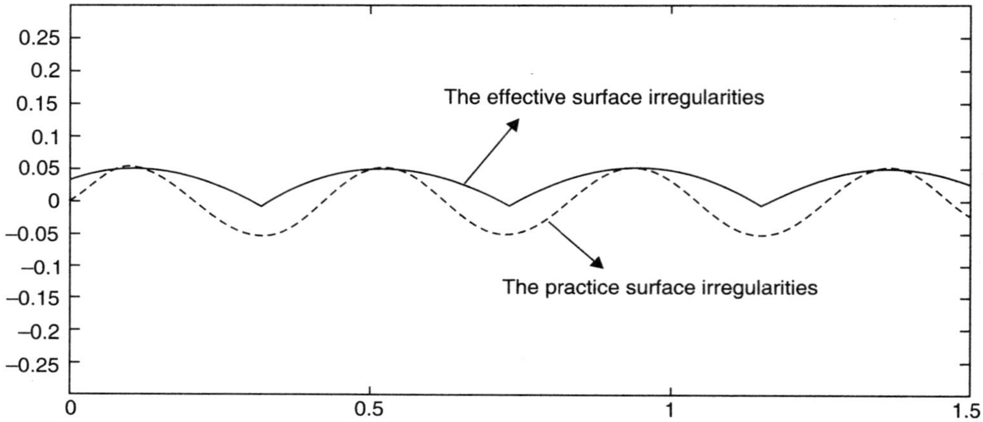

Fig. P. 3.3 The effective surface irregularities.

MATLAB Program:

```
R=0.3;
x=linspace(-R,R,100);
X=linspace(0,5*R,200);
h=0.05*sin(15*X);
for i=1:200,
    hq(i)=-R;
    for j=1:100,
        hj=0.05*sin(15*(X(i)+x(j)));
        hqt=hj-abs(R-sqrt(R^2-x(j).^2));
        if hqt>hq(i),
            hq(i)=hqt;
        end
    end
end
```

```
plot(X,hq,X,h,'-')
axis([0,1.5,-0.3,0.3])
```

Problem 3.4: *Please calculate the frequency response characteristic of the tire using the point contact tire model. Assume all the needed parameters are given.*

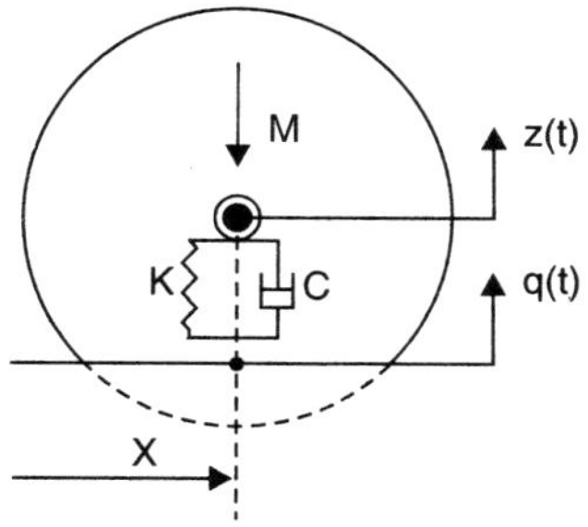

Fig. P. 3.4 (*a*) The point contact tire model.

Solution: As shown in the Fig. P. 3.4(*a*), the vertical dynamic equation of the tire is:

$$M \cdot \ddot{z} + C \cdot (\dot{z} - \dot{q}) + K \cdot (z - q) = 0 \tag{1}$$

Does Laplace transformation get:

$$-\omega^2 M \cdot Z + j\omega C \cdot (Z - Q) + K \cdot (Z - Q) = 0 \tag{2}$$

Solve the above equation, we get:

$$H(j\omega) = \frac{Z}{Q} = \frac{K + j\omega C}{K - \omega^2 M + j\omega C} \tag{3}$$

Define:

$$\omega_0^2 = \frac{K}{M}, n = \frac{C}{2M}, \zeta = \frac{n}{\omega_0}$$

Then: $$H(j\omega) = \frac{\omega_0^2 + 2j\zeta\omega_0\omega}{\omega_0^2 - \omega^2 + 2j\zeta\omega_0\omega} \tag{4}$$

Define frequency ratio: $$\lambda = \frac{\omega}{\omega_0}$$

Then it is module: $$|H(j\omega)| = \sqrt{\frac{1 + (2\zeta\lambda)^2}{(1-\lambda^2)^2 + (2\zeta\lambda)^2}} \tag{5}$$

It's phase: $$\varphi = \arctan \frac{2\zeta\lambda^3}{1 - \lambda^2 + (2\zeta\lambda^2)} \tag{6}$$

Here, if $M = 300\text{kg}$, $K = 3 \times 10^5$ N/m, $C = 100\text{N} \cdot \text{s/m}^2$, then we got the bode diagram as follow:

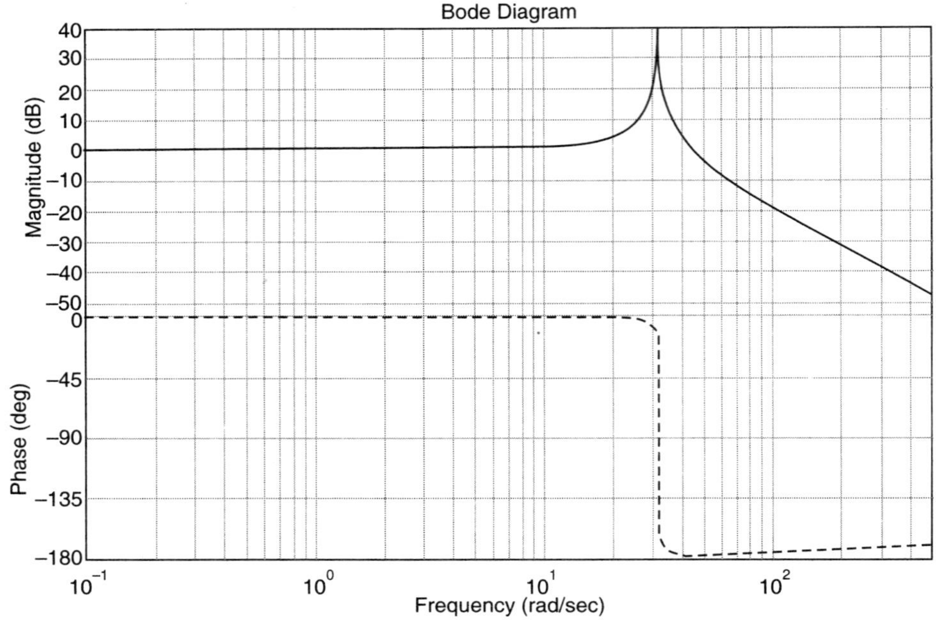

Fig. P. 3.4 (*b*) The bode diagram.

Problem 3.5: *Given that the radius of the tire is 0.3m, the camber angle is 1º, and the slip angle is 5º. Try to collect the relative literature concerning "H.B.Pacejka Magic Formula Model of tire" and write a piece of MATLAB program which can generate a two-dimensional curve representing the relationship between the longitudinal force (Fx) and the lateral force (Fy) of the tire in a condition of combined slipping. In additional, analyze the change trend of the curve with the change of gross vehicle weight.*

Solution: MATLAB program:

```
Fx=zeros(4,2001);
Fy=zeros(4,2001);
for fz=2000:2000:8000
vcx=40;
 fz0=fz; r0=0.3;
 re=r0;
 ydeg=1;
 adeg=5;

pcx1=1.65; pdx1=1; pdx2=0; pex1=-0.5; pex2=0; pex3=0; pex4=0;
pkx1=12; pkx2=10; pkx3=-0.6; phx1=0; phx2=0; pvx1=0; pvx2=0;
pcy1=1.3; pdy1=1; pdy2=0; pdy3=0;
pey1=-1; pey2=0; pey3=0; pey4=0;
```

```
pky1=10; pky2=1.5; pky3=0; phy1=0; phy2=0; phy3=0.25;
pvy1=0; pvy2=0; pvy3=0.15; pvy4=0;

rbx1=5; rbx2=8; rcx1=1; rhx1=0;
rby1=7; rby2=2.5; rby3=0; rcy1=1; rhy1=0.02; rhy2=0;
rex1=-0.5; rex2=0; rey1=-1; rey2=0;
rvy1=0; rvy2=0; rvy3=-0.2; rvy4=14; rvy5=1.9; rvy6=10;

%Longitudinal Force (Pure longitudinal slip)
dfz=(fz-fz0)./fz0;
k=0:0.0005:1;
shx=phx1+phx2.*dfz;
kx=k+shx;
cx=pcx1;
ux=pdx1+pdx2.*dfz;
dx=ux.*fz;
ex=(pex1+pex2.*dfz+pex3.*dfz.^2).*(1-pex4.*sign(kx));
kxk=fz.*(pkx1+pkx2.*dfz).*exp(pkx3.*dfz);
bx=kxk./(cx.*dx);
svx=fz.*(pvx1+pvx2.*dfz);
fx0=dx.*sin(cx.*atan(bx.*kx-ex.*(bx.*kx-atan(bx.*kx))))+svx;

%Lateral Force (Pure side slip)
a=tan(adeg*pi/180).*sign(vcx);
y=sin(ydeg*pi/180);
shy=phy1+phy2.*dfz+phy3.*y;
ay=a+shy;
cy=pcy1;
uy=(pdy1+pdy2.*dfz).*(1-pdy3.*y.^2);
dy=uy.*fz;
ey=(pey1+pey2.*dfz).*(1-(pey3+pey4.*y).*sign(ay));
kya0=pky1.*fz0.*sin(2*atan(fz./(pky2.*fz0)));
kya=kya0.*(1-pky3.*y.^2);
by=kya./(cy.*dy);
svy=fz.*(pvy1+pvy2.*dfz+(pvy3+pvy4.*dfz).*y);
```

```
fy0=dy.*sin(cy.*atan(by.*ay-ey.*(by.*ay-atan(by.*ay))))+svy;

%Longitudinal Force (Combined slip)
shxa=rhx1;
as=a+shxa;
bxa=rbx1.*cos(atan(rbx2.*k));
cxa=rcx1;
exa=rex1+rex2.*dfz;
gxa0=cos(cxa.*atan(bxa.*shxa-exa.*(bxa.*shxa-atan(bxa.*shxa))));
gxa=cos(cxa.*atan(bxa.*as-exa.*(bxa.*as-atan(bxa.*as))))./gxa0;
fx=gxa.*fx0;

%Lateral Force (Combined slip)
shyk=rhy1+rhy2.*dfz;
ks=k+shyk;
byk=rby1.*cos(atan(rby2.*(a-rby3)));
cyk=rcy1;
eyk=rey1+rey2.*dfz;
gyk0=cos(cyk.*atan(byk.*shyk-eyk.*(byk.*shyk-atan(byk.*shyk))));
gyk=cos(cyk.*atan(byk.*ks-eyk.*(byk.*ks-atan(byk.*ks))))./gyk0;
dvyk=uy.*fz.*(rvy1+rvy2.*dfz+rvy3.*y).*cos(atan(rvy4.*a));
svyk=dvyk.*sin(rvy5.*atan(rvy6.*k));
fy=gyk.*fy0+svyk;
cixu=fz/2000;
Fx(cixu,:)=fx;
Fy(cixu,:)=fy;
end
plot(Fx(1,:),Fy(1,:),
Fx(2,:),Fy(2,:),Fx(3,:),Fy(3,:),Fx(4,:),Fy(4,:));
```

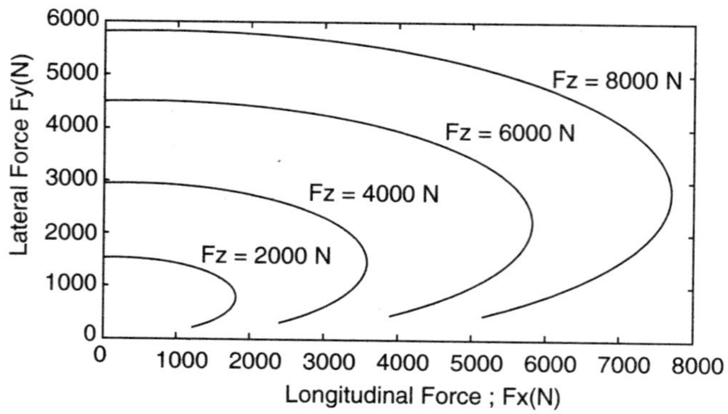

Fig. P. 3.5 The relationship between the longitudinal force and the lateral force.

Problem 3.6: *In a damped vibration locus of a certain tire with a mass of $m = 20kg$, the first peak appears at the time of $t_0 = 0s$ with a crest height of $x_1 = 0.05m$, and the second peak appears at the time of $t_1 = 0.03s$ with a crest height of $x_2 = 0.01m$. Try to calculate the natural frequency ω_d, the equivalent viscous damping coefficient C_{eq}, and the dynamic stiffness K_z of the tire, and use MATLAB to draw the damped vibration locus.*

Solution: The vibration period is:

$$\tau = t_1 - t_0 = (0.03 - 0)s = 0.03s \tag{1}$$

The attenuation rate is:

$$\delta = \ln\frac{x_1}{x_2} = \ln\frac{0.05}{0.01} \approx 1.61 \tag{2}$$

Set a temporary variable as:

$$\beta = \frac{\delta^2}{\delta^2 + 4\pi^2} = 0.0616 \tag{3}$$

Then, the natural frequency of the tire is:

$$\omega_d = \frac{2\pi}{\tau} = \frac{2\pi}{0.03}\text{rad/s} \approx 209.44 \text{ rad/s} \tag{4}$$

The equivalent viscous damping coefficient of the tire is:

$$C_{eq} = \sqrt{\frac{4m^2\omega_d^2\beta}{1-\beta}} = \sqrt{\frac{4\times 20^2 \times 209.44^2 \times 0.0616}{1-0.0616}}\text{N}\cdot\text{s/m} \approx 2.15\times 10^3\,\text{N}\cdot\text{s/m} \tag{5}$$

The dynamic stiffness of the tire is:

$$K_z = \frac{m\omega_d^2}{1-\beta} = \frac{20\times 209.44^2}{1-0.0616}\text{N/m} \approx 9.35\times 10^5\,\text{N/m} \tag{6}$$

In accordance with the damped vibration equation:

$$m\ddot{x} + C_{eq}\dot{x} + K_z x = 0 \tag{7}$$

Building a simulation model in MATLAB/SIMULINK as follows

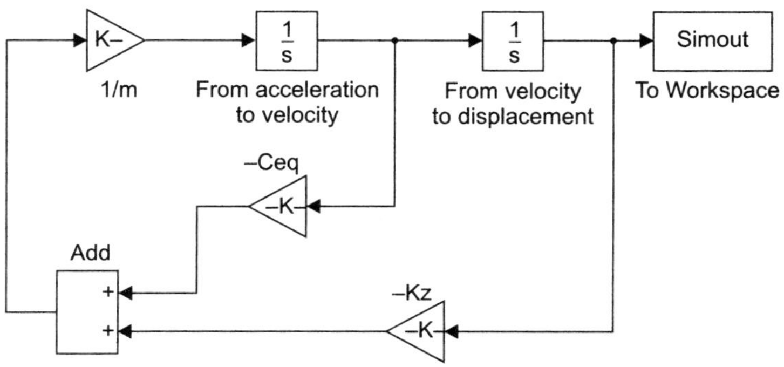

Setting the simulation time for 0.1s, then we get the certain damped vibration locus as follows:

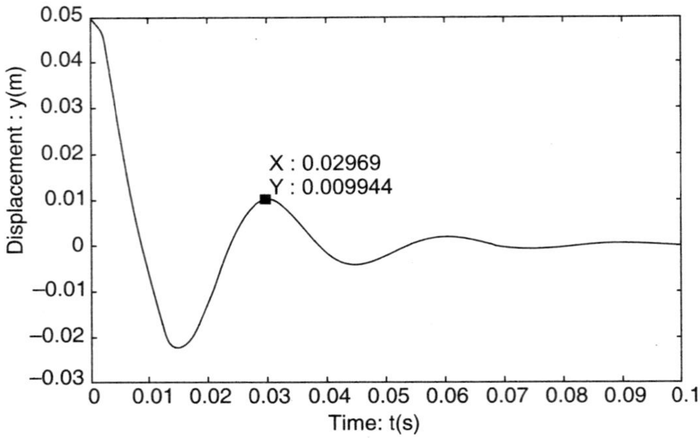

Fig. P. 3.6 The damped vibration locus.

Problem 3.7: *As we all know, we have limited the lateral accelerated speed under 0.4g, in the Motorcycle model, so as to keep the property of tire in the linear region. After learning this chapter, here we try to apply the Gim tire model into the Motorcycle model and then use the software of Matlab to do the simulation to help the readers know how to use the Gim tire model.*

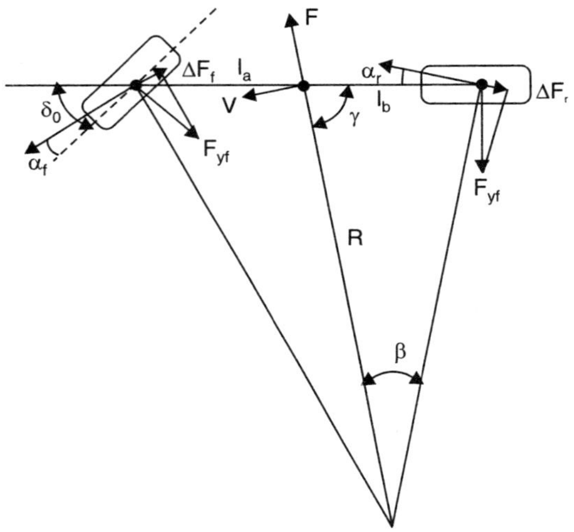

Fig. P. 3.7 Motorcycle model.

Solution: Firstly, we re-derive the function.

In the analysis, we have ignored the effect of steering system and taken the steering angle as the direct input, and have ignored the function of suspension system and suppose that the vehicle is moving along the trajectory that is parallel to the ground. In other words, the displacement along z axis, the pitching angle around y axis and the camber angle around x axis are zero. In addition, we suppose that the vehicle is moving at a constant speed, in this specific statement as shown in the Fig. P. 3.7.

As shown in the Fig. Centroid slip angle:

$$\beta = \frac{v}{Uc} \quad (1)$$

Slip angle of the front wheel:

$$\alpha_1 = \beta + \frac{aWr}{Uc} - \delta \quad (2)$$

Slip angle of the rear wheel:

$$\alpha_2 = \beta - \frac{bWr}{Uc} \tag{3}$$

Where: v is the lateral speed.

Uc is the longitudinal speed.

Wr is the yaw angular speed.

a is the la as shown in the figure.

b is the lb as shown in the figure.

δ is the steering angle (input value)

According to the Newton's law of balance, we can get the equation as follows:

$$\begin{cases} F_{y1} + F_{y2} = m(\dot{v} + U_c W_r) \\ aF_{y1} + bF_{y2} = I_z \dot{W}_r \end{cases} \tag{4}$$

Then we apply the tire model into this Motorcycle model, firstly take Gim model as an example.

Referring to the chapter 3, we know that F_{y1} and F_{y2} are determined by the longitudinal speed, slip angle, vertical force.

$$\begin{cases} F_{y1} = \text{Gim}(U_c, \alpha_1, F_z) \\ F_{y2} = \text{Gim}(U_c, \alpha_2, F_z) \end{cases} \tag{5}$$

Under this specific statement, we suppose the vertical force Fz, and longitudinal speed Uc are constant values, so Fy is only determined by slip angle as follows.

$$\begin{cases} F_{y1} = \text{Gim}(\alpha_1) \\ F_{y2} = \text{Gim}(\alpha_2) \end{cases} \tag{6}$$

Problem 3.8: *According to the Gim theory described above in the chapter 3, we construct the tire model of Gim in Matlab with the help of Simulink as follows.*

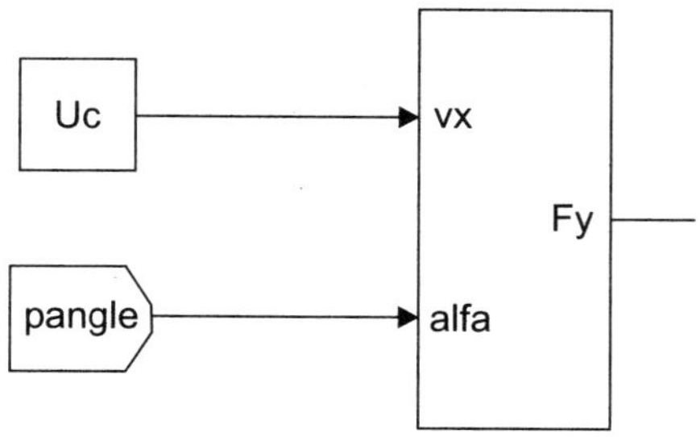

Fig. P. 3.8 Gim tire model.

Look under mask as follows:

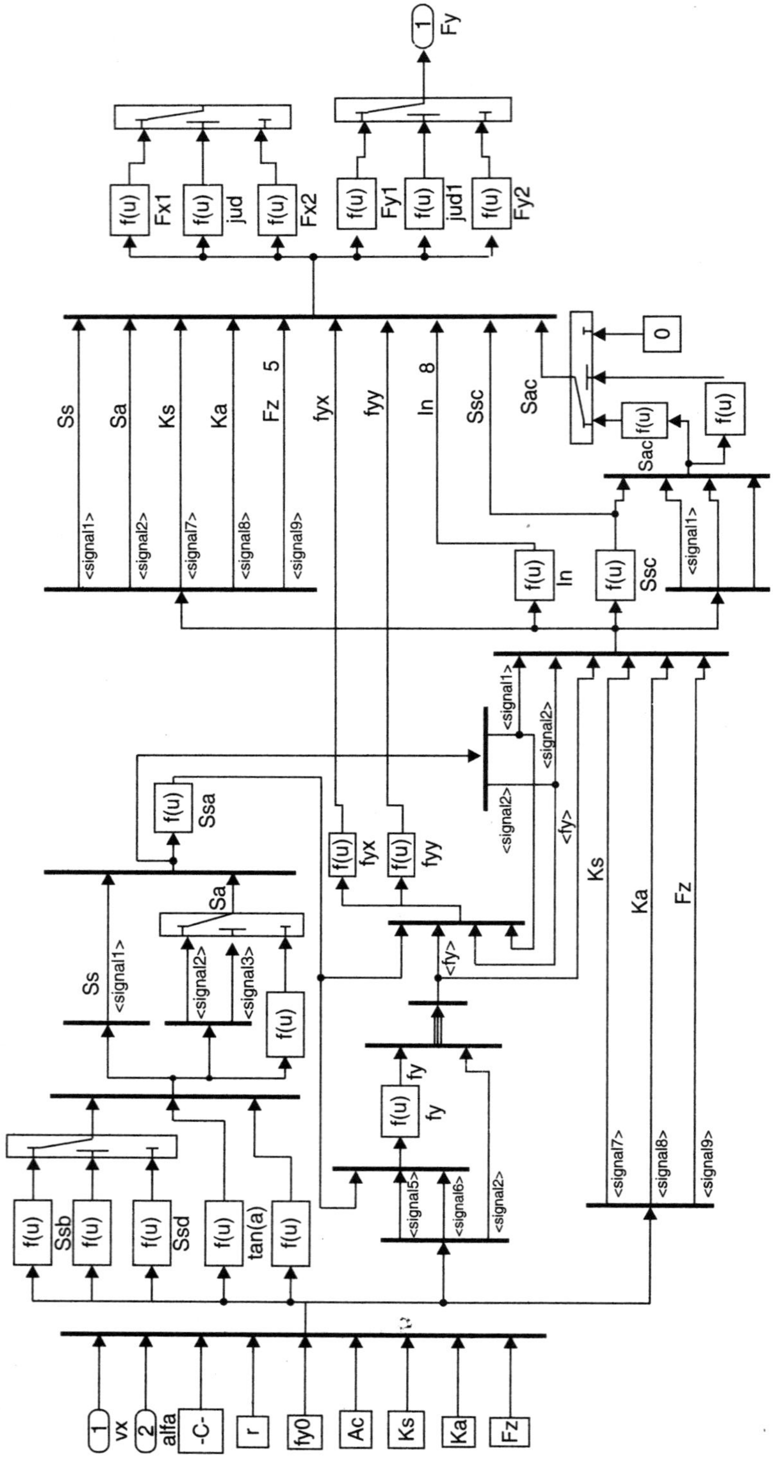

Input values are Uc and slip angle, and Uc is a constant value.

Finally, we apply this Gim model into the Motorcycle model and do the simulation with the help of Matlab as follows:

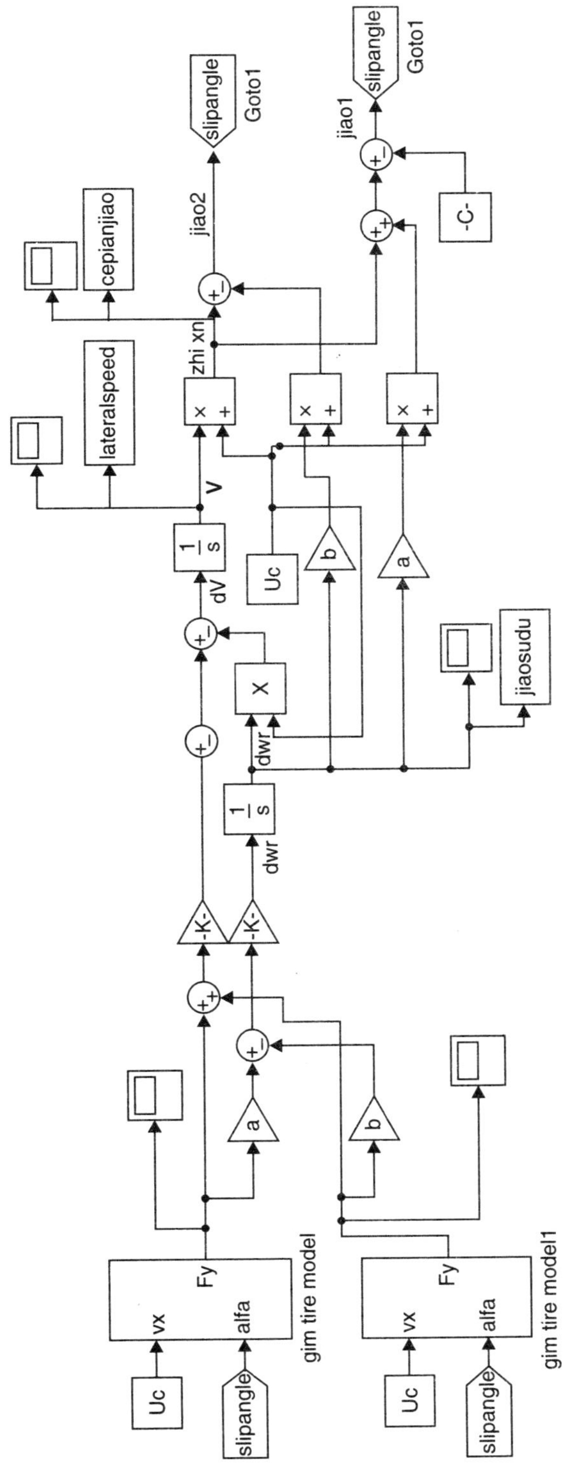

Result:

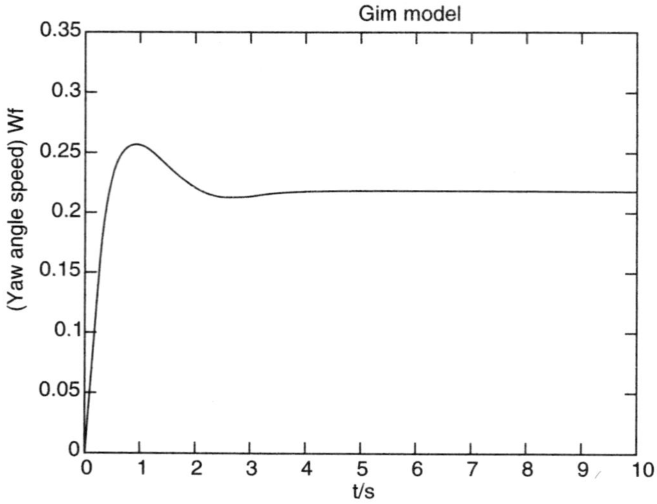

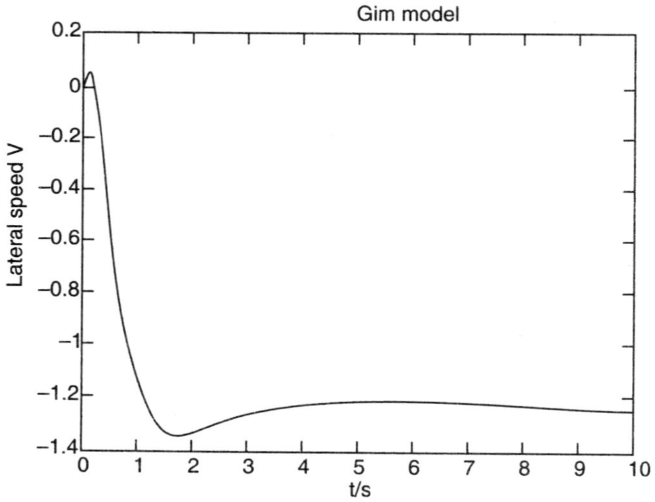

As we all know, we have limited the lateral accelerated speed under 0.4g, in the Motorcycle model, so as to keep the property of tire in the linear region. After learning this chapter, here we try to apply the Magic Formula Model tire model into the Motorcycle model and then use the software of Matlab to do the simulation to help the readers to know how to use the Magic Formula Model.

CHAPTER

4

Collision Analysis

4.1 INTRODUCTION

Road vehicle accident collision analysis involves the determination as far as possible of *what* happened during the collision, *where* it happened, and *how* it happened. The process is a reconstruction mode based on the damage, the debris, the highway marks, and the points of rest of the vehicles after the accident. In collision analysis, we are concerned in establishing three things:

(*a*) how the vehicles collided

(*b*) where they collided

(*c*) the speeds involved

4.2 ELEMENTS OF A COLLISION

An impact collision is a whole series of events in which two objects are in contact with each other. There are three events that are worthwhile noting.

1. First contact: First contact is the beginning of a collision.
2. Maximum engagement: Lynn B. Frickle [1990] points out that from first contact, penetration and force increase to maximum engagement. At maximum with a fixed object engagement instant, the vehicle, in contact with the object struck, stops. At that instant, the vehicle and object are at the same speed zero. After that, the vehicle moves backward due to the elasticity of vehicle material. Penetration and force decrease and velocity backward increases until the last contact, when the objects in collision disengage and separate. See Fig. 4.1.
3. Separation: Separation or stopping if the vehicles or other objects remain engaged occurs when force between the colliding or impacting bodies becomes zero.

Collisions may be divided into five general types. They are the following:

1. The *rear-end collision* and its variation, *the stopped vehicle collision*.
2. The *head-on collision*, where the vehicles strike each other dead center.

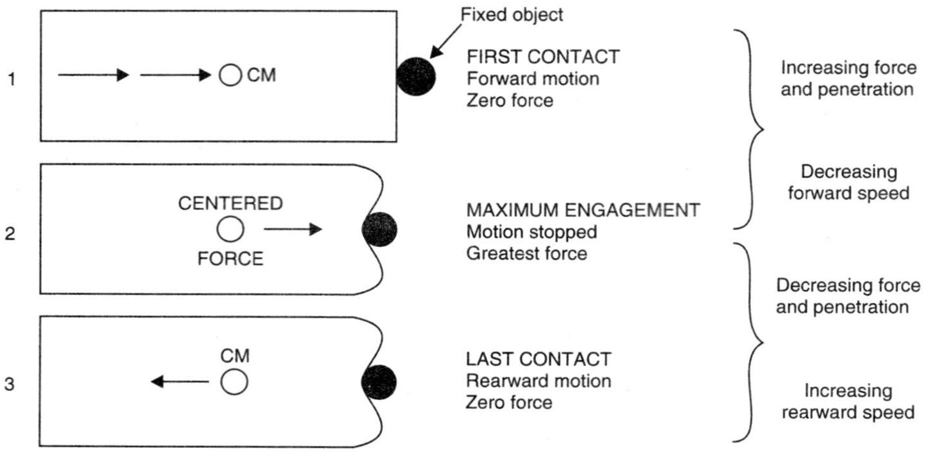

Fig. 4.1 Impact of a vehicle or other object in collision begins with first contact, progresses to maximum engagement, and ends with last contact at separation. A complete impact is illustrated here. In a complete impact, motion between the areas in contact momentarily ceases. This is a centered impact [Frickle (1990)].

3. The *corner-to-corner* collision, where the overlap is greater than six inches.
4. The *sideswipe*, where the vehicle overlap was less than six inches.
5. The *intersection collision* where the vehicles come together at right angles.

The determination of collision type is based on damage analysis. Damage analysis will give us the positional relationship of the vehicle at impact.

4.3 BASIC TERMS AND DEFINITIONS

A Run-off-Road Collision: A run-off-road collision is a type of single-vehicle accident that occurs when a vehicle leaves the roadway. Contributing factors often include loss of control or misjudging a curve, or attempting to avoid colliding with another road user or an animal. If the vehicle strikes a fixed object (an object that will move very little when struck, such as a tree, bridge structure or utility pole) or rolls over, the crash is likely to result in injuries or fatalities. In fact 2005 statistics from the US show that run off road crashes resulted in 31% of fatal crashes, but were only 16% of all crashes.

Angle Between Vectors: The angle between two nonzero vectors A and B is the angle $\theta (0 \leq \theta \leq \pi)$ between the directed segments representing both vectors.

Angle Collision: A collision between two traffic units approaching on separate roadways or other paths that intersect.

Angular Impact: When two vehicles collide in such a manner that their respective directions of force are not parallel.

Angular Velocity Change: The angular velocity change is the rotational speed given to the vehicle by the force created by the collision. This force usually creates a moment about the center of gravity. The velocity change at the location the force is different than that at the center of gravity. The difference in the two velocities changes is a function of the angular velocity change and the distance from the center of crush to the center of gravity. Refer to a dynamics text on the relationships which apply. The angular velocity change has the same sign (direction) as the moment arm.

Automobile Collision Analysis: Automobile collision analysis constitutes the determination as far as possible of what happened during the collision, where it happened, and how it happened. It is a reconstruction mode on the basis of the damage, the debris, the highway marks, and the point of rest of the vehicles after the accident.

Barrier Equivalent Velocity (BEV): The forward speed and corresponding kinetic energy with which a vehicle contacts a flat fixed rigid barrier at 90° with no rebound. See also Equivalent Energy Speed (EES).

Broadside Collisions: Broadside collisions commonly known as T-bone collisions, are where the side of one vehicle is impacted by the front or rear of another vehicle or a fixed object. Vehicle damage and occupant injury are more likely to be severe, but severity varies based on the part of the vehicle that is struck, safety features present, the speeds of both vehicles, and vehicle weight and construction. When a vehicle is hit on the side by another vehicle, the crumple zones of the striking vehicle will absorb some of the kinetic energy of the collision. The crumple zones of the struck vehicle may also absorb some of the collision's energy, particularly if the vehicle is not struck on its passenger compartment. Both vehicles are frequently turned from their original directions of travel. If the collision is severe, the struck vehicle may be spun or rolled over, potentially causing it to strike other vehicles, objects, or pedestrians. After the collision, the involved vehicles may be stuck together by the folding of their parts around each other. An occupant on the struck side of a vehicle may sustain far more severe injuries than an otherwise similar front or rear collision crash. Side-impact airbags can protect vehicle occupants during side collisions, but they face the same limitations as other airbags. Additionally, side impact wrecks are more likely to involve multiple individual collisions or sudden speed changes before motion ceases. Since the airbag can only provide protection during the first collision, it may leave occupants unprotected during subsequent collisions in the crash. However, the first collision in a crash typically has the most severe forces, so an effective airbag provides maximum benefit during the most severe portion of a crash. Broadside collisions are frequently caused by a failure to yield right of way. As with any crash, increased speed may increase crash severity.

Car Accident: A car accident is a road traffic incident which usually involves one road vehicle colliding with, another vehicle, another road user, or a stationary roadside object, and which may result in injury or property damage, or possibly death. Phrases used to describe accidents include: *auto accident*, *car crash*, *car smash*, *car wreck*, *fender bender*, *motor vehicle accident (MVA)*, *personal injury collision (PIC)*, *road accident*, *road traffic accident (RTA)*, *road traffic collision (RTC)*, *road traffic incident (RTI)*, *smash-up*, *bingle* and *traffic collision*.

Center of Mass (CM): The center of mass of a vehicle is considered to be located at a point midway between the vehicle's sides, midway along the vehicle's wheelbase and at 1/3rd the vehicle's height above the ground.

Change in Momentum: The difference of the momentum (product of mass and velocity) of a mass from one time to another; the difference of the momentum of a mass between the beginning and end of contact with another mass; the difference in the momentum of a system of bodies. See also Conservation of Momentum.

Closing Speed: The relative approach velocity is the sum of the two speed changes as calculated at the point where the force is applied to both vehicles. The approach velocity is along the common line which defines the direction of principal force. In some circumstances it will be useful as an estimate of one of the vehicle's pre-impact speed. Unless the collisions were central, the relative approach velocity is always greater than the sum of the speed changes at the centers of gravity.

Closing Velocity: The magnitude of the relative velocity between two vehicles at a given point in time as they approach each other; the magnitude of the relative velocity between two vehicles at the beginning of a crash; the vector difference between the velocity of the vehicle and the vehicle/object struck immediately before impact.

Coefficient of Restitution: The ratio of the relative normal velocity at the time of separation to the relative normal velocity at the time of initial contact between the impact centers of two colliding bodies.

Collinear Collision: A collision between two vehicles in which their respective directions of travel are parallel to one another, either as a rear - end or head - on collision.

Collinear Forces: If two or more forces have the same action line, they are collinear forces.

Collinear Impact: A collinear impact occurs when the angle of separation between the momentum vectors of the approaching vehicles is equal to or less than 5 degrees.

Collision Analysis: Automobile collision analysis constitutes the determination as far as possible of *what* happened during the collision, *where* it happened, and *how* it happened. It is a reconstruction mode on the basis of damage, debris, the highway marks, and the point of rest of the vehicles after the accident. Collision analysis is a study of engineering dynamics. Both the geometry of the motion of a vehicle and the forces which cause that motion are subjects of dynamics.

Collision Course: Motion of a *traffic unit* or other object which, if continued without modification, will result in collision with another *traffic unit* or other object. Objects are on a collision course if the space between them is diminishing and the angle between the path of either and a line between them is constant.

Collision Damage Marks: Collision damage marks are divided into direct and indirect damage, including: deformation, scrape marks, bending, twisting, buckling, folding, tearing, fracture, melting, burning, breaking and cracking.

Collision Deformation Classification (CDC): A classification of the extent of deformation to an automobile, utility vehicle, pickup, or van from a crash.

Collision Impact: A *central impact* occurs where the centers of mass of the two colliding bodies are aligned with each other when the vehicles first come into contact. However, this alignment does not make the collision collinear.

A *collinear impact* occurs when the angle of separation between the momentum vectors of a approaching vehicles is equal to or less than 5 degrees. There is no requirement that the centers of mass of the vehicles be aligned. Collinear impact requires the initial momentum vectors be *parallel* or *nearly parallel* to each other at the time of first contact. A collision can be collinear and not be a central impact collision. It can also be central impact collision but not collinear, (Brill [2004]). The collision could also be a central collinear impact.

Figure 4.2 shows a central impact where the centers of mass are aligned. The impact is not collinear. The angle of separation between momentum vectors is 90 degrees. Figure 4.3 shows a central impact where the centers of mass are aligned and the impact is collinear. The angle of separation between momentum vectors is zero degrees. The vehicle is stopped so the magnitude of its momentum vector is equal to zero. Figure. 4.4 shows a collinear impact but not a central impact. The angle of separation between momentum vectors is 3 degrees. Figure 4.5 illustrates a collinear impact and a central impact. The angle of separation between momentum vectors is 5 degrees and the centers of mass are aligned.

Figure 4.6 depicts a central impact but not collinear. The angle of separation between momentum vectors is 20 degrees but the centers of mass are aligned.

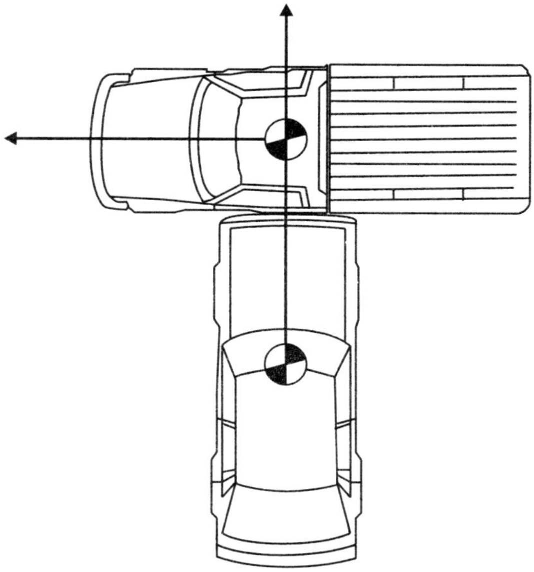

Fig. 4.2 Central impact where the centers of mass are aligned. The impact is not collinear. The angle of separation between momentum vectors is 90 degrees. Brill (2004).

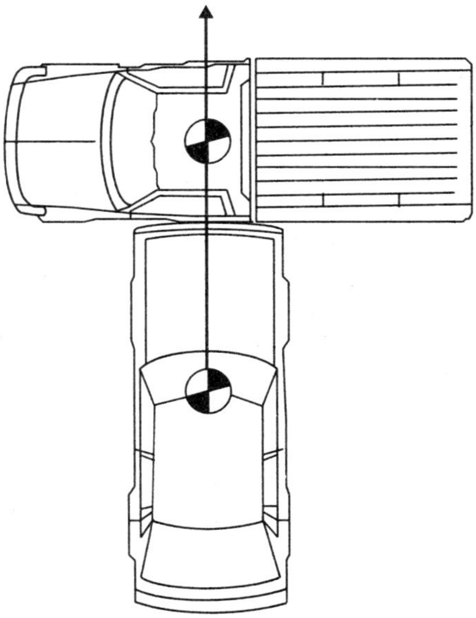

Fig. 4.3 Central impact - where the centers of mass are aligned and the impact is collinear. The angle of separation between momentum vectors is 0 degrees. The vehicle is stopped so the magnitude of its momentum vector is equal to zero. Brill (2004).

Fig. 4.4 Collinear impact but not a central impact. The angle of separation between momentum vectors is 3 degrees. Brill (2004).

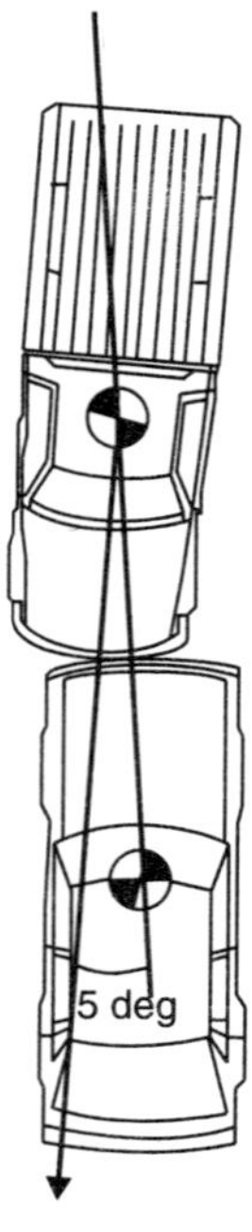

Fig. 4.5 Collinear impact and a central impact. The angle of separation between momentum vectors is 5 degrees and the centers of mass are aligned. Brill (2004).

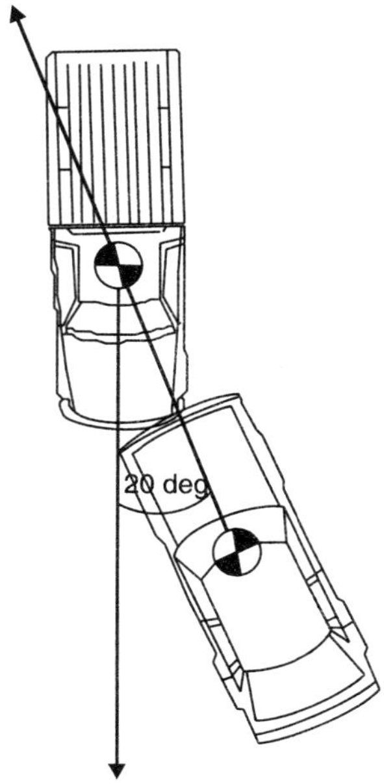

Fig. 4.6 Central impact but not collinear. The angle of separation between momentum vectors is 20 degrees but the centers of mass are aligned. Brill (2004).

Collision Scrub: A short, usually broad, *skidmark* or *yawmark* made during engagement of the vehicles in a collision. A *collision scrub* may be connected to *skidmarks* or *yaw-marks* made before or after collision, but have characteristics that differentiate it from such marks.

Collision Speed Change: The speed change is the magnitude of the velocity change vector for the center of gravity of the vehicle during the collision. It is often called Delta-V. The speed change is always in the direction of the force. The ratio of the vehicle speed changes is always equal to the inverse ratio of the vehicle masses.

Collision: A collision is an isolated event in which two or more bodies (colliding bodies) exert relatively strong forces on each other for a relatively short time. A collision is also a dynamic event consisting of the contact between two or more bodies resulting in a change of momentum of a least one participating body. Collisions involve forces (there is a change in velocity). Collisions can be elastic, meaning they conserve energy and momentum, inelastic, meaning they conserve momentum but not energy, or totally inelastic (or plastic), meaning they conserve momentum and the two objects stick together. See also Accident, Crash; Impact.

Collision-Primary: The impact between a vehicle and another vehicle or a vehicle and a fixed object.

Collision-Secondary: The impact between the occupant and some interior component of the vehicle.

Conservation of Angular Momentum: The total angular momentum of a system of masses about any one axis remains constant unless acted upon by an external torque about that axis.

Conservation of Energy: The principle of physics states that the amount of energy in a closed system is constant regardless of the changes in the form of that energy. Used in accident investigation in estimating vehicle speed from the distance required to slide to a stop and other applications, such as speed from damage. See Momentum.

Conservation of Linear Momentum: The total momentum of a system of masses in any one direction remains constant unless acted upon by an external force in that direction.

Conservation of Mass: The mass of a body remains unchanged by ordinary physical changes to which it may be subjected.

Conservation of Momentum: The linear momentum of a system of bodies remains unchanged if there is no resultant external force on the system. The angular momentum of a system of bodies about a fixed axis is unchanged if there is no resultant external momentum about that axis.

Conservation: Conservation is the act or process of retaining.

Conservation: The idea that the total amount of certain quantities in nature always stays the same. Energy, linear momentum, and angular momentum are examples of quantities that obey the law of conservation.

Conservative Forces: Conservative forces are forces which do work that are stored as potential energy. If the energy is released it will do as much work as was done to store the energy.

Conservative System: In a conservative system, there is no mechanism for dissipating or adding energy. Systems which contain only conservative forces are called conservative system. A linear spring is a conservative element of any system.

Conserve: Stay the same or constant. To say that something is conserved is to say that the amount of it stays the same and obeys conservation.

Constant Velocity: If a vehicle is traveling at constant velocity, then $v = d/t$, where d is the distance and t is the time.

Contact Damage: Damage to a *vehicle* resulting from direct pressure of some foreign object in a collision or roll over. It is usually indicated by *striations,* rub-off of material, or puncture. Compare with *induced damage.*

Contact Point: The point of intersection of the resultant contact impulse with the intervehicular contact surface of each of two colliding vehicles.

Corner-to-Corner Collisions: The corner-to-corner collisions usually involve rotation of the colliding vehicles. Since the stopping force due to impact is not along the center line of the vehicles, each vehicle tends to turn a different way. Each vehicle is deflected from the point of impact towards its own side of the road.

Crash Rates: The safety performance of roadways is almost always reported as rates. That is, some measure of harm (deaths, injuries, or property damage) divided by some indicator of exposure to the risk of this harm. Common rates related to road traffic fatalities include the number of deaths per capita, per registered vehicle, per licensed driver, or per vehicle mile traveled. Simple counts are almost never used. The annual count of fatalities is a rate, namely, the number of fatalities per year. There is no one rate that is superior to others in any general sense. The rate to be selected depends on the question being asked – and often also on what data is available. What is important is to specify exactly what rate is measured and how it relates to the problem being addressed.

Crash Test: A crash test is a form of destructive testing usually performed in order to ensure safe design standards in Crashworthiness and crash compatibility for automobiles or related components.

Crashworthiness: Crashworthiness is the ability of a structure to protect its occupants during an impact. This is commonly tested when investigating the safety of vehicles. Depending on the nature of the impact and the vehicle involved, different criteria are used to determine the crashworthiness of the structure. Crashworthiness may be assessed either prospectively, using computer models or experiments, or retrospectively by analyzing crash outcomes. Several criteria are used to assess crashworthiness prospectively, including the deformation patterns of the vehicle structure, the acceleration experienced by the vehicle during an impact, and the probability of injury predicted by human body models. Injury probability is defined using criteria, which are mechanical parameters (e.g., force, acceleration, or deformation) that correlate with injury risk. A common injury criterion is the Head Impact Criterion (HIC). Crashworthiness is assessed retrospectively by analyzing injury risk in real-world crashes, often using regression or other statistical techniques to control for the myriad of confounders that are present in crashes.

Crush Area: Area defined by the original vehicle exterior and a crush profile, crush equivalent speed: See Energy Equivalent Speed (EES).

Crush Distance: The permanent damage or deformation a vehicle sustains as a result of impact. "*C*" measurements are taken from the original bodyline at frame height and from left to right and from rear to front. To determine the five space intervals for *C*1-*C*6 measurements, double the damage width (in inches) and divide by 10.

Crush Energy: A number of crash tests were conducted to examine the crush behavior of vehicles. The crush energy, E_A, per unit width, w, of a crushed chassis can be expressed as a linear function of crush, C, as

$$\sqrt{\frac{2E_A}{w}} = d_0 + d_1 C$$

where the residual crush, C is measured normal (perpendicular) to and from the nominal undeformed surface. d_0 and d_1 are constants known as the crush stiffness coefficients. d_0 and d_1 are obtained experimentally for a particular vehicle. They differ from vehicle to vehicle and between the front, side, and rear of each vehicle. The crush energy is equal to the crush force multiplied by the distance during which the crush force was acting.

Crush Profile: The geometric shape in a specified plane (e.g., vertical, horizontal) that describes the vehicle damage resulting from an impact.

Crush Stiffness Coefficient: An empirical quantity used in the calculation of the energy dissipated in a collision and associated with each vehicle's velocity change, V. See CRASH3.

Damage Analysis: Damage analysis can predict what happened during a collision by what remains after in the form of damage.

Damage Closing Velocity: A damage related closing velocity is defined to be the sum of the speed changes at the point of common velocity or

$$V_C = \Delta V_1 + \Delta V_2$$

This closing velocity is in the direction of the line of action of the collision forces.

Damage Model: The damage model assumes that each vehicle consists of a series of radial springs which can record the damage. These springs are also referred to as force vectors as they transmit a radial force in a particular direction.

Damage Vector: A damage vector is generally assumed to be a linear spring anchored at the center of mass of the vehicle extending to the vehicle surface.

Damaging Stress: Is the lowest level of stress of any kind that will cause a part to not achieve its design performance be it measured as a maximum force or a time to failure.

Delta-*t* (Δt): A time interval associated with an event such as vehicle-to-vehicle contact; the time duration of impulse.

Delta-*V*: The velocity change experienced by a vehicle during impact is commonly called delta-V (ΔV). It is an important consideration for occupant injury severity. Delta-*V* is a function of the velocity before and after impact. It is a vector, and as such, requires magnitude and angular orientation for proper identification.

Direct Central Impact: If two inelastic bodies collide in a direct central impact, they have a common velocity after their impact.

Direct Damage: Damage or deformation of the vehicle that was directly caused by the collision itself.

Direct Eccentric Impact: Direct eccentric impact occurs if the initial velocities are parallel to the normal to the striking surfaces but are not collinear.

Direct Impact: If the velocities of the two bodies are directed along the line of impact, the impact is said to be direct impact. Direct central impact occurs if the mass centers of the two bodies are also along the line of impact.

Dynamic Crush: The deformation formed by the external surface of the vehicle at any time during an impact, usually measured relative to the corresponding as manufactured unreformed surface. See Crush Area; Crush Profile; Residual Crush.

Dynamic Event: A dynamic event changes one or more of the state variables: position, velocity, and time.

Elastic and Nonelastic Collisions: An *elastic collision* between two bodies is one in which no permanent deformation takes place in either body and both momentum and kinetic energy are conserved. In a perfectly elastic collision, the sum of the translational kinetic energies of the objects is unchanged.

A *nonelastic collision*, also referred to as an *inelastic collision*, is a collision between two bodies in which there is a permanent deformation, *i.e.*, the objects lose their original shape. In this type of collision, some kinetic energy is transformed into *internal energy*. Also, the linear momentum is conserved. When two bodies collide and stick and travel together with a common final velocity after the collision, the collision is called *completely inelastic*. However, some momentum is conserved, once the resultant components are found. The conservation laws can be applied in each direction.

Elastic Collision: An elastic collision is a collision in which the total kinetic energy of the colliding bodies after collision is equal to their total kinetic energy before collision. Elastic collisions occur only if there is no net conversion of kinetic energy into other forms. During the collision kinetic energy is first converted to potential energy associated with a repulsive force between the particles (when the particles move against this force, *i.e.* the angle between the force and the relative velocity is obtuse), then this potential energy is converted back to kinetic energy (when the particles move with this force, *i.e.* the angle between the force and the relative velocity is acute). Thus, it is a collision between two bodies in which no permanent (plastic) deformation takes place and both momentum and kinetic energy are conserved. When the amount of force between two objects in collision is sufficient to cause permanent deformation, the damage caused produces a loss of energy. Both momentum and kinetic energy are conserved. In a perfectly elastic collision the sum of the translational kinetic energies of the object is not changed.

Elastic Impact: An idealized impact where the kinetic energy at separation equals the kinetic energy at the initiation of contact; a fully elastic impact is an impact where the coefficient of restitution is equal to one.

Energy Equivalent Speed (EES): The speed and corresponding kinetic energy with which the vehicle must contact a fixed rigid object with no rebound for equivalence to conditions of another collision; for example, the energy may be equal to a specified level of residual crush; EES is the preferred term, broader than barrier equivalent velocity (BEV), equivalent barrier speed (EBS), and equivalent test speed (ETS).

Force: From Newton's second law of motion, we know that the rate of change of momentum is proportional to the force producing the change and takes place in the direction of applied force. In keeping with this law, a force acting on a vehicle will be equal to the mass of the vehicle times its acceleration.

Full Impact: An impact during which motion momentarily ceases between some areas of the colliding bodies while they are in contact with each other. When the colliding bodies do not separate after collision, the impact is complete.

Head-on Collision: A head on collision occurs when the front ends of two vehicles hit each other, these accidents result in the highest number of fatal crashes because the combined forces are so high. If two vehicles collide head-on, with each traveling at 50 mph the combined speed would equal 100mph. The reason for this is a law of physics called the Momentum conservation: In a collision occurring between car 1 and car 2, the total momentum of the two vehicles before the collision is equal to the total momentum of the two vehicles after the collision. Head on crashes often occur when passing, but can also be caused by driver inattention. See *Opposite Direction Collision.*

Head-on Impact: Frontal impact where the PDOF is at or near zero degrees.

Highway Collision: Highway collisions are the four types of collisions: the *rear-end* or *stopped vehicle*, the *head-on*, the *corner-to-corner*, and the *sideswipe*.

Impact Aspersion: Occurs when a fluid container is ruptured under extreme pressures of impact and then aspersed onto the roadway.

Impact Center: The point of intersection of the contact impulse and the inter vehicular contact surface for an impact. See Contact Point.

Impact Plane: The impact plane is a straight line set through the calculated damage profile of vehicle one.

Impact Point: The impact point is defined by the coordinates *x*, *y* on the impact plane in car coordinates with respect to the center of mass.

Impact Stresses: Impact stresses occur in regard to loads that are transient in respect to time. The duration of the load application is of the same order of magnitude as the natural period of vibration of the specimen.

Impact Velocity: The velocity of an object's center of gravity relative to a coordinate system fixed in the earth during an impact. See Preimpact Velocity, Postimpact Velocity.

Impact: An impact is a single collision of one mass in motion with a second mass, which may be either in motion or at rest. A collision between two bodies which occurs in a very small interval of time during which the two bodies exert relatively large forces on each other is called an impact. The common normal to the surfaces in contact during the impact is called the line of impact. The impact is the transmission of force from one body to another within a very short interval of time. The colliding bodies experience an elastic and/or plastic deformation. Generally interchangeable with collision.

Impulse Moment Arm: The perpendicular distance from an object's center of gravity to the line of action of an impulse. See Fig. 4.7 shown below:

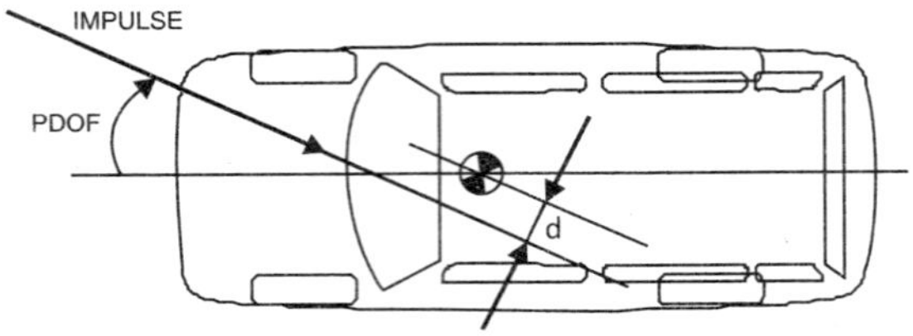

Fig. 4.7

Impulse Ratio: The ratio of the tangential and normal impulse components in planar impact mechanics. See Impulse.

Impulse: Impulse is a constant force acting at a time on a vehicle, or other object, having a certain mass and changing the vehicle's velocity from an initial value to a final value. Impulse units are those of mass multiplied by those of velocity, or those of force multiplied by those of time. Impulse is a vector quantity.

Impulsive Force: In non elastic collisions (collisions wherein the objects do not retain their original shape), the force is known as *impulsive force*.

Initial Path Heading: A vehicle as it comes into a collision has a path defined by the motion of its center of gravity. The direction of this path is called the initial path heading. The vehicle heading is normally the same as the initial path heading.

Kinematic Restitution: The kinematic definition of restitution states that the ratio of the relative velocity of the objects after the collision to the relative velocities of the objects before the collision is the coefficient of restitution.

Kinetic energy: Kinetic energy and momentum are both properties of moving objects. Whereas momentum is defined as mass times velocity, kinetic energy is defined as one-half times mass times velocity squared.

Kinetic Energy: Symbol *K.E.*: This is energy associated with movement; the energy of motion. It is a measure of the combination of how big and how fast. For example, a bullet fired from a gun (small, but fast) may have the same kinetic energy as a bowling ball tossed from a catapult, the ball is much larger but much slower. The formula of kinetic energy is one-half the mass of the object times its velocity squared. In standard equation form, kinetic energy is

$$KE = \frac{1}{2}mv^2$$

Law of Conservation of Energy: The law of conservation of energy states that when a rigid body or a system of rigid bodies moves under the action of conservation forces, the sum of the kinetic energy and the potential energy of the system remains constant.

Law of conservation of linear momentum: The law of conservation of linear momentum states that the total momentum of any system of bodies is unchanged by any interactions between the different members of the system.

Line of Impact: The line of impact is defined as the line of interaction between two colliding vehicles. It represents the actual force vector which each vehicle imposes on the other during the collision. It is the line along which the vehicle collides. The line of impact is a line pointing in the direction in which the parts were pushed. It is a line which indicates the direction of movement of the damaged metal. It is also the line which indicates the direction in which one vehicle was traveling relative to the other during that last inch before impact. Figures 4.8 (*a*) and (*b*) show two moving vehicles colliding. The line drawn on both vehicles shows the relative direction of travel of each, and which coincides with the direction in which the metal on each vehicle will be pushed. In Fig. 4.6, vehicle 2 has struck the right front corner of vehicle 1. Note that at the time of collision there is only one line of impact and the direction of damage on both vehicles will exactly coincide. At impact, the two lines of impact are one line. After impact, the vehicles generally separate and bounce away. With the line of impact determined, the motion of the vehicles just prior to impact is known, since they traveled along the line of impact, and the motion of the vehicles after the impact can be summarized.

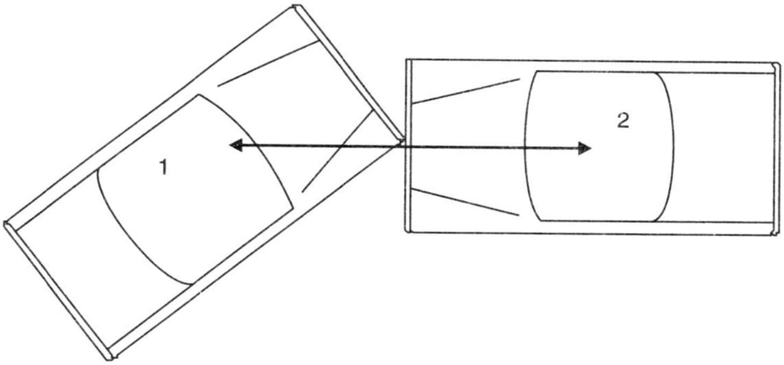

Fig. 4.8 (*a*)

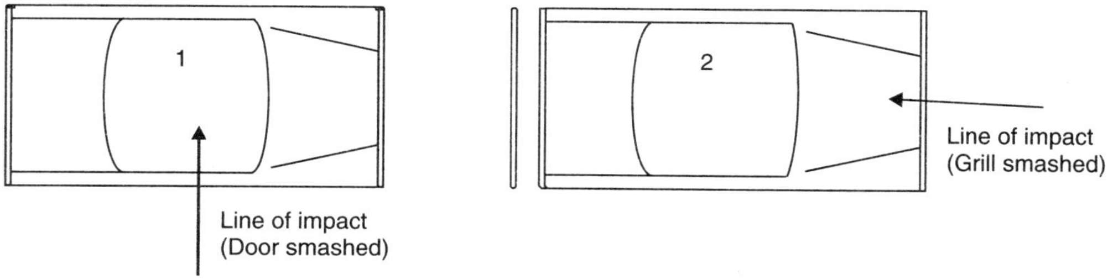

Fig. 4.8 (*b*)

Figure 4.8(*c*) shows one vehicle has a smashed-in front end and the other a smashed-in passenger door. At the time of collision, the two lines of impact were one line and the same line, so the vehicles must have collided as shown in Fig. 4.8(*c*).

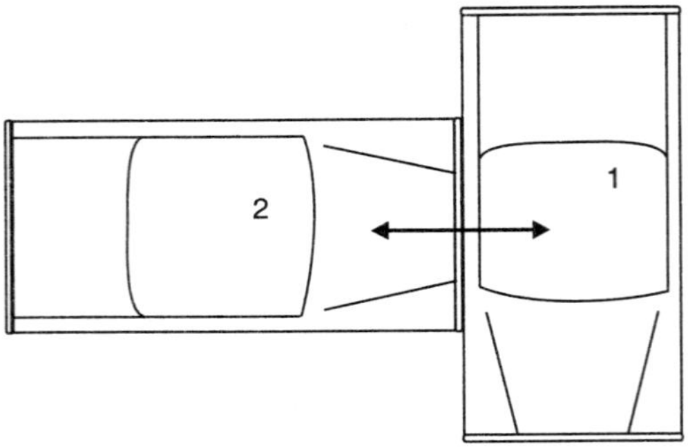

Fig. 4.8 (*c*)

When a vehicle is struck by another vehicle, the struck vehicle will have some motion imparted to it. This motion may be both linear and rotational. If the line of impact passes through the center of gravity of the vehicle no rotation will result. The vehicle will move as shown in Fig. 4.9.

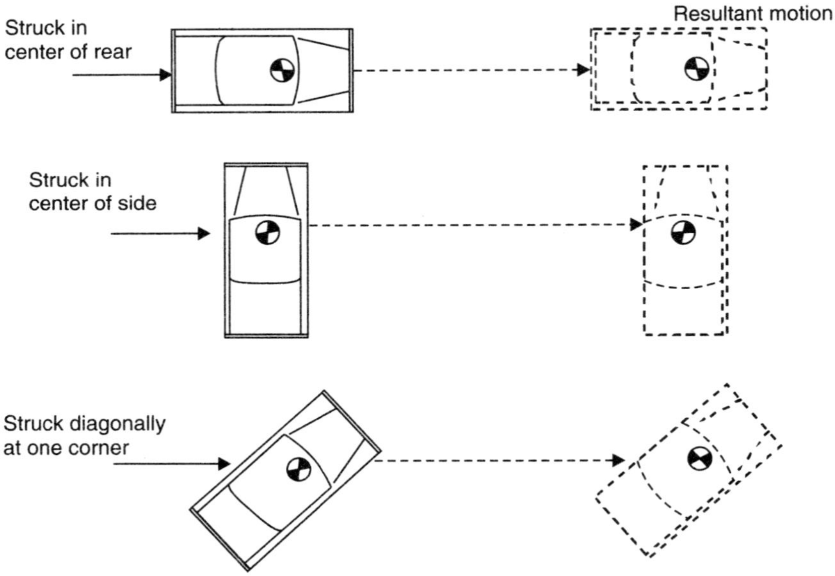

Fig. 4.9 (*a*)

The vehicle will move in a straight line along the line of impact. The vehicle's course does not deviate, and the vehicle does not rotate in any manner. If the line of impact does not pass through the center of gravity, the vehicle will also rotate during the skid. A vehicle struck from the right rear will turn counterclockwise. That is, front wheels to the left and rear wheels to the right. A vehicle struck in the left rear will rotate clockwise, front wheels to the right and rear wheels to the left. These motions occur as shown in Fig. 4.9 (*b*).

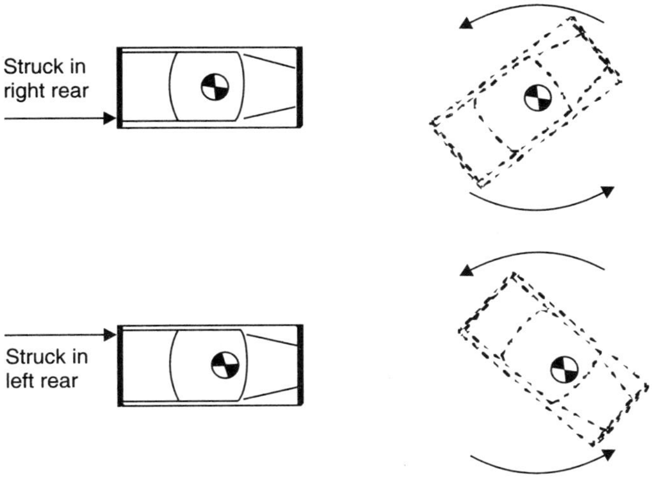

Fig. 4.9 (*b*)

Line of Impact: The line of Impact is defined as the line of interaction between two colliding automobiles. It represents the actual force vector which each vehicle imposes on the other during the collision. It is the line along which the vehicle collides.

Linear Momentum: A quantity associated with the motion of an object along a straight path. The linear momentum of an object is defined to be equal to its mass times its velocity.

Linear: Being in a straight line.

Mass Center of Gravity (CG): That point of a body from which it could be suspended or supported and be in the balance.

Mass Centre: Mass centre is that point in relation to a body where any plane passing through it divides the body into two parts of equal weight, and through which acts the resultant force due to the body's inertia when it is accelerated.

Mass Ratio: The effective mass ratio is defined as $KG/(KG + h^2)$. It is a dynamic measure of the effect of mass on the eccentricity of the force. A mass ratio of 1 indicates a central collision. The greater the force offset, the smaller the value of the effective mass ratio.

Mass: Mass is defined as the amount of matter in a body.

Mass: The mass of a body is determined by comparison with a standard mass, using a beam-type balance.

Momentum: Momentum is expressed by the three Newton's laws of motion, and is defined as the mass (or weight of an object) times its velocity. Momentum is the property of an object in motion that tends to keep it moving with the same velocity. Momentum is a vector quantity.

Multiple Collisions: In multiple collisions, there is an interaction of more than two vehicles. These situations are the kind where a vehicle may sideswipe another vehicle, be forced across the centerline and then strike an oncoming vehicle. The major issue in this case is to sort out the skid marks, dents and transfers so that they are assigned to the proper vehicle. In case of multiple impacts, there are clearly multiple lines of impact, one from each collision. Point transfers are useful in determining which vehicle was responsible for which impact.

Multiple Collisions: In the multiple collision situations the accident investigator must deal with the interactions of more than two vehicles. These situations are the kind where a vehicle may sideswipe another vehicle, be forced across the centerline and then strike an oncoming vehicle.

Non-Central Collision: In non-central collision of two colliding vehicles, the lines of the action of the collision forces are collinear but offset from the center of mass of each vehicle.

Nonelastic Collision: A nonelastic collision, also referred to as an inelastic collision, is a collision between two bodies in which there is permanent deformation, *i.e.*, the objects lose their original shape. Some kinetic energy is transformed into internal energy in this type of collision. When two bodies collide and stick and travel together with a common final velocity after the collision, the collision is said to be completely inelastic. Hence, an inelastic collision is a *collision* in which kinetic energy is not conserved. Inelastic collisions may not conserve kinetic energy, but they do obey conservation of momentum. Simple ballistic pendulum problems obey the conservation of kinetic energy *only* when the block swings to its largest angle.

Offset: The distance between the longitudinal heading axes of two vehicles in frontal contact; see Fig. 4.10 shown below.

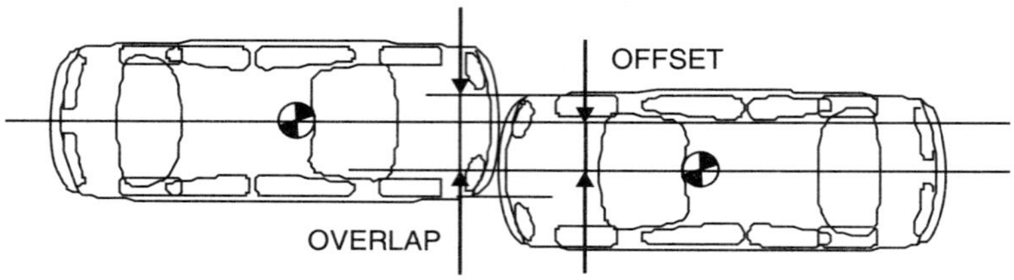

Fig. 4.10

Override: A condition in a collision where the main structural members, such as a bumper, of the striking vehicle are above the main structural members, such as frame rails, of the struck vehicle; See Fig. 4.11 shown below.

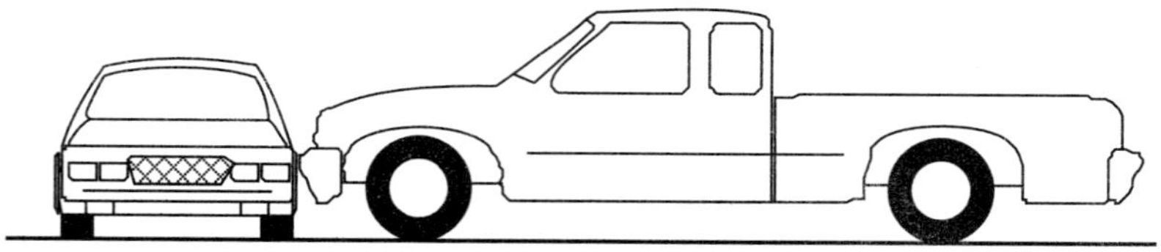

Fig. 4.11

Partial Impact: An impact in which motion is continuous between the parts of colliding objects which are in contact with each other; sideswipe.

Planar Impact: An impact in which all forces, moments, and motion take place in a plane.

Plastic Impact: An impact with little or no rebound at the end of impact; a perfectly plastic impact is where the coefficient of restitution is equal to zero.

Point of Contact: The point of intersection of the contact impulse and the inter vehicular contact surface during an impact. See also Direction Of Principal Force (DOPF); Final Contact Position; Impact Center; Principal Direction Of Force (PDOF).

Point of Impact (POI): The collision sequence where involved objects first come into one another or surface. Can be referred to as area of impact (AOI) and initial contact. The point of impact is the location on the roadway where two vehicles come together. This determination is often made by noting gouge marks in the pavement, the position of debris on the roadway or similar items.

Point of Impact: The point of impact is the location on the roadway where two vehicles come together. In the case of head-on collision, they are generally reliable guides, but if the two vehicles have sideswiped each other, the gouge marks and debris may be located at some distance from the actual point of impact. In the case of intersection collisions, the point of impact can usually be located by means of the skid marks leading to collision and determining the line of impact. If one vehicle strikes the side of the other, the skid marks of the struck vehicle will cross in front of the marks left by the other vehicle.

In a head-on collision, the vehicles stop at the point of impact or rebound equally if the weights and speeds are equal. If the weight and speeds are unequal, the lighter and slower vehicle will be pushed backward from the point of impact. The two vehicles do not spin and the debris is confined to a small area.

The corner-to-corner shown in Fig. 4.12(*b*) and sideswipe collisions shown in Fig. 4.14 are the most common roadway collision and the objective lies in determining which side of the roadway center line the accident occurred.

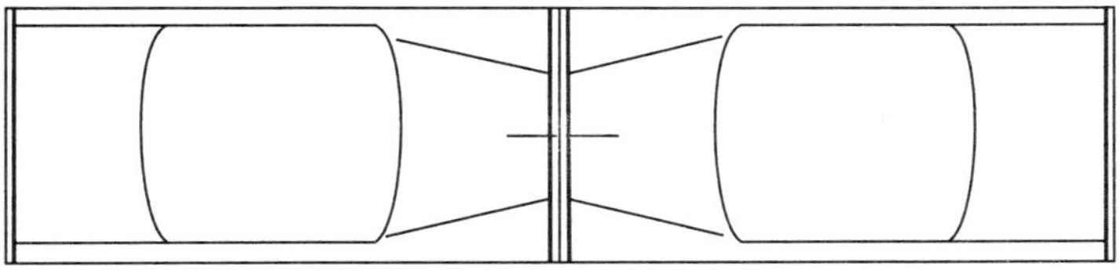

Fig. 4.12 (*a*) Head-on collision.

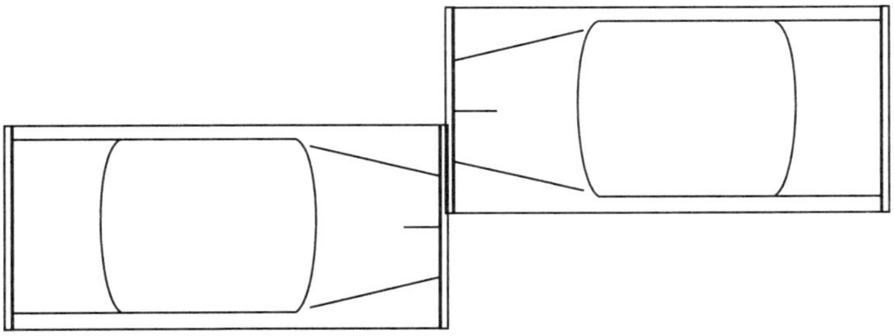

Fig. 4.12 (*b*) Corner-to-corner collision.

The corner-to-corner collisions usually involve rotation of the colliding vehicles. Since the stopping force due to collision is not along the centerline of the vehicles, each vehicle tends to turn as shown in Fig. 4.13(*a*). For left corner collisions both vehicles rotate counterclockwise as seen in Fig. 4.13(*a*). Similarly, the vehicles will tend to bounce away from each other, such that after coming to rest, the system will look as shown in Fig. 4.13(*b*). Rotation and lateral displacement are always present in this case. After collision of this kind, each vehicle is deflected from the point of impact towards its own side of the road.

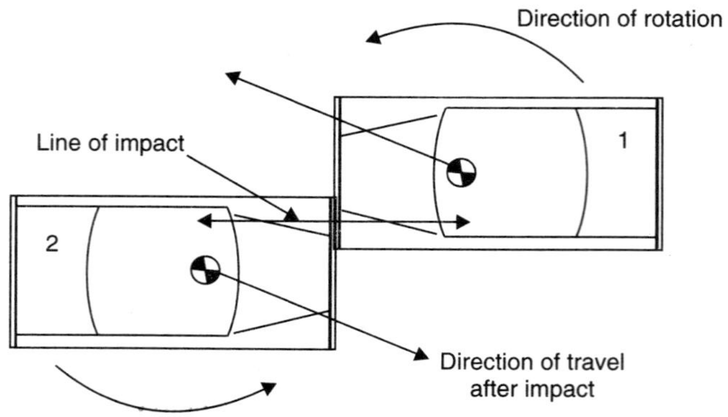

Fig. 4.13 (*a*) The line of impact and probable post-collision rebound.

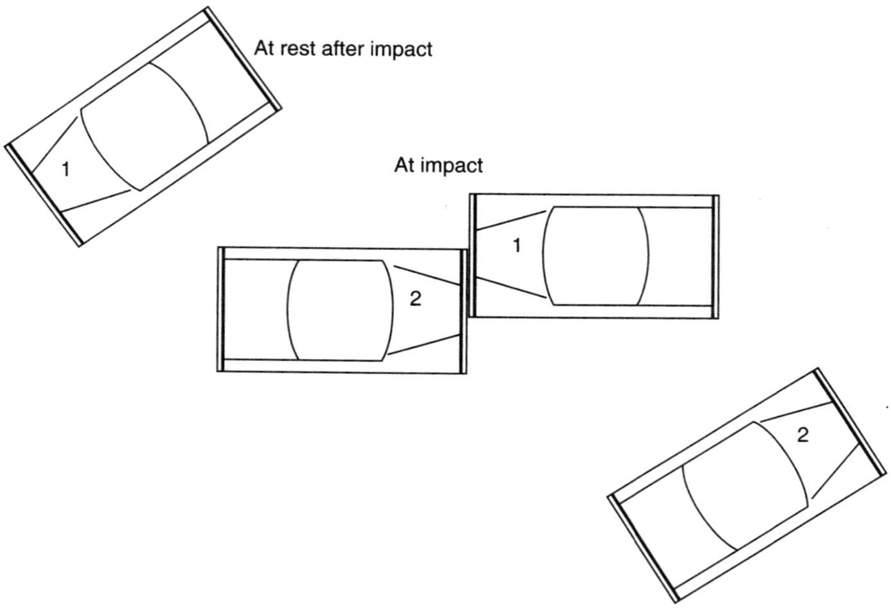

Fig. 4.13 (*b*) Relative final resting places often associated with a corner to corner collision.

The sideswipe collision is similar to the corner-to-corner but generally does not involve rotation of the vehicles. The vehicles are generally stopped by the drivers, and the physical damage to vehicles and occupants is minimum. The indicators of the point of impact are the mud, broken glass, and the skid marks of the vehicles.

The intersection accident is the collision case where two vehicles collide essentially at right angles. In this case, we are interested in determining the speed of each vehicle at collision, and if the vehicles skidded before colliding, the speed of each vehicle at the start of the skid marks. We are also interested in finding which vehicle entered the intersection first. These can be determined by manipulating the skid marks and momentum equation.

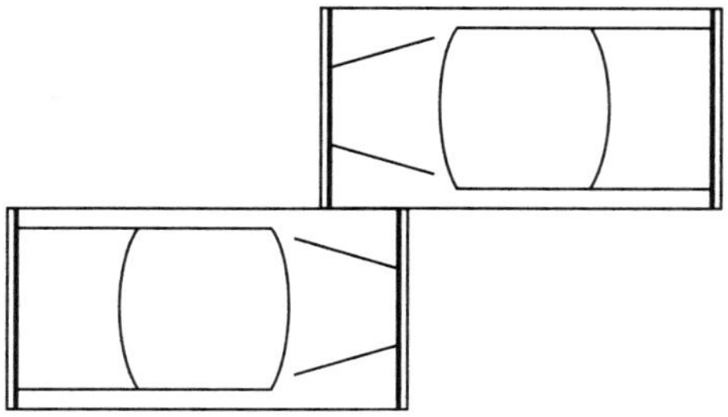

Fig. 4.14 Sideswipe.

Point of No Escape (by Braking): Consists of the perception/reaction distance plus the slide-to-stop distance for the specified velocity.

Point of Possible Perception: The place and time at which the hazard could have been perceived **by** a normal person. It precedes actual *perception* and is the beginning of *perception* delay.

Point of Rest: Controlled or uncontrolled post-collision position of vehicles or persons.

Point of Separation: to obtain this, each vehicle should be drawn on the scale diagram as it was positioned at the point of impact and also as it was positioned at the point of separation. The PS can usually be established from roadway marks such as skid or scrub marks, chips and gouges. It should be noted that the departure angle for a vehicle is the angle formed with the *X*-axis, or with a baseline drawn parallel to that axis, by the vehicle's path of travel after separation. This path is plotted by drawing a line from the center-of-mass location of the vehicle at time of impact through its center-of-mass location where the vehicle separated from the other vehicle or object.

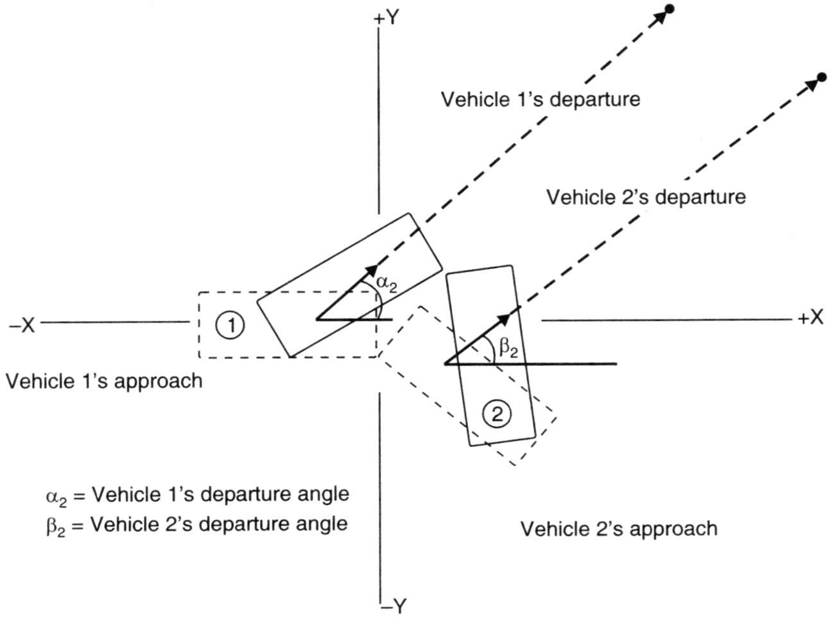

Fig. 4.15 (*a*)

Figure 4.15 (*a*) A diagram showing departure angles measured from the *X*-axis, or parallel baseline, to the line drawn from the center-of-mass of each vehicle at impact (broken-line rectangle) through the center-of-mass of the vehicle at separation, [Rivers (1997)]

Post Crash Damage: Damage existing to a vehicle after it comes to rest, including damage that may result during rescue, towing, and salvage operations.

Post Crash: The time in a collision sequence from the point of separation to the last event that significantly influences the vehicles or occupants of those vehicles.

Postimpact Skid Distance: The distance a vehicle skids after an impact or collision.

Post Impact Speed: The magnitude of the velocity of an object in a collision at the time of separation or end of contact. See Post Impact Velocity; Separation Speed.

Post Impact Trajectory: The path of the center of gravity of a vehicle from impact to rest.

Post Impact Velocity: The velocity of an object in a collision at the time of separation or end of contact.

Potential Energy: Potential energy is the capacity of a mass or body to do work due to the body's position. The American Heritage dictionary defines potential energy as the energy of a particle or system of particles, derived from position, or condition, rather than motion.

Pre Crash: The period of time in a collision sequence from the first significant event associated with a collision to initial contact.

Preimpact Skid Distance: The distance a vehicle skids prior to an impact or collision.

Preimpact Velocity: The velocity of a vehicle in a collision at the instant of its initial contact.

Principal Direction of Force (PDOF): Determined from vehicle damage, indicating the magnitude and direction of the opposing force acting on a vehicle during collision. Also known as thrust and direction of principal force. The direction of the line of action of the contact impulse in a planar collision expressed in degrees, measured positive clockwise from the longitudinal axis of a vehicle.**:** The principle of conservation of linear momentum assumes that the only forces that act on the vehicles during the collision are the collision forces between the vehicles. From Newton's Third Law, the collision forces that act on the vehicles are equal in magnitude but opposite in direction. The direction of the force can be reported as a global angle or as an angle measured relative to the vehicle. When measured relative to the vehicle, it is generally known as the Principal Direction of Force or PDOF.

The global force angles α_1 and α_2 give the direction of the force angles measured relative to the Global *X* coordinate axis which is fixed to the ground. $Pdof_1$ and $Pdof_2$ give the direction of the force angle relative to each vehicle's Local *x* coordinate axis. See Fig. 4.15 (*b*).

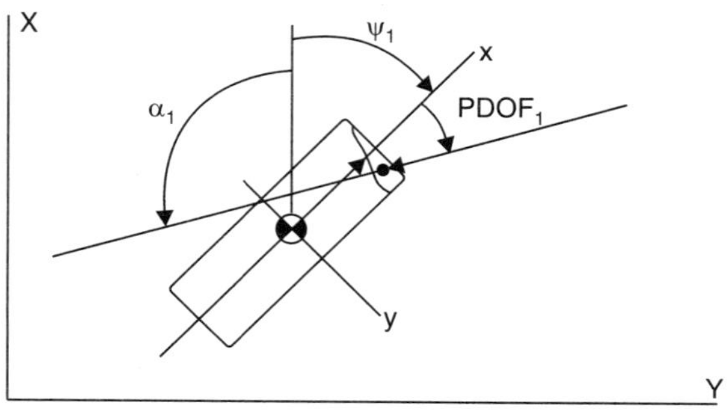

Fig. 4.15 (*b*) PDOF angle measurement.

On a vehicle's local coordinate system, the x-axis is directed towards the front of the wheel. As is always the case, angles are measured from the positive x-axis. The y-axis forms an angle of +90 degrees and determines the direction for positive angles. For a right handed coordinate system, the y-axis is directed to the right side of the vehicle.

The PDOF is defined by the tail of the vector consistent with the coordinate system. Refer to Fig. 4.15(b). From the geometry, it can be seen that the PODF can be calculated from the global force angle α and the entry angle ψ as

$$\text{PDOF} = \alpha - \psi + 180$$

For example, in a frontal collision where the force is directed to the rear of the vehicle, the PDOF would be zero degrees. With the right handed coordinate system shown in Fig. 4.15(b) (which is the normal system when discussing PDOFs), the PDOF will be reported as +30 degrees.

Real Impact: During a *real impact*, a portion of the deformation work is returned elastically, while same is used to crush vehicle parts permanently. For elastic impact, $e = 1$. For nonelastic or plastic impact, $e = 0$.

Rear-End Collision: A rear-end collision (often called simply rear-end or in the UK a shunt) is a traffic accident wherein a vehicle (usually an automobile or a truck) crashes into the vehicle in front of it, so called because it hits its rear. It may also be a rail accident wherein a train runs into the rear of a preceding train. Typical scenarios for rear-ends are a sudden deceleration by the first car (for example, to avoid someone crossing the street) the following car that does not have the time to brake and collides with the first at a road junction. The following car accelerates more rapidly than the leading. As a rule of thumb, crashing into another car is equivalent to crashing into a rigid surface (like a wall) at half of the speed. This means that rear-ending a still car while going at 30 mph is equivalent, in terms of forces, to crashing into a wall at 15 mph. The same is true for the vehicle crashed into. A typical medical consequence of rear-ends, even in case of collisions at moderate speed, is whiplash. In more severe cases permanent injuries, e.g. herniation, may occur. For purposes of insurance and policing, the driver of the car that rear-ends the other car is almost always considered to be at fault due to not being within stopping distance or lack of attention. An exception to this rule comes into play if the rear-ended vehicle is in reverse gear. If the driver of the car that was rear-ended files a claim against the driver who hit him, said driver could be responsible for all damages to the other driver's car.

Residual Crush: The permanent deformation formed by the nominal external surface of a vehicle caused by an impact, usually measured relative to the corresponding as-manufactured undeformed surface.

Same-Direction Collision: A collision between two *traffic units* moving in the same direction on the same *roadway*. Sometimes called *rear-end collisions*.

Second Impact: An impact between an occupant and an interior surface of a vehicle caused by and following an impact between the vehicle and another object. A second or subsequent impact between the same two vehicles during a crash.

Secondary Collision: When two vehicles collide at an acute angle, and then rotate into each other, a secondary impact occurs which redirects the paths of the centers of mass of both vehicles. This is called a secondary collision.

Secondary Impact: A second or subsequent impact between the same two vehicles during a crash. See Secondary Collision.

Sensation: A mental process due to immediate bodily stimulation whether consciously processed or not (Before the detection threshold is met the stimulus may not be consciously processed).

Sensor: Device in a control system that provides information from an input, as for example of rotational speed or line pressure.

Separation Speed: The speed at the time of loss of contact of two vehicles in a collision; it can refer to the speed of the centers of gravity or of the contact point.

Separation V: The vehicle speed at separation. Also known as the post-impact speed.

Separation Velocity: The vector velocity at the time of loss of contact of two vehicles in a collision; it can refer to the speed of the centers of gravity or at the contact point.

Separation: The point in the collision sequence where collision forces have been dissipated and the objects in contact have the opportunity to physically separate from each other, whether they actually separate or not.

Sideswipe Collisions: Sideswipe collisions are characterized by prolonged sliding contact and with very little structural deformation. A sideswipe collision is considered as a collision in which the angle of impact is shallow and the vehicle interaction in the longitudinal direction is frictional. Sideswipes are characterized by prolonged sliding between the contact surfaces.

Sustained Contact: Sustained contact is a dynamic situation where one vehicle pushes the other to path end. A sustained contact trajectory is possible when both vehicle trajectories are similar.

Sustained contact will only occur in almost central impacts. Sustained contact is unlikely for oblique collisions with distinct paths to rest.

Traffic Collision: When a road vehicle collides with another vehicle, pedestrian, animal, or geographical or architectural obstacle. Traffic collisions can result in injury, property damage, and death.

Underride: A condition in a collision in which the main structural components of one vehicle are below the main structural components of the other vehicle.

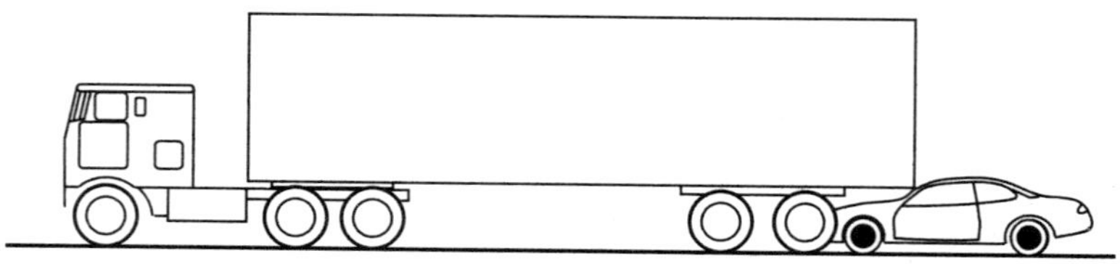

Fig. 4.16

Example 4.1: *The following information is given for an accident involving two vehicles:*

	Vehicle – 1	*Vehicle - 2*
Curb weight	*2086.56 kg (4600 lbs)*	*1451.52 kg (3200 lbs)*
Driver weight	*99.79 kg (220 lbs)*	*90.72 kg (200 lbs)*
Cargo weight	*22.68 kg (50 lbs)*	*18.14 kg (40 lbs)*
Travel direction at impact	*West*	*South*
Travel direction after impact	*W 18° S*	*W 40° S*
Post impact travel distance	*13.716 m (45 ft)*	*10.668 m (35 ft)*
Post impact drag factor	*0.35*	*0.40*

Determine:

(a) the impact speed of vehicle 1.

(b) the impact speed of vehicle 2.

(c) the delta V of vehicle 1.

(d) the delta V of vehicle 2.

Solution: We will begin by considering that West is our 0 degree line. All angles measured counterclockwise will be considered positive.

	Direction	θ	sin θ	cos θ
α_1	West	0	0	1
α_2	South	90°	1	0
β_1	W 18° S	18°	0.309	0.951
β_2	W 40° S	40°	0.643	0.766

Weight ratios:

Let $w_1 = 1$

then $$w_2 = \frac{1560.38}{2209.03} = 0.7064$$

where w_1 = weight of vehicle 1 = 2086.56 + 99.79 + 22.68 = 2209.03 kg

w_2 = weight of vehicle 2 = 1451.52 + 90.72 + 18.14 = 1560.38 kg

Post impact speed of vehicle 1:

$$s_1 = \sqrt{254 d_1 f_1} = \sqrt{254(13.716)(0.35)} = 34.92 \text{ km/h (mph)}$$

Post impact speed of *V*-2:

$$s_2 = \sqrt{254 d_2 f_2} = \sqrt{254(10.668)(0.40)} = 32.92 \text{ km/h (mph)}$$

The law of conservation of momentum along the east west axis:

$$S_1 w_1 \cos \alpha_1 + S_2 w_2 \cos \alpha_2 = s_1 w_1 \cos \beta_1 + s_2 w_2 \cos \beta_2$$

$$S_1 = (1)(1) + 0 = 34.92(1)(0.951) + 32.92(0.7064)(0.766)$$

$$S_1 = 33.209 + 17.813 = 51.02 \text{ km/h (31.71 mph)}$$

The law of conservation of momentum along the north-south axis:

$$S_1 w_1 \sin \alpha_1 + S_2 w_2 \sin \alpha_2 = s_1 w_1 \sin \beta_1 + s_2 w_2 \sin \beta_2$$

$$0 + S_2\,(0.7064)\,(1) = 34.92(1)\,(0.309) + 32.92(0.7064)\,(0.643) = 10.79 + 14.95$$

or

$$S_2 = 36.44 \text{ km/h } (22.65 \text{ mph})$$

Delta V calculation:

V-1 speed/direction at impact: West @ 51.02 km/hr.

V-1 speed/direction after impact: W 18° S @ 34.92 km/hr.

Breaking post impact moment into components:

West component: 34.92 cos 18° = 33.21

South component: 34.92 sin 18° = 10.79

Delta V West: 51.02 – 33.21 = 17.81 km/h (11.07 mph)

Delta V South: 0 + 10.79 = 10.79 km/h (6.71 mph)

V-1 Delta V: $\sqrt{17.81^2 + 10.79^2} = 20.82$ km/h (12.94 mph)

V-2 speed/ direction at impact: South @ 36.44 km/h

V-2 speed/ direction after impact: W 40° S @ 32.92 km/h

Breaking post impact moment into components:

West component: 32.92 cos 40° = 25.22

South component: 32.92 sin 40° = 21.16

Delta V West: 0 + 25.22 = 25.22 km/h (mph)

Delta V South: 36.44 – 21.16 = 15.28 km/h (mph)

V-2 Delta V: $\sqrt{22.22^2 + 15.28^2} = 29.47$ km/h (18.32 mph)

Check: The delta V's should be in inverse proportion to the weights.

Hence V-1 delta V / V-2 delta V $\quad V = \dfrac{w_2}{w_1}$

$$\frac{w_2}{w_1} = \frac{1560.38}{2209.03} = 0.7064$$

$$\frac{\text{V-1 Delta V}}{\text{V-2 Delta V}} = \frac{20.82}{29.47} = 0.7064$$

Example 4.2: *A car (V-1) makes a left turn into the path of a pickup truck (V-2). The pickup sees the car turning in front of him but cannot avoid striking the rear quarter panel of the car. The impact spins the car out in a clockwise fashion. Use the following data to calculate the impact speed of both vehicles and the speed of the pickup truck at the start of precrash skid.*

	V-1	*V-2*
Curb weight:	*3019 lb [1369 kg]*	*4550 lb [2064 kg]*
Driver weight:	*117 lb [53 kg]*	*217 lb [98 kg]*
Cargo weight:	*65 lb [29 kg]*	*120 lb [54 kg]*
Precrash skid dist.:	————	*53 ft [16.2 m]*
Precrash skid drag factor:	————	*0.75*

Direction at impact:	*S 70° E*	*North*
Direction after impact:	*E 45° N*	*N 2° W*
Distance after impact:	*30 ft [9.1 m]*	*41 ft [12.5 m]*
Drag factor after impact:	*0.55*	*0.75*

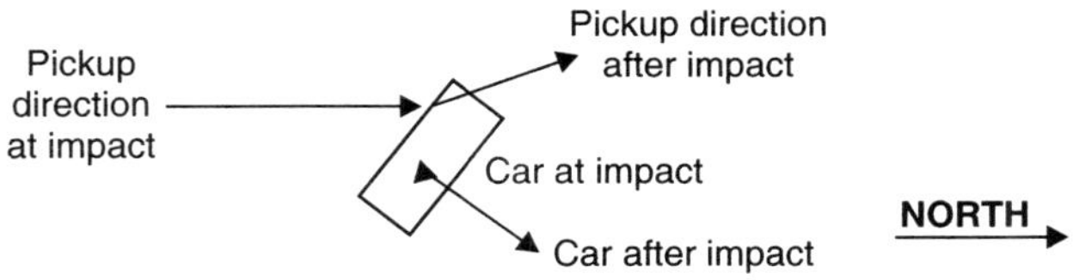

Solution: We have a conservation of momentum problem on our hands. The variables:

a_1 = Travel angle of vehicle one at impact
a_2 = Travel angle of vehicle two at impact
b_1 = Travel angle of vehicle one after impact
b_2 = Travel angle of vehicle two operator after impact
d_1 = Post impact travel distance of V-1 = 30 feet
d_2 = Post impact travel distance of V-2 = 41 feet
d_{2PRE} = Preimpact skid distance of V-2 = 53 feet
f_1 = Post impact drag factor of V-1 = 0.55
f_2 = Post impact drag factor of V-2 = 0.75
f_{2PRE} = Preimpact braking drag factor of V-2 = 0.75
S = Speed of V-2 at start of precrash braking
S_1 = Speed of V-l at impact
S_2 = Speed of V-2 at impact
s_1 = Speed of V-l after impact
s_2 = Speed of V-2 after impact
W_1 = Weight of V-l = 3019 + 117 + 65 = 3201 1b.
W_2 = Weight of V-2 = 4550 + 217 + 120 = 4887 1b.

We will start setting by assuming that North is our 0 degree line. All angles measured counter clockwise will be considered positive. We will tabulate our angles and then determine the trigonometric values:

	Direction	q	$\sin q$	$\cos q$
a_1:	S 70° E	250°	–0.940	–0.342
a_2:	North	0°	0	1
b_1:	E 45° N	315°	–0.707	0.707
b_2:	N 2° W	2°	0.035	0.999

The weight ratios: If $W_1 = 1$, $W_2 = 4887/3201. = 1.527$

Post impact speed of V-l:

$$s_1 = \sqrt{30 * d_1 * f_1} = \sqrt{30 * 30 * 0.55} = 22.2 \text{ mph}$$

Post impact speed of V-2:

$$s_2 = \sqrt{30 * d_2 * f_2} = \sqrt{30 * 41 * 75} = 30.4 \text{ mph}$$

The conservation of momentum formula along east/west axis:

$$S_1 * W_1 * \sin a_1 + S_2 * W_2 * \sin a_2 = s_1 * W_1 * \sin b_1 + s_2 * W_2 * \sin b_2$$

$$S_1 * 1 * -0.940 + S_2 * 1.527 * 0 = 22.2 * 1 * -0.707 + 30.4 * 1.527 * 0.035$$

$$-0.940 * S_1 = -15.7 + 1.9 = -13.8$$

$$S_1 = 14.7 \text{ or } 15 \text{ mph [24 km/hr]}$$

The conservation of momentum along the north-south axis:

$$S_1 * W_1 * \cos a_1 + S_2 * W_2 * \cos a_2 = s_1 * W_1 * \cos b_1 + s_2 * W_2 * \cos b_2$$

$$14.7 * 1 * -0.342 + S_2 * 1.527 * 1 = 22.1 * 1 * 0.707 + 30.4 * 1.527 * 0.999$$

$$-5.03 + 1.527 * S_2 = 15.7 + 46.4 = 62.1$$

$$1.527 * S_2 = 62.1 + 5.03 = 67.1$$

$$S_2 = 43.9 \text{ or } 44 \text{ mph [71 km/hr]}$$

The speed of V-2 at the start of precrash braking:

$$S = \sqrt{S_2^2 + 30 * d_{2PRE} * f_{2PRE}}$$

$$S = \sqrt{43.9^2 + 30 * 53 * 0.75}$$

$$S = \sqrt{927 + 1192} = \sqrt{3119}$$

$$S = 55.85 \text{ or } 56 \text{ mph [90 km/hr]}$$

Example 4.3: *Eastbound V-1 is struck broadside at an intersection by southbound V-2. Both vehicles spin out of control and strike a curb on the southeast side of the intersection. V-1 continues over the curb and strikes a pole rear-first and comes to rest. There is little damage to the pole, but the rear of V-1 sustains a maximum of 21 inches [56 cm] of crush. Given the data below, calculate V-1's and V-2's speed at impact. Use the equation from Problem 4 to determine V-1's speed at impact with the pole. Use 10 mph [16 km/hr] equivalent energy speed for each vehicle to account for the curb impacts.*

	V-1	*V-2*
Curb Weight:	*2874 lb [1304 kg]*	*3719 lb [1687 kg]*
Driver Weight:	*2101b [95 kg]*	*160 lb [73 kg]*
Passengers' Weight:	———	*70 lb [32 kg]*
Cargo Weight:	*55 lb [25 kg]*	*100 lb [45 kg]*
Travel direction at impact:	*East*	*South*
Travel direction after impact:	*S45°E*	*S37°E*
Travel dist. post impact:	*68 ft [20.7 m]*	*106 ft [32.3 m]*
Drag factor:	*0.53*	*0.38*

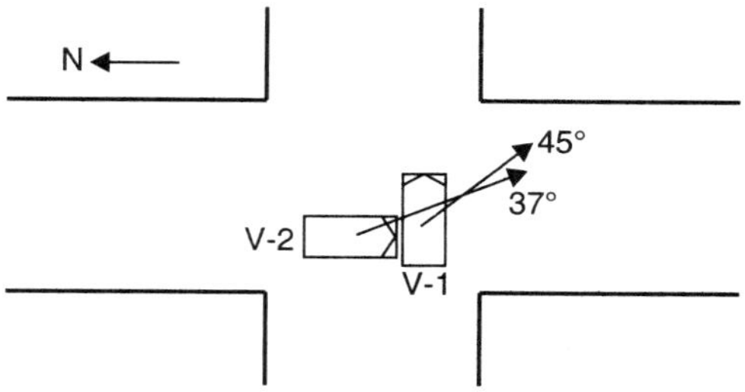

Solution: We have a conservation of momentum problem on our hands. The variables:

C_{MAX} = Maximum crush to rear of V-1 = 21 inches
d_1 = Post impact travel distance of V-l = 68 feet
d_2 = Post impact travel distance of V-2 = 106 feet
f_1 = Post impact drag factor of V-1 = 0.53
f_2 = Post impact drag factor of V-2 on guardrail = 0.38
S_1 = Speed of V-l at impact with V-2
S_2 = Speed of V-2 at impact with V-l
S_c = Equivalent energy speed loss of vehicle impacts into curb = 10 mph
S_p = Speed of V-l at impact with pole
s_1 = Speed of V-l after impact
s_2 = Speed of V-2 after impact
W_1 = Weight of V-l = 2874 + 210 + 55 = 3139 lb.
W_2 = Weight of V-2 = 3719 + 160 + 70 + 100 = 4049 lb.
α_1 = Travel angle of vehicle one at impact
α_2 = Travel angle of vehicle two at impact
β_1 = Travel angle of vehicle one after impact
β_2 = Travel angle of vehicle two after impact

We will start setting by assuming that South is our 0 degree line. All angles measured counter clockwise will be considered positive. We will tabulate our angles and then determine the trigonometric values:

	Direction	θ	$\sin\theta$	$\cos\theta$
α_1:	East	90°	1	0
α_2:	South	0°	0	1
β_1:	S45°E	45°	0.707	0.707
β_2:	S37°E	37°	0.602	0.799

The weight ratios: If W_1 =1, W_2 = 4049/3139 = 1.29

Since there is no significant post impact movement to consider, finding V-l's impact speed with the pole is easy. Simply substitute the 21 inches for C_{MAX} in the equation and crunch the numbers:

$$S_p = 0.55*C_{MAX} + 3.9$$

$$S_p = 0.55*21 + 3.9 = 11.55 + 3.9 = 15.4 \text{ mph}$$

The speed at the start of post impact skid can be solved by using the combined speed formula:

$$s_1 = \sqrt{S_p^2 + S_C^2 + 30 * d_1 * f_1} = \sqrt{15.4^2 + 10^2 + 30 * 68 * 0.53} = 37.7 \text{ mph}$$

We can use a similar approach for the post impact speed of V-2:

$$s_2 = \sqrt{S_C^2 + 30 * d_2 * f_2} = \sqrt{10^2 + 30 * 106 * 0.38} = 36.2 \text{ mph}$$

The conservation of momentum formula along east/west axis:

$$S_1 * W_1 * \sin \alpha_1 + S_2 * W_2 * \sin \alpha_2 = s_1 * W_1 * \sin \beta_1 + s_2 * W_2 * \sin \beta_2$$

$$S_1 * 1 * 1 + S_2 * 1.29 * 0 = 37.7 * 1 * 0.707 + 36.2 * 1.29 * 0.602$$

$$S_1 = 26.65 + 28.11$$

$$S_1 = 54.76 \text{ or } 55 \text{ mph } [88 \text{ km/hr}]$$

The conservation of momentum along the north-south axis:

$$S_1 * W_1 * \cos \alpha_1 + S_2 * W_2 * \cos \alpha_2 = s_1 * W_1 * \cos \beta_1 + s_2 * W_2 * \cos \beta_2$$

$$54.8 * 1 * 0 + S_2 * 1.29 * 1 = 37.7 * 1 * .707 + 36.2 * 1.29 * 0.799$$

$$0 + 1.29 * S_2 = 26.65 + 37.31 = 63.96$$

$$S_2 = 49.58 \text{ or } 50 \text{ mph } [80 \text{ km/hr}]$$

Vector: A vector is a quantity that has both magnitude and direction. Since velocity is defined as the speed of an object in a given direction, velocity is a vector. *Speed* is velocity's magnitude property, and the direction in which the speed occurs is velocity's directional property.

Velocity: Velocity is defined as a change in distance with respect to time and direction while the unit is in motion. Velocity is a vector quantity.

Weight: The property of an object that we know as weight is the result of the Euler's gravitational attraction acting on the object's mass. We can assume that the force of gravity is uniform everywhere on the Earth's surface. It follows that the mass of an object and its weight may be regarded as perfectly correlated, so that in a practical sense we can consider mass and weight synonymous. mass = W/g, where W is the weight of the object and g is the acceleration due to gravity (32.2 ft/sec^2 or 9.81 m/sec^2).

4.4 PRINCIPLE OF WORK AND ENERGY

Here, we derive the principle of work and energy or *law of conservation of energy* from the Newton's second law of motion. The work by a force is shown in Fig. 4.11. In Fig. 4.17, the normal force ΣF_n along the direction n is perpendicular to the path. The normal force does no work since it does not produce any displacement of mass m along n.

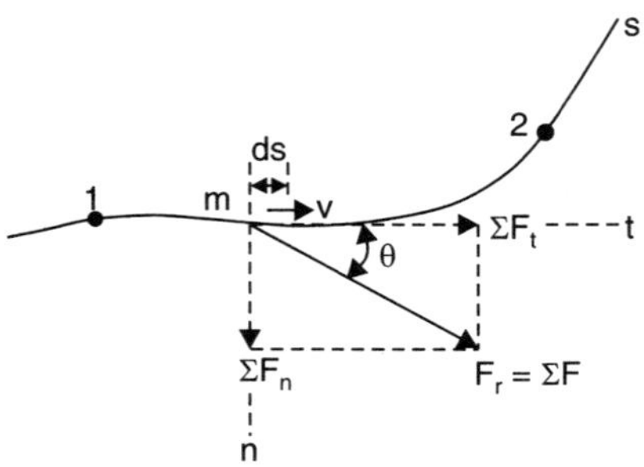

Fig. 4.17 Work by a force.

In tangential direction, we can write

$$\Sigma F_t = ma_t \tag{4.1}$$

$$\Sigma F \cos \theta = ma_t \tag{4.2}$$

where

$$a_t = v\frac{dv}{ds} \tag{4.3}$$

Thus

$$\Sigma F \cos \theta \, ds = mvdv$$

and

$$\int_{s_1}^{s_2} F \cos\theta ds = \int_{v_1}^{v_2} m v \, dv = \frac{1}{2} m v_2^2 - \frac{1}{2} m v_1^2 \tag{4.4}$$

The kinetic energy at position i is given by

$$T_i = \frac{1}{2} m v_i^2 \tag{4.5}$$

The work done by all the forces from position 1 to position 2 in Fig. 4.1 is given by

$$\Sigma U_{1-2} = T_2 - T_1 \tag{4.6}$$

or

$$T_1 + \Sigma U_{1-2} = T_2 \tag{4.7}$$

We can generalize Eq. (4.7) as follows:

$$\Sigma U_{1-2} = W_f + E_g + E_e + E_f \tag{4.8}$$

where W_f = work done by the external force

E_g = gravitational potential energy

E_e = elastic potential energy

E_f = frictional energy loss

Now, if W_f and E_f are absent, then

$$\Sigma U_{1-2} = U_1 - U_2 = \Delta E_g + \Delta E_e \tag{4.9}$$

or

$$T_1 + \Sigma U_{1-2} = T_1 + U_1 - U_2 = T_2 \tag{4.10}$$

or

$$T_1 + U_1 = T_2 + U_2 \tag{4.11}$$

The relationship given by Eq. (4.11) is referred to as the *law of conservation of energy*. Note that there is no energy loss between events 1 and 2. Hence, the energy at the ith position, $T_1 + U_1$, is conserved.

Example 4.4: *Derive an expression to determine the stopping distance, x, a vehicle skids to a stop, with an initial velocity of v on a pavement with a kinetic coefficient of friction, μ. Determine the stopping distance of a vehicle whose initial velocity is 90 km/h. The mass of the vehicle is 950 kg and the coefficient of kinetic friction, μ is 0.6.*

Solution: Here, we note that there are no changes in elastic and gravitational energies and no work done by external forces. Applying Eq. (4.10) gives the required relationship between the kinetic energy and the energy lost due to frictional force.

$$T_1 = \frac{1}{2}mv^2 \tag{E.1}$$

$$\Sigma U_{1-2} = -fx_s = -\mu mg(x_s) \tag{E.2}$$

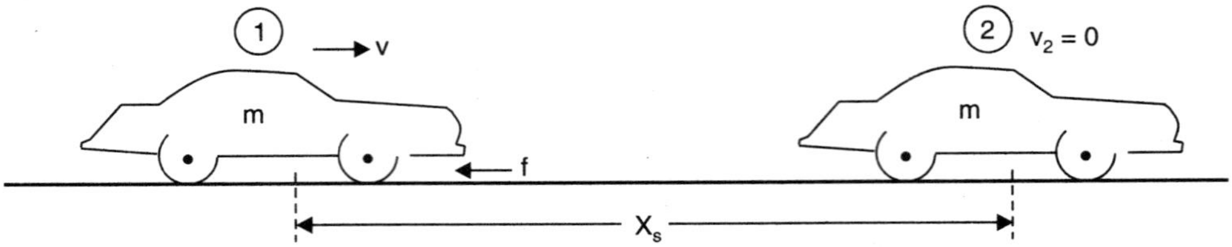

Fig. E4.4 Vehicle skidding to stop.

$$T_2 = \frac{1}{2} m v_2^2 = 0 \tag{E.3}$$

Hence,
$$\frac{1}{2} m v^2 - \mu m g\, x_s = 0 \tag{E.4}$$

or
$$x_s = \frac{v^2}{2\mu g} \tag{E.5}$$

For $v = 90$ km/h (25 m/s), $m = 900$ kg, and $\mu = 0.6$, Eq. (E.5) gives

$$x_s = \frac{v^2}{2\mu g} = \frac{25^2}{2(0.6)(9.81)} = 53.09 \text{ m.}$$

Example 4.5: *Figure E4.5 shows two vehicles A and B each having a weight of 4186 lb that collide on an icy pavement at an intersection. Vehicle A was traveling a speed of 30 mph just before the impact and skidded 10 ft after applying the brakes in the direction shown in the Fig. The coefficient of kinetic friction is 0.15. Determine the speed of vehicle B just before the collision or impact with vehicle A.*

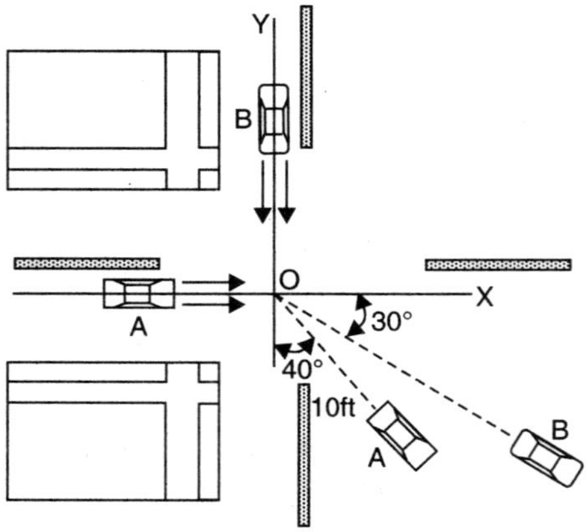

Fig. E4.5 Collision at an intersection.

Solution: The collision happened on an icy surface and therefore the external impulse to the system of two vehicles is negligible. Hence, by applying the law of conservation of momentum from event 0 to event 1, the velocity of vehicle B, v_{B0}, before impact or collision can be determined.

Let event 0 → 1 represents collision and event 1 → 2 represents skidding

v_{A0} = speed of vehicle A before the collision = 30 mph = 40 ft/sec

v_{A1} = separation speed of vehicle A after the collision

v_{B0} = speed of vehicle B before the collision

v_{B1} = separation speed of vehicle B after the collision

The masses of vehicles A and B are

$$m_A = m_B = \frac{4186\ \text{lb}}{32.2\ \text{ft/sec}} = 130\,\text{slugs}$$

Vehicle A: energy $1 \rightarrow 2$ (skidding)

$$T_1 + \Sigma U_{1\text{-}2} = T_2$$

or

$$\frac{1}{2}(130)v_{A1}^2 - 4186(0.15)(10) = 0$$

or

$$v_{A1} = 9.83\ \text{ft/sec} = 6.70\ \text{mph}$$

Vehicles A and B (in momentum): $0 \rightarrow 1$ (collision)

$$x\text{: } 130(44) + 0 = 130(9.83 \sin 40^\circ) + 130(v_{B1} \cos 30^\circ)$$

or

$$v_{B1} = 43.51\ \text{ft/sec} = 29.67\ \text{mph}$$

$$y\text{: } 0 + 130(v_{B0}) = 130(9.83 \cos 40^\circ) + 130(43.51 \sin 30^\circ)$$

or

$$v_{B0} = 29.29\ \text{ft/sec} = 19.99\ \text{mph}$$

4.5 IMPULSE, MOMENTUM AND ENERGY

Momentum is a vector quantity concerning the motion of an object. The magnitude of momentum is the mass of the object times its speed while its direction is the direction of motion. The units of momentum are kg-m/s or lb-ft/sec. In this section, we use p as the symbol for momentum. Impulse is a vector quantity pertaining to a force. Impulse is the integral of the force over a specified period of time. The units of impulse are the same as for momentum (kg-m/s) or (lb-ft/sec). In this section, we use the symbol I for impulse. The change in momentum is also referred to as an impulse, which is the force that will redirect the object. The effect of an impulse on a body is a change in the momentum of the body. That is $\Delta p = I$. Therefore, for the momentum before and after an impulse, we can write

$$\vec{p_B} + \vec{I} = \vec{p_A} \tag{4.12}$$

4.5.1 Impulse and Momentum of a Single Body

Figure 4.18 shows a rigid body acted on by several forces $F_1, F_2, \ldots, F_n$. The resultant force is obtained by adding these forces $F_1, F_2, \ldots, F_n$ vectorially. The system in Fig. 4.18 produces an acceleration in the direction of the resultant force.

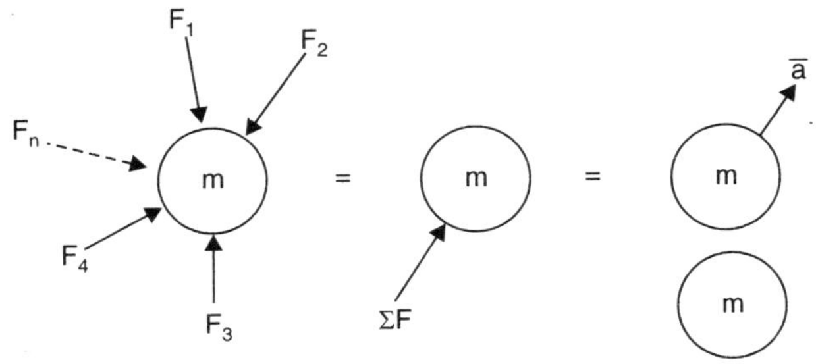

Fig. 4.18 Resultant force and acceleration.

Applying Newton's second law of motion to the system in Fig. 4.18, we can write

$$\Sigma F_i = M\vec{a} = m\frac{d\vec{v}}{dt} \tag{4.13}$$

or

$$\Sigma F_i dt = m d\vec{v} \tag{4.14}$$

Integrating Eq. (4.14), we get

$$\Sigma \int_{t_1}^{t_2} \vec{F}_t dt = m \int_{v_1}^{v_2} d\vec{v} = m\vec{v}_2 - m\vec{v}_1 \tag{4.15}$$

where v_i is the velocity at time t_i.

For a single body, with an impulse acting on it, we have $p_1 + I = p_2$. That is

$$m\vec{v}_1 + \Sigma \int_{t_1}^{t_2} \vec{F}_t dt = m\vec{v}_2 \tag{4.16}$$

Graphical representation of Eq. (4.16) is shown in Fig. 4.19.

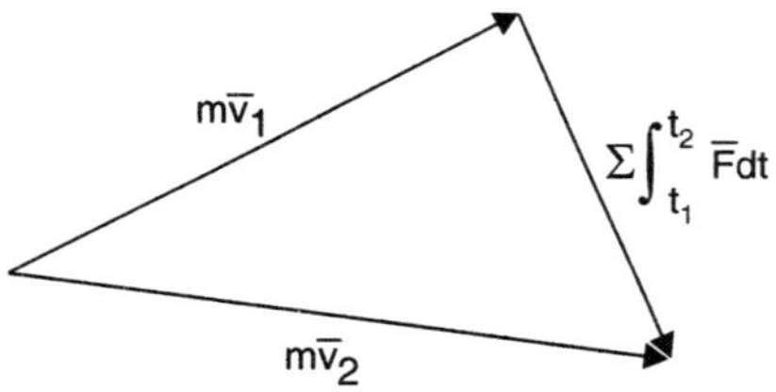

Fig. 4.19 Vector addition of momentum and impulse.

4.5.2 Impulse and Momentum of a System of Bodies

A rigid body acted upon by external and internal forces is shown in Fig. 4.20. Internal forces correspond to those exerted by other particles in the system while the external forces are those produced by elements that are not part of the system. For a road vehicle, the external forces are the forces of gravity, impact forces due to a collision or impact with an obstacle, frictional forces due to the contact of the tires with the roadway, etc.

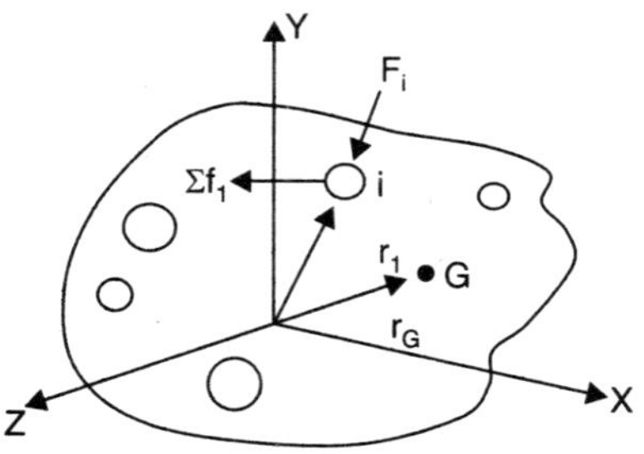

Fig. 4.20 External and internal forces acting on a rigid body.

Applying Newton's second law of motion for the system in Fig. 4.21, the equation of motion for particle *i* in the system can be written as

$$\vec{F}_i + \sum_j \vec{f}_{ij} = m_i \vec{a}_i \tag{4.17}$$

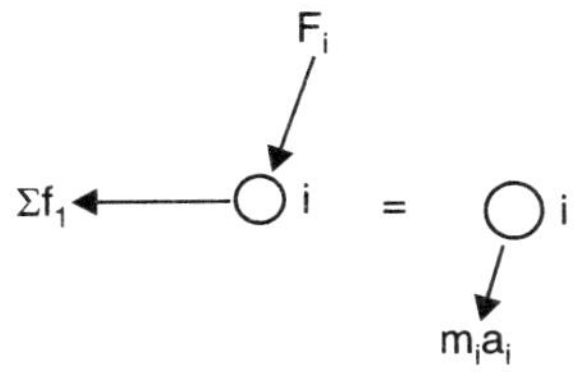

Fig. 4.21 External and internal forces acting on particle *i*.

In Eq. (4.17), the sum over j is for the internal forces associated with other particles. As a result of Newton's third law of motion, the internal forces all cancel out. This is due to the fact that for every internal force given by (i, j) there is an internal force given by (j, i) that is of equal magnitude but opposite in direction.

Hence, summing up over all particles in the system, Eq. (4.17) can be written as

$$\Sigma F_i = \Sigma m_i a_i \tag{4.18}$$

4.6 CENTER OF GRAVITY AND MOTION THEOREM

The center of mass (CM) or the center of gravity (CG). Motion of the system of particles with and without collision or impact is the same if the external impulse to the system of particles is negligible.

4.6.1 Location and Motion of Center of Mass (CM) or Center of Gravity (CG)

The vector position of the center of mass (CM) can be written as

$$m\vec{r}_G = \sum_i^n m_i \vec{r}_i \tag{4.19}$$

where $$m = \sum_i^n m_i$$

$$\vec{r}_G = \text{position vector of CG} = x_G\vec{i} + y_G\vec{j} + z_G\vec{k}$$

$$\vec{r}_i = \text{position vector of particle } i = x_i\vec{i} + y_i\vec{j} + z_i\vec{k} \tag{4.20}$$

and $\vec{i}, \vec{j}, \vec{k}$ are the unit vectors along the axes x, y and z respectively. Substituting Eq. (4.20) into Eq. (4.19), we obtain the components of the position vector of the center of gravity (CG) as follows:

$$x_G = \frac{\Sigma m_i x_i}{m}$$

$$y_G = \frac{\Sigma m_i y_i}{m}$$

$$z_G = \frac{\Sigma m_i z_i}{m} \tag{4.21}$$

Taking the first and second derivative of Eq. (4.19) with respect to time, t, we obtain the velocity and acceleration as

$$m\vec{v}_G = \Sigma m_i \vec{v}_i \tag{4.22}$$

$$m\vec{a}_G = \Sigma m_i \vec{a}_i = \Sigma m_i \frac{d\vec{v}_i}{dt} \tag{4.23}$$

Eq. (4.23) can be written as

$$\Sigma \vec{F}\, dt = \Sigma m_i d\vec{v}_i \tag{4.24}$$

Integrating Eq. (4.24) from t_1 to t_2 gives

$$\Sigma \int_{t_1}^{t_2} \vec{F} dt = \Sigma m_i \int_{t_1}^{t_2} d\vec{v}_i = \left[\Sigma m_i \vec{v}_i\right]_{t_1}^{t_2} \tag{4.25}$$

Now, substituting $m\vec{v}_G$ for $\Sigma m_i \vec{v}_i$ in Eq. (4.25) gives

$$m\vec{v}_G\Big|_{t_1} + \Sigma \int_{t_1}^{t_2} \vec{F}\, dt = m\vec{v}_G\Big|_{t_2} \tag{4.26}$$

Eq. (4.26) states that the final momentum is equal to the initial momentum plus the sum of all the impulses. Eq. (4.26) is shown graphically in Fig. 4.22.

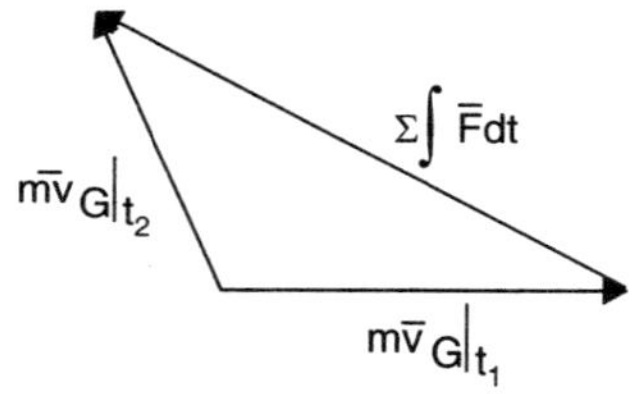

Fig. 4.22 Vector operation of linear impulse and momentum changes.

When the sum of all impulses is zero, we can write

$$m\vec{v}_G\Big|_{t_1} = m\vec{v}_G\Big|_{t_2} \tag{4.27}$$

and

$$\sum_{i=1}^{n} m_i \vec{v}_i \Bigg|_{t_1} = \sum_{i=1}^{n} m_i \vec{v}_i \Bigg|_{t_2} \tag{4.28}$$

Eq. (4.27) is the classical conservation of momentum law in terms of the center of mass (CM) or center of gravity (CG). Similarly, Eq. (4.28) represents the classical conservation of momentum law in terms of a system of n particles.

4.6.2 Conservation of Momentum and CG Formula

Figure 4.23 shows the CG motion of two particles m_1 and m_2. Also, if there is an inelastic collision or impact between m_1 and m_2, then the distance traveled by both particles in a given time is equal to $v_c t$. Here, we assume that the collision or impact of a system of particles or bodies is totally inelastic and the external frictional impulse is negligible, then the movement of the center of mass (CM) or the center of gravity (CG) of the system of particles or bodies after collision is the same as that without collision.

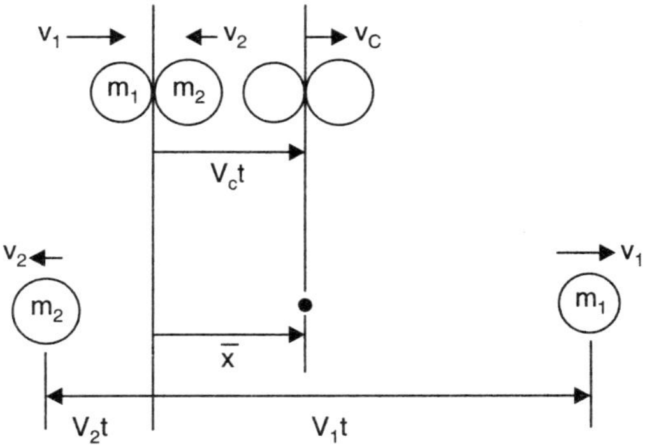

Fig. 4.23 CG motion of two particles.

Noting that the initial momentum of the particles is the same as that at the time of common velocity, the momentum equation can be written as

$$\sum_{i=1}^{2} m_i v_i = \left(\sum_{i=1}^{2} m_i\right) v_c \tag{4.29}$$

where m_i = mass of particle i

v_i = velocity of mass i

v_c = common velocity of total mass

(*a*) One-dimensional Analysis

The common velocity of the two particles after impact can be written as for conservation of momentum before and after collision.

$$v_c \text{ (common velocity)} = \frac{m_1 v_1 - m_2 v_2}{m_1 + m_2} \tag{4.30}$$

The x coordinate of the CG or CM position of a system of particles at time t is given by

$$x = \frac{\sum_{i=1}^{2} m_i x_i}{\sum_{i=1}^{2} m_i} = \frac{m_1(v_1 t) - m_2(v_2 t)}{m_1 + m_2} = \frac{(m_1 v_1 - m_2 v_2)}{m_1 + m_2} t = v_c t \tag{4.31}$$

Eq. (4.31) shows that the displacement of the CG of a system of particles with collision is the same as that without collision.

(*b*) Multi-Dimensional Analysis

The position vector relationship between the particles and the CG of the system of particles is given by

$$\sum_{i=1}^{2} m_i \vec{r}_i = \left(\sum_{i=1}^{2} m_i\right) \vec{r}_c \tag{4.32}$$

where $\vec{r}_i$ = position vector of mass m_i

$\vec{r}_c$ = position vector of mass CG

The relationships for the conservation of momentum for a system of two particles is given by the first derivative with respect to time of Eq. (4.32):

$$\sum_{i=1}^{2} m_i \frac{d\vec{r}_i}{dt} = \sum_{i=1}^{2} m_i \frac{d\vec{r}_c}{dt} \tag{4.33}$$

or

$$\sum_{i=1}^{2} m_i \vec{v}_i = \left(\sum_{i=1}^{2} m_i\right) \vec{v}_c \tag{4.34}$$

4.6.3 CG Motion Theorem

The CG motion theorem states that in the absence of any external forces, the center of mass (CM) or CG of a system of particles will move in a straight line at a constant speed. Any collision or impact between particles, elastic or inelastic will have no effect on this motion.

Proof of the theorem:

The position vector of CG or CM is given by

$$\vec{r}_c = \vec{r}_G + \vec{r}_{CM} \tag{4.35}$$

and

$$m\vec{r}_{CM} = m_1\vec{r}_1 + m_2\vec{r}_2 + ... + m_n\vec{r}_n \tag{4.36}$$

where

$$m = m_1 + m_2 + ... + m_n = \sum_{i=1}^{n} m_i \tag{4.37}$$

Differentiating Eq. (4.36) w.r.t. time gives

$$m\frac{d\vec{r}_{CM}}{dt} = m_1\frac{d\vec{r}_1}{dt} + m_2\frac{d\vec{r}_2}{dt} + ... + m_n\frac{d\vec{r}_n}{dt} \tag{4.38}$$

Note that in Eq. (4.38) each term on the right side is the momentum of a particle. Hence

$$m\frac{d\vec{r}_{CM}}{dt} = \vec{p}_1 + \vec{p}_2 + ... + \vec{p}_n = \sum_{i=1}^{n} \vec{p}_i \tag{4.39}$$

Eq. (4.39) states that the total momentum of the system is equal to the product of the total mass and velocity of the center of mass. This is called the *momentum of the center of mass.*

Differentiating Eq. (4.38), we obtain

$$m\frac{d^2\vec{r}_{CM}}{dt^2} = m_1\frac{d^2\vec{r}_1}{dt^2} + m_2\frac{d^2\vec{r}_2}{dt^2} + ... + m_n\frac{d^2\vec{r}_n}{dt^2} \tag{4.40}$$

Applying Newton's second law of motion to each term on the right of Eq. (4.40), we obtain

$$m_1\frac{d^2\vec{r}_i}{dt^2} = \sum_{k=1}^{m} \vec{F}_{ik} + \sum_{j=1}^{m} \vec{f}_{ij} = \vec{F}_i + f_i \tag{4.41}$$

where F_i = net external forces acting on particle i

f_i = net internal forces acting on it

Now, summing up over all particles, Eq. (4.30) gives

$$m\frac{d^2\vec{r}_{CM}}{dt^2} = \sum_{i=1}^{n} \vec{F}_i + \sum_{i}\sum_{j} f_{ij} \tag{4.42}$$

In Eq. (4.42), we note that f_{ij}'s are internal forces. Based on Newton's third law of motion for each of these, there is an equal and opposite reaction force acting on another particle. Therefore, in summing over i and j, all of these forces cancel out, leaving zero net internal force.

Hence, Eq. (4.42) becomes

$$m\frac{d^2\vec{r}_{CM}}{dt^2} = \sum_{i=1}^{n}\vec{F}_i \tag{4.43}$$

If there are no external forces acting on the system, then

$$m\frac{d^2\vec{r}_{CM}}{dt^2} = 0 \tag{4.44}$$

Solving Eq. (4.44) gives

$$\vec{r}_{CM} = \vec{r}_{CM}(0) + \vec{v}_{CM}(0)t \tag{4.45}$$

Eq. (4.45) represents the equation of a straight line whose coefficient is a constant speed. Hence, in the absence of external forces, the center of mass of a system of particles will move in a straight line at constant speed, independently of any collision of the particles with each other.

Example 4.6: *Two vehicles having masses m_1 and m_2 are traveling on a horizontal icy surface with velocities v_1 and v_2 respectively. Assume that they collide, show that the velocities of the center of mass (CM or CG) before and after collision are equal.*

Solution: The two vehicles with and without collision is shown in Fig. E4.6.

Fig. E4.6 Two vehicles with and without collision

Let v_1' and v_2' are the velocities of the vehicles after impact.

Momentum before impact = momentum after impact

or
$$m_1v_1 + m_2v_2 = m_1v_1' + m_2v_2'$$

The position and velocity vectors of CG before impact are given by

$$\vec{r} = \frac{m_1x_1 + m_2x_2}{m_1 + m_2}$$

$$\frac{d\vec{r}}{dt} = \frac{m_1v_1 + m_2v_2}{m_1 + m_2}$$

The velocity vector of CG after impact is

$$\frac{d\vec{r}'}{dt} = \frac{m_1v_1' + m_2v_2'}{m_1 + m_2}$$

Now, the frictional impulse is negligible. Therefore

$$m_1v_1 + m_2v_2 = m_1v_1' + m_2\,v_2'$$

or
$$\frac{d\vec{r}}{dt} = \frac{d\vec{r}'}{dt}$$

Hence, the velocities of CG before and after collision are equal.

4.7 IMPULSE AND CIRCLE OF CONSTANT ACCELERATION (COCA)

Figure 4.24 shows an impulse acting on a rigid body. Here, we analyze the instantaneous motion of the rigid body up to the moment at which the impulse ends. The rigid body will have both linear and angular accelerations, both lying in the plane of the body during the impulse. (See Hwang (2002)).

Notation:

The following notation will be used in the derivation of circle of constant acceleration (COCA).

m = mass of the rigid body

G = center of gravity or center of mass of rigid body located at the origin of the coordinate system

T = impulse duration (time)

$F(t)$ = force acting in the positive y direction from $t_0 = 0$ to $t_1 = T$

$\int F(t)dt$ = impulse

p = momentum arm of F about the center of mass perpendicular from G to the line of force

M = moment = force times the moment arm $[M(t) = pF(t)]$

Q = any given point on the rigid body defined by the polar coordinates r and θ and by Cartesian coordinates x and y

N = point along the line of force which lies on the x-axis

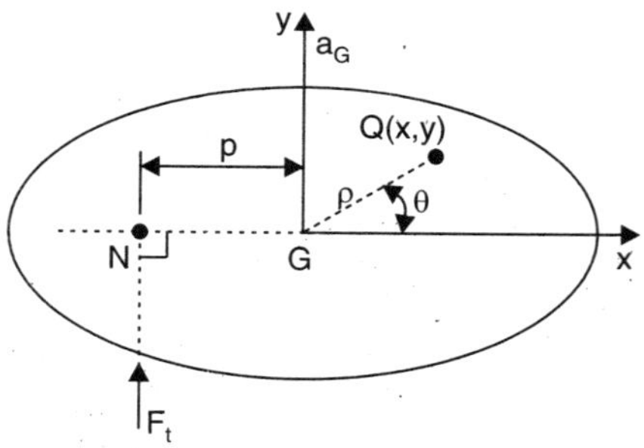

Fig. 4.24 An impulse acting on a rigid body.

Here, we conduct the analysis to

(a) derive the equations of motion of the rigid body.

(b) show that certain set of points on the rigid body will have the same magnitude of acceleration.

(c) derive the equations defining the above sets of points.

In all the above cases (a), (b), and (c), the sets will be obtained to be in the form of circles or arcs of circles as shown in Fig. 4.25. Here, we note that once the impulse is removed, the rigid body will have a constant linear speed along the y-axis plus a constant angular speed about the center of mass.

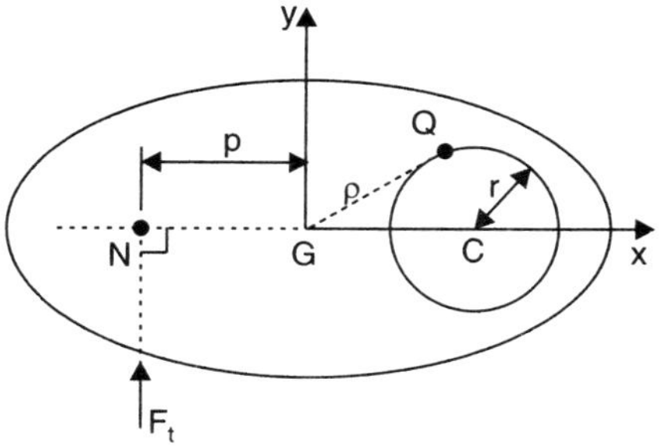

Fig. 4.25 A point Q on a COCA.

We assume here that the center of mass, G, is located at the origin of the coordinate system and that the body is oriented such that the impulsive force is parallel to the y-axis as shown in Fig. 4.26.

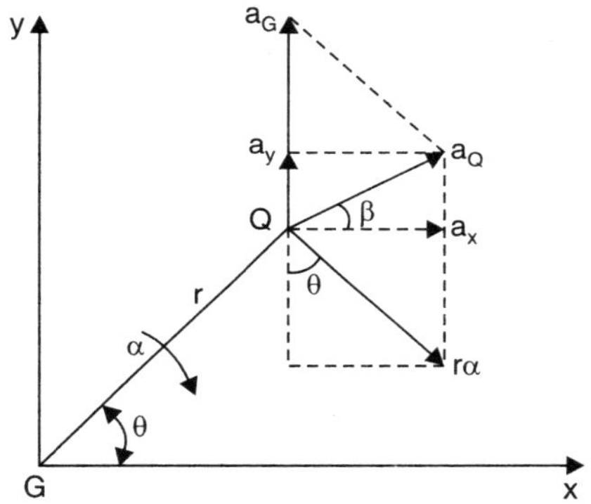

Fig. 4.26 Acceleration components at point Q.

4.7.1 Acceleration at Point Q

We assume that the impulse duration is very short such that all points on the rigid body suffer negligible displacement during this time. The acceleration at point Q can be stated as

$$\vec{a}_Q = \vec{a}_G + r\vec{\alpha} \tag{4.46}$$

where $\vec{a}_G$ = acceleration of the center of mass

r = radial distance from point G to Q

α = angular acceleration of point Q

The moment M_G about the center of mass G can be written as

$$\vec{M}_G = I_G\vec{\alpha} \tag{4.47}$$

where I_G = the moment of inertia of the body about the center of mass G.

Now setting

$$I_G = mk^2\alpha \tag{4.48}$$

where k = the radius of gyration about the center of mass G.

The moment

$$\vec{M}_G = F_p\text{(force times moment arm)} \tag{4.49}$$

The force equation in the y-direction is given by

$$ma_G = F \tag{4.50}$$

From Eq. (4.48), we can write α as

$$\alpha = \frac{F\,p}{mk^2} = \frac{a_G\,p}{k^2} \tag{4.51}$$

Now, using the standard relationship between the Cartesian and polar coordinates, the x and y components of acceleration are given by

$$a_x = \alpha r \sin\theta = \frac{p\,a_G}{k^2} r \sin\theta$$

$$a_y = a_G - \alpha r \cos\theta = a_G\left[1 - \frac{p}{k^2} r\cos\theta\right] \tag{4.52}$$

Since $a_Q^2 = a_x^2 + a_y^2$, we have

$$a_Q^2 = a_G^2\left[1 + \left(\frac{pr}{k^2}\right)^2 - 2\frac{p}{k^2} r\cos\theta\right] \tag{4.53}$$

where a_Q = the magnitude of the acceleration of point Q due to the impulse.

We note here that the force F contributes to this acceleration, a_Q through its effect on the translational motion of the center of mass and the effect on the angular motion of the body about the center of mass. Also, if p is zero, then all points Q have the same acceleration, a_G.

Special case: If Q lies at the intersection of F and the x-axis, then $r = p$ and $\theta \simeq 180°$. Figure 4.27 shows the special case when N and Q coincide.

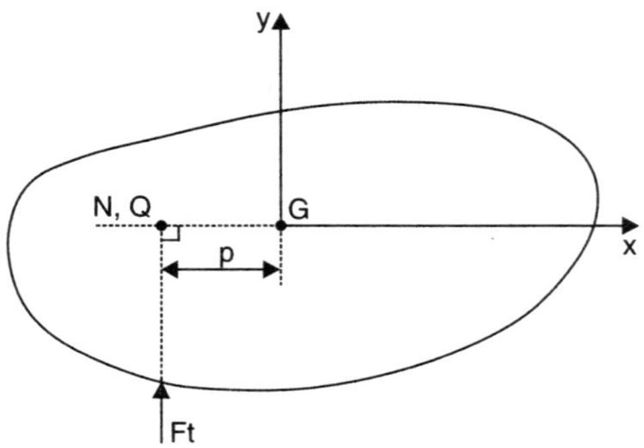

Fig. 4.27 Special case when N and Q coincides.

The acceleration a_Q given by Eq. (4.42) becomes

$$a_Q^2 = a_G^2\left[1 - \left(\frac{p^2}{k^2}\right)^2 + 2\frac{p^2}{k^2}\right] = a_G^2\left[1 + \left(\frac{p}{k}\right)^2\right]^2 \tag{4.54}$$

or $$a_Q = a_G\left[1+\left(\frac{p}{k}\right)^2\right] \quad \text{and} \quad a_G = a_Q\left[\frac{k^2}{k^2+p^2}\right] \tag{4.55}$$

Now, we have

$$\gamma = \frac{k^2}{k^2+p^2} \tag{4.56}$$

where γ is known as the *mass reduction factor* of a rigid body at point Q. For an offset impact or collision, γ is < 1 and $1/\gamma > 1$. The acceleration at point Q is $1/\gamma$ times greater than that at point G. Hence, the effect of the offset collision or impact on the acceleration at point Q is equivalent to the reduction of mass by $1/\gamma$ times in the direction of impact or collision. Therefore, the effective mass is reduced by $1/\gamma$ and the acceleration is increased by $1/\gamma$ at point Q.

Example 4.7: *A mid-size truck has a yaw radius of gyration, k, of 60 inches and a frontal width of 82 inches. The offset distance, p, of the truck is 18 inches. Determine the acceleration at the impacting side of the truck compared to that at the CG of the truck.*

Solution: From Eq. (4.45)

$$\gamma = \frac{k^2}{k^2+p^2} = \frac{60^2}{60^2+18^2} = 0.917$$

and $$\frac{1}{\gamma} = \frac{1}{0.917} = 1.0905$$

Since $1/\gamma = 1.0905$, the acceleration at the impacting side of the truck will be about 9.05% higher than that at the CG of the truck.

4.7.2 Circle of Constant Acceleration (COCA)

Considering the set of points for which the magnitude of acceleration, a_Q, is the same, then from Eq. (4.54) we have

$$a_Q^2 = a_G^2\left[1+\left(\frac{rp}{k^2}\right)^2 - 2\frac{rp}{k^2}\cos\theta\right] \tag{4.57}$$

Substituting x for $[r\cos\theta]$, and $[x^2+y^2]$ for r^2 in Eq. (4.57) gives

$$c^2 = \left[1+(x^2+y^2)\left(\frac{p}{k^2}\right)^2 - 2x\frac{p}{k^2}\right] \tag{4.58}$$

where $$c = \frac{a_Q}{a_G}$$

Multiplying both sides of Eq. (4.58) by $(k^2/p)^2$ and rearranging terms gives

$$\left(x-\frac{k^2}{p}\right)^2 + y^2 = \left(c\frac{k^2}{p}\right)^2 \tag{4.59}$$

Eq. (4.59) is the equation of a circle, namely,

$$(x - x_c)^2 + y^2 = p^2 \tag{4.60}$$

where x_c = the x-coordinate of center of circle = GC = $\dfrac{k^2}{p}$ (4.61)

ρ = the radius of the circle = $c\dfrac{k^2}{p}$ (4.62)

From Eqs. (4.61) and (4.62), we can write

$$\rho = c\ \mathrm{GC} \tag{4.63}$$

Eq. (4.59) represents the equation of a circle with center lying on the x-axis and at a distance GC $= k^2\rho$ from the center of mass. The moment is given by $M = F\rho$

The moment M

$$a_Q = c\ a_G = \left(\frac{\rho}{GC}\right)a_G = \left(\frac{\rho p}{k^2}\right)a_G \tag{4.64}$$

Noting that $a_G = F/m$ and $M = Fp$, Eq. (5.64) becomes

$$a_Q = \frac{\rho M}{k^2 m} \tag{4.65}$$

4.7.3 Direction of Acceleration, a_Q

Figure 4.28 shows the direction of acceleration, a_Q which is perpendicular to the radius of the circle and also tangent to the COCA at point Q.

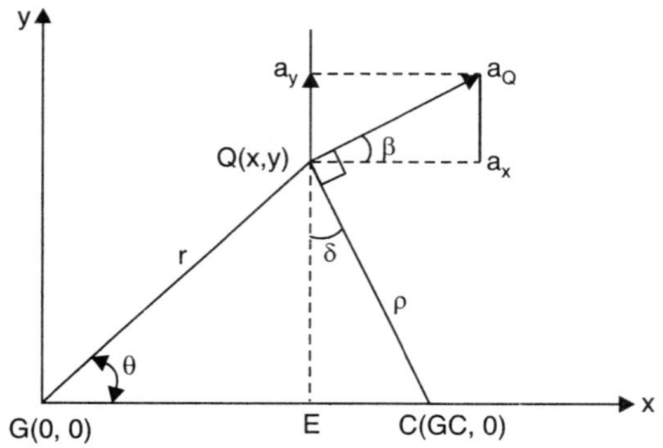

Fig. 4.28 Direction of acceleration, a_Q.

Let β = horizontal angle of a_Q

Then $$\tan\beta = \frac{a_y}{a_x} \tag{4.66}$$

Substituting for ax and ay from Eq. (4.52) and using k^2/p = GC, $x = r\cos\theta$, $y = r\sin\theta$ in Eq. (4.66), we obtain

$$\tan\beta = \frac{GC - x}{y}$$

where $$\tan\delta = \frac{GC - x}{y}, \ \beta = \delta \tag{4.67}$$

Since a_x is perpendicular to QE, a_Q will be perpendicular to QC, the radius of the circle.

4.8 ONE-DIMENSIONAL COLLISION

The principle of conservation of linear momentum is a powerful tool for applications in traffic accident reconstructions. It permits us to accurately determine the speed of a vehicle at impact with another vehicle. The law of conservation of linear momentum states that the vector sum of the linear momentum of a system of particles or bodies is the same both before and after the collision.

A collision happens when two or more particles or bodies come together in such a way that the forces acting on those bodies are relatively large but act for only a relatively short time. This results in an abrupt change in the motion of the colliding bodies that permits us to clearly distinguish the time and motion before the collision from the time and motion after the collision.

There are basically two types of collisions.

Elastic collision: If the kinetic energy is *conserved* in the collision, then the collision is classified as *elastic*.

Inelastic collision: If the kinetic energy is *not conserved,* then the collision is classified as *inelastic*. The collision is completely inelastic, if the two bodies in collision stick together after collision. In a completely inelastic collision, all kinetic energy is not *lost*. Instead, the loss of kinetic energy is consistent within the constraints imposed by the law of conservation of linear momentum.

4.8.1 One-dimensional Elastic Collision

Applying the law of conservation of linear momentum for the system in Fig. 4.29, we can write

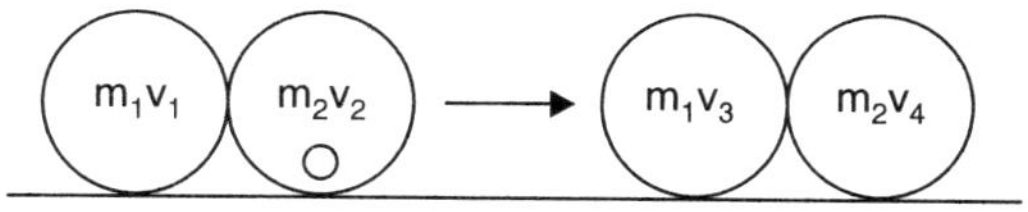

Fig. 4.29 Elastic collision.

$$m_1v_1 + m_2v_2 = m_1v_3 + m_2v_4 \tag{4.68}$$

Since this collision is elastic, kinetic energy is conserved. Therefore,

$$\frac{1}{2}m_1v_1^2 + \frac{1}{2}m_2v_2^2 = \frac{1}{2}m_1v_3^2 + \frac{1}{2}m_2v_4^2 \tag{4.69}$$

Notation:

m_1 = mass of vehicle 1

m_2 = mass of vehicle 2

v_1 = speed of vehicle 1 at impact

v_2 = speed of vehicle 2 at impact

v_3 = speed of vehicle 1 after impact

v_4 = speed of vehicle 2 after impact

v_5 = speed of vehicle mass $(m_1 + m_2)$ after impact
α_1 = approach angle (travel angle) of vehicle 1
α_2 = approach angle (travel angle) of vehicle 2
β_1 = departure angle (travel angle) of vehicle 1
β_2 = departure angle (travel angle) of vehicle 2

Eqs. (4.68) and (4.69) can be rearranged as

$$m_1(v_1 - v_3) = m_2(v_4 - v_2) \tag{4.70}$$

and

$$\frac{1}{2}m_1v_1^2 - \frac{1}{2}m_1v_3^2 = \frac{1}{2}m_4v_4^2 - \frac{1}{2}m_2v_2^2 \tag{4.71}$$

or

$$m_1(v_1^2 - v_3^2) = m_2(v_4^2 - v_2^2) \tag{4.72}$$

Dividing Eq. (4.72) by Eq. (4.70), we obtain

$$v_1 + v_3 = v_4 + v_2 \tag{4.73}$$

or

$$v_1 - v_2 = v_4 - v_3 \tag{4.74}$$

Eq. (4.74) shows that in an elastic collision, the difference in the velocities of approach equals the difference in the velocities of separation of the two bodies or vehicles.

From Eq. (4.74), we have

$$v_1 = v_4 - v_3 + v_2 \tag{4.75}$$

Substituting Eq. (4.75) in Eq. (4.70), we obtain v_2 as

$$m_1(v_4 - v_3 + v_2 - v_3) = m_2(v_4 - v_2) \tag{4.76}$$

or

$$m_1v_4 - 2m_1v_3 + m_1v_2 = m_2v_4 - m_2v_2$$

or

$$v_2 = \frac{2m_1v_3}{m_1 + m_2} + \frac{(m_2 - m_1)v_4}{m_1 + m_2} \tag{4.77}$$

From Eq. (4.74),

$$v_2 = v_1 + v_3 - v_4 \tag{4.78}$$

Substituting v_2 from Eq. (4.77) into Eq. (4.70), we get

$$m_1(v_1 - v_3) = m_2[v_4 - (v_1 + v_3 - v_4)]$$

or

$$v_1 = \frac{(m_1 - m_2)v_3}{(m_1 + m_2)} + \frac{2m_2m_4}{(m_1 + m_2)} \tag{4.79}$$

Eqs. (4.77) and (4.79) give the general solutions for impact velocities if the mass and final velocities are known for one-dimensional elastic impacts in which the kinetic energy is conserved.

4.8.2 Coefficient of Restitution (Elasticity)

The total linear momentum of the system in Fig. 4.30 before the collision is given by $m_1v_1 + m_2v_2$. If we replace two masses m_1 and m_2 with a single mass having a value, equal to the sum of m_1 and m_2, then this mass will have a velocity of v_5, which will be different from either v_1 or v_2.

Hence, $$m_1 v_1 + m_2 v_2 = (m_1 + m_2) v_5 \tag{4.80}$$

or $$v_5 = \frac{m_1 v_1 + m_2 v_2}{m_1 + m_2} \tag{4.81}$$

The kinetic energy of the mass $(m_1 + m_2)$ is given by

$$KE = \left(\frac{m_1 + m_2}{2}\right)\left(\frac{m_1 v_1 + m_2 v_2}{m_1 + m_2}\right)^2 \tag{4.82}$$

Eq. (4.82) represents the kinetic energy after the collision if the collision were totally inelastic. The kinetic energy available to deform the masses is obtained by subtracting the KE after collision from the sum of the individual kinetic energies of the two masses before collision.

$$KE = \left(\frac{1}{2} m_1 v_1^2 + \frac{1}{2} m_2 v_2^2\right) - \frac{1}{2}(m_1 + m_2)\left(\frac{m_1 v_1 + m_2 v_2}{m_1 + m_2}\right)^2 \tag{4.83}$$

or $$KE = \frac{1}{2}\left(\frac{m_1 m_2}{m_1 + m_2}\right)(v_1 - v_2)^2 \tag{4.84}$$

Eq. (4.84) gives the maximum amount of lost energy. When the collisions are not totally inelastic, some of this kinetic energy will be stored in a spring and will be released as the metal rebounds.

Expressing in terms of the post-impact velocities v_3 and v_4, we have

$$KE = \frac{1}{2}\left(\frac{m_1 m_2}{m_1 + m_2}\right)(v_3 - v_4)^2 \tag{4.85}$$

Eq. (4.85) can be related to the equation for maximum loss of energy by multiplying the former by the square of the coefficient of restitution (elasticity) E. In other words, we relate the unused KE to the maximum amount of KE that can be lost.

Hence, $$E^2 \frac{1}{2}\left(\frac{m_1 m_2}{m_1 + m_2}\right)(v_1 - v_2)^2 = \frac{1}{2}\left(\frac{m_1 m_2}{m_1 + m_2}\right)(v_3 - v_4)^2 \tag{4.86}$$

or $$E^2 (v_1^2 - v_2^2) = (v_3 - v_4)^2 \tag{4.87}$$

4.8.3 Inelastic Collision

If the collision is completely inelastic, the two masses will stick together, so that

$$v_3 = v_4 \tag{4.88}$$

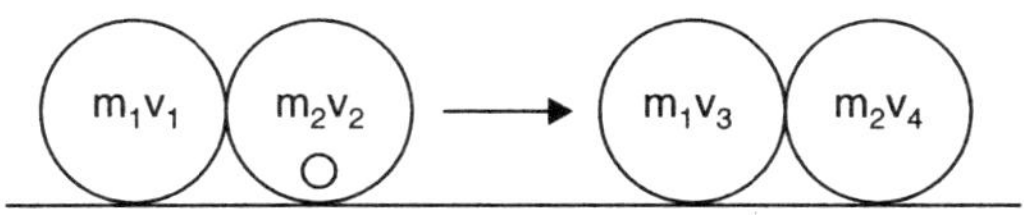

Fig. 4.30 Inelastic collision.

From the law of conservation of linear momentum, we can write

$$m_1 v_1 + m_2 v_2 = m_1 v_3 + m_2 v_4 \quad \text{(Note: } v_3 = v_4\text{)} \tag{4.89}$$

or
$$m_1v_1 = (m_1 + m_2)v_4 - m_2v_2 \tag{4.90}$$

Hence
$$v_1 = \frac{(m_1 + m_2)v_4}{m_1} - \left(\frac{m_2}{m_1}\right)v_2 \tag{4.91}$$

Similarly, we can write for v_2 as
$$m_1v_1 + m_2v_2 = (m_1 + m_2)v_4 \tag{4.92}$$

or
$$v_2 = \frac{(m_1 + m_2)}{m_2}v_4 - \left(\frac{m_1}{m_2}\right)v_1 \tag{4.93}$$

Eqs. (4.91) and (4.93) describe the inelastic collision in one dimension. If two vehicles collide with 100% overlap and thrust lines on the center of mass, they will generally cling together and slide to a stop. The post-impact velocity will be the same for each vehicle. The vector sum of linear momentum is the same after the collision as it was before the collision.

Example 4.8: *A vehicle weighing 4000 lbs hits a parked vehicle weighing 3000 lbs in a 100% overlap front to rear collision. The two vehicles move clinging together. The drag factor of the roadway is 0.80 and the two vehicles skid 100 ft. The front wheels of the striking vehicle and the rear wheels of the struck vehicle are locked by the collision. Determine the impact speed.*

Solution: This is recognized as a one-dimensional collision problem with all pre-impact momentum on the x-axis. The general conservation of linear momentum equation is
$$m_1v_1 + m_2v_2 = m_1v_3 + m_2v_4 \tag{E.1}$$

Since vehicle 2 is parked, $v_2 = 0$. Also since the vehicles move off together after the collision, $v_3 = v_4$. Eq. (E.1) becomes
$$m_1v_1 = (m_1 + m_2)v_4 \tag{E.2}$$

or
$$v_1 = \left(\frac{m_1 + m_2}{m_1}\right)v_4 \tag{E.3}$$

Now
$$v_4 = \sqrt{30 d f n} \tag{E.4}$$

Assume $n = 50\%$ [8 wheels supporting the load and only 4 wheels retarding it] From Eq. (E.4)
$$v_4 = \sqrt{30(100)(0.80)(0.50)} = 34.64 \text{ mph}$$

From Eq. (E.3)
$$v_1 = \left(\frac{4000 + 3000}{4000}\right)(34.64) = 60.62 \text{ mph.}$$

Example 4.9: *A vehicle weighing 3500 lbs impacts a parked vehicle weighing 2500 lbs. The drag factor for the roadway is 0.70. The striking vehicle leaves 60 ft of pre-impact skids with all four wheels. After the collision, the vehicle travels 80 ft with all four wheels locked. The struck vehicle, w_2, leaves 60 ft of skid marks from the rear wheel.*

(*a*) *determine the impact speed of w_1*

(*b*) *determine the impact speed of w_1 at the beginning of the skid.*

Solution: (*a*) The conservation of linear momentum equation is

$$w_1v_1 + w_2v_2 = w_1v_3 + w_2v_4 \tag{E.1}$$

In this problem, $v_2 = 0$ and $v_3 = v_4$.

Hence Eq. (E.1) gives

$$v_1 = v_3 + \left(\frac{w_2}{w_1}\right)v_4 \tag{E.2}$$

$$v_3 = \sqrt{30\,f\,d} \tag{E.3}$$

where d = skidding distance after the collision.

Therefore

$$v_3 = \sqrt{30(0.70)(80)} = 40.98 \text{ mph}$$

Similarly,

$$v_4 = \sqrt{30\,d\,f\,n} \qquad \text{(all four wheels locked, } n = 0.40\text{)}$$

$$= \sqrt{30(60)(0.70)(0.40)} = 22.45 \text{ mph}$$

Eq. (E.2) gives

$$v_1 = v_3 + \frac{w_2}{w_1}v_4 = 40.98 + \frac{2500}{3500}(22.45) = 57.01\,\text{mph}$$

The impact speed of w_1 = 57.01 mph.

(*b*) In order to determine the speed of w_1 at the beginning of the skid, we combine the pre-impact skid speed with the impact speed.

Substituting the value for v_1 in the combined speed formula, we have

$$S_1 = \sqrt{30\,df + S_i^2} = \sqrt{30(60)(0.70) + 57.01^2} = 67.15 \text{ mph}$$

Hence, the speed of w_1 at the beginning of the skid = 67.15 mph.

Example 4.10: (*a*) *Develop an expression for the linear momentum analysis of a rear-end collision where $v_2 \neq 0$ and $v_3 \neq v_4$.*

(*b*) *Develop a linear momentum equation for a head-on collision where $v_2 \neq 0$ and $v_3 \neq v_4$.*

Solution: (*a*) The linear momentum equation is

$$w_1v_1 + w_2v_2 = w_1v_3 + w_2v_4$$

or

$$v_1 = v_3 + \frac{w_2}{w_1}v_4 - \frac{w_2}{w_1}v_2 = v_3 + \frac{w_2}{w_1}(v_4 - v_2)$$

(*b*) The linear momentum equation is

$$w_1v_1 + w_2v_2 = w_1v_3 + w_2v_4$$

Since this is a head-on collision (opposite direction collision)

$$w_1v_1 - w_2v_2 = w_1v_3 + w_2v_4$$

or

$$v_1 = v_3 + \left(\frac{w_2}{w_1}\right)(v_4 + v_2)$$

4.9 ANGULAR COLLISIONS (QUADRANT METHOD)

Most collisions happen when the vehicles are moving at some angle to one another. Therefore, there is an angle of approach and also an angle of departure. Here, we use a two-dimensional coordinate system that consists of a horizontal *x*-axis and a vertical *y*-axis. These two axes are at right angle to each other. Figure 4.31 shows the vector components for the linear momentum problem.

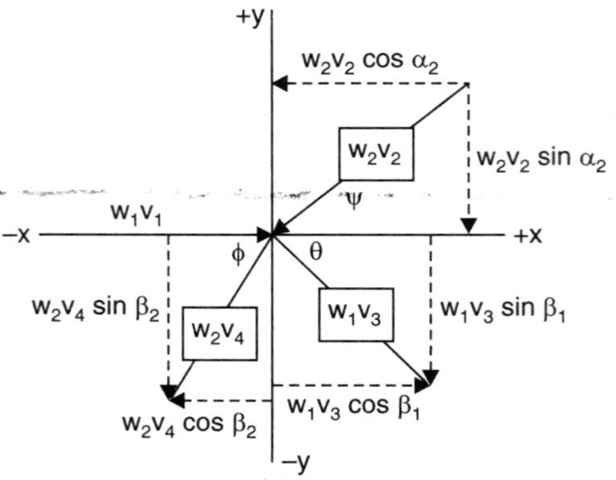

Fig. 4.31

4.9.1 90 Degree Collision Problem

Figure 4.32 shows the vector components for this problem.

We apply here the law of conservation of linear momentum for the system in Fig. 4.33 as follows:

In the *y*-direction:

$$w_1v_1 + w_2v_2 = w_1v_3 \sin \beta_1 + w_2v_4 \sin \beta_2 \quad (4.94)$$

and
$$w_2v_2 = w_1v_3 \sin \beta_1 + w_2v_4 \sin \beta_2 \quad (4.95)$$

or
$$v_2 = \left(\frac{w_1}{w_2}\right) v_3 \sin \beta_1 + v_4 \sin \beta_2 \quad (4.96)$$

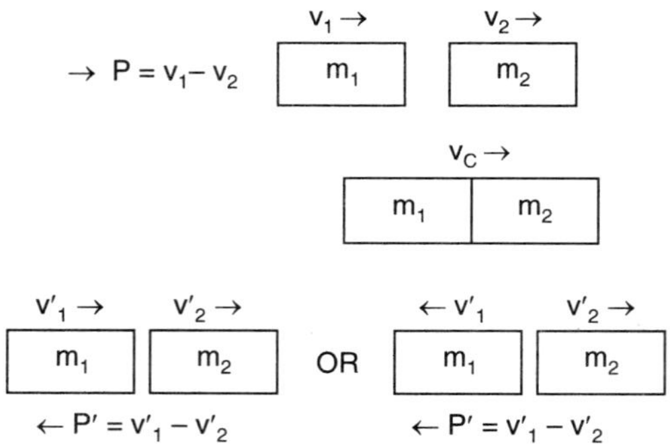

Fig. 4.32

In the *x*-direction:

$$w_1v_1 + w_2v_2 = w_1v_3 \cos \beta_1 + w_2v_4 \cos \beta_2 \quad (4.97)$$

and
$$w_1v_1 = w_1v_3 \cos \beta_1 + w_2v_4 \cos \beta_2 \quad (4.98)$$

or
$$v_1 = v_3 \cos \beta_1 + \left(\frac{w_2}{w_1}\right) v_4 \cos \beta_2 \quad (4.99)$$

4.9.2 Collisions other than 90 Degrees

Refer to Fig. 4.27 which shows the vector components for the problem. We apply the law of conservation of linear momentum for the problem referring to Fig. 4.31 as follows:

In the *y*-direction:

$$-w_2v_2 \sin \alpha_2 = -w_1v_3 \sin \beta_1 - w_2v_4 \sin \beta_2 \quad (4.100)$$

and $$-w_2v_2 \sin \alpha_2 = -w_1v_3 \sin \beta_1 - w_2v_4 \sin \beta_2 \tag{4.101}$$

or $$v_2 = \frac{w_1v_3 \sin\beta_1}{w_2 \sin\alpha_2} + \frac{v_4 \sin\beta_2}{\sin\alpha_2} \tag{4.102}$$

In the x-direction:

$$w_1v_1 - w_2v_2 \cos \alpha_2 = w_1v_3 \cos \beta_1 - w_2v_4 \cos \beta_2 \tag{4.103}$$

and $$w_1v_1 = w_1v_3 \cos \beta_1 - w_2v_4 \cos \beta_2 + w_2v_2 \cos \alpha_2 \tag{4.104}$$

or $$v_1 = v_3 \cos\beta_1 - \left(\frac{w_2}{w_1}\right) v_4 \cos\beta_2 + \left(\frac{w_2}{w_1}\right) v_2 \cos\alpha_2 \tag{4.105}$$

4.9.3 360 Degree Method

This method allows to measure the angles of approach and departure from the x-axis as either acute or obtuse angles. Figure 4.33 shows the vector components of this problem.

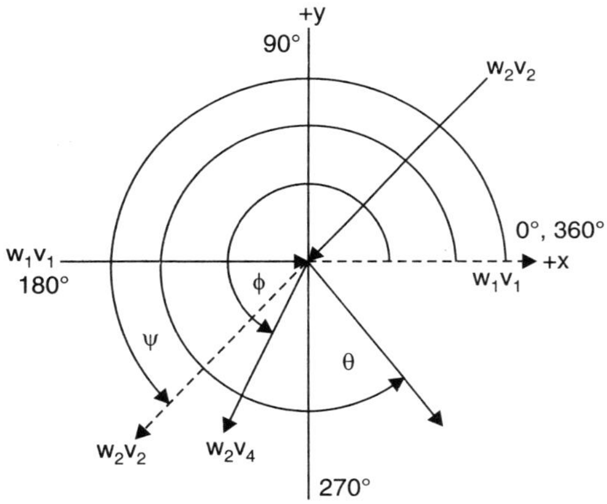

Fig. 4.33 360 degree coordinate system.

Using a 360 degree coordinate system greatly simplifies the analysis of oblique impact collisions. Since momentum is a vector, plus and minus signs must be used for a correct momentum analysis. A 360 degree coordinate system will automatically consider the proper signs associated with trigonometric values.

In the y-direction:

$$w_1v_1 \sin \alpha_1 + w_2v_2 \sin \alpha_2 = w_1v_3 \sin \beta_1 + w_2v_4 \sin \beta_2 \tag{4.106}$$

Since $\alpha_1 = 0$, we have

$$v_2 = \frac{w_1v_3 \sin\beta_1}{w_2 \sin\alpha_2} + \frac{v_4 \sin\beta_2}{\sin\alpha_2} \tag{4.107}$$

In the x-direction:

$$w_1v_1 \cos \alpha_1 + w_2v_2 \cos \alpha_2 = w_1v_3 \cos \beta_1 + w_2v_4 \cos \beta_2 \tag{4.108}$$

Since $\cos \alpha_1 = \cos 0° = 1$, we have from Eq. (4.108)

$$w_1v_1 + w_2v_2 \cos \alpha_2 = w_1v_3 \cos \beta_1 + w_2v_4 \cos \beta_2 \tag{4.109}$$

or
$$v_1 = v_3 \cos\beta_1 + \left(\frac{w_2}{w_1}\right) v_4 \cos\beta_2 - \left(\frac{w_2}{w_1}\right) v_2 \cos\alpha_2 \tag{4.110}$$

The 360 degree method always produces the same result as the quadrant method.

4.10 DAMAGE ANALYSIS

The dynamic model which relates the damage to a vehicle to the velocity changes is based on well established physical principles. This dynamic model is developed based on Newton's laws of motion and an application of the principle of conservation of energy.

4.10.1 Central Collision

Figure 4.34 shows the basic lumped mass model for central collision.

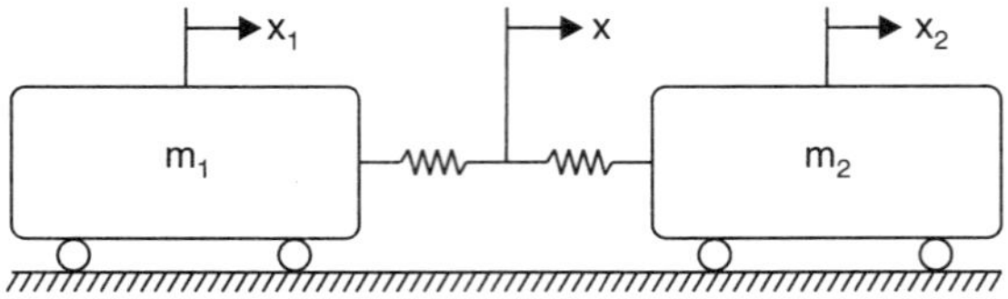

Fig. 4.34 Schematic representation of central collision.

The equation of motion for vehicle 1 is given by

$$m_1 a_1 = -\frac{k_1 k_2}{(k_1 + k_2)}(x_1 - x_2) \tag{4.111}$$

where a_1 is the acceleration of vehicle 1. A similar expression is written for vehicle 2. The energy absorbed by vehicle 1 in the crush of that spring assuming it as a linear spring is given by

$$E_1 = \frac{1}{2} k_1 (x_1 - x)^2 \tag{4.112}$$

where $(x_1 - x)$ is the deformation of vehicle 1 to the point of common velocity v_c for the vehicles. A similar equation is written for the energy for vehicle 2, E_2.

Let $\delta = (x_1 - x_2)$ = the relative separation of the centers of gravity

a_δ = relative acceleration.

The equation of motion for the relative displacement is given by

$$a_\delta = -\left[\frac{k_1 k_2}{k_1 + k_2}\right]\left[\frac{m_1 + m_2}{m_1 m_2}\right]\delta \tag{4.113}$$

Defining a constant k as

$$k^2 = \left[\frac{k_1 k_2}{k_1 + k_2}\right]\left[\frac{m_1 + m_2}{m_1 m_2}\right] \tag{4.114}$$

We can write Eq. (4.113) as

$$a\delta + k^2\delta = 0 \tag{4.115}$$

The solution of Eq. (4.115) becomes

$$\delta = A \sin kt + B \cos kt \tag{4.116}$$

At $t = 0$, $\delta = 0$, hence, $B = 0$. The closing velocity between the two masses m_1 and m_2 is defined by

$$v_c = \frac{d\delta}{dt} = k\,A\cos kt \tag{4.117}$$

At

$$t = 0,\ v_c = v_1 - v_2 \tag{4.118}$$

where v_1 and v_2 are the original speeds of vehicles 1 and 2 respectively before impact. Hence, Eq. (4.116) becomes

$$\delta = \left(\frac{v_1 - v_2}{k}\right)\sin kt \tag{4.119}$$

The maximum displacement occurs when $\sin kt = 1$

or

$$\delta_{max} = \left(\frac{v_1 - v_2}{k}\right) \tag{4.120}$$

Let

$\delta_1 = (x_1 - x)$ = the deformation of vehicle 1

$\delta_2 = (x - x_2)$ = the deformation of vehicle 2

Then, by Newton's third law of motion, the magnitude of the forces acting on each vehicle must be equal and opposite, or

$$k_1\delta_1 = -k_2\delta_2 \tag{4.121}$$

Let

$$\delta_1 + \delta_2 = \delta_{max} \tag{4.122}$$

Eliminating δ_2 from Eqs. (4.121) and (4.122), we have

$$\delta_1 = \frac{k_2}{k_1 + k_2}\delta_{max} \tag{4.123}$$

A similar result can be obtained for δ_2.

Now, Eq. (4.120) can be rearranged as

$$v_1 - v_2 = k\delta_{max} \tag{4.124}$$

Using Eq. (4.114) and substitution for δ_{max} for Eqs. (4.122) and (4.123) gives the closing speed as

$$v_1 - v_2 = \sqrt{\frac{(m_1 + m_2)(k_1\delta_1^2 + k_2\delta_2^2)}{m_1 m_2}} \tag{4.125}$$

Substitution of the energy expressions, defined by Eq. (4.112), gives the closing velocity of the centers of mass between the two vehicle masses m_1 and m_2 as

$$v_1 - v_2 = \sqrt{\frac{2(m_1 + m_2)(E_1 + E_2)}{m_1 m_2}} \tag{4.126}$$

Since the linear momentum is conserved for the two masses during the collision, assuming the collision to be plastic (coefficient of restitution, $e = 0$), the common velocity is given by

$$v_c = \frac{m_1 v_1 + m_2 v_2}{(m_1 + m_2)} \tag{4.127}$$

From Eqs. (4.126) and (4.127), the speed change of each vehicle can be determined as

$$\Delta v_1 = v_1 - v_c = \frac{m_2}{(m_1 + m_2)}(v_1 - v_2) = \sqrt{\frac{2(E_1 + E_2)}{m_1\left(1 + \frac{m_1}{m_2}\right)}} \tag{4.128}$$

$$\Delta v_2 = v_c - v_2 = \frac{m_1}{(m_1 + m_2)}(v_1 - v_2) = \sqrt{\frac{2(E_1 + E_2)}{m_2\left(1 + \frac{m_2}{m_1}\right)}} \tag{4.129}$$

Eqs. (4.128) and (4.129) can also be written as

$$\Delta v_1 = \sqrt{\frac{2m_2 \Delta E}{m_1(m_1 + m_2)}} \tag{4.128a}$$

$$\Delta v_2 = \sqrt{\frac{2m_1 \Delta E}{m_2(m_1 + m_2)}} \tag{4.129a}$$

where $\Delta E = E_1 + E_2$

Example 4.11 (Central collision): *Vehicle 1 and vehicle 2 suffer a head-on central collision. Vehicle 2 has a weight of 3000 kg and a speed of 10 km/h to the left. Vehicle 1 has a weight of 2000 kg and a speed to the right of 20 km/h. Determine the velocity change of vehicle 1 and 2 at common velocity.*

Solution: Applying Eq. (4.127) for v_c, we obtain

$$v_c = \frac{m_1 v_1 + m_2 v_2}{m_1 + m_2} = \frac{2000(-20) + 3000(10)}{(2000 + 3000)} = -2 \text{ km/h (to the right)}$$

where $m_1 = 2000$ kg, $m_2 = 3000$ kg, $v_1 = -20$ km/h, and $v_2 = 10$ km/h.

Next, applying the law of conservation of energy:

Total energy before collision = Total energy after collision

$$\frac{m_1 v_1^2}{2} + \frac{m_2 v_2^2}{2} = \frac{(m_1 + m_2)v_c^2}{2} + \Delta E$$

$$\frac{2000(-20)^2}{2} + \frac{3000(10)^2}{2} = \frac{(2000 + 3000)(-2)^2}{2} + \Delta E$$

or ΔE = energy dissipated in the collision = 540 kg (km/h)^2

By applying Eqs. (4.128*a*) and (4.129*a*), the change in velocity for each vehicle can be obtained. The velocity change of v_1 at common velocity is given by

$$\Delta v_1 = \sqrt{\frac{2m_2 \Delta E}{m_1(m_1 + m_2)}} = \sqrt{\frac{2(3000)(540)}{2000(2000 + 3000)}} = 18 \text{ km/h}$$

The velocity change of v_2 at common velocity is given by

$$\Delta v_2 = \sqrt{\frac{2m_1 \Delta E}{m_1(m_1 + m_2)}} = \sqrt{\frac{2(2000)(540)}{2000(2000 + 3000)}} = 12 \text{ km/h}$$

Hence, vehicle 2 (large vehicle) changes in speed from 10 to –2 km/h with a Δv_2 of –12 km/h, and vehicle 1 (small vehicle) changes in speed from –20 km/h (to the right) to –2 km/hr with a Δv_1 of 18 km/h. As expected, the vector sum of the two velocity changes is the same as the closing speed which is 30 km/h.

4.10.2 Non-Central Collision

Figure 4.35 shows the offset frontal collision. Here, we extend the central collision analysis to the non-central collision. The analysis for a non-central impact involves that the vehicles reach a common velocity at the common crush zone of the vehicles rather than at the center-of-gravity. In Fig. 4.35, point *C* is the centroid of the crush of the respective vehicles and we assume that the common velocity occurs at point *C*.

The relationship between the speed at the crush center and the speed of the center of gravity for vehicle 1 is given by

$$v_{c1} = v_1 + h_1 w_1 \tag{4.130}$$

In component form, the relative acceleration for vehicle 1 is given by

$$a_{c1} = a_1 + h_1 \alpha_1 \tag{4.131}$$

where a_{c1} = linear acceleration of point C.
a_1 = linear acceleration of the center of gravity.
h_1 = the offset of point C from the *x*-axis.
α_1 = angular acceleration of the vehicle.

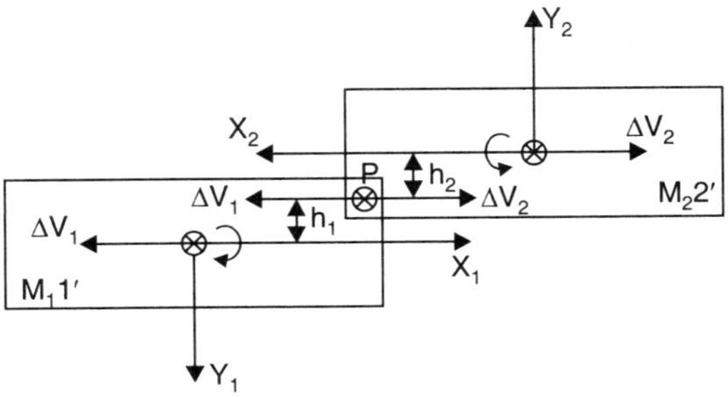

Fig. 4.35

The equation of motion for vehicle 1 is given by

$$F = -m_1 a_1 = -m_1(a_{c1} - h_1\alpha_1) \tag{4.132}$$

The moment acting on vehicle 1 is written as

$$M = -I\alpha_1 = -m_1 k^2 \alpha_1 \tag{4.133}$$

where k = the radius of gyration of the vehicle about the *z*-axis (yaw).

Solving a_{c1} from Eqs. (4.132) and (4.133) gives

$$ac_1 = -\frac{F}{m_1}\left[\frac{k_1^2 + h_1^2}{k_1^2}\right] \tag{4.134}$$

Therefore, the acceleration of the center of gravity is given by

$$a_1 = -\frac{F}{m_1} = \left[\frac{k_1^2}{k_1^2 + h_1^2}\right] a_{c1} \tag{4.135}$$

Effective Mass γ_m: The effective mass coefficient of a vehicle referred to a point C is defined here as

$$\gamma_{m1} = \frac{k_1^2}{k_1^2 + h_1^2} \tag{4.136}$$

then

$$a_1 = \gamma m_1 a_{c1} \tag{4.137}$$

From Eq. (4.132), the effective mass of vehicle 1 acting at point C is given by $\gamma_{m1} m_1$.

When Eq. (4.137) is integrated over the time interval of the collision until the common velocity is reached, then, we have

$$\Delta v_1 = \gamma_{m1} \Delta v_{c1} \tag{4.138}$$

where Δv_1 = speed change at the center of gravity.

Δv_{c1} = speed change at point C.

Note here that the derivation for the speed change of point C of the two vehicles can employ Eqs. (4.133) and (4.134) for central collisions if the effective mass of each vehicle is substituted for the vehicle mass.

Hence, for vehicle 1:

$$\Delta v_{c1} = \sqrt{\frac{2(E_1 + E_2)}{\gamma_{m1} m_1 \left(1 + \dfrac{\gamma_{m1}}{\gamma_{m2} m_2}\right)}} \tag{4.139}$$

From Eq. (4.138), taking γ_m inside the square root gives the speed change for the center of gravity as

$$\Delta v_1 = \sqrt{\frac{2\gamma_{m1}(E_1 + E_2)}{m_1 \left(1 + \dfrac{\gamma_{m1} m_1}{\gamma_{m2} m_2}\right)}} \tag{4.140}$$

Since the momentum is assumed to be conserved in a collision, the speed change of the vehicle 2 can be written as

$$\Delta v_2 = \Delta v_1 \left(\frac{m_1}{m_2}\right) \tag{4.141}$$

We can also write the relationships as follows:

For vehicle 1 at the center of mass C:

$$\Delta v_{c1} = \sqrt{\frac{2\gamma_{m1}\gamma_{m2} m_2 \Delta E}{m_1(\gamma_{m1} m_1 + \gamma_{m2} m_2)}} \tag{4.142}$$

For vehicle 2 at the center of mass C:

$$\Delta v_{c2} = \sqrt{\frac{2\gamma_{m1}\gamma_{m2} m_1 \Delta E}{m_2(\gamma_{m1} m_1 + \gamma_{m2} m_2)}} \tag{4.143}$$

For vehicle 1 at damage centroid, C:

$$\Delta v_p = \sqrt{\frac{2\gamma_{m2} m_2 \Delta E}{\gamma_{m1} m_1(\gamma_{m1} m_1 + \gamma_{m2} m_2)}} \tag{4.144}$$

For vehicle 2 at damage centroid, C:

$$\Delta v_p = \sqrt{\frac{2\gamma_{m1} m_1 \Delta E_1}{\gamma_{m2} m_2 (\gamma_{m1} m_1 + \gamma_{m2} m_2)}} \quad (4.145)$$

Equivalent Barrier Speed (EBS): A special case is a fixed barrier collision, where vehicle 1 impacts a barrier which absorbs no energy. The equivalent barrier speed (EBS) is obtained by setting $E_2 = 0$ and $m_2 = \infty$ in Eq. (4.140).

That is

$$\text{EBS} = \sqrt{\frac{2\gamma_{m1} E_1}{m_1}} \quad (4.146)$$

The EBS is a convenient concept to compare the energy absorbed in crushing vehicles. It is not equivalent to the speed change during a collision between two vehicles except in very special cases.

Example 4.12 (Non-central collision): *The offset of vehicle 1 (h_1 = lateral distance from the central line of the vehicle to the line through the collision point) is 20 cm. The value of k_1 is 4040 cm² (full size vehicle). For vehicle 2, the offset is 95% of h_1 or 19 cm and that k_2^2 is 91% of that of k_1^2 or 3676 cm². Determine the change in speed of the center of mass for the offset collision.*

Solution:

$$\gamma_{m1} = \frac{k_1^2}{k_1^2 + h_1^2} = 0.91 \text{ and } \gamma_{m2} = 0.91 \text{ also.}$$

Δv_{c1}, Δv_{c2}, and Δv_p for vehicles 1 and 2 are obtained by applying Eqs. (4.145) to (4.148).

For vehicle 1 at the center of mass C:

$$\Delta v_c = \sqrt{\frac{2\gamma_1\gamma_2 m_2}{m_1(\gamma_1 m_1 + \gamma_2 m_2)}\Delta E} = 17.1 \text{ km/h}$$

For vehicle 2 at the center of mass C:

$$\Delta v_c = \sqrt{\frac{2\gamma_1\gamma_2 m_1}{m_2(\gamma_1 m_1 + \gamma_2 m_2)}\Delta E} = 11.4 \text{ km/h}$$

For vehicle 1 at damage centroid, P:

$$\Delta v_p = \sqrt{\frac{2\gamma_2 m_2}{\gamma_1 m_1(\gamma_1 m_1 + \gamma_2 m_2)}\Delta E} = 19.0 \text{ km/h}$$

For vehicle 2 at damage centroid, P:

$$\Delta v_p = \sqrt{\frac{2\gamma_1 m_1}{\gamma_2 m_2(\gamma_1 m_1 + \gamma_2 m_2)}\Delta E} = 12.7 \text{ km/h}$$

Comparing the two cases, it can be seen that the change in speed of the center of mass for the offset collision is slightly less than for the central collision while the change in speed in the collision region is a little greater than for the central collision.

ΔV Computations: The velocity change experienced by a vehicle during impact is commonly called delta-V (ΔV). It is important consideration for occupant injury severity. Delta-V is a function of the velocity before and after impact. It is a vector, and as such, requires magnitude and angular orientation for proper identification.

Closing Velocity: The closing velocity is defined to be the sum of the speed changes at the point of common velocity or

$$V_C = \Delta v_{c1} + \Delta v_{c2} \tag{4.146a}$$

From the Eq. (4.132), the ΔV at the common velocity center C is related to that at the center of gravity as a function of γ_{m1} therefore, the closing velocity is defined in terms of the velocity changes of the center of gravity as

$$v_c = \frac{\Delta v_1}{\gamma_{m1}} + \frac{\Delta v_2}{\gamma_{m2}} \tag{4.146b}$$

The closing velocity is in the direction of the line of action of the collision forces. It can be employed to give some idea of the velocity of one of the two vehicles if something is known about the other velocity.

4.11 VEHICLE-TO-BARRIER (VTB) IMPACT

Here, we derive the relationship between the dynamic crush, barrier impact speed, and the circular natural frequency of the vehicle-to-barrier (VTB) impact by applying the principle of work and energy or the law of conservation of energy. Fig. 4.36 shows a simplified model of a spring-mass impact model for the system. Let v_0 be the velocity of the vehicle just before it contacts the rigid barrier. During the collision (from position 1 to position 2), there are no changes in the gravitational potential energy and no frictional energy loss.

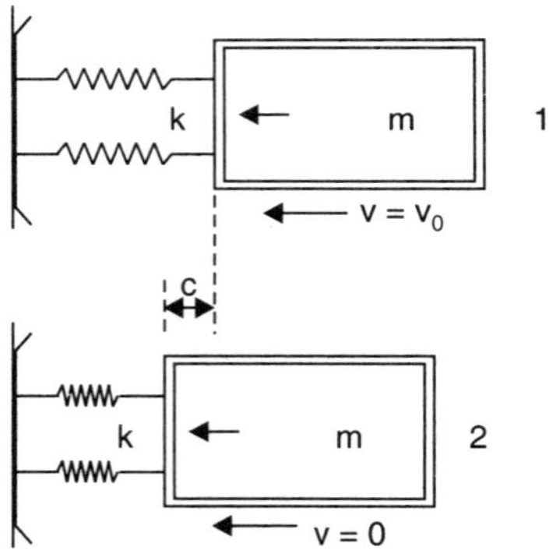

Fig. 4.36 A spring-mass impact model.

At position 2, the vehicle stops momentarily against the rigid barrier before it rebounds. Applying the law of conservation of energy, we have

$$T_1 + U_1 = T_2 + U_2 \tag{4.147}$$

or

$$\frac{1}{2}mv_0^2 + 0 = 0 + \frac{1}{2}kc^2 \tag{4.148}$$

or

$$v_0 = \sqrt{\frac{k}{m}}\, c \tag{4.149}$$

Denoting ω_n = the natural frequency of the system = $\sqrt{\frac{k}{m}}$, Eq. (4.149) becomes

$$v_0 = c\,\omega_n \quad (4.150)$$

Now
$$F = kc \quad (4.151)$$

where
F = the peak force

c = dynamic crush

Hence
$$\frac{1}{2}mv_0^2 = \frac{1}{2}kc^2 = \frac{1}{2}Fc \quad (4.152)$$

or
$$mv_0^2 = Fc \quad (4.153)$$

Therefore, the peak acceleration, a is given by

$$a = \frac{F}{m} = \frac{v_0^2}{c} = v_0\omega_n \quad (4.154)$$

4.11.1 Rebound Criterion in a Two-vehicle Impact

Here, we derive a relationship between the mass ratio and the coefficients of restitution such that immediately after separation, both vehicles move in the same direction.

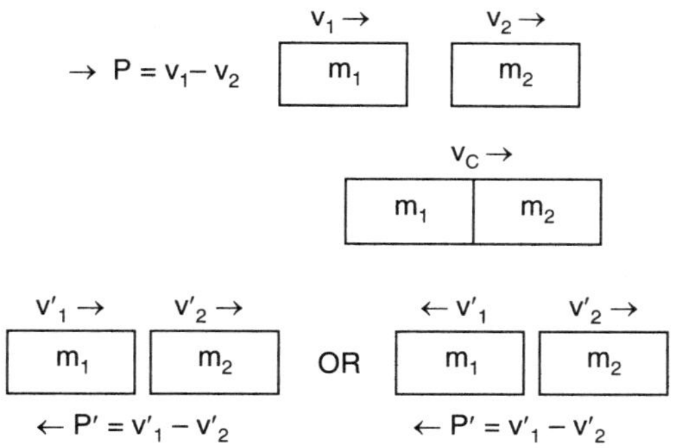

Fig. 4.37 Two vehicle collision and separation.

Figure 4.37 shows a two vehicle collision and separation.

Let

e = efficient of restitution (EoR)

v_1 = initial impact velocity

v_c = common velocity of m_1 and m_2, after impact

v_1', v_2' = rebound velocities of m_1 and m_2, respectively

The momentum equation is

$$m_1\bar{v}_1 = (m_1 + m_2)\bar{v}_c \quad (4.155)$$

Hence
$$v_c = \frac{m_1}{m_1 + m_2}v_1 = \frac{r_m}{1 + r_m}v_1 \quad (4.156)$$

where
$$v_1 = \sqrt{2gh}$$

$$r_m = \frac{m_1}{m_2} \quad (4.157)$$

The coefficient of restitution, e is defined by

$$e = \frac{\bar{v}_1' - \bar{v}_c}{\bar{v}_c - \bar{v}_1} = \frac{v_1' - v_c}{v_c - v_1} \tag{4.158}$$

Hence $$v_1' = v_c + e(v_c - v_1) \tag{4.159}$$

From Eqs. (4.156) and (4.159), we can write

$$\frac{v_1'}{v_1} = \frac{r_m - e}{1 + r_m} \tag{4.160}$$

For rebounding m_1:

$$\frac{v_1'}{v_1} < 0 \text{ or } r_m < e \tag{4.161}$$

For m_1 not rebounding backward:

$$r_m \geq e \tag{4.162}$$

Figure 4.38 shows the separation velocity ratio *vs.* r_m given by Eq. (4.160). The shaded portion in Fig. 4.38 is bounded by both the mass ratio, r_m and the coefficient of restitution, e, ranging from 0 to 1. In this shaded portion, the separation velocity ratio is negative.

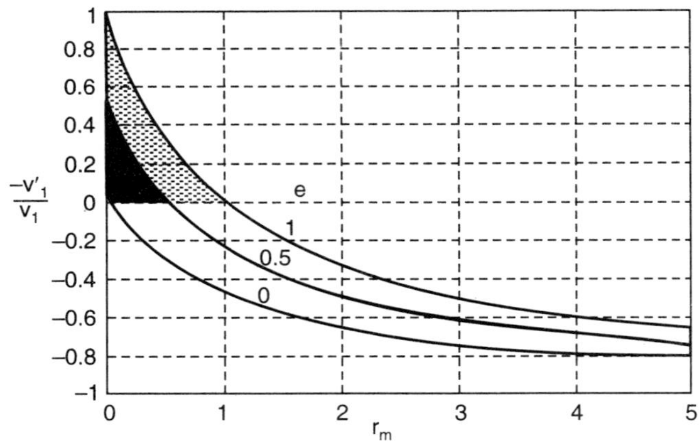

Fig. 4.38 Two-particle impact rebound condition.

Example 4.13: *Derive a relationship between the two coefficients of restitution such that at least the middle vehicle receives double whiplashes in a three-vehicle collision. Apply the equations of momentum and the coefficient of restitution at the times of impact and separation for each of the impacts.*

Solution: The momentum equation is given by

$$m_1\bar{v}_1 = m_1\bar{v}_1' + m_2\bar{v}_2' \tag{E.1}$$

The coefficient of restitution

$$e_1 = \frac{\bar{v}_1' - \bar{v}_2'}{\bar{v}_2 - \bar{v}_1} \tag{E.2}$$

Let $r_1 = m_1/m_2$ and $r_2 = m_2/m_3$.

For the first impact between m_1 and m_2, we have

$$v_1' = \frac{r_1 - e_1}{1 + r_1} v_1 \tag{E.3}$$

and

$$v_2' = \frac{r_1(1 + e_1)}{1 + r_1} v_1 \tag{E.4}$$

Similarly, for the second impact, between m_2 and m_3, we can write

$$v_2'' = \frac{r_3 - e_2}{1 + r_2} v_2' \tag{E.5}$$

$$v_3'' = \frac{r_2(1 + e_2)}{1 + r_2} v_2' \tag{E.6}$$

For vehicle 2 to be rear-ended twice, $v'_1 \geq v''_2$. Rewriting Eqs. (E.5) and (E.6), we obtain

$$e_2 > r_2 - \left[\frac{r_1 - e_1}{1 + e_1}\right]\left[\frac{1 + r_2}{r_1}\right] \tag{E.7}$$

Eq. (E.7) gives the relationship between the two coefficients of restitution such that the second vehicle is subjected to multiple rear impacts. This condition does not depend on the impact speed.

4.11.2 Coefficient of Restitution

Here, an expression for the coefficient of restitution, e, based on the relative separation velocity and relative approach velocity is derived assuming the two timings obtained from a crash test, t_m and t_f are given.

Let a = constant deceleration of Average Square Wave (ASW)

e = coefficient of restitution

t_f = final separation time

t_m = time of maximum dynamic crush

v = velocity change up to t_m

Figure 4.39 shows the velocity changes in the deformation and rebound phases. Figure 4.39 shows that based on the vehicle-to-barrier (VTB) test or the vehicle-to-vehicle (VTV) test, the relative acceleration of two masses in the deformation phase can be approximated by an average square wave (ASW). As shown in Fig. 4.39, the area under the square wave is the relative approach velocity. The vehicle deceleration in the restitution phase is approximated by a triangle, and the area is the rebound or relative separation velocity.

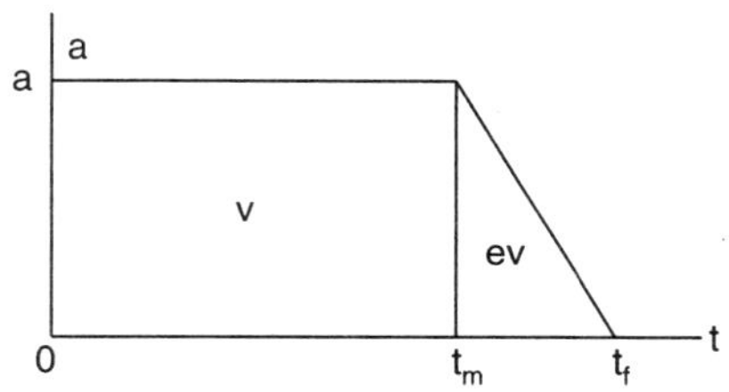

Fig. 4.39 Velocity changes in the deformation and rebound phases.

From Fig. 4.39, we can write

$$v = a\,t_m \tag{4.163}$$

$$ev = \frac{(t_f - t_m)a}{2} \tag{4.164}$$

Dividing Eq. (4.164) by Eq. (4.163), we obtain

$$e = \frac{t_f - t_m}{2t_m} \tag{4.165}$$

and

$$t_f = t_m(1 + 2e) \tag{4.166}$$

Based on the available crash results for vehicles and trucks in rigid barrier and vehicle-to-vehicle crashes, the two timings and the coefficient of restitution are summarized in Table 4.1.

Table 4.1 Relationship between t_f, t_m, and e

Crash mode	*e*	$t_f/t_m = 1 + 2e$	t_f/t_m *from crash tests*	t_f*, ms, from crash tests*
Vehicle-to-Barrier	0.15	1.3	1.3 to 1.5	80 to 100
Vehicle-to-Vehicle in line	0.10	1.2	1.2 to 1.4	100 to 140

4.11.3 Coefficient of Restitution in Vehicle Crashes (Huang (2002))

The coefficient of restitution, e, is defined as

$$e = \frac{\text{Relative separation speed}}{\text{Relative approach speed}} = \frac{v_2' - v_1'}{v_1 - v_2} \tag{4.167}$$

where v_1 = initial speed of vehicle 1

v'_1 = speed of vehicle 1 after collision

v_2 = speed of vehicle 2 or barrier

v'_2 = speed of vehicle 2 or barrier after collision

Prasad ( ) investigated the coefficient of restitution for vehicle structures and its application in estimating velocity changes in vehicle collisions. The coefficient of restitution, e, for a two-vehicle central collision can be determined using the data from vehicle-barrier collisions. The value of e obtained is only approximate since it assumes that the amount of energy of each vehicle losses is the same in both the vehicle-to-vehicle and the barrier collisions. The slope of the force-deflection curve, k, for each barrier crash must be known.

At some time during the collision, the two vehicles will have the same speed, called the *common speed*, v_c. At this point, the maximum deformation between the two vehicles occurs and it is at the end of deformation phase (or the loading phase). Beyond this point, the two vehicles will start to unload and separate, releasing any elastic rebound (potential) energy until the two vehicles are totally apart from each other.

If the mass of the barrier is assumed to be infinite, for a vehicle-fixed barrier collision (assuming a coordinate system at rest with respect to the barrier), v_2 is zero and v'_2 will also be zero.

Hence,

$$v_1^1 = -ev_1 \tag{4.168}$$

and
$$\frac{m_1}{2}(v_1^1)^2 = e^2\left(\frac{m_1}{2}v_1^2\right) \tag{4.169}$$

That is, kinetic energy after = $e^2 \times$ (kinetic energy before)

If $e = 1.0$, the kinetic energy after collision equals that before (there is no loss).

If $e = 0.0$, the kinetic energy after collision equals zero (car is at rest, resting against the barrier). It should be noted also here that the initial kinetic energy is equal to the stored, that is, available or elastic rebound, energy obtained during the collision.

4.11.4 Two-Vehicle Central Collision

Here, we assume a moving reference system which moves at the common velocity, v_c. At the time the speeds of the two vehicles are equal, their speed will be the common speed. Their speed will be zero in the common speed reference system used. Their speeds are v_1' and v_2' after separation.

Applying the law of conservation of momentum, we can write

$$0 = m_1(v_1^1 - v_c) + m_2(v_2^1 - v_c) \tag{4.170}$$

or
$$(v_2^1 - v_c) = -\frac{m_1}{m_2}(v_1' - v_c) \tag{4.171}$$

The kinetic energy after separation will be equal to the rebound energy. This rebound energy is retrievable upon separation of the two vehicles. Now, assuming the rebound energy in each vehicle in the case of vehicle-to-vehicle collisions is equal to that obtained when the vehicle collides with a fixed barrier, the rebound energy in the vehicle-barrier crash and the final kinetic energy in the vehicle-to-vehicle crash can be equated.

Thus
$$e_1^2\Delta E_1 + e_2^2\Delta E_2 = \frac{m_1}{2}(v_1^1 - v_c)^2 + \frac{m_2}{2}(v_2^1 - v_c)^2 \tag{4.172}$$

Note that the values of ΔE_1 and ΔE_2 are obtained from

$$\Delta E_1 = \frac{m_1 m_2}{2(m_1 + m_2)}\left(\frac{k_2}{k_1 + k_2}\right)\left(v_1^2 + v_2^2\right) \tag{4.173}$$

and
$$\Delta E_2 = \frac{m_1 m_2}{2(m_1 + m_2)}\left(\frac{k_1}{k_1 + k_2}\right)\left(v_1^2 + v_2^2\right) \tag{4.174}$$

Substituting ΔE_1 and ΔE_2 from Eqs. (4.173) and (4.174) into Eq. (4.172), we obtain

$$\left(v_1^1 - v_c\right) = \frac{m_2}{m_1 + m_2}(v_1 + v_2)\sqrt{\frac{e_1^2 k_2 + e_2^2 k_1}{k_1 + k_2}} \tag{4.175}$$

$$\left(v_2^1 - v_c\right) = \frac{m_2}{m_1 + m_2}(v_1 + v_2)\sqrt{\frac{e_1^2 k_1 + e_2^2 k_2}{k_1 + k_2}} \tag{4.176}$$

Now, substituting the velocity terms in the equation for e in Eq. (4.167), we obtain

$$e\sqrt{\frac{e_1^2 k_2 + e_2^2 k_1}{k_1 + k_2}} \tag{4.177}$$

In Eq. (4.177), e is the equivalent coefficient of restitution or combined COR for the two-vehicle-crash but expressed using the e values and the force deflection constants, k_i, for the two vehicle-barrier crashes. Note that when k_2 is extremely large, the value of the equivalent e becomes e_1 (as in vehicle-barrier case). Similarly, if the $e_1 = e_2$ and $k_1 = k_2$, then the value of e becomes the same as the e for either vehicle.

Eq. (4.177) can be normalized to the coefficient of restitution of either vehicle, say, vehicle 2. Dividing Eq. (4.177) by e_2 yields the following normalized equation:

$$\frac{e}{e_2} = \sqrt{\frac{\left(\frac{e_1}{e_2}\right)^2 + \left(\frac{k_1}{k_2}\right)}{\left(1 + \frac{k_1}{k_2}\right)}} \qquad (4.178)$$

Figure 4.35 shows the 3-D plot of $\left(\frac{e}{e_2}\right)$ vs. $\left(\frac{k_1}{k_2}\right)$ and $\left(\frac{e_1}{e_2}\right)$. It can be seen from Fig. 4.40 that for a given stiffness ratio $\left(\frac{k_1}{k_2}\right)$, the combined normalized coefficient of restitution ratio $\left(\frac{e}{e_2}\right)$ of two vehicles increases as the ratio of $\left(\frac{e_1}{e_2}\right)$ increases. Also, when $\frac{e_1}{e_2} = \frac{k_1}{k_2} = 1$, then $\frac{e}{e_2} = 1$.

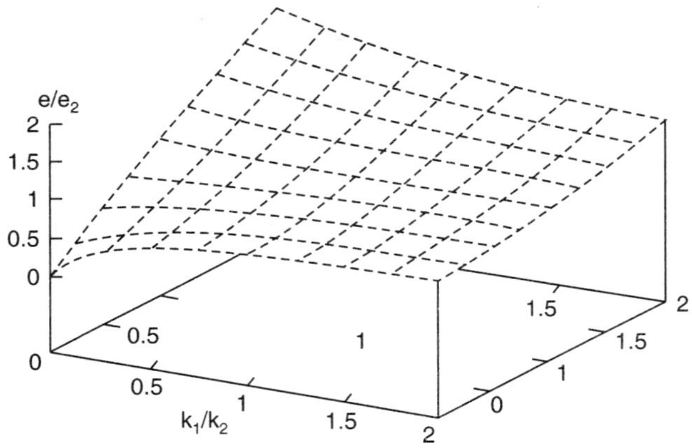

Fig. 4.40 Normalized e/e_2 as a function of k_1/k_2, e_1/e_2

4.12 COLLISION FORCES

The principle of defining a relationship between the collision damage and the velocity change of a vehicle during a collision was first developed in detail by Campbell (1974). Campbell reports results for the frontal collision of 1974 GM products into barriers and related the speed change during the collision to the stiffness of the vehicle and the amount of permanent crush. Work performed at Calspan sponsored by the National Highway Traffic Safety Administration (NHTSA) enlarged on Campbell's pioneer work and they developed a series of programs released to the public called CRASH. The last public version of CRASH3 was revised and released in 1982. The term CRASH is an acronym for CALSPAN Reconstruction of Accident Speeds on Highways.

Campbell (1974) has developed a method for utilizing energy methods to evaluate crash phenomena. The residual crush is the crush distance measured after the impact. It is also called static crush, the amount of crush after the elastic rebound from the dynamic crush.

Campbell proposed for a vehicle-fixed barrier collision, a linear relationship between the speed of the vehicle just before impact and the residual crush distance:

$$v = b_0 + b_1 x \tag{4.179}$$

where v = vehicle speed at impact (v must equal or exceed the bumper standard speed).
b_0, b_1 = vehicle structure dependent constants.
x or C = residual crush distance.

In addition to Eq. (4.179), Campbell proposed a linear force versus deflection (residual crush) equation to relate the velocity equation in terms of the energy relationship:

$$\frac{F}{W} = A + Bx \tag{4.180}$$

where x = residual crush.
F = crush force at a given x.
A, B = stiffness coefficients which are vehicle structure dependent.
W = width of section of body being crushed.

The force per unit width of damage section versus residual crush relationship is shown in Fig. 4.41. Similar to the barrier impact velocity versus residual crush, every point on the straight line is a data point approximated by the regression to the sets of data.

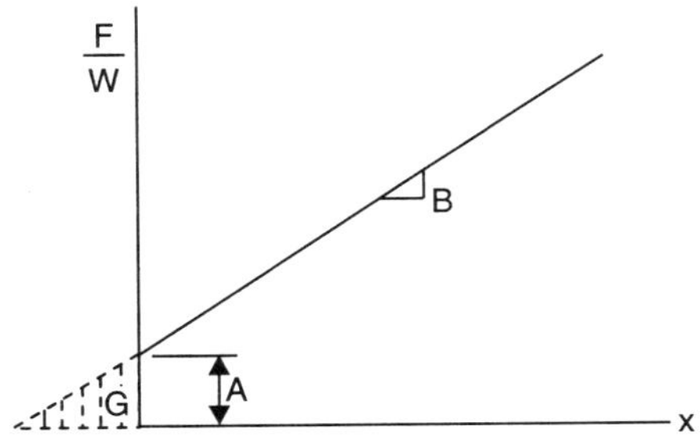

Fig. 4.41 Unit crush force versus residual crush.

Here zero restitution, $e = 0$ is assumed. In other words, the final velocity of the impacting vehicle is assumed to be equal to that of the barrier.

Applying the law of conservation of momentum and the law of conservation of energy we have

Work done on vehicle = Integral of force function over distance traversed

or
$$\int F\,dx = \frac{m}{2}v^2 \tag{4.181}$$

To this work must be added a constant term, GW, to account for work done on an energy-absorbing bumper, for which the crush distance is held at zero. Then, the work done (or crush energy) and the initial kinetic energy are given as follows:

Use C as residual crush, then

$$\text{Crush energy} = W\left(G + AC + \frac{B}{2}C^2\right) \tag{4.182}$$

$$T = \frac{1}{2}mv^2 = \frac{M}{2}(b_0 + b_1C)^2 = \frac{m}{2}(b_0^2 + 2b_0b_1C + b_1^2C^2) \tag{4.183}$$

Equating the like terms in Eq. (4.185), we get

$$A = \frac{mb_0b_1}{W}$$

$$B = \frac{mb_1^2}{W} \tag{4.184}$$

$$G = \frac{mb_0^2}{W}$$

By combining terms the coefficient G can be expressed in terms of A and B as

$$G = \frac{A^2}{2B} \tag{4.185}$$

The crush energy per unit width of vehicle in Eq. (4.183) is plotted and shown in Fig. 4.42. Note that the value of G when residual crush is zero depends solely on the square of b_0. Since b_0 is the intercept of the line on the v vs. x plot shown in Fig. 4.41, it affects the accuracy of the speed estimate. If b_0 is higher than what it should be, the actual speed of a subject vehicle involved in a low speed accident would be overestimated.

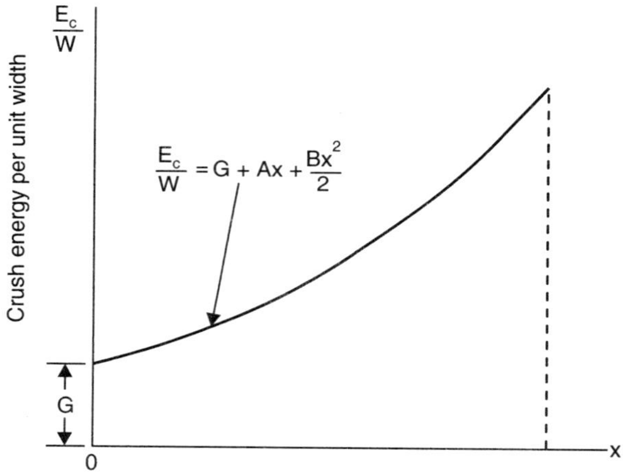

Fig. 4.42 Crush energy vs. residual crush.

The equations for A, B, and G become

$$A = 0.8\frac{mb_0b_1}{W}$$

$$B = 0.8\frac{mb_1^2}{W} \tag{4.186}$$

$$G = 0.4\frac{mb_0^2}{W}$$

Rearranging Eq. (4.186) and solving for b_0 and b_1, we obtain

$$b_0 = 1.12A\sqrt{\frac{W}{Bw}}$$

$$b_2 = 1.12\sqrt{\frac{W}{Bw}} \tag{4.187}$$

where b_0 = mph, b_1 = mph/in, mg = w = lb, W = in, A = lb/in, B = lb/in^2, and G = lb.

The NHTSA crash table has been developed based on vehicles collected in groups based on wheelbase. Table 4.2 gives vehicle stiffness categories by wheelbase and weight and default data. Table 4.3 gives vehicle stiffness categories as supplied by NHTSA.

The coefficients A and B are called stiffness or crush coefficients and have been developed from the results of collision testing. It is understood that they have been derived based on the following important assumptions:

- The force F_n is assumed to act normal to the car surface.
- The crush depth is measured perpendicular to the car surface.
- The crush depth profile extends approximately uniformly from the bottom to the top of the impact surface.
- The impacting surface reached a common velocity at maximum crush.
- The static crush dimensions were used as the basis for the derivation of A and B.

If the force is not perpendicular to the surface in the real application, an equivalent normal force is defined as,

$$F_n = F_\alpha / \cos \alpha$$

where α is measured from the normal to the surface. The direction of the force is defined from the positive local x axis of the vehicle.

Table 4.2 Vehicle stiffness Categories by wheelbase and weight and default data

		(1) *MI*	(2) *SU*	(3) *CO*	(4) *IN*	(5) *FU*
Wheelbase (in)		80.9-94.8	94.8-101.6	101.6-110.4	110.4-117.5	117.5-123.2
Track (in)		51.1	54.6	58.9	61.8	63.7
Length (in)		159.8	174.9	196.2	212.8	223.7
Width (in)		60.8	67.2	72.6	77	79.8
a (in)		45.1	46.3	51.3	54.7	58.1
b (in)		48.1	50.1	55.5	59.2	63
XF (in)		76	83.3	89.8	98.8	101.8
XR (in)		83.8	91.6	106.4	114	121.9
YS (in)		30.4	33.6	36.3	38.5	39.9
Kg (in^2)		2006	2951	3324	3741	4040
M (lb-sec^2/in)		5.7	7.9	9.18	10.99	12.59
Curb weight (lb)		2469	2753	3247	3947	4565
			Stiffness coefficients			
Front (F)	A	302 lb/in	259	317	356	325
	B	47 lb/in^2	43	56	34	37
Rear (B)	A	366	391	410	357	297
	B	38	41	44	13	70
Side (R, L)	A	77	140	173	143	177
	B	37	67	57	50	47

Definitions:

a = distance from c.g. to front axle.
b = distance from c.g. to rear axle.
XF = distance from c.g. to front of vehicle.
XR = distance from c.g. to rear of vehicle (measured negative direction).
YS = distance from c.g. to side of vehicle.
Kg = radius of gyration, squared.
M = vehicle mass, including 2 passenger loading.

Table 4.2 Vehicle stiffness Categories by wheelbase and weight and default data

(contd..)

		(6) LG	*(7) VN*	*(8) PU*	*(9) FW*	*(10) MB*
Wheelbase (in)		123.2-150	109-130			120
Track (in)		63.7	67.6			60
Length (in)		229.4	183.6	Select category 1 to 6		100
Width (in)		79.8	79	according to wheelbase		78
a (in)		60.1	48.5			54
b (in)		65.1	68.5			66
XF (in)		104.2	75.6			84
XR (in)		125.2	107			96
YS (in)		39.9	39.5			50
Kg (in^2)		4229	3713			4024
M (lb-sec^2/in)		13.74	11.2			10.35
Curb weight (lb)		5009	4000			4000
			Stiffness coefficients			
Front (F)	A	325	383	480	373	
	B	37	126	50	38	
Rear (B)	A	297	300	346		
	B	70	55	25		
Side (R, L)	A	177				
	B	47				

Definitions:

a = distance from c.g. to front axle.
b = distance from c.g. to rear axle.
XF = distance from c.g. to front of vehicle.
XR = distance from c.g. to rear of vehicle (measured negative direction).
YS = distance from c.g. to side of vehicle.
Kg = radius of gyration, squared.
M = vehicle mass, including 2 passenger loading.

Table 4.3 Vehicle stiffness categories as supplied by NHTSA

Stiffness category	*(1) MI*	*(2) SU*	*(3) CO*	*(4) IN*	*(5, 6) FU/LG*	*(7) VN*	*(8) PU*	*(9) FW*	*(10) MB*
	Pinto (FRONT)	Pinto (REAR)	Celica Supra	Chevelle (-76)	LeSabre (-76)	Vans***	Pick-ups***	Front Drive****	Movable
	Accord	Chev. Monza	Mustang (-73)	MonteCarlo (-77)	Chev V-8 (-76)	Econo E150	Courier	Citation	Barrier
V	Honda CVCC	Celica ST	AMC Concord	GrandPrix (-77)	Monaco (-76)	Dodge B-200	El Camino	Phoenix	
	Prelude	Celica GT	Malibu (78-)	Cutlass (-77)	Riviera (-77)	Chev G-20	Ford F150	Skylark	
M**	Corolla	Corona	Monarch	LeMans (-77)	Marquis (-76)	Ford F-500	Chev LUV	Omega	
	Chevette	Spirit	Zephyr	Phoenix	LTD (-76)	GMC G-35	Ford F250	Reliant	
E	Fiesta	Pacer	Fairmont	Chev V-8 (77-)	Eldorado (-78)	GMC G-1500	Dodge D-100	Aries	
O	Bobcat	Gremlin	Granada	Lesabre (77-)	Delta 88 (-76)	VW Vanagon	Ranchero F10	Escort	
	Datsum 210	VW Dasher	Firebird	Volare	T-Bird (-76)				
H	Datsum 310	Vega	Cressida	Monaco (77-)	Olds 98				
D	Arrow	Skyhawk	Datsun 810	Magnum	St. Regis				
	Champ	Omni	MonteCarlo (78-)	LeBaron	Newport				
I	Colt	Sunbird	Grand Prix (78-)	Riviera (77-)	Brghm. DeVille				
E	Porsche 924	Starfire	Cutlass (78-)	Marquis (77-)	Electra				
	Mazda GLC	Mustang (74-)	Regal	LTD (77-) 2-door	Fleetwood				
C	Fiat 124 Spider	Horizon	Aspen	Cordoba	Continental				
L	Fiat X1/9	Fiat 128 Sedan	Peugot 604L	Nova	Checker Cab				
	Datsun 280 ZX	Capri	BMW 528i	Eldorado (79-)	LTD (77-) 4-dr				
L	Opel	280 ZX 2+2	Volvo (all)	Delta 88 (77-)	Ford F100 1/2 ton	Lynx			
S	MG Midget	Challenger	Audi 5000	Diplomat	Other***				
	Tri Spitfire	BMW 320i	LeMans (78-)	T-Bird (77-)	Datsun p/u	GMC 1500			
E	VW Rabbit	Audi Fox	Century	Seville	Honcho4x4 p/u	Toy.SR5 long			
	VW Scirocco	Mazda Cosm.		Ventura	Wagoneer				
		Mazda RX		Cougar	Scout II				
		Renault LeCar			Chev. Blazer				
		Saab 900							
		Saab 99							
		Subaru							

** Includes all model years unless otherwise specified.
*** Front and rear crash modes only; for side damage pick a category (1 to 6) by wheelbase.
**** Front crash mode only; for side and rear pick a category (1 to 6) by wheelbase.

4.12.1 Energy Calculation

The energy absorbed by the vehicle during the collision is required to calculate the speed change. This energy is defined by the work done to crush the vehicle to the given profile.

For an applied force at an angle α measured from the normal to the surface, the collision energy absorbed by a vehicle would be

$$E = (1 + \tan^2 \alpha) \iint (A + Bx)\, dx\, dL \tag{4.188}$$

where x denotes the crush depth and L the crush width along the side of the vehicle. The units of energy are in-lbs.

To define the crush as a function of the crush width L, accident reconstruction software can accept a deformation profile with 2, 4, or 6 equally spaced crush dimensions as input by the user (see Fig. 4.43). The algorithm uses the crush dimensions to calculate the absorbed energy. For a two component profile, the following expanded formula is obtained through numerical integration of Eq. (4.188).

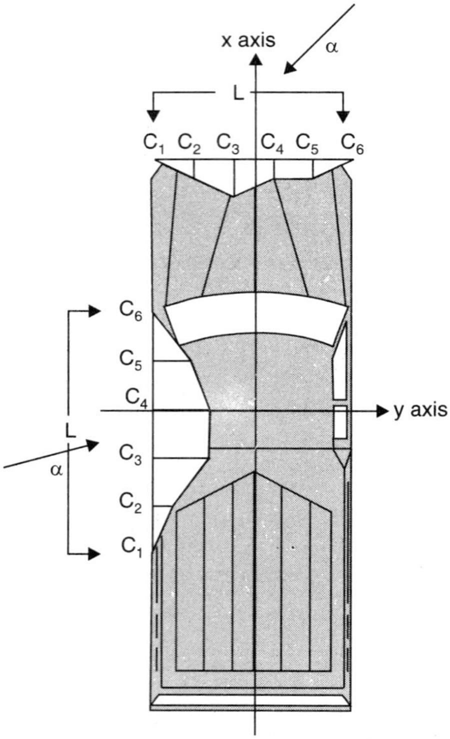

Fig. 4.43

$$E = L\left[\frac{A}{2}(C_1 + C_2) + \frac{B}{6}(C_1^2 + C_1C_2 + C_2^2) + \frac{A^2}{2B}\right] \tag{4.189}$$

For a four component profile the relation takes the form:

$$E = \frac{L}{3}\left[\frac{A}{2}(C_1 + 2C_2 + 2C_3 + C_4) + \frac{B}{6}(C_1^2 + 2C_2^2 + 2C_3^2 + C_4^2 + C_1C_2 + C_2C_3 + C_3C_4) + \frac{3A^2}{2B}\right] \tag{4.190}$$

For a six component profile the formula for energy takes the form:

$$E = \frac{L}{5}\left[\frac{A}{2}(C_1 + 2C_2 + 2C_3 + 2C_4 + 2C_5 + C_6) + \frac{B}{6}(C_1^2 + 2C_2^2 + 2C_3^2 + 2C_4^2 + 2C_5^2 + 2C_6^2 + C_1C_2 + C_2C_3 + C_3C_4 + C_4C_5 + C_5C_6) + \frac{5A^2}{2B}\right] \quad (4.191)$$

In Eqs. (4.190) and (4.191), the following notation is used.

E = energy absorbed by the vehicle.

C_i = residual crush of the vehicle measured normal to the undeformed surface at position i along the length.

L = width dimension of the damage region.

A, B, G = empirical coefficients of unit width properties obtained from crash test data.

The above equations for the energy absorbed are then multiplied by $(1 + \tan^2\alpha)$ for the total energy absorbed.

Once the energy is calculated for each vehicle, the velocity changes and equivalent barrier speeds can be calculated by substitution into the following equations:

$$\Delta v_1 = \sqrt{\frac{2\gamma_{m1}(E_1 + E_2)}{m_1\left(1 + \dfrac{\gamma_{m1}m_1}{\gamma_{m2}m_2}\right)}} \quad (4.192)$$

$$\Delta v_2 = \Delta v_1\left(\frac{m_1}{m_2}\right) \quad (4.193)$$

and

$$\text{EBS} = \sqrt{\frac{2\gamma_{m1}E_1}{m_1}} \quad (4.194)$$

4.13 LOW-SPEED SIDESWIPE COLLISION (FUNK, CORMIER, AND BAIN [2004])

Sideswipe collisions are characterized by prolonged sliding contact and with very little structural deformation. A sideswipe collision is considered as a collision in which the angle of impact is shallow and vehicle interaction in the longitudinal direction is frictional. Sideswipes are characterized by prolonged sliding between the contact surfaces.

The dynamics of sideswipe collisions were described by Bailey et al. [1995], who conducted eleven shallow angle (<30°) vehicle-to-vehicle sideswipe tests at approach speeds varying from 4.5 – 27 km/h. In some tests, vehicle sliding contact involved "snagging" of the two interacting surfaces at points in which the contact surface geometry was irregular. Snagging was recognized as a different contact mechanism from sliding. Damage from snagging involved forward or rearward deformation of a structure. They have in 2000 reported long contact durations (1.5 – 2 s) and generally low accelerations (≤ 3 g's in tests without snagging) in their tests. Levy [2000] presented a method for determining impact velocity in sideswipes in which a cycloidal truck tire mark has been imprinted on the side of the struck vehicle. This method is applicable to a small percentage of sideswipe collisions. Woolley [1987] modeled sideswipes in the "IMPAC" accident reconstruction computer program as frictional contact along a defined slip plane. SMAC [McHenry et al (1997)] and CRASHEX [Fonda (1987)] also incorporate frictional inter-vehicular sliding forces. All these models compute the inter-vehicular forces from stiffness coefficients based on vehicle crush profiles measured after crash tests. These stiffness coefficients may not be accurate when modeling sideswipe collisions with little or no crash. No crash test data has been presented to evaluate the accuracy of these models in analyzing collisions that are dominated by inter-vehicular sliding rather than crush.

Araszewski et al. [2002] calculated lateral inter-vehicular force based on damage using vehicle side stiffness coefficients. Longitudinal force was then calculated assuming an inter-vehicular sliding coefficient of friction of 0.6. Average accelerations were calculated based on these forces with additional analysis to determine regions where tire slippage occurred. Snagging was not considered, the vehicle was assumed to be a rigid body, all tires were assumed to be steered straight and free rolling, and inward crush and yaw rotations were assumed to be small. They reported on fourteen vehicle-to-barrier tests and four vehicle-to-vehicle tests used to validate the analytical method. The vehicle-to-barrier tests were conducted by steering the surface vehicle so that its side would contact a trunk lid mounted to a barrier. Similarly, in the vehicle-to-vehicle tests, the surface vehicle was steered so that its side would contact the bumper of the contact vehicle, which was also fixed to a barrier. Approach speeds varied from 4–10 km/h. Their method was reported to have accurately estimated the speed changes in their vehicle tests.

The method provides a useful starting point for evaluating sideswipe collision severity. The force calculation relies entirely on stiffness coefficients that are derived primarily from high speed crash tests with substantial crush. There is little or no data to validate the accuracy of the force calculation at the intended low levels of crush (0-4 cm). The method does not take into account the approach angle between the vehicles. The approach angle strongly influences the length of sliding contact damage, which in turn strongly influences the calculated velocity change and acceleration of the vehicles.

4.13.1 Funk-Cormier-Bain Model

This model is similar to SMAC [McHenry et al. (1997)] in that it calculates vehicle kinematics utilizing a forward timewise integration of the equations of rigid body motion. The model includes a novel non-crush-based method for calculating inter-vehicular force. An impact configuration is assumed in which the front corner of a bullet vehicle strikes the side of a stationary target vehicle at a given closing velocity and approach angle (Fig. 4.44). The bullet and target vehicles were modeled as rectangular rigid bodies having the dimensions, masses, and inertial properties of the subject vehicles.

The model was based on the following assumptions and limitations:

(*a*) The combined lateral compliance of both vehicles due to bumper and door panel deformation, suspension, tire sidewall stiffness, etc. was lumped into a single force-deflection relationship.

(*b*) Longitudinal interaction between the sliding surfaces was assumed to be frictional. The model may not be valid when snagging or substantial inward crush occurs, since this may cause the longitudinal force to exceed the force expected from pure friction.

(*c*) All tires on both vehicles were assumed to be steered straight and free rolling with lateral slippage governed by Coulomb friction.

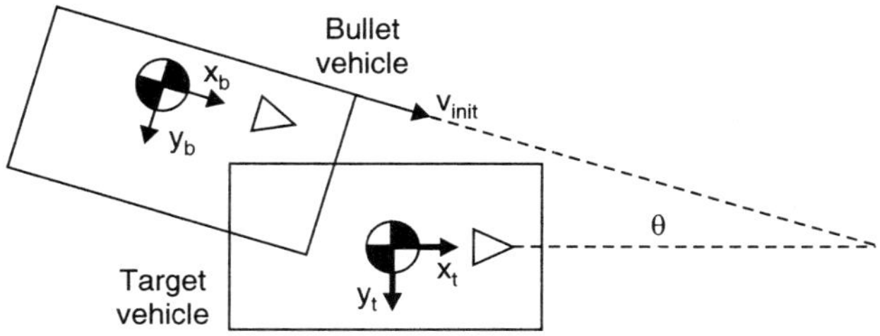

Fig. 4.44 Schematic of vehicle test setup.

An inertial coordinate frame was established to track the positions of both vehicles. The inertial frame was aligned with the initial position of the target vehicle frame such that its origin was coincident with the center of gravity (CG) of the target vehicle. The bullet vehicle was assigned an initial position based on its relative approach speed and angle such that its trajectory would cause the bullet vehicle bumper to strike the side of the target vehicle at a predetermined point of initial contact. Equations are presented assuming the right front corner of the forward-moving bullet vehicle strikes the left side of the initially stationary target vehicle. Vehicle positions were tracked with respect to an inertial coordinate frame. Vehicle forces, velocities, and accelerations were tracked in the local vehicle coordinate frame originating at the CG of each vehicle (Fig. 4.44). Vehicle velocities were coordinate transformed to the inertial frame before being integrated to obtain positions.

4.13.2 Modeling Procedure

Modeling of the vehicle dynamics was performed by solving the equations of motion at a given time point and successively incrementing in time to obtain results over a desired time duration. Since inter vehicular and tire slip conditions could change the governing equations of motion several times during a simulation, the model is considered as an initial value problem. Initial values for all parameters were established at the time in which vehicle contact initiated. At each subsequent time increment, the following calculations were performed:

1. Vehicle positions were calculated by integrating the vehicle velocity using the trapezoidal rule. Based on the vehicle dimensions, a lateral overlap distance (d_{lat}) was calculated as the perpendicular distance between the side of the target vehicle and the front corner of the bullet vehicle. The longitudinal point of contact was also calculated as the perpendicular distance between the rear axle of the target vehicle and the front corner of the bullet vehicle.
2. The lateral contact force applied to the target vehicle was calculated as a function of the lateral overlap distance (d_{lat}). The force-deflection relationship was modeled as a linear elastic spring:
$$F_{yt} = k \cdot d_{lat} \tag{4.195}$$
where F_{yt} was the y-axis force applied to the target vehicle and k was the spring stiffness.
3. The longitudinal contact force applied to the target vehicle was calculated assuming frictional contact:
$$F_{xt} = \mu_{slide} \cdot F_{yt} \tag{4.196}$$
where F_{xt} is the x-axis force applied to the target vehicle and μ_{slide} is the inter-vehicular sliding coefficient of friction. The maximum value for μ_{slide} was assumed to be 0.5. After a common contact velocity was reached, μ_{slide} was calculated to equilibrate the inter-vehicular

shear force such that no relative longitudinal acceleration occurred between the bumper of the bullet vehicle and the side of the target vehicle at the contact point. If, later in the event, this calculated shear force exceeded ±0.5. F_{yt}, then inter-vehicular sliding resumed.

4. The forces applied to the front corner of the bullet vehicle were calculated from Newton's third law by coordinate transforming the target vehicle forces to the bullet vehicle frame:

$$\begin{bmatrix} F_x \\ F_y \end{bmatrix}_b = -\begin{bmatrix} \cos\theta & \sin\theta \\ -\sin\theta & \cos\theta \end{bmatrix} \begin{bmatrix} F_x \\ F_y \end{bmatrix}_t \tag{4.197}$$

where F_{xb} and F_{yb} were the contact forces applied to the bullet vehicle along its x-axis and y-axis, respectively, and θ was the angle between the vehicles (Fig. 4.40).

5. Once the contact forces were calculated, the front and rear lateral tire forces (F_t and F_r), lateral acceleration (a_y), and angular acceleration (α) were calculated for each vehicle. Only two dynamic equilibrium equations are available to solve for these four unknowns (Fig. 4.45). These two equations may be obtained by summing the forces in the y-direction and summing the moments about the vehicle CG or the center of the front or rear axle (points f and r in Fig. 4.46. Two additional equations are available if the lateral slip conditions of the tires are known (Table 4.4). If no tires are slipping laterally, the lateral acceleration and angular acceleration of the vehicle are equal to zero. If the front and/or rear tires are slipping laterally, then the lateral tire force is equal to the maximum available tire force:

$$F_{f\,\max} = \mu_{\text{tire}} \cdot W \cdot \frac{d_{cg-r}}{d_{wb}} \tag{4.198}$$

$$F_{r\,\max} = \mu_{\text{tire}} \cdot W \cdot \frac{d_{cg-f}}{d_{wb}} \tag{4.199}$$

where μ_{tire} is the coefficient of friction between the tires and the ground, W is the weight of the vehicle, d_{cg-r} is the distance from the CG to the rear axle, d_{cg-f} is the distance from the CG to the front axle, and d_{wb} is the wheelbase.

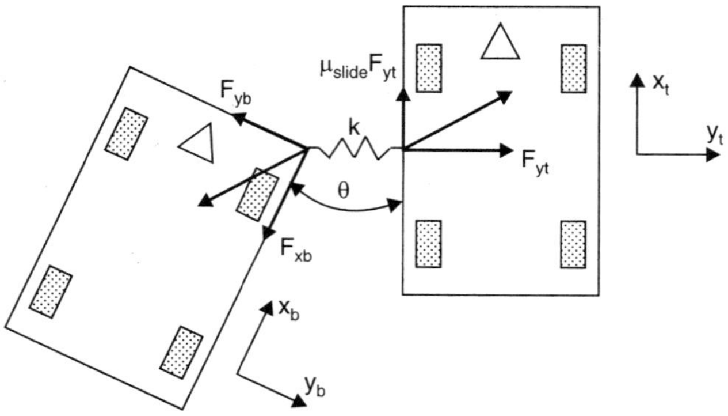

Fig. 4.45 Illustration of the inter-vehicular force interaction. For clarity, negative bullet vehicle forces are shown and the vehicles are drawn separated rather than overlapped.

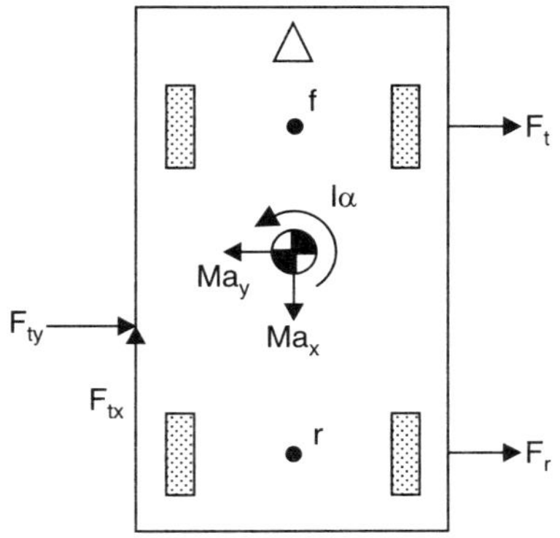

Fig. 4.46 Free body diagram of the target vehicle.

If the front tires are slipping laterally, but the rear tires are gripping, then the lateral acceleration at the CG and the angular acceleration of the vehicle are kinematically related as if the vehicle were rotating about the center of the rear axle:

$$a_y = \alpha\, d_{cg-r} \tag{4.200}$$

Similarly, when the rear tires are slipping laterally and the front tires are gripping, then the lateral acceleration at the CG and the angular acceleration of the vehicle are kinematically related as if the vehicle were rotating about the center of the front axle:

$$a_y = \alpha\, \mathrm{d}_{cg-f} \tag{4.201}$$

Table 4.4 Governing equations for the vehicle dynamics parameters that depend on the tire slip conditions.

Front tires	*Grip*	*Slip*	*Grip*	*Slip*
Rear tires	*Grip*	*Grip*	*Slip*	*Slip*
a_y	0	αd_{cg-r}	αd_{cg-r}	ΣF_y
α	0	ΣM_r	ΣM_f	ΣM_{cg}
F_t	ΣM_r	$F_{t\,\max}$	ΣF_y	$F_{t\,\max}$
F_r	ΣM_t	ΣF_y	$F_{r\,\max}$	$F_{r\,\max}$

6. After calculating the lateral tire forces and lateral and angular accelerations assuming a particular lateral slip condition for the tires, it was necessary to determine whether the assumed tire slip conditions were still valid. The calculation to determine whether the tire slip condition had changed depended on the previous tire slip condition. If a set of tires were not slipping, then slippage would initiate when the calculated lateral tire force exceeded the maximum allowable value Eqs. (4.198) and (4.199). If a set of tires were already slipping, then slippage would stop when its lateral velocity reached zero. Lateral tire velocities were obtained by first calculating the lateral accelerations of the vehicle at the center of the front (a_f) and rear (a_r) axles:

$$a_f = a_y + \alpha\, d_{cg-f} \tag{4.202}$$

$$a_r = a_y - \alpha\, d_{cg-r} \tag{4.203}$$

The lateral velocities of the front (*vf*) and rear (*vr*) tires were then calculated by numerically integrating the lateral accelerations. If the tire slip conditions were determined to have changed, it was necessary to repeat step 5 using a new set of governing equations determined by the new tire slip conditions (Table 4.4).

7. Next, all of the remaining parameters of interest describing the dynamics of the CG of each vehicle were calculated. The longitudinal acceleration (a_x) of the CG was calculated from Newton's second law. The linear velocity components (v_x and v_y) of the vehicle CG in the local vehicle coordinate frame were calculated by numerical integration of the acceleration components using the trapezoidal rule. Angular velocity and angular displacement were likewise calculated by integrating and double-integrating angular acceleration, respectively. The relative angle between the vehicles (θ) was then given by:

$$\theta = \theta_b - \theta_t \tag{4.204}$$

To calculate linear displacements, the velocity vector for each vehicle was first coordinate transformed to the inertial frame:

$$\begin{bmatrix} v_x \\ v_y \end{bmatrix}_{\text{inertial}} = \begin{bmatrix} \cos\theta_{\text{veh}} & -\sin\theta_{\text{veh}} \\ \sin\theta_{\text{veh}} & \cos\theta_{\text{veh}} \end{bmatrix} \begin{bmatrix} v_x \\ v_y \end{bmatrix}_{\text{veh}} \tag{4.205}$$

The position of the vehicle CG was then calculated by integrating the velocity vector for each direction in the inertial frame.

8. Once the calculations were complete, the data were output, the time was incremented, and steps 1-7 were repeated for the desired time duration.

In order to validate the model, seven low speed (3–10 km/h), shallow angle (15°) sideswipe collisions were staged with instrumented vehicles. These sideswipe collisions were characterized by long contact durations (~1 s) and low accelerations (< 0.4 g's). The experimental collisions were also simulated with EDSMAC [Day (1990)]. EDSMAC overpredicted peak longitudinal vehicle acceleration by an average of 83% and underpredicted the length of contact damage by an average of 50%. In contrast, the linear spring model accurately predicted the peak longitudinal vehicle acceleration (5% error) when the stiffness parameter was tuned to match the length of contact damage. These results suggest that a non-crush-based linear spring model for calculating inter-vehicular force could significantly improve the accuracy of reconstructions of low speed sideswipe collisions compared to existing methods such as SMAC. The analytical model developed here also demonstrated that it is not possible to fully reconstruct a sideswipe collision based on observations of vehicle damage alone.

4.14 EXAMPLE PROBLEMS AND SOLUTIONS

Example 4.14: *3700-pound [1678 kg] V-2 is stopped at a traffic signal when it is struck in the rear by 3800-pound [1724 kg] eastbound V-1. After impact V-1 travelled east 21 feet [6.4 m] at an average drag factor of 0.40. After impact V-2 travelled east 24 feet [7.3 m] at an average factor of 0.35. Calculate the impact speed of V-1.*

Solution: This is a simple application of the in-line (one dimensional) conservation of momentum formula. The variables:

S_1 = Speed of V-1 at impact

S_2 = Speed of V-2 at impact = 0 mph

s_1 = Speed of V-1 after impact

s_2 = Speed of V-2 after impact

d_1 = V-1 post impact skid distance = 21 feet

d_2 = V-2 post impact skid distance = 24 feet

f_1 = Drag factor of V-1 = 0.40

f_2 = Drag factor of V-2 = 0.35

W_1 = Weight of V-1 = 3800 lb

W_2 = Weight of V-2 = 3700 lb

The weight ratio:

If $W_1 = 1,\ W_2 = 3700/3800 = 0.974$

To find speeds after impact, we use the post impact skid distance and the basic skid formula. For vehicle One:

$$s_1 = \sqrt{30 * d_1 * f_1} = \sqrt{30 * 21 * 0.40} = 15.9 \text{ mph}$$

Speed after impact of vehicle Two:

$$s_2 = \sqrt{30 * d_2 * f_2} = \sqrt{30 * 24 * 0.35} = 15.9 \text{ mph}$$

The conservation of momentum formula:

$$S_1 * W_1 + S_2 * W_2 = s_1 * W_1 + s_2 * W_2$$

$$S_1 * 1 + 0 * 0.974 = 15.9 * 1 + 15.9 * 0.974$$

$$S_1 = 15.9 + 15.5$$

$$S_1 = 31.4 \text{ or } 31 \text{ mph } [50 \text{ km/hr}]$$

Example 4.15: *While crossing a public street, a golf cart containing the driver only is struck broadside by a 2006 Hyundai Sonata. The vehicles remain together after impact, traveling 96 feet [29.3 m] before coming to rest. The total weights of the vehicles, occupants and cargo are 1107 pounds [502 kg] for the golf cart and 3636 pounds [1649 kg] for the Hyundai. Use 0.65 for the post impact drag factor of the golf cart and 0.30 for the post impact drag factor of the Hyundai. Calculate the impact speed of the Hyundai.*

Solution: Since the much smaller golf cart is moving at low speed we may apply the in-line (one dimensional) conservation of momentum formula. We'll call the Hyundai vehicle one and the golf cart vehicle two. The variables:

$$S_1 = \text{Speed of V-1 at impact}$$
$$S_2 = \text{Speed of V-2 at impact} = 0 \text{ mph}$$
$$s_1 = \text{Speed of V-1 after impact}$$
$$s_2 = \text{Speed of V-2 after impact}$$
$$d_1 = \text{V-1 post impact skid distance} = 96 \text{ feet}$$
$$d_2 = \text{V-2 post impact skid distance} = 96 \text{ feet}$$
$$f_1 = \text{Drag factor of V-1} = 0.30$$
$$f_2 = \text{Drag factor of V-2} = 0.65$$
$$W_1 = \text{Weight of V-1} = 3636 \text{ lb}$$
$$W_2 = \text{Weight of V-2} = 1107 \text{ lb}$$

The weight ratio:

If $W_1 = 1,\ W_2 = 1107/3636 = 0.304$

Where we must be careful to use the weighted average drag factor. If we were to calculate independent slide-to-stop for both vehicles we would, in effect, be mixing momentum and energy and that would throw our results off somewhat. The weighted average drag factor:

$$f = (f_1 * W_1 + f_2 * W_2)/(W_1 + W_2)$$
$$f = (0.30*1 + 0.65*0.304)/(1 + 0.304)$$
$$f = 0.397$$

Now we can use the post impact skid distance and the basic skid formula. For both vehicles:

$$s_1 = \sqrt{30 * d * f} = \sqrt{30 * 96 * 0.397} = 33.8 \text{ mph}$$

The conservation of momentum formula:

$$S_1 * W_1 + S_2 * W_2 = s_1 * W_1 + s_2 * W_2$$
$$S_1 * 1 + 0*0.304 = 33.8*1 + 33.8*0.304$$
$$S_1 = 44.1 \text{ or } 44 \text{ mph [71 km/hr]}$$

Example 4.16: *In crash test MN8A-0, a 2467 lb [1119 kg] bogey vehicle impacted a portable stop sign at 63.0 mph [101.4 km/hr] and suffered a delta V of 4.0 mph [6.5 km/hr]. Determine the kinetic energy lost by the vehicle during the impact.*

Solution: Solution of this problem has four steps. First we subtract off speed loss (ΔV) from impact speed (S) to determine post impact speed:

$$s = S - \Delta V = 63.0 - 4.0 = 59.0 \text{ mph}$$

Second, we must convert the speeds of the vehicles from miles per hour to velocities feet per second.

Impact velocity of the vehicle: $V = 63.0*1.467 = 92.4$ ft/sec

Post-impact velocity of the vehicle: $v = 59.0*1.467 = 86.6$ ft/sec

Next we determine the kinetic energy of the vehicle before and after impact:

Impact kinetic energy: $E = 0.5*(W/g)*V^2 = 0.5*(2467/32.2)*92.4^2$

$= 0.5*(76.6)*8538 = 327{,}000$ ft*lb

Post-impact kinetic energy: $e = 0.5*(W/g)*v^2 = 0.5*(2467/32.2)*86.6^2$

$= 0.5*(76.6)*7500 = 287{,}000$ ft*lb

The loss of energy from the impact:

$$\Delta E = E - e = 327{,}000 - 287{,}000 = 40{,}000 \text{ ft*lb [54,000 Joules]}$$

Example 4.17: *A 2700 lb [kg] auto strikes a temporary stop sign. It then skids 75 feet [22.9 m] (drag factor = 0.70) through a construction site and then strikes a one-foot diameter fixed concrete pole head-on and stops. This results in 34 inches [86 cm] of maximum crush to the front of the vehicle. Use the data from Example E4.16 to estimate the speed/energy dissipated from the impact with the stop sign. Use one of the following equations to determine impact speed with the pole:*

S in mph, C_{MAX} in inches: V-1 front: $ebs = 1.05*C_{MAX} + 4$

S in km/hr, C_{MAX} in cm: V-1 front: $ebs = 0.67*C_{MAX} + 6$

Calculate the speed of the vehicle at impact with the stop sign.

Solution: The variables:

S = Impact speed with stop sign (start of accident)

s_p = Impact speed with pole

s_S = Equivalent energy speed from impact with stop sign

C_{MAX} = Maximum crush depth = 34 inches

d = Skid distance = 75 feet

f = Drag factor of auto = 0.70

ΔE = Energy loss from impact with stop sign = 40,000 ft*lb

W = Weight of the vehicle = 2700 lb

The auto/pole impact speed, from the given equation:

$$s_p = 1.05*C_{MAX} + 4 = 1.05*34 + 4 = 39.7 \text{ mph}$$

Equivalent energy speed from impact with stop sign:

$$s_s = \sqrt{30*\Delta E/W} = \sqrt{30*40000/2700} = 21.1 \text{ mph}$$

The initial speed portion of this problem can be solved by using the combined speed formula.

$$S = \sqrt{S_1^2 + 30*d*f + S_S^2}$$

$$S = \sqrt{39.7^2 + 30*75*0.70 + 21.1^2}$$

$$S = \sqrt{1576 + 1575 + 445}$$

$$S = 59.96 \text{ or } 60 \text{ mph [97 km/hr]}$$

Example 4.18: *A car (V-1) makes a left turn into the path of a pickup truck (V-2). The pickup sees the car turning in front of him but can not avoid striking the rear quarter panel of the car. The impact spins the car out in a clockwise fashion. Use the following data to calculate the impact speed of both vehicles and the speed of the pickup truck at the start of precrash skid.*

	V-1	*V-2*
Curb weight:	*3875 lb [1758 kg]*	*4550 lb [2064 kg]*
Driver weight:	*117 lb [53 kg]*	*217 lb [98 kg]*
Cargo weight:	*65 lb [29 kg]*	*120 lb [54 kg]*
Precrash skid dist.:	———	*53 ft [16.2 m]*
Precrash skid drag factor:	———	*0.70*
Direction at impact:	*S 75° E*	*North*
Direction after impact:	*E 60° N*	*N 5° W*
Distance after impact:	*41 ft [12.5 m]*	*45 ft [13.7 m]*
Drag factor after impact:	*0.55*	*0.70*

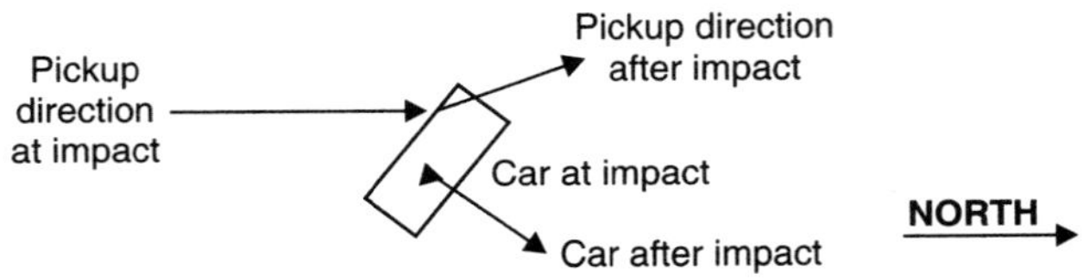

Solution: We have a conservation of momentum problem on our hands. The variables

d_1 = Post impact travel distance of V-1 = 41 feet

d_2 = Post impact travel distance of V-2 = 45 feet

d_{2PRE} = Preimpact skid distance of V-2 = 53 feet

f_1 = Post impact drag factor of V-1 = 0.55

f_2 = Post impact drag factor of V-2 = 0.70

f_{2PRE} = Preimpact braking drag factor of V-2 = 0.70

S = Speed of V-2 at start of precrash braking

S_1 = Speed of V-1 at impact

S_2 = Speed of V-2 at impact

s_1 = Speed of V-1 after impact

s_2 = Speed of V-2 after impact

W_1 = Weight of V-1 = 3875 + 117 + 65 = 4057 1b

W_2 = Weight of V-2 = 4550 + 217 +120 = 4887 1b

α_1 = Travel angle of vehicle one at impact

α_2 = Travel angle of vehicle two at impact

β_1 = Travel angle of vehicle one after impact

β_2 = Travel angle of vehicle two operator after impact

We will start setting by assuming that North is our 0 degree line. All angles measured counter clockwise will be considered positive. We will tabulate our angles and then determine the trigonometric values:

	Direction	θ	$\sin\theta$	$\cos\theta$
α_1:	S 75° E	255°	–0.966	–0.259
α_2:	North	0°	0	1
β_1:	E 60° N	330°	–0.500	0.866
β_2:	N 5° W	5°	0.087	0.996

The weight ratios:

If $W_1 = 1,\ W_2 = 4887/4057 = 1.205$

Post impact speed of V-1:

$$s_1 = \sqrt{30 * d_1 * f_1} = \sqrt{30 * 41 * 0.55} = 26.0 \text{ mph}$$

Post impact speed of V-2:

$$s_2 = \sqrt{30 * d_2 * f_2} = \sqrt{30 * 45 * 0.70} = 30.7 \text{ mph}$$

The conservation of momentum formula along east/west axis:

$$S_1 * W_1 * \sin\alpha_1 + S_2 * W_2 * \sin\alpha_2 = S_1 * W_1 * \sin\beta_1 + s_2 * W_2 * \sin\beta_2$$

$$S_1 * -0.966 + S_2 * 1.205 * 0 = 26.0 * 1 * -0.500 + 30.7 * 1.205 * 0.087$$

$$-0.966 * S_1 = -13.0 + 3.2 = -9.8$$

$$S_1 = 10.1 \text{ or } 10 \text{ mph [16 km/hr]}$$

The conservation of momentum along the north-south axis:

$$S_1 * W_1 * \cos\alpha_1 + S_2 * W_2 * \cos\alpha_2 = S_1 * W_1 * \cos\beta_1 + s_2 * W_2 * \cos\beta_2$$

$$10.1 * -0.259 + S_2 * 1.205 * 1 = 26.0 * 1 * 0.866 + 30.7 * 1.205 * 0.996$$

$$-2.61 + 1.205 * S_2 = 22.5 + 39.3 = 64.4$$

$$1.205 * S_2 = 61.8 + 2.61 = 64.4$$

$$S_2 = 53.4 \text{ or } 53 \text{ mph [85 km/hr]}$$

The speed of V-2 at the start of precrash braking:

$$S = \sqrt{S_2^2 + 30 * d_{2PRE} * f_{2PRE}}$$

$$S = \sqrt{53.4^2 + 30 * 53 * 0.70}$$

$$S = \sqrt{2852 + 1113} = \sqrt{3965}$$

$$S = 62.96 \text{ or } 63 \text{ mph [101 km/hr]}$$

Example 4.19: *A 2006 Ford F-250 pickup (V-1) collides head-on with 2002 Ford Focus (V-2). Neither vehicle moves significantly after impact. The F-250 has an average of 8.0 inches [20.3 cm] of crush on the top of its front bumper. V-2 has an average of 23.5 inches [59.7 cm] of crush on the top of its front bumper. The weights are 6408 lb [2907 kg] for the F-250 and 3425 lb [1554 kg] for the Focus. Use the following Campbell equations:*

S in mph, C_{AVG} in inches: V-l front: $ebs = 1.53*C_{MAX} + 7$

V-2 front: $ebs = 1.68*C_{MAX} + 7$

S in km/h, C_{AVG} in cm: V-l front: $ebs = 0.97C_{MAX} + 11$

V-2 front: $ebs = 1.06*C_{MAX} + 11$

Calculate the impact speeds of both vehicles.

Solution:

C_{AVG1} = Average crush depth to V-l front = 8.0 inches

C_{AVG2} = Average crush depth to V-2 front = 23.5 inches

ebs_1 = Equivalent barrier speed of V-l

ebs_2 = Equivalent barrier speed of V-2

s_1 = Speed of V-l after impact = 0 mph

s_2 = Speed of V-2 after impact = 0 mph

S_1 = Speed of V-l at impact

S_2 = Speed of V-2 at impact

W_1 = Weight of V-l = 6408 lb

W_2 = Weight of V-2 = 3425 lb

Weight Ratios:

If $W_1 = 1$, $W_2 = 3425/6408 = 0.534$

Equivalent barrier speed of V-l crush:

$$ebs_1 = 1.54*C_{1AVG} + 7 = 1.54*8.0 + 7 = 19.3 \text{ mph}$$

Equivalent barrier speed of V-2 crush:

$$ebs_2 = 1.68*C_{2AVG} + 7 = 1.68*23.5 + 7 = 46.5 \text{ mph}$$

We now substitute into the in-line conservation of momentum formula. Note that term for V-2's impact momentum is negative since it is heading in the opposite direction of V-1.

$$S_1*W_1 + S_2*W_2 = s_1*W_1 + s_2*W_2$$

$$S_1*1 - S_2*0.534 = 0*1 + 0*0.534$$

$$S_1 - S_2*0.534 = 0$$

$$S_1 = S_2*0.534$$

Substituting and solving the dissipation of energy equation:

$$S_1^2 * W_1 + S_2^2 * W_2 = s_1^2 * W_1 + s_2^2 * W_2 + ebs_1^2 * W_1 + ebs_2^2 * W_2$$

$$S_1^2 * 1 + S_2^2 * 0.534 = 0^2 * 1 + 0^2 * 0.534 + 19.3^2 * 1 + 46.5^2 * 0.534$$

$$S_1^2 + S_2^2 * 0.534 = 372 + 1155 = 1527$$

Substituting S_2*0.534 for S_1:

$$(0.534 * S_2)^2 + S_2^2 * 0.534 = 1527$$

$$0.258 * S_2^2 + S_2^2 * 0.534 = S_2^2 * 0.819 = 1527$$

$$S_2 = \sqrt{1864} = 43.17 \text{ or } 43 \text{ mph [69 km/h]}$$

Substituting this into the momentum equation:

$$S_1 = S_2 * 0.534 = 43.17 * 0.534 = 23.05 \text{ or } 23 \text{ mph [37 km/h]}$$

Example 4.20: *A 2003 Chevrolet Trailblazer, driven by an intoxicated driver, is approaching a T intersection. The driver fails to notice the stop sign in time. The truck brakes hard for 59 feet [18.0 m] on the asphalt road surface, 10 feet [3.0 m] grassy dirt surface, and collides head-on with a concrete retaining wall. The front of the truck is later measured and found to have an average crush of 21.7 inches [55 cm]. The drag factors were 0.80 for the asphalt surface and 0.48 for the dirt/grass surface. Estimate both the truck's impact speed and its speed at the start of precrash skid. The Trailblazer was crash tested at 29.7 and only suffered 9.0 inches of crush.*

Solution: We will derive the Campbell equation and then use it to estimate the impact speed into the brick wall. We will then use the combined speed formula to estimate speed at the start of skid. The Trailblazer was crash tested at 29.7 and only suffered 9.0 inches of crush.

The Campbell model, in equation form:

$$ebs = b_1 * C_{AVG} + b_0$$

where ebs = Equivalent barrier speed = 29.7 mph

b_1 = Stiffness constant, mph/in

C_{AVG} = Average crush depth = 9.0 inches

b_0 = Damage threshold constant, given 7 mph in this problem

The equation can also be generated for metric units. The b and b_0 would have to be different than the imperial unit equation.

Assume $b_0 = 7$ mph $ebs = b_1 * C_{AVG} + b_0$

$$29.7 = b_1 * 9.0 + 7$$

$$22.7 = b_1 * 9.0$$

$$2.52 = b_1$$

Therefore, the equation:

$$ebs = 2.52 * C_{AVG} + 7$$

Since the car hit a fixed object and stopped dead, ebs equals impact speed:

$$S_1 = 2.52 * C_{AVG} + 7 = 2.52 * 16.7 + 7 = 49.1 \text{ mph}$$

The speed at the start of skid can be solved by using the combined speed formula. The variables:

S_1 = Speed of V-l at impact = 49.1 mph

d_A = Skid distance on asphalt = 59 feet

f_A = Drag factor on asphalt = 0.80

d_G = Skid distance on grass/dirt = 10 feet

f_G = Drag factor on grass/dirt = 0.48

$$S = \sqrt{S_1^2 + 30 * d_A * f_A + 30 * d_G * f_G}$$

$$S = \sqrt{49.1^2 + 30 * 59 * 0.80 + 30 * 10 * 0.48}$$

$$S = \sqrt{2411 + 1416 + 144} = \sqrt{3971}$$

63.0 or 63 mph [101 km/h]

Example 4.21: *2900-pound [1315 kg] V-2 is stopped at a traffic signal when it is struck in the rear by 3330-pound [1510 kg] eastbound V-l. After impact V-l travelled east 53 feet [16.1 m] at an average drag factor of 0.38. After impact V-2 travelled east 28 feet [8.5 m] at an average drag factor of 0.72. Calculate the impact speed of V-l.*

Solution: This is a simple application of the in-line (one dimensional) conservation of momentum formula. The variables:

d_1 = V-l post impact skid distance = 53 feet

d_2 = V-2 post impact skid distance = 28 feet

f_1 = Drag factor of V-l = 0.38

f_2 = Drag factor of V-2 = 0.72

s_1 = Speed of V-l after impact

s_2 = Speed of V-2 after impact

S_1 = Speed of V-l at impact

S_2 = Speed of V-2 at impact = 0 mph

W_1 = Weight of V-l = 3330 lb.

W_2 = Weight of V-2 = 2900 lb.

The weight ratio: If $W_1 = 1$, $W_2 = 2900/3330 = 0.879$

To find speeds after impact, we use the post impact skid distance and the basic skid formula. For Vehicle One:

$$s_1 = \sqrt{30 * d_1 * f_1} = \sqrt{30 * 53 * 0.38} = 24.6 \text{ mph}$$

Speed after impact of Vehicle Two:

$$s_2 = \sqrt{30 * d_2 * f_2} = \sqrt{30 * 34 * 0.72} = 24.6 \text{ mph}$$

The conservation of momentum formula:

$$S_1 * W_1 + S_2 * W_2 = s_1 * W_1 + s_2 * W_2$$

$$S_1 * 1 + 0 * 0.879 = 24.6 * 1 + 24.6 * 0.879$$

$$S_1 = 24.6 + 21.6$$
$$S_1 = 46.2 \text{ or } 46 \text{ mph [74 km/h]}$$

Example 4.22: *Vehicle 1 is traveling west bound at 56.32 km/h (35 mph) and vehicle 2 is south bound traveling at 45.05 km/h (28 mph). They collide at an intersection. After impact vehicle 1 is traveling W 20° S at 43.44 km/h (27 mph). After impact vehicle 2 is traveling W 55° S at 32.18 km/h (20 mph). Determine the delta V's of the two vehicles.*

Weight of vehicle 1 = 1587.60 kg (3500 lbs)

Weight of vehicle 2 = 1270.08 kg (2800 lbs)

Solution:

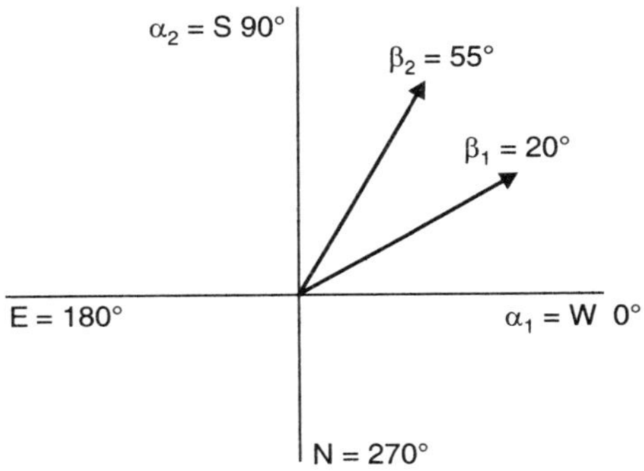

Fig. E4.22 Vehicle directions.

Vehicle 1 speed /direction at impact: West @ 56.32 km/h

Vehicle 1 speed /direction after impact: W 20° S @ 43.44 km/h

Breaking post impact movements into components:

West component:	$43.44 \cos 20° = 40.82$
South component:	$43.44 \sin 20° = 14.86$
Delta V West:	$56.32 - 40.82 = 15.5$ km/h (speed loss)
Delta V South:	$0 + 14.86 = 14.86$ km/h (speed gain)
V-1 delta V:	$\sqrt{15.5^2 + 14.86^2} = 21.47$ km/h (13.35 mph)

V-2 speed/direction at impact: South @ 45.05 km/h

V-2 speed/direction after impact: W 55° S @ 32.18 km/h

Breaking post impact movements into components:

West component:	$32.18 \cos 55° = 18.46$
South component:	$32.18 \sin 55° = 26.36$
Delta V West:	$0 + 18.46 = 18.46$ km/h (11.47 mph) speed gain
Delta V South:	$45.05 - 26.36 = 18.69$ km/h (11.62 mph) speed loss
V-2 delta V:	$\sqrt{18.46^2 + 18.69^2} = 26.27$ km/h (16.33 mph)

$$\frac{V-1\text{Delta V}}{V-2\text{ Delta V}} = \frac{21.47}{26.27} = 0.8173$$

$$\frac{w_2}{w_1} = \frac{1270.08}{1587.60} = 0.80$$

Example 4.23: *Vehicle 2 is slowing for a red traffic signal when it is struck squarely in the rear by vehicle 1. After impact, both vehicles were traveling 64.36 km/h (40 mph). Vehicle 2 received a delta V of 32.18 km/h (20 mph). Vehicle 2 received a delta V of 27.35 km/h (17 mph). Determine the impact speeds of the two vehicles.*

Weight of vehicle 1 = 2268 kg (5000 lbs)

Weight of vehicle 2 = 1587.60 kg (3500 lbs)

Solution: Since vehicle 2 was moving 64.36 km/h (40 mph) after impact and received a speed gain of 27.35 km/h (17 mph) from the rear impact, its impact speed was

$$S_2 + 27.35 = 64.36$$

or

$$S_2 = 37.01 \text{ km/h } (16.79 \text{ mph})$$

Here, we use the fact that the delta V's are going to be inversely proportional to the vehicle masses. Hence, vehicle 1's delta V was

$$\frac{\Delta V \text{ of vehicle 1}}{\Delta V \text{ of vehicle 2}} = \frac{\text{weight of vehicle 2}}{\text{weight of vehicle 1}}$$

$$\frac{\Delta V \text{ of vehicle 1}}{27.35} = \frac{1587.60}{2268} = 0.70$$

or

$$\Delta V \text{ of vehicle 1} = 0.70\,(27.35) = 19.15 \text{ km/h } (11.90 \text{ mph})$$

We now solve for vehicle 1's impact speed. Since it was going 64.36 km/h (40 mph) after impact and lost 19.15 km/h (11.90 mph), its impact speed was

$$S_2 - 19.15 = 64.36$$

or

$$S_2 = 83.51 \text{ km/h } (51.90 \text{ mph})$$

Example 4.24: *A 2006 Ford F-250 pickup (V-1) collides head-on with 2002 Ford Focus (A-2). Neither vehicle moves significantly after impact. The weights are 6408 lb [2907 kg] for the F-250 and 3425 lb [1554 kg] for the Focus. Use the averages of the following front crush data (various heights), as well as the following Campbell equations:*

V-1 Avg. Crush Depth:	*Hood:*	*0 inches (0 cm)*
	Top Bumper:	*4.3 inches (10.9 cm)*
	Bottom Bumper:	*7.2 inches (18.3 cm)*
V-2 Avg. Crush Depth:	*Hood:*	*41.9 inches (106.4 cm)*
	Top Bumper:	*20.9 inches (53.1 cm)*
	Bottom Bumper:	*17.3 inches (43.9 cm)*
S in mph, C_{AVG} in inches:	*V-1 front:*	*ebs = 1.54*C_{MAX} + 7*
	V-2 front:	*ebs = 1.68*C_{MAX} + 7*

S in km/h, C_{AVG} in cm: *V-1 front:* $ebs = 0.97*C_{MAX} + 11$

V-2 front: $ebs = 1.06*C_{MAX} + 11$

Calculate the impact speeds of both vehicles.

Solution: We will ultimately solve this problem using conservation of momentum and dissipation of energy. The parameters:

C_{AVG1} = Avg. crush depth to V-1 = (0 + 4.3 + 7.2)/3 = 3.8 inches

C_{AVG2} = Avg. crush depth to V-2 = (41.9 + 20.9 + 17.3)/3 = 26.7 inches

ebs_1 = Equivalent barrier speed of V-1

ebs_2 = Equivalent barrier speed of V-2

s_1 = Speed of V-1 after impact = 0 mph

s_2 = Speed of V-2 after impact = 0 mph

S_1 = Speed of V-1 at impact

S_2 = Speed of V-2 at impact

W_1 = Weight of V-1 - 6408 lb.

W_2 = Weight of V-2 = 3425 lb.

Weight Ratios: If $W_1 = 1$, $W_2 = 3425/6408 = 0.534$

Equivalent barrier speed of V-1 crush:

$$ebs_1 = 1.54*C_{1AVG} + 7 = 1.54*3.8 + 7 = 12.9 \text{ mph}$$

Equivalent barrier speed of V-2 crush:

$$ebs_2 = 1.68*C_{2AVG} + 7 = 1.68*26.7 + 7 = 51.9 \text{ mph}$$

We now substitute into the in-line conservation of momentum formula. Note that term for V-2's impact momentum is negative since it is heading in the opposite direction of V-1.

$$S_1*W_1 + S_2*W_2 = s_1*W_1 + s_2*W_2$$

$$S_1*1 - S_2*0.534 = 0*1 + 0*0.534$$

$$S_1 - S_2*0.534 = 0$$

$$S_1 = S_2*0.534$$

Substituting and solving the dissipation of energy equation:

$$S_1^2*W_1 + S_2^2*W_2 = s_1^2*W_1 + s_2^2*W_2 + ebs_1^2*W_1 + ebs_2^2*W_2$$

$$S_1^2*1 + S_2^2*0.534 = 0^2*1 + 0^2*0.534 + 12.9^2*1 + 51.9^2*0.534$$

$$S_1^2 + S_2^2*0.534 = 166 + 1438 = 1604$$

Substituting $S_2*0.534$ for S_1:

$$(0.534*S_2)^2 + S_2^2*0.534 = 1604$$

$$0.285*S_2^2 + S_2^2*0.534 = S_2^2*0.819 = 1604$$

$$S_2 = \sqrt{1958} = 44.25 \text{ or } 44 \text{ mph [71 km/h]}$$

Substituting this into the momentum equation:

$$S_1 - S_2*0.534 = 44.25*0.534 = 23.63 \text{ or } 24 \text{ mph [39 km/h]}$$

This scenario was NHTSA Crash Test #5683. The actual impact speeds were 24.0 mph for V-1 and 43.9 mph for V-2. The above Campbell equations were derived from the actual barrier tests for these year/make/model vehicles. The approach of averaging crush depths measured at top and bottom of bumper plus the hood, yielded much more accurate results for the partial underride impact than the two approaches evaluated.

Example 4.25: *While crossing a public street, a golf cart containing the driver only is struck broadside by a 2004 Chevrolet Cavalier. The vehicles remain together after impact, travelling 87 feet [26.5 m] before coming to rest. The total weights of the vehicles, occupants and cargo are 1107 pounds [502 kg] for the golf cart and 2953 pounds [1340 kg] for the Cavalier. Use 0.65 for the post impact drag factor of the golf cart and 0.30 for the post impact drag factor of the Cavalier. Calculate the impact speed of the Cavalier.*

Solution: Since the much smaller golf cart is moving at low speed we may apply the in-line (one dimensional) conservation of momentum formula. We'll call the Cavalier Vehicle One and the golf cart Vehicle Two. The variables:

d_1 = V-l post impact skid distance = 18 feet

d_2 = V-2 post impact skid distance = 22 feet

f_1 = Drag factor of V-l = 0.30

f_2 = Drag factor of V-2 = 0.65

s_1 = Speed of V-l after impact

S_1 = Speed of V-l at impact

s_2 = Speed of V-2 after impact

S_2 = Speed of V-2 at impact – 0 mph

W_2 = Weight of V-2 = 2700 lb.

W_1 = Weight of V-l = 3800 lb.

The weight ratio: If $W_1 = 1$, $W_2 = 1107/2953 = 0.375$

Where we must be careful to use the weighted average drag factor. If we were to calculate independent slide-to-stop for both vehicles we would, in effect, be mixing momentum and energy and that would throw our results off somewhat. The weighted average drag factor:

$$f = (f_1 * W_1 + f_2 * W_2)/(W_1 + W_2)$$

$$f = (0.30*1 + 0.65*0.375)/(1 + 0.375)$$

$$f = 0.395$$

Now we can use the post impact skid distance and the basic skid formula. For both vehicles:

$$s_1 = \sqrt{30 * d * f} = \sqrt{30 * 87 * 0.38} = 32.1 \text{ mph}$$

The conservation of momentum formula:

$$S_1 * W_1 + S_2 * W_2 = s_1 * W_1 + s_2 * W_2$$

$$S_1 * 1 + 0 * 0.375 = 32.1 * 1 + 32.1 * 0.375$$

$$S_1 = 44.1 \text{ or } 44 \text{ mph [71 km/hr]}$$

Example 4.26: *A 3650-lb [1656 kg] V-1 collides head-on with 3100-lb [1406 kg] V-2. Both vehicles bounce back a negligible distance and stop. The Campbell equations and crash measurements for the vehicles:*

	V-1	V-2
Campbell Eq. (mph, inch)	$ebs = 1.32*C_{AVG} + 7$	$ebs = 1.41*C_{AVG} + 7$
Campbell Eq. (km/h, cm)	$ebs = 0.84*C_{AVG} + 11$	$ebs = 0.89*C_{AVG} + 11$
C_1	20" [51 cm]	17" [43 cm]
C_2	21.5" [55 cm]	18" [46 cm]
C_3	22" [56 cm]	18.5" [47 cm]
C_4	22.5" [57 cm]	18" [46 cm]
C_5	22" [56 cm]	16.5" [42 cm]
C_6	18" [46 cm]	15" [38 cm]

Determine the impact speeds of both vehicles.

Solution: We will ultimately solve this problem using conservation of momentum and dissipation of energy. The parameters:

S_1 = Speed of V-1 at impact

s_1 = Speed of V-1 after impact = 0 mph

S_2 = Speed of V-2 at impact

s_2 = Speed of V-2 after impact = 0 mph

W_1 = Weight of V-1 = 3650 lb.

W_2 = Weight of V-2 = 3100 lb.

C_{AVE1} = Average crush depth to V-1 front

C_{AVE2} = Average crush depth to V-2 rear

ebs_1 = Equivalent barrier speed of V-1

ebs_2 = Equivalent barrier speed of V-2

Weight Ratios: If $W_1 = 1$, $W_2 = 3100/3650 = 0.849$

We must estimate the energies dissipated by crush via determination of their equivalent barrier impact speeds. Average crush to the front of Vehicle One:

$$C_{1avg} = (C_1/2 + C_2 + C_3 + C_4 + C_5 + C_6/2)/5$$

$$C_{1avg} = (20/2 + 21.5 + 22 + 22.5 + 22 + 18/2)/5 = 21.4 \text{ inches}$$

Equivalent barrier speed:

$$ebs_1 = 1.32*C_1 + 7 = 1.32*21.4 + 7 = 35.2 \text{ mph}$$

Average crush to the front of Vehicle Two:

$$C_{2avg} = (17/2 + 18 + 18.5 + 18 + 16.5 + 15/2)/5 = 17.4 \text{ inches}$$

Crush energy equivalent speed:

$$ebs_2 = 1.41*C_{2avg} + 7 = 1.41*17.4 + 7 = 31.5 \text{ mph}$$

We now substitute into the in-line conservation of momentum formula. Note that term for V-2's impact momentum is negative since it is heading in the opposite direction of V-1.

$$S_1*W_1 + S_2*W_2 = s_1*W_1 + s_2*W_2$$

$$S_1{*}1 - S_2{*}0.849 = 0{*}1 + 0{*}0.849$$
$$S_1 - S_2{*}0.849 = 0$$
$$S_1 = S_2{*}0.849$$

Substituting and solving the dissipation of energy equation:

$$S_1^{2}{*}W_1 + S_2^{2}{*}W_2 = s_1^{2}{*}W_1 + s_2^{2}{*}W_2 + \text{ebs}_1^{2}{*}W_1 + \text{ebs}_2^{2}{*}W_2$$
$$S_1^{2}{*}1 + S_2^{2}{*}0.849 = 0^{2}{*}1 + 0^{2}{*}0.849 + 35.2^{2}{*}1 + 31.5^{2}{*}0.849$$
$$S_1^{2} + S_2^{2}{*}0.849 = 1239 + 842 = 2081$$

Substituting S_2*.849 for S_1:

$$(0.849{*}S_2)^2 + S_2^{2}{*}0.849 = 2081$$
$$0.721{*}S_2^{2} + S_2^{2}{*}0.849 = S_2^{2}{*}1.570 = 2081$$
$$S_2 = \sqrt{1326} = 36.4 \text{ or } 36 \text{ mph [59 km/hr]}$$

Substituting this into the momentum equation

$$S_1 = S_2{*}0.849 = 36.4{*}0.849 = 30.9 \text{ or } 31 \text{ mph [50 km/hr]}$$

Example 4.27: *A 2000 Nisson Maxima skids 47 feet [14.3 m] on dry asphalt (drag factor = 0.82) and strikes the side of a stationary train blocking a grade crossing. The impact is to one of the solid iron 'trucks' (i.e., wheels, suspension) of a gondola car in the train. The ford rebounds a short distance and comes to rest. The front crush profile of the Ford was measured and compared to a NHTSA front barrier crash test:*

	NHTSA crash test	*Grade X-ing accident*
Impact speed:	*35.2 mph [56.4 km/h]*	
C_1	*18.3″ [46.5 cm]*	*20.7″ [52.6 cm]*
C_2	*20.0″ [50.8 cm]*	*27.9″ [70.9 cm]*
C_3	*20.0″ [50.8 cm]*	*25.5″ [64.8 cm]*
C_4	*20.5″ [52.1 cm]*	*25.5″ [64.8 cm]*
C_5	*20.5″ [52.1 cm]*	*23.3″ [59.2 cm]*
C_6	*18.4″ [46.7 cm]*	*20.5″ [52.1 cm]*

Treat the train as a fixed rigid barrier. Derive the Campbell equation for the Ford and use it to calculate its impact speed. Also, determine the Ford's speed at the start of its precrash skid.

Solution: We will ultimately solve this problem using the Campbell equation and the combined speed formula. We will use the following variables:

C_{AVG} = Average crush depth to Nissan front, inches
d = Precrash skid distance of Nissan = 47 feet
ebs = Equivalent barrier speed of Nissan, mph
f = Drag factor of Nissan during precrash skid = 0.82

We must estimate the energies dissipated by crush via determination of the Nissan's equivalent barrier impact speed. First we must calibrate the Campbell equation from the NHTSA NCAP crash test. Average crush to the front of the 2000 Nissan Maxima tested by NHTSA at 35.2 mph:

$$C_{AVG} = (C_1/2 + C_2 + C_3 + C_4 + C_5 + C_6/2)/5$$

$$C_{AVG} = (18.3/2 + 20.0 + 20.0 + 20.5 + 20.5 + 18.4/2)/5 = 19.9 \text{ inches}$$

Assume:

$$b_0 = 7 \text{ mph}$$
$$ebs = b_1{*}C_{AVG} + b_0$$
$$35.2 = b_1{*}19.9 + 7$$
$$28.2 = b_1{*}19.9$$
$$1.42 = b_1$$

The Campbell equation for a 2000 Nissan Maxima:

$$ebs = 1.42{*}C_{AVG} + 7$$

Average crush to the front of the Nissan from the train impact:

$$C = (20.7/2 + 27.3 + 25.5 + 25.5 + 23.3 + 20.5/2)/5 = 24.6 \text{ inches}$$

Since we are treating the stationary train as a fixed rigid barrier, and the Ford struck it head-on with negligible post impact movement, equivalent barrier speed will equal the car's impact speed (*S*).

$$S_1 = ebs = 1.42{*}C_{AVG} + 7 = 1.42{*}24.6 + 7 = 41.9 \text{ or } 42 \text{ mph [67 km/h]}$$

The speed of the car at the start of precrash skid (*S*) is determined by using the combined speed formula.

$$S = \sqrt{S_1^2 + 30 * d * f}$$
$$S = \sqrt{41.9^2 + 30 * 47 * 0.82}$$
$$S = \sqrt{1756 + 1156} = \sqrt{2912}$$
$$S = 53.96 \text{ or } 54 \text{ mph [87 km/h]}$$

Example 4.28: *A 3400-lb [1542 kg] V-1 collides head-on with 2750-lb [1247 kg] V-2. Both vehicles bounce back a negligible distance and stop. The Campbell equations and crush measurements for the vehicles:*

	V-1	*V-2*
Campbell Eq. (mph, inch)	$ebs = 1.32{*}C_{AVG} + 7$	$ebs = 1.41{*}C_{AVG} + 7$
Campbell Eq. (km/h, cm)	$ebs = 0.84{*}C_{AVG} + 11$	$ebs = 0.89{*}0{*}C_{AVG} + 11$
C_1	*30" [76 cm]*	*27" [69 cm]*
C_2	*31.5" [80 cm]*	*28" [71 cm]*
C_3	*32" [81 cm]*	*28.5" [72 cm]*
C_4	*32.5" [83 cm]*	*28" [71 cm]*
C_5	*32" [81 cm]*	*26.5" [67 cm]*
C_6	*28" [71 cm]*	*25.5" [64.8 cm]*

Determine the impact speeds of both vehicles.

Solution: We will ultimately solve this problem using conservation of momentum and dissipation of energy. The parameters:

C_{AVE1} = Average crush depth to V-1 front

C_{AVE2} = Average crush depth to V-2 rear

ebs_1 = Equivalent barrier speed of V-1

ebs_2 = Equivalent barrier speed of V-2

s_1 = Speed of V-1 after impact = 0 mph

S_1 = Speed of V-1 at impact

s_2 = Speed of V-2 after impact = 0 mph

S_2 = Speed of V-2 at impact

W_1 = Weight of V-1 = 3400 lb.

W_2 = Weight of V-2 = 2750 lb.

Weight Ratios: If $W_1 = 1$, $W_2 = 2750/3400 = 0.809$

We must estimate the energies dissipated by crush via determination of their equivalent barrier impact speeds. Average crush to the front of Vehicle One:

$$C_{1avg} = (C_1/2 + C_2 + C_3 + C_4 + C_5 + C_6/2)/5$$
$$C_{1avg} = (30/2 + 31.5 + 32 + 32.5 + 32 + 28/2)/5 = 31.4 \text{ inches}$$

Equivalent barrier speed:

$$\text{ebs}_1 = 1.32*C_{1avg} + 7 = 1.32*31.4 + 7 = 48.4 \text{ mph}$$

Average crush to the front of Vehicle Two:

$$C_{2avg} = (27/2 + 28 + 28.5 + 28 + 26.5 + 25/2)/5 = 27.4 \text{ inches}$$

Crush energy equivalent speed:

$$\text{ebs}_2 = 1.41 * C_{2avg} + 7 = 1.41*27.4 + 7 = 45.6 \text{ mph}$$

We now substitute into the in-line conservation of momentum formula. Note that term for V-2's impact momentum is negative since it is heading in the opposite direction of V-l.

$$S_1*W, + S_2*W_2 = s_1*W_1 + s_2*W_2$$
$$S_1* - S_2*0.809 = 0*1 + 0*0.809$$
$$S_1 - S_2*0.809 = 0 \quad S_1 = S_2*0.809$$

Substituting and solving the dissipation of energy equation:

$$S_1^2*W_1 + S_2^2*W_2 = s_1^2*W, + s_2^2*W_2 + \text{ebs}_1^2*W_1 + \text{ebs}_2^2*W_2$$
$$S_1^2*1 + S_2^2*0.809 = 0^2*1 + 0^2*0.809 + 48.4^2*1 + 45.6^2*0.809$$
$$S_1^2 + S_2^2*0.809 = 2343 + 1682 = 4025$$

Substituting $S_2*0.809$ for S_1:

$$(0.809*S_2)^2 + S_2^2*0.809 = 4025$$
$$0.654*S_2^2 + S_2^2*0.809 = S_2^2* 1.464 = 4025$$

$$S_2 = \sqrt{2749} = 52.44 \text{ or } 52 \text{ mph [84 km/hr]}$$

Substituting this into the momentum equation

$$S_1 = S_2*0.809 = 52.44*0.809 = 42.4 \text{ or } 42 \text{ mph [68 km/hr]}$$

Example 4.29: *An eastbound 2002 Dodge Ram 1500 collides head-on with a westbound 2004 Honda Accord and pushes the Accord straight backwards. Given the data below, determine the impact speeds of both vehicles.*

	Dodge	*Honda*
Curb Weight	*4952 lb [2246 kg]*	*3217 lb [1459 kg]*
Driver Weight	*264 lb [120 kg]*	*211 lb [96 kg]*
Cargo Weight	*300 lb [136 kg]*	*150 lb [68 kg]*
Damage Width	*Entire front*	*Entire front*
Crush Pt. C_1	*3.1" [8 cm]*	*17.4" 44 cm*
Crush Pt. C_2	*3.8" [9.5 cm]*	*25.7 [65 cm*
Crush Pt. C_3	*8.8" [22.5 cm]*	*27.6" [70 cm]*
Crush Pt. C_4	*8.5" [21.5 cm]*	*26.7" [68 cm]*
Crush Pt. C_5	*4.0" [10 cm]*	*24.8" [63 cm]*
Crush Pt. C_6	*7.5" [19 cm]*	*22.4" [59 cm]*
Post Crash Travel	*81′ [24.6 m] east*	*81′ [24.6 m] east*
Drag Factor	*0.50*	*0.50*

Solution: We were given that the Dodge pickup was subjected to a 35.1 mph impact and suffered 18.0 inches of crush. We also learn that the Honda was subjected to a 34.8 mph impact and suffered 20.7 inches of crush. We now derive the Campbell equations for both vehicles:

For the Dodge: $ebs = b_1 * C_{AVG} + b_0$

Assume $b_0 = 7$ mph $35.0 = b_1 * 18.0 + 7$

$$28.0 = b_1 * 18.0$$

$$1.56 = b_1$$

Therefore, the equation for the Dodge:

$$ebs_1 = 1.56 * C_{AVG1} + 7$$

For the Honda: $ebs = b_1 * C_{AVG2} + b_0$

Assume $b_0 = 7$ mph $34.8 = b_1 * 20.7 + 7$

$$27.8 = b_1 * 20.7$$

$$1.34 = b_1$$

Therefore, the equation for the Honda:

$$ebs_2 = 1.34 * C_{AVG2} + 7$$

We will ultimately solve this problem using conservation of momentum and dissipation of energy. The parameters:

C_{AVG1} = Average crush depth to Dodge front

C_{AVG2} = Average crush depth to Honda front

d = Post impact travel distance for both vehicles = 81 feet

ebs_1 = Equivalent barrier speed of V-1

ebs_2 = Equivalent barrier speed of V-2

f = Post impact drag factors for both vehicles = 0.50

s_1 = Speed of the Dodge after impact
s_2 = Speed of the Honda after impact
S_1 = Speed of the Dodge at impact
S_2 = Speed of the Honda at impact
W_1 = Weight of Dodge = 4952 + 264 + 300 = 5516 1b.
W_2 = Weight of Honda = 3217 + 211 + 150 = 3578 lb.

Weight Ratios: If $W_1 = 1$, $W_2 = 3578/5516 = 0.649$

We must estimate the energies dissipated by crush via determination of their equivalent barrier impact speeds. Average crush to the front of Vehicle One:

$$C_{AVG1} = (C_1/2 + C_2 + C_3 + C_4 + C_5 + C_6/2)/5$$
$$C_{AVG2} = (3.1/2 + 3.8 + 8.8 + 8.5 + 4.0 + 7.5/2)/5 = 6.1 \text{ inches}$$

Equivalent barrier speed:

$$\text{ebs}_1 = 1.56{*}\text{C}_{AVG1} + 7 = 1.56{*}6.1{+}7 = 16.5 \text{ mph}$$

Average crush to the front of Vehicle Two:

$$C_{AVG2} = (17.9/2 + 25.7 + 27.6 + 26.7 + 24.8 + 22.4/2)/5 = 249 \text{ inches}$$

Crush energy equivalent speed:

$$\text{ebs}_2 = 1.34{*}\text{C}_{AVG2} + 7 = 1.34{*}24.9 + 7 = 40.4 \text{ mph}$$

Speeds after impact:

$$s = \sqrt{30 * d * f} = \sqrt{30*81*0.5} = 40.4 \text{ mph}$$

We now substitute into the in-line conservation of momentum formula. Note that term for V-2's impact momentum is negative since it is heading in the opposite direction of V-l.

$$S_1{*}W_1 + S_2{*}W_2 = s_1{*}W_1 + s_2{*}W_2$$
$$S_1{*}1 - S_2{*}0.649 = 34.6{*}1 + 34.6{*}0.649$$
$$S_1 - S_2{*}0.649 = 57.0$$
$$S_1 = S_2{*}1.069 + 57.0$$

Substituting and solving the dissipation of energy equation:

$$S_1^2{*}W_1 + S_2^2{*}W_2 = s_1^2{*}W_1 + s_2^2{*}W_2 + \text{ebs}_1^2{*}W_1 + \text{ebs}_2^2{*}W_2$$
$$S_1^2{*}1 + S_2^2{*}0.649 = 34.6^2{*}1 + 34.6^2{*}0.649 + 16.5^2{*}1 + 40.4^2{*}0.649$$
$$S_1^2 + S_2^2{*}0.649 = 1497 + 778 + 272 + 1059 = 3306$$

If you solve both momentum and energy equations by trial and error, starting with $S_2 = 0$, S_1 = 57.0 mph for the momentum equation and 57.5 mph for the energy equation. This tells us V-1 was travelling approximately 57 mph at impact, while V-2 was stopped on the verge of stopping at impact.

In fact this problem was based on a NHTSA crash test in which a stopped Honda Accord was struck head-on by a Dodge Ram 1500 travelling 57.4 mph.

4.15 SUMMARY

Each and every vehicle collision reconstruction is quite unique and no two collisions are the same. Accident reconstruction of these vehicle collisions often requires the use of wide variety of methodologies and software based on the variation of physical evidence and accident investigative information. There is an ample room for ingenuity and insight for the application of methods and software. Each collision case needs to be evaluated on its own merits, although it is not uncommon to have similarities between some cases.

4.16 REFERENCES

Anon., "CRASH3 User's Guide and Technical Manual", DOT Report HS 805 732, NHTSA, Washington, DC, 1981.

Anon., *Automobile Handbook*, 2nd ed., Robert Bosch GmbH, Stuttgart, 1994.

Anon., "Laboratory Testing Machines for Measuring the Steady State Force and Moment Properties of Passenger Car Tyres", SAE J1107, Society of Automotive Engineers, Inc., Warrendale, PA, September 1975.

Anon., "CRASH3 User's Guide and Technical Manual", DOT Report DOT HS 805 732, Washington, DC, February 1981.

Anon., *Road Vehicles-Steady State Circular Test Procedure*, ISO 4138-1982 (E), International Organization for Standardization, 1982.

Anon., "Proposed Passenger Car and Light Truck Directional Control Response Test Procedures", SAE XJ 266, Society of Automotive Engineers, Warrendale, PA, 1985.

Anon., *Road Vehicles-Lateral Transient Response Test Methods,* ISO 7401-1988, International Organization for Standardization, 1988.

Anon., *Vehicle Dynamics Terminology,* SAE J670e, Society of Automotive Engineers, Warrendale, PA, July 1976. (See Appendix-E).

Araszewski, M., Toor, A., Overgaard, R. and Johal., R., "Lane Change Maneuver Modeling for Accident Reconstruction Applications", Paper No. 2002-01-0817,SAE International, Warrendale, PA, 2002.

Asay, A.F., Jewkes, D.B., and **Woolley, R.L.,** "Narrow Object Impact Analysis and Comparison with Flat Barrier Impacts", Paper 2002-01-0552, SAE International, Warrendale, PA, 2002.

Atkinson, D.R., Hennessy, C.J., and **Watts, A.J.,** "Low Speed Automobile Accidents and Occupant Kinematics, Dynamics, and Biomechanics", Lawyers and Judges Publishing Company, 1996.

Bailey, M.N., Wong, B.C., and **Lawrence, J.M.,** "Data and Methods for Estimating the Severity of Minor Impacts", Paper No. 950352, SAE International, Warrendale, PA, 1995.

Bastow, D., *Car Suspension and Handling*, 2nd ed., Pentech Press, London, 1990.

Bhat, R., and **Dukkipati, R.V.,** *Advanced Dynamics*, Narosa Publishing House, New Delhi, India, 2001.

Bohan, T.L., and **Damask, A.C.,** *Forensic Accident Investigation: Motor Vehicles,* Michie Butterworth law Publishers, Charlottesville, VA, 1995.

Brach, R.M., *Mechanical Impact Dynamics,* John Wiley & Sons, New York, NY, 1991.

Brach, R.M., "Impact Analysis of Two-Vehicle Collisions", Paper 830468, SAE International, Warrendale, PA, 1987.

Brach, R.M., "Energy Loss in Vehicle Collisions", Paper 871993, SAE International, Warrendale, PA, 1987.

Brach, R.M., and **Smith, R.A.,** "Re-Analysis of the RICSAC Car Crash Accelerometer Data", Paper 2002-01-1305, SAE International, Warrendale, PA, 2002.

Brach, R.M., "Energy Loss in Vehicle Collisions", Paper 871993, SAE International, Warrendale, PA, 1987.

Brach, Raymond M., and Matthew Brach, R., "Crush Energy and Planar Impact Mechanics for Accident Reconstruction", Paper 980025, SAE International, Warrendale, PA, 1998.

Brach, Raymond M., and **Matthew Brach, R.,** *Vehicle Accident Analysis and Reconstruction Methods*, SAE International, Warrendale, PA, 2005.

Brach, R.M., "Modeling of Low-Speed, Front-to-Rear Vehicle Impacts", Paper 2003-01-0491, SAE International, Warrendale, PA, 2003.

Brogliato, B., *Nonsmooth Impact Mechanics: Models, Dynamics and Control,* Springer, London, NY, 1996.

Brown, J.F., and **Obenski, K.S.,** *Forensic Engineering reconstruction of Accidents*, Charles C. Thomas Publisher, Springfield, IL, 1990.

Collins, J.C., and Morris, J.L., *Highway Collision Analysis*, Charles C. Thomas Publisher, Springfield, IL, 1974.

Campbell, C., *New Directions in Suspension Design,* Robert Bentley, Inc, Cambridge, MA, 1981.

Campbell, K., "Energy as a Basis for Accident Severity", Paper 74056, SAE International, Warrendale, PA, 1974.

Costin, M., and **Phillips, D.,** *Racing and Sports Car Chassis Design*, 2nd ed, Robert Bentley, Inc, Cambridge, MA, 1965.

Cheng, P.H., Sens, M.J., Weichel, J.F., and **Guenther, D.A.,** "An Overview of the Evolution of Computer Assisted Motor Vehicle Accident Reconstruction", SAE Paper 871991, in *Reconstruction of Motor Vehicle Accidents: A Technical Compendium,* PT-34, SAE International, Warrendale, PA, 1987.

Daily, J., *Fundamentals of Traffic Accident Reconstruction*, Institute of Police Technology and Management, University of North Florida, Jacksonville, FL, 1988.

Dukkipati, R.V., Ananda Rao M., and **Bhat, R.,** *Computer Aided Analysis and Design of Machine Elements*, New Age International (P) Limited, New Delhi, India, 2000.

Dukkipati, R.V., and **Amyot, J.R.,** *Computer Aided Simulation in Railway Dynamics,* Marcel Dekker Inc, New York, NY, 1988.

Dukkipati, R.V., Spatial *Mechanisms: Analysis and Synthesis*, Narosa Publishing House, New Delhi, India, 2001.

Dukkipati, R.V., *Vehicle Dynamics*, Narosa Publishing House, New Delhi, India, 2000.

Dukkipati, R.V., *Vibration Analysis,* Narosa Publishing House, New Delhi, India, 2004.

Dukkipati, R.V., *Engineering System Dynamics*, Narosa Publishing House, New Delhi, India, 2004.

Dukkipati, R.V., *Control Systems,* Narosa Publishing House, New Delhi, India, 2005.

Day, T.D., "Further Validation of EDSMAC using the RICSAC Staged Collisions," Paper 900102, SAE International, Warrendale, PA, 1990.

Ellis, J.R., *Road Vehicle Dynamics*, John R. Ellis, Inc., Akron, OH, 1988.

Ellis, J.R., *Vehicle Dynamics,* Business Books Limited, London, 1969.

Fonda, A.G., "Partially-Braked Impact and Trajectory Benchmarks, and Their Application to CRASH3 and CRASHEX, SAE International, Paper No. 2000-01-1315, Warrendale, PA, 2000.

Fonda, A.G., "CRASH Extended for Desk and Handheld Computers", Paper No. 870044, SAE International, Warrendale, PA, 1987.

Firch, J.W., *Motor Truck Engineering Handbook*, 3rd ed., James W. Fitch, Anacortes, WA, 1984.

Funk, F.R., Cormier, J.M., and **Bain, C.E.,** *Analytical Model for Investigating Low-Speed Sideswipe Collisions,* Paper 930654, SAE International, Warrendale, PA, 2004.

Garg, V.K., and **Dukkipati, R.V.,** *Dynamics of Railway Vehicle Systems*, Academic Press, New York, NY, 1984.

Giles, J.G., *Steering, Suspension and Tyres*, Iliffe Books Ltd., London, 1968.

Gillespie, T.D., *Fundamentals of Vehicle Dynamics*, Society of Automotive Engineers, Inc., Warrendale, PA, 1992.

Goodsell, D., *Dictionary of Automotive Engineering*, Butterworths, London, 1989.

Huang, M., *Vehicle Crash Mechanics*, CRC Press, Boca Raton, FL, 2002. Also SAE International, Warrendale, PA, 2002.

Ishikawa, H., "Computer Simulation of Automobile Collision–Reconstruction of Accidents", Paper 851729, SAE International, Warrendale, PA, 1985.

Ishikawa. H., "Impact Model for Accident Reconstruction–Normal and Tangential Restitution Coefficients", Paper 930654, SAE International, Warrendale, PA, 1993.

Ishikawa, H., "Impact Center and Restitution Coefficients for Accident Reconstruction", Paper 940564, SP1030, SAE International, Warrendale, PA, 1994.

Jones, I.S., and **Baum, A.S.,** "Research Input for Computer Simulation of Automobile Collisions (RICSAC)", vol. IV: "Staged Collision Reconstructions", DOT HS-805 040, Washington, DC, December 1978.

Jones, I.S., and **Baum, A.S.,** "Research Input for Computer Simulation of Automobile Collisions (RICSAC)", vol. IV: "Staged Collision Reconstructions", DOT HS-805 040, Washington, DC, 1978.

Kennedy, W.A., *Impact Velocity from Conservation of Linear Momentum for the Traffic Accident Investigator and Reconstructionist,* Institute of Police Technology and Management, University of North Florida, Jacksonville, FL, 1993.

Levy, R.A., Speed Determination in Car-Truck Sideswipe Collisions, SAE Paper No. 2000-01-0463, SAE International, Warrendale, PA, 2000.

Limpert, R., *Motor Vehicle Accident Reconstruction and Cause Analysis*, 4th ed, The Michie Company, Charlottesville, VA, 1994.

Lofgren, M.J., *Handbook for the Reconstructionist*, 3rd ed, Institute of Police Technology and Management, University of North Florida, Jacksonville, FL, 1983.

Magnus, K., ed., *Dynamics of Multibody Systems*, Springer-Verlag, Berlin-Heidelberg, 1978.

McHenry, B.G., and **McHenry, R.R.,** "SMAC-97-Refinement of the Collision Algorithm", Paper 970947, SAE International, Warrendale, PA, 1997.

McHenry, B.G., and **McHenry, R.R.,** "CRASH-97-Refinement of the Trajectory Solution Procedure", Paper 970949, SAE International, Warrendale, PA, 1997.

McHenry, B.G., and **McHenry, R.R.,** "Effects of Restitution in the Application of Crush Coefficients", Paper 970960, SAE International, Warrendale, PA, 1997.

McHenry, B.G., and **McHenry, R.R.,** "RICSAC-97- A Re-evaluation of the Reference Set of Full Scale Crash Tests", Paper 970961, SAE International, Warrendale, PA, 1997.

Milliken, W.F., and **Milliken, D.L.,** *Race Car Vehicle Dynamics*, Society of Automotive Engineers, Inc., Warrendale, PA, 1995.

Monk, M.W., and **Guenther, D.A.,** "Update of CrashII Computer Model Damage Tables – Volume I", Vehicle Research and Test center, National Highway Traffic Safety Administration (NHSTA), DOT HS 806-446, Washington, DC, 1983.

Monson, K.L., and **Germane, G.J.,** "Determination and Mechanisms of Motor Vehicle Structural Restitution from Crash Test Data", Paper 1999-01-0097, SAE International, Warrendale, PA, 1999.

Neptune, J.A., and **Flynn, J.E.,** "A Method for Determining Accident Specific Crush Stiffness Coefficients", Paper 940913, SAE International, Warrendale, PA, 1994.

Newton, K., Steeds, W., and **Garrett, T.K.,** *The Motor Vehicle*, 10th ed., Butterworths, London, 1983.

Noon, R.K., *Engineering Analysis of Vehicular Accidents,* CRC Press, Boca Raton, FL, 1994.

Pacejka, H.B., ed., The Dynamics of Vehicles on Roads and Railway Tracks, Swets & Zeitlinger, 1976.

Peters G.A., and **Peters, B.J.,** *Automotive Vehicle Safety*, SAE International, Warrendale, PA, 2002.

Prasad, A.K., "Coefficient of Restitution of Vehicle Structures and its Use in Estimating the Total ΔV in Automobile Collisions", AMD Vol. 126/BED Vol. 19, pp. 217-246, Crash-worthiness and Occupant Protection in Transportation Systems, ASME, New York, NY, 1991.

Prasad, A.K., "Missing Vehicle Algorithm (OLDMISS) Reformulation", Paper 910121, SAE International, Warrendale, PA, 1992.

Prasad, A.K., "Energy Absorbed by Vehicle Structures in Side Impacts", Paper 910599, SAE International, Warrendale, PA, 1991.

Prasad, A.K., "Energy Absorbing Properties of Vehicle Structures and Their Use in Estimating Impact Severity in Automobile Collisions", Paper 925209, SAE International, Warrendale, PA, 1992.

Puhn, F., *How to make your Car Handle,* H.P. Books, Los Angeles, CA, 1981.

Prasad, A.K., "CRASH 3 Damage Algorithm Reformulation for Front and Rear Collisions", Paper 900098, SAE International, Warrendale, PA, 1990.

Rao, J.S., and **Dukkipati, R.V.,** *Mechanism and Machine Theory*, 2nd ed, Wiley Eastern Limited, New Delhi, India, 1992.

Rivers, R.W., *Traffic Accident Analysis and Reconstruction Manual*, Charles C. Thomas Publisher, Springfield, IL, 1997.

Rivers, R.W., *Traffic Accident Investigators' Book of Formulae and Tables*, Charles C. Thomas Publisher, Springfield, IL, 1981.

Segel, L., "Research in the Fundamentals of Automobile Control and Stability", Transactions of the Society of Automotive Engineers, Warrendale, PA, 1985.

Slibar, A., and **Springer, H.,** *The Dynamics of Vehicles on Roads and Tracks*, Swets & Zeitlinger, 1978.

Steeds, W., *Mechanics of Road Vehicles*, Iliffe and Sons, Ltd., London, 1960.

Stronge, W.J., *Impact Mechanics,* Cambridge University Press, Cambridge, U.K., 2000.

Taborek, J.J., *Mechanics of Vehicles*, Towmotor Corporation, Cleveland, OH, 1957.

Terry, L., and **Baker, A.**, *Racing Car Design and Development*, Robert Bentley, Inc., Cambridge, MA, 1973.

Tumbas, N.S., and **Smith, R.A.,** "Measurement Protocol for Quantifying Vehicle Damage From an Energy Basis Point of View", Paper 880072, SAE International, Warrendale, PA, 1988.

Van Eldik, H.C.A., and **Pacejka, H.B.**, *The Tire as a Vehicle Component,* Delft University of Technology, Delft, The Netherlands, 1971.

Varat, M.S., and **Husher, S.E.,** "Vehicle Impact Response Analysis through the Use of Accelerometer Data", Paper 2000-01-0850, SAE International, Warrendale, PA, 2000.

Varat, M.S., and **Husher, S.E.,** "Vehicle Crash Severity Assessment in Lateral Pole Impacts", Paper 1999-01-0100, SAE International, Warrendale, PA, 1999.

Welsh, K.J., and **Struble, D.E.,** "Crush Energy & Structural Characterization", Paper 1999-01-0099, SAE International, Warrendale, PA, 1999.

Willumeit, H.P., ed., *The Dynamics of Vehicles on Roads and Tracks,* Swets & Zeitlinger, 1980.

Willke, D.T., and **Monk, M.W.,** "CRASH III Model Improvements: Derivation of New Side Stiffness Parameters from Crash Tests", *Accident Reconstruction Journal,* Waldorf, MD, Spring 1995.

Wong, J.Y., *Theory of Ground Vehicles*, 3rd ed., Wiley, New York, NY, 2001.

Woolley, R.L., "The 'IMPAC' Program for Collision Analysis", Paper 8700476, SAE International, Warrendale, PA, 1987.

Woolley, R.L., "Nonlinear Damage Analysis in Accident Reconstruction", Paper 2001-01-0504, SAE International, Warrendale, PA, 2001.

Wood, D.P, Ydenius, A., and **Adamson, D.,** "Velocity Changes, Mean Accelerations and Displacements of Some Car Types in Frontal Collisions", *International Journal of Crashworthiness,* vol. 8, no. 6., Woodhead Publishing, Ltd., Cambridge, U.K., 2003. pp. 591-603.

CHAPTER 5

Vehicle Rollover Analysis

5.1 INTRODUCTION

Vehicle vulnerability to rollover is one of the most important vehicle safety attributes. According to statistics reported on *www.autosafety.org,* rollover accidents kill 1 of every 5 people who die in motor vehicle accidents. Sports Utility Vehicles (SUVs) account for 17 of the 20 vehicles with the greatest risk of rollover, and a minivan can be expected to rollover 10% to 17.9% of the time it is in a single vehicle accident.

Rollover has been an important safety issue of the vehicles. Rollover is a major reason for severe and fatal injuries, especially for trucks and SUVs. National Automotive Sampling System (NASS) [92058] data shows that about 85~90 percent of the rollover occurred in single-vehicle accidents. The vehicles were out of control prior to overturning. Of these single vehicle crashes, more than 50 percent resulted in fatalities with vehicle being rolled over. Rollover crashes are the most dangerous type of collision for all classes of light vehicles. The rate of rollover accidents for SUVs and small trucks is higher than that of passenger cars.

Rollover accidents account for a small percentage of all vehicle accidents, but they account for a large percentage of fatal accidents. In 1997, NASS [970396] reported that only 8% car accidents resulted in rollover, but rollover accounted for 17% fatal injuries. In 1996, rollovers were 1.8 % of all crashes in the United States, but represented 9.9% of all fatal crashes [1999-01-0122]. Some reports showed that about 50% of single vehicle accidents fatalities resulted from rollover and about 10% multi-vehicle accident fatalities were caused by rollover [90315]. Almost 10,000 people die each year in the US in rollover accidents [920581].

Since rollover accidents tend to occur when vehicle enters a skid or leaves a roadway and the driver attempts to steer the vehicle to correct its path of travel, electronic stability control represents a highly significant feature in improving SUV, pickup and van safety. Electronic Stability Control (ESC) helps drive control a vehicle under extreme steering maneuvers, but less than one-third of 2006 vehicles had ESC. Until automakers reduce the risk of SUV rollover accidents through all available technology, including electronic stability control, the number of those injured and killed in rollover accidents will continue to climb.

In this Chapter we will introduce the rollover mechanism and present scenarios in which rollover could happen.

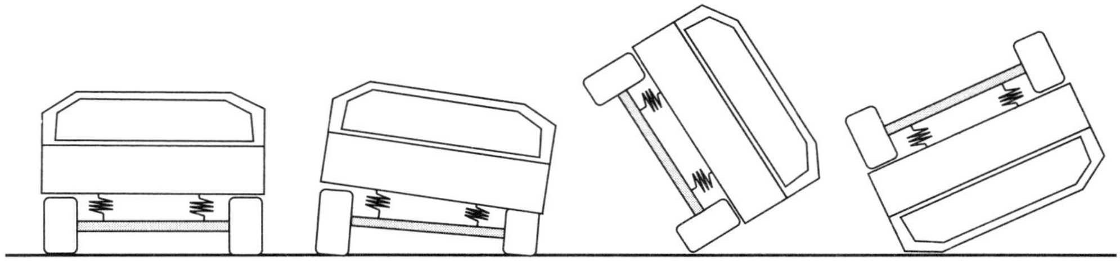

Fig. 5.1 A vehicle rollover scenario.

5.1.1 Rollover Scenario

Much attention has been paid to automotive rollover accidents during recent decades. Rollover accidents are a serious threat to the life of automobile occupants. Rollover is defined as a vehicle rotation of 90° or more around its longitudinal axis. The vehicle body will contact the ground, and the occupants will be injured. Figure 5.1 shows a vehicle rollover scenario. There are several factors that will affect the vehicle vulnerability to rollover. Among these factors are tire and vehicle characteristics, environmental conditions and drivers. A combination of vehicle and tire design is among the reasons of rollover. For example, vehicles with higher CG (like trucks), are more vulnerable to rollover when a simple but a severe steering input is given. Environmental conditions may also cause rollover. Such conditions may also cause rollover, which are mostly related to road conditions such as icy or bumpy roads. Drivers with 'dangerous' driving habits may cause the rollover. Rollover accidents may be the result of one or a combination of the above factors.

5.1.2 Importance of Rollover

Rollover has been an important safety issue of vehicles. Rollover is a major reason for severe and fatal injuries, especially for trucks and sport utility vehicles (SUVs). National Automotive Sampling System (NASS) data (Hinch *et al.*, 1992) shows that approximately 85 to 90% of rollover occurred in single-vehicle accidents. The vehicles were out of control prior to overturning. Of these single-vehicle crashes, more than 50% resulted in fatalities with the vehicles rolling over. Rollover crashes are the most dangerous type of collision for all classes of light vehicles. The rate of rollover accidents for SUVs and small trucks is higher than that of passenger cars.

Rollover accidents account for a small percentage of all vehicle accidents, but they account for a large percentage of fatal accidents. In 1997, NASS (James *et al.*, 1997) reported that only 8% of car accidents resulted in rollover, but rollover accounted for 17% of all fatal injuries. In 1996, rollovers were 1.8% of all crashes in the United States but represented 9.9% of all fatal crashes (SAE Paper No. 1999-01-0122). Some reports showed that approximately 50% of single-vehicle accident fatalities resulted from rollover, and approximately 10% of multi-vehicle accident fatalities were caused by rollover (SAE Paper No. 950315). Almost 10,000 people die each year in rollover accidents in the United States (Hinch *et al.*, 1992). More materials on rollover accidents and rollover control will be introduced in Section 5.7 of this chapter.

Several factors will affect the vulnerability of a vehicle to rollover. Among these factors are tire and vehicle characteristics, environmental conditions, and drivers. A combination of vehicle and tire design is among the reasons for rollover. For example, vehicles with a higher center of gravity (CG) such as trucks are more vulnerable to rollover when a simple but severe steering input is given. Environmental conditions also may cause rollover. Such conditions are related primarily to road conditions such as ice or bumpy roads. Drivers with dangerous driving habits likewise may cause rollover. Rollover accidents may result from one or a combination of these factors. Rollover can happen on a flat road, on a cross-slope road, or off road. Rollover can be divided into two categories: tripped rollover, and untripped rollover. Tripped rollover is caused by a vehicle hitting an obstacle. This rollover is caused by vehicle inclination where the center of gravity of the vehicle exceeds the stable point. National Highway and Traffic Safety Administration (NHTSA) reports show that only 10% or less of all rollover crashes occur under untripped conditions for the on-road environment. However, major rollovers occur in an off-road environment where the road conditions are unpredictable. Rollover for on-road accidents usually is caused by dangerous maneuvers that are induced either by high lateral acceleration or by yaw instability. The high lateral acceleration produces sufficient centrifugal force to pull the vehicle and cause it to rotate around its outside tire. During yaw motion, the tires produce saturated forces that cause tire sliding and rollover. Most rollover occurs under the following conditions:

Traveling at high speed on a curved road: When a vehicle travels on a curved road, lateral centrifugal force will pull it in an outboard motion, as shown in Fig. 5.2.

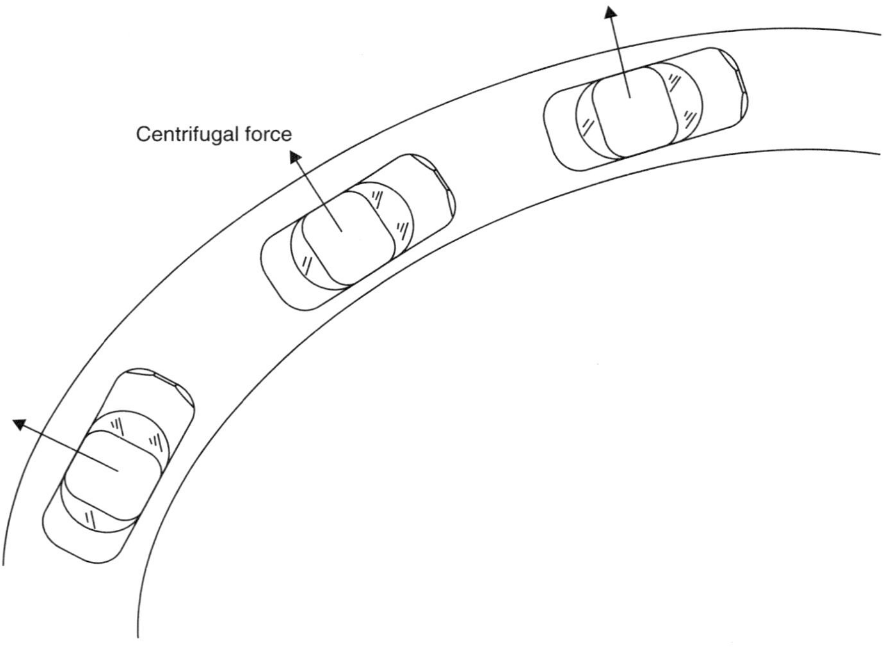

Fig. 5.2 A vehicle traveling on a curved road.

Severe cornering maneuver: This case is similar to the preceding case where the centrifugal force pulls the vehicle to rollover. For example, a driver avoiding an accident and steering rapidly can cause a yaw disturbance.

Traveling on a collapsing road and suddenly providing steering input for a vehicle with a low level of roll stability: This kind of rollover also is caused by a yaw disturbance. Figure 5.3 shows a vehicle rollover scenario caused by severe steering input.

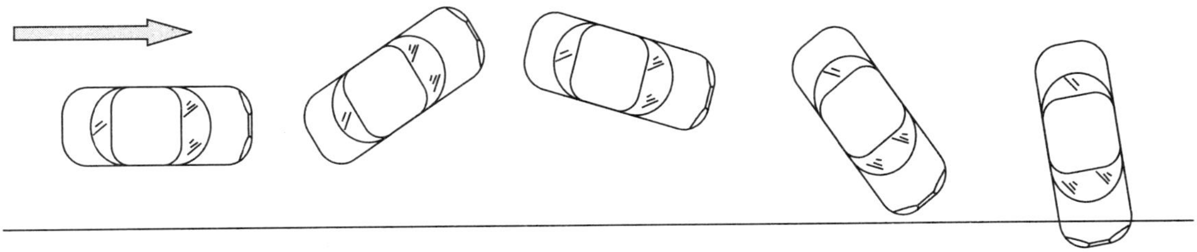

Fig. 5.3 A vehicle rollover scenario caused by severe steering input.

Losing control due to a rapid decrease of friction, such as driving on an icy road: Steering can cause yaw motion because forces on the tires in the lateral direction are strong enough to roll the vehicle. The forces also produce lateral acceleration on the vehicle center of gravity. The forces are produced by the friction between the tires and the road; therefore, these rollover scenarios are called friction rollovers. Figure 5.4 shows a vehicle losing control because of low friction on the road.

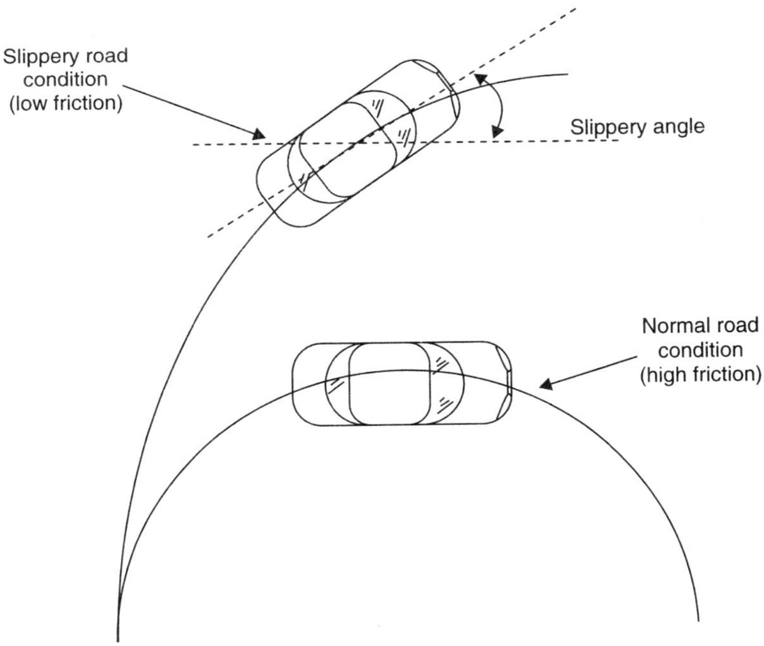

Fig. 5.4 A vehicle losing control because of low friction on the road.

Laterally sliding off the road: An example of this case would be a vehicle that can be decelerated by a barrier.

Sliding from a cliff: Figure 5.5 shows this type of scenario, in which a vehicle is sliding from a cliff.

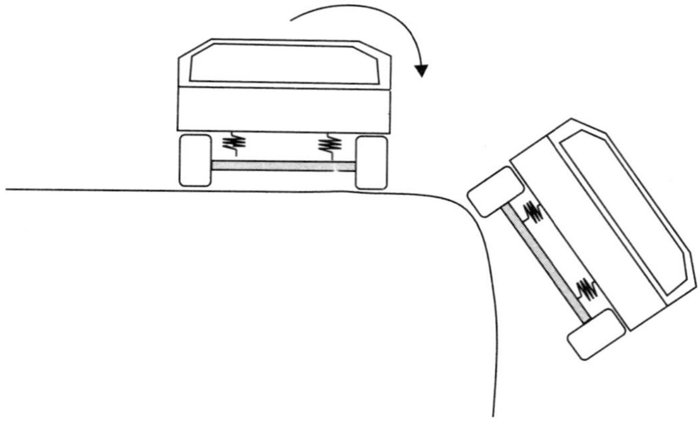

Fig. 5.5 A vehicle sliding from a cliff.

5.1.3 Research on Rollover

Because rollover is the major safety concern, it is important to develop testing and analytical methods to predict rollover and to improve design to control rollover. Rollover studies include field accident data, statistical analysis, occupant kinematics, injury mechanisms, rollover model analysis and testing, vehicle stability factors, rollover accident reconstruction, rollover control, and others. The study of vehicle rollover is a complicated topic. The phenomenon of rollover is difficult to predict. Rollover involves many vehicle factors and some nonvehicle factors, such as the road friction coefficient, obstacles on the road, the type of road shoulder, road and shoulder inclination angles, driver steering patterns, and others.

Vehicle Rollover Analysis

To simplify the complicated overall analysis, a steady-state rigid model and a steady-state suspended model are most commonly used. Some simple metrics have been developed to evaluate the static and dynamic stability of vehicles, such as the static stability factor (SSF), the tilt table ratio (TTR), the critical sliding velocity, the dynamic stability index, and others. Among the indicators, the static stability factor and the tilt table ratio are most commonly used.

A vehicle dynamic model involves many systems and parameters. Of particular interest are the steering and tire systems. Some simple dynamic models include only several degrees of freedom. The models can be used for quick review of rollover characteristics and parameter tuning. More complicated dynamic models include many components, and systems can be built using specific software such as ADAMS, NADSdyna, VDANL, and SIMPACK. Vehicle rollover scenarios can be animated on the computer screen. The taller and narrower a vehicle is, the easier it tends to roll over. The most effective way to keep the vehicle from rolling over is to make the center of gravity of the vehicle as low as possible and/or to make the vehicle wider (*i.e.*, increase the track width of the vehicle). The use of tires with less lateral force will help minimize the potential for rollover. Active control (as in active suspension) is one of the effective methods to prevent rollover. Some control algorithms are integrated with a dynamic model to simulate rollover and its control. Active control in rollover prevention is an active field of research. Research on rollover also includes parameter sensitivity analyses, dynamic stability studies, and correlation between vehicle design parameters and rollover statistical data.

Many studies have been done by the NHTSA on vehicle rollover since the 1970s. In 1973, NHSTA proposed an Advance Notice of Proposed Rule Making (ANPRM) to evaluate vehicle rollover tendencies on smooth and dry roads. Since then, several procedures have been proposed, but no standard procedures to evaluate either vehicle rollover stability or suitable rollover stability metric have been adopted by NHSTA. Other organizations, such as ISO, SAE, and CU (Customers Union), have developed some test procedures. The severe lane-change maneuver procedure to evaluate vehicle road-holding ability was developed by ISO. SAE proposed a procedure (J266) to evaluate both steady-state and transient behavior. CU developed an obstacle avoidance maneuver procedure, and ISO and CU developed full-scale vehicle handling tests to evaluate the dynamic response of a vehicle in an emergency situation.

5.1.4 Scope of This Chapter

This chapter introduces the concept of and possible scenarios for rollover. Static rollover models, including rigid and suspended models, are analyzed. In evaluating the static rollover threshold, the tilt table ratio and slide pull ratio are the most commonly used metrics. Many parameters influence the rollover threshold, such as the vehicle center-of-gravity height, track width, suspension stiffness, tire stiffness, tire lateral force and friction, and others. The influence of these parameters on the static rollover threshold is analyzed. The dynamic model is developed by extending the quasi-static model, which includes rotational and lateral types of motion. This chapter briefly introduces the rigid model, dependent suspension model, and independent suspension model. Dynamic rollover thresholds are based on the energy that is needed to cause the vehicle to rollover. The dynamic rollover thresholds include the dynamic stability index, rollover prevention energy reserve, rollover prevention metric and critical sliding velocity.

This chapter emphasizes the importance of rollover analysis and control, and it introduces the relation between rollover accidents and occupants. The rollover sensing system provides a warning when the vehicle is close to the rollover condition, and the control system activates an anti-rollover mechanism. Active anti-roll control has been widely used in modern vehicles, such as the active suspension, active roll-bar, active steering, anti-rollover braking system, and others.

5.2 RIGID VEHICLE ROLLOVER MODEL

Vulnerability to rollover using simple vehicle models is the subject of this section. The models are rigid body models and typically are used in the early stages of product development.

5.2.1 Rigid Vehicle Model

A vehicle is a complicated structure. In dynamic analysis, the vehicle usually is simplified as a combination of many rigid bodies. Spring, damping, and other connection components are used to connect these rigid bodies. To quickly understand basic rollover characteristics, assume that a vehicle is a rigid body model, as shown in Fig. 5.6. The road can either be flat or has a slope angle of θ. The deflection of the tires and suspension is neglected. When the vehicle undergoes a turn, centrifugal force caused by its body pulls it outward from the turning center. The centrifugal force equals the lateral acceleration of the vehicle multiplied by the weight of the vehicle.

Taking the moment at point B, the vehicle dynamic equation of Fig. 5.6 can be written as

$$I_b\ddot{\phi} = F_{1z}d - (mg\sin\theta)h - (mg\cos\theta)\frac{d}{2} + ma_yh \tag{5.1}$$

where I_b is the vehicle moment of inertia around point B, h and d are the vehicle center-of-gravity height and track width, respectively, and F_{1z} is the vertical force acting on the inner tires.

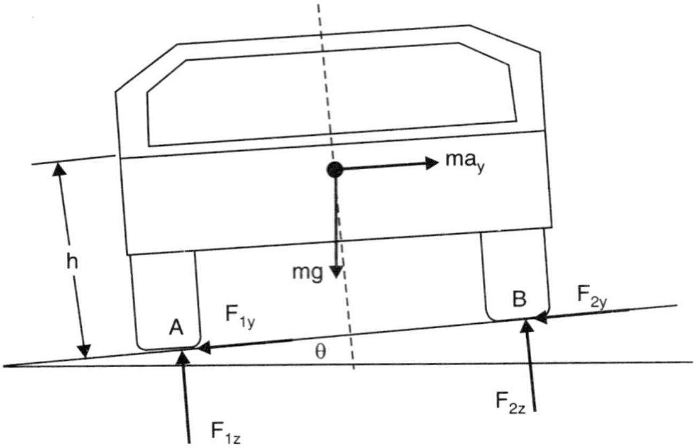

Fig. 5.6 A rigid vehicle model.

Usually, the slope angle θ is very small; therefore, sin θ ≈ θ and cos θ ≈ 1. Equation 5.1 can be simplified as

$$I_b\ddot{\varphi} = F_{1z}d - mgh\theta - mg\frac{d}{2} + ma_y h \tag{5.2}$$

At the beginning moment when rollover occurs, the rotation velocity and acceleration are neglected (*i.e.*, $\ddot{\phi} = 0$). This situation is called the quasi-static state (steady state). The discussions in this section and the next section are based on an assumption of the quasi-static state. Equation 5.2 can be simplified as

$$F_{1z}d - mgh\theta - mg\frac{d}{2} + ma_y h = 0 \tag{5.3}$$

5.2.2 Steady-State Rollover on a Flat Road

Rollover begins if the load on the inner tires is zero (*i.e.*, $F_{1z} = 0$). Equation 5.3 can be simplified as

$$\frac{a_y}{g} = \frac{d}{2h} + \theta \tag{5.4}$$

The lateral acceleration depends on the slope angle, the vehicle center-of-gravity height, and the vehicle track width. The larger the slope angle, the higher the rollover lateral acceleration. Usually, the slope angle θ is much small than $d/(2h)$. The first term, $d/(2h)$ often is called the first-order estimate of the static rollover threshold.

Example 5.1: *The center-of-gravity height of a truck is 1.8 m, and its tread width is 2 m. The truck travels on a road with a slope of 5°. Calculate the lateral acceleration of the vehicle at the moment rollover begins.*

Solution: According to Eq. (5.4), the lateral acceleration is

$$a_y = \left(\frac{d}{2h} + \theta\right)g = \left(\frac{1.8}{2*2} + \frac{5}{180}*\pi\right)g = 0.45g + 0.087g = 0.537g$$

The result shows that the lateral acceleration caused by the road angle is much smaller than that of the vehicle parameters h and d.

On a flat road surface, $\theta = 0$. Equation 5.4 can be simplified as

$$a_y = \frac{d}{2h} g \tag{5.5}$$

Equation 5.5 represents the relation between the lateral acceleration and the two basic dimensions of the vehicle: the center-of-gravity height (h), and the tread width (d). If $a_y > \frac{d}{2h} g$, the rollover will occur. The wider and the lower the center of gravity of the vehicle is, the more difficult it is for rollover to occur. Thus, Eq. (5.5) indicates the static roll stability and is called the rollover threshold for a flat road surface. The vehicle static rollover threshold is the maximum lateral acceleration that the vehicle can sustain in a steady-state turn on a level surface without rolling over.

Equation 5.5 also can be written as a nondimensional expression as

$$\frac{a_y}{g} = \frac{d}{2h} \tag{5.6}$$

Equation 5.6 is a ratio of half of the track width to the height of the center of gravity of the vehicle. This ratio is called the static stability factor (SSF). The static stability factor is a useful and simple metric for studying rollover propensity because only two basic properties of the vehicle are used for calculating the static stability factor. The static stability factor reflects the most fundamental dynamic relation of the vehicle. The static stability factor is obtained under the assumption that the vehicle is a rigid body, and deflections of the suspension and tire are neglected. When a vehicle is turning, the lateral force pulls the center of gravity of the vehicle to shift toward the outside of the tire and causes the outside tire and suspension deformation, which will significantly change the vehicle center-of-gravity movement and track width. All of these influencing factors reduce the vehicle rollover stability and the static stability factor. The estimation by Eq. (5.6) is not conservative, in that it virtually overestimates the actual static rollover threshold of real vehicles.

Thus, the static stability factor overestimates the rollover threshold. Although the vehicle is not a rigid body and the static stability factor is not an accurate metric to predict the rollover threshold, it provides the first guideline for a vehicle design and is useful for doing benchmark comparisons for different vehicles.

The inverse term of the roll threshold is called the rollover propensity (*i.e.*, $2h/d$). A higher number of the propensity means a lower rollover threshold.

5.2.3 Tilt Table Ratio

Figure 5.7 is a simple tilt test table, with a vehicle placed on its surface. The table is tilted from the horizontal by an angle φ. The vehicle is in steady state. The vehicle weight, in milligrams (mg), can be divided in two directions: one perpendicular to the tilt table surface, and the other parallel to the tilt table surface. The weight perpendicular to the tilt table surface is the actual weight acting on the vehicle, called the simulated weight, and the corresponding force can be expressed as

$$F_1 = m_s g = mg \cos \phi \tag{5.7}$$

The weight parallel to the tilt table surface is the simulated centrifugal force. Assume that a simulated lateral acceleration, a_s, is acted on the vehicle, and a centrifugal force pulls the vehicle into rollover. The centrifugal force can be expressed as

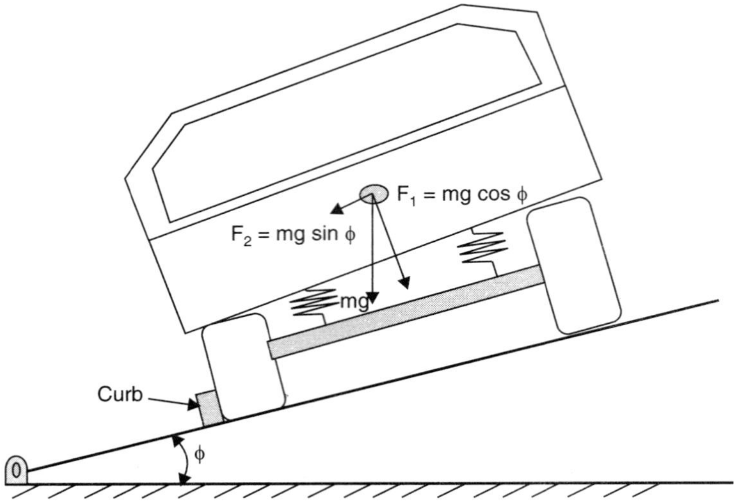

Fig. 5.7 Tilt test table with a vehicle placed on it.

$$F_2 = m_s a_s = mg \sin \phi \tag{5.8}$$

During the test, the tilt of the table is increased slowly until the tire at the high side loses contact with the surface of the table. The angle at which this occurs is called the tip-over angle. Dividing Eq. (5.7) by Eq. (5.8), the ratio of the simulated lateral acceleration over g is obtained and is called the tilt table ratio (TTR), which is expressed as

$$\text{TTR} = \frac{a_s}{g} = \tan \phi \tag{5.9}$$

The tilt table ratio is an estimator of the vehicle static rollover threshold. It is a simple test to estimate the vehicle rollover threshold because it is determined only by the tilt angle and does not need any additional measurement, such as the vehicle center of gravity or weight. Furthermore, the test is safe and nondestructive to the vehicle. The tilt table ratio is better used to estimate the rollover of unstable vehicles, where there is a small tilt angle, than stable vehicles. The tilt table ratio is more accurate in simulating a vehicle steady turn at high lateral acceleration or a vehicle on an embankment or side slope than the static stability factor, where a rigid vehicle model is used. The tilt table ratio replicates the physical road-vehicle condition, but it has some disadvantages. Because the real weight acting on the vehicle ($mg \cos \varphi$) is less than the vehicle weight, the lateral centrifugal force is less than the actual force. For most light-duty vehicles, the centrifugal force on a vehicle on the tilt table is approximately 60 to 80% of that while the vehicle turns on real roads. Thus, the vehicle rollover stability will be over-estimated. On the other hand, the reduced vertical load acting on the compliant tires and suspension of the vehicle may rise relative to the normal position, which results in a higher center-of-gravity position. Thus, the static rollover stability is underestimated. These two influencing factors tend to cancel each other.

Some vehicles could slide due to the limited surface-tire friction before rollover occurs. To prevent that, a small curb usually is placed next to the lower tires, as shown in Fig. 5.7. However, the presence of a curb can raise the effective side force of the center of the tire. That is, the effective center-of-gravity height is reduced, and the static rollover stability is overestimated.

5.2.4 Side Pull Ratio

The tilt table ratio is not a good estimator of rollover for more stable vehicles. To overcome this disadvantage, another static rollover estimator, the side pull ratio (SPR), is induced. Figure 5.8 shows the side pull ratio test setup. A vehicle is placed on a high-friction surface. Then a pull force through a belt is applied to the sprung mass at the vehicle center of gravity and slowly pulls laterally. The test is in a quasi-static state, and the belt is maintained in a horizontal line as the vehicle moves during the test.

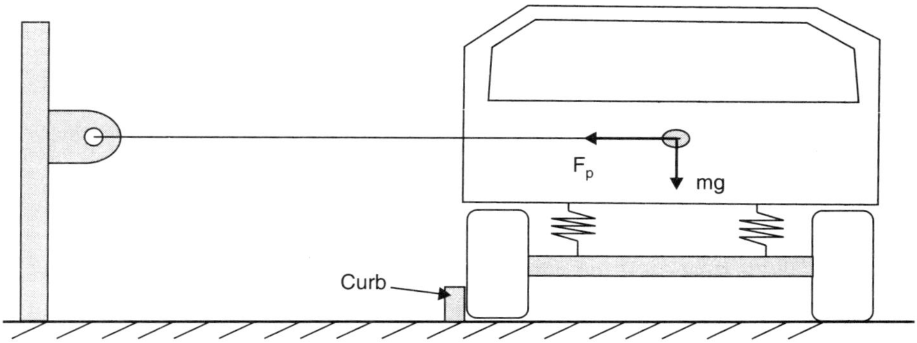

Fig. 5.8 A side pull ratio (SPR) test setup.

The side pull ratio is defined as the ratio of the minimum side pull force (*Fp*) required to lift all tires on one side of the vehicle from the ground to the vehicle weight and is expressed

$$\text{SPR} = \frac{F_p}{mg} \tag{5.10}$$

The test attempts to simulate lateral acceleration that acts on the vehicle in steady-state turns and reaches the rollover threshold. The side pull ratio uses the proper magnitude of vertical and lateral loads applied to the vehicle. The forces acting on the vehicle at the start of rollover are more accurate than those in the static stability factor and the tilt table ratio. Thus, the side pull ratio is a good estimator of the static rollover threshold. During the side pull test, the pulling force is acted upon the sprung mass at the vehicle center of gravity. However, the vehicle consists of many "rigid bodies," and each has its own center of gravity. The pull force represents the inertia forces of all of these rigid bodies; thus, the force transmitted from the sprung mass to the unsprung mass is higher than that in a real maneuver. The higher force causes a higher deflection of the suspension, which will lead to higher lateral displacement of the vehicle center of gravity. That is, the center-of-gravity height is reduced, and this causes the side pull ratio to be lower than the actual lateral acceleration at rollover. The other disadvantage is that the belt could damage the vehicle body when a force equal to the vehicle weight is applied. More belts or wider belts can reduce potential damage but add extra weight. The belt load is applied on a single point on the vehicle. This is different from the real distributed load. The single contact point is assumed as the center-of-gravity point. During the pulling procedure, the center-of-gravity height changes.

Similar to the tilt table test, a side curb also is used in the side pull ratio test because friction between the road surface and the tires is not enough to hold the vehicle before rollover occurs. This will influence the vehicle center-of-gravity height. The side pull test cannot give the absolute lateral acceleration when rollover occurs. It usually is used to study the static rollover stability of vehicles.

5.3 SUSPENDED VEHICLE ROLLOVER MODEL

In the rigid model, the deflections of the tires and suspension are neglected. However, these deflections will impact vehicle rollover stability. When rollover occurs, lateral force will pull the vehicle body outward. The outside suspension and tires will be compressed, which will cause the vehicle center of gravity to shift to the outside, accelerating the possibility of rollover. This section discusses the influence of suspension stiffness and tire stiffness on the rollover threshold.

5.3.1 Steady-State Rollover Model for a Suspended Vehicle

Figure 5.9 shows a suspended vehicle rollover model. The outside tire is compressed, and the inside tire is released, so there exists a deflection angle between the two tires, denoted by θ_1. The vehicle body is supported on the suspension. The centrifugal force pulls the body to the outside and compresses the right suspension spring. At the same time, the left suspension spring is extended. Another deflection angle exists between the two suspension springs and is denoted by θ_2.

The vehicle center of gravity rotates around an imaginary center, called the rollover center, which is point C in Fig. 5.9. When the vehicle is stationary on a flat road, a line connecting the center of gravity and the rollover center, called the rollover line, is in the vertical direction. During turning or cornering, the rollover line rotates an angle θ_1 due to the deflection of the tires and another angle θ_2 due to the suspension deflection. Thus, the total angle at which the rollover line rotates is $\theta_1 + \theta_2$, as shown in Fig. 5.9.

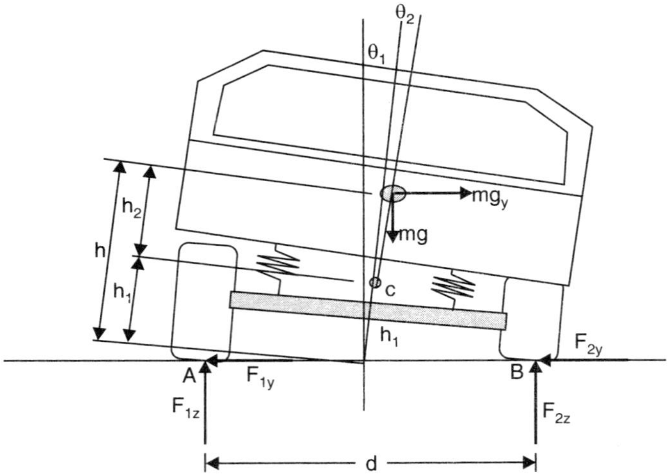

Fig. 5.9 Suspended rollover model.

Taking the moment about the contact point B between the outside wheel and the ground, the equilibrium equation is

$$F_{1z}d + ma_y[h_1 \cos\theta_1 + (h-h_1)\cos(\theta_1+\theta_2)] - mg[\frac{d}{2} - h_1 \sin\theta_1 - (h-h_1)\sin(\theta_1+\theta_2)] = 0 \quad (5.11)$$

where h_1 is the distance from the rollover center to the ground at the vehicle stationary condition, h_2 is the distance from the vehicle center of gravity to the rollover center, h is the vehicle center-of-gravity height, and $h = h_1 + h_2$

Equation 5.11 can be rewritten as

$$F_{1z}d + ma_y[h_1 \cos\theta_1 + h_2 \cos(\theta_1 + \theta_2)] - mg[\frac{d}{2} - h_1 \sin\theta_1 - h_2 \sin(\theta_1 + \theta_2)] = 0 \quad (5.12)$$

Because θ_1 and θ_2 are small numbers, it is assumed that $\cos\theta_1 \approx 1$, $\cos(\theta_1 + \theta_2) = 1$, $\sin\theta_1 \approx \theta_1$, and $\sin(\theta_1 + \theta_2) \approx \theta_1 + \theta_2$. When rollover occurs, the vertical force on the inside tire is zero (*i.e.*, $F_{1z} = 0$). Inserting these assumptions into Eq. (5.12), we obtain

$$\frac{a_y}{g} = \frac{d}{2h} - \theta_1 - \frac{h_2}{h}\theta_2 \quad (5.13)$$

Equation 5.13 is the rollover threshold for the suspended vehicle model. Compare Eq. (5.13) with Eq. (5.6), and it is clear that this rollover threshold is lower than that of the rigid vehicle model. The first item on the right side of Eq. (5.13) represents the rollover threshold for the rigid vehicle model. The second and third terms are the contributions from the tire deflection and suspension deflection, respectively. The contributions from the tire deflection and suspension deflection on the rollover threshold will be discussed separately.

5.3.2 Contribution from the Tire Deflection

In this section, the suspension is assumed to be rigid, and only the tire deflection is considered, as shown in Fig. 5.10. The outside tire is compressed Δr from the original height r. At the same time, the inside tire is released and is not subjected to any load. When the vehicle is at a straight stationary condition, each of the two tires takes half of the vehicle weight. However, when rollover occurs, the whole vehicle weight is supported by the outside tire. The increased weight supported by the tire is

$$\Delta F_{2z} = mg/2 \quad (5.14)$$

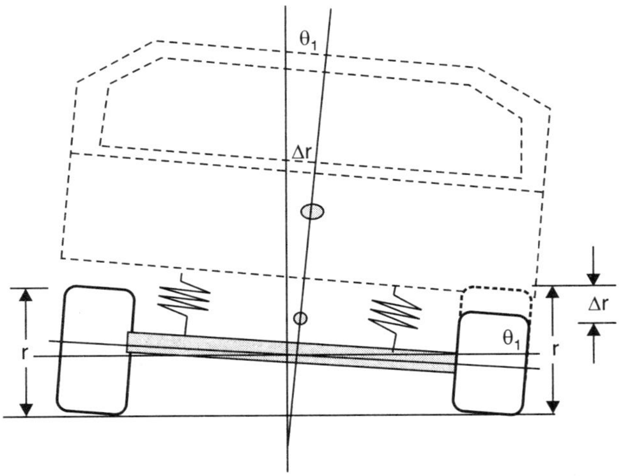

Fig. 5.10 Deflected tire and rigid body model.

The tire deflection, Δr, compared with the stationary condition, is

$$\Delta r = \frac{mg/2}{K_{tire}} \quad (5.15)$$

where K_{tire} is the tire stiffness.

Thus, the rotation angle at the rollover incipient caused by tire deflection is

$$\theta_1^R = \frac{\Delta r}{d/2} = \frac{mg}{dK_{\text{tire}}} \tag{5.16}$$

The rotation angle θ_1 is determined by the track width d, the tire stiffness k, and the vehicle weight. If the tire is rigid (*i.e.*, $k_{\text{tire}} \to \infty$), the tire deflection and angle will go to zero (*i.e.*,$\theta_1 \to 0$). The tire stiffness is the key element in controlling the deflection angle.

5.3.3 Contribution from the Suspension Deflection

In this section, the suspension deflection together with the tire deflection is considered, as shown in Fig. 5.11.

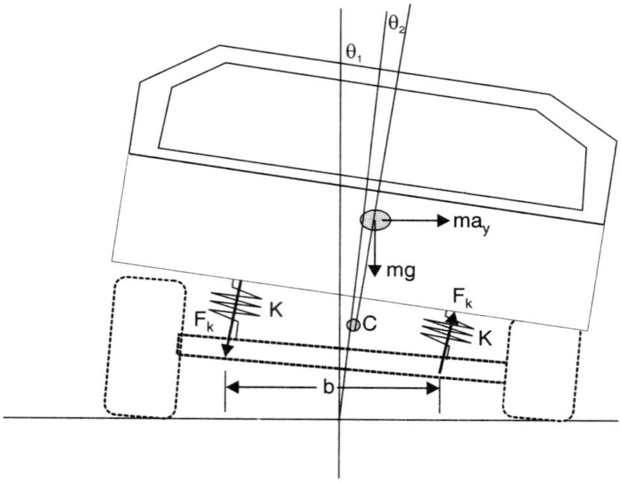

Fig. 5.11 Deflected suspension model.

During turning or cornering, the lateral force pulls the vehicle body toward the outside direction, and the suspension outside spring is compressed. The outside spring is compressed by Δz, and the inside spring is extended by Δz, which can be expressed as

$$\Delta z = \frac{b}{2} tg\theta_2 \approx \frac{b}{2}\theta_2 \tag{5.17}$$

where b is the distance between the two suspension springs, as shown in Fig. 5.11.

Taking the moment around the rollover center C, we obtain

$$ma_y h_2 \cos(\theta_1 + \theta_2) + mgh_2 \sin(\theta_1 + \theta_2) - F_K b = 0 \tag{5.18}$$

where F_K is the suspension spring force acting on the vehicle body, and $F_K = K\Delta z$. K is the suspension spring stiffness at one side.

For a small deflection angle, cos $(\theta_1 + \theta_2) \approx 1$ and sin $(\theta_1 + \theta_2) \approx \theta_1 + \theta_2$, Eq. (5.18) can be simplified as

$$ma_y h_2 + mgh_2(\theta_1 + \theta_2) - bK\Delta z = 0 \tag{5.19}$$

Substituting Eq. (5.17) into Eq. (5.19) gives

$$\theta_2 = \frac{a_y}{g}\frac{mgh_2}{\frac{b^2}{2}K - mgh_2} + \frac{mgh_2}{\frac{b^2}{2}K - mgh_2}\theta_1 \tag{5.20}$$

The rotation angle is determined by the suspension stiffness K, the distance between two suspension springs, the lateral centrifugal acceleration, the distance between the center of gravity and the rollover center, and the vehicle weight. If the suspension is rigid (*i.e.*, the stiffness $K \to \infty$), then the deflection angle caused by the suspension will be zero (*i.e.*, $\theta_2 \to 0$). The higher the location of the center of gravity is, the larger the deflection angle becomes, and the vehicle body becomes more unstable. During turning, the lateral force always pulls the vehicle body toward the outside, that is, the deflection angle is clockwise. Hence, $\theta_2 \geq 0$, so we obtain the relation

$$b \geq \sqrt{\frac{2mgh_2}{K}} \tag{5.21}$$

The distance between the two suspension springs should be long enough to satisfy Eq. (5.21); otherwise, the body will be unstable.

If Eqs. 5.16 and 5.20 are compared, the rotation angle θ_1 caused by the tire deflection is determined only by the vehicle parameters, but the rotation angle θ_2 depends not only on the vehicle parameter but also on the lateral acceleration and θ_1.

Example 5.2: *A truck weighs 12,000 kg. Its roll center height is 0.65 m, and the vehicle body center-of-gravity height is 1.7 m. The suspension stiffness is 80,000 N/m. Design a suspension springs' distance that makes the system stable.*

Solution: The height from the body center of gravity to the rollover center, h_2, is

$$h_2 = h - h_1 = 1.7 - 0.65 = 1.05 \text{ m}$$

According to Eq. (5.21), the minimal distance between the two springs is

$$b = \sqrt{\frac{2mgh_2}{K}} = \sqrt{\frac{2 \times 12{,}000 \times 9.8 \times 1.05}{80{,}000}} = 1.76 \text{ m}$$

5.3.4 Parameters Influencing the Suspended Rollover Model

Substituting Eqs. (5.16) and (5.20) into Eq. (5.13), the normalized lateral acceleration can be expressed as

$$\frac{a_y}{g} = \frac{d}{2h} - \frac{mg}{dK_{\text{tire}}} - \frac{a_y}{g}\frac{mgh_2^2}{h(\frac{b^2}{2}K - mgh_2)} - \frac{mgh_2^2}{h(\frac{b^2}{2}K - mgh_2)}\frac{mg}{dK_{\text{tire}}} \tag{5.22}$$

Equation (5.22) can be rewritten as

$$\frac{a_y}{g} = \frac{d}{2h}\frac{h(\frac{b^2}{2}K - mgh_2)}{\frac{b^2}{2}hK - mgh_2 + mgh_2^2} - \frac{mg}{dK_{\text{tire}}} \tag{5.23}$$

To reflect the roll moment influence on the lateral acceleration, the roll stiffness of the suspension, K_{θ_2}, is introduced. The two suspension springs shown in Fig. 5.11 provide the torque to the vehicle body, which causes the body roll moment. The torque T can be expressed as

$$T = F_k b = K\Delta z b = K\frac{b}{2}\theta_2 b = K_{\theta_2}\theta_2 \tag{5.24}$$

Subsequently, the roll stiffness is

$$K_{\theta_2} = \frac{b^2}{2}K \tag{5.25}$$

Substituting Eq. (5.25) into Eq. (5.23) gives

$$\frac{a_y}{g} = \frac{d}{2h}\frac{h(K_{\theta_1} - mgh_2)}{hK_{\theta_2} - mghh_2 + mgh_2^2} - \frac{mg}{dK_{\text{tire}}} \tag{5.26}$$

The right side of Eq. (5.26) can be extended into three independent terms as

$$\frac{a_y}{g} = \frac{d}{2h} - \frac{d}{2h}\frac{mgh_2^2}{hK_{\theta_2} - mghh_2 + mgh_2^2} - \frac{mg}{dK_{\text{tire}}} \tag{5.27}$$

To clearly express Eq. (5.27), the dimensionless stiffness is introduced for the tire stiffness and suspension stiffness.

Thus, the dimensionless stiffness for the tire is

$$K_{\text{tire}}^* = \frac{K_{\text{tire}}}{mg/d}$$

and the dimensionless stiffness for the suspension is

$$K_{\text{tire}}^* = \frac{K_{\text{tire}}}{mg/d}$$

Substitute these dimensionless stiffness values into Eq. (5.27), and the normalized lateral acceleration can be expressed as

$$\frac{a_y}{g} = \frac{d}{2h} - \frac{1}{K_{tire}^*} - \frac{d}{2h}\frac{1}{1 + \frac{h}{h_2}(K_{\theta_2}^* - 1)} \tag{5.28}$$

In the right side of Eq. (5.28), the three terms have their own physical meanings. The first depends only on the basic vehicle dimensions. The second has a relation with the tire dimensionless stiffness. The third term depends on the suspension dimensionless stiffness and the basic dimensions of the vehicle. The influences of the tire and suspension stiffness will be discussed in detail in the following case studies.

Case A: Assume the tire and suspension are infinitely stiff (*i.e.*, $K_{\text{tire}} \to \infty$ and $K_{\theta_2} \to \infty$). Equation (5.28) becomes

$$\frac{a_y}{g} = \frac{d}{2h} \tag{5.29}$$

Equation (5.29) is the same expression as Eq. (5.6). The suspended model becomes the rigid model. Or, we can say that the rigid rollover model is a special case of the suspension rollover model when the tire stiffness and suspension stiffness become infinite high. In this case, the rollover lateral acceleration is determined by only two vehicle dimensions: the center-of-gravity height, and the track width.

Case B: Assume the suspension is rigid, *i.e.*, $K_{\theta_2} \to \infty$, and the only flexibility considered is that of the tire. Eq.(5.28) becomes,

$$\frac{a_y}{g} = \frac{d}{2h} - \frac{1}{K^*_{\text{tire}}} \tag{5.30}$$

The rotation angle at the rollover incipient caused by the tire deflection, θ_1^R, can be expressed by the dimensionless stiffness as

$$\theta_1^R = \frac{1}{K^*_{\text{tire}}} \tag{5.31}$$

Substitute Eq. (5.31) into Eq. (5.30) to obtain the expression

$$\frac{a_y}{g} = \frac{d}{2h} - \theta_1^R \tag{5.32}$$

The normalized lateral acceleration of rollover is controlled by the basic dimensions of the vehicle as well as the tire dimensionless stiffness. In this case, the lateral acceleration for rollover to occur is reduced by $\theta_1^R g$ (*i.e.*, the rollover threshold is reduced). The higher the tire dimensionless stiffness parameter is, the smaller the rotational angle caused by the tire deflection. If the stiffness reaches infinitely high, the second item in Eq. (5.32) becomes zero. When the tire stiffness is high enough, the lateral acceleration caused by the tire deflection becomes negligible.

Example 5.3: *A loaded truck weighs 15,000 kg, and its tread width is 2 m. The roll center height is h_1 = 0.7 m, and the body center-of-gravity height is h = 1.8 m. The suspension is assumed to be rigid. Calculate the normalized lateral acceleration.*

Solution: According to Eq. (5.31), the normalized lateral acceleration is

$$\frac{a_y}{g} = \frac{2}{2 \times 1.8} - \frac{1}{K^*_{\text{tire}}} = 0.556 - \frac{1}{K^*_{\text{tire}}}$$

The tire stiffness should be high enough to support the whole vehicle. Because the lateral acceleration is larger than zero when the rollover occurs, the tire dimensional stiffness satisfies the expression

$$K^*_{tire} > 1.8$$

The relation between the normalized lateral acceleration and the tire dimensionless stiffness is plotted in Fig. 5.12.

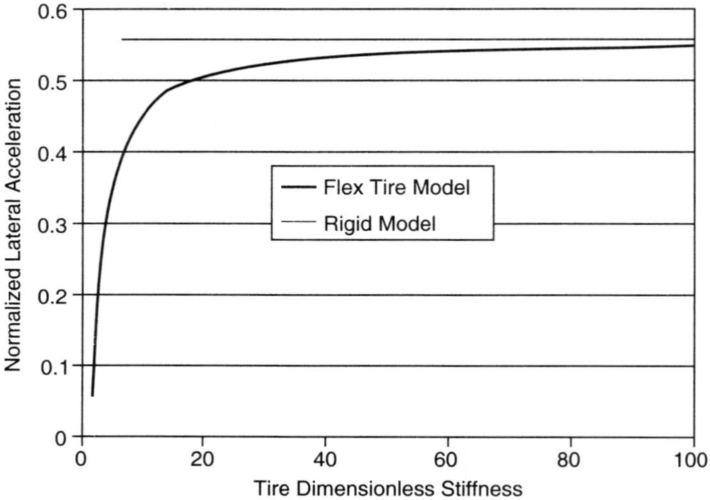

Fig. 5.12 The relation between the normalized lateral acceleration and the tire dimensionless stiffness.

When the tire dimensionless stiffness reaches 18, the lateral acceleration is 0.5g (*i.e.*, $a_y = 0.5$g), which reaches 90% of its maximum value of 0.556g. When the dimensionless stiffness reaches a certain value (18 in this example), the increase rate of the lateral acceleration gradually reduces. However, if the dimensional stiffness is below a certain value, the lateral acceleration is significantly influenced by the stiffness. If the stiffness is smaller than some value, rollover occurs easily at a very low lateral acceleration. Hence, to prevent vulnerability to rollover, it is necessary to increase the stiffness to a certain value.

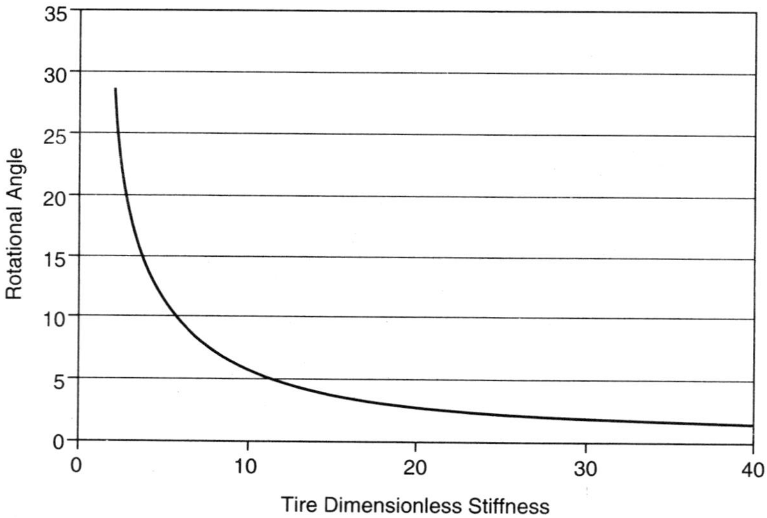

Fig. 5.13 The relation between the rollover angle and the tire dimensionless.

If rollover occurs at 0.5g lateral acceleration, the rollover angle of the vehicle becomes

$$\theta_1^R = \frac{1}{K_{\text{tire}}^*} = \frac{1}{18} * \frac{180}{3.14} = 3.18^\circ$$

The relation between the rollover angle and the tire dimensionless stiffness, is plotted in Fig. 5.13.

Case C: Assume a tire is rigid (*i.e.*, $K_{\text{tire}} \to \infty$), but the suspension deflection is finite. Equation 5.28 will be simplified as

$$\frac{a_y}{g} = \frac{d}{2h} - \frac{d}{2h} \frac{1}{1 + \frac{h}{h_2}(K_{\theta_2}^* - 1)} \tag{5.33}$$

Equation (5.33) is similar to Eq. (5.32). The second term on the right side of Eq. (5.32) is the rotational angle caused by the tire deflection. In Eq. (5.33), the second term on the right side is another rotational angle, θ_2^R caused by the suspension roll moment. Define this angle as

$$\theta_2^R = \frac{d}{2h} \frac{1}{1 + \frac{h}{h_2}(K_{\theta_2}^* - 1)} \tag{5.34}$$

Thus, Eq. (5.33) can be rewritten as

$$\frac{a_y}{g} = \frac{d}{2h} - \theta_2^R \tag{5.35}$$

The normalized lateral acceleration of rollover is controlled by the vehicle basic dimensions as well as the suspension dimensionless stiffness. In this case, the lateral acceleration needed for rollover to occur is reduced by θ_2^R g than that of the rigid vehicle model (*i.e.*, the rollover threshold is reduced). The higher the suspension dimensionless stiffness is, the smaller the rotational angle θ_2^R. If the stiffness becomes infinitely high, the second item in Eq. (5.35) goes to zero. When the suspension stiffness is high enough, the lateral acceleration caused by the suspension moment is negligible.

Example 5.4: *Assume that the tire is rigid, but the suspension stiffness is flexible. The remainder of the vehicle parameters are the same as those in example 5.3. Determine the required suspension roll stiffness for the lateral acceleration to be 85% of that of the rigid vehicle model. Also, calculate the normalized lateral acceleration.*

Solution: According to Eq. (5.33), the normalized lateral acceleration is

$$\frac{a_y}{g} = 0.556 - \frac{0.556}{1.636 K_{\theta_2}^* - 0.636}$$

The suspension stiffness should be high enough for the lateral acceleration to be larger than zero when rollover occurs. Thus,

$$0.556 - \frac{0.556}{1.636 K_{\theta_2}^* - 0.636} > 0$$

that is, $K_{\theta_2}^* > 1$

The relation between the normalized lateral acceleration and the suspension dimensionless stiffness is plotted in Fig. 5.14. The plot is similar to the one shown in Fig. 5.12.

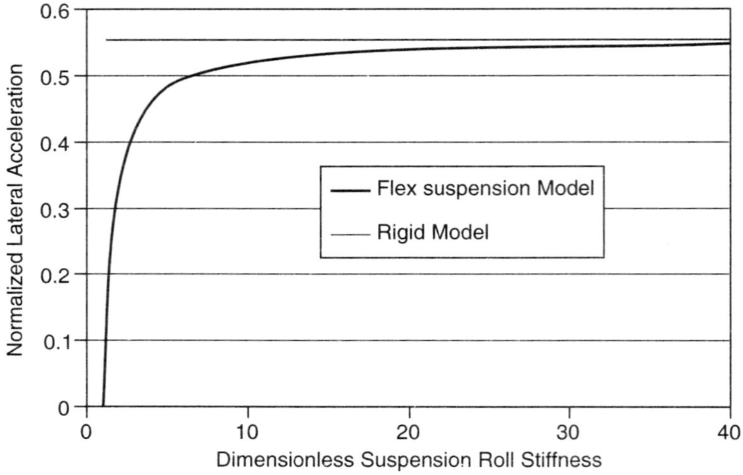

Fig. 5.14 The relation between the normalized lateral acceleration and the suspension dimensionless stiffness.

The lateral acceleration needs to be 85% that of the rigid model, that is,

$$\frac{a_y}{g} = 0.556 - \frac{0.556}{1.636K^*_{\theta_2} - 0.636} = 0.85 * 0.556$$

The dimensionless stiffness is $K^*_{\theta_2}$ = 4.46. Therefore, the roll stiffness is

$$K_{\theta_2} = mgh_2K^*_{\theta_2} = 15{,}000 * 9.8 * (1.8 - 0.7) * 4.46 \approx 721{,}000 \text{ Nm/rad}$$

The lateral acceleration corresponding to the stiffness is

$$a_y = 0.85 * 0.556g = 0.473g$$

Similar to Fig. 5.13, the relation between the rotational angle and the suspension roll dimensionless stiffness, $\theta_2^R = \frac{d}{2h}\frac{1}{1+\frac{h}{h_2}(K^*_{\theta_2} - 1)}$, can be plotted as well. The rotational angle corresponding to the lateral acceleration 0.473g is 0.0834 rad or 4.78°.

Case D: Both the tire stiffness and the suspension stiffness are flexible. The normalized lateral acceleration for this case can be written as

$$\frac{a_y}{g} = \frac{d}{2h} - \theta_1^R - \theta_2^R \tag{5.36}$$

The first term is the rollover threshold for the rigid vehicle model. The second and third terms are the corrected parameters of the rollover threshold for the tire deflection and suspension roll moment, respectively.

Example 5.5: *A loaded truck weighs 15,000 kg, and the tread width is 2 m. The roll center height is $h_1 = 0.7$ m, and the body center-of-gravity height is $h = 1.8$ m. Both the dimensionless stiffness for the tire and the suspension are 12. Calculate the rotational angle and the lateral acceleration at the moment rollover begins. Also compare the contribution of the tire and the suspension to the rotational angle and the lateral acceleration.*

Solution: The rotational angle due to the tire deflection is

$$\theta_1^R = \frac{1}{K^*_{\text{tire}}} = \frac{1}{12} = 0.083 \text{ rad} \quad \text{or} \quad 4.76^\circ$$

The rotational angle due to the suspension roll moment is

$$\theta_2^R = \frac{d}{2h}\frac{1}{1+\dfrac{h}{h_2}(K^*_{\theta_2}-1)} = \frac{2}{2*1.8}\frac{1}{1+\dfrac{1.8}{1.1}(12-1)} = 0.029 \text{ rad or } 1.68^\circ$$

For the same dimensionless stiffness, the rotational angle caused by the tire deflection is larger than that of the suspension roll moment.

The normalized lateral acceleration is

$$\frac{a_y}{g} = \frac{d}{2h} - \theta_1^R - \theta_2^R = \frac{2}{2*1.8} - 0.083 - 0.029 = 0.444$$

If only the tire deflection is considered, then the normalized acceleration is $0.556 - 0.083 = 0.473$. On the other hand, if only the suspension roll is considered, then the normalized acceleration is $0.556 - 0.029 = 0.527$.

5.4 DYNAMIC ROLLOVER MODEL

In static rollover analysis, the lateral acceleration is considered, and a vehicle is in a quasi-static situation. In addition to the lateral acceleration, the roll angular velocity and the roll angular acceleration should be considered in the vehicle rollover dynamic model. Dynamic models can be simple ones such as the rigid model, or more complicated ones such as the body-suspension model. These models can be treated analytically. In industry, dynamic models usually are built by some specific tools, including many degrees of freedom. The whole scenario of rollover motion can be simulated and animated.

5.4.1 Rigid Dynamic Model

Figure 5.15 shows a simple rigid dynamic model. Suppose the center of gravity is coincident with the roll axis. The dynamic equations can be written as

$$I_s\ddot{\varphi} = (F_{1z} - F_{2z})\frac{d}{2} + (F_{1y} + F_{2y})h \tag{5.37(a)}$$

$$ma_y = -F_{1y} - F_{2y} \tag{5.37(b)}$$

$$F_{1z} + F_{2z} = mg \tag{5.37(c)}$$

When rollover occurs, the left tire will leave the ground, so $F_{1y} = F_{2z} = 0$. Substitute Eqs. (5.37*b*) and (5.37*c*) into Eq. (5.37*a*) to obtain

$$I_s\ddot{\varphi} = mg\frac{d}{2} - ma_y h \tag{5.38}$$

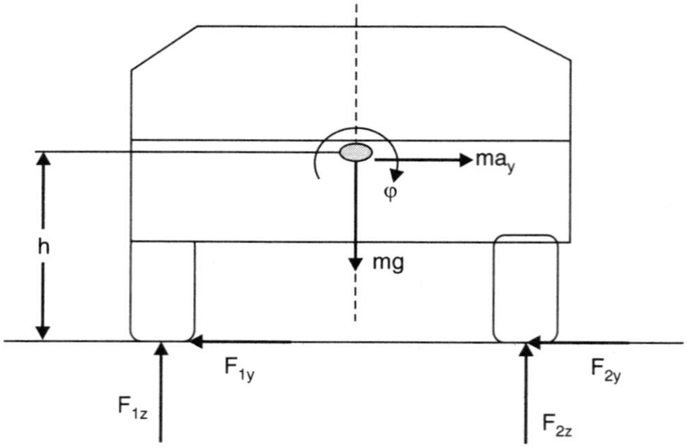

Fig. 5.15 A simple and rigid dynamic rollover model.

5.4.2 Dynamic Rollover Model for a Dependent Suspension Vehicle

Figure 5.9 is a dependent suspension model. The model can be divided into two parts: the body (sprung mass), and the suspension (unsprung mass). Force diagrams can be drawn separately, as shown in Fig. 5.16 and 5.17.

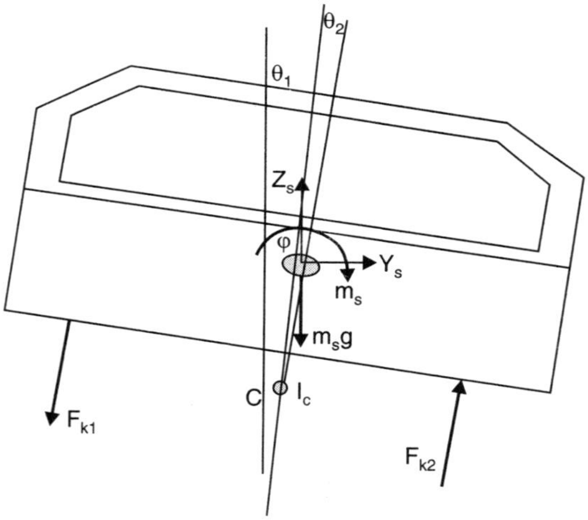

Fig. 5.16 Body (sprung mass) diagram of the suspended vehicle model.

The body is regarded as a rigid body, and its rotation center (C) and center of gravity are not coincident. The rollover dynamic equation for the body can be written as

$$m_s\ddot{z}_s = -mg + F_{k2}\cos(\theta_1 + \theta_2 + \varphi) - F_{k1}\cos(\theta_1 + \theta_2 + \varphi) \tag{5.39(a)}$$

$$m_s\ddot{y}_s = F_{k2}\sin(\theta_1 + \theta_2 + \varphi) - F_{k1}\sin(\theta_1 + \theta_2 + \varphi) \tag{5.39 (b)}$$

$$I_c\ddot{\varphi} = (m_s g - m_s\ddot{z}_s)h_2\sin(\theta_2 + \varphi) + m_s\ddot{y}_s h_2\cos(\theta_2 + \varphi) - (F_{k1} + F_{k2})\frac{b}{2} \tag{5.39(c)}$$

where I_c is the moment of inertia of the body around the roll center, and F_{kl} and F_{k2} are the spring forces.

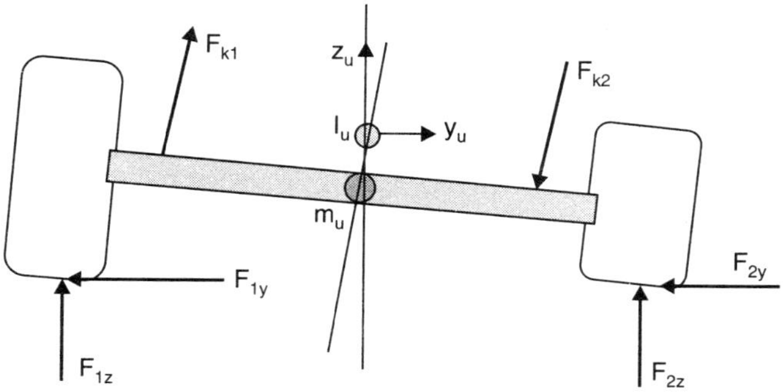

Fig. 5.17 Suspension (unsprung mass) diagram of the suspended vehicle model.

The suspension model is regarded as a rigid body, and its center of gravity is at point U. Its rotation center is at point C. When rollover starts, the forces between the inner tire and surface vanish (*i.e.*, $F_{1y} = F_{1z} = 0$). The dynamic equations for the dependent suspension are

$$m_u \ddot{z}_u = -m_u g + F_{k1}\cos(\theta_1+\theta_2+\varphi) - F_{k2}\cos(\theta_1+\theta_2+\varphi) + F_{2z} \quad (5.39(d))$$

$$m_u \ddot{y}_u = F_{k1}\sin(\theta_1+\theta_2+\varphi) - F_{k2}\sin(\theta_1+\theta_2+\varphi) - F_{2y} \quad (5.39(e))$$

$$I_u\ddot{\varphi} = (F_{k1}+F_{k2})\frac{b}{2}\cos(\theta_1+\theta_2+\varphi) + F_{2y}(h_1 - \frac{d}{2}\cos(\theta_1+\varphi) - F_{2z}\frac{d}{2}\cos(\theta_1+\varphi) \quad (5.39(f))$$

where I_u is the moment of inertia of the unsprung mass around the roll center C. If Eqs. (5.38*a*) through (5.38*d*) are combined, dynamic parameters will be obtained.

5.4.3 Dynamic Rollover Model for an Independent Suspension Vehicle

Figure 5.18 shows a dynamic rollover model for a vehicle with independent suspension. It is a half vehicle model, and the body is regarded as a rigid body. The tires are regarded as concentrated masses, and their moment of inertia is neglected. The model has seven degrees of freedom: body lateral motion, vertical motion and rotation, and vertical and lateral motion for each tire. Newton's equilibrium equations or the Lagrange method can be used to build the dynamic model.

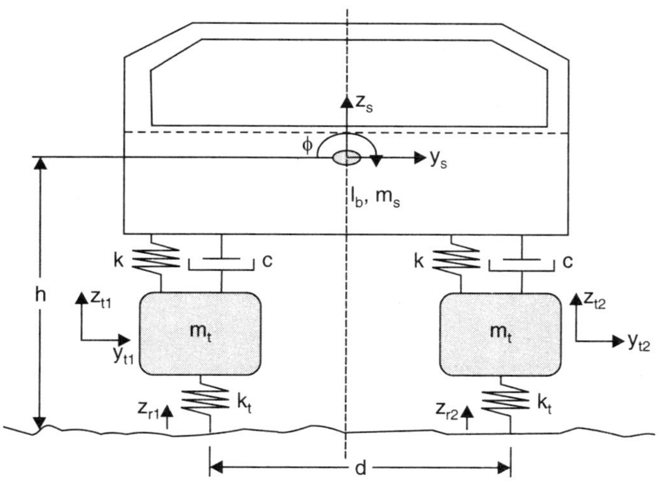

Fig. 5.18 A dynamic rollover model for a vehicle with independent suspension.

5.4.4 Rollover Simulation Tools

In this section, three dynamic models are introduced. These simple models have several degrees of freedom, which are useful for the analysis of base rollover characteristics. When more detailed information is needed, these models have their limitations, and more complicated rollover models must be developed. In industry, rollover models usually are built using specific software such as ADAMS, Carsim and NADSdyna.

With these tools, the whole virtual vehicle (Fig. 5.19) or system (a suspension as shown in Fig. 5.20) can be built. Rollover can be simulated by these virtual vehicle models. In addition, the models can be used for other dynamic attributes analysis, such as lane change, steering maneuver, accelerating, braking, cornering, handling, active vehicle dynamics, and others.

The virtual vehicle models provide many benefits for vehicle dynamics analysts. The models are especially valuable for parameter studies and optimization. The models can simulate different road terrains as well as forces and torques applied on the tires. Dynamic performance can be simulated, and the results can be animated on a computer screen. Figure 5.21 shows a truck ADAMS model on a designed road. Thus, the vehicle development time and cycle can be shortened, and the cost can be reduced. The risk caused by rollover testing also can be reduced significantly.

Fig. 5.19 A virtual vehicle dynamic model (from www.carsim.com).

Fig. 5.20 A virtual suspension dynamic model (from www.carsim.com).

The virtual vehicle models are extremely useful for situations where testing is dangerous (*e.g.*, rollover testing) and parameters are difficult to obtain experimentally. Rollover is sensitive to road conditions and transient disturbances. The tools are proven to simulate vehicle rollover well. Fig. 5.22 is an example of a rollover scenario called a fishhook maneuver. Fishhook is a difficult rollover test that requires a robotic programmable steering controller in place of the driver. However, with the help of the simulation tools, the fishhook procedure and performance can be simulated and animated on screen.

Fig. 5.21 A truck ADAMS model on a designed road (Elliott and Wheeler, SAE paper 2001-01-2734).

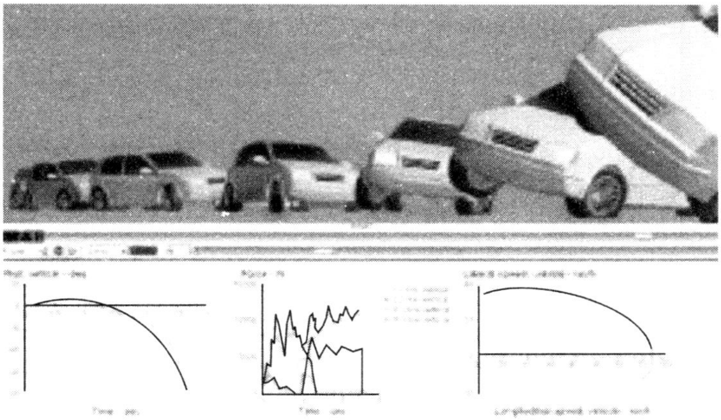

Fig. 5.22 Another virtual suspension dynamic model. (www.carsim.com.)

5.5 DYNAMIC ROLLOVER THRESHOLD

In the evaluation of the static rollover threshold, the vehicle is considered in a quasi-static state. Static stability factor, tilt table ratio, and side pull ratio are introduced. In addition to the definition of static rollover threshold, there is interest in studying the roll stability under transient and dynamic conditions. Several dynamic rollover thresholds are used, and some of them will be introduced in this section as follows:

- Dynamic stability index (DSI)
- Rollover prevention energy reserve (RPER)
- Rollover prevention metric (RPM)
- Critical sliding velocity (CSV)

These metrics are related to the vehicle lateral and rotational accelerations and/or velocities. The dynamic stability index includes rotational acceleration and velocity in addition to lateral acceleration. The last three thresholds are based on the concept of estimating the amount of energy needed to cause a vehicle to rollover. In these thresholds, the kinetic energy and potential energy are compared with the energy needed to raise the vehicle center of gravity and cause rollover.

5.5.1 Dynamic Stability Index

Equation 5.36 can be rewritten as

$$\frac{a_y}{g} + \frac{I\ddot{\varphi}}{mgh} = \frac{d}{2h} \tag{5.40}$$

Compared with Eq. 5.6, Eq. 5.40 has one more term on the left side, $\frac{I\ddot{\varphi}}{mgh}$. Because the rotational energy is considered, the left side of the equation is called the dynamic stability index (DSI), that is,

$$\text{DSI} = \frac{a_y}{g} + \frac{I\ddot{\varphi}}{mgh} \tag{5.41}$$

If the dynamic stability index is larger than the static stability factor, rollover will occur. In the analysis of the static rollover threshold, only lateral acceleration is considered as a cause of rollover. However, the roll velocity and acceleration also contribute to rollover. The dynamic stability index includes both. It provides a closer view of roll instability during dynamic testing, and it represents a practical metric to describe dynamic rollover propensity.

5.5.2 Rollover Prevention Energy Reserve

The rollover prevention energy reserve (RPER) function is defined as the difference between the energy required to bring a vehicle to its tip-over position, E_v, and the sum of the instantaneous rotational kinetic energy and potential energy, E_k. The rollover prevention energy reserve is expressed as

$$\text{RPER} = E_v - E_k \tag{5.42}$$

Prior to rollover of a vehicle, the motion of the vehicle includes both lateral and rotational motions. The vehicle kinetic energy of rollover motion consists of two parts: the kinetic energy caused by the lateral motion, and that caused by rotation, expressed as

$$E_K = \frac{1}{2}mV^2 + \frac{1}{2}I\dot{\varphi}^2 \tag{5.43}$$

where, m, V, and φ are the vehicle mass, lateral velocity, and rotational velocity, respectively.

If the rollover prevention energy reserve is positive, the kinetic energy is not enough to rollover the vehicle. If the rollover prevention energy reserve is negative, the kinetic energy exceeds the required energy to rollover the vehicle; thus, rollover will occur. During the rollover procedure, the rollover prevention energy reserve is changed from positive to negative. How soon the rollover prevention energy reserve is changed to negative and how rapidly the rollover prevention energy reserve is reduced below zero indicates the severity of the rollover. The rollover prevention energy reserve is a dynamic function of the vehicle motion, so it can be used to describe the dynamic rollover threshold. However, the measurement of energy is time consuming and expensive. Additionally, it is difficult to predict the storage of the tire and the dissipative energy.

5.5.3 Rollover Prevention Metric

The rollover prevention metric (RPM) is defined as the percentage difference between the lateral kinetic energy and the initial rotational energy as

$$\text{RPM} = 100\times\frac{T_0 - T_1}{T_0}\% \tag{5.44}$$

where T_0 and T_1 are the vehicle lateral kinetic energy and rotational energy, respectively. These terms are expressed as

$$T_0 = \frac{1}{2}mV^2 \tag{5.45}$$

$$T_1 = \frac{1}{2}I\dot{\varphi}_0^2 \tag{5.46}$$

where φ_0 is the initial rotational velocity.

According to the conservation of momentum, the momentum at impact is

$$mVH = I\dot{\varphi}_0 \tag{5.47}$$

Substituting Eq. (5.47) into Eq. (5.46), the initial rotational energy becomes

$$T_1 = \frac{1}{2}\frac{m^2V^2h^2}{I} \tag{5.48}$$

Substitute Eqs. (5.45) and (5.48) into Eq. (5.44) to obtain

$$T_1 = \frac{1}{2}\frac{m^2V^2h^2}{I} \tag{5.49}$$

The rollover prevention metric is not a function of lateral speed, and it depends only on the vehicle mass, the center-of-gravity height, and the moment of inertia.

5.5.4 Critical Sliding Velocity

When a vehicle slides on a road at a lateral velocity V and hits an object, it pivots about a hinge with an initial angular velocity $\dot{\varphi}$. The lateral kinetic energy depends on the lateral velocity. The lateral kinetic energy is transferred to the rotational kinetic energy. The rotational kinetic energy must be large enough to raise the vehicle center of gravity to the point where it is just above the pivot point about which the vehicle is rolling. Subsequently, we have

$$\frac{1}{2}I_B\dot{\varphi}_0^2 \geq mg\left(\sqrt{h^2 + \left(\frac{d}{2}\right)^2} - h\right) \tag{5.50}$$

where I_B is the moment of inertia at the hinge point B, as shown in Fig. 5.23. I_B can be expressed in terms of the moment of inertia at the center of gravity I by the expression

$$I_B = I + m(h^2 + d^2/4) \tag{5.51}$$

The critical sliding velocity (CSV) is defined as the minimum lateral velocity required to initiate rollover over an obstacle. Substitute Eq. (5.47) into Eq. (5.50) to obtain

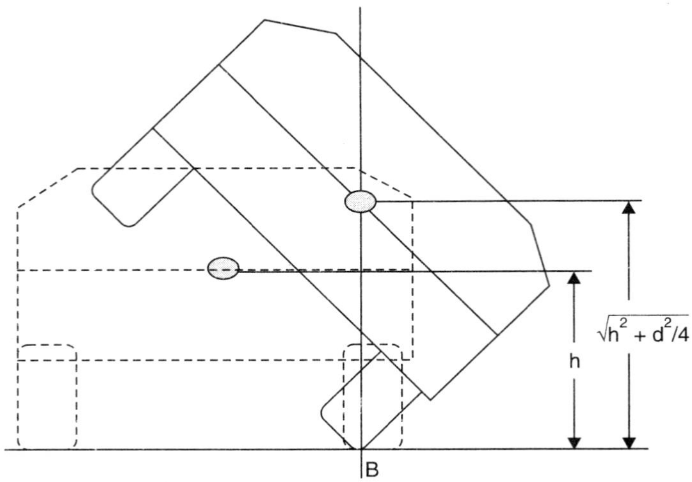

Fig. 5.23 The center of gravity of a vehicle is raised to the point where rollover occurs.

$$V^2 \geq \frac{2gI_B}{mh}\left(\sqrt{1+\left(\frac{d}{2h}\right)^2}-1\right) \tag{5.52}$$

The critical sliding velocity can be expressed as

$$V = \sqrt{\frac{2gI_B}{mh}\left(\sqrt{1+\left(\frac{d}{2h}\right)^2}-1\right)} \tag{5.53}$$

The critical sliding velocity depends on only the vehicle parameters, including its mass.

5.6 OCCUPANT IN ROLLOVER

5.6.1 Overview of the Occupant and Rollover

Rollover includes three phases: the tripping phase, the airborne phase and the ground-impact phase. In the tripping phase, occupants tend to tilt in the direction of the vehicle moment. In the airborne phase, the occupants tend to move upward and outward, away from the vehicle center of gravity. In the ground-impact phase, the occupants tend to continue moving until they impact the vehicle interior.

There are two types of injuries in rollover accidents. One is that the occupants are ejected from the vehicle when they do not wear seatbelts. The other is that occupants using seatbelts are injured by contacting with the vehicle body during airborne movement. This section focuses on the second case. The major purpose of the analysis of occupants' motion in rollover is to investigate the kinematics and forces experienced by them. The head and neck are the most vulnerable parts of the human body. Injury to these parts can be a result of contact with the roof and the upper vehicle interior during the ground-impact phase. The occupants' motion during the airborne phase is determined by restraint and centripetal forces that tend to project the occupant upward and outward relative to the compartment. Thus, it is important to understand the nature and extent of head and neck motion with respect to the interior vehicle structures.

Occupant kinematics in rollover accidents have been investigated experimentally and analytically using a variety of tools. Usually, anthropomorphic test dummies (ATDs), such as the Hybrid III ATD, are used for testing. The commonly used occupant models are the MADYMO ATD and the human facet computational model. The occupants' compartment used to investigate the motion of the occupants includes the human facet model and the vehicle interior model.

5.6.2 Testing of an Occupant Model

An occupant rollover test can be processed using test fixtures in a laboratory setting. The occupants used in rollover testing include ATDs and human volunteers. Human volunteers can be tested in a ground-based controlled rollover impact system (GB-CRIS). However, realistic rollover crash tests with people are forbidden because of the high risk of injury. Only dummies are used for frontal, side, or rear impacts in rollover crash tests. The 50% Hybrid III dummy probably is the most widely used surrogate for evaluating automotive safety in frontal crash testing.

An occupant's upward and outward motion and head excursion can be tested statically and dynamically. The static test is useful for estimating the magnitude of head excursion in a controlled acceleration ($1g$) environment but cannot quantify head excursion or head trajectory during a dynamic roll event. The dynamic test with a ground-based fixture can quantify head excursion in a dynamic environment.

The purpose of the rollover test is to investigate the occupants' kinematics in rollover crash scenarios, including head excursion and time histories. Linear accelerations in three directions (x, y and z) and three rotational velocities (roll, pitch and yaw) of the vehicle are measured. The vehicle kinematics can be obtained from these signals. Motion can be recorded by a camera, and the occupant excursion can be observed.

Figure 5.24 is a ground-based controlled rollover impact system experimental setup. A vehicle is mounted on a controlled rollover impact system and can be rotated at various rates about a principle axis that passes through its mass center. The vehicle is equipped with a continuous loop, a three-point seatbelt system configured with a sliding latch plate, an adjustable upper anchorage, and an emergency

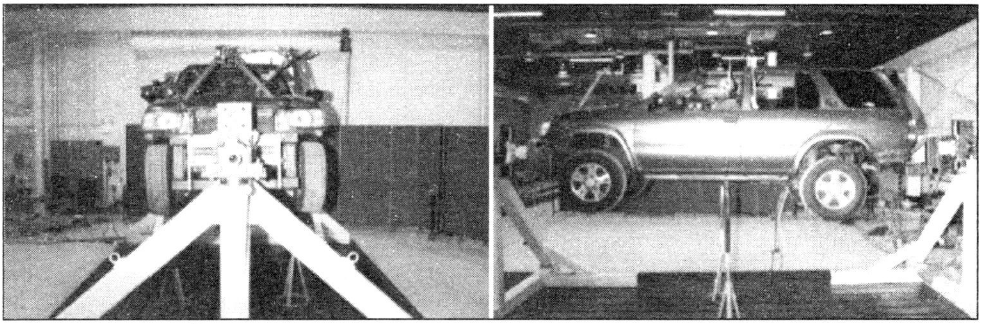

Fig. 5.24 A ground-based controlled rollover impact system experimental setup. (Newberry *et al.*, 2005.)

locking retractor mounted in the lower B-pillar. A Hybrid III ATD was positioned in the driver's seat. The kinematics of the occupant was measured by rotating the ground-based controlled rollover impact system fixture clockwise and counterclockwise. The fixture is rotated from the rest condition up to a steady-state roll rate. During the roll-up procedure, the vehicle is held for a period of several seconds at nominal constant rates to facilitate the steady-state evaluation of head excursion. Head excursion in both the outboard direction (*y* direction) and the upward direction (*z* direction) is recorded by cameras mounted inside the vehicle and is quantified by two-dimensional digitization of the Hybrid III ATD head position.

5.6.3 Simulation of Occupant Rollover

Simulation of occupant rollover is a quick and cost-effective way to understand different crash configurations and design variants, especially during the development of new vehicles. The simulated models of the vehicle and dummies are used to calculate the kinematics and dynamics of the crash.

Many occupant models have been developed in past years based on cadaver data. MADYMO is the most widely used one in automotive engineering. MADYMO is a mathematical dynamic modeling software package developed by TNO Automotive. Unlike dummy models that are developed for a particular loading direction, the MADYMO occupant model is a multidirectional model; it is not designed for a specific impact direction. Therefore, it is suitable for complex vehicle accident scenarios such as rollovers. MADYMO has two types of occupant models: a finite element (FE) model, and a multi-body facet model. The finite element model includes both rigid bodies and deformable bodies. This model can be used to simulate global deformations and kinematics as well as local deformations of components and flesh or skin. The multi-body facet models use a rigid surface finite element method. Massless shell elements are used to model the outer surfaces. The multi-body skeleton model is constructed by connecting the outer surface and the rigid and deformable bodies. The skeleton is made up of chains of rigid bodies connected by kinematic joints. The joint characteristics, joint motion, and inertial properties are based on published biomechanical data. Kinematic joints combined with dynamic restraint models are used to model the deformation of flexible components. The geometry of the model is represented by ellipsoids. Contact force characteristics for the ellipsoids are used to describe the interaction between the model and its environment and to represent the deformation of soft materials such as skin.

A multi-body facet model combined with a vehicle model forms an occupant compartment model, as shown in Figs. (5.25) and (5.26). This model is used to predict the occupants' kinematics and contact forces. The finite element model for the vehicle includes the vehicle bottom, roof, doors, windows, seats, belts, instrument panel and other components. The model can be used to calculate the kinematics and dynamics of linked bodies.

Fig. 5.25 An occupant and compartment model.
(Praxl *et al.*, 2003.)

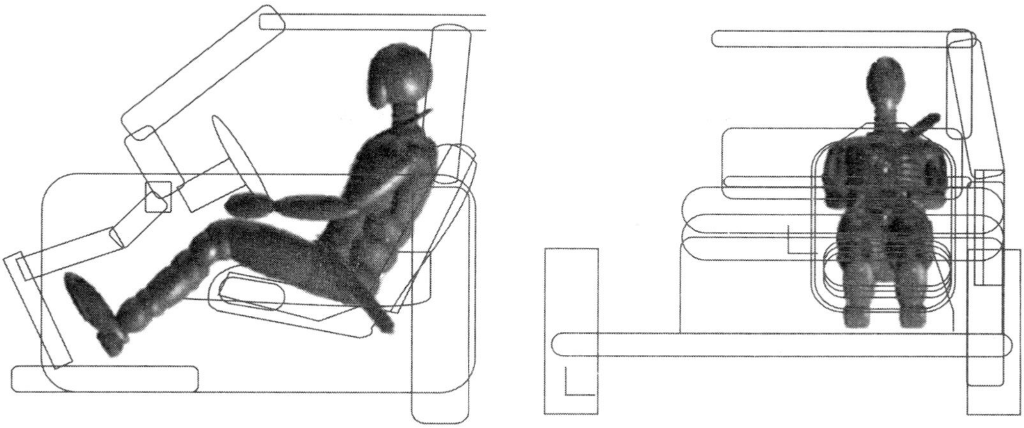

Fig. 5.26 A MADYMO model with the vehicle model.
(Newberry *et al.*, 2005.)

The occupant must be placed in the correct position in the vehicle to obtain accurate simulation results. Two major constraints of the occupant are the contact between the occupant and the seat and the contact with the seatbelt. The occupant is placed in the seat in such a way that equilibrium between the occupant's weight and the seat cushion force occurs. The seatbelt is modeled by a finite element method, as shown in Fig. 5.27. The finite element belt is placed near the outer geometry of the occupant model. The model is run until the end points of the belt move backward and the belt is wrapped around the facet.

The occupant kinematics is simulated, including the overall kinematics and the time histories of the head to contact with the side window. Figure 5.27 shows an occupant model position 30 ms after head-to-side-window contact.

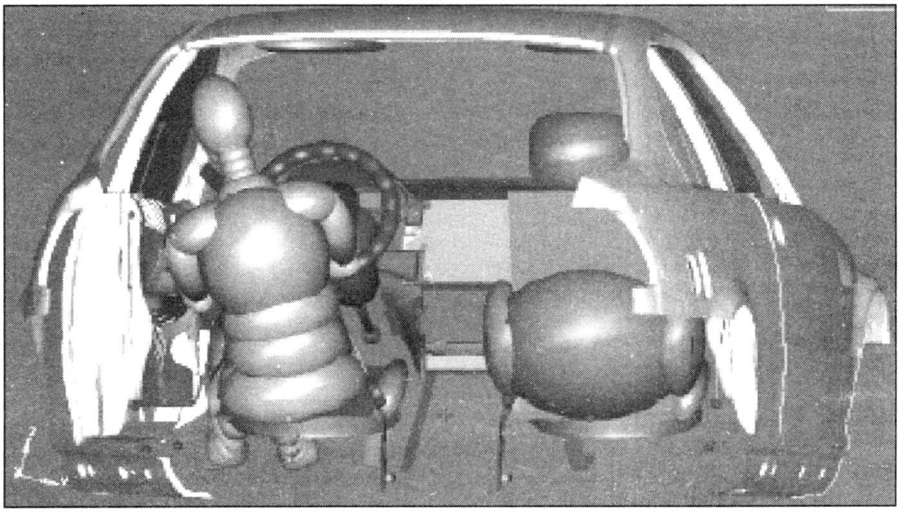

Fig. 5.27 An occupant model position 30 ms after head-to-side-window contact.
(Praxl *et al.*, 2003.)

5.7 SAFETY AND ROLLOVER CONTROL

5.7.1 Overview of Rollover Safety

According to either fatality statistics or incapacitate statistics, rollover is one of the most dangerous types of vehicle accidents. Rollover is second only to frontal crashes in the level of severity in terms of fatalities per registered vehicle. There are two types of injuries in rollover. One is the ejection of the vehicle occupants, and the other is the occupants' impact with the interior structure of the vehicle.

NASS data in 1991 show that there were approximately 200,000 rollovers. Among these accidents, about 56,000 cause serious incapacitating injuries, and about 9,000 are fatalities. The NHTSA (1999) data indicate that rollovers account for less than 5% of all vehicle crashes, but they account for about 15% of serious injuries and 25% of fatalities. Table 5.1 lists the accident distribution rate for different types of accidents and the corresponding harm rates and harm per occupant using 1988–1994 NASS and CDS (Crashworthiness Data System) data. Harm is a method developed by the NHTSA for weighting injuries according to their severity and frequency. Rollover accounts for only 8% of all accidents, but the injury or harm is 17%. The major reason for such a high injury rate is the higher number of incidents of ejection in rollover accidents.

Table 5.1 Distribution rate for different types of accidents and corresponding harm rates and harm per occupant (1988-1994 NASS and CDS, also SAE paper 970396)

	Occupants	*Harm*	*Harm Per Occupant*
Front	43%	37%	2.2
Side	23%	30%	3.4
Rear	8%	2%	0.8
Rollover	8%	17%	5.9
Other	5%	10%	5.5
Unknown	13%	4%	0.9
Total	100%	100%	

Single-vehicle rollover is the most frequent rollover accident of all vehicles. The 1989 NASS data shows that there were 137,600 rollover accidents of passenger cars. Among these, 124,800 were single-vehicle rollover accidents. The NASS-CDS database indicates that from 1995 to 1999, 81% of rollovers were single-vehicle crashes.

Single-vehicle rollover accidents are one of the major reasons for fatalities in vehicle accidents. According to 1990 FARS (Fatal Accident Reporting System) data, there were 15,901 fatalities in single-vehicle accidents. Among the accidents, 8,088 were caused by rollover, which occupies 51% of all fatalities. In 1999, 10,140 people were fatally injured in light-vehicle rollovers. Of that number, 8,348 were in single-vehicle rollover crashes. In 2001, 8,842 people lost their lives in single-vehicle rollover crashes. The number is 50% of all fatalities in single-vehicle crashes and 21% of all traffic fatalities.

From 1985 to 2000, the total number of single-vehicle crashes in the United States was reduced by 10%. However, the number of fatalities in rollover did not change much, as shown in Fig. 5.28.

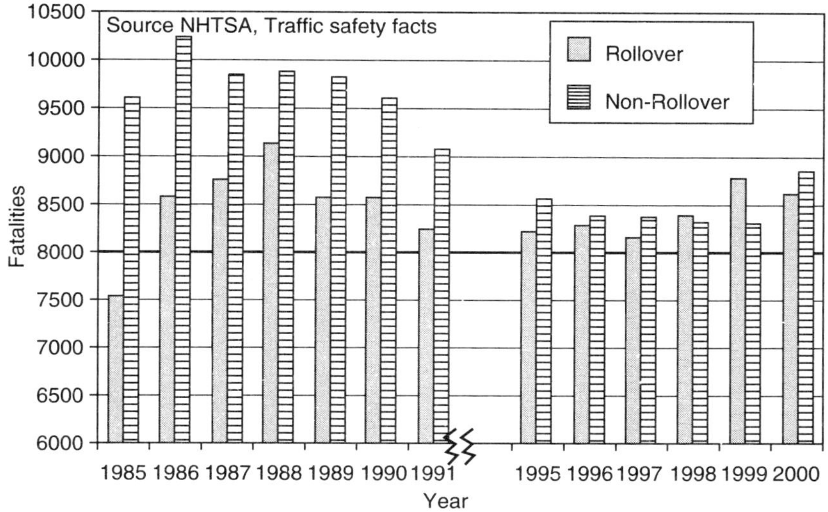

Fig. 5.28 Fatalities in single-vehicle crashes in the United States from 1985 to 2000. (Baumann and Eckstein, 2004.)

Some types of vehicles have more frequent rollover accidents than others. Figure 5.29 shows the number of rollover fatalities per million registered vehicles for different types of vehicles. Sport utility vehicles (SUVs) have the highest rates of rollover fatalities. Rollover fatalities for cars are much lower than those for SUVs and are lower than those for pickup trucks and vans.

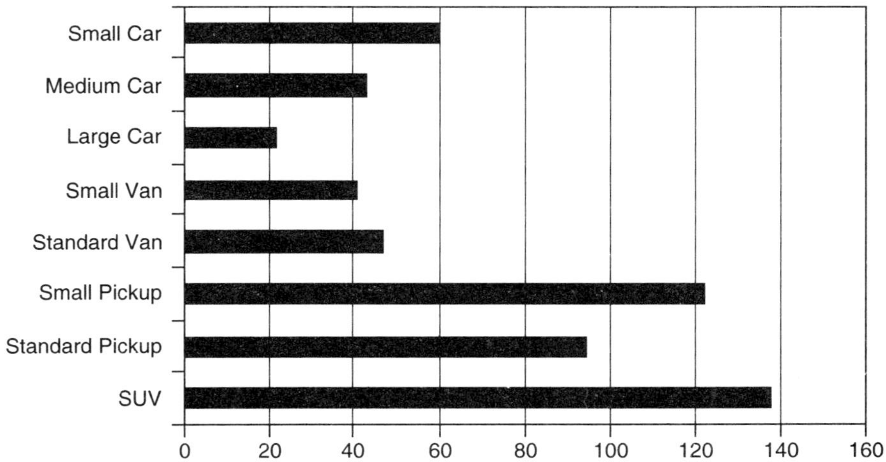

Fig. 5.29 The number of rollover fatalities per million registered vehicles for different types of vehicles, averaged from 1985 to 1990. (Hinch *et al.*, 1992.)

Table 5.2 lists rollover accident data collected by NASS in 1989. For light trucks and vans (LTVs), the number of single-vehicle rollover accidents was 65,800 from the overall 75,600 rollover accidents. Single-vehicle accidents account for about 90% of overall rollover accidents. Figure 5.30 shows statistics of fatal rollover crashes for different types of vehicles from 1988 to 2002 in the United States. This figure shows that the SUV rollover fatality has increased rapidly in recent years.

Table 5.2 Rollover accidents and single vehicle rollover accidents (NASS, 1989)

Vehicle	*Overall accidents*	*Single vehicle accidents*	*Rate of single vehicle accidents*
Passenger cars	137,600	124,800	90%
LTV	75,600	65,800	87%

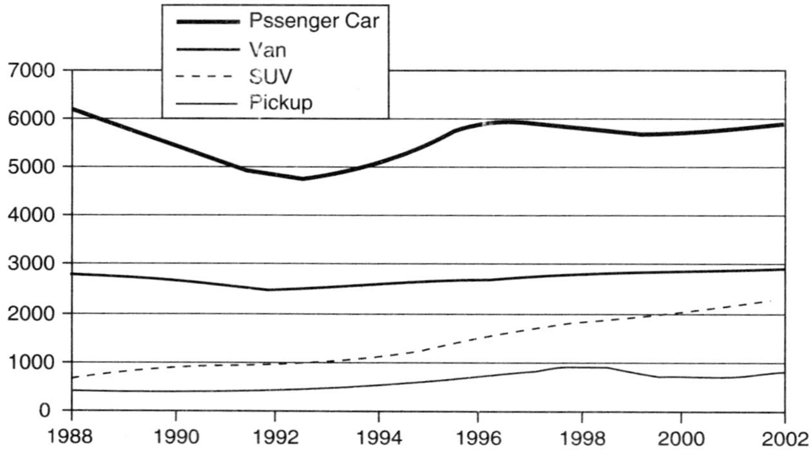

Fig. 5.30 Passenger vehicles involved in fatal crashes, by vehicle body type and year (DOT HS 809 586, 2003; also Baumann and Eckstein, 2004).

Most serious and fatal injuries in rollover accidents result from the ejection of occupants. The seatbelt is an effective tool to prevent such occupant ejection. The two functions of seatbelts are to prevent the ejection of the occupants from the vehicle and to balance the position of the occupants inside the vehicle. Unbelted occupants have a higher risk of ejection than those who are belted. The use of restraint reduces the risk of serious injury by more than 80% and fatal injuries by more than 65%. NASS and CDS data from 1982 to 1989 show that unrestrained occupants were 4.7 times more likely to be seriously injured than restrained occupants.

During the rollover phase, in addition to preventing ejection, seatbelt restraints provide substantial benefits to non-ejected occupants. Unrestrained and non-ejected occupants receive serious injuries six times more often than restrained occupants. Rollover fatalities can be reduced by as much as 75% if all occupants use seatbelts.

Table 5.3 shows that the injuries and fatalities from accidents can be reduced significantly by the use of seatbelts, especially for rollover accidents. Table 5.4 shows the distribution of occupants, harm percentage, and harm value per occupant for a vehicle that is involved in a rollover accident. Among the 39% of the occupants who did not use seatbelts, the harm is 73%, whereas the harm is only 27% for the 61% of occupants who were restrained.

Table 5.3 Seatbelt effectiveness "in fatality reduction", FARS data, Evans, 1991 (also from SAE paper 970396)

	Overall Fatality Reduction	*Reduction from Ejection Prevention*	*Reduction from Impacts with Vehicle Interior*
All Accidents	41%	18%	23%
Rollover Accidents	68%	50%	18%
Rollover As First Event	80%	64%	16%

Table 5.4 Distribution of occupants, Harm percentage and Harm value per occupant (1988-1994 NASS & CDS data, also from SAE paper 970396)

	Occupants	*Harm*	*Harm per Occupant*
Unrestrained	39%	73%	11.1
Restrained	61%	27%	2.6

For restrained occupants, the bodies move inside the vehicle and impact the interior structure. The head is the most frequently injured body region and is 4.5 times more likely to be injured than the neck during rollover.

5.7.2 Sensing of Rollover

Rollover is a complicated event. Many factors influence the initiation of rollover and the subsequent motion of the vehicle. Thus, it is important to develop rollover detection and prevent system. Active rollover control has focused mainly on two areas: rollover detection, and anti-rollover systems. A rollover detection system or sensing system consists of sensor architecture and the algorithm to activate the control system. The two major functions of a rollover sensing system are as follows:

- Accurate estimation of the vehicle roll rate, roll angle, lateral and vertical acceleration, and so forth
- Timely and accurate activation of the rollover restraint system

A tripped rollover is caused by a vehicle hitting against an obstacle or by lateral slipping. The vehicle is rotated around its longitudinal direction, as shown in Fig. 5.31(*a*). An untripped rollover is caused by the vehicle inclination where the vehicle center of gravity exceeds the stable point. The vehicle will be under vertical or lateral translation, as shown in Fig. 5.31(*b*). If rollover is caused by both of these cases, the rollover is a combination of a rotation moment and a translation force, as shown in Fig. 5.31(*c*).

Rollover includes the rotation and translation of the vehicle. Subsequently, signals that can be used as input to a rollover detector include the roll angle, roll angle rate, lateral and vertical accelerations, suspension deflection, and others. The roll angle is used as a signal because it directly correlates to the current roll stability of the vehicle. The roll angle rate is used because it is related to the rotation energy. Lateral and vertical acceleration are used because they are related to the dynamic energy. These signals are integrated in an algorithm as shown in Fig. 5.32 and then are summarized and calculated. The calculated results are compared with the rollover threshold, and then the control command is sent to deploy the rollover restraint system. The detector identifies signals using different sensors. Traditionally, there are two major sensors. One is a gravity-based sensor such as an inclinometer or tilt sensor. Gravity-based sensors respond slowly and can produce errors when exposed to linear acceleration.

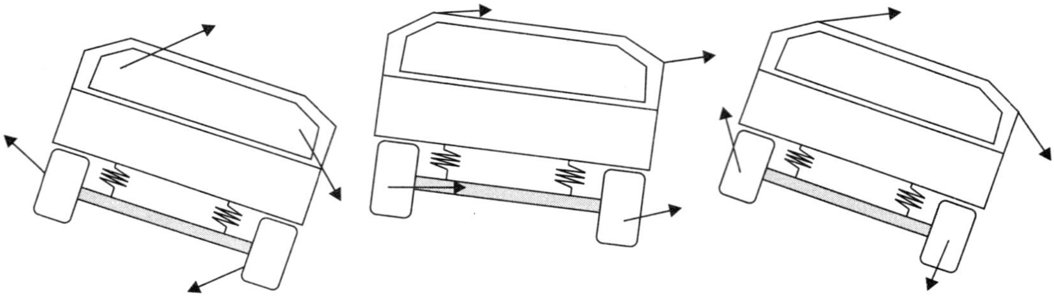

Fig. 5.31 Rollover motion: (*a*) rotation, (*b*) lateral translation, and (*c*) rotation and translation.

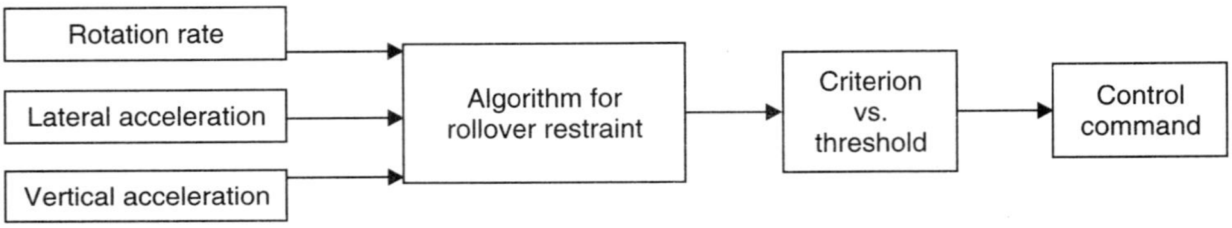

Fig. 5.32 Rollover detection system.

Likewise, they do not perform correctly during an airborne phase, on rough road, and during aggressive driving conditions. The other type of sensor is an angular rate sensor based on the Coriolis effect. The sensor is good for rollover detection but is not accurate for many other applications such as collision. Today, some special sensors have been designed, such as a satellite rollover sensor. Rollover sensors and algorithms are designed to be robust to environmental inputs such as vibration and shock.

5.7.3 Rollover Safety Control

The following methods are used to control vehicle rollover and to protect occupants:

- Seatbelts, including seatbelt pretensioners and electromechanical seatbelt retractor
- locks
- Inflatable head restraints/window curtains
- Active rollover controls
- Roll-bars
- Curtain airbags

In this section, only active rollover control will be introduced. It is a new field, and advanced technology is employed. In recent years with the development of advanced control technology and the cost reduction of electronic and control equipment, active control has been widely used in the automotive industry. Active rollover control is used to improve both comfort and handling. The following are the major anti-rollover control applications:

- Active suspensions
- Active roll-bars

- Active steering
- Anti-rollover braking systems

Active suspensions and active roll-bars directly control the vehicle roll motion. The active steering reduces vehicle oversteering and the vehicle yaw moment.

Active suspension: The suspension supports the vehicle body, controls handling, and provides ride comfort. In recent years, customers (especially SUV customers) demanded more comfortable ride and more stable handling. To satisfy their demands, active suspension usually is used. Active suspension is a regular suspension with a control system, as shown in Fig. 5.33. The control system uses electrohydraulic equipment to generate controlled forces, $F_{control}$, to react to rollover moments.

Figure 5.33 is a simplified half-vehicle model. The rollover index first must be defined. The signals, such as roll angle, roll angle rate, and lateral acceleration, are integrated in the algorithm. In critical situations where the rollover index indicates that the vehicle is under rollover danger, an emergency rollover prevention control in the suspension is made active.

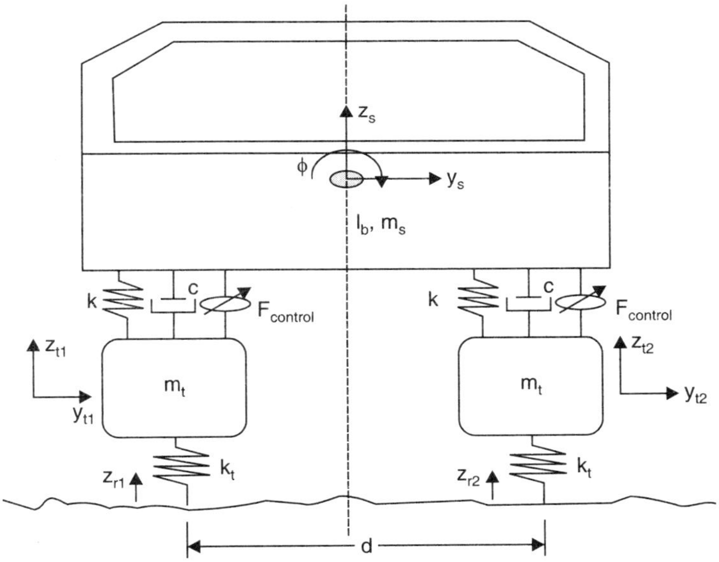

Fig. 5.33 An active suspension.

Active anti-roll-bar: An active anti-roll-bar system consists of stabilized bars, a hydraulic actuator, and an electronic control unit (ECU), as shown in Fig. 5.34. Each stabilized bar connects to a hydraulic actuator at one end. The ECU provides command to the hydraulic actuator according to the obtained lateral acceleration signal.

An active anti-roll-bar hydraulically determines the variation of the equivalent stiffness of the anti-roll-bars. The stiffness of the front and rear roll-bars also can be balanced. The vehicle load distribution is influenced by the roll-bar stiffness distribution. An active anti-roll-bar can effectively reduce the roll angle, control the roll motion, and improve ride comfort and handling by optimizing the ratio between the front and rear roll stiffness.

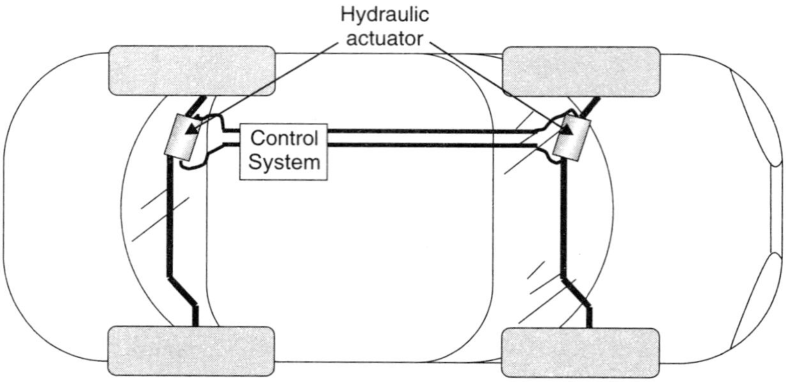

Fig. 5.34 An active anti-roll-bar system.

Active steering: Steering input significantly influences lateral vehicle dynamics. Excessive steering command may result in unstable vehicle motion. When a vehicle is at a severe left turn, its body rolls to the right in a clockwise direction, as shown in Fig. 5.35. To reduce or reverse the unstable roll movement, the steer angle must be reduced or reversed by an active steering system. Active steering with individual wheel torque control has the capability to recover a vehicle from an unstable roll state.

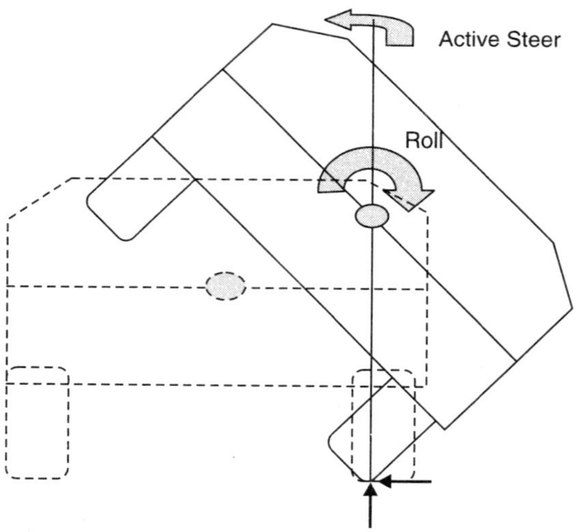

Fig. 5.35 Rollover and steering.

Anti-rollover braking system: An anti-rollover braking (ARB) system monitors vehicle motion. When a rollover is about to start, the ARB controls the front brakes. The brakes reduce the cornering capability of the front tires, which causes the vehicle to turn less sharply and reduces its speed. When the vehicle passes the rollover warning threshold, the ARB stops control, and the brakes are released.

5.8 SUMMARY

In this chapter, we reviewed the rollover scenario and its importance. Rigid models and the corresponding steady-static rollover thresholds were studied. This chapter also introduced suspension vehicle rollover models, and the influence of tires and the suspension stiffness was analyzed. In dynamic rollover models, roll acceleration/velocity in addition to lateral acceleration was considered. This chapter also provided dynamic rollover thresholds. The occupants' safety during rollover likewise was described. Finally, this chapter treated rollover sensing and anti-rollover systems.

5.9 REFERENCES

Bahling, G.S., **Bundorf, R.T.**, **Kaspzyk, G.S.**, **Moffatt, E.A.**, **Orlowski, K.F.**, and **Stocke, J.E.**, "Rollover and Drop Tests—The Influence of Roof Strength on Injury Mechanics Using Belted Dummies", SAE Paper No. 902314, SAE International, Warrendale, PA, 1990.

Baker, J.S., and **L.B. Fricke**, *The Traffic Accident Investigation Manual,* Northwestern University Traffic Institute, Evanston, IL, 1986.

Baumann, F.W., and **Eckstein, L.**, "Effects Causing Untripped Rollover of Light Passenger Vehicles in Evasive Maneuvers," SAE Paper No. 2004-01-1057, Society of Automotive Engineers, Warrendale, PA, 2004.

Bernard, J., **Shannon, J.**, and **Vanderploeg, M.**, "Vehicle Rollover on Smooth Surfaces", SAE Paper No. 891991, SAE International, Warrendale, PA, 1989.

Bready, J.E., **May, A.A.**, and **Allsop, D.**, "Physical Evidence Analysis and Roll Velocity Effects in Rollover Accident Reconstruction", SAE Paper No. 2001-01-1284, SAE International, Warrendale, PA, 2001.

Carter, J.W., **Habberstad, J.L.**, and **Croteau, J.**, "A Comparison of the Controlled Rollover Impact System (CRIS) with the J2114 Rollover Dolly", SAE Paper No. 2002-01-0694, SAE International, Warrendale, PA, 2002.

Chace, M.A., and **Wielenga, T.J.**, "A Test and Simulation Process to Improve Rollover Resistance", SAE Paper No. 1999-01-0125, SAE International, Warrendale, PA, 1999.

Chen, H.F., and **Guenther, D.**, "Modeling of Rollover Sequences", P SAE Paper No. 931976, SAE International, Warrendale, PA, 1993.

Chrstos, J.P., and **Guenther, D.A.**, "The Measurement of Static Rollover Metrics," SAE Paper No. 920582, Society of Automotive Engineers, Warrendale, PA, 1992.

Clark, S.K. [ed.], *Mechanics of Pneumatic Tires,* DOT HS 805 952, Washington, DC, August 1981.

Cliff, W.E., **Lawrence, Heinrichs, B.E.**, and **Fricker, T.R.**, "Yaw Testing of an Instrumented Vehicle with and without Braking", SAE Paper No. 2004-01-1187, SAE International, Warrendale, PA, 2004.

Cooperrider, N.K., **Hammoud, S.A.**, and **Colwell, J.**, "Characteristics of Soil-Tripped Rollovers", SAE Paper No. 980022, SAE International. Warrendale, PA, 1998.

Cooperrider, N.K., **Thomas, T.M.**, and **Hammoud, S.A.**, "Testing and Analysis of Vehicle Rollover Behavior", SAE Paper No. 900366, SAE International, Warrendale, PA, 1990.

D'Entremont, K.L., "The Effects of Light-Vehicle Design Parameters in Tripped-Rollover Maneuvers—A Statistical Approach Using an Experimentally Validated Computer Model", Paper 950315, SAE International, Warrendale, PA, 1995.

Dahlberg, E., "A Method Determining the Dynamic Rollover Threshold of Commercial Vehicles," SAE Paper No. 2000-01-3492, Society of Automotive Engineers, Warrendale, PA, 2000.

Danesin, D., Krief, P., Sorniotti, A., and **Velardocchia, M.**, "Active Roll Control to Increase Handling and Comfort," SAE Paper No. 2003-01-0962, Society of Automotive Engineers, Warrendale, PA, 2003.

Das, N.S., Suresh, B.A., and **Wambold, J.C.**, "Estimation of Dynamic Rollover Threshold of Commercial Vehicles Using Low Speed Experimental Data," SAE Paper No. 932949, Society of Automotive Engineers, Warrendale, PA, 1993.

Davis, M., Marting, P., Eger, R., Majjad, R., and **Nasr, N.**, "Rollover Simulation Based on a Nonlinear Model," SAE Paper No. 980208, Society of Automotive Engineers, Warrendale, PA, 1998.

Day, T.D., and **Garvey, J.T.**, "Applications and Limitations of 3-Dimensional Vehicle Rollover Simulation", Paper 2000-01-0852, SAE International, Warrendale, PA, 2000.

Dickerson, C.P., Arndt, S.M., Mowry, G.A., and **Arndt, M.W.**, "Effects of Outrigger Design on Vehicle Dynamics." Paper 9 40226, SAE International, Warrendale, PA, 1994.

Elliott, A.S., and **Wheeler, G.**, "Validation of ADAMS Models of Two USMC Heavy Logistic Vehicle Design Variants," SAE Paper No. 2001-01-2734, Society of Automotive Engineers, Warrendale, PA, 2001.

Erdogan, L., Guenther, D., and **Heydinger, G.**,"Suspension Parameter Measurement Using Side-Pull Test to Enhance Modeling of Vehicle Roll", SAE Paper No. 1999-01-1323, SAE International, Warrendale, PA, 1999.

Ervin, R., Winkler C., and **Karamihas, S.**, "A Centrifugal Concept for Measuring the Rollover Threshold of Light-Duty Vehicles," SAE Paper No. 2002-01-1603, Society of Automotive Engineers, Warrendale, PA, 2002.

Fay, R.J., and **J.D. Scott,** "New Dimensions in Rollover Analysis", SAE Paper No. 1999-01-0448, SAE International, Warrendale, PA, 1999.

Frimberger, M., Wolf, F., Scholpp, G., and **Schmidt, J.**, "Influences of Parameters at Vehicle Rollover," SAE Paper No. 2000-01-2669, Society of Automotive Engineers, Warrendale, PA, 2000.

Gertsch, J., and **Eichelhard, O.**, "Simulation of Dynamic Rollover Threshold for Heavy Trucks," SAE Paper No. 2003-01-3385, Society of Automotive Engineers, Warrendale, PA, 2003.

Gillespie, T., *Fundamentals of Vehicle Dynamics,* SAE International, Warrendale, PA, 1992.

Gillespie, T.D., *Fundamentals of Vehicle Dynamics*, Society of Automotive Engineers, Warrendale, PA, 1992.

Hac, A., "Influence of Chassis Characteristics on Sustained Roll, Heave, and Yaw Oscillations in Dynamic Rollover Testing," SAE Paper No. 2005-01-0398, Society of Automotive Engineers, Warrendale, PA, 2005.

Hac, A., "Rollover Stability Index Including Effects of Suspension Design", SAE Paper No. 2002-01-0965, SAE International, Warrendale, PA, 2002.

Hecker, F., ***et al.***, "Vehicle Dynamics Control for Commercial Vehicles," SAE Paper No. 973284, Society of Automotive Engineers, Warrendale, PA, 1997.

Heydinger, G., Durisek, N.J., Coovert, D.A., Guenther, D.A., and **Novak, S.J.,** "The Design of a Vehicle Inertia Measurement Facility", SAE Paper No. 950309, SAE International, Warrendale, PA, 1995.

Heydinger, G.J., and **Howe, J.G.,** "Analysis of Vehicle Response Data Measured During Severe Maneuvers," SAE Paper No. 2000-01-1644, Society of Automotive Engineers, Warrendale, PA, 2000.

Hinch, J., Shadle, S., and **Klein, T.M.,** "NHTSA's Rollover Rulemaking Program—Results of Testing and Analysis," SAE Paper No. 920581, Society of Automotive Engineers, Warrendale, PA, 1992.

James, M.B., Allsop, D.L., Nordhagen, R.P., and **Decker, R.,** "Injury Mechanisms and Field Accident Data Analysis in Rollover Accidents," SAE Paper No. 970396, Society of Automotive Engineers, Warrendale, PA, 1997.

Jialiang Le, Jerry, McCoy, Robert W., and **Chou, Clifford C.,** "Early Detection of Rollovers with Associated Test Development," SAE Paper No. 2005-01-0737, Society of Automotive Engineers, Warrendale, PA, 2005.

Jones, I.S., J.E., and **L.A. Wilson,** "Techniques for the Reconstruction of Rollover Accidents Involving Sport Utility Vehicles, Light Trucks and Minivans", SAE Paper No. 2000-01-0851, SAE International, Warrendale, PA, 2000.

Kaplan, M., Bilek, D., Kaplan, S., Vellos, D., and **Gilbert, M.G.,** "An Examination of Rim Gouging and Its Relation to On-Road Vehicle Rollover", SOARce, Society of Accident Reconstructionists, Wheat Ridge, CO, Spring 2004.

Kazemi, R., and **Soltani, K.,** "The Effects of Important Parameters on Vehicle Rollover with Sensitivity Analysis," SAE Paper No. 2003-01-0170, Society of Automotive Engineers, Warrendale, PA, 2003.

Lai, W., "Evaluation of Human Surrogate Models for Rollover," SAE Paper No. 2005-01-0941, Society of Automotive Engineers, Warrendale, PA, 2005.

Larson, R.E., Smith, J.W., Werner, S.M., and **Fowler, G.F.,** "Vehicle Rollover Testing in Recreating Rollover Collisions", SAE Paper No. 2000-01-1641, SAE International, Warrendale, PA, 2000.

Lund, Y.I., and **Bernard, J.E.,** "Analysis of Simple Rollover Metrics", SAE Paper No. 950306, SAE International, Warrendale, PA, 1995.

Lund, Y.I., and **Bernard, J.E.,** "Analysis of Simple Rollover Metrics," SAE Paper No. 950306, Society of Automotive Engineers, Warrendale, PA, 1995.

Marine, M.C., Wirth, J.L., and **Thomas, T.M.,** "Characteristics of On-Road Rollovers", SAE Paper No. 1999-01-0122, SAE International, Warrendale, PA, 1999.

Martinez, J.E., and **Schlueter, R.,** "A Primer on the Reconstruction and Presentation of Rollover Accidents", SAE Paper No. 960647, SAE International, Warrendale, PA, 1996.

Meyer, S.E., Davis, M., Forrest, S., Chng, D., and **Herbst, B.,** "Accident Reconstruction of Rollovers—A Methodology", SAE Paper No. 2000-01-0853, SAE International, Warrendale, PA, 2000.

Moffatt, E.A., and **Padmanaban, J.**, "The Relationship Between Vehicle Roof Strength and Occupant Injury in Rollover Crash Data", 39th Annual Proceedings, Association for the Advancement of Automotive Medicine, Chicago, IL, October 1995.

Nalecz, A.G., Lu, Z., and **d'Entermont, K.L.**, "An Investigation into Dynamic Measures of Vehicle Rollover Propensity," SAE Paper No. 930831, Society of Automotive Engineers, Warrendale, PA, 1993.

Newberry, W., Carhart, M., Lai, W., Corrigan, C.F., and **Croteau, J.**, "A Computational Analysis of the Airborne Phase of Vehicle Rollover: Occupant Head Excursion and Head-Neck Posture," SAE Paper No. 2005-01-0943, Society of Automotive Engineers, Warrendale, PA, 2005.

Orlowski, K.F., Bundorf, R.T., and **Moffatt, E.A.**, "Rollover Crash Tests–The Influence of Roof Strength on Injury Mechanics", SAE Paper No. 851734, SAE International, Warrendale, PA, 1985.

Orlowski, K.R., Moffatt, E.A., Bundorf, R.T., and **Holcomb, M.P.,** "Reconstruction of Rollover Collisions", SAE Paper No. 890857, SAE International, Warrendale, PA, 1989.

Pacejka, H.B., *Tire and Vehicle Dynamics,* Butterworth-Heinemann, Woburn, MA, 2002.

Padmanaban, Jeya., and **Stone, M.**, "Examination of Rollover Accident Characteristics Using Field Performance Data," SAE Paper No. 2002-01-2056, Society of Automotive Engineers, Warrendale, PA, 2002.

Ponticel, P., "Dynamic Rollover Testing On the Way", Automotive Engineering International, SAE International. Warrendale, PA, November 2003, pp. 26-28.

Praxl, N., Schönpflug, M., and **Adamec, J.,** "Application of Human Models in Vehicle Rollover," SAE Paper No. 2003-01-2188, Society of Automotive Engineers, Warrendale, PA, 2003.

Rains, G.C., and **Kanianthra, J.N.,** "Determination of the Significance of Roof Crush on Head and Neck Injury to Passenger Vehicle Occupants in Rollover Crashes", SAE Paper No. 950655, SAE International, Warrendale, PA, 1995.

Rosenthal, T.J., Szostak, H.T., and **Allen, R.W.,** "User's Guide and Program Description For Tripped Roll Over Vehicle Simulation", DOT HS 807 140 Final Report, Washington, DC, 1987.

Ruhl, R.L., and **Ruhl, R.A.,** "Prediction of Steady State Roll Threshold for Loaded Flat Bed Trailers—Theory and Calculation," SAE Paper No. 973261, Society of Automotive Engineers, Warrendale, PA, 1997.

Salaani, M.K., Heydinger, G.J., and **Grygier, P.A.,** "Heavy Tractor-Trailer Vehicle Dynamics Modeling for the National Advanced Driving Simulator," SAE Paper No. 2003-01-0965, Society of Automotive Engineers, Warrendale, PA, 2003.

Shim, T., and **Toomey, D.,** "Investigation of Active Steering/Wheel Torque Control at the Rollover Limit Maneuver," SAE Paper No. 2004-01-2097, Society of Automotive Engineers, Warrendale, PA, 2004.

Slaates, P.M., and **Coo, P.D.,** "Safety Restraint Systems in Heavy Truck Rollover Scenarios," SAE Paper No. 2003-01-3424, Society of Automotive Engineers, Warrendale, PA, 2003.

Stevens, D.C., "Passenger Vehicle Rollover Reconstruction: Investigation, Analysis and Presentation of Results", Passenger Vehicle Rollover TOPTEC: Causes, Prevention and Injury Prevalence, April 22-23, Scottsdale, AZ, SAE International, Warrendale, PA, 2002.

Tamny, S., "Friction Induced Rollover from Lift-Off to Launch", SAE Paper No. 2000-01-1649, SAE International, Warrendale, PA, 2000.

Viano, D.C., and **Parenteau, C.S.**, "Rollover Crash Sensing and Safety Overview," SAE Paper No. 2004-01-0342, Society of Automotive Engineers, Warrendale, PA, 2004.

Waller, E., and **Schiffmann, J.K.**, "Development of an Automotive Rollover Sensor," SAE Paper No. 2000-01-1651, Society of Automotive Engineers, Warrendale, PA, 2000.

Wielenga, T.J., and **Chace, M.A.**, "A Study in Rollover Prevention Using Anti-Rollover Braking," SAE Paper No. 2000-01-1642, Society of Automotive Engineers, Warrendale, PA, 2000.

Winkler, C.B., **Blower, D.**, **Ervin, R.D.**, and **Chalasani, R.M.**, "Rollover of Heavy Commercial Vehicles", RR-004, SAE International, Warrendale, PA, 2000.

www.adams.com, the ADAMS website.

www.carsim.com, the CarSim website.

www.madymo.com, the MADYMO website.

Yang, H., and **Liu, Y.**, "A Robust Active Suspension Controller with Rollover Prevention," SAE Paper No. 2003-01-0959, Society of Automotive Engineers, Warrendale, PA, 2003.

CHAPTER

6

Vehicle Dynamics

6.1 RIDE DYNAMICS

6.1.1 Introduction

Ride quality deals with passengers' comfort in the vehicle. The judgment of ride quality can vary from one customer to another according to their perception and experience. Several factors such as vibration, noise, heat, sitting posture, age, gender, and environment influence ride quality. Among these factors, vibration is considered the most important in determining the overall ride quality. Passengers in a moving vehicle are continuously exposed to a vibration environment caused by excitations from the road surface and from vehicle internal sources such as the powerplant, which is the combination of the engine and the transmission. Hence, this chapter discusses ride quality caused by vibration.

Ride comfort is one of the most important factors for a customer in determining whether or not to purchase a particular vehicle. Ride quality is an important attribute in modern car design and is a major contributor to customer satisfaction. For almost all vehicle OEMs (original equipment manufacturers) and suppliers, customer satisfaction has become the critical issue in vehicle product development and business strategy. Passengers provide the ride quality evaluation when they ride in the vehicles. The judgment of ride quality by passengers is a subjective evaluation. Ride quality also can be evaluated by objective methods, including simulation and testing. Ride quality is directly related to the seat comfort design. The seat is mounted on the vehicle; thus, the vehicle design also influences ride quality. The vehicle is subjected to external and internal vibration sources that determine the magnitude of ride discomfort. Therefore, research on ride quality and ride dynamics includes studies on vibration sources and modeling of the vehicle, seat, and human body. Figure 6.1 lists the ride dynamics system. The excitation to a passenger comes from both external and internal sources of the vehicle. The external sources include road input and wind excitation. Wind excitation has a significant impact on the vehicle body structure vibration that is transferred to interior sound. However, this contribution to passenger discomfort is less significant compared with other vibration sources. The major external excitation is from an uneven road profile. The power spectral density (PSD) of passenger vibration is roughly proportional to the power spectral density of the road surface profile. Ride vibration also depends on the vehicle speed. The faster the vehicle is moving, the more severe the vibration will be.

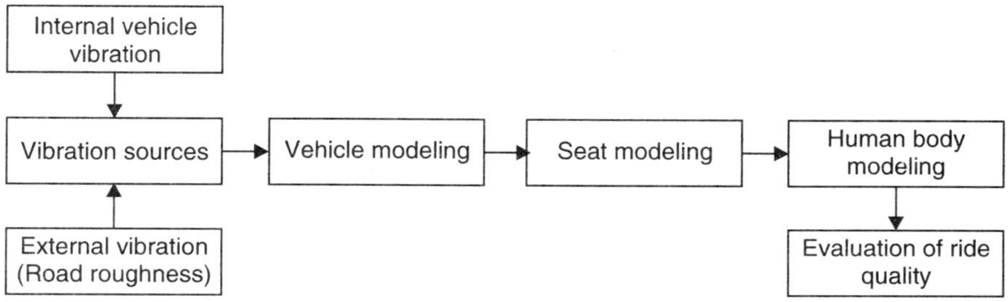

Fig. 6.1 The system of ride dynamics.

The major internal excitation is from the engine. Engine vibration is transferred to the body through the engine mounts, exhaust hangers, and driveshaft bearings. The mount system design is a key element in reducing vibration. Engine vibration consists mainly of two sources. The first is the engine firing pulse, and the second is the inertia force from the powerplant. Usually, the firing pulse excitation is much higher than the inertia force.

The driveshaft and exhaust are mounted on the body through bearings, hangers, and isolators. The vibration caused by the driveshaft and exhaust will be transferred to the body. Sometimes, the vibration sources from the driveshaft and exhaust are severe and can even reach the same level as the engine vibration.

The modeling of the body is important in the analysis of ride dynamics. Two models are widely used in ride dynamics analysis. One is the quarter car model, and the other is the bounce-pitch model. The quarter car model includes the unsprung mass system, which consists of the tire and spindle, and the sprung mass system, which consists of the suspension and body. Only vertical vibration is considered in this model. This two-degrees-of-freedom model predicts two major vertical modes. The bounce-pitch model also is a two-degrees-of-freedom model, but it considers both vertical motion and pitch motion. In addition to the two commonly used models, many other complicated models are studied. In automotive engineering, finite element method (FEM) models are used by more and more engineers. FEM models provide not only accurate results but also can be integrated with other applications.

Excitation is transferred to the seat through the vehicle structure. The SEAT (seat effectiveness amplitude transmissibility) value is used to evaluate the effectiveness of the seat vibration isolation. Traditionally, linear seat cushion models have been used for the calculation of seat transmissibility. However, the seat cushion is a nonlinear structure according to recent research results (Pang et al. 2003). This chapter describes both linear and nonlinear models.

Subjective and objective methods are used to evaluate ride quality. The VER (vehicle evaluation rating) is used for subjective evaluation. Passenger ride discomfort relates not only to the vibration amplitude but also to the vibration frequency. A weighting root-mean-square method and vibration dose value are used for objective evaluation. This chapter presents both linear and nonlinear seat-body models. Traditional passive control to reduce the vibration from the engine and from road input is not effective for all frequency ranges. Active and semi-active controls have found more and more applications in modern vehicles. Active control possesses the capability to reduce vibration at any frequency, but its cost is high and its structure is complicated. Semi-active control can reach some level of active control effectiveness with lower cost and a simpler structure. This chapter introduces the controls of the engine isolation system and the suspension system.

6.1.2 Vibration Environment in Road Vehicles

This section deals with the vehicle vibration environment: outside excitation, and internal excitation. The uneven road profile is the major source of outside vibration. The vibration will be transferred to occupants through the suspension and seating systems. The major internal sources of vibration include engine excitation, driveline excitation, and exhaust vibration.

6.1.2.1 Vibration Sources from the Road

Vibration contributed by the road is due to the interaction of the tire tread with the profile of a rough or coarse road surface. The level of road excitation depends on the road profile, vehicle speed, and tire dynamic properties. The road surface profile is one of the major excitation sources to ride dynamics. It has been found that the power spectral density of the interior vibration is roughly proportional to the power spectral density of the road surface profile. Speed is another important factor in ride quality. Increasing the vehicle speed biases the vibration spectrum to higher frequencies and usually increases the overall vibration level. The tire dynamic properties include the tire stiffness and pressure, the tire patch length and shape, and the width and size of the tire. These properties also influence the ride characteristics. The vibration levels increase as the tire pressure or stiffness increases. Shortening the tire patch length biases the vibration spectrum to higher frequencies. A wider tire typically involves a shorter tire patch, which will tend to bias frequencies upward.

6.1.2.1.1 Power Spectral Density in Spatial Frequency

Figure 6.2 shows a road profile. A vehicle travels on a road in the x direction, while the vehicle vibration is in the z direction. A road profile includes many waves. Spatial frequency, Ω, stands for the wavelength number per meter of the road, and the unit is m^{-1}. The spatial frequency is the inverse of the wavelength.

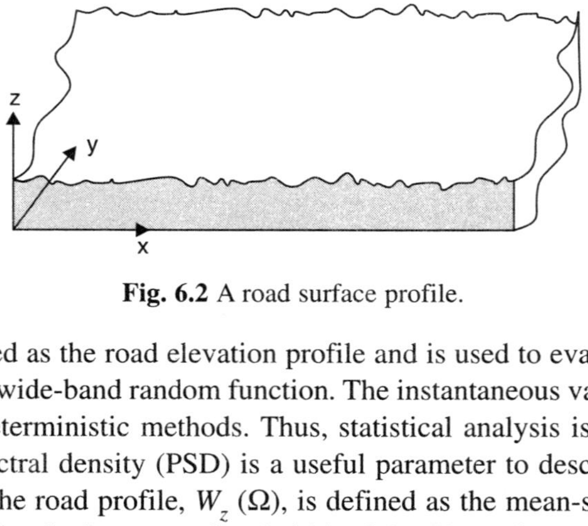

Fig. 6.2 A road surface profile.

Roughness is defined as the road elevation profile and is used to evaluate the road smoothness. Usually, a road profile is a wide-band random function. The instantaneous values for a random function cannot be expressed by deterministic methods. Thus, statistical analysis is used to describe the road vibration input. Power spectral density (PSD) is a useful parameter to describe random function. The power spectral density of the road profile, $W_z(\Omega)$, is defined as the mean-square response of an ideal narrow-band filter divided by the frequency bandwidth of the filter $\Delta\Omega$ as $\Delta\Omega \to 0$ at frequency Ω.

$$W_z(\Omega) = \lim_{\Delta\Omega \to 0} \frac{\overline{z_{\Delta\Omega}^2}}{\Delta\Omega} \tag{6.1}$$

where $\overline{z_{\Delta\Omega}^2}$ is the mean-square value within $\Delta\Omega$. Then, the mean-square value for the whole frequency range is

$$\overline{z^2} = \int_0^{\infty} W_z(\Omega)\, d\Omega \tag{6.2}$$

Or, the mean-square value during the wavelength L_i can be expressed as

$$\overline{zL_i^2} = \frac{1}{L_i}\int_0^{L_i} z_i^2 dx \tag{6.3}$$

The power spectral density is a function of the spatial frequency. Based on road test data, a relation between the power spectral density of the road profile and the spatial frequency was proposed by ISO as

$$W_z(\Omega) = W_z(\Omega_0)\left(\frac{\Omega}{\Omega_0}\right)^{-N} \tag{6.4}$$

where Ω_0 is the reference spatial frequency and $\Omega_0 = \frac{1}{2\pi}\mathrm{m}^{-1}$; $W_z(\Omega_0)$ is the profile spectral density corresponding to the reference spatial frequency. $W_z(\Omega_0)$ also is called the profile roughness coefficient. N is the frequency index and determines the frequency structure of the spectral density and the slope of the log-log curves.

The road roughness is divided into eight levels in the ISO standard from A to H. Table 6.1 lists the range and geometric mean values of the roughness coefficients.

Table 6.1 Roughness coefficients of road profiles

Road class	*Roughness coefficient $W_z(\Omega_0) \times 10^{-6}$*	
	Range	*Geometric mean*
A	8–32	16
B	32–128	64
C	128–512	256
D	512–2,048	1,024
E	2,048–8,192	4,096
F	8,192–32,768	16,384
G	32,768–131,072	65,536
H	131,072–524,288	262,144

The frequency index N can be chosen as several values for different road wavelengths. For most cases, N can be chosen as 2, that is, $N = 2$. For this case, Eq. (6.4) can be written as

$$W_z(\Omega) = \frac{W_z(\Omega_0)\Omega_0^2}{\Omega^2} \tag{6.5}$$

The power spectral density is inversely proportional to the square of the spatial frequency or is proportional to the square of the wavelength. Figure 6.3 plots the power spectral density of the road profile for the eight levels according to Eq. (6.5).

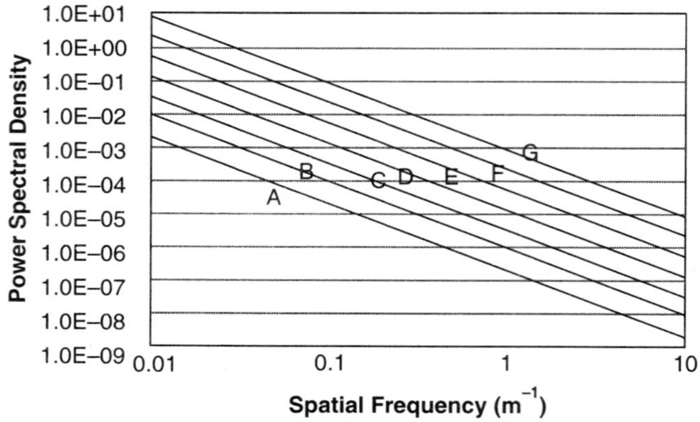

Fig. 6.3 Power spectral density for different

Usually, the elevation of the road surface along which a vehicle travels is used to describe the road roughness, which means that the vertical displacement input to the vehicle is the roughness. On the other hand, vibration is commonly expressed by acceleration and velocity. Hence, acceleration and velocity and their power spectral densities will be used as vibration input for vehicle ride dynamics analysis. The relation between displacement and acceleration can be obtained by harmonic analysis. Assume that the roughness displacement z of a road profile during a wavelength L can be expressed as

$$z(x) = A\sin\left(\frac{2\pi x}{L}\right) = A\sin(\Omega x) \tag{6.6}$$

where A is the amplitude of the road profile during the wavelength. The corresponding velocity and acceleration are given by

$$\dot{z}(x) = \frac{2\pi A}{L}\cos\left(\frac{2\pi x}{L}\right) = A\Omega\cos(\Omega x) \tag{6.7}$$

$$\ddot{z}(x) = -\left(\frac{2\pi}{L}\right)^2 A\sin\left(\frac{2\pi x}{L}\right) = -A\Omega^2\sin(\Omega x) \tag{6.8}$$

The mean square value of $\dot{z}(x)$ during the wavelength is obtained by

$$\overline{\dot{z}^2} = (2\pi\Omega)^2\,\overline{z^2} \tag{6.9}$$

The mean square value of $\ddot{z}(x)$ is given by

$$\overline{\ddot{z}^2} = (2\pi\Omega)^4\,\overline{z^2} \tag{6.10}$$

If Eqs. (6.9) and (6.10) are substituted into Eq. (6.1), the velocity and acceleration power spectral density can be obtained as

$$W_{\dot{z}}(\Omega) = (2\pi\Omega)^2 W_z(\Omega) \tag{6.11}$$

$$W_{\ddot{z}}(\Omega) = (2\pi\Omega)^4 W_z(\Omega) \tag{6.12}$$

For the special case where $N = 2$, substitute Eq. (6.5) into Eq. (6.12), and the velocity spectral density will be

$$W_{\ddot{z}}(\Omega) = (2\pi\Omega)^2 \frac{W_z(\Omega_0)\Omega_0^2}{\Omega^2} = 4\pi^2 W_z(\Omega_0)\Omega_0^2 \tag{6.13}$$

The velocity spectral density is a constant and is independent of the spatial frequency. That is, the spectral density is "white noise."

6.1.2.1.2 Power Spectral Density in Temporal Frequency

The power spectral density in spatial frequency has been discussed. However, vehicle vibration usually is described by temporal frequency instead of spatial frequency. Thus, it is more convenient to express the spectral density of surface profiles using temporal frequency in hertz than using spatial frequency in m^{-1}. The temporal frequency relates to the vehicle speed and spatial frequency. The relation between the temporal frequency and spatial frequency can be expressed as

$$f = V * \Omega \tag{6.14}$$

where f is the temporal frequency and V is the vehicle speed.

From Eq. (6.14), it can be seen that the temporal frequency is proportional to the vehicle speed for a given spatial frequency or a certain wavelength. The relation between the temporal frequency and spatial frequency can be plotted in Fig. 6.4.

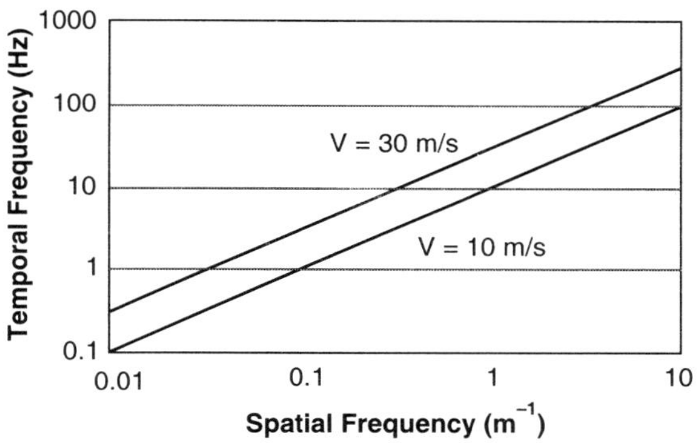

Fig. 6.4 Relation between temporal frequency and spatial frequency.

The statistical data shows that the spatial frequency for most road profiles is between 0.011 and 2.83 m^{-1}. The temporal frequency range is between 0.33 and 28.3 Hz for the corresponding spatial frequencies during the normal operation speed range of the vehicle (V = 10 ~ 30 m/s). Thus, the frequencies for the vehicle sprung mass (1 ~ 2 Hz) and the frequencies for the unsprung mass (10 ~ 15 Hz) fall into the road excitation frequency range, as shown in Fig. 6.4.

Similar to Eq. (6.1), the power spectral density in temporal frequency can be expressed as

$$W_z(f) = \lim_{\Delta f \to 0} \frac{\overline{z_{\Delta f}^2}}{\Delta f} \tag{6.15}$$

where Δf is the temporal frequency range. Substituting Eq. (6.14) into Eq. (6.15), the relation between the power spectral density in spatial frequency and that in temporal density can be obtained as

$$W_z(f) = \frac{1}{V} W_z(\Omega) \tag{6.16}$$

Substituting Eq. (6.4) into Eq. (6.16), the power spectral density in temporal frequency can be rewritten as

$$W_z(f) = \frac{W_z(\Omega_0)}{V}\left(\frac{\Omega}{\Omega_0}\right)^{-N} \tag{6.17}$$

For the special case where $N = 2$, the density is

$$W_z(f) = \frac{1}{V}W_z(\Omega_0)\left(\frac{\Omega}{\Omega_0}\right)^{-2} = \frac{\Omega_0^2 W_z(\Omega_0)}{V\Omega^2} = \frac{\Omega_0^2 W_z(\Omega_0)V}{f^2} \tag{6.18}$$

Performing similar manipulation on Eqs. (6.11) and (6.12), the velocity power spectral density and acceleration power spectral density are, respectively,

$$W_{\dot{z}}(f) = (2\pi f)^2 W_z(f) = 4\pi^2 W_z(\Omega_0)\Omega_0^2 V \tag{6.19}$$

$$W_{\ddot{z}}(f) = (2\pi f)^4 W_z(f) = 4\pi^4 W_z(\Omega_0)\Omega_0^2 V f^2 \tag{6.20}$$

Figure 6.5 shows the power density spectra of the road profile, velocity, and acceleration for road class *A* and *a* vehicle velocity of 10 m/s.

The velocity density spectrum is a constant for a given vehicle velocity. The road profile power spectral density decreases as the frequency increases. However, the acceleration power spectral density increases as the frequency increases. All three density spectra are proportional to the vehicle speed. The spectral densities increase as the speed increases. The acceleration density increases with the square of the temporal frequency, that is, the vibration transferred to the vehicle at high frequency is more severe than that transferred at low frequency.

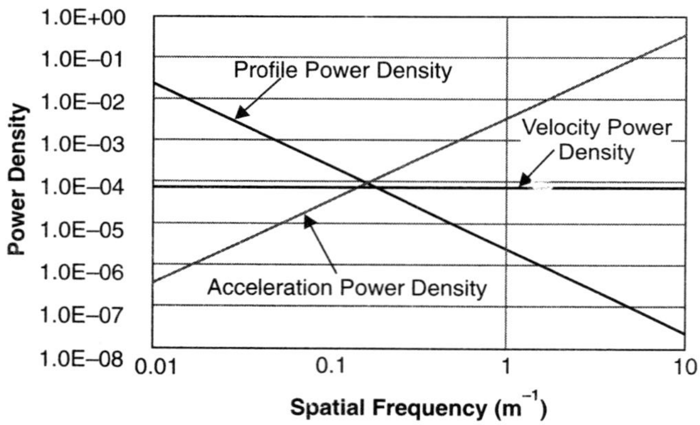

Fig. 6.5 Power density spectra of the road profile, velocity and acceleration.

6.1.2.2 Vehicle Internal Vibration Sources

Vehicle internal excitation sources come mainly from power train vibration. The power-train includes the engine, transmission, driveline system, exhaust system, and induction system. The powerplant is one of the major sources of vehicle vibration. The driveline, exhaust, and intake are connected with the powerplant. Vibration from the powerplant also is transferred to the vehicle body through these systems. Figure 6.6 shows the internal vibration sources. The induction system is connected with the engine using very soft tubing so that the engine vibration is almost isolated. This chapter describes in detail the excitation sources from the powerplant, driveline and exhaust.

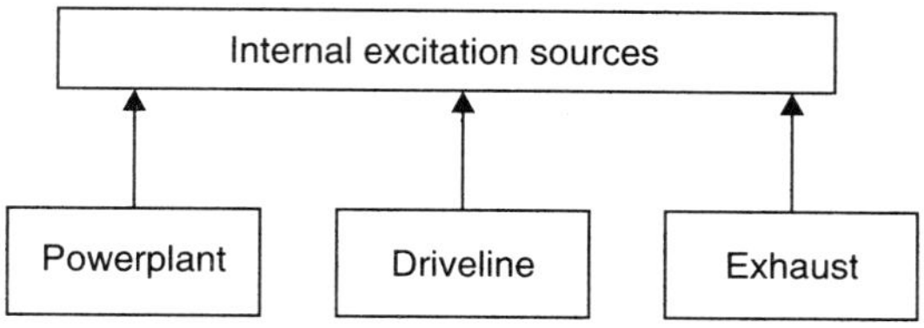

Fig. 6.6 Vehicle internal vibration sources.

6.1.2.2.1 Vibration Sources from the Powerplant

6.1.2.2.1.1 Coordinates and Powerplant Modes

The two major vibration sources from the engine are the pressure variation in the cylinder and inertial imbalance forces. The pressure variation is developed during the combustion process. The inertial imbalance force is generated by the reciprocating components. The responses excited by the sources depend not only on the engine structure but also on other structures attached to the engine, such as the transmission. Figure 6.7 shows the two basic configurations of powerplants. The powerplant x direction is always defined along the engine-transmission direction.

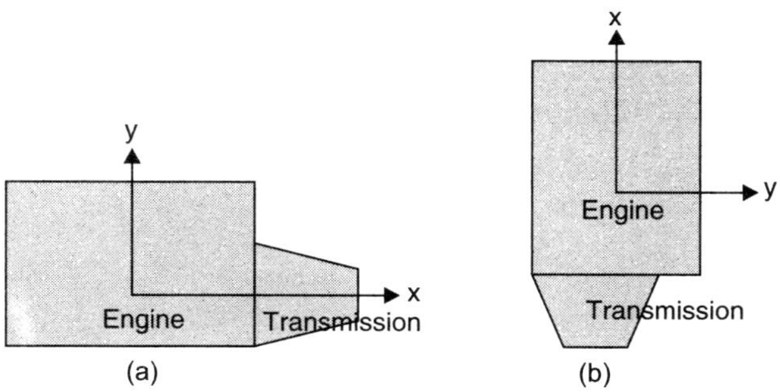

Fig. 6.7 Powerplant coordinates system:
(*a*) east-west direction, and (*b*) north-south direction.

Figure 6.8 is a vehicle drawing with its coordinates. The vehicle coordinates are defined as follows:

- x direction is defined from the vehicle front to the rear.
- y direction is defined from the driver to the passenger.
- z direction is defined from the bottom upward.

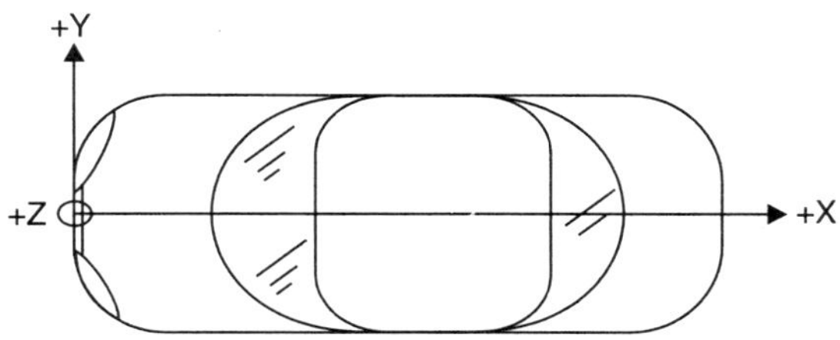

Fig. 6.8 Vehicle coordinates system.

There are six motions for a rigid body: bounce motion, lateral motion, fore-aft motion, roll motion, pitch motion, and yaw motion. The linear motion along the x axis is the fore-aft motion, the linear motion along the y axis is the lateral motion, and the linear motion along the z axis is the bounce motion. The rotation around the x axis is the roll motion, the rotation around the y axis is the pitch motion, and the rotation around the z axis is the yaw motion. Figure 6.9 shows the six motions.

When packaged in a vehicle, the powerplant could have the same coordinates as the vehicle, as shown in Fig. 6.10(a). This package of powerplant is called a north-south configuration. Most trucks and some large cars and sports cars have this kind of configuration. For this case, the six motions of the powerplant are the same as those for the vehicle body. However, the powerplant could be packaged in different coordinates, as shown in Fig. 6.10(b). This package is called an east-west configuration. This configuration can be seen in most passenger cars and some small sport utility vehicles (SUVs).

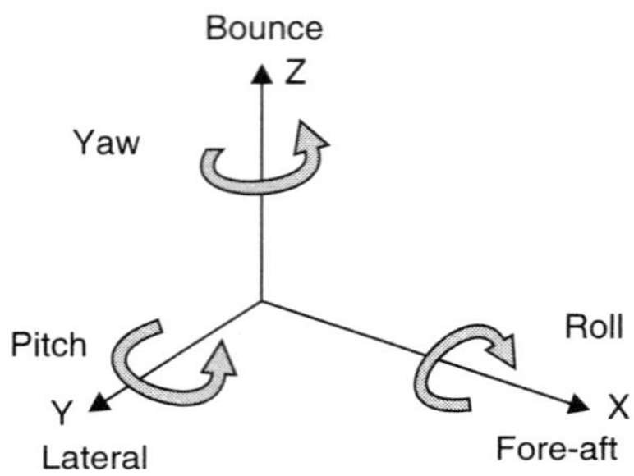

Fig. 6.9 Six-motion rotation of a rigid body.

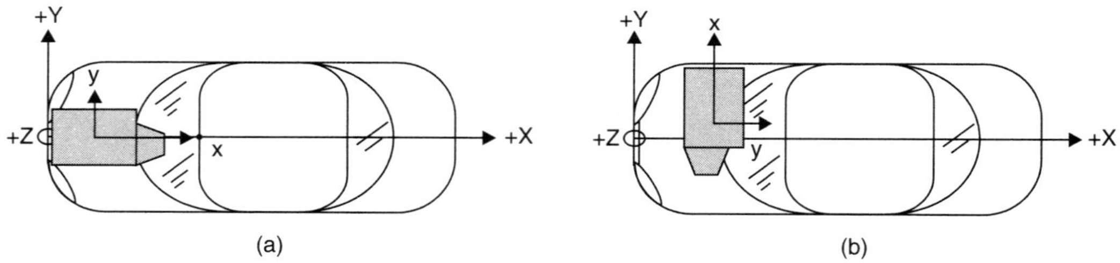

Fig. 6.10 Powerplant configurations:
(a) north-south configuration, and (b) east-west configuration.

The crankshaft of an engine is always along the x direction in the powerplant coordinates, so the most important engine excitations are around the x direction, that is, roll excitation. For a north-south configuration, the powerplant roll motion has the same direction as the vehicle roll motion. However, for an east-west configuration, the powerplant roll motion is different from the vehicle roll motion.

A powerplant usually is supported by several mounts and/or one or two roll restrictors. The configuration with three mounts and a roll restrictor is the most common case in passenger cars. The three mounts are installed on the front of the engine, on the rear of the engine, and on the transmission, respectively. These mounts have two major functions. The first is to support the powerplant weight, and the second is to reduce the powerplant vibration transferred to the vehicle body and to control the force transferred to the body from the road input. The roll restrictor usually is used to control the powerplant roll motion.

Figure 6.11 shows one possible configuration of these mounts, where a powerplant is supported on a subframe by three mounts and a roll restrictor. The subframe is a flexible structure, as is the powerplant. To primarily identify its modes and responses, the powerplant usually is assumed to be a rigid body, and the mounts are assumed to be fixed on the ground.

As a rigid body, the powerplant has six motions and six corresponding modes. The roll mode is the most critical because the engine excitation direction is the same as the mode. Two criteria guide the powerplant mount system design. One is that the roll mode frequency must be separated from the excitation frequency, and the powerplant vibration must be isolated effectively. The other is that the six modes must be decoupled.

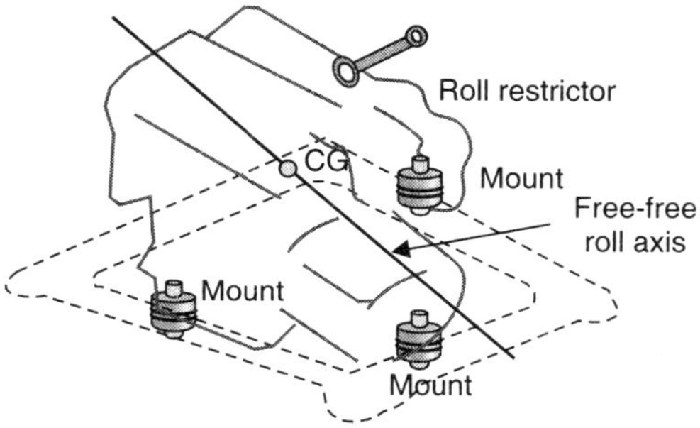

Fig. 6.11 A power-plant mount system on a subframe.

6.1.2.2.1.2 Vibration Sources from Engine Firing Pulsation

The purpose of this section is to identify the vibration sources transferred to the human body. We also will study how to isolate the excitation sources that transfer to the vehicle body and how to minimize the coupling of disturbances to the vehicle occupants. Engines used in most road vehicles are four-stroke engines. The engine finishes a working processing by four steps (strokes): air intake, air compression, combustion (explosion or firing), and exhaust. Figure 6.12 shows these steps. It takes 180° to process one step; therefore, the whole process takes 720° (*i.e.*, two cycles).

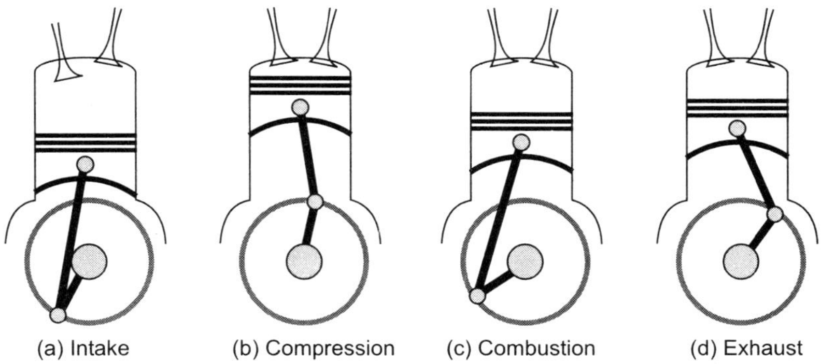

Fig. 6.12 Four steps of the engine working processing.

Among the four steps, the combustion phase produces the force because the gas and air are combined and fired. The firing pulse drives the engine crankshaft to move and then is transferred to the vehicle wheels through the driveline.

For a single-cylinder engine, a firing pulse appears in every two cycles, or there is a "half" firing pulse for each cycle. The firing order, sometimes simply called the order, is defined as the firing number in each cycle. Thus, the basic firing order for a single-cylinder engine is the half order. For an engine with two cylinders, there is a half firing order for each cylinder in one cycle, that is, there exists a whole firing order in one cycle for the two cylinders. Hence, the basic firing order for a two-cylinder engine is first order. Similarly, there are two, three, and four firing pulses for four , six-, and eight-cylinder engines, respectively, in one cycle. Thus, their corresponding firing orders are second order, third order, and fourth order, respectively.

The engine output provides the driving force to a vehicle, but it also causes vehicle vibration. The vibration usually is controlled by the engine mount isolation system. The design of an isolation system depends on the system natural frequency and the excitation frequency. To reduce the vibration excited by the firing pulse, it is necessary to understand the excitation frequency and the structure modes to be excited. The excitation frequency relates to the engine cylinders and speed. The engine speed range usually is from 600 to 6000 rpm. The idle speed range is from 600 to 1000 rpm, while the wide open throttle (WOT) speed range is from 1000 to 6000 rpm. The firing frequency f_{firing} for each engine speed can be calculated by

$$f_{\text{firing}} = \frac{\text{engine speed}}{60} * \text{firing order} = \frac{\text{engine speed}}{60} * \frac{\text{number of cylinders}}{2} \tag{6.21}$$

The relation among the engine speed, frequency, and order can be expressed clearly in a plot. Fig. 6.13 plots the first order and third order, but other orders also can be plotted.

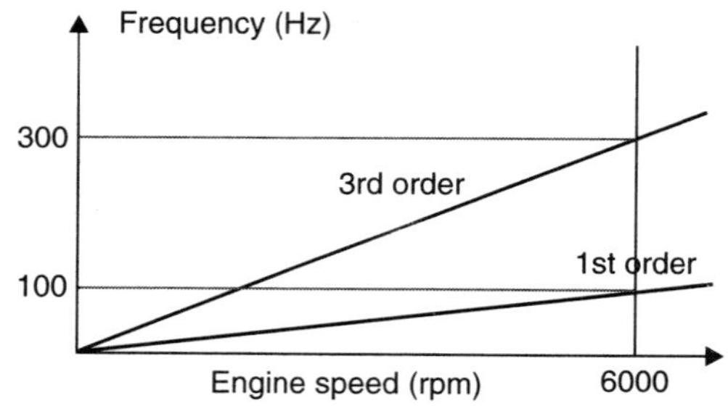

Fig. 6.13 Relation between engine speed, order and frequency.

The firing frequency varies with the engine speed. The lowest firing frequency is the most important because if vibration caused by this frequency is controlled, the higher excitation frequencies definitely will be controlled. Usually, the lowest engine working speed is 600 rpm. For a four-cylinder engine, second order is the dominant order; thus, its lowest firing frequency is 20 Hz. For a six-cylinder engine, third order is the dominant order; thus, the lowest firing frequency is 30 Hz. Table 6.2 lists the idle firing frequencies for different engines. The listed frequencies are the lowest excitation frequencies. These frequencies are important not only for powerplant isolation design but also for exhaust and driveline design.

Table 6.2 Relation of firing order and frequency for different engines

Engine cylinder number	*Firing order*	*Idle firing frequency (Assuming 600 rpm idle speed)*
2	First	10 Hz
4	Second	20 Hz
6	Third	30 Hz
8	Fourth	40 Hz

6.1.2.2.1.3 Vibration Sources from Powerplant Inertia Forces and Moments

The second source of engine excitation is from the inertial unbalance (or imbalance) forces generated by the reciprocating components (piston and rods) or rotating (crank-shaft and rod) masses within the engine. The directions of the inertia forces are both parallel to the piston axis and perpendicular to the crank and piston axes. The inertia torque acts about an axis, which is parallel to the crankshaft.

The inertia force is caused by system imbalance; thus, it increases with increasing engine speed because the force is proportional to the square of the operating speed. At high speeds, the inertia force could become the dominating excitation source.

For a multi-cylinder engine, the engine-unbalanced disturbance depends on the number and arrangement of cylinders in the engine. The engine inertia force can be canceled (or balanced) according to the engine cylinder number and crankshaft arrangement. For example, a four-cylinder engine has a vertical inertia force that acts on the engine block in addition to the oscillating torque about the crankshaft. Six- and eight-cylinder engines have no such inertia force, but they have torque oscillation. Generally, the forces become more balanced and are easier to handle with a greater number of cylinders. The inertia forces in six- and eight-cylinder engines are much smaller than those in two- and four-cylinder engines. Usually, the inertia imbalance force in a diesel engine is higher than that in a gasoline engine.

6.1.2.2.1.4 Powerplant Isolation Design

The most common method to reduce powerplant excitation is the use of an isolation system. The system analyzed with a six degrees-of-freedom model consists of the powerplant and supporting mounts. This is a complex spring-mass system. Many parameters, such as the powerplant moment of inertia, mass, center of gravity (CG), mount stiffness, and location, influence the effectiveness of isolation. The locations are always constrained by packaging limitations; thus, they must be chosen early in a vehicle development program and compete with other component package space.

The isolation design must perform the following functions:

- Location and support of the powertrain, which requires the mounts to be stiff.
- Isolation of the powerplant vibration transferred to the body, which requires that the mounts are soft.
- React to powerplant output torque and inertia unbalance.
- React to dynamic acceleration and deceleration forces due to starting, stopping, cornering, road inputs, and truck and rail shipment.
- Isolation of high-frequency vibration resulting from the powerplant structural elements.
- Absorption of road and tire/wheel-induced vibration.

The requirements usually conflict with each other. High mount stiffness is required for shock isolation at low frequency, but low stiffness is required for vibration isolation at high frequency. The dynamic stiffness of the mounts usually varies with the frequency. Well-designed hydraulic, semi-active, and active mounts can be closer to satisfying both conditions.

If an engine is chosen and an idle strategy is developed in a vehicle development, then the lowest firing frequency (firing) is known. The isolation design depends on the excitation frequency as well as the system natural frequencies. Thus, the first objective in vehicle design is to determine the powerplant rigid modes. In particular, the roll frequency is determined first because the roll motion is the most important one that must be isolated. Keep in mind that the firing pulses will cause torque to act on the engine block around the roll axle.

Vibration isolation theory can be used as a guide for the frequency choice. Let's consider a single-degree-of-freedom isolation system as shown in Fig. 6.14. Applying Newton's law, the dynamic equation can be written as

$$M\ddot{z} = -C(\dot{z} - \dot{z}_0) - K(z - z_0) \tag{6.22}$$

where M, C, and K are the mass, damping coefficient, and stiffness of the single-degree-of-freedom system, respectively.

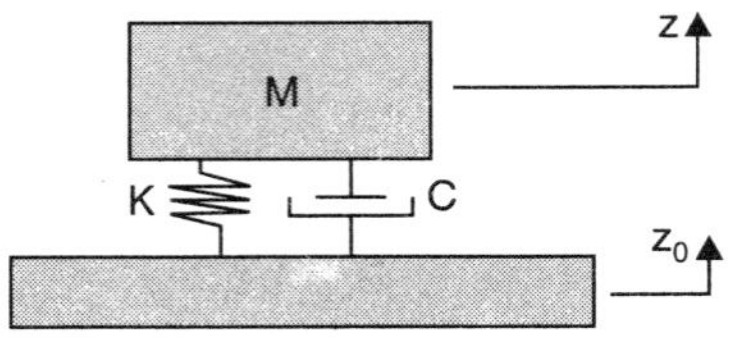

Fig. 6.14 A single-degree-of-freedom isolation system.

The transmissibility, T, between the response z and excitation z_0 is given by

$$T = \frac{Z}{Z_0} = \frac{1 + j2\xi\lambda}{1 - \lambda^2 + j2\xi\lambda} \tag{6.23}$$

where $\lambda = \dfrac{\omega}{\omega_n}$ is the frequency ratio, ω is the excitation frequency, $\omega_n = M$ is the natural frequency, and $\xi = \dfrac{C}{2\sqrt{KM}}$ is the damping ratio.

The magnitude of the transmissibility can be expressed as

$$|T| = \left|\frac{Z}{Z_0}\right| = \sqrt{\frac{1 + (2\xi\lambda)^2}{(1 - \lambda^2)^2 + (2\xi\lambda)^2}} \tag{6.24}$$

Figure 6.15 shows the transmissibility curves. The vibration can be reduced only in the region where the ratio between the excitation frequency and the natural frequency is greater than $\sqrt{2}$, that is, the isolation must meet the following condition

$$\lambda = \frac{\omega}{\omega_n} = \frac{f_{firing}}{f_{roll}} > \sqrt{2} \tag{6.25}$$

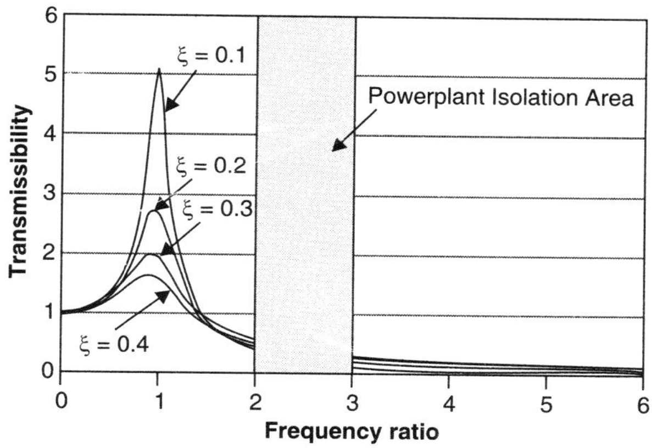

Fig. 6.15 Transmissibility of a single-degree-of-freedom isolation system.

The isolation of powerplant roll motion is similar to that of a single-degree-of-freedom system. If the roll frequency f_{roll} satisfies $f_{\text{roll}} < \dfrac{f_{\text{firing}}}{\sqrt{2}}$, the roll motion will be reduced. It is desirable to design the roll mode frequency as low as possible. The lower the roll frequency, the lower the transmissibility needed. However, a soft mounting system is impractical because the mounting system will have to support the powerplant weight, react to output torque, and resist dynamic acceleration and deceleration. Thus, determining the mount rate becomes one of the key steps in the design of the mount system. To effectively isolate the powerplant roll excitation, in engineering practice the frequency ratio usually is chosen between 2 and 3, as shown in Fig. 6.15, that is,

$$\frac{f_{\text{firing}}}{f_{\text{roll}}} = 2 \sim 3 \tag{6.26}$$

If the system is effectively isolated at the minimal firing frequency, it will satisfy all excitation frequency ranges. It is desirable to minimize the roll mode frequency because a higher frequency mode will result in higher vibration and noise levels to the occupants inside the vehicle. In fact, an increase of 1 Hz in this frequency will cause an equivalent proportional 4 to 6 dB(A) vibration and noise increase at the occupants' touch points (Qatu *et al.* 2002).

Example 6.1: *The idle speed of a six-cylinder engine is 600 rpm, and the firing order is third. The firing frequency is 30 Hz, that is,* $f_{firing} = 30$ *Hz. Determine the natural frequency of the powerplant roll mode that meets the isolation requirement.*

Solution: According to Eq. (6.26), if the firing frequency is twice the natural frequency, the roll frequency should be

$$f_{\text{roll}} = \frac{1}{2} f_{\text{firing}} = 15 \text{ Hz}$$

If the firing frequency is three times the natural frequency, the roll frequency should be

$$f_{\text{roll}} = \frac{1}{3} f_{\text{firing}} = 10 \text{ HZ}$$

Therefore, the roll mode frequency can be chosen between 10 and 15 Hz.

Note that in engineering practice, the roll mode frequency of a six-cylinder east-west engine usually is set between 10 and 12 Hz. In addition to the roll mode frequency requirement, the acceleration reduction ratio η is used to evaluate the isolation effectiveness of the mounts. The acceleration attenuation is defined as

$$\eta = \frac{a_{\text{active}}}{a_{\text{passive}}} \tag{6.27}$$

where a_{active} is the acceleration on the powerplant side (the active side), and a_{passive} is the acceleration on the vehicle frame or subframe (the passive side). For a good isolation system, the ratio should be larger than 10 that is, the vibration transferred to the vehicle should be less than one-tenth of the engine vibration. The criterion can be expressed in decibels as

$$20\lg\eta = 20\lg\left(\frac{a_{\text{active}}}{a_{\text{passive}}}\right) > 20\ \text{dB} \tag{6.28}$$

The roll mode frequency depends not only on the rate of its mounts but also on the locations of the mounts. Poor locations could introduce extra load to the system. To effectively design the isolation system, the mount locations should be chosen to allow the powerplant to move around its free-free torque roll axis. The free-free roll axis is the torque roll axis of the object when there are no external restraints to oppose its motion. This axis always goes through the center of gravity and depends only on the axis of the applied torque and the inertial properties of the object. If the mounts are located to allow the powerplant to move about the torque axis, no additional forces will be generated within the mounting system to cause it to move.

Figures 6.11 and 6.16 are two mounting systems. In Fig. 6.11, the powerplant is supported on a subframe by three mounts, and the mounts are not located at the roll axis. Thus, extra load will be induced in addition to the roll torque. However, for the design shown in Fig. 6.16, two mounts pass through the roll axis, and the extra load is avoided. The upper mount usually is fixed on the vehicle body, such as on the shock tower. This system is called a pendulum mounting system, which is the desirable system in mount design. However, it is not that easy to secure the desired pendulum mounting design because of package limitations and other requirements.

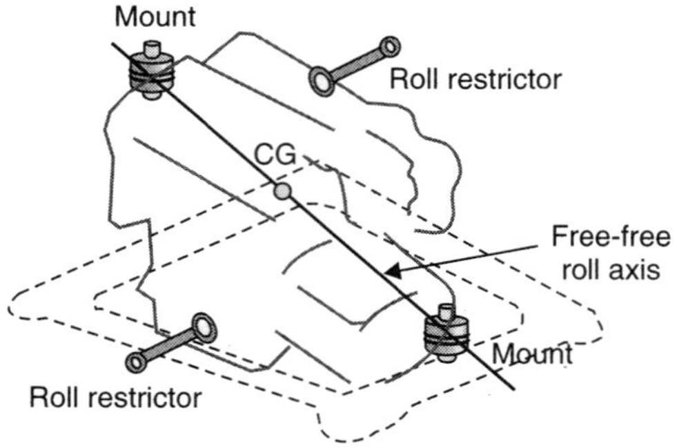

Fig. 6.16 A power-plant pendulum mounting system.

Modal decoupling: An excitation is applied on a powerplant to cause a motion in one mode, but other modes are excited, too. The two modes are coupled with each other. Some of the six modes of the powerplant could be coupled. If the modes are coupled, it will be hard to isolate a single mode, which makes the isolation problem rather complicated. It is desired to decouple these six modes. Decoupling makes each mode move independently of the other modes. Complete decoupling may not be possible, and the objective then becomes to minimize the coupling. The decoupling of powerplant modes depends on its mount rates, locations, center of gravity, and moment of inertia. Decoupling is another critical issue in powerplant isolation (in addition to the roll mode frequency determination). The desired decoupling is for all modes to be independent, no matter what the excitation is. However, the desired decoupling situation may not be possible in actual cases. Mode energy is used to measure the strength of the coupling between the modes.

A powerplant is assumed to be a rigid body and supported by the mounts, as shown in Fig. 6.17. The system has six degrees of freedom: three translation motions, and three rotational motions. The dynamic equations for a powerplant mounting system can be expressed as

$$\begin{cases} M\ddot{x}_c = F_x \\ M\ddot{y}_c = F_y \\ M\ddot{z}_c = F_z \\ I_{xx}\ddot{\alpha}_c - I_{xy}\ddot{\beta}_c - I_{xz}\ddot{\gamma}_c = M_x \\ I_{xy}\ddot{\alpha}_c - I_{yy}\ddot{\beta}_c - I_{yz}\ddot{\gamma}_c = M_y \\ I_{xz}\ddot{\alpha}_c - I_{yz}\ddot{\beta}_c - I_{zz}\ddot{\gamma}_c = M_z \end{cases} \tag{6.29}$$

where M is the powerplant mass; I_{xx}, I_{yy}, and I_{zz} are the moments of inertia; I_{xy}, I_{xz}, and I_{yz} are the products of inertia; x_c, y_c, and z_c are the coordinates of the powerplant center of gravity; α_c, β_c, and γ_c are the rotational angular acceleration around the x axis, y axis, and z axis, respectively; F_x, F_y, and F_z are the summation of all forces acting on the body in the x, y, and z directions, respectively; and M_x, M_y, and M_z are the moments acting around the x axis, y axis, and z axis, respectively.

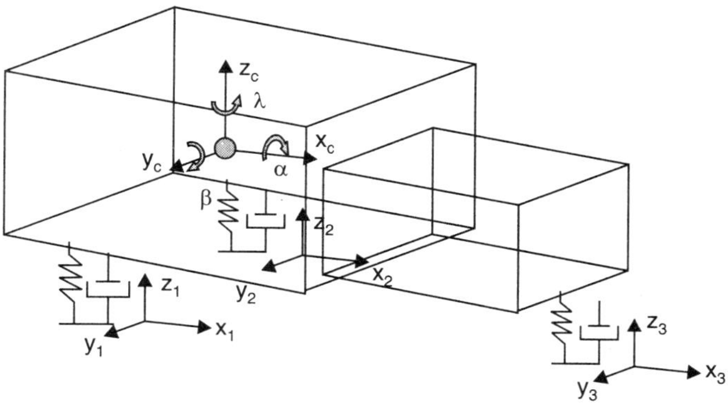

Fig. 6.17 A six-degrees-of-freedom powerplant-mounting system.

Assuming harmonic free motion and applying Laplace transforms, the equations will become

$$([K] - \omega^2[M])\,\Phi = \{0\} \tag{6.30}$$

where $[K]$ = stiffness matrix of the mounting system.

$[M]$ = mass matrix of the mounting system.

Φ = eigenvector matrix of the six degrees of freedom.

ω = natural frequency.

Setting the determinant of Eq. (6.30) to zero, the natural frequency and modal shape of the mounting system will be obtained. The matrix of actual mode Φ can be expressed as Ψ using the target modes Φ as a basis as

$$\Psi = \Phi^T M \Phi^E \tag{6.31}$$

The modal energy matrix E is defined as the matrix of modal energies expressed as a percent as

$$E = 100\ \Psi\Psi \tag{6.32}$$

E has the property that the sum of each column and row totals 100%. The percentage of modal energy for each mode presents its coupling and decoupling with other modes. For example, if the roll mode energy at its roll frequency is 95%, then the roll mode is almost decoupled from the other modes. If a modal energy is only 50% of a particular frequency, then the other modes have the other 50% modal energy at this frequency; thus, this mode is not well decoupled.

Example 6.2: *A powerplant with a V6 engine is supported by three mounts and one roll restrictor. Table E6.2 lists the location and stiffness of the mounts. The powerplant weighs 270 kg. Its center of gravity is located at* $x_c = 1700$ *mm,* $y_c = 0$ *mm, and* $z_c = 1200$ *mm. The moments of inertia are* $I_{xx} = 2.08e^7$ *kg mm²,* $I_{yy} = 1.05e^7$ *kg-mm², and* $I_{zz} = 1.95e^7$ *kg-mm². The products of inertia are* $I_{xy} = -1.6\ e^6$ *kg-mm²,* $I_{xz} = 6.5e^5$ *kg-mm, and* $I_{yz} = 3.8e^6$ *kg-mm². Determine the modal frequencies and the modal energy distribution.*

Table E 6.2 Mount location and stiffness of the powerplant mounting system

	Mount location (mm)			*Mount stiffness (N/mm)*		
	x	*y*	*z*	*Kx*	*Ky*	*Kz*
Mount 1	*1500*	*200*	*1100*	*150*	*150*	*300*
Mount 2	*1900*	*200*	*1150*	*150*	*150*	*300*
Mount 3	*1750*	*–500*	*1050*	*100*	*100*	*300*
Roll restrictor	*1750*	*–50*	*1600*	*60*	*1*	*1*

Solution: Substitute the parameters in this example into the modal equation and energy equation, and the natural frequency, modal shape, and modal energy can be obtained. Table E6.3 lists the calculation results.

Table E 6.3 Modal frequency and modal energy distribution

	Mode 1	*Mode 2*	*Mode 3*	*Mode 4*	*Mode 5*	*Mode 6*
Frequency (Hz)	6.5	5.97	9	10.51	7.69	11.42
Fore-aft	98.54	0	0.2	1.24	0	0.01
Lateral	0	97.06	0.4	0	0.85	1.69
Bounce	0.03	0.07	89.32	5.87	0.64	4.07
Roll	1.31	0.07	3.47	83.8	10.52	0.83
Pitch	0	2.08	4.86	0.42	0.09	92.55
Yaw	0.11	0.73	1.74	8.67	87.89	0.86

The results show that the modal energies for the fore-aft, lateral, and pitch modes are greater than 90%, which means that the modes are almost decoupled from the other modes. The bounce mode and yaw mode have 89.32 and 87.89% energy, respectively. The decoupling of these two modes from the other modes is acceptable. Usually, if the modal energy is greater than 85%, the decoupling of this mode is acceptable. Of course, the higher the modal energy, the better the decoupling. In this example, the roll modal energy is 83.80%. Therefore, the mounting system may need to be further optimized, although the roll mode frequency is as low as 10.51 Hz.

Bracket design: A mount is used to connect the powerplant and the body or subframe through two brackets, as shown in Fig. 6.18(*a*). The function of the brackets is to solidly hold the mount to the powerplant and the body. However, if the stiffness of the brackets is low, they will work in the same way as springs, as shown in Fig. 6.18(*b*).

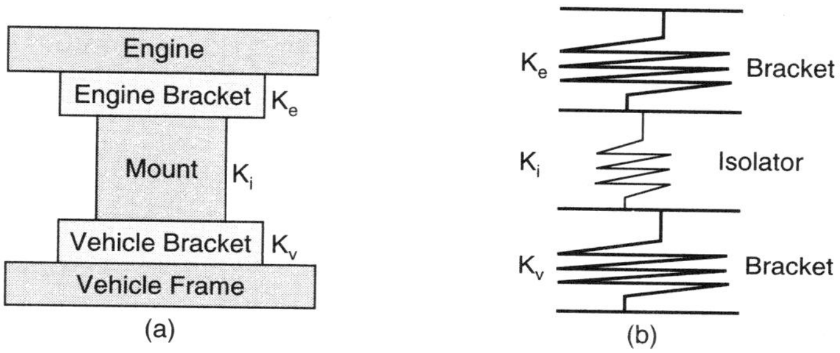

Fig. 6.18 (*a*) Mount and bracket system, and (*b*) the stiffness series of the mount-bracket system.

The equivalent stiffness for the series connection of the three springs is

$$\frac{1}{K} = \frac{1}{K_e} + \frac{1}{K_i} + \frac{1}{K_v} \tag{6.33}$$

where K_e, K_i and K_v are the stiffnesses of the engine bracket, the mount, and the body bracket, respectively.

If the stiffness of the brackets is too low, the equivalent stiffness of the assembly will be different from the designed stiffness of the mount K_i. Thus, the desirable roll mode frequency and mode decoupling cannot be met. The soft bracket in the mounting attachment also can result in resonance and the loss of isolation. Therefore, brackets on both mount sides must be sufficiently rigid. For rigid brackets, Eq. (6.33) becomes

$$K_e \cdot K_i \rightarrow \infty \quad K = K_i \tag{6.34}$$

Two rules are used to guide the bracket design. The first is that the stiffness of the brackets should be 6 to 10 times that of the stiffness of the mounts. The other is that the first natural frequency of the brackets should be above a certain value (500 Hz). The mount location should make the natural frequency in the effective vibration isolation range and make the powerplant modes decoupled. The mount brackets must possess adequate stiffness. In addition to these requirements, the following conditions should be considered when the mount locations are determined:

- Locations should be chosen to allow the powerplant to move around an axis as close as possible to its free-free torque roll axis.
- Mounts should be located at the points of low powertrain vibration.
- Mounts should be located at points of low body sensitivity.
- Mounts should be located near suspension attachment points for great absorption of suspension disturbances.
- Powerplant vibration modes should be decoupled.

6.1.2.2.2 Vibration Sources from the Driveline

The main purpose of the driveline is to transmit torque from the powerplant to the wheels. For rear-wheel-drive (RWD) and all-wheel-drive (AWD) vehicles, the driveline includes a front driveshaft, a rear driveshaft (if more than one shaft is used), a front axle, a rear axle, half shafts, and other components. In front-wheel-drive (FWD) applications, half shafts define most of the driveline system. Figure 6.19 shows an all-wheel-drive driveline configuration. All-wheel drive is widely used in cars and sport utility vehicles (SUVs) today. Thus, the vibration sources caused by the driveline become more important to ride quality.

The driveline is supported on the vehicle body by the transmission mount, central bearings and axle mounts. The driveline vibration will be transferred to the vehicle through the mounts and bearings. The mount and bearing locations are called the transmission plane, center bearing plane, and axle plane, respectively. The vibration transferred to the body will impact ride comfort. Most of the tactile vibration problems occur at frequencies below 100 Hz.

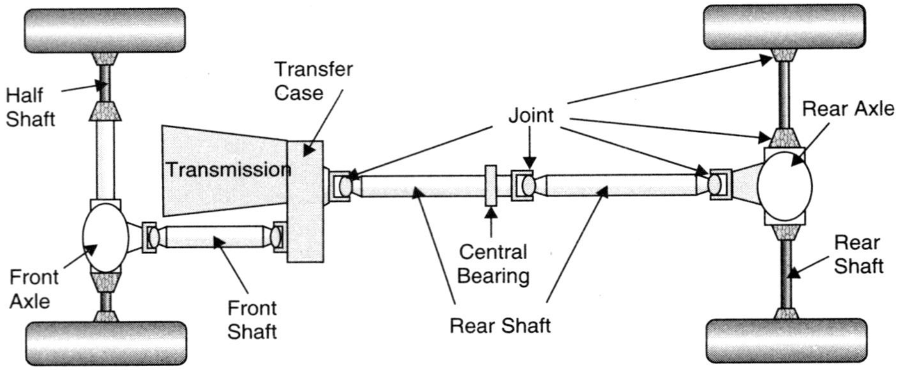

Fig. 6.19 A driveline system for rear-wheel-drive-based all-wheel-drive vehicles.

The driveline excitation sources can be described according to its components. The sources are the four generators: the transmission, driveshaft, front and rear axles, and universal joints. The mechanisms of these sources are divided into driveline imbalance, gear transmission error, and nonconstant velocity joint secondary couple excitation. Other excitations of constant velocity joints (CVJ) can appear similar to the plunging force of these joints. In addition to the component excitation sources, the driveshaft and the modes of the whole driveline and frequencies are important for driveline vibration.

6.1.2.2.2.1 Driveline Imbalance

The driveline imbalance forces are the results of individual imbalance effects and are related to the driveline rotation speed. The first order excitations are the results of driveline imbalance. The driveline will not produce disturbances on its bearing supports when the driveline is an axisymmetric structure and its rotation axis coincides with the principal axis. However, this ideal case is not achieved (unless a balancing process is used). Thus, driveline imbalance happens. The driveshaft is the most common component for driveline imbalance. The following reasons can cause the imbalance:

- The driveshaft and other rotating components are not axisymmetric. Figure 6.20(*a*) shows the center of gravity of a driveshaft that is offset from its geometric centerline.
- The driveshaft is supported off center on its supporting members.

- The driveshaft is not straight.
- The driveshaft itself deflects off its center (as a result of weight).
- Interface imbalance (runout) occurs. Figure 6.20(*b*) shows the centerlines of two connection shafts that are offset from each other.

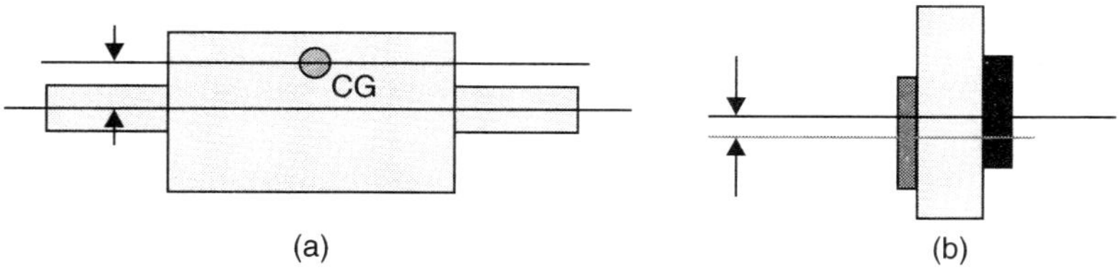

Fig. 6.20 (*a*) The center of gravity is offset from its geometric center, and (*b*) the centerlines of two connection shafts are offset.

The imbalance produces centrifugal force that is proportional to the square of the rotational speed. The imbalance force F can be expressed as

$$F = mr\omega^2 \tag{6.35}$$

where m = imbalance mass.
r = distance from the imbalance mass center to the shaft rotation center.
ω = shaft rotating angular speed.

The imbalance force is proportional to both the imbalance mass and the offset distance; therefore, the mass and the distance are considered simultaneously as a contribution of the imbalance force, called static imbalance. The unit for the static imbalance is gram per centimeter (g-cm). If the static imbalance is fixed, then the centrifugal force increases rapidly with the shaft rotating speed. The contribution by the rotation speed is dynamic imbalance. Figure 6.21 shows the centrifugal force varying with the shaft speed for different imbalance forces (g-cm).

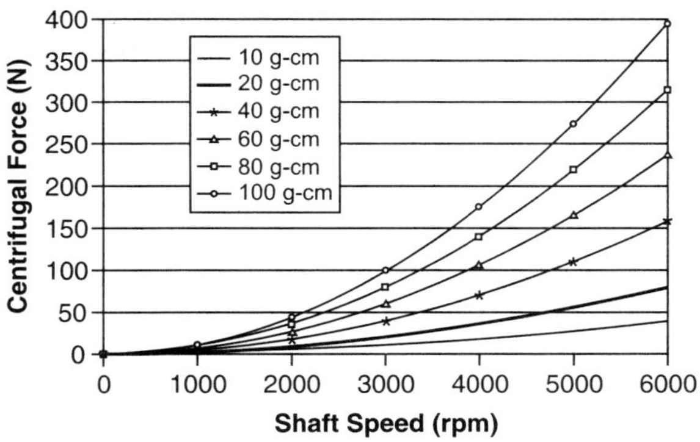

Fig. 6.21 The centrifugal force varying with the shaft speed for different static imbalance forces (g-cm).

Example 6.3: *Calculate the imbalance force for a driveshaft with an imbalance of 50 g-cm at 1000 and 6000 rpm, respectively.*

Solution: According to Eq. (6.35), the imbalance force for the driveshaft at 1000 rpm is

$$F_{1000} = 50\times10^{-3}\times10^{-2}\times\left(\frac{2\pi * 1000}{60}\right)^2 = 5.48 \text{ N}$$

The imbalance force at 6000 rpm is

$$F_{6000} = 50\times10^{-3}\times10^{-2}\times\left(\frac{2\pi * 6000}{60}\right)^2 = 197.4 \text{ N}$$

This example shows that the dynamic force due to imbalance is significantly different for different rotational speeds. The imbalance force at 6000 rpm is much higher than that at 1000 rpm.

6.1.2.2.2.2 Gear Transmission Error

Figure 6.22 shows a pair of the two meshing gears. If a pair of gears is perfectly connected, the following conditions must be satisfied for both gears:

- The gear circles are geometrically perfect.
- The gears are perfectly aligned.
- The gears are infinitely stiff.

For a perfect gear pair, the relation between the two gears can be established as

$$R_1\omega_1 = R_2\omega_2 \tag{6.36}$$

where R_1 and R_2 are the radii of the two gears, respectively, and ω_1 and ω_2 are the angular velocity of the two gears, respectively. In actual cases, the gears are not geometrically perfect, nor are they perfectly aligned or infinitely stiff. Thus, Eq. (6.36) is not always accurate. Instead, the actual meshing relation between the two gears is

$$R_1\omega_1 \neq R_2\omega_2 \tag{6.37}$$

The difference between the actual meshing position and the ideal meshing position of the two gears is called the transmission error (TE) and is expressed as

$$\text{TE} = \int_0^t (R_1\omega_1 - R_2\omega_2)\,dt \tag{6.38}$$

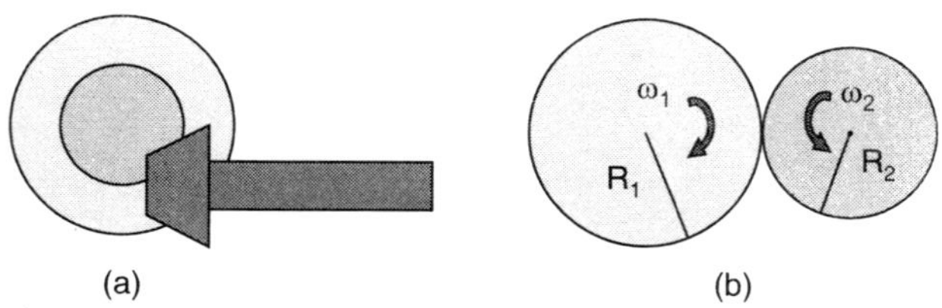

Fig. 6.22 Two gear meshing pairs.

This first order of gear meshing excitation is at the first order of the driveline excitation times the number of teeth in the driving pinion of the axle.

To reduce the excitation caused by transmission error, the gear teeth should be optimized, and the gear meshing or interaction should be designed optimally. The modal separation between the bending and torsional modes of the driveshafts also can reduce the excitation.

6.1.2.2.2.3 Second Order Excitation

The first order excitation sources are mainly from the driveshaft imbalance. Higher orders are from the transmission error. The excitations caused by second order driveshaft speed are related to the driveline angles and the use of nonconstant velocity joints in the driveshaft, such as Cardan universal joints as shown in Fig. 6.23. For a constant joint input speed, the output speeds go up or slow down twice per revolution. Thus, the system induces disturbances with frequencies that are twice the driveshaft rotation frequency. The magnitude of the disturbances increases with an increase in joint angle and torque load. These disturbances usually cause vehicle takeoff shudder or drive pinion flutter at low speeds (10 mph). Universal joints also can induce torsional and inertia vibrations.

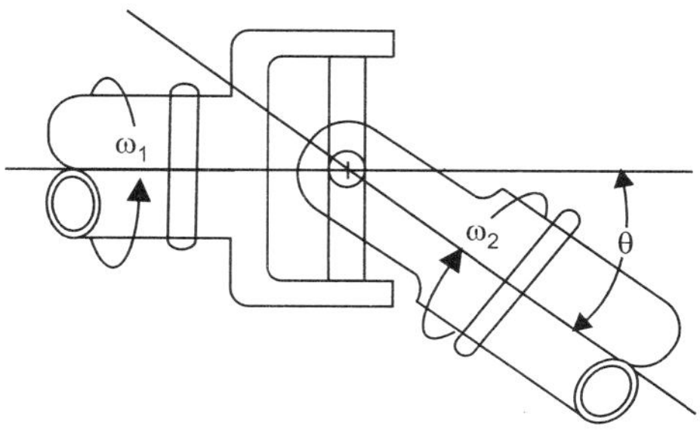

Fig. 6.23 Cardan universal joint system.

6.1.2.2.2.4 Driveshaft Modes and Driveline Modes

The vibration caused by components and component interactions is described. As a system, the design of the driveline modes and frequencies is another critical issue. The driveline includes several driveshafts and other components. The driveshaft modes are important in the whole driveline design. The driveshaft natural frequencies must be chosen above the driveline maximum excitation speed. If the driveshaft frequency is lower than the maximum excitation frequency, resonance of the driveline will occur. The driveshaft has a critical speed, which is defined as the shaft speed where the shaft rotating frequency matches the first bending mode/frequency of the shaft. Operation at or near critical speed can result in catastrophic failure of the driveshaft and/or the transmission or the transfer case. The critical speed can be avoided by the following suggestions:

- Increase the shaft stiffness (mostly by the use of a larger diameter) and reduce its weight.
- Use two- or three-piece shafts (reduce the length).
- Balance the shaft components.
- Separate the vertical and lateral P/T modes by at least 15%.

Figure 6.24 shows a simple driveline system. The driveline first bending frequency should be designed above the maximum excitation frequency.

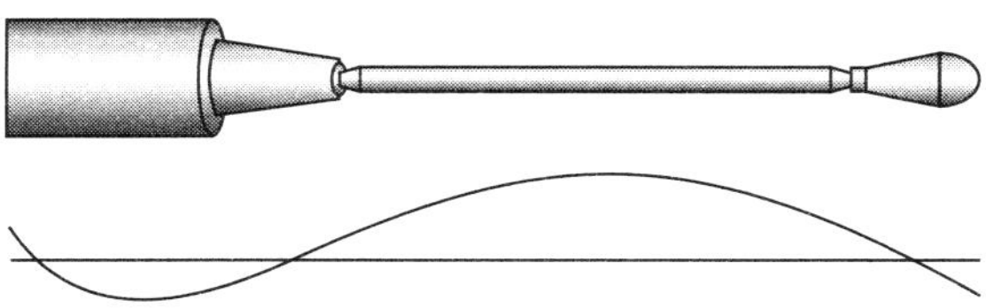

Fig. 6.24 A simple driveline system and its first bending mode.

6.1.2.2.3 Vibration Sources from the Exhaust System

An exhaust system, defined from the exhaust manifold to the tailpipe, consists of a down pipe or Y pipe, catalyst, resonator, muffler, tailpipe, hangers, and other components, as shown in Fig. 6.25. The fundamental purpose of the exhaust system is to route the exhaust gases from the engine and to keep the gases from entering the passenger compartment.

The exhaust system has interfaces with the engine and the vehicle body. The vibration is transferred from the powerplant to the exhaust system and then to the body structure through hangers, as shown in Fig. 6.25. To reduce the vibration levels, the transfer path is isolated. The flex decoupler is the key element to reducing the vibration transferred from the engine to the exhaust system. Hanger isolators are used to reduce the vibration transferred from the exhaust system to the body.

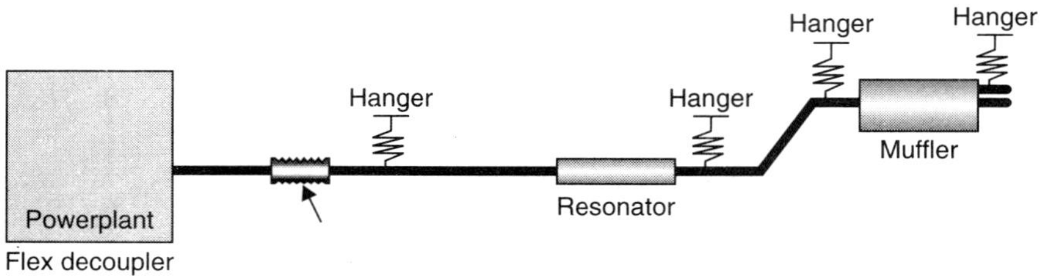

Fig. 6.25 An exhaust system.

The decoupler works similarly to an isolator that possesses three linear stiffness values and three rotational ones. To effectively isolate the engine vibration, the decoupler should be placed as close to the powerplant roll axis as possible. The isolation effectiveness is determined by the stiffness of the decoupler and can be evaluated from the transfer function between the vibration responses of the two sides of the decoupler. The stiffness of the flex decoupler also is very sensitive to the hanger force transferred to the body structure. Therefore, the transfer function between the vibration of the two sides of the decoupler and the sensitivity of the hanger force over the decoupler stiffness are critical for the decoupler design.

It is important to understand the exhaust system dynamic characteristics in order to reduce the vibrations coming from the engine and transferring to the body. The exhaust system is a complicated system, and it is almost impossible to solve its vibration problem using traditional analysis. The structure is divided into many finite segments, and finite element analysis (FEA) is used to implement the structure vibration analysis. In finite element analysis, the beam, shell, and solid elements usually are used to mimic the real structure. In some cases, only beam elements are used to analyze the exhaust system for a simple and quick analysis to determine the locations of the hangers. There are two major aspects in vibration analysis: modal analysis, and dynamic analysis. The exhaust system modes are critical for exhaust dynamic analysis. A normal modal analysis should be performed to ensure that the frequencies of the exhaust system do not line up with the frequencies of the powerplant and/or the body. Also, the same analysis is helpful in determining the nodal points on the exhaust system. These nodal points can be used as the locations where hanger rods can be attached to the exhaust system. This will minimize vibrational energy transferred to the body.

To mimic closely the real structure and engine excitation, we strongly suggest that a powerplant be included in the exhaust finite element model, as shown in Fig. 6.26. The powerplant in the model is represented by five rigid bar elements where the power-plant mass and the moments of inertia are added on the powerplant center of gravity. The powerplant is supported by engine mounts through three bar elements. The engine mounts are fixed (or grounded) to form the boundary condition. The exhaust is connected to the powerplant by two rigid bar elements. The engine load is applied on the location of the center of gravity. There are two advantages to modeling the exhaust with the powerplant: one is for load application, and the other is to check modal separation between the powerplant and exhaust modes. The model could be called the exhaust-powerplant model. Although some engineers use the "big mass method" in vibration analysis, the powerplant information will be lost, and a real lcad cannot be applied to system.

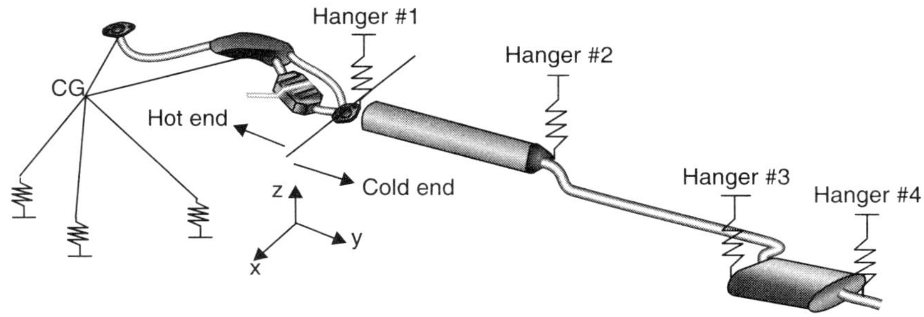

Fig. 6.26 An exhaust modeling and boundary condition.

The dynamic equation for the system can be expressed as

$$[M]\{\ddot{X}\}+[C]\{\dot{X}\}+[K]\{X\}=\{F\} \tag{6.39}$$

where $[M]$ $[C]$, and $[K]$ are the mass matrix, damping matrix, and stiffness matrix of the exhaust-powerplant system, respectively. $\{X\}$ and $\{F\}$ are the displacement and force vectors, respectively.

In the modal analysis, the right side of Eq. (6.39) is set to zero. Most of the important modes to be considered are below 150 Hz. The exhaust first vertical bending frequency and first lateral bending frequency must be 10% away from engine idle firing frequency. For example, the idle speed for a six-cylinder engine is 600 rpm, and the corresponding firing frequency is 30 Hz. Thus, the exhaust bending frequencies must be avoided between 27 and 33 Hz. A straight exhaust system has less modal density than a curved exhaust system. Therefore, a straight exhaust system generally is recommended, from a modal analysis point of view.

The lineup of the exhaust modes with the body modes must be avoided because the exhaust system and the body structure are connected through hangers. The choice of hanger location is made by modal analysis. Theoretically, the hangers should be located at or near the nodes of the exhaust system and at points of low body sensitivity. Nodes typically occur near masses such as mufflers, making them good candidates for hangers. Package is another problem limiting hanger location, and a careful tradeoff analysis between noise, vibration, and harshness (NVH) and package must be considered. The purpose of dynamic analysis is to determine the hanger force transferred to the body structure. The hanger forces depend on the engine excitation, as well as the number, stiffness, and locations of the hangers and the stiffness of the flex decoupler. The force should be as low as possible and within a certain target. The excitation is critical in the analysis. Several excitation methods of dynamic analysis are described next.

The first method is the large mass method. In this method, the powerplant is assumed to be a large mass, and only the roll rotation degree of freedom is set free, while the other five degrees of freedom remain fixed. A certain amount of acceleration is applied to the mass as system excitation. As mentioned, the method has a limitation because the powerplant information is lost, and its modes cannot be found. The second method is that the excitation is put on the powerplant center of gravity. In this method, the exhaust system is modeled with the powerplant, and the powerplant is supported with the engine mount. Thus, the excitation resource information and powerplant modes are retained. A torque around the roll axis, M_x, is applied on the powerplant center of gravity as an excitation. The generic criterion is that the dynamic force in hangers, F_{hanger}, must be less than 10 N, that is,

$$F_{hanger} \leq 10 \text{ N} \quad \text{under condition} \quad M_x = 100{,}000 \text{ N.mm} \tag{6.40}$$

The third method is the same as the second method in the structure, but the load is different. In this method, a real load exciting the exhaust will be established. For a certain vehicle, a hanger force target usually is set by benchmarking other vehicles or from a general acoustical transfer function equation. To simulate a certain target force response, an actual powerplant excitation is needed. However, it usually is difficult to measure the actual powerplant excitation.

Force response usually is considered from 15 to 150 Hz. Frequencies below 15 Hz will not be considered. For an engine with the lowest idle speed of 600 rpm, the firing frequency is 20 Hz for four-cylinder engines and 30 Hz for six-cylinder engines, respectively. The exhaust modes below the firing frequency are hard to excite. Also, it is not easy to excite high-frequency modes. The most prominent modes of an exhaust system are between 20 and 50 Hz. Figure 6.27 shows a typical hanger force response of an exhaust system.

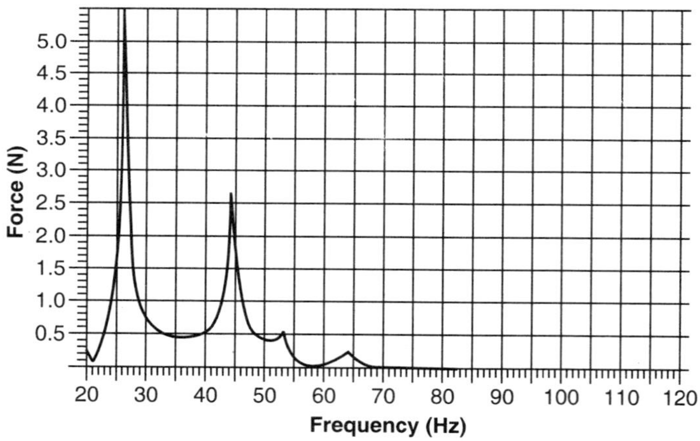

Fig. 6.27 A typical hanger force response of an exhaust system.

Usually, there are two brackets (sometimes called rods) for a hanger set: a body side bracket, and an exhaust side bracket. The body side bracket is attached to the body and is used to support the isolator. The exhaust side bracket is attached to the exhaust system. The bracket designer must avoid cantilevered brackets because such a design will result in a lower bracket natural frequency and higher stress in the bracket and force in the hanger. The bracket also should be stiff enough to avoid resonance with connected structures. Currently, the common criterion used to guarantee natural frequency to reach a satisfied target can be expressed as

$$\text{Natural frequency} = \frac{\text{maximum engine rpm * number of cylinders}}{120} \tag{6.41}$$

For example, for a six-cylinder engine with a maximum speed of 6000 rpm, the recommended minimum natural frequency is $6000 * \frac{6}{120} = 300$ Hz. This principle applies to both the exhaust and body side brackets.

The exhaust excitation contributed to the body comes not only from the engine but also from the exhaust acoustic excitation. When an exhaust acoustic mode (resonance of gas inside the exhaust system) is excited with its structural mode, the exhaust system likewise will be excited. Usually, this excitation is not as critical as that of the engine, but in some sensitive frequencies, the excitation should be considered.

6.1.3 Vehicle Ride Models

A vehicle model is very important in analyzing ride dynamics. The vehicle structure is a complicated system with many degrees of freedom. The vehicle can be modeled in many ways according to the type of response. For example, in automobile engineering, a finite element model usually is used to calculate the vehicle body modes and dynamic response. However, building and running the finite element model is complicated, time consuming, and costly. In a ride quality analysis, only a few low frequencies are considered. Approximate solutions are enough to determine vehicle parameters, such as suspension damping and stiffness. Thus, simplified models will satisfy the requirements.

In ride dynamics analysis, a vehicle can be treated as a body installed on its suspension system that is supported by the tires and axles. The body and suspension constitute the sprung mass. The tires and axles constitute the unsprung mass. The two simplified models used in this chapter are the quarter car model and the bounce-pitch model. These models will be described in detail.

The purpose of ride dynamics is to find the relation between the response at the occupants' touch points and the vibration sources. In this section, vehicle body response will be the focus. The vehicle response will be covered in later sections. As discussed in the previous section, two major vibration sources are input to the body. One is from the uneven road profile, and the other is from the vehicle, such as the engine, drive-line, and exhaust. Thus, the next section will analyze the relation between the body response and the road excitation, and the relation between the body response and the vehicle excitation.

6.1.3.1 Quarter Car Model

6.1.3.1.1 Modeling for the Quarter Car Model

A vehicle typically is a symmetric structure around the x axis of the vehicle. The vehicle body is supported by two front wheels and two rear wheels, and the body motions usually are coupled with the front and rear wheels. However, in many cases, the front and rear wheel motions are independent of each other. Thus, a quarter model, as shown in Fig. 6.28, usually is used to analyze the ride characteristics. In this model, only the vertical vibration is analyzed.

The quarter car model is a two-degrees-of-freedom model with two masses. The upper mass represents the vehicle body and the suspension mass, called sprung mass M. The lower mass is the weight of the tire and its axle, called the unsprung mass m. The quarter car model shows two frequencies associated with a vehicle suspension system. Dynamic equations for the quarter car model can be obtained by applying Newton's second law as

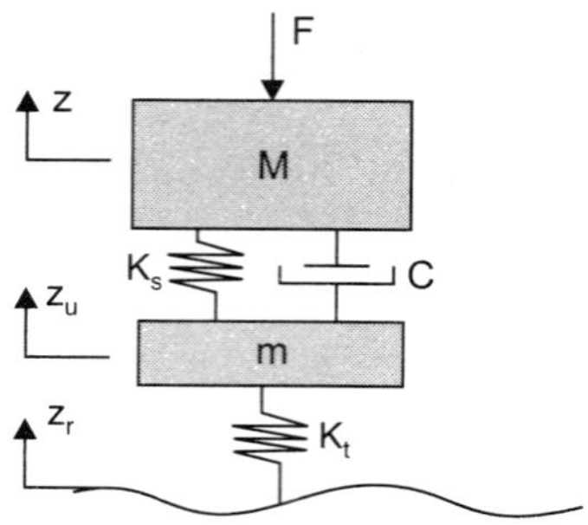

Fig. 6.28 A quarter car model.

$$M\ddot{z} + C\dot{z}_u + (K_s z \; = \; K_s)z_u + C\dot{z}_u + f \tag{6.42}$$

$$m\ddot{z}_u + C\dot{z}_u + (K_s + K_t)z_u \; = \; K_s z + C\dot{z} + K_t z_r \tag{6.43}$$

where M = sprung mass
m = unsprung mass
C = suspension damping coefficient
K_s = suspension stiffness
K_t = tire stiffness
z = sprung mass displacement
z_u = unsprung mass displacement
z_r = road displacement input
f = vehicle internal excitation force

The displacement solutions and force excitation can be assumed as

$$z = Ze^{j\omega t}$$
$$z_u = Z_u e^{j\omega t}$$
$$z_r = Z_r e^{j\omega t}$$
$$f = Fe^{j\omega t} \tag{6.44}$$

Substituting Eq. (6.44) into Eqs. (6.42) and (6.43), the equations are transferred into the frequency domain as

$$(K_s - M\omega^2 + jC\omega)Z = (K_s + jC\omega)\, Z_u + F \tag{6.45}$$

$$(K_s + K_t - m\omega^2 + jC\omega)Z_u = (K_s + jC\omega)Z_u + F \tag{6.46}$$

For a passenger car, the sprung mass is approximately an order higher than the unsprung mass, whereas the sprung stiffness is approximately an order lower than the unsprung stiffness. The unsprung mass is neglected, and the two springs are connected in series. The approximate frequency ω_s for the sprung system can be obtained by

$$\omega_s = \sqrt{\frac{K_s K_t}{(K_s + K_t)M}} \tag{6.47}$$

Usually, the sprung mass is about ten times the unsprung mass (*i.e.*, $M \approx 10$ m), and the unsprung stiffness is about nine times the sprung stiffness (*i.e.*, $K_t \approx 9K_s$). Thus, Eq. (6.47) can be simplified as

$$\omega_s = \sqrt{\frac{K_s K_t}{(K_s + K_t)M}} \approx \sqrt{\frac{K_s}{M}} = \omega_s' \tag{6.48}$$

where, ω_s' is the approximate value of ω_s.

The approximate frequency for the unsprung system can be calculated based on the assumption that the vehicle body is fixed and only the tire moves. The frequency is

$$\omega_u = \sqrt{\frac{K_s + K_t}{m}} \tag{6.49}$$

Usually, the sprung system frequency is around 1 Hz, whereas the unsprung system frequency is around 10 Hz. The two frequencies are so widely separated that the approximate frequencies are useful to understanding the system dynamic characteristics. The reason that the sprung system is designed for 1-Hz natural frequency is that this frequency is close to the walking pace frequency and generally does not excite resonances within the human body.

6.1.3.1.2 Modal Analysis for the Quarter Car Model

The damping ratio provided by the shock absorber is approximately 0.2 to 0.6. Thus, the damped natural frequency has little difference from the undamped natural frequency. To investigate the system natural frequencies and modal shapes, damping is neglected, and external excitation is not included. Substitute Eqs. (6.48) and (6.49) into Eqs. (6.45) and (6.46), and Eqs. (6.45) and (6.46) can be rewritten as

$$(\omega_s'^2 - \omega^2)Z - \omega_s'^2 Z_u = 0 \tag{6.50}$$

$$-\frac{K_s}{m} Z + (\omega_u^2 - \omega^2)Z_u = 0 \tag{6.51}$$

The characteristic equation of these equations can be written as

$$\omega^4 - \left(\omega_s'^2 + \omega_u^2\right)\omega^2 + \omega_s'^2\omega_u^2 - \omega_s'^2 \frac{K_s}{m} = 0 \tag{6.52}$$

The two real characteristic values of the solutions of these equation are the frequencies of the system, which are

$$\omega_{1,2}^2 = \frac{\omega_s'^2 + \omega_u^2}{2} \pm \sqrt{\frac{1}{4}\left(\omega_s'^2 - \omega_u^2\right)^2 + \omega_s'^2 \frac{K_s}{m}} \tag{6.53}$$

According to these relations between the sprung and unsprung masses, and between the sprung and unsprung stiffnesses, the approximate relation between the sprung and unsprung frequencies can be obtained as

$$\omega_u = \sqrt{\frac{K_s + K_t}{m}} = \sqrt{\frac{K_s + 9K_s}{\frac{M}{10}}} = 10\sqrt{\frac{K_s}{M}} 10\omega_s' \tag{6.54}$$

If Eq. (6.54) is substituted into Eq. (6.53), the two frequencies of the system are

$$\omega_1 = 0.95\omega_s'$$

$$\omega_2 = 10\omega_s' \tag{6.55}$$

The results show that the first frequency of the system is close to the sprung system frequency, and the second frequency is close to the unsprung system frequency. The mode shapes corresponding to these two natural frequencies can be obtained by substituting Eq. (6.55) into Eq. (6.50) or Eq. (6.51). The relation between Z and Z_u can be given by either Eq. (6.50) or Eq. (6.51). According to Eq. (6.50), the relation is

$$\frac{Z}{Z_u} = \frac{\omega_s'^2}{\omega_s'^2 - \omega^2} \tag{6.56}$$

Substitute Eq. (6.55) into Eq. (6.56), and the modal shapes of the two-degrees-of-freedom system will be obtained.

The first modal shape is

$$\left(\frac{Z}{Z_u}\right)_1 = \frac{\omega_s'^2}{\omega_s'^2 - (0.95\omega_s')^2} = 10.26$$

The second modal shape is

$$\left(\frac{Z}{Z_u}\right)_2 = \frac{\omega_s'^2}{\omega_s'^2 - (10\omega_s')^2} = -0.01$$

or

$$\left(\frac{Z_u}{Z}\right)_2 = -100$$

Figure 6.29 shows the two modal shapes for the two-degrees-of-freedom system. When the excitation frequency ω is close to the first natural frequency ω_1, the system will vibrate at frequency ω_1, and the mode shape is the first order mode shape. The body vibration is about ten times larger than that of the tires, as shown in Fig. 6.29(*b*). The apparent motion comes from the body; thus, the vibration of the tires can be neglected. The first mode is called the body mode. On the other hand, if the excitation frequency is close to the second natural frequency ω_2, the amplitude of the tires is about 100 times larger than motion of the body, as shown in Fig. 6.29(*c*). The body is almost at a static position, and the mode is called the tire mode. The separation of the two modes is significant for vibration isolation. One benefit is that the two modes can be controlled independently. The other benefit is that the body can be isolated effectively if the vehicle hits a bump and oscillates at the unsprung frequency.

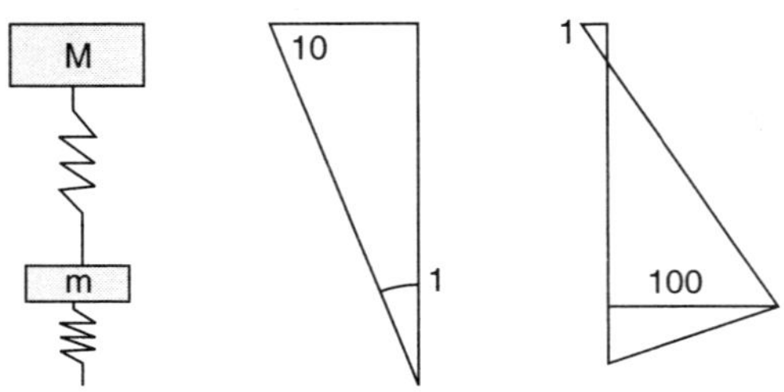

Fig. 6.29 The quarter car model and its two modal shapes: (*a*) the quarter car model, (*b*) first mode, and (*c*) second mode.

6.1.3.1.3 Dynamic Analysis for the Quarter Car Model

When the vehicle travels over an undulating road, the excitation includes multi-frequency input. As mentioned, high frequency can be isolated effectively by the suspension, but low-frequency excitation will be transferred to the body and may even be amplified. Thus, the dynamic response must be analyzed over a wide frequency spectrum. The vibration sources are either from road excitation or from vehicle excitation. The body responses caused by the two sources will be analyzed. Transmissibility between the body response and road input, and the transmissibility between the body response and vehicle excitation, will be described.

6.1.3.1.3.1 Transmissibility Between the Body Response and Road Excitation

If the vehicle excitation is neglected and only the road excitation is considered (*i.e.*, $F = 0$ in Eq. (6.45)), then the relation between the displacement at the sprung mass Z and that at the unsprung mass Z_u can be obtained as

$$\frac{Z_u}{Z} = \frac{(K_s - M\omega^2 + jC\omega)}{(K_s + jC\omega)} \tag{6.57}$$

Substituting Eq. (6.57) into Eq. (6.46), the transmissibility between the body response and road excitation Z_r is given as

$$\frac{Z}{Z_r} = \frac{K_t(K_s + jC\omega)}{(K_s + K_t - m\omega^2 + jC\omega)(K_s - M\omega^2 + jC\omega) - (K_s + jC\omega)^2} \tag{6.58}$$

The magnitude of the transmissibility can be expressed as

$$\left|\frac{Z}{Z_r}\right| = \frac{1}{P^2 + Q^2}\sqrt{(K_t K_s P + K_t CQ\omega)^2 + (K_t CP\omega - K_t K_s Q)^2} \tag{6.59}$$

where

$$P = mM\omega^2 - (K_s M + K_t M + K_s m)\omega^2 + K_s K_t$$

$$Q = K_t C\omega - C(m + M)\omega^3$$

The transmissibility must be controlled as low as possible, and the resonance should be avoided. Some factors, such as the damping of the shock absorber, the mass ratio, and the stiffness ratio between the sprung and unsprung systems, will influence the transmissibility. The influence of these parameters will be discussed.

Damping ratio: A system with the parameters of $M = 500$ kg, $K_t = 180{,}000$ N/m, $m = 1/10$ M, $K_t = 9K_s$ will be used as an example to analyze the influence of damping on transmissibility. The damping ratio $\xi = \dfrac{C}{2\sqrt{K_s M}}$ is set to 0.1, 0.2, and 0.3, respectively. Figure 6.30 shows the transmissibility of the system with different damping.

Figure 6.30 shows two peaks of the transmissibility magnitude. One peak is at 1 Hz, and the other is at 10 Hz. The 1-Hz resonant frequency comes from the sprung system. The sprung system is designed deliberately to have the specific frequency that is in the range of the comfortable frequencies of the occupants. The 10-Hz frequency coming from the unsprung system also avoids the range of 4 to 8 Hz in which the human body is uncomfortable.

The transmissibility is reduced significantly with the damping increase at the resonant frequency of the sprung system. However, the resonant frequencies do not change with the damping variation, which indicates that the damping has little impact on the frequency. In the range between the sprung resonant frequency and the unsprung resonant frequency, the lower the damping, the better the effectiveness of isolation, which is the same conclusion as that made for a single-degree-of-freedom isolation system.

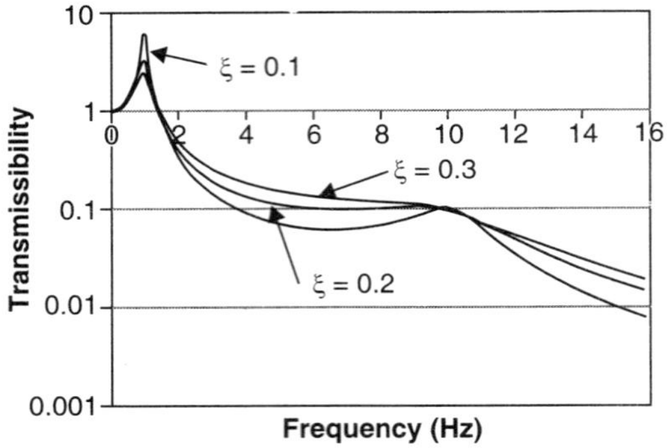

Fig. 6.30 Transmissibility between the body response and road excitation with different damping.

At the unsprung resonant frequency, the peaks are almost the same for the three different dampings, which means that the damping does not impact the response at this frequency. However, above the unsprung resonant frequency, the higher the damping, the better the isolation.

Mass ratio: Based on the previous example, the system parameters are set to M = 500 kg, K_t = 180,000 N/m, K_t = 9 K_s, ξ = 0.2. The effect of the mass ratio change on the transmissibility is studied. The mass ratio M/m is set to 5, 10, and 15, respectively.

Figure 6.31 shows the transmissibility changes with the mass ratio. The magnitude and resonant frequency for the sprung system are the same for different mass ratios. The change in the unsprung mass does not impact the performance of the sprung system. However, the response of the unsprung frequency shifts a lot. The lighter the mass, the higher the resonant frequency. Lighter mass also provides lower transmissibility magnitude between the sprung and unsprung frequencies, but the magnitude is increased above the unsprung frequency.

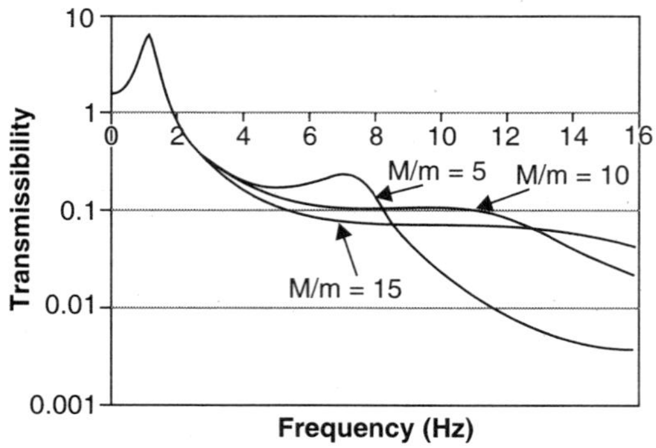

Fig. 6.31 Transmissibility between the body response and road excitation with different mass ratios.

Stiffness ratio: The system parameters are reset as M = 500 kg, K_t = 180,000 N/m, m = 1/10 M and ξ = 0.2. The influence of the stiffness ratio variation on transmissibility is studied. The stiffness ratio, Kt/Ks, is set to 7, 9, and 11, respectively.

Figure 6.32 shows the transmissibility change with the variation in stiffness ratio. At sprung frequencies, the higher the stiffness ratio, the lower the sprung frequency. This indicates that the suspension stiffness becomes relatively softer when the tire stiffness is increased. At the same time, the transmissibility magnitude decreases with the stiffness ratio increase. The higher stiffness ratio also has lower magnitude between the sprung and unsprung frequencies. Thus, the soft suspension will provide better isolation. However, at the unsprung frequency, the influence of the stiffness ratio on the magnitude and frequency is insignificant. Above the unsprung frequency, the magnitude decreases with the stiffness ratio increase.

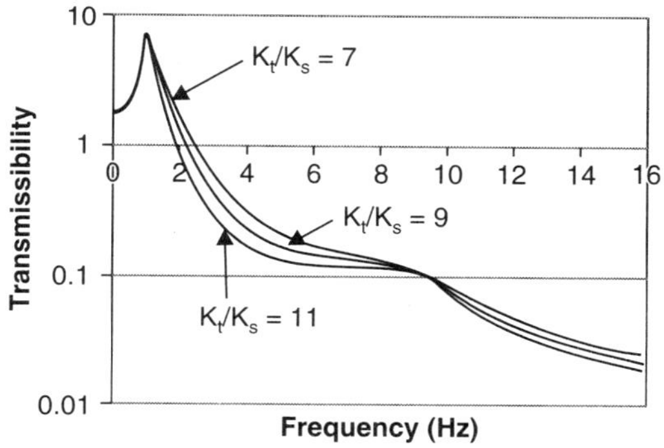

Fig. 6.32 Transmissibility between the body response and road excitation with different stiffness ratios.

Example 6.4: *A quarter car model is used for a vehicle that has the following parameters: M = 500 kg, K_t = 180,000 N/m, $K_t = 9\ K_s$ and $\xi = 0.3$. A single-degree-of-freedom system, shown in Fig. E6.4, has the same parameters as the sprung system of the quarter car model. Compare the transmissibility of the single-degree-of-freedom model with that of the quarter car model.*

Solution: The transmissibility of the single-degree-of-freedom model, as shown in Fig. 6.14, is

$$\frac{Z_s}{Z_{r_s}} = \frac{K_s + j\omega C}{K_s - M\omega^2 + jC\omega} \tag{6.60}$$

where Z_s and Z_{r_s} are the displacement magnitudes of the body and road systems, respectively. The transmissibility magnitude of the system is

$$\left|\frac{Z_s}{Z_{r_s}}\right| = \sqrt{\frac{K_s^2 + (C\omega^2)^2}{(K_s - M\omega^2)^2 + (C\omega)^2}} \tag{6.61}$$

Figure 6.4 plots the transmissibility of the single-degree-of-freedom model and the quarter car model. The two curves have the same trend. At the sprung frequency, the transmissibility magnitude of the quarter car model is 2.29, whereas the number is 1.99 for the single-degree-of-freedom model. The quarter car model has a second resonant peak at 10 Hz, whereas the single-degree-of-freedom model does not. The difference between the two models shows the coupling effectiveness between the sprung system and the unsprung system in the quarter car model. The coupling makes isolation effectiveness worse than the single-degree-of-freedom system.

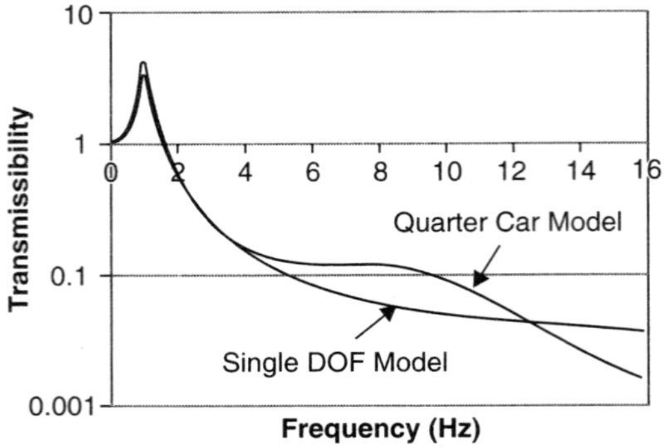

Fig. E6.4 Transmissibility of the single-degree-of-freedom model and the quarter-car model.

6.1.3.1.3.2 Transmissibility Between the Body Response and Vehicle Excitation

The transmissibility between the body response and road excitation has been analyzed. The powertrain is another important excitation source to the body response. It is essential to analyze the transmissibility between the body response and powertrain vibration. In this case, the road excitation is neglected (*i.e.*, $Z_r = 0$). The relation between the displacement at the sprung mass Z and the displacement at the unsprung mass Z_u can be obtained from Eq. (6.46) as

$$\frac{Z_u}{Z} = \frac{K_s + jC\omega}{K_s + K_t - m\omega^2 + jC\omega} \tag{6.62}$$

Substitute Eq. (6.62) into Eq. (6.45) and the ratio between the body displacement and the vehicle excitation force as shown in Fig. 6.28 is given as

$$\frac{Z}{F} = \frac{K_s + K_t - m\omega^2 + jC\omega}{(K_s - M\omega^2 + jC\omega)(K_s + K_t - m\omega^2 + jC\omega) - (K_s + jC\omega)^2} \tag{6.63}$$

The magnitude of the ratio can be expressed as

$$\left|\frac{Z}{F}\right| = \frac{1}{P^2 + Q^2}\sqrt{((K_s + K_t - m\omega^2)P + QC\omega)^2 + (CP\omega - Q(K_s + K_t - m\omega^2)^2} \tag{6.64}$$

Transmissibility usually is a non-unit value. However, the ratio of displacement over force in Eq. (6.64) has its unit. The ratio should be transferred to transmissibility to generally express the relation between response and excitation. The relation between the acceleration magnitude and the displacement magnitude can be expressed as

$$|\ddot{Z}| = |Z|\omega^2 \tag{6.65}$$

If the excitation force is divided by the mass, then the input acceleration can be obtained. Thus, the transmissibility can be written as

$$\left|\frac{\ddot{Z}}{F/M}\right| = \frac{M\omega^2}{P^2 + Q^2}\sqrt{((K_s + K_t - m\omega^2)P + QC\omega)^2 + (CP\omega - Q(K_s + K_t - m\omega^2)^2} \tag{6.66}$$

Some parameters, such as damping, stiffness ratio and mass ratio, will influence the transmissibility. Three cases with different damping ratio, different mass ratio and different stiffness ratio, respectively, will be discussed in following material just like in the section "6.3.1.3.1 Transmissibility between body response and road excitation".

Damping Ratio: Figure 6.33 is the transmissibility for three damping ratio, $\xi = 0.1, 0.2$ and 0.3. The results show that the damping has significant influence on the transmissibility magnitude at sprung frequency. The higher the damping, the lower the magnitude. The damping has almost no impact on the transmissibility beyond the sprung frequency range. Thus, the body response peak can be independently controlled by the suspension damping.

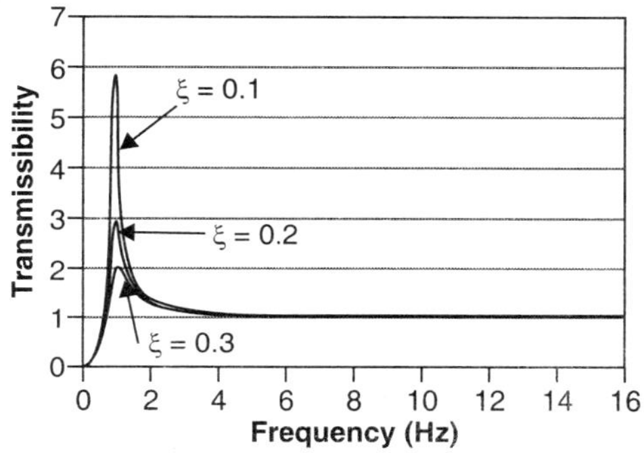

Fig. 6.33 Transmissibility between body response and vehicle excitation with different damping.

Mass Ratio: Figure 6.34 shows the transmissibility for three mass ratio, $M/m = 5$, 10 and 15. The variation of the mass ratio has a little impact on the transmissibility at the unsprung frequency. For the rest of the frequency range, transmissibility remains the same for different mass ratios. The result shows that the weight design of the tire and axles is independent of the ride response.

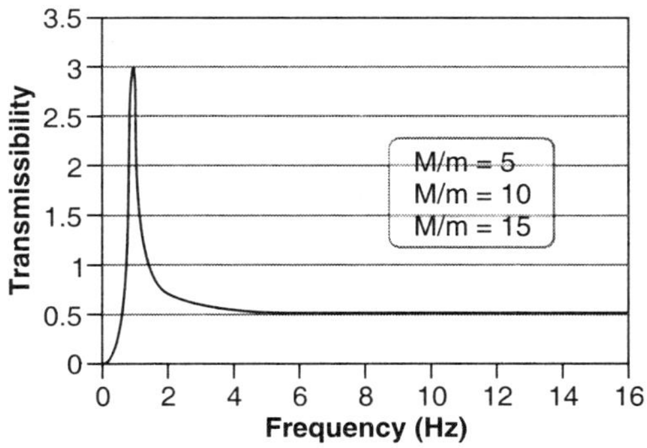

Fig. 6.34 Transmissibility between body response and vehicle excitation with various mass ratios.

Stiffness Ratio: Figure 6.35 shows the transmissibility for three stiffness ratio, $Kt/Ks = 7$, 9 and 11. The magnitude and sprung frequency become smaller with the increase in stiffness ratio, which indicates that softer suspensions are better for vibration isolation.

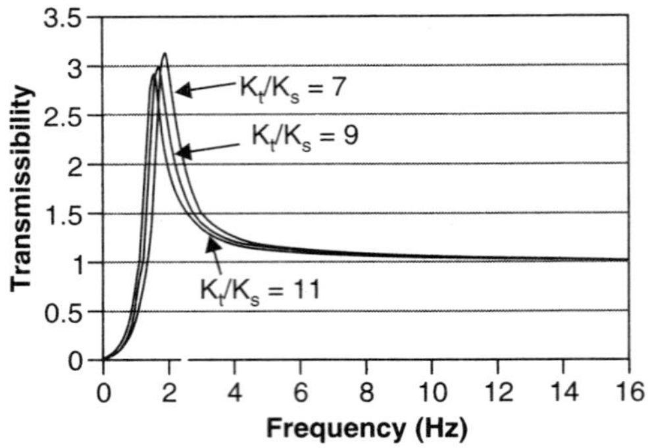

Fig. 6.35 Transmissibility between response and vehicle excitation with various stiffness ratios.

Analysis of the three cases shows that changing the damping, mass and stiffness ratios only influences the body response at the sprung frequency. The vibration control for the cases is similar to that of the single DOF system.

6.1.3.1.3.3 Dynamic Response at Random Input

Transmissibility, or transfer function, is used to describe the relation between the response and input. The transmissibility between the unsprung mass and road is

$$H_1(f) = \frac{Z_u}{Z_r} \tag{6.67}$$

The transmissibility between the sprung mass and unsprung mass is

$$H_2(f) = \frac{Z}{Z_u} \tag{6.68}$$

Thus, the response on the sprung mass will be,

$$z(t) = |H_1(f)||H_2(f)|z_r(t) \tag{6.69}$$

The mean square values of the vehicle response can be expressed by transmissibility and the mean square values of road input

$$\bar{z}^2 = |H_1(f)|^2|H_2(f)|^2\,\bar{z}_r^{\,2} \tag{6.70}$$

Eq. (6.15), the power spectral densities relation between vehicle response and road input can be written,

$$W(f) = |H_1(f)|^2|H_2(f)|^2\,W_z(f) \tag{6.71}$$

This formula shows that if the road input density is given, the vehicle response will be automatically obtained as long as the vehicle ride model is provided.

6.1.3.2 Pitch and Bounce Model

A vehicle has vertical as well as longitudinal dimensions, which means that the vehicle has a bounce motion as well as the pitch motion. As a vehicle travels on a road, the road excitations at the front and rear axles are not independent. The input to the rear axle is the same as that to the front axle, only in a delayed time. The delayed time is dependent on the vehicle speed and wheelbase. If the road wavelength is equal to or longer than the wheelbase, the vehicle will have bounce motion only. Otherwise, the pitch motion does exist. The pitch motions are important because they are the primary source of longitudinal vibrations at the location above the center of gravity. Any vertical and longitudinal vibration on the vehicle could be a combination of the bounce and the pitch motion.

The quarter car model provides important information on frequency and dynamic response at the vehicle body and axle that is useful for ride quality analysis. However, only bounce motion is considered in the quarter car model. Thus, it is necessary to establish a model that includes both bounce and pitch modes. In order to comprehensively understand the pitch and bounce vibration, a two-degree-of-freedom model is used, as shown in Fig. 6.36. In this model, the tires and suspensions are supported by two separated springs. One spring is at the front of the vehicle and the other one is at its rear. The body is assumed to be rigid. Damping and unsprung mass are neglected. The model is called the bounce-pitch model. By applying Newton second law, the dynamic equations of the system can be expressed as

$$M\ddot{z} = -k_f(z + l_1\theta) - k_r(z - l_2\theta) \tag{6.72}$$

$$I_y\ddot{\theta} = -k_f(z + l_1\theta)l_1 + k_r(z - l_2\theta)l_2 \tag{6.73}$$

where k_f: stiffness of front ride
k_r: stiffness of rear ride
l_1: distance between the front axle and the CG
l_2: distance between the rear axle and the CG
I_y: pitch moment of inertia

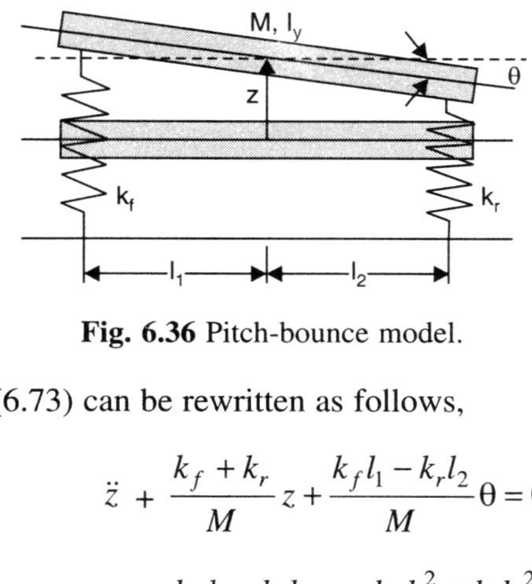

Fig. 6.36 Pitch-bounce model.

The Eq. (6.72) and Eq. (6.73) can be rewritten as follows,

$$\ddot{z} + \frac{k_f + k_r}{M}z + \frac{k_f l_1 - k_r l_2}{M}\theta = 0 \tag{6.74}$$

$$\ddot{\theta} + \frac{k_f l_1 - k_r l_2}{I_y}z + \frac{k_f l_1^2 + k_r l_2^2}{I_y}\theta = 0 \tag{6.75}$$

Assume $$A_1 = \frac{k_f + k_r}{M} \tag{6.76}$$

$$A_2 = \frac{k_f l_1 - k_r l_2}{M} \tag{6.77}$$

$$A_3 = \frac{k_f l_1^2 + k_r l_2^2}{I_y} = \frac{k_f l_1^2 + k_r l_2^2}{M r_y^2} \tag{6.78}$$

where, r_y^2 is radius of gyration and $r_y^2 = \sqrt{I_y / M}$

Substituting Eq. (6.76), (6.77) and (6.78) into Eq. (6.74) and Eq. (6.75), the two equations become,

$$\ddot{z} + A_1 z + A_2 \theta = 0 \tag{6.79}$$

$$\ddot{\theta} + \frac{A_2}{r_y^2} z + A_3 \theta = 0 \tag{6.80}$$

In the above equations, if θ is set to zero, *i.e.* $\theta = 0$, there will be no pitch motion. Only bounce motion exists. The equations become pure bounce vibration equation,

$$\ddot{z} + A_1 z = 0 \tag{6.81}$$

The natural frequency of the pure bounce motion is $\omega_{nz} = \sqrt{A_1}$.

If z is set to zero, *i.e.* $z = 0$, the system will have only pitch motion. The equations become,

$$\ddot{\theta} + A_3 \theta = 0 \tag{6.82}$$

The natural frequency of the pure pitch motion is $\omega_{n\theta} = \sqrt{A_3}$.

In most cases, neither θ nor z is zero. Thus, both pitch and pitch motions exist. The two motions are coupled. The coupling can be reflected on term A_2.

Example 6.5: *Calculate the pure bounce motion frequency and pure pitch motion frequency for the system with following parameters:*

Front Ride rate: $k_f = 35{,}000$ N/m

Rear Ride Rate: $k_r = 40{,}000$ N/m

Sprung mass: $M = 1600$ kg

Radius of gyration: $r_y = 1.2$ m

Distance between front axle and the vehicle CG: $l_1 = 1.5$ m

Distance between rear axle and the vehicle CG: $l_2 = 1.8$ m

Solution: According to the provided parameters, A_1 and A_3 can be calculated as follows,

$$A_1 = \frac{k_f + k_r}{M} = \frac{35000 + 40000}{1600} = 46.9\ s^{-2}$$

$$A_3 = \frac{k_f l_1^2 + k_r l_2^2}{I_y} = \frac{k_f l_1^2 + k_r l_2^2}{M r_y^2} = \frac{35000 * 1.5^2 + 40000 * 1.8^2}{1600 * 1.2^2} = 90.4\ s^{-2}$$

So, the pure bounce motion frequency is:

$$f_{nz} = \frac{\sqrt{A_1}}{2\pi} = \frac{\sqrt{46.9}}{2\pi} = 1.09 \text{ Hz}$$

The pure pitch motion frequency is:

$$f_{n\theta} = \frac{\sqrt{A_3}}{2\pi} = \frac{\sqrt{90.4}}{2\pi} = 1.51 \text{ Hz}$$

In order to obtain the coupled natural frequencies of the system, Eq. (6.79) and Eq. (6.80) must be solved. Assume,

$$z = \pi \sin \omega t \tag{6.83}$$

$$\theta = \Theta \sin \omega t \tag{6.84}$$

Substitute Eq. (6.83) and Eq. (6.84) into Eq. (6.79) and Eq.(6.80),

$$(A_1 - \omega^2)Z + A_2\Theta = 0 \tag{6.85}$$

$$\frac{A_2}{r_y^2}Z + (A_3 - \omega^2)\Theta = 0 \tag{6.86}$$

The eigenvalues for above equations can be solved by,

$$\omega^4 - (A_1 + A_3)\omega^2 + (A_1 A_3 - \frac{A_2^2}{r_y^2}) = 0 \tag{6.87}$$

The eigenvalues of Eq. (6.87), or the natural frequencies of the coupled system are,

$$\omega_{n1}^2 = \frac{A_1 + A_3}{2} - \sqrt{\frac{1}{4}(A_1 - A_3)^2 + \frac{A_2^2}{r_y^2}} \tag{6.88}$$

$$\omega_{n2}^2 = \frac{A_1 + A_3}{2} + \sqrt{\frac{1}{4}(A_1 - A_3)^2 + \frac{A_2^2}{r_y^2}} \tag{6.89}$$

The vibration modes, or the amplitude ratio between bounce and pitch modes can be obtained from Eq. (6.85) or Eq. (6.86). Using Eq. (6.85), the amplitude ratios corresponding to ω_{n1} and ω_{n2} are,

$$\left(\frac{Z}{\Theta}\right)_{\omega_{n1}} = \frac{A_2}{\omega_{n1}^2 - A_1} \tag{6.90}$$

$$\left(\frac{Z}{\Theta}\right)_{\omega_{n2}} = \frac{A_2}{\omega_{n2}^2 - A_1} \tag{6.91}$$

Among the two amplitude ratios, one is negative and the other is positive. In order to further understand the bounce mode and pitch mode, the concept of oscillation center is introduced. Oscillation center is a motion center for "pure" bounce motion or for "pure" pitch motion. The center depends on the two mode amplitude ratios.

If the amplitude ratio is positive, the oscillation center will be behind that of the vehicle body CG, *i.e.* the center is located on the right of the CG as shown in Fig. 6.5(*a*). The center location can be determined by,

$$L_{c1} = \frac{A_2}{\omega_{n1}^2 - A_1} \tag{6.92}$$

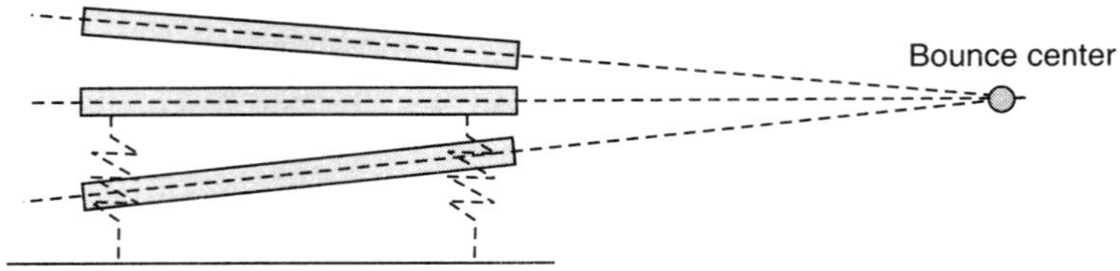

Fig. E 6.5 (*a*) The bounce center of the pitch-bounce model.

The center is usually large enough to fall outside wheelbase. For this case, the body motion is dominated by bounce mode, and the associated frequency is the bounce frequency ω_{n1}. The location is called the bounce center.

On the other hand, if the amplitude ratio is negative, the oscillation center will be ahead (left) of the CG, as shown in Fig. 6.5(*b*). The center location can be determined by,

$$L_{c2} = \frac{A_2}{\omega_{n2}^{\ 2} - A_1} \tag{6.93}$$

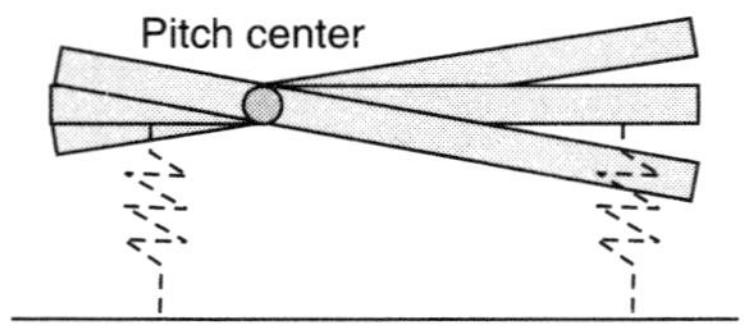

Fig. E 6.5 (*b*) The pitch center of the pitch-bounce model.

The center is usually small enough to fall within the wheelbase. For this case, the body motion is dominated by the pitch mode, and the associated frequency is the pitch frequency ω_{n2}. The location is called the pitch center.

Usually, the vehicle body motion includes both bounce and pitch modes. Only when $A_2 = 0$, *i.e.* $k_r l_2 = k_f l_1$, the body will have the pure bounce mode and pure pitch mode. For this special case, the bounce center will be at infinite distance from the CG and the pitch center will be at the CG. The bounce and the pitch oscillations are totally independent. Ride quality is poor for this case because the motions will be very irregular.

The other special case happens at $r_y^{\ 2} = l_1 l_2$. If the condition is substituted into Eq. (6.92) and Eq. (6.93), then the special locations are found, $L_{c1} = l_1$ and $L_{c2} = l_2$. In this case, one oscillation center will be at the front spring attached point and the other center is at the rear spring attached point. The vehicle model can be simplified as two independent models, one is at the front wheel and the other is at the rear wheel. The front wheel motion will have no influence on the rear wheel motion, and vice verse. This is a desirable condition for good ride quality.

Example 6.6: *Determine the bounce and pitch centers and their frequencies of the vehicle in Example 6.5.*

Solution: The three values A_1, A_2, A_3 used for calculating the frequencies are determined first. A_1 and A_3 were calculated in example 6.5. A_2 can be calculated from Eq. (6.77):

$$A_2 = \frac{k_f l_1 - k_r l_2}{M} = \frac{35000 * 1.5 - 40000 * 1.8}{1600} = -12.2$$

Substitute these values into Eq. (6.42) and Eq. (6.43), the frequencies will be given as follows,

$$\omega_{n1}^2 = \frac{A_1 + A_3}{2} - \sqrt{\frac{1}{4}(A_1 - A_3)^2 + \frac{A_2^2}{r_y^2}} = 44.65$$

$$\omega_{n2}^2 = \frac{A_1 + A_3}{2} + \sqrt{\frac{1}{4}(A_1 - A_3)^2 + \frac{A_2^2}{r_y^2}} = 92.65$$

Thus, $\omega_{n1} = 6.68,\quad \text{or}\quad f_{n1} = \omega_{n1}/(2\pi) = 1.06\ \text{Hz}$

$\omega_{n1} = 9.63,\quad \text{or}\quad f_{n1} = \omega_{n2}/(2\pi) = 1.53\ \text{Hz}$

Then, the locations of the bounce and pitch oscillation centers can be determined by Eq. (6.46) and Eq. (6.47), respectively

$$L_{c1} = \frac{A_2}{\omega_{n1}^2 - A_1} = \frac{-12.2}{44.65 - 46.9} = 5.4\ \text{m}$$

$$L_{c2} = \frac{A_2}{\omega_{n2}^2 - A_1} = \frac{-12.2}{92.65 - 46.9} = -0.27\ \text{m}$$

The bounce center is on the rear side and is 5.4 meters from the CG. The pitch center is within the wheelbase and is 0.27 meter front of the CG. For passenger car, the bounce frequency is usually between 1.0 to 1.5 Hz. The pitch mode frequency is a little higher than the bounce frequency.

6.2 HANDLING DYNAMICS

6.2.1 Introduction

Vehicle handling behavior is becoming increasingly important for today's discerning customers. The handling characteristics refer to the dynamic properties to steering input or environmental disturbance. Therefore, two essential issues associated with handling characteristics are direction control and disturbance stabilization. Vehicle dynamics plays an important role in the responsiveness of a vehicle. The performance of a vehicle depends on its handling characteristics. Handling is a measure of the directional response of a vehicle and is one of the most important characteristics from a vehicle dynamics point of view.

This section addresses the basic handling features of vehicles. A simplified model for vehicle dynamics will be presented. Based on the solutions of the model, steady-state characteristics of handling and the transient response of the vehicle as well as a frequency domain analysis will be discussed. The effects of suspension and the steering system on handling characteristics also are addressed. Overturning limit handling characteristics then are presented. Nonlinear models are introduced, and the simulation and testing of handling characteristics are presented. Traditional vehicle dynamics techniques view the vehicle as a system, which typically is simplified to a planar model in one of three planes: yaw, pitch, and roll. These planes are defined relative to the vehicle chassis at static equilibrium conditions. Figure 6.38 shows examples of these types of models. Several of the yaw plane models have been developed, with the most widely known being the bicycle model, as shown on the right side of Fig. 6.37. Yaw plane models often are used to predict vehicle handling response within the linear range because any additional degrees of freedom for the vehicle sprung mass have little effect. However, outside the linear handling range, additional degrees of freedom must be accounted for, because tire traction force saturation and load transfer effects play a predominant role.

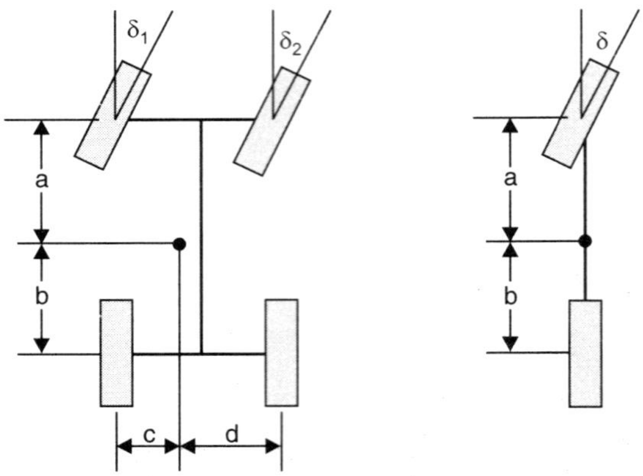

Fig. 6.37 Simplified yaw plane models.

6.2.1.1 Tire Cornering Forces

In addition to gravitational forces and aerodynamic disturbance, the dominant forces and moments influencing a vehicle are the tire forces. In the cornering process, the tire develops a lateral force and will experience lateral slip associated with a slip angle.

The slip angle is the angle between the tire's direction of heading and its direction of travel, as shown in Fig. 6.38. The lateral force, denoted by F_y, is named as the cornering force if the tire camber angle is ignored. Experiments show that cornering forces increase with the slip angle for a given tire load. Figure 6.39 shows the cornering force as a function of the slip angle. At a low slip angle, such as 5° or less, the relationship is linear and can be expressed as

$$F_y = C_\alpha \alpha \tag{6.94}$$

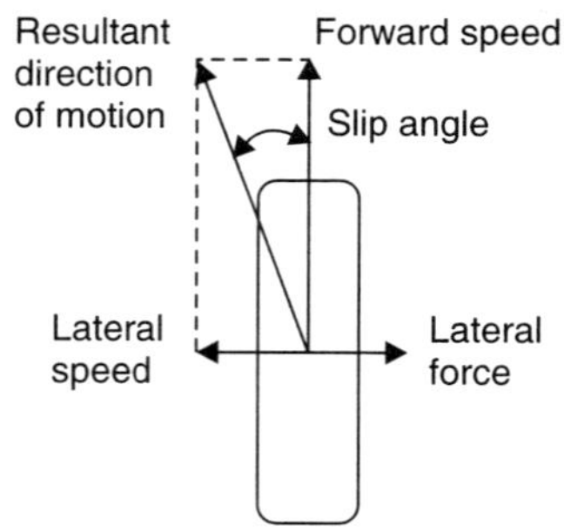

Fig. 6.38 The tire force.

The proportional constant C_α is known as the cornering stiffness, which is the slope of the curve of the cornering force versus the slip angle α at $\alpha = 0$.

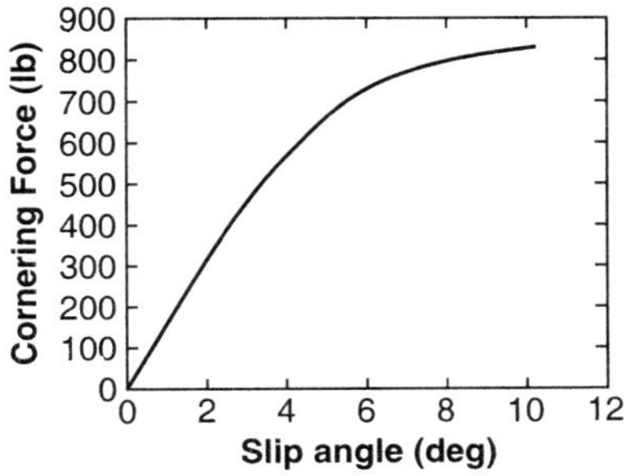

Fig. 6.39 The cornering force as a function of slip angle.

Example 6.7: *A typical car has* C_α *in the range of 30 to 80 kN/rad. Many factors affect cornering stiffness, including the tire size, type, number of plies, cord angles, and wheel width, as well as the tread. For a given tire, the inflation pressure and load are the main factors. Figures E6.7(a) and E6.7(b) show the effects of these factors.*

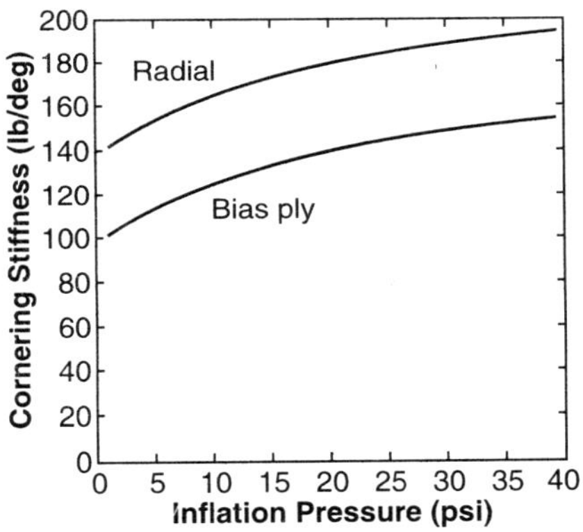

Fig. E6.7 (*a*) Cornering stiffness as a function of inflation pressure.

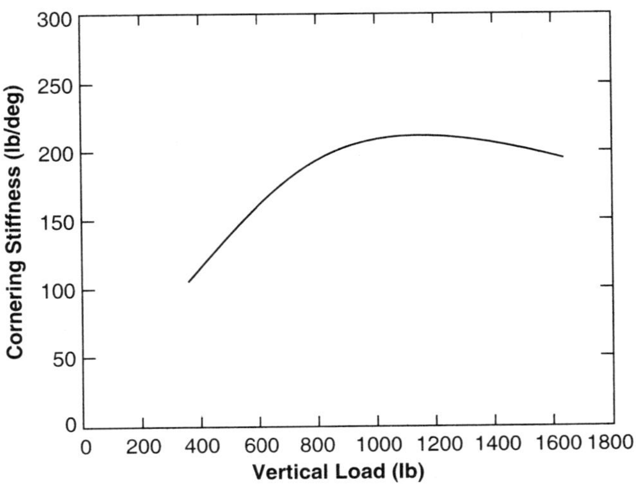

Fig. E6.7 (*b*) Cornering stiffness as a function of vertical load.

6.2.1.2 Forces and Torques in the Tire Contact Area

When a tire slips, in any point in the contact area between the tire and track, normal and frictional forces are developed, and some moment is associated with it. According to the profile design of the tire, the contact area forms an area that is not necessarily coherent. The effect of the contact forces can be described fully by a vector of force and a torque in reference to a point in the contact area. The vectors are described in a track-fixed coordinate system. The z axis is normal to the track, and the x axis is perpendicular to the z axis and perpendicular to the wheel rotation axis y. In Fig. 6.40, F_x is the longitudinal or circumferential force, F_y is the lateral force, F_z is the vertical force or wheel load, M_x is the tilting torque, M_y is the rolling resistance torque, and M_z is the self-aligning and bore torque. Nonsymmetric distributions of force in the contact patch cause torques around the x and y axes. The tilting torque M_x occurs when the tire is cambered. M_y also contains the rolling resistance of the tire. In particular, the torque around the z axis is relevant in vehicle dynamics. It consists of two parts:

$$M_z = M_b + M_s \tag{6.95}$$

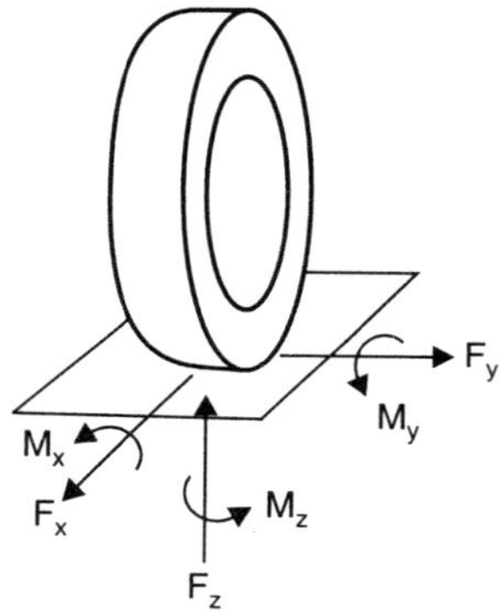

Fig. 6.40 Contact forces and torques.

Rotation of the tire around the z axis causes the bore torque M_b. The self-aligning torque M_s respects the fact that, in general, the resulting lateral force is not applied in the center of the contact patch.

6.3 THE SIMPLEST HANDLING MODELS—TWO-DEGREES-OF-FREEDOM

Yaw Plane Model

To study the basic handling dynamic characteristics, we make use of a two-degrees-of-freedom vehicle model, the bicycle model, which is the simplest one to describe the lateral motion of the vehicle.

The model considers the lateral and yaw dynamics of the vehicle, as shown in Fig. 6.41. At high speeds, the radius of turn is much larger than the wheelbase of the vehicle. The difference between the steer angles on the outside and inside front wheels is negligible. The two front wheels can be represented by one wheel at a steer angle δ, with a cornering force equivalent to both wheels. The same assumption is made for the rear wheels. The basic handling model has the following assumptions:

- The road surface is flat and level.
- The vehicle structure is rigid, including the suspension system.

- The steering system is either ignored or assumed to be rigid.
- The forward speed of the vehicle is constant.
- The vehicle undergoes only small perturbations.

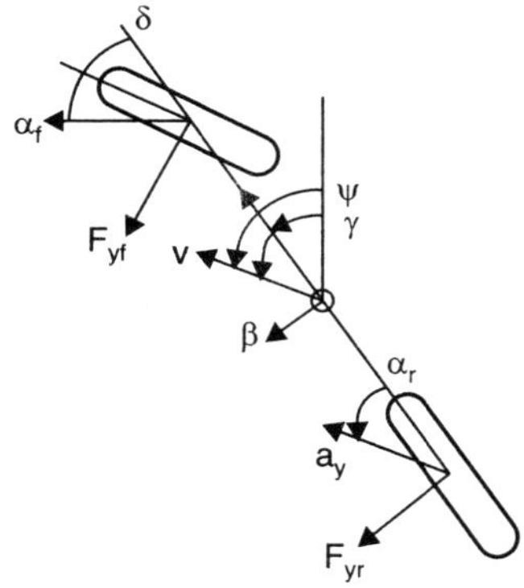

Fig. 6.41 Bicycle model.

The slip angle is less than 5°, and the cornering stiffness is constant. The front and rear cornering forces are represented by $F_{yf} = C_f\alpha_f$ and $F_{yr} = C_r\alpha_r$, where C_f and C_r are the cornering stiffness of the front and rear tires, respectively. Assume the sideslip angle β at the center of gravity (CG), which is defined as the angle between the longitudinal axis and the local direction of travel. The sideslip angle is defined by the difference between the vehicle heading angle γ and the direction angle of velocity ψ as

$$\beta = \psi - \gamma \tag{6.96}$$

In general, the sideslip angle will be different at every point on a car during cornering. Define the yaw rate or yaw velocity as $r = \dot{\psi}$. Denote that F_{xf} is the longitudinal force at the front axle, and F_{xr} is the longitudinal force at the rear axle, v is the forward velocity, R is the radius of the turn, δ is the steer angle at the front wheels (deg), L is the wheelbase (m), a_y is the lateral acceleration (g), W_f is the load on the front axle (N), W_r is the load on the rear axle (N), F_w is the wind gust excitation force, M_w is the wind gust excitation moment, and Vx and Vy are the longitudinal and lateral components of the center-of-gravity velocity, respectively. Assuming a constant longitudinal velocity of $Vx = v$, I is the moment of inertia of the vehicle about its yaw axis, m is the vehicle mass, and a and b are the distance of the front and rear axles from the center of gravity, respectively, where $L = a + b$. Apply Newton's second law at the center of gravity along the longitudinal and lateral directions, respectively, to obtain

$$F_{xr} + F_{xf}\cos\delta - F_{yf}\sin\delta = mv^2\sin\beta / R + m\dot{v}\cos\beta \tag{6.97}$$

$$F_{yr} + F_{yf}\cos\delta + F_{xf}\sin\delta - F_w = mv^2\cos\beta / R - m\dot{v}\sin\beta \tag{6.98}$$

Considering the moment equilibrium at the center of gravity, we obtain

$$F_{yf}a\cos\delta - F_{xf}a\sin\delta = I\dot{r} + M_w + F_{yr}b \tag{6.99}$$

and consider the relationship

$$F_{yf} = C_f\alpha_f, F_{yr} = C_r\alpha_r \tag{6.100}$$

For small angles, we can write

$$\sin\beta = \beta, \sin\delta = \delta, \cos\alpha = 1, \cos\delta = 1, \alpha_f \approx \beta + \delta - a\frac{r}{v}, \alpha_r \approx \beta + b\frac{r}{v} \tag{6.101}$$

Substituting above equations in Eq. (6.100), we get

$$F_{yf} = C_f\left(\beta + \delta - a\frac{r}{v}\right) \tag{6.102}$$

$$F_{yr} = C_r\left(\beta + b\frac{r}{v}\right) \tag{6.103}$$

Noting $\dot{\gamma} = v/R$, we obtain the following relationship among yaw angle, forward velocity and sideslip angle,

$$\psi = \beta + \gamma,\ r = \dot{\beta} + \dot{\gamma}, \tag{6.104}$$

Substituting Eqs. (6.102) through (6.104) into Eqs. (6.97) through (6.99), and ignoring the wind gust excitation, we obtain

$$F_{xr} + F_{xf} + \delta C_r(-\beta - \delta + a\frac{r}{v}) + mv\beta(\dot{\beta} - r) - m\dot{v} = 0 \tag{6.105a}$$

$$C_r(\beta + b\frac{r}{v}) + C_f(\beta + \delta - a\frac{r}{v}) + \delta F_{xr} + mv(\dot{\beta} - r) + m\dot{v}\beta = 0 \tag{6.105b}$$

$$C_f a\left(\beta + \delta - a\frac{r}{v}\right) + \delta F_{xr} a - I\dot{r} - C_r b\left(\beta + b\frac{r}{v}\right) = 0 \tag{6.105c}$$

Equation 6.105 is a set of nonlinear equations. By assuming forward velocity to be constant (namely $\dot{v} = 0$), we can rearrange these equations as

$$r\left(\delta C_f\frac{a}{v} - mv\beta\right) + \dot{\beta}mv\beta - \delta C_f\beta - \delta^2 C_f + F_{xf} + F_{xr} = 0 \tag{6.106a}$$

$$r\left(C_r\frac{b}{v} - C_f\frac{a}{v} - mv\right) + \dot{\beta}mv + \beta(C_f + C_r) + \delta(C_f + F_{xf}) = 0 \tag{6.106b}$$

$$\dot{r}I + r\left(C_f\frac{a^2}{v} + C_r\frac{b^2}{v}\right) + \beta(C_r b - C_f a) + \delta a(-C_r - F_{xf}) = 0 \tag{6.106c}$$

If the forward velocity v, the cornering stiffnesses C_f and C_r, and the parameters a, b, m, and I are known, with a given steer angle δ, the sideslip angle β, yaw or vehicle heading ψ, and the direction angle of velocity γ can be calculated. Generally, $C_f \gg F_{xf}$ because the longitudinal force will not affect the lateral motion. For the case of $F_{xf} = 0$, by ignoring Eq. (6.13a), we have

$$\dot{\beta} = -\left(\frac{C_r + C_f}{mv}\right)\beta + \left(\frac{C_f b - C_r a}{mv^2} - 1\right)r + \left(\frac{C_f + F_{xf}}{mv}\right)\delta \tag{6.107a}$$

$$\dot{r} = -\left(\frac{C_r b - C_f a}{I}\right)\beta - \left(\frac{C_r b^2 + C_f a^2}{I}\right)r + \frac{a(C_f + F_{xf})}{I}\delta \tag{6.107b}$$

or as a state equation, we have

$$P\dot{X} + QX = RU \tag{6.108}$$

where X is a state vector and U is the input vector

$$X = \begin{bmatrix} \beta \\ r \end{bmatrix}, U = [\delta]$$

$$P = \begin{bmatrix} m \\ I \end{bmatrix}, Q = \begin{bmatrix} \dfrac{C_r + C_f}{v} & \dfrac{-C_r b + C_f a}{v} + v \\ \dfrac{-C_r b + C_f a}{v}\beta & \dfrac{+C_r b^2 + C_f a^2}{v} \end{bmatrix}$$

$$R = \begin{bmatrix} \dfrac{C_f}{v} \\ \dfrac{C_f a}{I} \end{bmatrix} \tag{6.109}$$

By introducing following parameters

$$a_{11} = -\frac{C_r + C_f}{mv}$$

$$a_{12} = -\left(\frac{C_r b - C_f a}{mv^2} - 1\right)$$

$$a_{21} = -\frac{C_r b - C_f a}{I},$$

$$a_{22} = -\frac{C_r b^2 + C_f a^2}{Iv}$$

$$b_1 = \frac{C_f + F_{xf}}{mv},$$

$$b_2 = \frac{a(C_f + F_{xf})}{I} \tag{6.110}$$

Eq. (6.107) can be expressed as

$$\begin{aligned} \dot{\beta} &= a_{11}\beta + a_{12}r + b_1\delta \\ \dot{r} &= a_{21}\beta + a_{22}r + b_2\delta \end{aligned} \tag{6.111}$$

By eliminating one variable β, we obtain

$$\ddot{r} - (a_{11} + a_{22})\dot{r} + (a_{11}a_{22} - a_{12}a_{21})r = b_2\dot{\delta} + (a_{21}b_1 - a_{11}b_2)\delta \tag{6.112}$$

Similarly, we can get another as

$$\ddot{\beta} - (a_{11} + a_{22})\dot{\beta} + (a_{11}a_{22} - a_{12}a_{21})\beta = b_2\dot{\delta} + (a_{12}b_2 - a_{22}b_1)\delta \tag{6.113}$$

Note that both equations are similar to the equations for a single-degree-of-freedom system. The term $a_{11}a_{22} - a_{12}a_{21}$ corresponds to the square of the natural frequency, and the term $-a_{11} - a_{22}$ corresponds to the damping. The tire actually behaves similarly to a spring and damper, and the characteristics of the tire are similar to the stiffness and damping. The yaw rate and the sideslip angle as well as the directional angle have identical natural frequencies and damping. If the excitation is from the wind gust disturbance instead of the input of the steer angle, then the following equation governs the motion of the vehicle handling:

$$\dot{\beta} = a_{11}\beta + a_{12}r + \frac{F_w}{mv}$$

$$\dot{r} = a_{21}\beta + a_{22}r + \frac{M_w}{I} \tag{6.114}$$

6.3.1 Steady-State Handling Characteristics

Cornering is defined as the ability of the vehicle to travel a curved path. Many factors affect the cornering ability of a vehicle, such as the tire construction, tire tread, road surface, alignment, and tire loading. As a vehicle turns a corner, centrifugal force pushes outward on the center of gravity of the vehicle. Centrifugal force is resisted by the traction of the tires. The interaction of these two forces moves weight from the side of the vehicle on the inside of the turn to the outside of the vehicle, and the vehicle leans. As this occurs, weight leaves the springs on the inside, and that side of the vehicle rises. This weight goes to the springs on the outside, and that side of the vehicle lowers. This is what is known as body roll. When the cornering requirement of a particular maneuver is less than the traction that can be provided by the tires, the car will go where it is pointed and steered. However, if the cornering force exceeds the available traction from the tires, the tires will slip across the road surface and will skid. In this section, we will discuss the steady-state handling characteristics.

6.3.1.1 Yaw Velocity Gain and Understeer Gradient

Consider steady cornering, when a vehicle drives through a curve at low lateral acceleration and low lateral forces are needed. At the wheels, then hardly any lateral slip occurs. In the ideal case, with vanishing lateral slip, the wheels move only in the circumferential direction. Therefore,

$\ddot{\beta} = 0, \dot{\beta} = 0, \dot{\delta} = 0, \delta$ = constant, $F_{xf} = 0$, and Eq. (6.113) becomes

$$(a_{11}a_{22} - a_{12}a_{21})\beta = (a_{12}b_2 - a_{22}b_1)\delta \tag{6.115}$$

Then we have

$$\frac{\beta}{\delta} = \frac{a_{12}b_2 - a_{22}b_1}{a_{11}a_{22} - a_{12}a_{21}} = \frac{-C_f C_r b(a+b) + C_f a m v^2}{C_f C_r (a+b)^2 + (C_r b - C_f a) m v^2} \tag{6.116}$$

Similarly,

$$\frac{r}{\delta} = \frac{a_{21}b_1 - a_{11}b_2}{a_{11}a_{22} - a_{12}a_{21}} = \frac{v(a+b)C_f C_r}{(a+b)^2 C_f C_r + mv^2(bC_r - aC_f)} \tag{6.117}$$

This is defined as yaw velocity gain. We define the understeer gradient as

$$K = \frac{m(bC_r - aC_f)}{L^2 C_f C_r} \tag{6.118}$$

with the unit as deg/g or rad/m/s^2.

Example 6.8: *A typical understeer gradient for a car is K = 0.0024 s^2/m at 2.94 m/s^2, and K = 0.0026 s^2/m at 6.9 m/s^2.*

Example 6.9: *A typical North American car has an average $K = 0.45$ deg/m/s^2, and a typical European car has an average $K = 0.265$ deg/m/ s^2.*

From Eqs. (6.117) and (6.118), we obtain the yaw velocity gain as

$$\frac{r}{\delta} = \frac{v/L}{1 + Kv^2} \tag{6.119}$$

This ratio represents a gain that is proportional to velocity in the case of a neutral steer vehicle.

Example 6.10: *For $v = 22.35$ m/s, $a_y = 3.92$ m/ s^2, and $\frac{r}{\delta}$ is in the range of 0.16 to 0.33 s^{-1}.*

Larger lateral acceleration leads to nonlinearity of the tire slip angle. The steady-state response of the vehicle can be classified into three categories in terms of the value of K, as shown in Fig. E6.10.

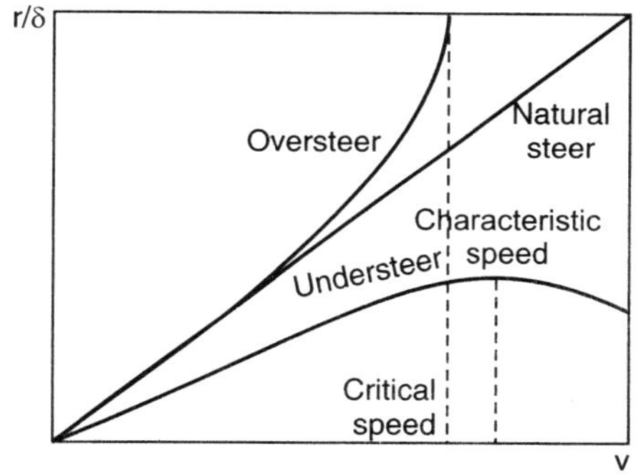

Fig. E6.10 Yaw velocity gain as a function of speed.

6.3.1.1.1 Neutral Steer

When $K = 0$, $r/\delta = v/L$, and the yaw velocity gain is proportional to the velocity of the vehicle, with a slope of $1/L$. It is named as neutral steer and is a pure rolling case with radius of $R \approx L/\delta$. Note that this relationship is the cornering characteristics of the vehicle when it travels with very low velocity and without sideslip angle. When the sideslip angle is $\beta = 0$, the steering angle $\delta \approx L/R$ is named as the Ackerman angle. The radius of turn is $R = L/\delta$, the yaw velocity is $r = (v/L)\delta$, and the yaw velocity gain is $\left.\frac{r}{\delta}\right|_s = v/L$.

On a constant-radius turn, no change in the steer angle will be required as the speed is varied.

6.3.1.1.2 Understeer

When $K > 0$, the denominator of Eq. (6.113) is larger than 1, and r/δ is not linearly proportional to the velocity of the vehicle. It is a downward bending curve. The larger the value of K, the lower the yaw velocity gain and, accordingly, the larger the under-steering. When the vehicle speed reaches $v_{ch} = \sqrt{\frac{1}{K}}$, the yaw velocity gain reaches its maximum value. Its value is exactly half of that of the neutral steer vehicle. The vehicle speed v_{ch} is defined as the characteristic speed, which is used to quantify the magnitude of understeering. When the value of the understeer increases, K increases, and v_{ch} decreases. The response is always stable and decreases with the increase in the speed of the vehicle.

6.3.1.1.3 Oversteer

When $K < 0$, the denominator of Eq. (6.119) is smaller than 1, and r/δ is not linearly proportional to the velocity of the vehicle. It is an upward bending curve. The smaller the value of K, the higher the yaw velocity gain and, accordingly, the larger the oversteering. When the vehicle speed reaches $v_{cr} = \sqrt{\frac{-1}{K}} = \sqrt{\frac{1^2 C_f C_r}{m(aC_f - bC_r)}}$, the yaw velocity gain reaches an infinite value. This speed characterizes a critical speed where the vehicle starts to lose stability. As approaches infinity, any trivial steering angle input results in an infinite yaw velocity, which means that the radius of turn is infinitely small; the vehicle has radical cornering and sideslip, or even rollover. Because the oversteer design of a vehicle has the trend of instability, vehicles generally are designed with appropriate understeer properties. For the understeer case, the yaw velocity increases with the speed of the vehicle until it reaches the characteristic speed. Then the yaw velocity begins to decrease thereafter. The characteristic speed is of significance as the speed at which the vehicle is most responsive in yaw. Figure 6.42 shows a typical curve of yaw rate gain as a function of vehicle speed.

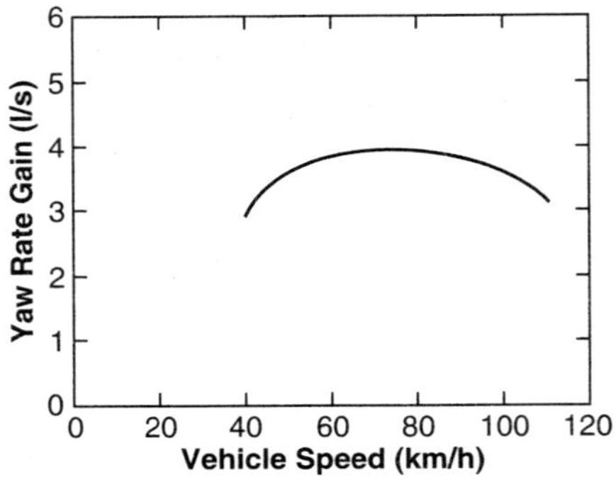

Fig. 6.42 Yaw rate gain as a function of vehicle speed.

6.3.1.2 Difference Between Slip Angles of the Front and Rear Wheels

The difference between the slip angles of the front and rear wheels on a constant-radius turn is another useful index. For a vehicle traveling forward with a speed of v, the sum of forces in the lateral direction from the tires must equal the mass times the centripetal acceleration

$$F_{yf} + F_{yr} = ma_y = mv^2/R \tag{6.120}$$

where R is the radius of the turn. Consider a vehicle with a moment equilibrium about the center of gravity. The sum of the moment from the front and rear lateral forces results in

$$F_{yf}a - F_{yr}b = 0 \tag{6.121}$$

From Eqs. (6.120) and (6.121), we can write

$$F_{yr} = mav^2/(LR)$$

$$F_{yf} = mbv^2/(LR) \tag{6.122}$$

By denoting

$$\alpha_f = \frac{mav^2}{(C_f LR)}$$

and
$$\alpha_r = \frac{mav^2}{(C_r LR)}$$

we obtain
$$K = \frac{1}{a_y L}(\alpha_f - \alpha_r) \tag{6.123}$$

$$(\alpha_f - \alpha_r) = Ka_y L \tag{6.124}$$

This relationship is plotted in Fig. 6.43. To measure the steady-state response of the vehicle, we apply a fixed value of the steering angle, which enables the vehicle to travel with a constant-radius turn of different speeds, measure the front and rear tire slip angles, and apply the difference between the front and rear tire slip angles versus lateral acceleration by curve to evaluate the steady response. When the lateral acceleration is larger than 3 to 4 m/s/s, $(\alpha_f - \alpha_r)$ is no longer linearly proportional to the lateral acceleration. This is due to the cornering stiffness, which is no longer in the linear range. From Eq. (6.124), we know that if $(\alpha_f - \alpha_r) > 0$, $K > 0$ and the vehicle is in understeer. Likewise, when $(\alpha_f - \alpha_r) = 0$, $K = 0$ and the vehicle is in neutral steer. When $(\alpha_f - \alpha_r) < 0$, $K < 0$ and the vehicle is in oversteer. Consider Eqs. (6.119) and (6.123). We have $\delta = \frac{L}{R} + LKa_y$. Therefore, we have $\delta = \frac{L}{R} + (\alpha_f - \alpha_f)$. This equation can be viewed from the model plot. Reorganizing the equation gives

$$R = \frac{L}{\delta - (\alpha_f - \alpha_r)} \tag{6.125}$$

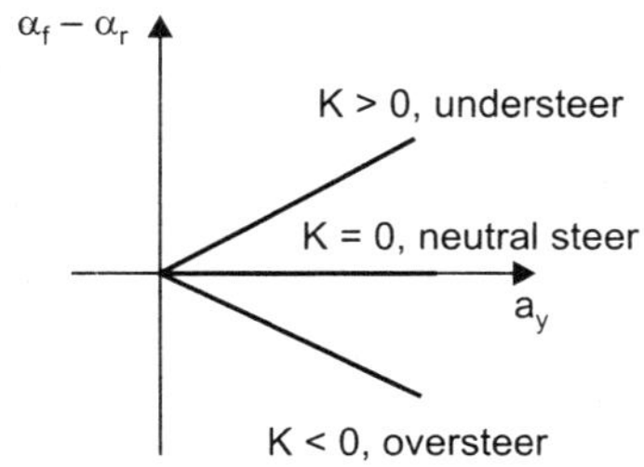

Fig. 6.43 As a function of acceleration.

If the slip angle can be ignored at very low speed, we get the radius of turn corresponding to the Ackerman angle as

$$R_0 = \frac{L}{\delta} \tag{6.126}$$

When the vehicle speed is raised and there is proper lateral acceleration, there is sufficient slip angle at the front and rear tires. If $(\alpha_f - \alpha_r) > 0$, then $R > R_0$ and it is understeer. If $(\alpha_f - \alpha_r) < 0$, then $R < R_0$.

6.3.1.3 Ratio of Radius of Turn

If the slip angle can be ignored at very low speed, we get the radius of turn R_0. When the vehicle speed is raised and there is proper lateral acceleration, there are enough slip angles at the front and rear tires, and then a different R, the ratio of the radius, can be applied. To characterize the steady-state response of vehicle, we derive

$$R = \frac{v}{r} = \frac{(1+Kv^2)L}{\delta} = (1+Kv^2)R_0 \tag{6.127}$$

Then

$$\frac{R}{R_0} = (1 + Kv^2) \tag{6.128}$$

When $K = 0$, $\frac{R}{R_0} = 1$, which means that the radius of turn of the vehicle with neutral steer does not vary with the vehicle speed and remains as R_0. When $K > 0$, $\frac{R}{R_0} > 1$, which means the turn radius of the understeer vehicle is always larger than R_0, and moreover, the radius of turn increases with the increase in the speed of the vehicle. When $K < 0$, $\frac{R}{R_0} < 1$, which means the turn radius of the understeer vehicle is always smaller than R_0, and moreover, the radius of turn decreases with the increase in the speed of the vehicle.

6.3.1.4 Static Margin

From Eq. (6.118), the sign of the understeer gradient depends on the sign of $bC_r - aC_f$, which is referred to as the static margin. The static margin is another measure of the steady-state handling behavior. It is determined by the point on the vehicle where a lateral force will not generate any steady-state yaw velocity (*i.e.*, the neutral steer point). We define a neutral steer line as shown in Fig. 6.44. The neutral steer line is the locus of points in the *x*-*y* plane along which external forces produce no steady-state yaw velocity. The static margin is defined as the distance the neutral steer point falls behind the center of gravity, normalized by the wheelbase, that is,

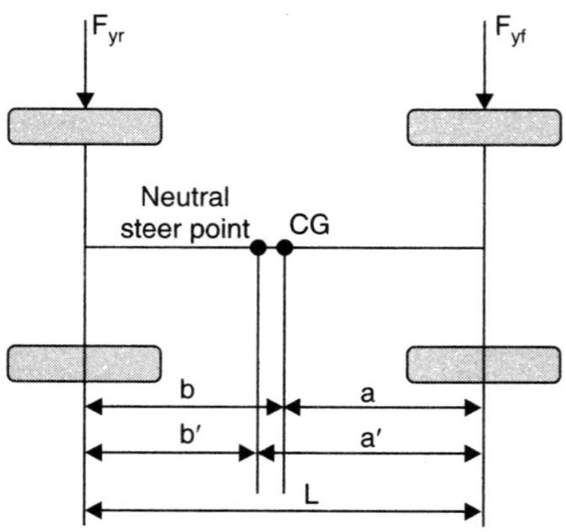

Fig. 6.44 Schematic of the static margin.

$$\text{Static margin} = \frac{e}{L}$$

When the point is behind the center of gravity, the static margin is positive and the vehicle is in understeer. At the center of gravity, the static margin is zero and the vehicle is in neutral steer. When the point is ahead of the center of gravity, the vehicle is in oversteering.

Example 6.11: *The static margin of a typical vehicle falls in the range of 0.05 to 0.07 behind the center of gravity. We can determine the location of the neutral steer point by the moment of equilibrium. When lateral forces are applied at the neutral steer point, the front and rear tires yield the same slip angle* α*, together with the lateral tire force* $F_{yf} = C_f\alpha$ *and* $F_{yr} = C_r\alpha$*. The distance between the neutral steer point and the front axis is*

$$a' = \frac{F_{yr}L}{F_{yf} + F_{yr}} = \frac{C_r}{C_f + C_r}L \tag{6.129}$$

Solution: The static margin is the difference between the distance of the neutral steer point to the front axis and the distance of the center of gravity to the front axis

$$\text{Static margin} = \frac{a' - a}{L} = \frac{C_r}{C_f + C_r} - \frac{a}{L} = \frac{bC_r - aC_f}{C_f + C_r} \tag{6.130}$$

When the neutral steer point coincides with the center of gravity, the static margin is zero. If the lateral force applied at the center of gravity causes the front and rear tires to have the same slip angle, the vehicle possesses neutral steer properties. When the center of gravity is in front of the neutral steer point, the static margin is negative. If the lateral force applied at the center of gravity causes the slip angle of the rear tire to be larger than that of the slip angle of the front tire, the vehicle has oversteer properties. When the center of gravity is behind the neutral steer point, the static margin is positive. Then the lateral force applied at the center of gravity causes the slip angle of the front tire to be larger than the slip angle of the rear tire, and the vehicle has understeer properties.

6.3.2 Dynamic Characteristics of Handling

In this section, we will discuss transient analysis and frequency domain analysis. We also will describe the stability analysis associated with the system eigenvalue.

6.3.2.1 Handling Damping and Natural Frequency

Next we discuss the free response characteristics of steering a vehicle. Consider Eqs. (6.112) and (6.113), where the system can be treated as a single-degree-of-freedom system. For Eqs. (6.112) and (6.113), we consider the free response case, namely,

$$b_2\dot{\delta} + (a_{21}b_1 - a_{11}b_2)\delta = 0 \text{ and } b_2\dot{\delta} + (a_{12}b_2 - a_{22}b_1)\delta = 0$$

By assuming $\omega_n^2 = a_{11}a_{22} - a_{12}a_{21}$ and $2\xi\omega_n = -(a_{11} + a_{22})$, we define ω_n as the natural circular frequency of the yaw velocity, and ξ as the damping factor of the yaw velocity. Hence, we obtain

$$\omega_n = \sqrt{\frac{mv(aC_f - bC_r) + C_fC_rL^2/v}{mvI}} = \frac{L}{v}\sqrt{\frac{C_fC_r(1 + Kv^2)}{mI}} \tag{6.131}$$

where ω_n is an important factor in assessing the transient response. It is desired to have a higher ω_n for vehicles.

The damping factor or ratio can be derived as

$$\xi = \frac{-[m(a^2C_f + b^2C_r) + I(C_f + C_r)]}{2mIL\sqrt{\dfrac{C_fC_r(1+Kv^2)}{mI}}} = \frac{-[m(a^2C_f + b^2C_r) + I(C_f + C_r)]}{2L\sqrt{mIC_fC_r(1+Kv^2)}} \quad (6.132)$$

Example 6.12: *A typical car has* $f_n = \dfrac{\omega_n}{2\pi}$ *about 1 Hz, and the damping is in the range of Figs. E6.12(a) and E6.12(b) respectively, show the damping ratio and the natural frequency as functions of the vehicle speed under different cornering stiffness differences.*

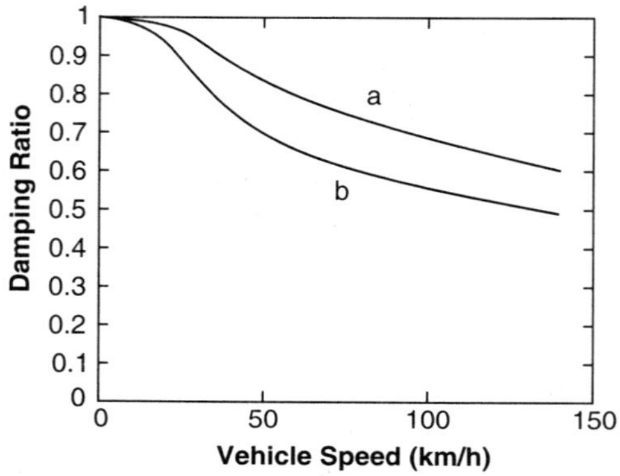

Fig. E6.12 (*a*) Damping as a function of speed for different cornering stiffness differences, where $\Delta C = C_r - C_f$: (*a*) ΔC_1, and (*b*) ΔC_2, $\Delta C_2 > \Delta C_1$.

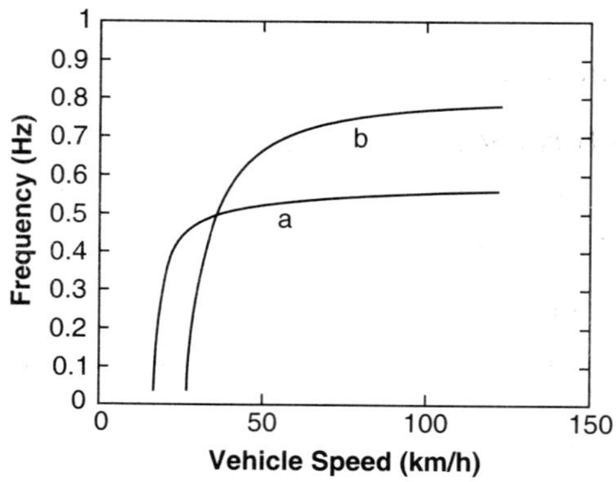

Fig. E6.12 (*b*) Natural frequency as a function of speed for different cornering stiffness differences, where $\Delta C = C_r - C_f$: (*a*) ΔC_1, and (*b*) ΔC_2, $\Delta C_2 > \Delta C_1$.

To find the transient response, Eq. (6.112) can be rewritten as

$$\dot{r}(t) + 2\xi\omega_n \dot{r}(t) + \omega_n^2 r(t) = 0 \quad (1)$$

To solve Eq. (1), we assume that

$$\dot{r}(t) = Ae^{st} \tag{2}$$

where A is a constant and s is a parameter that remains to be determined. By substituting Eq. (2) into Eq. (1), one obtains

$$(s^2 + 2\xi\omega_n s + \omega_n^2) Ae^{st} = 0 \tag{3}$$

Because $Ae_{st} \neq 0$, then

$$s^2 + 2\xi\omega_n s + \omega_n^2 = 0 \tag{4}$$

Equation (4) is known as the characteristic equation of the system. This equation has the following two roots

$$s_1, s_2 = \left(-\xi \pm \sqrt{\xi^2 - 1}\right)\omega_n \tag{5}$$

Solution (*a*): $\xi < 1$ (underdamped condition):

$$s_1, s_2 = \left(-\xi \pm i\sqrt{\xi^2 - 1}\right)\omega_n$$

Then the yaw velocity can be solved as

$$r(t) = A_1 \exp\left(-\xi + \sqrt{\xi^2 - 1}\right)\omega_n t + A_2 \exp\left(-\xi + \sqrt{\xi^2 - 1}\right)\omega_n t \tag{6}$$

or

$$r(t) = A \exp(-i\omega_n t) \cos(\omega_d t - \phi) \tag{7}$$

where $\omega_d = \omega_n\sqrt{1-\xi^2}$ is the damped frequency of the system. Constants A and φ are determined from the initial conditions.

Solution (*b*): $\xi > 1$ (overdamped condition):

$$s_1, s_2 = \left(-\xi \pm \sqrt{\xi^2 - 1}\right)\omega_n$$

Then the yaw velocity can be solved as

$$r(t) = A_1 \exp\left(-\xi + \sqrt{\xi^2 - 1}\right)\omega_n t + A_2 \exp\left(-\xi + \sqrt{\xi^2 - 1}\right)\omega_n t \tag{8}$$

The motion is a periodic and decays exponentially with time. Constants A_1 and A_2 are determined from the initial conditions.

Solution (*c*): $\xi = 1$ (critically damped condition):

$$s_1 = s_2 = -\omega_n$$

Then the yaw velocity can be solved as

$$r(t) = (A_1 + A_2) \exp(-\omega_n t) \tag{9}$$

Equation (9) represents an exponentially decaying response. The constants A_1 and A_2 depend on the initial conditions.

6.3.2.2 Step Steer Input Response

Yaw velocity properties can be characterized by the transient response to step function input or ramp function input from the steering angle. The frequency response of the yaw velocity can be obtained from the transient response test of the impulse input of the steering angle. Next we will derive the step input response by considering the step steer input to the following equation from Eq. (6.112)

$$\ddot{r}(t) + 2\xi\omega_n\,\dot{r}(t) + \omega_n^2\,r(t) = B_1\dot{\delta} + B_0\delta \tag{6.133}$$

where $B_1 = b_2$, $B_0 = (a_{21}b_1 - a_{11}b_2)$. Consider the following step input of the front tire steering angle as

$$\delta(t) = \begin{cases} 0, & t < 0 \\ \delta_0, & t \geq 0 \end{cases}, \quad \dot{\delta}(t) = \begin{cases} 0, & t < 0 \\ 0, & t \geq 0 \end{cases} \tag{6.134}$$

Then for $t \geq 0$, Eq. (6.133) can be simplified a

$$\ddot{r}(t) + 2\xi\omega_n\,\dot{r}(t) + \omega_n^2\,r(t) = B_0\delta_0 \tag{6.135}$$

The nonhomogeneous solution of Eq. (6.135) is

$$r = \frac{B_0\delta_0}{\omega_n^2} = \frac{v/L}{1+Kv^2}\delta_0 = \left.\frac{r}{\delta}\right|_s \delta_0 = r_0 \tag{6.136}$$

This is defined as the steady-state yaw velocity. Combining the homogenous solution given by Eq. (6) of Example E6.12 and the nonhomogeneous solution given by Eq. (6.136) yields

$$r(t) = \frac{B_0\delta_0}{\omega_n^2} + A\exp(-\xi\omega_n t)\cos(\omega_d t - \phi) \tag{6.137}$$

or

$$r(t) = \frac{B_0\delta_0}{\omega_n^2} + A_1\exp(-\xi\,\omega_n t)\cos\omega_d t + A_2\exp(-\xi\omega_n t)\sin\omega_d t \tag{6.138}$$

It can be shown that

$$A_1 = -\frac{B_0\delta_0}{\omega_n^2},\ A_2 = \frac{B_0\delta_0}{\omega_n^2}\left(\frac{B_1\omega_n^2}{B_0} - \xi\omega_n\right)\frac{1}{\omega_d} = r_0\left(\frac{-mva\omega_n}{LC_r} - \xi\right)\frac{1}{\sqrt{1-\xi^2}} \tag{6.139}$$

$$A = \sqrt{A_1^2 + A_2^2} = r_0\sqrt{[(\frac{-mva}{LC_r})^2\omega_n^2 + \frac{2mva\xi\omega_n}{LC_r} + 1]\frac{1}{1-\xi^2}} \tag{6.140}$$

Then

$$r(t) = r_0\sqrt{\left[\left(\frac{-mva}{LC_r}\right)^2\omega_n^2 + \frac{2mva\xi\omega_n}{LC_r} + 1\right]\frac{1}{1-\xi^2}}\exp(-\xi\omega_n t)$$

$$\cos\left(\omega_d t + \arctan\left[\frac{-\sqrt{1-\xi^2}}{\dfrac{-mva\omega_n}{LC_r} - \xi}\right]\right) \tag{6.141}$$

This is the yaw velocity transient response to step input of the front tire steer angle. From this equation, we know that when $t = \infty$, $r(\infty) = \left.\frac{r}{\delta}\right|_s \delta_0 = r_0$, which means the yaw velocity approaches a steady-state yaw velocity. For the transition process, the response is a damped oscillation, as shown in Fig. 6.46.

Example 6.13: *A typical overshoot ranges $r_{max}/r_0 = 1.12 - 1.65$ with a testing condition 31.3 m/s and $\alpha_y = 0.4g$. To further quantify the response, more dynamic factors can be defined as follows. Response time τ is defined as the time needed for yaw velocity to attain a steady value r_0 for the first time, after the step input is applied. Typically, it is desired to have smaller τ. To derive τ, from Eq. 6.53, when $t = \tau$, we have $A\exp(-\xi\omega_n\tau)\cos(\omega_d\tau - \phi) = 0$, so $\cos(\omega_d\tau - \phi) = 0$, then,*

$$\tau = -\frac{\Phi}{\omega_d} = \frac{\arctan\left[\dfrac{\sqrt{1-\xi^2}}{\dfrac{-mva\omega_n}{LC_r} - \xi}\right]}{\omega_n\sqrt{1-\xi^2}} \tag{1}$$

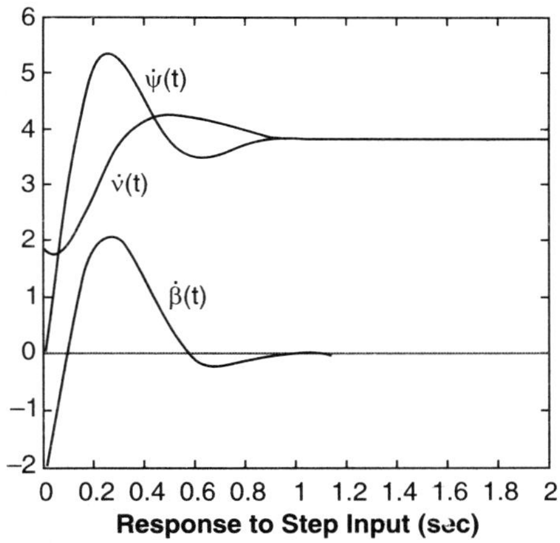

Fig. E6.13 Yaw rate, sideslip velocity, and directional velocity response.

Another dynamic factor is the time needed to attain the first peak ε, which is the peak response time. This is used to assess the transient response. Taking the first derivative of Eq. (6.137), we obtain

$$\frac{dr(t)}{dt} = A\exp(-\xi\omega_n t)\,[-\xi\omega_n \sin(\omega_d t + \phi) + \omega_d \cos(\omega_d t + \phi)] \tag{2}$$

Consider $\left.\dfrac{dr(t)}{dt}\right|_{t=\varepsilon} = 0$

We have $-\xi\omega_n \sin(\omega_d t + \phi) + \omega_d \cos(\omega_d t + \phi) = 0$, $\tan(\varepsilon\,\omega_d + \phi) = \dfrac{\omega_d}{\xi\omega_n}$

Therefore,

$$\varepsilon = -\frac{\arctan\dfrac{\omega_d}{\xi\omega_n} - \Phi}{\omega_d} = \frac{\arctan\left[\dfrac{\sqrt{1-\xi^2}}{\xi}\right]}{\omega_n\sqrt{1-\xi^2}} + \tau \tag{3}$$

Another dynamic factor is the vehicle factor T.B., which is defined as the product of the peak response time and the slip angle of the center of gravity, where T.B. = $\xi\beta$.

Example 6.14: *A typical car has the range of* $\varepsilon = 0.23$ *to 0.59 sec and T.B. = 0.25 to 1.45 sec, for tested condition of* $v = 31.3$ *m/s and* $a_y = 0.4g$.

6.3.2.3 Ramp Steer Input Response

Next we give the ramp input response by considering the ramp steer input of

$$\delta(t) = \begin{cases} 0, & t < 0 \\ kt, & t > 0 \end{cases}$$

$$\ddot{r} - (a_{11} + a_{22})\dot{r} + (a_{11}a_{22} - a_{12}a_{21})r = b_2\dot{\delta} + (a_{21}b_1 - a_{11}b_2)\delta \tag{6.142}$$

If the damping can be ignored, the solution can be derived as

$$r = \frac{k}{\omega_n^2}[b_2 + (a_{21}b_1 - a_{11}b_2)t - \frac{1}{\omega_n}(a_{21}b_1 - a_{12}b_2)\sin\omega_n t] \tag{6.143}$$

6.3.2.4 Impulse Input Excitation Response

Wind gust excitation can be represented as impulse input excitation. Next we will derive the impulse step response by considering the impulse steer input $\delta = \delta_0\delta(t)$ where $\delta(t)$ is the Dirac delta function, which is defined as

$$\delta(t) = 0, \text{ for } t \neq 0 \tag{6.144a}$$

and

$$\int_{-\infty}^{\infty} \delta(t)dt = 1 \tag{6.144b}$$

From the system equation

$$\ddot{r} - (a_{11} + a_{22})\dot{r} + (a_{11}a_{22} - a_{12}a_{21})r = b_2\dot{\delta} + (a_{21}b_1 - a_{11}b_2)\delta$$

the solution is derived as

$$r(t) = \frac{(a_{21}b_1 - a_{11}b_2)\delta_0}{\omega_d} \exp(-\xi\omega_n t) \sin \omega_d t \tag{6.145}$$

If we consider the wind gust impulse excitation, the system equation can be derived from Eq. (6.114) as

$$\ddot{r} - (a_{11} + a_{22})\dot{r} + (a_{11}a_{22} - a_{12}a_{21})r = a_{21}F_w/(mv) - a_{11}M_w \tag{6.146}$$

The solution is

$$r(t) = \frac{a_{21}F_w - a_{11}M_w mv}{\omega_d mv} \exp(-\xi\,\omega_n t) \sin \omega_d t \tag{6.147}$$

Figure 6.45 shows the yaw, sideslip, and directional angle response to impulse excitation.

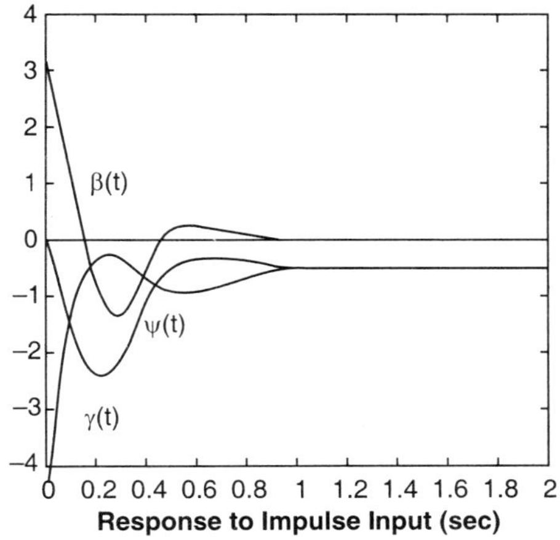

Fig. 6.45 Yaw, side-slip, and directional angle response to impulse excitation.

6.3.2.5 Frequency Response of Yaw Velocity

We now consider the response of a bicycle model to a harmonic excitation, based on the equation of system motion, which is Eq. (6.112). The approach uses a frequency domain analysis and is based on the concept of transfer and harmonic response functions. By taking the Laplace transformation of Eq. (6.112) or Eq. (6.133), we obtain

$$[s^2\bar{r}(s) - r(0)\,s - \dot{r}(0)] + 2\xi\omega_n\,[s\bar{r}(s) - r(0)] + \omega_n^2\bar{r}(s) = B_1 s\delta(s) + B_0\delta(s) \tag{6.148}$$

It follows that

$$\bar{r}(s) = \frac{B_1 s\delta(s) + B_0\delta(s)}{s^2 + 2\xi\omega_n s + \omega_n} + \frac{(s + 2\xi\omega_n)r(0) + \dot{r}(0)}{s^2 + 2\xi\omega_n s + \omega_n^2} \tag{6.149}$$

By ignoring the homogeneous solution, Eq. (6.149) can be written as

$$\bar{r}(s) = A(s)/B(s) \tag{6.150}$$

where $A(s)$ and $B(s)$ are polynomials, with $B(s)$ of a higher order. The response $r(t)$ is found by taking the inverse Laplace transformation. If only the forced solution is considered, we can define impedance transform as

$$Z(s) = [\,\bar{\delta}(s)/\bar{r}(s)\,] = (s^2 + 2\xi\omega_n\,s + \omega_n^2)/(B_1 s + B_0) \tag{6.151}$$

A transfer function $H(s)$ is defined

$$H(s) = 1/Z(s) = (B_1 s + B_0)\;(s^2 + 2\xi\omega_n\,s + \omega_n^2)^{-1} \tag{6.152}$$

The yaw rate and sideslip velocity transfer function can be solved, respectively, as

$$H_{r\delta}(s) = \frac{b_2 s + a_{21}b_1 - a_{11}b_2}{s^2 - (a_{22} + a_{11})s + (a_{11}a_{22} - a_{12}a_{21})} \tag{6.153}$$

$$H_{\beta\delta}(s) = \frac{b_1 s + a_{12}b_2 - a_{22}b_1}{s^2 - (a_{22} + a_{11})s + (a_{11}a_{22} - a_{12}a_{21})} \tag{6.154}$$

By using the definition of $B_1 = b_2,\ B_1' = b_1,\ B_0 = (a_{21}b_2 - a_{11}b_1),\ B_0' = a_{12}b_2 - a_{22}b_1$ we have

$$
\begin{aligned}
H_{r\delta}(i\omega) &= \frac{B_1 i\omega + B_0}{-\omega^2 + 2i\omega\xi\omega_n + \omega_n^2} \\
&= \frac{B_1 i\omega + B_0}{-\omega^2 + 2i\omega\xi\omega_n + \omega_n^2} \times \frac{-\omega^2 - 2i\omega\xi\omega_n + \omega_n^2}{-\omega^2 - 2i\omega\xi\omega_n + \omega_n^2} \qquad (6.155)\\
&= \frac{2B_1\xi\omega_n\omega^2 + B_0(\omega_n^2 - \omega^2)}{(\omega_n^2 - \omega^2)^2 - (2\xi\omega\omega_n)^2} + \frac{B_1\omega(\omega_n^2 - \omega^2) - 2B_0\xi\omega_n\omega}{(\omega_n^2 - \omega^2)^2 - (2\xi\omega\omega_n)^2} i \\
&= A(\omega) + B(\omega)i
\end{aligned}
$$

The amplitude characteristics are

$$D(\omega) = \sqrt{[A(\omega)]^2 + [B(\omega)]^2} \tag{6.156}$$

The phase characteristics are

$$\Phi(\omega) = \arctan\frac{B(\omega)}{A(\omega)} \tag{6.157}$$

Similarly,

$$
\begin{aligned}
H_{\dot{\beta}\delta}(i\omega) &= \frac{B_1' i\omega + B_0'}{-\omega^2 + 2i\omega\xi\omega_n + \omega_n^2} \\
&= \frac{B_1' i\omega + B_0'}{-\omega^2 + 2i\omega\xi\omega_n + \omega_n^2} \times \frac{-\omega^2 - 2i\omega\xi\omega_n + \omega_n^2}{-\omega^2 - 2i\omega\xi\omega_n + \omega_n^2} \qquad (6.158)\\
&= \frac{2B_1'\xi\omega_n\omega^2 + B_0'(\omega_0^2 - \omega^2)}{(\omega_n^2 - \omega^2)^2 - (2\xi\omega\omega_n)^2} + \frac{B_1'\omega(\omega_0^2 - \omega^2) - 2B_0'\xi\omega_n\omega}{(\omega_n^2 - \omega^2)^2 - (2\xi\omega\omega_n)^2} i \\
&= A'(\omega) + B'(\omega)i
\end{aligned}
$$

The amplitude characteristics are

$$D'(\omega) = \sqrt{[A'(\omega)]^2 + [B'(\omega)]^2} \tag{6.159}$$

The phase characteristics are

$$\Phi'(\omega) = \arctan\frac{B'(\omega)}{A'(\omega)} \tag{6.160}$$

Figure 6.46 shows the typical amplitude and phase frequency response of the yaw velocity. The x axis is the logarithmic of the input frequency. The gain is represented by level (unit decibels [dB]).

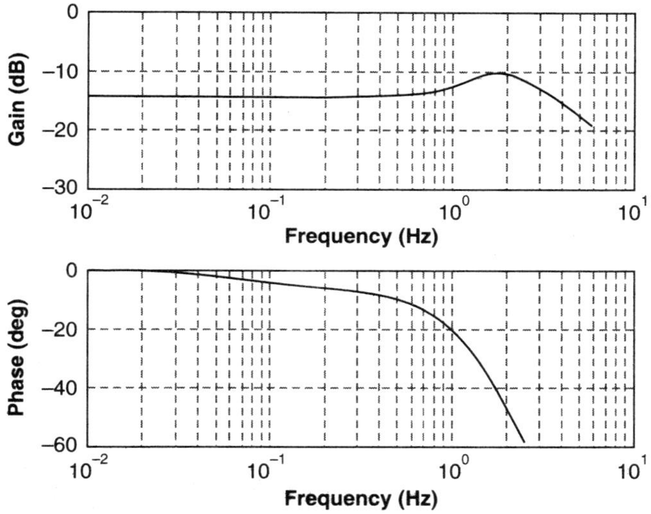

Fig. 6.46 Amplitude and phase frequency response of the yaw velocity.

The amplitude frequency response plot reflects the capability of the vehicle to follow steering orders with different frequencies. The amplitude frequency response curve levels off at the low frequency range, and the amplitude increases with the increase in frequency. The system tends toward resonance when a specific frequency is attained. If the frequency increases further, the amplitude reduces again and drops rapidly. From the viewpoint of steer stability, it is desirable to have a flatter amplitude frequency curve with a higher resonant frequency. Also, it is desirable to have a smaller phase difference to allow the vehicle to have a rapid response. Several parameters related to the frequency response curve have been used to quantify the steer stability:

- Steady gain, $H_{r\delta}(0)$, the amplitude ratio at zero frequency.
- Resonant frequency, where, the higher the f_r, the better the steer stability
- Amplitude ratio at the resonant point $H_{r\delta}(f_r)/H_{r\delta}(0)$, where it is desirable to have a smaller $H_{r\delta}(f_r)/H_{r\delta}(0)$.
- Phase delay angle at a frequency of 0.1 Hz, where it represents the response speed at low speed steering and is desirable for it to be zero.
- Phase delay angle at a frequency of 0.6 Hz, where it represents the response speed at fast speed steering and is desirable to be smaller.

Example 6.16: *A typical car has an f_r = 1.2 Hz, $H_{r\delta}(f_r) - H_{r\delta}(0)$ = 3.38 dB, and a phase delay of 26°.*

6.3.2.6 Stability Analysis

If the transient response $r(t)$ converges to a steady-state value, then the vehicle handling will be stable. When $r(t)$ approaches infinity, the vehicle handling exhibits instability. It is well known that the system natural properties will determine its stability. From Eq. (6.46), when $\xi \le 1$, as long as $\xi\omega_n > 0$, the solution converges; otherwise, it is divergent or unstable. It can be shown that

$$\xi\omega_n = \frac{-[m(a^2C_f + b^2C_r) + I(C_f + C_r)]}{2mvI} \tag{6.161}$$

Because C_f and C_r are negative, $\xi\omega_n > 0$. If $\xi < 1$, the solution will converge to zero. When $\xi > 1$, the characteristic roots should be negative for the solution to converge to zero, namely, $-\xi\omega_n \pm \sqrt{(\xi\omega_n)^2 - \omega_n{}^2}$ should be negative. In other words, ω_n^2 should be positive for the yaw velocity to converge. It can be shown that

$$\omega_n^2 = \frac{(aC_f - bC_r)}{I} + \frac{C_f C_r L}{mv^2 I} \tag{6.162}$$

The sign of the first term depends on the vehicle steady-state response. Because the understeer gradient is defined as

$$K = \frac{m(bC_r - aC_f)}{L^2 C_f C_r} \tag{6.163}$$

$$bC_r - aC_f = \frac{KL^2 C_f C_r}{m} \tag{6.164}$$

When the vehicle has an understeer feature, $K > 0$ and $bC_r - aC_f > 0$. The first term of Eq. (6.162) is positive when the vehicle has an understeer feature, $K < 0$ and $bC_r - aC_f < 0$. The first term of the above equation is negative. The second term of Eq. (6.162) is positive, and when the vehicle speed is slow, it is a very large value; therefore, ω_n^2 tends to be positive despite the sign of the first term. Then the yaw velocity $r(t)$ converges, and the vehicle is stable. With an increase in the vehicle speed, the second term becomes smaller and smaller. When $bC_r - aC_f < 0$ due to oversteer, ω_n^2 is likely to be negative, $r(t)$ diverges, and the vehicle is not stable. The velocity when $\omega_n^2 = 0$ with oversteer is defined as the critical speed v_{cr}. When the speed is larger than v_{cr} and $\omega_n^2 < 0$, the vehicle is unstable. By allowing ω_n^2 to equal zero, the critical speed can be derived as $v_{cr} = \sqrt{-1/K}$, which was given in Eq. (6.124).

6.3.2.7 Curvature Response

Curvature gain is another useful factor in some analyses. Consider $v = R\dot{\gamma} = \dot{\gamma}/\rho = (\dot{\psi} - \dot{\beta})/\rho$, where

$$\frac{\rho}{\delta}(i\omega) = \frac{\dot{\gamma}}{\delta v} = \frac{1}{v}\left(\frac{\dot{\psi}}{\delta} - \frac{\dot{\beta}}{\delta}\right) = \frac{1}{v}(H_{\gamma\delta} - H_{\beta\delta}) \tag{1.165}$$

Substituting the yaw velocity and sideslip transfer functions from Eqs. (6.155) and (6.158) into Eq. (6.165), we obtain

$$\frac{\rho}{\delta}(i\omega) = \frac{b_2(a_{11} - i\omega + a_{12}i\omega) - b_1(\omega^2 + a_{22}i\omega + a_{21})}{a_{12}v(\omega^2 + a_{22}i\omega + a_{21}) - (a_{11} - i\omega + a_{12}i\omega)(a_{22}v - vi\omega)} \tag{1.166}$$

Let us consider the following two cases.

In the first case, when $\omega = 0$ or δ is constant, Eq. (6.166) leads to

$$\frac{\rho}{\delta}(i\omega) = \frac{b_2 a_{11} - b_1 a_{21}}{a_{12}a_{21}v - a_{11}a_{22}v} = \frac{C_f C_r (a-b)}{C_f C_r (a+b)^2 - mv^2(C_f a - C_r b)} \tag{1.167}$$

When $v = 0$, the curvature gain becomes

$$\frac{\rho}{\delta}(i\omega) = \frac{1}{a+b} \tag{1.168}$$

If $C_f a = C_r b$, even $v \neq 0$ and Eq. (6.168) still holds.

In the second case, if ω is very big or $\omega^2 \gg \omega$, the curvature gain is given by

$$\frac{\rho}{\delta}(i\omega) = \frac{C_f a_{11}}{mv^2} = \frac{1}{v} b_1 \tag{1.169}$$

6.4 ACCELERATION

6.4.1 Introduction

Among the many attributes customers look for when purchasing a vehicle is its power. Power ranks as the top attribute with customers seeking sport vehicles, and it ranks among the top ten attributes for most other customers. Vehicle power affects the load-carrying capacity and acceleration of the vehicle. In this chapter, we will study the acceleration of vehicles under various driving and loading conditions.

Vehicle acceleration is limited by two main factors. The first is its power. In power-limited acceleration, the maximum acceleration is obtained when the vehicle is driven at its maximum power, provided that no slippage occurs at the tires. The second factor is traction. In traction-limited acceleration, the maximum acceleration is obtained when the tires start slipping. At low speeds and in low gears, acceleration typically is limited by traction. At high speeds and in high gears, acceleration is limited by the power generated by the powerplant. The term "powerplant" will refer to the engine and transmission.

6.4.1.1 Acceleration

If the vehicle is moving as a result of consistent and constant tractive forces at the tires, changing its velocity from an initial velocity V_i (which would be zero if one starts from an idle status) to a final velocity V_f, the acceleration can be calculated as

$$a = \frac{(V_f - V_i)}{t} \tag{6.170}$$

where t is time. Note that Eq. (6.170) is the same as that used in Chapter 7 Eq. (7.1) for braking and is repeated here for convenience. The only difference is that the acceleration will be a positive number when the vehicle is accelerating. If the total tractive force on both axles is Fb, then from Newton's second law of motion, the acceleration is F_b/M, where M is the mass of the vehicle and its load.

The distance traveled during acceleration (X_b) can be calculated by (repeated from Eq. (7.2) for convenience)

$$X_b = \frac{\left(V_f^{\,2} - V_i^2\right)}{2 \times a} \tag{6.171}$$

Example 6.16: *X_bA 2000-kg vehicle is to achieve a maximum speed of 100 km/hr in 6 seconds. Determine the tractive force needed between the tires and the road to achieve this speed and the distance it takes to reach that speed from an idle condition (i.e., complete stop).*

Solution: The final velocity can be written as

$$V_f = \frac{100\ \text{Km}}{\text{hr}} \times \frac{1000\ \text{m}}{\text{Km}} \times \frac{\text{hr}}{3600\ \text{sec}} = 27.78\ \text{m/sec}$$

The acceleration can be found from

$$a = \frac{V_f - V_i}{t} = \frac{27.78 - 0}{6} = 4.63\ \text{m/sec}^2$$

The force needed to achieve this acceleration

$$F = M \times a = 2000\ \text{kg} \times 4.63\ \text{m/sec}^2 = 9260\ \text{kg m/sec}^2 = 9260\ \text{N}$$

The distance needed to reach the maximum speed is

$$X_b = \frac{(V_f^2 - V_i^2)}{2 \times a} = \frac{(27.78)^2 - 0}{2 \times (4.63)} = 83.34\,\text{m}$$

In our quest for detailed understanding of vehicle acceleration, we need to understand the load transfer characteristics of an accelerating vehicle.

6.4.2 Load Transfer During Acceleration

The fundamental forces acting on an accelerating vehicle were discussed in Chapter 3 on Tire Forces. These equations and the treatment that followed are similar to that used in the previous chapter. Some of these forces will be reviewed briefly here. First we will consider acceleration of a vehicle on a horizontal road (*i.e.*, no grade).

6.4.2.1 Simple Acceleration on a Horizontal Road

Figure 6.47 gives the free body diagram of the vehicle that shows the forces acting on it under normal driving conditions. Note that we have neglected aerodynamic forces, rolling resistance, and powerplant resistance in this simple vehicle model. The mathematical manipulation of Newton's second law yields the equations derived in the previous chapter and will not be repeated here. The vertical forces at the wheels are

$$F_{z1} = Mg\left(\frac{c_2}{c_1 + c_2}\right) - \frac{h}{c_1 + c_2} Ma$$

$$F_{z2} = Mg\left(\frac{c_1}{c_1 + c_2}\right) + \frac{h}{c_1 + c_2} Ma \qquad (6.172)$$

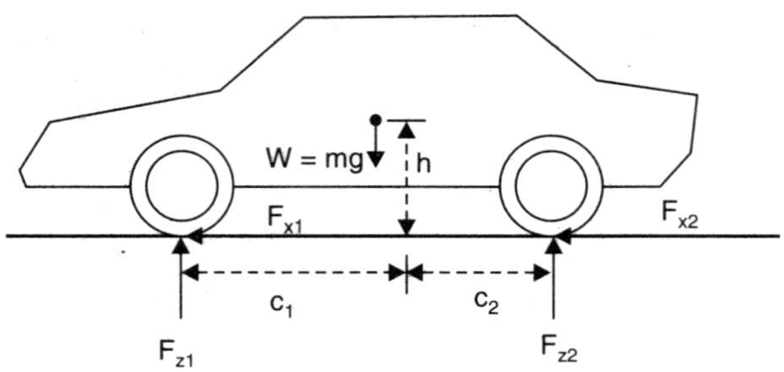

Fig. 6.47 Simple vehicle model on a horizontal plane.

These equations are the same as Eq. (7.25) from Chapter 7, repeated here for convenience. Note that we are assuming that the vehicle is accelerating; thus, a positive value for the acceleration (a) is applied.

When the vehicle is accelerating, the axle force on the front wheels will be smaller than those forces experienced under cruising driving conditions (*i.e.*, no acceleration or deceleration). As the vehicle accelerates, the load is transferred from the front axle to the rear axle. The total axle load of both axles remains equal to the weight of the vehicle.

Example 6.17: *Consider a midsize sedan that weighs 14,236.31 N (3200 lb) under a certain load. The wheelbase of the vehicle is 2.74 m (9 ft, or 108 in.), and the center of gravity is 1.067 m (3.5 ft) from the front axle and 0.508 m (20 in.) above the ground. Determine the load on each axle if the vehicle is under no acceleration, accelerating at 0.610 m/s² (2 ft/sec²) and 1.829 m/s² (6 ft/sec²).*

Solution: The following are given in the problem:

$$W = 3200 \text{ lb}$$

$$c_1 + c_2 = 108 \text{ in}$$

$$h = 20 \text{ in}$$

$$c_1 = 42 \text{ in}$$

First we will find the mass and c_2 as

$$M = 3200/32.2 = 99.38 \text{ slugs}$$

$$c_2 = 108 - 42 = 66 \text{ in}$$

Apply Eq. (8.3) (or Eq. (7.25) from Chapter 7) to obtain the loading under no acceleration (*i.e.*, $a = 0$) as

$$F_{z1} = Mg\left(\frac{c_2}{c_1 + c_2}\right) - \frac{h}{c_1 + c_2}Ma = 99.38 \times 32.2\left(\frac{66}{108}\right) - \frac{20}{108} \times 99.38 \times 0 = 1955.56 \text{ lb}$$

$$F_{z2} = Mg\left(\frac{c_1}{c_1 + c_2}\right) + \frac{h}{c_1 + c_2}Ma = 99.38 \times 32.2\left(\frac{42}{108}\right) + \frac{20}{108} \times 99.38 \times 0 = 1244.44 \text{ lb}$$

Similarly for $a = 2$ ft/sec²

$$F_{z1} = 99.38 \times 32.2\left(\frac{66}{108}\right) - \frac{20}{108} \times 99.38 \times (2) = 1918.76 \text{ lb}$$

$$F_{z2} = 99.38 \times 32.2\left(\frac{42}{108}\right) + \frac{20}{108} \times 99.38 \times (2) = 1281.24 \text{ lb}$$

and for $a = 6$ ft/sec²

$$F_{z1} = 99.38 \times 32.2\left(\frac{66}{108}\right) - \frac{20}{108} \times 99.38 \times (6) = 1845.14 \text{ lb}$$

$$F_{z2} = 99.38 \times 32.2\left(\frac{42}{108}\right) + \frac{20}{108} \times 99.38 \times (6) = 1354.86 \text{ lb}$$

The total load on the axles remains at 3200 lb, which is the weight of the vehicle. Load transfers from the front axle to the rear axle as the vehicle accelerates. Figure E6.17 shows the load distribution as a function of acceleration, using the parameters of Example 6.17. Of course, maximum acceleration is limited by the power of the vehicle.

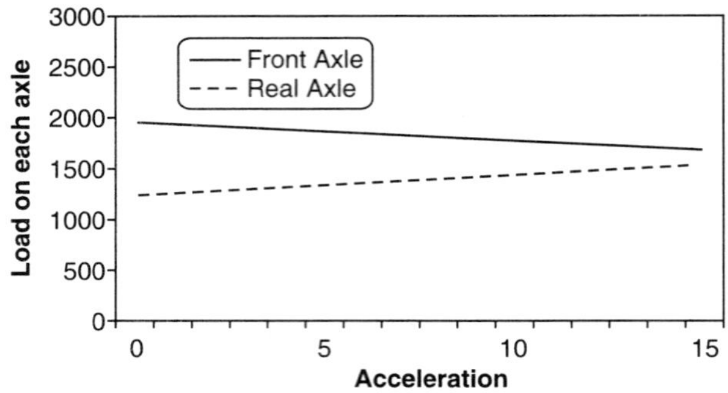

Fig. E6.17 Load transfer during acceleration.

6.4.2.2 Effect of Aerodynamic and Other Forces

In this section, other forces are considered in the accelerating vehicle model. Similar to Chapter 7 on braking, a more accurate accelerating vehicle model must take other factors into consideration. In particular, the rolling resistance of the tires, the aerodynamic drag resistance, and the driveline drag should be accounted for.

The rolling resistance is between the tires and the drive axle. If the rolling coefficient is distinguished by μ_r, then the rolling resistance between the tire and the drive axle can be written as in Eq. (7.26) of Chapter 7, or

$$R_r = \mu_r Mg$$

The coefficient of rolling resistance varies between 0.012 and 0.02 on concrete surfaces, with radial tires offering lower coefficients than bias ply tires. This coefficient reaches 0.1 when the vehicle is driven on hard soil, and it reaches 0.2 to 0.3 when the vehicle is driven on sand. The rolling resistance usually is less for tires with large diameters. The aerodynamic resistance is generated mainly by air flow around the vehicle and air flow through the vehicle. The aerodynamic resistance is estimated by Eq. (7.27) in Chapter 7 as

$$R_a = \frac{\rho}{2} C_D A_F V_r^2$$

where ρ is the density of the air, C_D is the drag coefficient, A_F is the projected area of the front of the vehicle and V_r is the relative velocity of the vehicle with respect to the wind.

The coefficient of drag ranges between 0.3 and 0.5 for passenger cars and can reach a value of 0.8 for vans. Similarly, aerodynamic lift and pitching can be treated (see Eqs. (7.28) and (7.29) in Chapter 7).

Now we will consider the aerodynamic drag and rolling resistance in developing the load transfer equations. Figure 6.48 (which is the same as Fig. 7.6 in Chapter 7) shows the vehicle model on a horizontal plane, including the forces of rolling resistance and aerodynamic drag.

Note that the resultant of the aerodynamic drag is represented here by the force R_a acting at some height ha. These are found either from wind tunnel testing or, more recently, from advanced computer models using computational fluid dynamics (CFD) software. The vertical forces (axle loads) can be written as (Eq. (7.30) from Chapter 7).

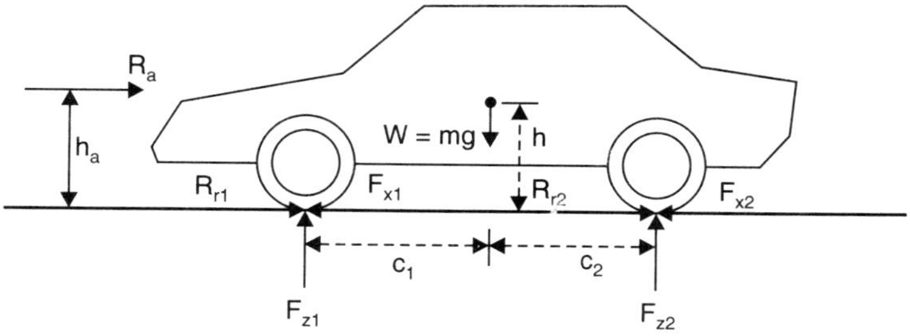

Fig. 6.48 Vehicle model on a horizontal plane.

$$F_{z1} = Mg\left(\frac{c_2}{c_1+c_2}\right) - \frac{h_a}{c_1+c_2}R_a - \frac{h}{c_1+c_2}Ma$$

$$F_{z2} = Mg\left(\frac{c_1}{c_1+c_2}\right) + \frac{h_a}{c_1+c_2}R_a + \frac{h}{c_1+c_2}Ma \tag{6.173}$$

Acceleration will be treated simply by using a positive value for a. Often, h_a is assumed to be the same as h.

6.4.2.3 Effect of Grade

Now we will consider the effect of grade as well as aerodynamic drag and rolling resistance forces in developing the load transfer equations. Figure 6.49 shows the vehicle model on an inclined surface, with consideration of rolling resistance and aerodynamic drag.

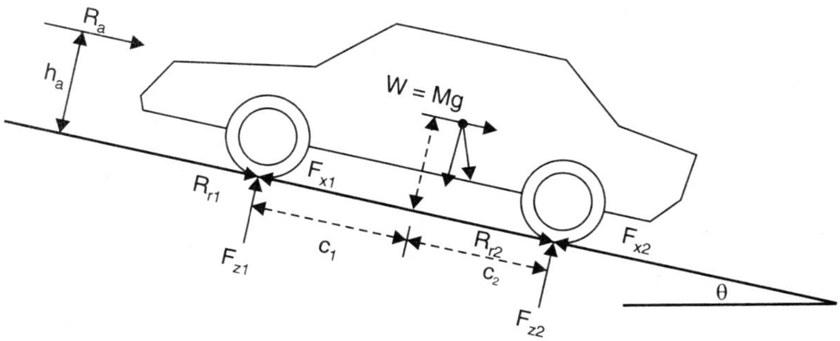

Fig. 6.49 Vehicle model on an incline (grade)
(*a*) front-wheel drive, and (*b*) all-wheel drive or four-by-four (front-wheel-drive based).

The vertical forces (axle loads) for a vehicle going uphill are (Eq. (7.31*b*) from Chapter 7 is repeated here for convenience)

$$F_{z1} = \frac{1}{c_1+c_2}(Mgc_2\cos\theta - h_a R_a - hMa - hMg\sin\theta)$$

$$F_{z2} = \frac{1}{c_1+c_2}(Mgc_1\cos\theta + h_a R_a + hMa + hMg\sin\theta) \tag{6.174}$$

Note that the sign of the last term should be reversed when the vehicle is going downhill. Acceleration will be treated simply by using a positive value for a. The total tractive force $F_x(F_{x1} + F_{x2})$ can be found as

$$F_x = R_a + R_r + Ma + Mg \sin\theta \tag{6.175}$$

or

$$F_x - R_r = R_a + Ma + Mg \sin\theta \tag{6.176}$$

where R_r is the total rolling resistance force of both axles ($R_r = R_{r1} + R_{r2} = Mg\,\mu r$). Assuming $h_a = h$ and substituting Eq. (6.176) into Eq. (6.174) yields the load on the front axle as

$$\begin{aligned} F_{z1} &= \frac{1}{c_1 + c_2}\left(Mgc_2 \cos\theta - h\left(R_a + Ma + Mg \sin\theta\right)\right) \\ &= \frac{1}{c_1 + c_2}\left(Mgc_2 \cos\theta - h\left(F_x - R_r\right)\right) \end{aligned} \tag{6.177}$$

The load on the rear axle is

$$\begin{aligned} F_{z2} &= \frac{1}{c_1 + c_2}\left(Mgc_1 \cos\theta + h\left(R_a + Ma + Mg \sin\theta\right)\right) \\ &= \frac{1}{c_1 + c_2}\left(Mgc_1 \cos\theta + h\left(F_x - R_r\right)\right) \end{aligned} \tag{6.178}$$

Example 6.18: *Consider a midsize sedan that weighs 14,236.3 N (3200 lb) under a certain load. The vehicle is going uphill at a 5° slope at a speed of 26.82 m/s (60 mph). The frontal area of the vehicle is 2.044 m² (22.0 ft²), and the coefficient of drag is 0.6. The aerodynamic resistance is acting on the center of gravity. The wheelbase of the vehicle is 2.74 m (9 ft), and the center of gravity is 1.372 m (6.5 ft, or 66 in.) from the front axle and 0.508 m (20 in.) above the ground. Determine the load on each axle if the vehicle is under no acceleration. What would the loads be if the vehicle is accelerating at 1.524 m/s² (5 ft/sec²)? The density of air is 1.226 kg/m³ (0.002378 slug/ft³). Calculate the axle forces when going downhill at a 5° slope with the same speed.*

Solution: The aerodynamic resistance first is calculated as

$$\begin{aligned} R_a &= \frac{\rho}{2} C_D A_F V_r^2 \\ &= \frac{0.002378}{2} \times 0.4 \times 22 \times (88)^2 = 81.03 \text{ lb} \end{aligned}$$

When traveling on a horizontal road, the forces on the axles are

$$\begin{aligned} F_{z1} &= \frac{1}{c_1 + c_2}\left(Mgc_2 \cos\theta - h_a R_a - hMa - hMg \sin\theta\right) \\ &= \frac{1}{c_1 + c_2}\left(Mgc_2 - h_a R_a - hMa\right) \\ &= \frac{1}{108}\left(3200 \times 66 - 20 \times 81 - 20 \times 99.38 \times 5\right) = 1848.5 \text{ lb} \end{aligned}$$

$$F_{z2} = \frac{1}{c_1 + c_2}\left(Mgc_1 \cos\theta + h_a R_a + hMa + hMg \sin\theta\right)$$

$$= \frac{1}{c_1 + c_2}(Mgc_1 + h_a R_a + hMa)$$

$$= \frac{1}{108}(3200 \times 42 + 20 \times 81 + 20 \times 99.38 \times 5) = 1351.5 \text{ lb}$$

When traveling uphill, the forces on the axles are

$$F_{z1} = \frac{1}{c_1 + c_2}(Mgc_2 \cos\theta - h_a R_a - hMa - hMg \sin\theta)$$

$$= \frac{1}{108}(3200 \times 66 \times \cos(5) - 20 \times 81 - 20 \times 99.38 \times 5 - 20 \times 3200 \times \sin(5))$$

$$= 1789.4 \text{ lb}$$

$$F_{z2} = \frac{1}{c_1 + c_2}(Mgc_1 \cos\theta + h_a R_a + hMa + hMg \sin\theta)$$

$$= \frac{1}{108}(3200 \times 42 \times \cos(5) + 20 \times 81 + 20 \times 99.38 \times 5 + 20 \times 3200 \times \sin(5))$$

$$= 1398.4 \text{ lb}$$

When the vehicle goes down hill, the forces are

$$F_{z1} = \frac{1}{c_1 + c_2}(Mgc_2 \cos\theta - h_a R_a - hMa + hMg \sin\theta)$$

$$= \frac{1}{108}(3200 \times 66 \times \cos(5) - 20 \times 81 - 20 \times 99.38 \times 5 + 20 \times 3200 \times \sin(5))$$

$$= 1892.7 \text{ lb}$$

$$F_{z2} = \frac{1}{c_1 + c_2}(Mgc_1 \cos\theta + h_a R_a + hMa - hMg \sin\theta)$$

$$= \frac{1}{108}(3200 \times 42 \times \cos(5) + 20 \times 81 + 20 \times 99.38 \times 5 - 20 \times 3200 \times \sin(5))$$

$$= 1295.1 \text{ lb}$$

Observe that uphill travel increased the forces on the rear wheels. On the other hand, downhill travel increased the forces on the front wheels. As will be shown in the next section, this will add to the advantages of rear-wheel-drive (RWD) vehicles in accelerating on uphill slopes.

The load described here represents the weight distribution at the initial acceleration from a startup speed of 60 mph. Aerodynamic forces will increase as the speed increases during acceleration. As was found earlier for vehicles under braking, the aerodynamic forces reduce the load on the front axle and increase it on the rear axle. Accurate analysis of the force distribution as the vehicle accelerates should take into consideration aerodynamic drag.

6.4.3 Traction-Limited Acceleration

One factor that limits acceleration is traction between the tires and the road. This is particularly felt by drivers of front-wheel-drive (FWD) vehicles under startup conditions and when accelerating on smooth surfaces. We will study the maximum attainable acceleration of various vehicles having different drive mechanisms.

6.4.3.1 Drivetrain Configurations

Many drive mechanisms are available in the marketplace for road vehicles. In particular, there are front-wheel-drive (FWD) systems, rear-wheel-drive (RWD) systems, and all-wheel-drive (AWD) or four-by-four (4×4) systems. In the latter system, different mechanisms exist for torque distribution between the front and rear axles.

A front-wheel-drive system usually consists of the engine configured such that the crankshaft direction is parallel to the axle direction (line connecting the front wheels). Such a configuration is called an east-west configuration. Figure 6.50 shows a typical front-wheel-drive configuration. The figure also shows how a front-wheel-drive system can be configured or adapted for an all-wheel-drive application. This can be done by adding several components. The first is a power takeoff unit (PTU) that consists of bevel gears to produce a line of motion perpendicular to that given at the output of the transmission. The second is the driveline, which may consist of a single-, two-, or three-piece driveshaft. Large cars and trucks tend to have a single-piece (relatively large in diameter) shaft. Small cars tend to have a two-piece or three-piece (smaller in diameter) driveline. The driveline also includes one or more center bearing (s) to mount the shaft (s) to the vehicle body. The third major unit is the coupler. In all-wheel-drive systems, the coupler is a unit of a clutch mechanism that engages the rear wheels (axle) only when a slip condition at the front wheels is observed. At this point, we will establish a difference between the all-wheel-drive and four-by-four systems. The all-wheel-drive systems are those systems in which the other axle (*i.e.*, the rear axle in this case) is engaged as needed, without input from the driver. In other words, the engagement happens when a slip condition of the front axle is met. The four-by-four system is a system in which the drive system is either front-wheel drive all the time or can engage the other axle (rear) with an on/off switch activated by the driver. The four-by-four system may not need a coupler. The last major unit in the front-wheel-drive-based all-wheel-drive system is the rear differential unit (RDU). This unit consists of a set of bevel gears to distribute the torque and motion from the driveshaft(s) to the rear wheels.

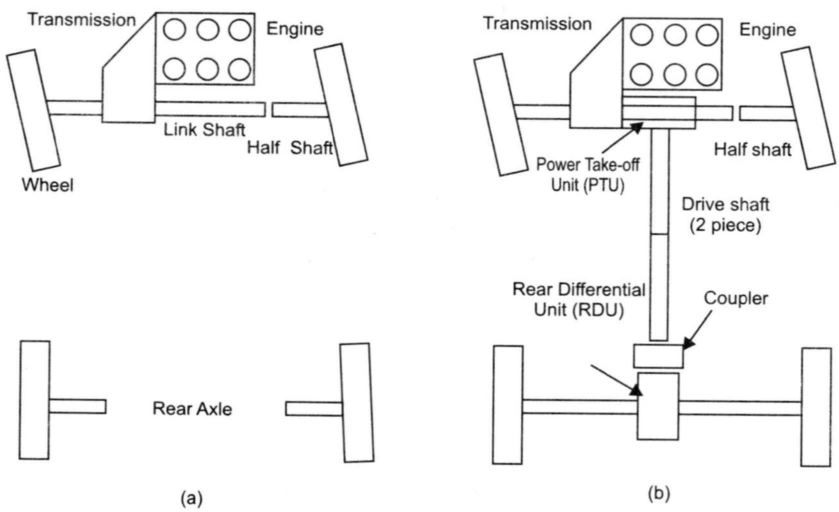

Fig. 6.50 (*a*) Front-wheel-drive (FWD) and (*b*) front-wheel-drive-based all-wheel-drive (AWD) systems.

Front-wheel-drive vehicles offer a major package advantage when compared with rear-wheel-drive ones. The power pack is compact, allowing more space to be used for the cabinet. Initially, front-wheel drive was used for light vehicles; however, one recently can find vehicles with V6 and V8 engines that are front-wheel drive delivering more than 200 hp. Because the engine is mounted on the front chassis systems, more load will be given to the front, or driving, wheels. This offers better traction capability and road handling, especially on icy and wet roads. Furthermore, front-wheel-drive vehicles are less sensitive to wind loads. This arrangement offers a shorter power flow from the engine to the driving wheels, thus minimizing power losses in the system.

Among the disadvantages of the front-wheel-drive system is its compact engine design, forcing shorter engines to be used. The gains made by offering higher loads to the driving front wheels are compromised with these wheels wearing off early. Also, this configuration presents challenges to the engineers designing the mount system. Frequently, poor mount design would lead to inferior noise and vibration performance. Because the transmission is packaged closer to the left (or right) wheels, it needs longer shafts to be connected with the right wheel than it does with the left wheel. This will allow the longer shaft to twist more under torque, causing the vehicle to steer unintentionally. This phenomenon is referred to as torque steer.

Torque distribution among front and rear axles is engineered to give maximum traction, as will be seen later. In certain applications, a distribution is determined *a priori* without an active control mechanism. Among common distributions are the 50/50 distribution, in which half the torques goes to each axle, and the 40/60 distribution for rear-wheel-drive-based all-wheel-drive and 60/40 for front-wheel-drive-based all-wheel-drive systems.

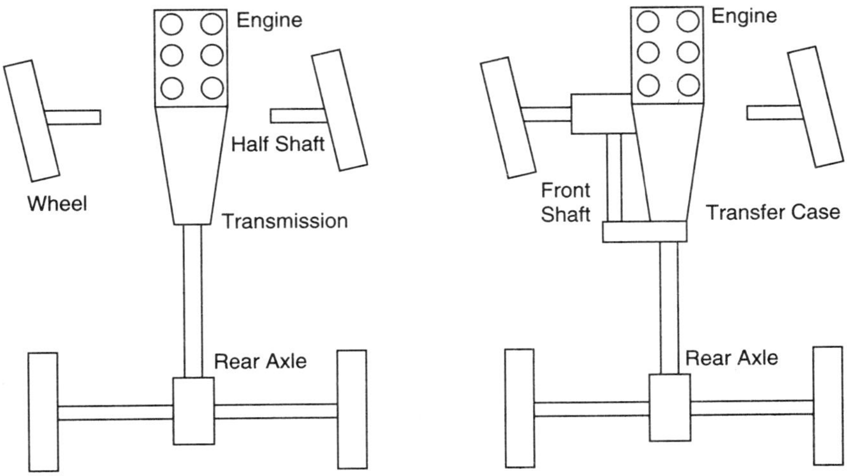

Fig. 6.51 (*a*) Rear-wheel-drive and (*b*) Rear-wheel-drive-based all-wheel-drive or four-by-four systems.

A rear-wheel-drive system consists of the engine configured such that the crankshaft direction is perpendicular to the axle direction. This configuration is called a north-south configuration. Figure 6.51 shows a typical rear-wheel-drive configuration. Similar to the front-wheel-drive configuration, the rear-wheel-drive system can be configured for all-wheel-drive and four-by-four applications. Several components must be added. The first is the transfer case (TC) that is added at the output end of the transmission. This unit consists of a chain and a set of spur gears. The purpose of the transfer case is to bring some torque back to the front wheels. The second major unit is the front driveshaft, and the third major unit is the coupler. The last major unit in the rear-wheel-drive-based all-wheel-drive system is the front differential unit (FDU).

Other changes take place as a result of making an all-wheel-drive (or four-by-four) version of a front-wheel-drive or rear-wheel-drive system. The major obstacle typically facing engineers in modern vehicles is the issue of package. All-wheel-drive and four-by-four systems are more challenging to package.

The rear-wheel-drive vehicle design offers certain important advantages. This design sets virtually no limit on the engine length. Thus, larger engines can be used (e.g., V8, V10, and V12). Insulation of the engine noise with this mount system is relatively easy and more robust when compared with the front-wheel-drive system. In addition, this system is more robust to torque steer. Rear-wheel-drive offers balanced weight distribution between the front and rear wheels, allowing longer life for the tires before they wear out.

Other configurations exist in the marketplace. These configurations include rear-engine vehicles. Such a configuration often is incorporated into sport vehicles used for racing purposes. In these vehicles, the engine is placed in the rear of the vehicle. Other configurations place the engine beneath the body of the vehicle. For more information on drive configurations, the reader is referred to the book by Reimpell and Stoll (1996). We will start by considering single-axle drive but will move later to consider all-wheel-drive systems.

6.4.3.2 Front-Wheel Drive

First let us consider a horizontal plane with no resistance or drag forces. For front-wheel-drive systems, the tractive forces are applied to the front wheels. These forces will equal the coefficient of friction multiplied by the normal force at those wheels

$$F_{x1} = \mu F_{z1} \tag{6.179}$$

The tractive force at the rear wheels will be zero. The maximum acceleration can be found from Newton's second law $F_{x1} + F_{x2} = Ma$ and by utilizing Eq. (8.3) as

$$Ma_F = \mu Mg\left(\frac{c_2}{c_1 + c_2}\right) - \mu\left(\frac{h}{c_1 + c_2}\right)Ma_F \tag{6.180}$$

These equations are similar to Eqs. (7.32) and (7.33) from Chapter 7. Note the sign changes due to tractive force being opposite in direction to the braking force. We replaced the acceleration symbol a by a_F to indicate that this equation is used to calculate the maximum acceleration when the tractive effort is applied to the front axle. This equation can be solved directly for the vehicle acceleration as

$$a_F = g\frac{\mu\left(\dfrac{c_2}{c_1 + c_2}\right)}{1 + \mu\left(\dfrac{h}{c_1 + c_2}\right)} \tag{6.181}$$

The maximum tractive force applied in a front-wheel-drive system is $F_{x1} = M \times a_F$. When the grade, rolling resistance, and aerodynamic drag are considered, Eq. (6.179) is substituted into Eq. (6.177). This yields the maximum tractive forces as

$$F_{z1} = \frac{1}{c_1 + c_2}\left(Mgc_2 \cos\theta - h\left(R_a + Ma + Mg \sin\theta\right)\right)$$
$$= \frac{1}{c_1 + c_2}\left(Mgc_2 \cos\theta - h\left(F_{x1} - R_r\right)\right) \tag{1.182}$$
$$F_{x1} = \frac{\mu}{c_1 + c_2}\left(Mgc_2 \cos\theta - h\left(F_{x1} - R_r\right)\right)$$

Substituting $Rr = \mu r$ Mg into Eq. (6.182) and rewriting that equation for F_x gives

$$F_{x1} + \frac{\mu h}{c_1 + c_2} F_{x1} = \frac{\mu Mg}{c_1 + c_2}(c_2 \cos\theta + h\mu_r)$$

or

$$F_{x1} = \frac{\frac{\mu Mg}{c_1 + c_2}(c_2 \cos\theta + h\mu_r)}{1 + \frac{\mu h}{c_1 + c_2}} \tag{6.183}$$

Note that F_{x2} is zero for a front-wheel-drive system. Equation (6.183) gives the maximum tractive force that can be applied to a front-wheel-drive system. The maximum acceleration can be obtained from Eq. (6.175).

6.4.3.3 Rear-Wheel Drive

Similar treatment can be made if the tractive forces are applied to only the rear wheels. In this case, the tractive force on the front wheels is zero. The maximum tractive force on the rear wheels is

$$F_{x2} = \mu F_{z2} \tag{6.184}$$

The maximum acceleration for rear-wheel-drive vehicles becomes

$$a_R = g \frac{\mu\left(\frac{c_1}{c_1 + c_2}\right)}{1 - \mu\left(\frac{h}{c_1 + c_2}\right)} \tag{6.185}$$

Figures (6.52) and (6.53) show the acceleration of a typical compact vehicle and the impact of the location of the center of gravity on the vehicle maximum traction-limited acceleration for both front-wheel-drive and rear-wheel-drive systems. In these figures, aerodynamic drag, rolling forces, and grade are not included.

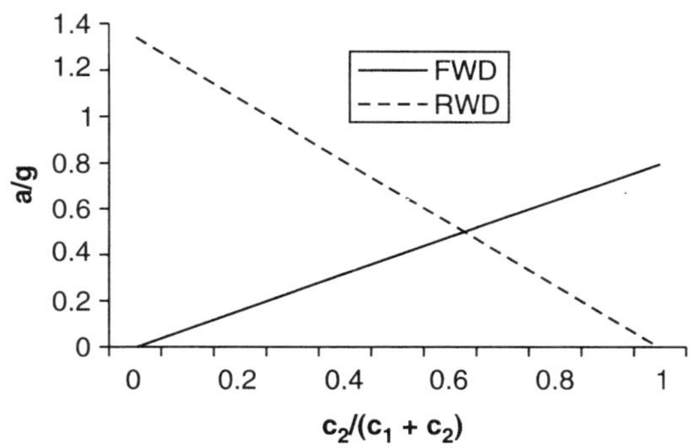

Fig. 6.52 Acceleration for front-wheel-drive and rear-wheel-drive vehicles, where $\mu = 1$, $h = 0.5$ m.

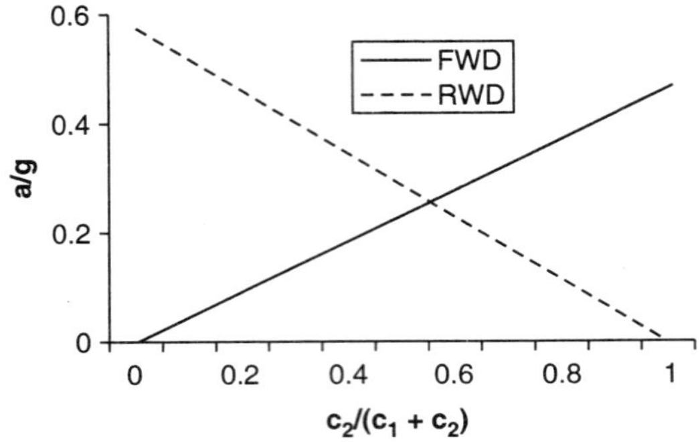

Fig. 6.53 Acceleration for front-wheel-drive and rear-wheel-drive vehicles, where $\mu = 0.5$, $h = 0.5$ m, and $c_1 + c_2 = 2.0$.

Note that for typical vehicle designs and loading conditions, the value of $c_2/(c_1 + c_2)$ ranges from 0.4 to 0.6. Within this range of application, the maximum acceleration that can be obtained for rear-wheel-drive systems is between the values of 0.22 and 0.34g when $\mu = 0.5$, and between 0.53 and 0.8g when $\mu = 1$. For front-wheel-drive systems, the maximum acceleration that can be achieved is between 0.18 and 0.27g when $\mu = 0.5$, and between 0.32 and 0.48g when $\mu = 1$. This shows that for normal acceleration conditions in a single drive, higher levels of maximum acceleration can be achieved with rear-wheel-drive. Thus, sport vehicles are almost always rear-wheel drive. In a rear-wheel-drive vehicle, grade, rolling resistance, and aerodynamic drag can be included in a fashion similar to that done for a front-wheel-drive system. This yields Eq. (6.186) for maximum tractive forces applied to the rear wheels as

$$F_{x2} = \frac{\dfrac{\mu M g}{c_1 + c_2}\left(c_1 \cos\theta - h\mu_r\right)}{1 - \dfrac{\mu h}{c_1 + c_2}} \tag{6.186}$$

Note that F_{x1} is zero for a rear-wheel-drive system. Equation (6.186) gives the maximum tractive force that can be applied to a rear-wheel-drive vehicle. The maximum acceleration can be obtained from Eq. (6.175).

6.4.3.4 All-Wheel-Drive and Four-by-Four Systems

In recent years, all-wheel-drive vehicles and four-by-four ones (also called four-wheel-drive) have received considerable attention. In four-by-four systems, the tractive forces of the vehicle are distributed on all its wheels under all driving conditions. The tractive forces may be distributed equally or have some bias toward a certain axis (front or rear). In all-wheel-drive systems, the vehicle normally is driven with one axis (either rear or front) until wheel spinning is detected. Only when one or more of the wheels is detected to be spinning is a certain portion of the torque distributed to the other axle.

Let us assume that the tractive torque is distributed such that a certain ratio K_f of the total torque goes to the front axle. The remainder that goes to the rear axle is $1 - K_f$ of the total torque. We then can generate Eq. (6.187) in a manner similar to that used in Eq. [3.37(*b*)] in Chapter 7

$$F_{x1} = \frac{K_f}{1 - K_f} F_{x2}$$

$$F_{x2} = \frac{1-K_f}{K_f} F_{x1} \tag{6.187}$$

Consider the free body diagram of the vehicle under direct tractive forces F_{x1} for the front axle and F_{x2} for the rear axle. The vehicle is still under the effects of aerodynamic forces as well as grade and rolling resistance. The total tractive force is $F_x(F_{x1} + F_{x2})$. The total rolling resistance force is $R_r\,(R_{r1} + R_{r2})$. The total weight is distributed as in Eq. (6.174). Equation (6.175) then can be written as

$$F_{x1} + F_{x2} - R_a - R_r - Mg\sin(\theta) = Ma \tag{6.188}$$

6.4.3.4.1 Front Skid

Under a front skid condition, $F_{x1} = \mu F_{z1}$. Substituting this and Eq. (6.174) into Eq. (6.188) yields

$$\mu\left(\frac{1}{c_1+c_2}(Mgc_2\cos\theta - h_aR_a - hMa_F - hMg\sin\theta)\right)\left(1+\frac{1-K_f}{K_f}\right)$$
$$-R_a - R_r - Mg\sin\theta = Ma_F \tag{6.189}$$

Rearranging for a_F gives

$$\mu\left(\frac{1}{c_1+c_2}(Mgc_2\cos\theta - h_aR_a - hMg\sin\theta)\right)\left(1+\frac{1-K_f}{K_f}\right) - R_a - R_r - Mg\sin\theta$$
$$= \left(\frac{\mu h}{c_1+c_2}\left(1+\frac{1-K_f}{K_f}\right)+1\right)Ma_F \tag{6.190}$$

or simply

$$a_F = \frac{\mu\left(\frac{1}{c_1+c_2}(Mgc_2\cos\theta - h_aR_a - hMg\sin\theta)\right)\left(1+\frac{1-K_f}{K_f}\right) - R_a - R_r - Mg\sin\theta}{\left(\frac{\mu h}{c_1+c_2}\left(1+\frac{1-K_f}{K_f}\right)+1\right)M} \tag{6.191}$$

Note that the acceleration symbol a has a subscript F to indicate that the acceleration obtained is the maximum acceleration achievable when the front tires are under the skid condition.

Ignoring aerodynamic drag, grade, and rolling resistance, the acceleration is

$$a_F = \frac{\mu\left(\frac{c_2}{c_1+c_2}\right)\left(1+\frac{1-K_f}{K_f}\right)}{\left(\frac{\mu h}{c_1+c_2}\left(1+\frac{1-K_f}{K_f}\right)+1\right)} g \tag{6.192}$$

Note that Eqs. (6.189) through (6.192) are similar to those of Eqs. (7.40) through (7.43) from Chapter 7 used earlier in braking. Eq. (6.192) will reduce to that of Eq. (6.181) when only a front axle is under loading in a front-wheel-drive system ($K_f = 1$).

6.4.3.4.2 Rear Skid

Equation (8.19) can be rewritten as

$$F_{x2}\left(1+\frac{K_f}{1-K_f}\right)-R_a-R_r-Mg\sin\theta=Ma \tag{6.193}$$

When under rear skid, $F_{x2}=\mu F_{z2}$. Substituting this and Eq. (8.5) into Eq. (6.193) yields Eq. (6.194).

$$\mu\left(\frac{1}{c_1+c_2}(Mgc_1\cos\theta+h_aR_a+hMa_R+Mg\sin\theta)\right)\left(1+\frac{K_f}{1-K_f}\right)$$
$$-R_a-R_r-Mg\sin\theta=Ma_R \tag{6.194}$$

Rearranging for a_R gives

$$\mu\left(\frac{1}{c_1+c_2}(Mgc_1\cos\theta+h_aR_a+Mg\sin\theta)\right)\left(1+\frac{K_f}{1-K_f}\right)-R_a-R_r-Mg\sin\theta$$
$$=\left(-\frac{\mu h}{c_1+c_2}\left(1+\frac{K_f}{1-K_f}\right)+1\right)Ma_R \tag{6.195}$$

Simplify as

$$a_R=\frac{\mu\left(\frac{1}{c_1+c_2}(Mgc_1+h_aR_a+Mg\sin\theta)\right)\left(1+\frac{K_f}{1-K_f}\right)-R_a-R_r-Mg\sin\theta}{\left(-\frac{\mu h}{c_1+c_2}\left(1+\frac{K_f}{1-K_f}\right)+1\right)M} \tag{6.196}$$

Ignoring aero-dynamic drag, grade, and rolling resistance, the acceleration is

$$a_R=\frac{\mu\left(\frac{c_1}{c_1+c_2}\right)\left(1+\frac{K_f}{1-K_f}\right)}{\left(-\frac{\mu h}{c_1+c_2}\left(1+\frac{K_f}{1-K_f}\right)+1\right)}g \tag{6.197}$$

Note that Eq. (8.28) will reduce to that of Eq. (8.16) in a rear-wheel-drive vehicle where $K_f=0$. Also note that in Eq. (8.28), the denominator comes closer to zero, which yields a higher value for a_R. This indicates the increased possibility that a front skid will happen first.

The front tires will skid first if $a_F < a_R$; otherwise, the rear tires will skid first.

Example 6.19: *A 21,351.46-N (4800-lb) passenger car has a wheelbase of 2846.80 mm (112 in.) and a center of gravity 1270.00 mm (50 in.) behind the front axle and 508 mm (20 in.) above the ground. The tractive effort distribution gives the front axle 50% of the total tractive force. The car is moving on a horizontal plane. Determine which tires will skid first if the car is moving on a surface with first a 0.75 coefficient of friction, and second a 0.25 coefficient of friction. Ignore drag and rolling resistance.*

Solution:

$$M = 4800 \text{ lbs} = 149.07 \text{ slugs}$$
$$c_1 + c_2 = 112 \text{ in}$$
$$c_1 = 50 \text{ in},$$
$$c_2 = 62 \text{ in},$$
$$K_f = 0.5$$
$$h = 20$$
$$\theta = 0$$
$$R_a = 0$$
$$R_r = 0$$

First surface: $\mu = 0.75$

Acceleration under front skid is

$$a_F = \frac{\mu\left(\dfrac{c_2}{c_1+c_2}\right)\left(1+\dfrac{1-K_f}{K_f}\right)}{\left(\dfrac{\mu h}{c_1+c_2}\left(1+\dfrac{1-K_f}{K_f}\right)+1\right)} g = \frac{0.75\left(\dfrac{62}{112}\right)\left(1+\dfrac{0.5}{0.5}\right)}{\left(\dfrac{0.75\times 20}{112}\left(1+\dfrac{0.5}{0.5}\right)+1\right)} g = 0.6549g$$

Second surface: $\mu = 0.25$

Acceleration under front skid is

$$a_R = \frac{\mu\left(\dfrac{c_1}{c_1+c_2}\right)\left(1+\dfrac{K_f}{1-K_f}\right)}{\left(-\dfrac{\mu h}{c_1+c_2}\left(1+\dfrac{K_f}{1-K_f}\right)+1\right)} g = \frac{0.75\left(\dfrac{50}{112}\right)\left(1+\dfrac{0.5}{0.5}\right)}{\left(-\dfrac{0.75\times 20}{112}\left(1+\dfrac{0.5}{0.5}\right)+1\right)} g = 0.9146g$$

Since $a_F < a_R$, front tires will skid first.

Second surface: $\mu = 0.25$

Acceleration under front skid

$$a_F = \frac{\mu\left(\dfrac{c_2}{c_1+c_2}\right)\left(1+\dfrac{1-K_f}{K_f}\right)}{\left(\dfrac{\mu h}{c_1+c_2}\left(1+\dfrac{1-K_f}{K_f}\right)+1\right)} g = \frac{0.25\left(\dfrac{62}{112}\right)\left(1+\dfrac{0.5}{0.5}\right)}{\left(\dfrac{0.25\times 20}{112}\left(1+\dfrac{0.5}{0.5}\right)+1\right)} g = 0.2541g$$

Acceleration under rear skid

$$a_R = \frac{\mu\left(\frac{c_1}{c_1+c_2}\right)\left(1+\frac{K_f}{1-K_f}\right)}{\left(-\frac{\mu h}{c_1+c_2}\left(1+\frac{K_f}{1-K_f}\right)+1\right)}g = \frac{0.25\left(\frac{50}{112}\right)\left(1+\frac{0.5}{0.5}\right)}{\left(-\frac{0.25\times 20}{112}\left(1+\frac{0.5}{0.5}\right)+1\right)}g = 0.2451g$$

Because $a_F > a_R$, the rear tires will skid first.

Similar to our finding under braking conditions, the preceding example shows the sensitivity of the tire skid (whether the front or rear tires skid first) to the coefficient of friction. The vehicle is to be driven on various types of surfaces with a high coefficient of friction (e.g., concrete surfaces) and a low coefficient of friction (e.g., icy or slippery roads). The drive system should be designed such that the maximum tractive force is obtained. Figure 6.19(*a*) shows how the acceleration, under front and rear skid conditions, changes with the coefficient of friction for the parameters described in Example 6.19. This figure clearly shows that rear skid is quite possible at surfaces with a lower coefficient of friction. Figure 6.19(*b*) shows that by changing only the tractive effort distribution such that 60% goes to the front axle, front axle skid will occur for all types of surfaces. Any factor higher than 55% will guarantee front skid conditions to occur first for all surfaces.

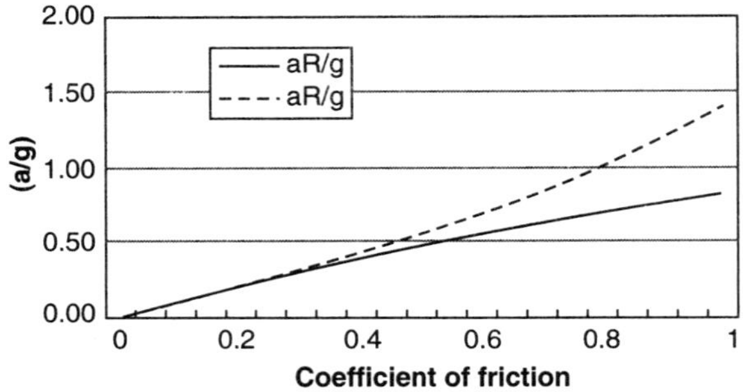

Fig. E6.19 (*a*) Acceleration under front and rear skid conditions for $c_1 = 50$ in, $c_2 = 62$ in, $h = 20$ and $K_f = 0.5$.

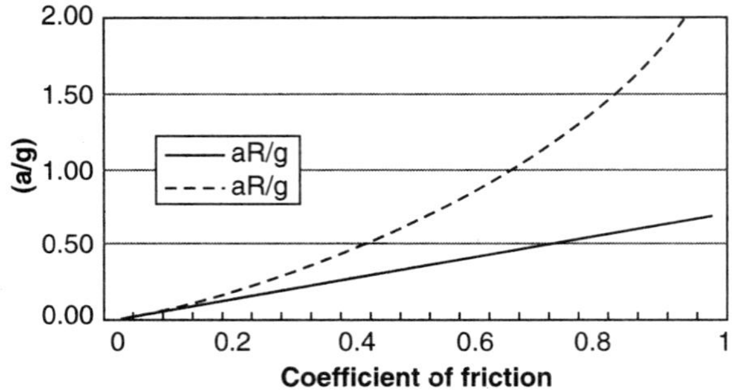

Fig. E6.19 (*b*) Acceleration under front and rear skid conditions for $c_1 = 50$ in, $c_2 = 62$ in, $h = 20$ and $K_f = 0.6$.

6.4.3.5 Optimal Tractive Effort

Maximum tractive effort is obtained when all wheels skid at the same time. To achieve this, torque distribution between the front and rear axle must follow the load on each axle, assuming that the wheels of both axles have the same coefficient of friction. In other words, if the front axle is experiencing 40% of the vehicle load under certain driving conditions, the maximum tractive effort is achieved if the motor torque is distributed such that 40% of it goes to the front axle. Similar to braking systems, three major factors affect the torque distribution between the front and rear axles. The first is that the loading conditions of the vehicle (fully loaded versus lightly loaded) will cause movement of the center of gravity of the vehicle. This actually means that a certain optimal distribution for a certain loading condition may not be the same as that for another loading condition. The second factor is the coefficient of friction. The coefficient of friction varies considerably between dry and wet surfaces and between various types of surface constructions and finishing. The third factor is safety considerations and/or vehicle handling objectives. This may dictate that one axle will experience skid conditions first (usually the front axle). Also similar to brakes, the drive system can be designed with a feedback control algorithm. Sensors can be attached to both axles to determine the axle load and to distribute motor torque (mechanically or hydraulically) to both axles according to the axle load distribution, with some bias toward achieving front skid conditions first (if desired).

Example 6.20: *Consider the vehicle treated in Example 6.19. The vehicle weighs 4800 lb (21,326 N) and has a wheelbase of 112 in. (2846.2 mm). Its center of gravity is 50 in. (1270.0 mm) behind the front axle and 20 in. (508 mm) above the ground and has a coefficient of friction of 0.75 with the road surface. The acceleration found for the assumptions of the front and rear lock-ups are given in Table 6.5 for various values of the torque distribution factor. The bold numbers are the smaller acceleration values. The skid condition that is associated with the smaller acceleration happens first. The maximum tractive force occurs when both F_{x1} and F_{x2} approach their skid (or maximum values) simultaneously. This can be achieved only if we set $F_{x1} = \mu F_{z1}$ and $F_{x2} = \mu F_{z2}$. The total friction force is*

$$F_x = F_{x1} + F_{x2} = \mu F_{z1} + \mu F_{z2} = \mu(F_{z1} + F_{z2}) = \mu Mg\cos\theta \tag{6.198}$$

The maximum tractive force is achieved at a point where a_R and a_F are close to each other. Note that at $K_f = 0.4196$, both acceleration values a_R and a_F are the same (0.75g), and maximum friction is achieved. From Eq. (6.188), the acceleration is

$$a = \frac{F_{x1} + F_{x2} - R_a - R_r - Mg\sin\theta}{M} = \frac{\mu Mg\cos\theta - R_a - R_r - Mg\sin\theta}{M} \tag{6.199}$$

The only way both the front and rear drives would skid at the same time is if the front torque distribution factor is equal to the ratio of the front axle load to the total load, or

$$K_{f\max} = \frac{F_{x1}}{F_x} = \frac{\mu F_{z1}}{\mu Mg\cos\theta} = \frac{\frac{\mu}{c_1 + c_2}\left(Mgc_2\cos\theta - h_a R_a - hMa - hMg\sin\theta\right)}{\mu Mg\cos\theta}$$

$$= \frac{\left(Mgc_2\cos\theta - h_a R_a - hMa - hMg\sin\theta\right)}{(c_1 + c_2)Mg\cos\theta} \tag{6.200}$$

or, by further manipulation;

$$K_{f\,\max} = \frac{\left(Mgc_2\cos\theta - h_a R_a - h\left(\mu Mg\cos\theta - R_a - R_r - Mg\sin\theta\right) - hMg\sin\theta\right)}{\left(c_1 + c_2\right)Mg\cos\theta} \tag{6.201}$$

Table E6.20 Acceleration and force distribution under front and rear skid assumptions for various front torque distribution factors, c_2 = 62 in (= 1574.8 mm), $c_1 + c_2$ = 112 in (= 2844.8 mm), μ = 0.75, h = 20 in (= 508 mm)

	Front skid						*Rear skid*					
K_f	aF/g	F_{z1}	F_{z2}	F_{x1}	F_{x2}	F_x	aR/g	F_{z1}	F_{z2}	F_{x1}	F_{x2}	F_x
0.0001	3.098						0.387	2326	2474	0	1856	1856
0.05	2.257						0.410	2305	2495	98	1871	1969
0.1	1.775						0.437	2283	2517	210	1888	2098
0.15	1.462						0.468	2256	2544	337	1908	2244
0.2	1.243						0.503	2226	2574	483	1930	2413
0.25	1.081						0.543	2191	2609	652	1957	2609
0.3	0.957						0.591	2150	2650	852	1987	2839
0.35	0.858						0.649	2101	2699	1090	2024	3114
0.4	0.778						0.718	2041	2759	1379	2069	3448
0.4196	0.75	2014	2786	1511	2090	3600	0.75	2014	2786	1510	2089	3600
0.45	0.711	2048	2752	1536	1877	3413	0.805					
0.5	0.655	2096	2704	1572	1572	3144	0.915					
0.55	0.607	2137	2663	1603	1311	2914	1.059					
0.6	0.566	2172	2628	1629	1086	2715	1.258					
0.65	0.530	2203	2597	1652	890	2542	1.550					
0.7	0.498	2230	2570	1673	717	2390	2.016					
0.75	0.470	2255	2545	1691	564	2255	2.885					
0.8	0.445	2276	2524	1707	427	2134	5.068					
0.85	0.422	2295	2505	1722	304	2025	20.833					
0.9	0.402	2313	2487	1735	193	1927	–9.868					
0.95	0.383	2329	2471	1747	92	1839	–3.989					
1	0.366	2343	2457	1757	0	1757	–2.500					

Ignoring aerodynamic forces, rolling resistance, and grade, the front torque distribution factor becomes

$$K_{f\,\max} = \frac{\left(c_2 - \mu h\right)}{\left(c_1 + c_2\right)} \tag{6.202}$$

Table E6.20 also displays the tractive forces on each axle for various front torque distribution factors. Figure E6.20 shows the total maximum tractive force.

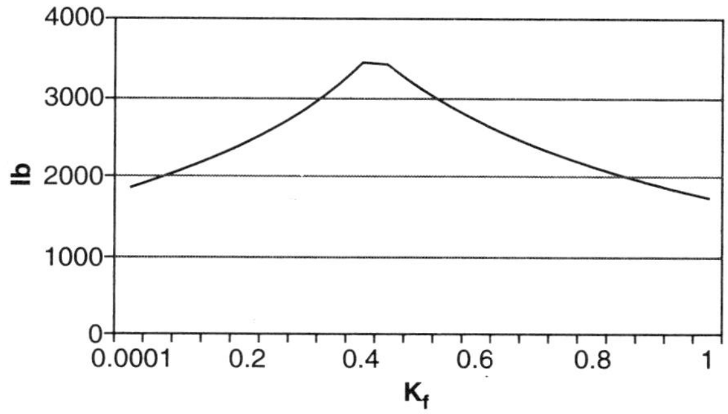

Fig. E6.20 Maximum tractive force and the front torque distribution factor.

Example 6.21: *Consider Example 8.4, where the 21,351.46-N (4800-lb) passenger car has a wheelbase of 2846.80 mm (112 in.) and a center of gravity 1270.00 mm (50 in.) behind the front axle and 508 mm (20 in.) above the ground. The car is moving on a horizontal plane, and drag and rolling resistance are negligible. Determine the maximum tractive force, maximum acceleration, and the distribution factor to achieve them. The coefficient of friction is 0.25.*

Solution: The front brake distribution factor to achieve the maximum braking force

$$K_f = \frac{(c_2 + \mu h)}{(c_1 + c_2)} = \frac{62 - 0.25 \times 20}{112} = 5089$$

The maximum braking force F_x is

$$F_x = \mu Mg \cos\theta = 0.25 \times 4800 = 1200 \text{ lb}$$

The maximum acceleration is

$$a = \frac{\mu Mg \cos\theta - R_a - R_r - Mg \sin\theta}{M} = \mu g = 8.05 \text{ ft/sec}^2$$

Note that the maximum acceleration is obtained when both a_R and a_F are equal. At that point, they both are proportional to the coefficient of friction times the axle load. To calculate the forces on both axles (including both friction and load forces), we need to develop a strategy to tackle the problem. The strategy is based on the following. Determine for the given problem a_F and a_R from Eqs. (6.192) and (6.197).

If $|a_F| < |a_R|$, front skid occurs first. Use the equations

$$F_{z1} = \frac{1}{c_1 + c_2}\left(Mgc_2 \cos\theta - h_a R_a - hMa_F - hMg \sin\theta\right)$$

$$F_{z2} = \frac{1}{c_1 + c_2}\left(Mgc_1 \cos\theta + h_a R_a + hMa_F + hMg \sin\theta\right)$$

Note that: $F_{z1} + F_{z2} = Mg \cos\theta$

$$F_{x1} = \mu F_{z1}$$

$$F_{x2} = \frac{1 - K_f}{K_f} F_{x1}$$

or

$$F_{x2} = Ma_F - F_{x1}$$

The total friction (tractive) force is

$$F_x = F_{x1} + F_{x2}$$

If rear skid occurs first. Use the equations

$$F_{z1} = \frac{1}{c_1 + c_2}\left(Mgc_2 \cos\theta - h_a R_a - hMa_R - hMg \sin\theta\right)$$

$$F_{z2} = \frac{1}{c_1 + c_2}\left(Mgc_1 \cos\theta + h_a R_a + hMa_R + hMg \sin\theta\right)$$

Note that: $F_{z1} + F_{z2} = Mg \cos\theta$

$$F_{x2} = \mu F_{z2}$$

$$F_{x1} = \frac{K_f}{1 - K_f} F_{x2}$$

or

$$F_{x1} = Ma_R - F_{x2}$$

The total friction (tractive) force is

$$F_x = F_{x1} + F_{x2}$$

Note the similarity between this strategy and the strategy developed for brakes.

Example 6.22: *Consider Example E8.5. The vehicle strategy requires a front axle skid on a dry surface, and the front torque distribution factor is chosen at 0.7. The vehicle is moving on a low friction surface with a 0.25 coefficient of friction. Determine the maximum tractive force and which axle would skid first.*

Solution: Because drag, rolling resistance, and grade are ignored, the acceleration under front skid is

$$a_F = \frac{\mu\left(\frac{c_2}{c_1 + c_2}\right)\left(1 + \frac{1 - K_f}{K_f}\right)}{\left(\frac{\mu h}{c_1 + c_2}\left(1 + \frac{1 - K_f}{K_f}\right) + 1\right)} g = \frac{0.25\left(\frac{62}{112}\right)\left(1 + \frac{0.3}{0.7}\right)}{\left(\frac{0.25 \times 20}{112}\left(1 + \frac{0.3}{0.7}\right) + 1\right)} g = 0.1368g$$

The acceleration under rear skid is

$$a_R = \frac{\mu\left(\frac{c_1}{c_1 + c_2}\right)\left(1 + \frac{K_f}{1 - K_f}\right)}{\left(-\frac{\mu h}{c_1 + c_2}\left(1 + \frac{K_f}{1 - K_f}\right) + 1\right)} g = \frac{0.25\left(\frac{50}{112}\right)\left(1 + \frac{0.7}{0.3}\right)}{\left(-\frac{0.25 \times 20}{112}\left(1 + \frac{0.7}{0.3}\right) + 1\right)} g = 0.7184g$$

Because $a_F < a_R$, the front skid occurs first. The normal forces are

$$F_{z1} = \frac{1}{c_1 + c_2}\left(Mgc_2 - hMa_F\right) = \frac{1}{112}\left(149 \times 32.2 \times 62 - 20 \times 149 \times (0.1368)\right) = 2517 \text{ lb}$$

$$F_{z2} = \frac{1}{c_1 + c_2}\left(Mgc_1 + hMa_F\right) = \frac{1}{112}\left(149 \times 32.2 \times 50 + 20 \times 149 \times (0.1368)\right) = 2283 \text{ lb}$$

Note that $F_{z1} + F_{z2} = 4800$ lb

$$F_{x1} = \mu F_{z1} = 0.25 \times 2517 = 629 \text{ lb}$$

$$F_{x2} = \frac{1-K_f}{K_f} F_{x1} = \frac{1-0.7}{0.7} \times (629) = 157 \text{ lb}$$

Total friction force is $F_x = F_{x1} + F_{x2} = 786$ lb

In a typical design problem, one needs to consider various surfaces and loading conditions. For example, let us consider the following problem where a vehicle is under different loading conditions and on various surfaces.

Example 6.23: *Consider a vehicle that weighs 14,235 N (3200 lb) when lightly loaded (driver only), with a center of gravity 1016.0 mm (40 in.) behind the front axle and 457.2 mm (18 in.) above the road, and 17,793 N (4000 lb) when loaded, with a center of gravity at 1295.4 mm (51 in.) behind the front axle and 508.0 mm (20 in.) above the road. The wheelbase is 2743.2 mm (108 in.). The vehicle is to achieve maximum possible tractive force and front skid under the following severe conditions:*

Vehicle is loaded and on a dry surface of $\mu = 0.8$.

Vehicle is loaded and on a wet surface of $\mu = 0.2$.

Vehicle is lightly loaded and on a dry surface of $\mu = 0.8$.

Vehicle is lightly loaded and on a wet surface of $\mu = 0.2$.

Determine the value of K_f *that you recommend, knowing that the drive system your company installs has a constant torque distribution value (no feedback control system). Determine the axle loads and the maximum tractive force.*

Solution: This problem is best solved by writing a spreadsheet with all possible tests. Table E6.23 shows the optimal value of K_f (Table E6.23A) and the associated acceleration and axle loads (Table E6.23B), as well as the maximum brake force for each case.

Table E6.23A Optimal value of K_f for example 6.23

Case	c_1	c_2	h	m	Mg	K_{fmax}	a_F/g	F_{z1}	F_{z2}	F_{x1}	F_{x2}	F_x
1	51	57	20	0.8	4000	0.3796	0.800	1519	2481	1215	1985	3200
2	51	57	20	0.2	4000	0.4907	0.200	1963	2037	393	407	800
3	40	68	18	0.8	3200	0.4963	0.800	1588	1612	1271	1289	2560
4	40	68	18	0.2	3200	0.5963	0.200	1908	1292	382	258	640

Table E6.23B Loads associated with $K_f = 0.65$ for example 6.23

Case	c_1	c_2	h	μ	Mg	K_f	a_F/g	F_{z1}	F_{z2}	F_{x1}	F_{x2}	F_x
1	51	57	20	0.8	4000	0.65	0.5290	1719	2281	1375	741	2116
2	51	57	20	0.2	4000	0.65	0.1536	1997	2003	399	215	615
3	40	68	18	0.8	3200	0.65	0.6430	1672	1528	1337	720	2058
4	40	68	18	0.2	3200	0.65	0.1843	1917	1283	383	206	590

To ensure front skid under all cases, the K_f must be chosen above the maximum value of 0.5963 determined for the fourth case. Thus, $K_f = 0.65$ is chosen. For this value, Table E6.23B shows the acceleration and loads obtained for each case. Note that with the choice made for K_f, front skid will occur first in all cases. To achieve this, major sacrifice is made for the maximum tractive force, especially for the first case, where the maximum tractive force is decreased from 3200 to 2116 lb.

In today's automotive drive systems, the front torque distribution factor K_f can be altered in a feedback control system. This will account for variations in the axle loads because of either different accelerations or different vehicle loading conditions (e.g., lightly loaded versus loaded conditions). This will enable the engineers to achieve higher tractive force at the surfaces with a low coefficient of friction. The K_f factor cannot be altered straightforwardly to account for the different coefficients of friction that various surfaces may offer. In that case, an advanced traction control system can be used.

6.4.4 Power-Limited Acceleration

The engine is the source of the power and energy of a vehicle. It transforms the energy from a source (typically chemical) into a kinetic energy, resulting in rotation of the crankshaft. The ideal powerplant puts forth constant power at all speeds. This power is defined as force times linear velocity, or torque times angular velocity, as

$$P = F \times V$$

or

$$P = T \times \omega \tag{6.203}$$

where ω is the angular velocity usually measured in radians per second (rad/s). In actual applications, the angular velocity is measured in engine revolutions per minute (rpm). The torque and power relation in an ideal powerplant can be determined based on this constant power assumption, as shown in Fig. 6.54. Note that for such an ideal powerplant, at low speeds, torque is needed more than power.

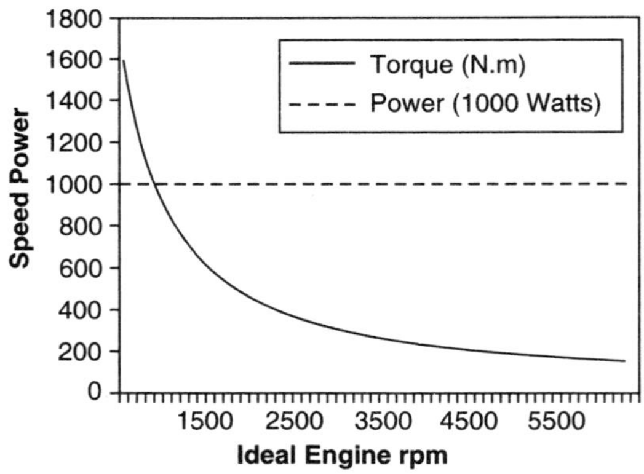

Fig. 6.54 Power and torque characteristics in an ideal engine.

Some electric motors can deliver this ideal output of the powerplant. On the other hand, the internal combustion engine performs differently. Despite this major shortcoming, it is still the most frequently used engine in automotive applications. Perhaps the abundance and affordability of the fuel are among the main reasons for that.

The powertrain consists of many components and systems. These systems include the engine cooling system, the drivetrain, the air conditioning, and others. In addition, some systems connect the powerplant to other systems. An example of these is the steering system, which has the power steering pump on the engine and the steering gear on the chassis of the vehicle. In addition to these systems (or subsystems), the engine carries other components such as the alternator. The arrangement of these subsystems and components will depend heavily on the powerplant configuration. Figure 6.55 shows a typical north-south rear-wheel-drive engine configuration. The key elements of the powerplant include its engine and transmission. Both of these will be discussed in detail in the next sections.

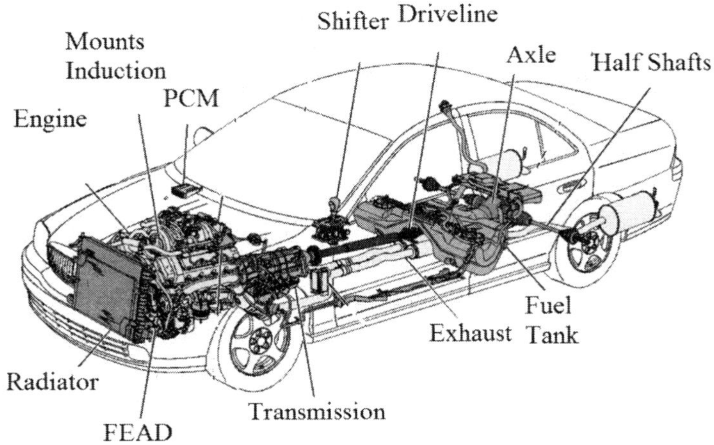

Fig. 6.55 Components and lay-out of a typical north-south RWD powertrain.

6.4.4.1 The Engine

Performance of an engine in the sense of torque and power versus engine revolutions per minute is only one of a few attributes that are critical in selecting an engine for a particular application. Automotive customers demand high torque at low speeds and high power in general. They also demand that the source of energy be economical and easy to replenish when consumed. To achieve high efficiency, the power-to-weight ratio is a factor considered by engineers and automotive customers. In addition, government regulations require fewer harmful emissions to the environment. Among the many engines considered or used for automotive applications are the following:

Gas turbine engine. This engine has a good power-to-weight ratio and lower emissions when compared with internal combustion diesel or gasoline engines. Tilagone *et al.* (2005) demonstrated the environmental and emission advantages of natural gas as a fuel source of engines for urban vehicles. However, this engine has low efficiency under no or little load conditions. This engine is not as popular as gasoline or diesel engines. One of the reasons could be the lack of an infrastructure that supports the delivery of this fuel to consumers (Unich, 1993).

Electric motor. This motor can be driven by batteries or fuel cells. Battery-driven road vehicles face the challenge of providing the customer demanded power and speed as well as driving range. The advancement of electric vehicles is completely dependent on that of battery technology. Fuel-cell-driven vehicles and fuel cell technology have received increased attention lately (e.g., Cantemir *et al.,* 2004). Fuel cells release the chemical energy and convert it directly to electrical energy. The fuel cells are filled with hydrogen that is combined with air to produce vapor, with little or no harm to the environment. Figure 6.56 shows a fuel-cell powertrain layout.

Stirling-cycle engine. In this engine, compressed helium and hydrogen are used to produce mechanical work. This engine was invented in 1816 by Robert Stirling, and its original use was to pump water out of mines. Small engines were built for fans and even were developed for an artificial heart that could be powered with an alpha-particle-emitting plutonium radioisotope (Oman, 1999). In the 1930s, a Dutch firm developed a small Stirling engine for driving generators that powered radios in remote regions. Transistor radios ended this use. General Motors subsequently developed this "Meijer" Stirling engine version for one application in the U.S. Army. The Stirling engine theoretically can convert the heat from burning hydrogen with Carnot-cycle efficiency. No other heat engine, including those based on cycles such as the diesel and Rankine engines, can achieve Carnot-cycle efficiency. A Stirling engine once was tested in a car. Although efficient, it was unsatisfactory because it did not give instant response to a driver's demand for acceleration power.

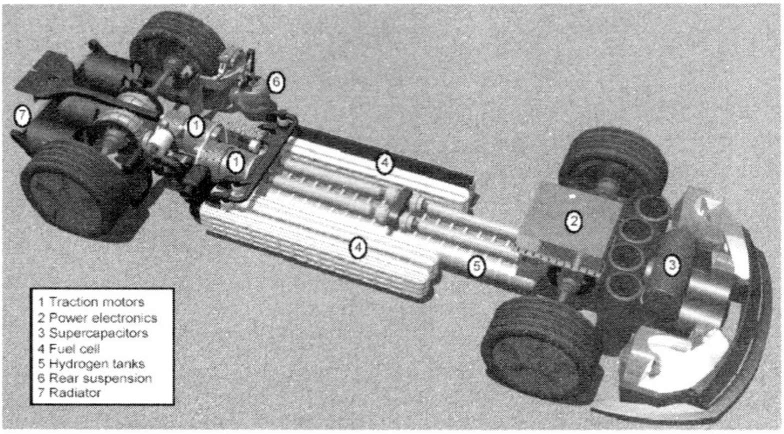

Fig. 6.56 Fuel Cell Powertrain Layout (Cantemir *et al*, 2004).

Rankine vapor cycle engine. This engine has many of the favorable characteristics of the ideal engine, including close-to-ideal torque delivery and low emissions. However, its low power-to-weight ratio is a major disadvantage.

Internal combustion engine. Gasoline or diesel is used in internal combustion engines as the source of energy. These fuels enter the combustion chamber and are given an electric charge to ignite the combustion process and release this energy. The high power-to-weight ratio of these engines makes them among the most frequently used engines in the automotive industry. Figure 6.55 shows a typical rear-wheel-drive powertrain layout with an internal combustion engine.

The main difference between a gasoline (petrol) and diesel engine is in how the ignition takes place. A gasoline engine draws a mixture of air and gasoline into the cylinder, compresses it, and ignites it with a spark to provide mechanical energy. On the other hand, a diesel brings in only air and compresses it, raising the temperature to the point where the fuel will ignite spontaneously when it is injected. This feature gives the diesel superior efficiency. The best car diesels have a thermal efficiency of approximately 43% compared with approximately 30% for gasoline engines. Because the process in a diesel engine needs air to be compressed to a higher degree than a gasoline engine, it causes more vibration. Thus, diesel engines must be stiffer and heavier to withstand those vibrations. This also results in the higher durability of diesel engines than that of gasoline engines. The higher compression ratio of diesel engines produces a higher torque output than that of equivalent gasoline engines. On the other hand, diesel power output usually is lower than that of an equivalent gasoline engine. This is the case because diesel engine speeds are lower than gasoline engine speeds. The diesel engine favored by the 1930s diesel pioneers, Citroën and Mercedes-Benz, was a similar type to that used in European passenger cars until the 1990s. The indirect injection (IDI) or pre-chamber design suited passenger cars best. Even now, indirect injection engines are still used in some European cars and light commercial vehicles.

Noise is the main problem of diesel engines. The fuel explodes on contact with the hot air in the engine, and the force of that explosion gives rise to the traditional diesel tick or rattle noise. To control that explosion, fuel is injected into a small chamber in the cylinder head, known as the pre-chamber, which is connected to the main combustion chamber directly over the piston. Combustion spreads rapidly into the main combustion chamber where the explosive force acts on the piston and forces it downward. The drawback of this controlled combustion is lower efficiency.

Efficiency could be further improved by injecting the fuel directly into the main combustion chamber and eliminating the pre-chamber. The introduction of control electronics made this type of engine appropriate to cars. Fiat introduced the first direct injection (DI) car engine in the Croma in 1988. The drawback to direct injection engines was greater combustion noise than indirect injection diesels. However, direct injection engines used up to 15% less fuel than indirect injection engines. This is one reason why premium brands such as Mercedes-Benz and BMW were late adopters of direct injection engines.

Electronic control systems meant that injection timing could be tightly controlled as required. Better diesel engines need high injection pressures to burn fuel more efficiently. The higher the fuel pressure, the better the fuel mixes with the air, completely burning more of the injected fuel in the process. That means more efficiency and cleaner emissions.

Diesel engines are believed to increase in popularity, especially when fuel economy becomes one of the more important vehicle attributes to consumers. Hybrid engines use an electric, relatively high-voltage battery to drive the vehicle when the demand for power is low, while an internal combustion engine is available when the demand for power is high. This type of powerplant is achieving favorable characteristics, particularly in the area of fuel economy. These engines now are installed on many small and medium-sized vehicles, but their initial cost remains a disadvantage. Figure 6.57 shows an exploded view of a Volkswagen hybrid drive (Kalberlah, 1991).

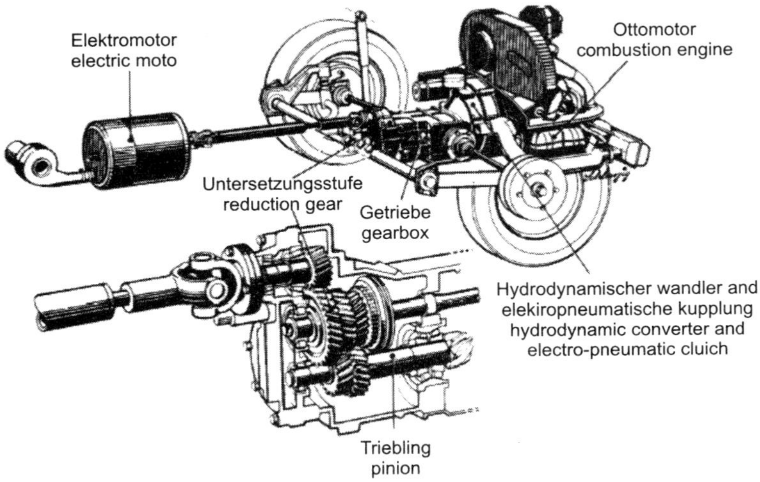

Fig. 6.57 Exploded view of a Volkswagen hybrid drive (Kalber-lah, 1991).

Nakamura *et al.* (1991) studied the potential of various engines in the future. The four-stroke engine of basically conventional structure will remain the mainstream powerplant up to the future. The engine must be improved in thermal efficiency and with significantly less tailpipe emissions. Future engines will turn toward downsizing, lighter weight, higher speed, and design diversity to meet diverse requirements from the standpoints of global energy saving, preservation of the environment and natural resources, and higher performance of the vehicle. Engine electronics and sensor technology must be improved for reliability and service-free operation. Progress is expected in catalytic technology to meet increasingly stringent emissions control requirements. Electric vehicles could be an alternative means of transportation for a limited application from the viewpoint of preserving urban environment and of utilizing possible excess electric energy supply in the future. Finally, various fuels that are alternatives to petroleum will continue to be investigated.

Next, we will spend more time reviewing the characteristics of the internal combustion engine because it is the most common engine used in automotive applications.

6.4.4.2 Internal Combustion Engines

The description of the combustion process and internal engine design are beyond the scope of this book. The focus here is on the engine output in terms of both torque and power that relate to vehicle dynamics.

The torque and power relations with the engine speed of typical gasoline and diesel engines are given in Figs. 6.58 and 6.59, respectively. In these engines, the torque and power are delivered at a lower speed (e.g., idle speed) but reach maximum values at an intermediate speed.

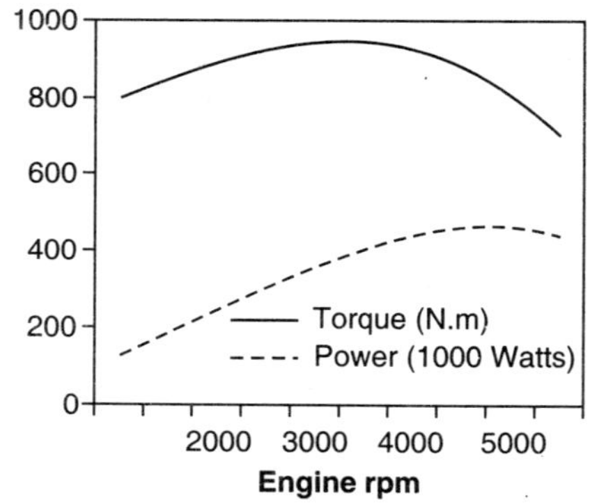

Fig. 6.58 Power and torque characteristics in a gasoline engine.

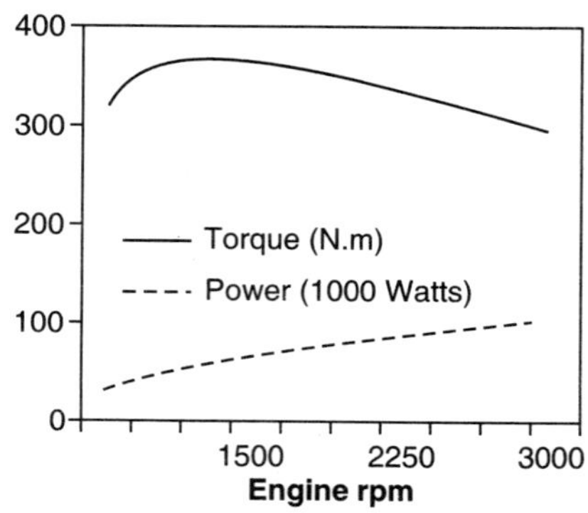

Fig. 6.59 Power and torque characteristics in a diesel engine.

The performance of these engines is not ideal. They do not produce constant power and hyperbolically reducing torque as a function of the engine revolutions per minute. Instead, the torque reaches a maximum value around 3000 rpm for gasoline engines and at a lower speed for diesel engines. After it reaches its maximum torque, losses within the engine reduce its output torque. Power reaches a maximum value at an engine speed higher than the one at which the maximum torque is obtained. The maximum engine speed is controlled to be slightly higher than that at which maximum power is achieved.

It should be mentioned that these torque and power characteristics are gross outputs. What the engine delivers in reality to the transmission is less than that. In particular, losses happen in the air handling systems (induction and exhaust), cooling systems (water pump and fan), front end accessory drive (FEAD) components (*i.e.*, power steering pump and alternator), and climate control systems (e.g., air conditioning). Air handling systems including exhaust and induction alone can consume approximately 10% of the total power. At approximately 3000 rpm, air conditioning consumes approximately 4 kW of power, the cooling system of the water pump and fan consume approximately 3 kW, and the alternator and power steering pump consume another 1.5 kW for each. This leads to approximately 10 kW of losses to the engine power before the power is delivered to the transmission input shaft. In total, these losses may add up to more than 20% of the engine power.

In addition to these factors, the output torque and power of the powerplant are functions of the ambient temperature and pressure. In general, low ambient pressure and/or high ambient temperature reduce the output power of the engine.

SAE International has published standards for finding the engine performance at various ambient pressure and temperature values (SAE J1995). In these standards are used a reference absolute pressure of Bo = 100 kPa or 16.5 psi, an absolute air dry pressure of B_{do} = 99 kPa or 16.36 psi, and an inlet temperature of T_o = 25°C. The power P under given temperature T in Celsius, dry pressure Bd in kilo Pascals (kPa), as compared with that found under the standard temperature and pressure values, are given using the formula

$$\frac{P}{P_o} = \frac{1}{1.18\left(\left(\frac{99}{B_d}\right)\left(\frac{T+273}{298}\right)^{1/2}\right) - 0.18} \tag{6.204}$$

Equation 6.204 is applicable to gasoline engines. Figure 6.60 gives the variation of the reasonable ambient variations, the output power of the engine may increase or decrease by approximately 10%.

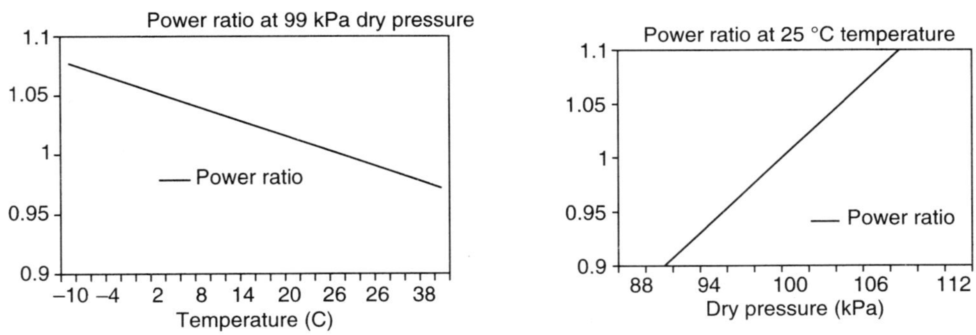

Fig. 6.60 Power variation as a function of (*a*) temperature and (*b*) dry pressure variations (SAE J1995).

The power prediction under various temperature and pressure values for diesel engines is given in the SAE J1995 standard, and readers are advised to refer to this standard. However, note that for diesel engines, the power ratio varies with $1/T$ rather than $1/\sqrt{T}$ as in gasoline engines.

One major difference between gasoline and diesel engines is the energy-to-weight ratio. At their highest efficiency, diesel engines seem to output twice the energy obtained by gasoline engines for the same weight.

Some of the useful relations in converting power among different units are

$$P\left(\frac{\text{ft}\times\text{lb}}{\text{sec}}\right) = \text{Torque}(\text{ft}\times\text{lb})\times\text{speed}\left(\frac{\text{rad}}{\text{sec}}\right)$$

$$P(\text{horsepower}) = \frac{\text{Torque}(\text{ft}\times\text{lb})\times\text{speed}\left(\frac{\text{rad}}{\text{sec}}\right)}{550} = \frac{\text{Torque}(\text{ft}\times\text{lb})\times\text{speed}(\text{rpm})}{5252} \tag{6.205}$$

$$\text{Power(kWatt)} = 0.746 \text{ Power (horse power)}$$

$$1 \text{ horsepower} = 550\frac{\text{ft}\times\text{lb}}{\text{sec}}$$

Before we study the transmission, let us consider an example on power calculations of a typical engine.

Example 6.24: *A V6 gasoline engine is outputting a maximum of 220 hp of gross power at 5000 rpm. Air induction losses are found to be 2.5% at that engine speed. The hot and cold end exhaust losses are found to be 7%. In addition, losses due to air conditioning, power steering, the alternator, and other accessories add up to 10 hp under specific driving conditions. The ambient temperature is 5°C, and the dry pressure is 95 kPa. Determine the power available to the transmission.*

Solution: The total losses of the induction and exhaust systems are

$$2.5\% + 7\% = 9.5\%$$

The power available to the accessories and transmission is

$$90.5\% \times 220 \text{ hp} = 199.1 \text{ hp}$$

The power available to the transmission under standard conditions is

$$199.1 - 10 = 189.1 \text{ hp}$$

$$\frac{P}{P_0} = \frac{1}{1.18\left(\left(\frac{99}{B_d}\right)\left(\frac{T+273}{298}\right)^{1/2}\right) - 0.18} = \frac{1}{1.18\left(\left(\frac{99}{95}\right)\left(\frac{5+273}{298}\right)^{1/2}\right) - 0.18} = 0.9924$$

The horsepower available to the transmission is

$$0.9924 \times 189.1 \text{ hp} = 187.6 \text{ hp}$$

6.4.4.3 The Transmission

As mentioned, an ideal powerplant delivers constant power throughout the engine speed range. However, the internal combustion engine delivers output torque that peaks at a certain speed, and power that peaks at a higher speed. Generally, the vehicle needs high torque at low speeds (for acceleration) and low torque at high speeds. To compensate for that shortcoming of internal combustion engines, a gearing mechanism is needed to deliver the required torque at the given speed. This gearing mechanism is the transmission.

Many types of transmissions are available in the marketplace. The automatic transmission is common in North America, whereas the manual transmission is widespread in Europe and the rest of the world. Recently, the continuously variable transmission (CVT) is finding applications in small vehicles.

In general, the transmission provides the required startup conditions, in forward and reverse, for a fully loaded vehicle on a relatively steep gradient (at approximately 30%). Under these conditions, the vehicle needs to accelerate, and high tractive forces at the tires are required. To produce these forces, higher torque is needed than that typically produced by the engine. Thus, the output engine speed should be reduced, and a lower gear (with a high gear reduction ratio) is needed. On the other hand, the higher gear (with the lower gear reduction ratio) is determined based on the desired maximum vehicle speed. To ensure smooth transition between these limits, other gear ratios are used in the transmission. For that purpose, typical transmissions for passenger road vehicles have three to six gear ratios.

At maximum vehicle speed, acceleration is not needed any further, and the engine should deliver enough power to maintain that speed and overcome the internal resistance of the transmission, rolling resistance of the tires, aerodynamic resistance, and reasonable grade (3 to 6%). Based on these factors, the gear ratio of the highest gear can be selected.

Thus, the transmission gear ratios are to be selected such that they bring the torque output of a typical engine (Fig. 6.58) to that of the ideal engine (Fig. 6.54). Figure 6.61 shows how the transmission output torque (and thus the tractive forces at the tires) changes with different gears.

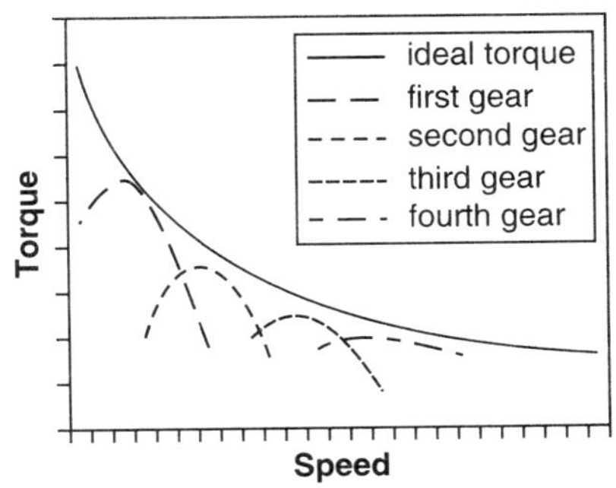

Fig. 6.61 Transmission torque output as a result of different gear engagement.

6.4.4.3.1 Manual Transmissions

A manual transmission consists of a clutch for engagement with the drive axle, a gear box that consists of the different gear sets that yield the desired gear ratios, a propeller shaft, and a drive axle. If there is no overdrive gear and as a general practice, the highest gear delivers no gear reduction. In other words, the speed of the engine is the same as that of the gear output shaft.

Common practice has been to select the maximum engine speed to be slightly higher (approximately 10% higher) than the speed at which the maximum power is obtained. The gear ratio of the highest gear (with the lowest gear reduction ratio) is found using the formula

$$\xi_n = \frac{n_e r(1-i)}{V_{max}\xi_x} \tag{6.206}$$

where n_e = engine speed corresponding to the maximum speed of the vehicle
r = radius of tire
i = tire slip
V_{max} = maximum speed of the vehicle
ξ_x = gear ratio in the drive axle

Note that if the lowest gear reduction ratio in the highest gear is selected to be 1 (*i.e.*, $\xi_n = 1$), then Eq. (6.206) can be used to find the gear ratio of the drive axle (*i.e.*, ξx).

The gear reduction ratio of the lowest gear is to be determined based on the ability of the vehicle to climb a certain grade (say, 30%) when fully loaded. Other criteria also may be used. One of these criteria states that the tires of the vehicle must not spin when climbing the maximum possible gradient.

Consider first rear-wheel-drive systems. At startup, drag can be neglected. Thus, Eq. (6.206) can be written as

$$F_x = F_{x2} = R_r + Ma + Mg\ \sin\ \theta = \mu_r Mg + Ma + Mg\ \sin\ \theta \tag{6.207}$$

The maximum slope (θ_{max}) at which the vehicle can withstand the load (without acceleration) can be calculated by equating Eq. (6.207) (with $a = 0$) with Eq. (6.186). This yields

$$Mg\ \sin(\theta_{max}) + \mu_r Mg = \frac{\dfrac{\mu Mg}{c_1 + c_2}(c_1 \cos\theta_{max} - h\mu_r)}{1 - \dfrac{\mu h}{c_1 + c_2}} \tag{6.208}$$

Simplifying further yields

$$\sin(\theta_{max}) = \frac{\dfrac{\mu}{c_1 + c_2}(c_1 \cos\theta_{max} - h\mu_r)}{1 - \dfrac{\mu h}{c_1 + c_2}} - \mu_r \tag{6.209}$$

Equation (6.209) is a nonlinear equation that can be solved for θ_{max}. At small angles of θ_{max}, the term cos (θ_{max}) can be approximated to 1, and the equation can be solved directly for θ_{max}. The value then can be substituted back into Eq. (6.209) to obtain a more accurate estimate. The process can be repeated until the desired accuracy is achieved.

Similar mathematical manipulation can be made for front-wheel-drive systems. This yields

$$\sin(\theta_{max}) = \frac{\dfrac{\mu}{c_1 + c_2}(c_2 \cos\theta_{max} + h\mu_r)}{1 + \dfrac{\mu h}{c_1 + c_2}} + \mu_r \tag{6.210}$$

Example 6.25: *Consider a vehicle that weighs 15,000 N when lightly loaded (driver only), with a center of gravity at 1.000 m behind the front axle and 0.460 m above the road. Consider a rolling resistance of 0.02. The wheelbase is 2.750 m. Find the maximum angle that can be withstood without losing traction for both front-wheel drive and rear-wheel drive when the coefficient of friction is 0.2.*

Solution: Given

$$c_1 + c_2 = 2.750 \text{ m}$$
$$c = 1.000 \text{ m}$$
$$\mu_r = 0.02$$
$$h = 0.46 \text{ m}$$
$$\mu = 0.2$$

For rear-wheel-drive vehicles:

First trial, $\theta_{max} = 0$

$$\sin(\theta_{max}) = \frac{\frac{\mu}{c_1+c_2}(c_2\cos\theta_{max} - h\mu_r)}{1-\frac{\mu h}{c_1+c_2}} - \mu_r$$

$$= \frac{\frac{0.2}{2.75}(1.75\times\cos(0) - 0.46\times 0.02)}{1-\frac{0.2\times 0.46}{2.75}} - 0.02 = 0.1110$$

Second trial, $\theta_{max} = \sin^{-1}(0.1110) = 0.1111$ rad

$$\sin(\theta_{max}) = \frac{\frac{0.2}{2.75}(1.75\times\cos(0.1111\text{ rad}) - 0.46\times 0.02)}{1-\frac{0.2\times 0.46}{2.75}} - 0.02 = 0.1101$$

Third trial, $\theta_{max} = \sin^{-1}(0.1101) = 0.1104$ rad

$$\sin(\theta_{max}) = \frac{\frac{0.2}{2.75}(1.75\times\cos(0.1101\text{ rad}) - 0.46\times 0.02)}{1-\frac{0.2\times 0.46}{2.75}} - 0.02 = 0.1102$$

Fourth trial, $\theta_{max} = \sin^{-1}(0.1102) = 0.1104$ rad $= 6.325$ degrees.

For front-wheel-drive vehicles

First trial, $\theta_{max} = 0$

$$\sin(\theta_{max}) = \frac{\frac{\mu}{c_1+c_2}(c_2\cos\theta_{max} + h\mu_r)}{1+\frac{\mu h}{c_1+c_2}} + \mu_r$$

$$= \frac{\frac{0.2}{2.75}(1.75\times\cos(0) + 0.46\times 0.02)}{1+\frac{0.2\times 0.46}{2.75}} + 0.02 = 0.1263$$

Second trial, $\theta_{max} = \sin^{-1}(0.1263) = 0.127$ rad

$$\sin(\theta_{max}) = \frac{\frac{0.2}{2.75}(1.75\times\cos(0.1267\text{rad})+0.46\times0.02)}{1+\frac{0.2\times0.46}{2.75}} + 0.02 = 0.1428$$

Third trial, $\theta_{max} = \sin^{-1}(0.1428) = 0.1433$ rad

$$\sin(\theta_{max}) = \frac{\frac{0.2}{2.75}(1.75\times\cos(0.1433\text{ rad})+0.46\times0.02)}{1+\frac{0.2\times0.46}{2.75}} + 0.02 = 0.1425$$

Fourth trial, $\theta_{max} = \sin^{-1}(0.1425) = 0.1430$ rad = 8.193 degrees.

Note that front-wheel-drive vehicles are more capable of achieving this low friction. For high friction, rear-wheel-drive vehicles have better capability for climbing slopes.

Assume T_{emax} = maximum engine torque
η_t = transmission efficiency
ξ_1 = gear ratio of the lowest gear

The total output torque available for the wheels is

$$T_{max} = T_{e\,max} \times \eta_t \times \xi_x \times \xi_1 \tag{6.211}$$

This must be equal to the total torque produced by the tractive force. This torque is calculated by using Eq. (6.176) for the tractive force and multiplying it by the rolling radius r. This gives (note $a = 0$)

$$T_{max} = (\mu_r Mg + Mg \sin\theta)\, r \tag{6.212}$$

Equating Eqs. (6.211) and (6.212) yields the gear ratio of the lowest gear as

$$\xi_1 = \frac{Mg(\sin(\theta_{max}) + \mu_r)r}{T_{e\,max}\,\xi_x\,\eta_t} \tag{6.213}$$

With the gear ratios of the high gear ξ_n and low gear $\xi_1$determined from Eqs. (6.175) and (6.182), respectively, we need to establish the gear ratios of the intermediate gears. For commercial vehicles, it is desired that the engine speed remains in the range that achieves the highest fuel economy. Thus, the gear ratios are set in a manner that follows geometric progression. To find these ratios, a general transmission ratio is to be found, based on the number of gear ratios (n), the low gear ratio ξ_1, and the high gear ratio ξ_n. This factor is

$$K_g = \sqrt[n-1]{\frac{\xi_n}{\xi_1}} \tag{6.214}$$

The gear ratios of the intermediate gears then are found by using

$$\xi_n = K_g \xi_{n-1} \tag{6.215}$$

For passenger vehicles, the gear ratios are found in a manner that is biased toward achieving the maximum vehicle speed in the shortest time. Also, higher gears are given closer ratios because shifting happens more frequently in these gears. Thus, the transmission factor K_g is not constant for passenger vehicles. Note that the lowest gear ratios (ξ_1/ξ_2) are approximately 10 to 15% higher than the value of K_g calculated by Eq. (6.214), and for the highest gear ratios are (ξ_{n-1}/ξ_n) 10 to 15% lower than the calculated value of K_g. x_n mes very close to that found using Eq. (6.214). Table 6.5 shows the gear ratios found in three small vehicles.

Table 6.3 Gear Ratios in Passenger Vehicles

	Gear Ratios										
	1st	*2nd*	*3rd*	*4th*	*5th*	*Kg*	ξ_1/ξ_2	ξ_2/ξ_3	ξ_3/ξ_4	ξ_4/ξ_5	*Average*
Audi A4 1.8	3.50	2.12	1.43	1.03	0.84	1.429	1.651	1.483	1.388	1.226	1.437
BMW 525i	4.20	2.49	1.67	1.24	1.00	1.432	1.687	1.491	1.347	1.240	1.441
Ford Escort 1.8	3.42	2.14	1.45	1.03	0.77	1.452	1.598	1.476	1.408	1.338	1.455

The mechanical efficiency of a transmission describes the losses due to friction between the engaging gears as well as friction within the bearings. This efficiency is a function of the speed and torque transmitted. The mechanical efficiency usually is between 90 and 98%, but it is lower for higher torque values. The transmission efficiency also is lower for lower gears and is higher for higher gears.

For a manual transmission at any engine speed, the tractive force available at the tires is calculated by

$$F_x = \frac{T_e \xi_o \eta_t}{r} \tag{6.216}$$

where T_e = engine output torque

ξ_o = overall reduction ratio

η_t = overall efficiency of the transmission

The engine speed is related to the vehicle speed as

$$V = \frac{n_e r(1-i)}{\xi_o} \tag{6.217}$$

where i is the slip (taken between 2 and 5%), and n_e is the engine speed.

Recently, automated manual transmissions (AMTs) have been receiving attention. These are different from manual transmissions because the driver does not actuate the power of the clutch and gear shifting. Clutch actuation and gear shifting occur by means of a hydraulic system or an electric motor and are operated electronically. The clutch pedal is obsolete; the clutch control is connected to the position and movement of the selector lever. The mechanical connection between the selector lever and the transmission is omitted; the transmission is controlled electronically (shift by wire). This offers more possibilities regarding the design of selector units as with the usual mechanical shifting systems. The driver usually activates the gearshift. With the implementation of a shift program in the transmission control unit, an automatic mode is easily established so that an AMT can execute gearshifts automatically. The automatic operation mode is selected by a separate switch.

6.4.4.3.2 Automatic Transmissions

Automatic transmissions are quite common in North America and less common in the rest of the world. Compared with the manual transmission, the automatic transmission offers more convenience to the driver because the driver does not need to manually shift at various speeds when the automatic transmission is used. However, the automatic transmission generally is more expensive and is not as economical with fuel efficiency. Figure 6.62 shows a typical automatic transmission used in a north-south powerplant configuration.

Fig. 6.62 A typical automatic transmission used in north-south powerplant configuration. ZF-6-speed automatic transmission (SAE 2001-01-0882).

The main components of an automatic transmission are the torque converter (Fig. 6.63) and the gear box. In automotive applications, the torque converter is a three-element hydrodynamic drive unit that links the engine to the automatic transmission. The torque converter decouples the engine from the transmission at idle conditions. It also increases the engine torque for input to the transmission during a launch condition. The first element of the torque converter is the impeller. It is a centrifugal pump that converts mechanical power from the engine to fluid power. The fluid power from the impeller is directed to the second element, a radial turbine that converts it back to mechanical power which then is transmitted to the transmission. The third element of the torque converter, the reactor or stator, redirects the working fluid back to the impeller.

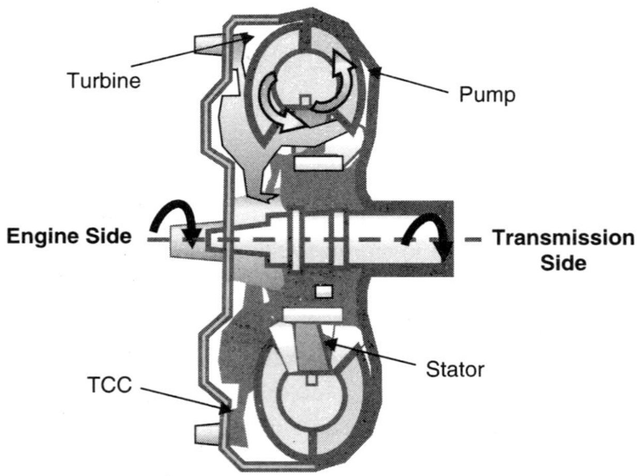

Fig. 6.63 Cross Sectional View of Torque Converter (Luk Inc., Lee and Lee, 2004).

The power output of the torque converter (Ptcout) is equal to the output torque (Ttcout) times the output speed (ωtcout), and the power input (Ptcin) is the input torque (Ttcin) times the input speed (ωtcin). The efficiency of the torque converter ηtc is defined as

$$\eta_{tc} = \frac{P_{out}}{P_{in}} = \frac{T_{tcout}\,\omega_{tcout}}{T_{tcin}\,\omega_{tcin}} \tag{6.218}$$

Note that the subscript *tc* stands for the torque converter.

The efficiency of the torque converter and the torque ratio are displayed in Fig. 6.64 as functions of the speed ratio. Note that at a speed ratio of zero (idle condition), the torque is the highest and the efficiency is the lowest.

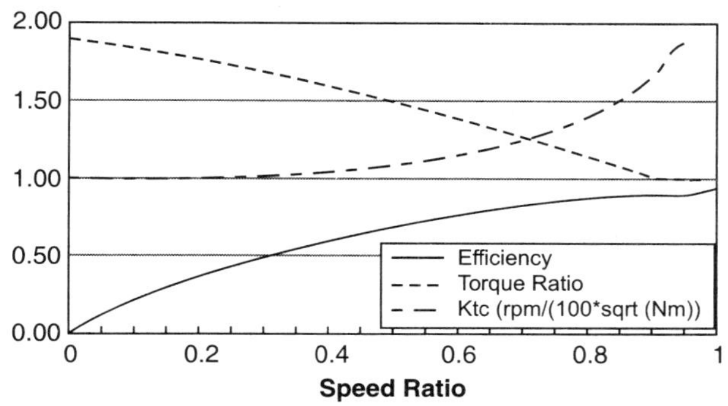

Fig. 6.64 Efficiency and torque ratios as functions of speed ratio for a torque converter.

At a maximum speed, the torque converter acts as a fluid coupling device, giving the highest efficiency and a torque ratio of 1.

Figure 6.65 shows the tractive force as a function of the vehicle speed and the various gear ratios. At the lowest gears, the vehicle speed is low, and acceleration can be high (under wide throttle). Hence, the tractive forces are high. At higher speeds, the need is mainly to maintain speed rather than the need for acceleration. This figure also shows how grade affects the gear ratio used. At no grade levels, the vehicle speed can be maintained with the highest gear, whereas at a certain grade, a lower gear must be chosen to maintain the vehicle speed. If the vehicle condition (speed and gear) are below the grade line given, the vehicle speed cannot be maintained (not enough torque or tractive forces).

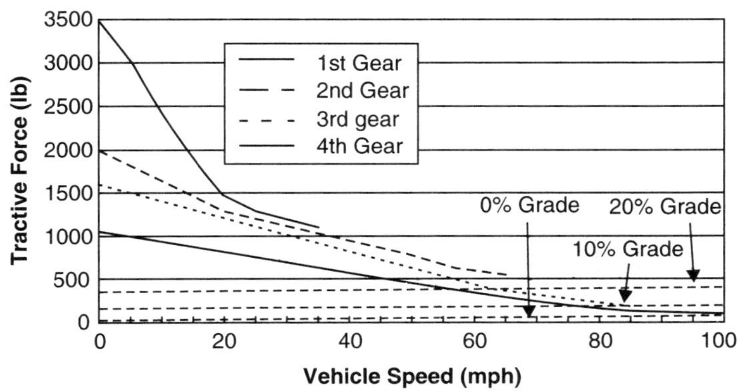

Fig. 6.65 Relation between tractive forces at various gears and vehicle speeds.

The introduction of new technology such as torque converter lock-up clutches and electronic transmission control units and an increase in the number of speeds and optimized components and parts all have contributed to the fact that the power losses of automatic transmissions were reduced drastically during the last decade. The fuel consumption of an automatic transmission today is approaching that of a vehicle with a manual transmission. Good shift quality and electronically controlled shift programs that adapt to driving conditions and driving behavior guarantee high driving comfort.

The introduction of five-speed transmissions since 1989, six-speed transmissions in general applications later, and the broadening numerous functions such as the manual shift mode using the lever to shift the gears (tiptronic) have added strongly to the image of improvement in automatic transmissions. The following are the advantages of high-speed (*i.e.*, five- or six-speed) automatic transmissions:

- Fuel consumption savings by approximately 5%.
- Reduced emissions.
- Improved acceleration by approximately 5%.
- Reduced weight by approximately 10%.

To determine the working output speed and torque of the torque converter, we need to specify a parameter called the capacity factor, defined as

$$K_{tc} = \frac{\omega_{tc}}{\sqrt{T_{tc}}} \tag{6.219}$$

This also is given in Fig. 6.64. The steps to be taken in finding the output speed and the torque of the torque converter are as follows:

- Determine the engine speed and torque (given).
- Find the engine capacity factor. The capacity factor of the torque converter is the same as that of the engine.
- Use a graph similar to Fig. 6.64 to determine the speed ratio, efficiency, and torque ratio.
- Calculate the output speed and torque as well as the efficiency.

Example 6.26: *An engine delivers 300 Nm torque at an engine speed of 2500 rpm. It is coupled to a torque converter and an automatic speed transmission. Determine the output speed and the output torque of the converter of the characteristics given in Fig. 6.66. Determine the efficiency of the torque converter.*

Solution: The engine speed and torque are given as

$$T_e = 300 \text{ Nm}$$
$$\eta_e = 2500 \text{ rpm}$$

Find the engine capacity factor. The capacity factor of the torque converter is the same as that of the engine

$$K_{tc} = K_e = \frac{\omega_e}{\sqrt{T_e}} = \frac{2500}{\sqrt{300}} = 144.3 \frac{\text{rpm}}{\sqrt{\text{Nm}}}$$

Use a graph similar to Fig. 6.66 to determine the speed and torque ratio.

The speed ratio is 0.81, and the torque ratio is 1.13. Then,

$$\eta_{tc} = 0.81 \times 2500 = 2025 \text{ rpm}$$
$$T_{tc} = 1.13 \times 300 = 339 \text{ Nm}$$

Using Eq. (8.49), the efficiency is

$$\eta_{tc} = \frac{P_{\text{out}}}{P_{\text{in}}} = \frac{T_{\text{tcout}}\omega_{\text{tcout}}}{T_{\text{tcin}}\omega_{\text{tcin}}} = \frac{2025 \times 339}{2500 \times 300} = 0.91$$

This also could be found from Fig. 6.66. Note that higher torque was given at a lower speed. This will help give a higher level of acceleration in startup condition (as desired by typical drivers).

For an automatic transmission, we need to modify Eqs. (6.216) and (6.217) to reflect the implementation of the torque converter. Instead of having the engine speed (n_e) and the engine torque (T_e) for a manual transmission, we will have the output speed of the torque converter (n_{tc}) and the output torque of the torque converter (T_{tc}). Hence, for the automatic transmission, the tractive force available at the tires is calculated by

$$F_x = \frac{T_{tc}\xi_o\eta_t}{r} \tag{6.220}$$

where ξ_o is the overall reduction ratio, and ηt is the overall efficiency of the transmission.

The vehicle speed is related to the engine speed (similar to Eq. (6.217)) as

$$V = \frac{n_{tc}r(1-i)}{\xi_o} \tag{6.221}$$

where i is the slip (taken between 2 and 5%).

6.4.4.3.3 Continuously Variable Transmissions (CVTs)

Continuously variable transmissions (CVTs) are transmissions where the creation of the ratio occurs by means of a variator with axial repositioning of a conical-shaped pair of discs. A chain or a belt between the discs transfers torque with friction contact. Figure 6.66 shows a typical continuously variable transmission used in an east-west powerplant configuration (Ford's Five Hundred and Freestyle).

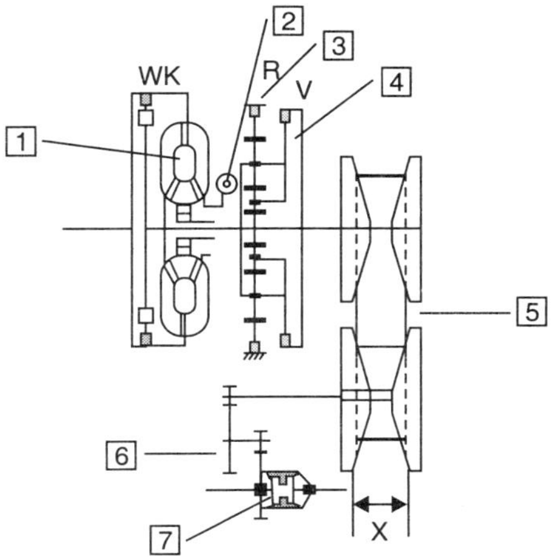

1. Torque Converter
2. Pump
3. Reverse Gear Set
4. Forward Clutch
5. Variator
6. Final Drive
7. Differential

Fig. 6.66 Typical continuously variable transmission used in an east-west power-plant configuration for a ZF CFT 30 continuously variable transmission for front-wheel drive. (Wagner *et al.*, 2006.)

Continuously variable transmissions are receiving interest for use in small and medium vehicles with low torque (300 Nm or less). Applications where high torque is needed (greater than 350 Nm) generally are not suitable for continuously variable transmissions.

The system shown in Fig. 6.66 is based on the Van Doorne belt concept. This system has two conically faced pulleys, with the effective radius of the pulleys (and hence the reduction ratio) changing as the distance x (Fig. 6.66) changes.

The other potential mechanism is the use of a toroidal (or Perby) system. Figure 6.67 shows the basic mechanism.

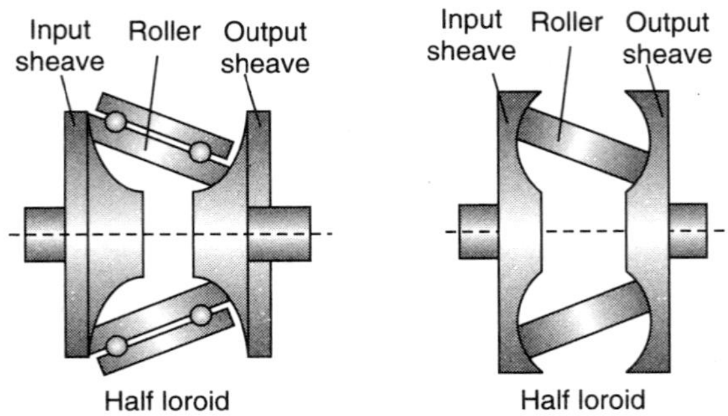

Fig. 6.67 The basic mechanism of a toroidal continuously variable transmission. (Wagner, 2001.)

Power is transferred through the first torus disc by friction on rollers and flows over a second sheave to the transmission output shaft. Twisting of the rollers changes the ratio. The fundamental advantage is the high torque capacity, where it is higher than with a Van Doorne continuously variable transmission. The toroidal continuously variable transmission also offers good driving comfort and the possibility of rapid ratio change. Disadvantages are the package requirements, the weight, and manufacturing costs. Furthermore, no real solution for traction oil has been found. Oils that currently are employed in development have not fulfilled satisfying functions over the whole temperature range of –40 and 150°C (Wagner, 2001).

The first toroidal transmissions went into mass production with General Motors in the 1930s, but the concept did not really catch on. Nissan undertook the second application in 1999 with a toroidal transmission for 350 Nm, a gear ratio of 6.36, and a weight of 105 kg for use in a standard driveline. These numbers are not competitive in comparison to a six-speed automatic transmission.

The three types of transmissions described here remain the ones used in the vast majority of vehicles today. In particular, the automated manual transmission (AMT) and the continuously variable transmission (CVT) are expected to receive increased interest, mainly due to the advantages they offer with fuel economy and shift quality. Conventional automatic transmissions are expected to gain higher efficiency levels, particularly with the introduction of six-speed transmissions.

6.4.4.4 Vehicle Acceleration

As stated, the tractive forces at the tires must be calculated to determine the vehicle acceleration and travel distance. In Section 6.4.3, we found the maximum possible acceleration based on the traction limitation. On the other hand, finite engine torque is the main factor that limits acceleration at off-idle conditions and on dry surfaces. The ultimate objective of the engine power and torque is to deliver tractive forces at the driving axles. These tractive forces result in the acceleration of the vehicle.

First we must account for the losses in the front end accessory drive, induction, and exhaust. These power losses are direct losses in the torque. Keep in mind that these losses are calculated at a specific engine speed (*i.e.*, one that corresponds to maximum power). These losses directly impact the torque delivered by the engine as the speed remains the same. Once the net engine power and torque are determined, we need to find the torque output of the transmission, taking into account the gear ratio of the transmission, the impact of the torque converter, and the system inefficiencies. Once the output torque of the transmission is determined, the axle gear ratio(s), if available, as well as driveline system inefficiencies must be considered to find the torque available for the driving axle(s) to accelerate the vehicle.

In addition to these losses, keep in mind that the resulting torque does more than simply accelerate the vehicle. This torque is used to rotate all rotating elements in the vehicle, including the wheels. To take into consideration the effect of inertia of the vehicle rotating elements, we will introduce the mass factor (γ_m). This is needed for application of Newton's second law to calculate vehicle acceleration as

$$F_{\text{net}} = F - \Sigma R = \gamma_m ma \tag{6.222}$$

where $\Sigma R = R_r + R_a + \ldots$

F_{net} is the net tractive force causing the vehicle to accelerate. The mass factor is

$$\gamma_m = 1 + \frac{1}{Mr^2}\left\{\sum I_w + \sum I_1\xi_1^2 + \sum I_2\xi_2^2 + \ldots + \sum I_n\xi_n^2\right\} \tag{6.223}$$

where M is the mass of the vehicle, r is the rolling radius of the wheels, I_w is the mass moment of inertia for a wheel, I_i (with $I = 1, 2, \ldots$) are the moments of inertia of the rotating components attached to the drive train. The following formula is suggested as an approximation for passenger cars (Taborek, 1957)

The following formula is suggested as an approximation for passenger cars (Taborek, 1957)

$$\gamma_m = 1.04 + 0.0025\xi_0^2 \tag{6.224}$$

where ξ_0 is the overall transmission ratio. Knowing that the acceleration is the first derivative of velocity with respect to time ($a = dv/dt$). Eq. (6.222) can be written as

$$F_{\text{net}} = F_x - \sum R = \gamma_m M\left(\frac{dV}{dt}\right) \tag{6.225}$$

Knowing that the net force (F_{net}) is a function of velocity (because of resistance forces, which are functions of velocity), this expression can be written to calculate time traveled between velocity V_i and V_f as

$$t = \gamma_m M \int_{V_i}^{V_f} \frac{dV}{F_{net}} \tag{6.226}$$

Similarly, the distance traveled (S) during this change in velocity would be

$$S = \gamma_m M \int_{V_i}^{V_f} \frac{dV}{F_{net}} \tag{6.227}$$

Keep in mind during acceleration the shifting time that takes place when going from a certain gear to another. Each shift is approximately 1 second, with the automatic transmission offering less shifting time than the manual transmission.

In general, Eqs. (6.226) and (6.227) should be used to calculate the key vehicle attributes of time and distance traveled to achieve a certain level of velocity. This is of particular and key interest to vehicles with a "sporty" image. The difficulty in these equations is in the integration that must be carried numerically. Realize that the engine speed (and torque) during acceleration changes. This results in a variable tractive force F_x. Also, the driver may shift the transmission gear during this acceleration, leading to a second source of variation in the tractive force. In addition, as the vehicle gains speed, the drag changes, which adds another reason for variation in the net force (F_{net}). All of these factors lead to a higher level of complexity when the acceleration time and/or distance are calculated.

Example 6.27: *A vehicle (including wheels) weighs 22,215 N (5000 lb). Each wheel weighs 266.6 N (60 lb), and the rolling radius of the wheels is 355.6 mm (14 in.). The radius of gyration is 279.4 mm (11 in.). The engine is driven at 3000 rpm and is delivering 406.27 Nm (300 ft-lb). The total drive train efficiency is 86%, and the transmission is in fourth gear with a ratio of 6.5 to 1. The components rotating at the engine speed have an inertia of 0.6 slug ft². Assume a slip of the running gear at 2% and a coefficient of rolling resistance of 2%. The front area of the vehicle is 1.858 m² (20 ft²), the aerodynamic drag coefficient is 0.4, and the speed of the wind is zero. Determine the acceleration at that point.*

Solution: Given

$$M = 5000/32.2 = 155.2800 \text{ slugs}$$
$$m_w = 60/32.2 = 1.8634 \text{ slugs}$$
$$r = 14 \text{ in}$$
$$r_g = 11 \text{ in}$$
$$n_e = 3000 \text{ rpm}$$
$$T_e = 300 \text{ ft.lb.}$$
$$\eta = 0.86$$
$$\xi_t = 4.5$$
$$\Sigma I = 0.6 \text{ slug.ft}^2 = 86.4 \text{ slug in}^2$$
$$i = 0.02$$
$$\mu_r = 0.02$$
$$C_D = 0.4$$
$$A_F = 20 \text{ ft}^2$$

1. Calculate the mass factor (γ_m)

$$I_w = m\,(r_g)^2 = 1.8634\,(11)^2 = 225.47 \text{ slug in}^2$$

$$\gamma_m = 1 + \frac{1}{Mr^2}\left\{\sum I_w + \sum I_1\xi_1^2 + \sum I_2\xi_2^2 + \ldots + \sum I_n\xi_n^2\right\}$$

$$= 1 + \frac{1}{155.28\times(14)^2}\left\{4\times 225.47 + 86.4\times 4.5^2\right\} = 1.0871$$

2. Calculate the tractive force Eq. (6.216)

$$F_x = \frac{T_e\xi_o\eta_t}{r} = \frac{36000\times 4.5\times 0.86}{14} = 995.14 \text{ lb}$$

3. Calculate the vehicle speed Eq. (6.217)

$$V = \frac{n_e r(1-i)}{\xi_o} = \frac{3000\times(2\pi)\times 14\times(1-0.02)}{4.5} = 57470.20 \text{ in/min} = 54.4225 \text{ mph}$$

Because the wind is at zero speed,

$$V_r = V = 54.4225 \text{ mph}$$

4. Calculate the rolling resistance force Eq. (7.26)

$$R_r = \mu_r mg = 0.02 \times 5000 = 100 \text{ lb}$$

5. Calculate the aerodynamic drag force Eq. (7.27)

$$R_a = \frac{\rho}{2} C_D A_F V_r^2 = \frac{0.002378}{2}\times 0.4\times 20\times\frac{(57470)^2}{(12)^2\times(60)^2} = 60.60 \ \frac{\text{slug.ft}}{\text{s}^2} = 60.60 \text{ lb}$$

Note unit conversion factors.

6. Calculate the acceleration Eq. (6.222)

$$a = \frac{F_{net}}{\gamma_m m} = \frac{F - \sum R}{\gamma_m m} = \frac{995.14 - 100.0 - 60.60}{1.0871\times 155.28} = 4.944 \text{ ft/s}^2$$

Example 6.28: *Consider Example 6.27. Assume that the engine torque remained constant during a takeoff acceleration. Also, acceleration is happening while the transmission is engaged in the same gear as in the previous example. Determine how long it would take the vehicle to reach a constant speed of 60 mph.*

Solution: 1. The unit conversion is

$$V_f = 60\frac{\text{mile}}{\text{hr}} = 60\times\frac{5280.0 \text{ ft}}{3600 \text{ s}} = 88.0 \text{ ft/s}$$

2. Determine the drag [Eq. (7.27)]

$$R_a = \frac{\rho}{2} C_D A_F V_r^2 = \frac{0.002378}{2}\frac{\text{slug}}{\text{ft}^3}\times 20\, ft^2 = 0.02378\frac{\text{slug}}{\text{ft}} V_r^2$$

3. Determine F_{net}

$$F_{\text{net}} = F - \sum R = 995.14 - 100.0 - 0.02378 V_r^2 = 895.14 - 0.02378 V_r^2$$

4. Determine time [Eq. (6.226)]

$$t = \gamma_m m \int_{V_i}^{V_f} \frac{dV}{F_{net}} = 1.0871\times 155.28\times\int_0^{88}\frac{dV}{895.14 - 0.02378 V_r^2} = 18.11 \text{ s}$$

The last example is a simple example that shows achievable acceleration with nonconstant net force (a function of velocity because of drag). Real problems would involve varying the engine revolutions per minute and varying the torque and gear ratio, as well as other complexities.

6.4.5 Safety Features

The most important safety concern that is encountered by customers trying to accelerate a vehicle is the loss of traction or slip condition. Although drivers generally possess more control while accelerating than braking (where drivers may lose control), traction is important to move the vehicle away from a possible accident or hazardous situation. Two features usually are used for that purpose. These are the use of a limited slip axle and traction control. The first is an axle that uses a clutch or locking device to make torque bias to a wheel that still has traction when the other wheel on the axle slips. Traction control applies brakes to slipping wheels.

6.4.5.1 Limited Slip Axle

While driving around a corner, the wheel on the outside of a given axle rotates more than the wheel on the inside. For this to occur, a differential gear set is incorporated in the axles. This differential unit receives torque from an input (the driveshaft) and distributes it between the axle shafts (or half shafts). It allows those outputs to rotate at different speeds.

A normal (or open) differential unit behaves such that if one output loses traction, little torque will be sent to the other output. In other words, neither of the wheels can receive more torque than the amount required slipping either one of them. If a front-wheel-drive vehicle has ice under the right front wheel, then that wheel will spin when torque is applied. If 25 ft-lb (33.87 Nm) caused that right front wheel to slip, the left front wheel will receive only 25 ft-lb (33.87 Nm).

Some axles incorporate a limited slip device to govern the differential to prevent this condition from occurring. This device directs some torque to the wheel that still has traction. Some of the devices that can be used as axle limited slip devices include the following:

1. **Plate clutch.** In this device, a series of plates is coated with friction material that provides resistance when the wheels move at different speeds. These have limited torque capacity. Fig. 6.68 shows one of these devices.
2. **Viscous coupling.** This device causes silicone fluid to be sheared between a series of clutch plates when the wheels move at different speeds. Cost is a drawback of these devices.
3. **Gerodisc/Visco-Lok devices.** In these devices, a difference in speed between the two wheels activates a pump that compresses a clutch. This, in turn, controls the slip. More information is available in a paper by Gassmann and Barlage, 1996.
4. **Torque multiplying differential.** This differential sends a multiple of the traction available at the slipping wheel to the wheel that still has traction. For example, if one wheel on an axle slips at 33.87 Nm (25 ft-lb) of torque, the other axle may receive 101.57 Nm (75 ft-lb).

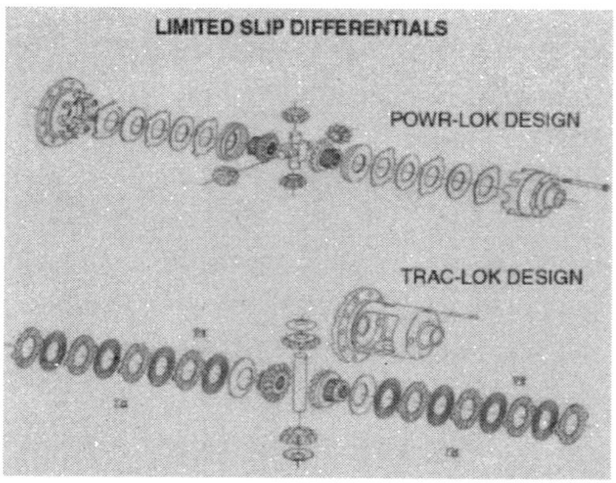

Fig. 6.68 Dana's limited slip axle design (Vettel and Bielik, 2003).

These devices are among those available on the market to control the amount of torque distributed at each wheel.

6.4.5.2 Traction Control

The traction control system works by applying the brake to the spinning wheel. This allows torque to be sent to the wheel that still has traction. For example, if the right wheel slips after 33.87 Nm (25 ft-lb) of torque, then a certain amount of brake torque (say, 75 ft-lb, or 101.57 Nm) is applied to that wheel. There is effectively a total of 135.4 Nm (100 ft-lb) of resistance at that right wheel—25 ft-lb from the limited traction, and 75 ft-lb from the brake. Hence, the left wheel will be able to receive 135.4 Nm (100 ft-lb). This torque is applied to the ground where traction is still present. Some traction control systems may reduce torque delivered by the engine to help control slip. One mechanism to perform that is to start in second gear when slippery roads are experienced.

Engineers have the objective of achieving the maximum tractive force. With advances in electronic controls and sensors, sensors are planted to assess skid, or slipping, conditions of the wheels. Traction control systems work in a manner opposite that of an anti-lock braking system (ABS). At slip during acceleration, the brakes are applied electronically and momentarily to bring the wheel to the nonskid condition. Throttle is applied again to reach the skid condition, gaining higher speed over a short period of time. This process is repeated until full traction at the desired speed is achieved. Traction control is now a standard safety feature on many vehicles and is available as an optional feature on most others. Similar to the advantages of anti-lock braking systems, traction control systems offer better treatment of complex surfaces frequently encountered in cold regions or in off-road maneuvers. Variations in the coefficient of friction between the two wheels of the same axle are possible. Thus, a condition of the wheels of the same axle having one side on a low coefficient of friction and the other wheel on a higher coefficient of friction can be encountered. Under such a condition, the maximum tractive force will be sacrificed if the same force is given to both wheels on the same axis.

When driving forces are applied to the tire, the tire experiences a condition of skid (slippage) over the road surface. The speed of the vehicle under a free-rolling condition (*i.e.*, no skid) can easily be related to the rotational speed of the wheels $V = r\omega$ as given in the last chapter. Here,

V = vehicle speed

r = rolling radius of the tire

ω = angular speed of the tire

Variation is observed between V and $r\omega$ under a slipping condition, where the angular speed of the tire is nonzero, while the vehicle is not moving ($V = 0$). Some skid is observed between the two cases of free rolling and the complete slippage. If we define the first as the case of 0% skid and the last as a 100% skid condition, Eq. (6.228) can be written to describe skid at any point between the two extreme cases as

$$i = \left(1 - \frac{V}{r\omega}\right) \times 100\% \tag{6.228}$$

As discussed, the coefficient of friction between the tires and the road is a function of many environmental parameters. This coefficient under given environmental conditions is a function of the skid between the tire and the ground, achieving a peak value only when a certain percentage of skid is achieved. The reader is encouraged to review the material on anti-lock braking systems to understand how traction control works.

6.5 TOTAL VEHICLE DYNAMICS

6.5.1 Introduction

In the previous chapters, the dynamics of various vehicle maneuvers was analyzed. These maneuvers or driving conditions included steering, handling, braking, and acceleration. Each of these conditions was discussed mostly as a stand-alone maneuver. In actual driving, one maneuver is experienced at a time for most drivers and/or driving conditions. However, for experienced drivers competing in car racing and for special events, more than one maneuver can be experienced at a time. Steering and braking to avoid an accident is one combination among these possible events. Steering and acceleration (while taking off in the parking lot or going around a curve in car racing) is another possible driving condition.

The purpose of this chapter is to bring to the reader the various driving conditions that can be experienced by a driver when driving a vehicle. In vehicle design and development, several extreme driving conditions are tested, and measurements are taken to express the driver's experience in technical terms.

Each driver has certain priorities in mind for vehicle attributes. For example, a young driver tends to prefer a small car with some power. He or she pays less attention to comfort and/or size of the car. On the other hand, an older driver tends to prefer a more comfortable and bigger vehicle. He or she pays less attention to power. Although ride, comfort, and safety play high on an older driver's list of attributes, style, power, and acceleration are the more important attributes to young drivers. There also is evidence that the customer's attitude toward the vehicle and its attributes is race and gender dependent, in addition to being age dependent.

6.5.1.1 Subjective and Objective Evaluations

The dynamics of a vehicle usually is expressed by the driver in various general subjective terms. Terms and phrases such as "fun to drive," "comfortable," "responsive," "sporty," and others frequently are used to describe the driver's experience with the vehicle. In fact, these phrases and terms can be used as metrics and can be rated subjectively, with 10 being the best to 0 being the worst. Both industry and customer advocacy groups as well as research centers perform what are called "customer clinics," where subjective assessments by customers are gathered. In most cases, these are actual customers, and the data are gathered and analyzed statistically. In other cases, experts who don the customer's hat deliver their assessments.

The engineering challenge is to develop objective metrics that can describe, or even predict, the terms used by the customer. For example, when the customer describes a vehicle as responsive, what does he or she exactly mean? Does he or she mean the vehicle accelerates with the desired or expected acceleration when it is given some throttle? Or does he or she mean that the vehicle responds instantly to steering? The first challenge faced by engineers in vehicle design is to establish the target customer and then to recognize the general subjective vehicle dynamics terms that describe the customer's wants and needs. This task requires a genuine multi-team effort, where marketing works hand in hand with engineering. The second challenge for the engineer is to describe the set of maneuvers and driving conditions that relate to the target customer's experience with the vehicle. In addition, the third challenge for the engineer is to find the actual objective metrics that should be used (*e.g.*, acceleration of the vehicle at a certain throttle, vibration level at a point such as seat track) and the actual values of these metrics that best describe the level desired by the customer. For example, at a certain steering torque and/or position, the customer expects the vehicle to steer at a certain radius. If the vehicle steers at a smaller radius, then it oversteers; if the steering radius is found to be more than the expected, the vehicle understeers. Needless to say, both oversteering and understeering are error states to the customer. In summary, engineers start their vehicle design by defining the target customer, describing his or her wants and needs in subjective terms, defining the testing condition or driving maneuvers that describe these wants and needs, deriving metrics that best relate to these wants and needs, and then determining the values that describe the customer's level of comfort with these metrics at these maneuvers.

6.5.1.2 Target Setting

Design targets can be set based only on the customer's wants and needs. The business case of the vehicle, which precedes the engineering case, is that the vehicle is built to be sold for a profit. Once the customer's wants and needs are established, the question of competitiveness in the marketplace is raised. In particular, questions should be asked about how competitors or potential competitors are addressing these wants and needs, how much success they have achieved, and where they would be by the time the engineered vehicles hit the market. The last question is of great importance because engineers need to know not only the status of the industry today but where it will be a number of years from now, when the product for which the targets are being set hits the market.

Knowing the customer provides the vision of the engineers and where the industry should be going. On the other hand, knowing the industry provides the strategy that should be followed to be competitive or move ahead of the competition. In addition, engineers should know what they can deliver, based on the current state of knowledge and available hardware.

Thus, targets should best describe the customer's wants and needs by taking into consideration the current market status. Targets should be measurable objectively. Subsequently, the hardware design process can be started. Using advanced engineering simulation techniques, the behavior of the vehicle with specific hardware can be assessed. It is understood that not all of the customer's wants and needs can be achieved with a certain vehicle package. At that point, "trade-off" studies are performed. These studies will compromise certain attributes in favor of others, depending on the customer's list of priorities.

Trade-off studies describe the level of attributes that can be achieved when a certain package, design limitations, or constraints are imposed on the engineering problem. It is understood that the intention is to give customers the best value for their wants and needs. Consideration of the target customer and/or group of customers and their priorities in the vehicle selection process is of great importance when conducting these trade-off studies. For example, a trade-off between "responsive" and "robust" may be needed. A vehicle that responds well to the driver's steering torque input may not be "forgiving" for a mother driving while talking to her children on their way to school. Also, trade-off studies among various attributes on one hand and between these attributes and cost in general will be performed. Indeed, most of the engineering task is to optimize the vehicle package such that the best value of these attributes is given to the customer.

6.5.1.3 Vehicle Dynamics Tests and Evaluations

During the target-setting process, knowledge of where the competition is with the vehicle dynamics attribute constitutes one of the basic pillars of vehicle design. To achieve such knowledge, competitors' products must be benchmarked or evaluated. This dictates the establishment of certain maneuvers that simulate the customer's behavior. In addition, engineers need to establish the measurable metrics that describe the customer's attitude toward the vehicle and the range of these metrics where a certain customer's comfort level is achieved.

Benchmarking occurs early in the design or product development process. In addition to benchmarking, engineers need to assess their own design at a later stage of the product development process, particularly when prototype hardware is available. Both of these actions (benchmarking and design assessment) necessitate establishing standardized tests for vehicle dynamics.

The challenge of coming up with these tests and procedures to describe the driver's interaction with the vehicle and its dynamics, as well as the metrics and the metric levels, constitutes the biggest challenge that the engineer faces. Based on such assessment of the customer, target values for certain metrics that can be measured objectively are established. The following is a general description of the many tests and evaluations that can be performed on the vehicle to describe the driver's interaction with it.

6.5.1.3.1 Ride

Ride was discussed in Chapter 6. Among the objectives of ride tests is to find the low-frequency movements of the vehicle. This includes assessment of vehicle movement on various roads, including smooth and rough roads. Assessments include the behavior of the vehicle under discrete events such as impact. These events cause substantial suspension movement. These tests typically describe the magnitude and characteristics of the vehicle motions relative to the road as a result of road inputs. They address the question of how the vehicle body movement relates to the road profile. Does it follow the profile exactly, or does it absorb it? They also address the delay of the vehicle response to road input, the nature of damping (overdamped or critically damped), and the relative motion of the front and rear (pitching) of the vehicle.

Disturbances to the driver and occupants as a result of the vehicle body motions also are evaluated. In particular, abruptness to occupants in the vertical direction, the lateral motions of the head and body, and impact of the suspension motions under various stop conditions are among the many tests and studies made on the vehicle.

Assessment of the vehicle ride includes its vibrations caused by road disturbances and felt by the occupants through the seat, steering wheel, floor pan, controls, and other customer touch points. Road-induced pitch and bounce motions of the body are among the examples. Another example is shake, which refers to the low-frequency (5- to 30-Hz) oscillations of the body, frame, suspension, powertrain, and seat that are felt by the driver or passengers.

6.5.1.3.2 Steering

Steering can be tested under different maneuvers. Some of these maneuvers are summarized as follows:

- **Parking**. Parking refers to very-low-speed driving and parking situations. Among the attributes considered are the steering effort, the uniformity of the effort, and the ability of the steering wheel to return to its original straight-ahead position. In addition, the ability to maneuver the vehicle into and out of restricted spaces is evaluated. This includes the steering wheel angle required for parking maneuvers or "angle demand."
- **Straight-ahead controllability**. This refers to the characteristics of steering around the straight-ahead position to allow the driver confident control of the steering. Tests cover the vehicle response to small steering corrections required to keep the vehicle in a straight path or lane changes. The acceleration in various directions due to small inputs on steering wheels

around the straight-ahead position is evaluated. The symmetry of the vehicle behavior for steering inputs to the right and left also is evaluated. Tests also include control of the vehicle motion around the longitudinal axis, as well as the relationship between small steering wheel torque or angle and the response of the vehicle (torque feedback).

- **Cornering controllability**. These tests are intended to describe driver confidence in the vehicle control while cornering, particularly at moderate and high speeds. The tests include the reaction of the vehicle to steering inputs, both when turning the vehicle into a corner and when keeping the vehicle on an intended curvature. They also include steering to return to the straight-ahead position.
- **Steering disturbances**. These refer to unwanted vehicle reactions, steering wheel motions, or torque feedback that can result from sources other than the driver's steering inputs. Such sources include insufficient steering system performance. An example is torque steer. This relates to deviations from the course during power takeoff, pedal tip-ins, gear changes, or others. The disturbance of these maneuvers to the stability of the straight-ahead is controllability tested. Nibble, or the vibration of the steering wheel around its rotational axis on smooth surfaces due to tire imbalance or force variations, also is studied.

6.5.1.3.3 Handling

Handling includes studies similar to those done under steering. Among these are the following:

- **Straight-ahead stability**. This is the ability of the vehicle to travel in a straight line without driver input. Tests include assessment for the directional stability of the vehicle when disturbed by road or other input. Also included is the sensitivity to side wind, passing, or being passed by other traffic (e.g., large trucks). Vehicle behavior including its roll, yaw, and deviation from straightness as well as corrective actions should be assessed.
- **Cornering stability**. Among these tests performed is whether the vehicle understeers or oversteers under constant throttle and varying steering input. The predictability and comfort level of steering are studied. Tests include studying whether braking on various types of surfaces changes the understeer or oversteer characteristics.
- **Transitional stability**. This is the control of the vehicle during lane changes or similar maneuvers. Stability and controllability of the vehicle under different steering inputs and levels are tested. Steering inputs can be continuous and smooth or abrupt.

6.5.1.3.4 Braking

Three types of braking maneuvers are considered. These are light, moderate, and maximum braking.

Light and moderate braking cover the low-to-moderate deceleration range of braking. Vehicle reaction to operation of the brakes, including the pedal effort, is tested. Measurement of pedal travel and pedal force before noticing deceleration is made. Controllability of the vehicle when under light braking is considered. The effect of the speed at which the braking force is applied on the vehicle behavior also is included. The smoothness and progression of the transition from light to moderate braking also is considered. Maximum braking is the operating point at which wheel lock-up is induced during straight-ahead driving. It should be evaluated on consistent friction surfaces. The time delay between the vehicle response and the pedal force engagement is evaluated. The travel of the braking pedal and its relation with the force applied are tested. Braking disturbances also are evaluated. These refer to any disturbance within or initiated by the braking system of the vehicle.

Finally, parking brake operation is evaluated. This refers to the overall function of the parking brake. This includes the force required to engage and hold the vehicle on a grade, the travel required to engage and hold it, the parking brake location and ease of use, and other attributes.

Other studies also are made on the relationship between the pedal effort and travel, the lateral free play of the brake pedal, pedal ergonomics, and others.

6.5.1.3.5 Performance

Performance relates to the powertrain and acceleration characteristics (basic powertrain power and torque, including gearing). Performance feel is the customer's perception of performance. In addition to power and acceleration, it includes the effects of shift response and ratio selection, sound quality, traction control, transmission, all-wheel-drive (AWD) performance, and accelerator control characteristics.

- **Standing start.** This includes part throttle and wide open throttle. Performance feel is tested from a standing start throughout the driving range into top gear.
- **Rolling start**. This includes testing or rating performance feel from a rolling 2- to 8-km/hr (1- to 5-mph) start, at steady part throttle (or wide open throttle), throughout the driving range into top gear. The time or distance to accelerate to various speeds also is considered, with an automatic transmission in overdrive and a manual transmission in low gear.

Other evaluations include city performance, traveling in the range of 30 to 80 km/hr (20 to 50 mph), and highway performance with speeds in the range of 65 to 110 km/hr (40 to 70 mph). Tests also can be performed on drivability and engine performance. Drivability relates to the calibration of the engine and the quality of the fuel charging system. It should be evaluated at low and high altitudes and during cold and hot operation. Tests include starting and stability of the engine speed at both idle and while driving. In addition, the performance of the transmission can be evaluated. Tests on the automatic transmission include shift quality. This is the customer's perception of how quickly, smoothly, quietly, consistently, and predictably the transmission events occur. It applies to torque converter clutch events as well as to gear ratio changes. Generally, the quicker, smoother, quieter, more consistent, and more predictable the transmission event is to the customer, the better. Evaluation of an automatic transmission includes testing up shifts, downshifts with or without braking, and shift timing (*i.e.*, the speed and vehicle conditions where shifting occurs). Assessment also includes the ergonomics of shift and static engagement. Static engagement refers to moving the gear from neutral or park conditions to drive or reverse conditions.

Shift feel of a manual or automatic transmission includes its ergonomics (*i.e.*, the location of the gear stick or lever with respect to the driver and the pattern or spacing between the gear position, shape, size, and other factors). Other evaluations include testing for the clutch pedal forces through the range of operation, shifting characteristics, smoothness of engagement and disengagement, the point of engagement (*i.e.*, the friction point), the motion required to engage and disengage the clutch, and others. Evaluations also can be made on the ease of locking or unlocking of the four-by-four (4×4) or all-wheel-drive system, smoothness and noise and vibration of the driveline systems, effectiveness on various types of surfaces (especially those with low coefficients of friction), and evaluation in off-road driving conditions. Additional evaluations include trailer tow performance.

6.5.2 Steering and Braking

We have noticed (see Chapter 6) that load transfers from the inside wheels to the outside wheels when the vehicle undergoes cornering or steering. On the other hand, we also have noticed (see Chapter 7) that load transfers from the rear to the front axles when braking. In many circumstances, braking happens while the vehicle is under steering conditions (not a straight-ahead condition). This will add significant complexity to the mechanics of the vehicle and load transfer characteristics. The issue of braking and steering was discussed as early as 1974 by Limpert *et al.* (1974) and in a publication by the U.S. Department of the Army in 1976. We will consider first the braking of a vehicle on a horizontal road (*i.e.*, no grade) and ignore the other forces (e.g., aerodynamic, powertrain resistance).

6.5.2.1 Simple Braking and Steering on a Horizontal Road

The equations derived in Chapter 6 that describe the load transfer from the inside to the outside wheels are repeated here

$$F_{z1i} = \frac{F_{z1}}{2} - \Delta F_{z1}$$

$$F_{z1o} = \frac{F_{z1}}{2} + \Delta F_{z1} \tag{6.229}$$

$$\Delta F_{z1} = \frac{F_{y1}h_r + K_\varphi \varphi}{t}$$

where, as was found in Section 6.2.

$$F_{y1} = F_{z1}\frac{V^2}{2Rg} \tag{6.230}$$

In these equations, F_{z1} is the total vertical force on the front axle, F_{y1} is the total lateral force on the front axle, R is the radius of the turn, $K\varphi$ is the roll stiffness, φ is the roll angle, t is the wheel tread or track width, h_r is the roll center height, and V is the vehicle speed.

Example 6.29: *Consider a midsize sedan that weighs 14,236.3 N (3200 lb) under a certain load. The wheelbase of the vehicle is 2.74 m (9 ft, or 108 in.), and the center of gravity is 1.067 m (3.5 ft) from the front axle and 0.508 m (20 in.) above the ground. The vehicle speed is 96.56 km/hr (60 mph), the turning radius is 182.88 m (600 ft), the wheel tread is 1.219 m (4 ft), K*φ *= 10157 Nm (90,000 in.-lb), the roll center height is 0.457 m (18 in.), and the roll angle is 5°. Determine the load on each wheel (*F_{z1i}, F_{z1o}, F_{z2i}, F_{z2o}*) if the vehicle is under no acceleration and is decelerating at 3.048 m/s*^{2} *(10 ft/sec*^{2}*). Given in the problem are the following:*

W = 3200 lb (Thus, M = 3200/32.2 = 99.38 slugs)

h = 20 in.

c_1 *= 42 in. (thus,* c_2 *= 108 – 42 = 66 in.)*

V = 60 mph (88 ft/sec)

R = 600 ft

t = 4 ft (48 in.)

K_φ *= 90,000 in.-lb*

h_r *= 18 in.*

φ *= 5° (0.08727 radians)*

Solution: Apply Eq. (7.25) to obtain the loading under no acceleration (*i.e.*, $a = 0$) as

$$F_{z1} = mg\left(\frac{c_2}{c_1+c_2}\right) - \frac{h}{c_1+c_2}ma = 99.38\times 32.2\left(\frac{66}{108}\right) - \frac{20}{108}\times 99.38\times 0 = 1955.56 \text{ lb}$$

$$F_{y1} = F_{z1}\frac{V^2}{2Rg} = 1955.56\times\frac{(88)^2}{2\times 600\times 32.2} = 391.92 \text{ lb}$$

$$\Delta F_{z1} = \frac{F_{y1}h_r + K_\varphi\varphi}{t} = \frac{391.92\times 18 + 90000\times 0.08727}{48} = 310.60 \text{ lb}$$

$$F_{z1i} = \frac{F_{z1}}{2} - \Delta F_{z1} = \frac{1955.56}{2} - 310.60 = 667.18 \text{ lb}$$

$$F_{z1o} = \frac{F_{z1}}{2} + \Delta F_{z1} = \frac{1955.56}{2} + 310.60 = 1288.38 \text{ lb}$$

Similarly for the rear axle

$$F_{z2} = mg\left(\frac{c_1}{c_1+c_2}\right) + \frac{h}{c_1+c_2}ma = 99.38\times 32.2\left(\frac{42}{108}\right) + \frac{20}{108}\times 99.38\times 0 = 1244.44 \text{ lb}$$

$$F_{y2} = F_{z2}\frac{V^2}{2Rg} = 1244.44\times\frac{(88)^2}{2\times 600\times 32.2} = 249.40 \text{ lb}$$

$$\Delta F_{z2} = \frac{F_{y2}h_r + K_\varphi\varphi}{t} = \frac{249.40\times 18 + 90000\times 0.08727}{48} = 257.16 \text{ lb}$$

$$F_{z2i} = \frac{F_{z2}}{2} - \Delta F_{z2} = \frac{1244.44}{2} - 257.16 = 365.06 \text{ lb}$$

$$F_{z2o} = \frac{F_{z2}}{2} + \Delta F_{z2} = \frac{1244.44}{2} + 257.16 = 879.38 \text{ lb}$$

For $a = -10$ ft/sec^2

$$F_{z1} = 99.38\times 32.2\left(\frac{66}{108}\right) - \frac{20}{108}\times 99.38\times(-10) = 2139.59 \text{ lb}$$

$$F_{y1} = F_{z1}\frac{V^2}{2Rg} = 2139.59\times\frac{(88)^2}{2\times 600\times 32.2} = 428.80 \text{ lb}$$

$$\Delta F_{z1} = \frac{F_{y1}h_r + K_\varphi\varphi}{t} = \frac{428.80\times 18 + 90000\times 0.08727}{48} = 324.43 \text{ lb}$$

$$F_{z1i} = \frac{F_{z1}}{2} - \Delta F_{z1} = \frac{2139.59}{2} - 324.43 = 745.36 \text{ lb}$$

$$F_{z1o} = \frac{F_{z1}}{2} + \Delta F_{z1} = \frac{2139.59}{2} + 324.43 = 1394.22 \text{ lb}$$

$$F_{z2} = 99.38\times 32.2\left(\frac{42}{108}\right) + \frac{20}{108}\times 99.38\times(-10) = 1060.41 \text{ lb}$$

$$F_{y2} = F_{z2}\frac{V^2}{2Rg} = 1060.41 \times \frac{(88)^2}{2 \times 600 \times 32.2} = 212.52 \text{ lb}$$

$$\Delta F_{z2} = \frac{F_{y2}h_r + K_\varphi \varphi}{t} = \frac{212.52 \times 18 + 90000 \times 0.08727}{48} = 243.32 \text{ lb}$$

$$F_{z2i} = \frac{F_{z2}}{2} - \Delta F_{z2} = \frac{1060.41}{2} - 243.32 = 286.88 \text{ lb}$$

$$F_{z2o} = \frac{F_{z2}}{2} + \Delta F_{z2} = \frac{1060.41}{2} + 243.32 = 773.52 \text{ lb}$$

Note that the total load on the axles remains at 3200 lb, which is the weight of the vehicle. However, the load transferred from the rear axle to the front axle as the vehicle was decelerating and from the inner wheels to the outer wheels. Figure 6.28(*a*) shows the load distribution as a function of deceleration. Figure 6.28(*b*) shows the load distribution on each wheel as a function of velocity. Note that at high speeds, the load reduces significantly at the inner wheels. Figure 6.28(*c*) shows the wheel load as a function of radius of the turn. Note here that at a low radius of 100 ft, the front and rear inner wheels have negative forces, which is not possible. At these drifting conditions, the forces at the wheels essentially become zero, and the possibility of rollover becomes real.

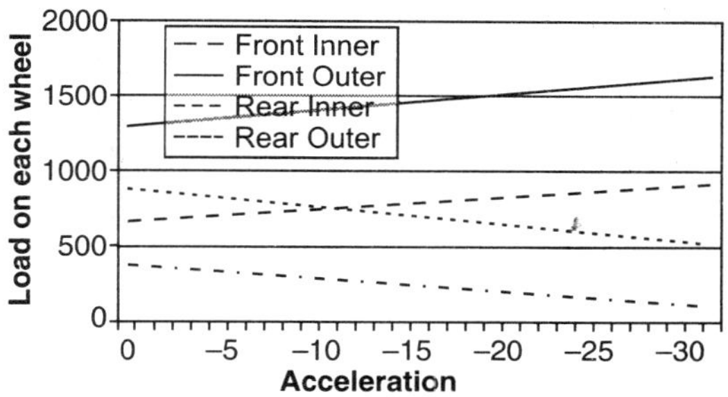

Fig. E6.28 (*a*) Load distribution under braking and cornering as a function of deceleration.

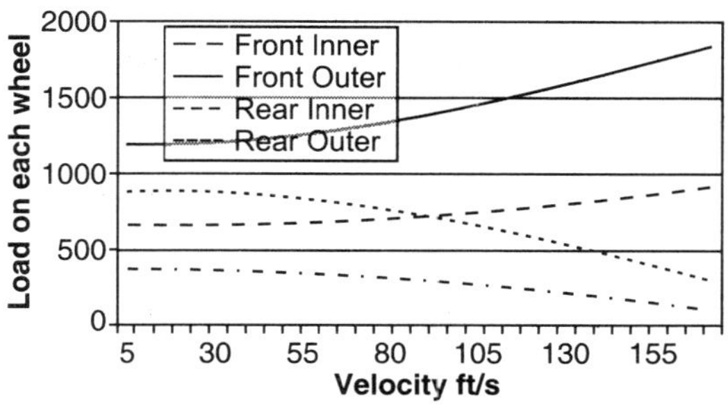

Fig. E6.28 (*b*) Load distribution under braking and cornering as a function of velocity.

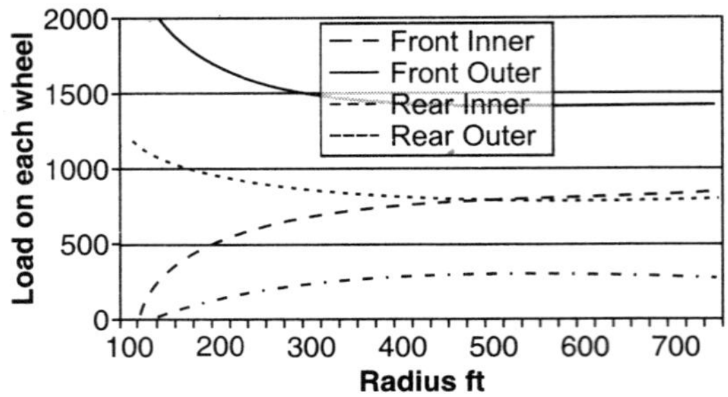

Fig. E6.28 (*c*) Load distribution under braking and cornering as a function of radius of the turn.

6.5.2.2 Optimal Braking Performance Under Steering

Braking performance under single axle braking conditions is not discussed here. We will consider braking with both axles, whereas single axle braking conditions are merely special cases. As mentioned, the brake engineer designs for the maximum braking force that can be obtained from the braking system at hand. Consideration is made for front lock-up to occur first under various loading and environmental conditions. Let us assume that, based on certain emergency conditions, the brake designer chooses to distribute the braking load such that a certain ratio (K_f) of the total braking force goes to the front brakes. The remainder that goes to the rear brakes is ($1 - K_f$) of the total braking force. Note here that in conventional braking systems, the force distributed to the front axle does not take into consideration that the wheels of the front axle will have different loads when steering or cornering. On the other hand, anti-lock braking systems (ABS) distribute the braking force according to the wheel load.

As noted in Chapter 7, the braking force of the front and rear axles can be written in terms of each other as

$$F_{x1} = \frac{K_f}{1-K_f} F_{x2}$$

$$F_{x2} = \frac{1-K_f}{K_f} F_{x1} \tag{6.231}$$

Note that front axle braking (single axle) is obtained when $K_f = 1$, and rear axle braking is obtained when $K_f = 0$. The total weight under no steering is distributed as

$$F_{z1} = \frac{1}{c_1 + c_2}\left(Mgc_2 \cos\theta - h_a R_a - hMa - hMg \sin\theta\right)$$

$$F_{z2} = \frac{1}{c_1 + c_2}\left(Mgc_1 \cos\theta + h_a R_a + hMa + hMg \sin\theta\right) \tag{6.232}$$

Manipulating equations 6.229 and 6.230

$$F_{z1i} = \frac{F_{z1}}{2} - \Delta F_{z1} = \frac{F_{z1}}{2} - \frac{\left(F_{z1}\dfrac{V^2}{2Rg}\right)h_r + K_\varphi \varphi}{t} = F_{z1}\left(\frac{1}{2} - \frac{V^2 h_r}{2Rgt}\right) - \frac{K_\varphi \varphi}{t}$$

$$= \left\{\frac{1}{c_1 + c_2}\left(Mgc_2 \cos\theta - h_a R_a - hMa - hMg \sin\theta\right)\right\}\left(\frac{1}{2} - \frac{V^2 h_r}{2Rgt}\right) - \frac{K_\varphi \varphi}{t} \tag{6.233}$$

$$F_{z1o} = \left\{ \frac{1}{c_1 + c_2} \left(Mgc_2 \cos\theta - h_a R_a - hMa - hMg \sin\theta \right) \right\} \left(\frac{1}{2} + \frac{V^2 h_r}{2Rgt} \right) + \frac{K_\varphi \varphi}{t}$$

Similar derivation can be made for the rear axle.

$$F_{z2i} = \left\{ \frac{1}{c_1 + c_2} \left(Mgc_1 \cos\theta + h_a R_a + hMa + hMg \sin\theta \right) \right\} \left(\frac{1}{2} - \frac{V^2 h_r}{2Rgt} \right) - \frac{K_\phi \phi}{t}$$

$$F_{z2o} = \left\{ \frac{1}{c_1 + c_2} \left(Mgc_1 \cos\theta + h_a R_a + hMa + hMg \sin\theta \right) \right\} \left(\frac{1}{2} + \frac{V^2 h_r}{2Rgt} \right) + \frac{K_\phi \phi}{t}$$

Note also from Chapter 7 that by summing the forces in the direction of motion, we obtain

$$F_{x1} + F_{x2} - R_a - R_r - Mg \sin\theta = Ma$$

or from Eq. (6.231)

$$\left(1 + \frac{K_f}{1 - K_f} \right) F_{x2} - R_a - R_r - Mg \sin\theta = Ma$$

$$\left(1 + \frac{1 - K_f}{K_f} \right) F_{x1} - R_a - R_r - Mg \sin\theta = Ma \tag{6.234}$$

in the above equation is longitudinal acceleration, to be distinguished from lateral acceleration that results from cornering. Observe the following simplifications

$$1 + \frac{K_f}{1 - K_f} = \frac{1 - K_f}{1 - K_f} + \frac{K_f}{1 - K_f} = \frac{1}{1 - K_f}$$

$$1 + \frac{1 - K_f}{K_f} = \frac{K_f}{K_f} + \frac{1 - K_f}{K_f} = \frac{1}{K_f} \tag{6.235}$$

6.5.2.2.1 Front Lock-Up

Under front lock-up, the maximum force at the front axle will be equal to twice that of the wheel where lock-up is introduced for conventional brakes. The inner wheel will always have less force; thus, the total axial force at the front axle will be $F_{x1} = -2\mu F_{z1i}$. Substituting this into the preceding equations yields

$$-\frac{2\mu}{K_f} \left(\left\{ \frac{1}{c_1 + c_2} \left(Mgc_2 \cos\theta - h_a R_a - hMa_F - hMg \sin\theta \right) \right\} \left(\frac{1}{2} - \frac{V^2 h_r}{2Rgt} \right) - \frac{K_\varphi \varphi}{t} \right)$$

$$-R_a - R_r - Mg \sin\theta = Ma_F \tag{6.236}$$

Rearranging for a_F

$$-\frac{2\mu}{K_f} \left(\left\{ \frac{1}{c_1 + c_2} \left(Mgc_2 \cos\theta - h_a R_a - hMg \sin\theta \right) \right\} \left(\frac{1}{2} - \frac{V^2 h_r}{2Rgt} \right) - \frac{K_\varphi \varphi}{t} \right)$$

$$-R_a - R_r - Mg \sin\theta = Ma_F \left[1 - \frac{\mu h}{(c_1 + c_2) K_f} \left(1 - \frac{V^2 h_r}{Rgt} \right) \right] \tag{6.237}$$

or simply

$$a_F = \frac{-\dfrac{2\mu}{K_f}\left(\left\{\dfrac{1}{c_1+c_2}\left(Mgc_2\cos\theta - h_aR_a - hMg\sin\theta\right)\right\}\left(\dfrac{1}{2}-\dfrac{V^2h_r}{2Rgt}\right)-\dfrac{K_\varphi\varphi}{t}\right)-R_a-R_r-Mg\sin\theta}{M\left[1-\dfrac{\mu h}{K_f\left(c_1+c_2\right)}\left(1-\dfrac{V^2h_r}{Rgt}\right)\right]} \tag{6.238}$$

Note that the acceleration a has a subscript F to indicate that the acceleration obtained by using Eq. (6.238) is valid only when the front tires are locked up. Also note the degree of complexity of this equation. Ignoring aerodynamic drag, grade, and rolling resistance, the deceleration is

$$a_R = \frac{-\dfrac{2\mu}{1-K_f}\left(\left\{\dfrac{Mgc_2}{c_1+c_2}\right\}\left(\dfrac{1}{2}-\dfrac{V^2h_r}{2Rgt}\right)-\dfrac{K_\phi\phi}{t}\right)}{M\left[1+\dfrac{\mu h}{\left(1-K_f\right)\left(c_1+c_2\right)}\left(1-\dfrac{V^2h_r}{Rgt}\right)\right]} \tag{6.239}$$

Note that this expression may yield a small denominator of zero. This will yield a high number for a_F, which will mean that rear lock-up must happen first. It is worth comparing this equation with Eq. (7.43).

6.5.2.2.2 Rear Lock-Up

Under rear lock-up, the maximum force at the rear axle will equal twice that of the wheel where lock-up is introduced for conventional brakes. The inner wheel will always have less force; thus, the total axial force at the rear axle will be $F_{x2} = -2\mu F_{z2i}$. Substituting this into the Eq. (6.234) yields

$$-2\mu\left(1+\frac{K_f}{1-K_f}\right)\left(\frac{1}{c_1+c_2}\left(Mgc_1\cos\theta + h_aR_a + hMa_R + hMg\sin\theta\right)\right)\left(\frac{1}{2}-\frac{V^2h_r}{2Rgt}\right)-\frac{K_\varphi\varphi}{t}$$
$$-R_a - R_r - Mg\sin\theta = Ma_R \tag{6.240}$$

Rearrange for a_R.

$$-\frac{2\mu}{1-K_f}\left(\left\{\frac{1}{c_1+c_2}\left(Mgc_2\cos\theta + h_aR_a + hMg\sin\theta\right)\right\}\left(\frac{1}{2}-\frac{V^2h_r}{2Rgt}\right)-\frac{K_\varphi\varphi}{t}\right)$$
$$-R_a - R_r - Mg\sin\theta = Ma_R\left[1+\frac{\mu h}{\left(c_1+c_2\right)\left(1-K_f\right)}\left(1-\frac{V^2h_r}{Rgt}\right)\right] \tag{6.241}$$

Simplify

$$a_R = \frac{-\dfrac{2\mu}{1-K_f}\left(\left\{\dfrac{1}{c_1+c_2}\left(Mgc_2\cos\theta + h_aR_a + hMg\sin\theta\right)\right\}\left(\dfrac{1}{2}-\dfrac{V^2h_r}{2Rgt}\right)-\dfrac{K_\varphi\varphi}{t}\right)-R_a-R_r-Mg\sin\theta}{M\left[1+\dfrac{\mu h}{\left(1-K_f\right)\left(c_1+c_2\right)}\left(1-\dfrac{V^2h_r}{Rgt}\right)\right]} \tag{6.242}$$

Ignoring aerodynamic drag, grade, and rolling resistance, the deceleration is

$$a_R = \frac{-\dfrac{2\mu}{1-K_f}\left(\left\{\dfrac{Mgc_2}{c_1+c_2}\right\}\left(\dfrac{1}{2}-\dfrac{V^2h_r}{2Rgt}\right)-\dfrac{K_\varphi\varphi}{t}\right)}{M\left[1+\dfrac{\mu h}{(1-K_f)(c_1+c_2)}\left(1-\dfrac{V^2h_r}{Rgt}\right)\right]} \tag{6.243}$$

It is worthwhile to compare Eq. (6.243) to Eq. (7.48) in Chapter 7.

The front tires will lock up first if $|a_F| < |a_R|$; otherwise, the rear tires will lock up first.

Example 6.30: *A 21,351.46-N (4800-lb) passenger car has a wheelbase of 2846.80 mm (112 in.) and a center of gravity 1270.00 mm (50 in.) behind the front axle and 508.00 mm (20 in.) above the ground. The braking effort distribution gives the front axle 60% of the total braking force. The car is moving on a horizontal plane. The car is going around a turning radius of 182.88 m (600 ft) at a speed of 96.56 km/hr (60 mph). The wheel tread is 1.219 m (4 ft), Kφ = 10,157 Nm (90,000 in.-lb), the roll center height is 0.457 m (18 in.), and the roll angle is 5°. Determine which tire will lock up first if the car is moving on a surface with first a 0.75 coefficient of friction and second a 0.25 coefficient of friction. Ignore drag and rolling resistance.*

Solution:

$M = 4800$ lbs $= 149.07$ slugs

$c_1 + c_2 = 112$ in

$c_1 = 50$ in,

$c_2 = 62$ in,

$K_f = 0.6$

$h = 20$

$\theta = 0$

$R_a = 0$

$R_r = 0$

$V = 60$ mph (= 88 ft/sec) = 1056 in/s

$R = 600$ ft = 7200 in

$t =$ 4 ft (= 48 in)

$K_\phi = 90{,}000$ in-lb

$h_r = 18$ in

$\phi = 5$ degrees (= 0.08727 radians)

$g = 32.2$ ft/s^2 = 386.4 in/s^2

First surface: $\mu = 0.75$

Acceleration under front-lock-up is

$$a_F = \frac{-\dfrac{2\mu}{K_f}\left[\left(\dfrac{Mgc_2}{c_1+c_2}\right)\left(\dfrac{1}{2}-\dfrac{V^2h_r}{2Rgt}\right)-\dfrac{K_\phi\phi}{t}\right]}{M\left[1-\dfrac{\mu h}{K_f(c_1+c_2)}\left(1-\dfrac{V^2h_r}{Rgt}\right)\right]}$$

$$a_F = \frac{-\dfrac{2\times 0.75}{0.6}\left[\left(\dfrac{149.07\times 386.4\times 62}{112}\right)\left(\dfrac{1}{2}-\dfrac{(1056)^2\times 18}{2\times 7200\times 386.4\times 48}\right)-\dfrac{90000\times 0.08727}{48}\right]}{149.07\left[1-\dfrac{0.75\times 20}{0.6\times 112}\left(1-\dfrac{(1056)^2\times 18}{7200\times 386.4\times 48}\right)\right]}$$

$$= -0.7168\ g$$

Acceleration under rear lock-up

$$a_R = \frac{-\dfrac{2\mu}{1-K_f}\left(\left\{\dfrac{Mgc_2}{c_1+c_2}\right\}\left(\dfrac{1}{2}-\dfrac{V^2h_r}{2Rgt}\right)-\dfrac{K_\phi\phi}{t}\right)}{M\left[1+\dfrac{\mu h}{(1-K_f)(c_1+c_2)}\left(1-\dfrac{V^2h_r}{Rgt}\right)\right]}$$

$$a_R = \frac{-\dfrac{2\times 0.75}{1-0.6}\left[\left(\dfrac{149.07\times 386.4\times 62}{112}\right)\left(\dfrac{1}{2}-\dfrac{(1056)^2\times 18}{2\times 7200\times 386.4\times 48}\right)-\dfrac{90000\times 0.08727}{48}\right]}{149.07\left[1+\dfrac{0.75\times 20}{(1-0.6)\times 112}\left(1-\dfrac{(1056)^2\times 18}{7200\times 386.4\times 48}\right)\right]}$$

$$= -0.6783g$$

Because $|a_F| > |a_R|$ the rear tires will lock up first.

Second surface: $\mu = \mathbf{0.25}$

Acceleration under front-lock-up is

$$a_F = \frac{-\dfrac{2\times 0.25}{0.6}\left[\left(\dfrac{149.07\times 386.4\times 62}{112}\right)\left(\dfrac{1}{2}-\dfrac{(1056)^2\times 18}{2\times 7200\times 386.4\times 48}\right)-\dfrac{90000\times 0.08727}{48}\right]}{149.07\left[1-\dfrac{0.25\times 20}{0.6\times 112}\left(1-\dfrac{(1056)^2\times 18}{7200\times 386.4\times 48}\right)\right]}$$

$$= -0.2067g$$

Acceleration under rear lock-up

$$a_R = \frac{-\dfrac{2\times 0.25}{1-0.6}\left[\left(\dfrac{149.07\times 386.4\times 62}{112}\right)\left(\dfrac{1}{2}-\dfrac{(1056)^2\times 18}{2\times 7200\times 386.4\times 48}\right)-\dfrac{90000\times 0.08727}{48}\right]}{149.07\left[1+\dfrac{0.25\times 20}{(1-0.6)\times 112}\left(1-\dfrac{(1056)^2\times 18}{7200\times 386.4\times 48}\right)\right]}$$

$$= -0.2653\ g$$

Because $|a_F| < |a_R|$, the front tires will lock up first.

Figure E6.30(a) shows how the acceleration under front and rear lock-up changes with the coefficient of friction for the parameters described in Example 6.30. It clearly shows that rear lock-up is quite possible at surfaces with high coefficients of friction. Fig. E6.30(b) shows that if the braking effort distribution is changed to 0.7, front brake lock-up will occur for all types of surfaces. However, this compromises the maximum braking force. The K_f factor cannot be altered straightforwardly to account for the different coefficients of friction that various surfaces may offer in conventional brakes. An advanced anti-lock braking system provides the maximum brake force per wheel without lock-up.

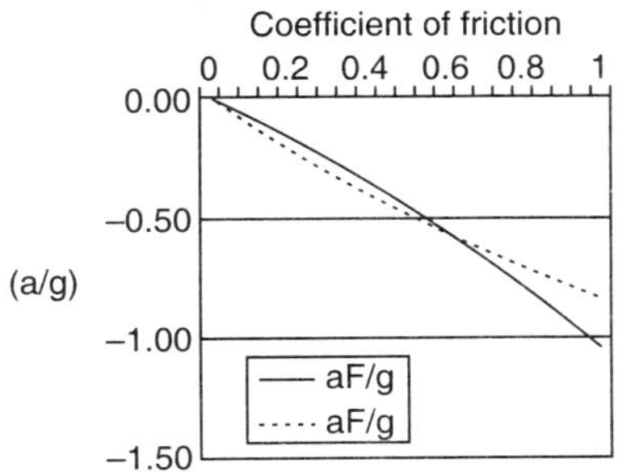

Fig. E6.30 (*a*) The values of a_F and a_R for different friction for the parameters of Example E6.29, $K_f = 0.6$

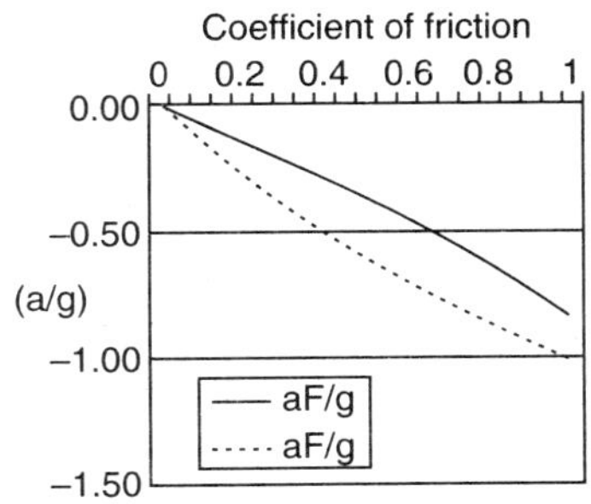

Fig. E6.30 (*b*) The values of a_F and a_R for different friction for the parameters of Example E9.29, $K_f = 0.7$

6.5.3 Steering and Acceleration

Load transfers from the inside wheels to the outside wheels when the vehicle undergoes cornering or steering, and it transfers from the front to the rear axles when the vehicle accelerates. Acceleration can happen while the vehicle is under steering conditions. First we will consider acceleration of a vehicle on a horizontal road (*i.e.*, no grade) and ignore the other forces (e.g., aerodynamic, powertrain resistance).

6.5.3.1 Simple Acceleration and Steering on a Horizontal Road

If we use Eq. (6.170) from Section 6.4 with Eq. (6.231), we can obtain the distribution of the weight of the vehicle under steering and acceleration. Consider the parameters of Example 6.29, and change the deceleration to an acceleration of 2 ft/s^2 (0.6096 m/s^2). Figure 6.69 shows the load distribution as a function of acceleration. Figure 6.70 shows the load distribution on each wheel as a function of velocity. Note that at high speeds, the load reduces significantly at the inner wheels. Figure 6.71 shows the wheel load as a function of radius of the turn. Note here that at a low radius of 100 ft, the front and rear inner wheels have negative forces, which is not possible. At these drifting conditions, the forces at the wheels essentially become zero, and the possibility of rollover becomes real.

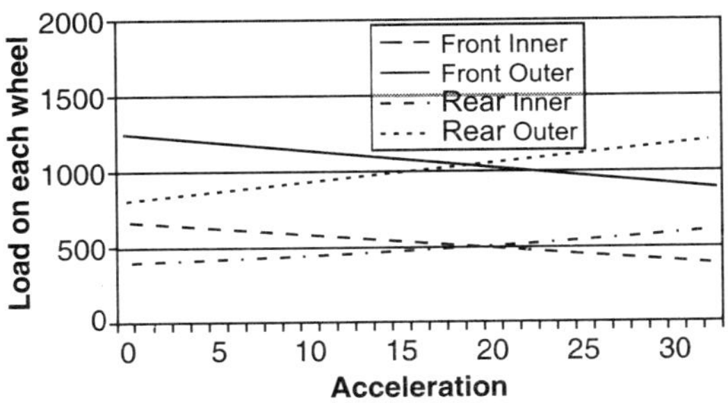

Fig. 6.69 Load distribution under acceleration and cornering as a function of acceleration.

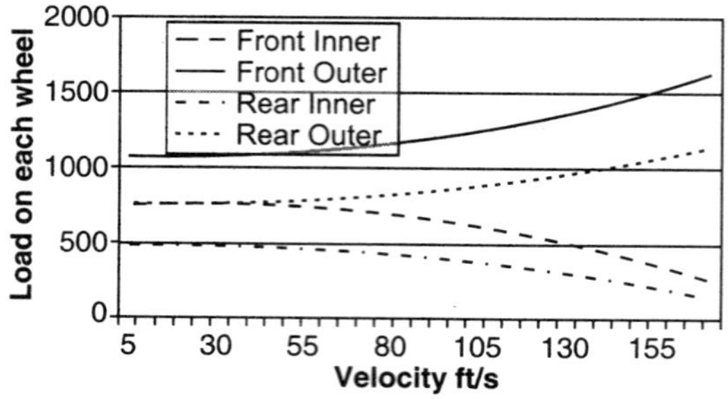

Fig. 6.70 Load distribution under acceleration and cornering as a function of velocity.

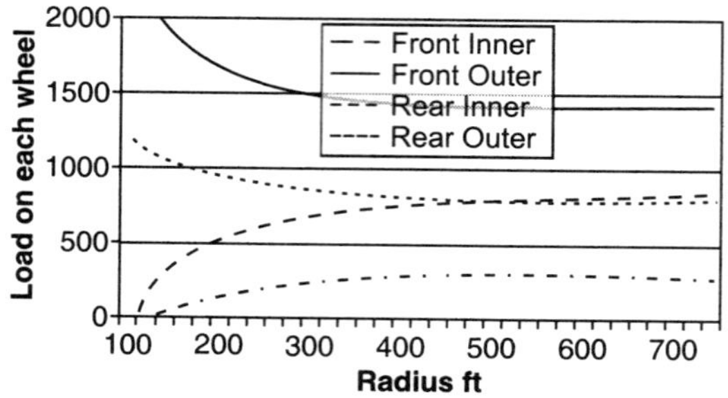

Fig. 6.71 Load distribution under acceleration and cornering as a function of radius.

6.5.3.2 Optimal Acceleration Performance Under Steering

Acceleration performance under a single axle drive condition (front-wheel drive or rear-wheel drive) is not discussed here. We will consider the cases of four-by-four or all-wheel-drive systems because the other front-wheel-drive and rear-wheel-drive conditions are special cases. The drive system is to be designed such that it delivers the maximum tractive forces. Let us distribute the tractive torque such that a certain ratio K_f of the total torque goes to the front axle, and the remainder goes to the rear axle. Note here that in conventional drive systems, the torque distributed to the front axle does not take into consideration the different loads on the right and left wheels when steering or cornering. Note that front-wheel drive (single axle) is obtained when $K_f = 1$, and rear-wheel drive is obtained when $K_f = 0$. The total weight is distributed to the front and rear axle according to Eq. (6.233). The wheel loads are distributed according to Eq. (6.233). Also note that by summing the forces in the direction of motion, we obtain Eq. (6.233).

6.5.3.2.1 Front Skid

The maximum force at the front axle will be equal to twice that of the wheel where it is first felt for conventional traction systems. The inner wheel will always have less force; thus, the total axial force at the front axle will be $F_{x1} = 2\mu F_{z1i}$. Substituting this into Eq. (6.233) and using Eqs. (6.232) and (6.234) yields

$$\frac{2\mu}{K_f}\left(\left\{\frac{1}{c_1+c_2}(Mgc_2\cos\theta - h_aR_a - hMa_F - hMg\sin\theta)\right\}\left(\frac{1}{2}-\frac{V^2h_r}{2Rgt}\right)-\frac{K_\varphi\varphi}{t}\right)$$
$$-R_a - R_r - Mg\sin\theta = Ma_F \tag{6.244}$$

Rearranging for a_F

$$\frac{2\mu}{K_f}\left(\left\{\frac{1}{c_1+c_2}\left(Mgc_2\cos\theta - h_a R_a - hMg\sin\theta\right)\right\}\left(\frac{1}{2}-\frac{V^2 h_r}{2Rgt}\right)-\frac{K_\varphi \varphi}{t}\right)$$

$$-R_a - R_r - Mg\sin\theta = Ma_F\left[1+\frac{\mu h}{(c_1+c_2)K_f}\left(1-\frac{V^2 h_r}{Rgt}\right)\right] \qquad (6.245)$$

or simply

$$a_F = \frac{\frac{2\mu}{K_f}\left(\left\{\frac{1}{c_1+c_2}\left(Mgc_2\cos\theta - h_a R_a - hMg\sin\theta\right)\right\}\left(\frac{1}{2}-\frac{V^2 h_r}{2Rgt}\right)-\frac{K_\phi \phi}{t}\right)-R_a - R_r - Mg\sin\theta}{M\left[1+\frac{\mu h}{K_f(c_1+c_2)}\left(1-\frac{V^2 h_r}{Rgt}\right)\right]}$$

(6.246)

The subscript F indicates that the acceleration obtained is valid when the front tires skid. Ignoring aerodynamic drag, grade, and rolling resistance, the deceleration is

$$a_F = \frac{\frac{2\mu}{K_f}\left[\left(\frac{Mgc_2}{c_1+c_2}\right)\left(\frac{1}{2}-\frac{V^2 h_r}{2Rgt}\right)-\frac{K_\varphi \varphi}{t}\right]}{M\left[1+\frac{\mu h}{K_f(c_1+c_2)}\left(1-\frac{V^2 h_r}{Rgt}\right)\right]} \qquad (6.247)$$

6.5.3.2.2 Rear Skid

The maximum force at the rear axle will be equal to twice that of the wheel where skid occurs first for conventional drive systems. Because the inner wheel has less force, the total axial force at the rear axle will be $F_{x2} = 2\mu F_{z2i}$. Substituting this into the Eq. (9.6) yields

$$2\mu\left(1+\frac{K_f}{1-K_f}\right)\left(\frac{1}{c_1+c_2}\left(Mgc_1\cos\theta + h_a R_a + hMa_R + hMg\sin\theta\right)\right)\left(\frac{1}{2}-\frac{V^2 h_r}{2Rgt}\right)-\frac{K_\varphi \varphi}{t}$$

$$-R_a - R_r - Mg\sin\theta = Ma_R \qquad (6.248)$$

Rearranging for a_R,

$$\frac{2\mu}{1-K_f}\left(\left\{\frac{1}{c_1+c_2}\left(Mgc_2\cos\theta + h_a R_a + hMg\sin\theta\right)\right\}\left(\frac{1}{2}-\frac{V^2 h_r}{2Rgt}\right)-\frac{K_\varphi \varphi}{t}\right)$$

$$-R_a - R_r - Mg\sin\theta = Ma_R\left[1-\frac{\mu h}{(c_1+c_2)(1-K_f)}\left(1-\frac{V^2 h_r}{Rgt}\right)\right] \qquad (6.249)$$

Simplify as

$$a_R = \frac{\dfrac{2\mu}{1-K_f}\left(\left\{\dfrac{1}{c_1+c_2}(Mgc_2\cos\theta + h_aR_a + hMg\sin\theta)\right\}\left(\dfrac{1}{2}-\dfrac{V^2h_r}{2Rgt}\right)-\dfrac{K_\varphi\varphi}{t}\right) - R_a - R_r - Mg\sin\theta}{M\left[1-\dfrac{\mu h}{(1-K_f)(c_1+c_2)}\left(1-\dfrac{V^2h_r}{Rgt}\right)\right]} \tag{6.250}$$

Ignoring aerodynamic drag, grade, and rolling resistance, the acceleration is

$$a_R = \frac{\dfrac{2\mu}{1-K_f}\left(\left\{\dfrac{Mgc_2}{c_1+c_2}\right\}\left(\dfrac{1}{2}-\dfrac{V^2h_r}{2Rgt}\right)-\dfrac{K_\varphi\varphi}{t}\right)}{M\left[1-\dfrac{\mu h}{(1-K_f)(c_1+c_2)}\left(1-\dfrac{V^2h_r}{Rgt}\right)\right]} \tag{6.251}$$

The front tires will skid first if $|a_F| < |a_R|$; otherwise, the rear tires will skid first.

Example 6.31: *Consider the vehicle in Example E6.30. The torque distribution gives the front axle 60% of the total torque delivered to the wheels. Determine which tire will skid first if the car is moving on a surface with first a 0.75 coefficient of friction and second a 0.25 coefficient of friction.*

Solution:

$M = 4800$ lbs $= 149.07$ slugs

$c_1 + c_2 = 112$ in

$c_1 = 50$ in

$c_2 = 62$ in

$K_f = 0.6$

$h = 20$ in

$\theta = 0$

$R_a = 0$

$R_r = 0$

$V = 60$ mph (= 88 ft/sec) = 1056 in/s

$R = 600$ ft = 7200 in

$t = 4$ ft (= 48 in)

$K_\phi = 90{,}000$ in-lb

$h_r = 18$ in

$\phi = 5$ degrees (= 0.08727 radians)

$g = 32.2$ ft/s^2 = 386.4 in/s^2

First surface: μ = 0.75

Acceleration under front skid is

$$a_F = \frac{\frac{2\times 0.75}{0.6}\left[\left(\frac{149.07\times 386.4\times 62}{112}\right)\left(\frac{1}{2}-\frac{(1056)^2\times 18}{2\times 7200\times 386.4\times 48}\right)-\frac{90000\times 0.08727}{48}\right]}{149.07\left[1+\frac{0.75\times 20}{0.6\times 112}\left(1-\frac{(1056)^2\times 18}{7200\times 386.4\times 48}\right)\right]} = 0.7815\ g$$

Acceleration under rear skid

$$a_R = \frac{\frac{2\times 0.75}{1-0.6}\left[\left(\frac{149.07\times 386.4\times 62}{112}\right)\left(\frac{1}{2}-\frac{(1056)^2\times 18}{2\times 7200\times 386.4\times 48}\right)-\frac{90000\times 0.08727}{48}\right]}{149.07\left[1-\frac{0.75\times 20}{(1-0.6)\times 112}\left(1-\frac{(1056)^2\times 18}{7200\times 386.4\times 48}\right)\right]} = 0.6290\ g$$

Since $|a_F| > |a_R|$, rear tires will skid first.

Second surface: μ = 0.25

Acceleration under front-lock-up is

$$a_F = \frac{\frac{2\times 0.25}{0.6}\left[\left(\frac{149.07\times 386.4\times 62}{112}\right)\left(\frac{1}{2}-\frac{(1056)^2\times 18}{2\times 7200\times 386.4\times 48}\right)-\frac{90000\times 0.08727}{48}\right]}{149.07\left[1+\frac{0.25\times 20}{0.6\times 112}\left(1-\frac{(1056)^2\times 18}{7200\times 386.4\times 48}\right)\right]} = 0.2117\ g$$

Acceleration under rear lock-up

$$a_R = \frac{\frac{2\times 0.25}{1-0.6}\left[\left(\frac{149.07\times 386.4\times 62}{112}\right)\left(\frac{1}{2}-\frac{(1056)^2\times 18}{2\times 7200\times 386.4\times 48}\right)-\frac{90000\times 0.08727}{48}\right]}{149.07\left[1-\frac{0.25\times 20}{(1-0.6)\times 112}\left(1-\frac{(1056)^2\times 18}{7200\times 386.4\times 48}\right)\right]} = 0.2574\ g$$

Because $|a_F| < |a_R|$, front tires will skid first.

Figure E6.31(*a*) shows how the acceleration under front and rear lock-up changes with the coefficient of friction for the parameters described in Example 6.30. Rear skid is possible at surfaces with high coefficients of friction. Figure E6.31(*b*) shows that by changing the torque distribution to 0.7, front skid will occur for all types of surfaces. The K_f factor cannot be altered straightforwardly in conventional drive systems. An advanced traction control provides the maximum force per wheel without skid.

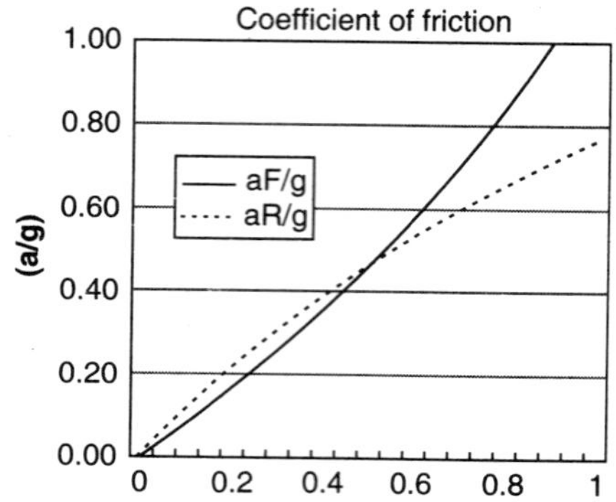

Fig. E6.31 (a) The values of a_F and a_R for different friction for the parameters of Example 6.30, where $K_f = 0.6$.

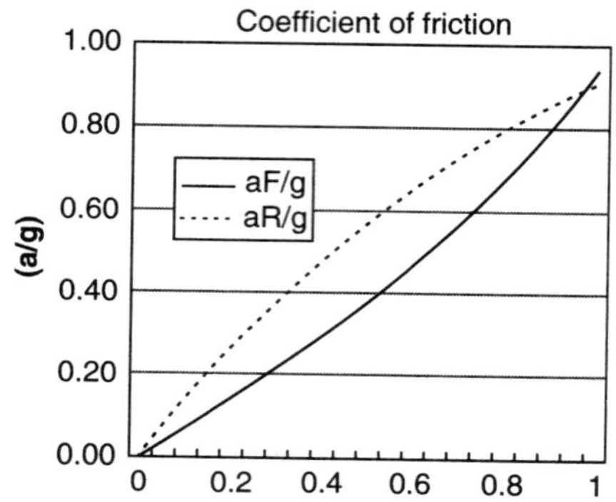

Fig. E6.31 (b) The values of a_F and a_R for different friction for the parameters of Example 6.30, where $K_f = 0.7$.

6.5.4 Vehicle Critical Speed

Vehicle critical speed (V_c) is the speed at which the tires of a cornering vehicle reach the limit of side friction and side slip begins. Slidge and Marshek (1997) reviewed formulas for estimating vehicle critical speed from yaw marks.

Figure 6.72 shows the vehicle forces in a banked plane. Noting that the lateral acceleration $a_y = V^2/R$, and applying Newton's laws, forces normal to the road surface are summed to give the normal force

$$F_z = Mg\cos\theta + \frac{MV^2 \sin\theta}{R} \tag{6.252}$$

Note that the angle θ is the embankment angle and not the grade. Summing the forces in the lateral direction yields the maximum friction force

$$F_f = \frac{MV^2 \cos\theta}{R} - Mg\sin\theta \tag{6.253}$$

Because $F_f = \mu F_z$ at the critical speed, further manipulation of Eq. (6.253) yields the equation of the critical speed (Slidge and Marshek, 1997).

$$V_c = \frac{\sqrt{gR(\mu + \tan\theta)}}{\sqrt{1 - \mu\tan\theta}} \tag{6.254}$$

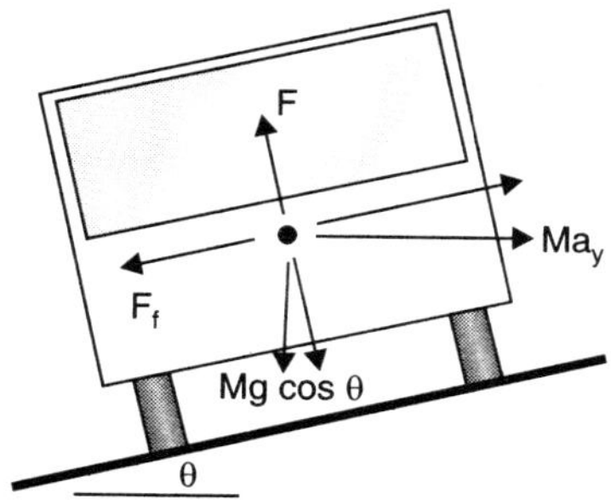

Fig. 6.72 Vehicle forces in a banked plane.

Example 6.32: *Determine the critical speed of a vehicle traveling on a 5% embankment at a radius of 50 m. The coefficient of friction between the tires and the road is 0.8. What would be the critical speed if the coefficient of friction is 0.2?*

Solution: Given $G = 9.81\ \text{m/s}^2$

$\tan\theta = 0.05$

$R = 50$ m

First, consider $\mu = 0.8$.

The critical speed is

$$V_c = \frac{\sqrt{gR(\mu + \tan\theta)}}{\sqrt{1 - \mu\tan\theta}} = \frac{\sqrt{9.81 \times 50(0.8 + 0.05)}}{\sqrt{1 - 0.8 \times 0.05}} = 20.84\ \text{m/s}$$

Second, consider $\mu = 0.2$.

The critical speed is

$$V_c = \frac{\sqrt{9.81 \times 50(0.2 + 0.05)}}{\sqrt{1 - 0.2 \times 0.05}} = 11.13\ \text{m/s}$$

Figure E6.32 shows the critical speed dependence on the coefficient of friction. Note that at lower friction values, the critical speed is relatively small and can be encountered in normal driving conditions.

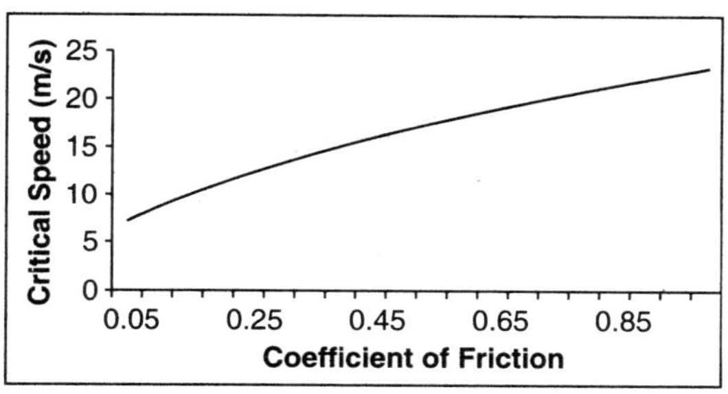

Fig. E6.32 Vehicle critical speed as a function of friction.

Calculation of the critical speed of a vehicle is of great importance in determining the steering characteristics of the vehicle. As soon as the critical speed is reached, slippage starts laterally at the wheels. This action will cause the vehicle to oversteer. In a vehicle without electronic stability systems, this behavior will cause the driver to countersteer, but the vehicle will not respond because of skidding. This leads to hazardous situations and may cause accidents.

Equations of critical speed also are presented for various other conditions. In particular, critical speed can be obtained for vehicles turning while braking or accelerating. Interested readers are encouraged to pursue these conditions as discussed in the paper by Slidge and Marshek (1997).

6.5.5 Vehicle Stability

We already learned that rear lock-up is not desirable because it introduces instability to the vehicle. Rear lock-up will cause the vehicle to spin around its vertical axis. Typical drivers apply a large brake force when faced with unexpected emergency conditions, which will force wheel lock-up to occur. Figure 6.73 illustrates the initiation of vehicle instability. Assume that rear wheel lock-up occurs first, as shown in Fig. 6.73(*a*), while the front wheels are still rolling (*i.e.*, no lock-up). Any lateral disturbance due to wind, road irregularity, grade, embankment, or others will cause the front wheels to develop side or lateral forces (F_y) to resist this disturbance. The rear wheels will not produce such a force because they are still sliding. Thus, the side forces developed at the front wheels will cause a yaw moment at the vehicle (the magnitude of which is $Fyc1$). This moment will cause the vehicle to rotate around its vertical axis. In turn, rotation causes the yawing moment to increase, thus causing instability.

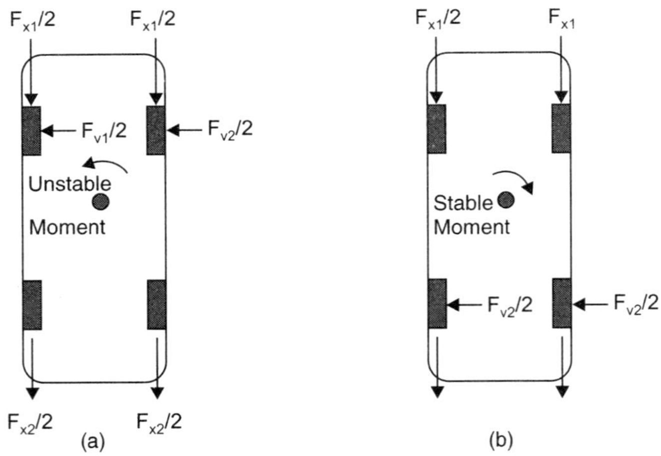

Fig. 6.73 Vehicle stability under rear and front lock-up: (*a*) rear wheel lock-up, and (*b*) front wheel lock-up.

On the other hand, if the front wheels are to lock up first, lateral forces will develop on the rear wheels, causing the yaw moment in an opposite direction to that developed when the rear lock-up happens first. This yaw moment is balanced with the vehicle friction forces, causing it to diminish and thus regaining vehicle stability.

Note here that applying front brakes that are different at each of the front wheels can be engineered to balance any of the yawing forces. This will enable the driver to regain control of the vehicle.

Advancement of anti-lock braking and traction control systems led to integrating these two systems into one electronic unit. Furthermore, it is found that the same unit can be used together with the engine control unit to enhance the vehicle stability, particularly when steering and cornering.

With electronic stability, the brakes can be used to enhance the steering of the vehicle to the point of stability. In situations where the vehicle starts to understeer or oversteer due to yaw instability, brakes on certain wheels (left or right) are applied to steer the vehicle in a direction that corrects the oversteering or understeering. Braking can be directed to the left rear wheel to counter understeer, or to the right front wheel to counter oversteer. Advanced electronic stability systems also with the engine electronic control unit can cause the vehicle to accelerate at a certain moment to regain stability. Situations that cause vehicle instability can be encountered in real driving. Rapid steering and counter steering can occur to avoid an obstacle that suddenly becomes apparent to the driver. Such behavior will generate substantial lateral forces on the front wheels. If the wheels were turned suddenly to the right, this would cause a clockwise rotation of the vehicle around its axis. Development of lateral forces at the rear wheel would be delayed, and typical drivers tend to countersteer; however, the vehicle will not respond. In this type of situation, the electronic stability will take charge and will apply braking on the left front wheel to counter the yawing moment produced by the lateral forces. Other actual maneuvers include fast lane changes, acceleration or deceleration during cornering, and rapid steering and counter steering required by certain roads (*e.g.*, S-shaped roads). The reader may refer to the book by Bosch (1999).

6.6 SUMMARY

In this chapter, the ride dynamics was divided into excitation sources, vehicle ride modeling, seating modeling and human body modeling. We reviewed the vehicle road surface uneven profile and powertrian excitation sources. The quarter-car model and bounce-pitch model were described in details. The active and semi-active control methods on engine isolation and suspension systems were also described. In this chapter, we also presented the simplest handling model and investigated the steady-state handling characteristics, which included the yaw rate gain and understeer gradient, the difference of the slip angles of the front and rear wheels, the ratio of the radius of turn and lateral acceleration gain, and the curvature gain. We also described the dynamic characteristics by using the simplest handling model, which included the damping and natural frequency, time response, and frequency response properties, as well as stability features. The effect of the chassis system on handling characteristics then was discussed, and some of the overturning limit handling characteristics also were described briefly. Finally, the nonlinear modeling and testing of handling characteristics were presented. This chapter overviewed aspects of vehicle dynamics from a customer's perspective. It covered the different tests and evaluations that can be made to assess the dynamics of a particular vehicle. Emphasis was made on evaluations that involve more than one maneuver by the vehicle driver occurring simultaneously. More detailed analysis was offered on the behavior of the vehicle under steering and braking conditions and under steering and acceleration conditions. Both treatments were made from a general perspective, where special cases could be derived from the general equations developed.

6.7 REFERENCES

Aoki, K., Shikata, T., Hyoudou, Y., Hirade, T., and **Aihara, T.,** "Application of an active control mount (ACM) for improved diesel engine vehicle quietness", SAE Paper No. 1999-01-0832, Automotive Engineers, Warrendale, PA, 1999.

Bosch Corporation, *Driving Safety Systems*, 2nd ed., Society of Automotive Engineers, Warrendale, PA, 1999.

Brunning, A.D., "The Relationship Between Objective Measurement and Perception of Low Levels of Sensory Inputs Experienced in a Motor Vehicle", M.Sc. Thesis, University of Bradford, Bradford, UK, 1996.

Casati, F. M., Berthevas, P. R., Herrington, R. M., and **Miyazaki, Y.,** "The Contribution of Molded polyurethane Foam Characteristics to Comfort and Durability of Car Seats", SAE Paper No. 1999-01-0585, Society of Automotive Engineers, Warrendale, PA, 1999.

Crede, C. E., *Vibration and Shock Isolation*, John Wiley & Sons, Inc., New York, NY, 1952.

Crolla, D., and Yu, V., *Vehicle Dynamics and Control*, China Communication Press, Beijing, China, 2006.

Demerly, J.D., and **Youcef-Toumi, Kamel.,** "Non-Linear Analysis of Vehicle Dynamics (NAVDyn): A Reduced Order Model for Vehicle Handling Analysis," SAE Paper No. 2000-01-1621, Society of Automotive Engineers, Warrendale, PA, 2000.

EI Demerdash, S. M., Selim, A. M., and **Corolla, D. A.,** "Vehicle body attitude using an electronically controlled active suspension", SAE Paper No. 1999-01-0724, Society of Automotive Engineers, Warrendale, PA, 1999.

Elliott, A.S., and **Wheeler, G.,** "Validation of ADAMS Models of Two USMC Heavy Logistic Vehicle Design Variants," SAE Paper No. 2001-01-2734, Society of Automotive Engineers, Warrendale, PA, 2001.

Fairley, T. E., and **Griffin, M. J.,** "The Apparent Mass of the Seated Human Body: Vertical Vibration," *Journal of Biomechanics*, Vol. 22, No. 2, pp. 81-96, 1989.

Foumani, M. S., and **Khajepour, A.,** "Application of Shape Memory Alloys to a New Adaptive Hydraulic Mount", ASE No. 2002-01-2163, New York, NY, 2002.

Genta, G., *Motor Vehicle Dynamics,* World Scientific Publishing, Singapore, 1997.

Gillespie, T.D., *Fundamentals of Vehicle Dynamics*, Society of Automotive Engineers, Warrendale, PA, 1992.

Gorder, K.V., Thomson, D., and **Basas, J.,** "Steering and Suspension Test and Analysis," SAE Paper No. 2000-01-1626, Society of Automotive Engineers, Warrendale, PA, 2000.

Griffin, M. J., "Evaluation of Vibration with Respect to Human Response", SAE paper No. 860047.

Harris, C. M., *Shock and Vibration Handbook*, 4th edition, McGraw-Hill, New York, NY, 1996.

Heydinger, G.J., Bixel, R.A., Durisek, N.J., Yu, E., and **Guenther, D.A.,** "Effects of Loading on Vehicle Handling," SAE Paper No. 980228, Society of Automotive Engineers, Warrendale, PA, 1998.

Kamel, S.M., and **Heydinger, G.J.,** "Model Validation of the 1997 Jeep Cherokee for the National Advanced Driving Simulator," SAE Paper No. 2000-01-0700, Society of Automotive Engineers, Warrendale, PA, 2000.

Kamel, S.M., Guenther, D.A., and **Heydinger, Gary J.,** "Vehicle Dynamics Modeling for the National Advanced Driving Simulator of a 1997 Jeep Cherokee," SAE Paper No. 1999-01-0121, Society of Automotive Engineers, Warrendale, PA, 1999.

Kazemi, R., Hamedi, B., and **Javadi, B.,** "Improving the Ride and Handling Qualities of a Passenger Car via Modification of Its Rear Suspension Mechanism," SAE Paper No. 2000-01-1630, Society of Automotive Engineers, Warrendale, PA, 2000.

Kim, T., Cho, Y., Yoon, Y., and **Park, S.,** "Dynamic Ride Quality Investigating and DB of Ride Values for Passenger and RV Cars", SAE paper No. 2001-01-0384, Society of Automotive Engineers, Warrendale, PA, 2001.

Kinkekaar, M. R., and **Neal, B. L.,** "The Influence of Polyurethane Foam Dynamics on the Vibration Isolation Character of Full Foam Seats", SAE Paper No. 980657, SAE International Congress and Exposition, Detroit, MI, 1989.

Kitazaki, S., and **Griffin, M. J.,** "A modal analysis of whole-body vertical vibration, using a finite element model of the human body", *J. of Sound and Vibration*, 1997, 200(1) 83-103.

Kluger, M.A., and **Long, D.M.,** "An Overview of Current Automatic, Manual and Continuously Variable Transmission Efficiencies and Their Projected Future Improvements," *SAE Transactions*, SAE Paper No. 1999-01-1259, Society of Automotive Engineers, Warrendale, PA, 1999.

Lewitzke, C., and **Lee, P.,** "Application of Elastomeric Components for Noise and Vibration Isolation in the Automotive Industry", SAE paper No. 2001-01-1447, Society of Automotive Engineers, Warrendale, PA, 2001.

Limpert, R., Gamero, F., and **Boyer, R.,** "An Investigation of Brake Force Distribution for Straight and Curved Braking," *SAE Transactions*, SAE Paper No. 741086, Society of Automotive Engineers, Warrendale, PA, 1976.

Matsumoto, Y., and **Griffin, M. J.,** "Movement of The Upper-Body of Seated Subjects Exposed to Vertical Whole-Body Vibration at the Principal Resonance Frequency", *J. of Sound and Vibration*, 1998, Vol. 215, No. 4

McManus, S. J., and **St. Clair, K. A.,** "Vibration and shock isolation performance of a commercial semi-active vehicle seat damper", SAE Paper No. 2000-01-3408, Society of Automotive Engineers, Warrendale, PA, 2000.

Milliken, W.F., and **Milliken, D.L.,** *Race Car Vehicle Dynamics,* Society of Automotive Engineers, Warrendale, PA, 1995.

Mo, C., Sunwoo, M., and **Patten, W. N.,** "Bistate control of a semiactive suspension", SAE Paper No. 1999-01-0725.

Pang, J., Kurrle, P., Qatu, M. and **Rebandt, R.,** "Attribute Analysis and Criteria for Automotive Exhaust Systems", SAE 2003 World Congress, Detroit, MI, USA, March, 2003.

Pang, J., Qatu, M., Dukkipati, R., Sheng, G., and **Patten, W. N.,** "Model Identification for Nonlinear Automotive Seat Cushion Structure", *International Journal of Vehicle Noise and Vibration*, Vol.1, No.1&2, 2003.

Park, S., Cheung, W., Cho, Y., and **Yoon, Y.,** "Dynamic Ride Quality Investigation for Passenger Car", SAE paper #980660.

Qatu, M., Sarifi, M., and **John F.,** "Robustness of powertrain mount system for noise, vibration and harshness at idle", *J. of Automobile Engineering*, Vol. 216, Society of Automotive Engineers, Warrendale, PA, 2002.

SAE., SAE J670e, "Vehicle Dynamics Terminology," Society of Automotive Engineers, Warrendale, PA, 1978.

Salaani, M.K., Christos, J.P., and **Guenther, D.A.,** "Parameter Measurement and Development of a NADSdyna Validation Data Set for a 1994 Ford Taurus," SAE Paper No. 970564, Society of Automotive Engineers, Warrendale, PA, 1997.

Salaani, M.K., Grygier, P.A., and **Heydinger, G.J.,** "Model Validation of the 1998 Chevrolet Malibu for the National Advanced Driving Simulator," SAE Paper No. 2001-01-0141, Society of Automotive Engineers, Warrendale, PA, 2001.

Sheng, G., Jiang, C.H., and **Yang, S.Z.,** "Dynamic Modeling of Pass-By Vehicle on Bridge," *Journal of Vibration, Measurement & Diagnosis*, Vol. 12, No. 2, 1992, pp. 20–26.

Shi, H.M., Sheng, G., and **Wu, Y.,** *Mechanical Vibration System: Analysis, Measurement, Modeling, and Control*, HUST Press, Wuhan, China, 1991.

Silvio, C., Data, Pascali, L., and **Santi, C.,** "Handling Objective Evaluation Using a Parametric Driver Model for ISO Lane Change Simulation," SAE Paper No. 2002-01-1569, Society of Automotive Engineers, Warrendale, PA, 2002.

Slidge, N., and **Marshek, K.,** "Formulas for Estimating Vehicle Critical Speed from Yaw Marks," *SAE Transactions*, SAE Paper No. 971147, Society of Automotive Engineers, Warrendale, PA, 1997.

Taborek, J.J., "Mechanics of Vehicles," *Machine Design,* May 30–December 26, ASME, New York, NY, 1957.

Tchernychouk, V., Rakheja, S., Stiharu, I., and **Boileau, P. E.,** "Study of Occupant-Seat Models for Vibration Comfort Analysis of Automotive Seats", SAE Paper No. 2000-01-2688, Society of Automotive Engineers, Warrendale, PA, 2000.

U.S. Department of the Army., *Engineering Design Handbook: Analysis and Design of Automotive Brake Systems*, DARCOM Pamphlet No. 706-358, Washington, DC, 1976.

Ushijima T., Takano, K., and **Kojima, H.,** "High Performance Hydraulic Mount for Improving Vehicle Noise and Vibration", SAE paper No. 880073, Society of Automotive Engineers, Warrendale, PA, 1988.

Vijayakumar, S., and **Barak, P.,** "Application of Bond Graph Technique and Computer Simulation to the Design of Passenger Car Steering System," SAE Paper No. 2002-01-0617, Society of Automotive Engineers, Warrendale, PA, 2002.

Wong, J.Y., *Theory of Ground Vehicle*, 3rd ed., John Wiley and Sons, New York, NY, 2001.

Yang, S.Z., Shi, H.M., Yang, K., Wu, Y., and **Sheng, G.,** *Modeling of Dynamic Mechanical Systems*, MEATU Press, Beijing, China, 1990.

Yu, Y., Peelamedu, S. M., Naganathan, N. G., and **Dukkipati, R. V.,** "Automotive Vehicle Engine Mounting System: A Survey", *J. of Dynamic Systems, Measurement, and Control*, Vol. 123, June 2001, ASME, New York, NY, 2001.

Yu, Z.S., *Auto Theory*, Machine Press, Beijing, China, 1989.

PROBLEMS

Ride Dynamics

6.1 Describe the factors influencing the ride quality.

6.2 Explain why the firing pulsation is the most important engine excitation. Please list the firing order for 4, 6, 8, 12 cylinder engines, respectively.

6.3 Analyze the dominated orders caused by inertia force.

6.4 For a four cylinder engine, the powerplant weight is 220 kg. The idle speed is 700 rpm. Assume the powerplant is uniformly supported by three mounts. The mounts' rate is the same. What is the practical stiffness range for the mounting system?

6.5 A vehicle has 6-cylinder engine and its body response is measured. The response spectrum is mainly dominated by 2nd order and 3rd order when the vehicle travels at WOT (Wide Open Throttle) condition. Explain the vibration sources to the body response.

6.6 Idle speed for a 4-cylinder engine is 600 rpm, please determine the natural frequency of powerplant roll mode.

6.7 Plot the relation among engine speed, frequency and order from 1st order to 4th order at increment of ½ order.

6.8 For a V8 engine, the maximum engine speed is 6500 rpm. Determine minimal exhaust hanger bracket frequency.

6.9 The driveshaft maximum imbalance force must be controlled less than 40 N. The maximum shaft speed is 3500 rpm. Determine the system static balance.

6.10 If only a disturbance force f_b is applied on the axle (unsprung mass) and other inputs are neglected, shown in Fig. P6.10, what is the transmissibility between the body response and the disturbed force?

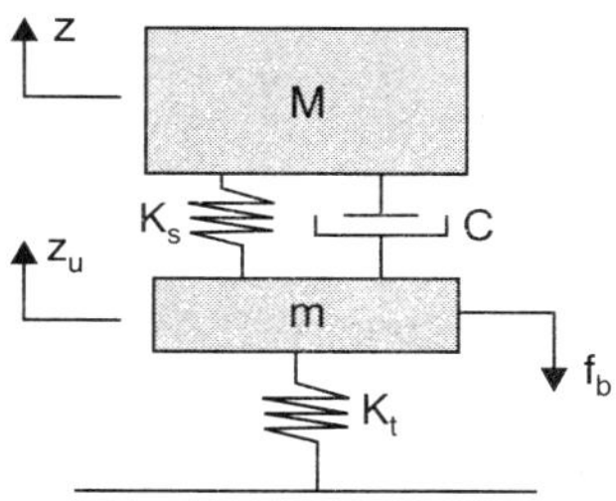

Fig. P6.10

6.11 Calculate the natural frequencies and modal shapes of the quarter model shown in Fig. 6.28 with following data:

Sprung mass: M = 1800 kg

Unsprung mass: m = 180 kg

Sprung stiffness: K_s = 80,000 N/m

Tire stiffness: K_t = 750,000 N/m

6.12 Calculate pure bounce and pitch frequencies and their oscillation centers for a vehicle with following characteristics.

Front Ride rate: k_f = 32,000 N/m

Rear Ride Rate: k_r = 38,000 N/m

Sprung mass: M = 1400 kg

Radius of gyration: r_y = 1.1 m

Distance between front axle and the vehicle CG: l_1 = 1.3 m

Distance between rear axle and the vehicle CG: l_2 = 1.7 m

6.13 A vehicle can be simplified as a 5 degree-of-freedom model shown in Fig. P6.13. Establish the dynamic equation of the model.

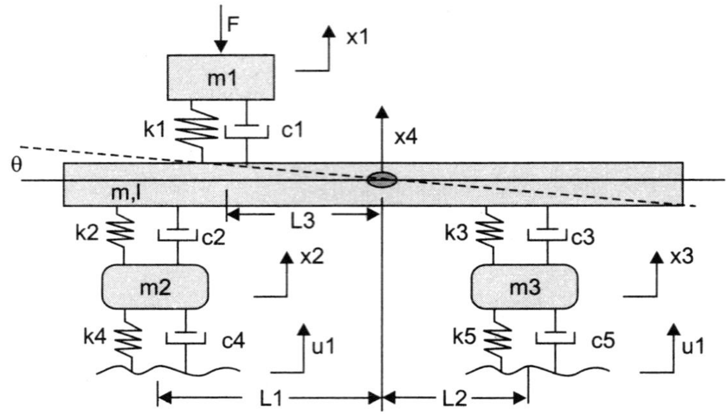

Fig. P6.13 A 5-DOF simplified vehicle model.

6.14 Describe the methods to evaluate ride quality and compare their advantage and disadvantage.

6.15 Use Newton's laws to derive a mathematical model of a vehicle similar to the bicycle model to describe the motion in the lateral direction. Simplify the model to be a steady state model, which means that the steering wheel angle is a constant.

6.16 Using Lagrange's equation of motion derive the model in Problem 6.1, and compare its features with the one obtained in Problem P6.1 using Newton's law.

6.17 Use the model obtained in P6.1 to determine the understeer coefficient (parameter, gradient).

6.18 Develop a mathematical model for the transient handling analysis of a passenger car under effects of lateral wind gust. The model should reflect speeds and positions for a vehicle moving in a horizontal ground plane. The speed in the longitudinal direction can be considered as constant. Propose quantities that are state variables and control variables.

6.19 Implement the model developed in problem P6.4 in MATLAB where the model can be a function of states and control as input and state derivatives as output.

6.20 Assume a car is driven on a road with a constant radius of 30 m at a velocity of 15 m/s, L = 2.8 m, determine the required steering angle.

6.21 Consider a car having following parameters: mass M = 2045 kg, inertia I = 5428 kg mm, distance from front axle to CG a = 1.49 m, distance from rear axle to CG a = 1.7 m, front cornering stiffness C_f = 77.9pkN/rad, rear cornering stiffness C_r = 76.5 kN/rad. Determine understeer gradient K.

6.22 For the P6.7, assume velocity = 20 m/s, lateral acceleration = 0.3 g, lateral angular velocity = 8.4 deg/s, and radius = 136 m, determine the stable response characteristic parameters ω_d, ξ.

6.23 Determine the function of eigenvalue of the above system as a function of velocity (from 20 m/s to 50 m/s).

6.24 Calculate and plot the response of the car in P 6.7 under step steer input and under the condition of velocity v = 50 m/s.

6.25 Calculate and plot the frequency response of the car in P6.7 under the condition of velocity = 20 m/s and 50 m/s.

6.26 Consider a car having following parameters: mass M = 1008 kg, inertia I = 1031 kg mm, distance from front axle to CG a = 1.2 m, distance from rear axle to CG a = 1 m, front cornering stiffness C_f = 117 kN/rad, rear cornering stiffness C_r = 145kN/rad. Determine understeer gradient K.

6.27 For the P6.12, assume velocity = 20 m/s, lateral acceleration = 0.3 g, lateral angular velocity = 8.4 deg/s, and radius = 136 m, determine the stable response characteristic parameters.

6.28 Determine the function of eigenvalue of the above system P6.12 as a function of velocity (from 20 m/s to 50 m/s).

6.29 Calculate and plot the response of the car in P6.12 under step steer input and under the condition of velocity v = 50 m/s.

6.30 Calculate and plot the frequency response of the car in P6.12 under the condition of velocity = 20 and 50 m/s.

Acceleration

6.31 A 18683 N (= 4200 lb) vehicle is to achieve a maximum speed of 104.6 km/hr (=65 mph) in 6 seconds. Assuming a constant engine torque, determine the traction force needed between the tires and the road to achieve this speed. Also, find the distance it takes to reach such a speed from an idle condition.

6.32 Consider a midsize sedan that has a mass of 1500 kg. The wheel base of the vehicle is 2.1 m and the center of gravity is 1.0 m from the front axle and 0.5 m above the ground. Determine the load on each axle if the vehicle is under no acceleration, accelerating at 0.6 m/sec^2 and 1.5 m/sec^2.

6.33 A midsize sedan that has a mass of 1500 kg. The vehicle is going at 100 km/hr. The front area of the vehicle is 2.0 m^2, and the coefficient of drag is 0.42. The aerodynamic resistance is acting on the C.G. The wheel base of the vehicle is at 2.1 m and the center of gravity is at 1.0 from the front axle and at 0.5 m above the ground. The density of air is 1.2256 kg/m^3. What would the loads be if the vehicle is accelerating at 0.6 m/s^2 on a horizontal plane. What would axle load be when the vehicle goes up hill and down hill at 6 degree slope with the same speed.

6.34 A 22215 N (=5000 lb) passenger car has 2743.2 mm (=108 in) wheel base and a CG at 1066.8 mm (42 in) behind the front axle and 457.2 mm (=18 in) above ground. The tractive effort distribution gives the front axle 90% of the total tractive force. The car is moving on a horizontal plane. Determine which tire will skid first if the car is moving on a surface with first a 0.85 coefficient of friction, and second, a 0.15 coefficient of friction. Ignore drag and rolling resistance.

6.35 Consider the previous problem. The car is moving on a horizontal plane and drag and rolling resistance are negligible. Determine the maximum tractive force, maximum acceleration and the distribution factor to achieve them. The coefficient of friction is 0.25.

6.36 Consider the previous problem. Vehicle strategy requires a front axle skid on a dry surface, the front torque distribution factor is chosen at 0.7. The vehicle is moving on a low friction surface with a 0.25 coefficient of friction. Determine the maximum tractive force and which axle would skid first.

6.37 A vehicle has a mass of 1600 kg when lightly loaded (driver only), with a C.G. at 0.8 m behind front axle and 0.45 m above road, and 2000 kg when loaded with a C.G. at 1.2 m behind front axle and 0.5 in above road. The wheel base is 2.6 m. The vehicle is to achieve maximum possible tractive force and front skid under the following severe conditions:

Vehicle is loaded, dry surface of μ = 0.85

Vehicle is loaded, wet surface of μ = 0.15

Vehicle is lightly loaded, a dry surface of μ = 0.85

Vehicle is lightly loaded, a wet road of μ = 0. 15

Determine the value of K_f that you recommend knowing that the drive system your company installs has a constant torque distribution value (no feedback control system). Determine the axle loads and the maximum tractive force.

6.38 A V6 gasoline engine is putting a maximum of 300 hp (horsepower) of gross power at 5000 rpm. Air induction losses are found to be 2.8 % at that engine speed. The hot and cold end exhaust losses are found to be 8%. In addition, losses due to air conditioning, power steering, alternator and other accessories add up to 15 horsepower under specific driving conditions. The ambient temperature is 35°C and dry pressure is 105 kPa. Determine the power available to the transmission.

6.39 An engine delivers 325 N m (240 ft lb) torque at 2500 rpm engine speed. It is coupled to a torque converter and an automatic speed transmission. Determine the out speed and output torque of the converter of the characteristics given in Fig. 6.22. Determine the efficiency of the torque converter.

6.40 A vehicle (including wheels) weighs 30,000 N. Each wheel weighs 300 N and has rolling radius of the wheels to the 0.35 m. and the radius of gyration is 0.28 m. The engine is driven at 3000 rpm and is delivering 400 N m. The total drivetrain efficiency is 84% and the transmission is in fourth gear with a 3.8 to 1 ratio. The components rotating at the engine speed have an inertia of 1.0 kg.m^2. Assume a slip of the running gear at 3% and a coefficient of rolling resistance of 2.5%. The front area of the vehicle is 2.2 m^2 and the aerodynamic drag coefficient is 0.38, and the speed of the wind is zero. Determine the acceleration at that point.

6.41 Consider the previous problem. Assume that the engine torque remained constant during take-off acceleration. Also, acceleration is happening while transmission is engaged in the same gear as the previous problem. Determine how long it would take the vehicle to reach a constant speed of 100 km/hr. Assume a wind speed of 20 km/hr against the vehicle direction.

6.42 Consider a light truck that weighs 20000 N under a certain load. The wheel base of the vehicle is at 3.0 m and the center of gravity is at 1.3 m from the front axle and at 0.75 m above the ground. Determine the load on each axle under maximum traction limited acceleration with a coefficient of friction of 0.8.

6.43 A 1300 kg compact passenger car with 2.0 m wheel base and a CG at 0.96 m behind the front axle and 0.6 m above ground. The car is moving on a horizontal plane. Determine which tire will skid first if the car is moving on a surface with a 0.9 coefficient of friction and if the front axle force distribution factor is 0.4, 0.6, and 0.8. Determine acceleration under front skid.

6.44 Consider a vehicle with a 1500 kg mass when lightly loaded (driver only), with a C.G. at 1.0 m behind front axle and 0.45 m above road. The mass is 2200 kg when loaded with a C.G. at 1.25 m behind front axle and 0.5 m above road. The wheel base is 2.7 m. The vehicle is to achieve maximum possible tractive force and front skid under the following severe conditions:

Vehicle is loaded, dry surface of $\mu = 0.9$

Vehicle is loaded, slippery surface of $\mu = 0.2$

Vehicle is lightly loaded, on a dry surface of $\mu = 0.9$

Vehicle is lightly loaded, on a slippery road of $\mu = 0.2$

Determine the value of the front force distribution factor K_f that you recommend knowing that the force system has a constant value. Determine the axle loads and the maximum tractive force.

6.45 The vehicle described in the previous example, has a traction control system which allows control of each wheel force distribution. Determine the maximum tractive force achieved under the following conditions:

Vehicle is loaded, dry surface of $\mu = 0.9$

Vehicle is loaded, slippery surface of $\mu = 0.2$

Vehicle is lightly loaded, on a dry surface of $\mu = 0.9$

Vehicle is lightly loaded, on a slippery road of $\mu = 0.2$.

6.46 For the vehicle of the previous two examples, determine the maximum tractive force if the vehicle is moving on a split mue surface with the right wheels on a 0.2 coefficient of friction surface and the left wheels on a 0.8 surface.

6.47 For the Problem P6.16, assume that control of the tractive force on each wheel is possible under an advanced traction control system. Determine the maximum tractive force that can be applied.

6.48 Derive Eq. (8.17).

6.49 Equations (8.40) and (8.41) give the maximum angle a RWD and FWD vehicle can climb at a constant speed. Derive a similar equation for AWD vehicles.

6.50 Consider a vehicle with a C.G. at 1.000 m behind front axle and 0.460 m above road. Consider a rolling resistance of 0.02. The wheel base is 2.750 m. Find the maximum angle that can be withstood without loosing traction for both FWD and RWD when the coefficient of friction is 1.0.

Total Vehicle Dynamics

6.51 Consider a vehicle 20000 N under a certain load. The wheel base of the vehicle is at 3 m and the center of gravity is at 1.20 m from the front axle and at 0.6 m above the ground. Vehicle speed is at 100 km/hr, the turning radius is of 180 m, wheel tread is of 1.25 m, K_ϕ = 10000 N m, roll center height is 0.45 m and a roll angle of 5 degrees. Determine the load on each wheel (F_{z1i}, F_{z1o}, F_{z2i}, F_{z2o}) if the vehicle is under no acceleration, and decelerating at 3.00 m/s^2.

6.52 Find the wheel forces for Problem P6.1 if the vehicle accelerates at a rate of 1.00 m/s^2.

6.53 A 2500 kg sport car with 2.5 m wheel base and a CG at 1.0 m behind the front axle and 0.35 m above ground. The braking effort distribution gives the front axle 60% of the total braking force. The car is moving on a horizontal plane and drag and rolling resistance are ignored. The car is going around a turning radius of 150 m at 120 km/hr. The wheel tread is of 1.40 m, K_ϕ = 12000 N m, roll center height is 0.4 m and a roll angle is of 8 degrees. Determine which tire will lock up first if the car is moving on a surface with first a 0.8 coefficient of friction, and second, a 0.2 coefficient of friction.

6.54 Consider the vehicle of the previous problem. The torque distribution gives the front axle 60% of the total torque delivered to the front wheels. Determine which tire will skid first if the car is moving on a surface with first a 0.8 coefficient of friction, and second, a 0.2 coefficient of friction.

6.55 Derive acceleration and tractive force expressions for skid of FWD systems going around a corner.

6.56 Derive acceleration and tractive force expressions for skid of RWD systems going around a corner.

6.57 Derive an expression for the brake force distribution factor so that lock-up occurs simultaneously at both front and rear wheels.

6.58 Derive an expression for the torque distribution factor so that skid occurs simultaneously at both front and rear wheels in AWD systems.

6.59 Consider the vehicle of Problem P6.1. The vehicle is going uphill at 6 degrees slope at 100 km/hr. The front area of the vehicle is 2.0 m^2, and the coefficient of drag is 0.42. The aerodynamic resistance is acting on the C.G. The density of air is 1.2256 kg/m^3. Determine the load on each axle if the vehicle is under no acceleration. What would the loads be if the vehicle is accelerating at 0.6 m/s^2. What would these forces be when the vehicle goes down hill at 6 degree slope with the same speed.

6.60 Repeat problem P6.9 when the vehicle is decelerating at 3.0 m/s^2 under braking effort.

6.61 Consider a vehicle with a 1500 kg mass when lightly loaded (driver only), with a C.G. at 1.0 m behind front axle and 0.45 m above road. The mass is 2200 kg when loaded with a C.G. at 1.25 m behind front axle and 0.5 m above road. The car is moving on a horizontal plane and drag and rolling resistance are ignored. The car is going around a turning radius of 150 m at 120 km/hr. The wheel base is 2.70 m and the wheel tread is 1.40 m, $K_\phi = 9000$ Nm, roll center height is 0.4 m and a roll angle is of 5 degrees. The vehicle is to achieve maximum possible braking force and front lock-up under the following conditions:

(*a*) Vehicle is loaded, dry surface of $\mu = 0.9$

(*b*) Vehicle is loaded, slippery surface of $\mu = 0.2$

(*c*) Vehicle is lightly loaded, on a dry surface of $\mu = 0.9$

(*d*) Vehicle is lightly loaded, on a slippery road of $\mu = 0.2$.

Determine the value of the front brake force distribution factor K_f that is to be recommended knowing that the brake system your company installs is conventional (no feedback control system). Determine the axle loads and the maximum braking force.

6.62 For the Problem P6.11 determine the maximum braking force if an ABS system is installed instead of a conventional braking system.

6.63 Prove Eq. (9.26).

6.64 Drive an expression of critical speed similar to Eq. (9.26) taking into consideration braking and weight transfer.

CHAPTER 7

Braking

7.1 INTRODUCTION

Braking performance of a vehicle is one of its most important safety features. This performance is judged by drivers under various conditions. One condition closely related to vehicle safety is emergency stop. Drivers expect certain behavior of the vehicle when under emergency. The direct expectation of the braking system under emergency conditions is simply avoiding an accident.

Performance of the braking system is also judged when under constant deceleration (*i.e.* normal stop) condition. Also, brakes can be used under normal driving conditions to adjust speed in traffic.

Pure braking is simple to understand. The input to the braking system is a direct force at the braking pedal by the driver with the direct response of deceleration in the longitudinal motion. The braking mechanism transfers the braking force applied by the driver at the braking pedal into a torque applied on the wheel, preventing the wheels from rotation, either on a continuous basis, or in a slip regulation fashion. This will result in large friction forces between the tires and the road, which reduce the speed of the vehicle.

Analysis of braking systems can be performed on various types of vehicles. These include on-road and off-road vehicles. The emphasis of this book is on road vehicles. Road vehicles can have one axle, or multi-axles. Also, they can have a trailer. The fundamental theory presented and equations derived are focused on single axle vehicles without a trailer. The same methods implemented here can be used to analyze multi-axle vehicles. They can also be made for vehicles with trailers. Such analysis is considered beyond the scope of this introductory book on vehicle dynamics.

Although much of the treatment made here is intended for passenger cars, the analysis and theory can be used for light trucks, sport utility vehicles (SUV) and other similar vehicles.

7.1.1 Types of Automotive Brakes

Automotive brakes used by standard road vehicles can be one of two types. These are drum and disk brakes. These types of brakes are shown in Fig. 7.1.

Each of the two types has advantages and disadvantages over the other brake type. In relation to braking performance, drum brakes require less actuation force (because of higher brake factor and better mechanical advantage) on the brake pedal when compared with disk brakes. Also, installation of parking brake features is easier with drum brakes. On the other hand, disk brakes offer better (*i.e.* more consistent) distribution of forces on the wheel when compared with disk brakes.

Fig. 7.1 Drum and disk brakes (Courtesy of Bosch Corporation).

Other motion control devices exist and are frequently used as auxiliary devices to help the braking action of the drum and/or disk brakes. Engine brakes are one version of such devices. In these brakes, the torque transmission can be interrupted by disengaging the clutch or selecting a neutral gear position. If the throttle is closed for a vehicle in motion, the engine will put a retarding force on the vehicle due to losses experienced to balance the kinetic energy of the vehicle. These forces are limited, and their effectiveness is increased if the compressor action of the engine is increased by shutting the exhaust. This type of retardation devices is referred to as exhaust brakes. It consists of a throttle in the exhaust duct system which can be controlled mechanically or electronically. The brake action that results from this will depend on the gearing and vehicle speed. The brake torque generated by shutting the exhaust can reach up to 70% of the drive torque. Other improvements can be made to the camshaft timing to increase the compression and subsequently the retardation. This leads to improvement of the engine brake torque to be up to 100 % of the drive torque. Brake torque devices have been used more on diesel engines. Usage of engine brakes has shown to reduce the usage of the foundation brakes by about 20% in normal driving conditions. Recent research on this auxiliary braking system includes the work of Moklegaard *et al.* (2000).

Another type of retardation devices is hydraulic. These use viscous damping as the mechanism to produce the retarding torque. This retarder is similar to the hydraulic clutch. The retarding torque is produced by the rotor which pumps a fluid to a stator, which reflects back against the rotor, and a continuous internal pumping cycle is developed. The reaction forces and hence the retarding torque are absorbed by the rotor which is connected to the wheels of the vehicle. The torque produced by hydraulic retarders becomes higher at higher speeds, a clear advantage. Also, these retarders are independent of the vehicle engine, drivetrain or electronic, and thus very robust in their action.

A third type of retardation is the electric retarders. These are based on producing eddy currents with a metal disc rotating between two electromagnets. This develops a retarding torque on the rotating disc. The eddy current results in heating the disc. The cooling of the disc is accomplished by means of convection heat transfer with ventilated rotors. These retarders need high temperatures to have effective cooling capacity. High temperatures, however, cause the retardation effectiveness to be reduced.

These retarding devices are installed in conjunction with foundation disk and drum brakes to help maximize the braking force. They are also used to reduce the use of the foundation braking devices especially when light braking is needed.

7.1.2 Braking Distance and Deceleration

Among the attributes in which braking performance is measured are the braking distance and deceleration. If the vehicle is moving as a result of a consistent and constant braking force and changing its velocity from V_i to V_f (which would be zero in the case of a complete stop), the acceleration, a, can be calculated as

$$a = \frac{\left(V_f - V_i\right)}{t} \tag{7.1}$$

where t is time; V_f is the final vehicle velocity and V_i is the initial vehicle velocity. Note that the above acceleration will be a negative number when the vehicle is decelerating. If the total braking force on both axles is F_b, then from Newton's second law of motion, the deceleration is F_b/M; where M is the mass of the vehicle.

The distance traveled during braking (X_b) can be calculated by

$$X_b = \frac{\left(V_f^{\,2} - V_i^2\right)}{2 \times a} \tag{7.2}$$

Example 7.1: *A 2000 kg vehicle traveling at 100 km/hr is stopped under a constant braking force of 10000 N. Find the time traveled during the deceleration and the stopping distance.*

Solution: The velocity can be written as

$$V_i = \frac{100\text{ km}}{\text{hr}} \times \frac{1000\text{ m}}{\text{km}} \times \frac{\text{hr}}{3600\text{ sec}} = 27.78\text{ m/sec}$$

The acceleration of the vehicle is calculated first

$$a = -F_b/M = -10000/20000 = -5\text{ m/sec}^2$$

The time needed to reach a complete stop is

$$t = (V_f - V_i)/a = (0 - 27.28)/-5 = 5.56\text{ sec}$$

The stopping distance is

$$X_b = \frac{\left(V_f^{\,2} - V_i^2\right)}{2 \times a} = \frac{0 - (27.78)^2}{2 \times (-5)} = 77.16\ \text{ m}$$

Requirements of stopping distance for various driving conditions have been established by regulatory institutions (e.g., FMVSS 121). More regulations can be found in the references used by Shaffer and Radlinski (2003). According to the later reference, the early regulations of 1936/1937 included only five brake-related requirements, paraphrased by Shaffer and Radlinski (2003) as follows:

1. Adequate brakes are required on all vehicles, with two separate, independent means of application. If the application means are connected, failure of one part shall not leave the vehicle without adequate brakes.

2. Brakes on one or more units of a combination vehicle must be adequate to stop and hold the combination.
3. Adequate brakes defined by stopping distances and decelerations are listed as:
   Vehicles with brakes on all wheels must be capable of stopping in 30 feet from 20 mph or sustained deceleration of 14 feet/sec^2. Vehicles with brakes on some wheels must be capable of stopping in 45 feet from 20 mph or sustained deceleration of 9.5 feet/sec^2.
4. For a combination vehicle, the trailer brakes must come on first, or at the same time as those on the towing vehicle.
5. The parking brake system must be able to hold the vehicle on any hill on which it is operated.

Later, in 1962, new regulations were established (e.g. Shaffer and Radlinski, 2003). In these regulations, the deceleration is no longer specified as sustained, but as "not less than", meaning peak, and the requirements for the new column, the percentage of the braking force (BF) with respect to the gross vehicle weight (GVW) are also "not less than". The stopping distance, which is listed as, "not more than the numbers listed in the Table, is now clarified as measured from the point of first movement of the brake pedal. There are 7 categories of commercial vehicles (CVs) listed. Vehicles must meet 3 requirements: braking force, required deceleration and stopping distance. Federal Motor Carrier Safety Regulation (FMCSR) has modified these later in 1971 to take the form of Table 7.1. Note that commercial vehicles are reduced to only six, three of which are classified as passenger carrying vehicles and the remaining three are property carrying vehicles.

Table 7.1 Safety requirements for different types of vehicles

Type of Motor Vehicles	*Service brake systems*			*Emergency brake systems*
	Braking Force as a percentage of gross vehicles or combination weight	*Deceleration in feet per second per second*	*Application and braking distance in feet from initial speed of 20 mph*	*Application and braking distance in feet from initial speed of 20 mph*
A Passenger-carrying vehicles				
1. Vehicles with a seating capacity of 10 persons or less. including driver and built on a passenger car chassis.	65.2	21	20	54
2. Vehicles with a seating capacity of more than 10 persons, including driver and built on a passenger car chassis, vehicles built on truck or bus chassis and having a manufactures' GVWR of 10,000 pounds or less.	52.8	17	25	66
3. All other passenger carrying vehicles.	43.5	14	35	85

(*Contd.*)

B Property-carrying Vehicles 1. Single unit vehicles having a manufacturer's GVWR of 10,000 pounds or less.	52.8	17	25	66
2. Single unit vehicles having a manufacturer's GVWR of more than 10,000 pounds, except truck tractors. Combinations of a 2-axle towing vehicle and trailer having a GVWR of 3,000 pounds or less. All combinations of 2 or less vehicles in drive away or towaway operation.	43.5	14	35	85
3. All other property carrying and combinations of property carrying vehicles.	43.5	14	40	90

7.2 BRAKE TORQUE DISTRIBUTION

One of the most important characteristics of the brake system is the distribution of torque on the rim of the wheel to assure certain deceleration behavior. The torque on the rim is a direct result of the actuation force. We will discuss this for both drum and disk brake systems.

7.2.1 Drum Brakes

The mechanical forces acting on the wheel as a result of applying a typical braking force in a drum brake are shown in Fig. 7.2.

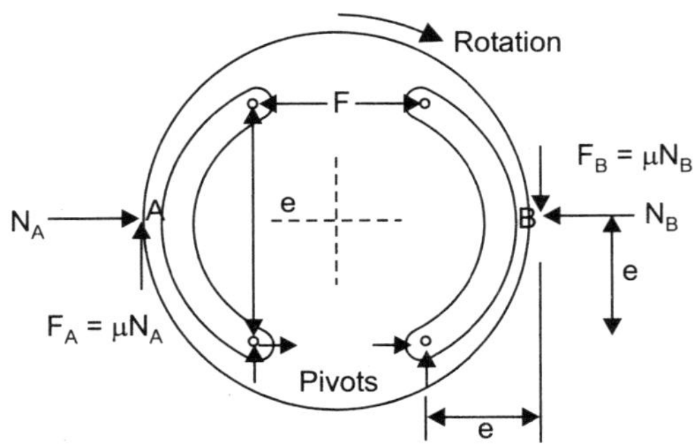

Fig. 7.2 Forces acting on the shoes of a drum brake.

Note that as the forces are applied on the pedal, they transfer directly through the brake system to the brake cylinders and translate into a force *F* applied on the brake shoes. Soon after the application of the brake force, the shoes are pushed against the rim. This generates the friction forces between the moving drum and the stationary shoe.

7.2.1.1 Mechanical Advantage

In actuality, friction will happen along an arc of contact between the shoes of the drum brake and the wheel rim. For simplicity, let us assume that the friction will actually happen at the points *A* and *B* during braking. One can then take the moment about the pivots (for shoe *B*) to obtain the following equation:

$$\sum M_{Pivot} = eF + e_2\mu N_B - e_1 N_B = 0 \tag{7.3}$$

where M_{Pivot} is the moment around the pivot, e is the vertical distance between the pivot and the contact with the brake cylinder, e_1 is the vertical distance between the pivot and the center of application of the friction force, e_2 is the horizontal distance between the pivot and the center of application of the friction force, μ is the coefficient of friction, and N_B is the normal force. Note that the friction force F_B at that point of contact is

$$F_B = \mu N_B \tag{7.4}$$

Mathematical treatment of the above equations gives the ratio of the friction force to the applied force:

$$\frac{F_B}{F} = \frac{\mu e}{e_1 - \mu e_2} \tag{7.5}$$

and similarly for the other shoe, one can find that

$$\frac{F_A}{F} = \frac{\mu e}{e_1 + \mu e_2} \tag{7.6}$$

The above two equations define the mechanical advantage obtained by using the drum shoe mechanism.

Example 7.2: *In a drum shoe system where e is equal to* $2e_1$ *and* $e_1 = e_2$*, determine the mechanical advantage of both shoes. Assume a coefficient of friction equal to 0.5.*

Solution:

$$\frac{F_B}{F} = \frac{\mu e}{e_1 - \mu e_2} = \frac{0.5 \times 2 \times e_1}{0.5 \times e_1} = 2$$

$$\frac{F_A}{F} = \frac{\mu e}{e_1 + \mu e_2} = \frac{0.5 \times 2 \times e_1}{1.5 \times e_1} = 0.67$$

This example shows that shoe B experiences higher mechanical advantage than shoe A. Also, if the coefficient of friction gets close to 1, the denominator gets closer to 0 and a high advantage is obtained. This will result in brake lock-up and wheel skid, which has a smaller friction coefficient and thus a smaller braking forces.

One important observation should be made. Assume a clockwise rotation and let us observe the force generation on the right hand shoe, which is frequently referred to as the leading shoe. As the force F is applied, this will result in the reaction Force N_B, which will result in the friction force ($F_B = \mu N_B$). This friction force will add to the total moment done around the pivot by the actuation force itself. In order to balance that moment, a higher value of N_B will be needed. This will create a higher friction force, a higher moment and a higher N_B again. This system is often referred to as 'self energizing' and soon the maximum pressure of the brake shoe is obtained.

7.2.1.2 Torque Calculations

A more detailed analysis of forces on the shoes will require treatment of the arc of contact between the shoe and the wheel rim. We will conduct this analysis for the leading shoe (the right shoe in a clockwise rotation). Assume a shoe width b, and that the pressure is linearly proportional from the distance to the pivot. In other words, the pressure at the pivot will remain at zero and the maximum pressure is achieved at an angle that is perpendicular to the line connecting the pivot with the center point (Fig. 7.3). The following equation then applies for the pressure distribution on the shoe:

$$p = p_{max} \sin \theta \tag{7.7}$$

where p is the pressure at any angle θ on the lining, and p_{max} is the maximum pressure over the range of contact.

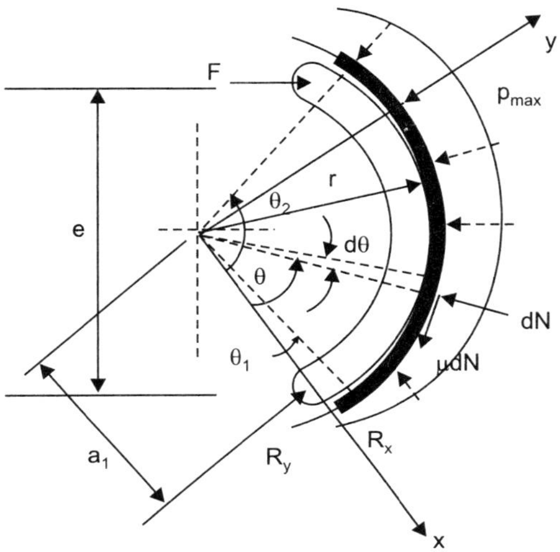

Fig. 7.3 Pressure distribution along the brake shoe.

At an angle θ, the normal force is

$$dN = pbrd\,\theta \tag{7.8}$$

where b is the width of the shoe and r is the radius to the lining. Proper mathematical manipulation will yield the moment resulting from the frictional forces (M_F) over the arc length of the shoe contact with the drum:

$$M_F = \mu p_{max} br \int_{\theta_1}^{\theta_2} \sin\theta \; (r - a_1 \cos\theta)\, d\theta \tag{7.9}$$

Note that the arm of the frictional force μdN is the expression $(r - a_1 \cos \theta)$. The moment produced by the normal forces (pressure) about the pivot is

$$M_N = p_{max} b\, r\, a_1 \int_{\theta_1}^{\theta_2} \sin^2 \theta \, d\theta \tag{7.10}$$

Applying the summation of moment's equation around the pivot will yield the actuation force equation

$$F = \frac{M_N - M_F}{e} \tag{7.11}$$

It should be noted here that a brake system can be designed where it will be self-locked (*i.e.* it does not need an actuation force). This is achieved by setting M_N to be equal to M_f. The brake must be designed such that $M_N > M_f$ to prevent self-locking.

The torque, T, applied to the drum as a result of the frictional forces can be obtained by integration product of the friction force and the radius to the drum of the friction area. This will yield:

$$T = \mu p_{max} b\, r^2 (\cos\theta_1 - \cos\theta_2) \tag{7.12}$$

The reaction forces at the pivot (hinge) can be obtained by summing the forces in the x and y directions.

$$R_x = p_{max} b\, r \left(\int_{\theta_1}^{\theta_2} \sin\theta \cos\theta \, d\theta - \mu \int_{\theta_1}^{\theta_2} \sin^2\theta \, d\theta \right) - F_x \tag{7.13}$$

$$R_y = p_{max} b\, r \left(\mu \int_{\theta_1}^{\theta_2} \sin\theta \cos\theta \, d\theta + \int_{\theta_1}^{\theta_2} \sin^2\theta \, d\theta \right) - F_y \tag{7.14}$$

Derivation can be made for the left shoe. The equations of M_f, M_N and the torque T will remain the same as those of the right shoe (note that the maximum pressure on the left shoe is usually less than that on the right shoe in a clock-wise rotation). The following are the resulting equations for the actuation force and the hinge reactions. The actuation force is:

$$F = \frac{M_N + M_F}{e} \tag{7.15}$$

Note here that the left shoe is not self-energizing and needs an actuation force to balance the moments. The hinge (*i.e.* pivot) reactions are:

$$R_x = p_{max} b\, r \left(\int_{\theta_1}^{\theta_2} \sin\theta \cos\theta \, d\theta + \mu \int_{\theta_1}^{\theta_2} \sin^2\theta \, d\theta \right) - F_x \tag{7.16}$$

$$R_y = p_{max} b\, r \left(-\mu \int_{\theta_1}^{\theta_2} \sin\theta \cos\theta d\theta + \int_{\theta_1}^{\theta_2} \sin^2\theta \, d\theta \right) - F_y \tag{7.17}$$

It should be noted here that the actuation force on both shoes is the same; the maximum pressure in the left shoe is not the same as that of the right shoe. The limiting maximum pressure is always in the right shoe (for clockwise rotation).

Example 7.3: *Consider the brake system shown below. The width of the shoe system is 30 mm, and the coefficient of friction is 0.3. The friction material used permits a maximum pressure value of 1000 kPa. Find the actuation force, the total torque applied by the brake system, and the reaction force at the pivots of both shoes.*

Solution: We will first analyze the right hand shoe. The distance (a_1) from the pivot to the center is:

$$a_1 = \sqrt{(100)^2 + (50)^2} = 111.80 \text{ mm}$$

The angles $\theta_1 = 0$ and θ_2 can be found as

$$\theta_2 = 180 - 2\tan^{-1}\left(\frac{50}{100}\right) = 180 - 2 \times 26.5 = 127 \text{ degrees}$$

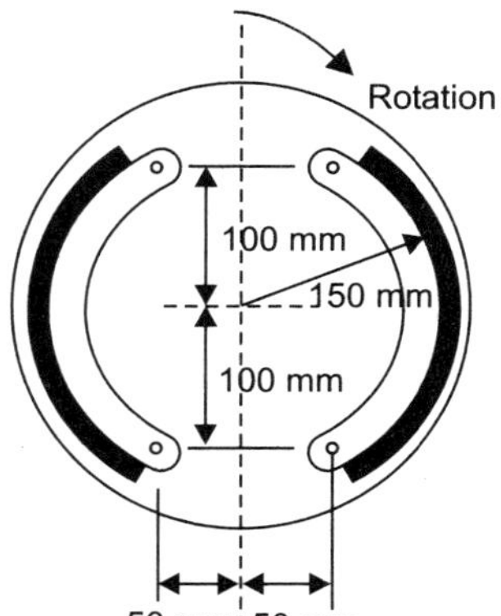

Fig. E.7.3 Brake described in Example 7.3.

Next, we need to find the moment from the friction forces:

$$M_F = \mu p_{\max} br\left[\left(-r\cos\theta\right)_0^{\theta_2} - a_1\left(\frac{\sin^2\theta}{2}\right)_0^{\theta_2}\right] = \mu p_{\max} br\left[r - r\cos\theta_2 - \frac{a_1}{2}\sin^2\theta_2\right]$$

$$= 0.3\times10^6\times0.03\times0.15\left[0.15 - 0.15\cos127° - \frac{0.118}{2}\sin^2 127°\right]$$

$$= 275.7 \quad \text{N.m}$$

The moment from the normal forces is

$$M_N = p_{\max} b\, r\, a_1\left[\frac{\theta}{2} - \frac{1}{4}\sin 2\theta\right]_0^{\theta_2} = p_{\max} b\, r\, a_1\left[\frac{\theta_2}{2} - \frac{1}{4}\sin 2\theta_2\right]$$

$$= 10^6\times0.03\times0.15\times0.1118\left[\frac{1}{2}\frac{\pi\times127}{180} - \frac{1}{4}\sin\left(2\times127°\right)\right]$$

$$= 677.8 \text{ N.m}$$

The applied or actuation force is

$$F = \frac{M_N - M_f}{e} = \frac{677.8 - 275.7}{0.2} = 2010.4 \text{ N}$$

Note that the above force is the resultant force which can be broken into x and y components (*i.e.* F_x and F_y). The reactions at the pivot of the right shoe are

$$R_x = p_{\max} b\, r\left(\int_{\theta_1}^{\theta_2}\sin\theta\cos\theta\, d\theta - \mu\int_{\theta_1}^{\theta_2}\sin^2\theta\, d\theta\right) - F_x$$

$$= 1000\times0.03\times0.15\left(\frac{1}{2}\sin^2 127 - 0.3\left(\frac{\pi\times127}{2\times180} - \frac{1}{4}\sin\left(2\times127\right)\right)\right) - 2.010\sin 26.5$$

$$= -1.2778 \text{ kN}$$

$$R_y = p_{max} b\, r\left(\mu\int_{\theta_1}^{\theta_2}\sin\theta\cos\theta\, d\theta + \int_{\theta_1}^{\theta_2}\sin^2\theta\, d\theta\right) - F_y$$

$$= 1000\times0.03\times0.15\left(0.3\left(\frac{1}{2}\sin^2 127°\right)+\left(\frac{\pi\times127}{2\times180}-\frac{1}{4}\sin(2\times127°)\right)\right)-2.010\cos 26.5$$

$$= 4.6961 \text{ kN}$$

The resultant force at the right pivot is

$$R = \sqrt{R_x^2 + R_y^2} = \sqrt{(-1.2778)^2 + (4.6961)^2} = 4.8668 \text{ kN}$$

Next, we will find the torque of the right hand shoe (T_R):

$$T_R = \mu p_{max} b\, r^2 (\cos\theta_1 - \cos\theta_2)$$

$$= 0.3\times1000\times0.032\times(0.15)^2(\cos 0° - \cos 127°)$$

$$= 0.3240 \text{ kNm}$$

Before finding any of the values of the left hand shoe, we will need to find the maximum pressure on that shoe. We will use the fact that the actuation force on both shoes is the same to determine the maximum pressure on the left shoe.

$$M_F = \mu p_{max} br\left[(-r\cos\theta)_0^{\theta_2} - a_1\left(\frac{\sin^2\theta}{2}\right)_0^{\theta_2}\right] = \mu p_{max} br\left[r - r\cos\theta_2 - \frac{a_1}{2}\sin^2\theta_2\right]$$

$$= 0.3\times p_{max}\times0.03\times0.15\left[0.15 - 0.15\cos127° - \frac{0.1118}{2}\sin^2 127°\right]$$

$$= 2.757\times10^{-4}\, p_{max}$$

The moment from the normal forces is

$$M_N = p_{max} b\, r\, a_1\left[\frac{\theta}{2}-\frac{1}{4}\sin 2\theta\right]_0^{\theta_2} = p_{max} b\, r\, a_1\left[\frac{\theta_2}{2}-\frac{1}{4}\sin 2\theta_2\right]$$

$$= p_{max}\times0.03\times0.15\times0.1118\left[\frac{1}{2}\frac{\pi\times127}{180}-\frac{1}{4}\sin(2\times127°)\right]$$

$$= 6.778\times10^{-4}\, p_{max}$$

The applied or actuation force is

$$F = \frac{M_N + M_f}{e} = \frac{(6.778+2.757)\times10^{-4}\, p_{max}}{0.2} = 2010.4 \text{ N}$$

or $$p_{max} = \frac{2010.4\times0.2}{(6.778+2.757)\times10^{-4}} = 4.21691\times10^5 \text{ N.m} = 421.7 \text{ kPa}$$

Thus the torque on the left hand shoe is

$$T_L = \mu p_{max} b\, r^2(\cos\theta_1 - \cos\theta_2)$$

$$= 0.3\times421\times0.032\times(0.15)^2(\cos 0° - \cos 127°)$$

$$= 0.1366 \text{ kN.m}$$

The total torque on the shoes is

$$T = T_L + T_R = 0.4606 \text{ kN.m}$$

Finally, we need to find the reaction forces on the pivot of the left shoe:

$$R_x = p_{max} b\, r\left(\int_{\theta_1}^{\theta_2} \sin\theta \cos\theta \, d\theta + \mu \int_{\theta_1}^{\theta_2} \sin^2\theta \, d\theta\right) - F_x$$

$$= 421.7\times0.03\times0.15\left(\frac{1}{2}\sin^2 127° + 0.3\left(\frac{\pi\times127}{2\times180} - \frac{1}{4}\sin(2\times127°)\right)\right) - 2.0104 \sin 26.5$$

$$= 0.4751 \text{ kN}$$

$$R_y = p_{max}\, b\, r\left(\mu \int_{\theta_1}^{\theta_2} \sin\theta \cos\theta \, d\theta - \int_{\theta_1}^{\theta_2} \sin^2\theta \, d\theta\right) - F_y$$

$$= 421.7\times0.03\times0.15\left(-0.3\left(\frac{1}{2}\sin^2 127°\right) + \left(\frac{\pi\times127}{2\times180} - \frac{1}{4}\sin(2\times127)\right)\right) - 2.0104 \cos 26.5$$

$$= 0.5761 \text{ kN}$$

The resultant force at the right pivot is

$$R = \sqrt{R_x^2 + R_y^2} = \sqrt{(0.4751)^2 + (0.5761)^2} = 0.7467 \text{ kN}$$

Note that the left shoe experiences less pressure and less reaction forces on its pivot than the right shoe.

Let us use the parameters used in the above example to find the total torque as a function of the coefficient of friction. The table below shows the variations in both the application force (needed to achieve a maximum pressure of 1000 kPa in the right hand shoe) and total torque.

Table 7.2 Relation between coefficient of friction actuation force and total torque

Coefficient of Friction	*Actuation Force (F)*	*Total Torque (T)*
0	3388.9	0
0.05	3159.1	101.14
0.1	2929.4	190.21
0.15	2699.6	269.24
0.2	2469.9	339.84
0.25	2240.0	403.29
0.3	2010.4	460.63
0.35	1780.6	512.69
0.4	1550.8	560.18
0.45	1321.1	603.66
0.5	1091.3	643.63

The most important observation is to note that the actuation force is linearly related to the coefficient of friction. The total torque, however, has a nonlinear relation with the coefficient of friction.

Customer's 'brake feel' is mostly related to the brake force needed to generate a braking action. The brake designer can actually choose two shoes for the B and one for A, if a higher advantage (or usually referred to as a brake factor) is desired. Brake designers should be careful as a higher advantage may result in a less robust design, particularly when the heat generated in the system alters the coefficient of friction slightly. Also, this may result in the undesired noise of brake squeal.

7.2.2 Disk Brakes

A more robust design in terms of the actuation force needed is that of a disk brake. In this brake system, the mechanical advantage is practically 1, and slight variation of the coefficient of friction will not result in magnified application force. For disk brakes, however, the application force is relatively high.

Disk brakes can be implemented by applying a frictional disk over the whole exposed area of the disk of the driven spindle. This is not very common in automotive engineering mainly because of the high energy that needs to be absorbed. For that reason, the disk brake shoes are only applied to a limited surface of the disk of the driven spindle. This will expose the remaining surface to air to dissipate heat and cool-off the surface. Caliper disk brakes are used extensively in the automotive industry, particularly for the front wheels. If the friction surface area of contact between the brake shoes and the driven disk is A, the applied force is F, then the pressure over the shoe pads is

$$p = F/A \tag{7.18a}$$

The torque applied at the wheel is

$$T = \mu F\, r_{ave} \tag{7.18b}$$

Where r_{ave} is the average radius from the center of the moving element to the contacting surface. Note that the torque applied is linearly dependent on the coefficient of friction, and thus variations in these coefficients due to heat will not be magnified as was observed for drum brakes due to the nonlinearity.

7.2.3 Consideration of Temperature

One important consideration that should be made when designing the braking system is that of heat generation and temperature rise. This is important because of two reasons. The first is that temperature rise affects the coefficient of friction considerably and thus affects the actuation force needed for a certain braking action. The second is that the generated heat is the main reason of accumulative damage in the braking system.

The energy that needs to be absorbed by the braking systems of the vehicle is the kinetic energy (K.E.) of the vehicle.

$$\text{K.E.} = \frac{1}{2} M \left(V_i^2 - V_f^{\,2}\right) \tag{7.19}$$

where M is the mass of the vehicle. V_f is zero if the vehicle is to get to a complete stop. This energy is to be absorbed by the braking system over the braking period. It will be transferred into heat energy at the braking system, resulting in a temperature rise.

The power absorption is a function of the braking force and the speed. If the braking force is constant over the braking action, the power absorption is highest at the beginning of the braking action, when the speed is maximum. The power at that point is the direct product of speed times the braking force. Over the interval time of the braking system, the average power can be calculated as the K.E. divided by the time interval.

Example 7.4: *A large sedan of 1779.29 N (4000 lb) traveling at 128.75 km/hr (80 mph) is to reach a complete in 10 seconds. Find the energy and average power.*

Solution:

$$M = \frac{4000 \text{ lb}}{32.2 \text{ ft/sec}^2} = 124.22 \text{ slug}$$

$$V_i = \frac{80 \text{ mile}}{\text{hr}} \times \frac{5280 \text{ ft}}{\text{mile}} \times \frac{\text{hr}}{3600 \text{ sec}} = 117.33 \text{ ft/sec}$$

$$\text{K.E.} = \frac{1}{2} M \left(V_i^2 - V_f^2\right) = \frac{1}{2} \times 124.22 \frac{\text{lb} \times \text{sec}^2}{\text{ft}} \times \left(117.33 \frac{\text{ft}}{\text{sec}}\right)^2 = 855075 \text{ ft} \cdot \text{lb}$$

$$\text{Power} = \frac{855075 \text{ ft} \cdot \text{lb}}{10 \text{ sec}} \frac{\text{HP}}{550 \frac{\text{ft} \cdot \text{lb}}{\text{sec}}} = 155 \text{ HP}$$

The temperature rise at the brake can be estimated using the following equation

$$\Delta\text{Temp} = \frac{K.E.}{Sm} \tag{7.20}$$

where S is the specific heat (500 J/kg.C° for steel) and m is the mass of the braking system.

Example 7.5: *Estimate the temperature rise of a vehicle of mass 2000 kg and traveling at 100 km/hr, to reach a complete stop in 10 seconds, the braking system has a mass of 10 kg. [Let S = 500 J/ (kg.C°)]*

Solution:

$$V_i = \frac{100 \text{ km}}{\text{hr}} \times \frac{1000 \text{ m}}{\text{km}} \times \frac{\text{hr}}{3600 \text{ sec}} = 27.78 \text{ m/sec}$$

$$\text{K.E.} = \frac{1}{2} M \left(V_i^2 - V_f^2\right) = \frac{1}{2} \times 2000 \text{ kg} \times \left(27.78 \frac{\text{m}}{\text{sec}}\right)^2 = 771605 \text{ J}$$

$$\Delta\text{Temp} = \frac{771605}{500 \times 10} = 154.3 \text{ C}°$$

The coefficient of friction is a function of various environmental factors, most notably are the velocity of the vehicle and the temperature of the shoes. The challenge to the application engineer is to try to find the value of the brake coefficient. Unfortunately, the application of a high brake force will result in high temperature (exceeding 600° F) and will result in permanent changes to the braking torque. Recently, advanced finite element packages have been used to predict the temperature and the consequent changes of the brake torque performance of the system.

7.3 LOAD TRANSFER DURING BRAKING

The fundamental forces acting on a vehicle when undergoing braking were discussed in Section 3.3.5 of Chapter 3. Some of these forces will be reviewed briefly here. We will consider first braking of a vehicle on a horizontal road (*i.e.* no grade).

7.3.1 Simple Braking on a Horizontal Road

The free body diagram of the vehicle that shows the forces acting on the vehicle under normal driving conditions is given in Fig. 7.4. Note that we have neglected aerodynamics forces, rolling resistance and powerplant resistance in this simple vehicle model. Applying Newton's second law in the horizontal direction yields:

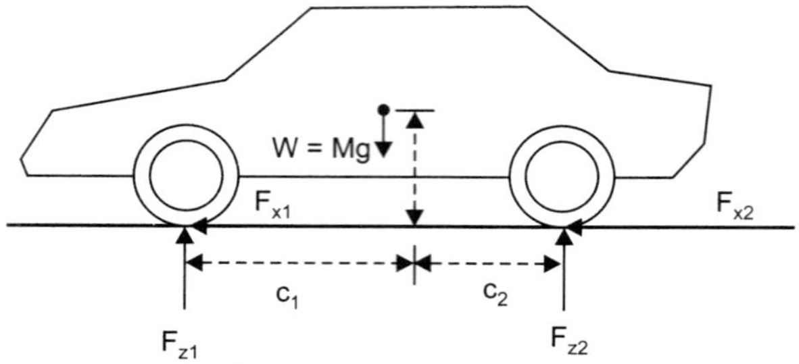

Fig. 7.4 Simple vehicle model on a horizontal plane.

$$F_{x1} + F_{x2} = Ma \tag{7.21}$$

where F_{x1} is the friction force on the front wheels, F_{x2} is the friction force on the rear wheels, c_1 is the horizontal (parallel to the road surface) distance between the contact point of the front wheel and the center of gravity, c_2 is the horizontal distance between the contact point of the rear wheel and the center of gravity, and a is its acceleration.

Applying Newton's second law in the vertical direction yields

$$F_{z1} + F_{z2} = Mg \tag{7.22}$$

where F_{z1} is the vertical force on the front wheels, F_{z2} is the vertical force on the rear wheels and g is the acceleration of gravity. Summing the moment about the center of gravity (CG) point yields

$$F_{z1}c_1 - F_{z2}\, c_2 + (F_{x1} + F_{x2})h = 0 \tag{7.23}$$

where c_1 is the horizontal (parallel to the road surface) distance between the contact point of the front wheel and the center of gravity, c_2 is the horizontal distance between the contact point of the rear wheel and the center of gravity, and h is the vertical (perpendicular to the road surface) distance between the contact points of the wheels contact with the ground and the CG. Note that the above three equations (Eq. (7.21-7.23) involve 4 unknowns: F_{x1}, F_{x2}, F_{z1}, and F_{z2}. This is an indeterminate system, where one needs one additional equation to solve for all the unknowns. Fortunately, two of the unknowns can be found by substituting the first equation into the third equation. This yields

$$F_{z1}c_1 - F_{z2}\, c_2 + Ma\, h = 0 \tag{7.24}$$

Equations (7.22) and (7.24) can be used directly to solve for the vertical forces (axle loads) independently:

$$F_{z1} = Mg\left(\frac{c_1}{c_1 + c_2}\right) - \frac{h}{c_1 + c_2} Ma$$

$$F_{z2} = Mg\left(\frac{c_1}{c_1 + c_2}\right) + \frac{h}{c_1 + c_2} Ma \tag{7.25}$$

Note that we are assuming that the vehicle is accelerating. Merely, deceleration will be treated simply by using a negative value for a.

It should also be noted that when the vehicle is braking, and negative a is used; the axle force on the front wheels will be larger than those forces experienced under cruising driving conditions (*i.e.*, no acceleration or deceleration). This is felt by drivers, particularly when the vehicle reaches a complete stop.

Example 7.6: *Consider a midsize sedan that weighs 14234.31 lb (3200 N) under a certain load. The wheel base of the vehicle is at 9 ft or 108 in (2.74 m) and the center of gravity is at 1.067 m (3.5 ft) from the front axle and at 0.508 m (20 in) above the ground. Determine the load on each axle if the brake is under no acceleration, decelerating at 1.524 m/s² (5 ft/sec²), 3.048 m/s² (10 ft/sec²) and 4.572 m/s² (15 ft/sec²).*

Solution: Given in the problem:

$$W = 3200 \text{ lb}$$
$$c_1 + c_2 = 108 \text{ in}$$
$$h = 20 \text{ in}$$
$$c_1 = 42 \text{ in}$$

We will first find the mass and c_2

$$M = 3200/32.2 = 99.38 \text{ slugs}$$
$$c_2 = 108 - 42 = 66 \text{ in}$$

Apply eqs. 7.25 to get the loading under no acceleration (*i.e.* $a = 0$)

$$F_{z1} = Mg\left(\frac{c_2}{c_1 + c_2}\right) - \frac{h}{c_1 + c_2} Ma = 99.38 \times 32.2\left(\frac{66}{108}\right) - \frac{20}{108} \times 99.38 \times 0 = 1955.56 \text{ lb}$$

$$F_{z2} = Mg\left(\frac{c_1}{c_1 + c_2}\right) + \frac{h}{c_1 + c_2} Ma = 99.38 \times 32.2\left(\frac{42}{108}\right) + \frac{20}{108} \times 99.38 \times 0 = 1244.44 \text{ lb}$$

Similarly for $a = -5$ ft/sec²

$$F_{z1} = 99.38 \times 32.2\left(\frac{66}{108}\right) - \frac{20}{108} \times 99.38 \times (-5) = 2047.57 \text{ lb}$$

$$F_{z2} = 99.38 \times 32.2\left(\frac{42}{108}\right) + \frac{20}{108} \times 99.38 \times (-5) = 1152.43 \text{ lb}$$

and for $a = -10$ ft/sec²

$$F_{z1} = 99.38 \times 32.2\left(\frac{66}{108}\right) - \frac{20}{108} \times 99.38 \times (-10) = 2139.59 \text{ lb}$$

$$F_{z2} = 99.38 \times 32.2\left(\frac{42}{108}\right) + \frac{20}{108} \times 99.38 \times (-10) = 1060.41 \text{ lb}$$

Finally, for $a = -15$ ft/sec²

$$F_{z1} = 99.38 \times 32.2\left(\frac{66}{108}\right) - \frac{20}{108} \times 99.38 \times (-15) = 2231.61 \text{ lb}$$

$$F_{z2} = 99.38 \times 32.2\left(\frac{42}{108}\right) + \frac{20}{108} \times 99.38 \times (-15) = 968.39 \text{ lb}$$

Note that in the previous example the total load on the axles remains at 3200 lb, which is the weight of the vehicle. What has happened in reality is that load has transferred from the rear axle to the front axle as the vehicle was decelerating. Figure 7.5 shows the load distribution as a function of the deceleration.

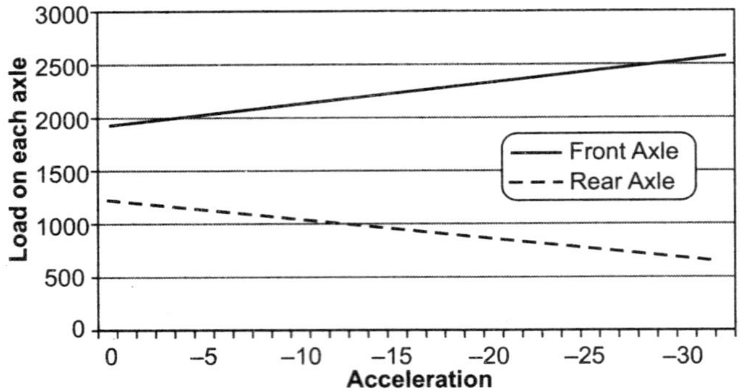

Fig. 7.5 Load distribution under braking.

7.3.2 Effect of Aerodynamic and Other Forces

A typical driver notices that speed reduction can be achieved without the braking action. In other words, if the vehicle is not cruising and the driver takes his foot off the gas pedal, deceleration is observed. This makes us conclude that there are other resistance forces that cause the vehicle to decelerate.

In the previous section, we studied a simple vehicle model where the vehicle is on a horizontal plane and under its own weight and braking forces without other loading conditions. There are other forces that should be accounted for in actual driving conditions. Even on a horizontal surface, the rolling resistance of the tires, the aerodynamic resistance and the driveline drag should be accounted for.

7.3.2.1 Rolling Resistance

The rolling resistance is between the tires and the drive axle. If the rolling coefficients are distinguished by μ_r then the rolling resistance can be estimated by:

$$R_r = \mu_r Mg \tag{7.26}$$

The rolling resistance is caused by the hysteresis in tire materials which results from the deflection of the tire while rolling. Friction between the tire and road when sliding and resistance due to air circulating inside the tire contribute minimally to the rolling resistance.

The tire structure including its construction and material as well as its operating conditions is found to be the major factor that affects the rolling resistance. Operation conditions include the road surface, temperature, pressure, and speed. The coefficient of rolling resistance varies between 0.012 and 0.02 on concrete surfaces with radial tires offering lower coefficients than bias tires. This coefficient reaches 0.1 when the vehicle is driven on hard soil and it reaches 0.2-0.3 when the vehicle is driven on sand. The rolling resistance is usually less for tires with larger diameters.

7.3.2.2 Aerodynamic Drag

Aerodynamic forces (drag) should also be considered when studying braking of a vehicle. The aerodynamic resistance is generated mainly by air flow around the vehicle and air flow through the vehicle. Air flow around the vehicle is mainly its flow over the exterior of the vehicle. Air flow through the vehicle is basically flow through the engine radiator for the purposes of heating, cooling and ventilation. The first of the two accounts for the main aerodynamic resistance. While skin friction forces are important, the normal pressure of the air on the vehicle acts against the motion of the vehicle and contributed the most to the aerodynamic drag.

Aerodynamic resistance is estimated using the following equation

$$R_a = \frac{\rho}{2} C_D A_F V_r^2 \tag{7.27}$$

where ρ is the density of the air, C_D is the drag coefficient, A_F is the projected area of the front of the vehicle and V_r is the relative velocity of the vehicle with respect to the wind. The air density is a function of the altitude and temperature. Normal operating temperatures can lead to approximately 20% variation in the density and similarly for altitude. Low temperatures at low altitudes give rise to the higher density. The coefficient of drag has historically been a subject of many studies. An aerodynamically designed vehicle can reach a coefficient of drag of as low as 0.3. It can also go as high as 0.5 if the car is not designed properly for aerodynamic resistance. This coefficient reaches the value of 0.8 for vans. The frontal area of the vehicle is the area that a person sitting in front of the vehicle sees. It can go from 1.7 m^2 for a mini car to a value of 2.2 m^2 for a luxury full size car. The most important observation of Eq. (7.27) is to note that the aerodynamic resistance is a function of square the speed.

It should be noted that equations for aerodynamic lift can be written in a fashion similar to that of Eq. (7.27), *i.e.*

$$R_L = \frac{\rho}{2} C_L A_F V_r^2 \tag{7.28}$$

The aerodynamic lift acts upward on the vehicle. The coefficient of aerodynamic lift (C_L) is usually determined in wind tunnel testing.

Similarly, the aerodynamic pitching can be calculated using the equation

$$M_a = \frac{\rho}{2} C_M L_C V_r^2 \tag{7.29}$$

where L_C is the characteristic length of the vehicle and C_M is the coefficient of aerodynamic pitching moment obtained from wind tunnel testing.

7.3.2.3 Powertrain Resistance

The powertrain of the vehicle consists of its powerplant (engine and transmission) and the driveline. These are the rotating components of the vehicle. Naturally, internal friction is present in these elements, mostly in the gearing actions of the transmission or power take-off or axle units in a 4 × 4 drive system.

The resistance of the powertrain is usually negligible in the consideration of the braking system of a typical vehicle.

7.3.2.4 Load Transfer on a Horizontal Plane

We will now consider the aerodynamic drag and rolling resistance in developing the load transfer equations. Figure 7.6 shows the vehicle model on a horizontal plane with consideration of rolling resistance and aerodynamic drag.

Note that the resultant of the aerodynamic drag is represented here by the force R_a acting at some height h_a. These are found either from wind tunnel testing or, more recently, from computational fluid dynamic (CFD) software packages. The vertical forces (axle loads) can be found in a manner similar to that used in Section 7.3.1:

$$F_{z1} = Mg\left(\frac{c_2}{c_1+c_2}\right) - \frac{h_a}{c_1+c_2}R_a - \frac{h}{c_1+c_2}Ma$$

$$F_{Z2} = Mg\left(\frac{c_1}{c_1+c_2}\right) + \frac{h_a}{c_1+c_2}R_a + \frac{h}{c_1+c_2}Ma \qquad (7.30)$$

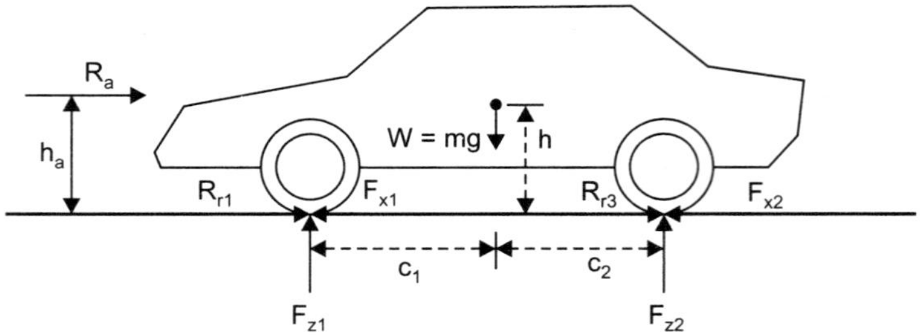

Fig. 7.6 Vehicle model on a horizontal plane.

Deceleration will be treated simply by using a negative value for a. Often h_a is assumed to be the same as h.

7.3.3 Effect of Grade

We will now consider the effect of grade as well as aerodynamic drag and rolling resistance in developing the load transfer equations. Figure 7.7 shows the vehicle model on an inclined surface with consideration of rolling resistance and aerodynamic drag.

The vertical forces (axle loads) can be found in a manner similar to that used in Section 7.3.1:

$$-F_{z1}(c_1+c_2) + Mgc_2\cos\theta - h_aR_a - hMg\sin\theta = hMa$$

$$F_{z2}(c_1+c_2) - Mgc_2\cos\theta - h_aR_a - hMg\sin\theta = hMa \qquad (7.31a)$$

$$F_{z1} = \frac{1}{c_1+c_2}\left(Mgc_2\cos\theta - h_aR_a - hMa - hMg\sin\theta\right)$$

or

$$F_{z2} = \frac{1}{c_1+c_2}\left(Mgc_1\cos\theta + h_aR_a + hMa + hMg\sin\theta\right) \qquad (7.31b)$$

where θ is the grade or slope angle, R_a is the aerodynamic drag, and h_a is the vertical distance (perpendicular to the road surface) between the road and the center of the aerodynamic force.

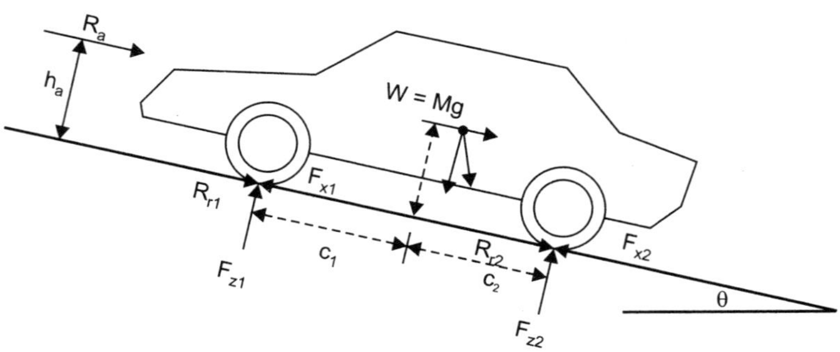

Fig. 7.7 Vehicle model on an incline (grade).

Note that when the vehicle is going uphill, the negative sign should be used for the last term. Deceleration will be treated simply by using a negative value for a.

Example 7.7: *Consider a midsize sedan that weighs 14234.31 N (3200 lb) under a certain load. The vehicle is going uphill a 5 degrees slope at 26.82 m/s (60 mph or 88 ft/sec). The front area of the vehicle is 2.044 m² (22.0 ft²) and the coefficient of drag is 0.4. The aerodynamic resistance is acting on the CG. The wheel base of the vehicle is at 2.74 m (9 ft) and the center of gravity is at 1.372 m (4.5 ft) from the front axle and at 0.508 m (20 in) above the ground. Determine the load on each axle if the vehicle is under no acceleration. What would the loads be if the vehicle is decelerating at 3.048 m/s² (10 ft/sec²). The density of air is 1.226 kg/m³ (0.002378 slug/ft³). Calculate forces when going down hill at 5 degree slope with the same speed.*

Solution: The aerodynamic resistance is first calculated

$$R_a = \frac{\rho}{2} C_D A_F V_r^2$$

$$= \frac{0.002378}{2} \times 0.4 \times 22 \times (88)^2 = 81.03 \text{ lb}$$

The forces on the axles are (when going on a horizontal road)

$$F_{z1} = \frac{1}{c_1 + c_2}\left(Mgc_2 \cos\theta - h_a R_a - hMa - hMg \sin\theta\right)$$

$$= \frac{1}{c_1 + c_2}\left(Mgc_2 - h_a R_a - hMa\right)$$

$$= \frac{1}{108}\left(3200 \times 66 - 20 \times 81 - 20 \times 99.38 \times (-10)\right) = 2124.6 \text{ lb}$$

$$F_{z2} = \frac{1}{c_1 + c_2}\left(Mgc_1 \cos\theta + h_a R_a + hMa + hMg \sin\theta\right)$$

$$= \frac{1}{c_1 + c_2}\left(Mgc_1 + h_a R_a + hMa\right)$$

$$= \frac{1}{108}\left(3200 \times 42 + 20 \times 81 + 20 \times 99.38 \times (-10)\right) = 1075.4 \text{ lb}$$

The forces on the axles are (when going uphill)

$$F_{z1} = \frac{1}{c_1 + c_2}\left(Mgc_2 \cos\theta - h_a R_a - hMa - hMg \sin\theta\right)$$

$$= \frac{1}{108}\left(3200 \times 66 \times \cos(5) - 20 \times 81 - 20 \times 99.38 \times (-10) - 20 \times 3200 \times \sin(5)\right) = 2065.5 \text{ lb}$$

$$F_{z2} = \frac{1}{c_1 + c_2}\left(Mgc_1 \cos\theta + h_a R_a + hMa + hMg \sin\theta\right)$$

$$= \frac{1}{108}\left(3200 \times 42 \times \cos(5) + 20 \times 81 + 20 \times 99.38 \times (-10) + 20 \times 3200 \times \sin(5)\right) = 1122.3 \text{ lb}$$

When the vehicle goes down hill, the forces are

$$F_{z1} = \frac{1}{c_1 + c_2}\left(Mgc_2 \cos\theta - h_a R_a - hMa + hMg \sin\theta\right)$$

$$= \frac{1}{108}\left(3200\times 66\times \cos(5) - 20\times 81 - 20\times 99.38\times(-10) + 20\times 3200\times \sin(5)\right) = 2168.8 \text{ lb}$$

$$F_{z2} = \frac{1}{c_1 + c_2}\left(Mgc_1 \cos\theta + h_a R_a + hMa - hMg \sin\theta\right)$$

$$= \frac{1}{108}\left(3200\times 42\times \cos(5) + 20\times 81 + 20\times 99.38\times(-10) - 20\times 3200\times \sin(5)\right) = 1019.0 \text{ lb}$$

The load described here represents the weight distribution at the initial braking. Aerodynamic forces will reduce as the speed starts decreasing and vanish at the full stop.

Comparing the results with those obtained in example 7.5, where aerodynamic loads were neglected, it is observed that the addition of aerodynamic forces has reduced the load on front axle, and increased it for rear axle. Also note that the above example represents normal operating conditions and not severe ones.

7.4 OPTIMAL BRAKING PERFORMANCE

Design consideration of automotive brake system, where 4 brakes exist on the four wheels, require considering vehicle safety and brake lock-up mechanism. In particular, vehicle stability and the ability to steer are of major concern, particularly when braking is conducted on surfaces with low coefficient of friction. Indeed, a braking strategy needs to be developed to achieve highest level of safety. We will start considering braking of single axis, but will move later to consider braking of both axes.

7.4.1 Braking at a Single Axle

We will first consider braking of a single axle, either front or rear. In order to get some insight of the vehicle behavior, we will only treat the braking force and the vehicle's weight on a horizontal plane.

7.4.1.1 Braking of Front Axle

Let us assume brakes are applied on the front axle. In this case, the braking force at the front wheels will be equal to the coefficient of friction multiplied by the normal force at those wheels:

$$F_{x1} = -\mu F_{z1} \tag{7.32}$$

The braking force at the rear wheels will be zero. The maximum deceleration then can be found from Eq. (7.21) by utilizing Eq. (7.25)

$$Ma_F = -\mu Mg\left(\frac{c_2}{c_1 + c_2}\right) + \mu\left(\frac{h}{c_1 + c_2}\right)Ma_F \tag{7.33}$$

Note that we replaced a by a_F to indicate that the above equation is only valid when only front brakes are applied. The above equation can be directly solved for the maximum deceleration:

$$a_F = -g\frac{\mu\left(\dfrac{c_2}{c_1 + c_2}\right)}{1 - \mu\left(\dfrac{h}{c_1 + c_2}\right)} \tag{7.34}$$

7.4.1.2 Braking of Rear Axle

Similar treatment can be made if the brakes are applied to the rear wheels only. In this case, the braking force on the front wheels is zero. The braking force on the rear wheels is

$$F_{x2} = \mu F_{z2} \tag{7.35}$$

One can develop an equation for the maximum deceleration when under rear wheel braking:

$$a_R = -g \frac{\mu\left(\dfrac{c_1}{c_1 + c_2}\right)}{1+\mu\left(\dfrac{h}{c_1 + c_2}\right)} \tag{7.36}$$

Figure 7.8 shows the deceleration of a typical compact vehicle.

Note that for typical vehicle designs and loading conditions in the vehicle, the value of $c_2/(c_1+ c_2)$ ranges from 0.4 to 0.6. Within this range of application, the maximum deceleration that can be obtained when the rear brakes only are activated is between the values –0.48g and –0.32g. This deceleration is between –0.53g and –0.8g when the vehicle is under front braking only. This shows that for normal braking conditions in a single drive braking, better deceleration can be achieved with front axle braking.

Figure 7.9 shows a similar graph when the coefficient of friction is 0.5 (instead of 1). Within the range of normal application, the maximum deceleration that can be obtained when the rear brakes only are activated is between the values –0.27g and –0.18g. This deceleration is between –0.23g and –0.34g when the vehicle is under front braking only. Note also here that, in general, front axle braking gives higher deceleration except in the vicinity of $c_2/(c_1+ c_2)$ of 0.4, where higher deceleration is obtained with rear axle braking.

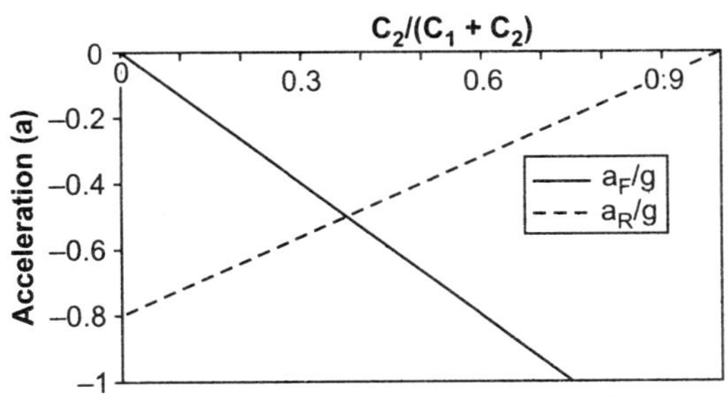

Fig. 7.8 Deceleration under front or rear braking, $\mu = 1$, $h = 0.5$ m, $c_1+ c_2 = 2.0$.

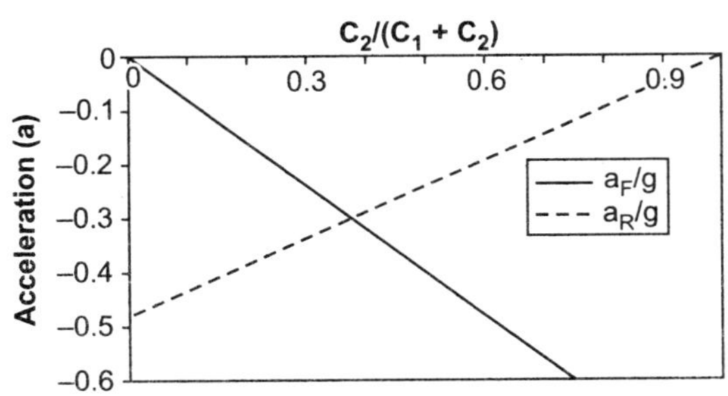

Fig. 7.9 Deceleration under front or rear braking, $\mu = 0.5$, $h = 0.5$ m, $c_1+ c_2 = 2.0$.

In real application brakes are supplied at all four wheels in order to achieve the maximum braking force before lock-up occurs. The weight is distributed at both axles differently depending on design and loading conditions as well as environmental factors (aerodynamics and grade). Assuming that the same coefficient of friction exists between all four wheels and the surface, the maximum braking force that can be achieved at each axle is directly proportional to the load on the axle. In order to achieve maximum braking force for the whole vehicle, the braking force distribution between the front and rear wheels should be proportional to that of the axle load. In other words, if the front axle carries 60% of the load and the rear axle carries the remaining 40% of the load, achieving maximum braking force will require that 60% of the total braking force should be provided by the front wheels and 40% provided by the rear wheels. This distribution of braking forces, which can be designed for, assures that lock-up happens at the same time for both axles. Lock-up, however, of any or both axles has significant consequences that should be discussed.

7.4.1.3 Safety Considerations

If the rear tires lock-up first, the vehicle will lose stability of the direction. Indeed, when driving on surfaces with low coefficient of friction (e.g. ice), rear lock-up of the brake system results in the vehicle rotating by an angle of 180 degrees. At rear lock-up, the rear axles can not resist any lateral movement. Slight forces, because of side winds or other factors, will cause the rear wheels to move laterally. This will result in a yawing moment of the inertia force about the yaw center, causing further lateral motion of the rear wheels. This will in turn result in a larger arm of the inertia forces and larger moment, which causes again further lateral motion. This self-energizing moment will not stop until the vehicle turns 180 degrees. Thus the directional stability of the vehicle is completely lost and serious safety consequences are possible.

On the other hand, if the lock-up starts at the front axle, directional control of the vehicle is lost. The driver will not be able to control the steering of the vehicle. Front tire lock-up does not cause the kind of directional instability noticed for rear wheel lock-up.

It should be noted here that when directional stability is lost, the driver can not regain it. In other words, the vehicle will continue with the self-energizing yawing moment until the vehicle reaches a complete stop with the 180 degree turn. The driver however can regain the ability to steer if the brakes are released slightly to prevent the front lock-up. Because of these consideration, braking requirement often calls for front lock-up to happen ahead of the rear lock-up. Although this will not allow us to achieve maximum braking force with both lock-ups happening at the same time, but it is worth it to design the braking system such that front lock-up happens ahead of rear lock-up for the mentioned safety reasons. The reader is referred to the book of Limpert (1992) and the paper by Dahlberg (1999) for more detailed analysis of the problem.

7.4.2 Braking at Both Axles

In today's vehicles, both axles have a braking system. Disk brakes are commonly used for the front axle and drum brakes used for the rear axle. As mentioned earlier, the brake designer needs the maximum braking force that can be obtained from the braking system at hand and needs to assure that front lock-up occurs first under various loading and environmental conditions. Let us assume that the brake designer chooses, based on certain emergency conditions, to distribute the braking load such that a certain ratio (K_f) of the total braking force goes to the front brakes. The remainder that goes to the rear brakes is $(1 - K_f)$ of the total braking force.

$$F_{x1} = K_f F_x$$

$$F_{x2} = (1 - K_f)\, F_x \tag{7.37a}$$

or

$$F_{x1} = \frac{K_f}{1 - K_f} F_{x2}$$

$$F_{x2} = \frac{1 - K_f}{K_f} F_{x1} \tag{7.37b}$$

where F_x is the total friction force. Consider the free-body-diagram (FBD) of the vehicle under direct braking forces $-F_{x1}$ for the front axle and $-F_{x2}$ for the rear axle. The vehicle is still under effects of aerodynamic forces as well as grade and rolling resistance. The total braking force is $(-F_x = -F_{x1} - F_{x2})$. Note the negative sign indicates that the direction of the force is opposite to that shown in Fig. 7.7. The total rolling resistance force is $(R_r = R_{r1} + R_{r2})$. The total weight is distributed as follows

$$F_{z1} = \frac{1}{c_1 + c_2}\left(Mgc_2 \cos\theta - h_a R_a - hMa - hMg \sin\theta\right) \tag{7.31b}$$

$$F_{z2} = \frac{1}{c_1 + c_2}\left(Mgc_1 \cos\theta + h_a R_a + hMa + hMg \sin\theta\right)$$

Note that $F_{z1} + F_{z2}$ will always be

$$F_{z1} + F_{z2} = \frac{1}{c_1 + c_2}\left(Mgc_2 \cos\theta\right) + \frac{1}{c_1 + c_2}\left(Mgc_1 \cos\theta\right) = Mg \cos\theta \tag{7.38}$$

Considering Fig. 7.7 again and summing the forces in the direction of motion yields

$$F_{x1} + F_{x2} - R_a - R_r - Mg \sin(\theta) = Ma \tag{7.39}$$

7.4.2.1 Front Lock-up

Under front lock-up, $F_{x1} = -\mu F_{z1}$, substituting this and Eqs. 7.31 into 7.39 yields:

$$-\mu\left(\frac{1}{c_1 + c_2}\left(Mgc_2 \cos\theta - h_a R_a - hMa_F - hMg \sin\theta\right)\right)\left(1 + \frac{1 - K_f}{K_f}\right)$$

$$-R_a - R_r - Mg \sin(\theta) = Ma_F \tag{7.40}$$

Rearranging for a_F

$$-\mu\left(\frac{1}{c_1 + c_2}\left(Mgc_2 \cos\theta - h_a R_a - hMg \sin\theta\right)\right)\left(1 + \frac{1 - K_f}{K_f}\right) - R_a - R_r - Mg \sin\theta$$

$$= \left(-\frac{\mu h}{c_1 + c_2}\left(1 + \frac{1 - K_f}{K_f}\right) + 1\right) Ma_F \tag{7.41}$$

or simply

$$a_F = \frac{-\mu\left(\frac{1}{c_1 + c_2}\left(Mgc_2 \cos\theta - h_a R_a - hMg \sin\theta\right)\right)\left(1 + \frac{1 - K_f}{K_f}\right) - R_a - R_r - Mg \sin\theta}{\left(-\frac{\mu h}{c_1 + c_2}\left(1 + \frac{1 - K_f}{K_f}\right) + 1\right) M} \tag{7.42}$$

Note that the acceleration a has a subscript F to indicate that the acceleration obtained by using the above equation is only valid when the front tires are locked-up.

Ignoring aerodynamic drag, grade and rolling resistance, and using the following simplification

$$1+\frac{K_f}{1-K_f}=\frac{1-K_f}{1-K_f}+\frac{K_f}{1-K_f}=\frac{1}{1-K_f} \quad \text{and} \quad 1+\frac{1-K_f}{K_f}=\frac{K_f}{K_f}+\frac{1-K_f}{K_f}=\frac{1}{K_f}$$

the deceleration becomes

$$a_F=-\frac{\mu\left(\frac{c_2}{c_1+c_2}\right)\left(1+\frac{1-K_f}{K_f}\right)}{\left(-\frac{\mu h}{c_1+c_2}\left(1+\frac{1-K_f}{K_f}\right)+1\right)}g=\frac{\frac{\mu}{K_f}\left(\frac{c_2}{c_1+c_2}\right)}{\left(1-\frac{\mu h}{K_f\left(c_1+c_2\right)}\right)}g \tag{7.43}$$

Note that the above expression may yield a denominator of zero. This will yield a very high number for a_F which will mean that rear lock-up must happen first in these circumstances. Eq. (7.43) will reduce to that of (7.34) when a front axle only is under breaking ($K_f = 1$).

7.4.2.2 Rear Lock-up

Eq. (7.38) can be re-written as

$$F_{x2}\left(1+\frac{K_f}{1-K_f}\right)-R_a-R_r-Mg\sin(\theta)=Ma \tag{7.44}$$

When under rear lock-up, $F_{x2}=-\mu F_{z2}$, substituting this and Eqs. 7.31 into 7.44 yields

$$-\mu\left(\frac{1}{c_1+c_2}\left(Mgc_1\cos\theta+h_aR_a+hMa_R+Mg\sin\theta\right)\right)\left(1+\frac{K_f}{1-K_f}\right)$$

$$-R_a-R_r-Mg\sin\theta=Ma_R \tag{7.45}$$

Rearranging for a_R

$$-\mu\left(\frac{1}{c_1+c_2}\left(Mgc_1\cos\theta+h_aR_a+Mg\sin\theta\right)\right)\left(1+\frac{K_f}{1-K_f}\right)-R_a-R_r-Mg\sin\theta$$

$$=\left(+\frac{\mu h}{c_1+c_2}\left(1+\frac{K_f}{1-K_f}\right)+1\right)Ma_R \tag{7.46}$$

Simplify

$$a_R=\frac{-\mu\left(\frac{1}{c_1+c_2}\left(Mgc_1+h_aR_a+Mg\sin\theta\right)\right)\left(1+\frac{K_f}{1-K_f}\right)-R_a-R_r-Mg\sin\theta}{\left(\frac{\mu h}{c_1+c_2}\left(1+\frac{K_f}{1-K_f}\right)+1\right)M} \tag{7.47}$$

Ignoring aerodynamic drag, grade and rolling resistance. The deceleration is

$$a_R = -\frac{\mu\left(\frac{c_1}{c_1+c_2}\right)\left(1+\frac{K_f}{1-K_f}\right)}{\left(\frac{\mu h}{c_1+c_2}\left(1+\frac{K_f}{1-K_f}\right)+1\right)}g \tag{7.48}$$

Note that Eq. (7.48) will reduce to that of (7.36) when a rear axle only is under breaking ($K_f = 0$). Front tires will lock-up first if $|a_F| < |a_R|$, otherwise rear tires will lock-up first.

Example 7.8: *A 21351.46 N (4800 lb) passenger car with 2844.80 mm (112 in) wheel base and a CG at 1270.00 mm (50 in) behind the front axle and 508 mm (20 in) above ground. The braking effort distribution gives the front axle 60% of the total braking force. The car is moving on a horizontal plane. Determine which tire will lock up first if the car is moving on a surface with first a 0.75 coefficient of friction, and second, a 0.25 coefficient of friction. Ignore drag and rolling resistance.*

Solution: $M = 4800 \text{ lbs} = 149.07 \text{ slugs}$

$c_1 + c_2 = 112$ in

$c_1 = 50$ in,

$c_2 = 62$ in,

$K_f = 0.6$

$h = 20$

$\theta = 0$

$R_a = 0$

$R_r = 0$

First surface: $\mu = 0.75$

Acceleration under front-lock-up

$$a_F = -\frac{\mu\left(\frac{c_2}{c_1+c_2}\right)\left(1+\frac{1-K_f}{K_f}\right)}{\left(-\frac{\mu h}{c_1+c_2}\left(1+\frac{1-K_f}{K_f}\right)+1\right)}g = -\frac{0.75\left(\frac{62}{112}\right)\left(1+\frac{0.4}{0.6}\right)}{\left(-\frac{0.75\times 20}{112}\left(1+\frac{0.4}{0.6}\right)+1\right)}g = -0.8908\, g$$

Acceleration under rear lock-up

$$a_R = -\frac{\mu\left(\frac{c_1}{c_1+c_2}\right)\left(1+\frac{K_f}{1-K_f}\right)}{\left(\frac{\mu h}{c_1+c_2}\left(1+\frac{K_f}{1-K_f}\right)+1\right)}g = -\frac{0.75\left(\frac{50}{112}\right)\left(1+\frac{0.6}{0.4}\right)}{\left(\frac{0.75\times 20}{112}\left(1+\frac{0.6}{0.4}\right)+1\right)}g = -0.2510\, g$$

Since $|a_F| > |a_R|$, rear tires will lock-up first.

Second surface: $\mu = 0.25$

Acceleration under front-lock-ups

$$a_F = -\frac{\mu\left(\frac{c_2}{c_1+c_2}\right)\left(1+\frac{1-K_f}{K_f}\right)}{\left(-\frac{\mu h}{c_1+c_2}\left(1+\frac{1-K_f}{K_f}\right)+1\right)}g = -\frac{0.25\left(\frac{62}{112}\right)\left(1+\frac{0.4}{0.6}\right)}{\left(-\frac{0.25\times 20}{112}\left(1+\frac{0.4}{0.6}\right)+1\right)}g = -0.2492\ g$$

Acceleration under rear lock-up

$$a_R = -\frac{\mu\left(\frac{c_1}{c_1+c_2}\right)\left(1+\frac{K_f}{1-K_f}\right)}{\left(\frac{\mu h}{c_1+c_2}\left(1+\frac{K_f}{1-K_f}\right)+1\right)}g = -\frac{0.25\left(\frac{50}{112}\right)\left(1+\frac{0.6}{0.4}\right)}{\left(\frac{0.25\times 20}{112}\left(1+\frac{0.6}{0.4}\right)+1\right)}g = -0.2510\ g$$

Since $|a_F| < |a_R|$, front tires will lock-up first.

The above example shows how sensitive the tire locks up (whether front or rear tires locks up first) to the coefficient of friction. The vehicle is to be driven on various types of surfaces with high coefficient of friction like concrete surfaces and low coefficient of friction like icy or slippery roads. For safety reasons, the brakes should be designed such that front lock-up occurs first under a wide-range of surfaces. Also, the brakes should be designed such that maximum braking force will be obtained.

Figure 7.10 shows how the acceleration under front and rear lock-ups changes with the coefficient of friction for the parameters described in Example 7.8. It shows clearly that rear lock-up is quite possible at surfaces with higher coefficient of friction. Figure 7.11 shows that only with changing the braking effort distribution such that 80% goes to the front brake, front brake lock-up will occur for all types of surfaces. This, however, is only done at the expense of maximum braking force. In other words, a higher braking force can be achieved with a braking effort factor of 60% than that of an 80% factor. This must be sacrificed to assure front lock-up happens first for all surfaces.

7.4.3 Optimal Braking Performance

Maximum braking performance is achieved when a maximum braking force is obtained. It is becoming clear from the previous examples that maximum braking force is achieved when both front and rear approach lock-up at the same time. In order to achieve that the braking force at each axle should be proportional to the axle load. In other words, if the front axle experiences 60% of the load, then maximum braking force is achieved when the braking force is distributed such that 60% of the braking force is directed towards the front axle. There are three major factors that affect the braking force distribution. The first is that the loading conditions of the vehicle (fully loaded versus lightly loaded) will cause movement of the center of gravity of the vehicle. This actually means that a certain optimal distribution for a certain loading condition may not be the same as that for another loading condition. The second factor is the coefficient of friction. Load distribution on front and rear axles is directly related to the coefficient of friction. The coefficient of friction varies considerably between dry and wet surfaces and between various types of surface constructions and finishing. The third factor is safety considerations which require the front brakes to lock-up first. In today's automobiles, the braking system is designed with a feed-back control system. Sensors are attached to the both axles to determine the axle load and distribute braking force hydraulically to both axis according to the axle load distribution with some bias toward achieving front lock-up first.

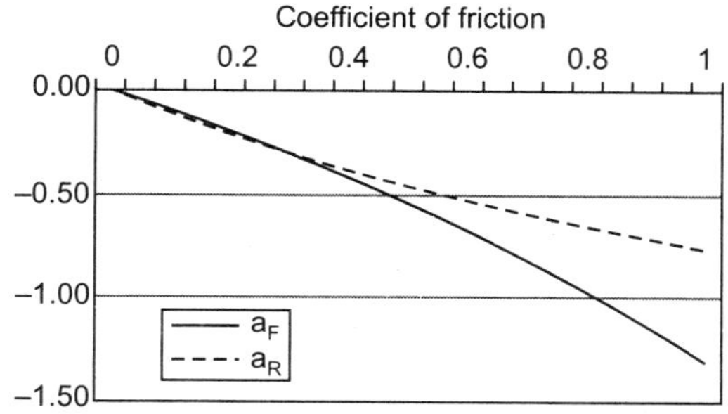

Fig. 7.10 Acceleration under front and rear lock-up conditions for c_1 = 50 in, c_2 = 62 in, h = 20 and K_f = 0.5.

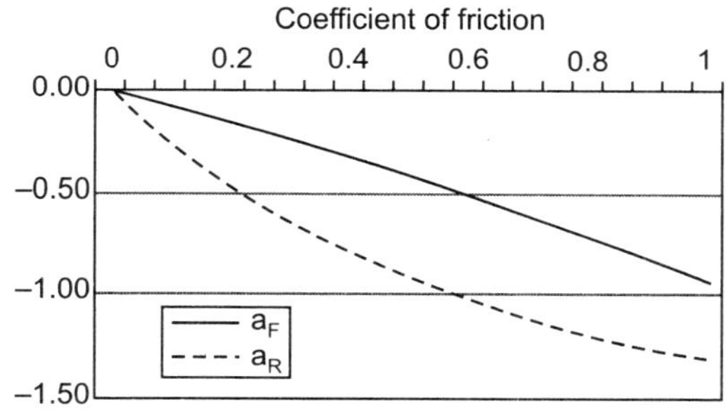

Fig. 7.11 Acceleration under front and rear lock-up conditions for c_1 = 50 in, c_2 = 62 in, h = 20 and K_f = 0.4.

A robust braking design will thus require the involvement of the braking engineers early in the vehicle development to determine the center of gravity of the vehicle under various loads. As shown by the previous examples in this chapter, this is a key input to the braking design and its optimal performance.

In addition, various countries had established different criteria and requirements for the braking system to assure its safety and acceptable performance. Such requirements take into consideration the initial speed at which braking is to occur, the condition of the braking system (new versus worn) and the loading of the vehicle (fully loaded versus lightly loaded vehicles). Most requirements are written for dry surfaces with a relatively high coefficient of friction. On such surfaces, the requirements call for an average deceleration in the range of 0.5 g to 0.65 g, depending on the vehicle initial speed, condition of the brake system and loading conditions. Optimal (*i.e.* maximum) average braking deceleration is noticed to be achieved when the starting speed is around 60 mph (96 km/hr).

Consider a vehicle that weighs 4800 lb (21351.46 N) and has a wheel base of 112 in (2844.80 mm). Its CG is at 50 in (1270.00 mm) behind the front axle and 20 in (508.00 mm) above ground and a coefficient of friction of 0.75. The accelerations found for the assumptions of front and rear lockups are given in Table 7.3 for various values of brake force distribution factor. The bold numbers show which deceleration is smaller. The lock-up associated with the smaller deceleration happens first.

The observation is made that, with only controlling the brake force distribution factor, we can obtain the maximum braking force. This is achieved at a point where a_R and a_F are close to each other. Note that at $K_f = 0.6875$, both acceleration values a_R and a_F are the same (–0.75g), and maximum friction is achieved.

The maximum braking force occurs when both F_{x1} and F_{x2} approach their lock-up (or maximum values) simultaneously. This can be achieved only if we set $F_{x1} = -\mu F_{z1}$ and $F_{x2} = -\mu F_{z2}$. The total friction force is

$$F_x = F_{x1} + F_{x2} = -\mu F_{z1} - \mu F_{z2} = -\mu\left(F_{z1} + F_{z2}\right) = -\mu Mg\cos\theta \tag{7.49}$$

Maximum deceleration:

$$a = \frac{F_{x1} + F_{x2} - R_a - R_r - Mg\sin\theta}{M} = \frac{-\mu Mg\cos\theta - R_a - R_r - Mg\sin\theta}{M} \tag{7.50}$$

The only way both front and rear brakes would lock up at the same time is if the front brake distribution factor is equal to the ratio of the front axle load to the total load, or

$$K_{f\max} = \frac{F_{x1}}{F_x} = \frac{-\mu F_{z1}}{-\mu Mg\cos\theta} = \frac{-\dfrac{\mu}{c_1 + c_2}\left(Mgc_2\cos\theta - h_a R_a - hMa - hMg\sin\theta\right)}{-\mu Mg\cos\theta} \tag{7.51}$$

$$K_{f\max} = \frac{\left(Mgc_2\cos\theta - h_a R_a - hMa - hMg\sin\theta\right)}{\left(c_1 + c_2\right)Mg\cos\theta}$$

$$K_{f\max} = \frac{\left(Mgc_2\cos\theta - h_a R_a - h\left(-\mu Mg\cos\theta - R_a - R_r - Mg\sin\theta\right) - hMg\sin\theta\right)}{\left(c_1 + c_2\right)Mg\cos\theta} \tag{7.52}$$

Ignoring aerodynamic forces, rolling resistance and grade, the front brake distribution factor becomes

$$K_{f\max} = \frac{\left(c_2 + \mu h\right)}{\left(c_1 + c_2\right)} \tag{7.53}$$

Table 7.3 Acceleration under front and rear lock-up assumptions for various front brake force distribution factors, c_2 = 62 in, $c_1 + c_2$ = 112 in, μ = 0.75, h = 20 in

K_f	a_F/g	a_R/g
0.0001	3.102	**–0.295**
0.05	4.947	**–0.309**
0.1	12.237	**–0.324**
0.15	–25.833	**–0.340**
0.2	–6.284	**–0.359**

(Contd.)

K_f	a_F/g	a_R/g
0.25	–3.577	**–0.379**
0.3	–2.500	**–0.401**
0.35	–1.921	**–0.427**
0.4	–1.560	**–0.456**
0.45	–1.314	**–0.490**
0.5	–1.134	**–0.528**
0.55	–0.998	**–0.573**
0.6	–0.891	**–0.627**
0.65	–0.804	**–0.692**
0.6875	**–0.750**	**–0.750**
0.7	**–0.733**	–0.772
0.75	**–0.674**	–0.872
0.8	**–0.623**	–1.003
0.85	**–0.580**	–1.179
0.9	**–0.542**	–1.431
0.95	**–0.509**	–1.820
1	**–0.479**	–2.500

Example 7.9: *Consider example 7.8, where the 21351.46 N (4800 lb) passenger car with 2844.80 mm (112 in) wheel base and a CG at 1270.00 mm (50 in) behind the front axle and 508 mm (20 in) above ground. The car is moving on a horizontal plane and drag and rolling resistance are negligible. Determine the maximum braking force, maximum deceleration and the distribution factor to achieve them. Consider first a 0.75 coefficient of friction, and second, a 0.25 coefficient of friction.*

Solution: For the first surface with 0.75 coefficient of friction:

The front brake distribution factor to achieve the maximum braking force is

$$K_f = \frac{(c_2 + \mu h)}{(c_1 + c_2)} = \frac{60 + 0.75 \times 20}{112} = 0.6875$$

Maximum braking force F_x

$$F_x = -\mu\, Mg \cos\theta = -0.75 \times 4800 = -3600 \text{ lb}$$

Maximum deceleration is

$$a = \frac{-\mu Mg\cos\theta - R_a - R_r - Mg\sin\theta}{M} = -\mu g = -24.15 \text{ ft/sec}^2$$

Note that maximum deceleration is obtained when both a_R and a_F are equal. At that point they are both equal to the coefficient of friction times the acceleration of gravity.

For the second surface with 0.25 coefficient of friction

The front brake distribution factor to achieve the maximum braking force is

$$K_f = \frac{60 + 0.25 \times 20}{112} = 0.5804$$

Maximum braking force F_x

$$F_x = -\mu Mg \cos\theta = -0.25 \times 4800 = -1200 \text{ lb}$$

Maximum deceleration is

$$a = -\mu g = -8.05 \text{ ft/sec}^2$$

In order to calculate the forces on both axles (including both friction and load forces), we need to develop a strategy to tackle the problem. The strategy is based on:

1. For the given problem, determine a_F and a_R from Eq. (7.42) and (7.47).
2. If $|a_F| < |a_R|$, front lock-up occurs first. The following equations should be used:

$$F_{z1} = \frac{1}{c_1 + c_2}\left(Mgc_2\cos\theta - h_a R_a - hMa_F - hMg\sin\theta\right)$$

$$F_{z2} = \frac{1}{c_1 + c_2}\left(Mgc_1\cos\theta + h_a R_a + hMa_F + hMg\sin\theta\right)$$

Note that: $F_{z1} + F_{z2} = Mg\cos\theta$

$$F_{x1} = -\mu F_{z1}$$

$$F_{x2} = \frac{1-K_f}{K_f}F_{x1}, \text{ or } F_{x2} = Ma_F - F_{x1}$$

Total friction force is

$$F_x = F_{x1} + F_{x2}$$

3. If $|a_F| > |a_R|$, rear lock-up occurs first. The following equations should be used:

$$F_{z1} = \frac{1}{c_1 + c_2}\left(Mgc_2\cos\theta - h_a R_a - hMa_R - hMg\sin\theta\right)$$

$$F_{z2} = \frac{1}{c_1 + c_2}\left(Mgc_1\cos\theta + h_a R_a + hMa_R + hMg\sin\theta\right)$$

Note that: $F_{z1} + F_{z2} = Mg\cos\theta$

$$F_{x2} = -\mu F_{z2}$$

$$F_{x1} = \frac{K_f}{1-K_f}F_{x2}, \text{ or } F_{x1} = Ma_R - F_{x2}$$

Total friction force is

$$F_x = F_{x1} + F_{x2}$$

Example 7.10: *Consider example 7.9. To assure front axle lock-up on a dry surface, the front brake distribution factor is chosen at 0.7. The vehicle is moving on a low friction surface with a 0.25 coefficient of friction. Determine the maximum braking force and which axle brake would lock-up first.*

Solution: Since drag, rolling resistance and grade are ignored, acceleration under front-lock-up

$$a_F = -\frac{\mu\left(\frac{c_2}{c_1+c_2}\right)\left(1+\frac{1-K_f}{K_f}\right)}{\left(-\frac{\mu h}{c_1+c_2}\left(1+\frac{1-K_f}{K_f}\right)+1\right)} g = -\frac{0.25\left(\frac{62}{112}\right)\left(1+\frac{0.3}{0.7}\right)}{\left(-\frac{0.25\times 20}{112}\left(1+\frac{0.3}{0.7}\right)+1\right)} g = -0.2112\, g$$

Acceleration under rear lock-up

$$a_R = -\frac{\mu\left(\frac{c_1}{c_1+c_2}\right)\left(1+\frac{K_f}{1-K_f}\right)}{\left(\frac{\mu h}{c_1+c_2}\left(1+\frac{K_f}{1-K_f}\right)+1\right)} g = -\frac{0.25\left(\frac{50}{112}\right)\left(1+\frac{0.7}{0.3}\right)}{\left(\frac{0.25\times 20}{112}\left(1+\frac{0.7}{0.3}\right)+1\right)} g = -0.3238\, g$$

since $|a_F| < |a_R|$, front lock-up occurs first. The normal forces are

$$F_{z1} = \frac{1}{c_1+c_2}\left(Mgc_2 - hMa_F\right) = \frac{1}{112}\left(149\times 32.2\times 62 - 20\times 149\times(-0.2112)\right) = 2838 \text{ lb}$$

$$F_{z2} = \frac{1}{c_1+c_2}\left(Mgc_1 + hMa_F\right) = \frac{1}{112}\left(149\times 32.2\times 50 + 20\times 149\times(-0.2112)\right) = 1962 \text{ lb}$$

Note that: $F_{z1} + F_{z2} = 4800$ lb

$$F_{x1} = -\mu F_{z1} = -0.25\times 2838 = -710 \text{ lb}$$

$$F_{x2} = \frac{1-K_f}{K_f} F_{x1} = \frac{1-0.7}{0.7}\times(-710) = -304 \text{ lb}$$

Total friction force is

$$F_x = F_{x1} + F_{x2} = -1014 \text{ lb}$$

Table 7.4 shows the acceleration and force distribution on each axle for the case of example 7.9 and a coefficient of friction of 0.75. Note that the maximum braking forces occur when the $K_{f\max}$ of 0.6875 found in example 7.9 is used. Also note that at that value, a_F and a_R are equal.

The maximum braking force is plotted in Fig. 7.12 as a function of the front brake distribution factor.

In a typical design problem, one needs to consider various surfaces and loading conditions. For example, let us consider the following problem where a vehicle is under different loading conditions and on various surfaces.

Table 7.4 Force distribution for different values of k_f under front and rear lock-up assumptions, c_2 = 62 in, $c_1 + c_2$ = 112 in, μ = 0.75, h = 20 in.

K_f	a_F/g	Fz_1	Fz_2	Fx_1	Fx_2	Fx	a_R/g	Fz_1	Fz_2	Fx_1	Fx_2	Fx
0.0001	3.102						–0.295	2910	1890	0	–1417	–1417
0.05	4.947						–0.309	2922	1878	–74	–1409	–1483
0.1	12.237						–0.324	2935	1865	–155	–1399	–1554
0.15	–25.833						–0.34	2949	1851	–245	–1388	–1633
0.2	–6.284						–0.359	2964	1836	–344	–1377	–1721
0.25	–3.577						–0.379	2982	1818	–455	–1364	–1818
0.3	–2.5						–0.401	3001	1799	–578	–1349	–1927
0.35	–1.921						–0.427	3023	1777	–718	–1333	–2050
0.4	–1.56						–0.456	3048	1752	–876	–1314	–2190
0.45	–1.314						–0.49	3077	1723	–1057	–1292	–2350
0.5	–1.134						–0.528	3110	1690	–1268	–1268	–2535
0.55	–0.998						–0.573	3149	1651	–1514	–1239	–2752
0.6	–0.891						–0.627	3195	1605	–1806	–1204	–3010
0.65	–0.804						–0.692	3250	1550	–2159	–1162	–3321
0.6875	–0.75	3300	1500	–2475	–1125	–3600	–0.75	3300	1500	–2475	–1125	–3600
0.7	–0.733	3286	1514	–2464	–1056	–3521	–0.772					
0.75	–0.674	3235	1565	–2426	–809	–3235	–0.872					
0.8	–0.623	3191	1609	–2394	–598	–2992	–1.003					
0.85	–0.58	3154	1646	–2366	–417	–2783	–1.179					
0.9	–0.542	3122	1678	–2341	–260	–2601	–1.431					
0.95	–0.509	3093	1707	–2320	–122	–2442	–1.82					
1	–0.479	3068	1732	–2301	0	–2301	–2.5					

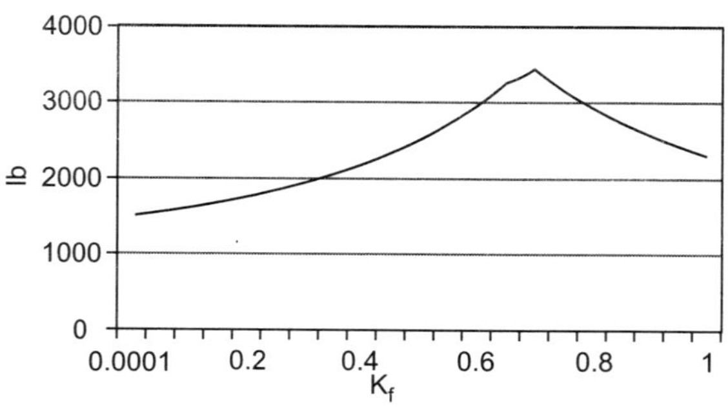

Fig. 7.12 Braking force as a function of the front brake force distribution factor.

Example 7.11: *Consider a vehicle that weighs 3200 lb (14235 N) when lightly loaded (driver only), with a CG at 40 in (1016.0 mm) behind front axle and 18 in (457.2 mm) above road, and 4000 lb (17793 N) when loaded with a CG at 51 in (1295.4 mm) behind front axle and 20 in (508.0 mm) above road.*

The wheel base is 108 in (2743.2 mm). The vehicle is to achieve maximum possible braking force and front lock-up under the following severe conditions:

1. *Vehicle is loaded, dry surface of μ= 0.8*
2. *Vehicle is loaded, slippery surface of μ = 0.2*
3. *Vehicle is lightly loaded, on a dry surface of μ = 0.8*
4. *Vehicle is lightly loaded, on a slippery road of μ = 0.2*

Determine the value of K_f that you recommend knowing that the brake system your company installs has a constant value (no feedback control system). Determine the axle loads and the maximum braking force.

Solution: This problem is best solved by writing a spreadsheet with all possible tests. Table E7.11 shows the optimal value of K_f (Table E7.11a) and the associated acceleration and axle loads (Table E7.11b), as well as maximum brake force for each of the cases.

Table E7.11a Optimal value of k_f for example E 7.11

Case	c_1	c_2	h	μ	*Mg*	K_{fmax}	a_F/g	Fz_1	Fz_2	Fx_1	Fx_2	Fx_2	*Fx*
1	51	57	20	0.8	4000	0.6759	–0.800	2704	1296	–2163	–1037	–1037	–3200
2	51	57	20	0.2	4000	0.5648	–0.200	2259	1741	–452	–348	–348	–800
3	40	68	18	0.8	3200	0.763	–0.800	2441	759	–1953	–607	–607	–2560
4	40	68	18	0.2	3200	0.663	–0.200	2121	1079	–424	–216	–216	–640

In order to avoid rear lock-up under all cases, the K_f must be chosen above the maximum value of 0.763 determined for case 3. Thus K_f = 0.8 is chosen. For this value, the following acceleration and loads are obtained for each of the cases mentioned (Table E7.11b).

Table E 7.11b Load distribution for k_f = 0.8 in example E 7.11

Case	c_1	c_2	h	μ	*Mg*	K_f	a_F/g	Fz_1	Fz_2	Fx_1	Fx_2	Fx_2	*Fx*	a_R/g
1	51	57	20	0.8	4000	0.8	–0.6477	2591	1409	–2073	–518	–518	–2591	–1.0851
2	51	57	20	0.2	4000	0.8	–0.1383	2214	1786	–443	–111	–111	–553	–0.3984
3	40	68	18	0.8	3200	0.8	–0.7556	2418	782	–1934	–484	–484	–2418	–0.8889
4	40	68	18	0.2	3200	0.8	–0.1643	2102	1098	–420	–105	–105	–526	–0.3175

Note that with the choice made for K_f all the cases will have front lock–up to occur first. In order to achieve this, major sacrifice is made for maximum braking force, specially for case 2, where maximum braking force is decreased from 800 lb to 553 lb.

As was stated earlier, in today's automotive brakes, the front brake force distribution factor K_f can be altered depending on the particular application at hand. This can be done straightforwardly to account for variations in the axle loads either because of different decelerations or because of different vehicle loading conditions (lightly loaded versus loaded conditions). In the above example, a factor of 0.8 can be used for the lightly loaded cases and 0.7 for the loaded cases. This will enable us to achieve higher braking force at the surfaces with low coefficient of friction. The K_f factor cannot be altered straightforwardly to account for the different coefficients of friction various surface may offer. In that case, an advanced anti-lock brake system (ABS) can be used.

European specifications calls for the following regulations with vehicles that do not have ABS (Buschman *et al.* 1992). First, the acceleration should always remain higher than 0.3 g, where g is the acceleration of gravity. Second, the braking force in the front should be lower than that of the rear when the vehicle acceleration is between 0.5 and 0.8 g.

Modern braking systems have electronic baking force distributions systems (EBD). These are proven to be more robust as the input signal describes current vehicle and driving conditions. EBD takes into account friction variations of the lining and road surfaces (Buschman *et al.* 1992).

7.5 CONSIDERATIONS OF VEHICLE SAFETY

Brakes are fundamentally a safety feature that will enhance motion control of the vehicle in a manner that will enable the driver to avoid a potential accident. The ability of brakes to deliver its function of motion control is dependent on the traction between the tires and the road. If this traction is lost, this safety feature is compromised. Hence, the main objective of additional safety features of brakes is to maximize traction effort between the tires and the road.

As was stated earlier, lock-up conditions will change the rolling action of the tires with a sliding action. This reduces the coefficient of friction considerably. It will also cause the vehicle to lose the ability to steer (under front lock-up conditions) or direction stability (under rear lock-up conditions). In addition, the distance traveled during braking will be far larger than that without lock-up conditions. This will cause the vehicle to behave in an unpredictable manner to the automotive customer, reducing his control over the vehicle and, more importantly, not give the required performance to avoid a potential accident. Thus, preventing lock-up was the challenge that faced engineer's overtime. Engineers had the objective of achieving the maximum force out of the brake system and minimize the possibility of, lock-up conditions. With advances in electronic controls and sensors, a system has been used effectively on vehicles to avoid lock-up conditions to happen. The key to this is for sensors planted for this purpose to assess skid, or slipping, conditions of the wheels. Brakes are then electronically and momentarily released to bring the wheel to the non-skid condition. Brakes are applied again to reach the skid condition, gaining higher speed reduction over a short period of time. This process is repeated until a complete stop, or the desired speed, is achieved. This system is referred to as anti-lock brake system and is becoming a standard safety feature on many vehicles and available as an optional feature on most others.

The anti-lock braking system offers better treatment of complex surfaces frequently encountered in cold regions. In area where temperature reaches the subfreezing point at night will cause pockets of water to freeze. This causes variations in the coefficient of friction between the two wheels of the same axle. Thus a condition of the wheels of the same axle having one side on a low coefficient of friction and the other wheel on a higher coefficient of friction can be encountered. Under such condition, maximum braking force will be sacrificed if the same braking force is given to both wheels on the same axis.

There are three major components of the ABS. These are the sensors to monitor wheel speeds and compare it with the vehicle speed, the electronic control unit (ECU), and finally the solenoid which is used to release and re-apply the pressure to the brakes. It is important to understand the relationship between the skid (or slip) condition of the tires and the braking force.

7.5.1 Skid (Slip) Condition and Braking

When forces (driving and braking) are applied to the tire, the tire experiences a condition of skid (slippage) over the road surface. The speed of the vehicle under free-rolling condition (*i.e.* no skid) can be easily related to the rotational speed of the wheels:

$$V = r_t\,\omega \tag{7.54}$$

where V is the vehicle speed, r_t is the radius of the tire and ω is the angular speed of the tire. In real applications and when a braking force is applied, variation is observed between V and $r_t\omega$. This can be observed when the wheels are under lock-up conditions. Under lock-up condition, the angular speed of the tire is zero ($\omega = 0$), while the vehicle is still moving ($V \neq 0$). Total skid and zero rolling is observed at that point. Some skid is observed between the two cases of free rolling and the wheel lock-up. If we define the first as the case of 0% skid and the last as a 100% skid condition, the following equation can be written to describe skid at any point between the two extreme cases:

$$i = \left(1 - \frac{V}{r_t\omega}\right) \times 100\% \tag{7.55}$$

Note that different definitions can be used for the skid of the tire. In fact, SAE offers the following definition (SAE J670*e*):

$$i_{\text{SAE}} = \left(\frac{r_t\omega}{V} - 1\right) \times 100\% \tag{7.56}$$

Note that i_{SAE} gives the value of 0% skid when the vehicle is under free-rolling condition and –100% when the tires are locked-up.

The phenomenon of tire-road interaction is a complex one. The coefficient of friction between the tires and the road is a function of many environmental parameters including the surface structure, the type of tire, the tire pressure, dry versus wet, wet versus icy, and many other factors. The coefficient of friction under given environmental conditions is found to be a function of the skid between the tire and the ground, achieving a peak value only when a certain percentage of skid is achieved (Fig. 7.13).

As the braking torque is applied to the wheel, a proportional braking force is developed between the tire and the ground. The skid condition is a function of the braking force applied. The peak value of the braking force is usually achieved with a skid of 15-20 % happening at the tire. Any additional skid after that (as is the case when the tires are locked up) will reduce the friction to the skid value (almost null on icy roads) and about 70%-80% the peak value on dry roads. This will reduce the braking efficiency (in the sense that longer distance is needed when wheels lock-up than when they do not).

Note that a portion of the curve shown in Fig. 7.13 is linear. This portion starts at the origin and ends up at a critical value F_{xc} and has a slope of C. For that portion, the braking force is

$$F_x = iC \tag{7.57}$$

where $F_x < F_{xc}$. The critical value is

$$F_{xc} = \frac{\mu_p W}{2} \tag{7.58}$$

where W is the total weight. Note that when F_x is below F_{xc}, the relationship between the braking effort and the skid is linear, indicating that the apparent skid is only due to the longitudinal stiffness of the tire. When F_x exceeds F_{xc}, actual slipping between the tread of the tire and the ground takes place. The braking effort is (Wong, 2001)

$$F_x = \mu_p W\left(1 - \frac{\mu_p W}{4iC}\right) \tag{7.59}$$

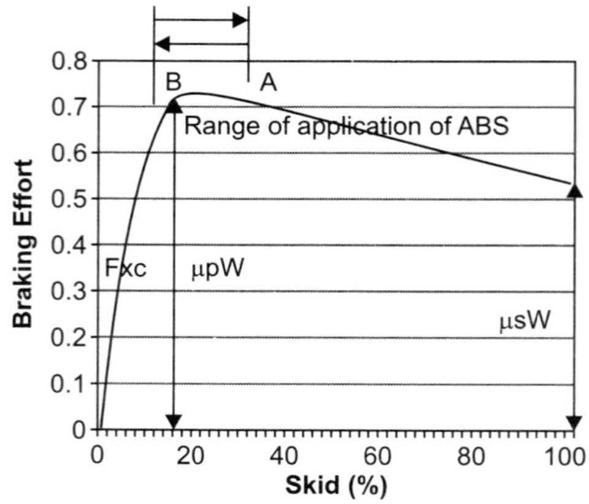

Fig. 7.13 Braking coefficient versus skid.

Note the nonlinearity between the baking effort and the skid. Eq. (7.59) applies until the maximum peak braking effort value is obtained. At that point, lock-up happens and skid moves rapidly from the 20% or so skid to 100% almost instantly with a reduced coefficient of friction.

7.5.2 Anti-Lock Braking System

Going back to Fig. 7.13, it is clear that maximum braking can be achieved if the brake efforts remain within the vicinity of the peak value. The danger in conventional braking systems is that as soon as the braking force gets to that peak value excessive skid occurs almost instantly. At that point, pressure is released off the brake in ABS with excessive skid, which brings the braking effort back to a value below that of the peak. Effort is then applied again to reach the maximum and so on.

This is best described with the help of Fig. 7.14. Under normal braking with braking effort below its peak value, speed reduction is noticed both for the vehicle and the wheels. When the braking force reaches its maximum value and excessive skid is observed (by noticing that the vehicle speed is reducing at a constant rate while wheel speed is reducing rapidly), the brake pressure is released. The wheel will then gain speed and an optimal value of skid is retained. The pressure is then applied again to bring the braking force back to its maximum value, and so on.

Comparing Fig. 7.13 with Fig. 7.14, point A on Fig. 7.14 is the point at which skid starts to occur excessively (i >25% or so). This is associated with point A on Fig. 7.13. At that point the ABS recognizes that the tire is locked-up and the pressure is released off the brakes. This will bring the system to point B where the wheel is gaining angular speed and lock-up condition is freed. The pressure is then applied to maximize braking action. This system will ensure that the braking effort will remain at its maximum range (Fig. 7.13) giving best braking performance and avoiding lock-up and the associated safety concerns of losing direction stability of the vehicle or the ability to steer it.

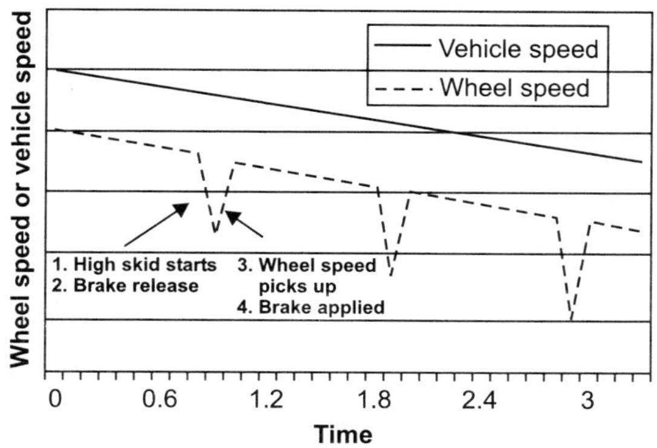

Fig. 7.14 Wheel speed and vehicle speed under ABS activation.

Angular speed of the tires can be measured easily with electromagnetic pulse pickups mounted on the tires or the driveline system. One needs to mount such pickups on each wheel to give more control over each of the wheels and observe their skid status. The angular speed is first found as a direct measurement and the angular deceleration is derived from that. A challenging issue of the ABS is to determine the vehicle speed. Instead, a linear accelerometer is mounted on the vehicle to find its linear deceleration. The comparison between vehicle deceleration and angular deceleration will help determine if the tires are locked-up or not.

The control system of the ABS consists of three main components. The first is a signal processing unit which measures the angular deceleration of the tire and the linear deceleration of the vehicle. The second is a module that calculates skid based on the signals processed and determines if skid can be avoided when certain criteria are met. The third unit is a module for sending the needed signal for the pressure control system of the brakes.

Variations exist in different ABS on how to determine lock-up conditions. One of which is the direct comparison between the linear acceleration of the vehicle and the angular deceleration of the tire. In the second, the angular deceleration of the vehicle is monitored and if it exceeds a certain value, it is held in memory for a certain time. During this time, both linear deceleration of the vehicle and angular deceleration of the tire are monitored closely. If a certain criterion is met, the ABS will be activated. In another control system, skid of the vehicle is estimated and angular speed of the tire is measured. If both meet a certain criterion, the ABS is activated. At low speeds, the ABS is found to run into errors in determining skid conditions and linear deceleration of the vehicle. The threshold of skid for determining lock-up is increased at low speeds. The re-application of pressure is modulated either by continuous monitoring of the angular tire deceleration and vehicle linear deceleration or by a fixed time-delay mechanism.

Different lay-outs of ABS are used in different applications. They all depend on the number of channels for the application and/or release of the pressure and sensors used to determine tire angular deceleration. For the front tires, each tire is monitored by its independent sensor and an independent channel is used for pressure release and application. For the rear tires, some ABS use one sensor and one pressure regulator, two sensors and one pressure regulator or two sensors and two pressure regulators.

7.6 PITCH PLANE MODELS

Vehicle braking may induce some of the fundamental vehicle modes, which will result in uncomfortable ride. Automotive customers experience the bounce and/or pitch types of motion when driving. Bounce is basically a mode in which the vehicle acts as a one-degree-of-freedom system, where its mass is simply setting on the 'spring' tires. This motion is experienced particularly if some road input is adding to the excitation. Pitch mode is the other mode that is experienced by customers when the vehicle is undergoing acceleration or deceleration. Deceleration is mostly due to a braking action.

A representative model of the vehicle would have to account for both bounce and pitch modes. If the whole vehicle is represented as a mass, setting on two different springs representing front tires and rear tires, it is possible to obtain a reasonable representation of the vehicle. Figure 7.15 shows a two-degree-of freedom model in which both bounce and pitch modes can be accounted for.

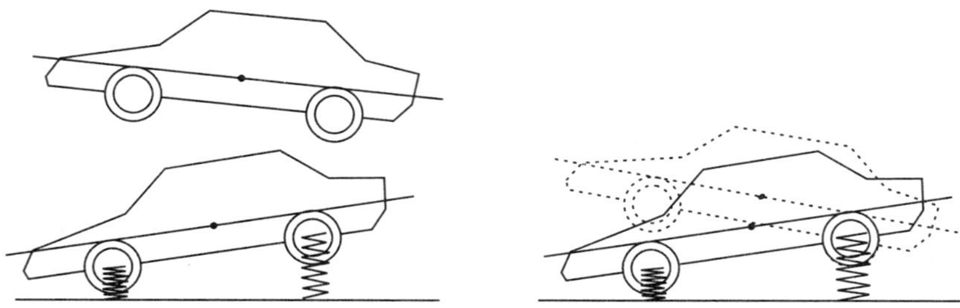

Fig. 7.15 Pitch and bounce modes.

Pitch and bounce modes are generally coupled. The simple vehicle model shown in Fig. 7.16 is a 2-degree-of-freedom model. Such model can be used for assessment of vehicle bounce and pitch modes. Note that normal loads are affected by the load transfer that occurs during braking. More detailed analysis was given in Chapter 4.

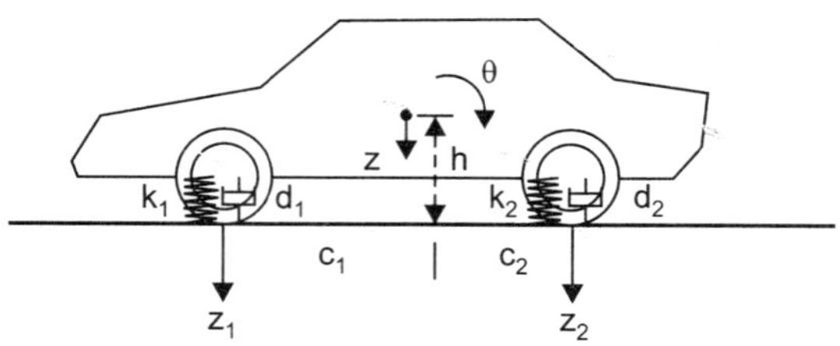

Fig. 7.16 Simple two-degree-of-freedom (pitch plane) vehicle model.

7.7 RECENT ADVANCES IN AUTOMOTIVE BRAKING

The major attributes that drive today's braking technology are their safety and their noise and vibration, particularly, brake squeal. Other aspects of new evolving braking technology relate to brake by wire, re-use of braking energy and heat (for fuel economy improvement).

Among recent technologies related to brakes is the use of brake assist. These are also referred to as 'power brakes'. Among the recent research on brake assist is the work of Tamura *et al.* (2001) who presented work on a brake assist system with a preview function. Such assist can be electrical or mechanical. The work of Feigel and Schonlau (1999) should also be mentioned. Figure 7.17 shows a cross section for a mechanical power assist.

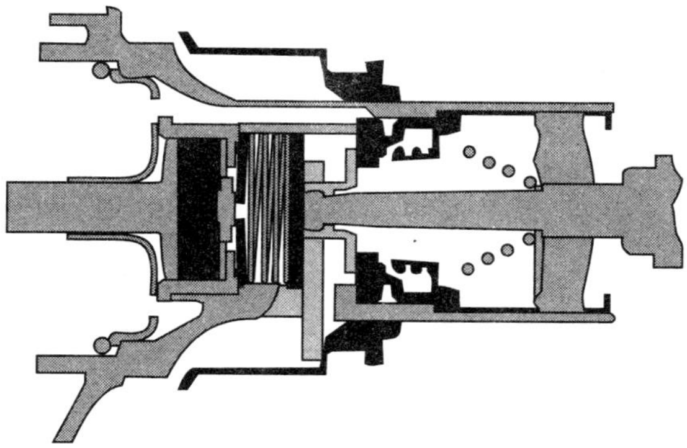

Fig. 7.17 Mechanical power assist (Feigel and Schonlau, 1999).

Utilizing vehicle braking force to achieve higher level of vehicle stability was the subject of various research. Estimation of the actual coefficient of friction between the tire and the road and adjusting the brake force distribution to achieve the desired dynamics was the subject of the study by Yoon et al (2004). Vehicle yaw stability during braking was studied by Seluga *et al.* (2004). Such studies are important to the discussion of accident reconstruction (Chapter 9).

Noise and vibrations induced by braking action is another corner of research receiving attention. In particular, brake squeal has been the subject of considerable research. Among this research is the work of Ouyang *et al.* (2004) on a review of numerical analysis of automotive disc brake squeal. Test and evaluation of automotive disc brake squeal was also discussed by Chen *et al* (2003). Brakes judder, another noise phenomenon induced by brake action, was the subject of several studies (e.g., Abdelhamid, 1997).

The possibility of recovering vehicle kinetic energy is one advantage of electric and hybrid electric vehicles. When a vehicle drives in heavy traffic, more than half of the total energy is dissipated in the brakes. Therefore, recovering braking energy is an effective approach for improving the driving range of electrical vehicles and the energy efficiency or fuel economy of hybrid vehicles. Gao *et al.* (1999) investigated the effectiveness of regenerative braking.

There has been good progress recently in using computational tools to estimate the braking performance. The use of advanced finite element methods (FEM) and codes as well as those of computational fluid dynamics (CFD) has become the standard engineering practice in the design and analysis of these tools. Qian (2002) recently studied the aerodynamic shape optimization using CFD Parametric model. The application was intended for brake cooling. Other applications can be found in various SAE transactions and the open literature on using computational aided engineering (CAE) tools. Figures 7.18 and 7.19 show results obtained in some of the recent publications using FEM and CFD tools.

Brake by wire is another technology rapidly evolving, particularly with the recent popularity of hybrid vehicles and electric cars, where higher voltage is now available. Among recent research on brake by wire are those of Yoon *et al.* (2004) and Langenwalter and Kelly (2003).

7.8 SUMMARY

The brake system constitutes an important safety feature in the vehicle that allows the driver to reduce speed either to allow for safe steering or to avoid accidents. Brake systems used in automotive applications are disk and drum brakes. Disk brakes are usually used for the front wheels and drum brakes used for the rear brakes. As the brakes are applied, a force will move the brake system closer to the wheels allowing touching the wheels and applying pressure. This results in a torque at each wheel against the angular motion if the *wheel and thus reducing speed.* In some cases, where excessive brake force is applied, the brakes will prevent the wheels from rotation (*i.e.* lock-up condition encountered). In these cases, the wheels will slide on the road with a less friction coefficient than that experienced with the wheels rolling (with some skid).

Some of the weight of the vehicle transfers to the front wheels during braking allowing higher friction forces for theses wheels. The brake force should be distributed proportional to the vertical load at each wheel to allow for maximum braking efforts. Also, when maximum braking effort is applied the brake design should allow for front wheel lock-up to occur ahead of the rear wheel lock-up. This will allow the loss of directional control (*i.e.* steeribility of the vehicle). While this is a safety hazard, it is of less danger than directional instability encountered with rear wheel lock-up. Newer vehicles are equipped with anti-lock braking systems (ABS). These systems will allow braking each wheel independently, allowing higher control of wheel motion and recovering the wheel spinning after a lock-up condition is encountered. They further allow for maximum braking force, even when different coefficients of friction are encountered by each wheel.

7.9 REFERENCES

Abdelhamid, M., "Brake judder analysis: case studies," *Society of Automotive Engineers Transactions,* Paper No 973018, 1997.

Buschmann, G., Ebner, H., and **Kuhn, W.,** "Electronic brake force distribution control a sophisticated addition to ABS," *Society of Automotive Engineers Transactions* SAE-920646, 1992.

Chen, F., Abdelhamid, M., Blaschke, P., and **Swayze1, J.,** "On Automotive Disc Brake Squeal Part III: Test and Evaluation," *Society of Automotive Engineers Transactions* 2003-01-1622, 2003.

Dahlberg, E., "Yaw Instability Due to Longitudinal Load Transfer During Braking in a Curve," *Society of Automotive Engineers Transactions* 1999-01-2952, 1999.

FMVSS 121., "Stability and Control of Medium and Heavy Duty Vehicles During Braking," 49 CFR Part 571, *Federal Register Final Rule,* v 60, n 47, pp.. 13216-13309, March 10, 1995.

Feigel, H. J., and **Schonlau, J.,** "Mechanical Brake Assist – A Potential New Standard Safety Feature," *Society of Automotive Engineers Transactions* 1999-01-0480, 1999.

Gao, Y., Chen, L., and **Ehsani, M.,** "Investigation of the Effectiveness of Regenerative Braking for EV and HEV," *Society of Automotive Engineers Transactions* 1999-01-2910, 1999.

Gillespe, T.D., *Fundamentals of Vehicle Dynamics*, Society of Automotive Engineers, Warrendale, PA, 1992.

Gillespe, T.D., "Heavy Truck Ride," *Society of Automotive Engineers Transactions* 850001, 1985.

Langenwalter, J., and **Kelly, B.,** "Virtual Design of a 42V Brake-by-Wire System," SAE Paper No. 2003-01-0305, *Society of Automotive Engineers Transactions*, Society of Automotive Engineers, Warrendale, PA, 2003.

Limbert, R., *Brake Design and Safety*, Society of Automotive Engineers, Warrendale, PA,1992.

Moklegaard, L, Stafanopoulou, A., and **Schmidt, J.,** "Transition from combustion to variable compression braking," SAE Paper No. 2000-01-1228, *Society of Automotive Engineers Transactions,* Society of Automotive Engineers, Warrendale, PA, 2000.

Ouyang, H., Nack, W., Yuan, Y., and **Chen, F.,** "Numerical Analysis of Automotive Aisc Brake Squeal: a Review," *International Journal of Vehicle Noise and Vibration,* Vol. 1, No. ¾, pp. 207 – 231, 2005.

Qian, C., "Aerodynamic Shape Optimization Using CFD Parametric Model with Brake Cooling Application," SAE Paper No. 2002-01-0599, Society *of Automotive Engineers Transactions,* 2002.

Seluga, K., Obert, R., and Ojalvo, I, "Articulated Vehicle Yaw Stability During Braking – A Parametric Study," SAE Paper No. 2004-01- 2630, *Society of Automotive Engineers Transactions,* Warrendale, PA, 2004.

Shaffer, S, and **Radlinski, R.,** "Braking Capability Requirements for In-Use Commercial Vehicles - A Chronology", SAE Paper No. 2003-01-3397, *Society of Automotive Engineers Transactions,* Warrendale, PA, 2003.

Shi, T., Dessouki, O., Warzecha, T., and **Chang, W.,** "Advances in Complex Eigenvalue Analysis for Brake Noise," SAE Paper No. 2001-01-1603*Society of Automotive Engineers Transactions*, Warrendale, PA, 2001.

Shigley, J.E., and Mischke, C.R., *Mechanical Engineering Design*, Fifth edition, McGraw Hill, New York, 1989.

Tamura, T., Inoue, H., Watanabe, T., and **Maruko, N.,** "Research on a Brake Assist System with a Preview Function," *Society of Automotive Engineers Transactions,* 2001-01-0357, 2001.

SAE., *Vehicle Dynamics Terminology*, SAE J670e, Society of Automotive Engineers, 1978.

Wong, J.Y., *Theory of Ground Vehicles,* Third edition, John Wiley and Sons, New York, 2001.

Yoon, P., Kang, H.J., and **Hwang, J.,** "Braking Status Monitoring for Brake-By-Wire Systems," *Society of Automotive Engineers Transactions,* 2004-01-0259, 2004.

PROBLEMS

7.1 A 18683 N (4200 lb) vehicle traveling at 104.61 km/hr (65 mile/hr) is stopped under a constant braking force. Find the braking force needed to bring the vehicle to a complete stop in 9 seconds, 7 seconds, and 5 seconds. Ignore drag, rolling resistance and grade.

7.2 For the problem P7.1, assume that the vehicle has a frontage area of 22 ft^2, and a coefficient of drag of 0.35. Knowing that the drag force itself changes as the velocity decreases (Eq. (7.27). Find the braking force needed to bring the vehicle to a complete stop in 5 seconds.

7.3 A 19572.2 N (4400 lb) vehicle traveling at 104.61 Km/hr (65 mile/hr) is stopped under a constant braking force of 889.6 N (200 lbs). Find the time needed for travel before the vehicle reaches a complete stop:

(*a*) when the vehicle is moving on a horizontal plane..

(*b*) when the vehicle is moving uphill a grade of 6%.

(*c*) when the vehicle is going down hill a grade of 6%.

7.4 Consider the brake system shown below. The width of the shoe system is 31.75 mm (1.25 in) and the coefficient of friction is 0.4. Vehicle specification calls for a braking force of 4448.2 N (1000 lb). Find the maximum pressure at the brake shoes to allow for achieving the desired braking force. Also find the total torque applied by the brake system, and the reaction force at the pivots of both shoes.

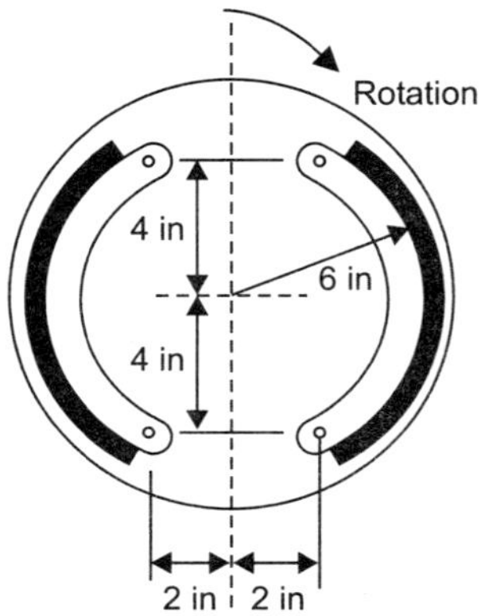

Fig. P7.4 Brake for Problem 7.4.

7.5 In Example 7.3, The width of the shoe system is 30 mm, and the coefficient of friction is 0.3. The friction material used permits a maximum pressure value of 1000 kPa. Find the actuation force, the total torque applied by the brake system, and the reaction force at the pivots of both shoes for the angles ψ = 20, 25, 30, 35, and 40 degrees. The distance between the center of the brake and the pivots is constant at 110 mm.

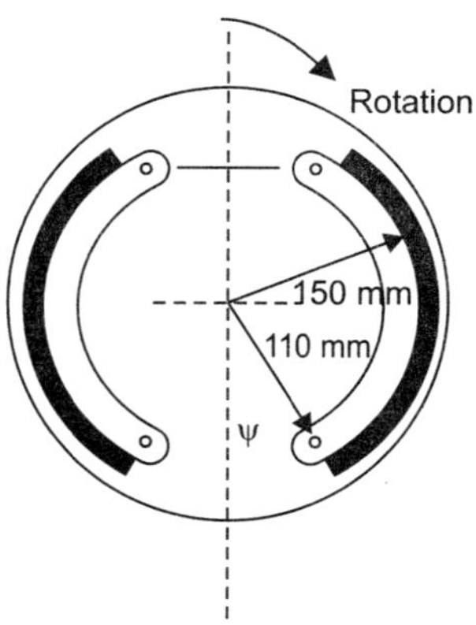

Fig. P7.5 Brake for problem 7.5.

7.6 A vehicle of mass 2200 kg and traveling at 120 km/hr, is to reach a complete stop in 10 seconds, the braking system has a mass of 10 kg. Let $C = 500$ J/ (kg.C^0). Determine the energy and average power. Also, estimate the temperature rise of the brakes.

7.7 Consider a light truck that weighs 17792.89 N (4000 lb) under a certain load. The wheel base of the vehicle is at 3.048 m (10 ft) and the center of gravity is at 1.2192 m (4 ft) from the front axle and at 0.6604 m (26 in) above the ground. Determine the load on each axle if the brakes are designed for all-wheel lock-up under these conditions with a coefficient of friction of 0.8.

7.8 Consider a midsize sedan that weighs 1600 kg under a certain load and going uphill a 5 degrees slope at 88 km/hr. The front area of the vehicle is 2.0 m^2, and the coefficient of drag is 0.4. The aerodynamic resistance is acting on the CG. The wheel base of the vehicle is at 2.8 m and the center of gravity is at 1.2 m from the front axle and at 0.5 m above the ground. Determine the load on each axle if the brake is under no acceleration, decelerating at 5 m/sec^2. The density of air is 1.225 kg/m^3.

7.9 For the problem described in 7.8, what the loads would be on each axle under no acceleration and a 5 m / sec^2 acceleration, if the vehicle was going down hill 5 degree slope.

7.10 A 13344.7 N (3000 lb) compact passenger car with 2.5908 m (102 in) wheel base and a CG at 1.016 m (40 in) behind the front axle and 0.4572 m (18 in) above ground. The car is moving on a horizontal plane. Determine which tire will lock up first if the car is moving on a surface with a 0.9 coefficient of friction, if the front brake force distribution factor is 0.4, 0.6, and 0.8.

Assumption: Ignore drag and rolling resistance.

7.11 Consider a vehicle with a 1500 kg mass when lightly loaded (driver only), with a CG at 1.0 m behind front axle and 0.45 m above road. The mass is 2200 kg when loaded with a CG at 1.25 m behind front axle and 0.5 m above road. The wheel base is 2.7 m. The vehicle is to achieve maximum possible braking force and front lock-up under the following severe conditions:

Case 1: Vehicle is loaded, dry surface of $\mu = 0.9$

Case 2: Vehicle is loaded, slippery surface of $\mu = 0.2$

Case 3: Vehicle is lightly loaded, on a dry surface of $\mu = 0.9$

Case 4: Vehicle is lightly loaded, on a slippery road of $\mu = 0.2$

Determine the value of the front brake force distribution factor K_f that is to be recommended knowing that the brake system your company installs has a constant value (no feedback control system). Determine the axle loads and the maximum braking force.

7.12 The vehicle described in Problem 7.11 has an ABS which allows control of each wheel force distribution. Determine the maximum braking force achieved under the four conditions described in Problem 7.11.

7.13 For the vehicle described in Problem 7.11, determine the maximum braking force that can be achieved if the vehicle is moving on a split-mue surface with the right wheels on a 0.2 coefficient of friction surface and the left wheels on a 0.8 surface.

7.14 For the Problem 7.13, an ABS is used in the braking action. Control of the braking force on each wheel is possible under this system. Determine the maximum braking force that can be achieved.

7.15 Perform the necessary manipulation of Eqs. (7.3) and (7.4) to establish Eq. (7.5).

7.16 Consider a vehicle with a 2000 kg mass when lightly loaded (driver only), with a CG at 1.2 m behind front axle and 0.5 m above road. The wheel base is 3.0 m. The vehicle is to achieve maximum possible braking force and front lock-up under the following severe conditions:

(*a*) front wheels are on the dry surface of $\mu = 0.9$, and the rear wheels are on a wet road of $\mu = 0.2$

(*b*) rear wheels are on the dry surface of $\mu = 0.9$, and the front wheels are on a wet road of $\mu = 0.2$

Determine the value of the front brake force distribution factor K_f that you recommend knowing that the brake system your company installs has a constant value (no feedback control system). Determine the axle loads and the maximum braking force.

7.17 Repeat Problem 7.16 with the vehicle loaded. The mass is 2500 kg when loaded with a CG at 1.4 m behind front axle and 0.6 m above road.

7.18 Verify that Eq. (7.47) reduces to (7.48) when aerodynamic drag, grade and rolling resistance are ignored.

7.19 A passenger carrying vehicle with less than 10 passengers, capacity (including the driver) is built on a passenger vehicle chassis system. The Gross Weight of the vehicle is 4000 lbs. Determine the minimum braking force required (use Table 7.1), the minimum deceleration required and the maximum stopping distance without emergency braking. Compare the minimum deceleration required found directly from Table 7.1 with that obtained from the stopping distance requirements.

7.20 Repeat Problem 7.19 for a property carrying vehicles weighing 35544 N (8,000 lb).
Assume the speed of 20 mph.

Vehicle-Pedestrian Collisions

CHAPTER

8

8.1 INTRODUCTION

A *pedestrian* is a person traveling on foot, whether walking or running. In some communities, those traveling using roller skates, skateboards, and similar devices are also considered to be pedestrians. In modern times, the term mostly refers to someone walking on a road or footpath, but this was not the case historically.

During the years 1994 to 1998 (see NHTSA, 1993, 1996, 1997), pedestrian injuries have continued to decline, whereas pedestrian fatalities remain about the same. In 1996, pedestrian-vehicle impacts resulted in 5,412 pedestrian fatalities and 82,000 pedestrians injuries (see NHTSA, 1997). These figures are a slight decrease from 1995 when 5,584 pedestrians were killed and 86,000 injured. Generally, since 1993, pedestrian fatalities have hovered about the 5,500 mark, whereas in terms of injuries, there has been a consistent and definite decrease from 94,000 injuries in 1993 to 82,000 injuries in 1996. Passenger car, van, and light truck involvements account for nearly 86 percent of the pedestrians injured. One pedestrian is killed by a motor vehicle every 97 minutes and injured every 6 minutes in the United Stages.

Data on pedestrian injuries from the 1982–1986 NASS data files (1982), the last pedestrian data files available for analyses, shows that approximately 40 percent of the pedestrian injuries resulted from contact with the vehicle, 32 percent resulted from contact with the ground and 26 percent of the injuries were from unknown contact sources.

The National Highway Traffic Safety Administration (NHTSA) has issued the latest statistics on accidents involving pedestrians and motor vehicles in the United States. The report details the number of pedestrian traffic related fatalities and injuries. The report shows that a pedestrian dies from a motor vehicle related collision once, on average, every 113 minutes and a pedestrian is injured in a traffic incident about every 8 minutes.

The pedestrian fatalities involving motor vehicles in the year 2005 are shown in Table 8.1.

Table 8.1 Pedestrian fatalities involving motor vehicles: 2005

State	*Total traffic fatalities*	*Pedestrians killed as percent of total*	*Pedestrian fatalities as percent of total*	*Population (thousands)*	*Pedestrian fatality as percent of total rate per 100,000 population*
Alabama	1,131	72	6.4	4,558	1.6
Alaska	72	7	9.7	664	1.1
Arizona	1,177	157	13.3	5,939	2.6
Arkansas	648	37	5.7	2,779	1.3
California	4,329	742	17.1	36,132	2.1
Colorado	606	48	7.9	4,665	1.0
Connecticut	274	34	12.4	3,510	1.0
Delaware	134	11	8.2	844	1.3
District of Columbia	48	16	33.3	551	2.9
Florida	3,543	576	16.3	17,790	3.2
Georgia	1,729	150	8.7	9,073	1.7
Hawaii	140	35	25.0	1,275	2.7
Idaho	275	9	3.3	1,429	0.6
Illinois	1,361	164	12.0	12,763	1.3
Indiana	938	63	6.7	6,272	1.0
Iowa	450	24	5.3	2,968	0.8
Kansas	428	24	5.6	2,745	0.9
Kentucky	985	54	5.5	4,173	1.3
Louisiana	955	109	11.4	4,524	2.4
Maine	169	9	5.3	1,322	0.7
Maryland	614	102	16.6	5,600	1.8
Massachusetts	442	76	17.2	6,399	1.2
Michigan	1,129	137	12.1	10,121	1.4
Minnesota	559	44	7.9	5,133	0.9
Mississippi	931	72	7.7	2,921	2.5
Missouri	1,257	88	7.0	5,800	1.5
Montana	251	13	5.2	936	1.4
Nebraska	276	8	2.9	1,759	0.5
Nevada	427	63	14.8	2,415	2.6
New Hampshire	166	5	3.0	1,310	0.4
New Jersey	748	154	20.6	8,718	1.8
New Mexico	488	61	12.5	1,928	3.2
New York	1,429	321	22.5	19,255	1.7
North Carolina	1,534	164	10.7	8,683	1.9

(Contd.)

North Dakota	123	9	7.3	637	1.4
Ohio	1,323	95	7.2	11,464	0.8
Oklahoma	802	50	6.2	3,548	1.4
Oregon	488	48	9.8	3,641	1.3
Pennsylvania	1,616	159	9.8	12,430	1.3
Rhode Island	87	14	16.1	1,076	1.3
South Carolina	1,093	98	9.0	4,255	2.3
South Dakota	186	14	7.5	776	1.8
Tennessee	1,270	70	5.5	5,963	1.2
Texas	3,504	419	12.0	22,860	1.8
Utah	282	20	7.1	2,470	0.8
Vermont	73	3	4.1	623	0.5
Virginia	947	88	9.3	7,587	1.2
Washington	647	71	11.0	6,288	1.1
West Virginia	374	23	6.1	1,817	1.3
Wisconsin	815	44	5.4	5,536	0.8
Wyoming	170	7	4.1	509	1.4
United States, total	43,443	4,481	11.2	296,410	1.6

SOURCE: U.S. Department of Transportation, National Highway Traffic Safety Administration, *Traffic Safety Facts 2005: Pedestrians*, Washington, DC: 2006, available at http://www-nrd.nhtsa.dot.gov/departments/nrd-30/insa/availinf.html as of Nov.30, 2006.

The totals reveal that in the year 2007: 4,654 pedestrians died in the United States and 70,000 injured in traffic accidents in the United States. The statistics also provide information about the location and conditions at the time of the traffic accidents. By far the majority (73%) of the deaths happened in urban areas and outside of crosswalks away from intersections (77%). Most happened in good weather (90%), and most at night (67%). The most alarming (but not surprising statistic) was that alcohol was found in a driver or the pedestrian in 49 percent of the fatal pedestrian traffic accidents. More children and elderly are involved in traffic crashes involving pedestrians than other victims.

In a pedestrian/motor vehicle collision, one of the primary objectives of a technical investigation is to determine the vehicle's impact speed.

Since the 1970's, this topic has received considerable attention in the published literature. Over 450 technical papers have been published in the last four decades on the topic of vehicle-pedestrian accidents (see Han and Brach, 2001).

A great number of these articles present and discuss statistics of pedestrian accidents, types and trends of injuries, types, trends and categories of vehicle frontal geometries, typical vehicle deformations, etc. Others present and discuss physical and mathematical models for the analysis and reconstruction of such pedestrian accidents. The form of these models varies widely, however and ranges from statistical and/or empirical equations to relatively sophisticated mathematical mechanics models. Examples of the former are Evans and Smith (1999), Happer, et al. (2000) and Limpert (1984, 1998). Examples of the latter include the simulation models of van Wijk, et al. (1983), Moser, et al. (2000) and the moment models of Wood, (1988). A compilation of numerous vehicle-pedestrian accident reconstruction formulas is given by Russell, (undated). A book by Eubanks (1994) gives broad coverage to pedestrian accidents including related topics such as tabulated pedestrian speeds. A compilation of technical papers representing basic studies and results is contained in volume edited by Backaitis (1990).

Limpert (1998) splits a vehicle-pedestrian collision into 3 phases: the impact phase, the flight phase and the sliding and/or rolling phase. Five categories of vehicle-pedestrian collisions are described by Ravani, et al. (1981), Brooks, et al. (1987) and others. These are wrap, forward projection, fender vault, roof vault and somersault collisions. A wrap collision is one where a significant portion of the pedestrian's body extends above the forward or frontal portion of the vehicle and the pedestrian's body "wraps" up onto another portion of the vehicle such as the hood (bonnet), windshield and/or roof. Wrap contact is accompanied by a rotation and bending of the pedestrian's body and often is followed by a secondary impact of the head and/or upper torso. In a fender vault, the pedestrian falls or rolls off to the side of the moving vehicle. For a roof vault, the vehicle passes under the pedestrian as it does for a somersault, where the pedestrian goes over the roof and rotates, typically to an inverted, head-down, position.

Confidence limits have been studied by other authors such as Evans and Smith (1999) and Happer, et al. (2000). Typically these limits are broad and may not apply directly to a particular data set or to a specific application of a model.

The size and frontal geometry of the vehicle are factors that influence the mechanics of throw distance. Many authors have studied these aspects of collisions; for example, see Ashton and Mackay (1979), Sturtz and Suren (1976) and Evans and Smith (1999). There appears to be some disagreement on the significance of frontal geometry. Limpert (1998) states that the frontal geometry and the identification and measurement of the location of the secondary impact can be critical in reconstruction to the extent that small errors can result in significant differences in speed reconstruction. Evans and Smith (1999) state that the effect of frontal geometry is overrated. In general, the frontal geometry probably does play a role in vehicle-pedestrian mechanics and reconstruction but has no special significance or sensitivity. It is possible that the vehicle frontal geometry, pedestrian height, weight and speed factors interact, or couple. For a given frontal geometry, the location and severity of the secondary impact depends on the speed and pedestrian geometry. The solutions proposed in the literature can be categorized as either empirical or mathematical. It should be noted that the true mathematical solutions are very difficult to apply to real world collisions, because the data necessary to solve the mathematical equations is only partly available from the real world collisions.

Empirical solutions overcome the problem of a lack of data that is required for each and every parameter necessary to solve the mathematical equations. The aim of the empirical solution is to work within the boundaries of the most commonly available data from real world collisions.

There are many mathematical and empirical models to facilitate technical analysis of vehicle-pedestrian collisions. The main objective of the technical analysis is to determine the vehicle impact speed. In the real world, post-impact data from vehicle/pedestrian collisions is often insufficient to use pure mathematical models. However mathematical models can be utilized successfully with the aid of some empirical data or reasonable assumptions. Conversely the empirical models overcome the difficulty of insufficient data, but must be applied within the model defined limits. This study compares the mathematical and empirical models. In addition, a slight variation of an earlier model [1] is presented to test the sensitivity of empirical models.

The aim of this study is to compare the mathematical models with the empirical models. In addition, this study compares two slightly different empirical models derived from the same data set. The criteria for developing the new empirical model are to keep it simple while maintaining a reasonable degree of accuracy (Toor et al. 2002).

8.2 FRONTAL IMPACT

Approximately 75% of all vehicle-pedestrian collisions involve the bumper and front of the vehicle.

8.2.1 Primary and Secondary Collisions

When a vehicle hits a pedestrian the movement of the pedestrian due to the impact can be divided into three phases:

(*a*) Contact phase or Impact phase

(*b*) Flight phase

(*c*) Sliding phase (ground contact phase) or rolling phase

The first phase is the contact phase, where one or more bodies of the pedestrian have contact with the vehicle and the pedestrian is accelerated. High contact forces can be seen during this phase (see Figure 8.1).

Fig. 8.1 Phases of the pedestrian movement (impacting vehicle not shown)

Depending on the deceleration of the vehicle the pedestrian separates from the car after the contact phase and travels through the air without any contacts neither with the car nor with the ground. During this phase there is only a minor speed loss of the pedestrian. As a result of gravity the pedestrian has ground contact after the flight phase. Depending on the pedestrian to ground friction the velocity of the pedestrian is reduced until the pedestrian reaches its final stop position.

The total trajectory used in this study is the distance from the point of impact to the final stop position of the pedestrian.

8.3 ACCIDENT INVESTIGATION

The accident investigation can be divided into three areas: (1) Technical data, (2) Witness statements and (3) Interpretation and analysis of technical data.

1. Technical Data

The first professionals to arrive at any serious traffic collisions are the emergency personal. Their main objectives are to attend to the accident victims. However, in the process they also contaminate technical evidence. Fortunately, considerable amount of the physical evidence survives to facilitate scientific analysis.

In a real world vehicle/pedestrian collision, the pedestrian trajectory data is very limited. In fact only the beginning and the end of the trajectory data are often available, i.e., only the point of impact and the final rest locations are known. Furthermore, in many cases even the point of impact is difficult to establish due to lack of physical evidence. The pedestrian airborne distance (i.e., launch to initial landing distance) is another parameter that often cannot be evaluated from the physical evidence.

In some cases secondary evidence may be available in the form of vehicle tire or skid marks. The skid marks can be used to determine the vehicle speed and provide an independent method to check the throw distance calculations. These real world cases are very special, as they provide rare opportunity to either enhance or develop empirical solutions.

2. Witness Statements

Witness statements can be mixture of opinions and observations, Baker et al. (1986). Therefore, witness statements must be interpreted with a reasonable degree of caution. However, reliable witness observations of pedestrian movement prior to the collision can establish pre-impact path of the pedestrian with some degree of accuracy.

3. Interpretation and Analysis of the Technical Data

In case where the physical evidence does not yield the point of impact, the point of impact can be assumed as the intersection of the pre-impact vehicle and pedestrian paths. The damage to the impacting vehicle's front end can locate the point of impact within the width of the vehicle's path.

Obviously the total pedestrian throw distance is the distance between the point of impact and the pedestrian's center of gravity at rest.

8.4 PEDESTRIAN TRAJECTORIES

During the motor vehicle accident, the pedestrian is often stuck by the leading edge of the vehicle on the lower limbs. The initial contact is often followed by a more pronounced pedestrian/vehicle interaction. The nature of the subsequent contacts is dependent on the vehicle geometry and the pedestrian stature.

8.4.1 Forward Projection Trajectory

If the pedestrian center of gravity (CG) is lower than the upper edge of the leading face of the vehicle, then a "forward projection trajectory" ensues. In the forward projection trajectory the upper body of the pedestrian may partially wrap over the hood of the vehicle; however, secondary contact between the pedestrian's head and the vehicle does not occur. If the CG of the pedestrian is propelled forward with respect to the vehicle then it is classified as a forward trajectory. Figure 8.2 illustrates the forward trajectory.

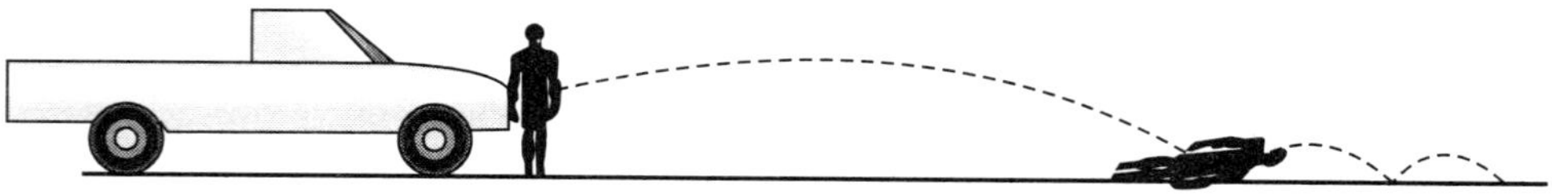

Fig. 8.2 Forward projection trajectory example

8.4.2 Wrap Trajectory

Conversely, if the pedestrian's center of gravity is above the upper leading edge of the vehicle then the wrap trajectory occurs. For a wrap trajectory the upper body will always wrap over the upper leading edge of the vehicle. After the initial impact the pedestrian CG moves rearwards with respect to the vehicle. If the pedestrian does not fall off to the side of the vehicle then a secondary contact with the windshield or the upper surface of the hood takes place. After the secondary contact the pedestrian's CG will begin to move forward with respect to the vehicle. Figure 8.3 illustrates the wrap trajectory.

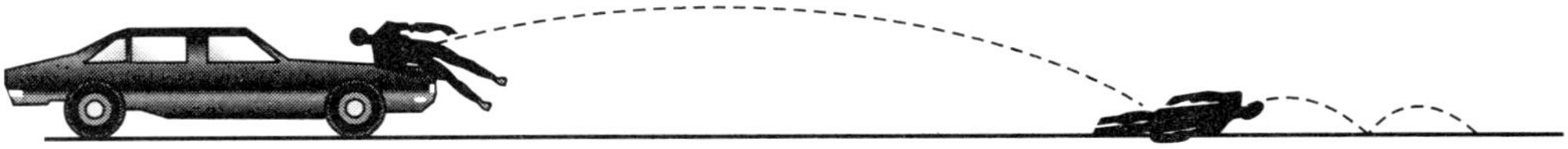

Fig. 8.3 Wrap trajectory example

During most pedestrian impacts, the impacting vehicle is decelerating prior to or immediately after impact. Upon impact the pedestrian is rapidly accelerated to a speed that is nearly equal to or less than the vehicle impact speed. At the same time the vehicle decelerates and continues decelerating, primarily due to braking. The vehicle and the pedestrian separate from each other when the pedestrian speed exceeds the vehicle speed. After the separation the pedestrian trajectory will include a free flight through the air at some launch angle, impact with the ground, and sliding or tumbling to rest. If the vehicle is not braking, then the pedestrian is more likely to be carried on the vehicle for some distance or vault over the roof of the vehicle.

If the pedestrian strikes near the corner of the vehicle and the pedestrian CG is above the upper leading edge of the vehicle, then after the initial contact the pedestrian can simply slide off the fender. This trajectory is termed "fender vault". The fender vault can also occur if the pedestrian is walking quickly and the impact occurs just prior to the pedestrian clearing the vehicle's path.

A "vault" trajectory occurs if the vehicle is traveling at a sufficient velocity and not decelerating. In this case the pedestrian vaults over the vehicle and comes to rest behind the vehicle, primarily due to the fact that the vehicle does not stop. Analysis of vault trajectories must be conducted with extreme caution.

An extensive review of the published literature pertaining to vehicle/pedestrian collisions lead to three methods for evaluating vehicle speeds. The pre-requisites for each method of assessing the impacting vehicles speed were knowledge of:

- Pedestrian injuries
- Vehicle damage
- Pedestrian throw distance

All three methods were based on empirical data. The assessment of the vehicle speed from pedestrian injuries and vehicle damage was subjective and was intended to serve as a check for the regression analysis based on pedestrian throw distance (Toor et al. 2000, 2002). Two tables were presented to facilitate the subjective assessment of the vehicle speed. The original tables are reproduced here (Toor et al. 2002) (Tables 8.2, 8.3 and 8.4).

Table 8.2 Vehicle damage summary for forward projection trajectories (Toor et al. 2002)

Approximate vehicle impact speed	*General damage summary*
< 20 km/h	Surface cleaning marks.
35 km/h	Leading edge of hood dented; deformation on front vehicle.
60 km/h	Middle of hood dented.

Table 8.3 Vehicle damage summary for wrap trajectories (Toor et al. 2002)

Approximate vehicle impact speed	*General damage summary*
< 20 km/h	Surface cleaning marks.
25 km/h	Head contact near bottom edge of windshield when pedestrian C.G. ~60 cm above low-fronted vehicle's bumper assembly; otherwise, head contact near middle of hood for average-sized vehicle and pedestrian.Body contact on roof when pedestrian C.G. ~85 cm above low-fronted vehicle's bumper assembly.
25 to 40 km/h	Head contact near trailing portion of hood or cowl; slight body pane deformation.
40 km/h	Head contact near bottom edge of windshield for impacts significantly below (~50 cm) pedestrian's C.G. (i.e., typical braking low-fronted vehicle).
40 to 50 km/h	Clearly defined dents on body panels.
50 km/h	Head contact near bottom edge of windshield when pedestrian C.G. ~40 cm above low-fronted vehicle's bumper assembly.Body contact on roof when pedestrian C.G. ~60 cm above low-fronted vehicle's bumper assembly.
50 to 55 km/h	Head contact near middle of windshield for typical braking low-fronted vehicle.
60 km/h	Head contact near bottom edge of windshield when vehicle's upper edge near pedestrian's C.G.
> 60 km/h	More probable body to roof contact.
70 km/h	Head contact near upper frame of windshield; significant deformation of body panels.
80 km/h	Pelvic contact with roof; roof deformation (unbraked vehicle).

Table 8.4 Pedestrian injury summary (Toor et al. 2002)

Approximate vehicle impact speed	*General injury description (subjectively assessed)*
< 20 to 25 km/h	Minor injuries.
30 to 50 km/h	Moderate or major injuries; however, injuries likely non-fatal.
> 50 km/h	Severe injuries which are likely fatal.

8.4.3 Vehicle Classification (Toor et al. 2002)

Several different types of vehicles were involved in the real world collisions and staged impact tests recorded in the available literature. Research revealed that the primary vehicular characteristic that affects the pedestrian's post-impact trajectory is the height of the upper edge at the front of the vehicle. Thus, the front profiles of vehicles were categorized into three distinct categories for the purpose of analyzing the available data:

1. *Blunt:* The front of the vehicle is almost flat and approximately vertical. The upper edge is above the center of gravity (C.G.) of a 50th percentile standing male (elevation > 1.05 m).
2. *High-fronted:* The upper edge of the front of the vehicle is above the knees, but at or below the C.G. of a 50th percentile standing male (0.55 m < elevation ≤ 1.05 m).
3. *Low-fronted:* The upper edge of the front of the vehicle is at or below the knees of a 50th percentile standing male (elevation < 0.55 m).

8.4.4 Pedestrian Trajectory Classification

In order to assess vehicle impact speed from the post-impact collision evidence, it is imperative to understand the post-impact kinematics of the struck pedestrian. Ravani (1981), Eubanks (1992, 1994) and Ashton (1981, 1982, 1989) provide adequate descriptions of the type of pedestrian trajectory one might expect from an impact with a vehicle: forward projection, wrap, fender vault, roof vault or somersault. In the latter three classifications, the pedestrian initially wraps onto the vehicle prior to vaulting over the fender, vaulting over the roof or performing a somersault. Thus, for simplifying the description of pedestrian post-impact motion in this paper, the pedestrian post-impact kinematics will be categorized as either a forward projection trajectory or wrap trajectory.

1. Forward projection trajectory—In a forward projection trajectory, the pedestrian is projected ahead of the vehicle such that the pedestrian's C.G. remains ahead of the vehicle. The vehicle's upper leading edge is above the pedestrian's C.G. in order to prevent rearward displacement of the pedestrian's C.G. with respect to the vehicle. Although the pedestrian's upper torso and head may wrap around the front portion of the vehicle, the trajectory will be classified as a forward projection due to the pedestrian's C.G. remaining ahead of the vehicle.

   When the pedestrian is contacted by the front of a vehicle, the pedestrian is accelerated in the direction of the vehicle's travel. As the height of the upper contact point on the front of the vehicle is higher than the pedestrian's C.G., the pedestrian will likely be vaulted forward primarily in the horizontal direction.

2. Wrap trajectory—In a wrap trajectory, the vehicle's upper leading edge is at or below the pedestrian's C.G. During the collision, the pedestrian's C.G. slides and/or rotates rearward of the front of the vehicle.

   After the initial contact, the pedestrian is accelerated forward and starts to rotate about the upper leading edge of the front of the vehicle. As the vehicle continues forward, the pedestrian shifts rearward with respect to the vehicle and continues to accelerate forward with respect to the ground. If the pedestrian has not moved off the side of the vehicle, then there will be secondary contact between the pedestrian and the vehicle; this contact typically consists of head contact with the hood or windshield. As the vehicle decelerates during the collision, the pedestrian will usually start to separate from the vehicle (i.e., shift forward with respect to the vehicle) after the secondary contact. The separation is due to the pedestrian's speed surpassing the decelerating vehicle's speed. The pedestrian will then be launched at a speed less than the initial impact speed of the vehicle.

   A vehicle must be decelerating in order for the pedestrian to be projected forward from a wrap trajectory. The forward motion of the pedestrian will be a vault at some angle with respect to the horizontal plane. After the vault through the air, the pedestrian impacts the ground and then tumbles/slides to rest. If the pedestrian's launch angle is zero, then there is no initial vertical velocity component; the pedestrian is thrown directly forward before impacting the ground and tumbling/sliding to rest (this specific scenario may occur in a wrap trajectory, but will always be the case in a forward projection).

   In some cases, the pedestrian may be contacted near the corner of the vehicle or the pedestrian may be traveling quickly. These collision circumstances in a wrap trajectory may cause the pedestrian to make only one contact with the vehicle. After being struck, the pedestrian may slide up the hood and off the side of the vehicle before reaching the windshield; this trajectory is classified as a fender vault. Due to the lack of available data for these types of collisions, it is difficult to assess impact speed in a fender vault scenario. Thus, these pedestrian post-impact trajectories are not considered in this paper.
3. Trajectory assessment—In the case of a real world collision reconstruction, both the vehicle's frontal profile and the pedestrian's size must be compared to assess the expected type of pedestrian post-impact trajectory. If the vehicle's upper leading edge impacts the pedestrian above the pedestrian's C.G., then a forward projection trajectory is expected. Otherwise, a wrap trajectory would likely result.

8.4.4.1 Vehicle Damage

The vehicle damage information reported in the available literature (Toor et al. 2000, 2002) was reviewed in order to ascertain trends in the damage for specific vehicle impact speeds and collision circumstances. A set of guidelines were derived for estimating approximate vehicle impact speed from the sustained vehicle damage for both forward projection and wrap trajectories.

The post-impact vehicle damage was reported for several collisions involving live humans, cadavers and dummies. The staged impacts primarily involved pedestrians that were about the height of an average-sized adult with the exception of four child-sized test dummies. The real world collisions with vehicle damage information also involved adult pedestrians with the exception of one child.

In the reviewed literature, the damage sustained from dummy collisions was compared to cadaver impacts. Specific tests were conducted by Kallieris (1988) to compare the wrap trajectories of dummies with those of cadavers. Kallieris noted that the depth of vehicular damage caused by head contact from the dummy or cadaver is similar; Kallieris also reported that the cadavers were observed to make head contact slightly further rearward than the dummies. In addition, the damage to the leading edge of the hood (caused by torso/leg contact) was found to be more significant in the dummy impacts than the cadaver collisions. Grosch (1989) confirmed the observations made by Kallieris and attributed the deeper hood damage to the dummy being less compliant than the cadaver; the cadaver was also found to rotate and slide up the vehicle more than the dummy.

Research literature also noted that vehicle's frontal profiles affect the pedestrian post-impact motion and the vehicle damage. Thus, for a specific impact speed and pedestrian size, vehicles with different frontal profiles will likely induce different damage patterns. In addition, vehicle damage will differ for similar sized and shaped vehicles with unlike structural characteristics. Thus, vehicle structural differences and other collision factors (e.g. pedestrian size, pedestrian pre-impact motion, etc.) preclude exact assessment of vehicle impact speed from the generalized post impact vehicle damage. However, a damage analysis does allow an order of magnitude estimate of the vehicle impact speed.

8.4.4.2 Forward Projection Trajectory Damage

The few considered collisions which involved children were forward projections; thus, this data was grouped with the few collisions involving adults who were also projected forward from vehicles. During the majority of forward projection trajectories, vehicles were braking prior to impact. Thus, Table 8.2 summarizes the damage expected for a braking vehicle that impacts an adult or child causing a forward projection trajectory.

Forward projections typically have damage isolated to the front of the vehicle extending to about the middle of the hood. Due to the limited extent of pedestrian motion over the vehicle and vehicle damage, it is more difficult to estimate an impact speed from observation of the vehicular damage resulting from a forward projection. However, as the impact speed increases, the extent of vehicle deformation is expected to increase correspondingly.

8.4.4.3 Wrap Trajectory Damage

Most vehicles were braking prior to impact in the wrap trajectory collisions that were considered; however, there was a smaller group of collisions in which there was no vehicle braking. In these latter collisions, pedestrians typically continue over the vehicle without achieving a common speed with the vehicle. It was found that the resulting vehicular damage patterns were similar to the damage patterns sustained by braking vehicles at similar impact speeds. Thus, Table 8.3 summarizes the damage expected for braking or non-braking high-fronted or low-fronted vehicles that impact an adult causing a wrap trajectory.

The wrap trajectory data reveals that the main indicator of vehicle impact speed is the location of secondary head contact. In wrap trajectories, the adult pedestrian's head will typically make contact with the vehicle somewhere between the hood and the upper frame of the windshield. In impacts with low-fronted vehicles, the average sized adult pedestrian will make head contact with the bottom portion of the windshield at an impact speed of around 40 km/h. If the pedestrian is struck near the C.G., then a vehicle impact speed of about 60 km/h is required for the pedestrian's head to strike the bottom portion of the windshield. Pedestrian contact with the roof becomes more likely at impact speeds of 60 km/h or greater. Furthermore, significant deformations of body panels will likely result at impact speeds greater than about 70 km/h.

For wrap trajectories, if an impacted pedestrian is much shorter or taller than the average height, then the expected vehicle damage will differ somewhat from that presented in Table 8.3 for a specific impact severity. For example, a shorter pedestrian may not impact the vehicle's windshield as high up as a 50th percentile male. The length of a vehicle's hood also affects the extent of pedestrian motion over the vehicle's frontal surfaces. If a vehicle has a much shorter hood than a typical sedan (e.g. compact vehicle), then the pedestrian would be expected to impact the vehicle further rearward than that presented in Table 8.3.

Vehicle damage summary: The differences in dummy kinematics from human kinematics were considered in the analysis of the post-impact vehicle damage. However, the data pertaining to the dummy collisions was not utilized as a basis for Table 8.2 and 8.3.

The reviewed data only includes collisions with impact speeds ranging up to about 80 km/h. Thus, the damage trends presented in Tables 8.2 and 8.3 may only be applied for vehicle impact speeds up to about 80 km/h. The impact speed ranges presented in Table 8.2 and 8.3 are not precise, but rather are approximate estimates based on the results of real world collisions and staged impacts. These speed ranges should only be utilized to assess an approximate vehicle impact speed. More uncertainty is associated with the higher speed ranges (i.e., above 60 km/h) as there is less data for these impact severities.

8.4.4.4 Pedestrian Injuries

The injuries sustained by a pedestrian involved in a vehicular collision may be utilized to aid in the reconstruction analysis. Several studies (Toor et al. 2000, 2002) have documented the injuries sustained by pedestrians in real world collisions and cadavers in staged impacts. Some research provides a detailed injury list of the struck pedestrian while other studies only provide partial lists or injury ratings corresponding to an injury scale.

The goal of this study was to correlate the specific injuries sustained by a struck pedestrian with a likely impact speed range. Unfortunately, there were no notable trends apparent to relate specific injuries sustained by the struck pedestrian with the vehicle impact speed.

The conglomeration and comparison of the referenced research studies yielded obvious trends and basic conclusions:

1. The severity of the pedestrian injury increased with impact speed.
2. The initial pedestrian/vehicle impact is more severe than the pedestrian/road impact.
3. The young and the elderly are at the highest risk to be injured.

Factors other than impact speed which influence the type of injury sustained may include the individual pedestrian characteristics and the vehicle structural characteristics. Although there was no detected relation between specific injuries and impact speed, some general injury trends were apparent. Table 8.5 provides a summary of the observed trends.

Table 8.5 Pedestrian injury summary

Approximate vehicle impact speed	*General injury description (subjectively assessed)*
< 20 to 25 km/h	Minor injuries.
30 to 50 km/h	Moderate or major injuries; however, injuries likely non-fatal.
> 55 km/h	Severe injuries which are likely fatal.

These injury trends are very generic and provide only an approximate estimate for the impact speed. Depending on the specific impact circumstances and the vehicle geometry, severe injuries (such as skull fractures) may result from relatively low-speed impacts (about 20 km/h); it is also possible for a pedestrian involved in a high speed impact to sustain relatively minor injuries. Injury data alone is insufficient to facilitate a precise estimate of impact speed; however, it may be used to support other analysis techniques.

8.4.4.5 Pedestrian Throw Distance

Review of the provided literature (Tool et al. 2002) yielded vehicle impact speeds with associated pedestrian throw distances for real world collisions and staged impact tests. The collisions involved adults, children, adult-sized dummies, adult-sized cadavers and child-sized dummies. Various blunt, high-fronted and low-fronted vehicles were either braked or unbraked at the moment of impact. Again, the data was categorized as either a forward projection or wrap trajectory.

(*a*) Forward projection trajectory throw distance—A total of 106 collisions were utilized in the analysis of vehicle impact speed and pedestrian throw distance for forward projection cases. Comparison of the adult and child trajectory data suggests that throw distances for a given impact speed are similar and thus may be considered in the same analysis for forward projection scenarios. 21 collisions involved adults and children projected forward from blunt vehicles. At least 10 of these impacts involved braking vehicles; there was no record of vehicle braking in the other 11 incidents. Data was present for 3 child impacts in which high-fronted vehicles were not braking. However, at least 36 of 82 collisions involved braking low or high-fronted vehicles and children.

(*b*) Wrap trajectory throw distance —The majority of the 202 wrap collisions considered were noted to have involved low or high-fronted braking vehicles. 10 collisions were noted to involve adults impacted by low or high-fronted vehicles that were not braking.

The data was categorized according to the front profile of the vehicle and its braking prior to impact. There is reduced correlation between impact speed and throw distance for non-braking vehicles. This variance is likely due to the pedestrians being carried on the vehicles for further distances before separation. Review of the provided literature confirms that there is no relationship between unbraked vehicle impact speeds and pedestrian throw distance. Consequently, the ten data points corresponding to unbraked vehicles were omitted for the remaining analysis.

(*c*) Combined throw distance data—During the reconstruction of a real world collision, there may be insufficient information to assess the type of pedestrian post-impact trajectory. Thus, a general relationship comparing vehicle impact speed and pedestrian throw distance is required. The throw distance data for both forward projection cases and wrap trajectories involving vehicle braking were compared.

(*d*) Regression analysis—The forward projection throw distance data in Figure B1 was categorized according to the size of pedestrian and frontal profile of the vehicle. The scatter of the various points in this chart suggests that the pedestrian size and vehicle frontal profile are not distinctly related to the pedestrian throw distance in a forward projection case. The wrap throw distance data (excluding non-braking vehicles) was categorized for both low and high-fronted vehicles. The distribution of the data in this chart shows that there is no determinable difference between the data for the two vehicle classifications.

Analysis of the forward projection, wrap and combined data indicates that there are second order relationships between vehicle impact speed and pedestrian throw distance. Thus, the square root of the pedestrian throw distance was plotted against the vehicle impact speed for each set of data.

A least squares linear regression was found to be a good fit for each set of transformed data. Statistical analyses revealed that there is a strong correlation in the data and that the calculated regression lines are useful predictors for vehicle impact speed.

Forward projection trajectory data:

$$V_v = 11.4(d_t)^{1/2} - 0.4 \quad [\pm 10.5 \text{ km/h}] \tag{8.1}$$

Wrap trajectory data:

$$V_v = 12.7(d_t)^{1/2} - 2.6 \quad [\pm 9.0 \text{ km/h}] \tag{8.2}$$

Combined throw distance data:

$$V_v = 12.3(d_t)^{1/2} - 1.9 \quad [\pm 9.5 \text{ km/h}] \tag{8.3}$$

where V_v = vehicle impact speed [km/h]

d_t = pedestrian throw distance [m]

As can be seen from the real world and staged collision data, the actual impact speed may vary for a specific pedestrian throw distance. Thus, a range for a vehicle's impact speed should be assessed for a specific collision scenario. Equations (8.1), (8.2) and (8.3) depict the magnitude of uncertainty associated with predicting a vehicle impact speed for a specific pedestrian throw distance. This uncertainty was derived from the 85th percentile prediction intervals which contain the majority of throw distance data. As there is limited data at the lower and upper ends of the throw distance data ranges, the reconstructionist should use caution when applying the derived regression equations near the limits of the data ranges.

The equations of the regression lines for the data approach the origin (within the uncertainty of the regression analysis). Comparison of Eqs. (8.1), (8.2) and (8.3) to the basic slide-to-stop formula allows the average drag factor for the entire pedestrian trajectory to be evaluated. Assuming the regression line intercepts the origin, the comparison becomes:

$$V^2 = 2\mu g d$$

$$V_v^2 = \begin{bmatrix} 11.4 \\ 12.7 \\ 12.3 \end{bmatrix}^2 d_t \tag{8.4}$$

An average drag factor for the entire pedestrian trajectory may be evaluated from these equations. Rearranging and solving for μ indicates that the average drag factor for a pedestrian in a forward projection case is about 0.5. Bratten (1989) evaluated a Tumble Number which was equivalent to the average drag factor of a body from the point of impact to its rest position. The body may vault through the air, tumble, bounce and/or slide to rest. The Bratten Tumble Number was derived to be about 0.5. The linear relationship obtained from the analysis of real world and staged forward projection impacts in this study yields a drag factor consistent with that derived in the Bratten paper. The average drag factor for a pedestrian in a wrap trajectory was evaluated to be about 0.6. The combined forward projection and wrap trajectory data also yielded an average drag factor of about 0.6.

8.4.4.6 Projection Efficiency

Pedestrian mathematical models typically assess the pedestrian throw speed (V_p)— the speed at which the pedestrian leaves the vehicle. In order to determine the vehicle impact speed from the pedestrian throw speed, the projection efficiency of a vehicle/pedestrian collision may be required. From the laws of physics, the vehicle impact speed (V_v) has to be greater than the pedestrian throw speed. The projection efficiency may be defined as:

$$\text{Projection efficiency} = \frac{V_p}{V_v} \times 100\% \tag{8.5}$$

A vehicle's frontal profile significantly affects the pedestrian's post-impact trajectory; thus, the projection efficiency will be discussed for both forward projection and wrap trajectories.

8.4.4.7 Forward Projection Trajectory Projection Efficiency

If it is assumed that the pedestrian's initial speed in the vehicle's travel direction is negligible, then the principle of conservation of linear momentum may be utilized to assess a relationship between the vehicle impact speed and the pedestrian throw speed. During the short duration of the impact, friction forces at the pedestrian's shoes and vehicle tiers are assumed to be negligible. Applying the momentum relationship to Eq. (8.5) yields the projection efficiency of the forward projection trajectory.

$$\text{Projection efficiency} = \frac{m_v}{(m_v + m_p)} \times 100\% \tag{8.6}$$

This formula illustrates that the pedestrian's throw speed may be related to the vehicle impact speed by the respective masses of the vehicle and pedestrian. The projection efficiency will only approach 100% when the vehicle mass becomes significantly larger than the pedestrian mass. A range for the projection efficiency for forward projection cases may be assessed by assuming reasonable maximum and minimum values for both the vehicle and pedestrian masses.

The results presented in Table 8.6 illustrate that the launch speed of the pedestrian would likely be 80 to 100% of the vehicle impact speed in a forward projection case. This range is consistent with the range presented in Limpert (1989) which indicates a projection efficiency of 80 to 100% for pedestrians impacted by large flat-fronted vehicles. Wood's (1991) mathematical model assesses the horizontal component of the pedestrian launch speed as a function of the vehicle impact speed for cases where there is only one contact between the two entities. This model considers the conservation of linear momentum, but also accounts for pedestrian rotation (as in fender vault scenario). In the case of a forward projection, Wood's formula simplifies to Eq. (8.6). This referenced publication derives the projection efficiency to be about 93% for cases where a 50th percentile male's C.G. is at or lower than the upper leading edge of the vehicle. This value is comparable to the approximate average value of 94% presented in Table 8.6.

Table 8.6 Projection efficiency for forward projection trajectories

Vehicle		*Pedestrian*		*Projection efficiency*
Description	*Approximate mass*	*Description*	*Approximate mass*	
Geo Metro	850 kg	Obese male	200 kg	81% (theoretical minimum)
Toyota Carmy	1,300 kg	50th% male	78 kg	94% (approximate average)
Tractor with Trailers	25,000 kg	5 year old	20 kg	99.9% (theoretical maximum)

Wrap trajectory projection efficiency—The mathematical model presented by Wood (1991) facilitates simple assessment of the projection efficiency for cases where there is only one contact between the vehicle and pedestrian (i.e., fender vault or forward projection). As described for forward projection trajectories, this model yields Eq. (8.6) when the pedestrian is impacted at or above the C.G. However, a fender vault trajectory will result when the pedestrian is impacted below the C.G. and only one contact is made with the vehicle. The efficiencies presented relate the horizontal component of the pedestrian throw speed to the vehicle impact speed. The actual pedestrian throw speed is greater than the horizontal component. Furthermore, the lack of secondary contact with the vehicle reduces the pedestrian's potential throw speed. Thus, the values likely underestimate the true projection efficiency values for wrap trajectories with secondary contact.

In order to assess the projection efficiency for collisions involving secondary contact, Wood derived further sets of equations. These more complicated formulae required detailed information of the vehicle geometry, vehicle contact points, pedestrian size measurements and extent of pedestrian rotation. This theory is relatively straight forward and appears to be useful for assessing projection efficiency. However, there are large uncertainties associated with several of the variables. Thus, Wood's model is not particular for making a direct comparison of pedestrian throw speed to vehicle impact speed.

In the more recent publication, Wood (1991) presents the "Impact Factor" for different collision circumstances: forward projection, fender vault and wrap that involve head contact. The "Impact Factor" appears to be comparable to the projection efficiency of the vehicle/pedestrian impact. However, the formulas used to evaluate this expression are dependent on several parameters including pedestrian drag factor, pedestrian C.G. height and extent of pedestrian rotation. Thus, the "Impact Factor" is difficult to accurately assess in the reconstruction of a real world collision.

Eubanks (1994) presents a relationship for assessing the impact projection efficiency for wrap trajectories based on the relative heights of the pedestrian's C.G. (h_p) and the upper leading edge of the vehicle (h_v).

$$\text{Projection efficiency} = \frac{h_v}{h_p} \times 100\% \tag{8.7}$$

This relationship indicates that the projection efficiency will approach 100% when the upper leading edge of the vehicle impacts the pedestrian near the C.G. (i.e., forward projection). It has already been hypothesized that a 100% projection efficiency occurs in a forward projection when the vehicle mass is much larger than the pedestrian mass. Thus, a true projection efficiency relationship for wrap trajectories may incorporate both mass and height parameters. A range for the projection efficiency for wrap cases may be assessed by using Eq. (8.7) and assuming reasonable maximum and minimum values for both the vehicle and pedestrian heights (see Table 8.7).

Table 8.7 Projection efficiency for wrap trajectories

Vehicle		*Pedestrian*		*Projection efficiency*
Description	*Approximate mass*	*Description*	*Approximate mass*	
Low-fronted	0.55 m (bumper)	$95^{th}\%$ male	1.12 m [54]	50% (theoretical minimum)
High-fronted	0.85 m (hood)	$5^{th}\%$ male	0.92 m	92% (theoretical minimum)

Grandel (1986) reported pedestrian throw speeds that were 90 and 95% of the vehicle impact speed in staged collisions involving dummies and high-fronted vehicles. Limpert (1989) indicated projection efficiencies of about 60 and 75% for impacts involving low-fronted and high-fronted vehicles, respectively. Thus, the reported and calculated value illustrate that a tall pedestrian impacted by a low-fronted vehicle will typically be accelerated to approximately 50 to 60% of the vehicle's impact speed (*i.e.*, when the vehicle's hood slopes down almost to the level of the bumper). A pedestrian impacted by a high-fronted vehicle just below the C.G. will likely be accelerated to about 90 to 95% of the vehicle's impact speed. Consequently, a vehicle with a leading edge situated between these two extreme elevations will likely impart projection efficiency between 60 and 90%.

8.4.4.8 Air Drag

When a pedestrian is vaulted into the air, there is some speed lost due to the resistance of the air around the pedestrian. Basic fluid theory dictates that the drag coefficient is directly proportional to the relative speed of the body through the fluid stream. Thus, up to a critical velocity, as the velocity increases, so does the drag on the body.

Air drag is neglected in the derivation of some of the reviewed models used to assess pedestrian launch speed from throw distance. However, this parameter is described in Collins (1979). Collins indicates that wind drag may be ignored for pedestrian launch speeds less than 40 km/h. He constructed a table of correction factors which should be added to specific pedestrian throw speeds in order to account for any speed loss due to air drag. The reported values increase with an increase in pedestrian speed; this result is consistent with the theory of fluid dynamics. Collins reported that typical pedestrian trajectory heights range from 1.5 to 2.1 m. Thus, the presented table of correction factors also has a dependency on the height of the pedestrian's trajectory.

Aronberg (1990) presents a graphical relationship between the speed loss due to air drag and the time spent in the air. This figure was based upon measurements recorded during the free-fall of a skydiver. The author correlates the speed loss of the skydiver to the duration of the free-fall and applies the relationship to the duration of the pedestrian's vault. As the speed of a vaulted pedestrian is much lower than that of a free-fall skydiver and the body dynamics are different, this data may not be directly applicable to a pedestrian's trajectory.

8.4.4.9 Drag Factor

Before evaluating vehicle impact speed from pedestrian throw distance, an understanding of the pedestrian drag factor is required. This parameter is factored into almost every mathematical model that addresses the post-impact kinematics of the struck pedestrian.

Newton's laws of motion can be utilized to assess a mathematical relationship for the deceleration rate, a, of an object sliding across a level surface:

$$a = \mu g \tag{8.8}$$

8.5 MATHEMATICAL MODELS

There are many published mathematical models. The traditional mathematical approach has been to treat the pedestrian as a projectile. Thus, either the projectile theory or some variation of the projectile theory is applied to develop mathematical models. There are simply too many published models to consider all models in a short study such as this. The mathematical models considered in this study were chosen for their "user friendliness". The "user friendliness" of the models was subjectively assessed.

8.5.1 List of Variables

a_v vehicle deceleration rate [g].
b_w half width of pedestrian's body.
d_h distance from pedestrian C.G. to top of head.
d_{hood} lateral distance of pedestrian travel from primary contact to secondary contact.
d_o vertical distance from pedestrian's C.G. to rest position on ground.
d_t pedestrian throw distance.
g gravitational acceleration.
h distance between pedestrian C.G. and upper contact point at front of vehicle.
h_a additional height pedestrian C.G. travels to top of trajectory.
h_{hood} highest point of pedestrian C.G. on vehicle.
I_p pedestrian's moment of inertia about horizontal axis.
k radius of gyration of pedestrian about horizontal axis.
m_p mass of pedestrian.
m_v mass of vehicle.
t_{hc} time between primary and secondary impacts, [sec].
V_p pedestrian throw speed.
V_{ph} horizontal component of the pedestrian throw speed.
V_{pi} initial traveling speed of pedestrian.
V_{pv} vertical component of the pedestrian throw speed.
V_v velocity of vehicle at primary impact.
α angle between upper contact point on front of vehicle and head contact point.
ϕ angle of pedestrian's initial path with respect to the vehicle's path (90° for perpendicular paths).
Θ pedestrian launch angle.
μ_p effective drag factor between pedestrian and ground.
Θ_p rotation angle of pedestrian's body from primary to secondary contacts.
μ_v vehicle braking coefficient.

8.5.2 Collin's Model

Collins (1979) took a simple approach by neglecting the pedestrian launch angle. Collins assumed that at impact the pedestrian trajectory is horizontal in the direction of the impacting vehicle. The pedestrian's vertical acceleration is only due to gravity. Upon landing, the vertical velocity is exhausted and the entire horizontal velocity is dissipated in the form of work done due to friction. This approach yielded the following quadratic equation.

$$S = V_p\sqrt{\frac{2h}{g}} + \frac{V_p^2}{2\mu g} \tag{8.9}$$

Although this is a simple model, it does incorporate the pedestrian C.G. height. This equation has also been incorporated in a book by Fricke, et al. (1990).

8.5.3 Searle's Model

Searle (1993) was critical of Collins for "incomplete consideration of the physics" involved in a vehicle/pedestrian collision. Searle also pointed out that the Collins method ignores any loss of horizontal speed due to impact with the ground on landing. The net result of this incomplete consideration is that Eq. (8.9) will slightly under-predict the average pedestrian launch speed. In addition, Collins indicates that a projection efficiency of 100% is applicable below impact speeds of about 50 km/h and references a pedestrian to road friction value of 0.8. Based on the data set available for this current study, both of these values are likely over-estimates. As the vehicle speed is inversely proportionate to the projection efficiency but proportionate to the frictional value, these two over-estimates tend to cancel each other to some degree. It is also noted that since the Collins model considers a horizontal throw angle and 100% projection efficiency, this suggests that the Collins model may be more appropriate for forward projection trajectories. If an appropriate friction value and lower projection efficiency value are incorporated, then Eq. (8.9) may also be appropriate for wrap trajectories.

Searle (1993) considered an idealized particle and evaluated the following relationship for the particle launch speed.

$$V_p = \frac{\sqrt{2\mu g(S + \mu h^+)}}{\cos\theta + \mu\sin\theta} \tag{8.10}$$

where h is the gain in elevation; for a pedestrian on level surface, this value will be negative.

In real world collisions, the data pertaining to launch angle is not normally available. In a study by Fugger et al. (2000), it was stated that at speeds between about 32 to 42 km/h, a wrap trajectory yielded typical throw angles of about 28 degrees. Searle (1993) eliminated the throw angle term by further developing Eq. (8.10) to yield a minimum pedestrian impact speed. The minimum value is given by differentiating with respect to the throw angle and equating this term to zero $\left(i.e., \frac{\partial v}{\partial \theta} = 0\right)$. Thus, the minimum pedestrian throw speed is given by

$$V_{p(\min)}\sqrt{\frac{2\mu g(S + \mu h^+)}{1 + \mu^2}} \tag{8.11}$$

Searle reported that the pedestrian throw speed would be slightly lower than the ideal particle speed. To yield more accurate throw speed values, Searle suggests the incorporation of a correction factor that is dependent on the average coefficient of friction. Searle indicates that for $\mu = 0.5$ to 0.7, the correction factor is approximately minus 5%.

Since Eq. (8.10) yields the minimum throw speed, factors must also be incorporated to yield the average vehicle impact speed. The ratio of the pedestrian launch speed and the vehicle impact speed is defined as the "projection efficiency". Searle indicates that the pedestrian throw speed is about 20% and 10% below the vehicle impact speed for adult and child pedestrians, respectively. The corresponding projection efficiency factors suggest that a wrap trajectory was referenced by Searle. Searle also indicates that the average vehicle impact speed is about 6% higher than the minimum.

8.5.4 Eubank's Model

Eubanks et al. (1994) published a 3-phase model: pedestrian wrap on the hood, fall to the ground, and slide to rest. The Eubanks approach is similar to Collins.

$$S = \frac{V_P^2}{2\mu g S} + \left(\sqrt{\frac{2d_{hood}}{g}} + \frac{d_{hood}}{V_P \sin \Phi} \right) \tag{8.12}$$

If the hood height is substituted for the pedestrian C.G. height and pedestrian speed is considered to be negligible, then the Eubanks and Collins equations are equivalent.

Another theoretical model was presented by Han and Brach (2001). The Han-Brach model also considered distance traveled over the three phases of the pedestrian collision: the initial vehicle/ pedestrian contact interaction, the pedestrian airborne trajectory, and the pedestrian slide/tumble/roll to rest. The Han-Brach model incorporates the launch angle, sliding friction, elevation gain/loss to determine the pedestrian launch speed. In order to assess the vehicle impact speed, the masses and a velocity ratio factor (to account for projection efficiency) are required. The airborne trajectory and the pedestrian slide/tumble/roll to rest components of the pedestrian impact are theoretically considered thoroughly; however, the initial impact phase and the projection efficiency factor need to be explored further. For a road grade angle of zero, the Han-Brach model equations are as follows:

$$V_v = A_P \left[\frac{m_v + m_e}{a m_c} \right] \sqrt{S - (x_L + \mu h)} \tag{8.13}$$

where

$$A_p = \sqrt{\frac{2\mu g}{\mu^2 \sin^2 \theta + \mu \sin 2\theta + \cos^2 \theta}} \tag{8.14}$$

Eubanks (1994) model is for a three phase pedestrian trajectory: travel on the vehicle, fall to the ground, and slide to rest. 100% projection efficiency is assumed, although the pedestrian wraps onto the hood. Thus, the use of this formula is limited.

The following coefficients are used in the standard quadratic equation to solve for vehicle impact speed:

$$a = \frac{1}{2\mu_p g}$$

$$b = \sqrt{\frac{2h_{hood}}{g}} + \frac{d_{hood}}{V_{pi} \sin \Phi}$$

$$c = -dc \tag{8.15}$$

8.5.5 Schmidt's Model

This model is for a vaulted pedestrian that slides to rest. Two preliminary parameters (e, f) are defined prior to assessment of the main variables.

$$e = 2\mu_p g d_t$$

$$f = \mu_p \sqrt{2g}\left[\sqrt{h_a} + \sqrt{h_a + d_o}\right] \tag{8.16}$$

The horizontal and vertical components of the pedestrian throw speed may then be assessed:

$$V_{ph} = -f + \sqrt{f^2 + e}$$

$$V_{pv} = \sqrt{2gh_a} \tag{8.17}$$

The pedestrian throw speed and launch angle are then determinable:

$$V_p = \sqrt{V_{ph}^2 + V_{pv}^2}$$

$$\Theta = \arctan\left(V_{pv} / V_{ph}\right) \tag{8.18}$$

In order to assess the vehicle impact speed from the pedestrian throw speed, the projection efficiency must be considered. Furthermore, any air drag effects should be considered as well.

8.5.6 Limpert's Model

In Limpert's model (1989), the vehicle impact speed [km/h] is assessed for a specific vehicle deceleration rate (a_v):

$$V_v = 10.62\sqrt{8.4a_v^4 + \frac{a_v d_t}{0.3048}} - 32.19a_v^2 \pm 4.02 \text{ [km/h]} \tag{8.19}$$

8.5.7 Wood's Model

Wood (1988, 2000 and 2002) has presented various models, both theoretical and empirical in origin. The Wood studies are generally thorough and very comprehensive; however, the discussion of the Wood models will be limited for this study. The Wood formulae simplifies considerably for the single contact impacts that represent vault or forward projection trajectories. For wrap trajectories with secondary head contact, the formulae remains complex and requires careful consideration of parameters.

8.5.8 Stcherbatcheff et al. Model

Stcherbatcheff et al. (1975) also published a quadratic model similar to Collins; however, Stcherbatcheff uses the vehicle deceleration rate rather than the pedestrian/road friction and vehicle impact speed rather than pedestrian launch speed as parameters. Stcherbatcheff also introduced an empirical coefficient based on test data. Their final equation took the following form:

$$S = \frac{V_v^2}{2\left|\mu_{v(mean)}\right|} + 0.03\mu_{v(mean)}V_v \tag{8.20}$$

where $\mu_{v\,(mean)}$ is the mean vehicle deceleration.

The inclusion of the empirical coefficient may overcome the deficiencies of the Collins model. This model is only useful if the vehicle deceleration can be established with a reasonable degree of accuracy. In other words, unless some form the tire marks are present at the impact scene, the rate of vehicle deceleration cannot be independently established.

For a vehicle implementing full wheel lock-up on dry roads, an average deceleration rate of 0.7 g may be appropriate. For a deceleration rate of 0.7 g, the Stcherbatcheff model may be simplified as:

$$S = \frac{V_v^2}{1.4g} + 0.02gV_v \quad (8.21)$$

or

$$V_v = 6.81\sqrt{0.0424 + 0.2918} - 1.4 \quad (8.21a)$$

8.5.9 Pedestrian Equations (Gregory Russell, 1999)

The equations in the following section are used to determine the speed of a vehicle after striking a pedestrian. The variables used by these equations are defined below (see Figs. 8.4, 8.5 and 8.6).

d_f = horizontal distance traveled from impact with vehicle to impact with ground
d_h = distance from pedestrian's impact with front of vehicle to exit off the side
d_o = vertical distance from pedestrian's center mass to final rest
d_t = horizontal distance traveled from impact with vehicle to final rest
f = vehicle drag factor
g = acceleration of gravity
h = height of pedestrian's center of mass
H_h = height of highest major contact between pedestrian and striking vehicle
h_t = maximum vertical height achieved by pedestrian's center of mass
L = distance from impact with front of vehicle to exit off hood
L_h = height of vehicle's leading edge
V_p = pedestrian's velocity (pre-impact)
w_p = pedestrian weight
w_v = vehicle weight
θ = pedestrian's take-off angle (when vaulted off of the striking vehicle)
ϕ = difference (angle) between pedestrian's and vehicle's approaches
μ = drag factor for pedestrian sliding on road surface

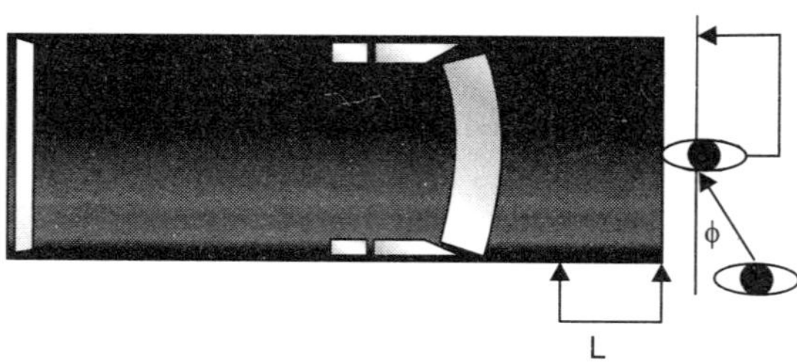

Fig. 8.4

Fig. 8.5

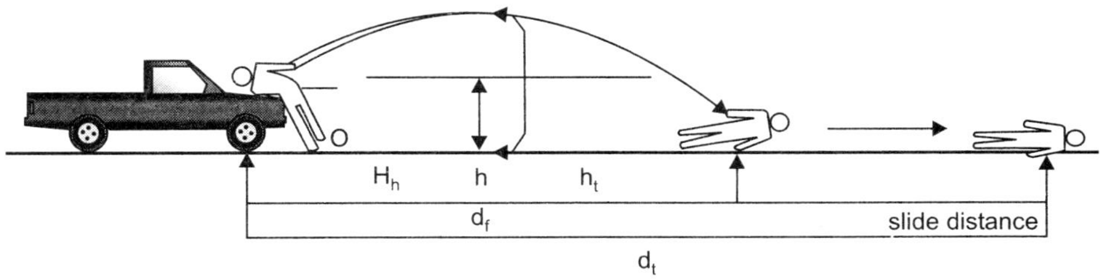

Fig. 8.6

Table 8.8

No.	*Model*	*Equation* — *Imperial*	*Equation* — *Metric*	*Comments*
1.	Aronberg and Snider	$V=\sqrt{\dfrac{2g(h_t-h)}{\sin\theta}}$ θ = launch angle of pedestrian $\theta=\tan^{-1}\left[\dfrac{\sqrt{h_t-h}}{\frac{g}{2}(\mu t)^2+\mu d_t-\sqrt{\frac{g}{2}}\mu t}\right]$ T = time pedestrian is in flight $T=\sqrt{\dfrac{2(h_t-h)}{g}}+\sqrt{\dfrac{2h_t}{g}}$	$V=\sqrt{\dfrac{2g(h_t-h)}{\sin\theta}}$ θ = launch angle of pedestrian $\theta=\tan^{-1}\left[\dfrac{\sqrt{h_t-h}}{\frac{g}{2}(\mu t)^2+\mu d_t-\sqrt{\frac{g}{2}}\mu t}\right]$ T = time pedestrian is in flight $T=\sqrt{\dfrac{2(h_t-h)}{g}}+\sqrt{\dfrac{2h_t}{g}}$	Solves for velocity of striking vehicle, when h_t, h, μ, and d_t are known. T and θ are solved as shown. This equation cannot be used for forward projections. μ = 0.4 to 1.2 (suggested by the authors, Aronberg and Snider).
2.	Limpent (1989)	$S=6.6\left[\sqrt{8.4\mu^4+\mu d_t}\right]-20\mu^2$	$S=1.6093\left[6.6\left\{\sqrt{8.4\mu^2}+3.281\mu d_t\right\}\right]-20\mu^2$	Solves for impact speed of striking vehicle, when d_t and μ are known. Requires pre-impact braking. μ = 0.7 to 1.2 suggested.
3.	Barzeley	$S=\sqrt{58+24d_t}-7.6$	$S=\sqrt{149+203d_t}-12$	Solves for impact speed of striking vehicle, when d_t is known. Equations assume no pre impact braking. Barzeley's equation is a derivation of Collin's equation that incorporates a value of 0.8 for μ and 3.0 feet/0.914 meters for the pedestrian's center of mass.

(Contd.)

4.	McHenry	$V = -\sqrt{2hg} + \sqrt{2\mu g(\mu h + d_t)}$	$V = -\sqrt{2hg} + \sqrt{2\mu g(\mu h + d_t)}$	Solves for impact velocity of striking vehicle, when g, d_t, μ and h are known. Can be used only in forward projections.
5.	Pultar	$V = g\left[-\sqrt{\frac{2h}{g}} + \sqrt{\left(\frac{2h}{g}\right)^2 + \left(\frac{2d_t}{g}\right)}\right]$	$V = g\left[-\sqrt{\frac{2h}{g}} + \sqrt{\left(\frac{2h}{g}\right)^2 + \left(\frac{2d_t}{g}\right)}\right]$	Solves for impact velocity of striking vehicle, when h, g, and d_t are known. Requires pre-impact braking, and pedestrian must come to rest in front of the vehicle.
6.	Schmidt and Nagel	$V = \sqrt{\mu\sqrt{2g}\left(\sqrt{h_t} + \sqrt{h_t - h}\right)^2 + 2\mu g d_t} - \left[\mu\sqrt{2g}\left(\sqrt{h_t} + \sqrt{h_t - h}\right)\right]$	$V = \sqrt{\mu\sqrt{2g}\left(\sqrt{h_t} + \sqrt{h_t - h}\right)^2 + 2\mu g d_t} - \left[\mu\sqrt{2g}\left(\sqrt{h_t} + \sqrt{h_t - h}\right)\right]$	Solves for impact velocity of striking vehicle, when d_t, g, h, h_t and μ are known. Cannot be used for forward projections. μ = 0.2 to 0.7 are suggested.
7.	Appel (1975)	$V = \sqrt{\frac{d_t}{K}}$	$V = \sqrt{\frac{d_t}{K}}$	Solves for impact speed of striking vehicle, when K and d_t are known.
8.	Wood (1988, 2000)	$V = \sqrt{\dfrac{[d_t - (\mu d_o)]2\mu g\left[\frac{w_v}{g} + \frac{w_p}{g}\right]^2}{\left(\frac{w_v}{g}\right)^2}}$	$V = \sqrt{\dfrac{[d_t - (\mu d_o)]2\mu g\left[\frac{w_v}{g} + \frac{w_p}{g}\right]^2}{\left(\frac{w_v}{g}\right)^2}}$	Solves for impact velocity of striking vehicle, when d_t, d_o, μ, w_v, w_p, and g are known. μ = 0.66 suggested.

Values of K	*Imperial*	*Metric*
Forward projection	0.0256	0.084
Wrap trajectories	0.0198	0.065
Adults (15 and older)	0.0213	0.070
Children (under 15)	0.0268	0.088

(Contd.)

9.	Wood's new (2002)	$S = K\sqrt{d_t}$ K (constant values)	$S = K\sqrt{d_t}$ K (constant values)	Solves for impact speed of striking vehicle when K and d_t are known.

Description	Imperial	Metric
Wrap trajectories (low values)	3.30	9.62
Wrap trajectories (high values)	4.70	13.20
Forward projections (low values)	3.01	8.77
Forward projections (high values)	4.72	13.76

10.	Northwestern University	$V_s = \sqrt{2g\mu(d_t - d_f)}$ $V_f = d_f\sqrt{\frac{-g}{2(-h)}}$ $d_f = -2\mu h + 2h\sqrt{\mu^2 + \frac{\mu d_t}{h}}$	$V_s = \sqrt{2g\mu(d_t - d_f)}$ $V_f = d_f\sqrt{\frac{-g}{2(-h)}}$ $d_f = -2\mu h + 2h\sqrt{\mu^2 + \frac{\mu d_t}{h}}$	Solves for impact speed of striking vehicle, basis on fall or slide speed of pedestrian, assuming a take-off angel of 0 degree. Assumes d_t, g, h, μ and d_f are known. μ = 0.45 to 0.60 suggested. Equation used for only forward projection (take-off angle = 0 degree) V_s and V_f must be equal.
11.	Searle (1993)	Maximum speed using 0° take-off angle: $V = \frac{\sqrt{2\mu g h_t}}{\cos\theta + \mu\sin\theta}$ when take-off angle equals 0° (forward projections): Highest value: $V = \sqrt{2\mu g d_t}$ Lowest value: $V = \sqrt{\frac{2\mu g d_t}{1+\mu^2}}$	Maximum speed using 0° take-off angle: $V = \frac{\sqrt{2\mu g h_t}}{\cos\theta + \mu\sin\theta}$ when take-off angle equals 0° (forward projections): Highest value: $V = \sqrt{2\mu g d_t}$ Lowest value: $V = \sqrt{\frac{2\mu g d_t}{1+\mu^2}}$	Solves for impact speed of striking vehicle, when d_t, g, θ and μ are known. μ = 0.66 for pedestrian sliding in a roadway and μ = 0.79 for pedestrian sliding on grass are suggested.Minimum speed calculated using approximately a 33.4° take-off angle with μ = 0.66 and 38.3° take-off angle with μ = 0.79.

(Contd.)

12.	Eubank (1994)	$a = \frac{1}{2g\mu}$ $b = \frac{d_h}{V_p \sin\phi} + \sqrt{\frac{H_h}{0.5g}}$ $c = -d_t$ $V = \left[\frac{-b + \sqrt{b^2 - 4ac}}{2a}\right] \Big/ \left(\frac{L_h}{h}\right)$	$a = \frac{1}{2g\mu}$ $b = \frac{d_h}{V_p \sin\phi} + \sqrt{\frac{H_h}{0.5g}}$ $c = -d_t$ $V = \left[\frac{-b + \sqrt{b^2 - 4ac}}{2a}\right] \Big/ \left(\frac{L_h}{h}\right)$	Solves for impact velocity of striking vehicle using the quadratic equation, when H_h, L_h, d_t, g, μ, ϕ, and V_p are known. $\mu = 1.14$ to 1.7 times the values of ϕ suggested. Equations should not be used for forward projections.
13.	Collins (1979)	$a = \frac{1}{2g\mu}$ $b = \frac{\sqrt{h}}{\sqrt{0.5g}}$ $c = -d_t$ $V = \frac{-b \pm \sqrt{b^2 - 4ac}}{2a}$	$a = \frac{1}{2g\mu}$ $b = \frac{\sqrt{h}}{\sqrt{0.5g}}$ $c = -d_t$ $V = \frac{-b \pm \sqrt{b^2 - 4ac}}{2a}$	Solves for impact velocity of striking vehicle in fps using the quadratic equation when h, d_t, and μ are known. $\mu = 0.8$ suggested.
14.	Fall and Slide	$a = \frac{1}{2g\mu}$ $b = \sqrt{\frac{2h}{g}}$ $c = -d_t$ $V = \frac{-b + \sqrt{b^2 - 4ac}}{2a}$	$a = \frac{1}{2g\mu}$ $b = \sqrt{\frac{2h}{g}}$ $c = -d_t$ $V = \frac{-b + \sqrt{b^2 - 4ac}}{2a}$	Solves for impact velocity of striking vehicle in fps using the quadratic equation when h, d_t, g, and μ are known. μ = 1.14 to 1.7 times the value of f suggested.
15.	Stcherbatcheff (1975)	$K = 0.0091$ $a = \frac{1}{2g\mu}$ $b = Kg\mu$ $c = -d_t$ $V = \frac{-b + \sqrt{b^2 - 4ac}}{2a}$	$K = 0.0091$ $a = \frac{1}{2g\mu}$ $b = Kg\mu$ $c = -d_t$ $V = \frac{-b + \sqrt{b^2 - 4ac}}{2a}$	Solves for impact velocity of striking vehicle using the quadratic equation when g, d_t, μ, and constant values are known. μ = 0.4 to 0.71 suggested.

(*Contd.*)

16.	Rich	$a = \frac{0.25}{\mu \cos^2 \theta} + \tan\theta$ $b = -(h + d_t \tan\theta)$ $c = d_t h$ $d = \frac{-b + \sqrt{b^2 - 4ac}}{2a}$	$a = \frac{0.25}{\mu \cos^2 \theta} + \tan\theta$ $b = -(h + d_t \tan\theta)$ $c = d_t h$ $d = \frac{-b + \sqrt{b^2 - 4ac}}{2a}$	Solves for distance pedestrian when air borne, when d_t, h, θ, and μ are known. h should be expressed as negative number in the equation. For wrap trajectories, μ = 1.1 to 1.7 times the value of f recommended. In forward projection trajectories use value of f form.
17.	IPTM	$V = \frac{-h}{4} + g\mu\sqrt{\left(\frac{h}{4}\right)^2 + \left(\frac{2d_t}{g\mu}\right)}$	$V = \frac{-h}{4} + g\mu\sqrt{\left(\frac{h}{4}\right)^2 + \left(\frac{2d_t}{g\mu}\right)}$	Solves for impact velocity of striking vehicle in fps, when h, d_t, g, and μ are known.
18.	Northwestern Fall distance	$f_d = 2fh - 2h\sqrt{\frac{f^2 + fd_t}{h}}$ $V_f = f_d\sqrt{\frac{g}{2h}}$		Solves for impact speed V_f, when g, f_d, f, h, and d_t are known.
19.	Northwestern slide distance	$s_d = d_t - \text{abs}(f_d)$ $f_d = 2fh - 2h\sqrt{\frac{f^2 + fd_t}{h}}$ $V_s = \sqrt{2fgs_d}$		Solves for impact speed V_s, when f_d, d_t, s_d, f, and g are known.

8.5.10 Pedestrian Walking Speeds

Walking speeds are needed frequently when analyzing pedestrian collisions. Table 8.9 includes the data for pedestrian walking speeds for person's different age groups.

Table 8.9

Age and Sex	*Walking speed*		
	mph	*ft/sec*	*m/s*
Men < 55 years	3.7	5.4	1.7
Men > 55 years	3.4	5.0	1.5
Women < 50 years	3.1	4.5	1.4
Women > 50 years	2.9	4.3	1.3
Women with small children	1.6	2.3	0.7
Children > 6 and < 10 years	2.5	3.7	1.1
Adolescents	4.0	5.9	1.8

Source: Ashton, S.J., "Pedestrian Accident Investigation and Reconstruction", paper presented at the seminar on Special Problems in Traffic Accident Reconstruction, Institute of Police Technology and Management (IPTM), Jacksonville, FL, 10-14, April, 1989.

8.6 EMPIRICAL MODELS

In a real world pedestrian collision a human body is not an ideal projectile. The pedestrian's trajectory during and after the collision is very complex. Many researchers have made excellent efforts to mathematically describe this behavior; however, the mathematical equations are often difficult to apply to real world collisions. In other cases additional assumptions must be made to solve these equations. The assumptions are typically based on empirical data.

Empirical models consider a "black box" approach, fitting a simple mathematical expression to the real data. These models incorporate many parameters necessary for solving mathematical equations into one or more constants. The outcome is usually a simple and practical mathematical expression. However, the disadvantage of the empirical models is that they typically model the average value of the considered data. If a limited sample size is considered, or if the data is skewed due to testing or recording techniques, then the empirical results may not be representative of real world collisions.

Appel et al. (1975) derived vehicle speed/throw distance relationship from the analysis of 137 real world accidents. This study categorized separate equation for vehicle front-end geometry, as low and high fronted and the pedestrian size as adults and children. By definition, the leading edge of the hood of a low and high fronted vehicle are below the C.G. of a 50th percentile adult; the leading edge of the hood of a blunt vehicle is above the C.G. of a 50th percentile adult. It is noted that a child or an adult impacted by a "blunt" vehicle would likely produce a forward projection type trajectory. The Appeal paper published the following relationships.

	Equation	
High fronted vehicle	$S = 0.084\, V_V^2$	(8.22)
Low fronted vehicle	$S = 0.065\, V_V^2$	(8.23)
Adults	$S = 0.070\, V_V^2$	(8.24)
Children	$S = 0.088\, V_V^2$	(8.25)

Sturtz, et al. (1976) re-analyzed the Appel data and defined the children as persons aged 15 years or less. They regrouped the data and found that the following cubic relationships were the best fit to the Appel data.

High fronted vehicles:

Children $$S = 0.6 + 0.76V_V + 0.0021\, V_V^2 \qquad (8.26)$$

Adults $$S = 1.0 + 0.61\,V_V + 0.0018\, V_V^2 \qquad (8.27)$$

Low fronted vehicles:

Children $$S = 0.6 + 0.0665\, V_V^2 \qquad (8.28)$$

Adults $$S = 1.0 + 0.654V_V + 0.00145\, V_V^2 \qquad (8.29)$$

Although the Sturtz, et al. (1976) equations are cubic equations, they simplify easily due to the lack of the square term. Thus, the Sturtz, et al. (1976) equations can be easily solved for vehicle impact speed in terms of throw distance.

Wood (2002) presented a hybrid model to yield minimum, maximum and average vehicle impact speeds from pedestrian throw distance. Wood's equations are:

$$V_{v(\min)} = 2.5\sqrt{S} \quad (8.30)$$

$$V_{v(\min)} = 3.6\sqrt{S} \quad (8.31)$$

$$V_{v(\min)} = 4.5\sqrt{S} \quad (8.32)$$

Wood has also published numerous other excellent papers which extensively considered and contributed to the theory of vehicle/pedestrian collisions. The above empirical model was chosen for its ease of use.

Fugger (2000) presents an empirical model that considers a power law relation between throw distance and vehicle impact speed. The Fugger model is applicable for wrap trajectories and is defined as:

$$V_V = 8.3604\ S^{0.6046} \quad (8.33)$$

where S is throw distance measured in [m] and V_V is the vehicle impact speed in [km/h].

Toor, et al. (2002) revised and validated an earlier empirical model published by the same authors (2000). The Tool empirical model was based on an extensive data set of 359 collisions from a variety of sources. This model was separated according to wrap or forward projection trajectories. In addition a combined trajectory model was also presented for that case where the trajectory is not known.

Forward projection trajectory model:

$$V_V = 9.19\ S^{0.59} \quad (8.34)$$

15th and 85th percentile prediction interval: ±10.5 km/h

5th and 95th percentile prediction interval: ±14.0 km/h

Wrap trajectory model:

$$V_V = 11.3\sqrt{S} - 0.3 \quad (8.35)$$

15th and 85th percentile prediction interval: ±9 km/h

5th and 95th percentile prediction interval: ±12 km/h

Combined throw distance model:

$$V_V = 12.8\sqrt{S} - 3.6 \quad (8.36)$$

15th and 85th percentile prediction interval: ±9.5 km/h

5th and 95th percentile prediction interval: ±12.8 km/h

The new model:

The vehicle/pedestrian collision data used to formulate the Eqs. (8.34), (8.35) and (8.36) were revaluated to test the sensitivity of empirical models.

In the pervious publication, Happer et al. (2000), a linear relationship between the square root of throw distance and the impact speed was assumed. In this new model a simple power law relationship has been considered. The following relationships were found to provide best correlation:

Forward projection trajectory model:

$$V_V = 8.25\ S^{0.61} \quad (8.37)$$

15th and 85th percentile prediction interval: ±7.7 km/h

5th and 95th percentile prediction interval: ±12.2 km/h

Wrap trajectory model:

$$V_V = 9.84\ S^{0.57} \tag{8.38}$$

15th and 85th percentile prediction interval: ±5.8 km/h

5th and 95th percentile prediction interval: ±9.2 km/h

Combined (wrap and forward) throw distance model:

$$V_V = 9.19\ S^{0.59} \tag{8.39}$$

15th and 85th percentile prediction interval: ±6.5 km/h

5th and 95th percentile prediction interval: ±10.2 km/h

As was the case with the previous model, statistical analysis indicates that there is a strong correlation in the data and the calculated power law relationship.

The 85th percentile prediction intervals are for use in cases where balance of probabilities is acceptable. However, there may be cases for which 85th percentile prediction may not be suitable. Typically, these cases most likely result from data uncertainty. In addition, there will be cases (for example criminal cases) where the accident reconstruction engineer will be asked to provide an opinion with greater certainty. Thus, in keeping with the earlier model (Eqs. (8.34), (8.35) and (8.36)) 5th and 95th prediction intervals are also included.

Coefficient of Friction

In order to compare the theoretical models with the empirical models, the coefficient of friction for the pedestrian's post impact trajectory is required. The published literature (Tool et al. 2002) indicates that the coefficient of friction varies according to pedestrian sliding on the ground or tumbling.

8.7 ANALYTICAL MODELS

8.7.1 Simplified Vehicle-Pedestrian Impact Model

In this model, the pedestrian is modeled as a rigid rod as shown in Fig. 8.7. The objective is to determine the horizontal velocities of the vehicle and pedestrian, and the angular velocity of the pedestrian after impact. Applying the impulse equation to the vehicle-pedestrian system (Limpert, 1980), we can write the following four equations:

$$-P = m_1(V_{12x} - V_0) \tag{8.40}$$

$$+P = m_2(V_{22x} - V_{21x}) \tag{8.41}$$

$$P(h_2 - h_1) = I_2(w_{22y} - w_{21y}) \tag{8.42}$$

$$\frac{-[V_{12x} - \{V_{22x} + w_{22y}(h_2 - h_1)\}]}{V_0 - 0} = k \tag{8.43}$$

where h_1 = contact point height, (ft)

h_2 = center-of-gravity-height of pedestrian, (ft)

I_2 = mass moment of inertial of pedestrian, (lb ft sec^2)

k = coefficient of restitution

m_1 = mass of vehicle, (lb sec^2/ft)

m_2 = mass of pedestrian, (lb sec^2/ft)

V_0 = impact speed of vehicle, (ft/sec)

P = impulse

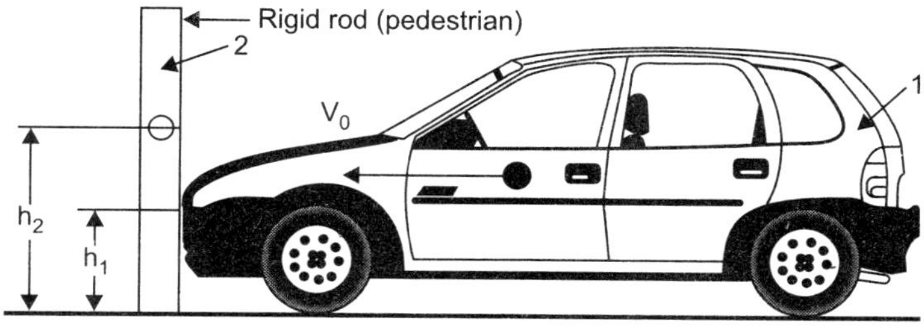

Fig. 8.7 Car-pedestrian impact model

Solving Eqs. (8.1) to (8.4) simultaneously for P, ω_{22y}, V_{12x}, and V_{22x}, we obtain:
The velocity of the car after impact:

$$V_{12x} = V_0 - \frac{P}{m_1} = V_0 - \frac{\left[(1+k)\left(\frac{m_2}{m_1}\right)V_0\right]}{\left(1+\frac{m_2}{m_1}\right)+\frac{(h_2-h_1)^2}{i_2^2}}, \text{ (ft/sec.)} \tag{8.44}$$

The velocity of pedestrian's cg after

$$V_{22x} = \frac{P}{m_2} = \frac{(1+k)V_0}{\left(1+\frac{m_2}{m_1}\right)+\frac{(h_2-h_1)^2}{i_2^2}}, \text{ (ft/sec.)} \tag{8.45}$$

Impulse P:

$$P = \frac{(1+k)m_2V_0}{\left(1+\frac{m_2}{m_1}\right)+\frac{m_2(h_2-h_1)^2}{I_2}}, \text{ (ft/sec.)} \tag{8.46}$$

The angular velocity of the pedestrian after impact:

$$\omega_{22y} = \frac{P(h_2-h_1)}{I_2} = \frac{(1+k)(h_2-h_1)V_0}{\left(1+\frac{m_2}{m_1}\right)+(h_2-h_1)^2}, \text{ (rad/sec.)} \tag{8.47}$$

where $$i_2^2 = \frac{I_2}{m_2}, \text{ (ft}^2\text{) [radius of gyration]}$$

Example 8.1: *Determine the velocity of the pedestrian's center-of-gravity after impact for the following data:*

$m_1 = 75$ *lb sec²/ft*, $m_2 = 6$ *lb sec²/ft*, $h_1 = 1.25$ *ft*, $h_2 = 2.75$ *ft*, $k = 0$, $V_0 = 62.4$ *ft/sec (42.5 mph)*

Solution: The radius of gyration for the pedestrian is approximately by that rod as

$$i_2 = \frac{h_2}{6} = \frac{2.75}{6} = 0.46 \text{ ft}^2$$

$$I_2 = i_2^2 m_2 = (0.46)^2(6) = 1.2696 \text{ lb ft sec}^2$$

The angular velocity of the pedestrian after impact is given by Eq. (8.47)

$$\omega_{22y} = \frac{(1+k)(h_2-h_1)V_0}{\left(1+\dfrac{m_2}{m_1}\right)+(h_2-h_1)^2} = \frac{(1+0)(2.75-1.25)(62.4)}{\left(1+\dfrac{6}{75}\right)+(2.75-1.25)^2} = 28.11 \text{ rad/sec.}$$

The velocity of the car after impact is given by Eq. (8.44):

$$V_{12x} = V_0 - \frac{\left[(1+k)\left(\dfrac{m_2}{m_1}\right)V_0\right]}{\left(1+\dfrac{m_2}{m_1}\right)+\dfrac{(h_2-h_1)^2}{i_2^2}} = 62.4 - \frac{\left[(1+0)\left(\dfrac{6}{75}\right)62.4\right]}{\left(1+\dfrac{6}{75}\right)+\dfrac{(2.75-1.25)}{(0.46)^2}} = 62.37 \text{ ft/sec.}$$

The velocity of the pedestrian's center-of-gravity after impact is given by Eq. (8.45):

$$V_{22x} = \frac{(1+k)V_0}{\left(1+\dfrac{m_2}{m_1}\right)+\dfrac{(h_2-h_1)^2}{i_2^2}} = \frac{(1+0)(62.4)}{\left(1+\dfrac{6}{75}\right)+\dfrac{(2.75-1.25)^2}{(0.46)^2}} = 5.3274 \text{ ft/sec.}$$

8.7.2 Brach and Han's Throw-Model for Frontal Pedestrian Collisions

In this section a planar model developed by Han and Brach (2001) for the mechanics of a vehicle-pedestrian collision is presented, analyzed and compared to experimental data. It takes into account the significant physical parameters of wrap and forward projection collisions and is suitable for solution using mathematics software or spreadsheets. Parameters related to the pedestrian and taken into account include horizontal distance traveled between primary and secondary impacts with the vehicle, launch angle, center-of-gravity height at launch, the relative forward speed of the pedestrian to the car at launch, distance from launch to a ground impact, distance from ground impact to rest and pedestrian-ground drag factor. Vehicle and roadway parameters include postimpact, constant-velocity vehicle travel distance, continued vehicle travel distance to rest with uniform deceleration and relative distance between rest positions of vehicle and pedestrian.

This model deals with throw distance, the distance along the road between the location of the initial pedestrian-vehicle contact and the pedestrian rest position. It deals with modeling of the relationship between throw distance and vehicle speed. Using mathematics and physics to model the throw distance presents a challenger, for several reasons, Han and Brach (2001). First and foremost is the fact that relative to the vehicle, the human body is articulated, flexible and soft, though the skeletal structure provides limited rigidity (but capable of fracture). For low vehicle frontal shapes, it is common to have a primary collision between the frontal portion of the vehicle (bumper, grill, etc.) and the lower part of the pedestrian followed by a secondary collision of an upper part of the pedestrian with the hood, windshield or roof as shown in Fig. 8.8. In low speed collisions, the secondary impact may be nonexistent or minor. Wide variations occur in human size, weight and form; wide variations exist in the size and frontal geometry of vehicles and wide variations occur in the initial conditions of contact between the two. For a given human form, size and weight and a given vehicle frontal geometry and compliance, the resulting motion of the pedestrian depends significantly on the initial speed of the vehicle. All of these effects couple since the location and severity of a secondary collision is dependent on the speed. In addition, a portion of the pedestrian's motion before coming to rest is with ground contact. Because of the articulated shape of the human body, this motion can consist of combinations of bouncing, tumbling, rolling and sliding, Han and Brach (2001). In fact, this type of motion has received considerable attention in its own right; for example see Bratten (1989) and Searle (1993). All of these variations are difficult to model. They also are difficult to control during experiments.

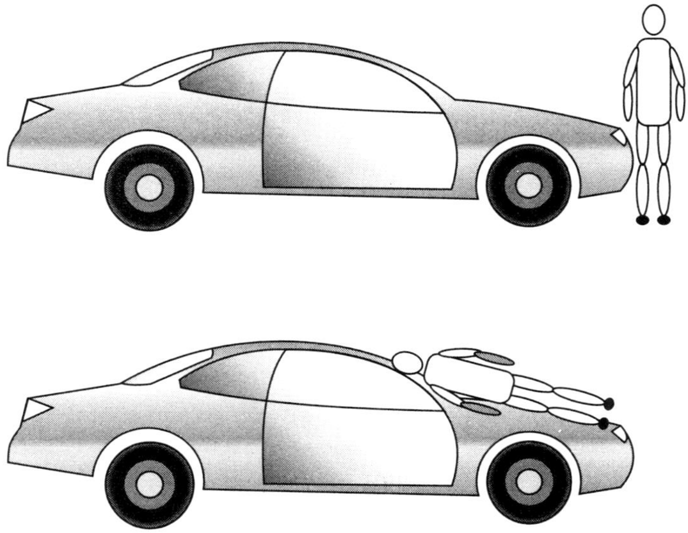

Fig. 8.8 Diagram of primary (top) and secondary (bottom) impacts between a pedestrian and a vehicle

Pedestrian Impact Model

Throw distance model:

Figure 8.9 shows a diagram that illustrates coordinates, parameters and some of the events of a vehicle-pedestrian collision. An objective is to determine the total throw distance, s_p, as a function of the initial vehicle speed, v_{c0}, with all the other quantities (such as launch angle, θ, launch height, *h*, road grade, φ, etc.) treated as known parameters. Another objective is to model the motion of the vehicle and relate it to the pedestrian throw distance.

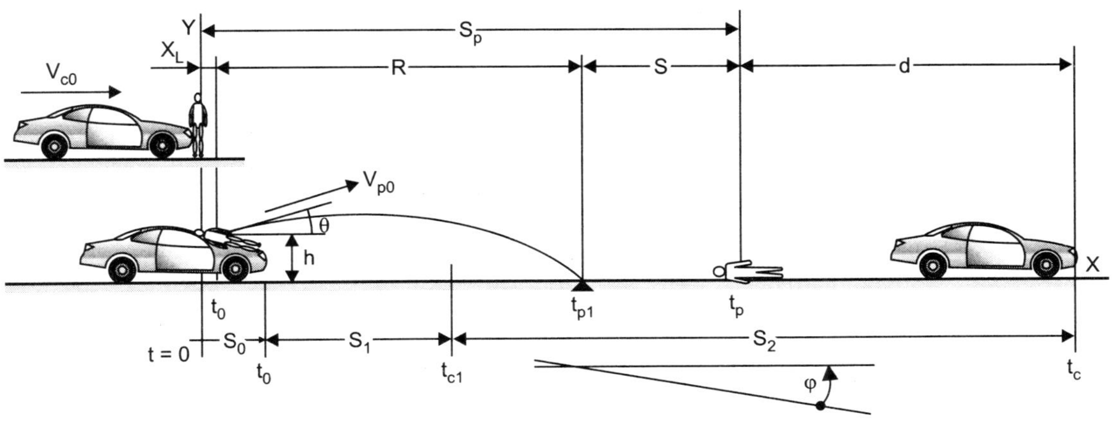

Fig. 8.9 Coordinates, variables and events corresponding to a vehicle-pedestrian collision

Table 8.9 Notation and list of variables [Han and Brach (2001)]

a_2	Deceleration of vehicle over distance s_2
d	Distance between rest positions of vehicle and pedestrian
f_p	Drag resistance coefficient of pedestrian over distance s
g	Acceleration of gravity
h	Height of pedestrian center of gravity at launch, τ_0
m_c	Mass of vehicle, weight/g
m_p	Mass of pedestrian, weight/g
R	Range of pedestrian throw, launch to ground impact
s	Pedestrian ground contact distance, impact to rest
s_0	Distance of travel of vehicle with pedestrian contact
s_1	Distance of travel of vehicle at uniform speed
s_2	Distance of travel of vehicle with uniform deceleration, a_2
s_p	Pedestrian throw distance, total distance, initial contact to rest
τ_c	Vehicle travel time, initial contact to rest
τ_{c1}	Time of travel of vehicle at steady speed
τ_p	Total time of travel of pedestrian, initial contact to rest
τ_{p1}	Time from impact to pedestrian initial contact with ground
υ'_{c0}	Velocity of vehicle after impact with pedestrian
υ_{c0}	Initial speed of vehicle
υ_{p0}	Initial speed of pedestrian
x_L	x-distance of pedestrian from initial contact to launch
α	Ratio of pedestrian speed to vehicle speed at time of launch
θ	Angle of launch of pedestrian relative to x axis
μ	Impulse ratio for pedestrian-ground impact
φ	Road grade angle

The throw distance can be written as:

$$S_p = x_L + R + s \tag{8.48}$$

where x_L is the distance the pedestrian moves in the time between the primary and secondary impacts. The parameter x_L is included to represent the combined influence of the vehicle frontal geometry, pedestrian center-of-gravity height at impact, and vehicle speed dependence of the secondary impact. R is the range, or ground distance, covered during the flight phase and s is the distance between the ground impact of the pedestrian and the rest position. Some analysis, such as Limpert (1998), represents the initial velocity of the pedestrian v_{p0}, as a fraction or multiple of the forward speed of the vehicle. For that reason (and to allow comparisons of this concept with experimental results) the initial launch speed of the pedestrian, v_{p0}, is expressed as

$$v_{p0} = \alpha v'_{c0} \tag{8.49}$$

where α is a constant and $\alpha v'_{c0}$ is the speed of the car at the time the pedestrian is launched. The pedestrian is launched following the secondary collision, or the primary, if no secondary collision takes place. Between initial contact and launch, the pedestrian is brought up to speed and the vehicle suffers a momentum loss. Based on conservation of momentum,

$$v'_{c0} = \frac{m_c}{m_c + m_p} v_{c0} \tag{8.50}$$

Where v_{c0} is the speed of the vehicle at initial contact, m_c is the mass of the vehicle and m_p is the mass of the pedestrian. The range, R can be expressed as

$$R = v_{p0} \cos\theta \, t_R - g \sin\varphi t_R^2 / 2 \tag{8.51}$$

The quantity t_R is the flight time from launch to ground contact and is

$$t_R = \frac{v_{p0} \sin\theta}{g \cos\varphi} + \frac{\sqrt{v_{p0}^2 \sin\theta + 2gh \cos\varphi}}{g \cos\varphi} \tag{8.52}$$

The velocity components of the pedestrian at the time of ground impact are:

$$v_{pRx} = v_{p0} \cos\theta - g \sin\varphi \, t_R \tag{8.53}$$

and

$$v_{pRy} = v_{p0} \sin\theta - g \cos\varphi \, t_R \tag{8.54}$$

These last two velocity components are velocity components just as the pedestrian impacts the ground. Following impact, vertical rebound is possible, even bouncing is possible. However, it can be shown that the horizontal travel distance is not highly dependent on multiple bouncing. Consequently, it is assumed that there is a single impact normal to the ground and that it is perfectly inelastic, so the postimpact rebound velocity is

$$v'_{pRy} = 0 \tag{8.55}$$

Further, it is assumed that Coulomb friction acts during the impact with the ground so the tangential component of the postimpact velocity is given by

$$v'_{pRx} = v_{pRx} + \mu \, v_{pRy} \tag{8.56}$$

The quantity, μ, is the ratio of the tangential impulse to the normal impulse during the impact. Throughout this paper, the pedestrian sliding drag factor is used for μ. Note that v_{pRy} is negative $v'_{pRx} < v_{pRx}$. The final part of the throw distance is the slide distance, s. This is calculated using a pedestrian drag factor, f_p, where:

$$s = \frac{(v'_{pRx})^2}{2g(f_p \cos\varphi + \sin\varphi)} \tag{8.57}$$

where $s_p = f(v_c)$ is given by using Eqs. (8.51) and (8.57) with Eq. (8.48).

Modeling of x_L:

It was mentioned above that x_L represents the distance the pedestrian (center of gravity) travels in the time between the initial and secondary impacts with the vehicle. In tests with dummies, Kallieris and Schmidt (1988) found that the location of the secondary impact depends on the speed of the car. Similar results were found by van Wijk, et al. (1983) from reconstructed accidents. In the following, a bilinear function is used to represent this distance as it varies with vehicle speed. Figure 8.10 shows the function where the values, $x_1 = 0.5$ m (1.6 ft), $x_2 = 1.5$ m (4.9 ft), $v_1 = 5$ m/s (11 mph) and $v_2 = 20$ m/s (45 mph) are used in this paper. By definition, $x_L = 0$ for forward projection collisions. To some extent, the nonlinear function in Fig.8.10 is arbitrary and requires experimental verification but it does allow roof vaults and somersaults to be modeled.

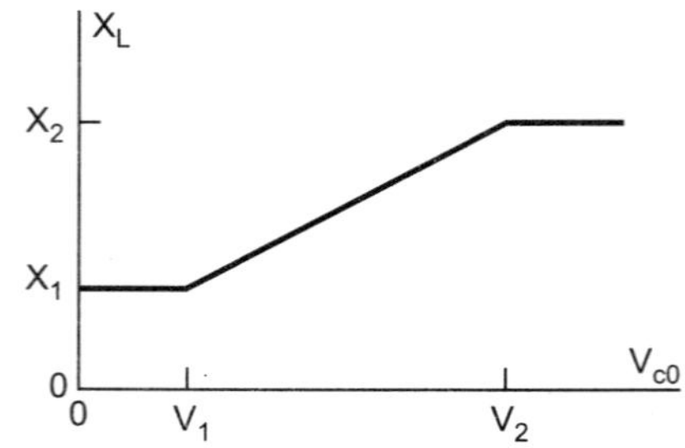

Fig. 8.10 Sample relationship between $x\hat{L}$ and the initial vehicle speed (Han and Brach (2001))

Vehicle Motion

Vehicle motion exists simultaneously with the pedestrian motion and is separated into 3 portions. The first is a distance, s_0. As a rough approximation, it is assumed that over the distance, x_L, the pedestrian travels at half its launch velocity so, $t_0 \approx x_L/(v'_{c0}/2)$ and $s_0 \approx v_{c0}\, t_0$. Next, the vehicle can travel a distance, s_1, at constant velocity and finally a subsequent portion of the motion, s_2, with constant deceleration, a_2. Of course, s_1 can be zero if deceleration exists from t_0 to rest. Consequently,

$$d = s_0 + s_1 + s_2 - s_p$$

$$d = v_{c0}t_0 + v'_{c0}(t_{c1} - t_0) = a_2(t_c - t_{c1})^2/2 - s_p \tag{8.58}$$

Throw Distance Index

Evans and Smith (1999) and Happer, et al. (2000), use curve fitting techniques to arrive at a relationship between the throw distance, s_p, and the initial vehicle speed, v_{c0}. A modified form of their relationship is:

$$s_p = c_1(v_{c0} + c_2)^2 \tag{8.59}$$

For wrap collisions, Evans and Smith choose $c_2 = 0$ and find (by fitting Eq. (8.59) to specific experimental data) that $c_1 = 1/3.58^2$ s^2/m. Happer, et al., find values of $c_1 = 1/3.53^2$ s^2/m and $c_2 = 0.72$ m/s for wrap collisions by fitting to a wide range of experimental data. As a means for comparison of experimental data sets later in this paper, a throw distance drag index is defined by an equation as Eq. (8.59) with $c_2 = 0$, representing equivalent constant acceleration motion:

$$f_{eq} = \frac{v_{c0}^2}{2g\, s_p} \tag{8.60}$$

Note that f_{eq} is an equivalent constant drag factor corresponding to the entire throw distance, s_p, and not just the distance, s, when the pedestrian is in contact with the ground. The quantity f_{eq} corresponds to a fictitious, average deceleration factor for the throw distance process from impact to rest.

Reconstruction Throw Model

If the road grade angle, φ, is close to zero, $\varphi = 0$, the pedestrian throw-distance model derived above (given by Eqs. (8.51), (8.57) and Eq. (8.48) can be put into a form more useful for reconstruction of the speed of the vehicle. Doing this, an equation for the initial launch speed of the pedestrian, v_{p0}, is:

$$V_{p0} = A_p\sqrt{s_p - B} \tag{8.61}$$

where

$$A_p = \sqrt{\frac{2f_p g}{f_p^2 \sin^2\theta + f_p \sin 2\theta + \cos^2\theta}} \tag{8.62}$$

and

$$B = x_L + f_p h \tag{8.63}$$

Using Eqs. (8.49), (8.50) and (8.61), the velocity of the car can be expressed as:

$$V_{co} = A\sqrt{s_p - B} \tag{8.64}$$

where

$$A = A_p \frac{m_c + m_p}{\alpha m_c} \tag{8.65}$$

With the exception that $\varphi \approx 0$, this model is a true inverse of the above throw-distance model. Recall that Eq.(8.59), solved for the velocity v_{c0}, also is an equation used for reconstruction. Eqs. (8.64) and (8.59) have subtle, but important differences. First, since it is derived analytically not empirically, the parameters contained in A and B have physical interpretation, making Eq.(8.64) more meaningful as well as more flexible. Admittedly, the values of the parameters in A and B are not always easy to estimate for specific reconstructions. Another difference between Eq. (8.64) and (8.59) concerns their form. Eq. (8.59) has an offset in the velocity whereas Eq.(8.64) has an offset in the throw distance. A deficiency with the throw distance index, Eq. (8.60), is that at low velocities, it tends to underestimate the throw distance. Allowing a shift in the s_p coordinate permits better matching of experimental results.

The method of least squares was used to find values of A and B that fit the model to experimental data. See Han and Brach (2001) for complete details.

Hill used the following equation to calculate the coefficient of friction from his test results.

$$f_p = \frac{v_{p0}^2}{2gs} \tag{8.66}$$

Hill found a coefficient for body to road surface of 0.8, on average, to adequately represent his test results and he used that value for calculation of vehicle speed from pedestrian throw in actual accidents.

Correction for Impact

However, Eq.(8.66) does not take account of the effect caused by the vertical impact which occurs when the dummy strikes the road surface. The impact may be a significant event that must be accounted for when calculating the coefficient of friction.

Assuming zero road grade, Eq. (8.67) can be obtained from Eq. (8.55) through (8.57) (where $\mu = f_p$),

$$s = \frac{1}{2f_p g}\left[v_{p0}\cos\theta - f_p\sqrt{v_{p0}^2 \sin^2\theta + 2gh}\right]^2 \tag{8.67}$$

Solving for f_p gives the following equation that provides the friction drag from the test results for a zero launch-angle (Hill's conditions):

$$f_p = \left[gs + v_{p0}\sqrt{2gh} - \sqrt{(gs + v_{p0}\sqrt{2gh})^2 - 2v_{p0}^2 gh}\right] \Big/ 2gh \tag{8.68}$$

Note that the measured distance in the tests is the sliding and/or rolling distance s, and does not include flight range R.

From Hill's tests, Eq. (8.68) yields an average value of friction drag of 0.74, less than 0.80 obtained by Hill. This shows that failure to account for the vertical impact with the ground results in a coefficient that is too high.

The coefficient of friction calculated using Eq.(8.68) was plotted by Han and Brach (2001) as a function of the horizontal release velocity for Hill's tests. Though the scatter in the data appears to vary with velocity, there seems to be no discernable trend between the releasing velocity and the coefficient of friction.

Wood and Simms (2000) carried out an analysis of the data of Hill (and others). Accounting for the impact gave average values of 0.702 and 0.734 for the coefficient using Hill's data.

Han and Brach (2001) compared the model results with the experimental data for more than fourteen selected sets of data corresponding to wrap and forward projection collisions. Han and Brach (2001) found an excellent agreement with the analysis model and the data trends. Some of the variables in Table 8, such as α, θ, R, and s could not be obtained directly from the experimental data due to the fact that these variables were not measured in the experiments. Values of α as high as 1.2 and 1.3 were found by Han and Brach (2001). A value of $\alpha = 1$ is expected for wrap collisions. For forward projection collisions, Wood and Walsh (2002) found values of α greater than 1.0, though smaller than Han and Brach (2001) values. Han and Brach (2001) found typical values of θ from about 4° to 13° and in some cases higher value, up to θ = 35°. Han and Brach (2001) recommend a range of $0 \leq \theta \leq 15°$ for reconstruction of typical wrap collisions.

An analysis, including a correction for impact effects, of Hill's direct measurement of pedestrian-ground friction yields an average value of 0.74 and indicates that values between 0.7 and 0.8 are appropriate for f_p (for an asphalt surface). Values of f_p greater than 0.7 also are expected since fitted values of the overall throw distance drag factor, f_{eq}, are as high as 0.70 and 0.74. These results are in contrast to considerably lower values of f_p by direct measurements of others. For example, Fricke (1990) finds values between 0.45 and 0.65 and Bratten (1989), found values in the vicinity of 0.5. Specific reasons for these differences were not identified.

Example 8.2: *The data for a vehicle-pedestrian wrap collision is given in Table E8.2 for a vehicle with low frontal geometry traveling at 25 mph (40 km/h) and a pedestrian who has a secondary impact with the windshield.*

(*a*) *determine the throw distance,* s_p

(*b*) *the effect on the throw distance of varying the launch angle,* θ, *from* $\theta = 0$ *to* $\theta = 10$ *degrees (Brach and Brach, 2005).*

Table E8.2

Symbol	*Value*	*Variable*
a_2	0.90	Deceleration of vehicle over distance s_2
f_p	0.80	Drag resistance coefficient of pedestrian over distance s
h	3.00 ft	Height of pedestrian center of gravity at launch, τ_0
s_1	0.00 ft	Distance for travel of vehicle at uniform speed
v_{c0}	36.670 ft/s	Initial speed of vehicle
	25 mph	Initial speed of vehicle
x_L	2.00 ft	x-distance of pedestrian from initial contact to launch
α	1.00	Ratio of pedestrian speed to vehicle speed at time of launch
θ	6.00	Angle of launch of pedestrian relative to x axis
φ	0.00	Road grade angle
μ	0.80	Impulse ratio for pedestrian-ground impact
m_c	100 lb-s²/ft	Mass of vehicle, weight/g
m_p	6 lb-s²/ft	Mass of pedestrian, weight/g

Solution:

(*a*) Analysis of Pedestrian Throw Distance from Initial Conditions version 2.0.1 pedthrow.xls

NOTATION, COORDINATES, UNITS AND VARIABLES:

x	-	coordinate parallel to ground
y	-	coordinate perpendicular to ground

INPUT INFORMATION (KNOWNS):

a_2	**0.90**		deceleration of vehicle over distance s_2, g's
f_p	**0.80**		drag resistance coefficient of pedestrian over distance s
g	**32.17**	ft/s^2	acceleration of gravity
h	**3.00**	ft	height of pedestrian center of gravity at launch, t_0
s_1	**0.00**	ft	distance of travel of vehicle at uniform speed
v_{c0}	**36.67**	ft/s	initial speed of vehicle
	25.0	mph	initial speed of vehicle
x_L	**2.00**	ft	x-distance of pedestrian from initial contact to launch
	1.00		ratio of pedestrian speed to vehicle speed at time of launch
	6.00	deg	angle of launch of pedestrian relative to x axis
	0.00	deg	road grade angle
	0.80		impulse ratio for pedestrian-ground impact
m_c	**100.00**	lb-s^2/ft	mass of vehicle, weight/g
m_p	**6.00**	lb-s^2/ft	mass of pedestrian, weight/g

OUTPUT INFORMATION (UNKNOWNS):

v'_{c0}	**34.59**	ft/s	velocity of vehicle after impact with pedestrian
v_{p0}	**34.59**	ft/s	initial speed of pedestrian
R	**19.22**	ft	range of pedestrian throw, launch to ground impact
t_{p1}	**0.67**	s	time from impact to pedestrian initial contact with ground
s	**10.20**	ft	pedestrian ground contact distance, impact to rest
s_p	**31.42**	ft	pedestrian throw distance; total distance, initial contact to rest
t_p	**1.56**	s	total time of travel of pedestrian, initial contact to rest
t_{c1}	**0.12**	s	time of travel of vehicle at steady speed
s_0	**4.24**	ft	distance of travel of vehicle with pedestrian contact
s_2	**20.66**	ft	distance of travel of vehicle with uniform deceleration, a_2

$s_0 + s_1 + s_2$	**24.90**	ft	total distance of travel of vehicle
t_c	**1.31**	s	vehicle travel time, initial contact to rest
d	**-6.52**	ft	distance between rest positions of vehicle and pedestrian

The throw distance *Sp* is 31.42 ft

Fig. E8.2 (*a*)

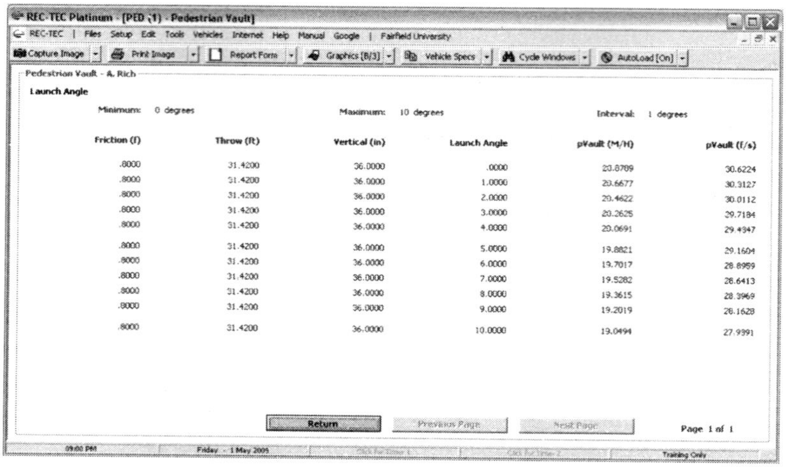

Friction (f)	Throw (ft)	Vertical (in)	Launch Angle	pVault (M/H)	pVault (f/s)
.8000	31.4200	36.0000	.0000	20.8789	30.6224
.8000	31.4200	36.0000	1.0000	20.6677	30.3127
.8000	31.4200	36.0000	2.0000	20.4622	30.0112
.8000	31.4200	36.0000	3.0000	20.2625	29.7184
.8000	31.4200	36.0000	4.0000	20.0691	29.4347
.8000	31.4200	36.0000	5.0000	19.8821	29.1604
.8000	31.4200	36.0000	6.0000	19.7017	28.8959
.8000	31.4200	36.0000	7.0000	19.5282	28.6413
.8000	31.4200	36.0000	8.0000	19.3615	28.3969
.8000	31.4200	36.0000	9.0000	19.2019	28.1628
.8000	31.4200	36.0000	10.0000	19.0494	27.9391

Fig. E8.2 (*b*)

Example 8.3: *Refer to Example E8.2 and the data in Table E8.2 with the speed of the vehicle unknown. Now, both the throw distance and the vehicle speed are unknown. Given that the distance traveled by the pedestrian was measured and found to exceed that of the vehicle by 15 ft (4.57m). Determine the speed of the vehicle.*

Solution:

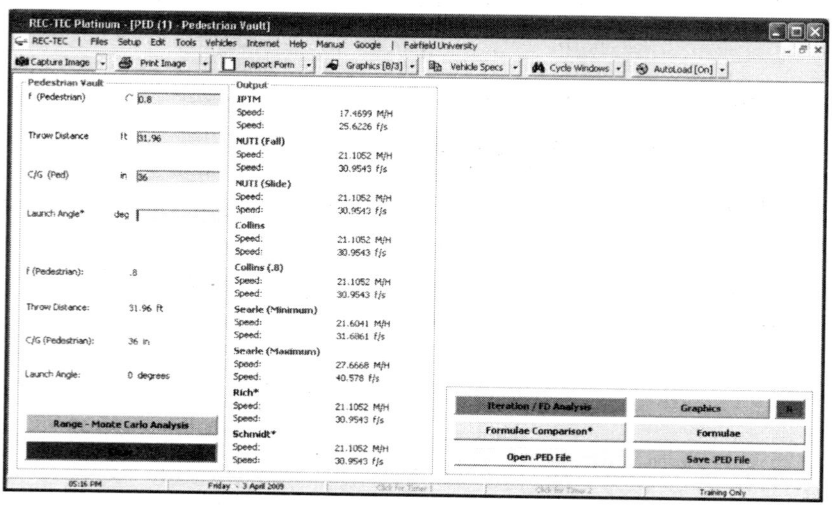

8.8 MODEL COMPARISON (TOOL ET AL. 2002)

The mathematical and empirical models discussed above were compared for forward trajection and wrap trajection by Toor et al. 2002. Since the various models used different parameters and present calculation methods for either pedestrian throw speed or vehicle impact speed, selection of appropriate projection efficiencies was required to ensure comparison of similar parameters. For a forward projection trajectory, an average projection efficiency of 95% was considered. For a wrap trajectory, an average projection efficiency of 80% was considered. In addition, the different studies list different friction values; thus a pedestrian coefficient of friction value of 0.5 was utilized for comparison purpose. Other parameters that were required for the mathematical model were the pedestrian displacement from primary to secondary contact (X_L in the Han-Brach model), pedestrian launch angle, pedestrian and vehicle masses, and the pedestrian center of gravity height. The average values for these parameters were used for the mathematical models and thus were defined as stated below:

Mass of pedestrian	:	$m_p = 70$ kg
Mass of vehicle	:	$m_v = 1300$ kg
Pedestrian CG height	:	$h = 1.0$ m
Pedestrian friction	:	$\mu = 0.5$
Pedestrian launch angle	:	$\theta = 30$ deg (wrap)
Pedestrian launch angle	:	$\theta = 0$ deg (forward)
Pedestrian displacement	:	$X_L = 1.5$ m (wrap)
Pedestrian displacement	:	$X_L = 0$ m (forward)
Projection efficiency	:	$PE = 80\%$ (wrap)
Projection efficiency	:	$PE = 95\%$ (forward)

It should be noted that the Searle (1993), Wood (1988) (hybrid model), Collins (1979) and Stcherbatcheff (1975) models do not explicitly differentiate between the wrap and forward projection trajectory. However, the discussions in the Searle study suggest that the correction factors are most applicable for wrap trajectories. The Stcherbatcheff model is based on a wrap trajectory with head to hood secondary contact. The Collins model assumes zero degree launch angle and 100% projection efficiency; both of these bases are characteristic of a forward projection trajectory. The Wood hybrid model is based on wrap trajectory tests which include head to hood secondary contact.

The Searle and Collins models were applied for both the wrap and forward projection trajectories by considering the appropriate projection efficiencies as listed above. The other models were only applied to their implied trajectory due to the explicit mathematical consideration of only one or two contacts between the pedestrian and the vehicle.

In the forward projection trajectory comparison, the Appel equation for children and the Sturtz equation for high fronted vehicles impacting children were plotted for comparison. It is noted that the definition of a high fronted vehicle is a vehicle with the leading edge of its hood between the knees and center of gravity of an adult pedestrian. Similarly for the wrap trajectory comparison, the Appel equation for low fronted and high fronted vehicles and the Strutz equation for low and high fronted vehicles impacting adults and low fronted vehicles impacting children are compared.

All the models (empirical, mathematical or hybrid) produced very reasonable results. In general, the models were in agreement with each other when appropriate parameters were considered. It appears that the correct selection and application of the parameters is more crucial than selecting a particular model. For almost any model, the results reasonably match the impact test data if proper factors are considered. In addition, there is considerable variance in the test data such that no single solution exactly models an impact.

(*a*) The Han-Brach Model: The Han-Brach model is slightly more detailed than most other models. It is in good agreement with the other models and impacts test data available. However, in order to facilitate the use of this model, it is necessary to quantify the distance that the pedestrian is carried from the initial point of contact to the point where the pedestrian separates from the vehicle as the pedestrian is launched forward. This distance must be assumed as the technical justification does not appear to be available or adequately presented in the Han-Brach study. The Han-Brach study suggests to use distances of about 1 to 2 meters for the pre-launch pedestrian carry distance; based on the empirical data, this appears to yield reasonable results.

The Han-Brach model incorporates an effective projection efficiency by incorporating the mass ratio factor and a velocity ratio. The velocity ratio is not thoroughly discussed, but the Han-Brach study suggests to apply a velocity ratio of 1.0. With a velocity ratio of 1.0, the effective projection efficiency then corresponds to 95% (when masses of 70 kg and 1300 kg) are considered for the pedestrian and vehicle masses, respectively; this is likely more consistent with forward projection trajectory projection efficiency. In order to yield an effective projection efficiency corresponding to a typical wrap trajectory, a velocity ratio of about 0.7 to 0.9 is likely appropriate. For a forward projection, a velocity ratio of 1.0 is likely appropriate.

A velocity ratio of 0.84 corresponds to an effective projection efficiency of 80% when masses of 70 kg and 1300 kg are considered for the pedestrian and vehicle masses, respectively. When a velocity ratio of 0.84 and a pedestrian friction value of 0.5 are applied, the Han-Brach model is in very good agreement with the available test data for the wrap trajectory scenario; however, if a velocity ratio of 1.0 is applied, then the Han-Brach model tends to under-estimate the wrap trajectory test data. Even if an increased pedestrian friction value ($\mu = 0.8$) is considered with a velocity ratio of 1.0, the Han-Brach model tends to under-estimate the wrap trajectory test data.

(*b*) The Collins Model: The methodology behind the Collins model suggests that the model would be best suited for cases involving forward trajectory. Even in the forward trajectory case, this model under-predicted compared to most other models considered in this paper; however, the results still fall well within the bounds of the test data. For collisions involving vehicle speed of about 65 km/h and above, the difference between this model and others become less. This model also consistently under-predicts the vehicle impact speed for the wrap trajectories as compared to other models, but is still within the variance of the test data.

Collins suggests that a pedestrian/ground coefficient of friction of 0.8 and projection efficiency of 100%. Although forward projection efficiency can be close to 100%, the coefficient of friction suggested by Collins is quite high. Although these two factors counteract one another, it is likely more technically appropriate to consider a lower friction value (depending on collision circumstances) and a projection efficiency selected, dependent on vehicle type and other collision circumstances.

(*c*) The Wood Model: The wood vault model is in good agreement with other models and the available throw distance data for forward projection trajectories. The Wood vault model slightly overpredicts the vehicle impact speed when applied for wrap trajectories; this is likely due to the fact that the model only considers one contact, and assumes secondary contact does not occur. This assumption of a single contact for a wrap trajectory is valid only for a specific collision scenario and should not be considered a typical event.

The Wood hybrid model predicts slightly lower impact speeds than the Wood vault model. Both of the Wood models fall well within the middle region of the test data.

***(d)* The Stcherbatcheff Model:** The vehicle impact speeds predicted by the Stcherbatcheff model were extremely close to the middle of the test data range when a vehicle braking coefficient of 0.7 was considered. This model is likely best applied if full wheel lock-up can be shown to have occurred.

In earlier discussion of this model, it was noted that to use the Stcherbatcheff model, vehicle deceleration rates must be known. For comparison purposes full wheel braking, with average deceleration of 0.7 g was assumed. The good agreement with this assumption reflects the fact that the majority of these data points had a fully braked vehicle on dry roads.

***(e)* The Fugger Model:** The Fugger model is very similar to the Toor model (2003), but tends to slightly under-predict the vehicle impact speed when compared to other models (including the model presented in this study) and the available throw distance data. In addition, the Fugger model data more closely resembles the forward projection trajectory model data from the Toor study.

***(f)* The Appel Model:** There was a good agreement between the Appel child model and the forward trajectory data. The Appel adult model was a slightly over-predictor when compared to the other models and the wrap trajectories data. The Appel high fronted model was the best predictor of the average vehicle impact speed as compared to the available wrap trajectory data. The Appel low fronted model predicted speeds near the upper end of the wrap data range; however, this is expected as the low fronted vehicles will correspond to a lower than average projection efficiency and thus a correspondingly higher impact speed. Therefore, the Appel models may be accurate when applied for a specific vehicle geometry and scenario.

***(g)* The Searle Model:** This theoretical model with the applied correction factors and projection efficiency slightly under-estimated the impact speed as compared to most other models and the available data for speeds below about 50 km/h. At speeds greater than about 50 km/h, the Searle model more significantly under-predicts the speed as compared to the other models and data. It is noted that the suggested correction factors for the Searle model are likely only valid to obtain average values for an average collision scenario. If the collision scenario involves a forward projection trajectory, then an increase in the projection efficiency should be considered. Likewise, if the vehicle is low fronted or the pedestrian taller than average, then correspondingly lower projection efficiencies should be considered.

***(h)* The Strutz Model:** The Strutz model for forward projection trajectories shows a reasonable agreement with other models for vehicle impact speed between about 20 and 40 km/h. Outside of this range the Strutz model significantly under-predicts and significantly over-predicts in the lower and upper ranges, respectively.

***(i)* The Two Toor Models:** The current and the previous Toor model yield similar vehicle impact speeds for given pedestrian throw distance. The maximum difference occurred in the forward trajectory cases. It should be noted that the region of largest difference between the two models coincides with region of least amount of data.

Both models predicted vehicle impact speeds with a reasonable degree of accuracy and were in good agreement with other published models and the test data.

There was a good agreement between the mathematical and empirical models discussed there. However, even the mathematical models which consider the projection theory in detail, often require empirical correction factors. The largest factors that affected the results obtained from the models were the parameters that were selected in order to apply the models. Therefore, it is crucial to select appropriate and accurate factors when applying the mathematical models.

A major advantage of the empirical or hybrid models is that they are easy to apply to real world collisions. The disadvantage is that the user must exercise caution that the case in hand meets the criteria of the model. In addition, the empirical models yield accurate results when the considered collision has similar conditions as the test data reference to derive the empirical relation. If the considered collision occurred on slippery roads with a significant downward slope, then the lower percentiles of the prediction intervals of the empirical model may be appropriate to consider. Alternatively, if the vehicle has an extremely low frontal profile, then the upper percentiles of prediction intervals of the empirical model may be appropriate to consider. However, proper and careful judgment is required when "customizing" the empirical model.

An older empirical model (Appel) predicted vehicle speeds higher than the recent model (Toor, 2001). This demonstrates a continued need for revising and validating empirical models. The geometry of the vehicle is continually changing. From year to year this change may not significantly alter vehicle/pedestrian interaction, but differences may be quite substantial from decade to decade. This is particularly true as the designs of newer vehicles improve toward "pedestrian friendliness".

The major advantage of the mathematical models is that they are not constrained by the "criteria" of the empirical model and are applicable to all vehicle/pedestrian collisions. The mathematical models are also advantageous when parameters different than the typical average values are known to be applicable to the collision circumstances. For example, if it is known that the collision occurred on icy roads with a significant slope and involved a child and a bus, then the specific model parameters may be selected to reflect to specific incident conditions. This may yield more accurate results than simply applying the 50th percentile prediction interval from the empirical models. However, the disadvantage of the mathematical models is the difficulty in establishing suitable values for all the variables necessary to solve the equations.

Additional mathematical model comparison was conducted by applying a range of pedestrian friction values. Although some authors suggest to implement the mathematical models by considering pedestrian friction values of about 0.7 or 0.8, the best correlation between the mathematical models and the available test data was achieved when a pedestrian friction value between the range of 0.45 and 0.55 was considered.

It is critical to select the proper and correct parameter values when applying the mathematical models. The parameters must be selected based on the specific collision circumstances and available research data. Therefore, there is an inherent empirical influence within the application of the mathematical models. However, it is also critical to apply the purely empirical models with proper judgment, particularly if the specific collision circumstances vary from typical or average values. When applied correctly, both empirical and mathematical models can provide good reference if compared against each other.

However, since these regions cannot be individually identified from real world collision data, the most useful factor is the average coefficient of friction. Based on the published literature, the average friction value varies between 0.37 and 0.8.

Equations (8.34), (8.35) and (8.36) were derived performing linear regression analysis to the vehicle/pedestrian data. The aim was to achieve the best fit to calculate vehicle impact speed. Using the same data, but forcing the intercept through the origin yields the following equations:

Forward trajectory $V_v = 11.23\sqrt{S}$ (8.69*a*)

Wrap trajectory $V_v = 12.17\sqrt{S}$ (8.69*b*)

Combined trajectory $V_v = 11.90\sqrt{S}$ (8.69*c*)

where S is in throw distance measured in [m] and V_v is the vehicle impact speed in [km/h].

Once the projection efficiency is incorporated to relate the pedestrian throw speed with the vehicle impact speed ($V_p = [PE]V_v$), these equations become directly comparable with slide to stop equation:

$$V_p^2 = 2\mu g S \tag{8.70a}$$

Correcting for units, the formula is expressed as:

$$V_p^2 = [PE_{\text{AVG}}]^2 \begin{bmatrix} 3.12 \\ 3.38 \\ 3.31 \end{bmatrix}^2 S \tag{8.70b}$$

$$\mu = [PE_{\text{AVG}}]^2 \begin{bmatrix} 0.50 \\ 0.58 \\ 0.56 \end{bmatrix} \tag{8.70c}$$

It is noted that the average projection efficiency is different for wrap and forward projection trajectories. The average projection efficiency for forward projection, wrap and combined trajectories can be defined as:

$$PE_{\text{AVG}} = \begin{bmatrix} 0.95 \\ 0.80 \\ 0.85 \end{bmatrix} \tag{8.70d}$$

Hence the average effective coefficient of friction calculated from the collision data for forward projection, wrap, and combined trajectory is as follows:

$$\mu = \begin{bmatrix} 0.45 \\ 0.37 \\ 0.40 \end{bmatrix} \tag{8.70e}$$

Thus, the average effective pedestrian coefficient of friction is calculated to be about 0.41; however, this is likely to be a lower bound based on the considered data due to its inclusion of the airborne portion of throw distance and the ground contact slide/tumble/roll distance.

The mathematical and empirical models predict vehicle impact with a reasonable degree of accuracy when each model is applied appropriately.

The advantage of empirical models is that they are simple to apply to real-world collision, particularly when the collision circumstances can be modeled by average or typical parameter values.

Example 8.4: *A pedestrian comes to rest 120 ft from the point of separation with the vehicle. The coefficient of friction from the pedestrian is determined to be 0.70. The cg height of the pedestrian is 2.6* *ft. The launch angle for the pedestrian is 7 degrees. Use the REC-TEC platinum software (described in Chapter 9).*

(*a*) *the launch of the pedestrian*

(*b*) *comparison of the formulae used*

(*c*) *graphical representation of the difference in speeds computed for the pedestrian*

(*d*) *comment on the difference in the computed speeds using the different formulae*

(*e*) *display the iteration/finite difference analysis menu*

(*f*) *display the table for the coefficient of friction for the Searle*

(*g*) *display the table for the coefficient of friction for the Rich formulae*

(*h*) *display the base value and the uncertainty level for the given ranges for part (g) above.*

(*i*) *change the launch angle to zero with all other inputs to zero and compute the launching speed of the pedestrian using different formulae.*

Solution:

(*a*) The launch speeds of the pedestrian computed using different formula using REC-TEC are shown in Fig. E8.4(*a*).

(*b*) and (*c*) Fig. E8.4(*b*) shows a comparison of the formulae used in REC-TEC software and a graphical representation of the difference in speeds computed for the pedestrian.

(*d*) Only the Rich and Schmidt formulae utilize the launch angle as part of the computations. The other formulae assume a launch angle, but may do not state what angle is assumed in the computation.

(*e*) Fig. E8.4(*c*) displays the iteration/finite difference analysis menu. The user must choose which specific formula is to be used for either the table or for the finite difference analysis.

(*f*) Fig. E8.4(*d*) shows a table for the coefficient of friction for the Searle (minimum) formula.

(*g*) and (*h*) Fig. E8.4(*e*) shows a finite difference analysis using Rich formula based on the ranges specified in Fig. E8.6(*d*) showing the base value and the uncertainty level for the given ranges.

(*h*) Fig. E8.4(*f*) shows the launch speeds for the pedestrian computed using different formula when the launch angle is zero.

8.9 SUMMARY

The mathematical and empirical models predict vehicle impact speed with a reasonable degree of accuracy when each model is appropriately applied.

The advantage of empirical models is that they are simple to apply to real world collision, particularly when the collision circumstances can be modeled by average or typical parameter values.

The mathematical models may provide useful insight when compared to empirical models and can provide trends if parameters vary significantly from the average or typical collision scenario values.

A simplified vehicle-pedestrian impact model developed by Limpert (1980) was presented. In this model, the pedestrian was modeled as a rigid rod. The model determines the horizontal velocities of the vehicle and pedestrian, and the angular velocity of the pedestrian after impact.

Brach and Han model for the mechanics of a vehicle pedestrian collision was presented in two forms: The first relates the throw distance, s_p, to the initial vehicle speed, v_{c0}, that is, $s_p = f(v_{c0})$. The second, intended for reconstruction, relates the vehicle speed, v_{c0}, to the pedestrian throw distance, *sp*, that is, $v_{c0} = f(s_p)$.

8.10 REFERENCES

Appel, H., Sturts, G., and **Gotzen, L.,** "Influence of Impact Speed and Vehicle Parameter on Injuries of Children and Adults in Pedestrian Accidents", Proceedings of the Second International IRCOBI Conference, p. 83-100, 1975.

Aronberg, R., "Airborne Trajectory Analysis Derivation for use in Accident Reconstruction", SAE Paper No. 900367, Society of Automotive Engineers Inc., Warrendale, PA, 1990.

Ashton, S.J., "Factors Associated with Pelvic and Knee Injuries in Pedestrians Struck by the Fronts of Cars", SAE Paper No.811026, Society of Automotive Engineers Inc., Warrendale, PA, 1981.

Ashton, S.J., "Pedestrian Accident Investigation and Reconstruction", Seminar on Special Problems in Traffic Accident Reconstruction, Institute of Police Technology and Management, University of North Florida, p.1-37, 1989.

Ashton, S.J., "Vehicle Design and Pedestrian Injuries", Pedestrian Accidents, edited by A.J. Chapman, F.M. Wade and H.C. Foot, John Wiley & Sons Ltd., Chichester, 1982.

Ashton, S.J., and **Mackay, G.M.,** "Car Design for Pedestrian Injury Minimization", Proc. Of the Seventh International Technical Conference on Experimental Safety Vehicles, U.S. DOT, Washington, D.C., 1979b, p. 630, also SAE paper No. 796057.

Backaitis, Stanley, H., "Accident Reconstruction Technologies – Pedestrians and Motorcycles in Automotive Collisions", PT-35, SAE International, Warrendale, PA, 1990.

Baker, J. Standard, et al., "Traffic Accident Reconstruction", Volume-II of the Traffic Accident Investigation Manual, Northwestern University Traffic Institute, Evanston, IL, 1989.

Baker, J.S., and **Fricke, L.B.,** "The Traffic Accident Investigation Manual", Northwestern University Traffic Institute, Evanston, IL, 1986.

Bratten, T.A., "Development of a Tumble Number for Use in Accident Reconstruction", SAE Paper No. 890859, Society of Automotive Engineers Inc., Warrendale, PA, 1989.

Brooks, D., Wiechel, J., Sens, M., and **Guenther, D.A.,** "A Comprehensive Review of Pedestrian Impact Reconstruction", SAE paper No. 870605, SAE International, Warrendale, PA, 1987.

Casteel, D.A., and **Moss, S.D.,** "Basic Collision Analysis and Scene Documentation", 2nd ed, Lawyers & Judge Publishing Company, Tucson, AZ, 1999.

Collins, J.C., *Accident Reconstruction*, Charles C. Thomas Publisher, Springfield, Illinois, 1979.

Daily, J., *Fundamental of Traffic Accident Reconstruction*, Institute of Police Technology and Management, University of North Florida, Jacsonville, FL, 1988.

Dettinger, J., "Methods of Improving the Reconstruction of Pedestrian Accidents: Development Differential, Impact factor, Longitudinal Forward Trajectory, Position of Glass Splinters" (in German), Verkehrsunfall and Fahrzeugtechnik, Dec. 1996, 324-330 and Jan. 1997, 25-30.

Eubanks, J.J., *Pedestrian Accident Reconstruction*, Lawyers & Judge Publishing Company, Tucson, AZ, 1994.

Eubanks, J.J., and **Haight, W.R.,** "Pedestrian Involved Traffic Collision Reconstruction Methodology", SAE Paper No. 921591, Society of Automotive Engineers Inc., Warrendale, PA, 1992.

Evans, A.K., and **Smith, R.,** "Vehicle Speed Calculation from Pedestrian Throw Distance", Proc. Instn Mech Engrs, Part D, Vol. 213, No. D5, p.441-448, 1999.

Fricke, L.B., "Traffic Accident Reconstruction", Traffic Accident Investigation Manual, Vol. 2, Northwestern University Traffic Institute, Evanston, IL, 1990.

Fricke, L.B., "Vehicle-Pedestrian Accident Reconstruction", Topic 877 of the Traffic Accident Investigation Manual, Northwestern University, Traffic Institute, 1990.

Fugger, T.F., Randies, D.C., Wobrock, J.L., Stein, A.C., and **Whiting, W.C.,** "Pedestrian Behavior at Signal-Controlled Crosswalks", SAE Paper No.2001-01-0896, SAE International, Warrendale, PA, 2001.

Fugger, T.F., and **Randles, B.C.,** "Comparison of Pedestrian Accident Reconstruction Models to Experimental Test Data for Wrap Trajectories", C567/031/2000, ImechE 2000.

Grandel, J., and **Zeisberger, H.,** "Kinematics and Head Injuries in Vehicle/Pedestrian Accidents at Speeds above 50 kph", Proceedings of the International IRCOBI, Conference on the Biomechanics of Impacts, p. 189-204, Germany, 1986.

Grosch, L., and **Hochgeschwender, J.,** "Experimental Simulation of Car/Pedestrian and Car/Cyclist Collisions and Application of Findings in Safety Features on the Vehicle", SAE Paper No. 890751, Society of Automotive Engineers Inc., Warrendale, PA, 1989.

Han, I., and **Brach, R.M.,** "Throw Model for Frontal Pedestrian Collisions", SAE Paper No.2001-01-0897, SAE International, Warrendale, PA, 2001.

Han, J., and **Brach, R.M.,** "Throw Model for Frontal Pedestrian Collisions", SAE Paper 2001-01-0898, SAE International, Warrendale, PA, 2001.

Happer, A., Araszewski, M., Toor, A, Overgaard, A, and **Johal, R.,** "Comprehensive Analysis Method for Vehicle/Pedestrian Collision", SAE Paper No.2000-10-0846, SAE International, Warrendale, PA, 2000.

Hill, G.S., "Calculations of Vehicle Speed from Pedestrian Throw Impact", J. Inst. Traffic Accident Investigation, p. 18-20, Spring 1994.

Isenberg, R., Chidester, C., and **Kaufman, R.,** "Pedestrian Crash Data Study (PCDS)-An Interim Evaluation", ESV paper, June 1996.

Kallieris, D., and **Schmidt, G.,** "New Aspects of Pedestrian Protection Loading and Injury Pattern in Simulated Pedestrian Accidents", SAE paper No.881725, 1988.

Kramer, Martin., "Pedestrian Vehicle Accident Simulation through Dummy Tests", Nineteenth Stapp Car Crash Conference, SAE paper No. 751165, 1975.

Kuhnel, A., "Vehicle-pedestrian Collision Experiments with the use of a Moving Dummy", Proc. of Eighteenth Conference of American Association of Automotive Medicine, p.223-245, Toronto, Sept. 1974.

Limpert, R., *Motor Vehicle Accident Reconstruction and Cause Analysis*, 4th ed., The Michie Company, Suppl., 1998.

Lucchini, E., and **Weissner, R.,** "Difference between the Kinematics and Loadings of Impacted Adults and Children; Results from Dummy Tests", Proc. of the Fifth International IRCOBI Conference on the Biomechanics of Impacts, p. 165-179, Lyon, 1980.

McHenry, B.G., and **McHenry, R.R.,** "SMAC-97: Refinement of the Collision Algorithm", SAE Paper 970947, SAE International, Warrendale, PA, 1997.

McHenry, R.R., "A Comparison of Results obtained with Different Analytical Techniques for Reconstruction of Highway Accidents", SAE International, SAE Paper No. 750983, Warrendale.

Ravani, B., Brougham, D., and **Mason, R.T.,** "Pedestrian Post-impact Kinematics and Injury Patterns", SAE paper No. 811024, Society of Automotive Engineers Inc., Warrendale, PA, 1981.

Rivers, R.W., "Speed Analysis for the Traffic Accident Investigation Manual", IPTM, University of North Florida, Jacksonville, FL, 1997.

Russel, C.G., Equations and Formulas for the Traffic Accident Investigation and Reconstruction, Lawyers & Judges Publishing Company Inc., Tucson, AZ, 1999.

Schmidt, D.N., and **Nagel, D.A.,** "Pedestrian Impact Case Study", Proceedings of the Fifteenth Conference of the Association for Automotive Medicine, p. 151-167, 1971.

Schneider, H., and **Beier, G.,** "Experiment and Accident: Comparison of Dummy Test Results and Real Pedestrian Accident", SAE paper No..741177, Society of Automotive Engineers Inc., Warrendale, PA,1974.

Searle, J., and **Searle, A.,** "The Trajectories of Pedestrians, Motorcycles, Motorcyclists, etc., Following a Road Accident", 831622, Society of Automotive Engineers, Inc., Warrendale, PA, 1983.

Searle, J.A., "The Physics of Throw Distance in Accident Reconstruction", 930659, Society of Automotive Engineers Inc., Warrendale, PA, 1993.

Severy, D.J., "Auto-Pedestrian Collision Experiments", SAE paper No. 660080, 1966.

Stcherbatcheff, G., Tarriere, C., Duclos, P., and **Fayon, A.,** "Simulation of Collisions Between Pedestrians and Vehicles Using Adult and Child Dummies", SAE Paper No. 751167, Society of Automotive Engineers, Inc., Warrendale, PA, 1975.

Sturtz, G., and **Suren, E.G.,** "Kinematic of Real Pedestrian and Two Wheel Rider Accidents, and Special Aspects of the Pedestrian Accident", Proc. Of IRCOBI Meeting on Biomechanics of Injury to Pedestrians, Cyclists and Motorcyclists, Amsterdam, 7-8, Sept. 1976.

Sturtz, G., Suren, E.G., Gotzen, L., Behrens, S., and **Richter, K.,** "Biomechanics of Real Child Pedestrian Accidents", SAE paper No. 760814, SAE International, Warrendale, PA, 1976.

Toor, A., "Theoretical Versus Empirical Solutions for Vehicle/Pedestrian Collisions", SAE Paper No. 2003-01-0883, SAE International, Warrendale, PA, 2003.

Toor, A., Araszewski, M., Johal, R., Overgaard, R., and **Happer, A.,** "Revision and Validation of Vehicle/Pedestrian Collision Analysis Method", SAE paper No. 2002-01-0550, SAE International, Warrendale, PA, 2002.

Traffic Safety Facts 1996, "A Compilation of Motor Vehicle Crash Data from the Fatal Accident Reporting System and the General Estimates System", U.S. Department of Transportation, NHTSA, Washington, DC, 1997.

van Wijk, J., Wismans, J., Maltha, J., and **Wittebrood, L.,** "MADYMO Pedestrian Simulations", International Congress and Exposition, SAE Paper No. 830060, SAE International, Warrendale, PA, 1983.

Vaughan, R., and **Bain, J.,** "Acceleration and Speeds of Young Pedestrians Phase II", Paper No.2000-01-0845, SAE International, Warrendale, PA, 2000.

Vaughan, R., and **Bain, J.,** "Acceleration and Speeds of Young Pedestrians", Paper No.1999-01-0440, SAE International, Warrendale, PA, 1999.

Wood, D., and **Simms, C.K.,** "A Hybrid Model for Pedestrian Impact and Projection", International journal of Crashworthiness, Vol.5, No.4, p.393-403, Woodhead Publishing Ltd., Cambridge, UK, 2000.

Wood, D.P., and **Walsh, D.G.,** "Pedestrian Forward Projection Impact", International journal of Crashworthiness, Vol. 7, No. 3, Woodhead Publishing Ltd., Cambridge, UK, 2002.

Wood, D.P., and **Simms, C.K.,** "Vehicle Speed Calculation from Pedestrian Throw Distance Communication", Proceedings, Institute of Mechanical Engineers, Vol. 214, pt D, pp. 462-469, 1999.

Wood, D.P., "Application of a Pedestrian Impact Model to the Determination of Impact Speed", 910814, Society of Automotive Engineers, Inc., Warrendale, PA, 1991.

Wood, D.P., "Impact and Movement of Pedestrians in Frontal Collisions with Vehicles", Proceedings of the Institution of Mechanical Engineers, Vol. 202, No. D2, p.101-110, London, UK, 1988.

Wood, D.P., and **Simms, C.K.,** "A Hybrid Model for Pedestrian Impact and Projection", International Journal of Crashworthiness, Vol. 5, No. 4, p.12-15, Woodhead Publishing Ltd., Cambridge, UK, 2000.

Wood, D.P., and **Simms, C.K.,** "Coefficient of Friction in Pedestrian Throw", Impact, Vol. 9, No.1, p.12-15, Institute of Traffic Accident Investigators, Shrewsbury, UK, 2000.

Appendix

A

Vector Algebra

A VECTOR ALGEBRA

A.1 Real and Complex Vectors

An n-dimensional complex *vector* **A** is an ordered n-tuple of complex numbers $(a_1, a_2, ..., a_n)$ which form an n-dimensional vector space. If the ordered n-tuple numbers $(a_1, a_2, ..., a_n)$ admit only real numbers, then we define an n-dimensional real vector **A.**

The vector **A** can be represented as

$$\mathbf{A} = \begin{Bmatrix} a_1 \\ a_2 \\ \cdot \\ \cdot \\ \cdot \\ a_n \end{Bmatrix} \tag{A.1}$$

in the matrix sense in which the components are distinguished by their positions in the column.

The *vectors*, $\mathbf{A} = (a_1, a_2, ..., a_n)$ and $\mathbf{B} = (b_1, b_2, ..., b_n)$, are said to be equal if and only if $a_1 = b_1$, $a_2 = b_2$, ..., $a_n = b_n$.

Vector Addition and Subtraction

The sum of the *vectors*, $\mathbf{A} = (a_1, a_2, ..., a_n)$ and $\mathbf{B} = (b_1, b_2, ..., b_n)$, is the vector $\mathbf{C} = \mathbf{A} + \mathbf{B} = (a_1 + b_1, a_2 + b_2, ..., a_n + b_n)$.

Vector Multiplication by a Scalar

The product of a vector $\mathbf{A} = (a_1, a_2,..., a_n)$ and a scalar number c is the vector
$\mathbf{B} = c\mathbf{A} = (ca_1, ca_2, ..., ca_n)$.

A.2 Laws of Vector Operation

Vector operations satisfy the following rules:

1. *Commutative law:* **A** + **B** = **B** + **A**.
2. *Associative law:* (**A** + **B**) + **C** = **A** + (**B** + **C**).
3. *Distributive Law:* $c(\mathbf{A} + \mathbf{B}) = c\mathbf{A} + c\mathbf{B}$ and $(c + d)\,\mathbf{A} = c\mathbf{A} + d\mathbf{A}$, where c and d are numbers.

$c(d\mathbf{A}) = (cd)\mathbf{A}$; $0\mathbf{A} = \mathbf{0}$, where **0** is called the zero vector or null vector; $c\mathbf{0} = \mathbf{0}$.

The equality $c\mathbf{A} = \mathbf{0}$ holds, if and only if $c = 0$ or $\mathbf{A} = \mathbf{0}$.

$-(c\mathbf{A}) = (-c)\mathbf{A} = c(-\mathbf{A})$.

A.3 Linear Dependence

Consider a vector **U**, in which $u_1, u_2, ..., u_k$ are the components and $c_1, c_2, ..., c_k$ are k scalars. Then, the vector **U** given by

$$\mathbf{U} = c_1u_1 + c_2u_2 + ... + c_ku_k \tag{A.2}$$

is called a linear combination of $u_1, u_2, ..., u_k$ with coefficients $c_1, c_2, ..., c_k$. The totality of linear combination of $u_1, u_2, ..., u_k$, obtained by letting $c_1, c_2, ..., c_k$ vary over R, is a vector space. The space of all linear combinations of $u_1, u_2, ..., u_k$. If the relation

$$c_1u_1 + c_2u_2 + ... + c_ku_k = 0 \tag{A.3}$$

can be satisfied only for the trivial case, namely, when all the coefficients $c_1, c_2, ..., c_k$ are identically zero, then the vectors $u_1, u_2, ..., u_k$ are said to be linearly independent. If at least one of the coefficients $c_1, c_2, ..., c_k$ is different from zero, then the vectors $u_1, u_2, ..., u_k$ are said to be linearly dependent, implying that one vector is a linear combination of the remaining vectors.

Linear Combination

A vector **A** is said to be a linear combination of vectors $\mathbf{A}_1, \mathbf{A}_2... , \mathbf{A}_k$ if numbers $c_1, c_2... , c_k$ exist such that

$$\mathbf{A} = c_1\mathbf{A}_1 + c_2\mathbf{A}_2 + ... + c_k\,\mathbf{A}_k \tag{A.4}$$

Vectors $\mathbf{A}_1, \mathbf{A}_2, ..., \mathbf{A}_k$ are linearly dependent if and only if at least one of them can be expressed as a linear combination of the others.

Length of Vector

The length of a vector $\mathbf{A} = (a_1, a_2, ..., a_n)$ is the non-negative number $[a_1^2 + a_2^2 + \cdots + a_n^2]^{1/2}$ which is denoted by $|\,\mathbf{A}\,|$ or a. A vector whose length equals unity is called a unit vector. The terms, modulus, magnitude, norm, or absolute value of a vector, are also used for the length of a vector.

A.4 Three-Dimensional Vectors

In rectangular Cartesian coordinate system, we employ x, y, and z to represent the three coordinates. The linearly independent unit vectors, $\mathbf{I} = (1, 0, 0)$, $\mathbf{J} = (0, 1, 0)$, and $\mathbf{K} = (0, 0, 1)$ in the directions x, y, and z, respectively, as shown in Fig. A.1 are called coordinate vectors.

In three-dimensional space, any four vectors are linearly dependent. Thus, every vector $\mathbf{A} = (a_1, a_2, a_3)$ can be expressed as a linear combination of the coordinate vectors **I**, **J**, and **K** as

$$\mathbf{A} = a_1\,\mathbf{I} + a_2\,\mathbf{J} + a_3\,\mathbf{K} \tag{A.5}$$

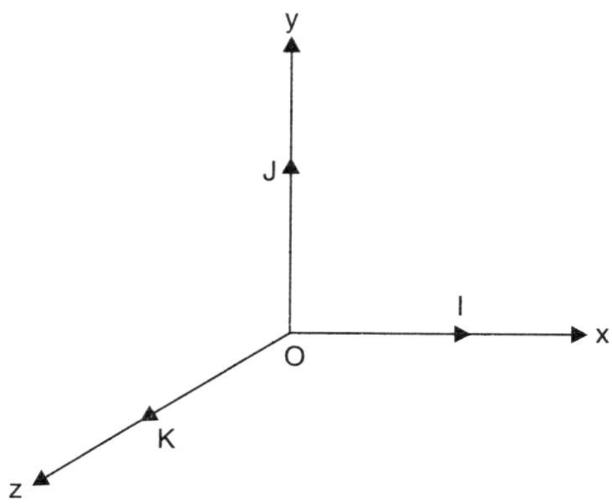

Fig. A.1 Coordinate vectors.

Two linearly dependent vectors are called collinear (parallel). Two vectors **A** and **B** are linearly dependent (parallel) if and only if one of them is a multiple of the other; that is, there is a number c such that $\mathbf{A} = c\mathbf{B}$. Three linearly dependent vectors are called coplanar.

Angle Between Vectors

The angle between two nonzero vectors **A** and **B** is the angle θ ($0 \le \theta \le \pi$) between the directed segments representing the vectors.

Scalar Product of Two Vectors

The scalar product or inner product or dot product of two vectors $\mathbf{A} = (a_1, a_2, a_3)$ and $\mathbf{B} = (b_1, b_2, b_3)$, written as $\mathbf{A} \cdot \mathbf{B}$, is defined by

$$\mathbf{A} \cdot \mathbf{B} = a_1 b_1 + a_2 b_2 + a_3 b_3 \qquad \text{(A.6)}$$

It can also be shown that

$$\mathbf{A} \cdot \mathbf{B} = |\mathbf{A}|\,|\mathbf{B}| \cos \theta \qquad \text{(A.7)}$$

where $|\mathbf{A}|$ and $|\mathbf{B}|$ are the magnitudes of vectors **A** and **B**, respectively, and θ is the angle between them.

Perpendicular Vectors

Two nonzero vectors **A** and **B** are perpendicular if and only if $\mathbf{A} \cdot \mathbf{B} = 0$.

A.5 Properties of Scalar Product of Vectors

The scalar product of vectors satisfies the relation:

1. $\mathbf{A} \cdot \mathbf{B} = \mathbf{B}.\mathbf{A}$ (i.e., it is commutative)
2. $(\mathbf{A} + \mathbf{B}) \cdot \mathbf{C} = \mathbf{A} \cdot \mathbf{C} + \mathbf{B} \cdot \mathbf{C}$ (i.e., it is distributive)
3. $\mathbf{A} \cdot \mathbf{A} = |\mathbf{A}|^2$.

A.6 Direction Angles

The angles that a nonzero vector makes with the coordinate vectors and thus with the coordinate axes are called the direction angles, and their cosines the direction cosines of the given vector.

Let α, β and γ be the direction angles of a nonzero vector $\mathbf{A} = (a_1, a_2, a_3)$. The direction cosines obey the rules

1. $\cos \alpha = a_1/|\mathbf{A}|$, $\cos \beta = a_2/|\mathbf{A}|$, $\cos \gamma = a_3/|\mathbf{A}|$
2. $\cos^2 \alpha + \cos^2 \beta + \cos^2 \gamma = 1$

A.7 Vector Product

The vector product or cross product or outer product of vectors $\mathbf{A} = (a_1, a_2, a_3)$ and $\mathbf{B} = (b_1, b_2, b_3)$, denoted by $\mathbf{A} \times \mathbf{B}$, is a vector $\mathbf{C}$ defined by

$$\mathbf{A} \times \mathbf{B} = \mathbf{C} \tag{A.8}$$

where $\mathbf{C}$ is a vector perpendicular to both $\mathbf{A}$ and $\mathbf{B}$, in the sense of a right-hand screw, and has magnitude $|\mathbf{A}||\mathbf{B}| \sin \theta$, as shown in Figure A.2. Here, θ is the angle between $\mathbf{A}$ and $\mathbf{B}$. The vector product $\mathbf{A} \times \mathbf{B}$ may be written as

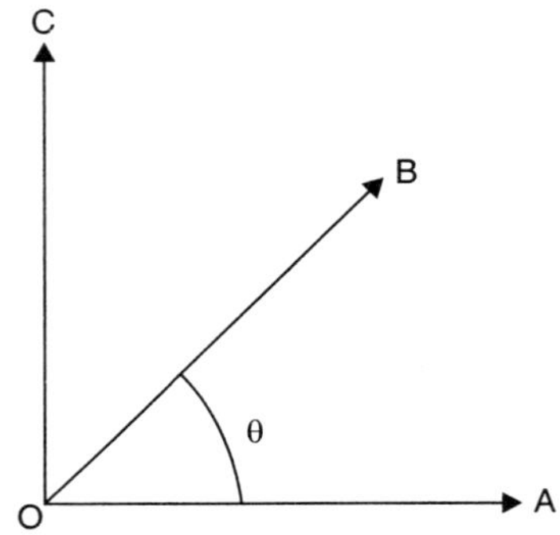

Fig. A.2 Vector product of two vectors

$$\mathbf{A} \times \mathbf{B} = (a_2b_3 - a_3b_2)\,\mathbf{I} + (a_3b_1 - a_1b_3)\,\mathbf{J} + (a_1b_2 - a_2b_1)\,\mathbf{K} \tag{A.9}$$

This result is also obtained by the expansion of the determinant

$$\mathbf{A} \times \mathbf{B} = \begin{vmatrix} I & J & K \\ a_1 & a_2 & a_3 \\ b_1 & b_2 & b_3 \end{vmatrix} \tag{A.10}$$

Properties of Vector Product

The vector product satisfies the following relations:

1. It is not commutative, *i.e.*, $\mathbf{A} \times \mathbf{B} \neq \mathbf{B} \times \mathbf{A}$ but $\mathbf{A} \times \mathbf{B} = -\mathbf{B} \times \mathbf{A}$.
2. It is distributive, *i.e*, $\mathbf{A} \times (\mathbf{B} + \mathbf{C}) = \mathbf{A} \times \mathbf{B} + \mathbf{A} \times \mathbf{C}$.
3. It is not associative, *i.e.* $\mathbf{A} \times (\mathbf{B} \times \mathbf{C}) \neq (\mathbf{A} \times \mathbf{B}) \times \mathbf{C}$.
4. $k\,\mathbf{A} \times \mathbf{B} = k\,(\mathbf{A} \times \mathbf{B})$, where k is a scalar number.

Mixed Product

The mixed product or triple scalar product of three vectors **A**, **B**, **C** is a scalar denoted by $\mathbf{A} \cdot (\mathbf{B} \times \mathbf{C})$. The mixed product of three vectors is also sometimes called a trivector.

Properties of Mixed Product

The mixed product satisfies the following relation:

$$\mathbf{A} \cdot (\mathbf{B} \times \mathbf{C}) = \mathbf{B} \cdot (\mathbf{C} \times \mathbf{A}) = \mathbf{C} \cdot (\mathbf{A} \times \mathbf{B})$$

$$= -\mathbf{A}.\ (\mathbf{C} \times \mathbf{B}) = -\mathbf{C} \cdot (\mathbf{B} \times \mathbf{A}) = -\mathbf{B} \cdot (\mathbf{A} \times \mathbf{C})$$

$$= \begin{vmatrix} a_1 & a_2 & a_3 \\ b_1 & b_2 & b_3 \\ c_1 & c_2 & c_3 \end{vmatrix} \qquad \text{(A.11)}$$

A.8 Derivative of a Vector

The components of vectors are functions of a scalar variable t. Thus, for every value of t in the domain under consideration, we get, in general, a different vector. We deal with a vector field and denote the vector by $\mathbf{A}(t)$. The derivative of $\mathbf{A}(t)$ with respect to t is defined by

$$\frac{d}{dt}\mathbf{A}(t) = \lim_{\Delta t \to 0} \frac{A(t+\Delta t) - A(t)}{\Delta t} = \dot{A}(t) \qquad \text{(A.12)}$$

Similarly, we can define higher derivatives. For example,

$$\frac{d^2}{dt^2}A(t) = \lim_{\Delta t \to 0} \frac{\dot{A}(t+\Delta t) - \dot{A}(t)}{\Delta t} = \ddot{A}(t) \qquad \text{(A.13)}$$

The components of a vector can in general be functions of several variables. The corresponding partial derivatives can be defined in a similar manner.

Rules of Vector Differentiation

It can be shown that

1. $\frac{d}{dt}(c\mathbf{A}) = c\dot{A} + \dot{c}\mathbf{A}$, where c is a scalar function of t.

2. $\frac{d}{dt}(\mathbf{A} + \mathbf{B}) = \dot{A} + \dot{B}$

3. $\frac{d}{dt}(\mathbf{A} \cdot \mathbf{B}) = \dot{A} \cdot B + A \cdot \dot{B}$

4. $\frac{d}{dt}(\mathbf{A} \times \mathbf{B}) = \dot{A} \times B + A \times \dot{B}$

REFERENCES

Dukkipati, R.V., *Advanced Engineering Analysis,* Narosa Publishing House, New Delhi, India, 2005.

Dukkipati, R.V., and **Cupic, K.,** *Applied Numerical Methods using MATLAB,* New Age International Publishing Ltd., New Delhi, India, 2010.

Dukkipati, R.V., *Applied Numerical Methods,* New Age International Publishing Ltd., New Delhi, India, 2009.

Dukkipati, R.V., *Mathematical Methods,* New Age International Publishing Ltd., New Delhi, India, 2009.

Greenberg, M.D., *Foundations of Applied Mathematics*, Prentice Hall, Englewood Cliffs, NJ, 1978.

Hildebrand, F.B., *Advanced Calculus for Applications*, Prentice-Hall, Englewood Cliffs, NJ, 1978.

Kaplan, W., *Advanced Calculus*, Addison-Wesley, Reading, 1952.

Lass, H., *Vector and Tensor Analysis*, McGraw-Hill, New York, 1950.

Rektorys, K., *Survey of Applicable Mathematics*, The MIT Press, 1962.

Strang, G., *Linear Algebra and its Applications*, Academic Press, New York, 1980.

Wylie, C.R., and **Barret, L.C.,** *Advanced Engineering Mathematics*, 5th edition, McGraw-Hill, New York, 1982.

Appendix

B

Basic Engineering Mathematics

B.1 ALGEBRA

B.1.1 Basic Laws

Cumulative law:

$$a + b = b + a \quad ab = ba$$

Associative law:

$$a + (b + c) = (a + b) + c \quad a(bc) = (abc)$$

Distributive law:

$$a(b + c) = ab + ac$$

B.1.2 Sums of Numbers

The sum of the first n numbers:

$$\sum_{1}^{n}(n) = \frac{n(n+1)}{2}$$

The sum of the squares of the first n numbers:

$$\sum_{1}^{n}(n^2) = \frac{n(n+1)(2n+1)}{6}$$

The sum of the cubes of the first n numbers:

$$\sum_{1}^{n}(n^3) = \frac{n^2(n+1)^2}{4}$$

B.1.3 Progressions

Arithmetic Progression

$$a, a + d, a + 2d, a + 3d, \ldots$$

where

a = first term

d = common difference

n = number of terms

S = sum of n terms

ℓ = last term

$$\ell = a + (n - 1)d$$

$$S = (n/2)\,(a + l)$$

$$(a + b)/2 = \text{arithmetic mean of } a \text{ and } b.$$

Geometric Progression

$$a,\ ar,\ ar^2,\ ar^3,\ \ldots$$

where

a = first term

r = common ratio

n = number of terms

S = sum of n terms

ℓ = last term

$$\ell = ar^{n-1}$$

$$S = a\frac{r^n - 1}{r - 1} = \frac{r\ell - a}{r - 1}$$

$$S = \frac{a}{1 - r} \quad \text{for } |r| < 1 \text{ and } n = \infty$$

$$\sqrt{ab} = \text{geometric mean of } a \text{ and } b.$$

B.1.4 Powers and Roots

$$a^x a^y = a^{x+y}$$

$$\frac{a^x}{a^y} = a^{x-y}$$

$$(ab)^x = a^x b^x$$

$$(a^x)^y = a^{xy}$$

$$a^0 = 1 \text{ if } a \neq 0$$

$$a^{-x} = 1/a^x$$

$$a^{x/y} = \sqrt[y]{a^x}$$

$$a^{1/y} = \sqrt[y]{a}$$

$$\sqrt[x]{ab} = \sqrt[x]{a}\sqrt[x]{b}$$

$$\sqrt[x]{a/b} = \sqrt[x]{a}/\sqrt[x]{b}$$

B.1.5 Binomial Theorem

$$(a \pm b)^n = a^n + (\pm 1)^1 na^{n-1}b + (\pm 1)^2 \frac{n(n-1)}{2!} a^{n-2} b^2 + (\pm 1)^3 \frac{n(n-1)(n-2)}{3!} a^{n-3} b^3 + \dots$$

$$+ (\pm 1)^m \frac{n(n-1)\dots(n-m+1)}{m!} a^{n-m} b^m + \dots$$

where $m! = 1.2.3. \dots (m-1)m$

The series is finite if n is a positive integer. If n is negative or fractional, the series is infinite and will converge for $|b| < |a|$ only.

B.1.6 Absolute Values

The numerical or absolute value of a number n is denoted by $|n|$ and represents the magnitude of the number without regard to its algebraic sign. For example,

$$|-5| = |+5| = 5$$

B.1.7 Logarithms

Laws of Logarithms

$$\log_b MN = \log_b M + \log_b N$$

$$\log_b 1 = 0$$

$$\log_b \frac{M}{N} = \log_b M - \log_b N$$

$$\log_b b = 1$$

$$\log_b N^m = m \log_b N$$

$$\log_b 0 = +\infty,\ 0 < b < 1$$

$$\log_b \sqrt[r]{N^m} = m/r \log_b N$$

$$\log_b 0 = -\infty,\ 1 < b < \infty$$

Important Constants

$$\log_{10} e = 0.434294481\ 9$$

$$\log_{10} x = 0.4343 \log_e x = 0.4343 \ln x$$

$$\ln 10 = \log_e 10 = 2.302585093\ 0$$

$$\ln x = \log_e x = 2.3026 \log_{10} x$$

B.2 TRIGONOMETRY

$\sin\theta = 0 \quad \Rightarrow \quad \theta = n\pi$, n is an integer

$\cos\theta = 0 \quad \Rightarrow \quad \theta = (4n+1)\frac{\pi}{2}$, n is an integer

$\sin\theta = \sin\alpha \quad \Rightarrow \quad \theta = n\pi + (-1)^n \alpha$, n is an integer

$\cos\theta = \cos\alpha \quad \Rightarrow \quad \theta = 2n\pi \pm \alpha$, n is an integer

$\tan\theta = 0 \quad \Rightarrow \quad \theta = n\pi$, n is an integer

$\tan\theta = \tan\alpha \quad \Rightarrow \quad \theta = n\pi + \alpha$, n is an integer

$$\cosh x = \frac{e^x + e^{-x}}{2}$$

$$\sinh x = \frac{e^x - e^{-x}}{2}$$

$$\tanh x = \frac{\sinh x}{\cosh x} = \frac{e^x - e^{-x}}{e^x + e^{-x}}$$

$$\coth x = \frac{e^x + e^{-x}}{e^x - e^{-x}} = \frac{1}{\tanh x} = \frac{\cosh x}{\sinh x}$$

$$\operatorname{sech} x = \frac{1}{\cosh x} = \frac{2}{e^x + e^{-x}}$$

$$\operatorname{cosech} x = \frac{1}{\sinh x} = \frac{2}{e^x - e^{-x}}$$

B.2.1 Trigonometric Identities

$$\sin^2 \alpha + \cos^2 \alpha = 1$$

$$1 + \tan^2 \alpha = \frac{1}{\cos^2 \alpha}$$

$$\tan \alpha \cot \alpha = 1$$

$$1 + \cot^2 \alpha = \frac{1}{\sin^2 \alpha}$$

$$\sin (\alpha \pm \beta) = \sin \alpha \cos \beta \pm \cos \alpha \sin \beta$$

$$\cos (\alpha \pm \beta) = \cos \alpha \cos \beta \mp \sin \alpha \sin \beta$$

$$\tan (\alpha \pm \beta) = \frac{\tan \alpha \pm \tan \beta}{1 \mp \tan \alpha \tan \beta}$$

$$\cot (\alpha \pm \beta) = \frac{\cot \alpha \cot \beta \mp 1}{1 \pm \cot \alpha \cot \beta}$$

$$\sin \alpha + \sin \beta = 2 \sin \frac{\alpha + \beta}{2} \cos \frac{\alpha - \beta}{2}$$

$$\sin \alpha - \sin \beta = 2 \cos \frac{\alpha + \beta}{2} \sin \frac{\alpha - \beta}{2}$$

$$\cos \alpha + \cos \beta = 2 \cos \frac{\alpha + \beta}{2} \cos \frac{\alpha - \beta}{2}$$

$$\cos \alpha - \cos \beta = -2 \sin \frac{\alpha + \beta}{2} \sin \frac{\alpha - \beta}{2}$$

$$\tan \alpha \pm \tan \beta = \frac{\sin (\alpha \pm \beta)}{\cos \alpha \cos \beta}$$

$$\cot\alpha \pm \cot\beta = \frac{\sin(\beta \pm \alpha)}{\sin\alpha \sin\beta}$$

$$\sin\alpha \cos\beta = \frac{1}{2}\sin(\alpha + \beta) + \frac{1}{2}\sin(\alpha - \beta)$$

$$\cos\alpha \cos\beta = \frac{1}{2}\cos(\alpha + \beta) + \frac{1}{2}\cos(\alpha - \beta)$$

$$\sin\alpha \sin\beta = \frac{1}{2}\cos(\alpha - \beta) - \frac{1}{2}\cos(\alpha + \beta)$$

$$\tan\alpha \tan\beta = \frac{\tan\alpha + \tan\beta}{\cot\alpha + \cot\beta} = -\frac{\tan\alpha - \tan\beta}{\cot\alpha - \cot\beta}$$

$$\cot\alpha \cot\beta = \frac{\cot\alpha + \cot\beta}{\tan\alpha + \tan\beta} = -\frac{\cot\alpha - \cot\beta}{\tan\alpha - \tan\beta}$$

$$\cot\alpha \tan\beta = \frac{\cot\alpha + \tan\beta}{\tan\alpha + \cot\beta} = -\frac{\cot\alpha - \tan\beta}{\tan\alpha - \cot\beta}$$

B.2.2 Cosine Law (Law of Cosines)

In any triangle, the square of any side is equal to the sum of the squares of the other two sides minus twice the product of those two sides multiplied by the cosine of their included angle, as

$$b^2 = a^2 + c^2 - 2ac\cos B$$
$$\cos B = (a^2 + c^2 - b^2)/2ac$$

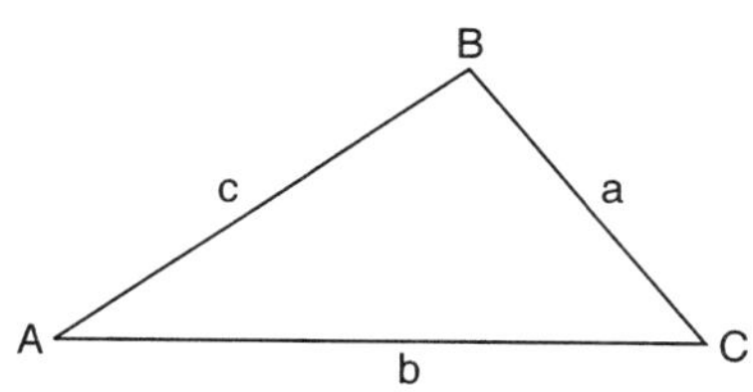

Formulas:

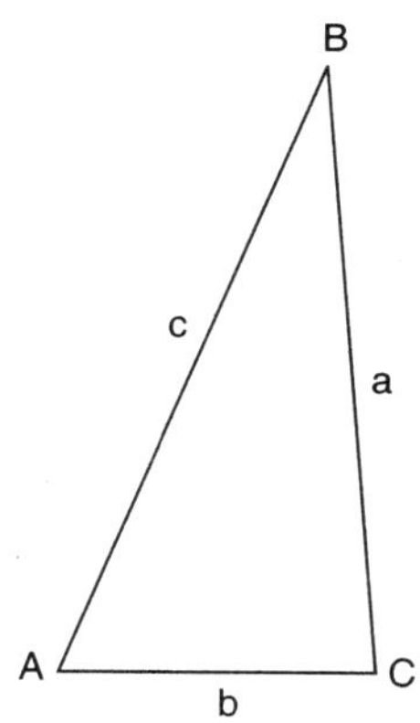

$\cos A = b/c = 1/\sec A = \sin A/\tan A$

$\cot A = b/a = 1/\tan A = \cos A/\sin A = \cos A \operatorname{cosec} A$

$\operatorname{cosec} A = c/a = 1/\sin A$

Double angle:

$$\cos 2A = 2\cos^2 A - 1 = \cos^2 A - \sin^2 A = 1 - 2\sin^2 A$$

$$\cot 2A = (\cot^2 A - 1)/2\cot A$$

$$\sin 2A = 2\sin A\cos A$$

$$\tan 2A = 2\tan A/(1 - \tan^2 A) = 2/(\cot A - \tan A)$$

$$\sin(A/2) = 0.707(1 - \cos A)^{1/2}$$

$$\tan(A/2) = \sin A/(1 + \cos A) = [(1 - \cos A)/(1 + \cos A)]^{1/2} = (1 - \cos A)/\sin A$$

$$\sec A = c/b = 1/\cos A$$

side:

$$a = c\sin A = b\tan A$$

$$\sin A = a/c = 1/\mathrm{cosec}\, A = \cos A/\cot A$$

$$\sin^2 A + \cos^2 A = 1$$

$$\tan A = a/b = 1/\cot A = \sin A/\cos A$$

B.2.3 Sine Law (Law of Sines)

In any triangle, the sides are proportional to the sines of the opposite angles, as

$$a/\sin A = b/\sin B = c/\sin C$$

$$a/b = \sin A/\sin B$$

$$b/c = \sin B/\sin C$$

$$a/c = \sin A/\sin C$$

B.3 DIFFERENTIAL CALCULUS

B.3.1 List of Derivatives

(Note: u, v and w are functions of x)

$$\frac{d}{dx}(a) = 0, \quad a = \text{constant}$$

$$\frac{d}{dx}(x) = 1$$

$$\frac{dy}{dx} = \frac{dy}{dv}\frac{dv}{dx}, \quad y = y(v)$$

$$\frac{d}{dx}(av) = a\frac{dv}{dx}$$

$$\frac{dy}{dx} = \frac{1}{dx/dy} \quad \text{if} \quad \frac{dx}{dy} \neq 0$$

$$\frac{d}{dx}(\pm u \pm v \pm \ldots) = \pm\frac{du}{dx} \pm \frac{dv}{dx} - \ldots$$

$$\frac{d}{dx}(u^n) = nu^{n-1}\frac{du}{dx}$$

$$\frac{d}{dx}(uv) = u\frac{dv}{dx} + v\frac{du}{dx}$$

$$\frac{d}{dx}\frac{u}{v} = \frac{v\,du/dx - u\,dv/dx}{v^2}$$

$$\frac{d}{dx}(u^v) = vu^{v-1}\frac{du}{dx} + u^v \ln u\frac{dv}{dx}$$

$$\frac{d}{dx}(a^u) = a^u \ln a\frac{du}{dx}$$

$$\frac{d}{dx}(e^u) = e^u\frac{du}{dx}$$

$$\frac{d}{dx}(\ln u) = \frac{1}{u}\frac{du}{dx}$$

$$\frac{d}{dx}(\log_a u) = \frac{\log_a e}{u}\frac{du}{dx}$$

$$\frac{d}{dx}(\sin u) = \cos u\frac{du}{dx}$$

$$\frac{d}{dx}(\cos u) = -sin\,u\frac{du}{dx}$$

$$\frac{d}{dx}(\tan u) = \sec^2 u\frac{du}{dx}$$

$$\frac{d}{dx}(\mathrm{cosec}\,u) = -\mathrm{cosec}\,u\cot u\frac{du}{dx}$$

$$\frac{d}{dx}(\sec u) = \sec u\tan u\frac{du}{dx}$$

$$\frac{d}{dx}(\cot u) = -\mathrm{cosec}^2 u\frac{du}{dx}$$

$$\frac{d}{dx}(\mathrm{vers}\,u) = \sin u\frac{du}{dx}$$

$$\frac{d}{dx}\sin^{-1} u = \frac{1}{\sqrt{1-u^2}}\frac{du}{dx} \qquad \frac{-\pi}{2} = \leqq \sin^{-1} u \leqq \frac{\pi}{2}$$

$$\frac{d}{dx}\cos^{-1} u = -\frac{1}{\sqrt{1-u^2}}\frac{du}{dx} \qquad 0 \leqq \cos^{-1} u \leqq \pi$$

$$\frac{d}{dx}\tan^{-1} u = \frac{1}{1+u^2}\frac{du}{dx}$$

$$\frac{d}{dx}\sinh^{-1} u = \frac{1}{\sqrt{u^2+1}}\frac{du}{dx}$$

$$\frac{d}{dx}\cosh^{-1} u = \frac{1}{\sqrt{u^2-1}}\frac{du}{dx} \qquad u > 1$$

$$\frac{d}{dx}\tanh^{-1} u = \frac{1}{1-u^2}\frac{du}{dx}$$

$$\frac{d}{dx}\operatorname{cosec h}^{-1} u = \frac{-1}{u\sqrt{u^2+1}}\frac{du}{dx}$$

$$\frac{d}{dx}\operatorname{sech}^{-1} u = -\frac{1}{\sqrt{1-u^2}}\frac{du}{dx} \qquad u > 0$$

$$\frac{d}{dx}\coth^{-1} u = \frac{1}{1-u^2}\frac{du}{dx}$$

$$\frac{d}{dx}\operatorname{cosec}^{-1} u = -\frac{1}{u\sqrt{u^2-1}}\frac{du}{dx} \qquad -\pi < \operatorname{cosec}^{-1} u \leqq -\frac{\pi}{2},\ 0 < \operatorname{cosec}^{-1} u \leqq \frac{\pi}{2}$$

$$\frac{d}{dx}\sec^{-1} u = \frac{1}{u\sqrt{u^2-1}}\frac{du}{dx} \qquad -\pi \leqq \sec^{-1} u < -\frac{\pi}{2},\ 0 \leqq \sec^{-1} u < \frac{\pi}{2}$$

$$\frac{d}{dx}\cot^{-1} u = \frac{-1}{1+u^2}\frac{du}{dx}$$

$$\frac{d}{dx}\operatorname{vers}^{-1} u = \frac{1}{\sqrt{2u-u^2}}\frac{du}{dx} \qquad 0 \leqq \operatorname{vers}^{-1} u \leqq \pi$$

$$\frac{d}{dx}\sinh u = \cosh u \frac{du}{dx}$$

$$\frac{d}{dx}\cosh u = \sinh u \frac{du}{dx}$$

$$\frac{d}{dx}\tanh u = \operatorname{sech}^2 u \frac{du}{dx}$$

$$\frac{d}{dx}\operatorname{cosech} u = -\operatorname{cosech} u \coth u \frac{du}{dx}$$

$$\frac{d}{dx}\operatorname{sech} u = -\operatorname{sech} u \tanh u \frac{du}{dx}$$

$$\frac{d}{dx}\coth u = -\operatorname{cosech}^2 u \frac{du}{dx}$$

B.3.2 Expansion in Series

Exponential and Logarithmic Series

$$e^x = 1 + \frac{x}{1!} + \frac{x^2}{2!} + \frac{x^3}{3!} + \frac{x^4}{4!} + \ldots, \qquad [-\infty < x < +\infty]$$

$$a^x = e^{mx} = 1 + \frac{m}{1!}x + \frac{m^2}{2!}x^2 + \frac{m^3}{3!}x^3 + \ldots, \quad [a > 0, -\infty < x < +\infty]$$

where $m = \ln a = (2.302\,6 \times \log_{10} a)$

$$\ln(1 + x) = x - \frac{x^2}{2} + \frac{x^3}{3} - \frac{x^4}{4} + \frac{x^5}{5} + \ldots, \quad [-1 < x < +1]$$

$$\ln(1 - x) = -x - \frac{x^2}{2} - \frac{x^3}{3} - \frac{x^4}{4} - \frac{x^5}{5} - \ldots, \quad [-1 < x < +1]$$

$$\ln\left(\frac{1+x}{1-x}\right) = 2\left(x + \frac{x^3}{3} + \frac{x^5}{5} + \frac{x^7}{7} + \ldots\right), [-1 < x < +1]$$

$$\ln\left(\frac{x+1}{x-1}\right) = 2\left(\frac{1}{x} + \frac{1}{3x^3} + \frac{1}{5x^5} + \frac{1}{7x^7} + \ldots\right), \quad [x < -1 \text{ or } +1 < x]$$

$$(1 + x)^n = 1 + nx + \frac{n(n-1)}{2!}x^2 + \frac{n(n-1)\,(n-2)}{3!}x^3 + \ldots\infty, |x| < 1$$

$$(1 + x)^{-n} = 1 - nx + \frac{n(n+1)}{2!}x^2 - \frac{n(n+1)\,(n+2)}{3!}x^3 + \ldots\infty, \quad |x| < 1$$

$$(1 - x)^{-n} = 1 + nx + \frac{n(n+1)}{2!}x^2 + \frac{n(n+1)(n+2)}{3!}x^3 + \ldots\infty, \quad |x| < 1$$

$$(1 + x)^{-1} = 1 - x + x^2 - x^3 + \ldots\infty, |x| < 1$$

$$(1 + x)^{-2} = 1 - 2x + 3x^2 - 4x^3 + \ldots\infty, |x| < 1$$

$$(1 - x)^{-1} = 1 + x + x^2 + x^3 + \ldots\infty, |x| < 1$$

$$(1 - x)^{-2} = 1 + 2x + 3x^3 + 4x^2 + \ldots\infty, |x| < 1$$

$$(1 - x)^{-1/2} = 1 + \frac{1}{2}x + \frac{1.3}{2.4}x^2 + \frac{1.3.5}{2.4.6}x^3 + \ldots.\infty, |x| < 1$$

$$\frac{1}{1^2} + \frac{1}{2^2} + \frac{1}{3^2} + \ldots\infty = \frac{\pi^2}{6}$$

$$\frac{1}{1^2} + \frac{1}{3^2} + \frac{1}{5^2} + \ldots\infty = \frac{\pi^2}{8}$$

$$\frac{1}{1^4} + \frac{1}{2^4} + \frac{1}{3^4} + \ldots\infty = \frac{\pi^4}{90}$$

$$\frac{1}{1^4} + \frac{1}{3^4} + \frac{1}{5^4} + \ldots\infty = \frac{\pi^4}{96}$$

$$\ln x = 2\left[\frac{x-1}{x+1} + \frac{1}{3}\left(\frac{x-1}{x+1}\right)^3 + \frac{1}{5}\left(\frac{x-1}{x+1}\right)^5 + \ldots.\right], \quad [0 < x < \infty]$$

$$\ln(a + x) = \ln a + 2\left[\frac{x}{2a+x} + \frac{1}{3}\left(\frac{x}{2a+x}\right)^3 + \frac{1}{5}\left(\frac{x}{2a+x}\right)^5 + \ldots.\right], \quad [0 < a < +\infty, -a < x < +\infty]$$

Series for the Trigonometric Functions

$$\sin x = x - \frac{x^3}{3!} + \frac{x^5}{5!} - \frac{x^7}{7!} + \ldots, \quad [-\infty < x < +\infty]$$

$$\cos x = 1 - \frac{x^2}{2!} + \frac{x^4}{4!} - \frac{x^6}{6!} + \frac{x^8}{8!} - \ldots, \quad [-\infty < x < +\infty]$$

$$\tan x = x + \frac{x^3}{3} + \frac{2x^5}{15} + \frac{17x^7}{315} + \frac{62x^9}{2835} + \ldots, \quad [-\pi/2 < x < +\pi/2]$$

$$\cot x = \frac{1}{x} - \frac{x}{3} - \frac{x^3}{45} - \frac{2x^5}{945} - \frac{x^7}{4725} - \ldots, \quad [-\pi < x < +\pi]$$

$$\sin^{-1} y = y + \frac{y^3}{6} + \frac{3y^5}{40} + \frac{5y^7}{112} + \ldots, \quad [-1 \le y \le +1]$$

$$\tan^{-1} y = y - \frac{y^3}{3} + \frac{y^5}{5} - \frac{y^7}{7} + \ldots, \quad [-1 \le y \le +1]$$

$$\cos^{-1} y = \frac{1}{2}\pi - \sin^{-1} y$$

$$\cot^{-1} y = \frac{1}{2}\pi - \tan^{-1} y$$

In these formulas, all angles are expressed in radians.

Series for the Hyperbolic Functions

$$\sinh x = x + \frac{x^3}{3!} + \frac{x^5}{5!} + \frac{x^7}{7!} + \ldots, \quad [-\infty < x < +\infty]$$

$$\cosh x = 1 + \frac{x^2}{2!} + \frac{x^4}{4!} + \frac{x^6}{6!} + \ldots, \quad [-\infty < x < +\infty]$$

$$\sinh^{-1} y = y - \frac{y^3}{6} + \frac{3y^5}{40} - \frac{5y^7}{112} + \ldots, \quad [-1 < y < +1]$$

$$\tanh^{-1} y = y + \frac{y^3}{3} + \frac{y^5}{5} + \frac{y^7}{7} + \ldots, \quad [-1 < y < +1]$$

General Formulas of Maclaurin and Taylor

If $f(x)$ and all its derivatives are continuous in the neighborhood of the point $x = 0$ (or $x = a$), then, for any value of x in this neighborhood, the function $f(x)$ may be expressed as a power series arranged according to ascending powers of x (or $x - a$), as follows:

$$f(x) = f(0) + \frac{f'(0)}{1!}x + \frac{f''(0)}{2!}x^2 + \frac{f'''(0)}{3!}x^3 + \ldots + \frac{f^{(n-1)}(0)}{(n-1)!}x^{n-1} + (P_n)x^n \quad \text{(Maclaurin)}$$

$$f(x) = f(a) + \frac{f'(a)}{1!}(x-a) + \frac{f''(a)}{2!}(x-a)^2 + \frac{f'''(a)}{3!}(x-a)^3 + \ldots$$

$$+ \frac{f^{(n-1)}(a)}{(n-1)!}(x-a)^{n-1} + (Q_n)(x-a)^n \quad \text{(Taylor)}$$

Here $(P_n)\, x^n$, or $(Q_n)\,(x-a)^n$, is called the remainder term; the values of the coefficients P_n and Q_n may be expressed as follows:

$$P_n = \frac{[f^{(n)}(sx)]}{n!} = \frac{(1-t)^{n-1} f^{(n)}(tx)}{(n-1)!}$$

$$Q_n = \frac{f^{(n)}[a+s(x-a)]}{n!} = \frac{(1-t)^{n-1} f^{(n)}[a+t(x-a)]}{(n-1)!}$$

where s and t are certain unknown numbers between 0 and 1.

B.4 INTEGRAL CALCULUS

B.4.1 List of Most Common Integrals

$$\int a\, du = a\int du = au + C$$

$$\int (u+v)\, dx = \int u\, dx + \int v\, dx$$

$$\int u\, dv = uv - \int u\, du \text{ (integration by parts)}$$

$$\int dy \int f(x, y)\, dx = \int dx \int f(x, y)\, dy$$

$$\int x^n dx = \frac{x^{n+1}}{n+1} + C, \quad \text{when } n \neq -1$$

$$\int \frac{dx}{x} = \ln x + C = \ln cx$$

$$\int e^x dx = e^x + C$$

$$\int \sin x\, dx = -\cos x + C$$

$$\int \cos x\, dx = \sin x + C$$

$$\int \frac{dx}{\sin^2 x} = -\cot x + C$$

$$\int \frac{dx}{\cos^2 x} = \tan x + C$$

$$\int \frac{dx}{\sqrt{1-x^2}} = \sin^{-1} x + C = -\cos^{-1} x + c$$

$$\int \frac{dx}{1+x^2} = \tan^{-1} x + C = -\cot^{-1} x + c$$

$$\int (a+bx)^n dx = \frac{(a+bx)^{n+1}}{(n+1)b} + C$$

$$\int \frac{dx}{a+bx} = \frac{1}{b}\ln(a+bx) + C = \frac{1}{b}\ln c(a+bx)$$

$$\int \frac{dx}{(a+bx)^2} = \frac{-1}{b(a+bx)} + C$$

$$\int \frac{dx}{1-x^2} = \frac{1}{2}\ln\frac{1+x}{1-x} + C = \tanh^{-1} x + C, \quad \text{when } x < 1$$

$$\int \frac{dx}{x^2-1} = \frac{1}{2}\ln\frac{x-1}{x+1} + C = -\coth^{-1} x + C, \quad \text{when } x > 1$$

$$\int \frac{dx}{a+2bx+cx^2} = \frac{1}{\sqrt{ac-b^2}}\tan^{-1}\frac{b+cx}{\sqrt{ac-b^2}} + C, \quad [ac-b^2 > 0]$$

$$= \frac{1}{2\sqrt{b^2-ac}}\ln\frac{\sqrt{b^2-ac}-b-cx}{\sqrt{b^2-ac}+b+cx} + C$$

$$= -\frac{1}{\sqrt{b^2-ac}}\tanh^{-1}\frac{b+cx}{\sqrt{b^2-ac}} + C \quad [b^2-ac > 0]$$

$$\int \sqrt{a+bx}\,dx = \frac{2}{3b}\left(\sqrt{a+bx}\right)^3 + C$$

$$\int \frac{dx}{\sqrt{a+bx}} = \frac{2}{b}\sqrt{a+bx} + C$$

$$\int \frac{(m+nx)\,dx}{\sqrt{a+bx}} = \frac{2}{3b^2}(3mb-2an+nbx)\sqrt{a+bx} + C$$

$$\int \sqrt{a+2bx+cx^2}\,dx = \frac{b+cx}{2c}\sqrt{a+2bx+cx^2} + \frac{ac-b^2}{2c}\int\frac{dx}{\sqrt{a+2bx+cx^2}} + C$$

$$\int a^x\,dx = \frac{a^x}{\ln a} + C$$

$$\int x^n e^{ax}\,dx = \frac{x^n e^{ax}}{a}\left[1-\frac{n}{ax}+\frac{n(n-1)}{a^2x^2}-\ldots\pm\frac{n!}{a^n x^n}\right] + C$$

$$\int \ln x\,dx = x\ln x - x + C$$

$$\int \frac{\ln x}{x^2}\,dx = -\frac{\ln x}{x} - \frac{1}{x} + C$$

$$\int \frac{(\ln x)^n}{x}\,dx = \frac{1}{n+1}(\ln x)^{n+1} + C$$

$$\int \sin^2 x\,dx = -\frac{1}{4}\sin 2x + \frac{1}{2}x + C = -\frac{1}{2}\sin x\cos x + \frac{1}{2}x + C$$

$$\int \cos^2 x\,dx = \frac{1}{4}\sin 2x + \frac{1}{2}x + C = \frac{1}{2}\sin x\cos x + \frac{1}{2}x + C$$

$$\int \sin mx\,dx = -\frac{\cos mx}{m} + C$$

$$\int \cos mx\,dx = \frac{\sin mx}{m} + C$$

$$\int \sin mx \cos nx \, dx = \frac{\cos(m+n)x}{2(m+n)} - \frac{\cos(m-n)x}{2(m-n)} + C$$

$$\int \sin mx \sin nx \, dx = \frac{\sin(m-n)x}{2(m-n)} - \frac{\sin(m+n)x}{2(m+n)} + C$$

$$\int \cos mx \cos nx \, dx = \frac{\sin(m-n)x}{2(m-n)} + \frac{\sin(m+n)x}{2(m+n)} + C$$

$$\int \tan x \, dx = -\ln \cos x + C$$

$$\int \cot x \, dx = \ln \sin x + C$$

$$\int \frac{dx}{\sin x} = \ln \tan \frac{x}{2} + C$$

$$\int \frac{dx}{\cos x} = \ln \tan\left(\frac{\pi}{4} + \frac{\pi}{2}\right) + C$$

$$\int \frac{dx}{1+\cos x} = \tan \frac{x}{2} + C$$

$$\int \frac{dx}{1-\cos x} = -\cot \frac{x}{2} + C$$

$$\int \sin x \cos x \, dx = \frac{1}{2}\sin^2 x + C$$

$$\int \frac{dx}{\sin x \cos x} = \ln \tan x + C$$

$$\int \frac{\cos x \, dx}{a + b\cos x} = \frac{x}{b} - \frac{a}{b}\int \frac{dx}{a + b\cos x} + C$$

$$\int \frac{\sin x \, dx}{a + b\cos x} = -\frac{1}{b}\ln(a + b\cos x) + C$$

$$\int \frac{A + B\cos x + C\sin x}{a + b\cos x + c\sin x} dx = A\int \frac{dy}{a + p\cos y} + (B\cos u + C\sin u)\int \frac{\cos y \, dy}{a + p\cos y}$$

$$-(B\sin u - C\cos u)\int \frac{\sin y \, dy}{a + p\cos y},$$ where $b = p\cos u$, $c = p\sin u$ and $x - u = y$.

$$\int e^{ax} \sin bx \, dx = \frac{a\sin bx - b\cos bx}{a^2 + b^2} e^{ax} + C$$

$$\int e^{ax} \cos bx \, dx = \frac{a\cos bx + b\sin bx}{a^2 + b^2} e^{ax} + C$$

$$\int \sinh x \, dx = \cosh x + C$$

$$\int \tanh x \, dx = \ln \cosh x + C$$

$$\int \cosh x \, dx = \sinh x + C$$

$$\int \coth x \, dx = \ln \sinh x + C$$

Glossary of Terms

Appendix C

Terminology, used frequently in the field of Road Vehicle Dynamics and Accident Reconstruction, is compiled here from various sources. For a more comprehensive list of glossary of terms and acronyms, the author recommends the readers to refer to author's companion book: *Dictionary of road Vehicle Dynamics and Accident Reconstruction*, Product Code: R-xxx, SAE International, Warrendale, PA, 2009.

A, B, C, D, Pillars and Posts: The vertical pillars and posts of a light vehicle form the major vertical structural members of a body. Pillars typically are at the window height; posts are below the window height. From front to rear, the A post/pillar is the most forward number, the B post/pillar is the second most forward vertical member, etc. See Figure A.1.

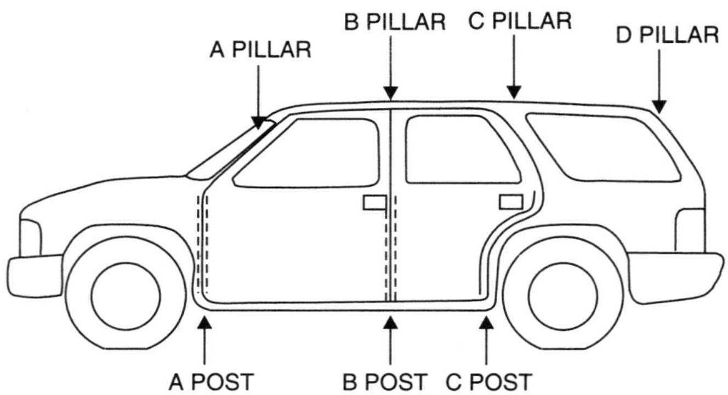

Fig. A.1 *A*, *B*, *C*, *D*, pillars and posts.

ABS Antilock Braking System: ABS antilock braking systems are closed loop control devices within the braking system which prevent wheel lock-up during braking and, as a result, retain the vehicle's steerability and stability.

ABS: Antilock braking system. This is a braking system designed to brake the vehicle in the most effective manner, without locking up the tires and causing them to go into a skid. This system is used differently than standard brakes while braking in an emergency situation. The brake pedal is stepped on and held in place while the electronic system takes over and modulates or applies the brakes until the wheels want to lock up, and then releases and brakes again. The driver does not have to do anything else. The old system required the driver to pump his/her brakes, thus modulating the brakes and doing the same thing as the ABS system.

Absorber: Absorber is an auxiliary vibrating system for modifying the vibrational characteristics of the main system, either damped or undamped.

Acceleration Drag Factor: See Drag Factor.

Acceleration due To Gravity: Ratio between the weight of a body and the mass of the body. The earth's gravitational constant is 32.2 feet/sec/sec (32.22) or 9.81meter/sec/sec (9.812).

Acceleration Factor: The acceleration factor θ is a measure of the freedom with which a wheel turns about its axle. Limpert has determined an empirical formula for rolling resistance as a function of tire type, inflation pressure, empirical coefficients based on the road surface and the speed of the vehicle when the wheel rolls. For example, on paved roadways, the rolling resistance for the ith wheel is given by

$$\theta_i = -\left[a_i + \frac{0.15}{p_i} + \frac{b_i}{p_i}\left(\frac{V}{100}\right)^2\right]$$

where V is the speed of the vehicle in mph, p_i is the tire inflation pressure in *psi*, and a_i and b_i are coefficients based on tire type. For radial tires, the coefficients are reported to be 0.005 and 0.67 respectively. For bias ply tires, they are 0.009 and 1.0. The velocity V is only a factor at high speed (> 75 mph) when standing waves are created.

Acceleration Marks: Marks that are just the opposite of skid marks. The tires are being rotated by the axle and when done fast enough, the outside of the tire, tread takes time to catch up to the rest of the tire, which is being held by the coefficient of friction between the tires and the road surface. Therefore, the acceleration marks are heavy at the beginning and lighten up as the tire tread catches up. See Skid Mark.

Gravitational Acceleration: The acceleration of a falling body due to the force of gravity (its weight); a standard of comparison for acceleration; usually taken as 32.2 ft per sec^2 or 9.81 m per sec^2.

Acceleration Scuff: Friction marks that occur when the tires on the drive wheels of the vehicle turn faster than they would normally turn, given the speed of the vehicle.

Acceleration Skid: Comes when sufficient power is applied to the drive wheels.

Acceleration Spin Marks: Acceleration spin marks occur when the tread speed is greater than the forward speed of the vehicle. They show pronounced longitudinal striation at the beginning of the accelerative skid marks. The acceleration mark stops when the tread speed equals the forward speed of the vehicle (when the tire is no longer spinning).

Acceleration: Time rate of change of velocity; change of velocity divided by time; a vector quantity measured in feet per second per second (ft per sec^2) or meters per second per second (m per sec^2) or expressed as a decimal fraction of acceleration due to gravity; $a = fg$. Positive acceleration increases the velocity of a body; negative acceleration (deceleration) reduces the velocity of a body. If a vehicle's velocity is constant, then its acceleration is zero. Acceleration results if an object speeds up, slows down, or changes direction. The definition of acceleration is given by $a = \Delta V/\Delta t$.

Accelerator: Pedal by which the fuel flows to the engine is controlled, depression of the pedal causing the vehicle to accelerate.

Accident Analysis: See Cause Analysis.

Accident Avoidance Analysis: With the overall accident circumstances established, such as pre-impact and impact speeds, as well as distance prior to impact, the accident avoidance analysis is conducted to determine if and how the accident could have been avoided or made less severe.

Accident Investigation: The detailed process of observation, acquisition, and documentation of physical evidence and other information involving an accident or crash is known as accident investigation.

Accident Reconstruction Software: Accident Reconstruction Software is used to recreate crash and crime scenes and is regularly used by law enforcement and accident reconstructionists in court to show what occurred in an accident.

Accident Reconstruction: A procedure carried out with the specific purpose of estimating in both a qualitative and quantitative manner how an accident occurred using engineering, scientific, and mathematical principles, based on evidence obtained through accident investigation. In other words, accident reconstruction is a process conducted with the specific objective of estimating in both a qualitative and quantitative manner how an accident occurred using engineering, scientific, and mathematical principles also based on evidences obtained through accident investigation. The general objective of accident reconstruction is to describe the events of the accident and to determine what happened at a particular point in time. It includes the investigation of the circumstances of an accident. These circumstances include the effect of human, vehicle, and environmental factors on the pre-impact, impact, and post-impact events. Specific objectives of accident reconstruction include the following: collision analysis, injury analysis, accident avoidance analysis, injury avoidance analysis, and accident causation analysis. The ultimate objective of accident reconstruction is product or process improvement for future risk reduction.

Accident Reconstructionist: The accident reconstructionist takes the evidence along with all the other data from various sources and rebuilds the accident. The data involves the objective data considered as facts, the engineering analysis and the witness statements. The objective data includes facts such as vehicles, drivers, and roadway involved in the accident, scene photographs and measurements. Webster's New Universal Unabridged Dictionary defines a reconstructionist as a person who builds up from remaining parts and other evidence an image of what was in its original and complete form. The definition does not use the word recreate. In general, it is impossible to recreate even the simplest of accidents.

Accident Scene: Accident scene is the place where a vehicular accident occurs, both during and immediately following the accident, and before vehicles and participants have departed. The accident scene can be defined as the central location of the accident, the final resting place of the vehicles involved, the skid marks and the accident debris. Taken all together, these factors constitute the accident scene.

Accident Site: Accident site refers to a place where an accident occurred, after vehicles and participants or persons involved have departed from the accident scene. See Accident Scene.

Accident: Accident is an event in which one or more vehicles undergo unexpected action(s), generally involving physical contact with another vehicle or other object, producing injury, death, and/or property damage. An unintended event that results in damage or injury. See also Crash and Collision.

Adhesion: Adhesion is a molecular mechanism attempting to pull two surfaces together, similar to water adhering to a surface. In road vehicle dynamics, adhesion is the grip between road and tire. It is proportional to the static coefficient of friction.

Aerodynamic Resistance: The aerodynamic resistance is generated mainly by air flow around the vehicle and air flow through the vehicle. Aerodynamic resistance is estimated by

$$R_a = \frac{\rho}{2} C_D A_F V_r^2,$$

where ρ is the density of the air, C_{D} is the drag coefficient, A_F is the projected area of the front of the vehicle and V_r is the relative velocity of the vehicle with respect to the wind. The coefficient of drag ranges between 0.3 and 0.5 for passenger cars and can reach the value of 0.8 for vans.

Aerodynamics: Aerodynamics deals with all processes which are observed as airflows through and around a vehicle.

After Accident Situation Map: A scale drawing of the scene after a traffic-accident situation has stabilized; a graphical summary of results of an accident without indication of any assumed events of the accident or other inferential elements.

Air Resistance: The force opposing movement of a body through the air surrounding it. Due to the relatively low speeds and the short airborne time periods dealt within vehicle collisions, the effects of wind and air resistance are negligible.

Aligning Stiffness (Aligning Torque Stiffness): The rate of change of aligning torque with respect to change in slip angle, usually evaluated at zero slip angle.

Aligning Stiffness Coefficient (Aligning Torque Coefficient): The ratio of aligning stiffness of a straight free-rolling tire to the vertical load.

Aligning Torque (Aligning Moment) (M_z): The component of the tire moment vector tending to rotate the tire about the Z' axis, positive clockwise when looking in the positive direction of the Z' axis. See Figure S.5.

Alignment Angles: Camber, caster, toe, turning radius, and steering axis inclination. All these angles must be set correctly to ensure proper handling and tire wear.

Alignment: The process of adjusting the position of the tires and steering axis to bring them to a specified predetermined position.

Alley: A street or highway intended to provide access to the rear or side of lots or buildings in urban districts and not intended for the purpose of thorough vehicular traffic.

Ambient Conditions: Included are weather conditions such as sun, rain, snow, or dust, wind gust, noise (for example, noise at airport), temperature, time of day, moon position, fog, and cloudiness.

Ambient Wind Angle (ν_a): The angle between the X axis of the earth-fixed axis system and the ambient wind velocity vector. A positive ambient wind angle is shown in Fig. A.2.

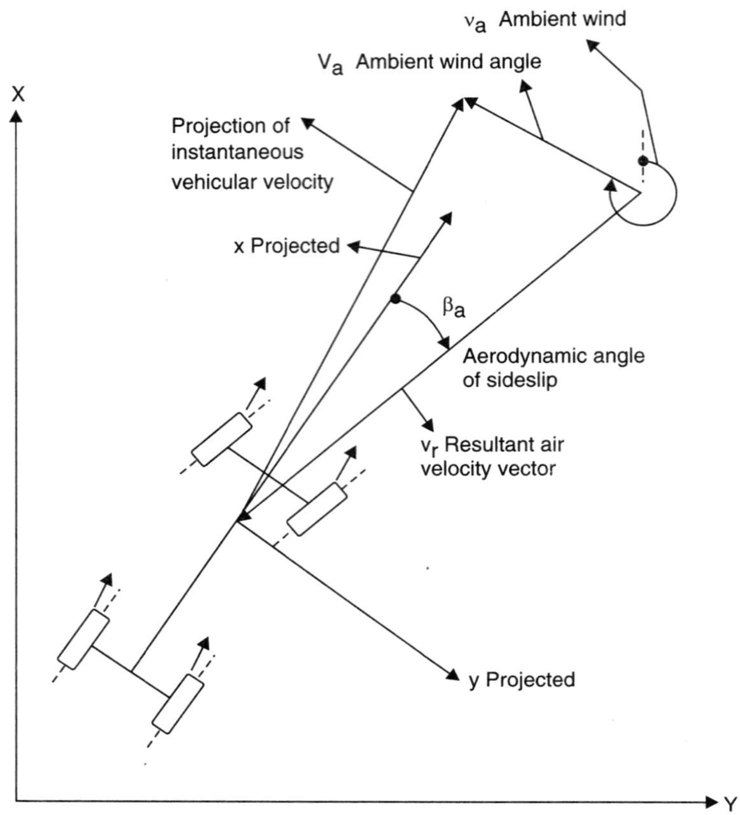

Fig. A.2 Wind vectors.

Ambient Wind Velocity: The horizontal component of the air mass velocity relative to the earth-fixed axis system in the vicinity of the vehicle.

Amplification Factor: See Magnification Ratio.

Amplitude Ratio (Relative Magnification Factor): The ratio of forced vibration amplitude to the static amplitude. Amplitude ratio or magnification factor is the ratio of the maximum force developed in the spring of a mass-spring-dashpot system to the maximum value of the exerting force.

Amplitude: Amplitude is the peak (maximum) value of a periodically varying quantity.

Angle Collision: A collision between two traffic units approaching on separate roadways or other paths that intersect.

Angular Acceleration: It is the rate of increase of rotational (angular) velocity, generally expressed in radians per second per second. The time rate of change of angular velocity.

Angular Displacement: The angle turned through by a body about a given axis, generally measured in degrees.

Angular Frequency (Circular Frequency): The angular frequency of a periodic quantity, in radians per unit time, is the frequency multiplied by 2π.

Angular Impact: When two vehicles collide in such a manner that their respective directions of force are not parallel.

Angular Impulse: Angular impulse is the product of the impulse and the distance from the force to the center of the center of moments.

Angular Momentum: A quantity associated with how an object moves around a reference point. It is often used to describe rotating objects. The angular momentum of an object is defined to be equal to its mass times its velocity about the point times its distance from that point.

Angular Orientation: The orientation of the vehicle axis system (x, y, z) with respect to the earth-fixed axis system (X, Y, Z) is given by a sequence of three angular rotations. The following sequence of rotations starting from a condition, in which the two sets of axes are initially aligned, is defined to be the standard:

1. A yaw rotation, ψ, about the aligned z- and Z-axis
2. A pitch rotation, Θ, about the vehicle y-axis
3. A roll rotation, ϕ, about the vehicle x-axis

Angular rotations are positive clockwise when looking in the positive direction of the axis about which rotation occurs.

Angular Velocity Change: The angular velocity change is the rotational speed given to the vehicle by the force created by the collision. This force usually creates a moment about the center of gravity. The velocity change at the location the force is different than that at the center of gravity. The difference in the two velocities changes is a function of the angular velocity change and the distance from the center of crush to the center of gravity. Refer to a dynamics text on the relationships which apply. The angular velocity change has the same sign (direction) as the moment arm.

Angular Velocity: The rate of change of angular displacement, expressed in radians per second or revolutions per minute.

Animation: A computer program that allows the reconstructionist to develop a videotape of an accident sequence to be used as an exhibition of his/her opinion of the accident scenario.

Anticipation: The amount of information available to the driver. Objectively defined here by whether the subject was aware of the stimulus, and/or appropriate response and whether the subject responded to one event or had to respond multiple times.

Antilock Braking System (ABS): Aggregate of all devices within the service-braking system which provide automatic slip control (in the direction of rotation) for one or several wheels under braking conditions. The braking force at wheels featuring direct control is controlled using data provided by the wheel's own wheel-speed sensor, while indirectly controlled wheels rely on data provided by the wheel-speed sensor(s) at one or more of the other wheels. An ABS system with "Select High" control has directly and indirectly controlled wheels. With an ABS system with "Select Low" control, all wheels with speed sensors are directly controlled.

Anti-Rollover Braking (ARB) System: Anti-rollover braking system monitors vehicle motion. When rollover is about to start, ARB controls the front brake. The brakes reduce the cornering capability of the front tires, which causes the vehicle to turn less sharply and also reduces its speed. When the vehicle passes the rollover warning threshold, the ARB stops control and the brakes are released.

Apogee: The maximum altitude a projectile will reach when shot in the air. Used here to define the maximum height the motorcycle driver or rider will reach when impacting an object, usually a vehicle.

A-Post: See *A, B, C, D*, Pillars and Posts. See Figure A.1.

Application Force: In friction brakes, the total force applied to a brake lining, which causes the. braking force by friction effect.

Applied Force: A force that is produced by direct contact between two bodies is an applied force.

Approach Angle: As a vehicle approaches a grade, it's the angle of slope which it can climb up without any part of the front (bumper, shackle etc.) hitting the ground/rock.

Approach Speed: Speed of a vehicle just prior to the first significant event such as contact in an accident. See Closing Speed.

Aquaplaning: See Hydroplaning.

Arc: Part of a curve, especially part of a circle, between two points on the curve.

Area of Impact: Area encompassed by the interface between colliding objects projected onto the road. See Impact Center.

Articulated Vehicle: A vehicle jointed or joined together permanently or semi permanently by means of a pivot connection for operating separate segments as a unit.

Articulation: The extent of movement between one tire and the other on the same axle.

Asphalt: See Bituminous Pavement.

Asphalted Concrete: Concrete cemented with a bituminous material such as tar or asphalt.

At-Scene Inquiry: Questioning informants at the scene of a traffic accident for the purpose of obtaining facts required for the official accident report and such other information as may be readily available.

At-Scene Investigation: Examining and recording results of the accident and obtaining additional information at the scene of a traffic accident which may not be available later and which supplements data obtained for the accident report. The information obtained is factual as far as possible.

Attitudes: More or less permanent ways of thinking and feeling about things.

Attribute: An attribute is any inherent characteristic of a road, a vehicle, or a person that affects the probability of a traffic accident.

Automatic Braking System: All the elements which automatically brake the towed vehicle as a result of intended or accidental separation from the towing vehicle.

Automobile Safety: Automobile safety is the avoidance of automobile accidents or the minimization of harmful effects of accidents, in particular as pertaining to human life and health. Numerous safety features have been built into cars for years, some for the safety of car's occupants only, some for the safety of others.

Auxiliary Brake: An auxiliary brake is a continuous brake in which the retarding torque is not generated by friction between two sliding surfaces, such as linings and a drum.

Available Path: The entire area in a traffic way in which a traffic unit may maneuver without interfering with other traffic units. Available path is not the same as the road space which a vehicle is legally entitled to travel.

Average Acceleration: Calculated by dividing the change in velocity of an object by the time that is required to gain that change in velocity.

Average Friction: This is the average of the specified friction range.

Average Theta: The lockup θ is the average of the four values centered for the vehicle. The minimum value for θ is 0.01 as the frictionless wheel has not been invented yet. Fully locked up wheels give θ = 1.0.

Axis of Rotation: In a rotating object, the axis of rotation is the line about which the object turns. The axis of rotation for a spinning record is in its middle. The reference point for angular momentum is often chosen to be along the axis of rotation.

Axis: A straight line around which an object rotates. Also a reference line from which or along which distances or angles are measured in a system of co-ordinates.

Axle Loads: The vertical forces (axle loads) are given independently by

$$F_{z1} = Mg\left(\frac{c_2}{c_1 + c_2}\right) - \frac{h}{c_1 + c_2} ma$$

$$F_{z2} = Mg\left(\frac{c_1}{c_1 + c_2}\right) + \frac{h}{c_1 + c_2} ma$$

where F_{x1} is the friction force on the front wheels, F_{x2} is the friction force on the rear wheels, c_1 is the horizontal (parallel to the road surface) distance between the contact point of the front wheel and the center of gravity, c_2 is the horizontal distance between the contact point of the rear wheel and the center of gravity, and a is its acceleration.

Note that we are assuming that the vehicle is accelerating. Merely deceleration will be treated simply by using a negative value for a. It should also be noted that when the vehicle is braking, negative a is used; the axle force on the front wheels will be larger than those forces experienced under cruising driving conditions (*i.e.*, no acceleration or deceleration). This is felt by drivers, particularly when the vehicle reaches a complete stop. See Figure A.3.

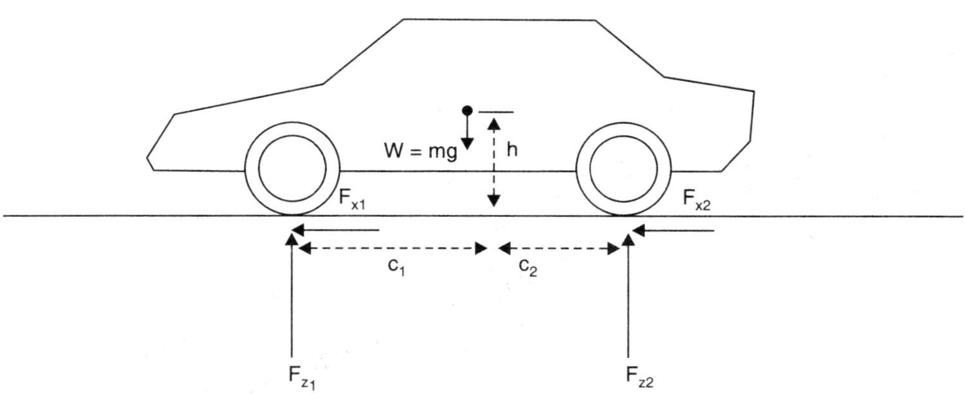

Fig. A.3 Simple vehicle model on a horizontal plane.

Axle Side Shake: Oscillatory motion of an axle, which consists of transverse displacement.

Axle Windup: Oscillatory motion of an axle about the horizontal transverse axis through its center of gravity.

Axle Yaw: Oscillatory motion of an axle around the vertical axis through its center of gravity.

Axle: A cross support that is designed to carry the weight of the car. Axle is a cross-shaft which carries the driving or freely mounted wheels of a vehicle.

Backlite Header: The structural body member which connects the upper portions of the rearmost driver and passenger pillars and forms the top edge of the backlite (back window).

Backlite: The rear or back window, which spans from the driver's to passenger's side of the vehicle.

Backup Collisions: Backup collisions happen when a driver reverses the vehicle into an object, person, other vehicle.

Balance Tire: A condition in which the tire can spin without causing a vibration of the suspension or car.

Balance: All forces generated by or acting on the rotating element of a machine-train are in a state of equilibrium.

Ball Joint: A connector consisting of a ball and socket. This configuration allows for angular and rotating motion at the same time.

Bank: The degree to which the outside edge of a *roadway* is higher than the inside edge at a specified point on a curve; the change in elevation per unit distance across the *roadway* from the inside of a curve to the outside edge; superelevation. Bank is designated in feet (meters) of rise, positive bank, or fall, negative bank, per foot (meter) of level distance from the inside to the outside of the curve. *Bank* may also be designated by rise as a percent of level distance. See Superelevation.

Barrier Equivalent Velocity (BEV): The forward speed and corresponding kinetic energy with which a vehicle contacts a flat fixed rigid barrier at 90° with no rebound. See also Equivalent Energy Speed (EES).

Barrier Line: A line which, when placed parallel to a *center* or *lane line,* indicates that all *traffic* must not cross the line for purposes of overtaking and passing; a double line consisting of two normal solid yellow lines delineating the separation between travel paths in opposite directions where overtaking and passing is prohibited in both directions; a two-direction no-passing marking. Crossing this marking with care is permitted only as a left turn maneuver. It is frequently used as a channelizing line in advance of an obstruction which must be passed on the right and to form a channelizing island separating traffic in counter directions.

Barrier: Massive or firmly located vertical obstruction for test evaluation of vehicle impact.

Bituminous Concrete: Concrete cemented with a bituminous material such as tar or asphalt.

Bituminous Pavement: A pavement comprising an upper layer or layers of aggregate with a bituminous binder (asphalt, coal tars, natural tars, etc.) and surface treatments such as chip seals, slurry seals, sand seals, and cape seals.

Brake Drum: A drum or pulley attached to a wheel (or shaft), to which is applied an external band or internal brake shoes.

Brake Fade: Repeated application of the brakes causes a loss of friction ability due to heat, resulting in the impairment of braking efficiency. If a vehicle is subjected to a series of severe stops in rapid succession, in many cases each successive stop requires a higher pedal force to maintain a specified deceleration level. This phenomenon is called *brake fade*.

Brake Feel: A customer's "brake feel" is related mostly to the brake force to generate a braking action.

Brake Hop: An oscillatory hopping motion of a single wheel or of a pair of wheels which occurs when brakes are applied in forward or reverse motion of the vehicle.

Brake Reaction Time: The time from the moment the stimulus is perceived until the moment the brake is applied in response to it.

Brake Slip: See Wheel Slip.

Brake: Parts of a braking system in which the forces opposing the vehicle's movement, or its tendency towards movement, are developed, such as friction brakes (disk or drum) or retarders (hydrodynamic or electrodynamic retarders, engine brakes).

Braking Efficiency: Braking efficiency, n, is defined as the ratio between the actual braking force generated, F, to the maximum amount of force that could be applied (μW), so $n = F/\mu W$.

Brake-Hysteresis: Difference in application force between application and release for the same braking torque.

Braking (Driving) Squeal: The squeal resulting from longitudinal slip.

Braking (Driving) Stiffness Coefficient: The ratio of braking (driving) stiffness of a straight free-rolling tire to the vertical load.

Braking (Driving) Stiffness: The rate of change of longitudinal force with respect to change in longitudinal slip, usually evaluated at zero longitudinal slip.

Braking Deceleration: Reduction of speed obtained by the braking system in the considered time, t. The following can be identified:

Braking Distance and Deceleration: Among the attributes in which braking performance is measured are the braking distance and deceleration. If the vehicle is moving as a result of a consistent braking force changing its velocity from V_i to V_f (which would be zero in the case of a complete stop), the acceleration can be calculated as: $a = (V_f - V_i)/t$, where t is time. Note that the above acceleration will have a negative value when the vehicle is decelerating. If the total braking force on both axles is F_b, then from Newton's second law of motion, the deceleration is F_b/M; where M is the mass of the vehicle. The distance traveled during braking (X_b) can be calculated by

$$X_b = \frac{(V_f^{\,2} - V_i^2)}{2 \times a}$$

Braking Distance: The distance through which brakes are applied to slow a vehicle; the shortest distance in which a particular vehicle can be stopped by braking from a specified speed on a particular surface; the distance from application of brakes to collision. See Stopping Distance.

Braking Disturbances: Braking disturbances refer to any disturbances within or initiated by the braking system of the vehicle.

Braking Effort: The braking effort is given by

$$F_x = \mu_p W \left(1 - \frac{\mu_p W}{4iC}\right)$$

Note the nonlinearity between the baking effort and the skid. The above equation applies until the maximum peak braking effort value is obtained. At that point, lock-up happens and skid moves rapidly from the 20% or so skid to 100% almost instantly with a reduced coefficient of friction.

Braking Equipment: All the braking systems fitted to a vehicle, whose function is to reduce its speed or bring it to a halt, or to hold the vehicle stationary if already halted.

Braking Factor z**:** Ratio between the total braking force, F_t, and the static weight, G_s, on the axle or the axles of the vehicle:

$$z = F_t / G_s$$

Braking Force Coefficient (Braking Coefficient): The ratio of the braking force to the vertical load.

Peak Braking Force: The largest force that can be developed during brake application as wheel slip is varied over the range of free rolling slip to locked wheel slip.

Braking Force: The force over the contact surface between a tire and a road in the direction of heading of the braked wheel that develops as a result of brake application. The negative longitudinal force results from braking torque application.

Braking Skidmark: See Skidmark.

Braking Time: Time required in traversing the braking distance.

Braking Torque: Product of the frictional forces resulting from the application force and the distance between the points of application of these forces and the axis of rotation of the wheels.

Braking Traction Coefficient: The maximum of the braking force coefficient, which can be reached without locking a wheel on a given tire and road surface for a given environment and operating condition.

Braking Work W: Integral of the product of the instantaneous total braking force, F_t and the elementary movement, d_s, over the distance travelled during braking, s.

$$W = \int_0^s F_t . ds$$

Braking-Force Distribution: Specification of braking force according to axle, given as % of the total braking force F_t.

Braking-System Hysteresis: Difference in control forces between application and release for the same braking torque.

Break-In: To run new or reconditioned machinery under light load for a preliminary period.

Breaking Load: See Rupture Load.

Breakover Angle: When a vehicle crests a hill, it's the angle at which the front tires still remain on the ground as the rear wheels continue to climb (See High Centered).

Bump Steer: A directional change in steering caused by road surface irregularities. As the suspension moves through jounce and rebound, changes in alignment at the front or rear wheels may alter the vehicle's path.

Bump Stop: An elastic member which increases the spring rate near the end of the compression and extension travel to reduce the effects of bottoming and/or topping.

Bus: Any motor vehicle designed for carrying more than 10 passengers and used for transporting persons; any motor vehicle other than a taxicab, designed and used for transporting persons for compensation.

Calculation Drag: Depending on the type of trajectory, the drag factor is either assigned by the program or calculated. For rollout, the drag factor is $\mu\theta$. For non-spinning skids, it is based on a friction circle which relates the heading of the vehicle to the path and modifies θ accordingly. For spinning skids, the drag factor is calculated from the change in speed between separations to skid end.

Camber Angle: Camber angle is defined as the inclination of a wheel outward from the body. Camber on a wheel will produce a lateral force known as "*camber thrust*. Typically, camber angle generates much less lateral force than slip angle. About 4 to 6 degrees of camber only produce the same lateral force as 1 degree of slip angle on a bias-ply tire. It is considered positive when the wheel leans outward at the top and negative when it leans inward. The inclination of the wheel plane to the vertical is considered positive when the wheel leans outward at the top and negative when it leans inward.

Camber Stiffness: The rate of change of lateral force with respect to change in inclination angle, usually evaluated at zero inclination angles.

Camber: When a tire is tilted so that it is not straight up and down, it is said to have *camber*. Camber is defined as negative, if the top of the tire tilts towards the center of the vehicle. Most vehicles in use today have little or no camber except for certain solid axle and swing axle designs.

Car Accident: A car accident is a road traffic incident which usually involves one road vehicle colliding with another vehicle, another road user, or a stationary roadside object, and which may result in injury or property damage, or possibly death. Phrases used to describe accidents include: *auto accident, car crash, car smash, car wreck, fender bender, motor vehicle accident (MVA), personal injury collision (PIC), road accident, road traffic accident (RTA), road traffic collision (RTC), road traffic incident (RTI), smash-up, bingle* and *traffic collision.*

Car Crash: See Car Accident.

Car Handling: Car handling and vehicle handling is a description of the way wheeled vehicles perform, transverse to their direction of motion, particularly during cornering and swerving. It also includes their stability when moving in a straight line. Handling and braking are the major components of a vehicle's "*active*" safety. The maximum lateral acceleration is sometimes discussed separately as "*road holding*". Handling is an esoteric performance area because rapid and violent maneuvers are often only used in unforeseen circumstances. This discussion is directed at road vehicles with at least three wheels, but some of it may apply to other ground vehicles.

Causation Analysis: In the causation analysis, all accident causation factors are studied based upon reasonable probability considering driver(s), vehicle(s), road, and environment.

Cause Analysis: Cause analysis is defined as the effort to determine, from whatever information is available, why an accident occurred, that is, the complete combination of factors that caused the highway transportation system to malfunction at the time and place of the accident with resultant injury or damage.

Cause: The combination of simultaneous and sequential factors without any one of which the result could not have occurred. The result might be a motor-vehicle traffic accident, or a successful trip.

CDC Number: The Collision Damage Classification (CDC) is an excellent tool for statistical studies which relate injury potential to collision severity when the resources are not available to completely reconstruct the accident. An SAE recommended practice- SAE J224 MAR80– defines the CDC number and how to create and interpret it.

As an example, "02FZEW4" would be a possible CDC code.

The first two columns represent the clock direction of the Principal Direction of Force- PDOF. In this case, the force is coming from the 2 o'clock direction.

Column 3 is the Area of Damage- In this case, *F* is for frontal damage.

Column 4 is the specific longitudinal or lateral area- In this case, *Z* indicates the right two thirds of the front are damaged.

Column 5 is the specific vertical or lateral area- In this case *E* refers to damage restricted to below the beltline.

Column 6 is type of damage distribution- In this case *W* indicates a wide distribution in the specific area.

Column 7 is the damage extent. In this case the vehicle is divided into nine zones with damage occurring to zone 4.

Center of Gravity (*cg*): That point of a body through which the resultant force of gravity (weight) acts irrespective of the orientation of the body.

Center of Impact: See Impact Center.

Center of Mass: See Center of Gravity.

Center of Tire Contact: The intersection of the wheel plane and the vertical projection of the spin axis of the wheel onto the road plane. The center of tire contact may not be the geometric center of the tire contact area due to distortion of the tire produced by applied forces.

Center Wear: Type of irregular tire wear characterized by increasing wear from shoulder to center of tread band.

Centered Force: A force generated in a collision that is directed through the center of gravity of vehicle. See also Eccentric Force.

Centered Thrust: A thrust directed toward the center of mass of the vehicle or other object in collision.

Central Force: The component of the tire force vector in the direction perpendicular to the direction of travel of the center of tire contact. Central force is equal to lateral force times cosine of slip angle minus longitudinal force times sine of slip angle.

Central Impact: A central impact is one where the collision force is approximately directed on a line through the center of gravity of the two vehicles.

Centrifugal Force: The force of a body in motion which tends to keep it continuing in the same direction rather than following a curved path. Force acting on a turning body that pushes the body outward. If a body rotates at the end of an arm, the force is provided by the tension in the arm. The reaction to this force acts at the center of rotation and is called the centrifugal force. It represents the inertia force of the body, resisting the change in the direction of its motion.

Chain-Reaction Accident: A series of two or more traffic accidents occurring one after another in the same vicinity in such a manner that, after one accident has reached a stabilized accident situation, its results contribute to the occurrence of a subsequent traffic accident. If there is no clear indication that a stabilized accident situation has been reached, the series of events is considered to be a single accident, although possibly a multiunit accident.

Change in Momentum: The difference of the momentum (product of mass and velocity) of a mass from one time to another. See also Conservation of Momentum.

Change of Velocity: The difference between velocity vectors at two points in time. See also Delta-v.

Characteristic Speed: That forward speed for an understeer vehicle at which the steering sensitivity at zero lateral acceleration trim is one-half the steering sensitivity of a neutral steer vehicle.

Chassis: The functions of the chassis are to produce the forces between the tires and road, to carry the vehicle weight, to drive, accelerate or brake the vehicle, and to direct the vehicle in turns. The chassis provides the basic unit carrying the road loads. The chassis consists of a frame or similar structure containing the engine, clutch, transmission drive shaft and differential, the drive axle, the wheels, and the steering and braking systems.

Chip: A short, deep gouge; a hole in pavement made by a strong, sharp, pointed metal object under great pressure, usually without striations.

Chop Mark: A broad shallow gauge even and regular on the deeper side and terminating in scratches and striations on the opposite side.

Chop: A broad shallow gouge in a road surface, beginning with an even, regular, deeper side and terminating in scratches and striations on the opposite shallower side; a depression in pavement made by strong, sharp metal edge moving under heavy pressure, more commonly occurring at an impact event as opposed to post impact trajectory.

Chord: A straight line connecting the ends of an arc or two points on a curve. A chord is never greater than the diameter of a circle of which the arc is a part.

Circular Frequency: See Angular Frequency.

Circular Motion: See Rotational Particle Motion.

Circular Vibration: The motion of a body, of which any point moves in a circle and in which any line maintains a constant direction; the motion is resolvable into two components of equal magnitude in space and time quadrature.

Closing Speed: The relative approach velocity is the sum of the two speed changes as calculated at the point where the force is applied to both vehicles. The approach velocity is along the common line which defines the direction of principal force. In some circumstances, it will be useful as an estimate of one of the vehicle's pre-impact speed. Unless the collisions were central, the relative approach velocity is always greater than the sum of the speed changes at the centers of gravity.

Closing Velocity: The magnitude of the relative velocity between two vehicles at a given point in time as they approach each other; the magnitude of the relative velocity between two vehicles at the beginning of a crash; the vector difference between the velocity of the vehicle and the vehicle/object struck immediately before impact.

Coefficient of Friction: A number representing the resistance to sliding of two surfaces in contact; the drag factor of a vehicle or other object sliding on a roadway or other level surface; the force parallel to a level surface required to keep an object sliding on that surface; the force parallel to a surface required to keep an object sliding on that surface in motion, divided by the force of the object against that surface; measured in pounds per pound; often designated by the Greek letter mu (μ). See Friction Force.

Coefficient of Kinetic Friction: Ratio of the magnitude of the force of kinetic friction to the magnitude of the normal force.

Coefficient of Restitution: The ratio of the relative normal velocity at the time of separation to the relative normal velocity at the time of initial contact between the impact center of two colliding bodies.

Coefficient of Rolling Resistance: The ratio of the force of resistance to rolling with zero slip to the vertical load of a wheel or vehicle. See Rolling Resistance.

Coefficient of Sliding Friction: The sliding coefficient of friction μ is defined as the dimensionless ratio of the force generated on a sliding object on a level surface to the weight of the object. For an object sliding on a flat surface, μ can also represent the acceleration or deceleration of the object, expressed as a decimal fraction of the gravitational acceleration constant, g (32.2 ft/sec^2), if the frictional force is the only unbalanced force acting on the object.

Coil Spring: A spring steel wire in the shape of a coil used as the springing medium for the suspension.

Compliance: The ability to yield elastically to change position.

Collinear Collision: A collision between two vehicles in which their respective directions of travel are parallel to one another, either as a rear - end or head - on collision.

Collinear Forces: If two or more forces have the same action line, they are collinear forces.

Collision Analysis: Automobile collision analysis constitutes the determination as far as possible of *what* happened during the collision, *where* it happened, and *how* it happened. It is a reconstruction made on the basis of damage, debris, the highway marks, and the point of rest of the vehicles after the accident. Collision analysis is a study of engineering dynamics. Both the geometry of the motion of a vehicle and the forces which cause that motion are subjects of dynamics.

Collision Course: Motion of a traffic unit or other object which, if continued without modification, will result in collision with another traffic unit or other object. Objects are on a collision course if the space between them is diminishing and the angle between the path of either and a line between them is constant.

Collision Damage Marks: Collision damage marks are divided into direct and indirect damage, including: deformation, scrape marks, bending, twisting, buckling, folding, tearing, fracture, melting, burning, breaking, and cracking.

Collision Scrub: A short, usually broad, skid mark or yaw mark made during engagement of the vehicles in a collision. A collision scrub may be connected to skid marks or yaw-marks made before or after collision, but has characteristics that differentiate it from such marks.

Collision Speed Change: The speed change is the magnitude of the velocity change vector for the center of gravity of the vehicle during the collision. It is often called Delta-*V*. The speed change is always in the direction of the force. The ratio of the vehicle speed changes is always equal to the inverse ratio of the vehicle masses.

Collision: A collision is an isolated event in which two or more bodies (colliding bodies) exert relatively strong forces on each other for a relatively short time. A collision is also a dynamic event consisting of the contact between two or more bodies resulting in a change of momentum of at least one participating body. Collisions involve forces (there is a change in velocity). Collisions can be elastic, meaning they conserve energy and momentum, inelastic, meaning they conserve momentum but not energy, or totally inelastic (or plastic), meaning they conserve momentum and the two objects stick together. See also Accident, Crash; Impact.

Collision-Primary: The impact between a vehicle and another vehicle or a vehicle and a fixed object.

Collision-Secondary: The impact between the occupant and some interior component of the vehicle.

Collision-Tertiary: The impact between the occupant and the restraint system.

Combination Friction: The base object and the opposed body are in mutual contact at surface peaks. The load is shared by the fluid/lubricating film and by the objects in contact.

Combined Speed-Analysis: In the combined speed-analysis, the velocity or speed at the beginning of a motion sequence is computed from several individual speeds at various points during the process.

Common Contact Point: See Impact Center.

Common Shaft: The individual shafts that exist in all machine-trains.

Common Velocity Conditions: Two independent conditions applicable to a collision, where at the time of separation, the relative normal velocity component is zero (no restitution) and the relative tangential component of velocity is zero (sliding has ended).

Complete Impact: See Full Impact.

Complex Reaction Time: The time to respond to more than one stimulus with one or more possible responses or while performing an additional task. Therefore, every response by a driver is a complex reaction time.

Complex Reaction: A choice between two or more possible reactions to an unexpected hazard. Time 1 sec, more or less.

Complex Vibration: Complex vibration is vibration whose components are sinusoids not harmonically related to one another. See Harmonic.

Compliance Oversteer: Compliance steer, which decreases vehicle under steer or increases vehicle over steer.

Compliance Steer Coefficient: The rate of change in compliance steer with respect to change in forces or moments applied at the tire-road contact.

Compliance Steer: The change in steer angle of front or rear wheels resulting from compliance in suspension and steering linkages and produced by forces and/or moments applied at the tire-road contact.

Compliance Understeer: Compliance steer, which increases vehicle under steer or decreases vehicle over steer.

Compliance: Compliance is the reciprocal of stiffness.

Compression: The relative displacement of sprung and unsprung masses in the suspension system in which the distance between the masses decreases from that at static condition.

Computer Simulation: A computer program that, through a mathematical means and a series of data input into the computer program, allows the solution to the equations of motion and renders a computer-generated drawing of the motion of the vehicle just prior to then through the collision, and then onto its point of rest.

Conclusions or Opinions: See Opinions or Conclusions.

Concrete Pavement: A solidified pavement with an upper layer of aggregate (such as sand and stone) mixed with Portland cement paste binder.

Concrete: A paving material consisting of an aggregate of stones of assorted sizes held together with a cement or binder; bituminous concrete; Portland cement concrete.

Concurrent Forces: The classification of a system of forces is made according to the arrangement of forces' action lines; if they pass through the same point, the system is concurrent.

Condition Factors: Deficiencies in basic attributes of roads, vehicles or people, as related to highway transportation and permanently or temporarily modified, that contribute to operational factors.

Conditions of Equilibrium: A body is in equilibrium when both the resultant force and the sum of the moments of all external forces about any random point are equal to zero.

Configuration: A configuration is defined as the geometric location of all the masses of the system.

Conicity Force: The component of lateral force offset, which changes sign (with respect to the Tire Axis System) with a change in direction of rotation (positive away from the serial number or toward the whitewall). The force is positive when it is directed away from the serial number on the right side tire and negative when it is directed toward the serial number on the left side tire.

Conservation of Angular Momentum: The total angular momentum of a system of masses about any one axis remains constant unless acted upon by an external torque about that axis.

Conservation of Energy: The principle of physics stating that the amount of energy in a closed system is constant regardless of the changes in the form of that energy. Used in accident investigation in estimating vehicle speed from the distance required to slide to a stop and other applications, such as speed from damage. See Momentum.

Conservation of Linear Momentum: The total momentum of a system of masses in any one direction remains constant unless acted upon by an external force in that direction.

Conservation of Mass: The mass of a body remains unchanged by ordinary physical changes to which it may be subjected.

Conservation of Momentum: The linear momentum of a system of bodies remains unchanged if there is no resultant external force on the system. The angular momentum of a system of bodies about a fixed axis is unchanged if there is no resultant external momentum about that axis.

Conservation: The idea that the total amount of certain quantities in nature always stays the same. Energy, linear momentum, and angular momentum are examples of quantities that obey the law of conservation.

Conservative System: In a conservative system, there is no mechanism for dissipating or adding energy. Systems which contain only conservative forces are called conservative system. A linear spring is a conservative element of any system.

Conservative Forces: Conservative forces are forces which do work that are stored as potential energy. If the energy is released, it will do as much work as was done to store the energy.

Conserve: Stay the same or constant. To say that something is conserved means to say that the amount of it stays the same and obeys conservation.

Constant Velocity: If a vehicle is travelling at constant velocity, then $v = d/t$, where d is the distance and t is the time.

Contact Damage: Damage to a vehicle resulting from direct pressure of some foreign object in a collision or roll over. It is usually indicated by striations, rub-off of material, or puncture. Compare with induced damage.

Contact Patch: Contact patch is the portion of a vehicle's tire that is in actual contact with the road surface. It is most commonly used in the discussion of pneumatic tires, (*i.e.* pressurized tires), where the term is strictly used to describe the portion of the tire's tread that touches the road surface. The term "footprint" is used almost synonymously.

Contact Path: The area or region of mutual contact between a tire and the surface over which it moves or rests.

Contact Point: The point of intersection of the resultant contact impulse with the intervehicular contact surface of each of two colliding vehicles. See Impact Center.

Contact Surface: See Intervehicular Contact Surface.

Continuous Braking System: Combination of braking systems for vehicles form a vehicle combination.

Contributing Factor: See Factor.

Control: The term *control* is used not only to denote the control process itself but also the entire system in which the control function takes place.

Control Force: Force exerted on the control device.

Control Line: A control line is a special pilot line by which the energy essential for control is transmitted from the towing vehicle to the towed vehicle.

Controlled Final Position: A final position reached because of conscious effort of some person to modify the motion of a traffic unit after an accident.

Coordinate Measurements: A method of locating any spot in an area by taking two measurements from the nearest point to the spot on a specified reference line: 1) the distance and direction from that point to the spot, and 2) the distance and direction from that point to a specified reference point on the reference line.

Coordinate Method: Technique of scene documentation where items of evidence are located by measuring the distance from two (2) reference lines. Similar in concept to locating points in the Cartesian plane.

Coordinate System of Vehicle: See Three-Axis Vehicle Coordinate System, Cornering Coefficient: See Sideslip Coefficient.

Coordinates: A method of locating a spot in an area by measurements along and at right angles to a reference line or by measurements of the shortest distances to each of two intercepting reference lines, the locations of which are identified for future reference. Compare with triangulation.

Cornering: Cornering is defined as the ability of the vehicle to travel a curved path. Many factors affect the cornering ability of a vehicle, such as the tire construction, tire tread, road surface, alignment, and tire loading.

Cornering Coefficient: See Cornering Stiffness Coefficient.

Cornering Controllability: These tests are to describe the driver confidence in vehicle control while cornering, particularly at moderate and high speeds. These include the reaction of the vehicle to steering inputs both when turning the vehicle into a corner and when keeping the vehicle on an intended curvature. They also include steering to return to straight-ahead.

Cornering Effect: See Sideslip Coefficient.

Cornering Force: The traction force in a lateral direction that is generated by a tire.

Cornering Limit: The maximum vehicular lateral acceleration of a vehicle.

Cornering Resistance: The cornering resistance F_K is given by $F_K = f_K G$, where f_K is the coefficient of cornering resistance, $G = mg$ = weight of vehicle, m is the vehicle mass, and g is the gravitational acceleration = 9.81 m/s^2.

Cornering Squeal: The squeal produced by a free-rolling tire resulting from slip angle.

Cornering Stability: Among the tests performed is whether the vehicle understeers or oversteers under constant throttle and varying steering input. The predictability and comfort level of steering are studied. Tests include studying whether braking on various types of surfaces changes the under/oversteer characteristics.

Cornering Stiffness Coefficient (Cornering Coefficient): The ratio of cornering stiffness of a straight free-rolling tire to the vertical load.

Cornering Stiffness: The negative of the rate of change of lateral force with respect to change in slip angle, usually evaluated at zero slip angle.

Corner-to-Corner Collisions: The corner-to-corner collisions usually involve rotation of the colliding vehicles. Since the stopping force due to impact is not along the center line of the vehicles, each vehicle tends to turn a different way. Each vehicle is deflected from the point of impact towards its own side of the road.

Coulomb Damping: Coulomb or dry-friction damping occurs when sliding contact exists between surfaces in contact which are dry or have insufficient lubrication. In this case, the damping force is constant in magnitude but opposite in direction to that of the motion. In dry-friction damping, energy is dissipated as heat.

Couple: A system of two equal forces, parallel, opposite in sense, and having different action lines is a couple.

Coupled Modes: Coupled modes are modes of vibration that are not independent but which influence one another because of energy transfer from one mode to the other. (See Mode of Vibration.)

Coupling: Coupling is the term used in mechanical vibration to indicate a connection between equations of motion.

Course Angle (υ): The angle between the trace of the vehicle velocity vector on the *X-Y* plane and *X*-axis of the earth-fixed axis system. A positive course angle is shown in Fig. H.1. Course angle is the sum of heading angle and sideslip angle ($\upsilon = \psi + \beta$).

Crash: An event in which one or more vehicles make unintended contact with another vehicle or other object producing injury, death, and/or property damage. See Accident, Vehicle; Collision; Impact.

Crash Reconstruction: See Accident Reconstruction.

Crash Test: A crash test is a form of destructive testing usually performed in order to ensure safe design standards in Crashworthiness and crash compatibility for automobiles or related components.

Crashworthiness: Crashworthiness is the ability of a structure to protect its occupants during an impact. This is commonly tested when investigating the safety of vehicles.

Critical Damping Coefficient: The smallest value of the damping coefficient required to prevent vibration.

Critical Damping: The minimum amount of viscous damping required in a linear system to prevent the displacement of the system from passing the equilibrium position upon returning from an initial displacement.

Critical Speed Formula: A formula: $v_{cr} = \sqrt{fgR}$ that calculates the speed of a vehicle from its radius of curvature, R, frictional drag coefficient, f, and acceleration of gravity, g.

Critical Speed Scuff: A tire mark left by a rotating wheel that is sliding or slipping sideways due to the centrifugal force exceeding peak lateral friction. The skid mark will normally have striation marks. Also referred to as yaw, centrifugal skid, sideslip or inertial mark.

Critical Speed: The maximum speed at which a vehicle can traverse a path with a specific radius of curvature without loss of directional control; the speed of a vehicle undergoing a sudden turn maneuver at which the tires leave visible sideslip marks.

Critical Velocity: The minimum velocity at the highest point of the loop in order to complete a cycle is called the critical velocity.

Critically Damped System: The system is said to be critically damped if the amount of damping is such that the resulting motion is on the border between the two cases of underdamped and over damped systems.

Critical-Speed Scuffs: See Yaw-Marks.

Crook: An abrupt change of direction of a tire mark due to collision forces. The crook often indicates the position of a tire at first contact.

Cross Slope: See Superelevation.

Crown: The amount by which the center of a roadway is higher than the edges, usually for drainage purposes.

Crucial Event: See First Harmful Event.

Cruise Control: Sometimes known as speed control or autocruise is a system that automatically controls the rate of motion of a motor vehicle. The driver sets the speed and the system will take over the throttle of the car to maintain the same speed.

Crumple Zone: That portion of the front or rear of a vehicle designed to absorb energy of a collision for the protection of the occupants.

Crush Area: Area defined by the original vehicle exterior and a crush profile, crush equivalent speed: See Energy Equivalent Speed (EES).

Crush Distance: The permanent damage or deformation a vehicle sustains as a result of impact. "*C*" measurements are taken from the original bodyline at frame height and from left to right and from rear to front. To determine the five space intervals for $C_1 - C_6$ measurements, double the damage width (in inches) and divide by 10.

Crush Energy: A number of crash tests were conducted to examine the crush behavior of vehicles. The crush energy, E_A, per unit width, w, of a crushed chassis can be expressed as a linear function of crush, C, as

$$\sqrt{\frac{2E_A}{w}} = d_0 + d_1 C$$

where the residual crush, C, is measured normal (perpendicular) to and from the nominal undeformed surface. d_0 and d_1 are constants known as the crush stiffness coefficients. d_0 and d_1 are obtained experimentally for a particular vehicle. They differ from vehicle to vehicle and between the front, side, and rear of each vehicle. The crush energy is equal to the crush force multiplied by the distance during which the crush force acts.

Crush Equivalent Speed: See Energy Equivalent Speed (EES).

Crush Profile: The geometric shape in a specified plane (*e.g.*, vertical, horizontal) that describes the vehicle damage resulting from an impact.

Crush Stiffness Coefficient: An empirical quantity used in the calculation of the energy dissipated in a collision and associated with each vehicle's velocity change, V. See CRASH3.

Crush Stiffness: See Crush Stiffness Coefficient.

Curb Rib: Raised circumferential rib on sidewall of tire, to prevent damage to sidewall in contact with curb.

Curb Weight: The weight of a motor vehicle with standard equipment and maximum fuel capacity.

Curvilinear Motion: Motion along a curved path which may be either planar or skewed.

Cycle: Cycle of oscillation is the complete sequence of variations in displacement, which occur during a period.

$$D = \text{Sqrt (width} \times \text{Closing Speed/0.0045)}$$

D'Alembert's Principle: The generalized principle of D'Alembert states that the virtual work performed by the effective forces through infinitesimal virtual displacements compatible with the system constraints is zero.

Damage Analysis: Damage analysis can predict what happened during a collision by what remains after in the form of damage.

Damage Closing Velocity: A damage related closing velocity is defined to be the sum of the speed changes at the point of common velocity or

$$V_C = \Delta V_1 + \Delta V_2$$

This closing velocity is in the direction of the line of action of the collision forces.

Damage Model: The damage model assumes that each vehicle consists of a series of radial springs which can record the damage. These springs are also referred to as force vectors as they transmit a radial force in a particular direction.

Damage Vector: A damage vector is generally assumed to be a linear spring anchored at the center of mass of the vehicle extending to the vehicle surface.

Damped: Damped systems are those in which energy is dissipated by forces opposing the vibratory motion.

Damped Natural Frequency: The damped natural frequency is the frequency of free vibration of a damped linear system. The free vibration of a damped system may be considered periodic in the limited sense that the time interval between zero crossings in the same direction is constant, even though successive amplitudes decrease progressively. The frequency of the vibration is the reciprocal of this time interval.

Damped, Dampened: A force or action opposing a vibrating motion to reduce the amount of vibration.

Damping Coefficient: The constant coefficient of the velocity term, $\dot{x}$, in a motion defined by the differential equation $m\ddot{x} + c\dot{x} + kx = 0$, where m is the mass and k is the stiffness of the system.

Damping Devices: As distinct from specific types of damping, damping devices refer to the actual mechanisms used to obtain damping of suspension systems.

Damping Factor: See Decay Factor.

Damping Ratio: The ratio of the amount of viscous damping present in a system to that required for critical damping.

Damping Slipper: A device restricting lateral thrash of a belt or chain especially a timing chain.

Damping: The process of energy dissipation is generally referred to, in the study of vibrations, as damping.

Danger: Danger is the likelihood for a hazard to be involved in causing an accident, and thus injury and harm.

Dashpot: (*a*) A damping device consisting of a piston and cylinder whose relative motion is opposed by the fluid friction of a liquid or of air. It provides forces proportional to the rate of movement of the piston when a spring is added to the device. A one-way valve may be incorporated to give a differential damping action. (*b*) A cylinder employed in steam engines fitted with trip gears for closing the admission valves suddenly as soon as they are released by the trip.

Datum: (*a*) A point from which all measurements are made, (*b*) A line from which all measurements are made, (*c*) A horizontal plane from which all vertical measurements are made. See Datum Line; Datum Plane; Datum Point.

Debris: Loose material strewn on the road as a result of a traffic accident; dirt, liquids, vehicle parts, clothing, and other materials from the involved traffic units at the site of the traffic accident.

Deceleration: The rate of decrease in the speed of a vehicle or a moving part.

Decision: Determination of what action to take after examination of a situation; choice of driving strategies or evasive tactics.

Deformation: The alteration of form or shape as a result of a collision.

Degree of Curve: The number of degrees of central angle subtended by an arc of 100 ft on the circumference; 5730 ft divided by the degree of a curve equals the radius of the curve.

Degree of Danger: The degree of danger for a particular condition may be expressed by the following relationship: *zx* Danger = Hazard consequence × Hazard frequency

Degree of Freedom: The number of degrees of freedom of a vibrating system is the sum total of all ways in which the masses of the system can be independently displaced from their respective equilibrium positions.

Delineators: Those devices that direct, channel and separate vehicular traffic.

Delta-t (Δt): A time interval associated with an event such as vehicle-to-vehicle contact; the time duration of impulse.

Delta-V: The velocity change experienced by a vehicle during impact is commonly called delta-V (ΔV). It is an important consideration for occupant injury severity. Delta-V is a function of the velocity before and after impact. It is a vector, and as such, requires magnitude and angular orientation for proper identification.

Density: Density is the ratio of the mass to the volume of a specific amount of substance.

Dent: Dent is a mark which tends to go into the metal rather than run along it. Here most of the metal movement is along the line of impact of the striping object.

Departure Angle: The angle between your rear tires and the first rear protrusion that would scrape the ground as you descend off of a hill.

Departure Velocity: See Separation Velocity.

Dependent Variables: Dependent variables are the variables that describe the physical behavior of the system.

Dependent Wheel Suspension: In the dependent wheel suspension, both wheels are connected rigidly by a solid beam and thus, remain parallel to each other regardless of the body or wheel motion.

Deposit: An imprint consisting of material deposited on the road surface from tire grooves. Sometimes found where a tire stops sliding. See Imprint.

Deposition: Pre-trial questioning of a witness, under oath, subject to cross examination for the purpose of discovering evidence or perpetuating testimony; a document in which a deposition is recorded.

Detailed Statement: A concise written summary of what a particular person has to say about a traffic accident or circumstances connected with it. A detailed statement is usually not a verbatim report of the information obtained from an informant.

Detection Task: Detection task is sensing a stimulus without identification.

Detection Threshold: See Absolute Threshold

Detection Time: The time from the point that a stimulus is first visible to the moment (start of when) that the stimulus is perceived by the observer.

Differential: The differential consists of a set of gears for changing the driveshaft's rotation into axle rotation. The rear axles can be solid or tubular shafts; these connect to the rear wheels which (finally) connect to the roadway.

Direct Cause: See Operational Factor.

Direct Central Impact: If two inelastic bodies collide in a direct central impact, they have a common velocity after their impact.

Direct Damage: Damage or deformation of the vehicle that was directly caused by the collision itself.

Direct Delta Function: See Unit Impulse Function.

Direct Eccentric Impact: Direct eccentric impact occurs if the initial velocities are parallel to the normal to the striking surfaces but are not collinear.

Direct Impact: If the velocities of the two bodies are directed along the line of impact, the impact is said to be direct impact. Direct central impact occurs if the mass centers of the two bodies are also along the line of impact.

Direction Angles: The angles that a nonzero vector makes with the coordinate vectors and with the coordinate axes are called direction cycles, and their cosines are the direction cosines of the given vector.

Direction Cosines: See Direction Angles.

Direction of Principal Force (DOPF): See Principal Direction of Force (PDOF).

Directional Stability: Ability of a car to travel in a straight line with a minimum of correction from the driver.

Displacement: The change in distance or position of an object relative to a reference point; the actual distance, off-centerline, of a rotating shaft as compared to a stationary reference, usually the machine housing. The term displacement defines a relative movement during a time interval.

Distance: Distance is a linear measurement from some point. For accident reconstruction problems, distance is measured relative to a coordinate system fixed on the earth.

Dive: A pitching motion of the sprung mass of a car downward at the front, which usually occurs during braking.

Drag Factor: A number representing the acceleration or deceleration of a vehicle or other body as a decimal fraction of the acceleration of gravity; $f = a/g$; the horizontal force needed to produce acceleration in the same direction divided by the weight of the body to which the force is applied. When a vehicle slides with all wheels locked, the coefficient of friction and the drag factor have the same value.

Drag Force: The negative tractive force.

Drag Link: A tube or rod used as connection between the pitman arm and steering knuckle. Normally used on cars and light trucks.

Drag Sled: A weighted device (whose bottom surface is covered with a portion of tire thread), which is pulled along a roadway surface and provides a sliding friction coefficient of that device and roadway surface by computing the ratio of pull force to its weight.

Drag: Drag is defined as a coefficient which represents the effective retardation of the linear velocity in a trajectory.

Dribble: Liquid debris, from a vehicle or its cargo, that drops to the ground, often leaving a trail, if the vehicle is moving, from a spatter area to a puddle.

Driver Response Time: A term used to describe research that measures a driver's response, regardless of the transition components included.

Driver: Any person who drives or who is in physical control of a vehicle.

Driver's Side: The left side (from the inside looking out) of the vehicle.

Driving Traction Coefficient: The maximum value of driving force coefficient which can be reached on a given tire and road surface for a given environment and operating condition.

Drum and Disc Brakes: Pure braking is simple to understand. The input to the braking system is a direct force at the braking pedal by the driver with the direct response of deceleration in the longitudinal motion. The braking mechanism transfers the braking force applied by the driver at the braking pedal into a torque applied on the wheel, preventing the wheels from rotation, either on a continuous basis, or in a slip regulation fashion. This will result in large friction forces between the tires and the road, which reduce the speed of the vehicle. Automotive brakes used by standard road vehicles are of two types. These are drum and disk brakes. These types of brakes are shown in Fig. D.1.

Drum Brakes

Disk Brakes

Fig. D.1 Drum and disk brakes (Courtesy of Bosch Corporation).

Dry Friction Damping: See Coulomb Damping.

Dry or Boundary Friction: The lubricating film no longer performs any support function, but residue from previously absorbed lubricants continues to exercise a tribological effect.

Dynamic Crush: The deformation formed by the external surface of the vehicle at any time during an impact, usually measured relative to the corresponding as manufactured unreformed surface. See Crush Area; Crush Profile; Residual Crush.

Dynamic Event: A dynamic event changes one or more of the state variables: position, velocity, and time.

Dynamic Index: (k^2/ab ratio) is the square of the radius of gyration (k) of the sprung mass about a transverse axis through the center of gravity, divided by the product of the two longitudinal distances (a and b) from the center of gravity to the front and rear wheel centers.

Dynamic Load: A rolling or moving load.

Dynamic Model: A model having the correct proportions for the simulation of the full scale dynamic conditions.

Dynamics: Dynamics is the study of the position of objects and how that position or state changes in time.

Dynamic Rate: Dynamic rate of an elastic member is the rate measured during rapid deflection where the member is not allowed to reach static equilibrium.

Dynamic Stiffness: The stiffness of a massless constraint which, substituted for a part of a system at a given frequency, will give the same vibrating force or torque for the same vibrating amplitude of the original part.

Dynamics: The branch of mechanics that studies the motion of bodies (kinematics) and action of forces that produce or change the motion.

Earth-Fixed Axis System (*X*, *Y*, *Z*): This system is a right-hand orthogonal axis system fixed on the earth. The trajectory of the vehicle is described with respect to this earth-fixed axis system. The *X*- and *Y*-axis are in a horizontal plane and the *Z*-axis is directed downward.

East-West Configuration: The power plant could also be packaged in different coordinate as shown in Figure E.1. This package is called east-west configuration. This configuration can be seen in most passenger cars and some small Sport Utility Vehicles (SUVs).

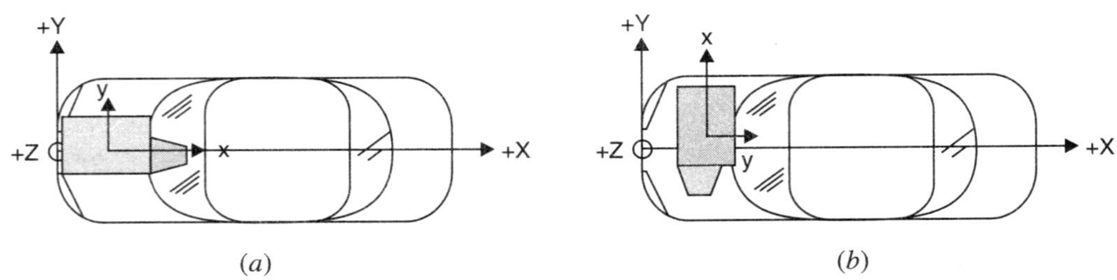

Fig. E.1 Power plant configurations (*a*) north south configuration (*b*) east west configuration.

Edge Line: A line which indicates the edge of the roadway.

Edge Ruts: They are developed on a soft shoulder when a vehicle runs off the roadbed surface onto the shoulder and then presses the soil, forming a rut at least 2 to 4 in. deep.

Effective Drag Factor: The ratio of the velocity squared to the quantity of the skidding distance multiplied by twice the gravitational constant.

Effective Inertia: The moment of inertia of a mass which, if substituted for a part of a system at a given frequency, would have the same vibratory acceleration under the same vibratory torque at the point of separation as the original part. It is measured by the value of the torque.

Effective Mass: That mass, which, if substituted for a part of a system at a given frequency, will have the same vibratory acceleration under the same vibratory force at the point of separation as the original part. It is measured by the value of the force.

Effective Rolling Radius: The ratio of the linear velocity of the wheel center in the x direction to the spin velocity.

Effective Static Deflection: Effective static deflection of a loaded suspension system equals the static load divided by the spring rate of the system at that load. Total static deflection and effective static deflection are equal when the spring rate is constant.

Elastic Collision: An elastic collision is a collision in which the total kinetic energy of the colliding bodies after collision is equal to their total kinetic energy before collision. Elastic collisions occur only if there is no net conversion of kinetic energy into other forms. During the collision, kinetic energy is first converted to potential energy associated with a repulsive force between the particles (when the particles move against this force, *i.e.* the angle between the force and the relative velocity is obtuse), then this potential energy is converted back to kinetic energy (when the particles move with this force, *i.e.* the angle between the force and the relative velocity is acute). Thus, it is a collision between two bodies in which no permanent (plastic) deformation takes place and both momentum and kinetic energy are conserved. When the amount of force between two objects in collision is sufficient to cause permanent deformation, the damage caused produces a loss of energy. Both momentum and kinetic energy are conserved. In a perfectly elastic collision, the sum of the translational kinetic energies of the object is not changed.

Energy Equivalent Speed (EES): The speed and corresponding kinetic energy with which the vehicle must contact a fixed rigid object with no rebound for equivalence to conditions of another collision; for example, the energy may be equal to a specified level of residual crush; EES is the preferred term, broader than barrier equivalent velocity (BEV), equivalent barrier speed (EBS), and equivalent test speed (ETS).

Energy Equivalent Velocity: See Energy Equivalent Speed (EES); Equivalent Barrier Speed (EBS).

Energy: Energy (potential or kinetic) of a body is the amount of work it can do by virtue of its position or its motion against forces applied to it.

Equilibrium: The state of a body where the body has no net force acting on it.

Equivalent Barrier Speed (EBS): A vehicle velocity at which the kinetic energy of the vehicle would equal the energy that was absorbed in plastic (permanent) deformation.

Equivalent Braking Force: Ratio of the total braking force of a vehicle to the gross weight.

Event: An event is some operation which takes a vehicle from one state to another with the passage of time. Events cause changes in the state variables, position, velocity, and time. Events occur because of forces or changes in forces that act on the vehicle with time.

Expert: The expert may be defined as an individual who has training and skill in a particular scope of endeavor, experience being an essential element of this definition. This definition implies that the expert is considerably more knowledgeable on his topic than the layman and has the ability to correlate facts, from which he can formulate logical conclusions.

External Forces: Actions of other bodies on a rigid body are known as external forces. Sea internal forces.

Fact: Information that corresponds to reality.

Factor: Each contributing circumstance is a factor or element of the cause, but none, by itself, is capable of producing the result. Factor is any circumstance contributing to a result without which the result could not have occurred. Each factor is an element which is necessary to produce the result, but not, by itself, sufficient.

Fade: Reduction of braking effort resulting from overheating or other transient effect.

Fall: A downward and onward movement in the air under the force of gravity after forward momentum carries an object beyond its supporting surface; rotation during a fall is gradual and the object usually lands right side up.

A fall occurs when a vehicle is travelling forward and is no longer in contact with the surface it is moving over. The velocity of a vehicle at the point of takeoff is given by $v = d\sqrt{\dfrac{g}{2\cos\theta(d\sin\theta - h\cos\theta)}}$, where v is the velocity of the vehicle at takeoff, m/s(ft/s), d is the horizontal distance travelled by the vehicle's center of mass from takeoff to landing, m(ft), g is the acceleration due to gravity, 9.81 m/s^2 (32.2 ft/s^2), h is the distance travelled vertically by the vehicle's center of mass from takeoff to landing (m or ft). h is positive if the landing point is above the takeoff point, negative if the landing point in below the takeoff point in the percent grade, and θ is the angle of take off as measured relative to a horizontal plane in degrees. For small angles, $v = d\sqrt{\dfrac{g}{2(dG-h)}}$ where $G = \sin\theta$ and $\cos\theta \approx 1$.

Federal Bumper: Bumper specially designed to meet the US Federal Safety Regulations which require the absorption of all the energy of a 5 mph impact.

Fender Bender: See Car Accident.

Fender Vault: A fender vault collision occurs when the pedestrian is struck by the front of a vehicle near one of its sides, wraps up onto the fender and sloughs off to the side of the moving vehicle without a significant secondary impact with a pillar or windshield.

Fender: Narrow curved panel mounted over or behind a wheel, to deflect tire-generated spray and roadstones.

Final Position: The location of a vehicle or body when it comes to rest after a collision; final position may be controlled or uncontrolled.

Final Rest Position: Controlled or uncontrolled post-collision resting position of vehicles, pedestrians, debris, etc. See Point of Rest.

Fineness Ratio: Ratio of length to thickness, of a vehicle.

First Contact Position: The position or location at an accident scene (measured relative to a coordinate system fixed to the earth) of a vehicle, pedestrian, or other object at the time it first has contact with another body in a collision.

First Contact Velocity: The velocity of the center of gravity of a vehicle, pedestrian, or other object at its first contact position.

First Contact: The initial touching of objects in a collision; the time and place on a traffic unit or traffic way where this touching occurs.

First Mode: The slightly eccentric rotation (off-center) of a shaft will generate a low-level frequency component that coincides with the rotating speed of the shaft.

First Order Lateral Force Variation: The peak-to-peak amplitude of the fundamental frequency component of the Fourier series-representing lateral force variation. Its frequency is equal to the rotational frequency of the tire.

Fixed Contact Patch Model of Tires: The fixed contact patch model of tires is created on the basis of the point contact tire model by converting the equivalent concentrated stiffness of the tires into the equivalent distributed stiffness within the fixed contact patch length of the tires.

Flip: Flip is the movement of a vehicle from a place where the forward velocity of a part of the vehicle suddenly is stopped by an object below its center of gravity such as curb, rail, or furrow, with the result that the ensuing rotational motion lifts the vehicle from the ground. Flips occur when vehicles are moving sideways and the resistance at the tires is sufficient to cause the vehicle to rise and move through the air.

Flips or Vaults: Flips or vaults occur when the moving vehicle hits something that stops the wheels suddenly. The vehicle then pivots upward and leaves the ground. It nearly always lands bottom up and then may continue on rolling or sliding. Vaults are endorsing flips. They occur only when front wheels, sliding or rolling, are stopped by an obstacle, usually a curb, high enough and vertical so that the wheel does not roll over it.

Footprint: The shape of the contact interface of a loaded tire with the ground. See Tire Contact Area or Tire Print.

Force: The action of one body on another. It is a vector quantity defined by its magnitude, direction, and point of application.

Forward Projection Pedestrian Collision: A frontal collision of vehicle and pedestrian or cyclist in which the initial contact area is at or above the height of the center of gravity of the pedestrian or cyclist and in which a single impact with the frontal geometry of the vehicle causes the pedestrian or cyclist to be projected forward relative to the vehicle.

Forward Speed: See Fig. T.3.

Four-Wheel Drive: Transmission system in which engine power is delivered to front and rear wheels of a vehicle.

Free Vibration: Free vibration of a system is the vibration during which no variable force is externally applied to the system. A vibration with periods depends solely on the properties of the disturbed system.

Friction Drag Coefficient: An average, uniform (constant) value of a friction coefficient applied to a specific sliding event such as when an object slides from an initial speed to stop over a distance, d, or during a speed change, delta V.

Friction Drag: Aerodynamic drag resulting from the friction between the moving air and the surface of a vehicle in motion.

Friction Force: The amount of force with which a motion is resisted. See Coefficient of Friction.

Friction Mark: A tire mark made when a slipping or sliding tire rubs the surface of the road or other surface.

Friction: The resistance to movement of an object across a surface. A way that a form of energy, such as kinetic energy, is changed into another form, usually heat energy. Friction causes you to slow down when you swing. It also causes a spinning top to eventually fall over.

Frictional Drag Coefficient: An average, uniform (constant) value of a friction coefficient applied to a specific sliding event such as when an object slides from an initial speed to a stop over a distance, d, or during a speed change, ΔV.

Frictional Drag Factor: See Frictional Drag Coefficient.

Front Lock-Up: Under front lock-up, the maximum force at the front axle will be equal to twice that of the wheel where lock-up is introduced for conventional brakes.

Front Skid: The maximum force at the front axle will be equal to twice that at the wheel where first felt for conventional traction systems.

Fundamental Frequency: The natural frequencies can be arranged in order of increasing magnitude and the lowest frequency is referred to as the fundamental frequency.

Fundamental Mode of Vibration: The fundamental mode of vibration of a system is the mode having the lowest natural frequency.

Fundamental or First Critical Speed: The lowest critical speed.

Furrow: A channel in a loose or soft material, such as snow or soil, made by a vehicle tire or some other part of a moving vehicle.

Furrow-In: A deep depression at the end of a furrow where a tire or other part of a vehicle stops sliding. A furrow-in often marks the beginning of a flip or roll.

G Force, g: A measurement of the amount of acceleration, braking, or cornering force that a car can generate.

GAW: Gross axle weight is the total weight carried by an individual axle (front or rear) including the vehicle weight and cargo.

GAWR: Gross axle weight rating is the maximum allowable weight that can be carried by a single axle (front or rear).

GCW: Gross combined weight is the weight of a loaded vehicle plus the weight of a fully loaded semitrailer.

GCWR: Gross combined weight rating is the maximum allowable weight of a vehicle and loaded semi trailer.

Generalized Coordinates: A set of independent coordinates, which properly and completely defines the configuration of a system and whose number is equal to the number of degrees of freedom, is called the generalized coordinates. The independent spatial coordinates describing the configuration of a dynamic system are also called generalized coordinates. Thus the number of degrees of a dynamic system is equal to the number of *generalized coordinates*.

Gouge Marks: Gouge marks are made by metal components that are forced into the pavement during impact or as the vehicle travels to its rest position. Gouge marks are extremely important since they, to a large extent, determine the point or area of impact between two vehicles. Gouge marks made during impact generally are deeper at the leading end of the mark than at the end of the mark.

Gouge: A pavement scar deep enough to be easily felt with the fingers; chip, chop, or groove. See Gouge Mark.

Grade Factor: The decimal equivalent of the grade of a slope by dividing the vertical distance (rise) by the horizontal distance (run).

Grade: The change in elevation in unit distance in a specified direction along the center line of a roadway or the path of a vehicle; the difference in level of two points divided by the level distance between the points. Grade is designated in feet per foot (meters per meter) of rise or fall per foot (meter) of level distance or in rise or fall as a percent of the level distance. Grade is positive (+) if the surface rises in the specified direction and negative (–) if it falls in that direction. The road grade, m, is the slope of the road in the direction of travel. It is specified as the ratio of rise over run or, $m = \text{Rise/Run} = Y/X$.

Gravitation: The force of the earth's attraction directed towards the earth's center.

Gravitational Acceleration (g): The constant describing the acceleration of any object falling toward the earth. Near the earth, gravitational acceleration is approximately equal to 9.8 meters per second per second (m/s^2).

Gravitational Potential Energy: A type of potential energy. Gravitational potential energy is associated in the interaction of an object with the earth. It is defined to be equal to the mass of the object times the gravitational acceleration times the distance of the object from the ground.

Gravity: The gravitational attraction at the surface of a planet or other celestial body. See Acceleration due to Gravity.

Groove: A long, narrow, pavement gouge; a channel in the pavement made by a small, strong, metal part being forced some distance along the surface while under great pressure. Narrow void in tire tread pattern.

Gross Axle Weight: Specified maximum carrying capacity of an axle, measured at tire-road interface.

Gross Combination Weight: Total weight of towing vehicle and trailer (or semi-trailer) with equipment, fuel, driver and passengers, payload, and trailer.

Gross Contact Area: The total area enclosing the pattern of the tire tread in contact with a flat surface, including the area of grooves or voids.

Gross Vehicle Weight: Maximum legal weight at which a vehicle can be operated. Curb weight plus payload. Combined weight of a truck and its load.

Handling Characteristics: The handling characteristics refer to the dynamic properties of steering input or environmental disturbance. The two essential issues associated with handling characteristics are direction control and disturbance stabilization.

Handling: The relative ability of a vehicle to maneuver through turns and go where the driver wants it to go.

Harmonic Excitations: If the excitation force is periodic the excitation is said to be harmonic.

Harmonic Response: Harmonic response is the periodic response of a vibrating system exhibiting the characteristics of resonance at a frequency that is a multiple of the excitation frequency.

Harmonics: A sinusoidal component of a periodic wave, having a frequency that is an integral multiple of the fundamental frequency. Also known as harmonic component.

Harshness: Vibrations that can be felt and/or heard that are caused by interaction between the tire and the road surface. It can be caused by tire or road irregularities.

Hazard: Hazard is the potential for causing injury or loss. Hazards may be associated with the driver, vehicle, road, or environment.

Heading Angle: The angle between the trace on the *X-Y* plane of the vehicle *x*-axis and the *X*-axis of the earth-fixed axis system. (See Fig. H.1). Angle by which the longitudinal axis of a moving vehicle deviates from its true direction of motion.

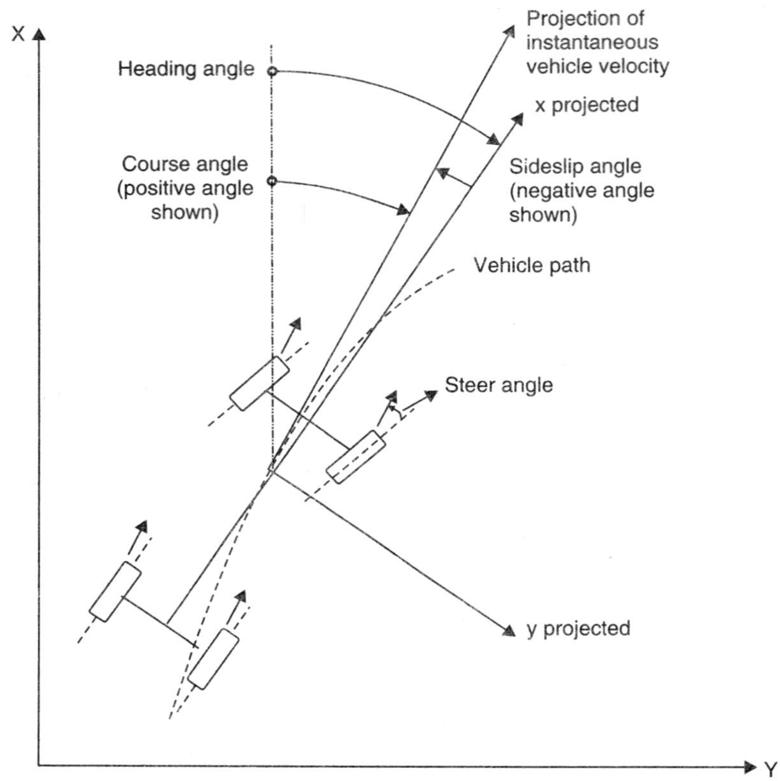

Fig. H.1 Heading sideslip, and course angles.

Head-on Collision: A head-on collision occurs when the front end of two vehicles hit each other. These accidents result in the highest number of fatal crashes because the combined forces are so high. If two vehicles collide head-on, with each travelling at 50 mph, the combined speed would equal 100 mph. The reason for this is a law of physics called the Momentum conservation: In a collision occurring between car 1 and car 2, the total momentum of the two vehicles before the collision is equal to the total momentum of the two vehicles after the collision. Head on crashes often occur when passing, but can also be caused by driver inattention. See Opposite Direction Collision.

Head-on Impact: Frontal impact where the PDOF is at or near zero degrees.

Headway: The distance that a vehicle follows behind another vehicle in seconds. For example, the measured time between the moments the lead car passes a landmark and the following car passes that same landmark. (Measured in seconds).

Highway Collision: Highway collisions are of four types: the *rear-end* or *stopped vehicle*, the *head-on*, the *corner-to-corner*, and the *sideswipe*.

Highway: The entire width between the boundary lines of every way publicly maintained when any part thereof is open to the use by the public for purposes of vehicular travel; a street; a publicly maintained traffic way.

Horsepower: The customary non-metric unit of power, equivalent to 0.7457 kilowatt, defined by a rate of working of 33,000 foot-pounds per minute.

Human Factors: Human factors involve all aspects of occupant influence on accident avoidance, interim collision damage, injury production, and mitigation.

Hydroplaning: A phenomenon in which a layer of fluid (usually water) on a roadway separates the load-bearing surface of one or more tires of a moving vehicle from the road surface and causes a full loss of traction (longitudinal) and steering (transverse) force components.

Impact Aspersion: Occurs when a fluid container is ruptured under extreme pressures of impact and then aspersed onto the roadway.

Impact Center: The point of intersection of the contact impulse and the inter vehicular contact surface for an impact. See Contact Point.

Impact Force (Lever Arm) Moment Arm: See Impulse Moment Arm.

Impact Plane: The impact plane is a straight line set through the calculated damage profile of vehicle one.

Impact Point: The impact point is defined by the coordinates x, y on the impact plane in car coordinates with respect to the center of mass.

Impact Velocity: The velocity of an object's center of gravity relative to a coordinate system fixed in the earth during an impact. See Preimpact Velocity, Postimpact Velocity.

Impact: An impact is a single collision of one mass in motion with a second mass, which may be either in motion or at rest. A collision between two bodies which occurs in a very small interval of time and during which the two bodies exert relatively large forces on each other is called an impact. The common normal to the surfaces in contact during the impact is called the line of impact. The impact is the transmission of force from one body to another within a very short interval of time. The colliding bodies experience an elastic and/or plastic deformation.

Imprint: A mark on the road or other surface made without sliding by a rolling tire or a person's foot. An imprint usually shows the pattern of the tire tread or shoe that made it.

Impulse Moment Arm: The perpendicular distance from an object's center of gravity to the line of action of an impulse; See Fig. I.1.

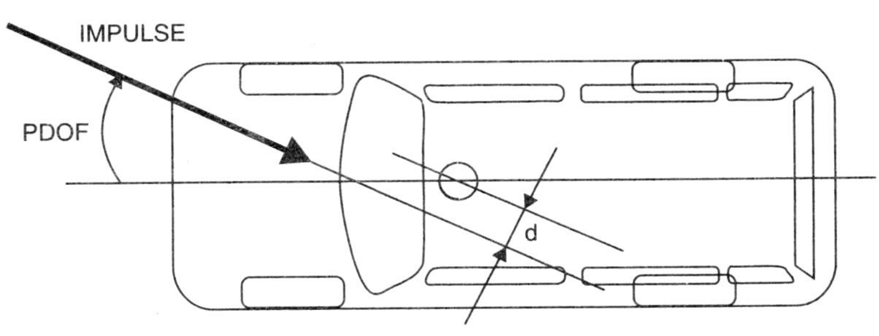

Fig. I.1

Impulse: Impulse is a vector quantity pertaining to a force; it is the integral of the force over a specified period of time. Usually, the time period is very short, as the impact occurs during a crash. The units of impulse are the same as for momentum (kg m/s).

Impulsive Force: An impulsive force is defined as a force, which has a large magnitude, and acts during a very short time duration such that the time integral of the force is finite.

Independent Variables: Independent variables are the variables with which the dependent variable changes.

Independent Wheel Suspension: With independent wheel suspensions, each wheel is connected rigidly by a solid beam and, thus, remain parallel to each other regardless of the body or wheel motion.

Inline Sideswipe Collisions: The approach velocity vectors of both vehicles are approximately parallel. The front corner or side of one vehicle slides along the side of the other vehicle without reaching a common velocity. Another requirement is that departure velocity vectors are approximately in line with the approach directions.

Kinematic Restitution: The kinematic definition of restitution states that the ratio of the relative velocity of the objects after the collision to the relative velocities of the objects before the collision is the coefficient of restitution.

Kinematics: The branch of engineering mechanics which deals with the motion of particles, lines and bodies without consideration of the forces required to produce or maintain motion is called kinematics. The study of the geometry of motion.

Kinematic Relationships: The kinematic relationships determine the relationship between the changes in motion.

Kinetic Energy of the Vehicle: The energy that needs to be absorbed by the braking systems of the vehicle is the kinetic energy (K.E.) of the vehicle. Thus, $\text{K.E.} = \frac{1}{2} M (V_i^2 - V_f^{\,2})$, where M is the mass of the vehicle, V_i is the initial velocity of the vehicle and V_f is the final velocity of the vehicle, which is zero if the vehicle comes to a complete stop. This energy is to be absorbed by the braking system over the braking period. It will be transferred into heat energy at the braking system, resulting in a temperature rise.

Kinetic Energy of Translation and Rotation: A body which has a combined motion of translation and rotation has both kinetic energy of translation and a kinetic energy of rotation. If v is the velocity of the center of gravity at any instant and ω is the angular velocity, the kinetic energy of the body is

$$E_k = \frac{1}{2}(mv^2 + I\omega^2)\,.$$

Kinetic Energy of Translation: If a body has a motion of translation, each particle of the body moves with the same velocity, so at each instant of velocity v, the kinetic energy of the body is

$$E_k = \frac{1}{2} mv^2\,.$$

Kinetic Energy: The kinetic energy of a body is the energy it possesses due to its velocity. If a body of mass m attains a velocity v from rest under the influence of a force P and moves a distance s, then, work done by P is Ps and the kinetic energy of the body is [1/2 mv^2].

Kinetic Friction: Describes the frictional forces present between surfaces in relative motion.

Kinetics: Kinetics is the study of motion and the forces that cause motion. Time for both the SI and English system is measured in seconds.

Lane: Lane is a term used only when estimating the response of a driver who responds to a vehicle changing lanes. Lanes refer to the number of lanes away that the intruding vehicle originally emerged from (From next lane, From one lane over (2 lanes away)).

Lateral Acceleration: The "acceleration" that gives rise to the sideways force on a vehicle (and its occupant) when cornering. In steady-state condition, lateral acceleration is equal to the product of centripetal acceleration times the cosine of the vehicle's sideslip angle. Since in most test conditions, the sideslip angle is small, for practical purposes, the lateral acceleration can be considered equal to centripetal acceleration.

Lateral Axis: The axis that extends from the left to the right and through the center of mass of a vehicle, perpendicular to the longitudinal axis. Also known as pitch axis.

Lateral Force (F_y): That portion of the friction force on the tires that is directed perpendicular to the vehicle centerline. The component of the tire force vector in the Y' direction. See Fig. T.3 and S.5.

Lateral Force Coefficient: The ratio of the lateral force to the vertical load.

Lateral Force Offset: The average lateral force of a straight free-rolling tire.

Lateral Forces on the Tires: Lateral forces on the tires are developed from the attitude of the tire to its path (slip angle) and the cornering stiffness characteristics of the tire.

Lateral Run out: Oscillation of the plane of symmetry of a rotating road wheel as a result of static misalignment.

Lateral Speed: See Fig. T.3.

Lateral Traction Coefficient: The maximum value of lateral force coefficient which can be reached on a free-rolling tire for a given road surface, environment and operating condition.

Lateral Velocity: Lateral Velocity of a point in the vehicle is the component of the vector velocity perpendicular to the *x*-axis and parallel to the road plane. See Fig. L.1.

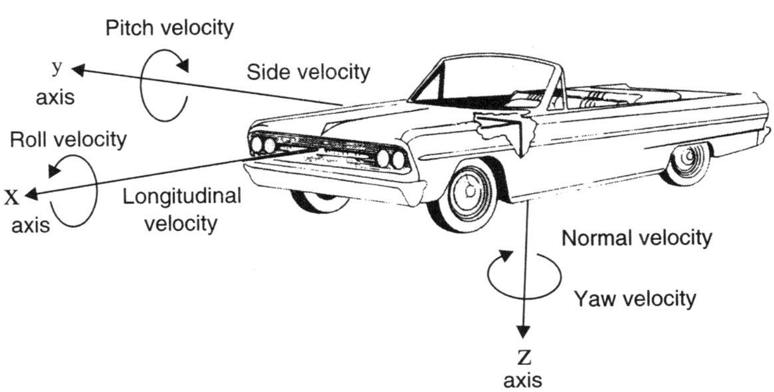

Fig. L.1 Directional control axis system.

Lateral: A direction that is to the side.

Law of Conservation of Energy: The law of conservation of energy states that when a rigid body or a system of rigid bodies moves under the action of conservative forces, the sum of the kinetic energy and the potential energy of the system remains constant.

Linear Momentum: A quantity associated with the motion of an object along a straight path. The linear momentum of an object is defined to be equal to its mass times its velocity.

Linear Spring: A linear spring is a flexible mechanical link between two particles in a mechanical system that is assumed to have negligible mass and damping.

Linear System: A linear system is one in which particles move only in straight line. Another name is rectilinear system.

Linear: Linear vibrating systems are those in which all the variable forces are directly proportional to the displacement, or to the derivatives of the displacement, with respect to time.

Liquid Debris: Debris consisting of liquids from a vehicle or its cargo; spatter, dribble, puddle, runoff and soak-in.

Liquid Marks: Liquid marks are made by liquids leaving the vehicle during and after a collision. Liquid marks are frequently made by radiator fluid. Others are battery acid, steering pump oil, or washing fluid.

Load Transfer: Effective increase or decrease in axle load due to acceleration or braking, or lateral transfer across axle on cornering.

Locked Wheel: Non-rotating wheel of a moving vehicle traversing the road surfaces without rotation, as in a *skid.*

Locked-Wheel Skid mark: A skid mark left by a braked, nonrotational wheel sliding in contact with a road surface.

Lockup Factor: See Wheel Lockup Factor.

Lock-up: Direct drive mode in which the driving and driven elements of a torque converter are mechanically coupled so that the complete unit, with casing, rotates at engine speed, thus eliminating fluid friction losses in the converter and increasing fuel economy.

Longitudinal Acceleration: The component of the vector acceleration of a point in the vehicle in the x-direction.

Longitudinal Axis: The axis extending from the rear to the front and through the center of mass of a vehicle, parallel to its sides. Also known as the roll axis or centerline.

Longitudinal Force: The component of the force vector in the x-direction.

Longitudinal Lane Change Distance: Considering the lateral lane change distance that corresponds to a constant vehicle speed and the average vehicle acceleration over the first half and latter half of the lane change maneuver derived the lane change model and resulted in formulae for the longitudinal lane change distance and the lane change time as follows:

$$d = 2V\sqrt{\frac{S}{a_{\mathrm{AVG}}}} \quad \text{and} \quad t = 2\sqrt{\frac{S}{a_{\mathrm{AVG}}}},$$

where d is the longitudinal lane change distance, t is the longitudinal lane change time, a_{AVG} is the average acceleration in [m/s^2], and a_{PEAK} is the peak acceleration in [m/s^2].

Longitudinal Slip (Percent Slip): The ratio of the longitudinal slip velocity to the spin velocity of the straight free-rolling tire expressed as a percentage.

Longitudinal Slip Velocity: The difference between the spin velocity of the driven or braked tire and the spin velocity of the straight free-rolling tire. Both spin velocities are measured at the same linear velocity at the wheel center in the X' direction. A positive value results from driving torque.

Longitudinal Slip: Tendency of a wheel's tire to slip slightly in normal drive or braking mode.

Longitudinal Velocity: Longitudinal Velocity of a point in the vehicle is the component of the vector velocity in the x-direction. See Fig. L.1.

Magic Formula: The "magic formula" presented by professor Pacejka in Holland Delft Industrial University has been shown to suitably match the experimental data of the dynamic tire characteristics by the combinational formula of the trigonometric function. Pacejka obtained a set of formulas of tire models that can describe the longitudinal and lateral forces and the self-aligning moment comprehensively, with the same form. The formula is called the "magic formula".

Magic Model: The magic model is empirical, requiring specification of a number of parameters determined from experimental measurements of the tire forces and moments. Such measurements require sophisticated test equipment, which makes the magic model impractical for organizations with modest tire-testing capabilities. Also, the magic model is not well suited to parametric studies of the impacts of snow- and ice-induced changes in the surface friction on vehicle performance. A static friction coefficient appears in the model, but it determines only the peak forces and moments. Its adjustment alone does not properly account for the variation in the surface friction commonly observed at different tire/road slip conditions. A full complement of force and moment measurements seems to be required instead for each road surface condition examined. Hence, an alternative tire model is needed that (1) requires only a limited set of easily measured tire input parameters; (2) is analytical in nature, with physically meaningful parameters that relate directly to tire characteristics and road surface conditions; (3) realistically accounts for changes in surface friction; and (4) is numerically simple and practical for use in complex vehicle simulations. A steady-state form of the solution is adequate because the problem being addressed is low-frequency vehicle maneuvering on smooth nondeformable road surfaces.

Mass Center of Gravity (CG): That point of a body from which it could be suspended or supported and be in the balance.

Mass Centre: Mass centre is that point in relation to a body where any plane passing through it divides the body into two parts of equal weight, and through which acts the resultant force due to the body's inertia when it is accelerated. See Fig. L.1.

Mass Ratio: The effective mass ratio is defined as $KG/\,(KG + h^2)$. It is a dynamic measure of the effect of mass on the eccentricity of the force. A mass ratio of 1 indicates a central collision. The greater the force offset, the smaller the value of the effective mass ratio.

Mass: The mass of a body is determined by comparison with a standard mass, using a beam-type balance.

Maximum Load: The greatest load a bar will hold is called the maximum load, and the corresponding unit stress is the ultimate strength.

Mechanical Efficiency: Mechanical efficiency η of a machine is the ratio of useful output to total input of work. Hence, $\eta = \dfrac{W_u}{W_a}$, where $W_a = W_u + W_f$, in which W_a is the work applied to the machine, W_u is the useful work performed, and W_f is the work required to overcome friction or other types of resistance.

Middle Ordinate: The distance of the center point of the chord, to the outer edge of the inertial scuffmark. Used in determining the radius of the circle that most closely approximates the shape of the arc. The perpendicular distance between an arc and its chord at the middle of the chord.

Mode of Vibration: In a system undergoing vibration, a mode of vibration is a characteristic pattern assumed by the system in which the motion of every particle is simple harmonic with the same frequency. Two or more modes may exist concurrently in a multiple degree-of-freedom system.

Modeling: Modeling refers to the determination of a quantitative picture of the important system characteristics.

Moment of a Force: The moment of a force M is defined as the product of an applied force, F, and the length of the moment arm, d. In equation form: $M = Fd$.

Moment of a Momentum: It is the product of the momentum and the distance from the center of mass to the center of moments. Also known as the angular momentum.

Moment of Inertia: A physical property of a body that represents its resistance to rotational acceleration.

Moment: The moment of a force F about a point O is $M = F\ell$, where ℓ is the perpendicular distance (called level armℓ) from the point O to the line of action of force F.

Momentary Deceleration: $a = d\upsilon/dt$

Moments: The external moments acting on the vehicle can be summed into one moment vector having the following components:

Momentum Attraction (Gravitation): Two particles attract each other with a force F proportional to their masses m_1 and m_2 and inversely proportional to the square of the distance r between them; or $F = Km_1m_2/r^2$, where $K = 6.673 \times 10^{-11}$ m^3/kg s^2 or 3.44×10^{-8} ft^3/lb s^2 is the gravitational constant.

Momentum: Momentum is a vector quantity pertaining to the motion of an object. Its magnitude is the mass of the object times its speed, and its direction is the direction of motion. The unit of momentum is mass times speed (kg m/s).

Motion: Motion of a particle with respect to other particles or objects is its state of continual changing of position with respect to them.

Motor Vehicle: Every vehicle which is self-propelled and every vehicle which is propelled by electric power obtained from overhead trolley wires, but not operated upon rails.

Motor: A prime mover. The engine of a vehicle.

Motor-Vehicle Accident: An accident involving a motor vehicle in transport, but not involving aircraft or watercraft.

Motor-Vehicle Traffic Accident: Any motor-vehicle accident that occurs on a traffic way or that occurs after the motor vehicle runs off roadway but before events are stabilized.

Multiple Collisions: In the multiple collision situations, the accident investigator must deal with the interactions of more than two vehicles. These situations are the kind where a vehicle may sideswipe another vehicle, be forced across the centerline and then strike with an oncoming vehicle.

Mu-split: Sudden change in road or ground surface friction, relevant to its effect on braking or cornering.

Natural Frequency: The natural frequency of free vibration of a system is the frequency at which an undamped system with a single degree of freedom will oscillate upon momentary displacement from its rest position. For a multiple degree-of-freedom system, the natural frequencies are the frequencies of the normal modes of vibration.

Natural Mode of Vibration: The manner of vibration associated with each natural frequency. The fundamental natural frequency.

Natural Modes: The eigenvectors are also referred to as modal vectors and represent physically so-called natural modes.

Natural Vibration: If the oscillating motion about an equilibrium point is the result of a disturbing force that is applied once and then removed, the motion is known as natural (or free) vibration.

Near Side: The side of the vehicle that is being impacted.

Net Contact Area: The area enclosing the pattern of the tire tread in contact with a flat surface, excluding the area of grooves or other depressions.

Neutral Stability: Neutral stability exists at a prescribed trim if, for any small temporary change in disturbance or control input, the resulting motion of the vehicle remains close to, but does not return to, the motion defined by the trim.

Neutral Steer Line: The set of points in the x-z plane at which external lateral forces applied to the sprung mass produce no steady-state yaw velocity.

Neutral Steer: A vehicle is neutral steer at a given trim if the ratio of the steering wheel angle gradient to the overall steering ratio equals the Ackerman steer angle gradient.

Neutral Turn: Turn executed by a track laying vehicle in which one track only is driven, and the drive to the second track is no neural.

Neutral: Transmission idling position when no gear is engaged.

Newton's First Law of Motion (Law of Inertia): An object at rest or in uniform motion will remain at rest or in uniform motion until acted upon by an unbalanced or external force.

Newton's Second Law of Motion (Law of Acceleration): The acceleration (a) of a body is directly proportional to the force (F) acting upon the body and inversely proportional to the mass (m) of the body $(F = ma)$. Also $a = F/m$.

Newton's Third Law of Motion (Law of Action and Reaction): For every action, there is an equal and opposite reaction. Or, whenever one body exerts force on another, the second body always exerts a force on the first body which is equal in magnitude, opposite in direction, and has the same line of action.

Nibbling: Effect of lateral force on a *tire* resulting from glancing contact with a rut or shallow ridge.

Nonelastic Collision: A nonelastic collision, also referred to as an inelastic collision, is a collision between two bodies in which there is permanent deformation, *i.e.*, the objects lose their original shape. Some kinetic energy is transformed into internal energy in this type of collision. When two bodies collide and stick and travel together with a common final velocity after the collision, the collision is said to be completely inelastic. Hence, an inelastic collision is a collision in which kinetic energy is not conserved. Inelastic collisions may not conserve kinetic energy, but they do obey conservation of momentum. Simple ballistic pendulum problems obey the conservation of kinetic energy *only* when the block swings to its largest angle.

Normal Force (F_z): The force that a body exerts on another body at right angles to their mutual interface. The component of the tire force vector in the Z' direction.

Normal Mode of Vibration: A normal mode of vibration is a mode of vibration that is uncoupled from (*i.e.*, can exist independently of) other modes of vibration of a system. When vibration of the system is defined as an eigenvalue problem, the normal modes are the eigenvectors and the normal mode frequencies are the eigenvalues. The term "classical normal mode" is sometimes applied to the normal modes of a vibrating system characterized by vibration of each element of the system at the same frequency and phase. In general, classical normal modes exist only in systems having no damping or having particular types of damping.

Normal Modes: The process of adjusting the elements of the natural modes to render their amplitude is called normalization, and the resulting vectors are referred to as normal modes.

Normal Velocity: Normal Velocity of a point in the vehicle is the component of the vector velocity in the z-direction. See Fig. L.1.

North-South Configuration: When packaged in a vehicle, the power plant could have the same coordinates as the vehicle. This package of power plant is called north-south configuration. Most trucks and some large and sport cars have this kind of configuration. For this case, the six motions of the power plant are the same as the vehicle body. See Figure D.4.

Number of Degrees of freedom: The number of degrees of freedom is equal to the number of coordinates required to completely specify the state of an object.

Oblique Collision: A collision between two vehicles in which the direction of the vehicles prior to the collision were neither parallel nor perpendicular. See also Angular Impact.

Oblique Impact: An impact in which the contact impulse does not pass through the center of gravity. See central impact.

Occupant Kinematics: An analysis of the motion of occupants during a collision for correlating occupant injuries to interior vehicle contact damage, and determination of relative seating position.

Occupant: Can be one of the passengers or the driver of a vehicle.

Offset: The distance between the longitudinal heading axes of two vehicles in frontal contact; see Fig. O.1. See also Overlap, Print, Intercept.

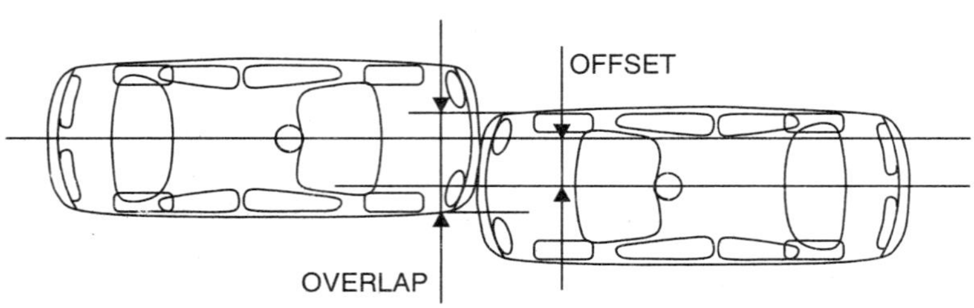

Fig. O.1

Override: A condition in a collision where the main structural members, such as a bumper, of the striking vehicle are above the main structural members, such as frame rails, of the struck vehicle; See Fig. O.2. See also Underride.

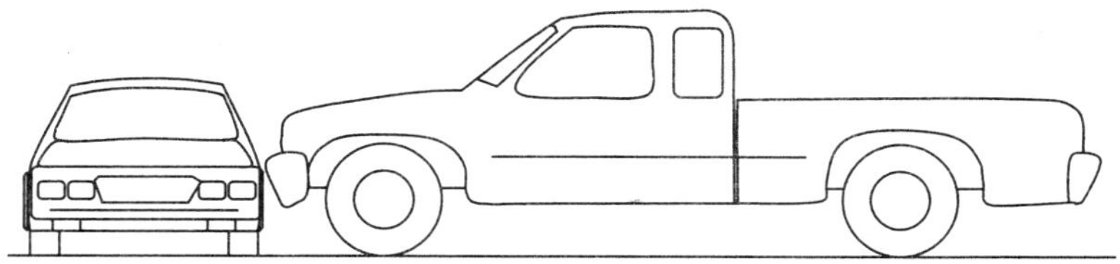

Fig. O.2

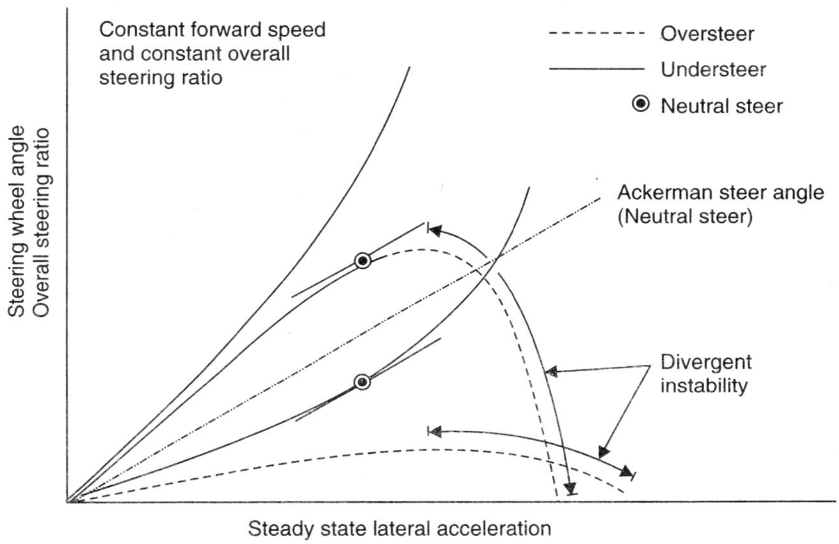

Fig. O.3 Steer properties.

Oversteer: When a vehicle, travelling on a circular path at constant speed and a constant front wheel steer angle, is accelerated, it will remain on a path with the same radius, tend to increase its radius, or tend to decrease its radius; these are defined as neutral steer, understeer, and oversteer, respectively. See Neutral steer; Understeer.

Overturning Moment (M_z): The component of the tire moment vector tending to rotate the tire about the X' axis, positive clockwise when looking in the positive direction of the X' axis. See Fig. S.5.

Partial Impact: An impact in which motion is continuous between the parts of colliding objects which are in contact with each other; sideswipe. Compare with full impact. 61-5

Part-Time Case: Four-wheel-drive transfer gear case that permits disengagement of drive to one axle.

Passenger Car: Vehicle with motive power designed for carrying usually ten persons or less, and not for hire or reward.

Passenger: Includes everyone except the driver of a vehicle.

Passenger's Side: Position opposite the driver's side.

Passive Restraint: Device to restrain vehicle occupant in event of an impact, and which does not require any action of the occupant, such as fastening or adjustment.

Passive Safety: Safety qualities of a vehicle assuming an impact has taken place, for example structural crashworthiness, energy absorption of steering column, etc.

Passive Vehicle: Vehicle constructed for carriage of passengers and their effects.

Patch: Area of contact of *tire* with ground.

Path Heading: A vehicle, as it comes into a collision, has a path defined by the motion of its center of gravity. The direction of this path is called the path heading.

Pavement: That part of a *roadway* having a constructed surface for the facilitation of vehicular *traffic.*

Payload: The cargo of a goods vehicle, or the weight of the cargo, as subject to local definition; effectively.

Payout: The term given to a condition where the length of the webbing of the seat belt increases due to a malfunction of the retractor mechanism.

PDOF: Principal direction of force. Many computer simulation programs today require an input telling it the principal direction from which the force was being applied to the vehicle. The clock directions usually do this, with the front of the vehicle being 12 o'clock.

PDR Time: See Perception-Decision-Reaction Time.

Peak Brake Power: Highest power developed within speed range of an engine. Most commercial diesel engines are governed to operate below peak brake power speeds.

Peak Power Engine Speed: Engine speed at which peak power occurs.

Peak torque Speed: Engine speed at which peak torque occurs.

Peak Value: The peak value generally refers to the maximum stress that the vibrating part undergoes.

Peak-to-Peak (Total) Lateral Force Variation: The difference between the maximum and minimum values of the lateral force during one revolution of the tire.

Peak-to-Peak (Total) Radial Force Variation: The difference between maximum and minimum values of the normal force during one revolution of the tire.

Peak-to-Peak Amplitude (Double Amplitude): Peak-to-Peak amplitude of displacement at a point in a vibrating system is the sum of the extreme values of displacement in both directions from the equilibrium position.

Peak-to-Peak Lateral Tire Run-Out: The difference between maximum and minimum indicator readings, measured parallel to the spin axis at the point of maximum tire section, on a true running wheel (measured separately for each sidewall).

Peak-to-Peak Lateral Wheel Run-Out: The difference between maximum and minimum indicator readings, measured parallel to the spin axis on the inside vertical portion of a rim flange (measured separately for each flange).

Peak-to-Peak Lateral Wheel Run-Out: The difference between maximum and minimum indicator readings, measured parallel to the spin axis on the inside vertical portion of a rim flange (measured separately for each flange).

Peak-to-Peak Loaded Radial Tire Run-Out: The difference between maximum and minimum values of the loaded radius on a true running wheel.

Peak-to-Peak Radial Wheel Run-Out: The difference between the maximum and minimum values of the wheel bead seat radius, measured in a plane perpendicular to the spin axis (measured separately for each bead seat).

Peak-to-Peak Unloaded Radial Tire Run-Out: The difference between maximum and minimum undeflected values of the tire radius, measured in plane perpendicular to the spin axis on a true running wheel.

Peak-to-Peak Value: The peak-to-peak value of a vibrating quantity is the algebraic difference between the extremes of the quantity.

Peak-to-Peak: Amplitude of an alternating quantity measured from positive peak to negative peak.

Pedestrian c.g.: To locate the center of gravity for an adult pedestrian, multiply height (in inches, without shoes) by 0.57, and than add shoe heel height.

Pedestrian Manifestations: Those items of evidence deposited at the collision scene by the pedestrian during or following impact.

Pedestrian: Any person afoot; any person not in or upon a motor vehicle or other road vehicle.

Perception Time: The time it takes the human body to perceive that a situation is about to take place, or is taking place on the highway ahead, that requires some action.

Perception: Comprehending a *situation,* such as a *hazard,* by means of the senses and the mind; in *highway* transpörtation, comprehending the risk of starting or continuing a movement; the time and place at which perception occurs.

Perception-Decision-Reaction Time: The time required by a person to complete a response to an event or stimulus.

Perception-Reaction: The time from the moment a stimulus is first perceived until the moment the observer first moves in response to it.

Perception-Response Time: It is the time from the moment the stimulus is perceived until the moment that the vehicle begins to respond. The perception-response time is the perception-reaction time plus the transition time. Or perception-response time is the perception-reaction time plus the movement time plus the mechanical latency.

Perfectly Inelastic Collision: In a perfectly inelastic collision, *i.e.*, a zero coefficient of restitution, the colliding particles stick together.

Period: Period of an oscillation is the smallest increment of time in which one complete sequence of variation in displacement occurs.

Pitch and Bounce Modes: A representative model of the vehicle would have to account for both bounce and pitch modes. If the whole vehicle is represented as a mass, setting on two different springs representing front tires and rear tires, it is possible to obtain a reasonable representation of the vehicle. Fig. P.1 shows a two-degree-of freedom model in which both bounce and pitch modes can be accounted for. Pitch and bounce modes are generally coupled.

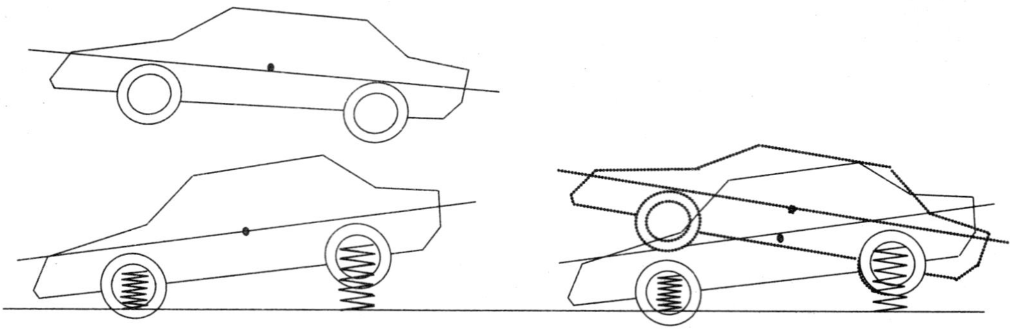

Fig. P.1 Pitch and bounce modes

Pitch Axis: See Lateral Axis.

Pitch Velocity: The angular velocity about the *y*-axis. See Fig. L.1.

Pitch, Roll, Yaw: Terms that distinguish rotations of a vehicle about three perpendicular axes with origin at the vehicle's center of gravity; pitch is rotation about the horizontal side-to-side axis, roll is rotation about a horizontal front-to-rear axis, and yaw is rotation about the vertical axis; See Fig. P.2. See also Yaw Angle.

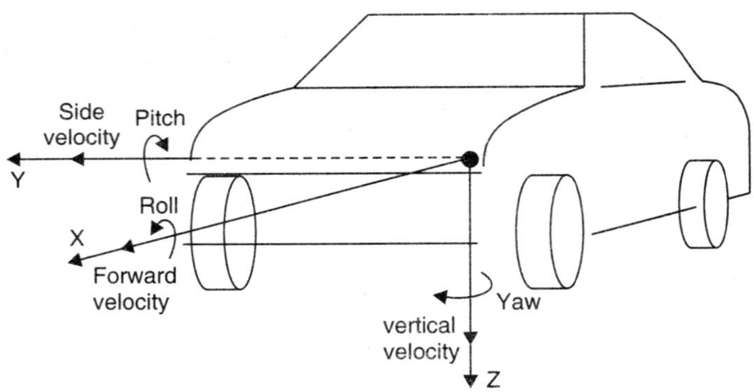

Fig. P.2

Pitch: Rotary motions of the sprung mass of a car around the transverse axis. The front end will rise while the rear lowers and vice versa. See Fig. P.2.

Pitching Moment: The component of the moment vector tending to rotate the vehicle about the *y*-axis, positive clockwise when looking in the positive direction of the *y*-axis.

Planar Impact: An impact in which all forces, moments, and motion take place in a plane.

Plastic Damage: Permanent damage to a vehicle that occurs during a collision.

Plastic Impact: An impact with little or no rebound at the end of impact; a perfectly plastic impact is where the coefficient of restitution is equal to zero.

Plasticity: This property is the exact opposite of elasticity. It is the state of a material that has been loaded beyond its elastic limit under a load beyond that required to cause elastic deformation (the elastic limit). A material possessing the property of plasticity deforms permanently.

Ply: Layer of rubberized fabric from which a tire carcass is made.

Pneumatic Tire: Hollow rubber toroid inflated by air pressure and mounted on the rim of a wheel.

Point Contact Model of Tires: In the point contact model of tires, it is assumed that the contact region between the tire and the ground is a point that is the projection of the center of the tire onto the ground

Point Mass: An idealized concept from mechanics in which an object is considered to have mass but no extent, no finite dimensions, and as a consequence, its rotation is irrelevant. See Rigid Body.

Point of Actual Perception: The point at which an object is detected and recognized as a potential hazard.

Point of Collision: See First Contact Point.

Point of Contact: The point of intersection of the contact impulse and the inter vehicular contact surface during an impact. See also Direction Of Principal Force (DOPF); Final Contact Position; Impact Center; Principal Direction Of Force (PDOF).

Point of First Contact: See First Contact.

Point of Impact (POI): The collision sequence where involved objects first come into one another or surface. Can be referred to as area of impact (AOI) and initial contact. The point of impact is the location on the roadway where two vehicles come together. This determination is often made by noting gouge marks in the pavement, the position of debris on the roadway or similar items.

Point of Intersection: See Intercept.

Point of Maximum Engagement: See Maximum Engagement.

Point of Perception: See Perception.

Point of Possible Perception: The place and time at which the hazard could have been perceived by a normal person. It precedes actual perception and is the beginning of perception delay.

Point of Rest: Controlled or uncontrolled post-collision position of vehicles or persons.

Point of Separation: See Separation. **Point of separation:** to obtain this, each vehicle should be drawn on the scale diagram as it was positioned at the point of impact and also as it was positioned at the point of separation. The PS can usually be established from roadway marks such as skid or scrub marks, chips and gouges. It should be noted that the departure angle for a vehicle is the angle formed with the *X*-axis, or with a baseline drawn parallel to that axis, by the vehicle's path of travel after separation. This path is plotted by drawing a line form the center-of-mass location of the vehicle at time of impact through its center-of-mass location where the vehicle separated from the other vehicle or object, as shown in Fig. P.3.

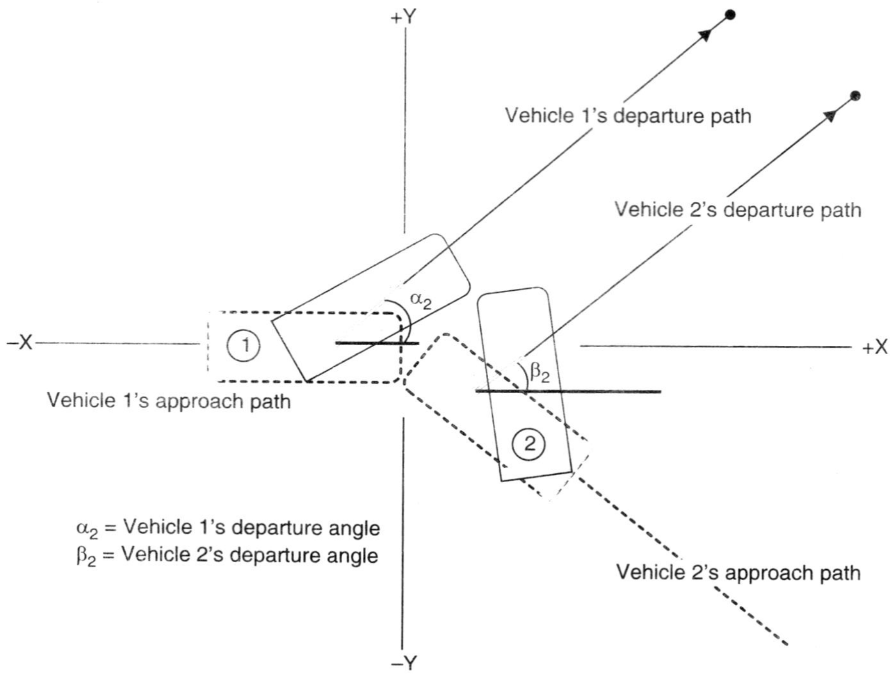

Fig. P.3 A diagram showing departure angles measured from the *X*-axis, or parallel baseline, to the line drawn from the center-of-mass of each vehicle at impact (broken-line rectangle) through the center-of-mass of the vehicle at separation, [Rivers (1997)]

Positive Offset Steering: Steering/suspension geometry in which point of intersection of *kingpin axis* with the *ground plane* lies inboard of the center of the *tire contact patch* or plane of the wheel.

Possible Perception: See Point of Possible Perception.

Post Collision Trajectory: See Post Impact Trajectory.

Post Crash Damage: Damage existing to a vehicle after it comes to rest, including damage that may result during rescue, towing, and salvage operations.

Post Crash: The time in a collision sequence from the point of separation to the last event that significantly influences the vehicles or occupants of those vehicles.

Post Impact Speed: The magnitude of the velocity of an object in a collision at the time of separation or end of contact. See Post Impact Velocity; Separation Speed.

Post Impact Trajectory: The path of the center of gravity of a vehicle from impact to rest.

Post Impact velocity: The velocity of an object in a collision at the time of separation or end of contact.

Post: Nominally upright structural member, particularly for supporting the roof and doors of a vehicle.

Postcollision Trajectory: See Postimpact Trajectory.

Postcrash Damage: Damage coming to a vehicle after it comes to rest, including damage that may result during rescue, towing, and salvage operations.

Postimpact Skid Distance: The distance a vehicle skids after an impact or collision.

Postimpact Speed: The magnitude of the velocity of an object in a collision at the time of separation or end of contact. See Postimpact Velocity; Separation Speed.

Postimpact Trajectory: The path of the center of gravity of a vehicle from impact to rest.

Postimpact Velocity: The velocity of an object in a collision at the time of separation or end of contact. See Postimpact Speed; separation Speed.

Potential Energy: A type of energy having to do with how two or more objects are interacting. Potential energy is stored in the interaction of an object with another object. Gravitational potential energy is a type of potential energy. Potential energy can be thought of as stored energy which, if released, can do work. An example is the energy stored in a linear spring when it is compressed.

Power Brakes: Brakes operated by mechanical, hydraulic or pneumatic power rather than by driver effort.

Power Hop: An oscillatory hopping motion of a single wheel or of a pair of wheels, which occurs when tractive force is applied in forward or reverse motion of the vehicle.

Position: Position involves the idea of measuring location in space.

Power Steering: Steering in which angular movement of the steering wheel actuates a powered steering system, there being no mechanical feedback between road and driver.

Power: Power is defined by the ratio of work or energy divided by time.

$$P = \frac{E}{t} \text{ (lb ft/sec.)}$$

where E = work done with a constant force, lb ft

t = time during which work is done, sec

Power: Power is the rate at which work is being done. If a force F is acting upon a body and moving it in the direction of the force with a velocity v, the power is $P = Fv$. The power of a torque T at any instant is $P = T\omega$, where ω is the instantaneous angular velocity of the body.

Pre Crash: The period of time in a collision sequence from the first significant event associated with a collision to initial contact.

Preimpact Skid Distance: The distance a vehicle skids prior to an impact or collision.

Preimpact Velocity: The velocity of a vehicle in a collision at the instant of its initial contact.

Pressure: A force per unit area.

Primary Cause: A misnomer loosely applied to the most obvious or easily explained factor in the cause of an accident or the most easily modified condition factor.

Principal Direction of Force (PDOF): Determined from vehicle damage, indicating of the magnitude and direction of the opposing force acting on a vehicle during collision. Also known as thrust and direction of principal force. The direction of the line of action of the contact impulse in a planar collision expressed in degrees, measured positive clockwise from the longitudinal axis of a vehicle.**:** The principle of conservation of linear momentum assumes that the only forces that act on the vehicles during the collision are the collision forces between the vehicles. From Newton's Third Law, the collision forces that act on the vehicles are equal in magnitude but opposite in direction. The direction of the force can be reported as a global angle or as an angle measured relative to the vehicle. When measured relative to the vehicle, it is generally known as the Principal Direction of Force or PDOF. The global force angles α_1 and α_2 give the direction of the force angles measured relative to the Global X coordinate axis which is fixed to the ground. $Pdof_1$ and $Pdof_2$ give the direction of the force angle relative to each vehicle's Local x coordinate axis.

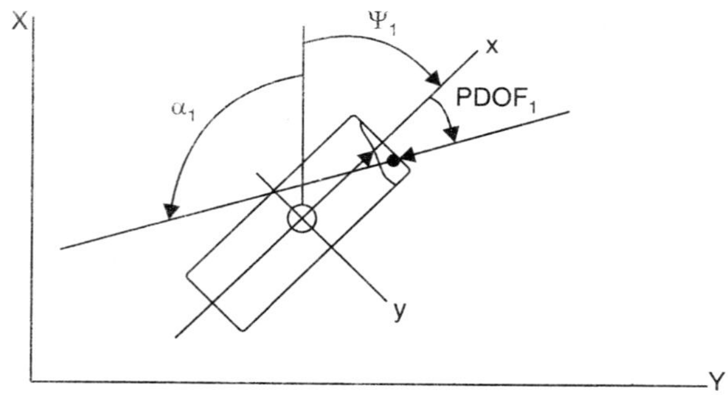

Fig. P.4 PDOF angle measurement

On a vehicle's local coordinate system, the x-axis is directed towards the front of the wheel. As is always the case, angles are measured from the positive x-axis. The y-axis forms an angle of +90 degrees and determines the direction for positive angles. For a right handed coordinate system the y-axis is directed to the right side of the vehicle.

The PDOF is defined by the tail of the vector consistent with the coordinate system. Refer to Fig.P.4. From the geometry, it can be seen that the PODF can be calculated from the global force angle α and the entry angle ψ as

$$\text{PDOF} = \alpha - \psi + 180$$

For example, in a frontal collision where the force is directed to the rear of the vehicle, the PDOF would be zero degrees. With the right handed coordinate system shown in Fig.P.4 (which is the normal system when discussing PDOFs), the PDOF will be reported as +30 degrees. See Thrust.

Print Marks: Print marks are made by a particular pattern on the opposite vehicle or object, including textiles or fabrics in car/pedestrian accidents.

Print: An imprint of liquid or fine dust picked up by a tire or shoe sole at one place and left at another, usually showing the pattern of the tire or shoe that made it.

Probability: The probability of an event occurring in an experiment equals the proportion of times the event occurs if the experiment is repeated a very large (infinite) number of times.

Professional Reconstruction: The effort to determine, from whatever information is available, how the accident happened. Reconstruction used to be referred to as determining behavioral or mediate causes of an accident. Reconstruction is the fourth level of traffic-accident investigation. It involves studying results of the accident, considering other circumstances, and applying scientific principles to form opinions relative to events of the accident which are otherwise unknown or are a matter of dispute.

Profile Drag: The aerodynamic drag of a vehicle attributable to the vehicle's external shape, and not to its surface texture or total surface area.

Projectile: A projectile is any object propelled through space by the exertion of a force, which ceases after launch.

Puddle: A wet area on the road or roadside where dribble accumulates after a vehicle has come to rest. A puddle often marks the final position of a vehicle after an accident.

Pull: A tendency of a car to steer toward one side.

Quantity: A characteristic of a system that can be well described and possibly assigned values. Energy, linear momentum, and angular momentum are examples of quantities associated with all systems.

Quarter Light: Rear side window of a passenger car or truck cab, particularly when set after the door.

Radius of Gyration: Radius of gyration is the square root of the ratio of the moment of inertia of a body about a given axis divided by its mass. The yaw radius of gyration can be related to vehicle wheel base and center of gravity position:

$$k = (1.023 \pm 0.085)\sqrt{a.b}$$

where k = yaw radius of gyration

a = center of gravity location to front axle

b = center of gravity location to rear axle

and the coefficient of fit $R^2 = 0.791$. The coefficient of fit R^2 is a measure of how well the data fits the assumed relationship and can range from zero to one. A value of one indicates the data fits perfectly.

Alternatively, the yaw radius of gyration is related to the overall length:

$$k = (0.298 \pm 0.030)L$$

where k = yaw radius of gyration

L = overall length

and the coefficient of fit $R^2 = 0.737$.

With these formulas, the unit of k will be the same as those used for a and b or for L.

Radius: The distance from the center of a circle to a point on its circumference; the distance from a point on an arc to the center of the circle of which the *arc* is part.

Raised Pavement Markers (R.P.M.): Non-reflectorized ceramic pavement markers, also known as *Bott's Dotts*, are most often used to delineate travel lane lines. Reflectorized RPM's come in many different colors and are used to delineate travel lane lines and outside boundaries.

Random Variable: A random variable is a variable whose value is determined by the outcome of a random experiment. Random variables are numerical values whose observed values are governed by the laws of probability. A random variable can be discrete or continuous.

Random Vibration: In the absence of any obvious pattern in a vibration record, the vibration is called a random vibration.

Rate: The amount of change of some quantity during a time interval divided by the length of the time interval.

Reaction Distance: The distance moved or travelled by a *vehicle* or other *traffic unit* during *reaction time.*

Reaction Time: The time taken by the human body to react to a perception that requires some action. It could be a brake application, or turning the steering wheel to move the vehicle out of the line of the action taking place ahead. See Perception-Decision-Reaction Time.

Reaction: A person's voluntary or involuntary response to a hazard or other situation that has been perceived; the response to a sensory stimulus. Reactions may be reflex, simple, complex, or discriminatory.

Real Impact: During *real impact*, a portion of the deformation work is returned elastically, while some is used to crush vehicle parts partially.

Rear Axle: The back axle.

Rear Drive System: A RWD system consists of the engine configured such that the crank shaft direction is perpendicular to the axle direction. This configuration is called a north-south configuration. Figure R.1 shows a typical RWD configuration. Similar to the FWD configuration, the RWD system can be configured for AWD and 4 × 4 applications. Several components must be added. The first is the transfer case (TC) that is added at the output end of the transmission. This unit consists of a chain and a set of spur gears. The purpose of the TC is to bring some torque back to the front wheels. The second major unit is the front drive shaft. The third major unit is the coupler.

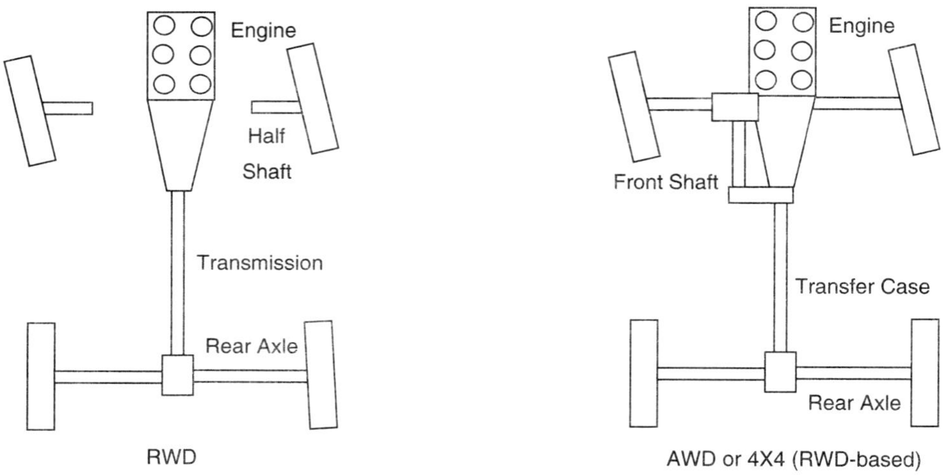

Fig. R.1 Rear wheel drive (RWD) and RWD-based all-wheel drive (AWD) systems.

Rear Fender: Body panel of passenger car between rear door and end of vehicle, and enclosing rear wheel.

Rear Lamp: Red lamp at extreme sides of back of vehicle to indicate presence and width.

Rear Lock-Up: Under rear lock-up, the maximum force at the rear axle will be equal to twice that of the wheel where lock-up is introduced for conventional brakes.

Rear Skid: The maximum force at the rear axle will be equal to twice that of the wheel where skid happens first for conventional drive systems.

Rear-End-Collision: In rear-end-collisions, both vehicles are moving in the same direction, in which the lead vehicle is stopped. In rear-end-collisions, both vehicles move forward down the highway. See Same Direction Collision.

Rear-View Mirror: Internal mirror to provide driver with rearward view through the backlight.

Rear-Wheel Drive: Transmission system in which the engine power is delivered to the rear wheels of a vehicle.

Rebound: A bounce motion during which the tire travels downward relative to the car and the spring and shock absorber extend.

Reconstruction: A systematic process of evaluating evidence associated with a collision sequence and applying accepted physical principles to determine how the collision occurred.

Recreational Vehicle: Vehicle intended primarily for off-highway leisure use. Also *RV.*

REC-TEC: A computer program used for basic accident reconstruction.

Rectilinear Motion: Motion along a straight path.

Reference Line: A line, often the edge of a roadway for which measurements are made. An extension of the reference point (RP) from which the *X* or *Y* direction measurement is taken. Generally speaking, the RP is a pole or hydrant set back from the street or road. A straight line is drawn from the RP to the curb or edge of the road. The line from the pole to the curb is the vertical reference line. A second line is then drawn on the outside of the curb and is the horizontal reference line.

Reference Point or Zero Point: The place where the two reference lines cross is a reference point, sometimes called a zero point, because it is the zero distance for each reference line. A point from which measurements are made to locate spots in an area; sometimes the intercept of two reference lines; RP. A reference point is described in terms of its relation to permanent landmarks.

Reflex Reaction: An involuntary response to a stimulus; an instinctive act resulting from perception of an imminently hazardous situation. Time: 0.2 sec, more or less. 64-14.

Regenerative Braking: Braking in which the loss of kinetic energy from braking is stored and subsequently fed back to provide tractive effort.

Regrooved Tire: Used tire on which a new *tread* has been cut, either into the original tread or a retread, but to a greater depth than that of the original molded groove.

Relative Motion: The displacement, velocity, and acceleration of a body with respect to a fixed point on the earth are called *absolute* displacement, velocity, and acceleration respectively. The displacement, velocity and acceleration of one body with respect to another which may also have displacement, velocity and acceleration with respect to the earth are called its *relative* displacement, velocity and acceleration respectively.

Relative Velocity: Drivers are not capable of detecting relative speed in a vehicle following situation until the subtended arc angle of the lead vehicle expands at a rate of 0.003 to 0.007 radians per second.

Release Time: Elapsed time between the moment when the actuating time for release starts and the moment when the braking force ceases.

Reliability: Reliability is concerned with failure-free performance of a system and the duration of time over which this failure-free performance is maintained. It is defined as the probability that a given system will perform its intended function satisfactorily for a specified period of time under specified operating conditions.

Residual Crush: The permanent deformation formed by the nominal external surface of a vehicle caused by an impact, usually measured relative to the corresponding as-manufactured undeformed surface. See Crush Area; Crush Profile.

Resonance Frequency (Resonant Frequency): A frequency at which resonance exists.

Resonance: A forced vibration phenomenon which exists if any small change in frequency of the applied force causes a decrease in the amplitude of the vibrating system.

Resonant Frequency: Frequency at which resonance exists.

Resonator: Acoustic chamber, generally of fixed volume, with a specific resonant frequency.

Response: The response of a device or system is the motion (or other output) resulting from an excitation (stimulus) under specified conditions.

Rest Position: The location of the center of gravity of a vehicle following an accident measured relative to a coordinate system fixed in the earth.

Restitution: See Coefficient of Restitution.

Resultant Force: Combined force acting on colliding vehicles to give a single resultant force.

Results from accident reconstructions are also useful in developing recommendations for making roads and highways safer, as well as improving safety aspects of motor vehicle designs.

Retarders: Any device that supplements a vehicle's *service brakes* by dissipating energy of motion, as by electrical or hydraulic means. Hydrodynamic or electrodynamic retarders are non-wearing auxiliary brakes for reducing the thermal load on the road wheel brakes under continuous braking.

Retreaded Tire: Used tire to which a new tread has been affixed by vulcanization to extend its useful life.

Returnability: The tendency of the front wheels to return to a straight-ahead position.

Reverse Gear: Reverse gear is the gear by which an engine provides power for movement in a direction opposite to the normal direction.

Rib Wheel: Open structure fabricated *tractor* wheel with broad annular rims usually fitted with removable lugs or "spades".

Rib: Raised circumferential feature of a tire tread pattern.

Ride Clearance: The maximum displacement in compression of the sprung mass relative to the wheel center permitted by the suspension system, from the normal load position.

Ride Comfort: Ride comfort is one of the most important factors for a customer in determining whether or not to purchase a particular vehicle. Ride quality is an important attribute in modern car design and is a major contributor to customer satisfaction. For almost all vehicle OEMs (original equipment manufacturers) and suppliers, customer satisfaction has become the critical issue in vehicle product development and business strategy.

Ride Control: Achieving the objectives of the suspension while a vehicle is in motion.

Ride Dynamics Analysis: In ride dynamics analysis, a vehicle is treated as a body installed on its suspension system that is supported by the tires and axles. The body and suspension constitute the sprung mass. The tires and axles constitute the unsprung mass.

Ride Height: The distance from a specific point on a vehicle to level ground. The actual measurement should be equal to vehicle specifications before correcting alignment.

Ride Quality: Ride quality deals with passengers' comfort in the vehicle. The judgment of ride quality can vary from one customer to another according to their perception and experience. Several factors such as vibration, noise, heat, sitting posture, age, gender, and environment influence ride quality. Among these factors, vibration is considered the most important in determining the overall ride quality. Passengers in a moving vehicle are continuously exposed to a vibration environment caused by excitations from the road surface and from vehicle internal sources such as the powerplant, which is the combination of the engine and the transmission. Hence, this chapter discusses ride quality caused by vibration.

Ride Rate: The change of wheel load, at the center of tire contact, per unit vertical displacement of the sprung mass relative to the ground at a specified load.

Ride Time: Period of a time when the bumper of a vehicle strikes a pedestrian or bicyclist and ending when the pedestrian or bicyclist strikes the ground. See also Through Distance.

Ride: Qualitative and quantitative nature of the response of a vehicle to road perturbations, particularly in relation to the performance of the suspension system and its influence on the degree of comfort experienced by the occupants.

Right Angle: An angle of 90 degrees formed by two lines perpendicular to each other.

Right Front Passenger: The seat opposite to the driver in the front seating area of the vehicle.

Right-Hand Drive: Controlled from the right-hand side (starboard) of the vehicle.

Rigid Axle: A *beam axle*.

Rigid Body: A rigid body does not deform when loaded and can be considered a combination of two or more particles that remain at, as dices, finite distance from each other.

Rigid Truck: A non-articulated vehicle, particularly a commercial vehicle carrying the payload on its own axles.

Rigid Vehicle: Vehicle which serves its function as a unit and not as a part of an articulated vehicle or train. Thus the tractor unit of an articulated vehicle is not, by many definitions, a rigid vehicle.

Rim Diameter: The diameter at the intersection of the bead seat and the flange.

Rim Flange: The retaining flange of a wheel rim, which supports and retains the *tire bead*.

Rim Pull: Tractive effort of a vehicle as measured at the wheel/road interface, or as calculated from torque and transmission ratio data.

Rim Width: The width across a wheel, measured between the rims.

Rim: The rim is the connecting component between vehicle and tire and is subjected to high forces and moments.

Ring Flash: A special photographic flash unit that attaches to the lens of the camera for close-up pictures in dark areas or for concentrated light.

Ringing: Method for exciting natural frequencies is to strike or excite a machine or structure with a timber or hammer.

Risks: Risk is a person's planned or inadvertent operation of a vehicle in such a manner that injury or harm may occur.

Road Accident: See Car Accident.

Road Feel: Feedback of front suspension and steering input to *steering wheel*, and to driver.

Road Geometry: A description of the collision site including name and other identification of the road, road radius, slope, crown, width of pavement, pavement markings, superior elevation, off-road surfaces, guard rails, pavement edge height, drainage, vegetation, and culverts.

Road Plane: Reference plane on which a vehicle stands. This may be inclined to represent road camber or surface curvature, subject to local definition.

Road Shock: A harsh force transmitted from the tires through the suspension or steering linkage.

Road Space: See Available Path.

Road Surface Conditions: A description of the road surface in terms of materials, such as concrete, asphalt, and gravel, and its condition, such as coarse or polished, pot holes, bumps, or frost breakups, is provided.

Road Traffic Incident: See Car Accident.

Road Type: Rural, urban, highway, or arterial.

Road: The part of a traffic way which includes both the roadway, which is the traveled part, and any shoulder alongside the roadway, where there are unmountable curbs.

Roadside: That part of a highway or street not occupied by a road or sidewalk.

Roadster: Passenger car of sporting appearance.

Roadway: That portion of a highway improved, designed, or ordinarily used for vehicular travel, exclusive of the beam or shoulder. In the event a highway includes two or more separate roadways the term 'roadway' as used herein shall refer to any such roadway separately but not to all such roadways collectively. If there is a curb, the roadway is considered to extend to the curb; hence roadway may include lanes commonly used for parking. If there is a paved shoulder, the roadway may be distinguished from the shoulder mainly by a painted line marking the edge of the roadway.

Roll Axis: The line joining the *roll centers* of the front and rear suspension of a vehicle. The axis about which a vehicle rolls in response to a roll perturbation.

Roll Camber: The camber displacement of a wheel resulting from suspension roll.

Roll Center: Instantaneous geometric center about which a vehicle body rolls, as determined by the suspension geometry. The roll center defined in 9.4.28 constitutes an idealized concept and does not necessarily represent a true instantaneous center of rotation of the sprung mass.

Roll Distance: The rollout distance is that portion of the trajectory when the vehicle tracks its rear and front wheels. If there is a skid end point, the vehicle is assumed to rollout to path end.

Roll Out: Part or all of a postimpact trajectory in which little or no sideslip of a vehicle's wheels occurs. See Spinout.

Roll Oversteer: Roll steer which decreases vehicle understeer or increases vehicle oversteer.

Roll Rate: Suspension roll stiffness.

Roll Steer Coefficient: The rate of change in roll steers with respect to change in suspension roll angle at a given trim.

Roll Steer: Roll steer is referred to as the steering motion of the wheels with respect to the sprung mass that is due to the rolling motion of the sprung mass. Roll steer could lag the steer input of the vehicle in handling.

Roll Stiffness: Resistance of a vehicle to rolling expressed as moment per unit angular displacement in roll.

Roll Understeer: Roll steer, which increases vehicle under steer or decreases vehicle over steer.

Roll Velocity: Angular velocity about the roll axis. See Figure L.1.

Roll: A rotary motion of the sprung mass of a car around the longitudinal (lengthwise) axis that results in body roll. See Figure P.2.

Rolling Friction: Resistance to motion of the rolling of one body over another.

Rolling Moment: The component of the moment vector tending to rotate the vehicle about the x-axis, positive clockwise when looking in the positive direction of the x-axis.

Rolling Radius: Radius of loaded tire from axis to center of *tire footprint*.

Rolling Resistance Coefficient: The rolling resistance coefficient f_r is computed by

$$f_r = \frac{F_{\text{pull}}}{W}$$

where F_{pull} = pulling force, lb
W = tire weight, lb

Rolling Resistance Force: The negative longitudinal force resulting from energy losses due to deformations of a rolling tire.

Rolling Resistance Moment (M_y): The component of the tire moment vector tending to rotate the tire about the Y' axis, positive clockwise when looking in the positive direction of the Y' axis. See Figure S.5.

Rolling Resistance of a Vehicle: The rolling resistance of a vehicle is the summation of all individual tire rolling resistances, computed by multiplying rolling resistance coefficient by vehicle weight.

Rolling Resistance of the Tire: The rolling resistance of the tire, which is applied to the wheel, is due to the deformation of the tire at the tire/road interface.

Rolling Resistance: Horizontal force required to keep a vehicle in motion on a level surface with the engine disconnected from the wheels and with no brake application; drag factor produced by friction within the vehicle and deformation of the tires and road surface. Compare with wind resistance.

Rollover Accident: Rollover accidents are unique, since in nearly all cases they involve a single vehicle, although the initial loss of vehicle control may have been caused by a prior impact with another vehicle or object. Rollover of a vehicle occurs when both inner wheels lift off the ground and the subsequent roll motion is sufficient to turn the vehicle on its side, roof, or into a multiple rollover. Rollover is directly related to lateral load transfer.

Rollover Preventive Metric: The Rollover Preventive Metric (RPM) is defined as the percentage difference between the lateral kinetic energy and the initial rotational energy as

$$RPM = 100\left[\frac{T_0 - T_1}{T_0}\right]$$

where T_0 and T_1 are the vehicle lateral kinetic energy and rotational energy, respectively.

Rollover Scenario: Rollover can happen on flat road, on cross-slope road, or on off-road. Much attention has been paid to automotive rollover accidents over past decades. Rollover accidents are a serious threat to the life of automobile occupants. Rollover is defined as a vehicle rotation of 90 degree or more around its longitudinal axis. The vehicle body will contact ground and occupants will be hurt. Fig. R.3 shows a vehicle rollover scenario.

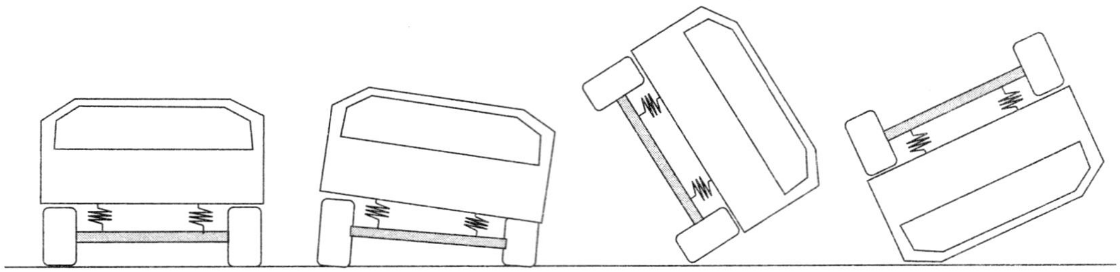

Fig. R.3 A vehicle rollover scenario.

Rollover Threshold: The lateral acceleration rate at which the vehicle will begin rotation about its longitudinal axis.

Rollover: The complete capsizing of a vehicle. See *Roll.* Vehicle motion in which its wheels leave the road surface and at least one side or top of the vehicle contacts the ground. See Flip; Vault.

Roof Vault: A roof vault begins as a wrap collision but, usually because of the high speed of the vehicle, the pedestrian continues up onto the roof. The pedestrian continues rearwards relative to the vehicle and may hit a rear surface such as a trans lid before reaching the ground.

Rotational Particle Motion: Also known as angular motion and circular motion is motion of a particle around a circular path.

Roughness: Roughness is defined as road elevation profile. Roughness is used to evaluate the road smoothness. Usually, a road profile is a wide-band random function.

RP (Reference Point): Chosen by the on-scene investigator, from which to take all measurements, in an *X* and *Y* or north and south direction.

Run Off: When fluid flows from a ruptured vehicle reservoir down a sloping surface.

Run-Flat Tire: Pneumatic tire capable of limited operation without air inflation.

Runoff: Rivulets of liquid debris from a puddle area downhill toward soak-in at the edge of the pavement; frequently the source of tire prints after a collision.

Run-On Tire: Pneumatic tire designed for use with special rims to prevent bead unseating in the event of a *puncture* or *blow-out*, thus enabling a moving vehicle to be brought safely to rest.

SAE Coordinate System: See Three-Axis Vehicle Coordinate System. See Fig. S.5.

Safe Approach Speed: A safe approach speed is the safe speed in preparation for entering or crossing a road or intersection.

Safe Driving: Safe driving is a form of training for motor vehicle drivers that goes beyond mastery of the rules of the road and the basic mechanics of driving. Its aim is to reduce the risk of driving by anticipating dangerous situations, despite adverse conditions or the mistakes of others. This can be achieved through adherence to a variety of general rules, as well as the practice of specific driving techniques.

Safe Speed: A safe speed is one adjusted to the potential or possible hazards of the road and traffic situation ahead.

Safe Stopping Distance (SSD): The safe stopping distance for vehicles knowing the actual time required for the driver to perceive and react to a hazard is given by

$$SSD = vt_p + \frac{v^2}{2a},$$

where SSD is the safe stopping distance for the vehicle (ft or m), t_p is the perception/reaction time (seconds), v is the velocity, m/s (ft/s), and a is the acceleration factor (the effective drag factor times the gravitational constant). Most design standards for rural and urban roadways use a 2.5 second perception/reaction time to include nearly all drivers and conditions for computing a safe stopping sight distance.

Safety Bumper: Bumper designed to protect vehicle on impact or reduce injury on impact with other road users.

Safety: Safety is a measure of probability that a hazard or danger does not exist. Safety is increased by minimizing or even eliminating, if possible, the hazard associated with the driver, vehicle, road, and environment.

Same-Direction Collision: A collision between two traffic units moving in the same direction on the same roadway. Sometimes called *rear-end collisions.*

Sandshoe: Trailer *landing leg* foot for use on soft or uneven terrain.

Scalar: Any quantity that requires only magnitude, irrespective of direction. A scalar is a quantity that does not have a direction. Examples are mass, time, and moment of inertia.

S-cam Brake: Drum brake with S-shaped cam to actuate the *shoes* and compensate for *lining* wear.

Scar: Any mark on a road, on the ground, or on a fixed object such that surface material is removed or displaced; durable sign on the road at the location of a traffic accident.

Scene: The location of a traffic accident before vehicles and people involved have left. Compare with site.

Scientific Method: 1) make observations of available facts; 2) develop a hypothesis; and 3) test the hypothesis.

Scouring: Form of *tire* wear caused by inaccurate tracking.

Scrape Marks: Scrape marks are made when vehicle components are sliding over the road surface without excessive downward force resulting in minimal groove depth. Large body surfaces such as the roof or side of a vehicle typically produce scrape marks.

Scrape: A broad area of a hard surface covered with many scratches or striations made by a sliding metal part without great pressure.

Scratch: A light and usually irregular scar mode on a hard surface, such as paving, by a sliding metal part without great pressure. Scratches are visible but not normally distinguishable to the touch.

Screen Wiper: Windscreen or windshield wiper.

Scrub: Skid marks left by tires pushed sideways or kept from rotating by forces of a collision,

Scuff Mark: A friction mark on a pavement made by a tire which is both rotating and slipping; acceleration scuffs, yaw marks, flat-tire marks.

Scuff: Scuff is a rough or irregular surface not requested in the design.

Scuttle: The lower, forward part of a driver's cab or passenger compartment that accommodates the legs of front-seated occupants and forms a *bulkhead* or *firewall* with the engine compartment.

Seat Belt Grabber: Device that prevents the unwinding or paying-out of a *seat belt* as it tightens on the reel or drum on impact.

Seat Belt: Occupant restraint system consisting of webbing and latching hardware, and firmly anchored to the vehicle structure.

Seat: The seat of a driver or passenger.

Second Impact: An impact between an occupant and an interior surface of a vehicle caused by and following an impact between the vehicle and another object. A second or subsequent impact between the same two vehicles during a crash.

Secondary Braking System: All the elements, the action of which may be modulated, allowing the driver to reduce, directly, or indirectly, the speed of a vehicle or to bring the vehicle to a halt in case of failure of the service braking system.

Secondary Collision: When two vehicles collide at an acute angle, and then rotate into each other, a secondary impact occurs which redirects the paths of the centers of mass of both vehicles. This is called a secondary collision.

Secondary Impact: A second or subsequent impact between the same two vehicles during a crash. See Secondary Collision.

Semi-Tracked Vehicle: Vehicle, particularly a military vehicle, with traction provided by track-lying apparatus, but steered by conventional wheels.

Semi-Tractor: Commercial vehicle tractor unit for hauling a *semi-trailer*.

Semitrailer: Any vehicle with or without motive power, other than a pole trailer, designed for carrying persons or property and for being drawn by a motor vehicle, and so constructed that some part of its weight and that of its load rests upon or is carried by another vehicle. See Truck.

Separation Speed: The speed at the time of loss of contact of two vehicles in a collision; it can refer to the speed of the centers of gravity or of the contact point.

Separation V: The vehicle speed at separation. Also known as the post-impact speed.

Separation Velocity: The vector velocity at the time of loss of contact of two vehicles in a collision; it can refer to the speed of the centers of gravity or at the contact point.

Separation: The point in the collision sequence where collision forces have been dissipated and the objects in contact have the opportunity to physically separate from each other, whether they actually separate or not.

Service Brake: Primary brake used for retarding and stopping a vehicle.

Service Braking System: All the elements, the action of which may be modulated, allowing the driver to reduce, directly or indirectly, the speed of a vehicle during normal driving or to bring the vehicle to halt. The primary brake system used for slowing and stopping a vehicle.

Severely Cornering Maneuver: Here the centrifugal force pulls the vehicle to rollover. For example, a driver avoiding an accident and steering rapidly causes a yaw disturbance. Traveling on collapsing road, and suddenly providing steering input for a vehicle with low level of roll stability. This kind of rollover is also caused by yaw disturbance. Fig. S.1 shows a vehicle rollover scenario because of severe steering input.

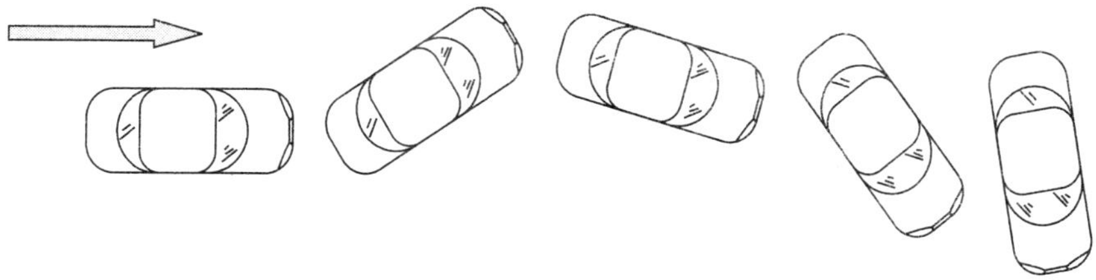

Fig. S.1 A vehicle rollover scenario because of severely steering put

Shake: Shake refers to the low-frequency (5 to 30 Hz) oscillations of the body, frame, suspension, powertrain, and seat that are felt by the driver or passengers. The intermediate frequency (5-25 Hz) vibrations of the sprung mass as a flexible body.

Shallow Angle Collision: A collision between two objects in which the angle between the directions of travel is less than 15 degrees.

Shimmy: A self-excited oscillation of a pair of steerable wheels about their steering axes, accompanied by appreciable tramp. A violent shake of the front wheels transmitted up to the steering wheel.

Shock Absorber: A generic term, which is commonly applied to hydraulic mechanisms for producing damping of suspension systems. A device, usually hydraulic, used to dampen or reduce the amount of spring oscillation after a bump.

Shock Impulse: See Shock Pulse.

Shock Isolator: (Shock Mount.): A shock isolator is a resilient support that tends to isolate a system from a shock motion.

Shock Motion: Shock motion is an excitation involving motion of a foundation. (See Foundation and Mechanical Shock.)

Shock Mount: (See Shock Isolator.)

Shock Pulse: A shock pulse (shock impulse) is a disturbing force characterized by a rise and subsequent delay of acceleration in a very short period of time.

Shock: Shock is a transient phenomenon. Shock results in a sharp, nearly sudden change in velocity.

Short/Long Wheel Base: The manufacturers wheel base dimension measured from the center point of a front wheel to the center point of the same side rear wheel (wheelbase from front axle to rear axle), *short* refers to *Wrangler* type lengths, long refers to *Grand Cherokee* type measurements.

Should Wear: Tire wear pattern where wear is greatest at the tire's *shoulder*.

Shoulder Rib: Rib or squared-off rim at the outer edge of a tire tread.

Shoulder: That portion of the *road* contiguous with the roadway for accommodation of stopped vehicles, for emergency use, and for lateral support of the roadway structure. The line between the roadway and the shoulder may be a painted edge line, a change in surface color or material, or a curb. On some modern traffic ways, there may be a surfaced shoulder on the right side, and frequently a narrower shoulder on the left side of a one-way roadway.

Side Acceleration: The component of the vector acceleration of a point in the vehicle in the y-direction.

Side Dumper: Dump truck capable of discharging to the side rather than (or as well as) to the rear.

Side Force: The component of the force vector in the *y*-direction.

Side Force: The component of the force vector in the *y*-direction.

Side Pull Ratio (SPR): The side pull ratio is defined as the ratio of the minimum side pull force (F_p) required to lift all tires on one side of the vehicle from the ground to the vehicle weight and is expressed as

$$SPR = \frac{F_p}{mg}$$

where m is the mass of the vehicle and g is the acceleration due to gravity (32.2 ft/s^2 or 9.81 m/s^2).

Side Rail: The outermost edge on the side of a vehicle's roof connecting the upper ends of the *A*, *B*, *C* and *D* pillars.

Side Scuffs: See Yaw marks.

Side Skid Mark: See Scrub.

Side Velocity: Side Velocity of a point in the vehicle is the component of the vector velocity in the *y*-direction. See Fig. L.1 and P.2.

Side Wall: The portion of either side of the tire which connects the head with the tread.

Sideboard: See *Side Panel.*

Sideslip Angle (Attitude Angle): The angle between the traces on the *X-Y* plane of the vehicle. *X*-axis and the vehicle velocity vector at some specified point in the vehicle. Sideslip angle is shown in Fig. 3 as a negative angle.

Sideslip Angle Gradient: The rate of change of sideslip angle with respect to change in steady-state lateral acceleration on a level road at a given trim and test conditions.

Sideslip Angle, Tire: See Wheel Sideslip Angle.

Sideslip Angle: A vehicle as it comes into a collision has a path defined by the motion of its center of gravity. The direction of this path is called the initial path heading. The vehicle heading is normally the same as the initial path heading. If the vehicle is sliding, slipping or yawing as it moves along the path before impact then the vehicle heading is not the same as the path heading. The difference is called the sideslip angle.

Sideslip Coefficient: The slope of the initial linear portion of the side force-sideslip curve of a tire.

Sideslip Mark: See Critical Speed Scuff.

Sideslip Stiffness: See Sideslip Coefficient.

Sideslip: Lateral/transverse translation of a vehicle perpendicular to its heading; See Fig. S.2. See also Wheel Slip.

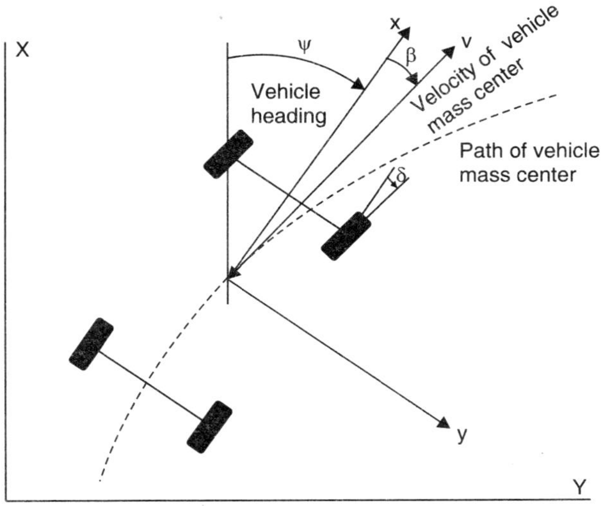

Fig. S.2

Sideswipe Collisions: Sideswipe collisions are characterized by prolonged sliding contact and with very little structural deformation. A sideswipe collision is considered as a collision in which the angle of impact is shallow and the vehicle interaction in the longitudinal direction is frictional. Sideswipes are characterized by prolonged sliding between the contact surfaces.

Sidewalk: That portion of a street between the curb lines, or the lateral lines of a roadway, and the adjacent property lines, intended for use by pedestrians.

Sidewall Rib: A raised circumferential rib located on the sidewall.

Sidewall: The portion of either side of the tire, which connects the bead with the tread.

Signature (FFT): Usually applied to the vibration frequency spectrum unique to a particular machine or machine component system, or subsystem at a specific location and point of time.

Simple Harmonic Vibration: Vibration at a point in a system is simple harmonic when the displacement with respect to time is described by a simple sine function.

Simple Reaction: A preplanned reaction to an expected hazard or other stimulii. Time: 0.5 sec, more or less.

Simulation: The development and use of computer models for the study of actual or postulated dynamic systems is known as simulation.

Simultaneous Factors: Factors which must be present at the same time to contribute to the cause of an accident; generally, condition factors.

Sine Wave: Sine wave is a wave in which points or particles move in transverse vibrations of a simple harmonic type.

Single Axle Weight: Total weight transmitted to the road by wheels whose axes are 40 inches or less apart.

Single Degree-of-freedom System: A single degree-of-freedom system is one for which only one coordinate is required to define completely the configuration of the system at any instant.

Single Pivot Steering: Steering in which a beam axle is pivoted at its mid-point.

Single Track Vehicle: Vehicle, such as a bicycle or motorcycle, that leaves only one track.

Single Vehicle Collisions: Single vehicle collisions are those collisions which involve a vehicle and a fixed barrier. The fixed barrier is assumed to not move nor absorb energy.

Single-Line Braking System: Assembly in which the braking systems of the individual vehicles act in such a way that the single line is used both for the energy supply to, and for the control of, the braking system of the towed vehicle.

Sinusoidal: Sinusoidal is the trace of an alternating quantity plotted to a time base in the form of a sine wave.

Sipe: Usually shallow groove or channel in a *tire tread pattern* that may serve to reduce tire noise, dissipate heat or increase flexibility of the tread pattern.

Site: The place where a traffic accident occurred after vehicles and people involved have left. Compare with scene.

Situation Hazard: A circumstance that more or less endangers a traffic unit on a trip and which must be avoided to prevent an accident.

Size: Size is a term denoting magnitude, particularly geometrical magnitudes but also includes weights, horsepower, etc.

Skid Distance: The skid distance is the length along the path for which the vehicle is assumed to be skidding. It is the arc length of a circular path if the user had specified a curved path upon data entry. It is calculated from the point of impact to the path end or the skid end points if specified.

Skid End V: The speed when the skidding ended.

Skid Mark: A friction mark left on road or any surface by a braked or locked tire. Skid marks are divided into "acceleration marks" created on acceleration, if the engine provides more power than the tire can transmit, "braking marks," if the brakes "lock up" and cause the tire to slide, or "yaw marks", if the tire slides sideways. Each has characteristic appearance, and an experienced accident reconstructor or forensic engineer can often determine what the vehicle was doing by examining the marks left by the tire.Skid marks are marks made by tires that are not rolling. The simplest form of computing vehicle speed is that of a vehicle coming to a complete stop at the end of the braking maneuver while the deceleration remains constant.

Skid Number: A number representing tire-pavement frictional drag determined by measurements made according to standard equipment, conditions, and procedures and usually stated as 100 times a friction coefficient.

Skid of the Tire: SAE offers the following definition (SAE J670e) for the skid of the tire:

$$i_{SAE} = \left(\frac{r\omega}{V} - 1\right) \times 100\%$$

Note that i_{SAE} gives the value of 0% skid when the vehicle is under free-rolling condition and –100% when the tires are locked-up.

Skid Steer: Steering of a vehicle, particularly an off-highway and earthmoving vehicle, by braking of the wheels on one side and driving on the other so that direction is changed by skidding around.

Skid Trajectory: A skid trajectory assumes the vehicle is not spinning and that the rear wheels are not tracking the path of the front wheels. A skidding trajectory can change to a rollout at any point before the path ends.

Skid: Motion of a vehicle when the wheels are locked or partially locked, or when the vehicle moves sideways as in a side slip.

Skidding Trajectory: When the rear wheels track outside the front wheels or if the vehicle is spinning then that portion of the trajectory is called a skidding trajectory.

Skidmark: A friction mark on a pavement made by a tire that is sliding without rotation. Sliding of wheels may be due to braking, to collision damage, or, rarely, to other circumstances. Dark deposits of tire and roadway material normally deposited on a roadway surface by a tire that is not free to rotate.

Skids, Yaw Marks, Prints, etc. Tire Sideslip Angle: See Wheel Sideslip Angle.

Sliding Braking Fraction Coefficient: The value of the braking force coefficient of a tire obtained on a locked wheel on a given tire and road surface for a given environment and operating condition.

Sliding Coefficient of Friction: The sliding coefficient,μ, of friction is defined as the dimensionless ratio of the force generated on a sliding object on a level surface to the weight of the object. For an object sliding on a flat surface, μ can also represent the acceleration or deceleration of the object, expressed as a decimal fraction of the gravitational acceleration constant, g(32.2 ft/s^2), if the frictional force is the only unbalanced force acting on the object.

Sliding Friction Coefficient: See Coefficient of Friction.

Slip Angle: The angle between the *X*' axis and direction of travel of the center of tire contact. See Figures T.3 and S.5.

Slip Angle Force: The lateral force when the inclination angle is zero and ply steer and conicity forces have been subtracted.

Slip Angle: The lateral or steering forces are generated when the tire tries to roll at an angle to its path. This angle is called the slip angle and consists of two parts: the steer angle of the wheel and yaw angle of the vehicle.

Slip Stiffness: See Wheel Slip Coefficient.

Slip: See Sideslip; Wheel Slip.

Slippage: Sliding movement between *tire*, or track of a *track-lying vehicle*, and the ground.

Slope: The ratio of a vertical distance change to the horizontal distance change. See Grade.

Smash Up: See Car Accident.

Smudge Marks: Smudge marks are marks made by objects depositing material onto the vehicle surface without deformation or crushing. Rubber transfer is a smudge mark, for example.

Snag: In a snag, parts of two vehicles hook together and create a change in the momentum of the two vehicles while can be completely unrelated to the damage profiles of the vehicles. That is to say that the coefficients derived from the classical collisions with a barrier or between cars have no relation to the actual accident or the damage profiles of snagged vehicles.

Snubber: A generic term which is commonly applied to mechanisms which employ dry friction to produce damping of suspension systems.

Soak-In: An area on the shoulder or roadside saturated with liquid debris either at the end of runoff or as a puddle marking the rest position of a vehicle after a collision.

Solid Axle Suspension: A suspension system consisting of one steel or aluminum 1-bearn extending the width of the vehicle.

Solid-Body Friction: Direct contact between the opposed surfaces.

Spatter: The collection of marks on the road made by liquid from the *vehicle* or its cargo squirted from containers on the *vehicle* by force of collision. Spatter areas are irregular in shape and often consist of many small spots.

Specific Brake Torque or Brake Factor: The specific brake torque or brake factor is defined as the ratio of drum drag to application force of one shoe. It is a general indication of brake's ability to produce torque for different lining/drum friction coefficients.

Specific Heat Capacity: Specific heat capacity (specific heat) is the quantity of heat in J required to raise the temperature of 1 kg of a substance by 1 K.

Spectral Luminous Efficiency: See Luminosity Function.

Spectrum: A spectrum is a definition of the magnitude of the frequency components that constitute a quantity.

Speed Control: See Cruise Control.

Speed Estimates: See Falls, Flips, Vaults, Conservation of Momentum, Velocity and Pedestrian Accidents.

Speed Reconstruction: Speed reconstruction is the determination of vehicle speed from the physical evidence from the accident.

Speed Sensitivity Analysis: The objective of a speed sensitivity analysis is to show to what extent the impact speeds predicted by an analysis or computer software are sensitive to input changes of critical accident parameters.

Speed to Slide Slip of a Vehicle: In order to calculate the speed to slide slip of a vehicle, one must know the radius of the curved path followed by the vehicle. The radius of curvature of the yaw mark is calculated using the following formula:

$$R = \frac{C^2}{8M} + \frac{M}{2}$$

where R = radius, m (ft.)

C = chord, m (ft)

M = middle ordinate, m (ft)

A chord measurement of 30.48 m (100 ft) is recommended. The vehicle's speed is higher at the beginning of the yaw mark since the vehicle will decelerate as its side slips. The critical speed is given by

$$V_{cr} = \sqrt{gRf}$$

where V_{cr} = critical speed, m/s (ft/sec)

g = acceleration due to gravity, 9.81 m/s^2 (32.2 ft/s^2)

f = coefficient of friction

We also have

$$S = \sqrt{15Rf}$$

where S = speed (mph)

R = radius, ft

f = coefficient of friction

Also in SI units, $S = \sqrt{127Rf}$

where S = speed (km/h)

R = radius, m

f = coefficient of friction.

For a roadway with super elevation, the critical speed is given by

$$V_{cr} = \sqrt{\frac{gR(f + \tan\gamma)}{(1 - f\tan\gamma)}}$$

where g = acceleration due to gravity, 9.81 m/s^2 (32.2 ft/s^2)

f = coefficient of friction

R = radius, m (ft)

γ = bank angle (super elevation)

Speed: A change in distance with respect to time regardless of direction.

Speedometer: Speedometer is a tachometer fitted to the gearbox or propeller shaft of a motor or other road vehicle and usually graduated to indicate the speed in miles per hour or km/hour.

Spin Axis: The axis of rotation of the wheel. (See Fig. S.3.)

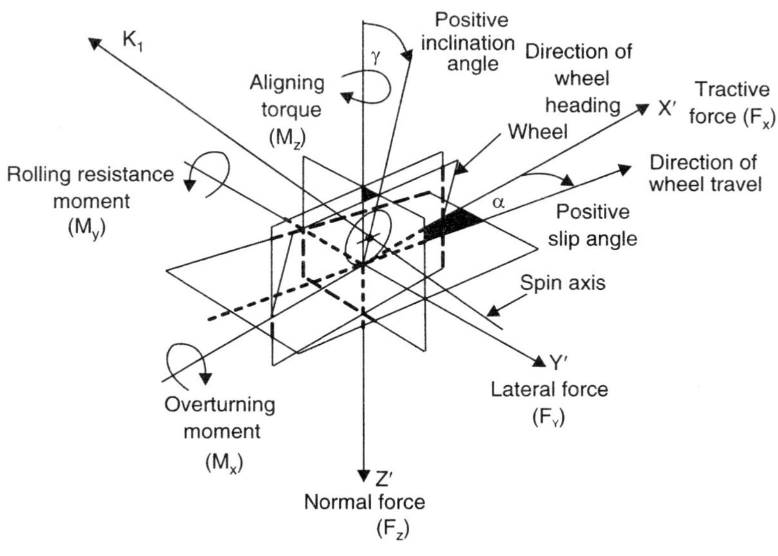

Fig. S.3 Tire axis system.

Spin Marks: Spin marks are tire marks made by a vehicle that is out of directional control, most likely spinning about its vertical axis. A typical characteristic of spin marks is one or more tire crossover points.

Spin Velocity: Angular velocity of a wheel about its *spin axis*.

Spinout: A descriptive term for postimpact vehicle motion including significant yaw rotation.

See Postimpact Trajectory.

Splash Marks: Splash marks are made by liquids, including blood. Detailed examination of the exterior surface marks will, in most cases, determine which of the vehicles or objects was sliding in what direction and which was moving faster.

Sporadic Trickle: The fluid trail that usually starts after impact aspersion and continues to the vehicle's final rest position or until the fluid container is empty.

Spot-type Disc Brake: A conventional *caliper disc brake.*

Spread Tandem Axle: Widely spaced *tandem axle* arrangement of a heavy vehicle, usually employed to circumvent axle weight restrictions applying to conventional close-coupled tandem arrangement.

Spring Brake: Pneumatic parking or *secondary brake* in which a helical spring is compressed by action of air pressure on a piston when the brake is not required, the pressure then being reduced to apply the brake.

Spring Constant: Spring constant or spring rate is the ratio of load to deflection for a spring.

Spring Rate: Load per unit deflection, or sometimes deflection per unit load, of a spring of a sprung system.

Spring Stiffness: A linear spring obeys a force-displacement law of $F = *x$ where * is called the spring stiffness or spring constant and has dimensions of force per length, and x is the displacement of the spring.

Spring: A spring is a flexible mechanical line between two particles in a mechanical system.

Sprung Mass: Considered to be a rigid body having equal mass, the same center of gravity, and the same moments of inertia about identical axes as the total sprung weight.

Sprung Weight: All weight which is supported by the suspension, including portions of the weight of the suspension members. In the case of most vehicles, the sprung weight is commonly defined as the total weight less the weight of unsprung parts.

Squat: A pitching motion of the sprung mass downward at the rear that often occurs during acceleration.

Squeal: Narrow band airborne tire noise (150-800 Hz) resulting from either longitudinal slip or slip angle or both.

Stability: The property of a system that causes it to return to a condition of equilibrium or steady motion when disturbed. Passenger vehicles exhibit varying characteristics depending upon test conditions and trim. Test conditions refer to vehicle conditions such as wheel loads, front wheel alignment, tire inflation pressure, and also atmospheric and road conditions which affect vehicle parameters. For example, temperature may change shock absorber damping characteristics and a slippery road surface may change tire cornering properties. Trim has been previously defined as the vehicle operating condition within a given environment, and may be specified in part by steer angle, forward velocity, and lateral acceleration. Since all these factors change the vehicle behavior, the vehicle stability must be examined separately for each environment and trim.

For a given set of vehicle parameters and particular test conditions, the vehicle may be examined for each theoretically attainable trim. The conditions which affect most stability are the steady-state values of forward velocity and lateral acceleration. In practice, it is possible for a vehicle to be stable under one set of operating conditions and unstable in another.

Stabilized Accident Situation: The condition prevailing after motion and other actions constituting the events of an accident has ceased and no further harm will ensue unless a new series of events is initiated by some means.

Standard Brakes: See Manual Brakes.

Statement: See Detailed Statement, Testimony.

Static Axle Load Distribution: The static axle load distribution is defined as the ratio of static rear axle load divided by the total vehicle weight, ψ:

$$\psi = F_{ZR}/W$$

where F_{ZR} = static rear axle load, lb
W = vehicle weight, lb

Static Crush: See Residual Crush.

Static Deflection: Static deflection is the deflection of a mechanical system due to gravitational force alone.

Static Friction: The frictional force exerted on a stationary body is known as static friction, Coulomb friction, and fluid friction.

Static Loaded Radius: The loaded radius of a stationary tire inflated to normal recommended pressure.

Static Margin: Static margin is another measure of the steady-state handling behavior. It is determined by the point on the vehicle where a lateral force will not generate any steady-state yaw velocity (*i.e.*, the neutral steer point). We define a neutral steer line as shown in Fig.S.4. The neutral steer line is the locus of points in the x-y plane along which external forces produce no steady-state yaw velocity. The static margin is defined as the distance the neutral steer point falls behind the CG, normalized by the wheelbase, that is,

$$\text{SM} = e/L$$

When the point is behind the CG, the static margin is positive and the vehicle is understeer. At the CG, the margin is zero and the vehicle is neutral steer. When ahead of the CG, the vehicle is oversteering. SM is the difference between the distance of neutral steer point to front axis and the distance of CG to front axis. When neutral steer point coincides with CG, SM = 0. If the lateral force applied at CG causes front tire and rear tire to have same slip angle, the vehicle possesses neutral steer properties. When CG is in front of neutral steer point, S.M. is negative. When the lateral force applied at CG causes that the rear tire slip angle is larger than that of front tire slip angle, the vehicle has oversteer property. When CG is behind the neutral steer point, S.M. is positive. When the lateral force applied at CG causes that the front slip angle is larger than rear slip angle, the vehicle has understeer property.

Static Rate: Static rate of an elastic member is the rate measured between successive stationary positions at which the member has settled to substantially equilibrium condition.

Static Stability Factor: A dimensionless number meant to indicate the resistance of vehicle to roll over. The number is calculated using the formula SSF = $T/2h$, where T is the vehicle track width and h is the height of the center of gravity of the vehicle.

Static Stiffness: Static stiffness is the change in deflection with change in dead load.

Static Tire Deflection: Effective decrease in wheel radius from *center of tire contact* to *spin axis*.

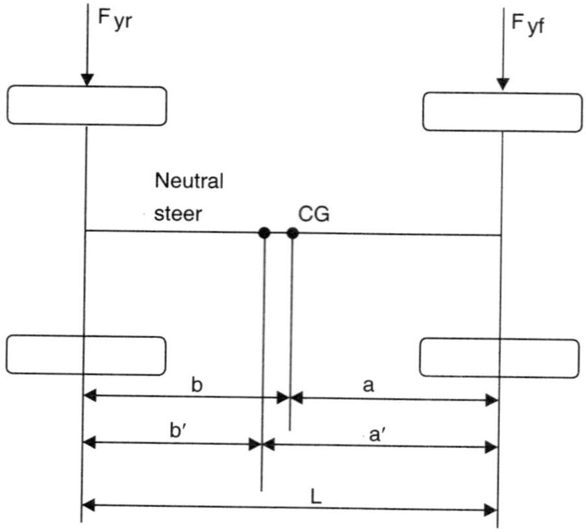

Fig. S.4 Schematic of static margin.

Static Toe: Static toe-in or toe-out of a pair of wheels, at a specified wheel load or relative position of the wheel center with respect to the sprung mass, is the difference in the transverse distances between the wheel planes, taken at the extreme rear and front points of the tire treads. When the distance at the rear is greater, the wheels are "toed-in" by this amount; and where smaller, the wheels are "toed-out."

Static Toe Angle (Deg): The static toe angle of a wheel, at a specified wheel load or relative position of the wheel center with respect to the sprung mass, is the angle between a longitudinal axis of the vehicle and the line of intersection of the wheel plane and the road surface. The wheel is "toed-in" if the forward portion of the wheel is turned toward a central longitudinal axis of the vehicle, and "toed-out" if turned away.

Static Toe: Static toe-in or toe-out of a pair of wheels, at a specified wheel load or relative position of the wheel center with respect to the sprung mass, is the difference in the transverse distances between the wheel planes, taken at the extreme rear and front points of the tire treads. When the distance at the rear is greater, the wheels are "toed-in" by this amount; and where smaller, the wheels are "toed-out." The static toe (inches) is equal to the sum of the toe angles (degrees) of the left and right wheels multiplied by the ratio of tire diameter (inches) to 57.3.

If the toe angles on the left and right wheels are the same and the outside diameter of tire is 28.65 in (727.7 mm), the static toe (inches) is equal to static toe angle (degrees).

Statically Coupled: If an equation of motions contains cross products of coordinates, that equation of motion is statically coupled.

Statics: The branch of mechanics which deals with the equilibrium of forces on bodies at rest (or moving at a uniform velocity in a straight line).

Station Line: Reference line from where measurements are obtained from predetermined reference points and from which measurements are made at right angles.

Steady State Response: The steady-state response is the system response after the transient motion has decayed sufficiently.

Steady State Vibration: Steady-state vibration exists in a system if the displacement at each point recurs for equal increments of time.

Steady State: Steady state exists when periodic (or constant) vehicle responses to periodic (or constant) control and/or disturbance inputs do not change over an arbitrarily long time. The motion responses in steady-state are referred to as steady-state responses. This definition does not require the vehicle to be operating in a straight line or on a level road surface. It can also be in a turn of constant radius or on a cambered road surface.

Steady-State Vibration: Steady-state vibration exists in a system if the displacement at each point recurs for equal increments of time

Steer Angle: The angle between the projection of a longitudinal axis of the vehicle and the line of intersection of the wheel plane and the road surface.

Steered Wheel: Road wheel that responds to applied steering input and therefore influences the path taken by the vehicle.

Steering Angle: Angle between the projection of the longitudinal axis of a vehicle and plane of *steered wheel.*

Steering Arm: Lever that imparts the steering action to the *steering knuckle* or *stub axle* assembly, as from a *tie rod* or *track rod.*

Steering Axis: The axis about which *steered wheel* assembly pivots when steered.

Steering Column: Shaft that transmits the rotation of steering wheel to a steering box or rack and pinion.

Steering Damper: Device for damping vibrations or shock loads in the *steering system.*

Steering Disturbances: These refer to unwanted vehicle reactions, steering wheel motions or torque feedback that can result from sources other than the driver's steering inputs. Such sources include insufficient steering system performance. An example is torque steer. This relates to deviations from the course during power take-off, pedal tip-ins, gear changes or others. The disturbance of these maneuvers to the stability of the straight-ahead is tested. Nibble, or the vibration of the steering wheel around its rotational axis on smooth surfaces due to tire imbalance or force variations, is also studied.

Steering Mechanism: The steering mechanism converts the driver's rotational input at the steering wheel into a change in the steering angle of the vehicle's steering road wheels.

Steering Ratio: Steering ratio refers to the ratio between the turn of the steering wheel (in degrees) or handlebars and the turn of the wheels (in degrees). In motorcycles and bicycles, the steering ratio is always 1:1, while in most passenger cars, it is between 12 and 20:1. Example: If one complete turn of the steering wheel (360 degrees) causes the wheels to turn 24 degrees, then the ratio is 15:1 (360/24 = 15).

Steering Response: The vehicle motion resulting from an input to the steering (control) element. Although the steering wheel is the primary directional control element, it should be recognized that longitudinal forces at the wheels resulting from driver inputs to brakes or throttle can modify directional response.

Steering System: The angle at the steering wheel may not be the decisive factor for the driving behavior. Actually, the steer angle at the road wheels is the dominant factor, which can differ from the steering wheel angle because of elasticity, friction influences and a servo-support. At very fast steering movements, the dynamic rise of tire forces also plays an important role. As the understeer gradient is usually measured at steering wheel, the compliance in the steering system which allows the road wheel to deviate from the steering wheel input has important effect on the understeer gradient and other handling properties. Fig. S.5 shows the schematic of steering system and system stiffness effects.

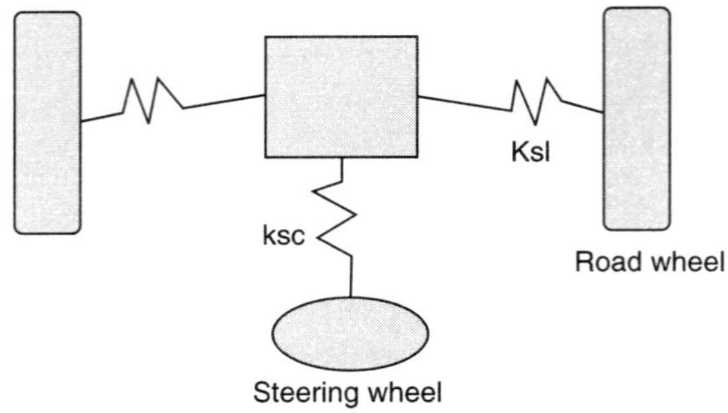

Fig. S.5 Schematic of steering system and system stiffness.

Steering Wheel Angle: Angular displacement of the steering wheel measured from the straight-ahead position (position corresponding to zero average steer angle of a pair of steered wheels).

Steering Wheel: Steering wheel is the spoked hand-wheel at the top of the steering column of an automobile which actuates the steering gear.

Steering: The steering (*i.e.*, change in travel direction of a vehicle) is accomplished by a rotation of the tires from their straight-ahead position relative to the vehicle.

Steering Forces: Steering forces or lateral forces are generated when the tire tries to roll at an angle to its path. This angle is called slip angle and consists of two parts: the steer angle of the wheel and the yaw angle of the vehicle.

Stiffness Coefficient: See Crush Stiffness Coefficient.

Stiffness: Stiffness is the ration of change of force (or torque) to the corresponding change in translational (or rotational) deflection of an elastic element.

Stopping Distance: Stopping distance is the distance between the moment when a hazard is recognized and the time when the vehicle comes to a complete stop. It is the sum of the distances traveled during the reaction time, the braking response time (at constant vehicle speed), and the braking time.

Straight Free-Rolling Tire: A free-rolling tire moving in a straight line at zero inclination angle and zero slip angle.

Straight-Ahead Controllability: This refers to the characteristics of steering around the straight-ahead position to allow the driver confident control of the steering. Tests cover the vehicle response to small steering corrections required to keep the vehicle in a straight path or lane changes. The acceleration in various directions due to small inputs on steering wheels around the straight-ahead position is evaluated. The symmetry of the vehicle behavior for steering inputs to the right and left is also evaluated. Tests also include control of the vehicle motion around the longitudinal axis. They also include the relationship between small steering wheel torque or angle and the vehicle's response (torque feedback).

Strategy: Adjusting speed, position on the road, and direction of motion, giving signals of intent to turn or to slow, or any other action in situations involving potential hazards; any maneuvers while on a trip which make evasive tactics easier or increase the chance of success in avoiding an actual hazard.

Striations: Narrow, light, parallel stripes or streaks usually made by friction or abrasion on the roadway or on vehicle parts.

Striking Vehicle: The vehicle initiating the collision. This is sometimes referred to as the bullet vehicle.

Struck Vehicle: The vehicle that was impacted by the other vehicle.

Structural Tire Performance: Structural performance of tire relates to the load-carrying function of the tire consisting of endurance involving fatigue life, resistance to wear, bond life, material insensitivity to chemical and thermal effects, and load and stress performance, involving static and dynamic load-carrying capacity, resistance to penetration, bead-unsteady resistance, and vibration performance.

Stub Axle: Short cantilevered *axle* on which a wheel is mounted.

Sub Harmonic Response: Sub harmonic response is a term sometimes used to denote a particular type of harmonic response which dominates the total response of the system. It frequently occurs when the excitation frequency is submultiples of the frequency of the fundamental resource.

Sub Harmonic: A sub harmonic is a sinusoidal quantity having a frequency that is integral submultiples of the fundamental frequency of a periodic quantity to which it is related.

Submarining: Action of the occupant sliding forward underneath the lap belt portion of the restraint system, with the lap belt webbing resting above the pelvic girdle.

Sustained Contact: Sustained contact is a dynamic situation where one vehicle pushes the other to path end. Sustained contact will only occur in almost central impacts.

Skidding Trajectory: When the rear wheels track outside the front wheels or if the vehicle is spinning then the portion of the trajectory is called a skidding trajectory. There are two types of skidding trajectories, a spin and a skid.

Skid Trajectory: The skid trajectory assumes the vehicle is not spinning and the rear wheels are not tracking the path of the front wheels.

Sustained Contact: Sustained contact is a dynamic situation where one vehicle pushes the other to path end. A sustained contact is possible when both vehicle trajectories are similar.

Superelevation: The degree to which the outside edge of a roadway is higher than the inside edge at a specified point on a curve; the change in elevation per unit distances across the roadway from the inside of a curve to the outside edge; bank.

Suspension Geometry: The geometric arrangement of the components and linkages of a suspension system and the numerical value of the lengths and angles thereof.

Suspension Rate: Change of wheel load at *center of tire contact* per unit displacement of the vehicle *sprung mass.*

Suspension Roll Angle: The angular displacement produced by suspension roll.

Suspension Roll Gradient: The rate of change in the suspension roll angle with respect to change in steady-state lateral acceleration on a level road at a given trim and test conditions.

Suspension Roll Stiffness: The rate of change in the restoring couple exerted by the suspension of a pair of wheels on the sprung mass of the vehicle with respect to change in suspension roll angle.

Suspension Roll: The rotation of the vehicle sprung mass about the x-axis with respect to a transverse axis joining a pair of wheel centers.

Suspension Travel: The length between the highest and lowest distance a wheel can *travel* from full compression to full hang or droop.

Suspension: Automobile suspensions have many variable characteristics, which are generally different in the front and rear and all of which affect handling. Some of these are: spring rate, damping, straight ahead camber angle, camber change with wheel travel, roll center height and the flexibility and vibration modes of the suspension elements. Suspension also affects unsprung weight. Many cars have suspension that connects the wheels on the two sides, either by a sway bar and/or by a solid axle. The Citroën 2CV has interaction between the front and rear suspension. It is possible for a vehicle to be understeer for small inputs and oversteer for large inputs (or the opposite), as shown in Fig. 4, since it is a nonlinear system and does not have the same characteristics at all trims. Consequently, it is necessary to specify the range of inputs and velocities when making a determination of the vehicle's steer characteristics. There is a set of equivalent definitions in terms of yaw velocity or curvature (reciprocal of radius of curvature), which can be used interchangeably with these definitions. These definitions only apply to two-axle vehicles, since the Ackerman steer angle only applies to two-axle vehicles.

System Response: Energy enters a dynamic system through the application of an excitation. The excitation varies in according with a prescribed function of time. The vibratory behavior of dynamic system is characterized by the motions caused by these excitations and is referred to as the system *response*.

Systems: Systems refers to an interrelated set of elements, components, or subsystems.

Tail Pipe: The rearmost pipe of an *exhaust system*, downstream of the rear *muffler (silencer)*. Occasionally used to describe intermediate pipes.

Tailgate: Hinged transverse rear door of a van or passenger car.

Takeoff Angle (Fall or Vault): The angle between the trajectory or path of the car and the horizontal in a fall or vault.

Takeoff Speed: The speed of a vehicle that leaves a surface and launches off a precipice, embankment or terminal point in the load and travels through the air.

Tare Weight: Weight of an unladen commercial vehicle without fuel, water and oil, but subject to local definition.

The Accident Investigator: The primary responsibility of the investigator is the collection and documentation of accident data.

The Accident Reconstruction Expert: The responsibility of an expert is the determination and establishment of facts, their analysis and evaluation, and the clear and understandable presentation of facts and factual relationships in the form of reports, or orally to judges and juries.

The Body: The body is the shell comprising the occupant compartment for passenger cars or buses and includes the shell-carrying payloads for trucks. The body provides support for many other functions, such as lighting, signals, and mirrors.

The Environment: The environment affects the driver, the road, and the vehicle. The driver is affected through factors such as fog, glare, traffic controls, weather, adjacent land use, buildings, and many more. The road is affected by the environment primarily through weather-related surface contamination. The vehicle is affected by the environment through fog reducing the head lights power, windshield contamination by mud, snow, or rain, and through water or ice affecting braking performance and/or tire traction.

The Longitudinal Force Coefficient (C_x) is based on the aerodynamic force acting on the vehicle in the x-direction (as established by 8.2) and is defined as:

$$C_x = F_x/(qA)$$

where q is dynamic pressure at any given relative air velocity as given by the formula

$$q = (pv^2)/2$$

The Maximum Force of a Tire: The maximum force F_ϕ of a tire on hard surfaces is defined as adhesion force and is proportional to the normal force applied on the wheel during driving or braking as

$$F_\phi = F_z\phi$$

where ϕ is the adhesion coefficient, which varies with the state of the tire rolling or slipping, and F_z is the normal force applied on the wheel during driving or braking.

The Normal Force Coefficient (C_z) is based on the aerodynamic force acting in the z-direction (as established by 8.2) and is defined as:

$$C_z = F_z/(qA)$$

The Pitching Moment Coefficient (C_{My}) is based on the pitching moment derived from the distribution of aerodynamic forces acting on the vehicle and is defined as:

$$C_{My} = M_y/(qA\ell)$$

The Rolling Moment Coefficient (C_{Mx}) is based on the rolling moment derived from the distribution of aerodynamic forces acting on the vehicle and is defined as:

$$C_{Mx} = M_x/(qA\ell)$$

The Yawing Moment Coefficient (C_{Mz}) is based on the yawing moment derived from the distribution of aerodynamic forces acting on the vehicle and is defined as:

$$C_{Mz} = M_z/(qA\ell)$$

Tilt Table Ratio: The Tilt Table Ratio (TTR) is an estimator of the vehicle static rollover threshold. It is a simple test to estimate the vehicle rollover threshold because it is determined only by the tilt angle and does not need any additional measurement, such as the vehicle center of gravity or weight.

Time: In dynamics, the position of the vehicle changes with time. Time has been called the fourth dimension of the characteristics of any body, it depends on its position and how that position changes with time. Time for both the SI and English system is measured in seconds.

Time-to-Contact: It is the time between the moment that a stimulus is first presented to the observer and the moment that the observer reaches that stimulus (the impact location typically).

Tire Aspect Ratio: The tire aspect ratio is given by $(H/W)100$ where H is the tire cross-section height and W is the tire cross-section width. The aspect ratio (H/W = height relative to width) on today's passenger car tires ranges between 80 and 50; the figure ranges down to 25 for sports car tires, and between 100 and 45 for the tires on heavy commercial vehicles.

Tire Axis System: The origin of the tire axis system is the center of tire contact. The X' axis is the intersection of the wheel plane and the road plane with a positive direction forward. The Z' axis is perpendicular to the road plane with a positive direction downward. The Y' axis is in the road plane, its direction being chosen to make the axis system orthogonal and right-hand. The origin of the tire axis system is the center of tire contact.

Tire Camber Angle: There is an angle between the wheel plane and the line perpendicular to the ground plane in the modern vehicle, which is called the camber angle of the tire. If the upper part of the wheel inclines outside the car body, the angle is positive. The front wheels of the car usually contain a tiny positive camber angle $g = 5' \sim 10'$ with zero load, which makes the tire perpendicular to the ground surface with a little vaulting and abrades uniformly with the decrease in rolling resistance. The angle turns out to be slightly negative under full load.

Tire Centrifugal Growth: A tire rotating at higher speeds will tend to develop a larger diameter, due to centrifugal forces that force the tread rubber away from the axis of rotation. As the tire diameter grows the tire width decreases. This centrifugal growth can cause rubbing of the tire against the vehicle at high speeds. Motorcycle tires are often designed with reinforcements aimed at minimizing centrifugal growth.

Tire Contact Area, Tire Print: The amount of tire tread that is in contact with the road surface. Also called the *footprint.*

Tire Cornering Forces: Due to the road bank, lateral wind, or the centrifugal force when driving, a side force F_y will be applied to the center of the tire. Correspondingly, a lateral reaction force will also be developed at the contact patch and is called the cornering force F_y. Besides, gravitational forces and aerodynamic disturbance, the dominant forces and moments influencing vehicle are the tire forces. In cornering process, tire develops a lateral force, and it will also experience lateral slip associated with a slip angle. The slip angle is the angle between its direction of heading and its direction of travel, as shown in Fig.T.3. The lateral force is named as the cornering force, if the tire camber angle is ignored. Experiments show that cornering forces increase with the slip angle for a given tire load. Fig.T.1 shows the cornering force as a function of slip angle. At low slip angle, like 5 degree or less, the relationship is linear, and it can be expressed by

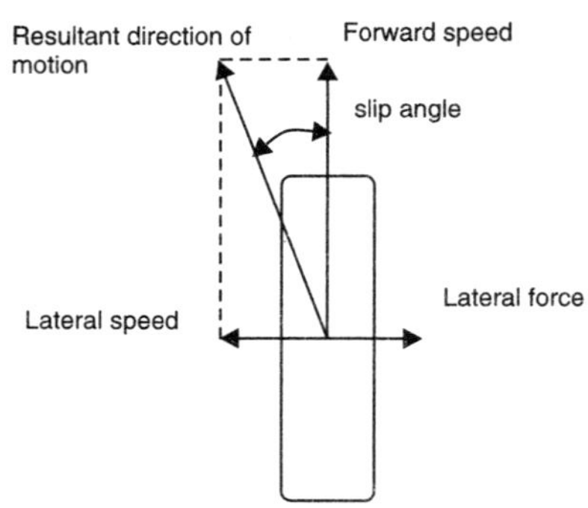

Fig. T.1 The tire force.

$$F_y = C_\alpha \alpha$$

The proportional constant C_α is defined as *cornering stiffness*, which is the slope of the curve of cornering force versus slip angle α at $\alpha = 0$.

Tire Face: Side of tire facing away from the vehicle.

Tire Force and Moment Stiffness: (may be evaluated at any set of operating conditions).

Tire Forces: The external force acting on the tire by the road having the following components.

Tire Friction Mark: A tire mark made when a slipping or skiding tire rubs the road or other surface; skidmarks; yawmarks; acceleration scuffs and flat tire marks.

Tire Lateral Dynamics: The tire lateral dynamics mainly discusses the cornering characteristics, roll characteristics, lateral forces at tire-ground contact patch, and self-aligning torque caused by slipping. It also analyzes the relationship among the slip angle, slip stiffness, cornering force, and self-aligning torque to obtain the precise lateral dynamics of a tire.

Tire Lateral Load Transfer Distribution: The distribution of the total tire lateral load transfer between front and rear tires expressed as the percentage of the total.

Tire Lateral Load Transfer: The vertical load transfer from one of the front tires (or rear tires) to the other that is due to acceleration, rotational, or inertial effects in the lateral direction.

Tire Load Rating: Maximum recommended load for a given tire at specified cold inflation pressure.

Tire Load: Load or weight supported by a tire.

Tire Longitudinal Dynamics: Tire longitudinal dynamics mainly presents tire rolling resistance, the slip coefficient, and the relationship between the tire traction and brake forces and the road surface condition.

Tire Longitudinal Load Transfer: The vertical load transferred from a front tire to the corresponding rear tire that is due to acceleration, rotational, or inertial effects in the longitudinal direction.

Tire Mark: A mark made on the surface of a road by a tire on a vehicle; friction marks and tire prints.

Tire Moments: The external moments acting on the tire by the road having the following components.

Tire or Tyre: Tires, or tyres (in American and British English, respectively), are ring-shaped parts, either pneumatic or solid (including rubber, metals and plastic composites), that fit around wheels to protect them and enhance their function. Pneumatic tires are used on many types of vehicles, such as bicycles, motorcycles, cars, trucks, earthmovers, and aircraft. Tires enable better vehicle performance by providing traction, braking, steering, and load support. Tires form a flexible cushion between the vehicle and the road, which smoothes out shock and makes for a comfortable ride. See Fig.T.2.

Fig. T.2 Firestone Pneumatic tire (Tyre).

Tire Overall Width: The width of the unloaded new tire, mounted on specified rim, inflated to the normal recommended pressure, including protective rib, bars, and decorations. (See Tire & Rim Association Yearbook.)

Tire Patch: That portion of the tire that rests on the road surface. It is oval shaped and has its longest axis along the normal direction of travel of the vehicle. When a vehicle is sliding sideways, the tire marks left on the road get almost twice as wide as the tire turns sideways and the short axis is in the direction of travel that makes the tire marks.

Tire Power Loss: Mechanical power input converted into heat by the tire.

Tire Print: See Tire Contact Area or Footprint.

Tire Pull Force: Tendency of a tire to pull to the left or right when running on a flat, even surface.

Tire Radial Run-out: Variation in extreme radius of the *tread* of a mounted tire in the *wheel plane*.

Tire Rate (Static): The static rate measured by the change of wheel load per unit vertical displacement of the wheel relative to the ground at a specified load and inflation pressure.

Tire Rolling Resistance: The rolling resistance of the tire, which is applied to the wheel, is due to the deformation of the tire at the tire/road interface. As is known from the tire experiment, tire deformation needs to consume energy, and the normal pressure distribution over the tire/road contact patch is not uniform because of the unequal force needed during the compression and elastic recovery. In the direction of travel, the normal force in the leading half of the contact patch is higher than that in the trailing half. The center point of the sum of the normal forces offsets a distance a to the centerline in the direction of rolling. Experiments show that the rolling resistance coefficient is proportional to the tire deformation, while being inversely proportional to the radius of a loaded tire.

Tire Rubber Prints: Tire rubber prints are made by the tire tread rolling over a relatively solid surface, and can be seen more easily on light colored road surfaces such as white fog lines.

Tire Scrub: Sliding movement of tire at right angles to the plane of the wheel, effectively the active component of *slip angle*.

Tire Section Height: Half the difference between the tire outside diameter and the nominal rim diameter.

Tire Section Width: The width of the unloaded new tire mounted on specified rim, inflated to the normal recommended pressure, including the normal sidewalls but not including protective rib, bars, and decorations. (See Tire & Rim Association Yearbook.)

Tire Sideslip angle: See Wheel Sideslip Angle.

Tire Slip: The tire slip may be computed by

$$S_T = \frac{V_V - V_T}{V_V} \times 100 \text{ (percent)}$$

where V_V = vehicle velocity, ft/sec

V_T = circumferential tire velocity, ft/sec

Tissue Marks: Tissue marks are made when human or animal body parts come in contact with the exterior surfaces of a vehicle during the accident.

Toe In: A condition where both tires of an axle are positioned so they are closer together at the front than the rear.

Toe Out: A condition where both tires of an axis are positioned so they are closer together at the rear than the front.

Toe: The difference in distance between the front and back of corresponding tires.

Top Speed: Maximum steady speed of a vehicle, usually measured on a flat surface and with no wind.

Topping: A noise and jolt when the extension cycle of the suspension travel ends at the bump stop.

Torque Applied at The Brake Wheel: The torque applied at the wheel is given by $T = F\, r_{ave}$, where r_{ave} is the average radius from the center of the moving element to the contacting surface. Note that the torque applied is linearly dependent on the coefficient of friction, μ and thus variations in these coefficients due to heat will not be magnified as was observed for drum brakes due to the nonlinearity.

Torque: A rotational effort produced by a force times a moment arm. A twisting or turning force. Simply defined as the twisting force from the power of the engine.

Torque-Arm Center in Braking: The instantaneous center in a vertical longitudinal plane through the wheel center about which the wheel moves relative to the sprung mass when the brake is locked.

Torsional Spring: A torsional spring is a link in a mechanical system where application of a torque leads to an angular displacement between the ends of the torsional spring.

Torsional Vibration Damper: A flywheel mounted on a shaft with the relative motion damped by viscous friction.

Torsional Vibration: Torsional vibration refers to vibration of a rigid body about a specific reference axis. The displacement is measured in terms of an angular coordinate.

Total Braking Force: Sum of the braking forces at the interfaces between all the wheels and the ground, produced by the effect of the braking system, which oppose the movement or the tendency to movement of the vehicle.

Total Distance: The total distance from impact to path end is the sum of the rollout and the skid distances.

Trace Evidence: Trace evidence contends that every contact, no matter how slight, will leave a trace. The trace is normally caused by objects or substances contacting one another, and leaving a minute sample on the contact surfaces. Material is often transferred by heat induced by contact friction.

Track Change: The change in wheel track resulting from vertical suspension displacements of both wheels in the same direction.

Track Width: Distance between the center of the lateral width of the wheel on one side of the vehicle to the center on the opposite side wheel.

Track: The distance on the ground between the center of the tire tread on one side of the vehicle, and the center of the tire tread on the opposite side.

Tracking Error: Misalignment of road wheels particularly where adversely affecting *steering* or tire wear.

Tracking: Marks made when tires roll through puddles, runoff or spatter and, become wet, leaving tire prints on the pavement as they roll on.

Traction Bar: Rod or strut attached to a *live axle* to react torque loads and prevent or minimize *axle wind-up* and *hop*.

Traction Coefficient: The traction coefficient is defined as the ratio of braking force between tire and ground of an axle to the normal force carried by that axle.

Traction Control: System that prevents the spinning of driven wheels of a vehicle when excess power is applied.

Traction Force: The adhesive friction of a body, such as a tire, on a surface, such as a roadway, that keeps the body from slipping on the surface. Traction breaks and slipping begins when another force, such as braking force or centrifugal force, exceeds traction force.

Traction Grade: Letter code of a *tire* representing the ability to stop a vehicle on a wet surface.

Traction: Driving force or effort of a motor vehicle, or its pulling power.

Tractive Force Variation: Variation of driving or tractive force exerted by a tire per revolution.

Tractive Effort: Quantitative measure of the ability of a vehicle to pull or haul.

Tractive Force (F_z): The component of the tire force vector in the direction of travel of the center of tire contact. Tractive force is equal to lateral force times sine of slip angle plus longitudinal force times cosine of slip angle. See also Lateral Force. See Fig. S.5.

Tractor Trailer: A truck tractor (cab) with two or more axles pulling a trailer.

Tractor, Semitrailer: A truck tractor (cab) with two or more axles pulling a semitrailer.

Traffic Accident: An accident involving travel or transportation on a traffic way.

Traffic Collision: See Car Accident.

Traffic Controls: A description of the controls used for directing vehicle traffic is provided including speed limit, location of warning and regulatory signs, traffic lane markings, raised medians, and traffic lights with time measurement for light cycles.

Traffic Unit: An element of traffic; a person using a traffic way for travel or transportation; vehicle, pedal-cycle, pedestrian, etc.

Traffic: Pedestrians, ridden or herded animals, vehicles, street cars, and other conveyances either singly or together while using any highway for purposes of travel.

Traffic way: The entire width between property lines, or other boundary lines, of every way or place of which any part is open to the public for purposes of vehicular travel as a matter of right or custom. All highways are traffic-ways, but trafficways include also some areas on private property such as shopping centers.

Trajectory: A typical representative point in the phase plane moves and traces a curve known as trajectory. The trajectory shows how the solution of the system varies with time. The term trajectory is used to describe the motion of a vehicle after the collision phase.

Transient Vibration: Transient vibration exists in a system when one or more component oscillations are discontinuous.

Transient: A non-steady motion impressed on a system which usually decays to a negligible value, but the lasting effect of the transient may be a different steady rate.

Transition Time: is the time between the first movements of the responder until the vehicle begins to respond. In essence it is the movement time plus the mechanical latency.

Transitional Stability: Transitional stability is the control of the vehicle during lane changes or similar maneuvers. Stability and controllability of the vehicle under different steering input and levels is tested. Steering inputs can be continuous and smooth or abrupt.

Translational Motion: Translational motion refers to the movement of a body's center of mass (that is, rotation about the center of mass not considered for translational motion).

Transmissibility: The ratio of the amplitude of vibration of a specified particle within a mechanical system to the amplitude of an applied vibration at a given frequency; normally the system is supported elastically, as by isolators.

Transmission: Transmissions are torque and speed converters. They provide the maximum engine power, which is produced over only a small range of engine speed, at any driving conditions, *i.e.*, at low and high speeds and when pulling heavy loads or climbing a grade. An ideal powerplant delivers constant power throughout the engine speed range. The internal combustion engine, however, delivers output torques that peak at a certain speed and power that peaks at a higher speed. Generally speaking, the vehicle needs the high torque at lower speeds (for acceleration) and lower torque at higher speeds. In order to compensate that shortcoming of internal combustion engines, a gearing mechanism is needed in order to deliver the needed torque at the given speed. This gearing mechanism is the transmission.

Transporter: Usually articulated commercial vehicle built for the transport of other vehicle, or of abnormally large loads.

Tread (Tire): The peripheral portion of the tire, the exterior of which is designed to contact the road surface.

Tread Arc Width: Peripheral distance between the two *shoulders* of a *tire*, measured along the tread contour.

Tread Chord Width: The distance measured parallel to the spin axis of an unloaded tire between one edge of the tread and the other. For tires with rounded tread edges, the point of measurement is that point in space which is at the intersection of the tread radius extended until it meets the prolongation of the upper sidewall contour.

Tread Contact Length: Length of a tire *contact patch* or *footprint*, usually measured in the wheel plane.

Tread Contact Width: The distance between the extreme edges of road contact at a specified load and pressure measured parallel to the Y' axis at zero slips angles and zero inclination angles.

Tread Contour: Cross-sectional shape of the tread surface of an inflated unloaded tire neglecting tread pattern depressions.

Tread Depth: Distance between the base of a tire tread *groove* and a line tangential to (or in simple measurements joining) the adjacent surface of the tire.

Tread Noise: Airborne sound (up to 5000 Hz) except squeal and slap produced by the interaction between the tire and the road surface.

Tread Pattern: The molded configuration on the face of the tread. It is generally composed of ribs, rows, grooves, bars, lugs, and the like.

Tread Profile: The general form of the tread of a *tire*, viewed in section. See also *Tread Radius*.

Tread Radius: Radius of a tire's tread measured in the plane of the *spin axis*.

Tread Separation: Separation of tire tread from the body of the tire.

Tread Shoulder: Extreme edge of a tire tread, particularly when coming to a distinct run.

Tread Width: The outside-edge-to-outside-edge width of the two tires on an axle.

Tread: Transverse distance between left and right side wheels on the same axle, measured between specified points, such as the *centers of tire contact*, or the centers of the widths of the tracks of a *track-laying vehicle*.

Treadwear Grade: Numerical code representing wear resistance of a tire.

Triangulation: A method of locating a spot in an area by measurements from two or more *reference points,* the locations of which are identified for future reference. Compare with *coordinates.*

Trim: The steady-state (that is, equilibrium) condition of the vehicle with constant input which is used as the reference point for analysis of dynamic vehicle stability and control characteristics.

Trip Point: That location along a ground surface at which the motion of a vehicle component is suddenly halted followed by a flip, rollover, or vault.

Trip Risk: The danger to a traffic unit in undertaking a particular trip; the combined effect of road, vehicle, driver, weather, traffic, and schedules on the possibility of maintaining adequate driving strategy and evasive tactics.

Tripped Rollover: The rollover can be divided into two categories: tripped rollover and untripped rollover. The tripped rollover is caused by a vehicle hitting against an obstacle. This rollover is caused by vehicle inclination where the vehicle CG exceeds the stable point. National Highway and Traffic Safety Administration (NHTSA) reports show that only 10% or less of all rollover crashes occur under untripped condition for the on-road environment. Major rollovers occur, however, when in off-road environment where the road conditions are unpredictable. The tripped rollover is caused by a vehicle hitting against an obstacle or by lateral slipping. The vehicle is rotated around its longitudinal direction, shown in Fig. T.3(*a*). The untripped rollover is caused by the vehicle inclination where the vehicle CG exceeds the stable point. The vehicle will be under vertical or lateral translation, as shown in Fig. T.3(*b*). If rollover is caused by the two cases, the rollover is a combination of a rotation moment and a translation force, as shown in Fig. T.3(*c*).

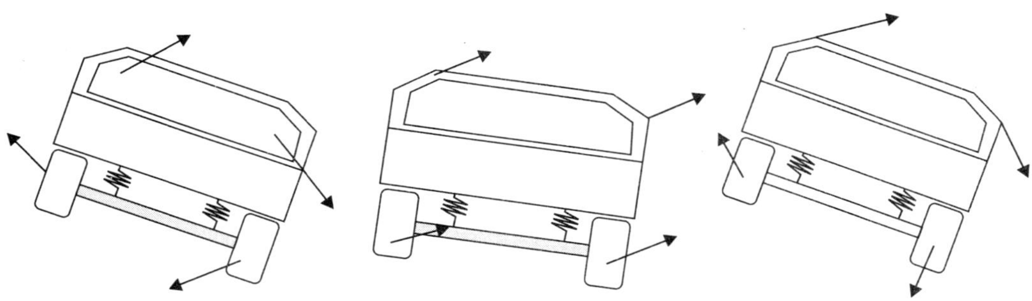

(*a*) (*b*) (*c*)

Fig. T.3 Rollover motion (*a*) Rotation (*b*) Lateral translation (*c*) Rotation and translation.

Tubeless Tire: Tire in which the air pressure is contained by the *carcass* of the tire and the *rim* of the wheel, there being no *inner tube*.

Turning Center: Center of the radius of the turn of a vehicle, at which the axes of all wheels ideally coincide.

Turning Radius: The radius of the *turning circle*, measured either to the nearest point on the plane of symmetry, or to the *center of tire contact* of the wheel describing the largest circle.

Turning Track: Radial width between centers of road contact (centers of contact patches) of innermost and outermost tires of a vehicle when negotiating a turn.

Two-Spread Axle: Axle with two reduction stages, one of which is a change-speed gear train with a choice of high or low ratios.

Two-Dimensional Motion: When there is motion in two directions at the same time a convention is required for the measurement of angles in the plane of the two dimensions. In collision analysis, the coordinate axes in a two-dimensional system are always perpendicular to each other.

Three-Dimensional Motion: Any body, including a vehicle, can have a very complex path in three dimensions. It is normal to assign three perpendicular axes or coordinates, one for each dimension. An origin is defined by the user from which all measurements are reference. The motion of a body is defined by its position with time on its path.

Uncontrolled Final Position: A final position reached by a traffic unit after an accident without conscious human intervention.

Under Steer: A characteristic of a vehicle that results in a tendency to steer towards the outside of a curve. See also Oversteer.

Underbody Debris: Debris consisting of mud, dust, rust, paint, snow, or road tar loosened by collision from the underside of the vehicle body, fenders, and other parts. Underbody debris is loosened by crumpling of vehicle parts in collapse and by collision shock.

Underbody: The body structure of the underside of a vehicle including floor, wheel wells and stiffening members.

Undercarriage: The wheels, axles and suspension system of a vehicle, but particularly of a commercial vehicle.

Underdamped System: For the motion of a system where the displacement is a harmonic function having amplitude which decays exponentially with time, the system is said to be underdamped, and the damping is below critical.

Underframe: Underframe is that part of a truck containing the axles and their bearing springs, the axle boxes, the buffer and draw bar spring and the wheels.

Underride: A condition in a collision in which the main structural components of one vehicle are below the main structural components of the other vehicle; See Fig. U.1. See also Override.

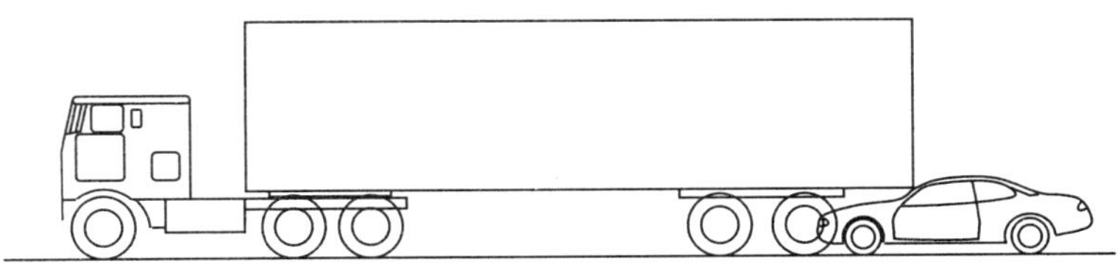

Fig. U.1

Understeer/Oversteer Gradient: The quantity obtained by subtracting the Ackerman steer angle gradient from the ratio of the steering wheel angle gradient to the overall steering ratio.

Understeer: A characteristic of a motor vehicle as loaded that results in a tendency to swerve toward the outside of a curve. Motor vehicles with more weight on front wheels than on rear wheels, or with too little pressure in front tires, are likely to understeer at high speed.

Undertread: Reinforcing *piles* laid beneath the *tread* of a *tire*.

Uniform Acceleration: Acceleration that changes at a constant rate.

Uniform Motion: The term uniform motion means uniform velocity.

Universal Joint: Rotating shaft coupling that permits angular axial displacement. Some types will also operate with linear misalignment of axes.

Unladen Weight: The maximum weight of a vehicle with no payload. Weight of vehicle driver and passenger and all service ancillaries, but without payload.

Unloaded Tire: Trailing tires that have less weight as a result of a weight transfer to the leading tires of a rotating vehicle. In this situation, some weight is shifted to those tires, but not enough to leave a gap between the tire surface and the ground, and not enough to allow downward force to produce a yaw mark.

Unrestrained Systems: An unrestrained system has a rigid body mode corresponding to a natural frequency of zero.

Unsprung Mass: Mass of a vehicle undercarriage below the springs. Vehicle components not carried by the suspension system.

Unsprung Weight: All weight which is not carried by the suspension system, but is supported directly by the tire or wheel, and considered to move with it.

Unstretched Length: The length of a spring when it is not subjected to external forces is called its unstretched length.

Untripped Rollover: See Tripped Rollover.

US Gravitational System: It is called gravitational system because force, length and time are considered absolute units in pounds, inches, and seconds.

Vacuum Brake: Conventional brake actuated by atmospheric pressure acting on a piston in a partially exhausted cylinder.

Vacuum-Assisted Brake: Hydraulic brake actuated or assisted by atmospheric pressure, usually through mechanical linkages.

Vacuum-Over-Hydraulic Brake: A vacuum-assisted *hydraulic brake* system.

Variable Acceleration: Acceleration that varies at a non-constant rate.

Variable Rate Coil Springs: Springs used to provide additional load bearing capacity for cars and light trucks. Unlike a standard coil spring, the coils of variable rate springs are not equally spaced. The bottom coils are more widely spaced than those at the top, which allows them to provide extra support when needed.

Variomatic Transmission: Infinite variable light car transmission using rubber belts on expanding pulleys.

Vault: Vault is a roll or pitch motion of a vehicle made following loss of ground contact. Vaults are similar to flips, except they are end-for-end flips. A vehicle or object launched in to the air where the takeoff is uphill. See also Flip and Fall.

Vehicle Acceleration: The vector quantity expressing the acceleration of a point in the vehicle relative to the earth-fixed axis system (X, Y, Z). The following motion variables are components of this vector, resolved with respect to the moving vehicle axis system.

Vehicle Area (*A*): The projected frontal area including tires and underbody parts.

Vehicle Axis System (*x*, *y*, *z*): This system is a right-hand orthogonal axis system fixed in a vehicle such that with the vehicle moving steadily in a straight line on a level road, the x-axis is substantially horizontal, points forward, and is in the longitudinal plane of symmetry. The y-axis points to driver's right and the z-axis points downward.

Vehicle Clearance Circle: Diameter of the smallest circle which will enclose the outermost points of projection of a vehicle when executing a turn at low speed and full lock.

Vehicle Coordinate System: See Three-Axis Vehicle Coordinate System.

Vehicle Critical Speed: Vehicle critical speed (V_c) is the speed at which the tires of a cornering vehicle reach the limit of side friction and slid-slip begins. The following equation yields the equation of the critical speed from yaw marks (Slidge and Marshek, 1997):

$$V_c = \frac{\sqrt{gR(\mu + \tan\theta)}}{\sqrt{1 - \mu\tan\theta}}$$

where R is the radius of the turn, g is the acceleration due to gravity, θ is the grade or slope of the road, and μ is the coefficient of friction.

Vehicle Handling: Vehicle handling describes those vehicle properties that allow a driver to negotiate maneuvers successfully. Since the term handling includes driver characteristics, the interpretation of handling qualities of a motor vehicle varies almost as much as an interpretation of driver skills.

Vehicle Heading: See Fig. L.1.

Vehicle Identification Number (VIN): Number assigned to a vehicle by the manufacturer primarily for registration and identification purposes. It often consists of numerals and letters.

Vehicle Length: The maximum dimension measured longitudinally between the foremost point and the rearmost point in the vehicle, including bumper, bumper guards, tow hooks, and/or rub strips, in standard equipment. Also known as overall length (OAL).

Vehicle Pitch Angle: The angle between the vehicle x-axis and the ground plane.

Vehicle Response: The vehicle motion resulting from some internal or external input to the vehicle. Response tests can be used to determine the stability and control characteristics of a vehicle.

Vehicle Roll Angle: The angle between the vehicle y-axis and the ground plane.

Vehicle Roll Gradient: The rate of change in vehicle roll angle with respect to change in steady-state lateral acceleration on a level road at a given trim and test conditions.

Vehicle Roll Stiffness: Sum of the separate suspension roll stiffnesses.

Vehicle Velocity: The vector quantity expressing velocity of a point in the vehicle relative to the earth-fixed axis system (X, Y, Z). The following motion variables are components of this vector resolved with respect to the moving vehicle axis system (x, y, z).

Vehicle Wheelbase (ℓ): The characteristic length upon which aerodynamic moment coefficients are based. The density of standard dry air shall be taken as 2378×10^{-6} slugs/ft at 59°F and 29.92 in Hg. The viscosity of standard dry air shall be taken as 373×10^{-9} slugs/ft-sec.

Vehicle Width: The maximum dimension measured between the widest points on the vehicle, excluding exterior mirrors, flexible mud flaps, and marker lamps, but including bumpers, moldings, sheet metal protrusions, or dual wheels in standard equipment. Also known as overall width (OAW).

Vehicle: Any device in, upon, or by which any person or property is or can be transported or drawn upon a highway, excepting devices moved by human power or those used exclusively upon stationary rails or tracks.

Vehicle-Pedestrian Interaction Dynamics: is aimed mostly at the determination of detailed pedestrian-vehicle interaction motion and forces in order to assess a specific injury or assess effects of changes in vehicle geometry.

Vehicular Accident Reconstructions: Vehicular accident reconstructions are often conducted by specialized units in law enforcement agencies, to answer questions about automobile accidents, such as who was driving, who was breaking the law, where were the victims seated, were they using seat belts? Through accident reconstruction, rigorous analysis is done, with expert witnesses that can present results in trial. Accident reconstructions are done in cases involving fatalities, and often when a school bus is involved. Factors looked at during accident reconstruction include steering angles, braking, use of lights, turn signals, speed, acceleration, engine rpm, cruise control, and anti-lock brakes. Witnesses are interviewed during accident reconstruction, and physical evidences such as tire marks are examined.

Velocity: The time rate of change of position of a body; it is a vector quantity having direction as well as magnitude.

Vertical (Bounce): The translational component of ride vibrations of the sprung mass in the direction of the vehicle *z*-axis. (See Fig. L.1)

Vertical Axis: The axis perpendicular to the plane of horizon and passing through the center of mass of the object. Also known as the yaw axis.

Vertical Movement: The up-and-down movement of the ball joint or tie rod end.

Vibration Damper, Vibration Isolater: A resilient mounting used to absorb the vibration of a fixed mechanical installation. A vibration damper is an auxiliary system composed of an inertia element and a viscous damper that is connected to a primary system as a means of vibration control. A vibration isolator is a resilient support that tends to isolate a system from steady-state excitation. See also Isolator; Damper.

Vibration: Vibration is the variation with time of the displacement of a body with respect to a specified reference dimension when the displacement is alternately greater and smaller than the reference. Thus, vibration is an oscillation where ($\propto$) in the quantity is a parameter that defines the motion of a mechanical system. See Oscillation.

Viscous Coupling: Friction shaft drive using the property of fluid friction within and *oil* or *thixotropic* fluid to transmit power.

Viscous Damper: Viscous damper, which is also referred to as a dashpot is characterized by the resistive force exerted on a body moving in a viscous fluid, and hence the name.

Viscous Friction Forces: Viscous friction forces are proportional to the velocity of a given particle and resist the motion because they act in a direction opposite to that of the velocity. Viscous friction forces are known as dissipative forces because the system loses energy due to their action.

Weight Distribution: Of a commercial vehicle, the percentage of the total weight distributed to each axle.

Weight Shift: When the weight of a vehicle shifts in reaction to a change in direction speed.

Weight Transfer: The amount of weight that moves laterally because of cornering forces or lengthwise, because of acceleration or braking forces.

Weight: Weight is the resultant force of attraction on the mass of a body when subjected to the gravitational field of another body. On Earth the units of weight are based on a gravitational acceleration of 9.81 m/s^2 (32.174 ft/sec^2). A mass of 1 kg exerts a force (weight) of 1 Newton when subjected to an acceleration of 1 m/s^2.

Well Base Wheel: Wheel with *one-piece rim* incorporating at or near its center a well to enable tire *beads* to be mounted over the *rim flanges*.

Wheel Alignment: When mounted on the vehicle, the wheel and tire may not be perfectly aligned to the direction of travel, and therefore may exhibit irregular wear. If the discrepancy in alignment is large, then the irregular wear will become quite substantial if left uncorrected. A wheel alignment is the procedure for checking and correcting this condition through adjustment of camber, caster and toe angles. These settings also affect the handling characteristics of the vehicle.

Wheel Base: The distance from the center of the front wheels to the center of the rear wheels or, if there is a tandem axle, the distance to the midpoint between the two tandem axles. The perpendicular distance between axes through front and rear wheel centerlines of a vehicle. In case of dual axles, the distance is to the midpoint of the centerlines of the dual axles.

Wheel Center: The point at which the spin axis of the wheel intersects the wheel plane.

Wheel Disc: The part of a wheel between *hub* and *rim*.

Wheel Fights: A rotary disturbance of the steering wheel produced by forces acting on the steerable wheels.

Wheel Flutter: Oscillation of *steered wheel* about its *steering axis*, usually at a frequency greater than that of rotation.

Wheel Force Factor: A force factor is defined as the ratio between the desired force and the largest force that can exist between the wheel and the ground. In other words, $\theta = F/F_{max}$, where $F_{max} = \mu W_{wheel}$, in which μ is the friction factor and W_{wheel} is the weight of the wheel.

Wheel Hop: Severe oscillation of a wheel in the vertical mode, in which the tire intermittently loses contact with the ground.

Wheel Imbalance: Wheel imbalance is the result of a nonuniform distribution of mass. The imbalance increases with the square of vehicle velocity, and may reach harmful effects when the misbalancing input frequency is in harmony with the natural frequency of the wheels/suspension system.

Wheel Lockup Factor: The wheel lockup factor is a measure of the freedom with which a wheel can turn about its axis. Limpert has determined an empirical formula for rolling resistance as a function of tire type, inflation pressure, and empirical coefficients based on the road surface and the speed of the vehicle when the wheel is rolling. The lockup factor θ is a measure of the freedom with which a wheel can turn about its axis. For example, on paved roadways, the rolling resistance for the ith wheel is given by

Wheel Plane: Central plane of the wheel, normal to the *spin axis*.

Wheel Rim: The periphery of a wheel on which the tire is mounted.

Wheel Sideslip Angle: The angle between a wheel's heading axis (x-axis) and direction of the velocity vector of the center of the wheel.

Wheel Skid: The occurrence of sliding between the tire and road interface which takes place within the entire contact area. Skid can result from braking, driving and/or cornering.

Wheel Slip Coefficient: The slope of the initial linear portion of the forward force-wheel slip curve of a tire.

Wheel Slip: The ratio of the forward velocity of a tire at the road contact patch to the forward velocity at the center of the wheel (for braking) or the ratio of the forward velocity of a tire at the center of the wheel to the forward velocity at the road contact patch (for traction or acceleration).

Wheel Spin: Rapid slipping rotation of a driven wheel resulting from the application of excessive torque for prevailing friction conditions. See also *Limited Slip Differential*.

Wheel Torque: The external torque applied to the tire from the vehicle about the spin axis.

Wheel Track (Wheel Tread): The lateral distance between the centers of tire contact of a pair of wheels. For vehicles with dual wheels, it is the distance between the points centrally located between the centers of tire contact of the inner and outer wheels.

Wheel Well: Cavity within a vehicle structure or coachwork to accommodate a wheel.

Wheel Wobble: A self-excited oscillation of steerable wheels about their steering axes, occurring without appreciable tramp.

Wheel: The tire-rim assembly is commonly called the wheel. See Fig. S.5.

Wheelbase: The distance from the center of the front wheels to the center of the rear wheels or, if there is a tandem axle, the distance to the midpoint between the two tandem .axles.

Wheelslip: Slipping of driven wheels when friction between tire and road surface is insufficient to react driving torque.

Wind Resistance: The force of the atmosphere against a moving vehicle; atmospheric drag factor; a combination of forces produced by motion of the vehicle through the air and motion of the air itself. Under certain circumstances, wind may assist rather than resist the motion of the vehicle. Compare with rolling resistance.

Wipe Marks: Wipe marks are generally made by soft objects wiping dirt or dust off without depositing paint, rubber, or metal.

Work and Power: The power required to maintain a consistent rate of acceleration (deceleration) varies according to vehicle speed. Power available for acceleration:

$$P_a = P\eta - PW$$

where P = engine output,

η = efficiency, and

PW = motive power.

Work of a Force: The work of a variable force in a moving body through the distance $\Delta s = s_2 - s_1$ is given by $W = \int_{s_1}^{s_2} F \cos\alpha \, ds$, where F is the variable force, ds is the elementary length of the path, and α is the angle between the force and the element ds.

Work: Work is the product of the average force and the distance moved in the direction of the force by its point of application.

Wrap Collisions: In wrap collisions, due to the low frontal contact surface of the wheel relative to the pedestrian's center of gravity as well as the pedestrian's body flexibility and freedom to rotate, the pedestrian's body wraps back on the hood following an initial contact.

Wrap Pedestrian Collision: A frontal collision of vehicle and pedestrian or cyclist in which initial contact occurs at a point below the center of gravity of the pedestrian or cyclist and where the frontal geometry of the vehicle allows the pedestrian to move rearward relative to the vehicle and strike another portion of the vehicle such as a windshield. The latter impact causes the pedestrian or cyclist to develop an airborne trajectory followed by an impact with the ground.

***X*-axis:** A horizontal axis in a system of rectangular coordinates.

Yaw Angle: Angle of a vehicle's plane of symmetry or heading to its actual direction of travel or course.

Yaw Mark: Yaw mark is a tire mark caused by a slide-slipping tire often showing a striped pattern called striations. Yaw marks are the result of an avoidance maneuver, where a driver perceives an object or potential hazard in front of the vehicle and swerves suddenly to the left or right or from a sudden yaw velocity. During extreme steering maneuvers the available tire friction can be saturated by the side forces on the tire, and the vehicle will then yaw and leave a peculiar type of tire mark. The speed at the beginning of the maneuver can often be calculated from yaw marks and is often referred to as critical speed. See also Skid Mark and Critical Speed Scuff.

Point of separation: to obtain this, each vehicle should be drawn on the scale diagram as it was positioned at the point of impact and also as it was positioned at the point of separation. The PS can usually be established from roadway marks such as skid or scrub marks, chips and gouges. It should be noted that the departure angle for a vehicle is the angle formed with the *X*-axis, or with a baseline drawn parallel to that axis, by the vehicle's path of travel after separation. This path is plotted by drawing a line from the center-of-mass location of the vehicle at time of impact through its center-of-mass location where the vehicle separated from the other vehicle or object, as shown in Fig. Y.1.

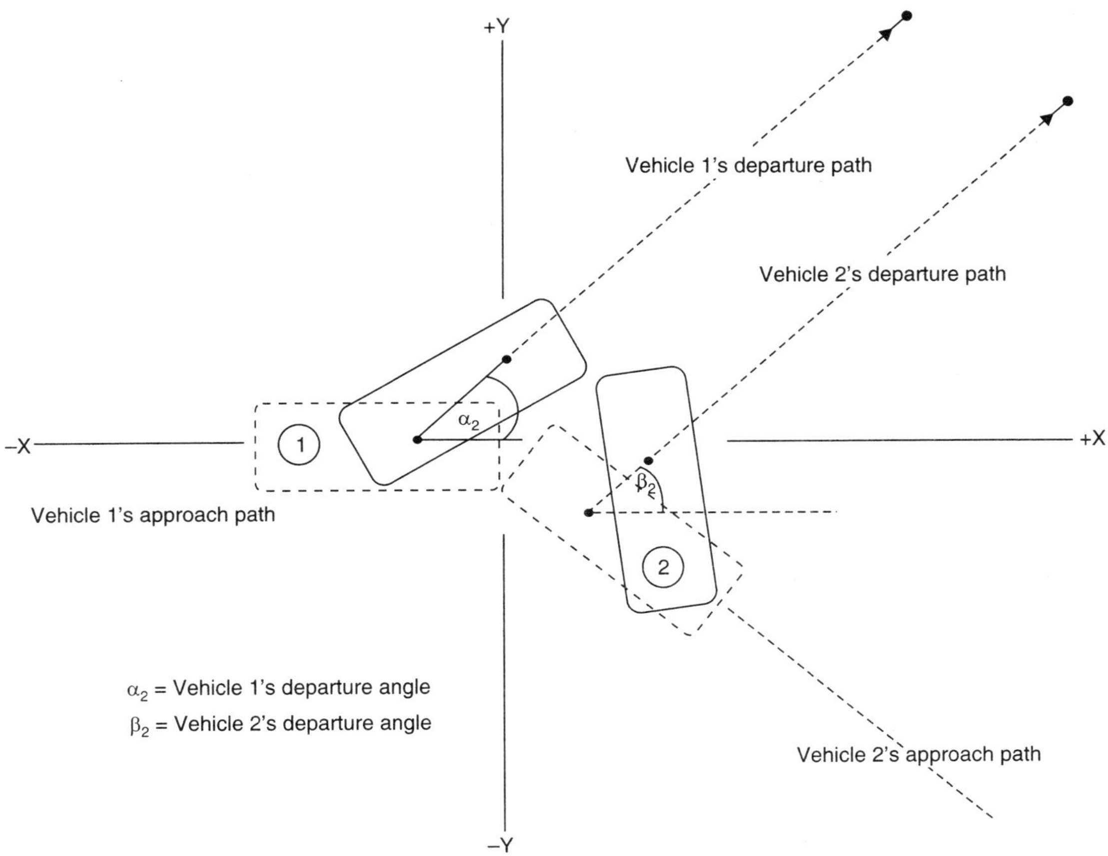

Fig. Y.1 A diagram showing departure angles measured from the *X*-axis, or parallel baseline, to the line drawn from the center-of-mass of each vehicle at impact (broken-line rectangle) through the center-of-mass of the vehicle at separation, [Rivers (1997)].

Yaw Moment of Inertia: The moment of inertia about a vertical axis of a vehicle. See Moment of Inertia; Radius of Gyration.

Yaw Rate: Angular velocity about the vertical axis.

Yaw Velocity: Angular velocity of a vehicle about its vertical axis. Also *yaw angular velocity*. See Fig. L.1.

Yaw: A sidewise movement of a vehicle in turning; movement of a vehicle in another direction than that in which it is headed; sidewise motion produced when centrifugal force exceeds traction force. Often the result of overreaction or exceeding the critical speed. Sometimes revealed by tire marks on the roadway. See Fig. P.2.

Yawing Moment Coefficient: Non-dimensional measure of the tendency of aerodynamic forces to displace a vehicle in the yaw sense.

Yawing Moment: The component of the moment vector tending to rotate the vehicle about the z-axis, positive clockwise when looking in the positive direction of the z-axis.

Yawing Rate: Frequency of oscillation of a vehicle in yaw. Yaw angular velocity.

Yawmark: A scuffmark made while a vehicle is yawing; the mark made on the road by a rotating tire which is slipping in a direction parallel to the axle of the wheel.

***Y*-axis:** A vertical axis in a system of rectangular coordinates.

Zero Point: See Reference Point.

Appendix

D

Units and Conversion

The SI System of Units

The International System of Units, whose name in French is Systeme International (SI) has been universally recognized and adopted by most of the countries of the world. The Unites States is still in the process of making the transition from the US/ English system to the SI system. Hence, both systems of units are followed throughout in this book, in worked examples and in exercise problems.

Units: A physical quantity can be measured only by a comparison with a like quantity. A distinct amount of a physical quantity is called a unit. Any physical quantity of the same kind can be compared with it, and its value can be stated in terms of a ratio number and the unit used.

Basic Units and Derived Units: The general unit of a physical quantity is defined as its dimension. A unit system can be developed by selecting the basic dimension of the system, a specific unit (for instance, meter for length, the kilogram for mass, and the second for time). Such a unit is called a basic unit and the corresponding physical quantity is called a basic quantity. Units that are not basic are called derived units.

Systematic Units: Systematic units are systematically derived units within a unit system. They can be obtained by replacing the general units (dimensions) by the basic units of the system. -

If we define the dimensions of length, mass, and time as $[L]$, $[M]$, and $[T]$, respectively, then the physical quantities may be expressed as $[L]^x\,[M]^y\,[T]^z$. For instance, the dimension of acceleration is $[L]\,[T]^{-2}$ and that of force is $[L]\,[M]\,[T]^{-2}$.

Absolute Systems of Units and Gravitational Systems of Units: Systems of units in which the mass is taken as a basic unit are called absolute systems of units, whereas those in which the force rather than the mass is taken as a basic unit are called gravitational systems of units.

British Engineering System of Units: The British engineering system of units is a gravitational system of units and is based on the foot, pound force, and second as basic units. This is the system used in the United States. The derived unit of mass is $lb_f\text{-}s^2/ft$ and is called a slug (1 slug = $1lb_f\text{-}s^2/ft$).

International System of Units (SI): The International System of units (abbreviated SI) is the internationally agreed system of units for expressing the values of physical quantities. (See Table D.1). In this system four basic units are added to the customary three basic units (meter, kilogram, second) of the metric (mks) absolute system of units. The four added basic units are the ampere as the unit of electric current, the kelvin as the unit of thermodynamic temperature, the candela as the unit of luminous intensity, and the mole as the unit of amount of substance. Thus in SI units the meter, kilogram, second, ampere, kelvin, candela, and mole constitute the seven basic units. There are two auxiliary units in the SI units: the radian, which is the unit of a plane angle, and the steradian, which is the unit of a solid angle. Table D.1 lists the seven basic units, two auxiliary units, and some of the derived units of the International System of units, (SI). (Multiples and submultiples of the units are indicated by a series of 16 prefixes for various powers of 10. See Table D.2.

In SI units the seven basic units and other units are defined as follows:

Base Units

Metre (m): The metre is the length equal to 1 650 763.73 wavelengths in vacuum of the radiation corresponding to the transition between the levels $2p_{10}$ and $5d_5$ of the krypton 86, the orange-red line.

Kilogram (kg): The kilogram is the unit of mass; it is equal to the mass of the international prototype of the kilogram.

Second (s): The second is the duration of 9 192 631 770 periods of the radiation corresponding to the transition between the two hyperfine levels of the ground state of the cesium-133 atom.

Ampere (A): The ampere is that constant current which, if maintained in two straight parallel conductors of infinite length, of negligible circular cross-section, and placed 1 m apart in vacuum, would produce between these conductors a force equal to 2×10^{-7} Newton per metre of length.

Kelvin (K): The Kelvin, unit of thermodynamic temperature, is the fraction 1/273.16 of the thermodynamic temperature of the triple point of water.

Candela (cd): The candela is the luminous intensity, in the perpendicular direction, of a surface of 1/600 000 m^2 of a black body at the temperature of freezing platium under a pressure of 101 325 N/ m^2 (Pa).

Mole (mol): The mole is the amount of substance of a system which contains as many elementary entities as there are atoms in 0.012 kg of carbon-12. The elementary entities must be specified and can be atoms, molecules, ions, electrons, other particles or specified groups of such particles.

Other Units

Bar: A unit of pressure = 105 N/m^2 (Pa).

Coulomb (C): The unit of electric charge; it is the quantity of electricity transported in one second by a current of one ampere.

Electron-Volt (eV): A unit of energy used in nuclear physics = 1.60206×10^{-19} J.

Degree (°): A unit of plane angle = $\pi/180$ rad.

Farad (F): The unit of electric capacitance; it is the capacitance of a capacitor between the plates of which there is a potential difference of one volt charged with one coulomb.

Henry (H): The unit of electric inductance; it is the inductance of a closed electrical circuit in which an emf of one volt is produced when the current varies uniformly at the rate of one ampere per second.

Hertz (Hz): The unit of frequency; it is equal to one cycle per second.

Joule (J): The unit of energy or work; it is the work done when a force of one Newton acts over a distance of one metre.

Kilowatt-hour (kWh): A unit of energy; it is the energy expended when a power of 1000 watts is supplied for one hour (= 3.6 MJ).

Knot (KN): A unit of speed; it is equal to one nautical mile per hour.

Litre (l): A unit of volume equal to one cubic decimetre.

Lumen (lm): The luminous flux emitted within a unit solid angle of one steradian by a point source having a uniform intensity of one candela.

Lux (lx): The unit of illumination = 1 lm/m^2.

Newton (N): The force required to accelerate a mass of one kilogram at one metre per second squared.

Ohm (Ω): The resistance between two points on a conductor at a potential difference of one volt when a current of one ampere is flowing.

Pascal (Pa): A unit of pressure = 1 N/m^2.

Poise (P): A unit of dynamic viscosity, usually quo ted in centipoise (cP); 1 cP = 10^{-3} N s/m^2.

Radian (rad): The angle subtended at the centre of a circle by an arc whose length is equal to the radius of the circle.

Siemen (S): The unit of conductance = l/Ω.

Steradian (sr): The solid angle subtended at the centre of a sphere by a cap with an area equal to the radius squared.

Stokes (St): A unit of kinematic viscosity, usually quoted in centistokes (cSt); 1 cSt = 10^{-6} m^2/s.

Tesla (T): The unit of magnetic flux density = 1 Wb/m^2.

Tonne (t): A unit of mass = 10^3 kg.

Volt (V): The unit of emf and potential difference; it is the difference in potential between two points on a conductor carrying a constant current of one ampere when the power dissipated is one watt.

Watt (W): The unit of power = 1 J/s.

Weber (Wb): The unit of magnetic flux; it is the flux which, linking a coil of one turn, produced in it an emf of one volt as it is reduced to zero at a uniform rate in one second.

The major difference between the two systems of units is attributed to the basic units and derived units established by Newton' second law of motion. They are summarized in Table D.1

Table D.1

International System of Units (SI)

	Quantity	*Unit*	*Symbol*	*Dimension*
Basic units	Length	meter	m	
	Mass	kilogram	kg	
	Time	second	s	
	Electric current	ampere	A	
	Temperature	kelvin	K	
	Luminous intensity	candela	cd	
	Amount of substance	mole	mol	
Auxiliary units	Plane angle	radian	rad	
	Solid angle	steradian	sr	
Derived units	Acceleration	meter per second squared	m/s^2	
	Activity (of radioactive source)	1 per second	s^{-1}	
	Angular acceleration	radian per second squared	rad/s^2	
	Angular velocity	radian per second	rad/s	
	Area	square meter	m^2	
	Density	kilogram per cubic meter	kg/m^3	
	Dynamic viscosity	Newton second per square meter	$N\text{-}s/m^2$	$m^{-1}\ kg\ s^{-1}$
	Electric capacitance	farad	F	$m^{-2}\ kg^{-1}\ s^4\ A^2$
	Electric charge	coulomb	C	As
Derived units	Electric field strength	volt per meter	V/m	$m\ kg\ s^{-3}\ A^{-1}$
	Electric resistance	ohm	Ω	$m^2\ kg\ s^{-3}\ A^{-2}$
	Entropy	joule per kelvin	J/K	$m^2\ kg\ s^{-2}\ K^{-1}$
	Force	Newton	N	$m\ kg\ s^{-2}$
	Frequency	Hertz	Hz	s^{-1}
	Illumination	lux	lx	m^{-2} cd sr
	Inductance	Henry	H	$m^2\ kg\ s^{-2}\ A^{-2}$
	Kinematic viscosity	square meter per second	m^2/s	
	Luminance	candela per square meter	cd/m^2	
	Luminous flux	lumen	1m	cd sr
	Magnetic field strength	ampere per meter	A/m	
	Magnetic flux	Weber	Wb	$m^2\ kg\ s^{-2}\ A^{-1}$

Table D.1 (*Continued*)

International System of Units (SI)

	Quantity	*Unit*	*Symbol*	*Dimension*
	Magnetic flux density	tesla	T	kg s^{-2} A^{-1}
	Magnetomotive force	ampere turn	A	A
	Power	Watt	W	m^2 kg s^{-3}
	Pressure	Pascal (Newton per square meter)	Pa(N/ m^2)	m^{-1} kg s^{-3}
	Radiant intensity	Watt per steradian	W/sr	m^2 kg s^{-3} sr^{-1}
	Specific heat	joule perkilogram kelvin	J/kg-K	m^2 $s^{-2}K^{-1}$
	Thermal conductivity	Watt per meter Kelvin	W/m-K	m kg s^{-3} K^{-1}
	Velocity	meter per second	m/s	
	Voltage	volt	V	m^2 kg s^{-3} A^{-1}
	Volume	cubic meter	m^3	
	Wave number	1 per meter	m^{-1}	
	Work, energy, quantity of heat	joule	J	m^2 kg s^{-2}

SI units Prefixes

The common prefixes for multiples and submultiples of SI units are given in Table D.2.

Table D.2 Prefixes and Abbreviated Prefixes for Multiples and Submultiples in Powers of 10

Multiplication Factor		*Prefix*	*Abbreviated Prefix*
1 000 000 000 000 000 000	$= 10^{18}$	exa	E
1 000 000 000 000 000	$= 10^{15}$	peta	P
1 000 000 000 000	$= 10^{12}$	Terra	T
1 000 000 000	$= 10^{9}$	giga	G
1 000 000	$= 10^{6}$	mega	M
1 000	$= 10^{3}$	kilo	k
100	$= 10^{2}$	hecto	h
10	$= 10^{1}$	deka	da
0.1	$= 10^{-1}$	deci	d
0.01	$= 10^{-2}$	centi	c
0.001	$= 10^{-3}$	milli	m
0.000 001	$= 10^{-6}$	micro	μ
0.000 000 001	$= 10^{-9}$	nano	n
0.000 000 000 001	$= 10^{-12}$	pico	p
0.000 000 000 000 001	$= 10^{-15}$	femto	f
0.000 000 000 000 000 001	$= 10^{-18}$	atto	a

SI Conversion

To convert the units of a given quantity from one system to another, the equivalence of units and conversion factors are given in Table D.3. A simple procedure is to write the desired SI unit equal to the US/English units, and put in canceling unit factors. The following example illustrates the procedure.

Example ED.1*: Express stress in SI units in terms of stress in US/ English units.*

Solution:

Stress: (Stress in SI Units) = (Stress in English units) × (Multiplication factor)

$$(\text{Pa}) = \left(\frac{\text{N}}{\text{m}^2}\right) = \left(\frac{\text{N}}{\text{lb}_\text{f}}\text{lb}_\text{f}\right)\left(\frac{1}{\frac{\text{m}}{\text{in}}}\text{in}\right)^2 = \left(\frac{\text{N}}{\text{lb}_\text{f}}\right)\left(\frac{1}{\frac{\text{m}}{\text{in}}}\right)^2\left(\frac{\text{lb}_\text{f}}{\text{in}^2}\right)$$

$$= \left(\frac{\text{Nper1lb}_\text{f}}{\text{mper1in}^2}\right)\left(\frac{\text{lb}_\text{f}}{\text{in}^2}\right)$$

$$= \left(\frac{4.4482}{0.0254^2}\right)\left(\frac{\text{lb}_\text{f}}{\text{in}^2}\right)$$

$$= 6894.757\left(\frac{\text{lb}_\text{f}}{\text{in}^2}\right)$$

Temperature Conversion

To convert from Fahrenheit t_f to Celsius t_c:

$$t_c = 5/9\,(t_f - 32)$$

To convert from Celsius t_c to Fahrenheit t_f:

$$t_f = 9/5 t_c + 32$$

REFERENCES

American National Standards for Metric Practice, ANSI/IEEE Std 268-1992, Institute of Electrical and Electronics Engineers, New York, NY, October 1992.

IBM., SI *Metric Reference Manual*, 3rd ed., New York, NY, March 1974.

IEEE/ASTM-SI-10., "Standard for use of the International System of Units (SI): The Modern Metric System", *ASTM International*, West Conshohocken, PA, 1992.

Mechtly, E.A., "The International System of Units", Second revision, *NASA SP-7012*, 1973.

National Bureau of Standards., "The International System of Units (SI)", July 1974.

National Institute for Science and Technology., "Guide for the Use of the International System of Units (SI), NIST Special Publication 811", abridged, *Gaithersburg*, MD,1995.

Taylor, B.N., (ed), The International System of Units (SI), National Institute Stand. Technol. Spec. Publ. 330, 1991 edition (U.S. Government Printing Office, Washington, DC, august 1991.

Wandmacher, C., *Metric Units in Engineering – Going SI*, Industrial Press, New York, NY, 1978.

D.1 PROBLEMS

D.1 Express torque, (SI) in US/English units.

D.2 Express mass moment of inertia, *I*, (SI) in US/English units.

D.3 Express modulus of elasticity, *E*, (SI) in US/English units.

D.4 Express spring constant, *k*, (SI) and mass (*M*) in US/English units

Table D.3 Conversion of Units

Quantity	*SI Equivalence*	*English* Equivalence
Acceleration	1 ft/s^2 = 0.3048 m/s^2	1 m/s^2 = 0.3281 ft/s^2
Angles	1 rad = 57.2958 degrees 1rpm = 0.1667 rev/sec = 0.1047 rad/sec	1 degree = 0.0175 rad 1 rad/sec = 9.5493 rpm
Angular Velocity	1 rpm = 0.1047 rad/sec	1 rad/sec = 9.5511 rpm
Area	1 in^2 = 645.16 × 10^{-6} m^2 1 ft^2 = 0.0929 m^2	1 m^2 = 1550 in^2 = 10.7639 ft^2
Area Moment of Inertia or Second Moment of Area	1 in^4 = 41.6231 × 10^{-8} m^4 1 ft^4 = 86.3097 × 10^{-4} m^4	1 m^4 = 240.251 × 10^4 in^4 = 115.862 ft^4
Damping Constant: Translatory Torsional	1 lb$_f$ .sec/in. = 175.1268 N s/m 1 in.lb$_f$ – sec/rad = 0.1130 m N s/rad	1 N.s/m = 5.7102 × 10^{-3} lb$_f$ .sec/in 1m.N.s/rad = 8.8507 lb$_f$.in.sec/rad
Force or Weight	1 lb$_f$ = 4.4482 N	1 N = 0.2248 lb$_f$
Length	1 in = 0.0254 m 1 ft = 0.3048 m 1 mile = 5280 ft = 1.6093 km	1 m = 39.370 in = 3.2808 ft 1 km = 3280.84 ft = 0.6214 mile
Mass	1lb$_f$ .sec^2/ft (slug) = 14.5939 kg = 32.174 lb$_m$ 1 lb$_m$ = 0.4536 kg 1 short ton=907.185 kg	1 kg = 2.2046 lb$_m$ = 0.0685 slugs (lb$_f$ .sec^2/ft)
Mass Density	1 lb$_m$/in^3 = 27.6799 × 10^3 kg/m^3 1 lb$_m$/ft^3 = 16.0185 kg/m^3	1 kg/m^3 = 36.127 × 10^{-6} lb$_m$/in^3 = 62.428 × 10^{-3} lb$_m$/ft^3
Mass Moment of Inertia	1 in.lb$_f$ – sec^2 = 0.1130 m^2 kg	1 m^2 kg = 8.8507 in.lb$_f$.sec^2
Moment	1 ft.lb = 1.3558 N m	1 N-m = 0.7376 ft-lb

Quantity	*SI Equivalence*	*English* Equivalence
Power	1 in.lb_f = 0.1129 W 1 ft.lb_f = 1.3558 W = 1818.2×10^{-6} hp 1 hp = 747.7 W	1 W = 8.8507 in.lb_f/sec 1 W = 0.7376 ft.lb_f/sec = 1.3410×10^{-3} hp
Spring constant: Translatory Torsional	1 lb_f/in = 175.1268 N/m 1 lb_f/ft = 14.5939 N/m 1 in.lb_f/rad = 0.1130 m N/rad	1 N/m = 5.7102×10^{-3} lb_f/in = 68.522×10^{-3} lb_f/ft 1 m N/rad = 8.8507 in.lb_f/rad = 0.7376 lb_f .ft/rad
Stress, Pressure, or Elastic Modulus	1 lb_f/in^2 (psi) = 6894.757 Pa 1 lb_f/ft^2 = 47.88 Pa	1 Pa = 1.4504×10^{-4} lb_f/in^2 (psi) = 208.8543×10^{-4} lb_f/ft^2
Torque or Moment	1 lb_f.in = 0.1130 N m 1 lb_f .ft = 1.3558 N m	1 N m = 8.8507 lb_f .in = 0.7376 lb_f – ft
Velocity	1 ft/sec = 0.682 mph = 0.3048 m/sec; 1 mph = 1.467 ft/sec = 1.609 km/h	1 m/s = 3.2808 ft/s = 2.237 mph 1 km/h = 0.622 mph
Volume	1 in^3 = 16.3871×10^{-6} m^3 1 ft^3 = 28.3168×10^{-3} m^3 1 US gallon = 3.7853 litres = 3.7853×10^{-3} m^3	1 m^3 = 61.0237×10^3 in^3 = 35.3147 ft^3 = 10^3 litres = 0.2642 US gallon
Work or Energy	1 lb_f.in = 0.1130 J_2 lb_f .ft = 1.3558 J 1 Btu = 1055.056 J 1 kWh = 3.6×10^6 J	1 J = 8.8507 in.lb_f 1 J = 0.7376 ft.lb_f = 0.9478×10^{-3} Btu = 0.2778 kWh

ACTAR Accreditation Commission for Traffic Accident Reconstruction

Appendix

E

E.1 ACTAR

In 1985, the National Highway Traffic Safety Administration (NHTS) provided a grant to develop national guidelines for the standardization of training in the field of traffic accident reconstruction. A task force of accident reconstructionists, engineers, police officers, educators and attorneys met and developed a report entitled Minimum Training Criteria for Police Traffic Accident Reconstructionists. In that report, the task force addressed certification of individuals in the field and recommended that "a certification board be formed" to accredit accident investigators and reconstructionists.

Five years later, twelve professional accident reconstruction associations with world-wide representation met to explore the possibility of forming an internationally recognized accreditation program open to both police and civilian accident reconstructionists. **The Accreditation Commission for Traffic Accident Reconstruction (ACTAR)** was the result of that coalition. The Governing Board of Directors, comprised of one representative from each participating association, has included police officers, engineers, educators and private consultants all working in the field of traffic accident investigation and reconstruction within the United States and Canada.

ACTAR was founded by and exists for the benefit of the traffic accident investigation and reconstruction community, as represented by the membership of the participating professional organizations. The Commission has not been obligated to nor controlled by any governmental body or agency. Since its incorporation in 1991, it has been the ongoing goal of ACTAR to promote, within the legal and scientific community, recognition of the minimum standards established by the NHTSA study, as well as those developed by an ongoing review of the latest technology and trends in the profession.

Minimum standards have been designed to advance the recognition of the ACTAR accreditation program, and in doing so, to encourage the integrity, consistency and professionalism of those involved in traffic accident investigation and reconstruction, to promote the professional and intellectual development of those individuals, organizations and institutions involved in traffic accident investigation and reconstruction, to assist the legal and scientific community in weighing the suitability of individuals offering their services as Accident Reconstructionists and to improve public awareness of the profession as it relates to the legal system.

By way of committee discussions of different aspects of accident investigation and reconstruction training programs, as well as review of other disciplines practicing in the field, the ACTAR Governing Board of Directors developed a formula for minimum training and experience requirements. Applying those minimum standards to a higher level of understanding and knowledge in the collision investigation and analysis field, the Governing Board of Directors created and refined a multi-part accreditation examination. That examination was reviewed by outside independent professionals in the testing field, as well as educators, to ensure an objective, clear and thorough examination.

Although participation in the accreditation program is voluntary, people who are properly trained and experienced in accident investigation and reconstruction can successfully complete the examination and achieve accreditation. Those accredited must obtain a minimum number of continuing educational units (C.E.U.'s) over a five year period from completion of the initial examination to maintain their status with ACTAR.

E.2 ACTAR PARTICIPATING ORGANIZATIONS

ASSE: American Society of Safety Engineers
CA²RS: California Association of Accident Reconstruction Specialists
CATAIR: Canadian Association of Technical Accident Investigators and Reconstructionists
CSP: Colorado State Patrol
DRI: Denver Research Institute
FARO: Forensic Accident Reconstructionists of Oregon
IAARS: International Association of Accident Reconstruction Specialists
IATAI: Illinois Association of Technical Accident Investigators
MdATAI: Maryland Association of Traffic Accident Investigators
UCMO: Central Missouri State University
MSP: Michigan State Police
MwATAI: Midwest Association of Technical Accident Investigators
NAPARS: National Association of Professional Accident Reconstruction Specialists
NATARI: National Association of Traffic Accident Reconstructionists and Investigators
NJAAR: New Jersey Association of Accident Reconstructionists
NYSTARS: New York Statewide Traffic Accident Reconstruction Society
OPP: Ontario Provincial Police
OSP: Oregon State Police
PSP: Pennsylvania State Police
SATAI: Southwestern Association of Technical Accident Investigators
SOAR: Society of Accident Reconstructionists
TAARS: Texas Association of Accident Reconstruction Specialists
WATAI: Washington Association of Technical Accident Investigators

E.3 WEB SITE LINKS

ACTAR does not endorse or support the following site link information. These sites are listed for the use of the reader. Please advise the webmaster of any broken or out-of-date links.

Accident Reconstruction and Investigation

The Accident Reconstruction Network (ARC Network): is the largest Internet portal web site for the industry of accident reconstruction and traffic accident investigation. This web site provides a research section, news article archive, discussion forum, expert witness directory, event calendar, products directory, book store, education services directory, police department directory and more.

Traffic Accident Reconstruction Origin (TARO): The premier non-commercial web location for Traffic Accident Reconstruction discussion. TARO is like any technical journal. It contains articles of a technical nature that report on the evolving state of traffic accident reconstruction.

- Traffic Accident Reconstruction Web Log
- Indiana Association of Certified Accident Investigators
- Institute of Traffic Accident Investigators (United Kingdom)
- European Association for Accident Research & Analysis

E.4 COLLEGES AND UNIVERSITIES

- Collision Safety Institute (CSI) - Training
- Institute of Police Technology and Management (IPTM) - University of North Florida
- TEEX Law Enforcement and Security Training Division (LESTD) - The Texas A&M University System
- Center for Public Safety (formerly the Traffic Institute) - Northwestern University
- University of Michigan - Transportation Research Institute
- United Transportation Training Inc.
- North Coast Polytechnic Institute
- Michigan State University- Highway Traffic Safety Program
- SRR Training

E.5 GOVERNMENT AGENCIES

- Bureau of Transportation Statistics
- Code of Federal Regulations
- Department of Transportation (DOT)
- Federal Highway Administration (FHWA)
- Office of Highway Safety (MUTCD)
- Federal Register
- Library of Congress
- National Crash Analysis Center

- National Highway Traffic Safety Administration (NHTSA)
- National Library of Medicine
- National Transportation Safety Board (NTSB)
- Transport Canada
- Turner-Fairbanks Highway Research Center

E.6 OTHER RELATED SITES

- All Things Automotive
- Applied Kinematics
- Association for the Advancement of Automotive Medicine (AAAM)
- Autoguide
- CRASH
- Highway Loss Data Institute
- Insurance Institute for Highway Safety (IIHS)
- Lawyers & Judges Publishing Company
- Northwestern University Transportation Library
- Transportation Research Board (TRB)
- Transport Research Laboratory (TRRL)
- Underride Network
- WorldWide Web - Virtual Library: Transportation

E.7 ADDITIONAL SITES OF INTEREST

- United States Post Office Information Center
- United States Naval Observatory (Sun/Moon Positions)
- Latitude and Longitude Search
- U.S. Gazetteer – Maps by Zipcode Area
- Universal Currency Converter
- Vehicle Identification Number (VIN) Lookup

E.8 TRAFFIC CRASH REPORTS & OVERLAY FORMS

The following table contains hyper-links to state traffic crash report forms and any overlay code sheets applicable to that state. All forms are in Adobe Portable Document File Format (PDF). In order to eliminate the fraudulent use of crash reports, all state report forms cannot be printed or modified. Overlay code sheets can be printed.

In cases where states have revised their reporting forms, both old and new formats are provided. ACTAR assumes no liability for the accuracy or use of these documents. They are provided as a service to the traffic crash reconstruction community.

State	*Crash Report*	*Overlay Code Sheet*	*Comments*
Alabama	Form AST-27 1/91	Codes on report	
Alaska	Form 12-200 9/12/01	Codes on report	
Arkansas	Form 5-2000	Code Sheet	
Arizona	Form 01-2704 3/01	Codes on report	
California	Form CHP555 1/88	Codes on report	Vehicle Code Data
	Form CHP555 7/03	Codes on report	
Colorado		Code Sheet	
	Form 2447 11/05	Code Sheet	
	Form 2447 2/06	Code Sheet	
Connecticut	Form PR1 12/94	Form PR1 12/94	
Delaware	Form 438 1/88	Codes on report	
District of Columbia	Form PD10 7/86	Codes on report	
Florida	Form HSMV-90003 1/01	Codes on report	
	Form HSMV-90003 1/02	Codes on report	
Georgia	Form DPS523 2/94	Code Sheet	
	Form DMVS 523 12/03	Code Sheet	
Hawaii		Form IMVSO 1-851	
Iowa	Form 433003 1/01	Form 433014 1/01	
	Form 433002		Civilian Report
Idaho	Form IDT-90 11/96	Codes on report	
Illinois	Form SR1050 12/01	Form SR1000A 3/01	
	SR1050 10/03	Form SR1000A&B 12/01	
	Form SR1050 10/06	Form SR1000A 10/06	
	Form SR1510 10/06	Form SR1000B 10/06	
Indiana	Form 23558 7/91	Code Sheet	
	Form 23558 9/01		
	Form 23558 5/03		

Kansas		Form 852 1/95	
Kentucky	Form KSP74 1/00	Code Sheet	
Louisiana	Form DSSSP3105-11	Codes on report	
Maine		Code Sheet	
Maryland	Form MSP-1 3/95	Code Sheet	
		Code Sheet (8/01)	
		Code Sheet (1/06)	
Massachusetts	Form CRA-65 9/01	Code Sheet	
Michigan	Form UD10	Code Sheet 11/98	
Minnesota	Form PS32003-07	Code Sheet	
	Form PS32003-10	Code Sheet	
Mississippi	Form	Code Sheet	
Missouri	Form 1996 version	Codes on report	
	Form SHP2P 1/02		
Montana		Code Sheet	
Nebraska	Form 40 10/94	Code Sheet 1994	
	Form 40 1/02	Code Sheet 2002	
Nevada	Form 1/04	Code Sheet	
New Hampshire		DSMV160 8/93	
New Jersey	Form NJTR1 8/95	Code Sheet	
	Form NJTR1 7/05	Code Sheet	
New Mexico	Form SH10074 6/97	Codes on report	
New York		MV104AW 7/96	
		MV104A 7/01	
North Carolina	Form DMV349 9/99	DMV349	
North Dakota		DOT2356 1/99	
		SFN2356 2/96	
		SFN2356 11/86	
		DOT2356 12/06	

Ohio	Form OH-1 10/99	Codes on report	
Oklahoma	Form DPS67-1 9/96	Codes on report	
	Form DPS0192-1 1/07	Codes on report	
Oregon	Form 73546-A 8/98	Codes on report	
Pennsylvania	Form AA-45	Code Sheet (1996)	
		Code Sheet (1998)	
	Form AA-500	Code Sheet (2005)	
Rhode Island	Form DMVSAF 1/83	Codes on report	
South Carolina	Form TR310 1/01	Codes on report	
South Dakota		Code Sheet	
		DPSHS621 1/83	
Tennessee	Form SF1203 1/00	HR04	
Texas	Form ST3 1/00		
	Form CBR-3 1/06	Codes on report	
Utah	Form DI9 10.93	DI9 1997	
		11/2005	
Virginia		FR300P 1/78	
		FR300T 9/03	
Vermont	Form	TA VA 01 11/01	
Washington	Form 3000345159 1/97	Code Sheet 1/97	
	Form 30034513 7/06	Code Sheet 7/06	
West Virginia	Form DMV17F 5/97		
Wisconsin	Form MV4000 12/96	MV4000	
Wyoming	Form PR802 4/91	PR802A 1/91	
	Form PR902 1/07	Codes on report	
Port Authority NY-NJ		Code Sheet	
CANADA			
Alberta	Form MTS284 1991	Code Sheet	
	Form MTS284A 1991		Walk-in report form

British Columbia	Form MV6020 2/98	MV6020A 2/98
		CL160 3/04
New Brunswick		Code Sheet
Ontario		Code Sheet PtA Code Sheet PtB
Quebec	Form 5185-2 10/2000	Code Sheet PtA Code Sheet PtB

E.9 ACCIDENT RECONSTRUCTION EDUCATION INFORMATION

Accident Reconstruction, Traffic Accident Investigation, Crash Data Retrieval, Law Enforcement Education, Crash Investigation, Forensic Mapping, Traffic, DUI/DWI, Criminal Investigation

Welcome to the Accident Reconstruction Education and Training section of the ARC Network. In this section you will find quality schools, colleges and training institutions dedicated to providing the best accident reconstruction education. Please use the accident reconstruction education directory below to find the accident reconstruction training you are looking for.

Educational Resources

Schools and Colleges

Training Institution	*Overview*	*Classes Offered*
Collision Safety Institute (CSI)	The Collision Safety Institute (CSI) is an independent traffic collision investigation, reconstruction training and research center.	• Crash Data Retrieval System • Pedestrian/bicycle Reconstruction • Collision Reconstruction • Motorcycle Reconstruction • Collision Investigation • Collision Trauma Biomechanics
FactualDiagrams.com	Our primary focus is to provide quality Forensic Mapping Training in California and the Western United States. Our instructors are all ACTAR accredited collision reconstructionists and are certified Forensic Mapping Specialists. They also are all members of the Professional Society of Forensic Mapping.	• ACTAR Test Prep. Course • Bosch/Vetronix Crash Data Retrieval Technician • CrashZoneTM/CrimeZoneTM- Basic • CrashZoneTM/CrimeZoneTM- Advanced • Forensic Mapping Course • Forensic Mapping – Advanced • We Sponsor Specialized Courses
IPTM (Institute of Police Technology and Management)	The mission of the Institute of Police Technology and Management (IPTM) is to provide the law enforcement community with the highest quality of training at competitive prices. By providing this service, IPTM continues to support law enforcement's efforts in building and maintaining safer communities.	• Crash Investigation Courses • Criminal Investigation Courses • Forensics and Technology Courses • DUI/DWI Courses • Traffic Courses

(Contd.)

MapScenes	At MapScenes we're interested in your Law Enforcement Officers efficiency. They will quickly become more productive after one of our forensic mapping software courses. By mastering our products and understanding simple to advanced software features. They will bring faster evidence collection to the field, and faster drawing to the desktop. MapScenes Systems has introduced the Course Training Standard (CTS). Map Scenes acknowledges its responsibility to the industry ensuring that all trainers and educational courses within the CTS are kept up to date through a base educational standard maintaining quality and integrity. Qualified students become knowledgeable using the Mapscenes System and therefore are designated a *MapScenes Certified Professional*.	• MapScenes 2006 Upgrade and Advanced Techniques Class • MapScenes System Upgrade Training Class • Forensic Mapping with Evidence Recorder 3.0 and MapScenes PRO 2006 • Evidence Recorder 3.0 and MapScenes PRO 2006 Upgrade Class
Northwestern University	The Center for Public Safety was founded by Northwestern University as the Traffic Institute in 1936 for the purpose of traffic safety. The Center serves local, county, state, federal, and foreign government agencies through programs of specialized training, continuing education, research and development, publications and direct assistance.	• Accident Reconstruction • Accident Investigation • Police Training • Management Consulting • Research and Development • Transportation Engineering
SRR Training	Law enforcement must be prepared to face numerous, ever-changing challenges. The training received should be ready to face these same challenges as well. While the foundation of our courses is based on proven methodologies, SRR Training is constantly modifying them to reflect the latest trends in crash and criminal investigatory procedures. If you are looking to provide your officers with a unique learning experience, or if you yourself wish to enhance your own professionalism, our courses and instructors will help you to meet your goal.	• ACTAR Prep Course • Crash Investigation • Crash Reconstruction • Cad Zone • Commercial Vehicle • Crash/Crime Scene Measuring/ Diagramming • Crush • Human Factors • Motorcycle, Bicycle, Pedestrian • Night-Time Investigation • Report Writing

(*Contd.*)

VS Visual Statement	Visual Statement recognizes that many users would enjoy the benefits of hands-on training found within training courses. We offer training sessions within major cities in North America based on demand and availability.	• Collision and Crime Scene Diagramming • Advanced Diagramming and 3D Animation
West Coast Accident Reconstruction Equipment & Education (WeCARE)	WeCARE offers a variety of traffic collision courses from the basics through advanced reconstruction. We also design specialized training seminars and courses for Law Enforcement Agencies, Insurance Companies, Law Firms, Professional Associations and other organizations. Please call to discuss your specific training needs.	• Accident Investigation • Accident Reconstruction • Speed from Crush • Auto-Pedestrian Investigation • Computer Assisted Diagramming

Other Training Resources

- Accident Reconstruction Dictionary (Glossary)
- Test Your Skill
- Equation Resource

Appendix

F

Miscellaneous Accident Reconstruction Formulae

F.1 CRITICAL SPEEDS OF DRIVING SHAFTS

The critical speed (the speed at which instability occurs) of hollow shafts, such as the drive shaft of a road vehicle is calculated by

$$n_c = \frac{4.8(10^6)\sqrt{D^2 + d^2}}{L^2}; \text{ (RPM)} \qquad \text{(F.1)}$$

where d = inner shaft diameter, (in)

D = outer shaft diameter, (in)

L = distance between universal joints, (in)

The recommended maximum shaft design speed for elastic shafts should be approximately 15 to 30% below the critical speed.

For example for a drive shaft with an outer diameter of 3 in., and inner diameter of 2.5 in., and a shaft length of 60 in., Eq. (F.1) gives a critical speed of 5206.84 rpm.

F.2 TIRE TREAD SEPARATION SPEED

The tire tread separation speed of the free wheel is calculated by

$$V_{\text{tread}} = \frac{dn_e}{7i}; \text{ (mph)} \qquad \text{(F.2)}$$

where d = tire diameter, (ft)

n_e = engine speed, (RPM)

i = final drive reduction gear ratio.

For example, for a tire diameter of 32 in., an engine speed of 4500 rpm, and a final gear ratio of 5.0, the tire tread speed is 343.29 mph.

F.3 SPEED ON SIDE SLOPES

The velocity V_0 at the beginning of the sliding process is calculated by

$$V_0 = 2f\sqrt{\frac{g\ell}{\sin\alpha}};\ (\text{ft/sec}) \qquad \text{(F.3)}$$

where g = gravitational constant or 32.2 ft/sec²
f = tire/road friction coefficient
ℓ = lateral sliding distance, (ft)
α = lateral or cross slope angle (deg) for ($\alpha < 10°$)

The path distance S traveled by the vehicle is given by

$$S = \frac{2f\ell}{\sin\alpha};\ (\text{ft}) \qquad \text{(F.4)}$$

Example F.1: *A car locks its brakes and slides on an icy road laterally for 15 ft, where it hits a child waiting at a bus stop. The cross slope of the highway was 8 degrees. Determine the speed when the brakes were locked and the distance by while the car would have missed the child had the cross slope been 3 degrees. Assume a coefficient of friction of 0.20.*

Solution: We apply Eq. (F.3)

$$V_0 = 2(0.20)\sqrt{\frac{(32.2)(15)}{\sin 8°}} = 23.56 \text{ ft/sec} = 16.07 \text{ mph}$$

The skidding distance

$$S = \frac{2(0.20)(15)}{\sin 8°} = 43.11 \text{ ft}$$

With a lateral distance of 15 ft and a cross slope of 3 deg., the skidding distance would be

$$S = \frac{2(0.20)(15)}{\sin 3°} = 114.64 \text{ ft}$$

The car would have missed the child by more than 71 ft. The speed would have been

$$V_0 = 2(0.20)\sqrt{\frac{32.2(15)}{\sin 3°}} = 38.43 \text{ ft/sec} = 26.21 \text{ mph.}$$

F.4 VEHICLE WEIGHT DISTRIBUTION

The distance of the center of gravity from the front wheel can be calculated by

$$\ell_f = \frac{F_s W_b}{W};(\text{ft}) \qquad \text{(F.5)}$$

where F_s = static rear axle wheel load, lb
W_b = wheelbase, ft
W = vehicle weight, lb.

F.5 VEHICLE CENTER-OF-GRAVITY HEIGHT

The vertical height of the center of gravity above ground h is given by

$$h = R_f + \frac{(F_{f,r} - F_f)(W_b)\left[W_b^2 - (a - R_f)^2\right]^{1/2}}{W(a - R_f)}; \text{(ft)} \tag{E.6}$$

where a = raised vertical distance of rear axle, ft
F_f = front axle load without raising rear axle, lb
$F_{f,r}$ = front axle load with rear axle raised, lb
W_b = wheelbase, ft
R_f = front tire radius, ft.

F.6 RADIUS OF YAW MARKS

The radius of yaw marks can be calculated by

$$R = \frac{C^2 + 4D^2}{8D}; \text{(ft)} \tag{F.7}$$

where C = cord length in ft (A straight line connecting beginning and end of scuff markings)
D = maximum perpendicular distance between cord and scuff markings, (ft).

F.7 ROAD CURVATURE

The radius of the road can be calculated by (see Fig.F.1)

$$R = \frac{57.3\ell}{\alpha}; \text{(ft)} \tag{F.8}$$

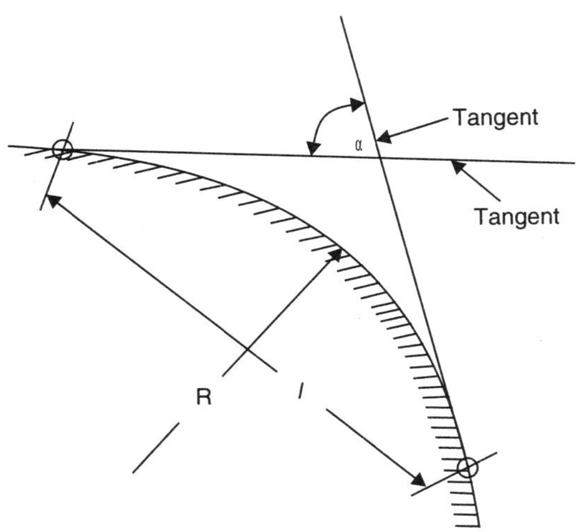

Fig. F.1 Road Curvature.

where ℓ = length of curve, ft

α = tangent angle of curvature, deg.

F.8 SAFE FOLLOWING DISTANCE OF TWO VEHICLES

The safety distance, S between two vehicles may be computed by

$$S = S_0 + V\left[t_r + \frac{V}{2}\left(\frac{1}{d_1} - \frac{1}{d_2}\right)\right];\ (\text{ft}) \tag{F.9}$$

where d_1 = deceleration of rear vehicle, ft/sec^2

d_2 = deceleration of lead vehicle, ft/sec^2

S_0 = distance between stopped vehicles, ft

V = velocity of vehicles, ft/sec

t_r = reaction time of driver in rear vehicle, sec.

F.9 CONSTANT SPEED PASSING

The passing maneuver distance S_p is computed by

$$S_p = V_1\left[\frac{\ell_1 + \ell_2 + \ell_s}{V_1 - V_2} + 0.9\sqrt{\frac{\ell}{a_y}}\right];\ (\text{ft}) \tag{F.10}$$

where a_y = average lateral acceleration of lane change and return lane change, g units

ℓ_1 = length of passing vehicle, ft

ℓ_2 = length of vehicle being passed, ft

ℓ_s = safe distance between vehicles before and after passing, ft

ℓ = lateral distance, determined by sum of 1/2 width of both vehicles and lateral safety distance, ft

V_1 = velocity of passing vehicle, ft/sec

V_2 = velocity of vehicle being passed, ft/sec

The time required for the passing maneuver is given by

$$t_p = \frac{S_p}{V_1};\ (\text{sec}) \tag{F.11}$$

Example F.2: *Determine the distance and time required for a passing maneuver by the use of the following given data: V_1 = 62 mph (90.89 ft/sec), V_2 = 52 mph (76.23 ft/sec), ℓ_1 = 18 ft, ℓ_2 = 59 ft, ℓ_s = 2(20) = 40 ft, ℓ = 12 ft and a_y = 0.2 g.*

Solution: Eq. (F.10) gives

$$S_p = 90.89\left[\frac{18 + 59 + 40}{90.89 - 76.23} + 0.9\sqrt{\frac{12}{0.2(32.2)}}\right] = 837.04 \text{ ft}$$

and

$$t_p = \frac{S_p}{V_1} = \frac{837.04}{90.89} = 9.21 \text{ sec.}$$

F.10 TIME TO AVOID HEAD-ON CRASH WHILE PASSING

(*a*) Constant speeds – both vehicles

The time available before impact is given by

$$t_a = \frac{\Delta S}{V_1 + V_2}\text{; (sec)} \qquad \text{(F.12)}$$

where ΔS = distance between approaching vehicles, ft

V_1 = speed of vehicle 1, ft/sec

V_2 = speed of vehicle 2, ft/sec

(*b*) One or both vehicles braking

The time available before braking is given by

$$t_b = \frac{V_1 + V_2}{a_1 + a_2}\sqrt{\left(\frac{V_1 + V_2}{a_1 + a_2}\right)^2 - \frac{2\Delta S}{a_1 + a_2}} \qquad \text{(F.13)}$$

where a_1 = deceleration of vehicle 1, ft/sec^2

a_2 = deceleration of vehicle 2, ft/sec^2

If one of the vehicles is not braking, then its deceleration is equal to zero.

F.11 BRAKING TO AVOID COLLISION

(*a*) Stopping distance

The stopping distance S_b is given by

$$S_b = \frac{V_2}{2(32.2)(f \pm \sin\alpha)} + t_r V\text{ ; (ft)} \qquad \text{(E.14)}$$

where S_b = the stopping distance of a vehicle consisting of the distance during which the vehicle driver reacts and applies the brakes, and the distance during which the vehicle slows from its travel speed until the vehicle stops

f = drag factor or deceleration, g units

V = vehicle velocity, ft/sec

t_r = driver reaction time, sec

α = road incline, deg. (+ sign for uphill and – sign for downhill).

(*b*) Maximum speed to stop at point of impact

The maximum noncollision speed to stop just at the point of impact is given by

$$V_{\max, g} = \sqrt{(at_r)^2 + 2aS_{r-c}} - at_r\text{; (ft/sec)} \qquad \text{(F.15)}$$

where a = deceleration, ft/sec^2

t_r = driver reaction time, sec

S_{r-c} = distance between points where reaction began in actual accident to point of impact, ft

The distance S_{r-c} is computed by

$$S_{r-c} = S_1 + t_r V_{1b} \tag{F.16}$$

where S_1 = pre-impact braking distance, ft

V_{1b} = speed of vehicle at moment of braking in actual accident, ft/sec.

(*c*) Maximum speed for other vehicle to clear impact area

The maximum noncollision speed for other vehicle to clear the impact area is given by

$$V_{\max,t} = \frac{\frac{a}{2}(t_{b-c} + t_a)^2 + S_{r-c}}{t_{r-c} + t_a} \tag{F.17}$$

where a = deceleration, ft/sec^2

S_{r-c} = distance between reaction beginning and collision point, ft

t_{b-c} = time from beginning of braking (skid marks) to point of impact, sec

t_{r-c} = time from beginning of reaction to point of impact, sec

t_a = time required by other vehicle (or pedestrian) to clear impact area, sec.

The time t_a is given by

$$t_a = \frac{S_{\text{safe}}}{V_{0v}}; \text{(sec)} \tag{F.18}$$

where S_{safe} = distance required to clear impact area, ft

V_{0v} = speed of other vehicle, ft/sec

$$t_{b-c} = \frac{-V_t}{a} + \sqrt{\left(\frac{V_t}{a}\right)^2 + \frac{2S}{a}}; \text{(sec)}$$

or

$$t_{b-c} = \frac{V_0}{a} - \sqrt{\left(\frac{V_0}{a}\right)^2 - \frac{2S}{a}}; \text{(sec)} \tag{F.19}$$

where V_0 = initial velocity at time $t = 0$ sec

V_t = final velocity at time t

a = acceleration, ft/sec^2

S = distance, ft

t_{b-c} is then computed using Eq. (F.19*a*) or (F.19*b*) given below:

$$t = -\frac{V_t}{a} + \sqrt{\left(\frac{V_t}{a}\right)^2 + \frac{2S}{a}}; \text{(sec)} \tag{F.19a}$$

$$t = -\frac{V_c}{a} - \sqrt{\left(\frac{V_0}{a}\right)^2 - \frac{2S}{a}}; \text{(sec)} \tag{F.19b}$$

Example F.3: *Vehicle A drives across an intersection at a constant speed of 10 mph. Vehicle B approaches the intersection at a speed of 48 mph and collides with vehicle A. The skid marks of vehicle B begin 24 ft from the point of impact. The speed on that roadway is 30 mph. Damage inspection showed that the left front corner of vehicle B impacted against the left rear wheel of vehicle A approximately 3.5 ft* *ahead of the rear bumper. Determine the maximum speeds possible for accident avoidance for a reaction time t_R = 1.1 sec. Assume a drag factor of 0.8.*

Solution: The stopping distance of vehicle *B* had it traveled at 30 mph is given by Eq. (F.14)

$$S_b = \frac{V_2}{2(32.2)(f \pm \sin\alpha} + t_r V$$

$$S_b = \frac{44^2}{2(32.2)(0.8)} + (44)(1) = 81.58 \qquad (1)$$

Now S_{r-c} is given by Eq. (F.16). This is in order to find how far vehicle *B* was from the point of impact when its driver began to react in the actual accident.

$$S_{r-c} = 24 + (1.0)\ (50)\ (1.466) = 97.3 \text{ ft.} \qquad (2)$$

Comparing Eqs. (1) and (2), we see that the accident could have been avoided had vehicle A traveled at the speed limit of 30 mph.

Now, we determine the maximum speed vehicle A could have traveled and stopped just at the point of impact.

Substituting in Eq. (F.15) gives

$$V_{\max,\,g} = \sqrt{[(25.76)(1.1)]^2 + 2(25.76)(97.3) - (25.76)(1.1)}$$

$$= 47.93 \text{ ft/sec or } 32.72 \text{ mph}$$

The time avoidance for vehicle *A* is given by

$$t_{AV} = \frac{3.5}{(10)(1.466)} = 0.239 \text{ sec.}$$

Hence, if vehicle *B* arrives 0.239 seconds later at the point of impact, vehicle *A* will have traveled 3.5 ft., it will have moved out of the danger zone.

The time t_{b-c} is obtained using equation given below,

$$t_{b-c} = \frac{V_0}{a} - \sqrt{\left(\frac{V_0}{a}\right)^2 - \frac{2S}{a}}\ ; \text{sec}$$

or

$$t_{b-c} = \frac{70.35}{25.76} - \sqrt{\left(\frac{70.32}{25.76}\right)^2 - \frac{2(24)}{25.76}} = 0.366 \text{ seconds}$$

Now, Eq. (F.17) gives

$$V_{\max,\,t} = \frac{\frac{a}{2}(t_{b-c} + t_a)^2 + S_{r-c}}{t_{r-c} + t_a} = \frac{\left(\frac{2.76}{2}\right)(0.366 + 0.239)^2 + 97.3}{(1.466 + 0.273)} = 58.66 \text{ ft/sec or } 39.98 \text{ mph.}$$

F.12 VEHICLE ROLLOVER ANALYSIS

(*a*) Simplified static rollover analysis

The limit lateral acceleration is computed by

$$a_{y\,\lim} = \frac{V^2}{Rg} \le f_s\,;\ g \text{ units} \tag{F.20}$$

where g = acceleration due to gravity, (32.2 ft/sec^2)
R = turn radius, ft
V = vehicle velocity, ft/sec
f_s = side or lateral friction coefficient

A vehicle may rollover only when the tire-road friction equals or exceeds the following value:

$$f_r \le \frac{t}{2h} \tag{F.21}$$

where h = center-of-gravity height, in.
t = truck width, in.

From Eqs. (F.20) and (F.21), the minimum velocity $V_{\min}$ at which rollover occurs is given by

$$V_{\min} = \sqrt{\frac{Rgt}{2h}}\,,\ \text{ft/sec} \tag{F.22}$$

(*b*) Rollover caused by side sliding and tripping

The minimum side sliding speed vs. required for rollover when striking a curb is given by

$$V_s > 45.5\sqrt{\left(\frac{I_{\text{roll}}}{Wh}\right)\left[\sqrt{1+\left(\frac{t}{2h}\right)^2}-1\right]}\,;\ \text{ft/sec} \tag{F.23}$$

where I_{roll} = mass moment of inertia of vehicle about point of rotation, ft-lb-sec^2
h = center-of-gravity, height, ft
t = track width, ft
W = vehicle weight, lb

and

$$I_{\text{roll}} = I_{\text{roll(cg)}} + \left(\frac{W}{32.2}\right)\left[h^2 + \left(\frac{t}{2}\right)^2\right];\ \text{ft-lb-sec}^2 \tag{F.24}$$

where $I_{\text{roll (cg)}}$ = mass moment of inertia about longitudinal axis of vehicle, ft-lb-sec^2

Example F.4: *Determine the minimum sliding velocity of a vehicle required to cause rollover when striking a curb. The dynamic vehicle parameters given are: W = 4800 lbs,* $I_{roll\,(cg)}$ *= 500 ft lb sec*^{2}*, t = 5.35 ft, and h = 2.05 ft.*

Solution:

$$I_{\text{roll}} = 500 + \left(\frac{4800}{32.2}\right)\left[(2.05)^2 + \left(\frac{5.32}{2}\right)^2\right] = 2193.14 \text{ ft lb sec}^2$$

The minimum velocity is then given by Eq. (F.23):

$$V_s = 45.5\sqrt{\frac{2193.14}{(4800)(2.05)}\left[\sqrt{1+\left(\frac{5.35}{(2)(2.05)}\right)^2} - 1\right]} = 17.24 \text{ ft/sec or } 11.76 \text{ mph.}$$

F.13 BRAKE FADE

(*a*) The energy absorbed by one foundation brake during downhill travel at a constant speed is calculated by

$$q = 6.8\; W\, V\, (G - f_{re})\; z;\ \text{BTU/hr} \tag{F.25}$$

where W = vehicle weight, lbs
V = speed, mph
G = road grade, percentage
f_{re} = summation of all retarding coefficients
z = fraction of total braking done by brake to be analyzed

(*b*) The brake temperature achieved by the raking during continued brake application is calculated by

$$T_b = \left(T_i - \left[T_a + \frac{q}{Ah}\right]\right)_e - \frac{Aht}{\rho cv} + \frac{q}{Ah} + T_a;\ \text{deg F} \tag{F.26}$$

where q = braking energy per brake, BTU/hour
T_i = initial brake temperature, °F
T_a = ambient air temperature, °F
A = brake drum cooling area, ft^2
v = brake drum cooling volume, ft^2
h = convective cooling coefficient, BTU/ (hr F ft^2)
ρ = drum density, lbm/ft^3
c = specific heat, BTU/lbm °*F*
t = brake time, hour

Typical values for a standard size 16.5 × 7 in foundation brake are: A = 3.5 ft^2, v = 0.25 ft^3, ρ = 455 lbm/ ft^3, c = 0.11 BTU(lbm °F), h = 5 BTU/(hr F ft^2) for a vehicle speed of 10 mph, 9 BTU/(hr F ft^2) for 20 mph, 13 BTU/hf F ft^2) for 30 mph, 16 BTU(hr F ft^2) for 40 mph, 19 BTU(hr F ft^2) for 50 mph, and 23 BTU/(hr F ft^2) for 60 mph. The braking time is obtained by dividing the interval distance by speed.

F.14 STEERING TO AVOID COLLISION

(*a*) Normal and limit lateral acceleration

In normal lane change maneuvers with a steering wheel angle less than 20 degrees, the lateral acceleration is determined by

$$a_{y,\,\text{normal}} = 0.1 - 0.0013V;\ g \text{ units} \tag{F.27}$$

For severe lane change maneuvers, the lateral acceleration is determined by

$$a_{y,\,\text{severe}} = 0.22 - 0.002V;\ g \text{ units} \tag{F.28}$$

where V = vehicle speed, mph.

(*b*) Lane change accident avoidance

The minimum distance S_l between vehicle and obstacle required by vehicle to attain a safe lane change maneuver from a given initial velocity is determined by (Fig.F.2)

$$S_\ell = 0.458V\sqrt{\frac{\ell}{a_y}};\ \text{ft} \tag{F.29}$$

where a_y = average lateral acceleration, g units

l = lateral displacement required for obstacle avoidance, ft

V = vehicle velocity, ft/sec.

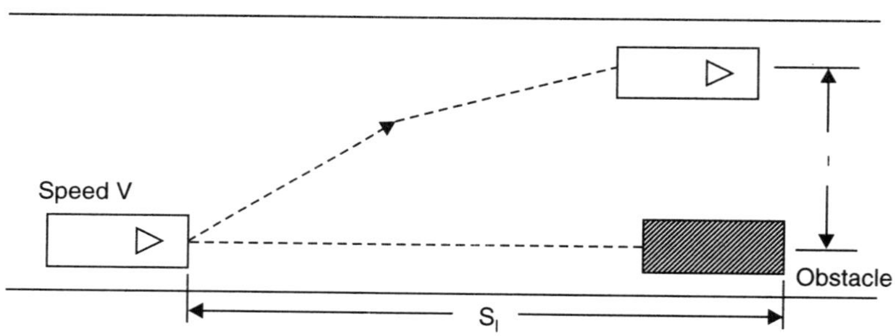

Fig. F.2 Lane changing accident avoidance on straight roadway.

(*c*) Lane change time

For constant velocity, the time required to change lanes is determined by

$$t_\ell = 0.458\sqrt{\frac{\ell}{a_y}};\ \text{sec} \tag{F.30}$$

(*d*) Minimum path radius

For a sine-shaped lane change, the minimum radius along the curved path is determined by

$$R_{\min} = \frac{S_\ell^2}{2\pi\ell};\ \text{ft} \tag{F.31}$$

(*e*) Path geometry

The path of the vehicle's center of gravity along the sine-shaped curve is determined by

$$y = \ell\left[\frac{x}{S_\ell} - \frac{1}{2\pi}\sin\left(\frac{2\pi x}{S_\ell}\right)\right]; \text{ ft} \tag{F.32}$$

where x = linear displacement in direction of vehicle travel, ft

y = lateral displacement, ft

(*f*) Normal and limit lateral acceleration in single direction steering

The upper level of lateral acceleration in a single direction steering maneuver is determined by

$$a_{y,\,\text{normal}} = 0.025V - \left(\frac{V}{44}\right)^{1.85}; \text{ g units} \tag{F.33}$$

where V = vehicle velocity, mph

(*g*) Turning dynamics

The minimum distance between vehicle and obstacle is determined by

$$S_\ell = 0.3V\sqrt{\frac{\ell}{a_y}}\,; \text{ ft} \tag{F.34}$$

Then time required to avoid is given by (Fig. F.3)

$$t_t = 0.3\sqrt{\frac{\ell}{a_y}}\,; \text{ ft} \tag{F.35}$$

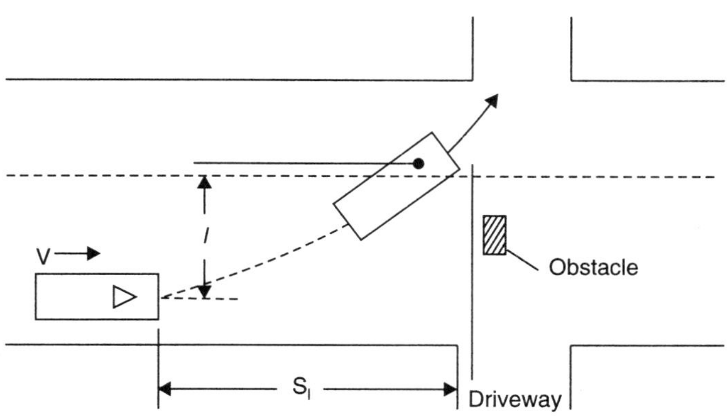

Fig. F.3 Basic steering avoidance scheme.

(*h*) Critical speed to avoid braking or steering

Steering maneuvers at higher speed show an increased potential for accident avoidance since braking distance increases with the speed squared while lane change distance increases linearly. From Eqs. (F.14) and (F.16), we obtain the critical speed V_c as

$$V_c = 29(f \pm \sin\alpha)\sqrt{\frac{\ell}{a_y}}\ ;\ \text{ft/sec} \tag{F.36}$$

where f = drag factor

α = slope of the roadway, deg.

For speeds < V_c, braking has a great potential for accident avoidance and for speeds > V_c, lane change has a great potential.

(*i*) Lane change and braking

The lane change distance when both steering and braking inputs are provided at the same time is given by

$$S_{\ell,b} = 0.458V\sqrt{\frac{\ell}{a_y}} - 2Ef\sqrt{\frac{\ell}{a_y}}\ ;\ \text{ft} \tag{F.37}$$

where E = braking efficiency.

Example F.5: *A vehicle travels at 65 mph (95.29 ft/sec) when it approaches a stopped vehicle in its lane. The lateral displacement required for avoidance is 13 ft. Determine the distances and times required for lane changing for normal turning.*

Solution:

(*a*) For normal turning, the lateral acceleration is given by Eq. (F.27):

$$a_{y,\,\text{normal}} = 0.1 - 0.0013(65) = 0.0155\ \text{g}$$

(*b*) The lane change distance is given by Eq. (F.29):

$$S_\ell = 0.458(95.29)\sqrt{\frac{13}{0.0155}} = 12.392\ \text{ft}$$

(*c*) The time required is given by Eq. (F.30):

$$t_\ell = 0.458\sqrt{\frac{\ell}{a_y}} = 0.458\sqrt{\frac{13}{0.0155}} = 13.26\ \text{sec.}$$

Example F.6: *A car traveling at 60 mph (87.96 ft/sec) and the driver tries to avoid a 12 ft obstacle by lane changing at 0.25 g and braking at E = 0.6 on a dry roadway with a tire-road friction coefficient of F = 0.7. Determine the lane change distance.*

Solution: The lane change distance is given by Eq. (F.37)

$$S_{\ell,b} = 0.458(87.96)\sqrt{\frac{12}{0.25}} - 2(0.6)\left(\frac{12}{0.25}\right) = 221.51\ \text{ft.}$$

Single steering and braking:

The required avoidance distance for a single direction steering wheel rotation and braking is given by

$$S_{\ell,b} = 0.3V\sqrt{\frac{\ell}{a_y}} - Ef\sqrt{\frac{\ell}{a_y}}\ ;\ \text{ft} \tag{F.38}$$

F.15 TIP AND ROLLOVER

F.15.1 Level Roadway

A vehicle may tip sideways and rollover when traveling on a curved roadway at an excessive speed. Vehicle tips and rollover depends upon various factors including: the drag factor, radius of the curved roadway, track width of the vehicle, and the location of the center of mass (CM) of the vehicle.

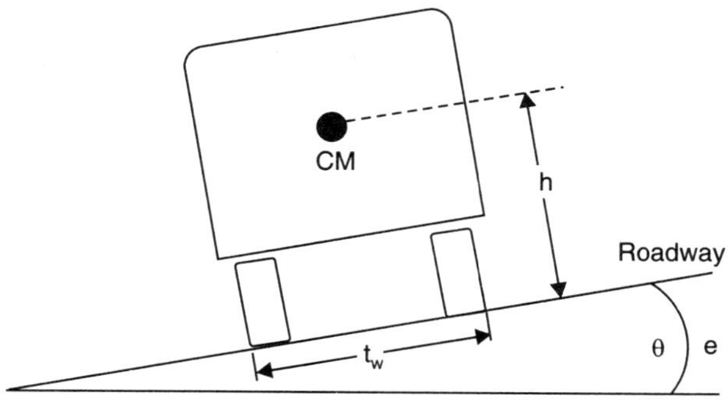

Fig. F.4 Tips and rollover phenomenon.

The drag factor must be high for a vehicle to tip and rollover in order to hold the vehicle from side slipping and getting into yaw.

A vehicle's speed (in mph) at the moment of twist or at the time at starts to tip (rotate around the center of mass (CM)) resulting in a rollover can be computed for a *level roadway* by the following formula:

$$S_{tr} = 0.682\sqrt{\frac{r\,g\,t_w}{2h}}\text{, (mph)} \tag{F.39}$$

where S_{tr} = vehicle speed in tip and rollover, mph
r = radius of the curved roadway, ft
g = acceleration due to gravity, 32.2 ft/sec^2
t_w = track width of vehicle, in
h = height of vehicle's center of mass (CM), in

In SI units:

$$S_{tr} = 3.61\sqrt{\frac{r\,g\,t_w}{2h}}\text{, (kph)} \tag{F.40}$$

where S_{tr} = vehicle speed in tip and rollover, km/h
r = radius of the curved roadway, m
g = acceleration due to gravity, 9.81 m/sec^2
t_w = track width of vehicle, cm
h = height of vehicle's center of mass (CM), cm

F.15.2 Banked or Superelevated Roadway

Bank: The degree to which the outside edge of a roadway is higher than the inside edge at a specified point on a curve; the change in elevation per unit distance across the roadway from the inside of a curve to the outside edge; superelevation. Bank is designated in feet (meters) of rise, positive bank, or fall, negative bank, per foot (meter) of level distance from the inside to the outside of the curve. Bank may also be designated by rise as a percent of level distance. See Superelevation.

Superelevation: The degree to which the outside edge of a roadway is higher than the inside edge at a specified point on a curve; the change in elevation per unit distances across the roadway from the inside of a curve to the outside edge; bank.

A vehicle's speed at the moment of twist or at the time it starts to tip (rotate around the center of mass (CM)) resulting in a rollover can be obtained for a *banked or superelevated* roadway using the following formula:

$$S_{bs} = 0.682\sqrt{\frac{r\,g(0.5t_w + he)}{(h - 0.5t_w\,e)}} \text{ (mph)} \tag{F.41}$$

where S_{bs} = vehicle's speed in banked or superelevated roadway, mph
r = radius of the curved roadway, ft
g = acceleration due to gravity, 32.2 ft/sec^2
t_w = vehicle's track width, in
h = height of the center of mass (CM) of the vehicle, in
e = bank angle or superelevation, percentage

In SI units:

$$S_{bs} = 3.61\sqrt{\frac{r\,g(0.5t_w + he)}{(h - 0.5t_w e)}} \text{ (kph)} \tag{F.42}$$

where S_{bs} = vehicle's speed in banked or superelevated roadway, km/h
r = radius of the curved roadway, m
g = acceleration due to gravity, 9.81 m/sec^2
t_w = track width of the vehicle, cm
h = height of the center of mass (CM) of the vehicle, cm

Example F.7: [Level Roadway] *A vehicle traveled into a curved roadway at an excessive speed and overturned. The roadway was level (no bank or superelevation). The following data is known:*

Radius of the curved roadway = 350 ft (106.68 m)
Track width of the vehicle = 102 in (259 cm)
Height of vehicle's center of mass (CM) = 100 in (253.92 cm)

Determine the vehicle's speed at the moment of twist or at the time it starts to tip (rotate around the center of mass (CM)) resulting in a overturning of the vehicle.

Solution: In SI units:

The vehicle's speed is given by Eq. (F.39)

$$S_{tr} = 0.682\sqrt{\frac{r\,g\,t_w}{2h}} \text{ (mph)}$$

Here $r = 350$ ft, $t_w = 102$ in, $g = 32.2$ ft/sec^2, and $h = 100$ in.

$$S_{tr} = 0.682\sqrt{\frac{350(32.2)(102)}{2(100)}} = 51.70 \text{ mph}$$

In SI units:

The vehicle's speed is given by Eq. (F.40):

$$S_{tr} = 3.61\sqrt{\frac{r\,g\,t_w}{2h}}\text{, km/h}$$

Here $r = 106.68$, $t_w = 259$ cm, $g = 9.81$ m/sec^2, and $h = 253.92$ cm.

$$S_{tr} = 3.610\sqrt{\frac{106.68(9.81)(259)}{2(253.92)}} = 83.40 \text{ km/h}$$

Example F.8 [Banked or superelevated roadway] *A four wheeled passenger vehicle traveled into a curved roadway at an excessive speed and overturned. The accident investigation revealed the following information:*

Radius of the curved roadway = 350 ft (106.68 m)

Track width of the vehicle = 102 in (259 cm)

Height of vehicle's center of mass (CM) = 100 in (253.92 cm)

Superelevation = 0.04.

Determine the speed of the vehicle at the moment of overturning.

Solution: In SI units:

The vehicle's speed is given by Eq. (F.41)

$$S_{bs} = 0.682\sqrt{\frac{r\,g(0.5t_w + h\,e)}{(h - 0.5t_w\,e)}} \text{ (mph)}$$

$$= 0.682\sqrt{\frac{350(32.2)[0.5(102)+100(0.04)]}{[100-0.5(102)(0.04)]}} = 54.25 \text{ mph}$$

In SI units: we use Eq. (F.42)

$$S_{bs} = 3.61\sqrt{\frac{r\,g(0.5t_w + h\,e)}{(h - 0.5t_w e)}}\text{, (km/h)}$$

$$= 3.61\sqrt{\frac{106.68(9.81)[0.5(259)+253.92(0.04)]}{[253.92-0.5(259)0.04]}} = 87.51 \text{ km/h}$$

Example F.9: [Banked or superelevated roadway] *A vehicle traveled into a curved roadway at an excessive speed and overturned. The accident investigation revealed the following information:*

Radius of the curved roadway = 400 ft (121.92 m)

Track width of the vehicle = 102 in (259 cm)

Height of vehicle's center of mass (CM) = 110 in (253.92 cm)

Superelevation = 0.04.

Determine the speed of the vehicle at the moment of overturning.

Solution: In SI units:

The vehicle's speed is given by Eq. (F.41)

$$S_{bs} = 0.682\sqrt{\frac{r\,g(0.5t_w + he)}{(h - 0.5t_w\,e)}},\ \text{(mph)}$$

Here $r = 400$ ft, $t_w = 110$ in, $g = 32.2$ ft/sec^2, $e = 0.04$ and $h = 110$ in.

$$= 0.682\sqrt{\frac{400(32.2)[0.5(110) + (110)(0.04)]}{[110 - 0.5(110)(0.04)]}} = 57.45 \text{ mph}$$

In SI units:

$$S_{bs} = 3.61\sqrt{\frac{r\,g(0.5t_w + he)}{(h - 0.5t_w e)}},\ \text{(kph)}$$

Here $r = 121.92$ cm, $t_w = 259$ cm, $g = 9.81$ m/sec^2, $e = 0.04$ and $h = 253.92$ cm.

$$S_{bs} = 3.61\sqrt{\frac{121.92(9.81)[0.5(259) + 253.92(0.04)]}{[253.92 - 0.5(259)0.04]}} = 93.55 \text{ (km/h)}$$

F.16 HYDROPLANING

Hydroplaning is a phenomenon in which a layer of fluid (usually water) on a roadway surface separates the load-bearing surface of one or more tires of a moving vehicle from the road surface and causes a full loss of traction (longitudinal) and steering (transverse) force components. Hydroplaning causes the front tires to slow down or even completely stop turning as they skim down the highway. This condition is known as *spin-down* and can be noticed on the highway. The hydroplaning has an adverse effect on the turning and stopping ability of a vehicle. Hydroplaning speeds are influenced by several factors. They include

(*a*) The depth of the fluid on the highway: The tires will not hydroplane when the fluid height is below certain values. Tests conducted by Horne and Drecher (1963) and others indicate that the minimum water depth is most likely between 0.2 inches and 0.3 inches (5.08 and 7.62 mm).

(*b*) Surface texture of the roadway: It appears a rough or open textured pavement surface will require a greater fluid depth than a smooth surface for hydroplaning to occur.

(*c*) Pavement crown: Crowned paving will allow the fluid to drain off quickly and pavement sufficient fluid buildup to allow hydroplaning.

(*d*) Transverse grooves: Transverse grooves usually found in concrete paving play the same part as the *hills* and *valleys* of the surface texture in allowing the water or fluid trapped under the tire to escape.

(*e*) Tire tread design: The tire tread design has two effects on hydroplaning speed: (1) circumferential ribs tend to require higher ground speeds for hydroplaning than smooth tread tires, (2) good tread design tends to increase the minimum water/fluid depth required for a tire to hydroplane.

(*f*) Tire inflation pressure: Low pressures result in hydroplaning at lower speeds. The normal tire pressures for passenger and other small-type vehicles are between 16 and 30 psi (110 and 206 kPa), critical hydroplaning speeds for these pressures are between 41 and 57 mph (65 and 91 km/h). Commercial-type vehicles (such as large trucks) use tire inflation pressures of 50 to 90 psi before they attain speeds of between 73 and 98 mph (117 to 157 km/h).

Experiments have shown that the following simplified critical hydroplaning speed formula can be employed as a guide in establishing the speed a vehicle would have to be traveling in order to hydroplane.

The critical hydroplaning speed (V_p) is given by

$$S_h = 10.35\sqrt{p} \tag{F.43}$$

where S_h = hydroplaning g speed, mph

p = tire inflation pressure, psi.

In SI units:

$$S_h = 6.36\sqrt{p} \tag{F.44}$$

where S_h = hydroplaning g speed, km/h

p = tire inflation pressure, kPa.

Eqs. (F.43) and (F.44) are applicable to smooth or closed patterned tires that do not allow water to escape, and for rib tires where the roadway water depth exceeds the depth of the tire tread groove.

Example F.10: *A four-wheeled passenger vehicle was traveling on a smooth surface highway that had a film of water measuring 0.29 in (7.366 mm) in depth. The front tires had a smooth tread and carried 25 psi (172.37 kPa) air pressure. The driver of the passenger vehicle lost steering and braking control on the front wheels due to hydroplaning. Determine the approximate speed required for the vehicle to hydroplane.*

Solution: The approximate speed required for the passenger vehicle to hydroplane is given by Eq. (F.43)

$$S_h = 10.35\sqrt{p} = 10.35\sqrt{25} = 51.75 \text{ mph.}$$

In SI units:

$$S_h = 6.35\sqrt{p} = 6.35\sqrt{172.37} = 83.37 \text{ km/h}$$

REFERENCES

Collins, J.C., and **Morris, J.L.**, *Highway Collision Analysis*, Charles, C. Thomas Publishers, Springfield, IL, 1979.

Dukkipati, R.V., Pang, J., Qatu, M., Sheng, G., and **Shuguang, Z.,** *Road Vehicle Dynamics*, Product Code: R-366, SAE International, Warrendale, PA, June 2008.

Dukkipati, R.V., Pang, J., Qatu, M., Sheng, G., and **Shuguang, Z.,** *Road Vehicle Dynamics: Problems and Solutions*, Product Code: R-386, SAE International, Warrendale, PA, September, 2009.

Horne, W.B., and Drecher, R.C., "Phenomena of Pneumatic Tire Hydroplaning", NASA-TN-D-2056, November, 1963.

Limpert, R., *Motor Vehicle Accident Reconstruction and Cause Analysis*, 5th ed, Lexis Publishing Charlottesville, VA, 1999.

Rivers, R.W., and **Lofgren, M.J.,** *Technical Traffic Accident Investigators' Handbook: A Level 3 Reference, Training and Investigation Manual*, 2nd ed., Charles C. Thomas Publishers, Springfield, IL, 1997.

Rivers, R.W., *Speed Analysis for Traffic Accident Investigation*, 2nd ed., Institute of Police Technology and Management (IPTM), University of North Florida, Jacksonville, FL, 1997.

Russel, C.G., *Equations and Formulas for the Traffic Accident Investigator and Reconstructionist*, Lawyers and Judges Publishing Company, Inc., Tucson, AZ, 1999.

PROBLEMS

Problem F.1: *A vehicle locks its brakes and slides on an icy road laterally for 17 ft, where it hits a child waiting at a bus stop. The cross slope of the roadway was 7 degrees. Determine the speed when the brakes were locked and the distance by which the vehicle would have missed the child had the cross slope been 2 degrees. Assume a coefficient of friction of 0.17.*

Solution: From Eq. (F.3)

$$V_0 = 2(0.17)\sqrt{\frac{(32.2)(17)}{\sin 7^\circ}} = 22.79 \text{ ft/sec} = 15.5 \text{ mph}$$

The skidding distance

$$S = \frac{2(0.17)(17)}{\sin 7^\circ} = 47.4 \text{ ft}$$

With a lateral distance of 17 ft and a cross slope of 2 deg., the skidding distance would be

$$S = \frac{2(0.17)(17)}{\sin 2^\circ} = 165.6 \text{ ft}$$

The car would have missed the child by more than 118 ft. The speed would have been

$$V_0 = 2(0.17)\sqrt{\frac{32.2(17)}{\sin 2^\circ}} = 42.58 \text{ ft/sec} = 29.43 \text{ mph.}$$

Problem F.2: *Determine the distance and time required for a passing maneuver by the use of the following given data:*

$$V_1 = 60 \text{ mph } (87.96 \text{ ft/sec})$$
$$V_2 = 53 \text{ mph } (73.30 \text{ ft/sec})$$
$$\ell_1 = 17 \text{ ft}, \ \ell_2 = 60 \text{ ft},$$
$$\ell_s = 40 \text{ ft}, \ \ell = 12 \text{ ft, and } a_y = 0.20\, g.$$

Solution: Eq. (F.10) gives

$$S_p = 87.96\left[\frac{17+60+40}{87.96-73.30} + 0.9\sqrt{\frac{12}{0.2(32.2)}}\right] = 810.06 \text{ ft}$$

and $$= \frac{S_p}{V_1} = \frac{810.06}{87.96} = 9.21 \text{ sec.}$$

Problem F.3: *Vehicle A drives across an intersection at a uniform speed of 10 mph. Vehicle B approaches the intersection at a speed of 48 mph and collides with vehicle A. The skid marks of vehicle B begin 23 ft from the point of impact. The roadway is dry and the speed limit is 30 mph. The damage shows that the left front corner of vehicle B impacted against the left rear wheel of vehicle A approximately 3.4 ft ahead of the rear bumper. Determine the maximum speeds possible for accident avoidance for a reaction time* $t_R = 1.0$ *sec.*

Solution: The stopping distance of vehicle B had it traveled at 30 mph is given by Eq. (F.14)

$$S_b = \frac{44^2}{2(32.2)(0.8)} + (44)(1)$$

Now S_{r-c} is given by Eq. (F.16). This is in order to find how far vehicle *B* was from the point of impact when its driver began to react in the actual accident.

$$S_{r-c} = 23 + (1.0)\,(48)\,(1.466) = 93.4 \text{ ft.}$$

Comparing Eqs. (1) and (2), we see that the accident could have been avoided had vehicle A traveled at the speed limit of 30 mph.

Now, we determine the maximum speed vehicle A could have traveled and stopped just at the point of impact.

Substituting in Eq. (F.15) gives

$$V_{\max,g} = \sqrt{[(25.76)(1.0)]^2 + 2(25.76)(93.4)} - (25.76)(1.0) \;=\; 48.24 \text{ ft/sec or } 32.9 \text{ mph}$$

The time avoidance for vehicle A is given by

$$t_{AV} = \frac{3.4}{(10)(1.466)} = 0.232 \text{ sec.}$$

Hence, if vehicle *B* arrives 0.232 seconds later at the point of impact, vehicle A will have traveled 3.5 ft., it will have moved out of the danger zone.

The time t_{b-c} is obtained using equation given below,

$$t_{b-c} = \frac{V_0}{a} - \sqrt{\left(\frac{V_0}{a}\right)^2 - \frac{2S}{a}}\text{; sec}$$

or $$t_{b-c} = \frac{70.4}{25.76} - \sqrt{\left(\frac{70.40}{25.76}\right)^2 - \frac{2(23)}{25.76}} = 0.349 \text{ seconds}$$

Now, Eq. (F.17) gives

$$V_{\max}, t = \frac{\frac{a}{2}(t_{b-c} + t_a)^2 + S_{r-c}}{t_{r-c} + t_a} = \frac{\left(\frac{25.76}{2}\right)(0.349 + 0.232)^2 + 93.4}{(1.349 + 0.204)} = 62.94 \text{ ft/sec or } 42.91 \text{ mph.}$$

Problem F.4: *Determine the minimum sliding velocity of a vehicle required to cause rollover when striking a curb. The dynamic vehicle parameters are:*

$$W = 1966\ lb;\ I_{roll\,(cg)} = 214\ ft\ lb\ sec^2;\ t = 4.5\ ft\ and\ h = 1.94\ ft.$$

Solution:

$$I_{\text{roll}} = 214 + \left(\frac{1966}{32.2}\right)\left[(1.94)^2 + \left(\frac{4.5}{2}\right)^2\right] = 752.9 \text{ ft lb sec}^2$$

The minimum velocity is then given by Eq. (F.23):

$$V_s = 45.5\sqrt{\frac{752.9}{(1966)(1.94)}\left[\sqrt{1+\left(\frac{4.5}{(2)(1.94)}\right)^2} - 1\right]} = 14.73 \text{ ft/sec or } 10.04 \text{ mph.}$$

Problem F.5: *A car traveling at 60 mph (87.96 ft/sec) when it approaches a stopped car in its lane. The lateral displacement required for avoidance is 12 ft. Determine the distances and times required for lane changing for normal turning.*

Solution: For normal turning, the lateral acceleration is given by Eq. (F.27):

$$a_{y,\,\text{normal}} = 0.1 - 0.0013(60) = 0.022 \text{ g}$$

The lane change distance is given by Eq. (F.29):

$$S_\ell = 0.458(87.96)\sqrt{\frac{12}{0.022}} = 941.3 \text{ ft}$$

The time required is given by Eq. (F.30):

$$t_\ell = 0.458\sqrt{\frac{\ell}{a_y}} = 0.458\sqrt{\frac{12}{0.022}} = 10.7 \text{ sec.}$$

Problem F.6: *A car traveling at 60 mph (87.96 ft/sec) when it approaches a stopped car in its lane. The lateral displacement required for avoidance is 12 ft. Determine the distances and times required for severe lane change maneuvers turning.*

Solution: For severe turning, the lateral acceleration is given by Eq. (F.28):

$$a_{y,\,\text{normal}} = 0.22 - 0.002(60) = 0.1 \text{ g}$$

The lane change distance is given by Eq. (F.37)

$$S_{\ell,\,b} = 0.458(87.96)\sqrt{\frac{12}{0.1}} - 2(0.6)\left(\frac{12}{0.1}\right) = 297.3 \text{ ft}$$

The time required is given by Eq. (F.30):

$$t_\ell = 0.458\sqrt{\frac{\ell}{a_y}} = 0.458\sqrt{\frac{12}{0.1}} = 5.01 \text{ sec.}$$

Problem F.7: *A vehicle traveling at 65 mph (95.29 ft/sec) and the driver attempts to avoid a 13 ft obstacle by lane changing at 0.2 g and braking at E = 0.55 on a dry roadway with a tire-road friction coefficient f = 0.8. Determine the lane change distance.*

Solution: From (F.38), the required avoidance distance for a single direction steering wheel rotation and braking is given by

$$S_{\ell,b} = 0.3V\sqrt{\frac{\ell}{a_y}} - Ef\sqrt{\frac{\ell}{a_y}} \text{ ft}$$

$$S_{\ell,b} = 0.3\times 95.29\sqrt{\frac{13}{0.2}} - 0.55\times 0.8\sqrt{\frac{13}{0.2}} = 226.93 \text{ ft}$$

Problem F.8: [Level roadway]. *A vehicle traveled into a curved roadway at an excessive speed and overturned. The roadway was level (no bank or superelevation). The accident investigation revealed the following data:*

Radius of the curved roadway = 400 ft (121.92 m)

Track width of the vehicle = 102 in (259 cm)

Height of vehicle's center of mass (CM) = 110 in (279.4 cm)

Determine the vehicle's speed at the moment of overturn.

Solution: In SI units:

The vehicle's speed is given by Eq. (F.39)

$$S_{tr} = 0.682\sqrt{\frac{r\,g\,t_w}{2h}}, \text{ (mph)}$$

Here r = 400 ft, t_w = 102 in, g = 32.2 ft/sec^2 and h = 110 in.

$$S_{tr} = 0.682\sqrt{\frac{400(32.2)(102)}{2(110)}} = 52.70 \text{ mph}$$

In SI units:

The vehicle's speed is given by Eq. (F.40):

$$S_{tr} = 3.61\sqrt{\frac{r\,g\,t_w}{2h}}, \text{ km/h}$$

Here r = 121.92, t_w = 259 cm, g = 9.81 m/sec^2 and h = 279.4 cm.

$$S_{tr} = 3.610\sqrt{\frac{121.92(9.81)(259)}{2(279.4)}} = 85.0 \text{ km/h}$$

Problem F.9: *A vehicle was traveling on a smooth-surface highway that had a film of water measuring 0.28 in (7.112 mm) in depth. The front tires had a smooth tread and carried 24 psi (165.47 kPa) air pressure. The driver of the vehicle lost steering and braking control on the front wheels due to hydroplaning. Determine the approximate speed required for the vehicle to hydroplane.*

Solution: The approximate speed required for the passenger vehicle to hydroplane is given by Eq. (F.43)

$$S_h = 10.35\sqrt{p} = 10.35\sqrt{24} = 50.7 \text{ mph.}$$

In SI units:

$$S_h = 6.35\sqrt{p} = 6.35\sqrt{165.47} = 81.68 \text{ kph.}$$

Appendix

G

Sensitivity Analysis of Accident Reconstruction Computations

The potential for the accident reconstructionist to be able to reconstruct any accident depends on the data and measurements available. Such data is subject to measurement and estimation errors. In this appendix, we present the sensitivity of standard computations to variations in the measured or estimated values of the independent variables needed for computation. Here, we adopt the methods proposed by Metz and Metz (1998).

G.1 SPEED FROM SLIDE-TO-STOP MEASUREMENTS

A skid length is measured from roadway marks and speed is determined from the measured skid length d and the tire/road coefficient of friction μ. The expression used for such computations is given by

$$v = 15.9\sqrt{d \times \mu} \tag{G.1}$$

with velocity v in km/h, d in meters, dimensionless μ and 15.9 = units conversion factor. Symbolically, we can write Eq. (G.1) as

$$v_{\text{slide-to-stop}} = f(d, \mu) \tag{G.2}$$

To determine the *sensitivity* of a calculation of speed from distance and tire/road coefficient of friction data, we need to calculate the partial derivative of the speed expression w.r.t. each independent variable:

$$S_d^v = \frac{\partial v}{\partial d} \tag{G.3}$$

$$S_\mu^v = \frac{\partial v}{\partial \mu} \tag{G.4}$$

It is clear that for the slide-to-stop calculation of Eq. (G.1), the resulting partial derivatives become:

$$S_d^v = \frac{7.95\mu}{\sqrt{d \times \mu}} \tag{G.5}$$

$$S_\mu^v = \frac{7.95d}{\sqrt{d \times \mu}} \tag{G.6}$$

Example G.1: *Determine the sensitivity of a calculation of speed from distance and tire/road coefficient of friction data for a vehicle which slides to stop in 60 m with a tire/road coefficient of friction of μ = 0.80. Given that* $V = 15.9\sqrt{d\mu}$, *where V is vehicle speed (km/h), d is the distance (m), and μ is the tire/road coefficient of friction (no units).*

Solution: The nominal value for the computed speed from the slide-to-stop measurements is given by

$$V = 15.9\sqrt{60(0.80)} = 110.19 \text{ km/h} \tag{E.1}$$

Here, we are interested in the variation in computed speeds produced by Eq. (E.1) when d and μ are allowed to vary away from their nominal values. At those nominal values:

$$S_d^v = \frac{7.95(0.80)}{\sqrt{60(0.80)}} = 0.92 \tag{E.2}$$

$$S^v_\mu = \frac{7.95(60)}{\sqrt{60(0.80)}} = 68.85 \tag{E.3}$$

Eqs. (E.1) and (E.2) show that every meter of error present in the measurement of skid mark length, d, and the calculated vehicle velocity obtained from Eq. (G.1) will be in error by 0.92 km/h, or almost a 1:1 sensitivity *in the units of the calculation*. For every unit change in tire/road coefficient of friction μ away from the nominal value, the calculated speed will be in error by 68.85 km/h.

It is necessary for the reconstructionist to compute *both raw and normalized sensitivities* in order to produce a complete picture of the sensitivities involved in a particular calculation. In all types of sensitivity analysis, it is customary to perform the *normalization* of the sensitivity computations by multiplying them by a ratio of the variable of interest divided by the calculated value:

$$S^v_{d \text{ norm}} = S_d^v \times \frac{d}{v} \tag{E.4}$$

$$S^v_{\mu \text{ norm}} = S_\mu^v \times \frac{\mu}{v} \tag{E.5}$$

With the nominal values given above, the normalized sensitivities become, respectively:

$$S^v_{d \text{ norm}} = 0.92 \times \frac{60}{110.19} = 0.50 \tag{E.6}$$

$$S^v_{\mu \text{ norm}} = 68.87 \times \frac{0.80}{110.19} = 0.50 \tag{E.7}$$

As expected, the two normalized sensitivities are identical. The symmetrical structure of Eq. (G.1) is such that d and μ occur in an identical way in the expression used for calculating speed from slide-to-stop distance and tire/road coefficient of friction.

G.2 SPEED FROM SLIDE-TO-STOP WITH SEVERAL DISTANCE MEASUREMENTS ON DIFFERING COEFFICIENTS OF FRICTION AND/OR SPILT SURFACES

Often speed from slide-to-stop calculations requires accommodation of slides which traverse several different μ-values. Taking as a typical case a slide across two distinct surfaces, the calculation for speed in SI units is given by:

$$v = 15.9\sqrt{d_1\mu_1 + d_2\mu_2} \tag{G.7}$$

Therefore, we have:

$$v_{\text{slide-to-stop}} = f(d_1, d_2, \mu_1, \mu_2) \tag{G.8}$$

Computing the sensitivities w.r.t. only the $(d, \mu)_1$ variable gives:

$$S_{d_1}^{v} = \frac{7.95 \times \mu_1}{\sqrt{d_1\mu_1 + d_2\mu_2}} \tag{G.9}$$

$$S_{\mu_1}^{v} = \frac{7.95 \times d_1}{\sqrt{d_1\mu_1 + d_2\mu_2}} \tag{G.10}$$

Example G.2: *A vehicle slides to a stop, having skidded 25 m over a surface with a tire/road coefficient of friction of* $\mu_1 = 0.65$ *and an additional 38 m over a surface with* $\mu_2 = 0.25$. *Determine the normalized sensitivities with respect to* μ_1, μ_2, d_1, *and* d_2. *The vehicle speed is computed from* $V = 15.9\sqrt{d_1\mu_1 + d_2\mu_2}$, *where V is the vehicle speed (km/h),* d_1 *and* d_2 *are the distances (m) and* μ_1 *and* μ_2 *are the coefficients of friction between tire/road (no units).*

Solution: The nominal speed is

$$V = 15.9\sqrt{25(0.65) + 38(0.25)} = 80.68 \text{ km/h (E.1)}$$

All the nominal values for the independently measured values d_1, d_2, μ_1 and μ_2, we have:

$$s_{d_1\,\text{norm}}^{v} = \frac{7.95 \times 0.65}{\sqrt{25(0.65) + 38(0.25)}} \times \frac{25}{80.68} = 0.316 \tag{E.2}$$

$$s_{\mu_1\,\text{norm}}^{v} = \frac{7.95 \times 25}{\sqrt{25(0.65) + 38(0.25)}} \times \frac{0.65}{80.86} = 0.316 \tag{E.3}$$

Again, as expected, the two normalized sensitivities are identical, as would be those for sensitivities calculated w.r.t. d_2 and μ_2. Examination of Eq. (G.7) makes this expectation obvious.

G.3 SPEED FROM YAW MEASUREMENTS

Calculation of vehicle speeds based on yaw marks proceeds in two distinct phases. First, the radius of the (assumed to be circular) arc traversed by one or more of the vehicle tires is determined from measurements of chord length and middle ordinate, Dukkipati, et al. (2008). Once the radius has been determined, the yaw velocity is calculated from a tire/road coefficient of friction and/or maximum lateral acceleration value for the vehicle.

For determining radius from measurements of chord length and middle ordinate, we have, from simple geometry:

$$r = \frac{c^2}{8m} + \frac{m}{2} \tag{G.11}$$

with r = curve radius, c = chord length, m = middle ordinate, all measured in consistent units.

With the curve radius r determined from Eq. (G.11), the cornering speed can be calculated from:

$$v = \sqrt{r \times \mu \times g} \tag{G.12}$$

with μ = tire/road coefficient of friction and g = gravitational constant, again in units consistent with those used in Eq. (G.11). Substituting Eq. (G.11) into Eq. (G.12) and using SI units gives:

$$v = 11.3\sqrt{\left(\frac{c^2}{8m} + \frac{m}{2}\right) \times \mu \times g} \tag{G.13}$$

with v in km/h, $g = 9.8$ m/sec^2 and c, m in meters. To compute the sensitivities, we again note that:

$$v_{\text{yaw}} = f(c, m, \mu) \tag{G.14}$$

$$S^v_{c\ \text{norm}} = \frac{1.41 \times c \times \mu}{m\sqrt{\left(\frac{c^2}{8m} + \frac{m}{2}\right) \times \mu}} \times \frac{c}{v} \tag{G.15}$$

$$S^v_{m\ \text{norm}} = \frac{5.63 \times \left(\frac{1}{2} - \frac{c^2}{8m^2}\right) \times \mu}{\sqrt{\left(\frac{c^2}{8m} + \frac{m}{2}\right) \times \mu}} \times \frac{m}{v} \tag{G.16}$$

$$S^v_{\mu\ \text{norm}} = \frac{5.63 \times \left(\frac{c^2}{8m} + \frac{m}{2}\right) \times \mu}{m\sqrt{\left(\frac{c^2}{8m} + \frac{m}{2}\right) \times \mu}} \times \frac{\mu}{v} \tag{G.17}$$

Obviously, for more complex expressions such as Eqs. (G.15) to (G.17), the derivatives become cumbersome. Use of a symbolic mathematical manipulation program (e.g., MATLAB or MATHEMATICA™) is helpful in computing the sensitivities for such expressions.

Example G.3: *The vehicle speed from yaw measurements is estimated by the equation*

$$V = 11.27\sqrt{\left(\frac{c^2}{8m} + \frac{m}{2}\right)\mu g}$$

where r = *curve radius (m)*

c = *chord length (m)*

m = *middle ordinate (m)*

μ = *tire/road coefficient of friction.*

A vehicle leaves a yaw marks on a roadway with a tire/road coefficient of friction $\mu = 0.75$. The yaw mark is assumed to be an arc of a circle with a measured chord length $c = 20$ m, and a measured middle ordinate $m = 0.60$ m. Determine the normalized sensitivities with respect to c, m, and μ.

Solution: The nominal computed speed to produce the roadway yaw mark is given by

$$V = 11.27\sqrt{\left(\frac{20^2}{8(0.60)} + \frac{0.60}{2}\right)(0.75)} = 89.25 \text{ km/h} \tag{E.1}$$

At these nominal values, the normalized sensitivities become, after some algebra,

$$S^v_{c\ \text{norm}} = \frac{1.41 \times c \times \mu}{m\sqrt{\left(\frac{c^2}{8m} + \frac{m}{2}\right) \times \mu}} \times \frac{c}{v} \tag{E.2}$$

$$S^v_{c\ \text{norm}} = \frac{1.41 \times 20 \times 0.75}{0.6\sqrt{\left(\frac{20^2}{8 \times 0.6} + \frac{0.6}{2}\right) \times 0.75}} \times \frac{20}{89.25} = 0.996$$

$$S^v_{m\ \text{norm}} = \frac{5.63 \times \left(\frac{1}{2} - \frac{c^2}{8m^2}\right) \times \mu}{\sqrt{\left(\frac{c^2}{8m} + \frac{m}{2}\right) \times \mu}} \times \frac{m}{v} \tag{E.3}$$

$$S^v_{\mu\ \text{norm}} = \frac{5.63 \times \left(\frac{20^2}{8 \times 0.6} + \frac{0.6}{2}\right) \times 0.75}{0.6\sqrt{\left(\frac{20^2}{8 \times 0.6} + \frac{0.6}{2}\right) \times 0.75}} \times \frac{0.75}{89.25} = 0.623$$

A number of insights become apparent from the results given in Eqs. (E.2 to E.4). Firstly, because Eq. (G.13) is *not symmetric* w.r.t. independent quantities c, m, and μ, we expect the normalized sensitivities to have *differing magnitudes*. At the nominal values, the speed-from-yaw calculation is almost twice as sensitive to measurement errors in chord length as it is to errors in middle ordinate and coefficient of friction measurements. Examination of the non-normalized values indicates that the middle ordinate measurements are, on an *absolute scale*, the most sensitive.

It is important for the reconstructionist to examine *both* normalized and absolute sensitivities in such a calculation. Measurements of the middle ordinate which are in error will have a strong effect on the result. This is especially true because the middle ordinate measurement will be << chord length, and will involve measuring a (usually) small distance, often < 1.0 m, to a yaw mark which is approximately as wide as a tire footprint (on the order of 0.1 m). Under those circumstances, middle ordinate measurement errors are extremely easy to make.

It is also possible for a normalized sensitivity value to be *negative*, indicating an inverse relationship between the calculated result and the independent variable being examined. In the case of middle ordinate length m, we expect this inverse relationship; a larger middle ordinate for a given chord length indicates a sharper cornering radius, and thus a lower calculated speed at nominal values of c and μ.

G.4 SPEED DURING A VAULT OR TRIPPED ROLLOVER

A vault occurs when a vehicle encounters an obstacle such as a curb and is launched into flight. If the vehicle is traveling laterally when the curb is struck, the maneuver is termed a tripped rollover. If the vehicle is traveling forward when the curb is struck, the maneuver is simply called a vault.

When a tripped rollover or vault occurs, the vehicle trajectory follows a parabolic arc under the influence of gravity. The length of the flight distance traveled and measurement of any ground differential enable a speed calculation to be made. Under these conditions, speed can be calculated as Dukkipati et al. (2008):

$$V = \frac{11.27 \times d}{\sqrt{d - H}} \tag{G.18}$$

where v = calculated velocity in km/h, d = distance traveled during vault maneuver in meters, H = height differential, if any, between landing point and launch point (H is < 0 if the vehicle lands on a surface lower than its launch surface and > 0 if the vehicle lands on a surface higher than its launch surface) in meters, and 11.27 = units conversion factor.

For the circumstances of a vault or tripped rollover, we have:

$$V = f(d, H) \tag{G.19}$$

As usual, we compute

$$S^{v}_{d\text{ norm}} = \frac{\partial v}{\partial d} = \left(\frac{-5.64 \times d}{(d - H)^{3/2}} + \frac{11.27}{\sqrt{d - H}} \right) \times \frac{d}{v} \tag{G.20}$$

$$S^{v}_{H\text{ norm}} = \frac{\partial v}{\partial H} = \left(\frac{5.64 \times d}{(d - H)^{3/2}} \right) \times \frac{H}{v} \tag{G.21}$$

Example G.4: *The vehicle speed during a vault or tripped rollover is given by*

$$V = \frac{11.27d}{\sqrt{d - H}}$$

where V = *calculated speed in km/h*

d = *distance traveled during vault maneuver in meters*

H = *height differential, if any, in meters between landing point and launch point (H is < 0, if the vehicle lands on a surface lower than its launch surface and > 0 if the vehicle lands on a surface higher than its launch surface).*

A vehicle sliding sideways encounters a curb and executes a tripped rollover, traveling a total distance of 15 m and landing on a spot 0.25 m lower than its launch point. Determine the raw and normalized sensitivities with respect to d and H respectively.

Solution: The nominal calculated vehicle speed at launch on the basis of this measurement is:

$$V = \frac{11.27 \times 15}{\sqrt{15 - 0.25}} = 44.02 \text{ km/h} \tag{E.1}$$

The resulting raw and normalized sensitivities are:

Raw

$$S^{v}_{d} = \left(\frac{-5.64 \times 15}{(15 - 0.25)^{3/2}} + \frac{11.27}{\sqrt{15 - 0.25}} \right) = 1.441 \tag{E.2}$$

$$S^{v}_{H} = \left(\frac{5.64 \times 15}{(15 - 0.25)^{3/2}} \right) = 1.493$$

Normalized

$$S^{v}_{d\text{ norm}} = \left(\frac{-5.64\times 15}{(15-0.25)^{3/2}} + \frac{11.27}{\sqrt{15-0.25}}\right)\times\frac{15}{44.02} = 0.49 \tag{E.3}$$

$$S^{v}_{H\text{ norm}} = \left(\frac{5.64\times 15}{(15-0.25)^{3/2}}\right)\times\frac{0.25}{44.02} = 0.0085$$

Again, it is obviously important for the reconstructionist to examine both raw and normalized values.

G.5 SPEED DURING A FALL

A fall is distinguished from a vault or tripped rollover in that there is no tripping mechanism to convert an external force to the vehicle D'Alembert inertial force and create a rotation couple. In a fall maneuver, the vehicle is simply driven off a structure and allowed to free-fall to contact, again under the influence of gravity. Under these circumstances, velocity is calculated from Dukkipati, et al. (2008):

$$v = \frac{7.97\times d}{\sqrt{m\times d - H}} \tag{G.22}$$

where v, d, and H are defined identically to the vault calculations and m = grade of the surface from which the vehicle is driven (Grade is defined as rise/run, so a grade of +1.0 = a 45° upward incline). Now we have:

$$v = f(d, H, m) \tag{G.23}$$

$$S^{v}_{d\text{ norm}} = \left(\frac{-3.98\times d\times m}{(-H+d\times m)^{3/2}} + \frac{7.97}{\sqrt{-H+d\times m}}\right)\times\frac{d}{v} \tag{G.24}$$

$$S^{v}_{m\text{ norm}} = \frac{-3.98\times d^2}{(-H+d\times m)^{3/2}} \tag{G.25}$$

$$S^{v}_{H\text{ norm}} = \frac{-3.98\times d}{(-H+d\times m)^{3/2}} \tag{G.26}$$

Example EG.5: *In a free-fall maneuver, the vehicle is simply driven off a structure and allowed to free-fall to contact under the influence of gravity. Under these circumstances, the vehicle velocity is computed from*

$$V = \frac{7.97\, d}{\sqrt{md - H}}$$

where V = *computed velocity in km/h*

d = *distance traveled during free-fall in meters*

m = *grade of the surface from which the vehicle is driven (Grade is defined as rise/run, so a grade of +1.0 = a 45 degree upward incline)*

H = *height differential, if any, between landing point and launch point (H is < 0, if the vehicle lands on a surface lower than its launch surface and > 0 if the vehicle lands on a surface higher than its launch surface) in meters.*

A vehicle is driven off a slight downhill grade (m = –0.011) and lands 20 m away from the launch point at a height of 1.75 m lower than the launch point. Determine the raw and normalized sensitivities with respect to the variables d, m, and H.

Solution: The nominal computed velocity for these measurements is:

$$v = \frac{7.97 \times 20}{\sqrt{-0.011 \times 20 - (-1.75)}} = 128.87 \text{ km/h} \tag{E.1}$$

The resulting raw and normalized sensitivities for these nominal values are:

$$S_d^v = \left(\frac{-3.98 \times 20 \times (-0.011)}{(-(-1.75) + 20 \times (-0.011))^{3/2}} + \frac{7.97}{\sqrt{-(-1.75) + 20 \times (-0.011)}} \right) \tag{E.2}$$

$$S_d^v = 6.91$$

$$S_m^v = \frac{-3.98 \times 20^2}{(-(-1.75) + 20 \times (-0.011))^{3/2}}$$

$$S_m^v = -842.33$$

$$S_H^v = \frac{-3.98 \times 20}{(-(-1.75) + 20 \times (-0.011))^{3/2}}$$

$$S_H^v = -42.6$$

$$S_{d\text{ norm}}^v = \left(\frac{-3.98 \times 20 \times (-0.011)}{(-(-1.75) + 20 \times (-0.011))^{3/2}} + \frac{7.97}{\sqrt{-(-1.75) + 20 \times (-0.011)}} \right) \times \frac{20}{128.87} \tag{E.3}$$

$$S_{d\text{ norm}}^v = 1.07$$

$$S_{m\text{ norm}}^v = \frac{-3.98 \times 20^2}{(-(-1.75) + 20 \times (-0.011))^{3/2}} \times \frac{-0.011}{128.87}$$

$$S_{m\text{ norm}}^v = 0.072$$

$$S_{H\text{ norm}}^v = \frac{-3.98 \times 20}{(-(-1.75) + 20 \times (-0.011))^{3/2}} \times \frac{-1.75}{128.87}$$

$$S_{H\text{ norm}}^v = 0.571$$

G.6 SPEED FROM CONSERVATION OF MOMENTUM

For collisions involving two vehicles which subsequently separate and come to rest at different positions, it is common to employ conservation of momentum calculations to determine the vehicle speeds at impact. The calculation is often employed in circumstances for which an appropriate post-impact drag factor can be determined. Under those conditions, the post-impact slide-to-stop speeds can be calculated as above, and the pre-impact speeds thus determined from conservation of momentum.

The momentum balance before and after impact is given by

$$m_1 v_{1B} + m_2 v_{2B} = m_1 v_{1A} + m_2 v_{2A} \tag{G.27}$$

Boldface type, *e.g.*, v_{1B} indicates a vector quantity. It is usual to separate the momentum components into coordinate directions and components. For example, using figure one below as an exemplar accident, Eq. (G.27) in component form becomes:

x-direction:

$$m_1 v_{1b} \sin \varphi_1 - m_2 v_{2b} \cos \varphi_2 = -m_1 v_{1a} \sin \psi_1 - m_2 v_{2a} \sin \psi_2 \qquad \text{(G.28)}$$

y-direction:

$$m_1 v_{1b} \cos \varphi_1 - m_2 v_{2b} \sin \varphi_2 = -m_1 v_{1a} \cos \psi_1 - m_2 v_{2a} \cos \psi_2 \qquad \text{(G.29)}$$

$(\,)_a$ and $(\,)_b$ imply quantities before and after the impact. We can combine Eqs. (G.29) and (G.30) with Eq. (G.1) to get the following set of equations which describe the momentum balance for the above accident:

x-direction:

$$m_1 v_{1b} \sin \varphi_1 - m_2 v_{2b} \cos \varphi_1 = 15.9 m_1 \sqrt{d_1 \mu_1} \sin \psi_1 - 15.9 m_2 \sqrt{d_2 \mu_2} \sin \psi_2 \qquad \text{(E.30)}$$

y-direction:

$$m_1 v_{1b} \cos \varphi_1 - m_2 v_{2b} \sin \varphi_2 = 15.9 m_1 \sqrt{d_1 \mu_1} \cos \psi_1 - 15.9 m_2 \sqrt{d_2 \mu_2} \cos \psi_2 \qquad \text{(E.31)}$$

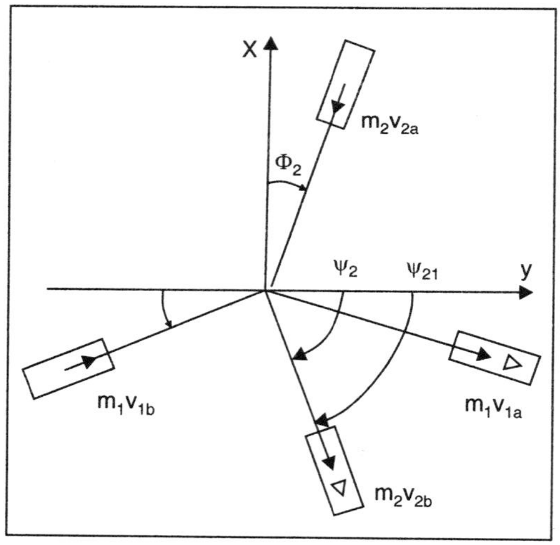

Fig. G.1 Accident analysis using momentum

Eqs. (G.31) and (G.32) represent a set of two simultaneous algebraic equations involving 10 measured quantities or parameters. Symbolically we can write that:

$$\left\{ \frac{v_{1b}}{v_{2b}} \right\} = f(m_1, m_2, \varphi_1, \varphi_2, \psi_1, \psi_2, d_1, d_2, \mu_1, \mu_2) \qquad \text{(G.32)}$$

In order to study the sensitivity of such a momentum calculation completely, we have to (*a*) simultaneously solve Eqs. (G.31) and (G.32) for the unknown v_{1b} and v_{2b}, and (*b*) calculate the 20 partial derivatives involved.

Using MATHEMATICA™ to solve Eqs. (G.31) and (G.32) simultaneously for v_{2b}, for example, gives:

$$v_{2b} = \frac{15.9}{m_1 m_2} \left\{ m_1^2 \sqrt{d_1 \mu_1} \sec[\varphi_1 + \varphi_2] \sin[\varphi_1 + \psi_1] \right\} + 15.9 \sqrt{d_2 \mu_2} \sin[\varphi_1 + \psi_2] \qquad \text{(E.33)}$$

And, as an example of calculation of just one of the 20 possible raw sensitivity calculations, we can calculate:

$$S_{d_1}^{v_{2b}} = \frac{7.95 m_1 \mu_1 \sec[\varphi_1 + \varphi_2]\sin[\varphi_1 + \psi_1]}{m_2 \sqrt{d_1 \mu_1}} \tag{E.34}$$

Eqs. (G.31) and (G.32) could readily be also solved for v_{1b}, and the remaining 19 sensitivity functions can be calculated for both v_{1b} and v_{2b}, either in raw form {as in Eq.(G.34) above} or in normalized form.

Example G.6: *For collisions involving two vehicles which subsequently separate and come to rest at different positions, it is common practice to employ conservation of momentum calculations to determine the vehicle speeds at impact. The calculation is often employed in circumstances for which an appropriate post-impact drag factor can be determined. Under those circumstances or conditions, the post-impact slide-to-stop speeds can be computed as above, and the pre-impact speeds thus determined from conservation of momentum. As shown in Fig. E2.5, ($)_a$ and ($)_b$ imply quantities before and after the impact. The notation used for the two vehicles is depicted in Fig. E.2.5.*

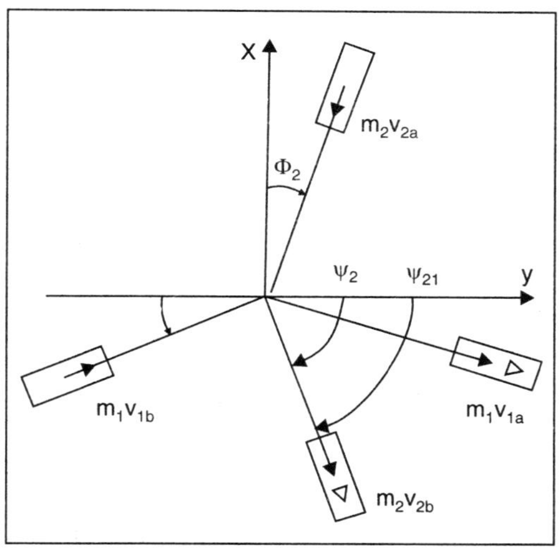

Fig. E2.5 Accident analysis using conservation of momentum.

The vehicle 2 speed after impact is given by

$$V_{2b} = \frac{15.9}{m_1 m_2}\left[m_1^2 \sqrt{d_1 \mu_1} \sec(\phi_1 + \phi_2)\sin(\phi_1 + \psi_1)\right] + 15.9\sqrt{d_2 \mu_2} \sin(\phi_1 + \psi_2)$$

Two vehicles are traveling toward one another at an intersection, and collide, as shown in Fig.E2.5. Both vehicles slide to a stop after the collision with all four wheels locked. The following values were measured by the investigating police officers and/or the accident reconstructionist:

$$m_1 = 2000,\ m_2 = 3100,\ \phi_1 = 0,\ \phi_2 = 0,\ \psi_1 = 80^\circ,\ \psi_2 = 50^\circ,$$
$$d_1 = 25 \text{ m},\ d_2 = 30 \text{ m},\ \mu_1 = 0.80, \text{ and } \mu_2 = 0.50$$

Determine:

(*a*) speed of vehicle 2 after impact, V_{2b}

(*b*) sensitivity of v_{2b} with respect to d_1

(*c*) normalized sensitivity of v_{2b} with respect to d_1

Solution: (*a*) The angular values given above are consistent with the definitions shown in Fig.G.1. For the above set of measurements, the nominal value of v_{2b} is given by

$$v_{2b} = \frac{15.9}{3182 \times 2055}\left[2055^2 \sqrt{24 \times 0.81}\sec(0^\circ)\sin(80^\circ)\right]$$

$$+15.9\sqrt{27 \times 0.49}\sin(52^\circ) = 90.16 \text{ km/h} \quad \text{(E.1)}$$

(*b*) The sensitivity of the calculation for v_{2b} w.r.t. d_1 is:

$$S_{d_1}^{v_{2b}} = \frac{7.95 \times 2055 \times 0.81 \times \sec(0^\circ)\sin(80^\circ)}{3182\sqrt{24 \times 0.81}} = 0.929 \quad \text{(E.2)}$$

In this non-normalized form, we can see that a change or measurement error in measured distance d_1 produces a corresponding error or change in the calculated value v_{2b} of 0.929 km/h. The units of Eq. (E.2) are thus km/h.

(*c*) In normalized form, and for the nominal values given above:

$$S_{d_1\,norm}^{v_{2b}} = 0.929 \times \frac{24}{90.16} = 0.247 \quad \text{(E.3)}$$

As noted, numerous other calculations for various sensitivities are possible for the momentum expression.

REFERENCES

Metz, L.D., and **Metz, L.G.**, "Sensitivity of Accident Reconstruction Calculations", SAE paper No.980375, SAE International, Warrendale, PA, 1998.

Baker, J.S., *Traffic Accident Investigation Manual*, The Traffic Institute, Northwestern University, Evanston, IL, 1975.

Reed, W.S., and **Keskin, A.T.**, "Vehicular Response to Emergency Braking", SAE paper No.87051, SAE International, Warrendale, PA, 1987.

Dukkipati, R.V., Pang, G., Qatu, M., Sheng, G., and **Shugant, Z.**, *Road Vehicle Dynamics*, Product code R-366, SAE International, Warrendale, PA, June 2008.

Dukkipati, R.V., Pang, G., Qatu, M., Sheng, G., and **Shugant, Z.**, *Road Vehicle Dynamics-Problems and Solutions Set*, Product code R-386, SAE International, Warrendale, PA, September 2009.

Appendix H

Graph of Minimum Stopping Distance versus Speed based on a Friction Coefficient of 0.7.

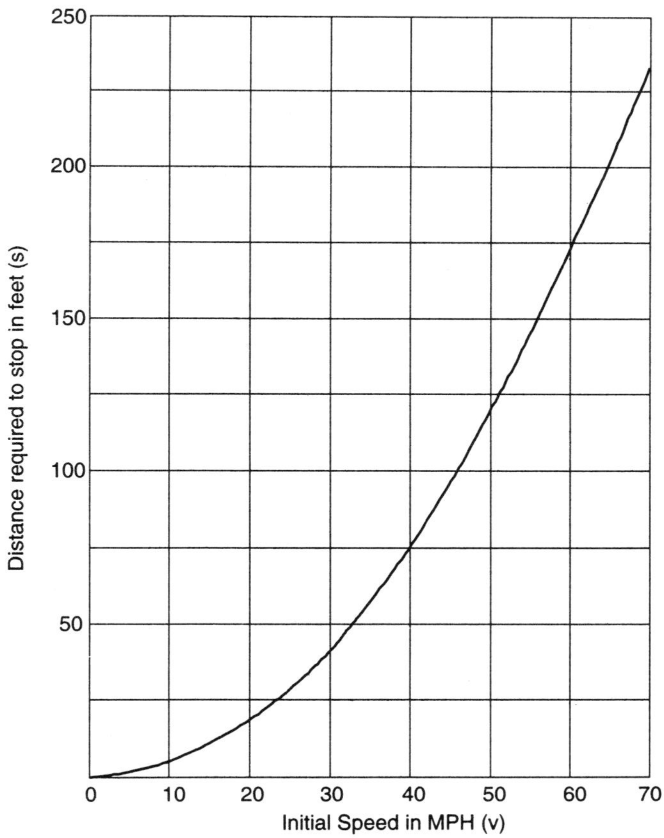

Source: J.C. Collins and J.L. Morris, *Highway collision Analysis*, Charles C. Thomas Publishing, Springfield, IL, 1979.

Table of Minimum Stopping Distance versus Speed

Appendix I

Based on Clean, Dry pavement $\mu = 0.7$
(Numbers rounded to nearest foot)

Speed mph	*¾ second perception distance Feet*	*¾ second reaction distance Feet*	*Skid distance Feet*	*Total distance Feet*
5	5.5	5.5	1	12
10	11	11	5	27
15	16	16	10	42
20	22	22	20	64
25	27	27	30	94
30	33	33	43	109
35	39	39	58	136
40	44	44	76	164
45	50	50	96	196
50	55	55	119	229
55	61	61	144	266
60	66	66	171	303
65	72	72	201	345
70	77	77	233	387
75	78	78	267	423
80	88	88	304	480

Appendix J

Maximum Safe Driving Speeds for Various Visibility Conditions.
[Chart based on a total reaction time of ¾ seconds]

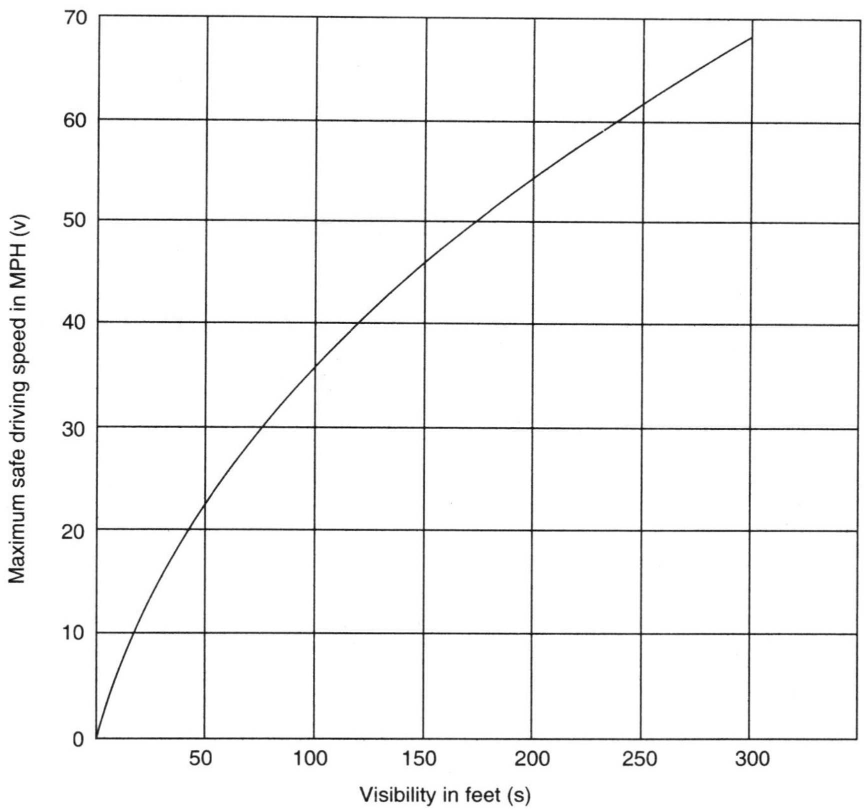

Source: J.C. Collins and J.L. Morris, *Highway collision Analysis*, Charles C. Thomas Publishing, Springfield, IL, 1979.

Appendix K

Speed in miles per hour required to slow to a stop for various distances and surfaces.

Length of slide to stop in meters	Average drag factor during slide to stop																			
	← ice →										← Clean, wet paving →									
		← Snow →				← Gravel →							← Clean dry paving →							
	← Illegal brakes←						← Fair brakes →				← Good brakes →				← Excellent brakes →					
meters	0.05	0.10	0.20	0.30	0.35	0.40	0.45	0.50	0.55	0.60	0.65	0.70	0.75	0.80	0.85	0.90	0.95	1.00	1.10	1.20
2	2	2	3	4	5	5	5	5	6	6	6	6	7	7	7	7	8	8	8	8
4	2	3	5	6	6	7	7	8	8	8	9	9	9	10	10	10	11	11	11	12
6	3	4	6	7	8	8	9	9	10	10	11	11	12	12	12	13	13	13	14	15
8	3	5	7	8	9	10	10	11	11	12	12	13	13	14	14	15	15	15	16	17
10	4	5	8	9	10	11	12	12	13	13	14	14	15	15	16	16	17	17	18	19
12	4	6	8	10	11	12	13	13	14	15	15	16	16	17	17	18	18	19	20	21
14	5	6	9	11	12	13	14	14	15	16	17	14	18	18	19	19	20	20	21	22
16	5	7	10	12	13	14	15	15	16	17	18	18	19	20	20	21	21	22	23	24
18	5	7	10	13	14	15	16	16	17	18	19	19	20	21	21	22	23	23	24	25
20	5	8	11	13	14	15	16	17	18	19	20	20	21	22	23	23	24	24	26	27
											Ice									
25	6	9	12	15	16	17	18	19	20	21	22	23	24	24	25	26	17	17	29	30
30	7	9	13	16	18	19	20	21	22	23	24	25	26	27	28	28	29	30	31	33
35	7	10	14	18	19	20	22	23	24	25	26	27	28	29	30	31	32	32	34	35
40	8	11	15	19	20	22	23	24	26	27	28	29	30	31	32	33	34	35	36	38
45	8	12	16	20	22	23	25	26	27	28	30	31	32	33	34	35	36	37	38	40
50	9	12	17	21	23	24	26	27	29	30	31	32	34	35	36	37	38	39	41	42
55	9	13	18	22	24	26	27	29	30	31	33	34	35	36	37	39	40	41	43	44
60	9	13	19	23	25	27	28	30	31	33	34	35	37	38	39	40	41	42	44	46
65	10	14	20	24	26	28	30	31	33	34	36	37	38	39	41	42	43	44	46	48
70	10	14	20	25	27	29	31	32	34	35	37	38	40	41	42	43	45	46	48	50
75	11	15	21	26	28	30	32	34	35	37	38	40	41	42	44	45	46	47	50	52
80	11	15	22	27	29	31	33	35	36	38	39	41	42	44	45	46	48	49	51	54
85	11	16	23	28	30	32	34	36	37	39	41	42	44	45	47	48	49	50	53	55
90	12	16	23	28	31	33	35	37	39	40	42	43	45	46	48	49	51	52	54	57
95	12	17	24	29	32	34	36	38	40	41	43	45	46	48	49	51	53	53	56	58

100	12	17	24	30	32	35	37	39	41	42	44	46	47	49	50	52	53	55	57	60
105	13	18	25	31	33	35	38	40	42	43	45	47	49	50	52	53	55	56	59	61
110	13	18	26	31	34	36	39	41	43	44	46	48	50	51	53	54	56	57	60	63
115	13	19	26	32	35	37	39	42	44	45	47	49	51	53	54	56	57	59	62	64
120	13	19	27	33	35	38	40	42	44	46	48	50	52	54	55	57	58	60	63	66
125	14	19	27	34	36	39	41	43	45	47	49	51	53	55	56	58	60	61	64	67
130	14	20	28	34	37	39	42	44	46	48	50	52	54	56	58	59	61	62	65	68
135	14	20	28	35	38	40	43	45	47	49	51	53	55	57	59	60	62	64	67	70
140	14	20	29	35	38	41	43	46	48	50	52	54	56	58	60	61	63	65	68	71
145	15	21	29	36	39	42	44	47	49	51	53	55	57	59	61	63	64	66	69	72
150	15	21	30	37	40	42	45	47	50	52	54	56	58	60	62	64	65	67	70	73
155	15	22	30	37	40	43	46	48	51	53	55	57	59	61	63	65	66	68	72	75
160	15	22	31	38	41	44	46	49	51	54	56	58	60	62	64	66	68	69	73	76
165	16	22	31	39	42	44	47	50	51	54	57	59	61	63	65	67	69	70	74	77
170	16	23	32	39	42	45	48	50	53	55	58	60	62	64	66	68	70	71	75	78
175	16	23	32	40	43	46	49	51	54	56	58	61	63	65	67	69	71	72	76	79
180	16	23	33	40	43	46	49	52	54	57	59	61	64	66	68	70	72	73	77	80
185	17	24	33	41	44	47	50	53	55	58	60	62	65	67	69	71	73	74	78	82
190	17	24	34	41	45	48	51	53	56	58	61	63	65	68	70	72	74	75	79	83
195	17	24	34	12	45	48	51	54	57	59	62	64	66	68	70	72	74	76	80	84
200	17	24	35	42	46	49	52	55	57	60	62	65	67	69	71	73	75	77	81	85
220	18	26	36	44	48	51	54	57	60	63	65	68	70	73	75	77	79	81	85	89
240	19	27	38	46	50	54	57	60	63	66	68	71	73	76	78	80	83	85	89	93
260	20	28	39	48	52	56	59	62	65	68	71	74	76	79	81	84	86	88	93	97
280	20	29	41	50	54	58	61	65	68	71	74	77	79	82	84	87	89	92	96	100
300	21	30	42	52	56	60	64	67	70	73	76	79	82	85	87	90	92	95	99	104
325	22	31	44	54	58	62	66	70	73	76	80	83	86	88	91	94	96	99	104	108
350	23	32	46	56	61	65	69	72	76	79	83	86	89	92	94	97	100	102	107	112
375	24	34	47	58	63	67	71	75	79	82	85	89	92	95	98	101	103	106	111	116
400	2]4	35	49	60	65	69	73	77	81	85	88	92	95	98	101	104	107	109	115	120

Source: Baker, J.S., and Fricke, L.B., "The Traffic Accident Investigation Manual", Northwestern University Traffic Institute, Evanston, IL, 1986.

Appendix L

Braking Efficiency of Cars

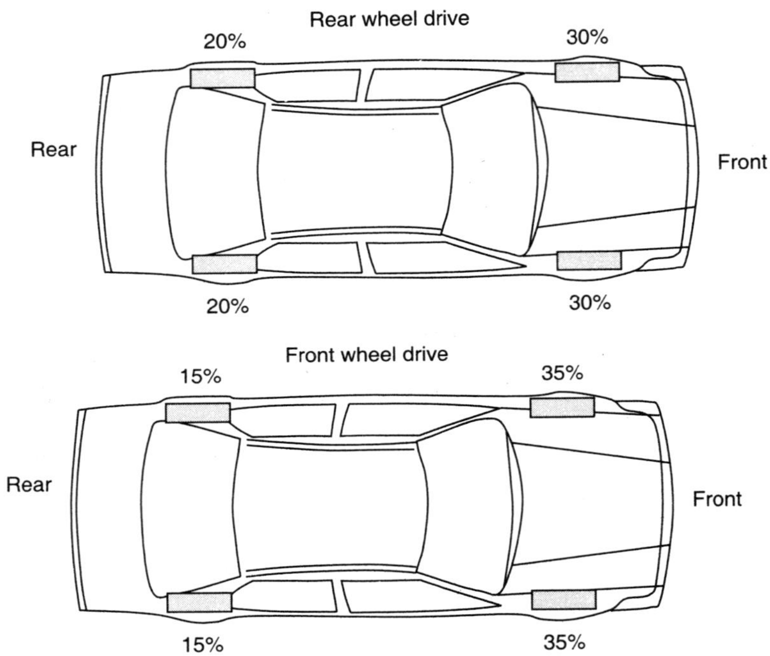